Optical Fiber Communications

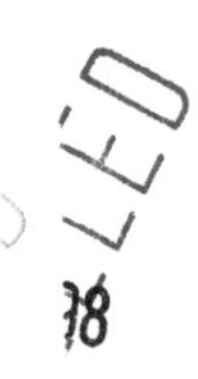

TURN

1997

Prentice Hall International Series in Optoelectronics

Consultant editors: John Midwinter, University College London, UK
Alan Snyder, Australian National University
Bernard Weiss, University of Surrey, UK

Fundamentals of Optical Fiber Communications
W. van Etten and J. van der Plaats

Light Emitting Diodes
K. Gillessen and W. Schairer

Optical Communication Systems
J. Gowar

Optical Sensing Techniques and Signal Processing
T. E. Jenkins

Optical Fiber Communications: Principles and Practice (Second Edition)
J. M. Senior

Lasers: Principles and Applications
J. Wilson and J. F. B. Hawkes

Optoelectronics: An Introduction (Second Edition)
J. Wilson and J. F. B. Hawkes

Optical Fiber Communications

Principles and Practice

John M. Senior

Second Edition

Prentice Hall

New York London Toronto Sydney Tokyo Singapore

First published 1985
This second edition published 1992 by
Prentice Hall International (UK) Ltd
Campus 400, Maylands Avenue
Hemel Hempstead
Hertfordshire HP2 7EZ
A division of
Simon & Schuster International Group

© Prentice Hall International (UK) Ltd, 1985, 1992

All rights reserved. No part of this publication may be reproduced, stored in a retrieval system, or transmitted, in any form, or by any means, electronic, mechanical, photocopying, recording or otherwise, without prior permission, in writing, from the publisher.
For permission within the United States of America contact Prentice Hall Inc., Englewood Cliffs, NJ 07632

Typeset in 10/12pt Times
by Mathematical Composition Setters, Salisbury, Wiltshire

Printed and bound in Great Britain at the University Press, Cambridge

Library of Congress Cataloging-in-Publication Data

Senior, John M., 1951–
Optical fiber communications : principles and practice / John M. Senior. — 2nd ed.
p. cm. – (Prentice Hall international series in optoelectronics)
Includes bibliographical references and index.
ISBN 0–13–635426–2 : $39.00
1. Optical communications. 2. Fiber optics. I. Title.
II. Series: Prentice Hall international series in optoelectronics.
TK5103.59.S46 1992
621.382′75—dc20 91-36019
CIP

British Library Cataloguing in Publication Data

Senior, John M.
Optical fibre communications: Principles and practice. – 2nd ed. – (Prentice Hall International series in optoelectronics)
I. Title II. Series
621.36

ISBN 0–13–635426–2

3 4 5 96 95 94

Contents

Preface to the second edition

It was suggested in the preface to the first edition that the relentless onslaught in the development and application of optical fiber communication technology would continue over the next decade. Now the greater part of that period is over, it may be observed that this was clearly an accurate statement as improvements and developments in the technology have occurred with tremendous rapidity in parallel with its increasingly widescale deployment. In particular, developments in relation to single-mode fibers, mid-infrared transmission, optical couplers, fiber lasers, narrow linewidth and frequency tunable lasers, superluminescent diodes, optical amplifiers, integrated optics and optical computation, together with coherent transmission techniques, have caused increases in the sophistication of the component technology whilst the trend towards greater transmission capacity and optical fiber networking (rather than point-to-point communications) has continued unabated. These substantial advances made the writing of a second edition both essential and urgent.

The above factors are particularly important when considering the major worldwide utilization of the first edition within academia and industry. In this context the question could also be asked as to why there has been a gap of seven years between the two editions. Although the continuing popularity of the first edition has provided a strong indication that the material it contains is still very relevant, it is not the answer to the aforementioned question. The answer is simply that at the commencement of the writing of the second edition, it was apparent that a substantial amount of additional material would need to be incorporated both to update the text and to take it into the new areas of technological development which have evolved since the mid-1980s. Hence this second edition constitutes a major revision which has necessitated the inclusion of much important new material whilst retaining the essential elements of the first edition. For example, the book now has three additional chapters as well as many new sections.

In common with the first edition, this edition has been developed from both teaching the subject to final-year undergraduates as well as from the continuation of a series of successful short courses on optical fiber communications conducted for professional engineers at the Manchester Metropolitan University (formerly Manchester Polytechnic). Furthermore it draws upon the diverse research activities of the Research Group which I lead in the area of optical fiber communications and networks. The book remains a comprehensive introductory text for use by both undergraduate and postgraduate engineers and scientists to provide them with a firm grounding in all significant aspects of the technology whilst providing strong insights into the potential future developments together with the growing areas of application.

The reader should therefore be in a position to appreciate such developments as they occur.

The enhanced treatment of the practical areas, together with the incorporation of the relevant standardization issues, will enable the book to continue to find major use as a reference text for practising engineers and scientists. Nevertheless, the book has been produced as a teaching/learning text and to this end it includes over 100 worked examples interspersed throughout in order to assist the learning process by illustrating the use of equations, by providing realistic values for parameters encountered and to aid the reader in aspects of design associated with optical fiber communication systems and networks. A total of 275 problems is also provided at the end of relevant chapters to examine the reader's understanding and to assist tutorial work. In a number of cases they also extend and elucidate the text, and in this context they should be considered as an integral part of the book. A *Solutions Manual* containing solutions to these problems may be obtained from the publisher.

In keeping with the status of an introductory text the fundamentals are included where necessary and there has been no attempt to cover the entire field in full mathematical rigour. However, selected proofs are developed in important areas throughout the text. It is assumed that the reader is conversant with differential and integral calculus and differential equations. In addition, the reader will find it useful to have a grounding in optics as well as a reasonable familiarity with the fundamentals of solid state physics.

Chapter 1 gives a short introduction to optical fiber communications by considering the historical development, the general system and the major advantages provided by this technology. In Chapter 2 the concept of the optical fiber as a transmission medium is introduced using a simple ray theory approach. This is followed by discussion of electromagnetic wave theory applied to optical fibers prior to consideration of light wave transmission within the various fiber types. The major transmission characteristics of optical fibers are then discussed in some detail in Chapter 3. A particular focus in this second edition within both Chapters 2 and 3 concerns the properties and characteristics of single-mode fibers.

Chapters 4 and 5 deal with the more practical aspects of optical fiber communications and therefore could be omitted from an initial teaching program. A number of these areas, however, are of crucial importance and thus should not be lightly overlooked. Chapter 4 deals with the manufacture and cabling of the various fiber types, whilst in Chapter 5 the different techniques to provide optical fiber connection are described. In this latter chapter both fiber to fiber joints (i.e. connectors and splices) are discussed as well as fiber branching devices, or couplers, which provide versatility within the configuration of optical fiber systems and networks.

Chapters 6 and 7 discuss the light sources employed in optical fiber communications. In Chapter 6 the fundamental physical principles of photoemission and laser action are covered prior to consideration of the various types of semiconductor and nonsemiconductor laser currently in use, or under investigation, for optical fiber communications. The other important semiconductor optical source, namely the light emitting diode, is dealt with in Chapter 7.

The next two chapters are devoted to the detection of the optical signal and the amplification of the electrical signal obtained. Chapter 8 discusses the basic principles of optical detection in semiconductors; this is followed by a description of the various types of photodetector currently utilized. The optical fiber direct detection receiver is then considered in Chapter 9, with particular emphasis on its performance characteristics.

Active optical devices and components are described in Chapter 10, which commences with detailed consideration of the various types of optical amplifier, followed by an account of the technology involved in integrated optics and opto-electronic integration. This is continued by discussion of optical bistability and digital optics which leads into an overview of optical computation.

Chapter 11 draws together the preceding material in a detailed discussion of the major current implementations of optical fiber communication systems (i.e. those using intensity modulation and the direct detection process) in order to give an insight into the design criteria and practices for all the main aspects of both digital and analog fiber systems. Both optical fiber distribution systems and advanced multiplexing strategies, together with the application of optical amplifiers within systems, are discussed to provide an understanding of these fast-growing areas.

It is apparent from the attention that has been devoted to coherent optical fiber communications over recent years that this is a future area of major exploitation of the technology. Hence coherent optical fiber systems are dealt with in some detail in Chapter 12, which covers all major aspects of this developing communications strategy in relation to both single and multicarrier systems.

Chapter 13 gives a general treatment of the major measurements which may be undertaken on optical fibers in both the laboratory and the field. This chapter, which occurred earlier in the book in the first edition, has been repositioned in order to enable the reader to obtain a more complete understanding of optical fiber subsystems and systems prior to consideration of these issues. Furthermore, it has been extended to include the measurements normally required to be taken on single-mode fibers and focused on to the measurement techniques which have been adopted as national and international standards.

Finally, Chapter 14 describes the many current and predicted application areas for optical fiber communications by drawing on practical examples from deployed systems as well as research and development activities. In particular, consideration is given to the possible developments in the telecommunication local access network together with the standardization associated with synchronous optical networks. The discussion is also expanded into the areas of optical fiber sensing together with optical fiber local area networking, both of which demonstrate major potential for the future application of optical fiber communication technology.

The book is referenced throughout to extensive end-of-chapter references which provide a guide for further reading and indicate a source for those equations which have been quoted without derivation. A complete glossary of symbols, together with a list of common abbreviations employed in the text, is also provided. SI units are used throughout the text.

I am very grateful for the many useful comments and suggestions provided by reviewers which have resulted in significant improvements to this text. Thanks must also be given to the authors of numerous papers, articles and books which I have referenced whilst preparing the text, and especially to those authors, publishers and companies who have kindly granted permission for the reproduction of diagrams and photographs. Further, I would like to thank the many readers of the first edition for their positive and helpful feedback which has assisted me greatly in the formulation of this second edition. I am also grateful to my family and friends who have continued to tolerate my infrequent appearances over the period of writing and revising the book. In particular I would like to dedicate this second edition to my late father, Ken, whose interest and encouragement in this work was never failing and who tragically did not quite see it completed. Finally, very special thanks are due to Judy for her patience and support in doing all the things that I should have done during the time I devoted to writing this edition.

Professor John M. Senior

Glossary of symbols and abbreviations

A	constant, area (cross-section, emission), far field pattern size, mode amplitude, wave amplitude (A_0)
A_{21}	Einstein coefficient of spontaneous emission
A_c	peak amplitude of the subcarrier waveform (analog transmission)
a	fiber core radius, parameter which defines the asymmetry of a planar guide (given by equation (10.21)), baseband message signal ($a(t)$)
$a_b(\lambda)$	effective fiber core radius
a_{eff}	bend attenuation fiber
a_k	integer 1 or 0
$a_m(\lambda)$	relative attenuation between optical powers launched into multimode and single-mode fibers
B	constant, electrical bandwidth (post detection), magnetic flux density, mode amplitude, wave amplitude (B_0)
B_{12}, B_{21}	Einstein coefficients of absorption, stimulated emission
B_F	modal birefringence
B_{fib}	fiber bandwidth
B_{FPA}	mode bandwidth (Fabry–Perot amplifier)
B_m	bandwidth of an intensity modulated optical signal $m(t)$, maximum 3 dB bandwidth (photodiode)
B_{opt}	optical bandwidth
B_r	recombination coefficient for electrons and holes
B_T	bit rate, when the system becomes dispersion limited ($B_T(DL)$)
b	normalized propagation constant for a fiber, ratio of luminance to composite video, linewidth broadening factor (injection laser)
C	constant, capacitance, crack depth (fiber), wave coupling coefficient per unit length, coefficient incorporating Einstein coefficients
C_a	effective input capacitance of an optical fiber receiver amplifier
C_d	optical detector capacitance
C_f	capacitance associated with the feedback resistor of a transimpedance optical fiber receiver amplifier
C_j	junction capacitance (photodiode)
C_L	total optical fiber channel loss in decibels, including the dispersion–equalization penalty (C_{LD})
C_0	wave amplitude
C_T	total capacitance
CT	polarization crosstalk

c	velocity of light in a vacuum, constant (c_1, c_2)
c_i	tap coefficients for a transversal equalizer
D	amplitude coefficient, electric flux density, distance, diffusion coefficient, corrugation period, decision threshold in digital optical fiber transmission, fiber dispersion parameters: material (D_M); profile (D_P); total first order (D_T); waveguide (D_W), detectivity (photodiode), specific detectivity (D^*)
D_c	minority carrier diffusion coefficient
D_f	frequency deviation ratio (subcarrier FM)
D_L	dispersion–equalization penalty in decibels
D_p	frequency deviation ratio (subcarrier PM)
d	fiber core diameter, distance, width of the absorption region (photodetector), thickness of recombination region (optical source), pin diameter (mode scrambler)
d_f	far field mode-field diameter (single-mode fiber)
d_n	near field mode-field diameter (single-mode fiber)
d_o	fiber outer (cladding) diameter
E	electric field, energy, Youngs modulus, expected value of a random variable, electron energy
E_a	activation energy of homogeneous degradation for an LED
E_F	Fermi level (energy), quasi-Fermi level located in the conduction band (E_{Fc}), valence band (E_{Fv}) of a semiconductor
E_g	separation energy between the valence and conduction bands in a semiconductor (bandgap energy)
$E_m(t)$	subcarrier electric field (analog transmission)
E_o	optical energy
E_q	separation energy of the quasi-Fermi levels
e	electronic charge, base for natural logarithms
F	probability of failure, transmission factor of a semiconductor–external interface, excess avalanche noise factor ($F(M)$), optical amplifier noise figure
$\mathcal{F}$	Fourier transformation
F_n	noise figure (electronic amplifier)
F_{to}	total noise figure for system of cascaded optical amplifiers
f	frequency
f_D	peak to peak frequency deviation (PFM–IM)
f_d	peak frequency deviation (subcarrier FM and PM)
f_o	Fabry–Perot resonant frequency (optical amplifier), pulse rate (PFM–IM)
G	open loop gain of an optical fiber receiver amplifier, photoconductive gain, cavity gain of a semiconductor laser amplifier
$G_i(r)$	amplitude function in the WKB method
G_o	optical gain (phototransistor)
G_R	Raman gain (fiber amplifier)

G_s	single pass gain of a semiconductor laser amplifier
Gsn	Gaussian (distribution)
g	degeneracy parameter
$\bar{g}$	gain coefficient per unit length (laser cavity)
g_m	transconductance of a field effect transistor, material gain coefficient
g_0	unsaturated material gain coefficient
g_R	power Raman gain coefficient
$\bar{g}_{th}$	threshold gain per unit length (laser cavity)
H	magnetic field
$H(\omega)$	optical power transfer function (fiber), circuit transfer function
$H_A(\omega)$	optical fiber receiver amplifier frequency response (including any equalization)
$H_{CL}(\omega)$	closed loop current to voltage transfer function (receiver amplifier)
$H_{eq}(\omega)$	equalizer transfer function (frequency response)
$H_{OL}(\omega)$	open loop current to voltage transfer function (receiver amplifier)
$H_{out}(\omega)$	output pulse spectrum from an optical fiber receiver
h	Planck's constant, thickness of a planar waveguide, power impulse response for optical fiber ($h(t)$), mode coupling parameter (PM fiber)
$h_A(t)$	optical fiber receiver amplifier impulse response (including any equalization)
h_{eff}	effective thickness of a planar waveguide
h_{FE}	common emitter current gain for a bipolar transistor
$h_f(t)$	optical fiber impulse response
$h_{out}(t)$	output pulse shape from an optical fiber receiver
$h_p(t)$	input pulse shape to an optical fiber receiver
$h_t(t)$	transmitted pulse shape on an optical fiber link
I	electrical current, optical intensity
I_b	background radiation induced photocurrent (optical receiver)
I_{bias}	bias current for an optical detector
I_c	collector current (phototransistor)
I_d	dark current (optical detector)
I_o	maximum optical intensity
I_p	photocurrent generated in an optical detector
I_S	output current from photodetector resulting from intermediate frequency in coherent receiver
I_{th}	threshold current (injection laser)
i	electrical current
i_a	optical receiver preamplifier shunt noise current
i_{amp}	optical receiver, preamplifier total noise current.
i_D	decision threshold current (digital transmission)
i_d	photodiode dark noise current
i_{det}	output current from an optical detector
i_f	noise current generated in the feedback resistor of an optical fiber receiver transimpedance preamplifier

i_N	total noise current at a digital optical fiber receiver
i_n	multiplied shot noise current at the output of an APD excluding dark noise current
i_s	shot noise current on the photocurrent for a photodiode
i_{SA}	multiplied shot noise current at the output of an APD including the noise current
i_{sig}	signal current obtained in an optical fiber receiver
i_t	thermal noise current generated in a resistor
i_{TS}	total shot noise current for a photodiode without internal gain
J	Bessel function, current density
J_{th}	threshold current density (injection laser)
j	$\sqrt{-1}$
K	Boltzmann's constant, constant, modified Bessel function
K_I	stress intensity factor, for an elliptical crack (K_{IC})
k	wave propagation constant in a vacuum (free space wave number), wave vector for an electron in a crystal, ratio of ionization rates for holes and electrons, integer, coupling coefficient for two interacting waveguide modes, constant
k_f	angular frequency deviation (subcarrier FM)
k_p	phase deviation constant (subcarrier PM)
L	length (fiber), distance between mirrors (laser), coupling length (waveguide modes)
L_{ac}	insertion loss of access coupler in distribution system
L_B	beat length in a monomode optical fiber
L_{bc}	coherence length in a monomode optical fiber
L_c	characteristic length (fiber)
L_D	diffusion length of charge carriers (LED)
L_{ex}	star coupler excess loss in distribution system
L_0	constant with dimensions of length
L_t	lateral misalignment loss at an optical fiber joint
L_{tr}	tap ratio loss in distribution system
$\mathscr{L}$	transmission loss factor (transmissivity) of an optical fiber
l	azimuthal mode number, distance, length
l_a	atomic spacing (bond distance)
l_0	wave coupling length
M	avalanche multiplication factor, material dispersion parameter, total number of guided modes or mode volume; for a multimode step index fiber (M_s); for multimode graded index fiber (M_g), mean value (M_1) and mean square value (M_2) of a random variable
M_a	safety margin in an optical power budget
M_{op}	optimum avalanche multiplication factor
M^x	excess avalanche noise factor, (also denoted as $F(M)$)
m	radial mode number, Weibull distribution parameter, intensity modulated optical signal ($m(t)$), mean value of a random variable,

	integer, optical modulation index (subcarrier amplitude modulation)
m_a	modulation index
N	integer, density of atoms in a particular energy level (eg N_1, N_2, N_3), minority carrier concentration in n type semiconductor material, number of input/output ports on a fiber star coupler, number of nodes on distribution system, noise current
NA	numerical aperture of an optical fiber
NEP	noise equivalent power
N_g	group index of an optical waveguide
N_{ge}	effective group index or group index of a single-mode waveguide
N_0	defined by equation (11.80)
N_p	number of photons per bit (coherent transmission)
n	refractive index (eg n_1, n_2, n_3), stress corrosion susceptibility, negative type semiconductor material, electron density
n_e	effective refractive index of a planar waveguide
n_{eff}	effective refractive index of a single-mode fiber
n_0	refractive index of air
n_{sp}	spontaneous emission factor (injection laser)
P	electrical power, minority carrier concentration in p type semiconductor material, probability of error (P(e)), of detecting a zero level ($P(0)$), of detecting a one level ($P(1)$), of detecting z photons in a particular time period ($P(z)$), conditional probability, of detecting a zero when a one is transmitted ($P(0/1)$), of detecting a one when a zero is transmitted ($P(1/0)$), optical power (P_1, P_2, etc.)
P_a	total power in a baseband message signal $a(t)$
P_B	threshold optical power for Brillouin scattering
P_b	backward travelling signal power (semiconductor laser amplifier), power transmitted through fiber sample
P_c	optical power coupled into a step index fiber, optical power level
P_D	optical power density
P_{dc}	dc optical output power
P_e	optical power emitted from an optical source
P_G	optical power in a guided mode
P_i	mean input (transmitted) optical power launched into a fiber
P_{in}	input signal power (semiconductor laser amplifier)
P_{int}	internally generated optical power (optical source)
P_L	optical power of local oscillator signal (coherent system)
P_m	total power in an intensity modulated optical signal $m(t)$
P_o	mean output (received) optical power from a fiber
P_{opt}	mean optical power travelling in a fiber
P_{out}	initial output optical (prior to degradation) power from an optical source
P_p	optical pump power (fiber amplifier)
P_{po}	peak received optical power

P_r reference optical power level, optical power level
P_R threshold optical power for Raman scattering
$P_{Ra}(t)$ backscattered optical power (Rayleigh) within a fiber
P_S optical power of incoming signal (coherent system)
P_s total power transmitted through a fiber sample
P_{sc} optical power scattered from a fiber
P_t optical transmitter power, launch power (P_{tx})
p crystal momentum, average photoelastic coefficient, positive type semiconductor material, probability density function ($p(x)$)
q integer, fringe shift
R photodiode responsivity, radius of curvature of a fiber bend, electrical resistance (eg R_{in}, R_{out}); facet reflectivity (R_1, R_2)
R_{12} upward transition rate for electrons from energy level 1 to level 2
R_{21} downward transition rate for electrons from energy level 2 to level 1
R_a effective input resistance of an optical fiber receiver preamplifier
R_b bias resistance, for optical fiber receiver preamplifier (R_{ba})
R_c critical radius of an optical fiber
R_D radiance of an optical source
RE_{dB} ratio of electrical output power to electrical input power in decibels for an optical fiber system
R_f feedback resistance in an optical fiber receiver transimpedance preamplifier
R_L load resistance associated with an optical fiber detector
RO_{dB} ratio of optical output power to optical input power in decibels for an optical fiber system
R_t total carrier recombination rate (semiconductor optical source)
R_{TL} total load resistance within an optical fiber receiver
r radial distance from the fiber axis, Fresnel reflection coefficient, mirror reflectivity, electro-optic coefficient.
r_e generated electron rate in an optical detector
r_{ER}, r_{ET} reflection and transmission coefficients, respectively, for the electric field at a planar, guide–cladding interface
r_{HR}, r_{HT} reflection and transmission coefficients respectively for the magnetic field at a planar, guide-cladding interface
r_{nr} nonradiative carrier recombination rate per unit volume
r_p incident photon rate at an optical detector
r_r radiative carrier recombination rate per unit volume
r_t total carrier recombination rate per unit volume
S fraction of captured optical power, macroscopic stress, dispersion slope (fiber), power spectral density $S(\omega)$
S_f fracture stress
$S_i(r)$ phase function in the WKB method
$S_m(\psi)$ spectral density of the intensity modulated optical signal $m(t)$
S/N peak signal power to rms noise power ratio, with peak to peak signal power $[(S/N)_{p-p}]$ with rms signal power $[(S/N)_{rms}]$

S_0	scale parameter; zero dispersion slope (fiber)
S_t	theoretical cohesive strength
s	pin spacing (mode scrambler)
T	temperature, time
T_a	insertion loss resulting from an angular offset between jointed optical fibers
T_c	10 to 90% rise time arising from intramodal dispersion on an optical fiber link
T_D	10 to 90% rise time for an optical detector
T_F	fictive temperature
T_l	insertion loss resulting from a lateral offset between jointed optical fibers
T_n	10 to 90% rise time arising from intermodal dispersion on an optical fiber link
T_0	threshold temperature (injection laser), nominal pulse period (PFM–IM)
T_R	10 to 90% rise time at the regenerator circuit input (PFM–IM)
T_S	10 to 90% rise time for an optical source
T_{syst}	total 10 to 90% rise time for an optical fiber system
T_T	total insertion loss at an optical fiber joint
T_t	temperature rise at time t
T_∞	maximum temperature rise
t	time, carrier transit time, slow(t_s), fast (t_f)
t_c	time constant
t_d	switch on delay (laser)
t_e	$1/e$ pulse width from the centre
t_r	10 to 90% rise time
U	eigenvalue of the fiber core
V	electrical voltage, normalized frequency for an optical fiber or planar waveguide
V_{bias}	bias voltage for a photodiode
V_c	cutoff value of normalized frequency (fiber)
V_{CC}	collector supply voltage
V_{CE}	collector–emitter voltage (bipolar transistor)
V_{EE}	emitter supply voltage
V_{eff}	effective normalized frequency (fiber)
V_{opt}	voltage reading corresponding to the total optical power in a fiber
V_{sc}	voltage reading corresponding to the scattered optical power in a fiber
v	electrical voltage
v_a	amplifier series noise voltage
$v_A(t)$	receiver amplifier output voltage
v_c	crack velocity
v_d	drift velocity of carriers (photodiode)
v_g	group velocity
$v_{out}(t)$	output voltage from an RC filter circuit

v_p phase velocity
W eigenvalue of the fiber cladding, random variable
W_e electric pulse width
W_o optical pulse width
w depletion layer width (photodiode)
X random variable
x coordinate, distance, constant, evanescent field penetration depth, slab thickness, grating line spacing
Y constant, shunt admittance, random variable
y coordinate, lateral offset at a fiber joint
Z random variable, constant
Z_0 electrical impedance
z coordinate, number of photons
z_m average or mean number of photons arriving at a detector in a time period τ
z_{md} average number of photons detected in a time period τ
α characteristic refractive index profile for fiber (profile parameter), optimum profile parameter (α_{op}), linewidth enhancement factor (injection laser), optical link loss
$\bar{\alpha}$ loss coefficient per unit length (laser cavity)
α_{cr} connector loss at transmitter and receiver in decibels
α_{dB} signal attenuation in decibels per unit length
α_{fc} fiber cable loss in decibels per kilometre
α_i internal wavelength loss per unit length (injection laser)
α_j fiber joint loss in decibels per kilometre
α_m mirror loss per unit length (injection laser)
α_N signal attenuation in nepers
α_0 absorption coefficient
α_p fiber transmission loss at the pump wavelength (fiber amplifier)
α_r radiation attenuation coefficient
β wave propagation constant
$\bar{\beta}$ gain factor (injection laser cavity)
β_c isothermal compressibility
β_0 proportionality constant
β_r degradation rate
Γ optical confinement factor (semiconductor laser amplifier)
γ angle, attenuation coefficient per unit length for a fiber
γ_p surface energy of a material
γ_R Rayleigh scattering coefficient for a fiber
Δ relative refractive index difference between the fiber core and cladding
Δf linewidth of single frequency injection laser
ΔG peak–trough ratio of the passband ripple (semiconductor laser amplifier)
Δn index difference between fiber core and cladding ($\Delta n/n_1$ fractional index difference)

δ_E	phase shift associated with transverse electric waves
δf	uncorrelated source frequency widths
δ_H	phase shift associated with transverse magnetic waves
$\delta\lambda$	optical source spectral width (linewidth), mode spacing (laser)
δT	intermodal dispersion time in an optical fiber
δT_g	delay difference between an extreme meridional ray and an axial ray for a graded index fiber
δT_s	delay difference between an extreme meridional ray and an axial ray for a step index fiber, with mode coupling (δT_{sc})
δT_g	polarization mode dispersion in fiber
ε	electric permittivity, of free space (ε_0), relative (ε_r), semiconductor (ε_s), extinction ratio (optical transmitter)
ζ	solid acceptance angle
η	quantum efficiency (optical detector)
η_{ang}	angular coupling efficiency (fiber joint)
η_c	coupling efficiency (optical source to fiber)
η_D	differential external quantum efficiency (optical source)
η_{ep}	external power efficiency (optical source)
η_i	internal quantum efficiency injection laser
η_{int}	internal quantum efficiency (LED)
η_{lat}	lateral coupling efficiency (fiber joint)
η_{pc}	overall power conversion efficiency (optical source)
η_T	total external quantum efficiency (optical source)
θ	angle, fiber acceptance angle (θ_a)
θ_B	Bragg diffraction angle, blaze angle diffraction grating
Λ	acoustic wavelength, period for perturbations in a fiber
Λ_c	cutoff period for perturbations in a fiber
λ	optical wavelength
λ_B	Bragg wavelength (DFB laser)
λ_c	long wavelength cutoff (photodiode), cutoff wavelength for single-mode fiber, effective cutoff wavelength (λ_{ce})
λ_0	wavelength at which first order dispersion is zero
μ	magnetic permeability, relative permeability, (μ_r), permeability of free space (μ_0)
ν	optical source bandwidth in gigahertz
ρ	polarization rotation in a monomode optical fiber
ρ_f	spectral density of the radiation energy at a transition frequency f
σ	standard deviation, (rms pulse width), variance (σ^2)
σ_c	rms pulse broadening resulting from intramodal dispersion in a fiber
σ_m	rms pulse broadening resulting from material dispersion in a fiber
σ_n	rms pulse broadening resulting from intermodal dispersion in a graded index fiber (σ_g), in a step index fiber (σ_s)
σ_T	total rms pulse broadening in a fiber or fiber link
τ	time period, bit period, signalling interval, pulse duration 3, dB pulse width (τ(3 dB))

τ_{21}	spontaneous transition lifetime between energy levels 2 and 1
τ_E	time delay in a transversal equalizer
τ_e	$1/e$ full width pulse broadening due to dispersion on an optical fiber link
τ_g	group delay
τ_i	injected (minority) carrier lifetime
τ_{ph}	photon lifetime (semiconductor laser)
τ_r	radiative minority carrier lifetime
τ_{sp}	spontaneous emission lifetime (equivalent to τ_{21})
Φ	linear retardation
ϕ	angle, critical angle (ϕ_c), photon density, phase shift
ψ	scalar quantity representing **E** or **H** field.
ω	angular frequency, of the subcarrier waveform in analog transmission (ω_c), of the modulating signal in analog transmission (ω_m), pump frequency (ω_p), Stokes component (ω_s) antistokes component (ω_a), intermediate frequency of coherent heterodyne receiver (ω_{IF}), normalized spot size of the fundamental mode
ω_0	spot size of the fundamental mode
∇	vector operator, Laplacian operator (∇^2)

A–D	analog to digital
ac	alternating current
AFC	automatic frequency control
AGC	automatic gain control
AM	amplitude modulation
APD	avalanche photodiode
AR	antireflection (surface, coating)
ARROW	antiresonant reflecting optical waveguide
ASK	amplitude shift keying
ATM	alternative test method (fiber), asynchronous transfer mode (multiplexing)
BAP	broadband access point
BER	bit error rate
BH	buried heterostructure (injection laser)
BIDS	broadband integrated distributed star (potential local access network)
BOD	bistable optical device
CAM	computer aided manufacture
CATV	common antenna television
CCITT	International Telephone and Telegraph Consultative Committee
CCTV	close circuit television
CDH	constricted double heterojunction (injection laser)

CMI	coded mark inversion
CMOS	complementary metal oxide silicon
CNR	carrier to noise ratio
CO	central office (telephone switching centre)
CPFSK	continuous phase frequency shift keying
CPU	central processing unit
CSMA/CD	carrier sense multiple access with collision detection
CSP	channelled substrate planar (injection laser)
CW	continuous wave or operation
D–A	digital to analog
dB	decibel
D–IM	direct intensity modulation
DC	depressed cladding (fiber design)
dc	direct current
DF	dispersion flattened (single-mode fiber)
DFB	distributed feedback (injection laser)
DH	double heterostructure or heterojunction (injection laser or LED)
DLC	digital loop carrier
DLD	dark line defect (semiconductor optical source)
DPSK	differential phase shift keying
DQDB	distributed queue dual bus (emerging standard for metropolitan area networks)
DS	dispersion shifted (single-mode fiber)
DSB	double sideband (amplitude modulation)
DSD	dark spot defect (semiconductor optical source)
ECL	emitter coupler logic
EH	traditional mode designation
EIA	Electronics Industries Association
ELED	edge-emitter light emitting diode
EMI	electromagnetic interference
EMP	electromagnetic pulse
erf	error function
erfc	complementary error function
ESI	equivalent step index (fiber)
FBT	fused biconical taper (fiber coupler)
FC	fiber connector
FDDI	fiber distributed data interface
FDM	frequency division multiplexing
FET	field effect transistor, junction (JFET)
FM	frequency modulation
FMS	flexible manufacturing systems
FOTP	fiber optic test procedure
FPA	Fabry–Perot amplifier
FSK	frequency shift keying

FWHP	full width half power
FWHM	full width half maximum (equivalent to FWHP)
GRIN	graded index (rod lens)
HB	high birefringence (fiber)
HBT	heterojunction bipolar transmitter
HDB	high density bipolar
HDTV	high definition television
HE	traditional mode designation
HEMT	high electron mobility transistor
He–Ne	helium-neon (laser)
HF	high frequency
HV	high voltage
IF	intermediate frequency
ILD	injection laser diode
IM	intensity modulation, with direct detection (IM/DD)
IO	integrated optics
I/O	input/output
I & Q	inphase and quadrature (coherent receiver),
ISDN	integrated services digital network, broadband (BISDN)
ISI	intersymbol interference
ISO	International Standardization Organization
LAN	local area network
LB	low birefringence (fiber)
LEC	long external cavity (laser)
LED	light emitting diode
LLC	logical link control (LAN)
LOC	large optical cavity (injection laser)
LP	linearly polarized (mode notation)
LPE	liquid phase epitaxy
MAC	medium access control (LAN), isochronous (I-MAC)
MAN	metropolitan area network
MAP	manufacturing automation protocol
MBE	molecular beam epitaxy
MC	matched cladding (fiber design)
MCVD	modified chemical vapour deposition
MESFET	metal Schottky field effect transistor
MFD	mode-field diameter (single-mode fiber)
MFSK	multilevel frequency shift keying
MISFET	metal integrated-semiconductor field effect transistor
MMF	multimode fiber
MOSFET	metal oxide semiconductor field effect transistor
MOVPE	metal oxide vapour-phase epitaxy
MQW	multiquantum-well
MUSE	multiple sub-Nyquist sampling encoding

Nd : YAG	neodymium-doped yttrium-aluminium-garnet (laser)
NRZ	non-return to zero
OC	optical carrier (SONET)
OCWR	optical continuous wave reflectometer
OEIC	optoelectronic integrated circuit
OFDM	optical frequency division multiplexing
OOK	on–off keying (equivalent to binary amplitude shift keying)
ORL	optical return loss
OSI	open systems interconnection
OTDM	optical time division multiplexing
OTDR	optical time domain reflectrometry
OVPO	outside vapour-phase oxidation
PAM	pulse amplitude modulation
PANDA	polarization maintaining and absorption reducing (fiber)
PC	physical contact (fiber connector)
PCM	pulse code modulation
PCS	plastic-clad silica (fiber)
PCVD	plasma-activated chemical vapour deposition
PCW	planar convex waveguide (injection laser)
PD	photodiode, photodetector
PDF	probability density function
PFM	pulse frequency modulation
PHY	physical layer protocol (FDDI)
PIN–FET	*p–i–n* photodiode followed by a field transistor
PLL	phase locked loop
PM	phase modulation, polarization maintaining (fiber)
PMD	physical medium dependent (FDDI)
PMMA	polymethyl methacrylate
PoLSK	polarization shift keying
PON	passive optical network, telephony (TPON), broadband (BPON), asynchronous (APON)
POTDR	polarization optical time domain reflectrometer
PPL	passive photonic loop (optical local access network proposal)
PPM	pulse position modulation
PSK	phase shift keying
PTT	post, telegraph and telecommunications
PWM	pulse width modulation
QAM	quadrature amplitude modulation
RAPD	reach through avalanche photodiode
RIN	relative intensity noise (injection laser)
rms	root mean square
RNF	refracted near field (method for fiber refractive index profile measurement)
RO	relaxation oscillation

RTM reference test method (fiber)
RZ return to zero
SAGM separate absorption, grading and multiplication (avalanche photodiode)
SAM separate absorption and multiplication (avalanche photodiode)
SAW surface acoustic wave
SBS stimulated Brillouin scattering
SC subscriber connector (fiber)
SCM subcarrier multiplexing
SDH synchronous digital hierarchy
SDM space division multiplexing
SEED self-electro-optic device, symmetric (S-SEED)
SHF super high frequency
SIU subscriber interface unit
SLA semiconductor laser amplifier
SLD superluminescent diode
SLED surface emitter light emitting diode
SMA subminiature assembly (fiber connector)
SMF single-mode fiber
SML separated multiclad layer (injection laser)
SMT station management (FDDI)
SNR signal to noise ratio
SONET synchronous optical network
SOP state of polarization
SPE synchronous payload envelope (SONET)
SQW single quantum-well
SRS stimulated Raman scattering
ST straight tip (fiber connector)
STM synchronous transport module (SDH)
STS synchronous transport signal (SONET)
TDM time division multiplexing
TDMA time division multiple access
TE transverse electric
TEM transverse electromagnetic
TJS transverse junction stripe (injection laser)
TM transverse magnetic
TTL transistor–transistor logic
TWA travelling wave amplifier
UHF ultra high frequency
VAD vapour axial deposition
VCO voltage controlled oscillator
VHF very high frequency
VPE vapour-phase epitaxy
VSB vestigial sideband (modulation)

VT	virtual tributary (SONET)
WDM	wavelength division multiplexing
WKB	Wentzel, Kramers, Brillouin (analysis technique for graded fiber)
WPS	wideband switch point
ZD	zenner diode
ZMD	zero material dispersion (fiber)

To my father, Ken

1

Introduction

Communication may be broadly defined as the transfer of information from one point to another. When the information is to be conveyed over any distance a communication system is usually required. Within a communication system the information transfer is frequently achieved by superimposing or modulating the information on to an electromagnetic wave which acts as a carrier for the information signal. This modulated carrier is then transmitted to the required destination where it is received and the original information signal is obtained by demodulation. Sophisticated techniques have been developed for this process using electromagnetic carrier waves operating at radio frequencies as well as microwave and millimetre wave frequencies. However, 'communication' may also be achieved using an electromagnetic carrier which is selected from the optical range of frequencies.

1.1 Historical development

The use of visible optical carrier waves or light for communication has been common for many years. Simple systems such as signal fires, reflecting mirrors and, more recently, signalling lamps have provided successful, if limited, information transfer. Moreover, as early as 1880 Alexander Graham Bell reported the transmission of speech using a light beam [Ref. 1]. The photophone proposed by Bell just four years after the invention of the telephone modulated sunlight with a diaphragm giving speech transmission over a distance of 200 m. However, although some investigation of optical communication continued in the early part of the twentieth century [Refs. 2 and 3] its use was limited to mobile, low capacity communication links. This was due to both the lack of suitable light sources and the problem that light transmission in the atmosphere is restricted to line of sight and is severely affected by disturbances such as rain, snow, fog, dust and atmospheric turbulence. Nevertheless lower frequency and hence longer wavelength electromagnetic waves* (i.e. radio and microwave) proved suitable carriers for information transfer in the atmosphere, being far less affected by these atmospheric conditions. Depending on their wavelengths these electromagnetic carriers can be transmitted over considerable distances but are limited in the amount of information they can convey by their frequencies (i.e. the information-carrying capacity is directly related to the bandwidth or frequency extent of the modulated carrier, which is generally limited to a fixed fraction of the carrier frequency). In theory, the greater the carrier frequency, the larger the available transmission bandwidth and thus the information-carrying capacity of the communication system. For this reason radio communication was developed to higher frequencies (i.e. VHF and UHF) leading to the introduction of the even higher frequency microwave and, latterly, millimetre wave transmission. The relative frequencies and wavelengths of these types of electromagnetic wave can be observed from the electromagnetic spectrum shown in Figure 1.1. In this context it may also be noted that communication at optical frequencies offers an increase in the potential usable bandwidth by a factor of around 10^4 over high frequency microwave transmission. An additional benefit of the use of high carrier frequencies is the general ability of the communication system to concentrate the available power within the transmitted electromagnetic wave, thus giving an improved system performance [Ref. 4].

A renewed interest in optical communication was stimulated in the early 1960s with the invention of the laser [Ref. 5]. This device provided a powerful coherent light source, together with the possibility of modulation at high frequency. In addition the low beam divergence of the laser made enhanced free space optical transmission a practical possibility. However, the previously mentioned constraints of light transmission in the atmosphere tended to restrict these systems to short distance applications. Nevertheless, despite the problems some modest free space optical communication links have been implemented for applications such as the linking of a television camera to a base vehicle and for data links of a few hundred

* For the propagation of electromagnetic waves in free space, the wavelength λ equals the velocity of light in a vacuum c times the reciprocal of the frequency f in hertz or $\lambda = c/f$.

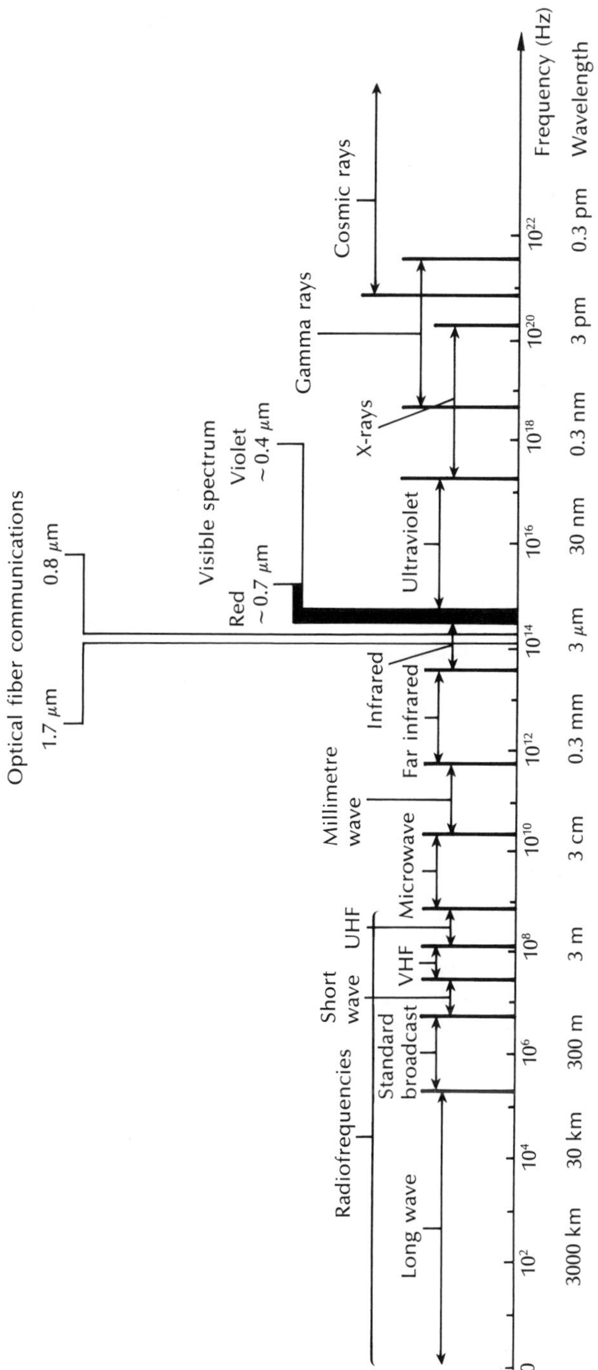

Figure 1.1 The electromagnetic spectrum showing the region used for optical fiber communications.

metres between buildings. There is also some interest in optical communication between satellites in outer space using similar techniques [Ref. 6].

Although the use of the laser for free space optical communication proved somewhat limited, the invention of the laser instigated a tremendous research effort into the study of optical components to achieve reliable information transfer using a lightwave carrier. The proposals for optical communication via dielectric waveguides or optical fibers fabricated from glass to avoid degradation of the optical signal by the atmosphere were made almost simultaneously in 1966 by Kao and Hockham [Ref. 7] and Werts [Ref. 8]. Such systems were viewed as a replacement for coaxial cable or carrier transmission systems. Initially the optical fibers exhibited very high attenuation (i.e. 1000 dB km^{-1}) and were therefore not comparable with the coaxial cables they were to replace (i.e. 5 to 10 dB km^{-1}). There were also serious problems involved in jointing the fiber cables in a satisfactory manner to achieve low loss and to enable the process to be performed relatively easily and repeatedly in the field. Nevertheless, within the space of ten years optical fiber losses were reduced to below 5 dB km^{-1} and suitable low loss jointing techniques were perfected.

In parallel with the development of the fiber waveguide, attention was also focused on the other optical components which would constitute the optical fiber communication system. Since optical frequencies are accompanied by extremely small wavelengths the development of all these optical components essentially required a new technology. Thus semiconductor optical sources (i.e. injection lasers and light emitting diodes) and detectors (i.e. photodiodes and to a certain extent phototransistors) compatible in size with optical fibers were designed and fabricated to enable successful implementation of the optical fiber system. Initially the semiconductor lasers exhibited very short lifetimes of at best a few hours, but significant advances in the device structure enabled lifetimes greater than 1000 hr [Ref. 9] and 7000 hr [Ref. 10] to be obtained by 1973 and 1977 respectively. These devices were originally fabricated from alloys of gallium arsenide (AlGaAs) which emitted in the near infrared between 0.8 and 0.9 μm.

Subsequently the above wavelength range was extended to include the 1.1 to 1.6 μm region by the use of other semiconductor alloys (see Section 6.3.6) to take advantage of the enhanced performance characteristics displayed by optical fibers over this range. In particular for this longer wavelength region, semiconductor lasers and also the simpler structured light emitting diodes based on the quarternary alloy InGaAsP are now available which have projected median lifetimes in excess of 25 years (when operated at 10 °C) for the former and 100 years (when operated at 70 °C) for the latter device type [Ref. 11]. Direct modulation of the commercial devices is also feasible at rates of several giga bit s^{-1} which is especially useful in the first longer wavelength window region around 1.3 μm where fiber intramodal dispersion is minimized and hence the transmission bandwidth is maximized, particularly for single-mode fibers. It is also noteworthy that this fiber type has quickly come to dominate system applications within telecommunications. Moreover, the lowest silica glass fiber losses of about 0.2 dB km^{-1} are obtained in the other longer wavelength window at 1.55 μm, but, unfortunately, intramodal dispersion is greater at this wavelength thus limiting the maximum bandwidth achievable with conventional single-mode fiber.

To obtain both the low loss and low dispersion at the same operating wavelength, new advanced single-mode fiber structures have been realized: namely, dispersion shifted and dispersion flattened fibers. Hence developments in fiber technology have continued rapidly over recent years, encompassing other specialist fiber types such as polarization maintaining fibers, as well as glass materials for even longer wavelength operation in the mid-infrared (2 to 5 μm) and far-infrared (8 to 12 μm) regions. In addition, the implementation of associated fiber components (splices, connectors, couplers, etc.) and active optoelectronic devices (sources, detectors, amplifiers, etc.) has also moved forward with such speed that optical fiber communication technology would seem to have reached a stage of maturity within its developmental path [Ref. 12]. Therefore, high performance, reliable optical fiber communication systems are now widely deployed both within telecommunications networks and many other more localized communication application areas.

1.2 The general system

An optical fiber communication system is similar in basic concept to any type of communication system. A block schematic of a general communication system is shown in Figure 1.2(a), the function of which is to convey the signal from the information source over the transmission medium to the destination. The communication system therefore consists of a transmitter or modulator linked to the information source, the transmission medium, and a receiver or demodulator at the destination point. In electrical communications the information source provides an electrical signal, usually derived from a message signal which is not electrical (e.g. sound), to a transmitter comprising electrical and electronic components which converts the signal into a suitable form for propagation over the transmission medium. This is often achieved by modulating a carrier, which, as mentioned previously, may be an electromagnetic wave. The transmission medium can consist of a pair of wires, a coaxial cable or a radio link through free space down which the signal is transmitted to the receiver, where it is transformed into the original electrical information signal (demodulated) before being passed to the destination. However, it must be noted that in any transmission medium the signal is attenuated, or suffers loss, and is subject to degradations due to contamination by random signals and noise, as well as possible distortions imposed by mechanisms within the medium itself. Therefore, in any communication system there is a maximum permitted distance between the transmitter and the receiver beyond which the system effectively ceases to give intelligible communication. For long-haul applications these factors necessitate the installation of repeaters or line amplifiers (see Section 11.4) at intervals, both to remove signal distortion and to increase signal level before transmission is continued down the link.

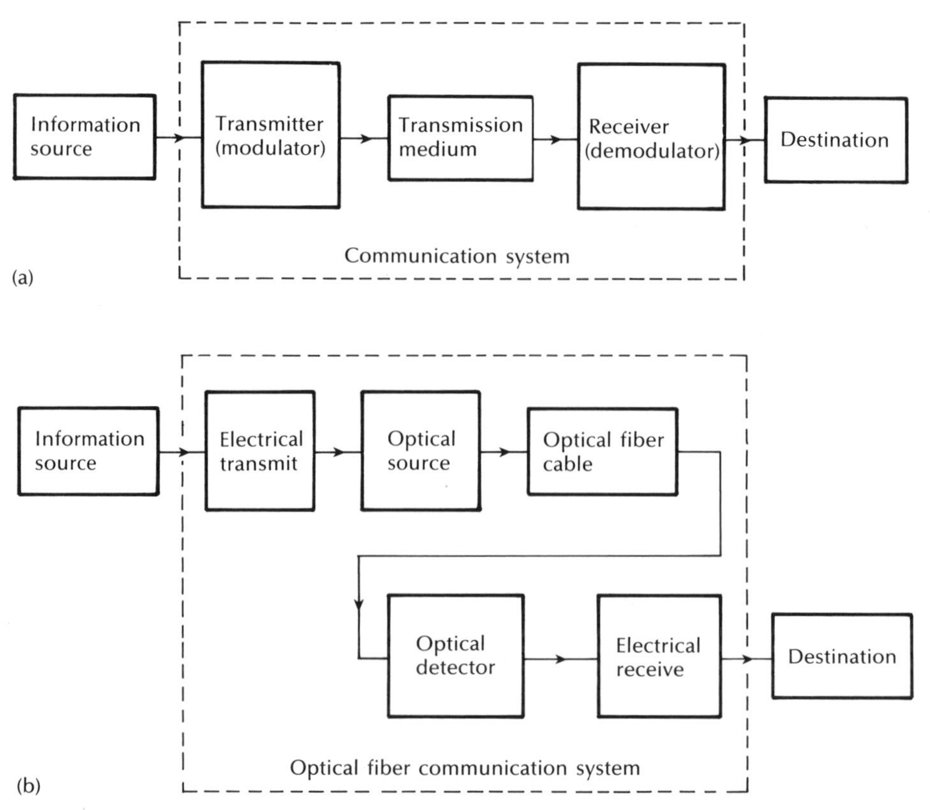

Figure 1.2 (a) The general communication system. (b) The optical fiber communication system.

For optical fiber communications the system shown in Figure 1.2(a) may be considered in slightly greater detail, as given in Figure 1.2(b). In this case the information source provides an electrical signal to a transmitter comprising an electrical stage which drives an optical source to give modulation of the lightwave carrier. The optical source which provides the electrical–optical conversion may be either a semiconductor laser or light emitting diode (LED). The transmission medium consists of an optical fiber cable and the receiver consists of an optical detector which drives a further electrical stage and hence provides demodulation of the optical carrier. Photodiodes (p–n, p–i–n or avalanche) and, in some instances, phototransistors and photoconductors are utilized for the detection of the optical signal and the optical–electrical conversion. Thus there is a requirement for electrical interfacing at either end of the optical link and at present the signal processing is usually performed electrically.*

* Significant developments have already taken place in devices for optical signal processing which may alter this situation in the future (see Chapter 10).

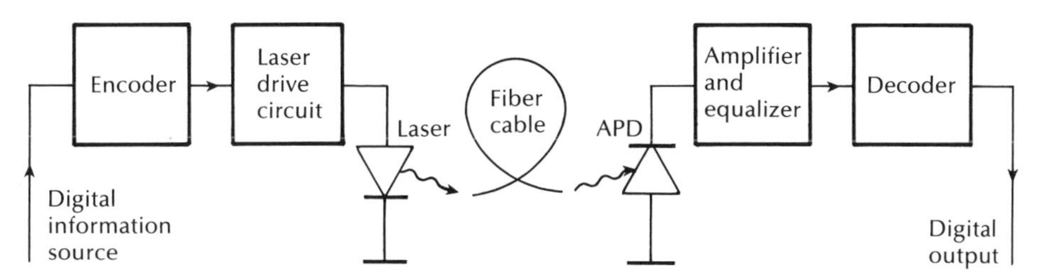

Figure 1.3 A digital optical fiber link using a semiconductor laser source and an avalanche photodiode (APD) detector.

The optical carrier may be modulated using either an analog or digital information signal. In the system shown in Figure 1.2(b) analog modulation involves the variation of the light emitted from the optical source in a continuous manner. With digital modulation, however, discrete changes in the light intensity are obtained (i.e. on–off pulses). Although often simpler to implement, analog modulation with an optical fiber communication system is less efficient, requiring a far higher signal to noise ratio at the receiver than digital modulation. Also, the linearity needed for analog modulation is not always provided by semiconductor optical sources, especially at high modulation frequencies. For these reasons, analog optical fiber communication links are generally limited to shorter distances and lower bandwidths than digital links.

Figure 1.3 shows a block schematic of a typical digital optical fiber link. Initially, the input digital signal from the information source is suitably encoded for optical transmission. The laser drive circuit directly modulates the intensity of the semiconductor laser with the encoded digital signal. Hence a digital optical signal is launched into the optical fiber cable. The avalanche photodiode (APD) detector is followed by a front-end amplifier and equalizer or filter to provide gain as well as linear signal processing and noise bandwidth reduction. Finally, the signal obtained is decoded to give the original digital information. The various elements of this and alternative optical fiber system configurations are discussed in detail in the following chapters. However, at this stage it is instructive to consider the advantages provided by lightwave communication via optical fibers in comparison with other forms of line and radio communication which have brought about the introduction of such systems in many areas throughout the world.

1.3 Advantages of optical fiber communication

Communication using an optical carrier wave guided along a glass fiber has a number of extremely attractive features, several of which were apparent when the technique was originally conceived. Furthermore, the advances in the technology to date have surpassed even the most optimistic predictions, creating additional advantages. Hence it is useful to consider the merits and special features offered by optical fiber communications over more conventional electrical communications. In this context we commence with the originally foreseen advantages and then consider additional features which have become apparent as the technology has been developed.

(a) Enormous potential bandwidth. The optical carrier frequency in the range 10^{13} to 10^{16} Hz (generally in the near infrared around 10^{14} Hz or 10^{5} GHz) yields a far greater potential transmission bandwidth than metallic cable systems (i.e. coaxial cable bandwidth up to around 500 MHz) or even millimetre wave radio systems (i.e. systems currently operating with modulation bandwidths of 700 MHz). At present, the bandwidth available to fiber systems is not fully utilized but modulation at several gigahertz over a hundred kilometres and hundreds of megahertz over three hundred kilometres without intervening electronics (repeaters) is possible. Therefore, the information-carrying capacity of optical fiber systems has proved far superior to the best copper cable systems. By comparison the losses in wideband coaxial cable systems restrict the transmission distance to only a few kilometres at bandwidths over one hundred megahertz.

Although the usable fiber bandwidth will be extended further towards the optical carrier frequency, it is clear that this parameter is limited by the use of a single optical carrier signal. Hence a much enhanced bandwidth utilization for an optical fiber can be achieved by transmitting several optical signals, each at different centre wavelengths, in parallel on the same fiber. This wavelength division multiplexed operation [Ref. 13], particularly with dense packing of the optical wavelengths (or, essentially, fine frequency spacing), offers the potential for a fiber information-carrying capacity which is many orders of magnitude in excess of that obtained using copper cables or a wideband radio system.

(b) Small size and weight. Optical fibers have very small diameters which are often no greater than the diameter of a human hair. Hence, even when such fibers are covered with protective coatings they are far smaller and much lighter than corresponding copper cables. This is a tremendous boon towards the alleviation of duct congestion in cities, as well as allowing for an expansion of signal transmission within mobiles such as aircraft, satellites and even ships.

(c) Electrical isolation. Optical fibers which are fabricated from glass, or sometimes a plastic polymer, are electrical insulators and therefore, unlike their metallic counterparts, they do not exhibit earth loop and interface problems. Furthermore, this property makes optical fiber transmission ideally suited for communication in electrically hazardous environments as the fibers create no arcing or spark hazard at abrasions or short circuits.

(d) Immunity to interference and crosstalk. Optical fibers form a dielectric waveguide and are therefore free from electromagnetic interference (EMI), radiofrequency interference (RFI), or switching transients giving electromagnetic

pulses (EMP). Hence the operation of an optical fiber communication system is unaffected by transmission through an electrically noisy environment and the fiber cable requires no shielding from EMI. The fiber cable is also not susceptible to lightning strikes if used overhead rather than underground. Moreover, it is fairly easy to ensure that there is no optical interference between fibers and hence, unlike communication using electrical conductors, crosstalk is negligible, even when many fibers are cabled together.

(e) Signal security. The light from optical fibers does not radiate significantly and therefore they provide a high degree of signal security. Unlike the situation with copper cables, a transmitted optical signal cannot be obtained from a fiber in a noninvasive manner (i.e. without drawing optical power from the fiber). Therefore, in theory, any attempt to acquire a message signal transmitted optically may be detected. This feature is obviously attractive for military, banking and general data transmission (i.e. computer network) applications.

(f) Low transmission loss. The development of optical fibers over the last twenty years has resulted in the production of optical fiber cables which exhibit very low attenuation or transmission loss in comparison with the best copper conductors. Fibers have been fabricated with losses as low as 0.2 dB km^{-1} (see Section 3.3.2) and this feature has become a major advantage of optical fiber communications. It facilitates the implementation of communication links with extremely wide repeater spacing (long transmission distances without intermediate electronics), thus reducing both system cost and complexity. Together with the already proven modulation bandwidth capability of fiber cable this property provides a totally compelling case for the adoption of optical fiber communication in the majority of long-haul telecommunication applications.

(g) Ruggedness and flexibility. Although protective coatings are essential, optical fibers may be manufactured with very high tensile strengths (see Section 4.7). Perhaps surprisingly for a glassy substance, the fibers may also be bent to quite small radii or twisted without damage. Furthermore. cable structures have been developed (see Section 4.9.4) which have proved flexible, compact and extremely rugged. Taking the size and weight advantage into account, these optical fiber cables are generally superior in terms of storage, transportation, handling and installation to corresponding copper cables, whilst exhibiting at least comparable strength and durability.

(h) System reliability and ease of maintenance. These features primarily stem from the low loss property of optical fiber cables which reduces the requirement for intermediate repeaters or line amplifiers to boost the transmitted signal strength. Hence with fewer repeaters, system reliability is generally enhanced in comparison with conventional electrical conductor systems. Furthermore, the reliability of the optical components is no longer a problem with predicted lifetimes of 20 to 30 years

now quite common. Both these factors also tend to reduce maintenance time and costs.

(i) Potential low cost. The glass which generally provides the optical fiber transmission medium is made from sand – not a scarce resource. So, in comparison with copper conductors, optical fibers offer the potential for low cost line communication. Although over recent years this potential has largely been realized in the costs of the optical fiber transmission medium which for bulk purchases is now becoming competitive with copper wires (i.e. twisted pairs), it has not yet been achieved in all the other component areas associated with optical fiber communications. For example, the costs of high performance semiconductor lasers and detector photodiodes are still relatively high, as well as some of those concerned with the connection technology (demountable connectors, couplers, etc.).

Overall system costs when utilizing optical fiber communication on long-haul links, however, are substantially less than those for equivalent electrical line systems because of the low loss and wideband properties of the optical transmission medium. As indicated in (f), the requirement for intermediate repeaters and the associated electronics is reduced, giving a substantial cost advantage. Although this cost benefit gives a net gain for long-haul links it is not always the case in short-haul applications where the additional cost incurred, due to the electrical–optical conversion (and vice versa), may be a deciding factor. Nevertheless, there are other possible cost advantages in relation to shipping, handling, installation and maintenance, as well as the features indicated in (c) and (d) which may prove significant in the system choice.

The reducing costs of optical fiber communications has not only provided strong competition with electrical line transmission systems, but also for microwave and millimetre wave radio transmission systems. Although these systems are reasonably wideband the relatively short span 'line of sight' transmission necessitates expensive aerial towers at intervals no greater than a few tens of kilometres. Hence optical fiber is fast becoming the dominant transmission medium within the major industrialized societies.

Many advantages are therefore provided by the use of a lightwave carrier within a transmission medium consisting of an optical fiber. The fundamental principles giving rise to these enhanced performance characteristics, together with their practical realization, are described in the following chapters. However, a general understanding of the basic nature and properties of light is assumed. If this is lacking, the reader is directed to the many excellent texts encompassing the topic, a few of which are indicated in Refs. 18 to 25.

References

[1] A. G. Bell, 'Selenium and the photophone', *The Electrician*, pp. 214, 215, 220, 221, 1880.

[2] W. S. Huxford and J. R. Platt,'Survey of near infra-red communication systems', *J. Opt. Soc. Am.*, **38**, pp. 253–268, 1948.

[3] N. C. Beese, 'Light sources for optical communication', *Infrared Phys.*, **1**, pp. 5–16, 1961.

[4] R. M. Gagliardi and S. Karp, *Optical Communications*, John Wiley, 1976.

[5] T. H. Maiman, 'Stimulated optical radiation in ruby', *Nature, Lond.*, **187**, pp. 493–494, 1960.

[6] A. R. Kraemer, 'Free-space optical communications', *Signal*, pp. 26–32, 1977.

[7] K. C. Kao and G. A. Hockham, 'Dielectric fiber surface waveguides for optical frequencies', *Proc. IEE*, **113**(7), pp. 1151–1158, 1966.

[8] A. Werts, 'Propagation de la lumière cohérente dans les fibres optiques', *L'Onde Electrique*, **46**, pp. 967–980, 1966.

[9] R. L. Hartman, J. C. Dyment, C. J. Hwang and H. Kuhn, 'Continuous operation of GaAs–$Ga_xAl_{1-x}As$, double heterostructure lasers with 30 °C half lives exceeding 1000 h', *Appl. Phys. Lett.*, **23**(4), pp. 181–183, 1973.

[10] A. R. Goodwin, J. F. Peters, M. Pion and W. O. Bourne, 'GaAs lasers with consistently low degradation rates at room temperature', *Appl. Phys. Lett.*, **30**(2), pp. 110–113, 1977.

[11] S. E. Miller, 'Overview and summary of progress', in S. E. Miller and I. P. Kaminow (Eds.), *Optical Fiber Telecommunications II*, Academic Press, pp. 1–27, 1988.

[12] J. M. Senior and T. E. Ray, 'Optical fibre communications: the formation of technological strategies in the UK and USA', *Internat. J. of Technol. Management*, **5**(1), pp. 71–88, 1990.

[13] J. M. Senior, 'Wavelength division in optical fibre networks', *Communications Internat.*, **15**(4), pp. 52–54, 59, 1988.

[14] P. Russer, 'Introduction to optical communication', M. J. Howes and D. V. Morgan (Eds.), *Optical Fiber Communications*, pp. 1–26, John Wiley, 1980.

[15] J. E. Midwinter, *Optical Fibres for Transmission*, John Wiley, 1979.

[16] B. Costa, 'Historical remarks', in *Optical Fibre Communication* by the Technical Staff of CSELT, McGraw-Hill, 1981.

[17] C. P. Sandbank (Ed.), *Optical Fibre Communication Systems*, John Wiley, 1980.

[18] H. F. Wolf (Ed.), *Handbook of Fiber Optics Theory and Applications*, Granada, 1981.

[19] F. A. Jenkins and H. E. White, *Fundamentals of Optics* (4th Edn), McGraw-Hill, 1976.

[20] E. Hecht and A. Zajac, *Optics*, Addison-Wesley, 1974.

[21] G. R. Fowles, *Introduction to Modern Optics* (2nd edn), Holt, Rinehart & Winston, 1975.

[22] R. S. Longhurst. *Geometrical and Physical Optics* (3rd edn), Longman, 1973.

[23] F. G. Smith and J. H. Thomson, *Optics*, John Wiley, 1980.

[24] S. G. Lipson and H. Lipson, *Optical Physics* (2nd edn), Cambridge University Press, 1981.

[25] M. Born and E. Wolf, *Principles of Optics* (6th edn), Pergamon Press, 1980.

2

Optical fiber waveguides

2.1 Introduction

The transmission of light via a dielectric waveguide structure was first proposed and investigated at the beginning of the twentieth century. In 1910 Hondros and Debye [Ref. 1] conducted a theoretical study, and experimental work was reported by Schriever in 1920 [Ref. 2]. However, a transparent dielectric rod, typically of silica glass with a refractive index of around 1.5, surrounded by air, proved to be an impractical waveguide due to its unsupported structure (especially when very thin waveguides were considered in order to limit the number of optical modes propagated) and the excessive losses at any discontinuities of the glass–air interface. Nevertheless, interest in the application of dielectric optical waveguides in such areas as optical imaging and medical diagnosis (e.g. endoscopes) led to proposals [Refs. 3 and 4] for a clad dielectric rod in the mid-1950s in order to overcome these problems. This structure is illustrated in Figure 2.1, which shows a

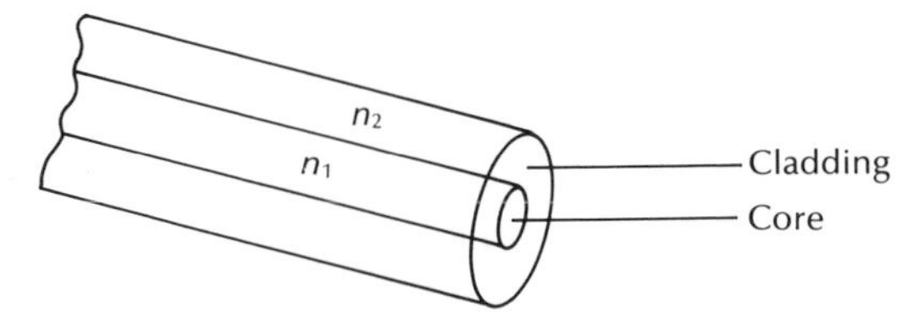

Figure 2.1 Optical fiber waveguide showing the core of refractive index n_1, surrounded by the cladding of slightly lower refractive index n_2.

transparent core with a refractive index n_1 surrounded by a transparent cladding of slightly lower refractive index n_2. The cladding supports the waveguide structure whilst also, when sufficiently thick, substantially reducing the radiation loss into the surrounding air. In essence, the light energy travels in both the core and the cladding allowing the associated fields to decay to a negligible value at the cladding–air interface.

The invention of the clad waveguide structure led to the first serious proposals by Kao and Hockham [Ref. 5] and Werts [Ref. 6], in 1966, to utilize optical fibers as a communications medium, even though they had losses in excess of 1000 dB km^{-1}. These proposals stimulated tremendous efforts to reduce the attenuation by purification of the materials. This has resulted in improved conventional glass refining techniques giving fibers with losses of around 4.2 dB km^{-1} [Ref. 7]. Also, progress in glass refining processes such as depositing vapour-phase reagents to form silica [Ref. 8] has allowed fibers with losses below 1 dB km^{-1} to be fabricated.

Most of this work was focused on the 0.8 to 0.9 μm wavelength band because the first generation optical sources fabricated from gallium aluminum arsenide alloys operated in this region. However, as silica fibers were studied in further detail it became apparent that transmission at longer wavelengths (1.1 to 1.6 μm) would result in lower losses and reduced signal dispersion. This produced a shift in optical fiber source and detector technology in order to provide operation at these longer wavelengths. Hence at longer wavelengths, especially around 1.55 μm, fibers with losses as low as 0.2 dB km^{-1} have been reported [Ref. 9].

Such losses, however, are very close to the theoretical lower limit for silicate glass fiber and, more recently, interest has grown in glass forming systems which can provide low loss transmission in the mid-infrared (2 to 5 μm) and also the far-infrared (8 to 12 μm) optical wavelength regions. At present the best developed of these systems which offers the potential for ultra-low-loss transmission of around 0.01 dB km^{-1} at a wavelength of 2.55 μm is based on fluoride glass [Ref. 10].

In order to appreciate the transmission mechanism of optical fibers with dimensions approximating to those of a human hair, it is necessary to consider the optical waveguiding of a cylindrical glass fiber. Such a fiber acts as an open optical waveguide, which may be analysed utilizing simple ray theory. However, the concepts of geometric optics are not sufficient when considering all types of optical fiber, and electromagnetic mode theory must be used to give a complete picture. The following sections will therefore outline the transmission of light in optical fibers prior to a more detailed discussion of the various types of fiber.

In Section 2.2 we continue the discussion of light propagation in optical fibers using the ray theory approach in order to develop some of the fundamental parameters associated with optical fiber transmission (acceptance angle, numerical aperture, etc.). Furthermore, this provides a basis for the discussion of electromagnetic wave propagation presented in Section 2.3, where the electromagnetic mode theory is developed for the planar (rectangular) waveguide. Following, in Section 2.4, we discuss the waveguiding mechanism within cylindrical fibers prior to consideration of both step and graded index fibers. Finally, in Section 2.5 the theoretical concepts and important parameters (cutoff wavelength, spot size, propagation constant, etc.) associated with optical propagation in single-mode fibers are introduced and approximate techniques to obtain values for these parameters are described.

2.2 Ray theory transmission

2.2.1 Total internal reflection

To consider the propagation of light within an optical fiber utilizing the ray theory model it is necessary to take account of the refractive index of the dielectric medium. The refractive index of a medium is defined as the ratio of the velocity of light in a vacuum to the velocity of light in the medium. A ray of light travels more slowly in an optically dense medium than in one that is less dense, and the refractive index gives a measure of this effect. When a ray is incident on the interface between two dielectrics of differing refractive indices (e.g. glass–air), refraction occurs, as illustrated in Figure 2.2(a). It may be observed that the ray approaching the interface is propagating in a dielectric of refractive index n_1 and is at an angle ϕ_1 to the normal at the surface of the interface. If the dielectric on the other side of the interface has a refractive index n_2 which is less than n_1, then the refraction is such that the ray path in this lower index medium is at an angle ϕ_2 to the normal, where ϕ_2 is greater than ϕ_1. The angles of incidence ϕ_1 and refraction ϕ_2 are related to each other and to the refractive indices of the dielectrics by Snell's law of refraction [Ref. 11], which states that:

$$n_1 \sin \phi_1 = n_2 \sin \phi_2$$

or

$$\frac{\sin \phi_1}{\sin \phi_2} = \frac{n_2}{n_1} \tag{2.1}$$

It may also be observed in Figure 2.2(a) that a small amount of light is reflected back into the originating dielectric medium (partial internal reflection). As n_1 is

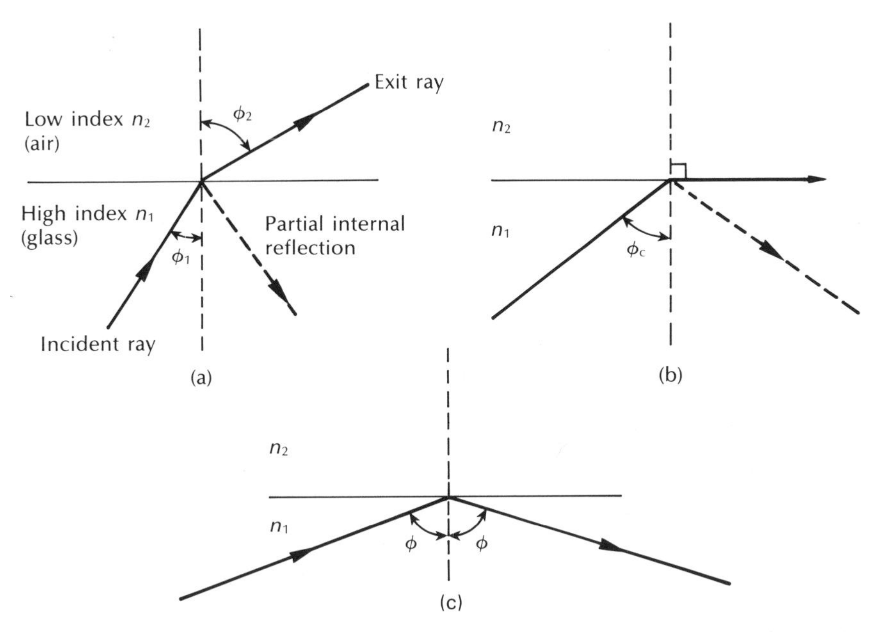

Figure 2.2 Light rays incident on high to low refractive index interface (e.g. glass–air): (a) refraction; (b) the limiting case of refraction showing the critical ray at an angle ϕ_c; (c) total internal reflection where $\phi > \phi_c$.

greater than n_2, the angle of refraction is always greater than the angle of incidence. Thus when the angle of refraction is 90° and the refracted ray emerges parallel to the interface between the dielectrics the angle of incidence must be less than 90°. This is the limiting case of refraction and the angle of incidence is now know as the critical angle ϕ_c, as shown in Figure 2.2(b). From Eq. (2.1) the value of the critical angle is given by:

$$\sin \phi_c = \frac{n_2}{n_1} \tag{2.2}$$

At angles of incidence greater than the critical angle the light is reflected back into the originating dielectric medium (total internal reflection) with high efficiency (around 99.9%). Hence, it may be observed in Figure 2.2(c) that total internal reflection occurs at the interface between two dielectrics of differing refractive indices when light is incident on the dielectric of lower index from the dielectric of higher index, and the angle of incidence of the ray exceeds the critical value. This is the mechanism by which light at a sufficiently shallow angle (less than $90° - \phi_c$) may be considered to propagate down an optical fiber with low loss. Figure 2.3 illustrates the transmission of a light ray in an optical fiber via a series of total internal reflections at the interface of the silica core and the slightly lower refractive index silica cladding. The ray has an angle of incidence ϕ at the interface which is greater than the critical angle and is reflected at the same angle to the normal.

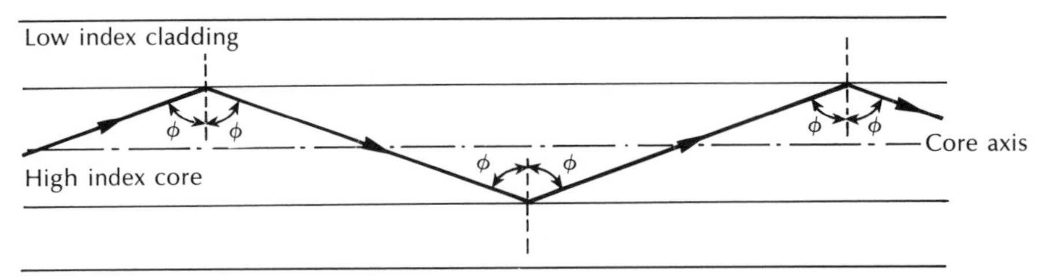

Figure 2.3 The transmission of a light ray in a perfect optical fiber.

The light ray shown in Figure 2.3 is known as a meridional ray as it passes through the axis of the fiber core. This type of ray is the simplest to describe and is generally used when illustrating the fundamental transmission properties of optical fibers. It must also be noted that the light transmission illustrated in Figure 2.3 assumes a perfect fiber, and that any discontinuities or imperfections at the core–cladding interface would probably result in refraction rather than total internal reflection, with the subsequent loss of the light ray into the cladding.

2.2.2 Acceptance angle

Having considered the propagation of light in an optical fiber through total internal reflection at the core–cladding interface, it is useful to enlarge upon the geometric optics approach with reference to light rays entering the fiber. Since only rays with a sufficiently shallow grazing angle (i.e. with an angle to the normal greater than ϕ_c) at the core–cladding interface are transmitted by total internal reflection, it is clear that not all rays entering the fiber core will continue to be propagated down its length.

The geometry concerned with launching a light ray into an optical fiber is shown in Figure 2.4, which illustrates a meridional ray A at the critical angle ϕ_c within the fiber at the core–cladding interface. It may be observed that this ray enters the fiber core at an angle θ_a to the fiber axis and is refracted at the air–core interface before transmission to the core–cladding interface at the critical angle. Hence, any rays which are incident into the fiber core at an angle greater than θ_a will be transmitted to the core–cladding interface at an angle less than ϕ_c, and will not be totally internally reflected. This situation is also illustrated in Figure 2.4, where the incident ray B at an angle greater than θ_a is refracted into the cladding and eventually lost by radiation. Thus for rays to be transmitted by total internal reflection within the fiber core they must be incident on the fiber core within an acceptance cone defined by the conical half angle θ_a. Hence θ_a is the maximum angle to the axis at which light may enter the fiber in order to be propagated, and is often referred to as the acceptance angle* for the fiber.

* θ_a is sometimes referred to as the maximum or total acceptance angle.

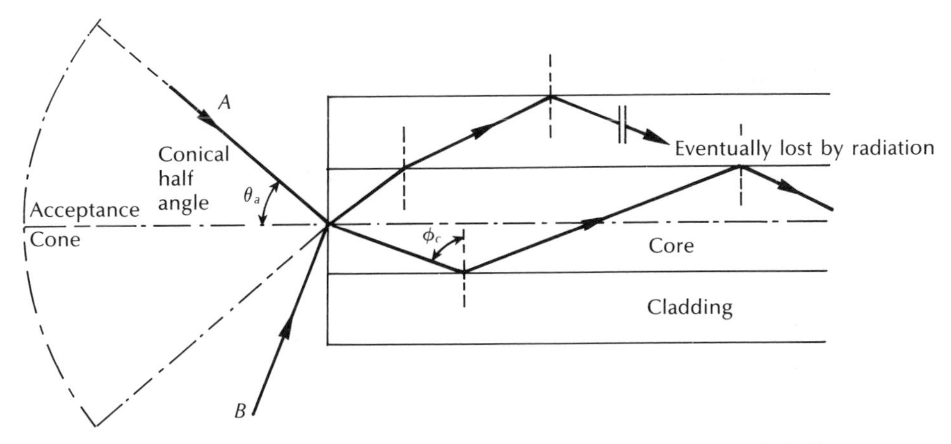

Figure 2.4 The acceptance angle θ_a when launching light into an optical fiber.

If the fiber has a regular cross section (i.e. the core–cladding interfaces are parallel and there are no discontinuities) an incident meridional ray at greater than the critical angle will continue to be reflected and will be transmitted through the fiber. From symmetry considerations it may be noted that the output angle to the axis will be equal to the input angle for the ray, assuming the ray emerges into a medium of the same refractive index from which it was input.

2.2.3 Numerical aperture

The acceptance angle for an optical fiber was defined in the preceding section. However, it is possible to continue the ray theory analysis to obtain a relationship between the acceptance angle and the refractive indices of the three media involved, namely the core, cladding and air. This leads to the definition of a more generally used term, the numerical aperture (NA) of the fiber. It must be noted that within this analysis, as with the preceding discussion of acceptance angle, we are concerned with meridional rays within the fiber.

Figure 2.5 shows a light ray incident on the fiber core at an angle θ_1 to the fiber axis which is less than the acceptance angle for the fiber θ_a. The ray enters the fiber from a medium (air) of refractive index n_0, and the fiber core has a refractive index n_1, which is slightly greater than the cladding refractive index n_2. Assuming the entrance face at the fiber core to be normal to the axis, then considering the refraction at the air–core interface and using Snell's law given by Eq. (2.1)

$$n_0 \sin \theta_1 = n_1 \sin \theta_2 \tag{2.3}$$

Considering the right-angled triangle ABC indicated in Figure 2.5, then:

$$\phi = \frac{\pi}{2} - \theta_2 \tag{2.4}$$

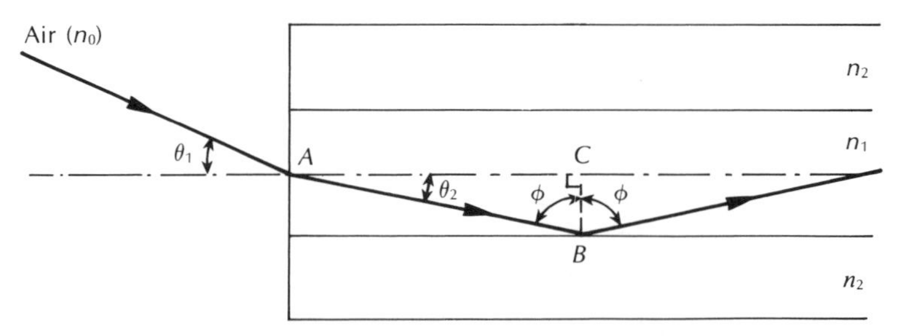

Figure 2.5 The ray path for a meridional ray launched into an optical fiber in air at an input angle less than the acceptance angle for the fiber.

where ϕ is greater than the critical angle at the core–cladding interface. Hence Eq. (2.3) becomes

$$n_0 \sin \theta_1 = n_1 \cos \phi \tag{2.5}$$

Using the trigonometrical relationship $\sin^2 \phi + \cos^2 \phi = 1$, Eq. (2.5) may be written in the form

$$n_0 \sin \theta_1 = n_1 (1 - \sin^2 \phi)^{\frac{1}{2}} \tag{2.6}$$

When the limiting case for total internal reflection is considered, ϕ becomes equal to the critical angle for the core–cladding interface and is given by Eq. (2.2). Also in this limiting case θ_1 becomes the acceptance angle for the fiber θ_a. Combining these limiting cases into Eq. (2.6) gives:

$$n_0 \sin \theta_a = (n_1^2 - n_2^2)^{\frac{1}{2}} \tag{2.7}$$

Equation (2.7), apart from relating the acceptance angle to the refractive indices, serves as the basis for the definition of the important optical fiber parameter, the numerical aperture (NA). Hence the NA is defined as:

$$NA = n_0 \sin \theta_a = (n_1^2 - n_2^2)^{\frac{1}{2}} \tag{2.8}$$

Since the NA is often used with the fiber in air where n_0 is unity, it is simply equal to $\sin \theta_a$. It may also be noted that incident meridional rays over the range $0 \leqslant \theta_1 \leqslant \theta_a$ will be propagated within the fiber.

The numerical aperture may also be given in terms of the relative refractive index difference Δ between the core and the cladding which is defined as:*

$$\Delta = \frac{n_1^2 - n_2^2}{2n_1^2}$$

$$\simeq \frac{n_1 - n_2}{n_1} \quad \text{for } \Delta \ll 1 \tag{2.9}$$

* Sometimes another parameter $\Delta n = n_1 - n_2$ is referred to as the index difference and $\Delta n / n_1$ as the fractional index difference. Hence Δ also approximates to the fractional index difference.

Hence combining Eq. (2.8) with Eq. (2.9) we can write:

$$NA = n_1(2\Delta)^{\frac{1}{2}} \tag{2.10}$$

The relationships given in Eqs. (2.8) and (2.10) for the numerical aperture are a very useful measure of the light-collecting ability of a fiber. They are independent of the fiber core diameter and will hold for diameters as small as 8 μm. However, for smaller diameters they break down as the geometric optics approach is invalid. This is because the ray theory model is only a partial description of the character of light. It describes the direction a plane wave component takes in the fiber but does not take into account interference between such components. When interference phenomena are considered it is found that only rays with certain discrete characteristics propagate in the fiber core. Thus the fiber will only support a discrete number of guided modes. This becomes critical in small core diameter fibers which only support one or a few modes. Hence electromagnetic mode theory must be applied in these cases [Ref. 12].

Example 2.1

A silica optical fiber with a core diameter large enough to be considered by ray theory analysis has a core refractive index of 1.50 and a cladding refractive index of 1.47.

Determine: (a) the critical angle at the core–cladding interface; (b) the NA for the fiber; (c) the acceptance angle in air for the fiber.

Solution: (a) The critical angle ϕ_c at the core–cladding interface is given by Eq. (2.2) where:

$$\phi_c = \sin^{-1}\frac{n_2}{n_1} = \sin^{-1}\frac{1.47}{1.50}$$
$$= 78.5^\circ$$

(b) From Eq. (2.8) the numerical aperture is:

$$NA = (n_1^2 - n_2^2)^{\frac{1}{2}} = (1.50^2 - 1.47^2)^{\frac{1}{2}}$$
$$= (2.25 - 2.16)^{\frac{1}{2}}$$
$$= 0.30$$

(c) Considering Eq. (2.8) the acceptance angle in air θ_a is given by:

$$\theta_a = \sin^{-1} NA = \sin^{-1} 0.30$$
$$= 17.4^\circ$$

Example 2.2

A typical relative refractive index difference for an optical fiber designed for long distance transmission is 1%. Estimate the NA and the solid acceptance angle in air for the fiber when the core index is 1.46. Further, calculate the critical angle at the core–cladding interface within the fiber. It may be assumed that the concepts of geometric optics hold for the fiber.

Solution: Using Eq. (2.10) with $\Delta = 0.01$ gives the numerical aperture as:

$$NA = n_1(2\Delta)^{\frac{1}{2}} = 1.46(0.02)^{\frac{1}{2}} = 0.21$$

For small angles the solid acceptance angle in air ζ is given by:

$$\zeta \simeq \pi\theta_a^2 = \pi \sin^2 \theta_a$$

Hence from Eq. (2.8):

$$\zeta \simeq \pi(NA)^2 = \pi\ 0.04 = 0.13 \text{ rad}$$

Using Eq. (2.9) for the relative refractive index difference Δ gives:

$$\Delta \simeq \frac{n_1 - n_2}{n_1} = 1 - \frac{n_2}{n_1}$$

Hence

$$\frac{n_2}{n_1} = 1 - \Delta = 1 - 0.01 = 0.99$$

From Eq. (2.2) the critical angle at the core–cladding interface is:

$$\phi_c = \sin^{-1}\frac{n_2}{n_1} = \sin^{-1} 0.99 = 81.9^\circ$$

2.2.4 Skew rays

In the preceding sections we have considered the propagation of meridional rays in the optical waveguide. However, another category of ray exists which is transmitted without passing through the fiber axis. These rays, which greatly outnumber the meridional rays, follow a helical path through the fiber, as illustrated in Figure 2.6, and are called skew rays. It is not easy to visualize the skew ray paths in two dimensions but it may be observed from Figure 2.6(b) that the helical path traced through the fiber gives a change in direction of 2γ at each reflection where γ is the angle between the projection of the ray in two dimensions and the radius of the fiber core at the point of reflection. Hence, unlike meridional rays, the point of emergence of skew rays from the fiber in air will depend upon the number of reflections they undergo rather than the input conditions to the fiber. When the light input to the fiber is nonuniform, skew rays will therefore tend to have a smoothing

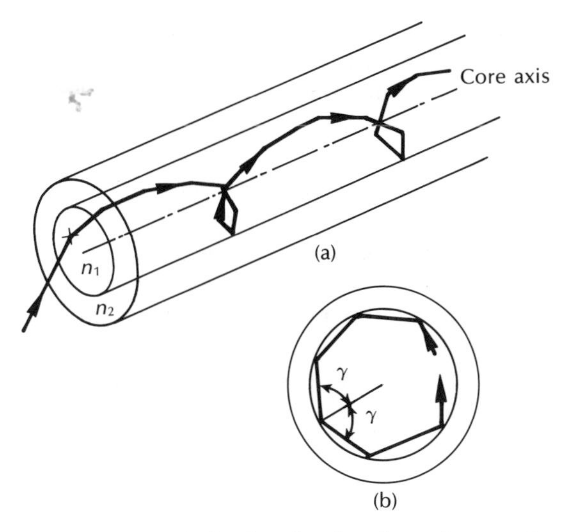

Figure 2.6 The helical path taken by a skew ray in an optical fiber: (a) skew ray path down the fiber; (b) cross-sectional view of the fiber.

effect on the distribution of the light as it is transmitted, giving a more uniform output. The amount of smoothing is dependent on the number of reflections encountered by the skew rays.

A further possible advantage of the transmission of skew rays becomes apparent when their acceptance conditions are considered. In order to calculate the acceptance angle for a skew ray it is necessary to define the direction of the ray in two perpendicular planes. The geometry of the situation is illustrated in Figure 2.7

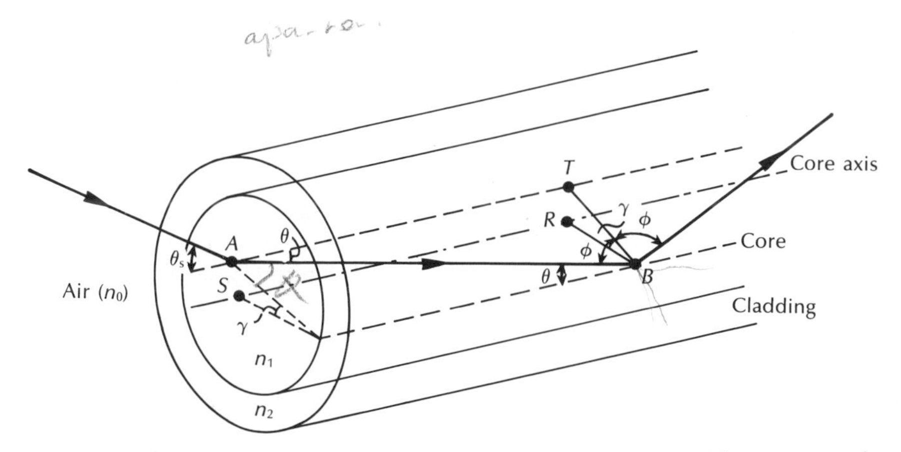

Figure 2.7 The ray path within the fiber core for a skew ray incident at an angle θ_s to the normal at the air–core interface.

where a skew ray is shown incident on the fiber core at the point A, at an angle θ_s to the normal at the fiber end face. The ray is refracted at the air–core interface before travelling to the point B in the same plane. The angles of incidence and reflection at the point B are ϕ, which is greater than the critical angle for the core–cladding interface.

When considering the ray between A and B it is necessary to resolve the direction of the ray path AB to the core radius at the point B. As the incident and reflected rays at the point B are in the same plane, this is simply $\cos\phi$. However, if the two perpendicular planes through which the ray path AB traverses are considered, then γ is the angle between the core radius and the projection of the ray on to a plane BRS normal to the core axis, and θ is the angle between the ray and a line AT drawn parallel to the core axis. Thus to resolve the ray path AB relative to the radius BR in these two perpendicular planes requires multiplication by $\cos\gamma$ and $\sin\theta$.

Hence, the reflection at point B at an angle ϕ may be given by:

$$\cos\gamma\sin\theta = \cos\phi \tag{2.11}$$

Using the trigonometrical relationship $\sin^2\phi + \cos^2\phi = 1$, Eq. (2.11) becomes

$$\cos\gamma\sin\theta = \cos\phi = (1 - \sin^2\phi)^{\frac{1}{2}} \tag{2.12}$$

If the limiting case for total internal reflection is now considered, then ϕ becomes equal to the critical angle ϕ_c for the core–cladding interface and, following Eq. (2.2), is given by $\sin\phi_c = n_2/n_1$. Hence, Eq. (2.12) may be written as:

$$\cos\gamma\sin\theta \leqslant \cos\phi_c = \left(1 - \frac{n_2^2}{n_1^2}\right)^{\frac{1}{2}} \tag{2.13}$$

Furthermore, using Snell's law at the point A, following Eq. (2.1) we can write:

$$n_0 \sin\theta_a = n_1 \sin\theta \tag{2.14}$$

where θ_a represents the maximum input axial angle for meridional rays, as expressed in Section 2.2.2, and θ is the internal axial angle. Hence substituting for $\sin\theta$ from Eq. (2.13) into Eq. (2.14) gives:

$$\sin\theta_{as} = \frac{n_1\cos\phi_c}{n_0\cos\gamma} = \frac{n_1}{n_0\cos\gamma}\left(1 - \frac{n_2^2}{n_1^2}\right)^{\frac{1}{2}} \tag{2.15}$$

where θ_{as} now represents the maximum input angle or acceptance angle for skew rays. It may be noted that the inequality shown in Eq. (2.13) is no longer necessary as all the terms in Eq. (2.15) are specified for the limiting case. Thus the acceptance conditions for skew rays are:

$$n_0\sin\theta_{as}\cos\gamma = (n_1^2 - n_2^2)^{\frac{1}{2}} = NA \tag{2.16}$$

and in the case of the fiber in air ($n_0 = 1$):

$$\sin\theta_{as}\cos\gamma = NA \tag{2.17}$$

Therefore by comparison with Eq. (2.8) derived for meridional rays, it may be noted that skew rays are accepted at larger axial angles in a given fiber than meridional rays, depending upon the value of $\cos \gamma$. In fact, for meridional rays $\cos \gamma$ is equal to unity and θ_{as} becomes equal to θ_a. Thus although θ_a is the maximum conical half angle for the acceptance of meridional rays, it defines the minimum input angle for skew rays. Hence, as may be observed from Figure 2.6, skew rays tend to propagate only in the annular region near the outer surface of the core, and do not fully utilize the core as a transmission medium. However, they are complementary to meridional rays and increase the light-gathering capacity of the fiber. This increased light-gathering ability may be significant for large NA fibers, but for most communication design purposes the expressions given in Eqs. (2.8) and (2.10) for meridional rays are considered adequate.

Example 2.3

An optical fiber in air has an *NA* of 0.4. Compare the acceptance angle for meridional rays with that for skew rays which change direction by 100° at each reflection.

Solution: The acceptance-angle for meridional rays is given by Eq. (2.8) with $n_0 = 1$ as

$$\theta_a = \sin^{-1} NA = \sin^{-1} 0.4 = 23.6^\circ$$

The skew rays change direction by 100° at each reflection, therefore $\gamma = 50^\circ$. Hence using Eq. (2.17) the acceptance angle for skew rays is:

$$\theta_{as} = \sin^{-1}\left(\frac{NA}{\cos \gamma}\right) = \sin^{-1}\left(\frac{0.4}{\cos 50^\circ}\right) = 38.5^\circ$$

In this example, the acceptance angle for the skew rays is about 15° greater than the corresponding angle for meridional rays. However, it must be noted that we have only compared the acceptance angle of one particular skew ray path. When the light input to the fiber is at an angle to the fiber axis, it is possible that γ will vary from zero for meridional rays to 90° for rays which enter the fiber at the core–cladding interface giving acceptance of skew rays over a conical half angle of $\pi/2$ radians.

2.3 Electromagnetic mode theory for optical propagation

2.3.1 Electromagnetic waves

In order to obtain an improved model for the propagation of light in an optical fiber, electromagnetic wave theory must be considered. The basis for the study of electromagnetic wave propagation is provided by Maxwell's equations [Ref. 13]. For a medium with zero conductivity these vector relationships may be written in terms of the electric field **E**, magnetic field **H**, electric flux density **D** and magnetic flux density **B** as the curl equations:

$$\nabla \times \mathbf{E} = -\frac{\partial \mathbf{B}}{\partial t} \tag{2.18}$$

$$\nabla \times \mathbf{H} = \frac{\partial \mathbf{D}}{\partial t} \tag{2.19}$$

and the divergence conditions:

$$\nabla \cdot \mathbf{D} = 0 \qquad \text{(no free charges)} \tag{2.20}$$

$$\nabla \cdot \mathbf{B} = 0 \qquad \text{(no free poles)} \tag{2.21}$$

where ∇ is a vector operator.

The four field vectors are related by the relations:

$$\begin{aligned} \mathbf{D} &= \varepsilon \mathbf{E} \\ \mathbf{B} &= \mu \mathbf{H} \end{aligned} \tag{2.22}$$

where ε is the dielectric permittivity and μ is the magnetic permeability of the medium.

Substituting for **D** and **B** and taking the curl of Eqs. (2.18) and (2.19) gives

$$\nabla \times (\nabla \times \mathbf{E}) = -\mu\varepsilon \frac{\partial^2 \mathbf{E}}{\partial t^2} \tag{2.23}$$

$$\nabla \times (\nabla \times \mathbf{H}) = -\mu\varepsilon \frac{\partial^2 \mathbf{H}}{\partial t^2} \tag{2.24}$$

Then using the divergence conditions of Eqs. (2.20) and (2.21) with the vector identity

$$\nabla \times (\nabla \times \mathbf{Y}) = \nabla(\nabla \cdot \mathbf{Y}) - \nabla^2(\mathbf{Y})$$

we obtain the nondispersive wave equations:

$$\nabla^2 \mathbf{E} = \mu\varepsilon \frac{\partial^2 \mathbf{E}}{\partial t^2} \tag{2.25}$$

and

$$\nabla^2 \mathbf{H} = \mu\varepsilon \frac{\partial^2 \mathbf{H}}{\partial t^2} \tag{2.26}$$

where ∇^2 is the Laplacian operator. For rectangular Cartesian and cylindrical polar coordinates the above wave equations hold for each component of the field vector, every component satisfying the scalar wave equation:

$$\nabla^2\psi = \frac{1}{v_p^2}\frac{\partial^2\psi}{\partial t^2} \tag{2.27}$$

where ψ may represent a component of the **E** or **H** field and v_p is the phase velocity (velocity of propagation of a point of constant phase in the wave) in the dielectric medium. It follows that

$$v_p = \frac{1}{(\mu\varepsilon)^{\frac{1}{2}}} = \frac{1}{(\mu_r\mu_0\varepsilon_r\varepsilon_0)^{\frac{1}{2}}} \tag{2.28}$$

where μ_r and ε_r are the relative permeability and permittivity for the dielectric medium and μ_0 and ε_0 are the permeability and permittivity of free space. The velocity of light in free space c is therefore

$$c = \frac{1}{(\mu_0\varepsilon_0)^{\frac{1}{2}}} \tag{2.29}$$

If planar waveguides, described by rectangular Cartesian coordinates (x, y, z), or circular fibers, described by cylindrical polar coordinates (r, ϕ, z), are considered, then the Laplacian operator takes the form:

$$\nabla^2\psi = \frac{\partial^2\psi}{\partial x^2} + \frac{\partial^2\psi}{\partial y^2} + \frac{\partial^2\psi}{\partial z^2} \tag{2.30}$$

or

$$\nabla^2\psi = \frac{\partial^2\psi}{\partial r^2} + \frac{1}{r}\frac{\partial\psi}{\partial r} + \frac{1}{r^2}\frac{\partial^2\psi}{\partial\phi^2} + \frac{\partial^2\psi}{\partial z^2} \tag{2.31}$$

respectively. It is necessary to consider both these forms for a complete treatment of optical propagation in the fiber, although many of the properties of interest may be dealt with using Cartesian coordinates.

The basic solution of the wave equation is a sinusoidal wave, the most important form of which is a uniform plane wave given by:

$$\psi = \psi_0 \exp \mathrm{j}(\omega t - \mathbf{k}\cdot\mathbf{r}) \tag{2.32}$$

where ω is the angular frequency of the field, t is the time, $\mathbf{k}$ is the propagation vector which gives the direction of propagation and the rate of change of phase with distance, whilst the components of $\mathbf{r}$ specify the coordinate point at which the field is observed. When λ is the optical wavelength in a vacuum, the magnitude of the propagation vector or the vacuum phase propagation constant k (where $k = |\mathbf{k}|$) is given by:

$$k = \frac{2\pi}{\lambda} \tag{2.33}$$

It should be noted that in this case k is also referred to as the free space wave number.

2.3.2 Modes in a planar guide

The planar guide is the simplest form of optical waveguide. We may assume it consists of a slab of dielectric with refractive index n_1 sandwiched between two regions of lower refractive index n_2. In order to obtain an improved model for optical propagation it is useful to consider the interference of plane wave components within this dielectric waveguide.

The conceptual transition from ray to wave theory may be aided by consideration of a plane monochromatic wave propagating in the direction of the ray path within the guide (see Figure 2.8(a)). As the refractive index within the guide is n_1, the optical wavelength in this region is reduced to λ/n_1, whilst the vacuum propagation constant is increased to $n_1 k$. When θ is the angle between the wave propagation vector or the equivalent ray and the guide axis, the plane wave can be resolved into two component plane waves propagating in the z and x directions, as shown in Figure 2.8(a). The component of the phase propagation constant in the z direction β_z is given by:

$$\beta_z = n_1 k \cos\theta \tag{2.34}$$

The component of the phase propagation constant in the x direction β_x is:

$$\beta_x = n_1 k \sin\theta \tag{2.35}$$

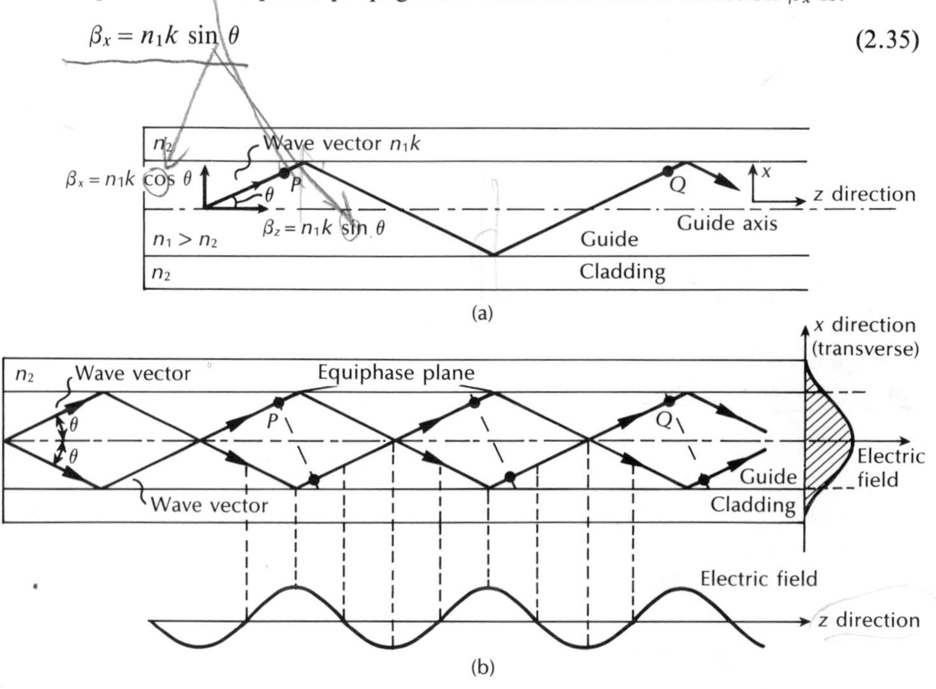

Figure 2.8 The formation of a mode in a planar dielectric guide: (a) a plane wave propagating in the guide shown by its wave vector or equivalent ray – the wave vector is resolved into components in the z and x directions; (b) the interference of plane waves in the guide forming the lowest order mode ($m = 0$).

The component of the plane wave in the x direction is reflected at the interface between the higher and lower refractive index media. When the total phase change* after two successive reflections at the upper and lower interfaces (between the points P and Q) is equal to $2m\pi$ radians, where m is an integer, then constructive interference occurs and a standing wave is obtained in the x direction. This situation is illustrated in Figure 2.8(b), where the interference of two plane waves is shown. In this illustration it is assumed that the interference forms the lowest order (where $m = 0$) standing wave, where the electric field is a maximum at the centre of the guide decaying towards zero at the boundary between the guide and cladding. However, it may be observed from Figure 2.8(b) that the electric field penetrates some distance into the cladding, a phenomenon which is discussed in Section 2.3.4.

Nevertheless, the optical wave is effectively confined within the guide and the electric field distribution in the x direction does not change as the wave propagates in the z direction. The sinusoidally varying electric field in the z direction is also shown in Figure 2.8(b). The stable field distribution in the x direction with only a periodic z dependence is known as a mode. A specific mode is obtained only when the angle between the propagation vectors or the rays and the interface have a particular value, as indicated in Figure 2.8(b). In effect, Eqs. (2.34) and (2.35) define a group or congruence of rays which in the case described represents the lowest order mode. Hence the light propagating within the guide is formed into discrete modes, each typified by a distinct value of θ. These modes have a periodic z dependence of the form $\exp(-j\beta_z z)$ where β_z becomes the propagation constant for the mode as the modal field pattern is invariant except for a periodic z dependence. Hence, for notational simplicity, and in common with accepted practice, we denote the mode propagation constant by β, where $\beta = \beta_z$. If we now assume a time dependence for the monochromatic electromagnetic light field with angular frequency ω of $\exp(j\omega t)$, then the combined factor $\exp j(\omega t - \beta z)$ describes a mode propagating in the z direction.

To visualize the dominant modes propagating in the z direction we may consider plane waves corresponding to rays at different specific angles in the planar guide. These plane waves give constructive interference to form standing wave patterns across the guide following a sine or cosine formula. Figure 2.9 shows examples of such rays for $m = 1, 2, 3$, together with the electric field distributions in the x direction. It may be observed that m denotes the number of zeros in this transverse field pattern. In this way m signifies the order of the mode and is known as the mode number.

When light is described as an electromagnetic wave it consists of a periodically varying electric field **E** and magnetic field **H** which are orientated at right angles to each other. The transverse modes shown in Figure 2.9 illustrate the case when the

* It should be noted that there is a phase shift on reflection of the plane wave at the interface as well as a phase change with distance travelled. The phase shift on reflection at a dielectric interface is dealt with in Section 2.3.4.

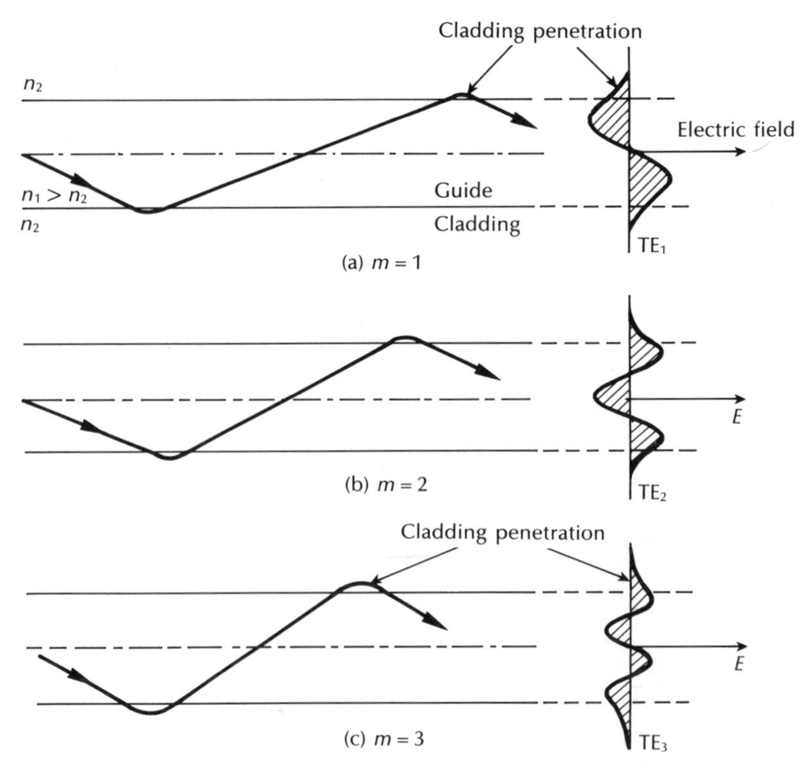

Figure 2.9 Physical model showing the ray propagation and the corresponding transverse electric (TE) field patterns of three lower order models ($m = 1, 2, 3$) in the planar dielectric guide.

electric field is perpendicular to the direction of propagation and hence $E_z = 0$, but a corresponding component of the magnetic field **H** is in the direction of propagation. In this instance the modes are said to be transverse electric (TE). Alternatively, when a component of the **E** field is in the direction of propagation, but $H_z = 0$, the modes formed are called transverse magnetic (TM). The mode numbers are incorporated into this nomenclature by referring to the TE_m and TM_m modes, as illustrated for the transverse electric modes shown in Figure 2.9. When the total field lies in the transverse plane, transverse electromagnetic (TEM) waves exist where both E_z and H_z are zero. However, although TEM waves occur in metallic conductors (e.g. coaxial cables) they are seldom found in optical waveguides.

2.3.3 Phase and group velocity

Within all electromagnetic waves, whether plane or otherwise, there are points of constant phase. For plane waves these constant phase points form a surface which is referred to as a wavefront. As a monochromatic light wave propagates along a waveguide in the z direction these points of constant phase travel at a phase velocity v_p given by:

$$v_p = \frac{\omega}{\beta} \tag{2.36}$$

where ω is the angular frequency of the wave. However, it is impossible in practice to produce perfectly monochromatic light waves, and light energy is generally composed of a sum of plane wave components of different frequencies. Often the situation exists where a group of waves with closely similar frequencies propagate so that their resultant forms a packet of waves. The formation of such a wave packet resulting from the combination of two waves of slightly different frequency propagating together is illustrated in Figure 2.10. This wave packet does not travel at the phase velocity of the individual waves but is observed to move at a group velocity v_g given by

$$v_g = \frac{\delta\omega}{\delta\beta} \tag{2.37}$$

The group velocity is of greatest importance in the study of the transmission characteristics of optical fibers as it relates to the propagation characteristics of observable wave groups or packets of light.

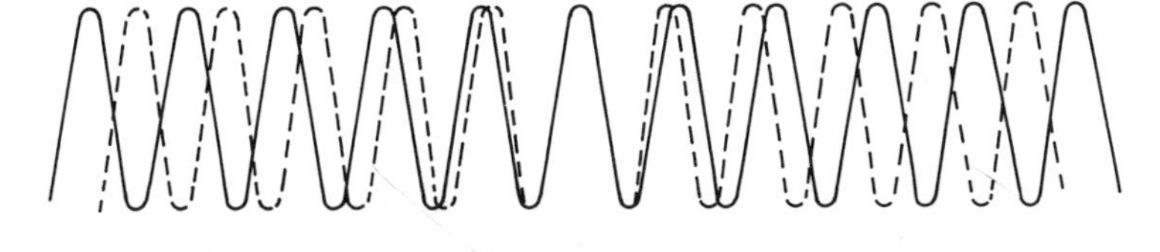

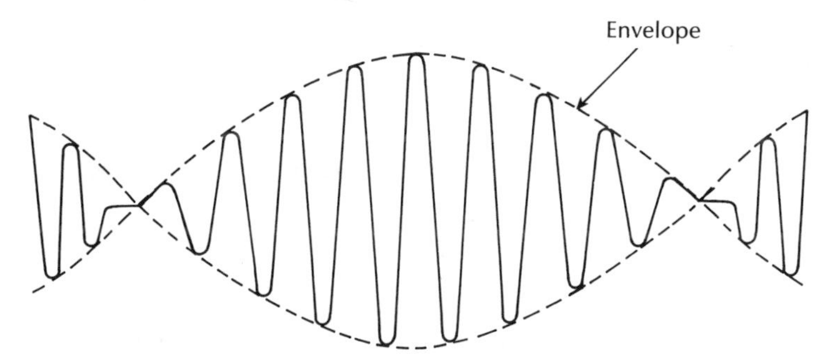

Figure 2.10 The formation of a wave packet from the combination of two waves with nearly equal frequencies. The envelope of the wave package or group of waves travels at a group velocity v_g.

If propagation in an infinite medium of refractive index n_1 is considered, then the propagation constant may be written as:

$$\beta = n_1 \frac{2\pi}{\lambda} = \frac{n_1 \omega}{c} \tag{2.38}$$

where c is the velocity of light in free space. Equation (2.38) follows from Eqs. (2.33) and (2.34) where we assume propagation in the z direction only and hence $\cos\theta$ is equal to unity. Using Eq. (2.36) we obtain the following relationship for the phase velocity

$$v_p = \frac{c}{n_1} \tag{2.39}$$

Similarly, employing Eq. (2.37), where in the limit $\delta\omega/\delta\beta$ becomes $d\omega/d\beta$, the group velocity:

$$\begin{aligned} v_g &= \frac{d\lambda}{d\beta} \cdot \frac{d\omega}{d\lambda} = \frac{d}{d\lambda}\left(n_1 \frac{2\pi}{\lambda}\right)^{-1}\left(\frac{-\omega}{\lambda}\right) \\ &= \frac{-\omega}{2\pi\lambda}\left(\frac{1}{\lambda}\frac{dn_1}{d\lambda} - \frac{n_1}{\lambda^2}\right)^{-1} \\ &= \frac{c}{\left(n_1 - \lambda \dfrac{dn_1}{d\lambda}\right)} = \frac{c}{N_g} \end{aligned} \tag{2.40}$$

The parameter N_g is known as the group index of the guide.

2.3.4 Phase shift with total internal reflection and the evanescent field

The discussion of electromagnetic wave propagation in the planar waveguide given in Section 2.3.2 drew attention to certain phenomena that occur at the guide–cladding interface which are not apparent from ray theory considerations of optical propagation. In order to appreciate these phenomena it is necessary to use the wave theory model for total internal reflection at a planar interface. This is illustrated in Figure 2.11, where the arrowed lines represent wave propagation vectors and a component of the wave energy is shown to be transmitted through the interface into the cladding. The wave equation in Cartesian coordinates for the electric field in a lossless medium is:

$$\nabla^2 \mathbf{E} = \mu\varepsilon \frac{\partial^2 \mathbf{E}}{\partial t^2} = \frac{\partial^2 \mathbf{E}}{\partial x^2} + \frac{\partial^2 \mathbf{E}}{\partial y^2} + \frac{\partial^2 \mathbf{E}}{\partial z^2} \tag{2.41}$$

As the guide–cladding interface lies in the y–z plane and the wave is incident in the x–z plane on to the interface, then $\partial/\partial y$ may be assumed to be zero. Since the phase fronts must match all points along the interface in the z direction, the three waves shown in Figure 2.11 will have the same propagation constant β in this

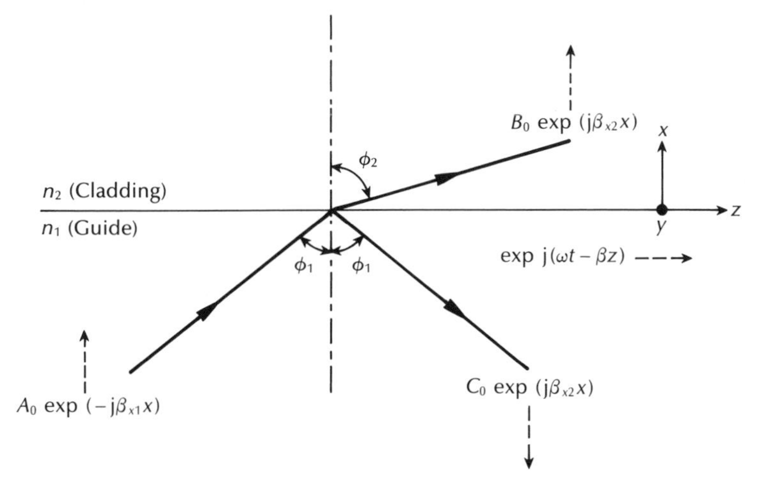

Figure 2.11 A wave incident on the guide–cladding interface of a planar dielectric waveguide. The wave vectors of the incident, transmitted and reflected waves are indicated (solid arrowed lines) together with their components in the z and x directions (dashed arrowed lines).

direction. Therefore from the discussion of Section 2.3.2 the wave propagation in the z direction may be described by exp $j(\omega t - \beta z)$. In addition, there will also be propagation in the x direction. When the components are resolved in this plane:

$$\beta_{x1} = n_1 k \cos \phi_1 \tag{2.42}$$

$$\beta_{x2} = n_2 k \cos \phi_2 \tag{2.43}$$

where β_{x1} and β_{x2} are propagation constants in the x direction for the guide and cladding respectively. Thus the three waves in the waveguide indicated in Figure 2.11, the incident, the transmitted and the reflected, with amplitudes A, B and C, respectively, will have the forms:

$$A = A_0 \exp - (j\beta_{x1} x) \exp j(\omega t - \beta z) \tag{2.44}$$

$$B = B_0 \exp - (j\beta_{x2} x) \exp j(\omega t - \beta z) \tag{2.45}$$

$$C = C_0 \exp (j\beta_{x1} x) \exp j(\omega t - \beta z) \tag{2.46}$$

Using the simple trigonometrical relationship $\cos^2 \phi + \sin^2 \phi = 1$:

$$\beta_{x1}^2 = (n_1^2 k^2 - \beta^2) = -\xi_1^2 \tag{2.47}$$

and

$$\beta_{x2}^2 = (n_2^2 k^2 - \beta^2) = -\xi_2^2 \tag{2.48}$$

When an electromagnetic wave is incident upon an interface between two dielectric media, Maxwell's equations require that both the tangential components of **E** and **H** and the normal components of **D** ($=\varepsilon\mathbf{E}$) and **B** ($=\mu\mathbf{H}$) are continuous across the boundary. If the boundary is defined at $x=0$ we may consider the cases of the transverse electric (TE) and transverse magnetic (TM) modes.

Initially, let us consider the TE field at the boundary. When Eqs. (2.44) and (2.46) are used to represent the electric field components in the y direction E_y and the boundary conditions are applied, then the normal components of the **E** and **H** fields at the interface may be equated giving

$$A_0 + C_0 = B_0 \tag{2.49}$$

Furthermore it can be shown (see Appendix A) that an electric field component in the y direction is related to the tangential magnetic field component H_z following

$$H_z = \frac{\mathrm{j}}{\mu_r\ \mu_0\omega}\frac{\partial E_y}{\partial x} \tag{2.50}$$

Applying the tangential boundary conditions and equating H_z by differentiating E_y gives:

$$-\beta_{x1}A_0 + \beta_{x2}C_0 = -\beta_{x2}B_0 \tag{2.51}$$

Algebraic manipulation of Eqs. (2.49) and (2.51) provides the following results:

$$C_0 = A_0\left(\frac{\beta_{x1} - \beta_{x2}}{\beta_{x1} + \beta_{x2}}\right) = A_0 r_{ER} \tag{2.52}$$

$$B_0 = A_0\left(\frac{2\beta_{x1}}{\beta_{x1} + \beta_{x2}}\right) = A_0 r_{ET} \tag{2.53}$$

where r_{ER} and r_{ET} are the reflection and transmission coefficients for the **E** field at the interface respectively. The expressions obtained in Eqs. (2.52) and (2.53) correspond to the Fresnel relationships [Ref. 11] for radiation polarized perpendicular to the interface (**E** polarization).

When both β_{x1} and β_{x2} are real it is clear that the reflected wave C is in phase with the incident wave A. This corresponds to partial reflection of the incident beam. However, as ϕ_1 is increased the component β_z (i.e. β) increases and, following Eqs. (2.47) and (2.48), the components β_{x1} and β_{x2} decrease. Continuation of this process results in β_{x2} passing through zero, a point which is signified by ϕ_1 reaching the critical angle for total internal reflection. If ϕ_1 is further increased the component β_{x2} becomes imaginary and we may write it in the form $-\mathrm{j}\xi_2$. During this process β_{x1} remains real because we have assumed that $n_1 > n_2$. Under the conditions of total internal reflection Eq. (2.52) may therefore be written as:

$$C_0 = A_0\left(\frac{\beta_{x1} + \mathrm{j}\xi_2}{\beta_{x2} - \mathrm{j}\xi_2}\right) = A_0 \exp 2\mathrm{j}\delta_E \tag{2.54}$$

where we observe there is a phase shift of the reflected wave relative to the incident wave. This is signified by δ_E which is given by:

$$\tan \delta_E = \frac{\xi_2^2}{\beta_{x1}} \tag{2.55}$$

Furthermore, the modulus of the reflected wave is identical to the modulus of the incident wave ($|C_0| = |A_0|$). The curves of the amplitude reflection coefficient $|r_{ER}|$ and phase shift on reflection, against angle of incidence ϕ_1, for TE waves incident on a glass–air interface are displayed in Figure 2.12 [Ref. 14]. These curves illustrate the above results, where under conditions of total internal reflection the reflected wave has an equal amplitude to the incident wave, but undergoes a phase shift corresponding to δ_E degrees.

A similar analysis may be applied to the TM modes at the interface, which leads to expressions for reflection and transmission of the form [Ref. 14]:

$$C_0 = A_0\left(\frac{\beta_{x1}n_2^2 - \beta_{x2}n_1^2}{\beta_{x1}n_2^2 + \beta_{x2}n_1^2}\right) = A_0 r_{HR} \tag{2.56}$$

and

$$B_0 = A_0\left(\frac{2\beta_{x1}n_2^2}{\beta_{x1}n_2^2 + \beta_{x2}n_1^2}\right) = A_0 r_{HT} \tag{2.57}$$

where r_{HR} and r_{HT} are, respectively, the reflection and transmission coefficients for the **H** field at the interface. Again, the expressions given in Eqs. (2.56) and (2.57) correspond to Fresnel relationships [Ref. 11], but in this case they apply to radiation polarized parallel to the interface (**H** polarization). Furthermore,

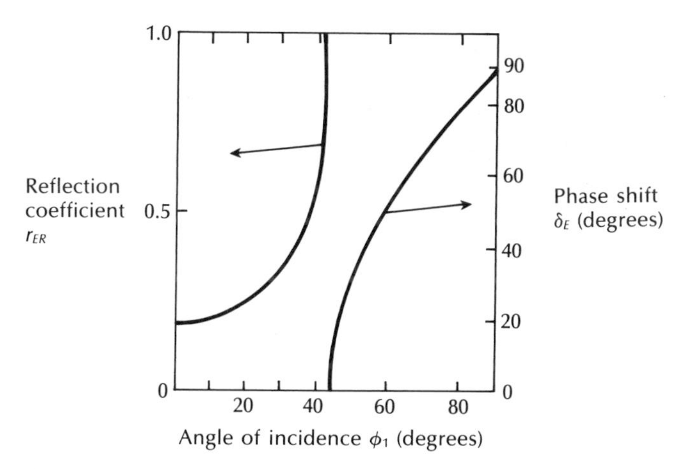

Figure 2.12 Curves showing the reflection coefficient and phase shift on reflection for transverse electric waves against the angle of incidence for a glass–air interface ($n_1 = 1.5$, $n_2 = 1.0$). Reproduced with permission from J. E. Midwinter, *Optical Fibers for Transmission*, John Wiley & Sons Inc., 1979.

considerations of an increasing angle of incidence ϕ_1, such that β_{x2} goes to zero and then becomes imaginary, again results in a phase shift when total internal reflection occurs. However, in this case a different phase shift is obtained corresponding to

$$C_0 = A_0 \exp(2j\delta_H) \tag{2.58}$$

where

$$\tan \delta_H = \left(\frac{n_1}{n_2}\right)^2 \tan \delta_E \tag{2.59}$$

Thus the phase shift obtained on total internal reflection is dependent upon both the angle of incidence and the polarization (either TE or TM) of the radiation.

The second phenomenon of interest under conditions of total internal reflection is the form of the electric field in the cladding of the guide. Before the critical angle for total internal reflection is reached, and hence when there is only partial reflection, the field in the cladding is of the form given by Eq. (2.45). However, as indicated previously, when total internal reflection occurs, β_{x2} becomes imaginary and may be written as $-j\xi_2$. Substituting for β_{x2} in Eq. (2.45) gives the transmitted wave in the cladding as:

$$B = B_0 \exp(-\xi_2 x) \exp j(\omega t - \beta z) \tag{2.60}$$

Thus the amplitude of the field in the cladding is observed to decay exponentially* in the x direction. Such a field, exhibiting an exponentially decaying amplitude, is often referred to as an evanescent field. Figure 2.13 shows a diagrammatic representation of the evanescent field. A field of this type stores energy and transports it in the direction of propagation (z) but does not transport energy in the

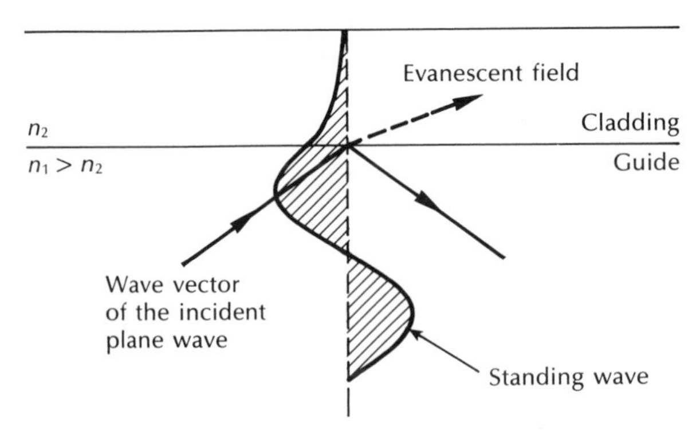

Figure 2.13 The exponentially decaying evanescent field in the cladding of the optical waveguide.

* It should be noted that we have chosen the sign of ξ_2 so that the exponential field decays rather than grows with distance into the cladding. In this case a growing exponential field is a physically improbable solution.

transverse direction (x). Nevertheless, the existence of an evanescent field beyond the plane of reflection in the lower index medium indicates that optical energy is transmitted into the cladding.

The penetration of energy into the cladding underlines the importance of the choice of cladding material. It gives rise to the following requirements:

1. The cladding should be transparent to light at the wavelengths over which the guide is to operate.
2. Ideally, the cladding should consist of a solid material in order to avoid both damage to the guide and the accumulation of foreign matter on the guide walls. These effects degrade the reflection process by interaction with the evanescent field. This in part explains the poor performance (high losses) of early optical waveguides with air cladding.
3. The cladding thickness must be sufficient to allow the evanescent field to decay to a low value or losses from the penetrating energy may be encountered. In many cases, however, the magnitude of the field falls off rapidly with distance from the guide–cladding interface. This may occur within distances equivalent to a few wavelengths of the transmitted light.

Therefore, the most widely used optical fibers consist of a core and cladding, both made of glass. The cladding refractive index is thus higher than would be the case with liquid or gaseous cladding giving a lower numerical aperture for the fiber, but it provides a far more practical solution.

2.3.5 Goos–Haenchen shift

The phase change incurred with the total internal reflection of a light beam on a planar dielectric interface may be understood from physical observation. Careful examination shows that the reflected beam is shifted laterally from the trajectory predicted by simple ray theory analysis, as illustrated in Figure 2.14. This lateral displacement is known as the Goos–Haenchen shift, after its first observers.

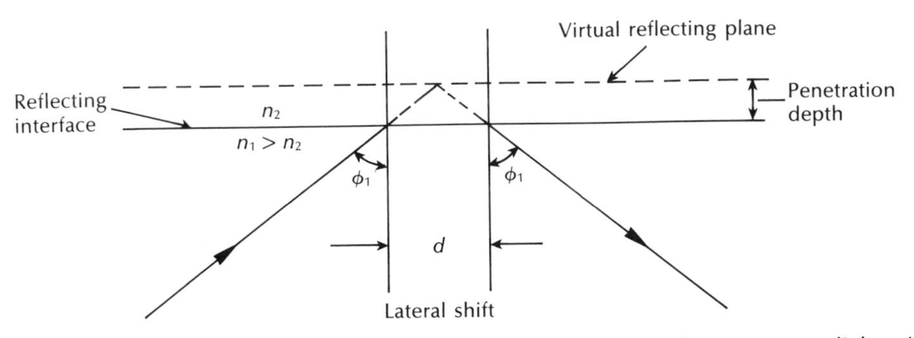

Figure 2.14 The lateral displacement of a light beam on reflection at a dielectric interface (Goos–Haenchen shift).

The geometric reflection appears to take place at a virtual reflecting plane which is parallel to the dielectric interface in the lower index medium, as indicated in Figure 2.14. Utilizing wave theory it is possible to determine this lateral shift [Ref. 14] although it is very small ($d \simeq 0.06$ to $0.10\ \mu m$ for a silvered glass interface at a wavelength of $0.55\ \mu m$) and difficult to observe. However, this concept provides an important insight into the guidance mechanism of dielectric optical waveguides.

2.4 Cylindrical fiber

2.4.1 Modes

The exact solution of Maxwell's equations for a cylindrical homogeneous core dielectric waveguide* involves much algebra and yields a complex result [Ref. 15]. Although the presentation of this mathematics is beyond the scope of this text, it is useful to consider the resulting modal fields. In common with the planar guide (Section 2.3.2), TE (where $E_z = 0$) and TM (where $H_z = 0$) modes are obtained within the dielectric cylinder. The cylindrical waveguide, however, is bounded in two dimensions rather than one. Thus two integers, l and m, are necessary in order to specify the modes, in contrast to the single integer (m) required for the planar guide. For the cylindrical waveguide we therefore refer to TE_{lm} and TM_{lm} modes. These modes correspond to meridional rays (see Section 2.2.1) travelling within the fiber. However, hybrid modes where E_z and H_z are nonzero also occur within the cylindrical waveguide. These modes which result from skew ray propagation (see Section 2.2.4) within the fiber are designated HE_{lm} and EH_{lm} depending upon whether the components of **H** or **E** make the larger contribution to the transverse (to the fiber axis) field. Thus an exact description of the modal fields in a step index fiber proves somewhat complicated.

Fortunately, the analysis may be simplified when considering optical fibers for communication purposes. These fibers satisfy the weakly guiding approximation [Ref. 16] where the relative index difference $\Delta \ll 1$. This corresponds to small grazing angles θ in Eq. (2.34). In fact Δ is usually less than 0.03 (3%) for optical communications fibers. For weakly guiding structures with dominant forward propagation, mode theory gives dominant transverse field components. Hence approximate solutions for the full set of HE, EH, TE and TM modes may be given by two linearly polarized components [Ref. 16]. These linearly polarized (LP) modes are not exact modes of the fiber except for the fundamental (lowest order) mode. However, as Δ in weakly guiding fibers is very small, then HE–EH mode pairs occur which have almost identical propagation constants. Such modes are said to be degenerate. The superpositions of these degenerating modes characterized by a common propagation constant correspond to particular LP modes regardless of

* This type of optical waveguide with a constant refractive index core is known as a step index fiber (see Section 2.4.3).

their HE, EH, TE or TM field configurations. This linear combination of degenerate modes obtained from the exact solution produces a useful simplification in the analysis of weakly guiding fibers.

The relationship between the traditional HE, EH, TE and TM mode designations and the LP_{lm} mode designations are shown in Table 2.1. The mode subscripts l and m are related to the electric field intensity profile for a particular LP mode (see Figure 2.15(d)). There are in general $2l$ field maxima around the circumference of the fiber core and m field maxima along a radius vector. Furthermore, it may be observed from Table 2.1 that the notation for labelling the HE and EH modes has changed from that specified for the exact solution in the cylindrical waveguide mentioned previously. The subscript l in the LP notation now corresponds to HE and EH modes with labels $l + 1$ and $l - 1$ respectively.

The electric field intensity profiles for the lowest three LP modes, together with the electric field distribution of their constituent exact modes, are shown in Figure 2.15. It may be observed from the field configurations of the exact modes that the field strength in the transverse direction (E_x or E_y) is identical for the modes which belong to the same LP mode. Hence the origin of the term 'linearly polarized'.

Using Eq. (2.31) for the cylindrical homogeneous core waveguide under the weak guidance conditions outlined above, the scalar wave equation can be written in the form [Ref. 17]:

$$\frac{d^2\psi}{dr^2} + \frac{1}{r}\frac{d\psi}{dr} + \frac{1}{r^2}\frac{d^2\psi}{d\phi^2} + (n_1^2k^2 - \beta^2)\psi = 0 \tag{2.61}$$

where ψ is the field (**E** or **H**), n_1 is the refractive index of the fiber core, k is the propagation constant for light in a vacuum, and r and ϕ are cylindrical coordinates. The propagation constants of the guided modes β lie in the range:

$$n_2k < \beta < n_1k \tag{2.62}$$

Table 2.1 Correspondence between the lower order in linearly polarized modes and the traditional exact modes from which they are formed

Linearly polarized	Exact
LP_{01}	HE_{11}
LP_{11}	HE_{21}, TE_{01}, TM_{01}
LP_{21}	HE_{31}, EH_{11}
LP_{02}	HE_{12}
LP_{31}	HE_{41}, EH_{21}
LP_{12}	HE_{22}, TE_{02}, TM_{02}
LP_{lm}	HE_{2m}, TE_{0m}, TM_{0m}
LP_{lm} ($l \neq 0$ or 1)	$HE_{l+1,m}$, $EH_{l-1,m}$

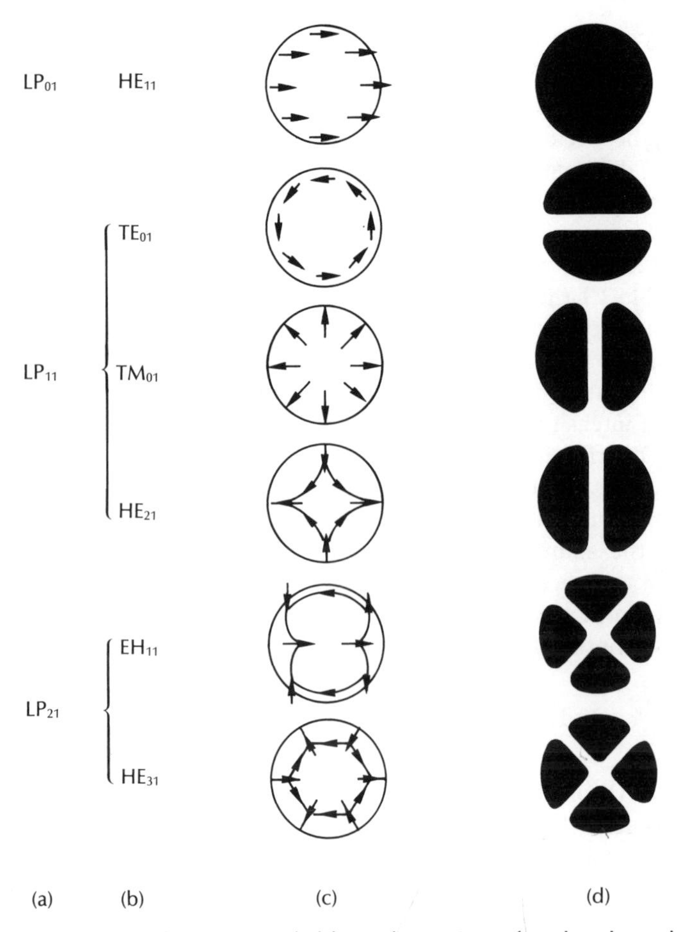

Figure 2.15 The electric field configurations for the three lowest LP modes illustrated in terms of their constituent exact modes: (a) LP mode designations; (b) exact mode designations; (c) electric field distribution of the exact modes; (d) intensity distribution of E_x for the exact modes indicating the electric field intensity profile for the corresponding LP modes.

where n_2 is the refractive index of the fiber cladding. Solutions of the wave equation for the cylindrical fiber are separable, having the form:

$$\psi = E(r)\begin{Bmatrix}\cos l\phi \\ \sin l\phi\end{Bmatrix} \exp\,(\omega t - \beta z) \tag{2.63}$$

where in this case ψ represents the dominant transverse electric field component. The periodic dependence on ϕ following $\cos l\phi$ or $\sin l\phi$ gives a mode of radial order l. Hence the fiber supports a finite number of guided modes of the form of Eq. (2.63).

Introducing the solutions given by Eq. (2.63) into Eq. (2.61) results in a differential equation of the form:

$$\frac{d^2\mathbf{E}}{dr^2} + \frac{1}{r}\frac{d\mathbf{E}}{dr} + \left[(n_1 k^2 - \beta^2) - \frac{l^2}{r^2}\right]\mathbf{E} = 0 \tag{2.64}$$

For a step index fiber with a constant refractive index core, Eq. (2.64) is a Bessel differential equation and the solutions are cylinder functions. In the core region the solutions are Bessel functions denoted by J_l. A graph of these gradually damped oscillatory functions (with respect to r) is shown in Figure 2.16(a). It may be noted

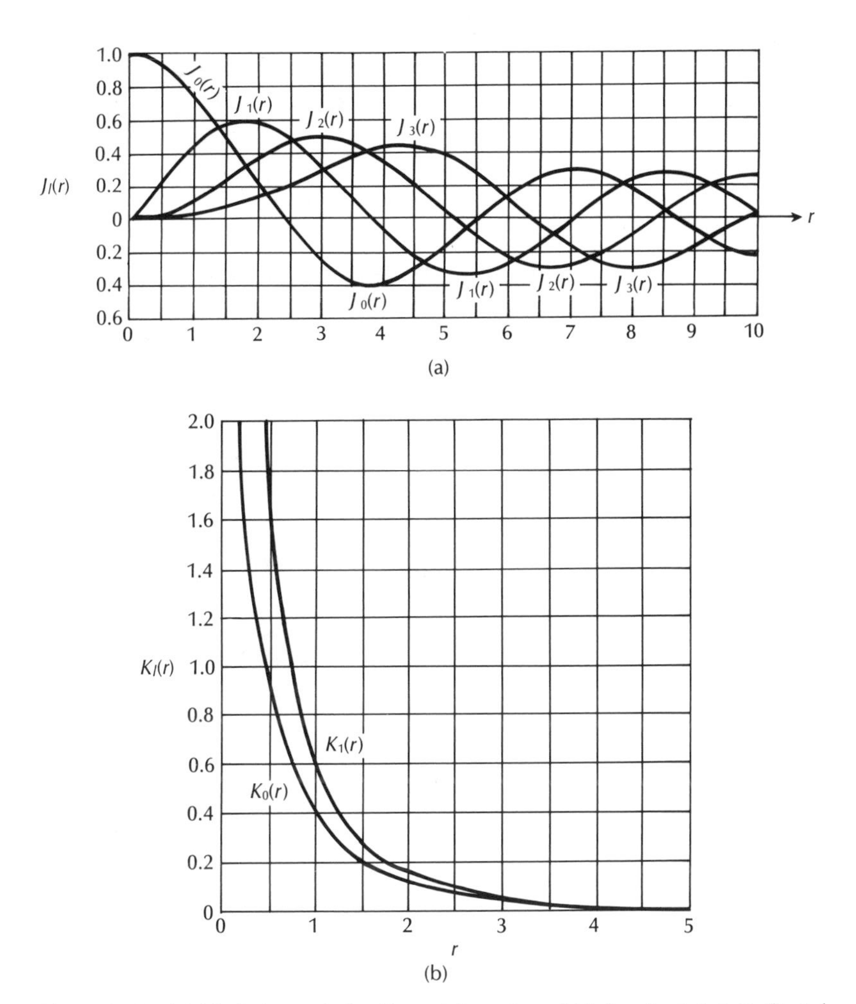

Figure 2.16 (a) Variation of the Bessel function $J_l(r)$ for $l = 0, 1, 2, 3$ (first four orders), plotted against r. (b) Graph of the modified Bessel function $K_l(r)$ against r for $l = 0, 1$.

that the field is finite at $r = 0$ and may be represented by the zero order Bessel function J_0. However, the field vanishes as r goes to infinity and the solutions in the cladding are therefore modified Bessel functions denoted by K_l. These modified functions decay exponentially with respect to r, as illustrated in Figure 2.16(b). The electric field may therefore be given by:

$$\begin{aligned} \mathbf{E}(r) &= GJ_l(UR) && \text{for} \quad R < 1 \text{ (core)} \\ &= GJ_l(U)\frac{K_l(WR)}{K_l(W)} && \text{for} \quad R > 1 \text{ (cladding)} \end{aligned} \tag{2.65}$$

where G is the amplitude coefficient and $R = r/a$ is the normalized radial coordinate when a is the radius of the fiber core; U and W which are the eigenvalues in the core and cladding respectively,* are defined as [Ref. 17]:

$$U = a(n_1^2k^2 - \beta^2)^{\frac{1}{2}} \tag{2.66}$$

$$W = a(\beta^2 - n_2^2k^2)^{\frac{1}{2}} \tag{2.67}$$

The sum of the squares of U and W defines a very useful quantity [Ref. 18] which is usually referred to as the normalized frequency† V where

$$V = (U^2 + W^2)^{\frac{1}{2}} = ka(n_1^2 - n_2^2)^{\frac{1}{2}} \tag{2.68}$$

It may be observed that the commonly used symbol for this parameter is the same as that normally adopted for voltage. However, within this chapter there should be no confusion over this point. Furthermore, using Eqs. (2.8) and (2.10) the normalized frequency may be expressed in terms of the numerical aperture NA and the relative refractive index difference Δ, respectively, as:

$$V = \frac{2\pi}{\lambda}a(NA) \tag{2.69}$$

$$V = \frac{2\pi}{\lambda}an_1(2\Delta)^{\frac{1}{2}} \tag{2.70}$$

The normalized frequency is a dimensionless parameter and hence is also sometimes simply called the V number or value of the fiber. It combines in a very useful manner the information about three important design variables for the fiber: namely, the core radius a, the relative refractive index difference Δ and the operating wavelength λ.

* U is also referred to as the radial phase parameter or the radial propagation constant, whereas W is known as the cladding decay parameter [Ref. 19].

† When used in the context of the planar waveguide, V is sometimes known as the normalized film thickness as it relates to the thickness of the guide layer (see Section 10.5.1).

It is also possible to define the normalized propagation constant b for a fiber in terms of the parameters of Eq. (2.68) so that:

$$b = 1 - \frac{U^2}{V^2} = \frac{(\beta/k)^2 - n_2^2}{n_1^2 - n_2^2}$$

$$= \frac{(\beta/k)^2 - n_2^2}{2n_1^2\Delta} \tag{2.71}$$

Referring to the expression for the guided modes given in Eq. (2.62), the limits of β are n_2k and n_1k, hence b must lie between 0 and 1.

In the weak guidance approximation the field matching conditions at the boundary require continuity of the transverse and tangential electrical field components at the core–cladding interface (at $r = a$). Therefore, using the Bessel function relations outlined previously, an eigenvalue equation for the LP modes may be written in the following form [Ref. 18]:

$$U \frac{J_{l\pm 1}(U)}{J_l(U)} = \pm W \frac{K_{l\pm 1}(W)}{K_l(W)} \tag{2.72}$$

Solving Eq. (2.72) with Eqs. (2.66) and (2.67) allows the eigenvalue U and hence β to be calculated as a function of the normalized frequency. In this way the propagation characteristics of the various modes, and their dependence on the optical wavelength and the fiber parameters may be determined.

Considering the limit of mode propagation when $\beta = n_2k$, then the mode phase velocity is equal to the velocity of light in the cladding and the mode is no longer properly guided. In this case the mode is said to be cut off and the eigenvalue $W = 0$ (Eq. 2.67). Unguided or radiation modes have frequencies below cutoff where $\beta < kn_2$, and hence W is imaginary. Nevertheless, wave propagation does not cease abruptly below cutoff. Modes exist where $\beta < kn_2$ but the difference is very small, such that some of the energy loss due to radiation is prevented by an angular momentum barrier [Ref. 21] formed near the core–cladding interface. Solutions of the wave equation giving these states are called leaky modes, and often behave as very lossy guided modes rather than radiation modes. Alternatively, as β is increased above n_2k, less power is propagated in the cladding until at $\beta = n_1k$ all the power is confined to the fiber core. As indicated previously, this range of values for β signifies the guided modes of the fiber.

The lower order modes obtained in a cylindrical homogeneous core waveguide are shown in Figure 2.17 [Ref. 16]. Both the LP notation and the corresponding traditional HE, EH, TE and TM mode notations are indicated. In addition, the Bessel functions J_0 and J_1 are plotted against the normalized frequency and where they cross the zero gives the cutoff point for the various modes. Hence, the cutoff point for a particular mode corresponds to a distinctive value of the normalized frequency (where $V = V_c$) for the fiber. It may be observed from Figure 2.17 that the value of V_c is different for different modes. For example, the first zero crossing J_1 occurs when the normalized frequency is 0 and this corresponds to the cutoff for

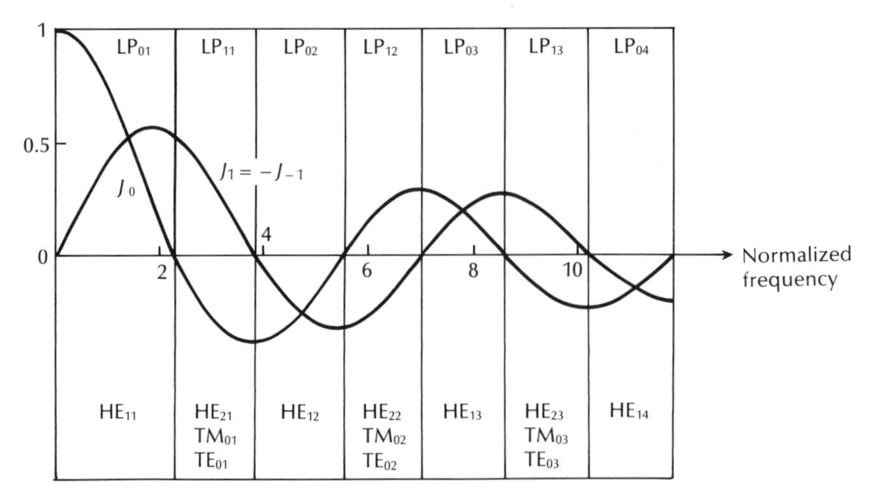

Figure 2.17 The allowed regions for the LP modes of order $l = 0, 1$ against normalized frequency (V) for a circular optical waveguide with a constant refractive index core (step index fiber). Reproduced with permission from D. Gloge. *Appl. Opt.*, **10**, p. 2552, 1971.

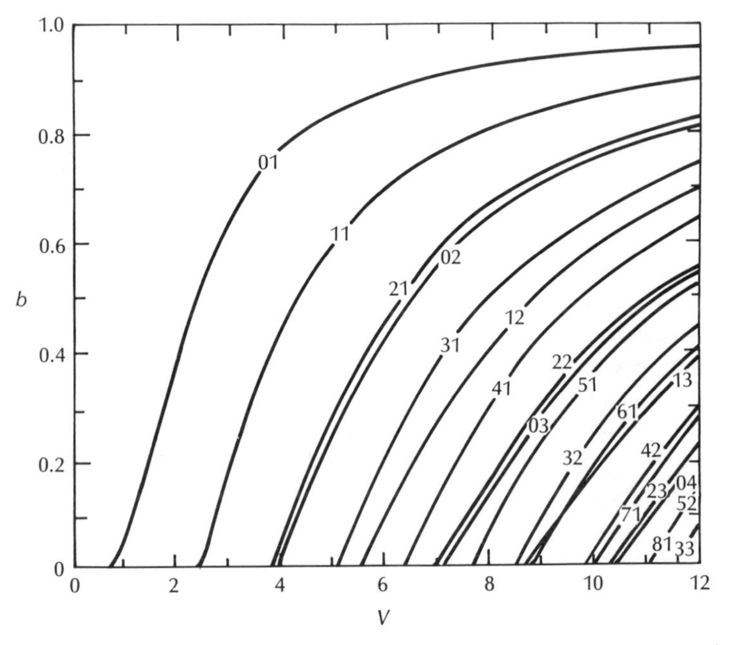

Figure 2.18 The normalized propagation constant b as a function of normalized frequency V for a number of LP modes. Reproduced with permission from D. Gloge. *Appl. Opt.*, **10**, p. 2552, 1971.

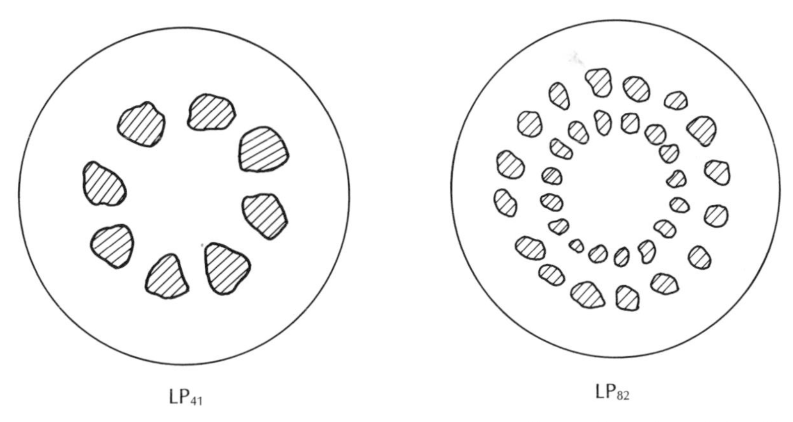

Figure 2.19 Sketches of fiber cross sections illustrating the distinctive light intensity distributions (mode patterns) generated by propagation of individual linearly polarized modes.

the LP_{01} mode. However, the first zero crossing for J_0 is when the normalized frequency is 2.405, giving a cutoff value V_c of 2.405 for the LP_{11} mode. Similarly, the second zero of J_1 corresponds to a normalized frequency of 3.83, giving a cutoff value V_c for the LP_{02} mode of 3.83. It is therefore apparent that fibers may be produced with particular values of normalized frequency which allow only certain modes to propagate. This is further illustrated in Figure 2.18 [Ref. 16] which shows the normalized propagation constant b for a number of LP modes as a function of V. It may be observed that the cutoff value of normalized frequency V_c which occurs when $\beta = n_2k$ corresponds to $b = 0$.

The propagation of particular modes within a fiber may also be confirmed through visual analysis. The electric field distribution of different modes gives similar distributions of light intensity within the fiber core. These waveguide patterns (often called mode patterns) may give an indication of the predominant modes propagating in the fiber. The field intensity distributions for the three lower order LP modes were shown in Figure 2.15. In Figure 2.19 we illustrate the mode patterns for two higher order LP modes. However, unless the fiber is designed for the propagation of a particular mode it is likely that the superposition of many modes will result in no distinctive pattern.

2.4.2 Mode coupling

We have thus far considered the propagation aspects of perfect dielectric waveguides. However, waveguide perturbations such as deviations of the fiber axis from straightness, variations in the core diameter, irregularities at the core–cladding interface and refractive index variations may change the propagation characteristics of the fiber. These will have the effect of coupling energy travelling in one mode to another depending on the specific perturbation.

Ray theory aids the understanding of this phenomenon, as shown in Figure 2.20, which illustrates two types of perturbation. It may be observed that in both cases the ray no longer maintains the same angle with the axis. In electromagnetic wave theory this corresponds to a change in the propagating mode for the light. Thus individual modes do not normally propagate throughout the length of the fiber without large energy transfers to adjacent modes, even when the fiber is exceptionally good quality and is not strained or bent by its surroundings. This mode conversion is known as mode coupling or mixing. It is usually analysed using coupled mode equations which can be obtained directly from Maxwell's equations. However, the theory is beyond the scope of this text and the reader is directed to Ref. 17 for a comprehensive treatment. Mode coupling affects the transmission properties of fibers in several important ways, a major one being in relation to the dispersive properties of fibers over long distances. This is pursued further in Sections 3.8–3.11.

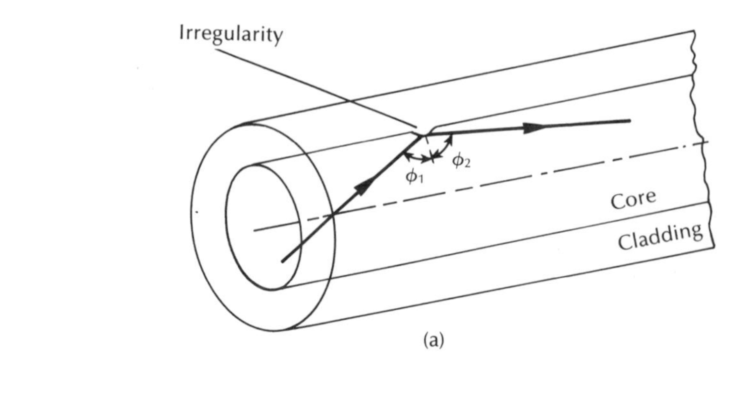

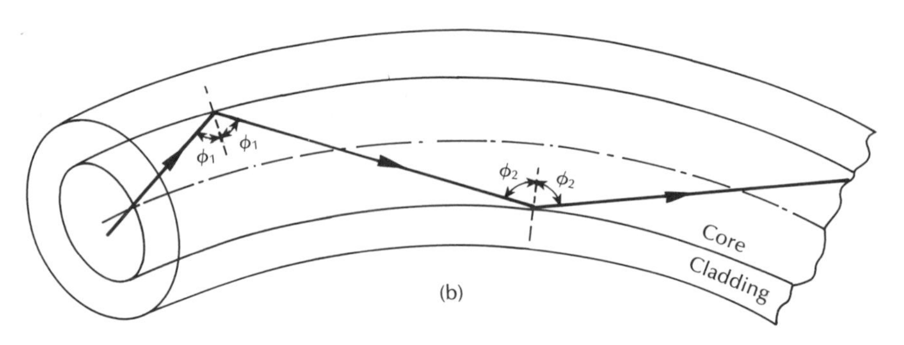

Figure 2.20 Ray theory illustrations showing two of the possible fiber perturbations which give mode coupling: (a) irregularity at the core–cladding interface; (b) fiber bend.

2.4.3 Step index fibers

The optical fiber considered in the preceding sections with a core of constant refractive index n_1 and a cladding of a slightly lower refractive index n_2 is known as step index fiber. This is because the refractive index profile for this type of fiber makes a step change at the core–cladding interface, as indicated in Figure 2.21, which illustrates the two major types of step index fiber. The refractive index profile may be defined as:

$$n(r) = \begin{cases} n_1 & r < a \quad \text{(core)} \\ n_2 & r \geqslant a \quad \text{(cladding)} \end{cases} \tag{2.73}$$

in both cases.

Figure 2.21(a) shows a multimode step index fiber with a core diameter of around 50 μm or greater, which is large enough to allow the propagation of many modes within the fiber core. This is illustrated in Figure 2.21(a) by the many different possible ray paths through the fiber. Figure 2.21(b) shows a single-mode or monomode step index fiber which allows the propagation of only one transverse electromagnetic mode (typically HE_{11}), and hence the core diameter must be of the order of 2 to 10 μm. The propagation of a single mode is illustrated in

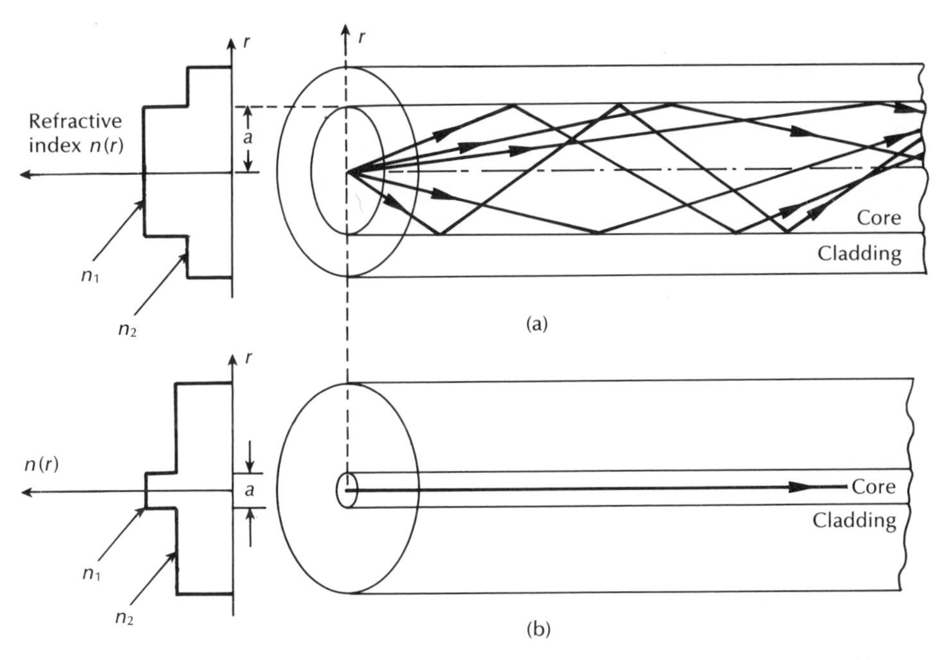

Figure 2.21 The refractive index profile and ray transmission in step index fibers: (a) multimode step index fiber; (b) single-mode step index fiber.

Figure 2.21(b) as corresponding to a single ray path only (usually shown as the axial ray) through the fiber.

The single-mode step index fiber has the distinct advantage of low intermodal dispersion (broadening of transmitted light pulses), as only one mode is transmitted, whereas with multimode step index fiber considerable dispersion may occur due to the differing group velocities of the propagating modes (see Section 3.10). This in turn restricts the maximum bandwidth attainable with multimode step index fibers, especially when compared with single-mode fibers. However, for lower bandwidth applications multimode fibers have several advantages over single-mode fibers. These are:

(a) the use of spatially incoherent optical sources (e.g. most light emitting diodes) which cannot be efficiently coupled to single-mode fibers;
(b) larger numerical apertures, as well as core diameters, facilitating easier coupling to optical sources;
(c) lower tolerance requirements on fiber connectors.

Multimode step index fibers allow the propagation of a finite number of guided modes along the channel. The number of guided modes is dependent upon the physical parameters (i.e. relative refractive index difference, core radius) of the fiber and the wavelengths of the transmitted light which are included in the normalized frequency V for the fiber. It was indicated in Section 2.4.1 that there is a cutoff value of normalized frequency V_c for guided modes below which they cannot exist. However, mode propagation does not entirely cease below cutoff. Modes may propagate as unguided or leaky modes which can travel considerable distances along the fiber. Nevertheless, it is the guided modes which are of paramount importance in optical fiber communications as these are confined to the fiber over its full length. It can be shown [Ref. 16] that the total number of guided modes or mode volume M_s for a step index fiber is related to the V value for the fiber by the approximate expression

$$M_s \simeq \frac{V^2}{2} \tag{2.74}$$

which allows an estimate of the number of guided modes propagating in a particular multimode step index fiber.

Example 2.4

A multimode step index fiber with a core diameter of 80 μm and a relative index difference of 1.5% is operating at a wavelength of 0.85 μm. If the core refractive index is 1.48, estimate: (a) the normalized frequency for the fiber; (b) the number of guided modes.

Solution: (a) The normalized frequency may be obtained from Eq. (2.70) where:

$$V \simeq \frac{2\pi}{\lambda} an_1(2\Delta)^{\frac{1}{2}} = \frac{2\pi \times 40 \times 10^{-6} \times 1.48}{0.85 \times 10^{-6}} (2 \times 0.015)^{\frac{1}{2}} = 75.8$$

(b) The total number of guided modes is given by Eq. (2.74) as:

$$M_s \simeq \frac{V^2}{2} = \frac{5745.6}{2}$$

$$= 2873$$

Hence this fiber has a V number of approximately 76, giving nearly 3000 guided modes.

Therefore, as illustrated in Example 2.4, the optical power is launched into a large number of guided modes, each having different spatial field distributions, propagation constants, etc. In an ideal multimode step index fiber with properties (i.e. relative index difference, core diameter) which are independent of distance, there is no mode coupling, and the optical power launched into a particular mode remains in that mode and travels independently of the power launched into the other guided modes. Also, the majority of these guided modes operate far from cutoff, and are well confined to the fiber core [Ref. 16]. Thus most of the optical power is carried in the core region and not in the cladding. The properties of the cladding (e.g. thickness) do not therefore significantly affect the propagation of these modes.

2.4.4 Graded index fibers

Graded index fibers do not have a constant refractive index in the core* but a decreasing core index $n(r)$ with radial distance from a maximum value of n_1 at the axis to a constant value n_2 beyond the core radius a in the cladding. This index variation may be represented as:

$$n(r) = \begin{cases} n_1(1 - 2\Delta(r/a)^{\alpha})^{\frac{1}{2}} & r < a \quad \text{(core)} \\ n_1(1 - 2\Delta)^{\frac{1}{2}} = n_2 & r \geqslant a \quad \text{(cladding)} \end{cases} \tag{2.75}$$

where Δ is the relative refractive index difference and α is the profile parameter which gives the characteristic refractive index profile of the fiber core. Equation (2.75) which is a convenient method of expressing the refractive index profile of the fiber core as a variation of α allows representation of the step index profile when $\alpha = \infty$, a parabolic profile when $\alpha = 2$ and a triangular profile when $\alpha = 1$. This range of refractive index profiles is illustrated in Figure 2.22.

* Graded index fibers are therefore sometimes referred to as inhomogeneous core fibers.

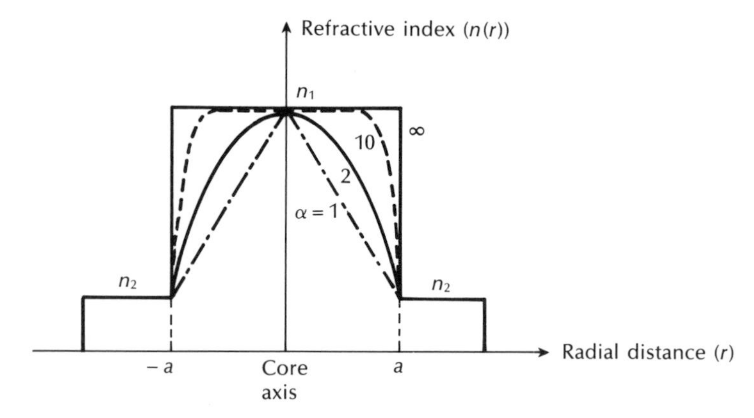

Figure 2.22 Possible fiber refractive index profiles for different values of α (given in Eq. (2.75)).

The graded index profiles which at present produce the best results for multimode optical propagation have a near parabolic refractive index profile core with $\alpha \approx 2$. Fibers with such core index profiles are well established and consequently when the term 'graded index' is used without qualification it usually refers to a fiber with this profile. For this reason in this section we consider the waveguiding properties of graded index fiber with a parabolic refractive index profile core.

A multimode graded index fiber with a parabolic index profile core is illustrated in Figure 2.23. It may be observed that the meridional rays shown appear to follow curved paths through the fiber core. Using the concepts of geometric optics, the gradual decrease in refractive index from the centre of the core creates many refractions of the rays as they are effectively incident on a large number of high to low index interfaces. This mechanism is illustrated in Figure 2.24 where a ray is

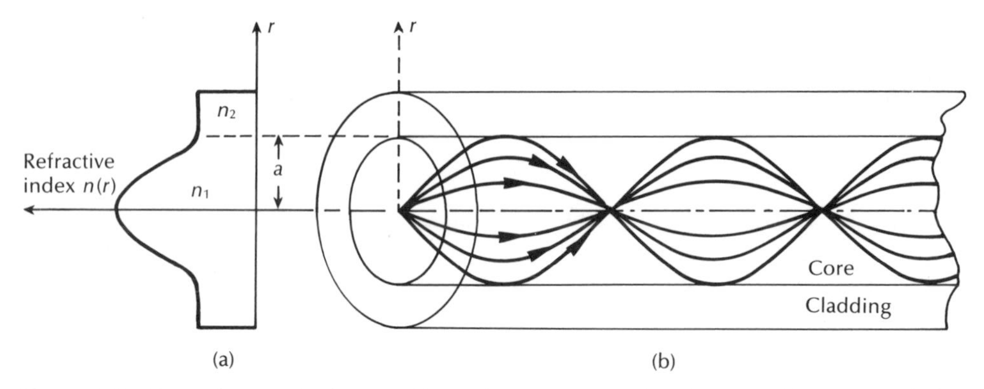

Figure 2.23 The refractive index profile and ray transmission in a multimode graded index fiber.

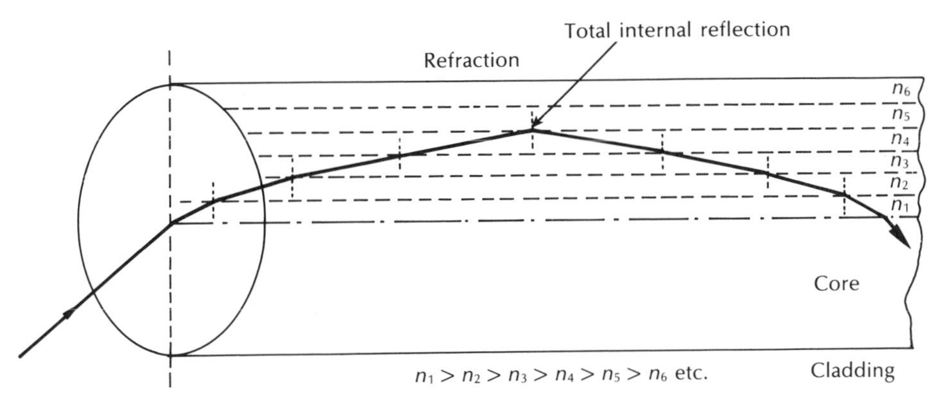

Figure 2.24 An expanded ray diagram showing refraction at the various high to low index interfaces within a graded index fiber, giving an overall curved ray path.

shown to be gradually curved, with an ever-increasing angle of incidence, until the conditions for total internal reflection are met, and the ray travels back toward the core axis, again being continuously refracted.

Multimode graded index fibers exhibit far less intermodal dispersion (see Section 3.10.2) than multimode step index fibers due to their refractive index profile. Although many different modes are excited in the graded index fiber, the different group velocities of the modes tend to be normalized by the index grading. Again considering ray theory, the rays travelling close to the fiber axis have shorter paths when compared with rays which travel into the outer regions of the core. However, the near axial rays are transmitted through a region of higher refractive index and therefore travel with a lower velocity than the more extreme rays. This compensates for the shorter path lengths and reduces dispersion in the fiber. A similar situation exists for skew rays which follow longer helical paths, as illustrated in Figure 2.25. These travel for the most part in the lower index region at greater speeds, thus giving the same mechanism of mode transit time equalization. Hence, multimode graded index fibers with parabolic or near parabolic index profile cores have

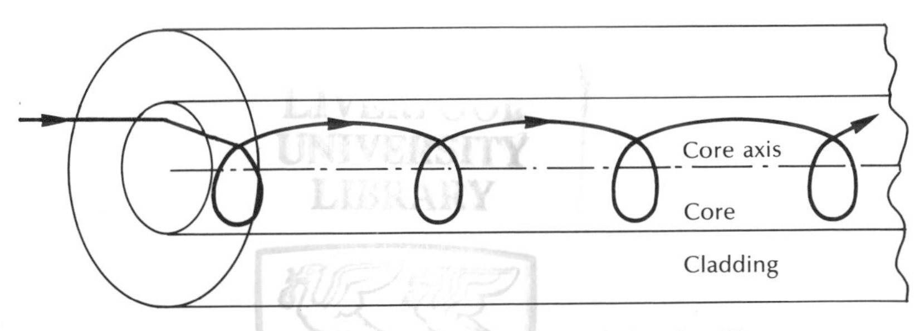

Figure 2.25 A helical skew ray path within a graded index fiber.

LIVERPOOL UNIVERSITY LIBRARY FIAT LVX

transmission bandwidths which may be orders of magnitude greater than multimode step index fiber bandwidths. Consequently, although they are not capable of the bandwidths attainable with single-mode fibers, such multimode graded index fibers have the advantage of large core diameters (greater than 30 μm) coupled with bandwidths suitable for long distance communication.

The parameters defined for step index fibers (i.e. NA, Δ, V) may be applied to graded index fibers and give a comparison between the two fiber types. However, it must be noted that for graded index fibers the situation is more complicated since the numerical aperture is a function of the radial distance from the fiber axis. Graded index fibers, therefore, accept less light than corresponding step index fibers with the same relative refractive index difference.

Electromagnetic mode theory may also be utilized with the graded profiles. Approximate field solutions of the same order as geometric optics are often obtained employing the WKB method from quantum mechanics after Wentzel, Kramers and Brillouin [Ref. 22]. Using the WKB method modal solutions of the guided wave are achieved by expressing the field in the form.

$$E_x = \tfrac{1}{2}[G_1(r)\exp[jS(r)] + G_2(r)\exp[-jS(r)]]\begin{pmatrix}\cos l\phi\\ \sin l\phi\end{pmatrix}\exp(j\beta z) \tag{2.76}$$

where G and S are assumed to be real functions of the radial distance r.

Substitution of Eq. (2.76) into the scalar wave equation of the form given by Eq. (2.61) (in which the constant refractive index of the fiber core n_1 is replaced by $n(r)$) and neglecting the second derivative of $G_i(r)$ with respect to r provides approximate solutions for the amplitude function $G_i(r)$ and the phase function $S(r)$. It may be observed from the ray diagram shown in Figure 2.23 that a light ray propagating in a graded index fiber does not necessarily reach every point within the fiber core. The ray is contained within two cylindrical caustic surfaces and for most rays a caustic does not coincide with the core–cladding interface. Hence the caustics define the classical turning points of the light ray within the graded fiber core. These turning points defined by the two caustics may be designated as occurring at $r = r_1$ and $r = r_2$.

The result of the WKB approximation yields an oscillatory field in the region $r_1 < r < r_2$ between the caustics where:

$$G_1(r) = G_2(r) = D/[(n^2(r)k^2 - \beta^2)r^2 - l^2]^{\frac{1}{4}} \tag{2.77}$$

(where D is an amplitude coefficient) and

$$S(r) = \int_{r_1}^{r_2} [(n^2(r)k^2 - \beta^2)r^2 - l^2]^{\frac{1}{2}} \frac{\mathrm{d}r}{r} - \frac{\pi}{4} \tag{2.78}$$

Outside the interval $r_1 < r < r_2$ the field solution must have an evanescent form. In the region inside the inner caustic defined by $r < r_1$ and assuming r_1 is not too close to $r = 0$, the field decays towards the fiber axis giving:

$$G_1(r) = D\exp(jmx)/[l^2 - (n^2(r)k^2 - \beta^2)r^2]^{\frac{1}{4}} \tag{2.79}$$

$$G_2(r) = 0 \tag{2.80}$$

where the integer m is the radial mode number and

$$S(r) = \mathrm{j} \int_{r}^{r_1} [l^2 - (n^2(r)k^2 - \beta^2)r^2]^{\frac{1}{2}} \frac{\mathrm{d}r}{r} \tag{2.81}$$

Also outside the outer caustic in the region $r > r_2$, the field decays away from the fiber axis and is described by the equations:

$$G_1(r) = D \exp(\mathrm{j}mx)/[l^2 - (n^2(r)k^2 - \beta^2)r^2]^{\frac{1}{4}} \tag{2.82}$$

$$G_2(r) = 0 \tag{2.83}$$

$$S(r) = \mathrm{j} \int_{r_2}^{r} [l^2 - (n^2(r)k^2 - \beta^2)r^2]^{\frac{1}{2}} \frac{\mathrm{d}r}{r} \tag{2.84}$$

The WKB method does not initially provide valid solutions of the wave equation in the vicinity of the turning points. Fortunately, this may be amended by replacing the actual refractive index profile by a linear approximation at the location of the caustics. The solutions at the turning points can then be expressed in terms of Hankel functions of the first and second kind of order $\frac{1}{3}$ [Ref. 23]. This facilitates the joining together of the two separate solutions described previously for inside and outside the interval $r_1 < r < r_2$. Thus the WKB theory provides an approximate eigenvalue equation for the propagation constant β of the guided modes which cannot be determined using ray theory. The WKB eigenvalue equation of which β is a solution is given by [Ref. 23]:

$$\int_{r_1}^{r_2} [(n^2(r)k^2 - \beta^2)r^2 - l^2]^{\frac{1}{2}} \frac{\mathrm{d}r}{r} = (2m - 1)\frac{\pi}{2} \tag{2.85}$$

where the radial mode number $m = 1, 2, 3 \ldots$ and determines the number of maxima of the oscillatory field in the radial direction. This eigenvalue equation can only be solved in a closed analytical form for a few simple refractive index profiles. Hence, in most cases it must be solved approximately or with the use of numerical techniques.

Finally the amplitude coefficient D may be expressed in terms of the total optical power P_G within the guided mode. Considering the power carried between the turning points r_1 and r_2 gives a geometric optics approximation of [Ref. 26]:

$$D = \frac{4(\mu_0/\varepsilon_0)^{\frac{1}{2}} P_G^{\frac{1}{2}}}{n_1 \pi a^2 I} \tag{2.86}$$

where

$$I = \int_{r_1/a}^{r_2/a} \frac{x\mathrm{d}x}{[(n^2(ax)k^2 - \beta^2)a^2x^2 - l^2]^{\frac{1}{2}}} \tag{2.87}$$

The properties of the WKB solution may by observed from a graphical representation of the integrand given in Eq. (2.78). This is shown in Figure 2.26, together with the corresponding WKB solution. Figure 2.26 illustrates the functions $(n^2(r)k^2 - \beta^2)$ and (l^2/r^2). The two curves intersect at the turning points $r = r_1$ and $r = r_2$. The oscillatory nature of the WKB solution between the turning points (i.e.

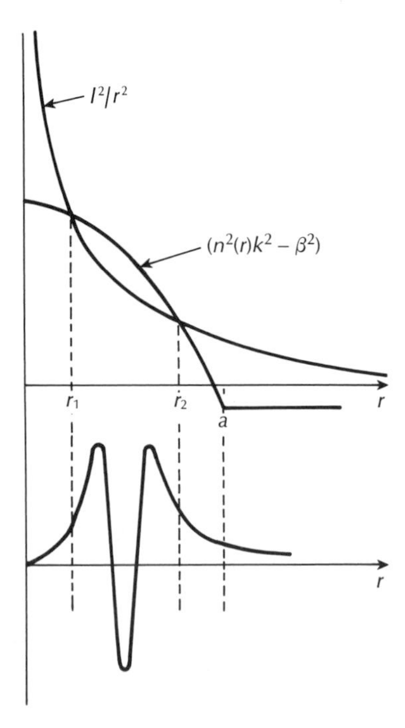

Figure 2.26 Graphical representation of the functions $(n^2(r)k^2 - \beta^2)$ and (l^2/r^2) that are important in the WKB solution and which define the turning points r_1 and r_2. Also shown is an example of the corresponding WKB solution for a guided mode where an oscillatory wave exists in the region between the turning points.

when $l^2/r^2 < n^2(r)k^2 - \beta^2)$ which changes into a decaying exponential (evanescent) form outside the interval $r_1 < r < r_2$ (i.e. when $l^2/r^2 > n^2(r)k^2 - \beta^2)$ can also be clearly seen.

It may be noted that as the azimuthal mode number l increases, the curve l^2/r^2 moves higher and the region between the two turning points becomes narrower. In addition, even when l is fixed the curve $(n^2(r)k^2 - \beta^2)$ is shifted up and down with alterations in the value of the propagation constant β. Therefore, modes far from cutoff which have large values of β exhibit more closely spaced turning points. As the value of β decreases below n_2k, $(n^2(r)k^2 - \beta^2)$ is no longer negative for large values of r and the guided mode situation depicted in Figure 2.26 changes to one corresponding to Figure 2.27. In this case a third turning point $r = r_3$ is created when at $r = a$ the curve $(n^2(r)k^2 - \beta^2)$ becomes constant, thus allowing the curve (l^2/r^2) to drop below it. Now the field displays an evanescent, exponentially decaying form in the region $r_2 < r < r_3$, as shown in Figure 2.27. Moreover, for $r > r_3$ the field resumes an oscillatory behaviour and therefore carries power away from the fiber core. Unless mode cutoff occurs at $\beta = n_2k$ the guided mode is no

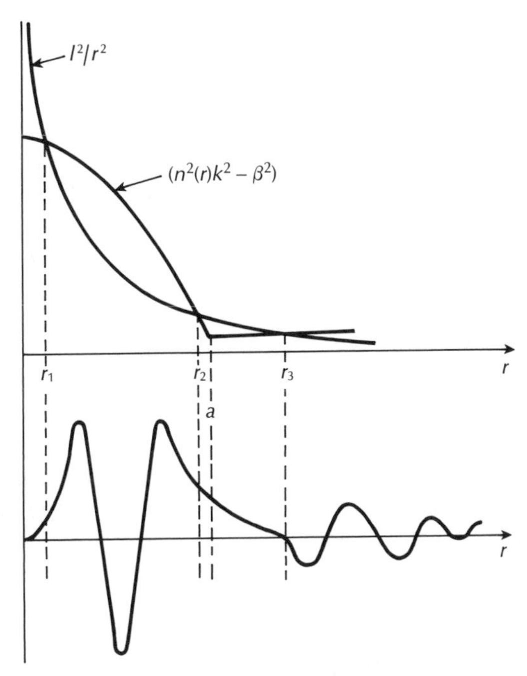

Figure 2.27 Similar graphical representation as that illustrated in Figure 2.26. Here the curve $(n^2(r)k^2 - \beta^2)$ no longer goes negative and a third turning point r_3 occurs. This corresponds to leaky mode solutions in the WKB method.

longer fully contained within the fiber core but loses power through leakage or tunnelling into the cladding. This situation corresponds to the leaky modes mentioned previously in Section 2.4.1.

The WKB method may be used to calculate the propagation constants for the modes in a parabolic refractive index profile core fiber where, following Eq. (2.75):

$$n^2(r) = n_1^2\left(1 - 2\left(\frac{r}{a}\right)^2 \Delta\right) \quad \text{for} \quad r < a \tag{2.88}$$

Substitution of Eq. (2.88) into Eq. (2.85) gives:

$$\int_{r_1}^{r_2} \left[n_1^2k^2 - \beta^2 - 2n_1^2k^2\left(\frac{r}{a}\right)^2 \Delta - \frac{l^2}{r^2}\right]^{\frac{1}{2}} \mathrm{d}r = (m + \tfrac{1}{2})\pi \tag{2.89}$$

The integral shown in Eq. (2.89) can be evaluated using a change of variable from r to $u = r^2$. The integral obtained may be found in a standard table of indefinite integrals [Ref. 27]. As the square root term in the resulting expression goes to zero at the turning points (i.e. $r = r_1$ and $r = r_2$), then we can write

$$\left[\frac{a(n_1k^2 - \beta^2)}{4n_1k\surd(2\Delta)} - \frac{l}{2}\right]\pi = (m + \tfrac{1}{2})\pi \tag{2.90}$$

Solving Eq. (2.90) for β^2 gives:

$$\beta^2 = n_1^2 k^2 \left[\frac{1 - 2\sqrt{(2\Delta)}}{n_1 ka} (2m + l + 1) \right] \tag{2.91}$$

It is interesting to note that the solution for the propagation constant for the various modes in a parabolic refractive index core fiber given in Eq. (2.91) is exact even though it was derived from the approximate WKB eigenvalue equation (Eq. 2.85). However, although Eq. (2.91) is an exact solution of the scalar wave equation for an infinitely extended parabolic profile medium, the wave equation is only an approximate representation of Maxwell's equation. Furthermore, practical parabolic refractive index profile core fibers exhibit a truncated parabolic distribution which merges into a constant refractive index at the cladding. Hence Eq. (2.91) is not exact for real fibers.

Equation (2.91) does, however, allow us to consider the mode number plane spanned by the radial and azimuthal mode numbers m and l. This plane is displayed in Figure 2.28, where each mode of the fiber described by a pair of mode numbers is represented as a point in the plane. The mode number plane contains guided, leaky and radiation modes. The mode boundary which separates the guided modes from the leaky and radiation modes is indicated by the solid line in Figure 2.28. It depicts a constant value of β following Eq. (2.91) and occurs when $\beta = n_2 k$. Therefore, all the points in the mode number plane lying below the line $\beta = n_2 k$ are associated with guided modes, whereas the region above the line is occupied by leaky and radiation modes. The concept of the mode plane allows us to count the total number of guided modes within the fiber. For each pair of mode numbers m and l the corresponding mode field can have azimuthal mode dependence cos $l\phi$ or sin $l\phi$ and can exist in two possible polarizations (see Section 3.13). Hence the

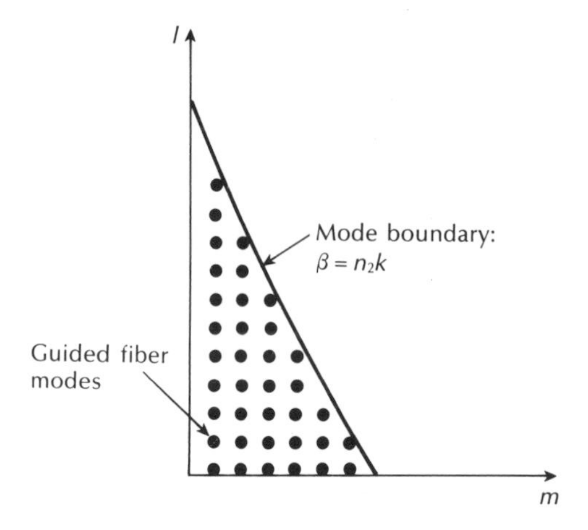

Figure 2.28 The mode number plane illustrating the mode boundary and the guided fiber modes.

modes are said to be fourfold degenerate.* If we define the mode boundary as the function $m = f(l)$, then the total number of guided modes M is given by

$$M = 4 \int_0^{l_{\max}} f(l)\, \mathrm{d}l \tag{2.92}$$

as each representation point corresponding to four modes occupies an element of unit area in the mode plane. Equation (2.92) allows the derivation of the total number of guided modes or mode volume M_g supported by the graded index fiber. It can be shown [Ref. 23] that:

$$M_g = \left(\frac{\alpha}{\alpha + 2}\right)(n_1 ka)^2 \Delta \tag{2.93}$$

Furthermore, utilizing Eq. (2.70), the normalized frequency V for the fiber when $\Delta \ll 1$ is approximately given by:

$$V = n_1 ka(2\Delta)^{\frac{1}{2}} \tag{2.94}$$

Substituting Eq. (2.94) into Eq. (2.93), we have:

$$M_g \simeq \left(\frac{\alpha}{\alpha + 2}\right)\left(\frac{V^2}{2}\right) \tag{2.95}$$

Hence for a parabolic refractive index profile core fiber ($\alpha = 2$), $M_g \approx V^2/4$, which is half the number supported by a step index fiber ($\alpha = \infty$) with the same V value.

Example 2.5

A graded index fiber has a core with a parabolic refractive index profile which has a diameter of 50 μm. The fiber has a numerical aperture of 0.2. Estimate the total number of guided modes propagating in the fiber when it is operating at a wavelength of 1 μm.

Solution: Using Eq. (2.69), the normalized frequency for the fiber is:

$$V = \frac{2\pi}{\lambda} a(NA) = \frac{2\pi \times 25 \times 10^{-6} \times 0.2}{1 \times 10^{-6}}$$

$$= 31.4$$

The mode volume may be obtained from Eq. (2.95) where for a parabolic profile:

$$M_g \simeq \frac{V^2}{4} = \frac{986}{4} = 247$$

Hence the fiber supports approximately 247 guided modes.

---

* An exception to this are the modes that occur when $l = 0$ which are only doubly degenerate as $\cos l\phi$ becomes unity and $\sin l\phi$ vanishes. However, these modes represent only a small minority and therefore may be neglected.

2.5 Single-mode fibers

The advantage of the propagation of a single mode within an optical fiber is that the signal dispersion caused by the delay differences, between different modes in a multimode fiber may be avoided (see Section 3.10). Multimode step index fibers do not lend themselves to the propagation of a single mode due to the difficulties of maintaining single-mode operation within the fiber when mode conversion (i.e. coupling) to other guided modes takes place at both input mismatches and fiber imperfections. Hence, for the transmission of a single mode the fiber must be designed to allow propagation of only one mode, whilst all other modes are attenuated by leakage or absorption.

Following the preceding discussion of multimode fibers, this may be achieved through choice of a suitable normalized frequency for the fiber. For single-mode operation, only the fundamental LP_{01} mode can exist. Hence the limit of single-mode operation depends on the lower limit of guided propagation for the LP_{11} mode. The cutoff normalized frequency for the LP_{11} mode in step index fibers occurs at $V_c = 2.405$ (see Section 2.4.1). Thus single-mode propagation of the LP_{01} mode in step index fibers is possible over the range:

$$0 \leqslant V < 2.405 \tag{2.96}$$

as there is no cutoff for the fundamental mode. It must be noted that there are in fact two modes with orthogonal polarization over this range, and the term single-mode applies to propagation of light of a particular polarization. Also, it is apparent that the normalized frequency for the fiber may be adjusted to within the range given in Eq. (2.96) by reduction of the core radius, and possibly the relative refractive index difference following Eq. (2.70) which, for single-mode fibers, is usually less than 1%.

Example 2.6

Estimate the maximum core diameter for an optical fiber with the same relative refractive index difference (1.5%) and core refractive index (1.48) as the fiber given in Example 2.4 in order that it may be suitable for single-mode operation. It may be assumed that the fiber is operating at the same wavelength (0.85 μm). Further, estimate the new maximum core diameter for single-mode operation when the relative refractive index difference is reduced by a factor of 10.

Solution: Considering the relationship given in Eq. (2.96), the maximum V value for a fiber which gives single-mode operation is 2.4. Hence, from Eq. (2.70) the core radius a is:

$$a = \frac{V\lambda}{2\pi n_1 (2\Delta)^{\frac{1}{2}}} = \frac{2.4 \times 0.85 \times 10^{-6}}{2\pi \times 1.48 \times (0.03)^{\frac{1}{2}}}$$

$$= 1.3\ \mu\text{m}$$

Therefore the maximum core diameter for single-mode operation is approximately 2.6 μm.

Reducing the relative refractive index difference by a factor of 10 and again using Eq. (2.70) gives:

$$a = \frac{2.4 \times 0.85 \times 10^{-6}}{2\pi \times 1.48 \times (0.003)^{\frac{1}{2}}} = 4.0\ \mu\text{m}.$$

Hence the maximum core diameter for single-mode operation is now approximately 8 μm.

It is clear from Example 2.6 that in order to obtain single-mode operation with a maximum V number of 2.4 the single-mode fiber must have a much smaller core diameter than the equivalent multimode step index fiber (in this case by a factor of 32). However, it is possible to achieve single-mode operation with a slightly larger core diameter, albeit still much less than the diameter of multimode step index fiber, by reducing the relative refractive index difference of the fiber.* Both these factors create difficulties with single-mode fibers. The small core diameters pose problems with launching light into the fiber and with field jointing, and the reduced relative refractive index difference presents difficulties in the fiber fabrication process.

Graded index fibers may also be designed for single-mode operation and some specialist fiber designs do adopt such non-step index profiles (see Section 3.12). However, it may be shown [Ref. 28] that the cutoff value of normalized frequency V_c to support a single mode in a graded index fiber is given by:

$$V_c = 2.405(1 + 2/\alpha)^{\frac{1}{2}} \tag{2.97}$$

Therefore, as in the step index case, it is possible to determine the fiber parameters which give single-mode operation.

Example 2.7

A graded index fiber with a parabolic refractive index profile core has a refractive index at the core axis of 1.5 and a relative index difference of 1%. Estimate the maximum possible core diameter which allows single-mode operation at a wavelength of 1.3 μm.

Solution: Using Eq. (2.97) the maximum value of normalized frequency for single-mode operation is

$$V = 2.4(1 + 2/\alpha)^{\frac{1}{2}} = 2.4(1 + 2/2)^{\frac{1}{2}}$$
$$= 2.4\sqrt{2}$$

* Practical values for single-mode step index fiber designed for operation at a wavelength of 1.3 μm are $\Delta \simeq 0.3\%$, giving $2a \simeq 8.5$ μm.

The maximum core radius may be obtained from Eq. (2.70) where:

$$a = \frac{V\lambda}{2\pi n_1 (2\Delta)^{\frac{1}{2}}} = \frac{2.4\sqrt{2} \times 1.3 \times 10^{-6}}{2\pi \times 1.5 \times (0.02)^{\frac{1}{2}}}$$

$$= 3.3\ \mu\text{m}$$

Hence the maximum core diameter which allows single-mode operation is approximately 6.6 μm.

It may be noted that the critical value of normalized frequency for the parabolic profile graded index fiber is increased by a factor of $\sqrt{2}$ on the step index case. This gives a core diameter increased by a similar factor for the graded index fiber over a step index fiber with the equivalent core refractive index (equivalent to the core axis index), and the same relative refractive index difference.

The maximum V number which permits single-mode operation can be increased still further when a graded index fiber with a triangular profile is employed. It is apparent from Eq. (2.97) that the increase in this case is by a factor of $\sqrt{3}$ over a comparable step index fiber. Hence, significantly larger core diameter single-mode fibers may be produced utilizing this index profile. Such advanced refractive index profiles, which came under serious investigation in the early 1980s [Ref. 29], have now been adopted, particularly in the area of dispersion modified fiber design (see Section 3.12).

A further problem with single-mode fibers with low relative refractive index differences and low V values is that the electromagnetic field associated with the LP_{10} mode extends appreciably into the cladding. For instance, with V values less than 1.4, over half the modal power propagates in the cladding [Ref. 21]. Thus the exponentially decaying evanescent field may extend significant distances into the cladding. It is therefore essential that the cladding is of a suitable thickness, and has low absorption and scattering losses in order to reduce attenuation of the mode. Estimates [Ref. 30] show that the necessary cladding thickness is of the order of 50 μm to avoid prohibitive losses (greater than 1 dB km^{-1}) in single-mode fibers, especially when additional losses resulting from microbending (see Section 4.8.1) are taken into account. Therefore, the total fiber cross section for single-mode fibers is of a comparable size to multimode fibers.

Another approach to single-mode fiber design which allows the V value to be increased above 2.405 is the W fiber [Ref. 32]. The refractive index profile for this fiber is illustrated in Figure 2.29 where two cladding regions may be observed. Use of such two step cladding allows the loss threshold between the desirable and undesirable modes to be substantially increased. The fundamental mode will be fully supported with small cladding loss when its propagation constant lies in the range $kn_3 < \beta < kn_1$.

If the undesirable higher order modes which are excited or converted to have values of propagation constant $\beta < kn_3$, they will leak through the barrier layer

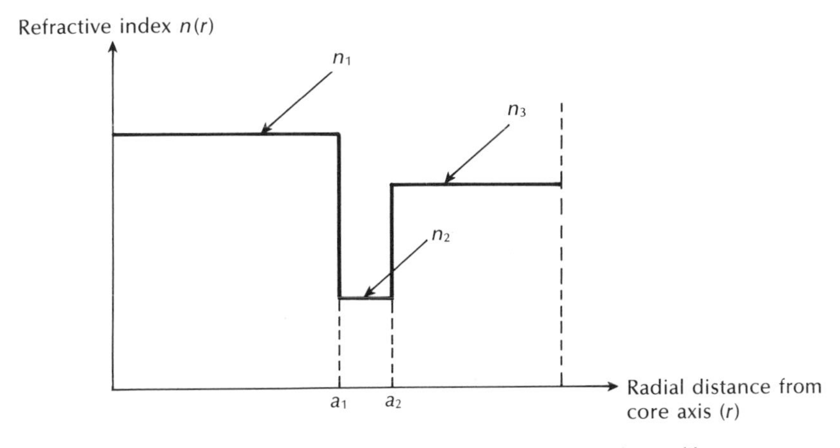

Figure 2.29 The refractive index profile for a single-mode *W* fiber.

between a_1 and a_2 (Figure 2.29) into the outer cladding region n_3. Consequently these modes will lose power by radiation into the lossy surroundings. This design can provide single-mode fibers with larger core diameters than can the conventional single cladding approach which proves useful for easing jointing difficulties; W fibers also tend to give reduced losses at bends in comparison with conventional single-mode fibers.

Although single-mode fibers have only relatively recently emerged (i.e. since 1983) as a viable optical communication medium they have quickly become the dominant and the most widely used fiber type within telecommunications.* Major reasons for this situation are as follows:

1. They currently exhibit the greatest transmission bandwidths and the lowest losses of the fiber transmission media (see Chapter 3).
2. They have a superior transmission quality over other fiber types because of the absence of modal noise (see Section 3.10.3).
3. They offer a substantial upgrade capability (i.e. future proofing) for future wide bandwidth services using either faster optical transmitters and receivers or advanced transmission techniques (e.g. coherent technology, see Chapter 12).
4. They are compatible with the developing integrated optics technology (see Chapter 10).
5. The above (1) to (4) provide a confidence that the installation of single-mode fiber will provide a transmission medium which will have adequate performance such that it will not require replacement over its twenty-plus-year anticipated lifetime.

At present the most commonly used single-mode fibers employ a step index (or near

* Multimode fibers are still finding extensive use within more localized communications (e.g. in data links and local area networks).

step index) profile design and are dispersion optimized (see Section 3.11.2) for operation in the 1.3 μm wavelength region. These fibers are either of a matched-cladding (MC) or a depressed-cladding (DC) design, as illustrated in Figure 2.30. In the conventional MC fibers, the region external to the core has a constant uniform refractive index which is slightly lower than the core region, typically consisting of pure silica. Alternatively when the core region comprises pure silica then the lower index cladding is obtained through fluorine doping. A mode-field diameter (MFD) (see Section 2.5.2) of 10 μm is typical for MC fibers with relative refractive index differences of around 0.3%. However, improved bend loss performance (see Section 3.6) has been achieved in the 1.55 μm wavelength region with reduced MFDs of about 9.5 μm and relative refractive index differences of 0.37%. [Ref. 40].

A more recent experimental MC fiber design employs a segmented core as shown in Figure 2.30(b) [Ref. 41]. Such a structure provides conventional single-mode dispersion optimized performance at wavelengths around 1.3 μm but is multimoded with a few modes (two or three) in the shorter wavelength region around 0.8 μm. The multimode operating region is intended to help relax both the tight tolerances involved when coupling LEDs to such single-mode fibers (see Section 7.3.6) and their connectorization. Thus segmented core fiber of this type could find use in

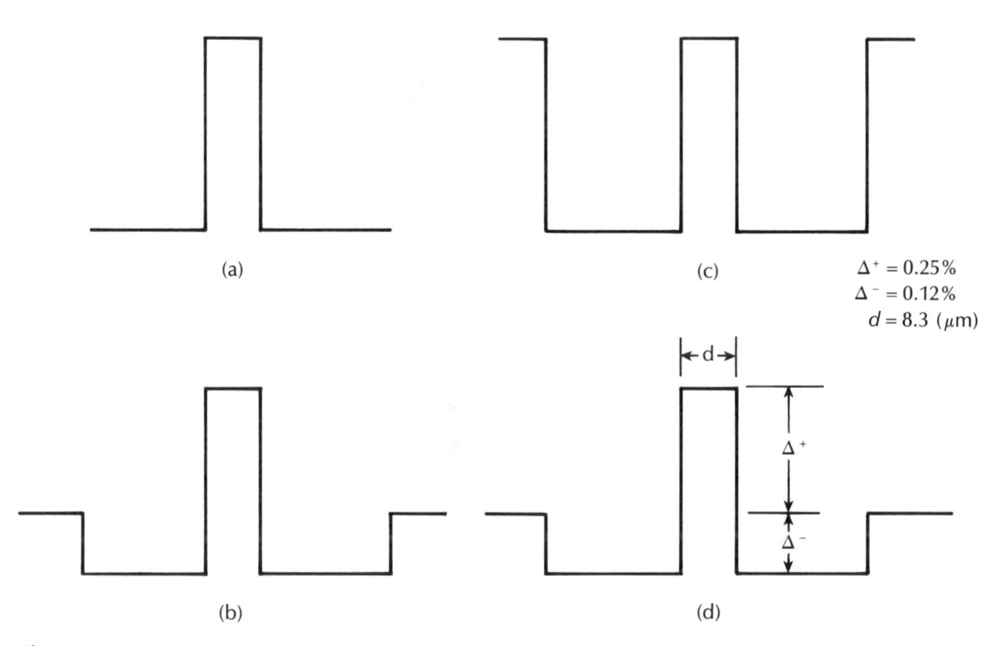

Figure 2.30 Single-mode fiber step index profiles optimized for operation at a wavelength of 1.3 μm: (a) conventional matched-cladding design; (b) segmented core matched-cladding design; (c) depressed-cladding design; (d) profile specifications of a depressed-cladding fiber [Ref. 42].

applications which require an inexpensive initial solution but upgradeability to conventional single-mode fiber performance at the 1.3 μm wavelength in the future.

In the DC fibers shown in Figure 2.30 the cladding region immediately adjacent to the core is of a lower refractive index than that of an outer cladding region. A typical MFD (see Section 2.5.2) of a DC fiber is 9 μm with positive and negative relative refractive index differences of 0.25% and 0.12% (see Figure 2.30(d)) [Ref. 42].

2.5.1 Cutoff wavelength

It may be noted by rearrangement of Eq. (2.70) that single-mode operation only occurs above a theoretical cutoff wavelength λ_c given by:

$$\lambda_c = \frac{2\pi a n_1}{V_c}(2\Delta)^{\frac{1}{2}} \tag{2.98}$$

where V_c is the cutoff normalized frequency. Hence λ_c is the wavelength above which a particular fiber becomes single-moded. Dividing Eq. (2.98) by Eq. (2.70) for the same fiber we obtain the inverse relationship:

$$\frac{\lambda_c}{\lambda} = \frac{V}{V_c} \tag{2.99}$$

Thus for step index fiber where $V_c = 2.405$, the cutoff wavelength is given by [Ref. 43]:

$$\lambda_c = \frac{V\lambda}{2.405} \tag{2.100}$$

An effective cutoff wavelength has been defined by the CCITT* [Ref. 44] which is obtained from a 2 m length of fiber containing a single 14 cm radius loop. This definition was produced because the first higher order LP_{11} mode is strongly affected by fiber length and curvature near cutoff. Recommended cutoff wavelength values for primary coated fiber range from 1.1 to 1.28 μm for single-mode fiber designed for operation in the 1.3 μm wavelength region in order to avoid modal noise and dispersion problems. Moreover, practical transmission systems are generally operated close to the effective cutoff wavelength in order to enhance the fundamental mode confinement, but sufficiently distant from cutoff so that no power is transmitted in the second order LP_{11} mode.

Example 2.8

Determine the cutoff wavelength for a step index fiber to exhibit single-mode operation when the core refractive index and radius are 1.46 and 4.5 μm, respectively, with the relative index difference being 0.25%.

* Recommendation G 652.

Solution: Using Eq. (2.98) with $V_c = 2.405$ gives:

$$\lambda_c = \frac{2\pi a n_1 (2\Delta)^{\frac{1}{2}}}{2.405} = \frac{2\ 4.5 \times 1.46(0.005)^{\frac{1}{2}}}{2.405}\ \mu\text{m}$$

$$= 1.214\ \mu\text{m}$$
$$= 1214\ \text{nm}$$

Hence the fiber is single-moded to a wavelength of 1214 nm.

2.5.2 Mode-field diameter and spot size

Many properties of the fundamental mode are determined by the radial extent of its electromagnetic field including losses at launching and jointing, microbend losses, waveguide dispersion and the width of the radiation pattern. Therefore, the mode-field diameter (MFD) is an important parameter for characterizing single-mode fiber properties which takes into account the wavelength dependent field penetration into the fiber cladding. In this context it is a better measure of the functional properties of single-mode fiber than the core diameter. For step index and graded (near parabolic profile) single-mode fibers operating near the cutoff wavelength λ_c, the field is well approximated by a Gaussian distribution (see Section 2.5.5). In this case the MFD is generally taken as the distance between the opposite $1/e = 0.37$ field amplitude points and the power $1/e^2 = 0.135$ points in relation to the corresponding values on the fiber axis, as shown in Figure 2.31.

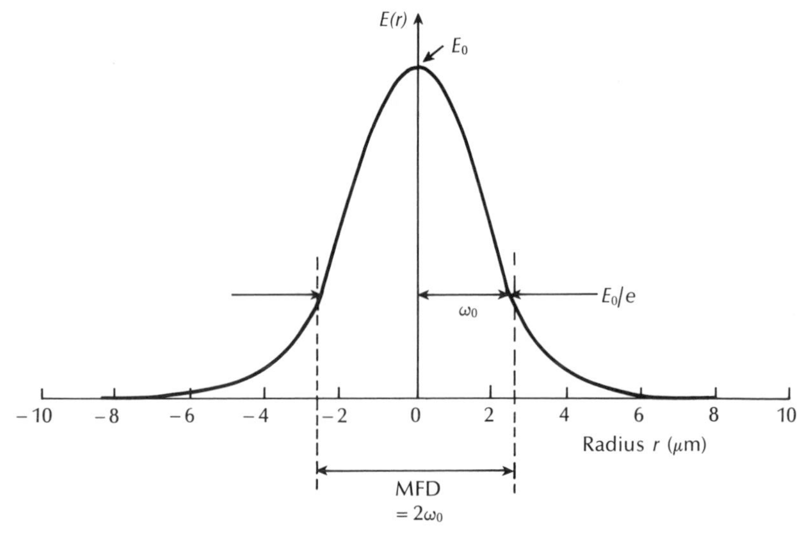

Figure 2.31 Field amplitude distribution $E(r)$ of the fundamental mode in a single-mode fiber illustrating the mode-field diameter (MFD) and spot size (ω_0).

Another parameter which is directly related to the mode-field diameter of a single-mode fiber is the spot size (or mode-field radius) ω_0. Hence $MFD = 2\,\omega_0$, where ω_0 is the nominal half width of the input excitation (see Figure 2.31). The MFD can therefore be regarded as the single-mode analog of the fiber core diameter in multimode fibers [Ref. 45]. However, for many refractive index profiles and at typical operating wavelengths the MFD is slightly larger than the single-mode fiber core diameter.

Often, for real fibers and those with arbitrary refractive index profiles the radial field distribution is not strictly Gaussian and hence alternative techniques have been proposed. However, the problem of defining the MFD and spot size for non-Gaussian field distributions is a difficult one and at least eight definitions exist [Ref. 19]. Nevertheless, a more general definition based on the second moment of the far field and known as the Petermann II definition [Ref. 46] is recommended by the CCITT. Moreover, good agreement has been obtained using this definition for the MFD using different measurement techniques on arbitrary index fibers [Ref. 47].

2.5.3 Effective refractive index

The rate of change of phase of the fundamental LP_{01} mode propagating along a straight fiber is determined by the phase propagation constant β (see Section 2.3.2). It is directly related to the wavelength of the LP_{01} mode λ_{01} by the factor 2π, since β gives the increase in phase angle per unit length. Hence:

$$\beta\lambda_{01} = 2\pi, \text{ or } \lambda_{01} = \frac{2\pi}{\beta} \tag{2.101}$$

Moreover, it is convenient to define an effective refractive index for single-mode fiber, sometimes referred to as a phase index or normalized phase change coefficient [Ref. 48] n_{eff}, by the ratio of the propagation constant of the fundamental mode to that of the vacuum propagation constant:

$$n_{\text{eff}} = \frac{\beta}{k} \tag{2.102}$$

Hence, the wavelength of the fundamental mode λ_{01} is smaller than the vacuum wavelength λ by the factor $1/n_{\text{eff}}$ where:

$$\lambda_{01} = \frac{\lambda}{n_{\text{eff}}} \tag{2.103}$$

It should be noted that the fundamental mode propagates in a medium with a refractive index $n(r)$ which is dependent on the distance r from the fiber axis. The effective refractive index can therefore be considered as an average over the refractive index of this medium [Ref. 19].

Within a normally clad fiber, not depressed-cladded fibers (see Section 2.5), at long wavelengths (i.e. small V values) the mode-field diameter is large compared to the core diameter and hence the electric field extends far into the cladding region.

In this case the propagation constant β will be approximately equal to n_2k (i.e. the cladding wavenumber) and the effective index will be similar to the refractive index of the cladding n_2. Physically, most of the power is transmitted in the cladding material. At short wavelengths, however, the field is concentrated in the core region and the propagation constant β approximates to the maximum wave-number n_1k. Following this discussion, and as indicated previously in Eq. (2.62), then the propagation constant in single-mode fiber varies over the interval $n_2k < \beta < n_1k$. Hence, the effective refractive index will vary over the range $n_2 < n_{\text{eff}} < n_1$.

In addition, a relationship between the effective refractive index and the normalized propagation constant b defined in Eq. (2.71) as:

$$b = \frac{(\beta/k)^2 - n_2^2}{n_1^2 - n_2^2} = \frac{\beta^2 - n_2^2k^2}{n_1k^2 - n_2^2k^2} \tag{2.104}$$

may be obtained. Making use of the mathematical relation, $A^2 - B^2 = (A + B)(A - B)$, Eq. (2.104) can be written in the form:

$$b = \frac{(\beta + n_2k)(\beta - n_2k)}{(n_1k + n_2k)(n_1k - n_2k)} \tag{2.105}$$

However, taking regard of the fact that $\beta \simeq n_1k$, then Eq. (2.105) becomes:

$$b \simeq \frac{\beta - n_2k}{n_1k - n_2k} = \frac{\beta/k - n_2}{n_1 - n_2}$$

Finally, in Eq. (2.102) n_{eff} is equal to β/k, therefore:

$$b \simeq \frac{n_{\text{eff}} - n_2}{n_1 - n_2} \tag{2.106}$$

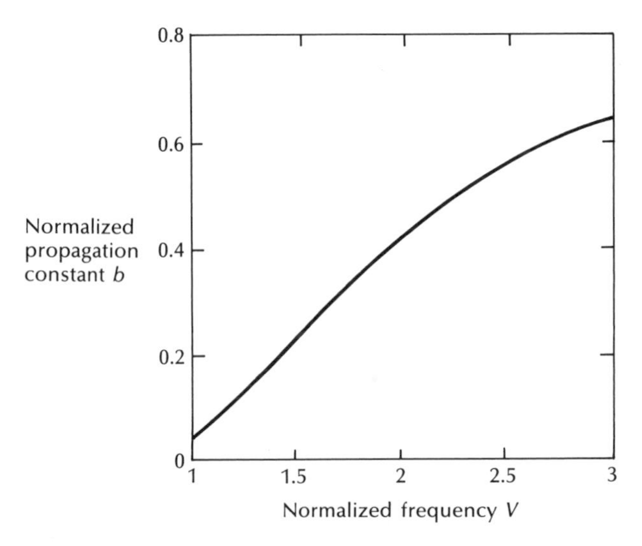

Figure 2.32 The normalized propagation constant (b) of the fundamental mode in a step index fiber shown as a function of the normalized frequency (V).

The dimensionless parameter b which varies between 0 and 1 is particularly useful in the theory of single-mode fibers because the relative refractive index difference is very small giving only a small range for β. Moreover, it allows a simple graphical representation of results to be presented as illustrated by the characteristic shown in Figure 2.32 of the normalized phase constant of β as a function of normalized frequency V in a step index fiber.* It should also be noted that $b(V)$ is a universal function which does not depend explicitly on other fiber parameters.

Example 2.9

Given that a useful approximation for the eigenvalue of the single-mode step index fiber cladding W is [Ref. 43]:

$$W(V) \simeq 1.1428\ \mathrm{V} - 0.9960$$

deduce an approximation for the normalized propagation constant $b(V)$.

Solution: Substituting from Eq. (2.68) into Eq. (2.71), the normalized propagation constant is given by:

$$b(V) = 1 - \frac{(V^2 - W^2)}{V^2} = \frac{W^2}{V^2}$$

Then substitution of the approximation above gives:

$$b(V) \simeq \frac{(1.1428\ \mathrm{V} - 0.9960)^2}{V^2}$$

$$= \left(1.1428 - \frac{0.9960}{V}\right)^2$$

The relative error on this approximation for $b(V)$ is less than 0.2% for $1.5 \leqslant V \leqslant 2.5$ and less than 2% for $1 \leqslant V \leqslant 3$ [Ref. 43].

2.5.4 Group delay and mode delay factor

The transit time or group delay τ_g for a light pulse propagating along a unit length of fiber is the inverse of the group velocity v_g (see Section 2.3.3). Hence:

$$\tau_g = \frac{1}{v_g} = \frac{d\beta}{d\omega} = \frac{1}{c}\frac{d\beta}{dk} \tag{2.107}$$

The group index of a uniform plane wave propagating in a homogeneous medium

* For step index fibers the eigenvalue U, which determines the radial field distribution in the core, can be obtained from the plot of b versus V because from Eq. (2.71), $U^2 = V^2(1 - b)$.

has been determined following Eq. (2.40) as:

$$N_g = \frac{c}{v_g}$$

However, for a single-mode fiber, it is usual to define an effective group index* N_{ge} [Ref. 48] by:

$$N_{ge} = \frac{c}{v_g} \tag{2.108}$$

where v_g is considered to be the group velocity of the fundamental fiber mode. Hence, the specific group delay of the fundamental fiber mode becomes:

$$\tau_g = \frac{N_{ge}}{c} \tag{2.109}$$

Moreover, the effective group index may be written in terms of the effective refractive index n_{eff} defined in Eq. (2.102) as:

$$N_{ge} = n_{eff} - \lambda \frac{dn_{eff}}{d\lambda} \tag{2.110}$$

It may be noted that Eq. (2.110) is of the same form as the denominator of Eq. (2.40) which gives the relationship between the group index and the refractive index in a transparent medium (planar guide).

Rearranging Eq. (2.71) β may be expressed in terms of the relative index difference Δ and the normalized propagation constant b by the following approximate expression:

$$\beta = k[(n_1^2 - n_2^2)b + n_2^2]^{\frac{1}{2}} \simeq kn_2[1 + b\Delta] \tag{2.111}$$

Furthermore, approximating the relative refractive index difference as $(n_1 - n_2)/n_2$, for a weakly guiding fiber where $\Delta \ll 1$, we can use the approximation [Ref. 16]:

$$\frac{n_1 - n_2}{n_2} \simeq \frac{N_{g1} - N_{g2}}{N_{g2}} \tag{2.112}$$

where N_{g1} and N_{g2} are the group indices for the fiber core and cladding regions respectively. Substituting Eq. (2.111) for β into Eq. (2.107) and using the approximate expression given in Eq. (2.112) we obtain the group delay per unit distance as:

$$\tau_g = \frac{1}{c}\left[N_{g2} + (N_{g1} - N_{g2})\frac{d(Vb)}{dV}\right] \tag{2.113}$$

The dispersive properties of the fiber core and the cladding are often about the same and therefore the wavelength dependence of Δ can be ignored [Ref. 19].

* N_{ge} may also be referred to as the group index of the single-mode waveguide.

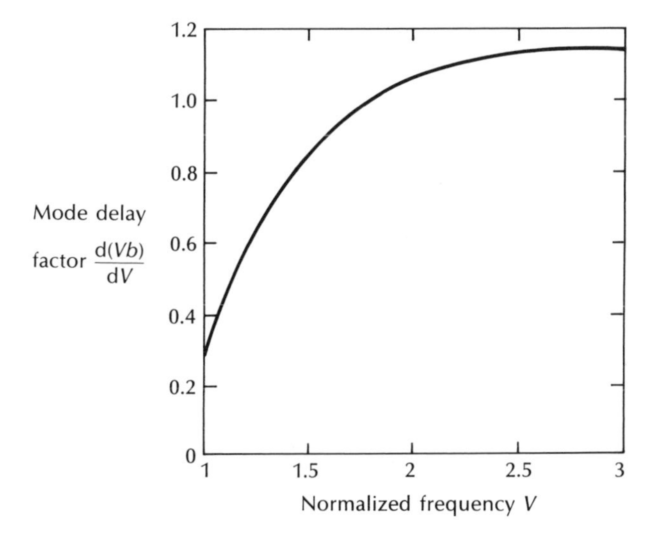

Figure 2.33 The mode delay factor ($d(Vb)/dV$) for the fundamental mode in a step index fiber shown as a function of normalized frequency (V).

Hence the group delay can be written as:

$$\tau_g = \frac{1}{c}\left[N_{g2} + n_2\Delta\,\frac{d(Vb)}{dV}\right] \tag{2.114}$$

The initial term in Eq. (2.114) gives the dependence of the group delay on wavelength caused when a uniform plane wave is propagating in an infinitely extended medium with a refractive index which is equivalent to that of the fiber cladding. However, the second term results from the waveguiding properties of the fiber only and is determined by the mode delay factor $d(Vb)/dV$, which describes the change in group delay caused by the changes in power distribution between the fiber core and cladding. The mode delay factor [Ref. 50] is a further universal parameter which plays a major part in the theory of single-mode fibers. Its variation with normalized frequency for the fundamental mode in a step index fiber is shown in Figure 2.33.

2.5.5 The Gaussian approximation

The field shape of the fundamental guided mode within a single-mode step index fiber for two values of normalized frequency is displayed in Figure 2.34. As may be expected, considering the discussion in Section 2.4.1, it has the form of a Bessel function ($J_0(r)$) in the core region matched to a modified Bessel function ($K_0(r)$) in the cladding. Depending on the value of the normalized frequency a significant proportion of the modal power is propagated in the cladding region, as mentioned

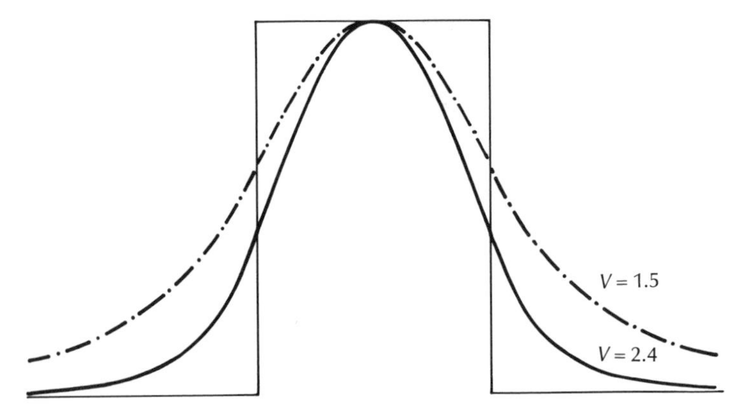

Figure 2.34 Field shape of the fundamental mode for normalized frequencies, $V = 1.5$ and $V = 2.4$.

earlier. Hence, even at the cutoff value (i.e. V_c) only about 80% of the power propagates within the fiber core.

It may be observed from Figure 2.34 that the shape of the fundamental LP_{01} mode is similar to a Gaussian shape, which allows an approximation of the exact field distribution by a Gaussian function.* The approximation may be investigated by writing the scalar wave equation Eq. (2.27) in the form:

$$\nabla^2\psi + n^2k^2\psi = 0 \tag{2.115}$$

where k is the propagation vector defined in Eq. (2.33) and $n(x, y)$ is the refractive index of the fiber, which does not generally depend on z, the coordinate along the fiber axis. It should be noted that the time dependence $\exp(j\omega t)$ has been omitted from the scalar wave equation to give the reduced wave equation† in Eq. (2.115) [Ref. 23]. This representation is valid since the guided modes of a fiber with a small refractive index difference (i.e. $\Delta \ll 1$) have one predominant transverse field component, for example E_y. By contrast E_x and the longitudinal component are very much smaller [Ref. 23].

The field of the fundamental guided mode may therefore be considered as a scalar quantity and need not be described by the full set of Maxwell's equations. Hence Eq. (2.115) may be written as:

$$\nabla^2\phi + n^2k^2\phi = 0 \tag{2.116}$$

where ϕ represents the dominant transverse electric field component.

The near Gaussian shape of the predominant transverse field component of the fundamental mode has been demonstrated [Ref. 51] for fibers with a wide range of refractive index distributions. This proves to be the case not only for the LP_{01}

* However, it should be noted that $K_0(r)$ decays as $\exp(-r)$ which is much slower than a true Gaussian.
† Eq. (2.115) is also known as the Helmholtz equation.

mode of the step index fiber but also for the modes with fibers displaying arbitrary graded refractive index distributions. Therefore, the predominant electric field component of the single guided mode may be written as the Gaussian function [Ref. 23]:

$$\phi = \left(\frac{2}{\pi}\right)^{\frac{1}{2}} \frac{1}{\omega_0} \exp\left(-r^2/\omega_0^2\right) \exp\left(-j\beta z\right) \tag{2.117}$$

where the radius parameter $r^2 = x^2 + y^2$, ω_0 is a width parameter which is often called the spot size or radius of the fundamental mode (see Section 2.5.2) and β is the propagation constant of the guided mode field.

The factor preceding the exponential function is arbitrary and is chosen for normalization purposes. If it is accepted that Eq. (2.117) is to a good approximation the correct shape [Ref. 26], then the parameters β and ω_0 may be obtained either by substitution [Ref. 52] or by using a variational principle [Ref. 26]. Using the latter technique solutions of the wave equation, Eq. (2.116), are claimed to be functions of the minimum integral:

$$J = \int_V \left[(\nabla\phi)\cdot(\nabla\phi^*) - n^2k^2\phi\phi^*\right] \mathrm{d}V = \min \tag{2.118}$$

where the asterisk indicates complex conjugation. The integration range in Eq. (2.118) extends over a large cylinder with the fiber at its axis. Moreover, the length of the cylinder L is arbitrary and its radius is assumed to tend towards infinity.

Use of variational calculus [Ref. 53] indicates that the wave equation Eq. (2.116) is the Euler equation of the variational expression given in Eq. (2.118). Hence, the functions that minimize J satisfy the wave equation. Firstly, it can be shown [Ref. 23] that the minimum value of J is zero if ϕ is a legitimate guided mode field. We do this by performing a partial integration Eq. (2.118) which can be written as:

$$J = \int_s \phi^*(\nabla\phi)\, \mathrm{d}s - \int_V \left[\nabla^2\phi + n^2k^2\phi\right]\phi^*\, \mathrm{d}V \tag{2.119}$$

where the surface element ds represents a vector in a direction normal to the outside of the cylinder. However, the function ϕ for a guided mode disappears on the curved cylindrical surface with infinite radius. In this case the guided mode field may be expressed as:

$$\phi = \hat{\phi}(x, y) \exp\left(-j\beta z\right) \tag{2.120}$$

It may be observed from Eq. (2.120) that the z dependence is limited to the exponential function and therefore the integrand of the surface integral in Eq. (2.119) is independent of z. This indicates that the contributions to the surface integral from the two end faces of the cylinder are equal in value, opposite in sign and independent of the cylinder length. Thus the entire surface integral goes to zero. Moreover, when the function ϕ is a solution of the wave equation, the volume integral in Eq. (2.119) is zero and hence J is also equal to zero.

The variational expression given in Eq. (2.118) can now be altered by substituting Eq. (2.120). In this case the volume integral becomes an integral over the infinite cross section of the cylinder (i.e. the fiber) which may be integrated over the length coordinate z. Integration over z effectively multiplies the remaining integral over the cross section by the cylinder length L because the integrand is independent of z. Hence dividing by L we can write:

$$\frac{J}{L} = \int_{-\infty}^{\infty} \int_{-\infty}^{\infty} \{(\nabla_t \hat{\phi})(\Delta_t \hat{\phi}^*) - [n^2(x, y)k^2 - \beta^2]\hat{\phi}\hat{\phi}^*\} \, dx \, dy \qquad (2.121)$$

where the operator V_t indicates the transverse part (i.e. the x and y derivatives) of Δ.

We have now obtained in Eq. (2.121) the required variational expression that will facilitate the determination of spot size and propagation constant for the guided mode field. The latter parameter may be obtained by solving Eq. (2.121) for β^2 with $J = 0$, as has been proven to be the case for solutions of the wave equation. Thus:

$$\beta^2 = \frac{\int_{-\infty}^{\infty} \int_{-\infty}^{\infty} [n^2k^2\hat{\phi}\hat{\phi}^* - (\nabla_t\hat{\phi})(\nabla_t\hat{\phi}^*)] \, dx \, dy}{\int_{-\infty}^{\infty} \int_{-\infty}^{\infty} \hat{\phi}\hat{\phi}^* \, dx \, dy} \qquad (2.122)$$

Equation (2.122) allows calculation of the propagation constant of the fundamental mode if the function ϕ is known. However, the integral expression in Eq. (2.122) exhibits a stationary value such that it remains unchanged to the first order when the exact mode function $\hat{\phi}$ is substituted by a slightly perturbed function. Hence a good approximation to the propagation constant can be obtained using a function that only reasonably approximates to the exact function. The Gaussian approximation given in Eq. (2.117) can therefore be substituted into Eq. (2.122) to obtain:

$$\beta^2 = \left[\frac{4k^2}{\omega_0^2} \int_0^{\infty} rn^2(r) \exp(-2r^2/\omega_0^2) \, dr\right] - \frac{2}{\omega_0^2} \qquad (2.123)$$

Two points should be noted in relation to Eq. (2.123). Firstly, following Marcuse [Ref. 23] the normalization was picked to bring the denominator of Eq. (2.122) to unity. Secondly, the stationary expression of Eq. (2.123) was obtained from Eq. (2.122) by assuming that the refractive index was dependent only upon the radial coordinate r. This condition is, however, satisfied by most common optical fibre types.

Finally, to derive an expression for the spot size ω_0 we again make use of the stationary property of Eqs. (2.122) and (2.123). Hence, if the Gaussian function of Eq. (2.117) is the correct mode function to give a value for ω_0, then β^2 will not alter if ω_0 is changed slightly. This indicates that the derivative of β^2 with respect to ω_0 becomes zero (i.e. $d\beta^2/d\omega_0 = 0$). Therefore, differentiation of Eq. (2.123) and setting the result to zero yields:

$$1 + 2k^2 \int_0^{\infty} r\left(\frac{2r^2}{\omega_0^2} - 1\right)n^2(r) \exp(-2r^2/\omega_0^2) \, dr = 0 \qquad (2.124)$$

Equation (2.124) allows the Gaussian approximation for the fundamental mode within single-mode fiber to be obtained by providing a value for the spot size ω_0. This value may be utilized in Eq. (2.123) to determine the propagation constant β.

For step index profiles it can be shown [Ref. 52] that an optimum value of the spot size ω_0 divided by the core radius is only a function of the normalized frequency V. The optimum values of ω_0/a can be approximated to better than 1% accuracy by the empirical formula [Ref. 52]:

$$\frac{\omega_0}{a} = 0.65 + 1.619\,\mathrm{V}^{-\frac{3}{2}} + 2.879\,\mathrm{V}^{-6} \tag{2.125}$$

$$= 0.65 + 1.619\left(2.40\,\frac{\lambda_c}{\lambda}\right)^{-\frac{3}{2}} + 2.879\left(2.405\,\frac{\lambda_c}{\lambda}\right)^{-6}$$

$$\omega_0 = a\left[0.65 + 0.434\left(\frac{\lambda}{\lambda_c}\right)^{\frac{3}{2}} + 0.0149\left(\frac{\lambda}{\lambda_c}\right)^{6}\right] \tag{2.126}$$

The approximate expression for spot size given in Eq. (2.126) is frequently used to determine the parameter for step index fibers over the usual range of λ/λ_c (i.e. 0.8 to 1.9) [Ref. 43].

Example 2.10

Estimate the fiber core diameter for a single-mode step index fiber which has a MFD of 11.6 μm when the normalized frequency is 2.2.

Solution: Using the Gaussian approximation, from Eq. (2.125) the fiber core radius is:

$$a = \frac{\omega_0}{0.65 + 1.619(\mathrm{V})^{-\frac{3}{2}} + 2.879(\mathrm{V})^{-6}}$$

$$= \frac{5.8 \times 10^{-6}}{0.65 + 1.619(2.2)^{-\frac{3}{2}} + 2.879(2.2)^{-6}}$$

$$= 4.95\ \mu\mathrm{m}$$

Hence the fiber core diameter is 9.9 μm.

The accuracy of the Gaussian approximation has also been demonstrated for graded index fibers [Ref. 54], having a refractive index profile given by Eq. (2.76) (i.e. power law profiles in the core region). When the near parabolic refractive index profile is considered (i.e. $\alpha = 2$) and the square law medium is assumed to extend to infinity rather than to the cladding where $n(r) = n_2$, for $r \geqslant a$ (Eq. (2.76)); then the Gaussian spot size given in Eq. (2.124) reduces to:

$$\omega_0^2 = \frac{a}{n_1 k}\left(\frac{2}{\Delta}\right)^{\frac{1}{2}} \tag{2.127}$$

Furthermore, the propagation constant becomes:

$$\beta^2 = n_1^2 k^2\left[1 - \frac{2(2\Delta)^{\frac{1}{2}}}{n_1 ka}\right] \tag{2.128}$$

It is interesting to note that the above relationships for ω_0 and β in this case are identical to the solutions obtained from exact analysis of the square law medium [Ref. 26].

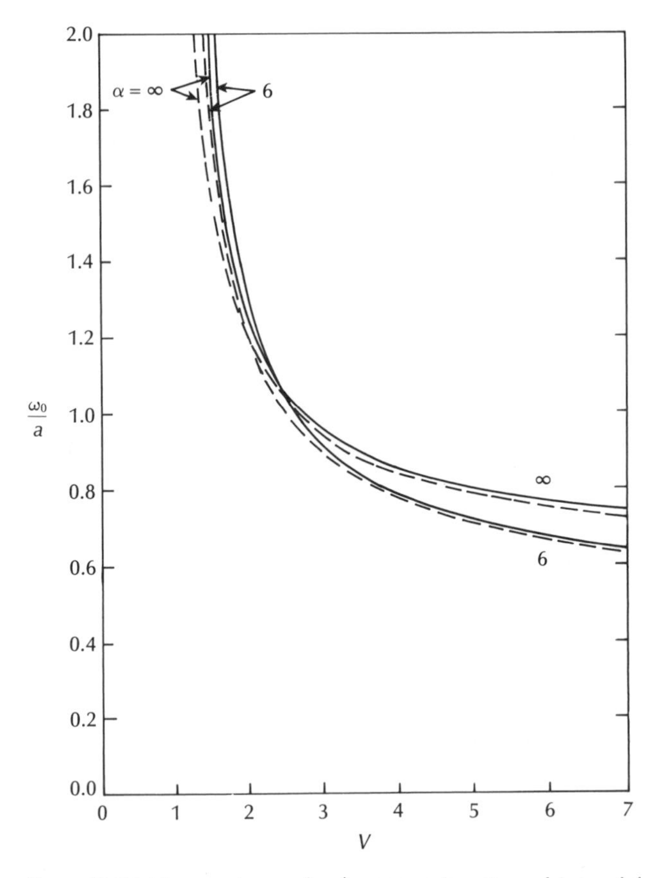

Figure 2.35 Comparison of ω_0/a approximation obtained from Eqs. (2.123) and (2.124) (broken lines) with values obtained from numerical integration of the wave equation and subsequent optimization of its width (solid lines). Reproduced with permission from D. Marcuse, 'Gaussian approximation of the fundamental modes of graded-index fibers', *J. Opt. Soc. Am.*, **68**, p. 103, 1978.

Numerical solutions of Eqs. (2.123) and (2.124) are shown in Figure 2.35 (broken lines) for values of α of 6 and ∞ for profiles with constant refractive indices in the cladding region [Ref. 51]. In this case Eqs. (2.123) and (2.124) cannot be solved analytically and computer solutions must be obtained. The solid lines in Figure 2.35 show the corresponding solutions of the wave equation, also obtained by a direct numerical technique. These results for the spot size and propagation constant are provided for comparison as they are not influenced by the prior assumption of Gaussian shape.

The Gaussian approximation for the transverse field distribution is very much simpler than the exact solution and is very useful for calculations involving both launching efficiency at the single-mode fiber input as well as coupling losses at splices or connectors. In this context it describes very well the field inside the fiber core and provides good approximate values for the guided mode propagation constant. It is a particularly good approximation for fibers operated near the cutoff wavelength of the second order mode [Ref. 26] but when the wavelength increases, the approximation becomes less accurate. In addition, for single-mode fibers with homogeneous cladding, the true field distribution is never exactly Gaussian since the evanescent field in the cladding tends to a more exponential function for which the Gaussian provides an underestimate.

However, for the calculations involving cladding absorption, bend losses, crosstalk between fibers and the properties of directional couplers, then the Gaussian approximation should not be utilized [Ref. 26]. Better approximations for the field profile in these cases can, however, be employed such as the exponential function [Ref. 55], or the modified Hankel function of zero order [Ref. 56], giving the Gaussian-exponential and the Gaussian–Hankel approximations respectively. Unfortunately, these approximations lose the major simplicity of the Gaussian approximation, in which essentially one parameter (the spot size) defines the radial amplitude distribution, because they necessitate two parameters to characterize the same distribution.

2.5.6 Equivalent step index methods

Another strategy to obtain approximate values for the cutoff wavelength and spot size in graded index single-mode fibers (or arbitrary refractive index profile fibers) is to define an equivalent step index (ESI) fiber on which to model the fiber to be investigated. Various methods have been proposed in the literature [e.g. Refs. 57 to 62] which commence from the observation that the fields in the core regions of graded index fibers often appear similar to the fields within step index fibers. Hence, as step index fiber characteristics are well known, it is convenient to replace the exact methods for graded index single-mode fibers [Refs. 63, 64] by approximate techniques based on step index fibers. In addition, such ESI methods allow the propagation characteristics of single-mode fibers to be represented by a few parameters.

Several different suggestions have been advanced for the choice of the core radius a_{ESI}, and the relative index difference Δ_{ESI}, of the ESI fiber which lead to good approximations for the spot size (and hence joint and bend losses) for the actual graded index fiber. They are all conceptually related to the Gaussian approximation (see Section 2.5.5) in that they utilize the close resemblance of the field distribution of the LP_{01} mode to the Gaussian distribution within single-mode fiber. An early proposal for the ESI method [Ref. 58] involved transformation of the basic fiber parameters following:

$$a_s = Xa, \qquad V_s = YV, \qquad NA_s = (Y/X)NA \tag{2.129}$$

where the subscript s is for the ESI fiber and X, Y are constants which must be determined. However, these ESI fiber representations are only valid for a particular value of normalized frequency V and hence there is a different X, Y pair for each wavelength. The transformation can be carried out either on the basis of compared radii or relative refractive index differences. Figure 2.36 compares the refractive index profiles and the electric field distributions for two graded index fibres ($\alpha = 2, 4$) and their ESI fibers. It may be observed that their fields differ slightly only near the axis.

An alternative ESI technique is to normalize the spot size ω_0 with respect to an optimum effective fiber core radius a_{eff} [Ref. 61]. This latter quantity is obtained from the experimental measurement of the first minimum (angle θ_{min}) in the diffraction pattern using transverse illumination of the fiber immersed in an index matching fluid. Hence:

$$a_{eff} = 3.832/k \sin \theta_{min} \tag{2.130}$$

where $k = 2\pi/\lambda$. In order to obtain the full comparison with single-mode step index fiber, the results may be expressed in terms of an effective normalized frequency V_{eff} which relates the cutoff frequencies/wavelengths for the two fibers:

$$V_{eff} = 2.405(V/V_c) = 2.405(\lambda_c/\lambda) \tag{2.131}$$

The technique provides a dependence of ω_0/a_{eff} on V_{eff} which is almost identical for a reasonably wide range of profiles which are of interest for minimizing dispersion (i.e. $1.5 < V_{eff} < 2.4$).

A good analytical approximation for this dependence is given by [Ref. 61]:

$$\frac{\omega_0}{a_{eff}} = 0.6043 + 1.755\ V_{eff}^{-\frac{3}{2}} + 2.78\ V_{eff}^{-6} \tag{2.132}$$

Refractive index profile dependent deviations from the relationship shown in Eq. (2.132) are within $\pm 2\%$ for general power law graded index profiles.

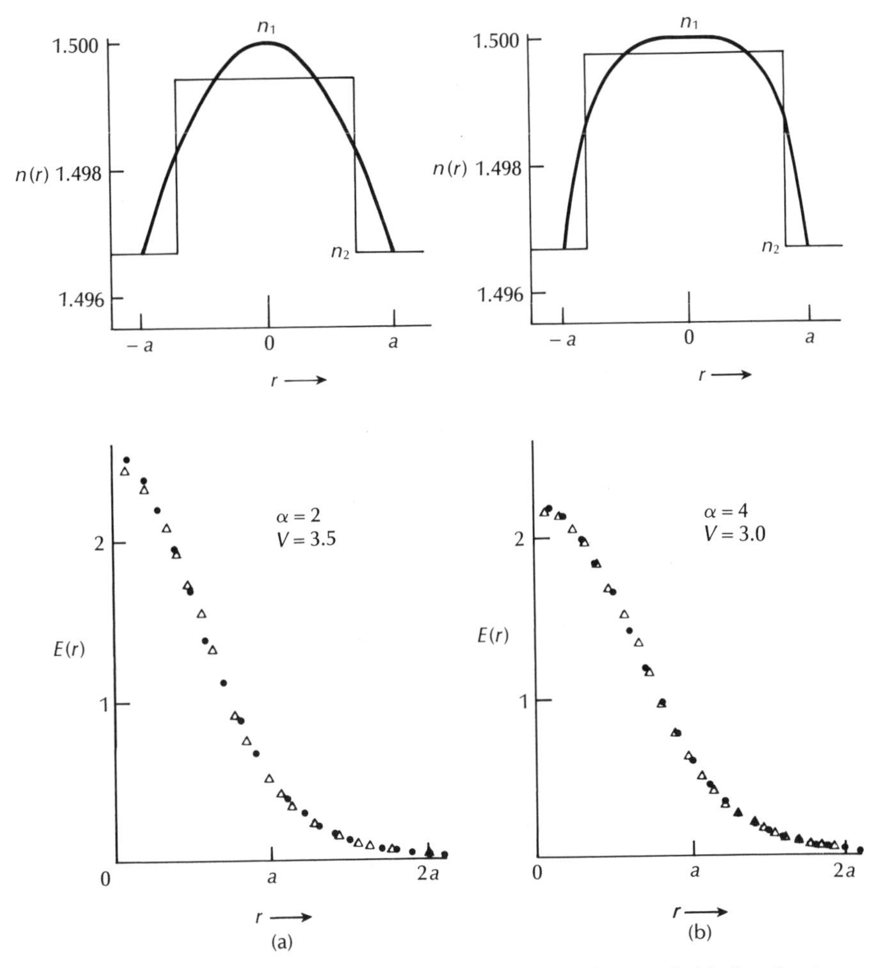

Figure 2.36 Refractive index distributions $n(r)$ and electric field distributions $E(r)$ for graded index fibers and their ESI fibers for: (a) $\alpha = 2$, $V = 3.5$; (b) $\alpha = 4$, $V = 3.0$. The field distributions for the graded index and corresponding ESI profiles are shown by solid circles and open triangles respectively. Reproduced with permission from H. Matsumura and T. Suganuma, *Appl. Opt.*, **19**, p. 3151, 1980.

Example 2.11

A parabolic profile graded index single-mode fiber designed for operation at a wavelength of 1.30 μm has a cutoff wavelength of 1.08 μm. From experimental measurement it is established that the first minimum in the diffraction pattern occurs at an angle of 12°. Using an ESI technique, determine the spot size at the operating wavelength.

Solution: Using Eq. (2.130), the effective core radius is:

$$a_{\text{eff}} = \frac{3.832\lambda}{2\pi \sin \theta_{\min}} = \frac{3.832 \times 1.30 \times 10^{-6}}{2\pi \sin 12^{\circ}}$$

$$= 3.81\ \mu\text{m}$$

The effective normalized frequency can be obtained from Eq. (2.131) as:

$$V_{\text{eff}} = 2.405\frac{\lambda_c}{\lambda} = 2.405\frac{1.08}{1.30} = 2.00$$

Hence the spot size is given by Eq. (2.132) as:

$$\omega_0 = 3.81 \times 10^{-6}[0.6043 + 1.755(2.00)^{-\frac{3}{2}} + 2.78(2.00)^{-6}]$$
$$= 4.83\ \mu\text{m}$$

Other ESI methods involve the determination of the equivalent parameters from experimental curves of spot size against wavelength [Ref. 62]. All require an empirical formula, relating spot size to the normalized frequency for a step index fiber, to be fitted by some means to the data. The usual empirical formula employed is that derived by Marcuse for the Gaussian approximation and given in Eq. (2.125). An alternative formula which is close to Eq. (2.125) is provided by Snyder [Ref. 65] as:

$$\omega_0 = a(\ln V)^{-\frac{1}{2}} \tag{2.133}$$

However, it is suggested [Ref. 62] that the expression given in Eq. (2.133) is probably less accurate than that provided by Eq. (2.125).

A cutoff method can also be utilized to obtain the ESI parameters [Ref. 66]. In this case the cutoff wavelength λ_c and spot size ω_0 are known. Therefore, substituting $V = 2.405$ into Eq. (2.125) gives:

$$\omega_0 = 1.099a_{\text{ESI}} \text{ or } 2a_{\text{ESI}} = 1.820\omega_0 \tag{2.134}$$

Then using Eq. (2.70) the ESI relative index difference is:

$$\Delta_{\text{ESI}} = (0.293/n_1^2)(\lambda_c/2a_{\text{ESI}})^2 \tag{2.135}$$

where n_1 is the maximum refractive index of the fiber core.

Example 2.12

Obtain the ESI relative refractive index difference for a graded index fiber which has a cutoff wavelength and spot size of 1.190 μm and 5.2 μm respectively. The maximum refractive index of the fiber core is 1.485.

Solution: The ESI core radius may be obtained from Eq. (2.134) where:

$$2a_{\text{ESI}} = 1.820 \times 5.2 \times 10^{-6} = 9.464\ \mu\text{m}$$

Using Eq. (2.135), the ESI relative index difference is given by:

$$\Delta_{\text{ESI}} = (0.293/1.485^2)(1.190/9.464)^2$$
$$= 2.101 \times 10^{-3} \text{ or } 0.21\%$$

Alternatively, performing a least squares fit on Eq. (2.125) provides 'best values' for the ESI diameter ($2a_{\text{ESI}}$) and relative index difference (Δ_{ESI}) [Ref. 62]. It must be noted, however, that these best values are dependent on the application and the least squares method appears most useful in estimating losses at fiber joints [Ref. 67]. In addition, recent work [Ref. 68] has attempted to provide a more consistent relationship between the ESI parameters and the fiber mode-field diameter. Overall, the concept of the ESI fiber has been relatively useful in the specification of standard matched-cladding and depressed-cladding fibers by their equivalent a_{ESI} and Δ_{ESI} values. Unfortunately, ESI methods are unable accurately to predict mode-field diameters and waveguide dispersion in dispersion shifted and dispersion flattened (see Section 3.12) fibers [Ref. 19].

Problems

2.1 Using simple ray theory, describe the mechanism for the transmission of light within an optical fiber. Briefly discuss with the aid of a suitable diagram what is meant by the acceptance angle for an optical fiber. Show how this is related to the fiber numerical aperture and the refractive indices for the fiber core and cladding.

An optical fiber has a numerical aperture of 0.20 and a cladding refractive index of 1.59. Determine:

(a) the acceptance angle for the fiber in water which has a refractive index of 1.33;
(b) the critical angle at the core–cladding interface.

Comment on any assumptions made about the fiber.

2.2 The velocity of light in the core of a step index fiber is 2.01×10^8 m s^{-1}, and the critical angle at the core–cladding interface is 80°. Determine the numerical aperture and the acceptance angle for the fiber in air, assuming it has a core diameter suitable for consideration by ray analysis. The velocity of light in a vacuum is 2.998×10^3 m s^{-1}.

2.3 Define the relative refractive index difference for an optical fiber and show how it may be related to the numerical aperture.

A step index fiber with a large core diameter compared with the wavelength of the transmitted light has an acceptance angle in air of 22° and a relative refractive index difference of 3%. Estimate the numerical aperture and the critical angle at the core–cladding interface for the fiber.

2.4 A step index fiber has a solid acceptance angle in air of 0.115 radians and a relative refractive index difference of 0.9%. Estimate the speed of light in the fiber core.

2.5 Briefly indicate with the aid of suitable diagrams the difference between meridional and skew ray paths in step index fibers.

Derive an expression for the acceptance angle for a skew ray which changes direction by an angle 2γ at each reflection in a step index fiber in terms of the fiber NA and γ. It may be assumed that ray theory holds for the fiber.

A step index fiber with a suitably large core diameter for ray theory considerations has core and cladding refractive indices of 1.44 and 1.42 respectively. Calculate the acceptance angle in air for skew rays which change direction by $150°$ at each reflection.

2.6 Skew rays are accepted into a large core diameter (compared to the wavelength of the transmitted light) step index fiber in air at a maximum axial angle of $42°$. Within the fiber they change direction by $90°$ at each reflection. Determine the acceptance angle for meridional rays for the fiber in air.

2.7 Explain the concept of electromagnetic modes in relation to a planar optical waveguide.

Discuss the modifications that may be made to electromagnetic mode theory in a planar waveguide in order to describe optical propagation in a cylindrical fiber.

2.8 Briefly discuss, with the aid of suitable diagrams, the following concepts in optical fiber transmission:

(a) the evanescent field;
(b) Goos–Haenchen shift;
(c) mode coupling.

Describe the effects of these phenomena on the propagation of light in optical fibers.

2.9 Define the normalized frequency for an optical fiber and explain its use in the determination of the number of guided modes propagating within a step index fiber.

A step index fiber in air has a numerical aperture of 0.16, a core refractive index of 1.45 and a core diameter of 60 μm. Determine the normalized frequency for the fiber when light at a wavelength of 0.9 μm is transmitted. Further, estimate the number of guided modes propagating in the fiber.

2.10 Describe with the aid of simple ray diagrams:

(a) the multimode step index fiber;
(b) the single-mode step index fiber.

Compare the advantages and disadvantages of these two types of fiber for use as an optical channel.

2.11 A multimode step index fiber has a relative refractive index difference of 1% and a core refractive index of 1.5. The number of modes propagating at a wavelength of 1.3 μm is 1100. Estimate the diameter of the fiber core.

2.12 Explain what is meant by a graded index optical fiber, giving an expression for the possible refractive index profile. Using simple ray theory concepts, discuss the transmission of light through the fiber. Indicate the major advantage of this type of fiber with regard to multimode propagation.

2.13 The relative refractive index difference between the core axis and the cladding of a graded index fiber is 0.7% when the refractive index at the core axis is 1.45. Estimate values for the numerical aperture of the fiber when:

(a) the index profile is not taken into account; and
(b) the index profile is assumed to be triangular.

Comment on the results.

2.14 A multimode graded index fiber has an acceptance angle in air of $8°$. Estimate the relative refractive index difference between the core axis and the cladding when the refractive index at the core axis is 1.52.

2.15 The WKB value for the propagation constant β given in Eq. (2.91) in a parabolic refractive index core fiber assumes an infinitely extended parabolic profile medium. When in a practical fiber the parabolic index profile is truncated, show that the mode numbers m and l are limited by the following condition:

$$2(2m + l + 1) \leqslant ka(n_1^2 - n_2^2)^{\frac{1}{2}}$$

2.16 A graded index fiber with a parabolic index profile supports the propagation of 742 guided modes. The fiber has a numerical aperture in air of 0.3 and a core diameter of 70 μm. Determine the wavelength of the light propagating in the fiber.

Further estimate the maximum diameter of the fiber which gives single-mode operation at the same wavelength.

2.17 A graded index fiber with a core axis refractive index of 1.5 has a characteristic index profile (α) of 1.90, a relative refractive index difference of 1.3% and a core diameter of 40 μm. Estimate the number of guided modes propagating in the fiber when the transmitted light has a wavelength of 1.55 μm, and determine the cutoff value of the normalized frequency for single-mode transmission in the fiber.

2.18 A single-mode step index fiber has a core diameter of 7 μm and a core refractive index of 1.49. Estimate the shortest wavelength of light which allows single-mode operation when the relative refractive index difference for the fiber is 1%.

2.19 In problem 2.18, it is required to increase the fiber core diameter to 10 μm whilst maintaining single-mode operation at the same wavelength. Estimate the maximum possible relative refractive index difference for the fiber.

2.20 Show that the maximum value of a/λ is approximately 1.4 times larger for a parabolic refractive index profile single-mode fiber than for a single-mode step index fiber. Hence, sketch the relationship between the maximum core diameter and the propagating optical wavelength which will facilitate single-mode transmission in the parabolic profile fiber.

2.21 A single-mode step index fiber which is designed for operation at a wavelength of 1.3 μm has core and cladding refractive indices of 1.447 and 1.442 respectively. When the core diameter is 7.2 μm, confirm that the fiber will permit single-mode transmission and estimate the range of wavelengths over which this will occur.

2.22 A single-mode step index fiber has core and cladding refractive indices of 1.498 and 1.495 respectively. Determine the core diameter required for the fiber to permit its operation over the wavelength range 1.48 to 1.60 μm. Calculate the new fiber core diameter to enable single-mode transmission at a wavelength of 1.30 μm.

2.23 A single-mode fiber has a core refractive index of 1.47. Sketch a design characteristic of relative refractive index difference Δ against core radius for the fiber to operate at a wavelength of 1.30 μm. Determine whether the fiber remains single-mode at a transmission wavelength of 0.85 μm when its core radius is 4.5 μm.

2.24 Convert the approximation for the normalized propagation constant of a single-mode step index fiber given in Example 2.9 into a relationship involving the normalized wavelength λ/λ_c in place of the normalized frequency. Hence, determine the range of values of this parameter over which the relative error in the approximation is between 0.2% and 2%.

2.25 Given that the Gaussian function for the electric field distribution of the fundamental mode in a single-mode fiber of Eq. (2.117) takes the form:

$$E(r) = E_0 \exp(-r^2/\omega_0^2)$$

where $E(r)$ and E_0 are shown in Figure 2.31, use the approximation of Eq. (2.125) to evaluate and sketch $E(r)/E_0$ against r/a over the range 0 to 3 for values of normalized frequency $V = 1.0, 1.5, 2.0, 2.5, 3.0$.

2.26 The approximate expression provided in Eq. (2.125) is valid over the range of normalized frequency $1.2 < V < 2.4$. Sketch ω_0/a against V over this range for the fundamental mode in a step index fiber. Comment on the magnitude of ω_0/a as the normalized frequency is reduced significantly below 2.4 and suggest what this indicates about the distribution of the light within the fiber.

2.27 The spot size in a parabolic profile graded index single-mode fiber is 11.0 μm at a transmission wavelength of 1.55 μm. In addition, the cutoff wavelength for the fiber is 1.22 μm. Using an ESI technique, determine the fiber effective core radius and hence estimate the angle at which the first minimum in the diffraction pattern from the fiber would occur.

2.28 The cutoff method is employed to obtain the ESI parameters for a graded index single-mode fiber. If the ESI relative index difference was found to be 0.30% when the spot size and cutoff wavelength were 4.6 μm and 1.29 μm, respectively, calculate the maximum refractive index of the fiber core.

Answers to numerical problems

2.1 (a) 8.6°; (b) 83.6°
2.2 0.263, 15.2°
2.3 0.375, 75.9°
2.4 2.11×10^8 m s^{-1}
2.5 34.6°
2.6 28.2°
2.9 33.5, 561
2.11 92 μm
2.13 (a) 0.172; (b) 0.171
2.14 0.42%
2.16 1.2 μm, 4.4 μm
2.17 94, 3.45
2.18 1.36 μm
2.19 0.24%
2.21 down to 1139 nm
2.22 12.0 μm, 10.5 μm
2.24 $0.8 \leqslant \lambda/\lambda_c \leqslant 1.0$ and $1.6 \leqslant \lambda/\lambda_c \leqslant 2.4$
2.27 3.0 μm, 18.4°
2.28 1.523

References

[1] D. Hondros and P. Debye, 'Electromagnetic waves along long cylinders of dielectric', *Annal. Physik*, **32**(3), pp. 465–476, 1910.

[2] O. Schriever, 'Electromagnetic waves in dielectric wires', *Annal. Physik*, **63**(7), pp. 645–673, 1920.

[3] A. C. S. van Heel, 'A new method of transporting optical images without aberrations', *Nature, Lond.*, **173**, p. 39, 1954.

[4] H. H. Hopkins and N. S. Kapany, 'A flexible fibrescope, using static scanning', *Nature, Lond.*, **113**, pp. 39–41, 1954.

[5] K. C. Kao and G. A. Hockham, 'Dielectric-fibre surface waveguides for optical frequencies', *Proc IEE*, **113**, pp. 1151–1158, 1966.

[6] A. Werts, 'Propagation de la lumière coherente dans les fibres optiques', *L'Onde Electrique*, **46**, pp. 967–980, 1966.

[7] S. Takahashi and T. Kawashima, 'Preparation of low loss multi-component glass fiber', *Tech. Dig. Int. Conf. Integr. Opt. and Opt. Fiber Commun.*, p. 621, 1977.

[8] J. B. MacChesney, P. B. O'Connor, F. W. DiMarcello, J. R. Simpson and P. D. Lazay, 'Preparation of low-loss optical fibres using simultaneous vapour phase deposition and fusion', *Proc. 10th Int. Conf. on Glass*, paper 6–40, 1974.
[9] T. Miya, Y. Terunuma, T. Hosaka and T. Miyashita, 'Ultimate low-loss single-mode fibre at 1.55 μm', *Electron Lett.*, **15**(4), pp. 106–108, 1979.
[10] S. Sakaguchi, 'Low loss optical fibers for midinfrared optical communication', *J. of Lightwave Technol.*, **LT-5**(9), pp. 1219–1228, 1987.
[11] M. Born and E. Wolf, *Principles of Optics*, 6th edn., Pergamon Press, 1980.
[12] D. C. Agarwal, 'Ray concepts in optical fibers', *Indian J. Theoret. Phys.*, **28**(1), pp. 41–54, 1980.
[13] R. P. Feyman, *The Feyman Lectures on Physics*, Vol. 2, Addison-Wesley, 1969.
[14] J. E. Midwinter, *Optical Fibers for Transmission*, John Wiley, 1979.
[15] E. Snitzer, 'Cylindrical dielectric waveguide modes', *J. Opt. Soc. Am.*, **51**, pp. 491–498, 1961.
[16] D. Gloge, 'Weakly guiding fibers', *Appl. Opt.*, **10**, pp. 2252–2258, 1971.
[17] D. Marcuse, *Theory of Dielectric Optical Waveguides*, Academic Press, New York, 1974.
[18] A. W. Snyder, 'Asymptotic expressions for eigenfunctions and eigenvalues of a dielectric or optical waveguide', *Trans IEEE Microwave Theory Tech.*, **MTT-17**, pp. 1130–1138, 1969.
[19] E. G. Neumann, *Single-Mode Fibers: Fundamentals*, Springer-Verlag, 1988.
[20] D. Gloge, 'Optical power flow in multimode fibers', *Bell Syst. Tech. J.*, **51**, pp. 1767–1783, 1972.
[21] R. Olshansky, 'Propagation in glass optical waveguides', *Rev. Mod. Phys.*, **51**(2), pp. 341–366, 1979.
[22] P. M. Morse and H. Fesbach, *Methods of Theoretical Physics*, Vol. II, McGraw-Hill, 1953.
[23] D. Marcuse, *Light Transmission Optics*, 2nd edn, Van Nostrand Reinhold, 1982.
[24] A. Ghatak and K. Thyagarajan, 'Graded index optical waveguides', in E. Wolf (ed.), *Progress in Optics Vol. XVIII*, pp. 3–128, North-Holland 1980.
[25] D. B. Beck, 'Optical fiber waveguides', in M. K. Barnoski (Ed.), *Fundamentals of Optical Fiber Communications*, pp. 1–58, Academic Press, 1976.
[26] D. Marcuse, D. Gloge, E. A. J. Marcatili, 'Guiding properties of fibers', in *Optical Fiber Telecommunications*, S. E. Miller and A. G. Chynoweth (Eds.), Academic Press, pp. 37–100, 1979.
[27] I. S. Gradshteyn and I. M. Ryzhik, *Tables of Integrals, Series and Products*, 4th edn, Academic Press, 1965.
[28] K. Okamoto and T. Okoshi, 'Analysis of wave propagation in optical fibers having core with α-power refractive-index distribution and uniform cladding', *IEEE Trans. Microwave Theory Tech.*, **MTT-24**, pp. 416–421, 1976.
[29] M. A. Saifi, 'Triangular index monomode fibres' in *Proc. SPIE Int. Soc. Opt. Eng. (USA)*, **374**, pp. 13–15, 1983.
[30] D. Gloge, 'The optical fibre as a transmission medium', *Rep. Prog. Phys.*, **42**, pp. 1777–1824, 1979.
[31] M. M. Ramsey and G. A. Hockham, 'Propagation in optical fibre waveguides' in C. P. Sandbank (Ed.) *Optical Fibre Communication Systems*, pp. 25–41, John Wiley, 1980.

[32] S. Kawakami and S. Nishida, 'Characteristics of a doubly clad optical fiber with a low index cladding', *IEEE J. Quantum Electron*, **QE-10**, pp. 879–887, 1974.

[33] C. W. Yeh, 'Optical waveguide theory', *IEEE Trans. Circuits and Syst.*, **CAS-26**(12), pp. 1011–1019, 1979.

[34] C. Pask and R. A. Sammut, 'Developments in the theory of fibre optics', *Proc. IREE Aust.*, **40**(3), pp. 89–101, 1979.

[35] W. A. Gambling, A. H. Hartog and C. M. Ragdale, 'Optical fibre transmission lines', *The Radio Electron. Eng.*, **51**(7/8), pp. 313–325, 1981.

[36] H. G. Unger, *Planar Optical Waveguides and Fibres*, Clarendon Press, 1977.

[37] M. J. Adams, *An Introduction to Optical Waveguides*, John Wiley, 1981.

[38] Y. Suematsu and K.-I. Iga, *Introduction to Optical Fibre Communications*, John Wiley, 1982.

[39] T. Okoshi, *Optical Fibers*, Academic Press, 1982.

[40] H. Kanamori, H. Yokota, G. Tanaka, M. Watanabe, Y. Ishiguro, I. Yoshida, T. Kakii, S. Itoh, Y. Asano and S. Takana, 'Transmission characteristics and reliability of silica-core single-mode fibers', *J. of Lightwave Technol.*, **LT-4**(8), pp. 1144–1150, 1986.

[41] V. A. Bhagavatula, J. C. Lapp, A. J. Morrow and J. E. Ritter, 'Segmented-core fiber for long-haul and local-area-network applications', *J. of Lightwave Technol.*, **6**(10), pp. 1466–1469, 1988.

[42] D. P. Jablonowski, 'Fiber manufacture at AT&T with the MCVD Process', *J. of Lightwave Technol.*, **LT-4**(8), pp. 1016–1019, 1986.

[43] L. B. Jeunhomme, *Single-Mode Fiber Optics*, Marcel Dekker Inc., 1983.

[44] CCITT, COM XV-83-E, USA, 'Cut-off wavelength of cabled single-mode fibers', 1985.

[45] K. I. White, 'Methods of measurements of optical fiber properties', *J. Phys. E. Sci. Instrum.*, **18**, pp. 813–821, 1985.

[46] K. Petermann, 'Constraints for the fundamental-mode spot size for broadband dispersion-compensated single-mode fibres', *Electron. Lett.*, **19**, pp. 712–714, 1983.

[47] W. T. Anderson, V. Shah, L. Curtis, A. J. Johnson and J. P. Kilmer, 'Mode-Field diameter measurements for single-mode fibers with non-Gaussian field profiles', *J. of Lightwave Technol.*, **LT-5**, pp. 211–217, 1987.

[48] H. Kogelnik and H. P. Weber, 'Rays, stored energy, and power flow in dielectric waveguides', *J. Opt. Soc. Am.*, **64**, pp. 174–185, 1974.

[49] A. W. Snyder and J. D. Love, *Optical Waveguide Theory*, Chapman and Hall, 1983.

[50] H. G. Unger, *Planar Optical Waveguides and Fibres*, Clarendon Press, Oxford, 1977.

[51] D. Marcuse, 'Gaussian approximation of the fundamental modes of graded-index fibers', *J. Opt. Soc. Am.*, **68**(1), pp. 103–109, 1978.

[52] D. Marcuse, 'Loss analysis of single-mode fiber splices', *Bell Syst. Tech. J.*, **56**(5), pp. 703–718, 1977.

[53] G. A. Bliss, *Lectures on the Calculus of Variations*, University of Chicago Press, Chicago, 1946.

[54] D. Marcuse, 'Excitation of the dominant mode of a round fiber by a Gaussian beam', *Bell. Syst. Tech. J.*, **49**, pp. 1695–1703, 1970.

[55] A. K. Ghatak, R. Srivastava, I. F. Faria, K. Thyagaranjan and R. Tiwari, 'Accurate method for characterizing single-mode fibers: theory and experiment', *Electron. Lett.*, **19**, pp. 97–99, 1983.

[56] E. K. Sharma and R. Tewari, 'Accurate estimation of single-mode fiber characteristics from near-field measurements', *Electron. Lett.*, **20**, pp. 805–806, 1984.

[57] A. W. Snyder and R. A. Sammut, 'Fundamental (HE) modes of graded optical fibers', *J. Opt. Soc. Am.*, **69**, pp. 1663–1671, 1979.

[58] H. Matsumura and T. Suganama, 'Normalization of single-mode fibers having arbitrary index profile', *Appl. Opt.*, **19**, pp. 3151–3158, 1980.

[59] R. A. Sammut and A. W. Snyder, 'Graded monomode fibres and planar waveguides', *Electron. Lett.*, **16**, pp. 32–34, 1980.

[60] C. Pask and R. A. Sammut, 'Experimental characterisation of graded-index single-mode fibres', *Electron. Lett.*, **16**, pp. 310–311, 1980.

[61] J. Streckert and E. Brinkmeyer, 'Characteristic parameters of monomode fibers', *Appl. Opt.*, **21**, pp. 1910–1915, 1982.

[62] M. Fox, 'Calculation of equivalent step-index parameters for single-mode fibres', *Opt. and Quantum Electron.*, **15**, pp. 451–455, 1983.

[63] W. A. Gambling and H. Matsumura, 'Propagation in radially-inhomogeneous single-mode fibre', *Opt. and Quantum Electron.*, **10**, pp. 31–40, 1978.

[64] W. A. Gambling, H. Matsumura and C. M. Ragdale, 'Wave propagation in a single-mode fibre with dip in the refractive index', *Opt. and Quantum Electron.*, **10**, pp. 301–309, 1978.

[65] A. W. Snyder, 'Understanding monomode optical fibers', *Proc. IEEE*, **69**(1), pp. 6–13, 1981.

[66] V. A. Bhagavatula, 'Estimation of single-mode waveguide dispersion using an equivalent–step–index approach', *Electron, Lett.*, **18**(8), pp. 319–320, 1982.

[67] D. Davidson, 'Single-mode wave propagation in cylindrical optical fibers', *Optical-Fiber Transmission*, E. E. Basch (Ed.), H. W. Sams & Co., pp. 27–64, 1987.

[68] F. Martinez and C. D. Hussey, '(E) ESI determination from mode-field diameter and refractive index profile measurements on single-mode fibers', *IEE Proc. Pt. J.*, **135**(3), pp. 202–210, 1988.

3

Transmission characteristics of optical fibers

3.1 Introduction

The basic transmission mechanisms of the various types of optical fiber waveguide have been discussed in Chapter 2. However, the factors which affect the performance of optical fibers as a transmission medium were not dealt with in detail. These transmission characteristics are of utmost importance when the suitability of optical fibers for communication purposes is investigated. The transmission characteristics of most interest are those of attenuation (or loss) and bandwidth.

The huge potential bandwidth of optical communications helped stimulate the birth of the idea that a dielectric waveguide made of glass could be used to carry wideband telecommunication signals. This occurred, as indicated in Section 2.1 in the celebrated papers by Kao and Hockham, and Werts, in 1966. However, at the time the idea may have seemed somewhat ludicrous as a typical block of glass could support optical transmission for at best a few tens of metres before it was attenuated to an unacceptable level. Nevertheless, careful investigation of the attenuation showed that it was largely due to absorption in the glass, caused by impurities such as iron, copper, manganese and other transition metals which occur in the third row of the periodic table. Hence, research was stimulated towards a new generation of 'pure' glasses for use in optical fiber communications.

A major breakthrough came in 1970 when the first fiber with an attenuation below 20 dB km^{-1} was reported [Ref. 1]. This level of attenuation was seen as the absolute minimum that had to be achieved before an optical fiber system could in any way compete economically with existing communication systems. Since 1970 tremendous improvements have been made, leading to silica-based glass fibers with losses of less than 0.2 dB km^{-1} in the laboratory [Ref. 2]. Hence, comparatively low loss fibers have been incorporated into optical communication systems throughout the world. Moreover, as the fundamental lower limits for attenuation in silicate glass fibers have virtually been achieved, activities are increasing in relation to the investigation of other material systems which may exhibit substantially lower losses when operated at longer wavelengths. Such mid-infrared (and possibly far-infrared) transmitting fibers could eventually provide for extremely long-haul repeaterless communication assuming that, in addition to the material considerations, the optical source and detector requirements can be satisfactorily met [Ref. 2].

The other characteristic of primary importance is the bandwidth of the fiber. This is limited by the signal dispersion within the fiber, which determines the number of bits of information transmitted in a given time period. Therefore, once the attenuation was reduced to acceptable levels attention was directed towards the dispersive properties of fibers. Again, this has led to substantial improvements, giving wideband fiber bandwidths of many tens of gigahertz over a number of kilometres.

In order to appreciate these advances and possible future developments, the optical transmission characteristics of fibers must be considered in greater depth. Therefore, in this chapter we discuss the mechanisms within optical fibers which give rise to the major transmission characteristics mentioned previously (attenuation and dispersion), whilst also considering other, perhaps less obvious, effects when light is propagating down an optical fiber (modal noise, polarization and nonlinear phenomena).

We begin the discussion of attenuation in Section 3.2 with calculation of the total losses incurred in optical fibers. The various attenuation mechanisms (material absorption, linear scattering, nonlinear scattering, fiber bends) are then considered in detail in Sections 3.3 to 3.6. The primary focus within these sections is on silica-based glass fibers. However, in Section 3.7 consideration is given to other material systems which may be employed for mid-infrared and far-infrared optical transmission. Dispersion in optical fibers is described in Section 3.8, together with the associated limitations on fiber bandwidth. Sections 3.9 and 3.10 deal with intramodal and intermodal dispersion mechanisms and included in the latter section is a discussion of the modal noise phenomenon associated with intermodal dispersion. Overall signal dispersion in both multimode and single-mode fibers is then considered in Section 3.11. This is followed in Section 3.12 by discussion of the modification of the dispersion characteristics within single-mode fibers in order to obtain dispersion shifted and dispersion flattened fibers. Section 3.13 presents a brief account of polarization within single-mode fibers which includes description of the salient features of polarization maintaining fibers.

Finally, nonlinear optical phenomena, which can occur at relatively high optical power levels within single-mode fibers, are dealt with in Section 3.14.

3.2 Attenuation

The attenuation or transmission loss of optical fibers has proved to be one of the most important factors in bringing about their wide acceptance in telecommunications. As channel attenuation largely determined the maximum transmission distance prior to signal restoration, optical fiber communications became especially attractive when the transmission losses of fibers were reduced below those of the competing metallic conductors (less than 5 dB km^{-1}).

Signal attenuation within optical fibers, as with metallic conductors, is usually expressed in the logarithmic unit of the decibel. The decibel, which is used for comparing two power levels, may be defined for a particular optical wavelength as the ratio of the input (transmitted) optical power P_i into a fiber to the output (received) optical power P_o from the fiber as:

$$\textit{Number of decibels } (\mathrm{dB}) = 10 \log_{10} \frac{P_i}{P_o} \tag{3.1}$$

This logarithmic unit has the advantage that the operations of multiplication and division reduce to addition and subtraction, whilst powers and roots reduce to multiplication and division. However, addition and subtraction require a conversion to numerical values which may be obtained using the relationship:

$$\frac{P_i}{P_o} = 10^{(\mathrm{dB}/10)} \tag{3.2}$$

In optical fiber communications the attenuation is usually expressed in decibels per unit length (i.e. dB km^{-1}) following:

$$\alpha_{\text{dB}} L = 10 \log_{10} \frac{P_i}{P_o} \tag{3.3}$$

where α_{dB} is the signal attenuation per unit length in decibels and L is the fiber length

---

Example 3.1

When the mean optical power launched into an 8 km length of fiber is 120 μW, the mean optical power at the fiber output is 3 μW.

Determine:

(a) the overall signal attenuation or loss in decibels through the fiber assuming there are no connectors or splices;
(b) the signal attenuation per kilometre for the fiber.
(c) the overall signal attenuation for a 10 km optical link using the same fiber with splices at 1 km intervals, each giving an attenuation of 1 dB;
(d) the numerical input/output power ratio in (c).

Solution: (a) Using Eq. (3.1), the overall signal attenuation in decibels through the fiber is:

$$\textit{Signal attenuation} = 10 \log_{10} \frac{P_i}{P_o} = 10 \log_{10} \frac{120 \times 10^{-6}}{3 \times 10^{-6}}$$

$$= 10 \log_{10} 40 = 16.0 \text{ dB}$$

(b) The signal attenuation per kilometre for the fiber may be simply obtained by dividing the result in (a) by the fiber length which corresponds to it using Eq. (3.3) where,

$$\alpha_{\text{dB}} L = 16.0 \text{ dB}$$

hence,

$$\alpha_{\text{dB}} = \frac{16.0}{8}$$

$$= 2.0 \text{ dB km}^{-1}$$

(c) As $\alpha_{\text{dB}} = 2 \text{ dB km}^{-1}$, the loss incurred along 10 km of the fiber is given by

$$\alpha_{\text{dB}} L = 2 \times 10 = 20 \text{ dB}$$

However, the link also has nine splices (at 1 km intervals) each with an attenuation of 1 dB. Therefore, the loss due to the splices is 9 dB.

Hence, the overall signal attenuation for the link is:

$$\begin{aligned} \textit{Signal attenuation} &= 20 + 9 \\ &= 29 \text{ dB} \end{aligned}$$

(d) To obtain a numerical value for the input/output power ratio, Eq. (3.2) may be used where:

$$\frac{P_i}{P_o} = 10^{29/10} = 794.3$$

A number of mechanisms are responsible for the signal attenuation within optical fibers. These mechanisms are influenced by the material composition, the preparation and purification technique, and the waveguide structure. They may be categorized within several major areas which include material absorption, material scattering (linear and nonlinear scattering), curve and microbending losses, mode coupling radiation losses and losses due to leaky modes. There are also losses at connectors and splices, as illustrated in Example 3.1. However, in this chapter we are interested solely in the characteristics of the fiber; connector and splice losses are dealt with in Section 5.2. It is instructive to consider in some detail the loss mechanisms within optical fibers in order to obtain an understanding of the problems associated with the design and fabrication of low loss waveguides.

3.3 Material absorption losses in silica glass fibers

Material absorption is a loss mechanism related to the material composition and the fabrication process for the fiber, which results in the dissipation of some of the transmitted optical power as heat in the waveguide. The absorption of the light may be intrinsic (caused by the interaction with one or more of the major components of the glass) or extrinsic (caused by impurities within the glass).

3.3.1 Intrinsic absorption

An absolutely pure silicate glass has little intrinsic absorption due to its basic material structure in the near-infrared region. However, it does have two major intrinsic absorption mechanisms at optical wavelengths which leave a low intrinsic absorption window over the 0.8 to 1.7 μm wavelength range, as illustrated in Figure 3.1, which shows a possible optical attenuation against wavelength characteristic for absolutely pure glass [Ref. 3]. It may be observed that there is a fundamental absorption edge, the peaks of which are centred in the ultraviolet wavelength region. This is due to the stimulation of electron transitions within the glass by higher energy excitations. The tail of this peak may extend into the window region at the shorter wavelengths, as illustrated in Figure 3.1. Also in the infrared

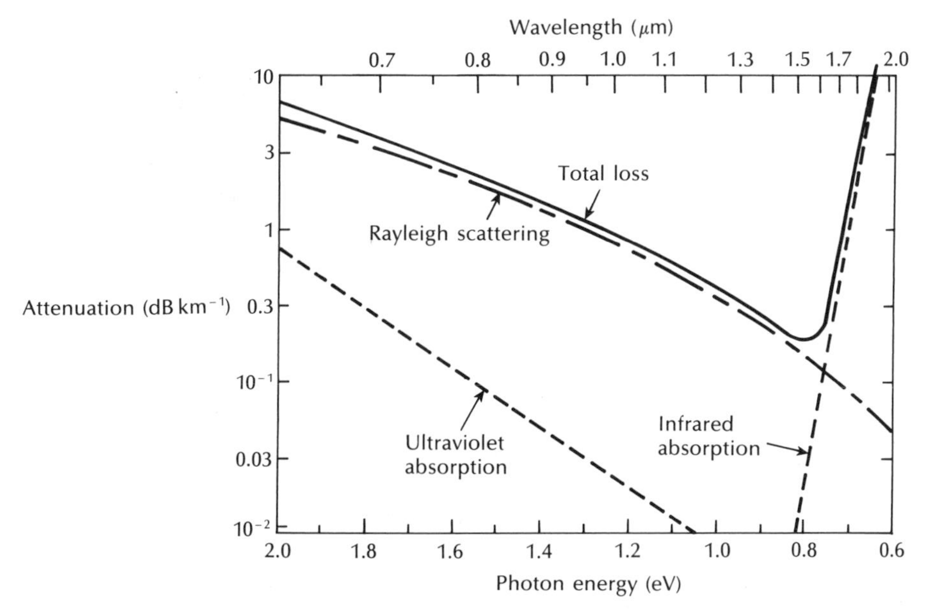

Figure 3.1 The attenuation spectra for the intrinsic loss mechanisms in pure GeO_2—SiO_2 glass [Ref. 3].

and far-infrared, normally at wavelengths above 7 μm, fundamentals of absorption bands from the interaction of photons with molecular vibrations within the glass occur. These give absorption peaks which again extend into the window region. The strong absorption bands occur due to oscillations of structural units such as Si—O (9.2 μm), P—O (8.1 μm), B—O (7.2 μm) and Ge—O (11.0 μm) within the glass. Hence, above 1.5 μm the tails of these largely far-infrared absorption peaks tend to cause most of the pure glass losses.

However, the effects of both these processes may be minimized by suitable choice of both core and cladding compositions. For instance, in some nonoxide glasses such as fluorides and chlorides, the infrared absorption peaks occur at much longer wavelengths which are well into the far-infrared (up to 50 μm), giving less attenuation to longer wavelength transmission compared with oxide glasses.

3.3.2 Extrinsic absorption

In practical optical fibers prepared by conventional melting techniques (see Section 4.3), a major source of signal attenuation is extrinsic absorption from transition metal element impurities. Some of the more common metallic impurities found in glasses are shown in the Table 3.1, together with the absorption losses caused by one part in 10^9 [Ref. 4]. It may be noted that certain of these impurities, namely chromium and copper, in their worst valence state can cause attenuation in excess of 1 dB km^{-1} in the near-infrared region. Transition element contamination may be

Table 3.1 Absorption losses caused by some of the more common metallic ion impurities in glasses, together with the absorption peak wavelength

	Peak wavelength (nm)	One part in 10^9 (dB km^{-1})
Cr^{3+}	625	1.6
C^{2+}	685	0.1
Cu^{2+}	850	1.1
Fe^{2+}	1100	0.68
Fe^{3+}	400	0.15
Ni^{2+}	650	0.1
Mn^{3+}	460	0.2
V^{4+}	725	2.7

reduced to acceptable levels (i.e. one part in 10^{10}) by glass refining techniques such as vapour-phase oxidation [Ref. 5] (see Section 4.4), which largely eliminates the effects of these metallic impurities.

However, another major extrinsic loss mechanism is caused by absorption due to water (as the hydroxyl or OH ion) dissolved in the glass. These hydroxyl groups are bonded into the glass structure and have fundamental stretching vibrations which occur at wavelengths between 2.7 and 4.2 μm depending on group position in the glass network. The fundamental vibrations give rise to overtones appearing almost harmonically at 1.38, 0.95 and 0.72 μm, as illustrated in Figure 3.2 [Ref. 6]. This

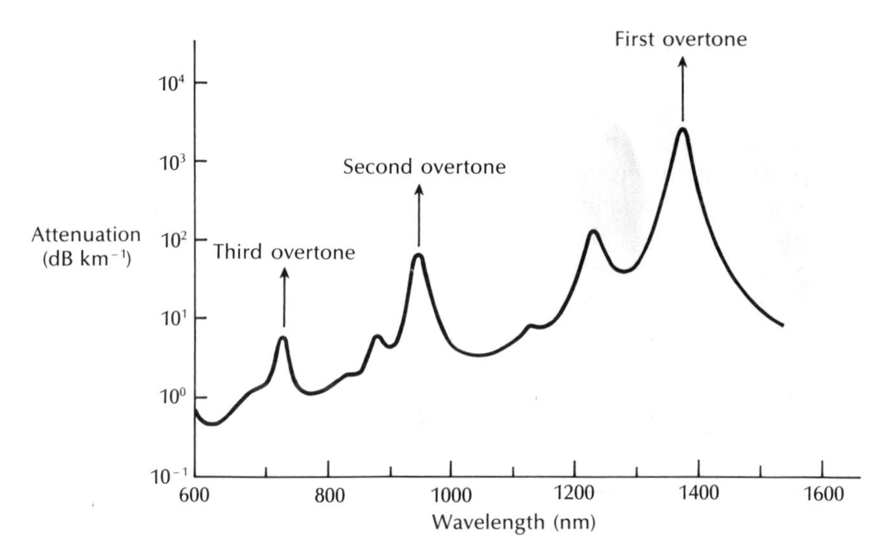

Figure 3.2 The absorption spectrum for the hydroxyl (OH) group in silica. Reproduced with permission from D. B. Keck, R. D. Maurer and P. C. Schultz, *Appl. Phys. Lett.*, **22**, p. 307, 1973.

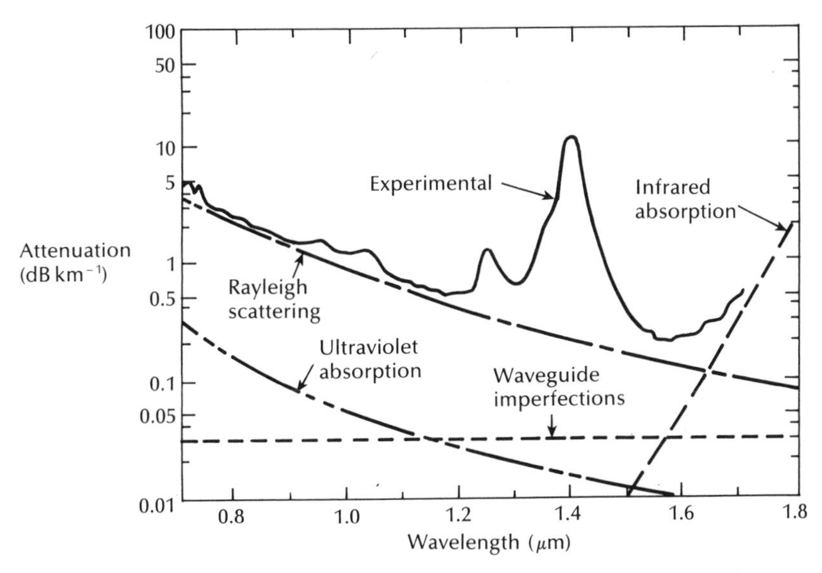

Figure 3.3 The measured attenuation spectrum for an ultra-low-loss single-mode fiber (solid line) with the calculated attenuation spectra for some of the loss mechanisms contributing to the overall fiber attenuation (dashed and dotted lines) [Ref. 3].

shows the absorption spectrum for the hydroxyl group in silica. Furthermore, combinations between the overtones and the fundamental SiO_2 vibration occur at 1.24, 1.13 and 0.88 μm, completing the absorption spectrum shown in Figure 3.2.

It may also be observed in Figure 3.2 that the only significant absorption band in the region below a wavelength of 1 μm is the second overtone at 0.95 μm which causes attenuation of about 1 dB km^{-1} for one part per million (ppm) of hydroxyl. At longer wavelengths the first overtone at 1.38 μm and its sideband at 1.24 μm are strong absorbers giving attenuation of about 2 dB km^{-1} ppm and 4 dB km^{-1} ppm respectively. Since most resonances are sharply peaked, narrow windows exist in the longer wavelength region around 1.3 and 1.55 μm which are essentially unaffected by OH absorption once the impurity level has been reduced below one part in 10^7. This situation is illustrated in Figure 3.3, which shows the attenuation spectrum of an ultra-low-loss single-mode fiber [Ref. 3]. It may be observed that the lowest attenuation for this fiber occurs at a wavelength of 1.55 μm and is 0.2 dB km^{-1}. This is approaching the minimum possible attenuation of around 0.18 dB km^{-1} at this wavelength [Ref. 8].

3.4 Linear scattering losses

Linear scattering mechanisms cause the transfer of some or all of the optical power contained within one propagating mode to be transferred linearly (proportionally to the mode power) into a different mode. This process tends to result in attenuation of the transmitted light as the transfer may be to a leaky or radiation mode which does not continue to propagate within the fiber core, but is radiated from the fiber. It must be noted that as with all linear processes there is no change of frequency on scattering.

Linear scattering may be categorized into two major types: Rayleigh and Mie scattering. Both result from the nonideal physical properties of the manufactured fiber which are difficult and, in certain cases, impossible to eradicate at present.

3.4.1 Rayleigh scattering

Rayleigh scattering is the dominant intrinsic loss mechanism in the low absorption window between the ultraviolet and infrared absorption tails. It results from inhomogeneities of a random nature occurring on a small scale compared with the wavelength of the light. These inhomogeneities manifest themselves as refractive index fluctuations and arise from density and compositional variations which are frozen into the glass lattice on cooling. The compositional variations may be reduced by improved fabrication, but the index fluctuations caused by the freezing-in of density inhomogeneities are fundamental and cannot be avoided. The subsequent scattering due to the density fluctuations, which is in almost all directions, produces an attenuation proportional to $1/\lambda^4$ following the Rayleigh scattering formula [Ref. 9]. For a single component glass this is given by:

$$\gamma_R = \frac{8\pi^3}{3\lambda^4} n^8 p^2 \beta_c K T_F \tag{3.4}$$

where γ_R is the Rayleigh scattering coefficient, λ is the optical wavelength, n is the refractive index of the medium, p is the average photoelastic coefficient, β_c is the isothermal compressibility at a fictive temperature T_F, and K is Boltzmann's constant. The fictive temperature is defined as the temperature at which the glass can reach a state of thermal equilibrium and is closely related to the anneal temperature. Furthermore, the Rayleigh scattering coefficient is related to the transmission loss factor (transmissivity) of the fiber $\mathscr{L}$ following the relation [Ref. 10]:

$$\mathscr{L} = \exp(-\gamma_R L) \tag{3.5}$$

where L is the length of the fiber. It is apparent from Eq. (3.4) that the fundamental component of Rayleigh scattering is strongly reduced by operating at the longest possible wavelength. This point is illustrated in Example 3.2.

Example 3.2

Silica has an estimated fictive temperature of 1400 K with an isothermal compressibility of 7×10^{-11} m^2 N^{-1} [Ref. 11]. The refractive index and the photoelastic coefficient for silica are 1.46 and 0.286 respectively [Ref. 11]. Determine the theoretical attenuation in decibels per kilometre due to the fundamental Rayleigh scattering in silica at optical wavelengths of 0.63, 1.00 and 1.30 μm. Boltzmann's constant is 1.381×10^{-23} J K^{-1}.

Solution: The Rayleigh scattering coefficient may be obtained from Eq. (3.4) for each wavelength. However, the only variable in each case is the wavelength, and therefore the constant of proportionality of Eq. (3.4) applies in all cases. Hence:

$$\gamma_R = \frac{8\pi^3 n^8 p^2 \beta_c K T_F}{3\lambda^4}$$

$$= \frac{248.15 \times 20.65 \times 0.082 \times 7 \times 10^{-11} \times 1.381 \times 10^{-23} \times 1400}{3 \times \lambda^4}$$

$$= \frac{1.895 \times 10^{-28}}{\lambda^4} \text{ m}^{-1}$$

At a wavelength of 0.63 μm:

$$\gamma_R = \frac{1.895 \times 10^{-28}}{0.158 \times 10^{-24}} = 1.199 \times 10^{-3} \text{ m}^{-1}$$

The transmission loss factor for one kilometre of fiber may be obtained using Eq. (3.5),

$$\mathscr{L}_{km} = \exp(-\gamma_R L) = \exp(-1.199 \times 10^{-3} \times 10^3)$$
$$= 0.301$$

The attenuation due to Rayleigh scattering in dB km^{-1} may be obtained from Eq. (3.1) where:

$$Attenuation = 10 \log_{10}(1/\mathscr{L}_{km}) = 10 \log_{10} 3.322$$
$$= 5.2 \text{ dB km}^{-1}$$

At a wavelength of 1.00 μm:

$$\gamma_R = \frac{1.895 \times 10^{-28}}{10^{-24}} = 1.895 \times 10^{-4} \text{ m}^{-1}$$

Using Eq. (3.5):

$$\mathscr{L}_{km} = \exp(-1.895 \times 10^{-4} \times 10^3) = \exp(-0.1895)$$
$$= 0.827$$

and Eq. (3.1):

$$Attenuation = 10 \log_{10} 1.209 = 0.8 \text{ dB km}^{-1}$$

At a wavelength of 1.30 μm:

$$\gamma_R = \frac{1.895 \times 10^{-28}}{2.856 \times 10^{-24}} = 0.664 \times 10^{-4}$$

Using Eq. (3.5):

$$\mathscr{L}_{\text{km}} = \exp(-0.664 \times 10^{-4} \times 10^{3}) = 0.936$$

and Eq. (3.1):

$$\textit{Attenuation} = 10 \log_{10} 1.069 = 0.3 \text{ dB km}^{-1}$$

The theoretical attenuation due to Rayleigh scattering in silica at wavelengths of 0.63, 1.00 and 1.30 μm, from Example 3.2, is 5.2, 0.8 and 0.3 dB km^{-1} respectively. These theoretical results are in reasonable agreement with experimental work. For instance, a low reported value for Rayleigh scattering in silica at a wavelength of 0.6328 μm is 3.9 dB km^{-1} [Ref. 11]. However, values of 4.8 dB km^{-1} [Ref. 12] and 5.4 dB km^{-1} [Ref. 13] have also been reported. The predicted attenuation due to Rayleigh scattering against wavelength is indicated by a broken line on the attenuation characteristics shown in Figures 3.1 and 3.3.

3.4.2 Mie scattering

Linear scattering may also occur at inhomogeneities which are comparable in size to the guided wavelength. These result from the nonperfect cylindrical structure of the waveguide and may be caused by fiber imperfections such as irregularities in the core–cladding interface, core–cladding refractive index differences along the fiber length, diameter fluctuations, strains and bubbles. When the scattering inhomogeneity size is greater than $\lambda/10$, the scattered intensity which has an angular dependence can be very large.

The scattering created by such inhomogeneities is mainly in the forward direction and is called Mie scattering. Depending upon the fiber material, design and manufacture, Mie scattering can cause significant losses. The inhomogeneities may be reduced by:

(a) removing imperfections due to the glass manufacturing process;
(b) carefully controlled extrusion and coating of the fiber;
(c) increasing the fiber guidance by increasing the relative refractive index difference.

By these means it is possible to reduce Mie scattering to insignificant levels.

3.5 Nonlinear scattering losses

Optical waveguides do not always behave as completely linear channels whose increase in output optical power is directly proportional to the input optical power. Several nonlinear effects occur, which in the case of scattering cause disproportionate attenuation, usually at high optical power levels. This nonlinear scattering causes the optical power from one mode to be transferred in either the forward or backward direction to the same, or other modes, at a different frequency. It depends critically upon the optical power density within the fiber and hence only becomes significant above threshold power levels.

The most important types of nonlinear scattering within optical fibers are stimulated Brillouin and Raman scattering, both of which are usually only observed at high optical power densities in long single-mode fibers. These scattering mechanisms in fact give optical gain but with a shift in frequency, thus contributing to attenuation for light transmission at a specific wavelength. However, it may be noted that such nonlinear phenomena can also be used to give optical amplification in the context of integrated optical techniques (see Section 10.8). In addition, these nonlinear processes are explored in further detail both following and in Section 3.14.

3.5.1 Stimulated Brillouin scattering

Stimulated Brillouin scattering (SBS) may be regarded as the modulation of light through thermal molecular vibrations within the fiber. The scattered light appears as upper and lower sidebands which are separated from the incident light by the modulation frequency. The incident photon in this scattering process produces a phonon* of acoustic frequency as well as a scattered photon. This produces an optical frequency shift which varies with the scattering angle because the frequency of the sound wave varies with acoustic wavelength. The frequency shift is a maximum in the backward direction reducing to zero in the forward direction making SBS a mainly backward process.

As indicated previously, Brillouin scattering is only significant above a threshold power density. Assuming that the polarization state of the transmitted light is not maintained (see Section 3.12), it may be shown [Ref 16] that the threshold power P_B is given by:

$$P_B = 4.4 \times 10^{-3} d^2 \lambda^2 \alpha_{dB} \nu \text{ watts} \tag{3.6}$$

where d and λ are the fiber core diameter and the operating wavelength, respectively, both measured in micrometres, α_{dB} is the fiber attenuation in decibels per kilometre and ν is the source bandwidth (i.e. injection laser) in gigahertz. The expression given in Eq. (3.6) allows the determination of the threshold optical power which must be launched into a single-mode optical fiber before SBS occurs (see Example 3.3).

3.5.2 Stimulated Raman scattering

Stimulated Raman scattering (SRS) is similar to stimulated Brillouin scattering except that a high frequency optical phonon rather than an acoustic phonon is

* The phonon is a quantum of an elastic wave in a crystal lattice. When the elastic wave has a frequency f, the quantized unit of the phonon has energy hf joules, where h is Planck's constant.

generated in the scattering process. Also, SRS can occur in both the forward and backward directions in an optical fiber, and may have an optical power threshold of up to three orders of magnitude higher than the Brillouin threshold in a particular fiber.

Using the same criteria as those specified for the Brillouin scattering threshold given in Eq. (3.6), it may be shown [Ref. 16] that the threshold optical power for SRS P_R in a long single-mode fiber is given by:

$$P_R = 5.9 \times 10^{-2} d^2 \lambda \alpha_{dB} \text{ watts} \quad (3.7)$$

where d, λ and α_{dB} are as specified for Eq. (3.6).

Example 3.3

A long single-mode optical fiber has an attenuation of 0.5 dB km^{-1} when operating at a wavelength of 1.3 μm. The fiber core diameter is 6 μm and the laser source bandwidth is 600 MHz. Compare the threshold optical powers for stimulated Brillouin and Raman scattering within the fiber at the wavelength specified.

Solution: The threshold optical power for SBS is given by Eq. (3.6) as

$$\begin{aligned} P_B &= 4.4 \times 10^{-3} d^2 \lambda^2 \alpha_{dB} \nu \\ &= 4.4 \times 10^{-3} \times 6^2 \times 1.3^2 \times 0.5 \times 0.6 \\ &= 80.3 \text{ mW} \end{aligned}$$

The threshold optical power for SRS may be obtained from Eq. (3.7), where:

$$\begin{aligned} P_R &= 5.9 \times 10^{-2} d^2 \lambda \alpha_{dB} \\ &= 5.9 \times 10^{-2} \times 6^2 \times 1.3 \times 0.5 \\ &= 1.38 \text{ W} \end{aligned}$$

In Example 3.3, the Brillouin threshold occurs at an optical power level of around 80 mW whilst the Raman threshold is approximately seventeen times larger. It is therefore apparent that the losses introduced by nonlinear scattering may be avoided by use of a suitable optical signal level (i.e. working below the threshold optical powers). However, it must be noted that the Brillouin threshold has been reported [Ref. 17] as occurring at optical powers as low as 10 mW in single-mode fibers. Nevertheless, this is still a high power level for optical communications and may be easily avoided. SBS and SRS are not usually observed in multimode fibers because their relatively large core diameters make the threshold optical power levels extremely high. Moreover, it should be noted that the threshold optical powers for both these scattering mechanisms may be increased by suitable adjustment of the other parameters in Eqs. (3.6) and (3.7). In this context, operation at the longest possible wavelength is advantageous although this may be offset by the reduced fiber attenuation (from Rayleigh scattering and material absorption) normally obtained.

3.6 Fiber bend loss

Optical fibers suffer radiation losses at bends or curves on their paths. This is due to the energy in the evanescent field at the bend exceeding the velocity of light in the cladding and hence the guidance mechanism is inhibited, which causes light energy to be radiated from the fiber. An illustration of this situation is shown in Figure 3.4. The part of the mode which is on the outside of the bend is required to travel faster than that on the inside so that a wavefront perpendicular to the direction of propagation is maintained. Hence, part of the mode in the cladding needs to travel faster than the velocity of light in that medium. As this is not possible, the energy associated with this part of the mode is lost through radiation. The loss can generally be represented by a radiation attenuation coefficient which has the form [Ref. 19]:

$$\alpha_r = c_1 \exp(-c_2 R)$$

where R is the radius of curvature of the fiber bend and c_1, c_2 are constants which are independent of R. Furthermore, large bending losses tend to occur in multimode fibers at a critical radius of curvature R_c which may be estimated from [Ref. 20]:

$$R_c \simeq \frac{3n_1^2\lambda}{4\pi(n_1^2 - n_2^2)^{\frac{3}{2}}} \tag{3.8}$$

It may be observed from the expression given in Eq. (3.8) that potential macrobending losses may be reduced by:

(a) designing fibers with large relative refractive index differences;
(b) operating at the shortest wavelength possible.

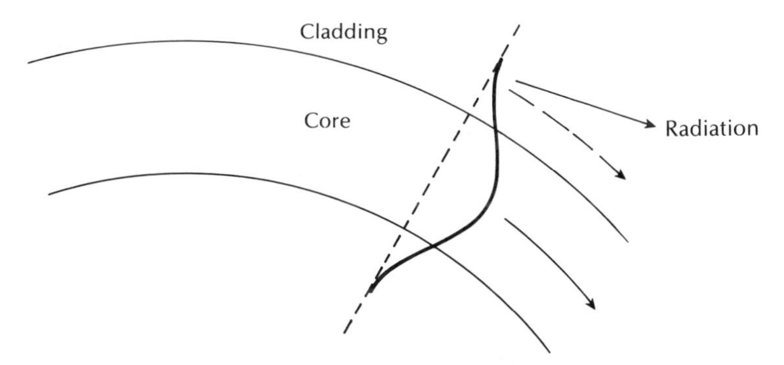

Figure 3.4 An illustration of the radiation loss at a fiber bend. The part of the mode in the cladding outside the dashed arrowed line may be required to travel faster than the velocity of light in order to maintain a plane wavefront. Since it cannot do this, the energy contained in this part of the mode is radiated away.

The above criteria for the reduction of bend losses also apply to single-mode fibers. One theory [Ref. 21], based on the concept of a single quasi-guided mode, provides an expression from which the critical radius of curvature for a single-mode fiber R_{cs} can be estimated as:

$$R_{cs} \simeq \frac{20\lambda}{(n_1 - n_2)^{\frac{3}{2}}}\left(2.748 - 0.996\frac{\lambda}{\lambda_c}\right)^{-3} \quad (3.9)$$

where λ_c is the cutoff wavelength for the single-mode fiber. Hence again, for a specific single-mode fiber (that is, a fixed relative index difference and cutoff wavelength), the critical wavelength of the radiated light becomes progressively shorter as the bend radius is decreased. The effect of this factor and that of the relative refractive index difference on the critical bending radius is demonstrated in the following example.

Example 3.4

Two step index fibers exhibit the following parameters:

(a) A multimode fiber with a core refractive index of 1.500, a relative refractive index difference of 3% and an operating wavelength of 0.82 μm.

(b) An 8 μm core diameter single-mode fiber with a core refractive index the same as (a), a relative refractive index difference of 0.3% and an operating wavelength of 1.55 μm.

Estimate the critical radius of curvature at which large bending losses occur in both cases.

Solution: (a) The relative refractive index difference is given by Eq. (2.9) as:

$$\Delta = \frac{n_1^2 - n_2^2}{2n_1^2}$$

Hence

$$n_2^2 = n_1^2 - 2\Delta n_1^2 = 2.250 - 0.06 \times 2.250 = 2.115$$

Using Eq. (3.8) for the multimode fiber critical radius of curvature:

$$R_c \simeq \frac{3n_1^2\lambda}{4\pi(n_1^2 - n_2^2)^{\frac{3}{2}}} = \frac{3 \times 2.250 \times 0.82 \times 10^{-6}}{4\pi \times (0.135)^{\frac{3}{2}}} = 9\ \mu\text{m}$$

(b) Again, from Eq. (2.9):

$$n_2^2 = n_1^2 - 2\Delta n_1^2 = 2.250 - (0.006 \times 2.250) = 2.237$$

The cutoff wavelength for the single-mode fiber is given by Eq. (2.98) as:

$$\lambda_c = \frac{2\pi a n_1 (2\Delta)^{\frac{1}{2}}}{2.405}$$

$$= \frac{2\pi \times 4 \times 10^{-6} \times 1.500\ (0.06)^{\frac{1}{2}}}{2.405}$$

$$= 1.214\ \mu\text{m}$$

Substituting into Eq. (3.9) for the critical radius of curvature for the single-mode fiber gives:

$$R_{cs} \simeq \frac{20 \times 1.55 \times 10^{-6}}{(0.043)^{\frac{3}{2}}} \left(2.748 - \frac{0.996 \times 1.55 \times 10^{-6}}{1.214 \times 10^{-6}}\right)^{-3}$$

$$= 34\ \text{mm}$$

Example 3.4 shows that the critical radius of curvature for guided modes can be made extremely small (e.g. 9 μm), although this may be in conflict with the preferred design and operational characteristics. Nevertheless, for most practical purposes, the critical radius of curvature is relatively small (even when considering the case of a long wavelength single-mode fiber, it was found to be around 34 mm) to avoid severe attenuation of the guided mode(s) at fiber bends. However, modes propagating close to cutoff, which are no longer fully guided within the fiber core, may radiate at substantially larger radii of curvature. Thus it is essential that sharp bends, with a radius of curvature approaching the critical radius, are avoided when optical fiber cables are installed. Finally, it is important that microscopic bends with radii of curvature approximating to the fiber radius are not produced in the fiber cabling process. These so-called microbends, which can cause significant losses from cabled fiber, are discussed further in Section 4.8.1.

3.7 Mid-infrared and far-infrared transmission

In the near-infrared region of the optical spectrum, fundamental silica fiber attenuation is dominated by Rayleigh scattering and multiphonon absorption from the infrared absorption edge (See Figure 3.2). Therefore, the total loss decreases as the operational transmission wavelength increases until a crossover point is reached around a wavelength of 1.55 μm where the total fiber loss again increases because at longer wavelengths the loss is dominated by the phonon absorption edge. Since the near fundamental attenuation limits for near-infrared silicate glass fibers have been achieved, more recently researchers have turned their attention to the mid-infrared (2 to 5 μm) and the far-infrared (8 to 12 μm) optical wavelengths.

In order to obtain lower loss fibers it is necessary to produce glasses exhibiting longer infrared cutoff wavelengths. Potentially, much lower losses can be achieved if the transmission window of the material can be extended further into the infrared by utilizing constituent atoms of higher atomic mass and if it can be drawn into fiber exhibiting suitable strength and chemical durability. The reason for this possible loss reduction is due to Rayleigh scattering which displays a λ^{-4} dependence and hence becomes much reduced as the wavelength is increased. For example, the scattering loss is reduced by a factor of 16 when the optical wavelength is doubled. Thus it may be possible to obtain losses of the order of 0.01 $dB\,km^{-1}$ at a wavelength of 2.55 μm, with even lower losses at wavelengths of between 3 μm and 5 μm [Ref. 23].

Candidate glass forming systems for mid-infrared transmission are fluoride, fluoride–chloride, chalcogenide and possibly oxide. In particular, heavy metal oxide glasses based on bismuth and gallium oxides offer a near equivalent transmittance range to many of the fluoride glasses and hence show promise if their scatter losses can be made acceptably low [Ref. 24]. Chalcogenide glasses, which generally comprise one or more of the elements S, Se and Te, together with one or more elements Ge, Si, As and Sb, are capable of optical transmission in both the mid-infrared and far-infrared regions.* However, research activities into far-infrared transmission using chalcogenide glasses, halide glasses and halide crystal fibers are at present mainly concerned with radiometry, infrared imaging and power transmission rather than telecommunications [Ref. 25].

The research activities into ultra-low-loss fibers for long-haul repeaterless communications have to date centred on the fluorozirconates, with zirconium fluoride (ZrF_4) as the major constituent and fluorides of barium, lanthanum, aluminium, gadolinium, sodium, lithium and occasionally lead added as modifiers and stabilizers [Ref. 26]. Such alkali additives improve the glass stability and working characteristics. Moreover, hafnium tetrafluoride (HfF_4) can be substituted for ZrF_4 to vary the refractive index and form fluorohafnate glasses. Both these glass systems offer transmittance to a wavelength of around 5.5 μm [Ref. 23].

Extensive work has been undertaken on two particular heavy metal fluoride systems, namely: zirconium–barium–lanthanum–aluminium fluoride (ZBLA) and zirconium–barium–gadolinium–aluminium fluoride (ZBGA). Furthermore, sodium fluoride is often added to ZBLA to increase its stability and form ZBLAN glass. In this case the core–cladding refractive index differences are obtained by varying the sodium fluoride level or by partially substituting hafnium tetrafluoride for the zirconium, whereas with ZBGA the aluminium fluoride content is usually varied. Typically, the above glasses contain approximately 50 to 60% ZrF_4 or HfF_4, together with 30 to 40% BaF_2, with the other alkali and rare earth fluorides completing the composition.

In order to fabricate low loss, long length fluoride fibers a basic problem concerned with reducing the extrinsic losses remains to be resolved [Ref. 27]. At

* A typical chalcogenide fiber glass is As_2S_3.

present the most critical and difficult problems are associated with the minimization of the scattering losses resulting from extrinsic factors such as defects, waveguide imperfections and radiation caused by mechanical deformation. The estimated losses of around 0.01 dB km^{-1} at a wavelength of 2.5 μm for ZrF_4-based fibers are derived from an assessment of the extrinsic losses due to ultraviolet and infrared absorptions together with Rayleigh scattering. However, experimental losses obtained so far remain significantly higher (i.e. one order of magnitude) than this estimated value. Nevertheless, it is useful to consider the theoretical characteristics for intrinsic losses obtained for a range of materials which are displayed in Figure 3.5 [Ref. 28]. The effect of increasing the atomic weight of the anion from oxide to fluoride may be clearly observed. Although the theoretical losses shown in Figure 3.5 cannot yet be achieved in practical fiber, progress has been made in relation to the fabrication of heavy metal fluoride glass fibers, particularly for single-mode operation [Ref. 29].

Materials such as $ZnCl_2$ and As_2S_3 are also being considered for mid-infrared transmission since this is their region of minimum loss, as can be observed in

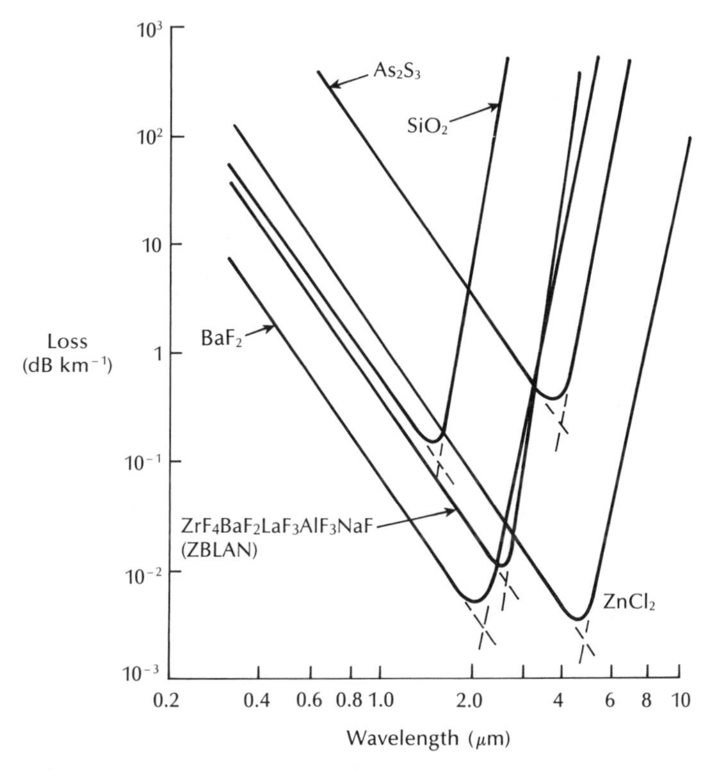

Figure 3.5 Theoretical intrinsic losses for a number of mid-infrared transmitting materials. Reproduced with permission from J. A. Savage, 'Materials for infrared fiber optics', Mat. Sci. Rep., **2**, pp. 99–138, 1987.

Figure 3.5. In addition these glasses offer potential for far-infrared transmission applications as fiber lengths of a few metres may prove sufficient for use with CO_2 laser radiation.* Moreover, both monocrystalline and polycrystalline halide fibers [Ref. 23] as well as hollow core glass fibers [Ref. 30] are being studied to assess their potential for high power transmission applications. A commercial example in the former case is Kristen 5 (KRS-5) fiber fabricated from TlBr and TlI produced by Horiba of Japan [Ref. 25].

3.8 Dispersion

Dispersion of the transmitted optical signal causes distortion for both digital and analog transmission along optical fibers. When considering the major implementation of optical fiber transmission which involves some form of digital modulation, then dispersion mechanisms within the fiber cause broadening of the transmitted light pulses as they travel along the channel. The phenomenon is illustrated in Figure 3.6, where it may be observed that each pulse broadens and overlaps with its neighbours, eventually becoming indistinguishable at the receiver input. The effect is known as intersymbol interference (ISI). Thus an increasing number of errors may be encountered on the digital optical channel as the ISI becomes more pronounced. The error rate is also a function of the signal attenuation on the link and the subsequent signal to noise ratio (SNR) at the receiver. This factor is not pursued further here but is considered in detail in Section 11.6.3. However, signal dispersion alone limits the maximum possible bandwidth attainable with a particular optical fiber to the point where individual symbols can no longer be distinguished.

For no overlapping of light pulses down on an optical fiber link the digital bit rate B_T must be less than the reciprocal of the broadened (through dispersion) pulse duration (2τ). Hence:

$$B_T \leqslant \frac{1}{2\tau} \tag{3.10}$$

This assumes that the pulse broadening due to dispersion on the channel is τ which dictates the input pulse duration which is also τ. Hence Eq. (3.10) gives a conservative estimate of the maximum bit rate that may be obtained on an optical fiber link as $1/2\tau$.

Another more accurate estimate of the maximum bit rate for an optical channel with dispersion may be obtained by considering the light pulses at the output to have a Gaussian shape with an rms width of σ. Unlike the relationship given in Eq. (3.10), this analysis allows for the existence of a certain amount of signal overlap on the channel, whilst avoiding any SNR penalty which occurs when intersymbol interference becomes pronounced. The maximum bit rate is given approximately by

* Operating at a wavelength of 10.6 μm.

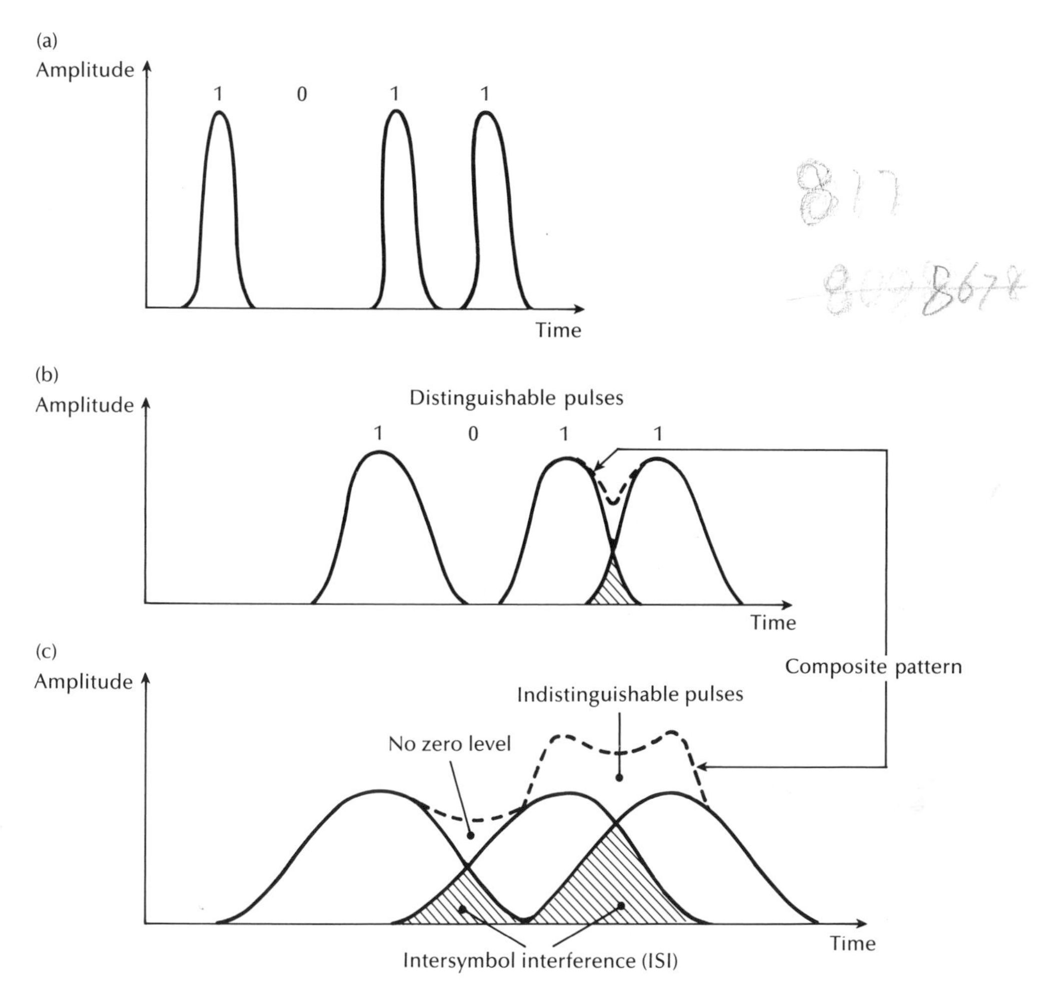

Figure 3.6 An illustration using the digital bit pattern 1011 of the broadening of light pulses as they are transmitted along a fiber: (a) fiber input; (b) fiber output at a distance L_1; (c) fiber output at a distance $L_2 > L_1$.

(see Appendix B):

$$B_T(\max) \simeq \frac{0.2}{\sigma} \text{ bit s}^{-1} \tag{3.11}$$

It must be noted that certain sources [Refs. 31, 32] give the constant term in the numerator of Eq. (3.11) as 0.25. However, we take the slightly more conservative estimate given, following Olshansky [Ref. 9] and Gambling *et al.* [Ref. 33]. Equation (3.11) gives a reasonably good approximation for other pulse shapes which may occur on the channel resulting from the various dispersive mechanisms within the fiber. Also, σ may be assumed to represent the rms impulse response for the channel, as discussed further in Section 3.10.1.

The conversion of bit rate to bandwidth in hertz depends on the digital coding format used. For metallic conductors when a nonreturn to zero code is employed, the binary one level is held for the whole bit period τ. In this case there are two bit periods in one wavelength (i.e. two bits per second per hertz), as illustrated in Figure 3.7(a). Hence the maximum bandwidth B is one half the maximum data rate or

$$B_T(\text{max}) = 2B \tag{3.12}$$

However, when a return code is considered, as shown in Figure 3.7(b), the binary one level is held for only part (usually half) the bit period. For this signalling scheme the data rate is equal to the bandwidth in hertz (i.e. one bit per second per hertz) and thus $B_T = B$. The bandwidth B for metallic conductors is also usually defined by the electrical 3 dB points (i.e. the frequencies at which the electrical power has dropped to one half of its constant maximum value). However, when the 3 dB optical bandwidth of a fiber is considered it is significantly larger than the corresponding 3 dB electrical bandwidth for the reasons discussed in Section 7.4.3. Hence, when the limitations in the bandwidth of a fiber due to dispersion are stated (i.e. optical bandwidth B_{opt}), it is usually with regard to a return to zero code where the bandwidth in hertz is considered equal to the digital bit rate. Within the context of dispersion the bandwidths expressed in this chapter will follow this general criterion unless otherwise stated. However, as is made clear in Section 7.4.3, when electro-optical devices and optical fiber systems are considered it is more usual to state the electrical 3 dB bandwidth, this being the more useful measurement when

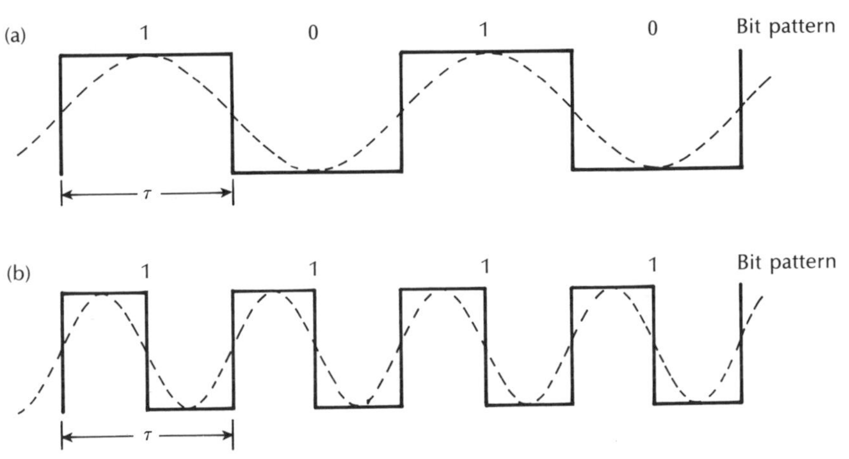

Figure 3.7 Schematic illustration of the relationships of the bit rate to wavelength for digital codes: (a) nonreturn to zero (NRZ); (b) return to zero (RZ).

interfacing an optical fiber link to electrical terminal equipment. Unfortunately, the terms of bandwidth measurement are not always made clear and the reader must be warned that this omission may lead to some confusion when specifying components and materials for optical fiber communication systems.

Figure 3.8 shows the three common optical fiber structures, multimode step index, multimode graded index and single-mode step index, whilst diagrammatically illustrating the respective pulse broadening associated with each fiber type. It may be observed that the multimode step index fiber exhibits the greatest dispersion of a transmitted light pulse and the multimode graded index fiber gives a considerably improved performance. Finally, the single-mode fiber gives the minimum pulse broadening and thus is capable of the greatest transmission bandwidths which are currently in the gigahertz range, whereas transmission via multimode step index fiber is usually limited to bandwidths of a few tens of megahertz. However, the amount of pulse broadening is dependent upon the distance the pulse travels within the fiber, and hence for a given optical fiber link

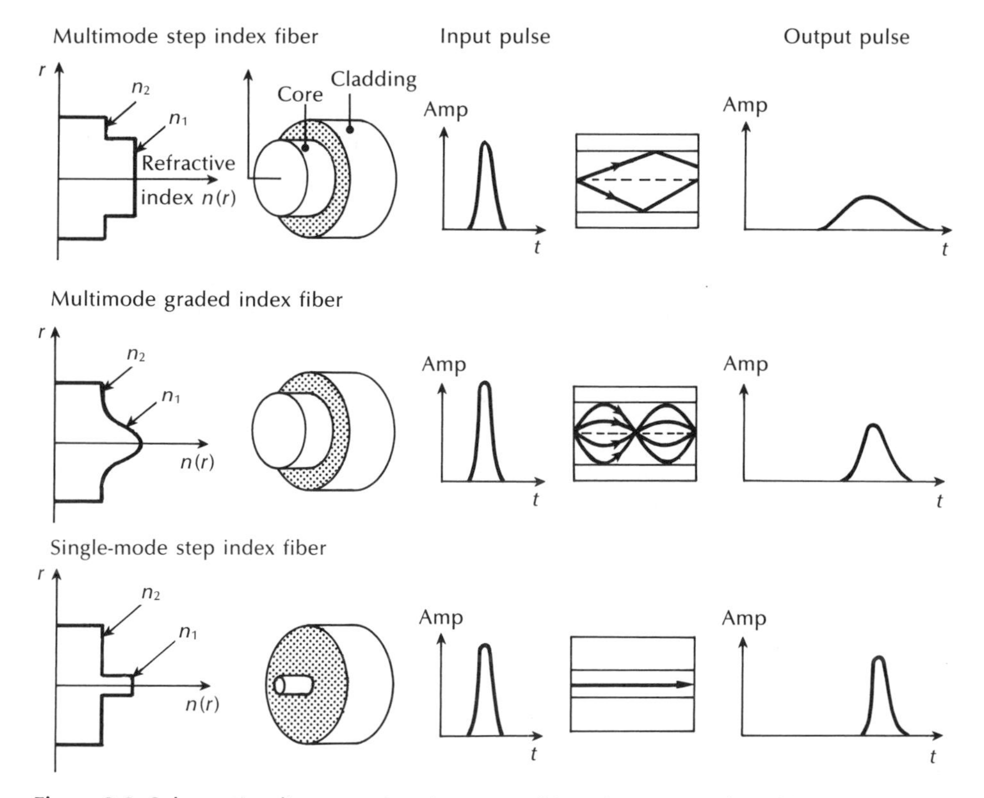

Figure 3.8 Schematic diagram showing a multimode step index fiber, multimode graded index fiber and single-mode step index fiber, and illustrating the pulse broadening due to intermodal dispersion in each fiber type.

the restriction on usable bandwidth is dictated by the distance between regenerative repeaters (i.e. the distance the light pulse travels before it is reconstituted). Thus the measurement of the dispersive properties of a particular fiber is usually stated as the pulse broadening in time over a unit length of the fiber (i.e. ns km^{-1}).

Hence, the number of optical signal pulses which may be transmitted in a given period, and therefore the information-carrying capacity of the fiber, is restricted by the amount of pulse dispersion per unit length. In the absence of mode coupling or filtering, the pulse broadening increases linearly with fiber length and thus the bandwidth is inversely proportional to distance. This leads to the adoption of a more useful parameter for the information-carrying capacity of an optical fiber which is known as the bandwidth–length product (i.e. $B_{opt} \times L$). The typical best bandwidth–length products for the three fibers shown in Figure 3.8, are 20 MHz km, 1 GHz km and 100 GHz km for multimode step index, multimode graded index and single-mode step index fibers respectively.

Example 3.5

A multimode graded index fiber exhibits total pulse broadening of 0.1 μs over a distance of 15 km. Estimate:

(a) the maximum possible bandwidth on the link assuming no intersymbol interference;
(b) the pulse dispersion per unit length;
(c) the bandwidth–length product for the fiber.

Solution: (a) The maximum possible optical bandwidth which is equivalent to the maximum possible bit rate (for return to zero pulses) assuming no ISI may be obtained from Eq. (3.10), where:

$$B_{opt} = B_T = \frac{1}{2\tau} = \frac{1}{0.2 \times 10^{-6}} = 5 \text{ MHz}$$

(b) The dispersion per unit length may be acquired simply by dividing the total dispersion by the total length of the fiber

$$\text{dispersion} = \frac{0.1 \times 10^{-6}}{15} = 6.67 \text{ ns km}^{-1}$$

(c) The bandwidth–length product may be obtained in two ways. Firstly by simply multiplying the maximum bandwidth for the fiber link by its length. Hence:

$$B_{opt}L = 5 \text{ MHz} \times 15 \text{ km} = 75 \text{ MHz km}$$

Alternatively, it may be obtained from the dispersion per unit length using Eq. (3.10) where:

$$B_{opt}L = \frac{1}{2 \times 6.67 \times 10^{-9}} = 75 \text{ MHz km}$$

In order to appreciate the reasons for the different amounts of pulse broadening within the various types of optical fiber, it is necessary to consider the dispersive mechanisms involved. These include material dispersion, waveguide dispersion, intermodal dispersion and profile dispersion which are considered in the following sections.

3.9 Intramodal dispersion

Intramodal or chromatic dispersion may occur in all types of optical fiber and results from the finite spectral linewidth of the optical source. Since optical sources do not emit just a single frequency but a band of frequencies (in the case of the injection laser corresponding to only a fraction of a per cent of the centre frequency, whereas for the LED it is likely to be a significant percentage), then there may be propagation delay differences between the different spectral components of the transmitted signal. This causes broadening of each transmitted mode and hence intramodal dispersion. The delay differences may be caused by the dispersive properties of the waveguide material (material dispersion) and also guidance effects within the fiber structure (waveguide dispersion).

3.9.1 Material dispersion

Pulse broadening due to material dispersion results from the different group velocities of the various spectral components launched into the fiber from the optical source. It occurs when the phase velocity of a plane wave propagating in the dielectric medium varies nonlinearly with wavelength, and a material is said to exhibit material dispersion when the second differential of the refractive index with respect to wavelength is not zero (i.e. $d^2n/d\lambda^2 \neq 0$). The pulse spread due to material dispersion may be obtained by considering the group delay τ_g in the optical fiber which is the reciprocal of the group velocity v_g defined by Eqs. (2.37) and (2.40). Hence the group delay is given by:

$$\tau_g = \frac{d\beta}{d\omega} = \frac{1}{c}\left(n_1 - \lambda \frac{dn_1}{d\lambda}\right) \tag{3.13}$$

where n_1 is the refractive index of the core material. The pulse delay τ_m due to material dispersion in a fiber of length L is therefore:

$$\tau_m = \frac{L}{c}\left(n_1 - \lambda \frac{dn_1}{d\lambda}\right) \tag{3.14}$$

For a source with rms spectral width σ_λ and a mean wavelength λ, the rms pulse broadening due to material dispersion σ_m may be obtained from the expansion of Eq. (3.14) in a Taylor series about λ where:

$$\sigma_m = \sigma_\lambda \frac{d\tau_m}{d\lambda} + \sigma_\lambda \frac{2d^2\tau_m}{d\lambda^2} + \cdots \tag{3.15}$$

As the first term in Eq. (3.15) usually dominates, especially for sources operating over the 0.8 to 0.9 μm wavelength range, then:

$$\sigma_m \simeq \sigma_\lambda \frac{d\tau_m}{d\lambda} \tag{3.16}$$

Hence the pulse spread may be evaluated by considering the dependence of τ_m on λ, where from Eq. (3.14):

$$\frac{d\tau_m}{d\lambda} = \frac{L\lambda}{c}\left[\frac{dn_1}{d\lambda} - \frac{d^2n_1}{d\lambda^2} - \frac{dn_1}{d\lambda}\right]$$

$$= \frac{-L\lambda}{c}\frac{d^2n_1}{d\lambda^2} \tag{3.17}$$

Therefore, substituting the expression obtained in Eq. (3.17) into Eq. (3.16), the rms pulse broadening due to material dispersion is given by:

$$\sigma_m \simeq \frac{\sigma_\lambda L}{c}\left|\lambda \frac{d^2n_1}{d\lambda^2}\right| \tag{3.18}$$

The material dispersion for optical fibers is sometimes quoted as a value for $|\lambda^2(d^2n_1/d\lambda^2)|$ or simply $|d^2n_1/d\lambda^2|$.

However, it may be given in terms of a material dispersion parameter M which is defined as:

$$M = \frac{1}{L}\frac{d\tau_m}{d\lambda} = \frac{\lambda}{c}\left|\frac{d^2n_1}{d\lambda^2}\right| \tag{3.19}$$

and which is often expressed in units of ps nm^{-1} km^{-1}.

Example 3.6

A glass fiber exhibits material dispersion given by $|\lambda^2(d^2n_1/d\lambda^2)|$ of 0.025. Determine the material dispersion parameter at a wavelength of 0.85 μm, and estimate the rms pulse broadening per kilometre for a good LED source with an rms spectral width of 20 nm at this wavelength.

Solution: The material dispersion parameter may be obtained from Eq. (3.19):

$$M = \frac{\lambda}{c}\left|\frac{d^2n_1}{d\lambda^2}\right| = \frac{1}{c\lambda}\left|\lambda^2\frac{d^2n_1}{d\lambda^2}\right|$$

$$= \frac{0.025}{2.998 \times 10^5 \times 850} \text{ s nm}^{-1} \text{ km}^{-1}$$

$$= 98.1 \text{ ps nm}^{-1} \text{ km}^{-1}$$

The rms pulse broadening is given by Eq. (3.18) as:

$$\sigma_m \simeq \frac{\sigma_\lambda L}{c} \left| \lambda \frac{d^2 n_1}{d\lambda^2} \right|$$

Therefore in terms of the material dispersion parameter M defined by Eq. (3.19):

$$\sigma_m \simeq \sigma_\lambda L M$$

Hence, the rms pulse broadening per kilometre due to material dispersion:

$$\sigma_m(1 \text{ km}) = 20 \times 1 \times 98.1 \times 10^{-12} = 1.96 \text{ ns km}^{-1}$$

Figure 3.9 shows the variation of the material dispersion parameter M with wavelength for pure silica [Ref. 34]. It may be observed that the material dispersion tends to zero in the longer wavelength region around 1.3 μm (for pure silica). This provides an additional incentive (other than low attenuation) for operation at longer wavelengths where the material dispersion may be minimized. Also, the use of an injection laser with a narrow spectral width rather than an LED as the optical source leads to a substantial reduction in the pulse broadening due to material dispersion, even in the shorter wavelength region.

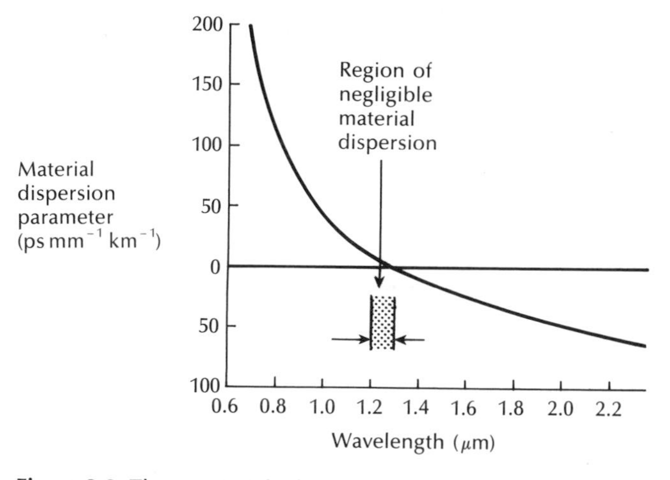

Figure 3.9 The material dispersion parameter for silica as a function of wavelength. Reproduced with permission from D. N. Payne and W. A. Gambling, *Electron. Lett.*, **11**, p. 176, 1975.

Example 3.7

Estimate the rms pulse broadening per kilometre for the fiber in Example 3.6 when the optical source used is an injection laser with a relative spectral width σ_λ/λ of 0.0012 at a wavelength of 0.85 μm.

Solution: The rms spectral width may be obtained from the relative spectral width by:

$$\sigma_\lambda = 0.0012\lambda = 0.0012 \times 0.85 \times 10^{-6}$$
$$= 1.02 \text{ nm}$$

The rms pulse broadening in terms of the material dispersion parameter following Example 3.6 is given by:

$$\sigma_m \simeq \sigma_\lambda LM$$

Therefore, the rms pulse broadening per kilometre due to material dispersion is:

$$\sigma_m \simeq 1.02 \times 1 \times 98.1 \times 10^{-12} = 0.10 \text{ ns km}^{-1}$$

Hence, in this example the rms pulse broadening is reduced by a factor of around 20 (i.e. equivalent to the reduced rms spectral width of the injection laser source) compared with that obtained with the LED source of Example 3.6.

3.9.2 Waveguide dispersion

The waveguiding of the fiber may also create intramodal dispersion. This results from the variation in group velocity with wavelength for a particular mode. Considering the ray theory approach it is equivalent to the angle between the ray and the fiber axis varying with wavelength which subsequently leads to a variation in the transmission times for the rays, and hence dispersion. For a single mode whose propagation constant is β, the fiber exhibits waveguide dispersion when $(d^2\beta)/(d\lambda^2) \neq 0$. Multimode fibers, where the majority of modes propagate far from cutoff, are almost free of waveguide dispersion and it is generally negligible compared with material dispersion (≈ 0.1 to 0.2 ns km^{-1}) [Ref. 34]. However, with single-mode fibers where the effects of the different dispersion mechanisms are not easy to separate, waveguide dispersion may be significant (see Section 3.11.2).

3.10 Intermodal dispersion

Pulse broadening due to intermodal dispersion (sometimes referred to simply as modal or mode dispersion) results from the propagation delay differences between modes within a multimode fiber. As the different modes which constitute a pulse in a multimode fiber travel along the channel at different group velocities, the pulse width at the output is dependent upon the transmission times of the slowest and fastest modes. This dispersion mechanism creates the fundamental difference in the overall dispersion for the three types of fiber shown in Figure 3.8. Thus multimode step index fibers exhibit a large amount of intermodal dispersion which gives the greatest pulse broadening. However, intermodal dispersion in multimode fibers may be reduced by adoption of an optimum refractive index profile which is provided by the near parabolic profile of most graded index fibers. Hence, the overall pulse broadening in multimode graded index fibers is far less than that obtained in multimode step index fibers (typically by a factor of 100). Thus graded index fibers used with a multimode source give a tremendous bandwidth advantage over multimode step index fibers.

Under purely single-mode operation there is no intermodal dispersion and therefore pulse broadening is solely due to the intramodal dispersion mechanisms. In theory, this is the case with single-mode step index fibers where only a single mode is allowed to propagate. Hence they exhibit the least pulse broadening and have the greatest possible bandwidths, but in general are only usefully operated with single-mode sources.

In order to obtain a simple comparison for intermodal pulse broadening between multimode step index and multimode graded index fibers it is useful to consider the geometric optics picture for the two types of fiber.

3.10.1 Multimode step index fiber

Using the ray theory model, the fastest and slowest modes propagating in the step index fiber may be represented by the axial ray and the extreme meridional ray (which is incident at the core–cladding interface at the critical angle ϕ_c) respectively. The paths taken by these two rays in a perfectly structured step index fiber are shown in Figure 3.10. The delay difference between these two rays when travelling in the fiber core allows estimation of the pulse broadening resulting from intermodal dispersion within the fiber. As both rays are travelling at the same velocity within the constant refractive index fiber core, then the delay difference is directly related to their respective path lengths within the fiber. Hence the time taken for the axial ray to travel along a fiber of length L gives the minimum delay time T_{Min} and:

$$T_{\text{Min}} = \frac{distance}{velocity} = \frac{L}{(c/n_1)} = \frac{Ln_1}{c} \tag{3.20}$$

where n_1 is the refractive index of the core and c is the velocity of light in a vacuum.

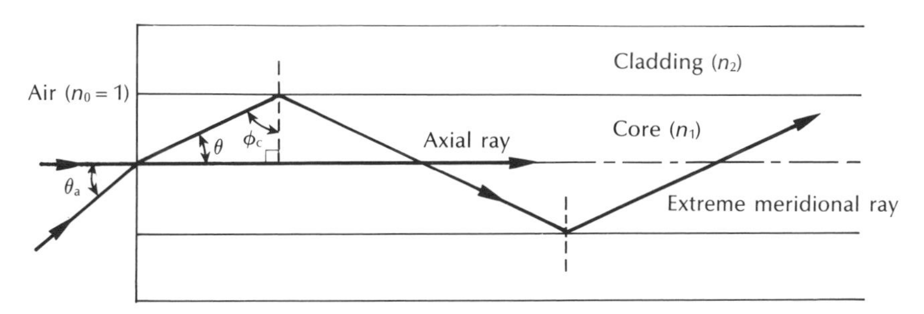

Figure 3.10 The paths taken by the axial and an extreme meridional ray in a perfect multimode step index fiber.

The extreme meridional ray exhibits the maximum delay time T_{Max} where:

$$T_{\text{Max}} = \frac{L/\cos\theta}{c/n_1} = \frac{Ln_1}{c\cos\theta} \tag{3.21}$$

Using Snell's law of refraction at the core–cladding interface following Eq. (2.2):

$$\sin\phi_c = \frac{n_2}{n_1} = \cos\theta \tag{3.22}$$

where n_2 is the refractive index of the cladding. Furthermore, substituting into Eq. (3.21) for $\cos\theta$ gives:

$$T_{\text{Max}} = \frac{Ln_1^2}{cn_2} \tag{3.23}$$

The delay difference δT_s between the extreme meridional ray and the axial ray may be obtained by subtracting Eq. (3.20) from Eq. (3.23). Hence:

$$\delta T_s = T_{\text{Max}} - T_{\text{Min}} = \frac{Ln_1^2}{cn_2} - \frac{Ln_1}{c}$$

$$= \frac{Ln_1^2}{cn_2}\left(\frac{n_1 - n_2}{n_1}\right) \tag{3.24}$$

$$\simeq \frac{Ln_1^2\Delta}{cn_2} \quad \text{when } \Delta \ll 1 \tag{3.25}$$

where Δ is the relative refractive index difference. However, when $\Delta \ll 1$, then from the definition given by Eq. (2.9), the relative refractive index difference may also be given approximately by:

$$\Delta \simeq \frac{n_1 - n_2}{n_2} \tag{3.26}$$

Hence rearranging Eq. (3.24):

$$\delta T_s = \frac{Ln_1}{c}\left(\frac{n_1 - n_2}{n_2}\right) \simeq \frac{Ln_1\Delta}{c} \tag{3.27}$$

Also substituting for Δ from Eq. (2.10) gives:

$$\delta T_s \simeq \frac{L(NA)^2}{2n_1 c} \tag{3.28}$$

where NA is the numerical aperture for the fiber. The approximate expressions for the delay difference given in Eq. (3.27) and (3.28) are usually employed to estimate the maximum pulse broadening in time due to intermodal dispersion in multimode step index fibers. It must be noted that this simple analysis only considers pulse broadening due to meridional rays and totally ignores skew rays with acceptance angles $\theta_{as} > \theta_a$ (see Section 2.2.4).

Again considering the perfect step index fiber, another useful quantity with regard to intermodal dispersion on an optical fiber link is the rms pulse broadening resulting from this dispersion mechanism along the fiber. When the optical input to the fiber is a pulse $p_i(t)$ of unit area, as illustrated in Figure 3.11, then [Ref. 37]:

$$\int_{-\infty}^{\infty} p_i(t)\, dt = 1 \tag{3.29}$$

It may be noted that $p_i(t)$ has a constant amplitude of $1/\delta T_s$ over the range

$$\frac{-\delta T_s}{2} \leqslant p(t) \leqslant \frac{\delta T_s}{2}$$

The rms pulse broadening at the fiber output due to intermodal dispersion for the multimode step index fiber σ_s (i.e. the standard deviation) may be given in terms of the variance σ_s^2 as (see Appendix C):

$$\sigma_s^2 = M_2 - M_1^2 \tag{3.30}$$

where M_1 is the first temporal moment which is equivalent to the mean value of the pulse and M_2, the second temporal moment, is equivalent to the mean square value of the pulse. Hence:

$$M_1 = \int_{-\infty}^{\infty} t p_i(t)\, dt \tag{3.31}$$

and

$$M_2 = \int_{-\infty}^{\infty} t^2 p_i(t)\, dt \tag{3.32}$$

The mean value M_1 for the unit input pulse of Figure 3.11 is zero, and assuming

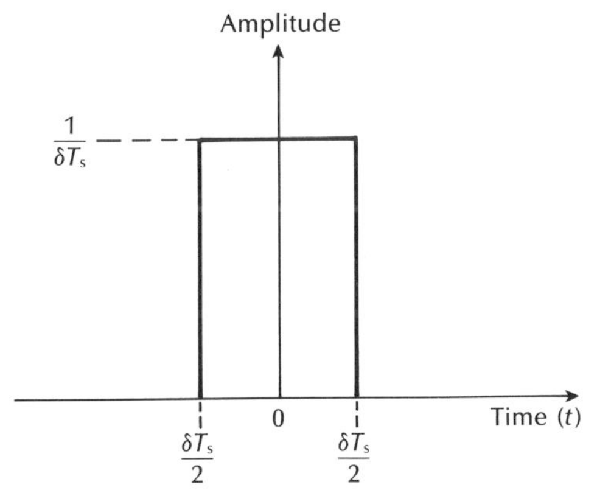

Figure 3.11 An illustration of the light input to the multimode step index fiber consisting of an ideal pulse or rectangular function with unit area.

this is maintained for the output pulse, then from Eqs. (3.30) and (3.32):

$$\sigma_s^2 = M_2 = \int_{-\infty}^{\infty} t^2 p_i(t)\, dt \tag{3.33}$$

Integrating over the limits of the input pulse (Figure 3.11) and substituting for $p_i(t)$ in Eq. (3.33) over this range gives:

$$\begin{aligned}\sigma_s^2 &= \int_{-\delta T_s/2}^{\delta T_s/2} \frac{1}{\delta T_s} t^2\, dt \\ &= \frac{1}{\delta T_s}\left[\frac{t^3}{3}\right]_{-\delta T_s/2}^{\delta T_s/2} = \frac{1}{3}\left(\frac{\delta T_s}{2}\right)^2\end{aligned} \tag{3.34}$$

Hence substituting from Eq. (3.27) for δT_s gives:

$$\sigma_s \simeq \frac{L n_1 \Delta}{2\sqrt{3}c} \simeq \frac{L(NA)^2}{4\sqrt{3}n_1 c} \tag{3.35}$$

Equation (3.35) allows estimation of the rms impulse response of a multimode step index fiber if it is assumed that intermodal dispersion dominates and there is a uniform distribution of light rays over the range $0 \leqslant \theta \leqslant \theta_a$. The pulse broadening is directly proportional to the relative refractive index difference Δ and the length of the fiber L. The latter emphasizes the bandwidth–length trade-off that exists, especially with multimode step index fibers, and which inhibits their use for wide-band long haul (between repeaters) systems. Furthermore, the pulse broadening is reduced by reduction of the relative refractive index difference Δ for the fiber. This suggests that weakly guiding fibers (see Section 2.4.1) with small Δ are best for low dispersion transmission. However, as may be seen from Eq. (3.35) this is also subject to a trade-off as a reduction in Δ reduces the acceptance angle θ_a and the NA, thus worsening the launch conditions.

Example 3.8

A 6 km optical link consists of multimode step index fiber with a core refractive index of 1.5 and a relative refractive index difference of 1%. Estimate:

(a) the delay difference between the slowest and fastest modes at the fiber output;
(b) the rms pulse broadening due to intermodal dispersion on the link;
(c) the maximum bit rate that may be obtained without substantial errors on the link assuming only intermodal dispersion;
(d) the bandwidth–length product corresponding to (c).

Solution: (a) The delay difference is given by Eq. (3.27) as:

$$\delta T_s \simeq \frac{L n_1 \Delta}{c} = \frac{6 \times 10^3 \times 1.5 \times 0.01}{2.998 \times 10^8}$$

$$= 300 \text{ ns}$$

(b) The rms pulse broadening due to intermodal dispersion may be obtained from Eq. (3.35) where:

$$\sigma_s = \frac{Ln_1\Delta}{2\sqrt{3}c} = \frac{1}{2\sqrt{3}} \frac{6 \times 10^3 \times 1.5 \times 0.01}{2.998 \times 10^8}$$

$$= 86.7 \text{ ns}$$

(c) The maximum bit rate may be estimated in two ways. Firstly, to get an idea of the maximum bit rate when assuming no pulse overlap Eq. (3.10) may be used where:

$$B_T(\max) = \frac{1}{2\tau} = \frac{1}{2\delta T_s} = \frac{1}{600 \times 10^{-9}}$$

$$= 1.7 \text{ Mbit s}^{-1}$$

Alternatively an improved estimate may be obtained using the calculated rms pulse broadening in Eq. (3.11) where

$$B_T(\max) = \frac{0.2}{\sigma_s} = \frac{0.2}{86.7 \times 10^{-9}}$$

$$= 2.3 \text{ Mbit s}^{-1}$$

(d) Using the most accurate estimate of the maximum bit rate from (c), and assuming return to zero pulses, the bandwidth–length product is

$$B_{opt} \times L = 2.3 \text{ MHz} \times 6 \text{ km} = 13.8 \text{ MHz km}$$

Intermodal dispersion may be reduced by propagation mechanisms within practical fibers. For instance, there is differential attenuation of the various modes in a step index fiber. This is due to the greater field penetration of the higher order modes into the cladding of the waveguide. These slower modes therefore exhibit larger losses at any core–cladding irregularities which tends to concentrate the transmitted optical power into the faster lower order modes. Thus the differential attenuation of modes reduces intermodal pulse broadening on a multimode optical link.

Another mechanism which reduces intermodal pulse broadening in nonperfect (i.e. practical) multimode fibers is the mode coupling or mixing discussed in Section 2.4.2. The coupling between guided modes transfers optical power from the slower to the faster modes, and vice versa. Hence, with strong coupling the optical power tends to be transmitted at an average speed, which is a mean of the various propagating modes. This reduces the intermodal dispersion on the link and makes it advantageous to encourage mode coupling within multimode fibers.

The expression for delay difference given in Eq. (3.27) for a perfect step index fiber may be modified for the fiber with mode coupling among all guided modes to

[Ref. 38]:

$$\delta T_{sc} \simeq \frac{n_1 \Delta}{c} (LL_c)^{\frac{1}{2}} \tag{3.36}$$

where L_c is a characteristic length for the fiber which is inversely proportional to the coupling strength. Hence, the delay difference increases at a slower rate proportional to $(LL_c)^{\frac{1}{2}}$ instead of the direct proportionality to L given in Eq. 3.27). However, the most successful technique for reducing intermodal dispersion in multimode fibers is by grading the core refractive index to follow a near parabolic profile. This has the effect of equalizing the transmission times of the various modes as discussed in the following section.

3.10.2 Multimode graded index fiber

Intermodal dispersion in multimode fibers is minimized with the use of graded index fibers. Hence, multimode graded index fibers show substantial bandwidth improvement over multimode step index fibers. The reason for the improved performance of graded index fibers may be observed by considering the ray diagram for a graded index fiber shown in Figure 3.12. The fiber shown has a parabolic index profile with a maximum at the core axis, as illustrated in Figure 3.12(a). Analytically, the index profile is given by Eq. (2.75) with $\alpha = 2$ as:

$$\begin{aligned} n(r) &= n_1(1 - 2\Delta(r/a)^2)^{\frac{1}{2}} \qquad r < a \text{ (core)} \\ &= n_1(1 - 2\Delta)^{\frac{1}{2}} = n_2 \qquad r \geqslant a \text{ (cladding)} \end{aligned} \tag{3.37}$$

Figure 3.12(b) shows several meridional ray paths within the fiber core. It may be observed that apart from the axial ray the meridional rays follow sinusoidal trajectories of different path lengths which result from the index grading, as was discussed in Section 2.4.4. However, following Eq. (2.40) the local group velocity

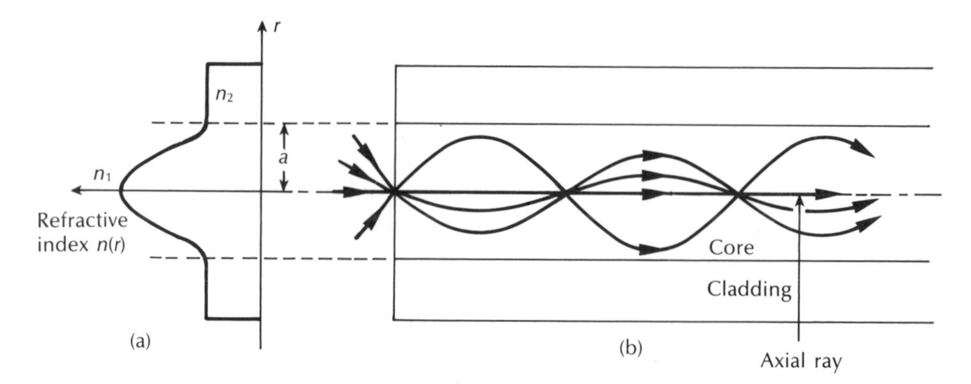

Figure 3.12 A multimode graded index fiber: (a) parabolic refractive index profile; (b) meridional ray paths within the fiber core.

is inversely proportional to the local refractive index and therefore the longer sinusoidal paths are compensated for by higher speeds in the lower index medium away from the axis. Hence there is an equalization of the transmission times of the various trajectories towards the transmission time of the axial ray which travels exclusively in the high index region at the core axis, and at the slowest speed. As these various ray paths may be considered to represent the different modes propagating in the fiber, then the graded profile reduces the disparity in the mode transit times.

The dramatic improvement in multimode fiber bandwidth achieved with a parabolic or near parabolic refractive index profile is highlighted by consideration of the reduced delay difference between the fastest and slowest modes for this graded index fiber δT_g. Using a ray theory approach the delay difference is given by [Ref. 39]:

$$\delta T_g \simeq \frac{L n_1 \Delta^2}{2c} \simeq \frac{(NA)^4}{8 n_1^3 c} \tag{3.38}$$

As in the step index case Eq. (2.10) is used for conversion between the two expressions shown.

However, a more rigorous analysis using electromagnetic mode theory gives an absolute temporal width at the fiber output of [Refs. 40, 41]:

$$\delta T_g = \frac{L n_1 \Delta^2}{8c} \tag{3.39}$$

which corresponds to an increase in transmission time for the slowest mode of $\Delta^2/8$ over the fastest mode. The expression given in Eq. (3.39) does not restrict the bandwidth to pulses with time slots corresponding to δT_g as 70% of the optical power is concentrated in the first half of the interval. Hence the rms pulse broadening is a useful parameter for assessment of intermodal dispersion in multimode graded index fibers. It may be shown [Ref. 41] that the rms pulse broadening of a near parabolic index profile graded index fiber σ_g is reduced compared to the similar broadening for the corresponding step index fiber σ_s (i.e. with the same relative refractive index difference) following:

$$\sigma_g = \frac{\Delta}{D} \sigma_s \tag{3.40}$$

where D is a constant between 4 and 10 depending on the precise evaluation and the exact optimum profile chosen.

The best minimum theoretical intermodal rms pulse broadening for a graded index fiber with an optimum characteristic refractive index profile for the core α_{op} of [Refs. 41, 42]:

$$\alpha_{op} = 2 - \frac{12\Delta}{5} \tag{3.41}$$

is given by combining Eqs. (3.27) and (3.40) as [Refs. 33, 42]:

$$\sigma_g = \frac{Ln_1\Delta^2}{20\sqrt{3}c} \tag{3.42}$$

Example 3.9

Compare the rms pulse broadening per kilometre due to intermodal dispersion for the multimode step index fiber of Example 3.8 with the corresponding rms pulse broadening for an optimum near parabolic profile graded index fiber with the same core axis refractive index and relative refractive index difference.

Solution: In Example 3.8, σ_s over 6 km of fiber is 86.7 ns. Hence the rms pulse broadening per kilometre for the multimode step index fiber is:

$$\frac{\sigma_s(1\text{ km})}{L} = \frac{86.7}{6} = 14.4\text{ ns km}^{-1}$$

Using Eq. (3.42), the rms pulse broadening per kilometre for the corresponding graded index fiber is:

$$\sigma_g(1\text{ km}) = \frac{Ln_1\Delta^2}{20\sqrt{3}c} = \frac{10^3 \times 1.5 \times (0.01)^2}{20\sqrt{3} \times 2.998 \times 10^8}$$

$$= 14.4\text{ ps km}^{-1}$$

Hence, from Example 3.9, the theoretical improvement factor of the graded index fiber in relation to intermodal rms pulse broadening is 1000. However, this level of improvement is not usually achieved in practice due to difficulties in controlling the refractive index profile radially over long lengths of fiber. Any deviation in the refractive index profile from the optimum results in increased intermodal pulse broadening. This may be observed from the curve shown in Figure 3.13, which gives the variation in intermodal pulse broadening (δT_g) as a function of the characteristic refractive index profile α for typical graded index fibers (where $\Delta = 1\%$). The curve displays a sharp minimum at a characteristic refractive index profile slightly less than 2 ($\alpha = 1.98$). This corresponds to the optimum value of α in order to minimize intermodal dispersion. Furthermore, the extreme sensitivity of the intermodal pulse broadening to slight variations in α from this optimum value is evident. Thus at present improvement factors for practical graded index fibers over corresponding step index fibers with regard to intermodal dispersion are around 100 [Ref. 40].

Another important factor in the determination of the optimum refractive index profile for a graded index fiber is the dispersion incurred due to the difference in refractive index between the fiber core and cladding. It results from a variation in the refractive index profile with optical wavelength in the graded fiber and is often given by a profile dispersion parameter $d\Delta/d\lambda$. Thus the optimized profile at a given

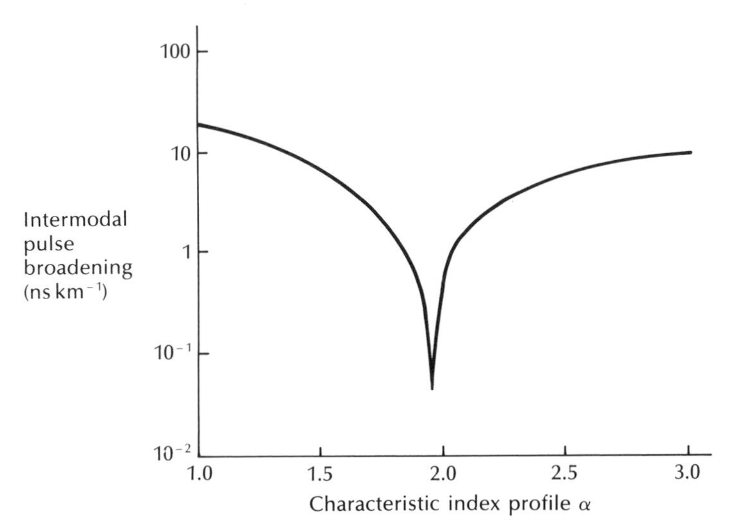

Figure 3.13 The intermodal pulse broadening δT_g for graded index fibers having $\Delta = 1\%$, versus the characteristic refractive index profile α.

wavelength is not necessarily optimized at another wavelength. As all optical fiber sources (e.g. injection lasers and light emitting diodes) have a finite spectral width, the profile shape must be altered to compensate for this dispersion mechanism. Moreover, the minimum overall dispersion for graded index fiber is also limited by the other intramodal dispersion mechanisms (i.e. material and waveguide dispersion). These give temporal pulse broadening of around 0.08 and 1 ns km^{-1} with injection lasers and light emitting diodes respectively. Therefore, practical pulse broadening values for graded index fibers lie in the range 0.2 to 1 ns km^{-1}. This gives bandwidth–length products of between 0.5 and 2.5 GHz km when using lasers and optimum profile fiber.

3.10.3 Modal noise

The intermodal dispersion properties of multimode optical fibers (see Sections 3.10.1 and 3.10.2) create another phenomenon which affects the transmitted signals on the optical channel. It is exhibited within the speckle patterns observed in multimode fiber as fluctuations which have characteristic times longer than the resolution time of the detector, and is known as modal or speckle noise. The speckle patterns are formed by the interference of the modes from a coherent source when the coherence time of the source is greater than the intermodal dispersion time δT within the fiber. The coherence time for a source with uncorrelated source frequency width δf is simply $1/\delta f$. Hence, modal noise occurs when:

$$\delta f \gg \frac{1}{\delta T} \qquad (3.43)$$

Disturbances along the fiber such as vibrations, discontinuities, connectors, splices and source/detector coupling may cause fluctuations in the speckle patterns and hence modal noise. It is generated when the correlation between two or more modes which gives the original interference is differentially delayed by these disturbances. The conditions which give rise to modal noise are therefore specified as:

(a) a coherent source with a narrow spectral width and long coherence length (propagation velocity multiplied by the coherence time);
(b) disturbances along the fiber which give differential mode delay or modal and spatial filtering;
(c) phase correlation between the modes.

Measurements [Ref. 43] of rms signal to modal noise ratio using good narrow linewidth injection lasers show large signal to noise ratio penalties under the previously mentioned conditions. The measurements were carried out by misaligning connectors to create disturbances. They gave carrier to noise ratios reduced by around 10 dB when the attenuation at each connector was 20 dB due to substantial axial misalignment.

Modal noise may be avoided by removing one of the conditions (they must all be present) which give rise to this degradation. Hence modal noise free transmission may be obtained by the following:

1. The use of a broad spectrum source in order to eliminate the modal interference effects. This may be achieved by either (a) increasing the width of the single longitudinal mode and hence decreasing its coherence time or (b) by increasing the number of longitudinal modes and averaging out of the interference patterns [Ref. 44].
2. In conjunction with 1(b) it is found that fibers with large numerical apertures support the transmission of a large number of modes giving a greater number of speckles, and hence reduce the modal noise generating effect of individual speckles [Ref. 45].
3. The use of single-mode fiber which does not support the transmission of different modes and thus there is no intermodal interference.
4. The removal of disturbances along the fiber. This has been investigated with regard to connector design [Ref. 46] in order to reduce the shift in speckle pattern induced by mechanical vibration and fiber misalignment.

Hence, modal noise may be prevented on an optical fiber link through suitable choice of the system components. However, this may not always be possible and then certain levels of modal noise must be tolerated. This tends to be the case on high quality analog optical fiber links where multimode injection lasers are frequently used. Analog transmission is also more susceptible to modal noise due to the higher optical power levels required at the receiver when quantum noise effects are considered (see Section 9.2.5). Therefore, it is important that modal noise is taken into account within the design considerations for these systems.

Modal noise, however, can be present in single-mode fiber links when propagation of the two fundamental modes with orthogonal polarization is allowed or, alternatively, when the second order modes* are not sufficiently attenuated. The former modal noise type, which is known as polarization modal noise, is outlined in Section 3.13.1. For the latter type, it is apparent that at shorter wavelengths, a nominally single-mode fiber can also guide four second order LP modes (see Section 2.4.1). Modal noise can therefore be introduced into single-mode fiber systems by time-varying interference between the LP_{01} and the LP_{11} modes when the fiber is operated at a wavelength which is smaller than the cutoff wavelength of the second order modes. The effect has been observed in overmoded single-mode fibers [Ref. 47] and may be caused by a number of conditions. In particular the insertion of a short jumper cable or repair section, with a lateral offset, in a long single-mode fiber can excite the second order LP_{11} mode [Ref. 41]. Moreover, such a repair section can also attenuate the fundamental LP_{01} mode if its operating wavelength is near the cutoff wavelength for this mode. Hence to reduce modal noise, repair sections should use special fibers with a lower value of cutoff wavelength than that in the long single-mode fiber link; also offsets at joints should be minimized.

3.11 Overall fiber dispersion

3.11.1 Multimode fibers

The overall dispersion in multimode fibers comprises both intramodal and intermodal terms. The total rms pulse broadening σ_T is given (see Appendix D) by:

$$\sigma_T = (\sigma_c^2 + \sigma_n^2)^{\frac{1}{2}} \tag{3.44}$$

where σ_c is the intramodal or chromatic broadening and σ_n is the intermodal broadening caused by delay differences between the modes (i.e. σ_s for multimode step index fiber and σ_g for multimode graded index fiber). The intramodal term σ_c consists of pulse broadening due to both material and waveguide dispersion. However, since waveguide dispersion is generally negligible compared with material dispersion in multimode fibers, then $\sigma_c \simeq \sigma_m$.

Example 3.10

A multimode step index fiber has a numerical aperture of 0.3 and a core refractive index of 1.45. The material dispersion parameter for the fiber is 250 ps nm^{-1} km^{-1} which makes material dispersion the totally dominating intramodal dispersion mechanism. Estimate (a) the total rms pulse broadening per kilometre when the

* In addition to the two orthogonal LP_{01} modes, at shorter wavelengths 'single-mode' fiber can propagate four LP_{11} modes.

fiber is used with an LED source of rms spectral width 50 nm and (b) the corresponding bandwidth–length product for the fiber.

Solution: (a) The rms pulse broadening per kilometre due to material dispersion may be obtained from Eq. (3.18), where

$$\sigma_{\mathrm{m}}(1\ \mathrm{km}) \simeq \frac{\sigma_\lambda L \lambda}{c}\left|\frac{\mathrm{d}^2 n_1}{\mathrm{d}\lambda^2}\right| = \sigma_\lambda L M = 50 \times 1 \times 250\ \mathrm{ps\,km^{-1}}$$

$$= 12.5\ \mathrm{ns\,km^{-1}}$$

The rms pulse broadening per kilometre due to intermodal dispersion for the step index fiber is given by Eq. (3.35) as:

$$\sigma_{\mathrm{s}}(1\ \mathrm{km}) \simeq \frac{L(NA)^2}{4\sqrt{3}n_1 c} = \frac{10^3 \times 0.09}{4\sqrt{3} \times 1.45 \times 2.998 \times 10^8}$$

$$= 29.9\ \mathrm{ns\,km^{-1}}$$

The total rms pulse broadening per kilometre may be obtained using Eq. (3.43), where $\sigma_{\mathrm{c}} \approx \sigma_{\mathrm{m}}$ as the waveguide dispersion is negligible and $\sigma_{\mathrm{n}} = \sigma_{\mathrm{s}}$ for the multimode step index fiber. Hence:

$$\sigma_{\mathrm{T}} = (\sigma_{\mathrm{m}}^2 + \sigma_{\mathrm{s}}^2)^{\frac{1}{2}} = (12.5^2 + 29.9^2)^{\frac{1}{2}}$$

$$= 32.4\ \mathrm{ns\,km^{-1}}$$

(b) The bandwidth–length product may be estimated from the relationship given in Eq. (3.11) where:

$$B_{\mathrm{opt}} \times L = \frac{0.2}{\sigma_{\mathrm{T}}} = \frac{0.2}{32.4 \times 10^{-9}}$$

$$= 6.2\ \mathrm{MHz\,km}$$

3.11.2 Single-mode fibers

The pulse broadening in single-mode fibers results almost entirely from intramodal or chromatic dispersion as only a single-mode is allowed to propagate.* Hence the bandwidth is limited by the finite spectral width of the source. Unlike the situation in multimode fibers, the mechanisms giving intramodal dispersion in single-mode fibers tend to be interrelated in a complex manner. The transit time or specific group delay τ_{g} for a light pulse propagating along a unit length of single-mode fiber may be given, following Eq. (2.107), as:

$$\tau_{\mathrm{g}} = \frac{1}{\mathrm{c}}\frac{\mathrm{d}\beta}{\mathrm{d}k} \tag{3.45}$$

where c is the velocity of light in a vacuum, β is the propagation constant for a

* Polarization mode dispersion can, however, occur in single-mode fibers (see Section 3.13.1).

mode within the fiber core of refractive index n_1 and k is the propagation constant for the mode in a vacuum.

The total first order dispersion parameter or the chromatic dispersion of a single-mode fiber, D_T, is given by the derivative of the specific group delay with respect to the vacuum wavelength λ as:

$$D_T = \frac{d\tau_g}{d\lambda} \tag{3.46}$$

In common with the material dispersion parameter it is usually expressed in units of $\text{ps nm}^{-1}\,\text{km}^{-1}$. When the variable λ is replaced by ω, then the total dispersion parameter becomes:

$$D_T = -\frac{\omega}{\lambda}\frac{d\tau_g}{d\omega} = -\frac{\omega}{\lambda}\frac{d^2\beta}{d\omega^2} \tag{3.47}$$

The fiber exhibits intramodal dispersion when β varies nonlinearly with wavelength. From Eq. (2.71) β may be expressed in terms of the relative refractive index difference Δ and the normalized propagation constant b as:

$$\beta = kn_1[1 - 2\Delta(1 - b)]^{\frac{1}{2}} \tag{3.48}$$

The rms pulse broadening caused by intramodal dispersion down a fiber of length L is given by the derivative of the group delay with respect to wavelength as [Ref. 27]:

$$\text{Total rms pulse broadening} = \sigma_\lambda L\left|\frac{d\tau_g}{d\lambda}\right|$$

$$= \frac{\sigma_\lambda L 2\pi}{c\lambda^2}\frac{d^2\beta}{dk^2} \tag{3.49}$$

where σ_λ is the source rms spectral linewidth centred at a wavelength λ.

When Eq. (3.44) is substituted into Eq. (3.45), detailed calculation of the first and second derivatives with respect to k gives the dependence of the pulse broadening on the fiber material's properties and the normalized propagation constant b. This gives rise to three interrelated effects which involve complicated cross-product terms. However, the final expression may be separated into three composite dispersion components in such a way that one of the effects dominates each term [Ref. 50]. The dominating effects are as follows:

1. The material dispersion parameter D_M defined by $\lambda/c\,|d^2n/d\lambda^2|$ where $n = n_1$ or n_2 for the core or cladding respectively.

2. The waveguide dispersion parameter D_W, which may be obtained from Eq. (3.47) by substitution from Eq. (2.114) for τ_g, is defined as:*

$$D_W = -\left(\frac{n_1 - n_2}{\lambda c}\right) V \frac{d^2(Vb)}{dV^2} \qquad (3.50)$$

   where V is the normalized frequency for the fiber. Since the normalized propagation constant b for a specific fiber is only dependent on V, then the normalized waveguide dispersion coefficient $V\,d^2(Vb)/dV^2$ also depends on V. This latter function is another universal parameter which plays a central role in the theory of single-mode fibers.
3. A profile dispersion parameter D_P which is proportional to $d\Delta/d\lambda$.

This situation is different from multimode fibers where the majority of modes propagate far from cutoff and hence most of the power is transmitted in the fiber core. In the multimode case the composite dispersion components may be simplified and separated into two intramodal terms which depend on either material or waveguide dispersion, as was discussed in Section 3.9. Also, especially when considering step index multimode fibers, the effect of profile dispersion is negligible. Although material and waveguide dispersion tend to be dominant in single-mode fibers, the composite profile should not be ignored. However, the profile dispersion parameter D_P can be quite small (e.g. less than 0.5 ps nm^{-1} km^{-1}), especially at long wavelengths and hence is often neglected in rough estimates of total dispersion within single-mode fibers.

Strictly speaking, in single-mode fiber with a power law refractive index profile the composite dispersion terms should be employed [Ref. 51]. Nevertheless, it is useful to consider the total first order dispersion D_T in a practical single-mode fiber as comprising:

$$D_T = D_M + D_W + D_P \qquad (\text{ps nm}^{-1}\,\text{km}^{-1}) \qquad (3.51)$$

which is simply the addition of the material dispersion D_M, the waveguide dispersion D_W and the profile dispersion D_P components. However, in standard single-mode fibers the total dispersion tends to be dominated by the material dispersion of fused silica. This parameter is shown plotted against wavelength in Figure 3.9. It may be observed that the characteristic goes through zero at a wavelength of 1.27 μm. This zero material dispersion (ZMD) point can be shifted anywhere in the wavelength range 1.2 to 1.4 μm by the addition of suitable dopants [Ref. 52]. For instance, the ZMD point shifts from 1.27 μm to approximately 1.37 μm as the GeO_2 dopant concentration is increased from 0 to 15%. However, the ZMD point alone does not represent a point of zero pulse broadening since the pulse dispersion is influenced by both waveguide and profile dispersion.

* Equation (3.50) does not provide the composite waveguide dispersion term (i.e. taking into account both the fiber core and the cladding) from which it differs by a factor near unity which contains $dn_2/d\lambda$ [Ref. 51].

With zero material dispersion the pulse spreading is dictated by the waveguide dispersion coefficient $V\,\mathrm{d}^2(Vb)/\mathrm{d}V^2$, which is illustrated in Figure 3.14 as a function of normalized frequency for the LP_{01} mode. It may be seen that in the single-mode region where the normalized frequency is less than 2.405 (see Section 2.5) the waveguide dispersion is always positive and has a maximum at $V = 1.15$. In this case the waveguide dispersion goes to zero outside the true single-mode region at $V = 3.0$. However, a change in the fiber parameters (such as core radius) or in the operating wavelength alters the normalized frequency and therefore the waveguide dispersion.

The total fiber dispersion, which depends on both the fiber material composition and dimensions, may be minimized by trading off material and waveguide dispersion whilst limiting the profile dispersion (i.e. restricting the variation in refractive index with wavelength). For wavelengths longer than the ZMD point, the material dispersion parameter is positive whereas the waveguide dispersion parameter is negative, as shown in Figure 3.15. However, the total dispersion D_T is approximately equal to the sum of the material dispersion D_M and the waveguide dispersion D_W following Eq. (3.50). Hence for a particular wavelength, designated λ_0, which is slightly larger than the ZMD point wavelength, the waveguide dispersion compensates for the material dispersion and the total first order dispersion parameter D_T becomes zero (See Figure 3.15). The wavelength at which the first order dispersion is zero λ_0 may be selected in the range 1.3 to 2 μm by careful control of the fiber core diameter and profile [Ref. 50]. This point is illustrated in Figure 3.16 where the total first order dispersion as a function of

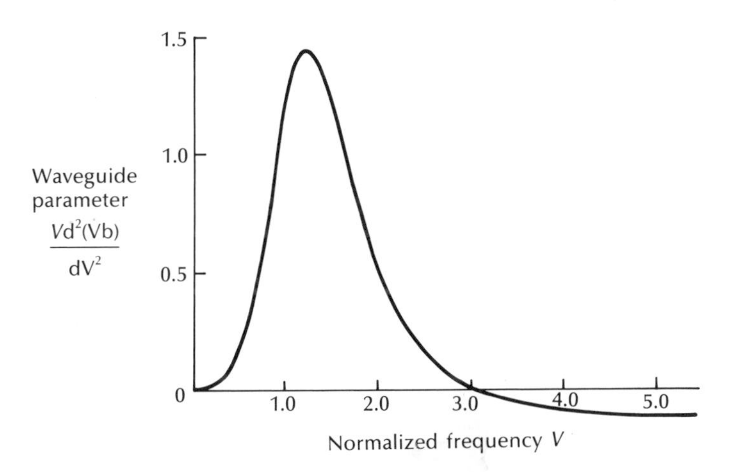

Figure 3.14 The waveguide parameter $Vd^2(Vb)/dV^2$ as a function of the normalized frequency V for the LP_{01} mode. Reproduced with permission from W. A. Gambling. A. H. Hartog and C. M. Ragdale, *The Radio and Electron. Eng.*, **51**, p. 313, 1981.

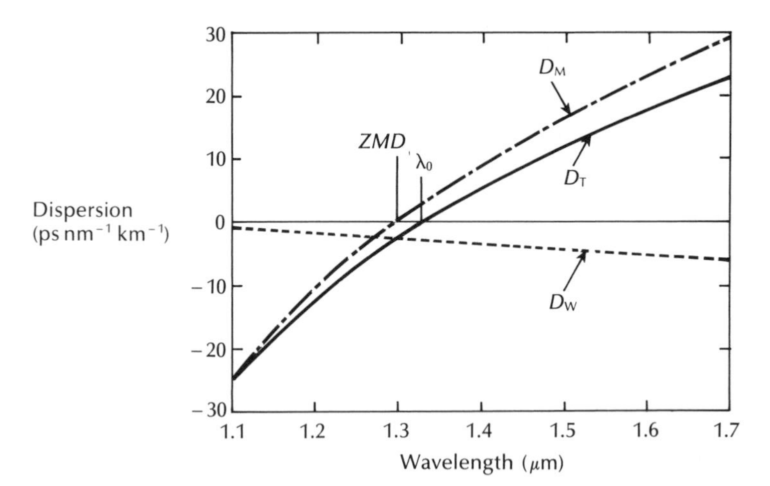

Figure 3.15 The material dispersion parameter (D_M), the waveguide dispersion parameter (D_W) and the total dispersion parameter (D_T) as functions of wavelength for a conventional single-mode fiber.

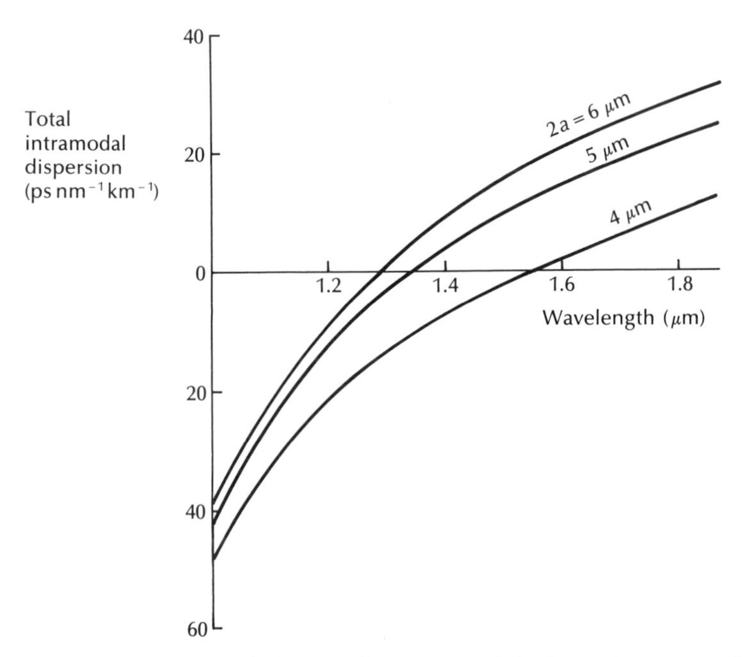

Figure 3.16 The total first order intramodal dispersion as a function of wavelength for single-mode fibers with core diameters of 4, 5, and 6 μm. Reproduced with permission from W. A. Gambling. A. H. Hartog, and C. M. Ragdale, *The Radio and Electron. Eng.*, **51**, p. 313, 1981.

wavelength is shown for three single-mode fibers with core diameters of 4, 5 and 6 μm.

The effect of the interaction of material and waveguide dispersion on λ_0 is also demonstrated in the dispersion against wavelength characteristics for a single-mode silica core fiber shown in Figure 3.17. It may be noted that the ZMD point occurs at a wavelength of 1.27 μm but that the influence of waveguide dispersion shifts the total dispersion minimum towards the longer wavelength giving a λ_0 of 1.32 μm.

The wavelength at which the first order dispersion is zero λ_0 may be extended to

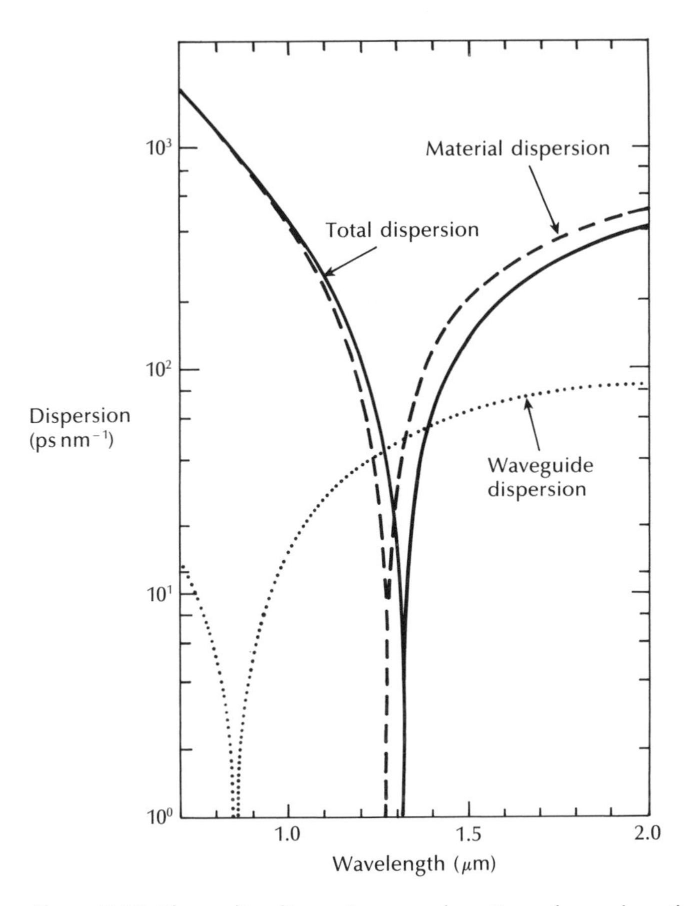

Figure 3.17 The pulse dispersion as a function of wavelength in 11 km single-mode fiber showing the major contributing dispersion mechanisms (dashed and dotted curves) and the overall dispersion (solid curve). Reproduced with permission from J. I. Yamada, M. Saruwatari, K. Asatani, H. Tsuchiya, A. Kawana, K. Sugiyama and T. Kumara, 'High speed optical pulse transmission at 1.29 μm wavelength using low-loss single-mode fibers' *IEEE J. Quantum Electron.*, **QE-14** p. 791, 1978. Copyright © 1980, IEEE.

wavelengths of 1.55 μm and beyond by a combination of three techniques. These are:

(a) lowering the normalized frequency (V value) for the fiber;
(b) increasing the relative refractive index difference Δ for the fiber;
(c) suitable doping of the silica with germanium.

This allows bandwidth–length products for such single-mode fibers to be in excess of 100 GHz km^{-1} [Ref. 53] at the slight disadvantage of increased attenuation due to Rayleigh scattering within the doped silica.

For single-mode fibers optimized for operation at a wavelength of 1.3 μm, the CCITT* [Ref. 54] recommends that the maximum value of chromatic dispersion D_T shall not exceed 3.5 ps nm^{-1} km^{-1} in the wavelength range 1.285 to 1.330 μm. Moreover, for the same fiber D_T should be less than 20 ps nm^{-1} km^{-1} at the wavelength of 1.55 [Ref. 55]. Hence, although the wavelength of zero first order chromatic dispersion (i.e. $D_T = 0$) is often called the zero-dispersion wavelength, it is more correct to refer to it as the wavelength of minimum dispersion because of the significant second order dispersion effects.

The variation of the intramodal dispersion with wavelength is usually characterized by the second order dispersion parameter or dispersion slope S which may be written as [Ref. 56]:

$$S = \frac{dD_T}{d\lambda} = \frac{d^2\tau_g}{d\lambda^2} \tag{3.52}$$

Whereas the first order dispersion parameter D_T may be seen to be related only to the second derivative of the propagation constant β with respect to angular frequency in Eq. (3.47), the dispersion slope can be shown to be related to both the second and third derivatives [Ref. 51] following:

$$S = \frac{(2\pi c)^3}{\lambda^4}\frac{d^3\beta}{d\omega^3} + \frac{4\pi c}{\lambda^3}\frac{d^2\beta}{d\omega^2} \tag{3.53}$$

It should be noted that although there is zero first order dispersion at λ_0, these higher order chromatic effects impose limitations on the possible bandwidths that may be achieved with single-mode fibers. For example, a fundamental lower limit to pulse spreading in silica-based fibers of around 2.50×10^{-2} ps nm^{-1} km^{-1} is suggested at a wavelength of 1.273 μm [Ref. 57]. These secondary effects such as birefringence arising from ellipticity or mechanical stress in the fiber core are considered further in Section 3.13. However, they may cause dispersion, especially in the case of mechanical stress of between 2 and 40 ps km^{-1}. If mechanical stress is avoided, pulse dispersion around the lower limit may be obtained in the longer wavelength region (i.e. 1.3 to 1.7 μm). By contrast the minimum pulse spread at a wavelength of 0.85 μm is around 100 ps nm^{-1} km^{-1} [Ref. 39].

* CCITT Recommendation G.652.

An important value of the dispersion slope $S(\lambda)$ is obtained at the wavelength of minimum intramodal dispersion λ_0 such that:

$$S_0 = S(\lambda_0) \tag{3.54}$$

where S_0 is called the zero-dispersion slope which, from Eqs. (3.46) and (3.52), is determined only by the third derivative of β. Typical values for the dispersion slope for standard single-mode fiber at λ_0 are in the region 0.085 to 0.092 ps nm^{-2} km^{-1}. Moreover, for such fibers the CCITT has recently proposed that λ_0 lies in the range 1.295 to 1.322 μm with S_0 less than 0.095 ps nm^{-2} km^{-1} [Ref. 58]. The total chromatic dispersion at an arbitrary wavelength can be estimated when the two parameters λ_0 and S_0 are specified according to [Ref. 55]:

$$D_T(\lambda) = \frac{\lambda S_0}{4}\left[1 - \left(\frac{\lambda_0}{\lambda}\right)^4\right] \tag{3.55}$$

Example 3.11

A typical single-mode fiber has a zero-dispersion wavelength of 1.31 μm with a dispersion slope of 0.09 ps nm^{-2} km^{-1}. Compare the total first order dispersion for the fiber at the wavelengths of 1.28 μm and 1.55 μm. When the material dispersion and profile dispersion at the latter wavelength are 13.5 ps nm^{-1} km^{-1} and 0.4 ps nm^{-1} km^{-1}, respectively, determine the waveguide dispersion at this wavelength.

Solution: The total first order dispersion for the fiber at the two wavelengths may be obtained from Eq. (3.55). Hence:

$$\begin{aligned} D_T(1280\text{ nm}) &= \frac{\lambda S_0}{4}\left[1 - \left(\frac{\lambda_0}{\lambda}\right)^4\right] \\ &= \frac{1280 \times 0.09 \times 10^{-12}}{4}\left[1 - \left(\frac{1310}{1280}\right)^4\right] \\ &= -2.8\text{ ps nm}^{-1}\text{ km}^{-1} \end{aligned}$$

and

$$\begin{aligned} D_T(1550\text{ nm}) &= \frac{1550 \times 0.09 \times 10^{-12}}{4}\left[1 - \left(\frac{1310}{1550}\right)^4\right] \\ &= 17.1\text{ ps nm}^{-1}\text{ km}^{-1} \end{aligned}$$

The total dispersion at the 1.28 μm wavelength exhibits a negative sign due to the influence of the waveguide dispersion. Furthermore, as anticipated the total dispersion at the longer wavelength (1.55 μm) is considerably greater than that obtained near the zero-dispersion wavelength.

The waveguide dispersion for the fiber at a wavelength of 1.55 μm is given by Eq. (3.51) where:

$$\begin{aligned} D_W &= D_T - (D_M + D_P) \\ &= 17.1 - (13.5 + 0.4) \\ &= 3.2 \text{ ps nm}^{-1}\text{ km}^{-1} \end{aligned}$$

3.12 Dispersion modified single-mode fibers

It was suggested in Section 3.11.2 that it is possible to modify the dispersion characteristics of single-mode fibers by the tailoring of specific fiber parameters. However, the major trade-off which occurs in this process between material dispersion (Eq. 3.19) and waveguide dispersion (Eq.3.50) may be expressed as:

$$D_T = D_M + D_W = \underbrace{\frac{\lambda}{c}\left|\frac{d^2 n_1}{d\lambda^2}\right|}_{\text{material dispersion}} \; - \underbrace{\left[\frac{n_1 - n_2}{\lambda c}\right]\frac{V d^2(Vb)}{dV^2}}_{\text{waveguide dispersion}} \tag{3.56}$$

At wavelengths longer than the zero material dispersion (ZMD) point in most common fiber designs, the D_M and D_W components are of opposite sign and can therefore be made to cancel at some longer wavelength. Hence the wavelength of zero first order chromatic dispersion can be shifted to the lowest loss wavelength for silicate glass fibers at 1.55 μm to provide both low dispersion and low loss fiber. This may be achieved by such mechanisms as a reduction in the fiber core diameter with an accompanying increase in the relative or fractional index difference to create so-called dispersion shifted (DS) single-mode fibers. However, the design flexibility required to obtain particular dispersion, attenuation, mode-field diameter and bend loss characteristics has resulted in specific, different refractive index profiles for these dispersion modified fibers.

An alternative modification of the dispersion characteristics of single-mode fibers involves the achievement of a low dispersion window over the low loss wavelength region between 1.3 μm and 1.6 μm. Such fibers, which relax the spectral requirements for optical sources and allow flexible wavelength division multiplexing (see Section 11.9.3) are known as dispersion flattened (DF) single-mode fibers. In order to obtain DF fibers multilayer index profiles are fabricated with increased waveguide dispersion which is tailored to provide overall dispersion (e.g. less than 2 ps nm^{-1} km^{-1}) over the entire wavelength range 1.3 to 1.6 μm [Ref. 59]. In effect these fibers exhibit two wavelengths of zero total chromatic dispersion. This factor may be observed in Figure 3.18 which shows the overall dispersion characteristics as a function of optical wavelength for standard single-mode fiber optimized for operation at 1.3 μm in comparison with both DS and DF fiber [Ref. 60].

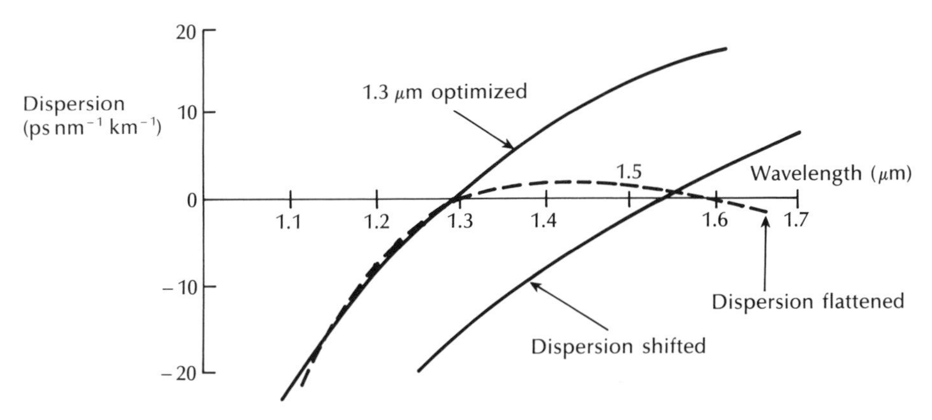

Figure 3.18 Total dispersion characteristics for the various types of single-mode fiber.

3.12.1 Dispersion shifted fibers

A wide variety of single-mode fiber refractive index profiles are capable of modification in order to tune the zero-dispersion wavelength point λ_0 to a specific wavelength within a region adjacent to the zero-material-dispersion (ZMD) point. In the simplest case, the step index profile illustrated in Figure 3.19 gives a shift to longer wavelength by reducing the core diameter and increasing the fractional index difference. Typical values for the two parameters are 4.4 μm and 0.012 respectively [Ref. 61]. For comparison, the standard nonshifted design is shown dotted in Figure 3.19.

It was indicated in Section 3.11.2 that λ_0 could be shifted to longer wavelength by altering the material composition of the single-mode fiber. For suitable power confinement of the fundamental mode, the normalized frequency V should be maintained in the range 1.5 to 2.4 μm and the fractional index difference must be increased as a square function whilst the core diameter is linearly reduced to keep V constant. This is normally achieved by substantially increasing the level of germanium doping in the fiber core. Figure 3.20 [Ref. 61] displays typical material and waveguide dispersion characteristics for single-mode step index fibers with various compositions and core radii. It may be observed that higher concentrations of the dopant cause a shift to longer wavelength which when coupled with a reduction in the mode-field diameter (MFD), giving a larger value (negative of waveguide dispersion), leads to the shifted fiber characteristic shown in Figure 3.20.

A problem that arises with the simple step index approach to dispersion shifting displayed in Figure 3.19 is that the fibers produced exhibit relatively high dopant dependent losses at operation wavelengths around 1.55 μm. This excess optical loss, which may be of the order of 2 dB km^{-1}, [Ref. 61] could be caused by stress-induced defects which occur in the region of the core–cladding interface [Ref. 62]. Alternatively, it may result from refractive index inhomogeneities associated with waveguide variations at the core–cladding interface [Ref. 63]. A logical assumption

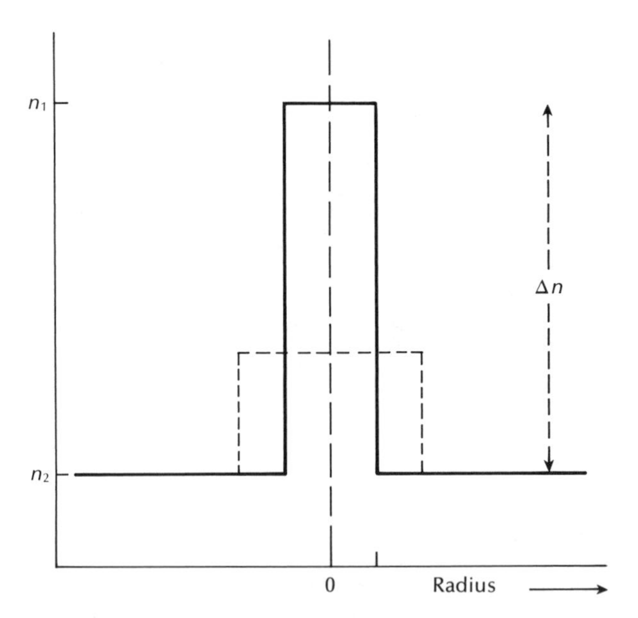

Figure 3.19 Refractive index profile of a step index dispersion shifted fiber (solid) with a conventional nonshifted profile design (dashed).

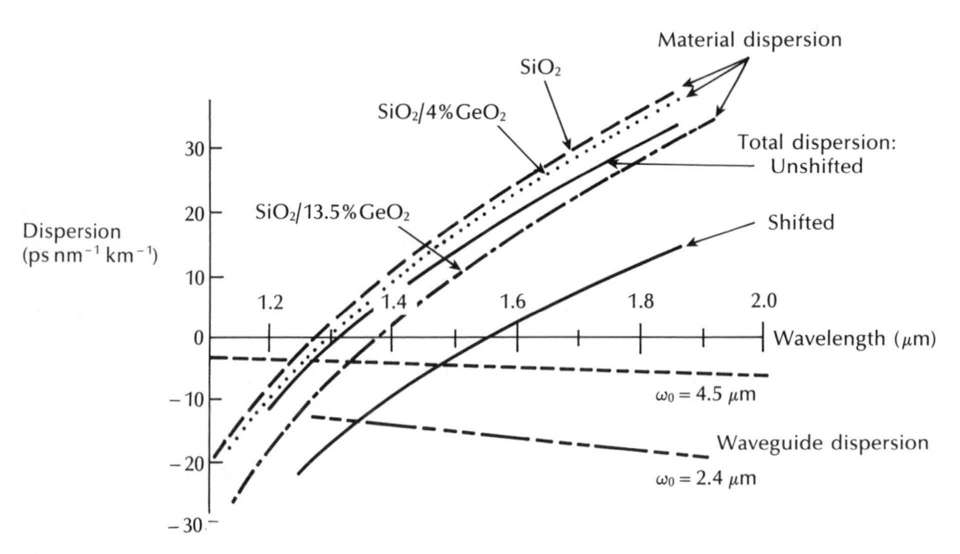

Figure 3.20 Material, waveguide and total dispersion characteristics for conventional and dispersion shifted step index single-mode fibers showing variation with composition and spot size (ω_0).

is that any stress occurring across the core–cladding interface might be reduced by grading the material composition and therefore an investigation of graded index single-mode fiber designs was undertaken.

Several of the graded refractive index profile DS fiber types are illustrated in Figure 3.21. The triangular profile shown in Figure 3.21(a) is the simplest and was the first to exhibit the same low loss (i.e. 0.24 dB km^{-1}) at a wavelength of 1.56 μm (i.e. λ_0) as conventional nonshifted single-mode fiber [Ref. 64]. Furthermore, such fiber designs also provide an increased MFD over equivalent step index structures which assists with fiber splicing [Ref. 61]. However, in the basic triangular profile design the optimum parameters giving low loss together with zero dispersion at a wavelength of 1.55 μm cause the LP_{11} mode to cutoff in the wavelength region 0.85 to 0.9 μm. Thus the fiber must be operated far from cutoff which produces

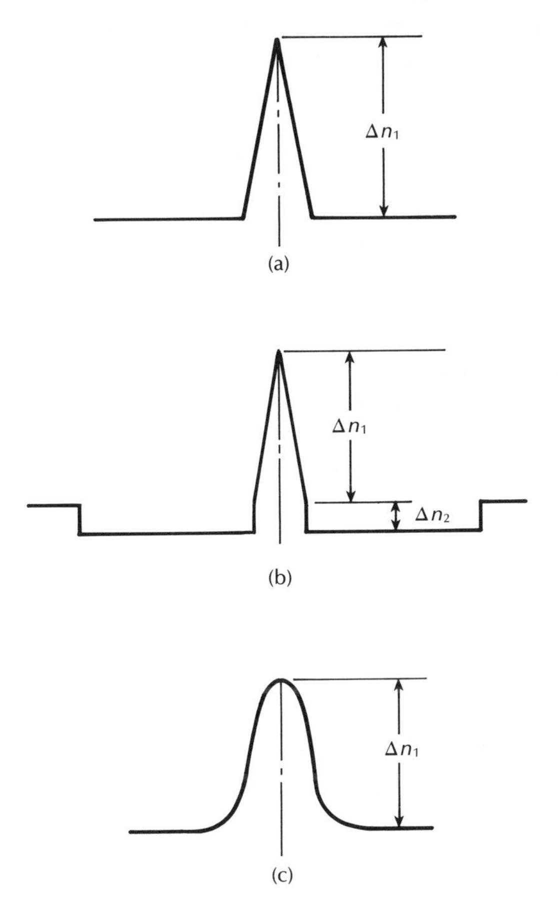

Figure 3.21 Refractive index profiles for graded index dispersion shifted fibers: (a) triangular profile; (b) depressed-cladding triangular profile; (c) Gaussian profile.

sensitivity to bend induced losses (in particular microbending) at the 1.55 μm wavelength [Ref. 65]. One method to overcome this drawback is to employ a triangular index profile combined with a depressed cladding index, as shown in Figure 3.21(b) [Ref. 66]. In this case the susceptibility to microbending losses is reduced through a shift of the LP_{11} cutoff wavelength to around 1.1 μm with a MFD of 7 μm at 1.55 μm.

Low losses and zero dispersion at a wavelength of 1.55 μm have also been obtained with a Gaussian refractive index profile, as illustrated in Figure 3.21(c). This profile, which was achieved using the vapour axial deposition fabrication process (see Section 4.4.2), produced losses of 0.21 dB km^{-1} at the λ_0 wavelength of 1.55 μm [Ref. 67].

The alternative approach for the production of DS single-mode fiber has involved the use of multiple index designs. One such fiber type which has been used to demonstrate dispersion shifting but which has been more widely employed for DF fibers (see Section 3.12.2) is the doubly clad or W fiber (see Section 2.5). However the multiple index triangular profile fibers [Ref. 68] and the segmented-core triangular profile designs [Ref. 69] which are shown in Figure 3.22(a) and (b), respectively, have reduced the sensitivity to microbending by shifting the LP_{11} mode cutoff to longer wavelength whilst maintaining a MFD of around 9 μm at a wavelength of 1.55 μm. The latter technique of introducing a ring of elevated index around the triangular core enhances the guidance of the LP_{11} mode towards longer wavelength. Such fibers may be obtained as commercial products and have been utilized within the telecommunication network [Ref. 70], exhibiting losses as low as 0.17 dB at 1.55 μm [Ref. 71].

More recently, dual-shaped core DS fibers have come under investigation in order to provide an improvement in bend loss performance over the 1.55 μm wavelength region [Refs. 72, 73]. A dual-shape core refractive index profile is shown in Figure 3.22(c), which illustrates a step index fiber design. However, several graded

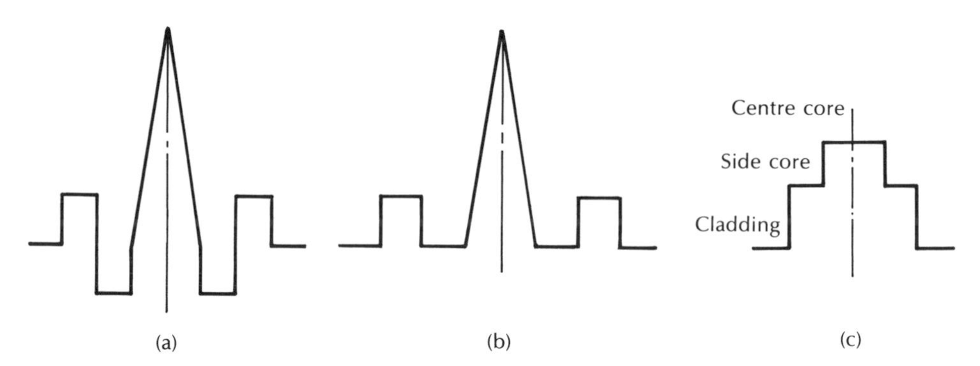

Figure 3.22 Advanced refractive index profiles for dispersion shifted fibers (a) triangular profile multiple index design; (b) segmented-core triangular profile design; (c) dual-shaped core design.

index profiles have also been studied in relation to improvements in bend loss characteristics without the incursion of increases in splice loss when the fibers are jointed [Ref. 73].

3.12.2 Dispersion flattened fibers

The original W fiber structure mentioned in Section 3.12.1 was initially employed to modify the dispersion characteristics of single-mode fibers in order to give two wavelengths of zero dispersion, as illustrated in Figure 3.18. A typical W fiber index profile (double clad) is shown in Figure 3.23(a). The first practical demonstration of dispersion flattening using the W structure was reported in 1981 [Ref. 74]. However, drawbacks with the W structural design included the requirement for a high degree of dimensional control so as to make reproducible DF fibers [Ref. 75], comparatively high overall fiber losses (around 0.3 dB km^{-1}), as well as a very high sensitivity to fiber bend losses. The latter factor results from operation very close to the cutoff (or leakage) of the fundamental mode in the long wavelength window in order to obtain a flat dispersion characteristic.

To reduce the sensitivity to bend losses associated with the W fiber structure the light which penetrates into the outer cladding area can be retrapped by introducing a further region of raised index into the structure. This approach has resulted in the triple clad (TC) and quadruple clad (QC) structures shown in Figure 3.23(b) and (c) [Refs. 76, 77]. An independent but similar program produced segmented-core DF fiber designs [Ref. 71]. Reports of low attenuation of 0.19 dB km^{-1} for DF single-mode fiber at a wavelength of 1.55 μm [Ref. 78] with significantly reduced bending losses [Ref. 79] have been made. In addition, mean splice losses of 0.04 to 0.05 dB have been achieved for MFDs of typically 6 μm and 7 μm at the 1.3 μm and 1.55 μm wavelengths respectively [Ref. 78]. However, although it is suggested [Ref. 58] that efforts remain directed towards improvements in the reduction of bend loss sensitivity and the identification of the optimum MFD, it is the case that certain DF single-mode fibers have been transferred in production [Ref. 73]. Nevertheless, it is likely that their eventual use within the telecommunications network will depend upon their performance, cost and compatibility in relation to conventional (i.e. 1.3 μm dispersion optimized) single-mode fiber.

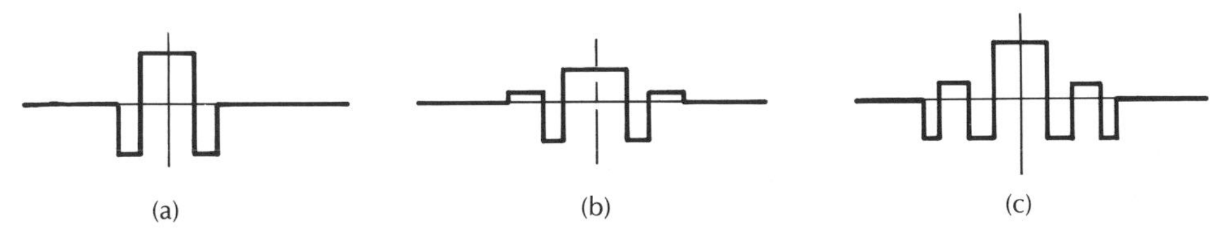

Figure 3.23 Dispersion flattened fiber refractive index profiles: (a) double clad fiber (*W* fiber); (b) triple clad fiber; (c) quadruple clad fiber.

3.13 Polarization

Cylindrical optical fibers do not generally maintain the polarization state of the light input for more than a few metres, and hence for many applications involving optical fiber transmission some form of intensity modulation (see Section 7.5) of the optical source is utilized. The optical signal is thus detected by a photodiode which is insensitive to optical polarization or phase of the light wave within the fiber. Nevertheless, systems and applications have been investigated [Ref. 81] (see Sections 12.1 and 14.5.1) which could require the polarization states of the input light to be maintained over significant distances, and fibers have been designed for this purpose. These fibers are single-mode and the maintenance of the polarization state is described in terms of a phenomenon known as modal birefringence.

3.13.1 Modal birefringence

Single-mode fibers with nominal circular symmetry about the core axis allow the propagation of two nearly degenerate modes with orthogonal polarizations. They are therefore bimodal supporting HE_{11}^{x} and HE_{11}^{y} modes where the principal axes x and y are determined by the symmetry elements of the fiber cross section. Thus the fiber behaves as a birefringent medium due to the difference in the effective refractive indices, and hence phase velocities, for these two orthogonally polarized modes. The modes therefore have different propagation constants β_x and β_y which are dictated by the anisotropy of the fiber cross section. When the fiber cross section is independent of the fiber length L in the z direction, then the modal birefringence B_F for the fiber is given by [Ref. 82]:

$$B_F = \frac{(\beta_x - \beta_y)}{(2\pi/\lambda)} \tag{3.57}$$

where λ is the optical wavelength. Light polarized along one of the principal axes will retain its polarization for all L.

The difference in phase velocities causes the fiber to exhibit a linear retardation $\Phi(z)$ which depends on the fiber length L in the z direction and is given by [Ref. 82]:

$$\Phi(z) = (\beta_x - \beta_y)L \tag{3.58}$$

assuming that the phase coherence of the two mode components is maintained. The phase coherence of the two mode components is achieved when the delay between the two transit times is less than the coherence time of the source. As indicated in Section 3.11 the coherence time for the source is equal to the reciprocal of the uncorrelated source frequency width $(1/\delta f)$.

It may be shown [Ref. 83] that birefringent coherence is maintained over a length of fiber L_{bc} (i.e. coherence length) when:

$$L_{bc} \simeq \frac{c}{B_F \delta f} = \frac{\lambda^2}{B_F \delta \lambda} \tag{3.59}$$

where c is the velocity of light in a vacuum and $\delta\lambda$ is the source linewidth.

However, when phase coherence is maintained (i.e. over the coherence length) Eq. (3.48) leads to a polarization state which is generally elliptical but which varies periodically along the fiber. This situation is illustrated in Figure 3.24(a) [Ref. 82] where the incident linear polarization which is at 45° with respect to the x axis becomes circular polarization at $\Phi = \pi/2$, and linear again at $\Phi = \pi$. The process continues through another circular polarization at $\Phi = 3\pi/2$ before returning to the initial linear polarization at $\Phi = 2\pi$. The characteristic length L_B corresponding to this process is known as the beat length. It is given by:

$$L_B = \frac{\lambda}{B_F} \tag{3.60}$$

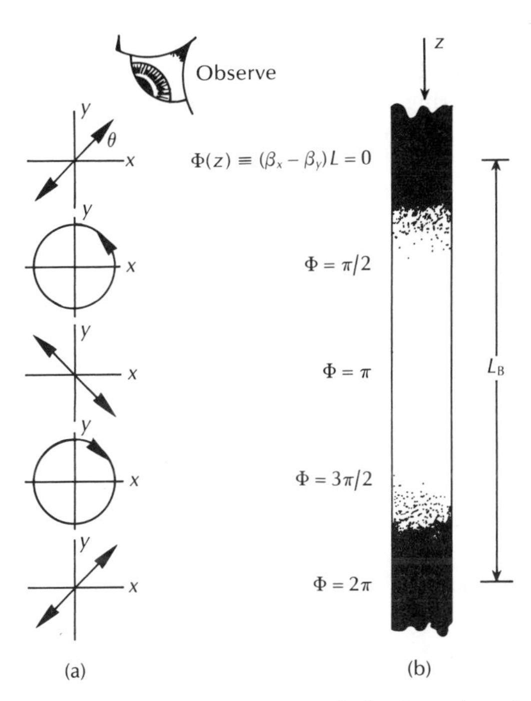

Figure 3.24 An illustration of the beat length in a single-mode optical fiber [Ref. 82]: (a) the polarization states against $\Phi(z)$; (b) the light intensity distribution over the beat length within the fiber.

Substituting for B_F from Eq. (3.47) gives:

$$L_B = \frac{2\pi}{(\beta_x - \beta_y)} \tag{3.61}$$

It may be noted that Eq. (3.61) may be obtained directly from Eq. (3.58) where:

$$\Phi(L_B) = (\beta_x - \beta_y)L_B = 2\pi \tag{3.62}$$

Typical single-mode fibers are found to have beat lengths of a few centimetres [Ref. 84], and the effect may be observed directly within a fiber via Rayleigh scattering with use of a suitable visible source (e.g. He–Ne laser) [Ref. 85]. It appears as a series of bright and dark bands with a period corresponding to the beat length, as shown in Figure 3.24(b). The modal birefringence B_F may be determined from these observations of beat length.

Example 3.12

The beat length in a single-mode optical fiber is 9 cm when light from an injection laser with a spectral linewidth of 1 nm and a peak wavelength of 0.9 μm is launched into it. Determine the modal birefringence and estimate the coherence length in this situation. In addition calculate the difference between the propagation constants for the two orthogonal modes and check the result.

Solution: To find the modal birefringence Eq. (3.60) may be used where:

$$B_F = \frac{\lambda}{L_B} = \frac{0.9 \times 10^{-6}}{0.09} = 1 \times 10^{-5}$$

Knowing B_F, Eq. (3.59) may be used to obtain the coherence length:

$$L_{bc} \simeq \frac{\lambda^2}{B_F \delta\lambda} = \frac{0.81 \times 10^{-12}}{10^{-5} \times 10^{-9}} = 81 \text{ m}$$

The difference between the propagation constant for the two orthogonal modes may be obtained from Eq. (3.61) where:

$$\beta_x - \beta_y = \frac{2\pi}{L_B} = \frac{2\pi}{0.09} = 69.8$$

The result may be checked by using Eq. (3.57) where:

$$\beta_x - \beta_y = \frac{2\pi B_F}{\lambda} = \frac{2\pi \times 10^{-5}}{0.9 \times 10^{-6}}$$

$$= 69.8$$

In a nonperfect fiber various perturbations along the fiber length such as strain or variations in the fiber geometry and composition lead to coupling of energy from one polarization to the other. These perturbations are difficult to eradicate as they may easily occur in the fiber manufacture and cabling. The energy transfer is at a maximum when the perturbations have a period Λ, corresponding to the beat length, and defined by [Ref. 81]:

$$\Lambda = \frac{\lambda}{B_F} \tag{3.63}$$

However, the cross polarizing effect may be minimized when the period of the perturbations is less than a cutoff period Λ_c (around 1 mm). Hence polarization maintaining fibers may be designed by either:

1. High (large) birefringence: the maximization of the modal birefringence, which, following Eq. (3.60), may be achieved by reducing the beat length L_B to around 1 mm or less; or
2. Low (small) birefringence: the minimization of the polarization coupling perturbations with a period of Λ. This may be achieved by increasing Λ_c giving a large beat length of around 50 m or more.

Example 3.13

Two polarization maintaining fibers operating at a wavelength of 1.3 μm have beat lengths of 0.7 mm and 80 m. Determine the modal birefringence in each case and comment on the results.

Solution: Using Eq. (3.60), the modal birefringence is given by:

$$B_F = \frac{\lambda}{L_B}$$

Hence, for a beat length of 0.7 mm:

$$B_F = \frac{1.3 \times 10^{-6}}{0.7 \times 10^{-3}} = 1.86 \times 10^{-3}$$

This typifies a high birefringence fiber.

For a beat length of 80 m:

$$B_F = \frac{1.3 \times 10^{-6}}{80} = 1.63 \times 10^{-8}$$

which indicates a low birefringence fiber.

In a uniformly birefringent fiber, as mentioned previously, the orthogonal fundamental modes have different phase propagation constants β_x and β_y. Hence the two modes exhibit different specific group delays (see Section 2.5.4) of τ_{gx} and τ_{gy}. A delay difference $\delta\tau_g$ therefore occurs between the two orthogonally polarized waves such that:

$$\delta\tau_g = \tau_{gx} - \tau_{gy} \tag{3.64}$$

where $\delta\tau_g$ is known as the polarization mode dispersion [Ref. 83]. Measured values of polarization mode dispersion range from significantly less than 1 ps km^{-1} in conventional single-mode fibers [Ref. 86] to greater than 1 ns km^{-1} in high birefringence polarization maintaining fibers [Ref. 87]. However, in specific low birefringence fibers, that is, spun fiber (see Section 3.13.2), polarization mode dispersion is negligible [Ref. 88].

Since the two fundamental modes generally launched into single-mode fiber have different group velocities, the output from a fiber length L will comprise two elements separated by a time interval $\delta\tau_g L$. For high birefringence fibers, the product $\delta\tau_g L$ provides a good estimate of pulse spreading in long fiber lengths. In this case the 3 dB bandwidth B is given by [Ref. 89]:

$$B = \frac{0.9}{(\delta\tau_g L)} \tag{3.65}$$

However, for short fiber lengths and fiber lengths longer than a characteristic coupling length L_c, the pulse spreading is proportional to $(LL_c)^{\frac{1}{2}}$ instead of simply L. Moreover, the maximum bit rate $B_T(\text{max})$ for digital transmission in relation to polarization mode dispersion may be obtained from [Ref. 90]:

$$B_T(\text{max}) = \frac{B}{0.55} \tag{3.66}$$

Example 3.14

The polarization mode dispersion in a uniformly birefringent single-mode fiber is 300 ps km^{-1}. Calculate the maximum bit rate that may be obtained on a 20 km repeaterless link assuming only polarization mode dispersion to occur.

Solution: Combining Eqs. (3.65) and (3.66), the maximum bit rate is:

$$B_T(\text{max}) = \frac{0.9}{0.55(\delta\tau_g)L} = \frac{0.9}{0.55 \times 300 \times 10^{-12} \times 20 \times 10^3}$$

$$= 273 \text{ kbit s}^{-1}$$

Although the maximum bit rate for the high birefringence fiber obtained in Example 3.14 is only 273 kbit s^{-1}, experimental results tend to indicate that polarization mode dispersion in long lengths of conventional single-mode fiber will not present a serious bandwidth limitation, even for systems operating to 1 Gbit s^{-1} with 100 km repeater spacings [Ref. 82]. For instance, measurement of polarization mode dispersion after cabling and jointing on an installed 30 km link was found to be less than 0.5 ps [Ref. 92]. Nevertheless, polarization mode dispersion can cause intersymbol interference in digital optical fiber communication systems, or signal distortion in analog systems, known as polarization mode distortion.

Although certain single-mode fibers can be fabricated to propagate only one polarization mode (see Section 3.13.2), fibers which transmit two orthogonally polarized fundamental modes can exhibit interference between the modes which may cause polarization modal noise. This phenomenon occurs when the fiber is slightly birefringent and there is a component with polarization dependent loss. Hence, when the fiber link contains an element whose insertion loss is dependent on the state of polarization, then the transmitted optical power will depend on the phase difference between the normal modes and it will fluctuate if the transmitted wavelength or the birefringence alters. Any polarization sensitive loss will therefore result in modal noise within single-mode fiber [Ref. 93].

Polarization modal noise is generally of larger amplitude than modal noise obtained within multimode fibers (see Section 3.10.3). It can therefore significantly degrade the performance of a communication system such that high quality analog transmission may prove impossible [Ref. 51]. Moreover, with digital transmission it is usually necessary to increase the system channel loss margin (see Section 10.6.4). It is therefore important to minimize the use of elements with polarization dependent insertion losses (e.g. beam splitters, polarization selective power dividers, couplers to single polarization optical components, bends in high birefringence fibers) on single-mode optical fiber links. However, other types of fiber perturbation such as bends in low birefringence fibers, splices and directional couplers do not appear to introduce significant polarization sensitive losses [Ref. 81].

Techniques have been developed to produce both high and low birefringence fibers, initially to facilitate coherent optical communication systems. Birefringence occurs when the circular symmetry in single-mode fibers is broken which can result from the effect of geometrical shape or stress. Alternatively, to design low birefringence fibers it is necessary to reduce the possible perturbations within the fiber manufacture. These fiber types are discussed in the following section.

3.13.2 Polarization maintaining fibers

Although the polarization state of the light arriving at a conventional photodetector is not distinguished and hence of little concern, it is of considerable importance in coherent lightwave systems in which the incident signal is superimposed on the field of a local oscillator (see Section 12.3). Moreover, interference and delay differences between the orthogonally polarized modes in birefringent fibers may cause polarization modal noise and polarization mode dispersion respectively (see Section 3.13.1). Finally, polarization is also of concern when a single-mode fiber is coupled to a modulator or other waveguide device (see Section 10.6.2) that can require the light to be linearly polarized for efficient operation. Hence, there are several reasons why it may be desirable to use fibers that will permit light to pass through whilst retaining its state of polarization. Such polarization maintaining (PM) fibers can be classified into two major groups: namely, high birefringence (HB) and low birefringence (LB) fibers.

The birefringence of conventional single-mode fibers is in the range $B_F = 10^{-6}$ to 10^{-5} [Ref. 88]. A HB fiber requires $B_F > 10^{-5}$ and a value better than 10^{-4} is a minimum for polarization maintenance [Ref. 94]. HB fibers can be separated into two types which are generally referred to as two-polarization fibers and single-polarization fibers. In the latter case in order to allow only one polarization mode to propagate through the fiber, a cutoff condition is imposed on the other mode by utilizing the difference in bending loss between the two polarization modes.

The various types of PM fiber, classified in terms of their linear polarization maintenance, are shown in Figure 3.25 [Ref. 95]. In addition, a selection of the most common structures is illustrated in Figure 3.26. The fiber types illustrated in Figure 3.26(a) and (b) employ geometrical shape birefringence, whilst Figure 3.26(c) to (g) utilize various stress effects. Geometrical birefringence is a somewhat weak effect and a large relative refractive index difference between the fiber core and cladding is required to produce high birefringence. Therefore, the elliptical core fiber of Figure 3.26(a) generally has high doping levels which tend to increase the optical losses as well as the polarization cross coupling [Ref. 96]. Alternatively, deep low refractive index side pits can be employed to produce HB fibers, as depicted in Figure 3.26(b).

Stress birefringence may be induced using an elliptical cladding (Figure 3.26(c)) with a high thermal-expansion coefficient. For example, borosilicate glass with some added germanium or phosphorus to provide index compensation can be utilized [Ref. 97]. The HB fibers shown in Figure 3.26(d) and (e) employ two distinct stress regions and are often referred to as the bow-tie [Ref. 98] and PANDA* [Ref. 99] fibers because of the shape of these regions. Alternatively, the flat cladding fiber design illustrated in Figure 3.26(f) has the outer edge of its elliptical cladding touching the fiber core which therefore divides the stressed cladding into two separate regions [Ref. 100].

In order to produce LB fibers attempts have been made to fabricate near-perfect, round-shaped core fibers. Ellipticity of less than 0.1% and modal birefringence of 4.5×10^{-9} has been achieved using the MCVD (see Section 4.4.3) fabricational technique [Refs. 101, 102]. Moreover, the residual birefringence within conventional single-mode fibers can be compensated for by twisting the fiber after manufacture, as shown in Figure 3.26(g). A twist rate of around five turns per metre is sufficient to reduce crosstalk significantly between the polarization modes [Ref. 95].

The reduction occurs because a high degree of circular birefringence is created by the twisting process. Hence, it is found that the propagation constants of the modes polarized in the left hand and right hand circular directions are different. This has the effect of averaging out the linear birefringence and thus produces a low

* The mnemonics PANDA, however, represent polarization maintaining and absorption reducing.

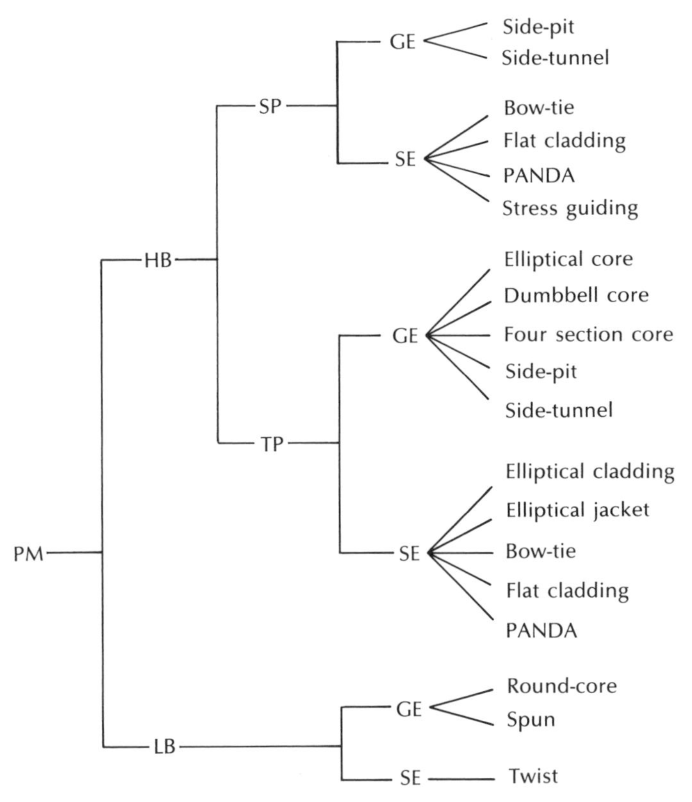

Figure 3.25 Polarization maintaining fiber types classified from linear polarization maintenance view point. PM: polarization maintaining, HB: high-birefringent, LB: low-birefringent, SP: single-polarization, TP: two-polarization modes, GE: geometrical effect, SE: stress effect [Ref. 95].

birefringence fiber. Unfortunately, the method has limitations as the fiber tends to break when beat lengths are reduced to around 10 cm [Ref. 104].

An alternative method of compensation for the residual birefringence in conventional circularly symmetric single-mode fibers is to rotate the glass preform during the fiber drawing process to produce spun fiber [Ref. 103]. This geometric effect also decreases the residual linear birefringence on average by introducing circular birefringence, but without introducing shear stress. The technique has produced fibers with modal birefringence as low as 4.3×10^{-9} [Ref. 95].

Another effective method of producing circularly birefringent fibers and thus reducing linear birefringence is to fabricate a fiber in which the core does not lie along the longitudinal fiber axis; instead the core follows a helical path about this axis [Ref. 105]. To obtain such fibers a normal MCVD (see Section 4.4.3) preform containing core and cladding glass is inserted into an off-axis hole drilled in a silica

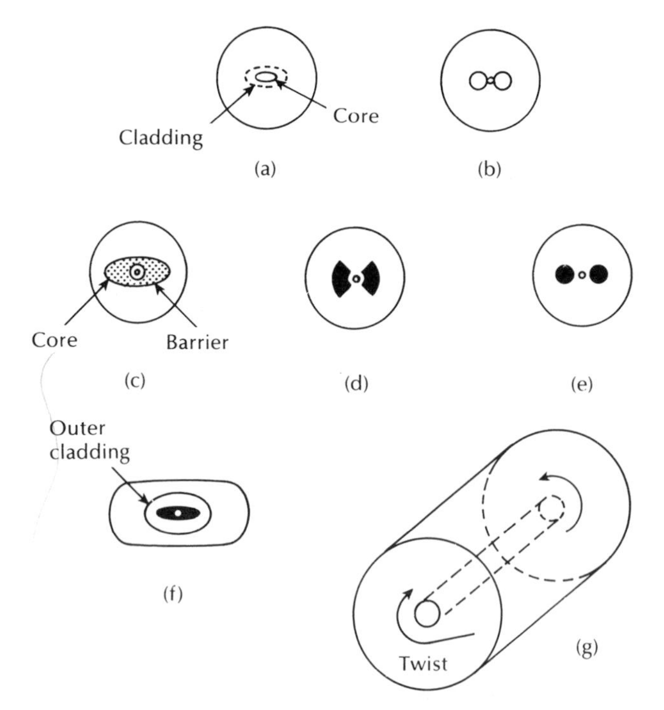

Figure 3.26 Polarization maintaining fiber structure; (a) elliptical core; (b) side-pit fiber; (c) elliptical stress-cladding; (d) bow-tie stress regions; (e) circular stress regions (PANDA fiber); (f) flat fiber; (g) twisted fiber.

rod. Then as the silica rod containing the offset core–cladding preform is in the process of being drawn into fiber, it is rotated about its longitudinal axis. The resulting fiber core forms a tight helix which has a pitch length of a few millimetres. In this case the degree of circular birefringence tends to be an order of magnitude or more greater than that achieved by twisting the fiber, giving beat lengths of around 5 mm or less.

The characteristics of the aforementioned PM fibers are not only described by the modal birefringence or beat length but also by the mode coupling parameters or polarization crosstalk as well as their transmission losses. The mode coupling parameter or coefficient h, which characterizes the polarization maintaining ability of fibers based on random mode coupling, proves useful in the comparison of different lengths of PM fiber. It is related to the polarization crosstalk* CT by [Ref. 95]:

$$CT = 10 \log_{10} \frac{P_y}{P_x} = 10 \log_{10} \tanh(hL) \qquad (3.67)$$

* The crosstalk is also referred to as the extinction ratio at the fiber output between the unwanted mode and the launch mode.

where P_x and P_y represent the optical power in the excited (i.e. unwanted) mode and the coupled (i.e. launch) mode, respectively, in an ensemble of fiber length L. However, it should be noted that the expression given in Eq. (3.67) applies with greater accuracy to two-polarization fibers because the crosstalk in a single polarization fiber becomes almost constant around -30 dB and is independent of the fiber length beyond 200 m [Ref. 106].

Example 3.15

A 3.5 km length of two polarization mode PM fiber has a polarization crosstalk of -27 dB at its output end. Determine the mode coupling parameter for the fiber.

Solution: Using Eq. (3.67) relating the mode coupling parameter h to the polarization crosstalk CT:

$$\log_{10} \tanh (hL) = \frac{CT}{10} = -2.7$$

Thus $\tanh (hL) = 2 \times 10^{-3}$ and $hL \simeq 2 \times 10^{-3}$

Hence $$h = \frac{2 \times 10^{-3}}{3.5 \times 10^{3}} = 5.7 \times 10^{-7}\ \text{m}^{-1}$$

The generally higher transmission losses exhibited by PM fibers over conventional single-mode fibers is a major consideration in their possible utilization within coherent optical fiber communication systems. This factor is, however, less important when dealing with the short fiber lengths employed in fiber devices (see Section 5.6). Nevertheless, care is required in the determination of the cutoff wavelength or the measurement of fiber loss at longer wavelengths than cutoff because the transmission losses of the HE_{11}^{x} and HE_{11}^{y} modes in HB fibers exhibit different wavelength dependencies. PM fibers with losses approaching those of conventional single-mode fiber have been fabricated recently. For example, optical losses of around 0.23 dB km^{-1} at a wavelength of 1.55 μm with polarization crosstalk of -36 dB km^{-1} have been obtained [Refs. 107, 108]. Such PM fibers could therefore eventually find application within long-haul coherent optical fiber transmission systems.

3.14 Nonlinear phenomena

Although the initial work concerned with nonlinear optical effects used relatively large core multimode fibers, more recently such phenomena have become very important within the development of low loss, single-mode fibers [Ref. 106]. The small core diameters, together with the long propagation distances that may be obtained with these fibers, has enabled the observation of certain nonlinear phenomena at power levels of a few milliwatts which are well within the capability of semiconductor lasers. Fiber attenuation associated with nonlinear scattering was discussed in Section 3.5 but these and other nonlinear processes may also be employed in important applications of single-mode fibers.

Interest has grown in the use of fibers as an interaction medium for stimulated Brillouin and Raman scattering as well as self phase modulation and four-wave mixing. The latter process, however, requires at least two propagating modes in order to fulfil the associated phase matching conditions, unless it takes place around the zero chromatic dispersion wavelength [Ref. 109]. In this section we will therefore concentrate on the other phenomena and their effects within single-mode fibers.

It was indicated in Section 3.5 that when an optical wave is within a fiber medium incident photons may be scattered, producing a phonon emitted at acoustic frequencies by exciting molecular vibrations, together with another photon at a shifted frequency. In quantum mechanical terms this process can be described as the molecule absorbing the photon at the original frequency whilst emitting a photon at the shifted frequency and simultaneously making a transition between vibrational states. The scattered photon therefore emerges at a frequency shifted below or above the incident photon frequency with the energy difference between the two photons being deposited or extracted from the scattering medium. An upshifted photon frequency is only possible if the material gives up quantum energy equal to the energy difference between the incident and scattered photon. The material must therefore be in a thermally excited state before the incident photon arrives, and at room temperature (i.e. 300 K) the upshifted scattering intensity is much weaker than the downshifted one. The former scattered wave is known as the Stokes component whereas the latter is referred to as the anti-Stokes component. In contrast to linear scattering (i.e. Rayleigh), which is said to be elastic because the scattered wave has the same frequency as the incident wave, these nonlinear scattering processes are clearly inelastic. A schematic of the spectrum obtained from these inelastic scattering processes is shown in Figure 3.27. It should be noted that the schematic depicts the spontaneous scattering spectrum rather than the stimulated one.

The frequency shifts associated with inelastic scattering can be small (less than 1 cm^{-1}), which typifies Brillouin scattering with an acoustic frequency phonon. Larger frequency shifts (greater than 100 cm^{-1}) characterize the Raman regime where the photon is scattered by local molecular vibrations or by optical frequency phonons. An interesting feature of these inelastic scattering processes is that they not only result in a frequency shift but for sufficiently high incident intensity they provide optical gain at the shifted frequency. The incident optical frequency is also known as the pump frequency ω_p, which gives the Stokes (ω_s) and anti-Stokes (ω_a) components of the scattered radiation (see Figure 3.27). For a typical fiber, a pump power of around one watt in 100 m of fiber results in a Raman gain of about a factor of 2 [Ref. 110]. By contrast, the peak Brillouin gain is more than two orders

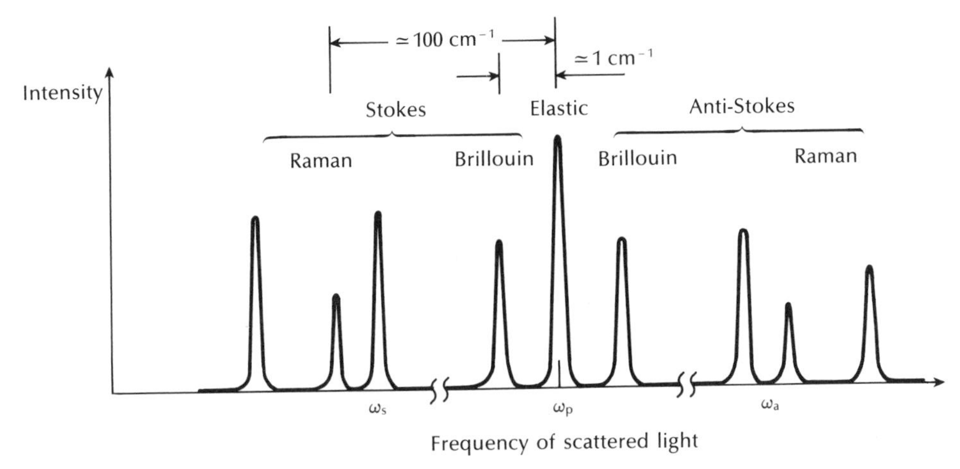

Figure 3.27 Spectrum of scattered light showing the inelastic scattering processes. Not drawn to scale as the intensities of the anti-Stokes Raman lines are far less than those of the Stokes Raman lines.

of magnitude greater than the Raman gain, but the Brillouin frequency shift and gain bandwidth are much smaller. Furthermore, Brillouin gain only exists for light propagation in the opposite direction to the pump light whilst Raman amplification will occur for light propagating in either direction.

Raman gain also extends over a substantial bandwidth, as may be observed in Figure 3.28 [Ref. 111]. Hence, with a suitable pump source, a fiber can function as a relatively high gain, broad bandwidth, bidirectional optical amplifier (see Section 10.4.2). Although given its much greater peak gain, it might be expected that Brillouin amplification would dominate over Raman amplification. At present this is not usually the case because of the narrow bandwidth associated with the Brillouin process which is often in the range 20 to 80 MHz. Pulsed semiconductor laser sources generally have much broader bandwidths and therefore prove inefficient pumps for such a narrow gain spectrum.

Nonlinear effects which can be readily described by the intensity dependent refractive index of the fiber are commonly referred to as Kerr nonlinearities. The refractive index of a medium results from the applied optical field perturbing the atoms or molecules of the medium to induce an oscillating polarization, which then radiates, producing an overall perturbed field. At low intensities the polarization is a linear function of the applied field and hence the resulting perturbation of the field can be realistically described by a constant refractive index. However, at higher optical intensities the perturbations do not remain linear functions of the applied field and Kerr nonlinearities may be observed. Typically, in the visible and infrared wavelength regions Kerr nonlinearities do not exhibit a strong dependence on the frequency of the incident light because the resonant frequencies of the oscillations tend to be in the ultraviolet region of the spectrum [Ref. 112].

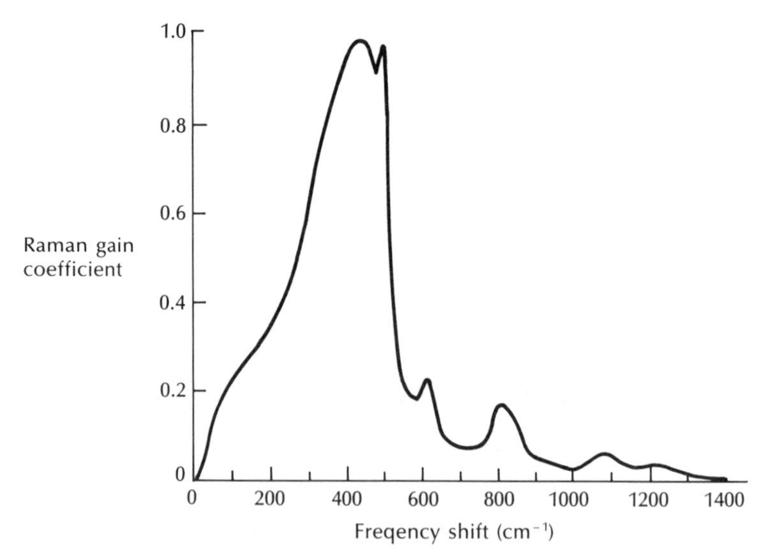

Figure 3.28 Raman gain spectrum for a silica core single-mode fiber. The peak gain occurred at 440 cm^{-1} with a pump wavelength of 0.532 μm. Reproduced with permission from W. J. Tomlinson and R. H. Stolen 'Nonlinear phenomena in optical fibers', *IEEE Commun. Mag.*, **26**, p. 36, 1988. Copyright © 1988 IEEE.

The intensity dependent refractive index causes an intensity dependent phase shift in the fiber. Hence, for a light pulse propagating in the fiber, Kerr nonlinearities result in a different transmission phase for the peak of the pulse compared to the leading and trailing pulse edges. This effect, which is known as self phase modulation, causes modifications to the pulse spectrum. As the instantaneous frequency of a wave is the time derivative of its phase, then a time-varying phase creates a time-varying frequency. Thus self phase modulation can alter and broaden the frequency spectrum of the pulse. In addition to self phase modulation, or the alteration of the pulse by itself, Kerr nonlinearities also allow a pulse to be modified by another pulse which can be at a different polarization or alternatively just a different mode of the fiber.

Although self phase modulation can simply be used for frequency shifting [Ref. 113], it has found major application for pulse compression within single-mode fiber transmission [Refs. 108 to 110]. In this context self phase modulation effectively imposes a chirp, or positive frequency sweep, on the pulse. This phenomenon combined with the group-velocity dispersion* occurring within the fiber allows

* It is usual to describe the group-velocity dispersion resulting from the frequency dependence of the group velocity (i.e. the different spectral components within a pulse exhibit a different group delay τ_g thus causing pulse spread) in terms of the intramodal or chromatic dispersion which for a unit length of fiber is defined by $d\tau_g/d\lambda$.

optical pulses to be compressed by employing, for example, a pair of diffraction gratings in which the longer wavelength light travelling at the front of the pulse follows a longer path length than the shorter wavelength light at the rear of the pulse. Hence, the rear of the pulse catches up with the front of the pulse and compression occurs. Furthermore, for critical pulse shapes and at high optical power levels, such pulse compression can be obtained in the fiber itself which forms the basis of so-called soliton propagation [Ref. 114]. Such nonlinear pulses can propagate without any dispersive changes (i.e. their shapes are self maintaining) and hence they are of great interest for communications [Ref. 110]. However, since soliton propagation is the result of a nonlinear phenomenon, it is critically dependent on the intensity of the pulse. Unfortunately, even with low loss fibers, a soliton propagating along a fiber will gradually lose energy and hence its special characteristics. Nevertheless, strategies to overcome this drawback are under investigation [e.g. Ref. 115].

Problems

3.1 The mean optical power launched into an optical fiber link is 1.5 mW and the fiber has an attenuation of 0.5 dB km^{-1}. Determine the maximum possible link length without repeaters (assuming lossless connectors) when the minimum mean optical power level required at the detector is 2 μW.

3.2 The numerical input/output mean optical power ratio in a 1 km length of optical fiber is found to be 2.5. Calculate the received mean optical power when a mean optical power of 1 mW is launched into a 5 km length of the fiber (assuming no joints or connectors).

3.3 A 15 km optical fiber link uses fiber with a loss of 1.5 dB km^{-1}. The fiber is jointed every kilometre with connectors which give an attenuation of 0.8 dB each. Determine the minimum mean optical power which must be launched into the fiber in order to maintain a mean optical power level of 0.3 μW at the detector.

3.4 Discuss absorption losses in optical fibers, comparing and contrasting the intrinsic and extrinsic absorption mechanisms.

3.5 Briefly describe linear scattering losses in optical fibers with regard to:

(a) Rayleigh scattering;
(b) Mie scattering.

The photoelastic coefficient and the refractive index for silica are 0.286 and 1.46 respectively. Silica has an isothermal compressibility of 7×10^{-11} $m^2 N^{-1}$ and an estimated fictive temperature of 1400 K. Determine the theoretical attenuation in decibels per kilometre due to the fundamental Rayleigh scattering in silica at optical wavelengths of 0.85 and 1.55 μm. Boltzmann's constant is 1.381×10^{-23} J K^{-1}.

3.6 A K_2O–SiO_2 glass core optical fiber has an attenuation resulting from Rayleigh scattering of 0.46 dB km^{-1} at a wavelength of 1 μm. The glass has an estimated fictive temperature of 758 K, isothermal compressibility of 8.4×10^{-11} $m^2 N^{-1}$, and a photoelastic coefficient of 0.245. Determine from theoretical considerations the refractive index of the glass.

3.7 Compare stimulated Brillouin and stimulated Raman scattering in optical fibers, and indicate the way in which they may be avoided in optical fiber communications.

The threshold optical powers for stimulated Brillouin and Raman scattering in a long 8 μm core diameter single-mode fiber are found to be 190 mW and 1.70 W, respectively, when using an injection laser source with a bandwidth of 1 GHz. Calculate the operating wavelength of the laser and the attenuation in decibels per kilometre of the fiber at this wavelength.

3.8 The threshold optical power for stimulated Brillouin scattering at a wavelength of 0.85 μm in a long single-mode fiber using an injection laser source with a bandwidth of 800 MHz is 127 mW. The fiber has an attenuation of 2 dB km^{-1} at this wavelength. Determine the threshold optical power for stimulated Raman scattering within the fiber at a wavelength of 0.9 μm assuming the fiber attenuation is reduced to 1.8 dB km^{-1} at this wavelength.

3.9 Explain what is meant by the critical bending radius for an optical fiber.

A multimode graded index fiber has a refractive index at the core axis of 1.46 with a cladding refractive index of 1.45. The critical radius of curvature which allows large bending losses to occur is 84 μm when the fiber is transmitting light of a particular wavelength. Determine the wavelength of the transmitted light.

3.10 A single-mode step index fiber with a core refractive index of 1.49 has a critical bending radius of 10.4 mm when illuminated with light at a wavelength of 1.30 μm. If the cutoff wavelength for the fiber is 1.15 μm calculate its relative refractive index difference.

3.11 (a) A multimode step index fiber gives a total pulse broadening of 95 ns over a 5 km length. Estimate the bandwidth–length product for the fiber when a nonreturn to zero digital code is used.

(b) A single-mode step index fiber has a bandwidth–length product of 10 GHz km. Estimate the rms pulse broadening over a 40 km digital optical link without repeaters consisting of the fiber, and using a return to zero code.

3.12 An 8 km optical fiber link without repeaters uses multimode graded index fiber which has a bandwidth–length product of 400 MHz km. Estimate:

(a) the total pulse broadening on the link;
(b) the rms pulse broadening on the link.

It may be assumed that a return to zero code is used.

3.13 Briefly explain the reasons for pulse broadening due to material dispersion in optical fibers.

The group delay τ_g in an optical fiber is given by:

$$\tau_g = \frac{1}{c}\left(n_1 - \frac{\lambda \mathrm{d}n_1}{\mathrm{d}\lambda}\right)$$

where c is the velocity of light in a vacuum, n_1 is the core refractive index and λ is the wavelength of the transmitted light. Derive an expression for the rms pulse broadening due to material dispersion in an optical fiber and define the material dispersion parameter.

The material dispersion parameter for a glass fiber is 20 ps nm^{-1} km^{-1} at a wavelength of 1.5 μm. Estimate the pulse broadening due to material dispersion within the fiber when light is launched from an injection laser source with a peak wavelength of 1.5 μm and an rms spectral width of 2 nm into a 30 km length of the fiber.

3.14 The material dispersion in an optical fiber defined by $|d^2n_1/d\lambda^2|$ is 4.0×10^{-2} μm^{-2}. Estimate the pulse broadening per kilometre due to material dispersion within the fiber when it is illuminated with an LED source with a peak wavelength of 0.9 μm and an rms spectral width of 45 nm.

3.15 Describe the mechanism of intermodal dispersion in a multimode step index fiber.

Show that the total broadening of a light pulse δT_s due to intermodal dispersion in a multimode step index fiber may be given by:

$$\delta T_s \simeq \frac{L(NA)^2}{2n_1c}$$

where L is the fiber length, NA is the numerical aperture of the fiber, n_1 is the core refractive index and c is the velocity of light in a vacuum.

A multimode step index fiber has a numerical aperture of 0.2 and a core refractive index of 1.47. Estimate the bandwidth–length product for the fiber assuming only intermodal dispersion and a return to zero code when:

(a) there is no mode coupling between the guided modes;
(b) mode coupling between the guided modes gives a characteristic length equivalent to 0.6 of the actual fiber length.

3.16 Using the relation for δT_s given in Problem 3.15, derive an expression for the rms pulse broadening due to intermodal dispersion in a multimode step index fiber. Compare this expression with a similar expression which may be obtained for an optimum near parabolic profile graded index fiber.

Estimate the bandwidth–length product for the step index fiber specified in Problem 3.15 considering the rms pulse broadening due to intermodal dispersion within the fiber and comment on the result. Indicate the possible improvement in the bandwidth–length product when an optimum near parabolic profile graded index fiber with the same relative refractive index difference and core axis refractive index is used. In both cases assume only intermodal dispersion within the fiber and the use of a return to zero code.

3.17 An 11 km optical fiber link consisting of optimum near parabolic profile graded index fiber exhibits rms intermodal pulse broadening of 346 ps over its length. If the fiber has a relative refractive index difference of 1.5%, estimate the core axis refractive index. Hence determine the numerical aperture for the fiber.

3.18 A multimode, optimum near parabolic profile graded index fiber has a material dispersion parameter of 30 ps nm^{-1} km^{-1} when used with a good LED source of rms spectral width 25 nm. The fiber has a numerical aperture of 0.4 and a core axis refractive index of 1.48. Estimate the total rms pulse broadening per kilometre within the fiber assuming waveguide dispersion to be negligible. Hence, estimate the bandwidth–length product for the fiber.

3.19 A multimode step index fiber has a relative refractive index difference of 1% and a core refractive index of 1.46. The maximum optical bandwidth that may be obtained with a particular source on a 4.5 km link is 3.1 MHz.

(a) Determine the rms pulse broadening per kilometre resulting from intramodal dispersion mechanisms.
(b) Assuming waveguide dispersion may be ignored, estimate the rms spectral width of the source used, if the material dispersion parameter for the fiber at the operating wavelength is 90 ps nm^{-1} km^{-1}.

3.20 Describe the phenomenon of modal noise in optical fibers and suggest how it may be avoided.

3.21 Discuss dispersion mechanisms with regard to single-mode fibers indicating the dominating effects. Hence, describe how intramodal dispersion may be minimized within the single-mode region.

3.22 An approximation for the normalized propagation constant in a single-mode step index fiber shown in Example 2.9 is:

$$b(V) \simeq \left(1.1428 - \frac{0.9960}{V}\right)^2$$

Obtain a corresponding approximation for the waveguide parameter $V\,\mathrm{d}^2(Vb)/\mathrm{d}V^2$ and hence write down an expression for the waveguide dispersion in the fiber.

Estimate the waveguide dispersion in a single-mode step index fiber at a wavelength 1.34 μm when the fiber core radius and refractive index are 4.4 μm and 1.48 respectively.

3.23 A single-mode step index fiber exhibits material dispersion of 7 ps nm^{-1} km^{-1} at an operating wavelength of 1.55 μm. Using the approximation obtained in Problem 3.22, estimate the fiber core diameter which will enable the waveguide dispersion to cancel the material dispersion so that zero intramodal dispersion is obtained at this wavelength. The refractive index of the fiber core is 1.45.

3.24 A single-mode step index fiber has a zero-dispersion wavelength of 1.29 μm and exhibits total first order dispersion of 3.5 ps nm^{-1} km^{-1} at a wavelength of 1.32 μm. Determine the total first order dispersion in the fiber at a wavelength of 1.54 μm.

3.25 Describe the techniques employed and the fiber structures utilized to provide:

(a) dispersion shifted single-mode fibers;
(b) dispersion flattened single-mode fibers.

3.26 Explain what is meant by:

(a) modal birefringence;
(b) the beat length;

in single-mode fibers.

The difference between the propagation constants for the two orthogonal modes in a single-mode fiber is 250. It is illuminated with light of peak wavelength 1.55 μm from an injection laser source with a spectral linewidth of 0.8 nm. Estimate the coherence length within the fiber.

3.27 The difference in the effective refractive indices $(n_x - n_y)$ for the two orthogonally polarized modes in conventional single-mode fibers are in the range $9.3 \times 10^{-7} < n_x - n_y < 1.1 \times 10^{-5}$. Determine the corresponding range for the beat lengths of the fibers when they are operating at a transmission wavelength of 1.3 μm. Hence obtain the range of the modal birefringence for the fibers.

3.28 A single-mode fiber maintains birefringent coherence over a length of 100 km when it is illuminated with an injection laser source with a spectral linewidth of 1.5 nm and a peak wavelength of 1.32 μm. Estimate the beat length within the fiber and comment on the result.

3.29 Provide a definition for polarization mode dispersion in single-mode optical fibers.

The maximum bit rate that can be achieved over a 6 km length of highly birefringent single-mode fiber is 400 kbit s^{-1}. Assuming polarization mode dispersion to be the dominant dispersive mechanism, calculate its value within this fiber.

3.30 Describe, with the aid of sketches, the techniques that can be employed to produce both high and low birefringence PM fibers.

A two polarization mode PM fiber has a mode coupling parameter of 2.3×10^{-5} m^{-1} when operating at a wavelength of 1.55 μm. Estimate the polarization crosstalk for the fiber at this wavelength.

3.31 Explain what is meant by self phase modulation.

Identify and discuss a major application area for this nonlinear phenomenon.

Answers to numerical problems

3.1 57.5 km
3.2 10.0 μW
3.3 703 μW
3.5 1.57 dB km^{-1}, 0.14 dB km^{-1}
3.6 1.49
3.7 1.50 μm, 0.30 dB km^{-1}
3.8 2.4 W
3.9 0.86 μm
3.10 0.47%
3.11 (a) 13.2 MHz km; (b) 800 ps
3.12 (a) 10 ns; (b) 4 ns
3.13 1.2 ns
3.14 5.4 ns km^{-1}
3.15 (a) 11.0 MHz km; (b) 14.2 MHz km
3.16 15.3 MHz km; improvement to 10.9 GHz km
3.17 1.45, 0.25
3.18 774 ps km^{-1}, 258 MHz km
3.19 (a) 2.82 ns km^{-1}; (b) 31 nm
3.22 -3.92 ps nm^{-1} km^{-1}
3.23 7.2 μm
3.24 23.6 ps nm^{-1} km^{-1}
3.26 48.6 m
3.27 12 cm $< L_B <$ 1.4 m; $9.3 \times 10^{-7} < B_F < 1.1 \times 10^{-5}$
3.28 113.6 m
3.29 682 ps km^{-1}
3.30 -16.4 dB km^{-1}

References

[1] F. P. Kapron, D. B. Keck and R. D. Maurer, 'Radiation losses in optical waveguides', *Appl. Phys. Lett.*, **10**, pp. 423–425, 1970.

[2] S. R. Nagel, 'Optical fiber – the expanding medium', *IEEE Commun. Mag.*, **25**(4), pp. 33–43, 1987.

[3] T. Miya, Y. Teramuna, Y. Hosaka and T. Miyashita, 'Ultimate low-loss single-mode fibre at 1.55 μm', *Electron. Lett.*, **15**(4), pp. 106–108, 1979.

[4] P. C. Schultz. 'Preparation of very low loss optical waveguides', *J. Am. Ceram. Soc.*, **52**(4), pp. 383–385, 1973.

[5] H. Osanai, T. Shioda, T. Morivama, S. Araki, M. Horiguchi, T. Izawa and H. Takata, 'Effect of dopants on transmission loss of low OH-content optical fibres', *Electron. Lett.*, **12**(21), pp. 549–550, 1976.

[6] D. B. Keck, R. D. Maurer and P. C. Schultz, 'On the ultimate lower limit of attenuation in glass optical waveguides', *Appl. Phys. Lett.*, **22**(7), pp. 307–309, 1973.

[7] A. R. Tynes, A. D. Pearson and D. L. Bisbee, 'Loss mechanisms and measurements in clad glass fibers and bulk glass', *J. Opt. Soc. Am.*, **61**, pp. 143–153, 1971.

[8] K. J. Beales and C. R. Day, 'A review of glass fibres for optical communications', *Phys. Chem. Glasses*, **21**(1), pp. 5–21, 1980.

[9] R. Olshansky, 'Propagation in glass optical waveguides', *Rev. Mod. Phys.*, **51**(2), pp. 341–367, 1979.

[10] R. M. Gagliardi and S. Karp, *Optical Communications*, John Wiley, 1976.

[11] J. Schroeder, R. Mohr, P. B. Macedo and C. J. Montrose, 'Rayleigh and Brillouin scattering in $K_2O–SiO_2$ glasses', *J. Am. Ceram. Soc.*, **56**, pp. 510–514, 1973.

[12] R. D. Maurer, 'Glass fibers for optical communications', *Proc. IEEE*, **61**, pp. 452–462, 1973.

[13] D. A. Pinnow, T. C. Rich, F. W. Ostermayer Jr and M. DiDomenico Jr, 'Fundamental optical attenuation limits in the liquid and glassy state with application to fiber optical waveguide materials', *App. Phys. Lett.*, **22**, pp. 527–529, 1973.

[14] E. A. J. Marcatili, 'Objectives of early fibers: evolution of fiber types', in S. E. Miller and A. G. Chynoweth (Eds.), *Optical Fiber Telecommunications*, pp. 1–35, Academic Press, 1979.

[15] D. Gloge, 'Propagation effects in optical fibers', *IEEE Trans. Microwave Theory Tech.*, **MTT-23**, pp. 106–120, 1975.

[16] R. H. Stolen, 'Nonlinearity in fiber transmission', *Proc. IEEE*, **68**(10), pp. 1232–1236, (1980).

[17] R. H. Stolen, 'Nonlinear properties of optical fibers', in S. E. Miller and A. G. Chynoweth (Eds.), *Optical Fiber Telecommunications*, pp. 125–150, Academic Press, 1979.

[18] Y. Ohmori, Y. Sasaki and T. Edahiro, 'Fiber-length dependence of critical power for stimulated Raman scattering', *Electron. Lett.*, **17**(17), pp. 593–594, 1981.

[19] M. M. Ramsay and G. A. Hockham, 'Propagation in optical fibre waveguides', in C. P. Sandbank (Ed.), *Optical Fibre Communication Systems*, pp. 25–41, John Wiley, 1980.

[20] H. F. Wolf, 'Optical waveguides', in H. F. Wolf (Ed.), *Handbook of Fiber Optics: Theory and Applications*, pp. 43–152, Granada, 1979.

[21] W. A. Gambling, H. Matsumura and C. M. Ragdale, 'Curvature and microbending losses in single mode optical fibres', *Opt. Quantum Electron.*, **11**, pp. 43–59, 1979.

[22] T. Li, 'Structures, parameters and transmission properties of optical fibers', *Proc. IEEE*, **68**(10), pp. 1175–1180, 1980.

[23] J. A. Savage, 'Materials for infrared fibre optics', *Materials Science Reports*, **2**, pp. 99–138, 1987.

[24] W. H. Dumbaugh, 'Oxide glasses with superior infrared transmission', *Proc. SPIE Int. Soc. Opt. Eng (USA)*, **505**, pp. 97–101, 1984.

[25] M. Saito, M. Takizawa and M. Miyagi, 'Optical and mechanical properties of infrared fibers', *J. of Lightwave Technol.*, **6**(2), pp. 233–239, 1988.

[26] S. R. Nagel, 'Fiber materials and fabrication methods', in *Optical Fiber Telecommunications II*, S. E. Miller and I. P. Kaminow (Eds.), pp. 121–215, Academic Press Inc., 1988.

[27] S. Sakaguchi and S. Takahashi, 'Low-loss fluoride optical fibers for midinfrared optical communication', *J. of Lightwave Technol.*, **LT-5**(9), pp. 1219–1228, 1987.

[28] M. Nishimura, 'The two modes of a "single-mode" fiber', *Photonics Spectra*, **20**(6), pp. 109–116, June 1986.

[29] Y. Ohishi and S. Takahashi, 'Low-dispersion fluoride glass single-mode fibres operating in two spectral ranges', *Electron. Lett.*, **24**(4), pp. 220–221, 1988.

[30] A. Bornstein and N. Croitoru, 'Experimental evaluation of a hollow glass fiber', *Appl. Opt.*, **25**(3), pp. 355–358, 1986.

[31] I. P. Kaminow, D. Marcuse and H. M. Presby, 'Multimode fiber bandwidth: theory and practice', *Proc. IEEE*, **68**(10), pp. 1209–1213, 1980.

[32] M. J. Adams, D. N. Payne, F. M. Sladen and A. H. Hartog, 'Optimum operating wavelength for chromatic equalisation in multimode optical fibres', *Electron. Lett.*, **14**(3), pp. 64–66, 1978.

[33] W. A. Gambling, A. H. Hartog and C. M. Ragdale, 'Optical fibre transmission lines', *Radio Electron. Eng. J. IERE*, **51**(7/8), pp. 313–325, 1981.

[34] D. N. Payne and W. A. Gambling, 'Zero material dispersion in optical fibres', *Electron. Lett.*, **11**(8), pp. 176–178, 1975.

[35] F. P. Kapron and D. B. Keck, 'Pulse transmission through a dielectric optical waveguide', *Appl. Opt.*, **10**(7), pp. 1519–1523, 1971.

[36] M. DiDomenico Jr, 'Material dispersion in optical fiber waveguides', *Appl. Opt.*, **11**, pp. 652–654, 1972.

[37] F. G. Stremler, *Introduction in Communication Systems*, 2nd edn, Addison-Wesley, 1982.

[38] D. Botez and G. J. Herkskowitz, 'Components for optical communication systems: a review', *Proc. IEEE*, **68**(6), pp. 689–730, 1980.

[39] A. Ghatak and K. Thyagarajan, 'Graded index optical waveguides: a review', in E. Wolf (Ed.), *Progress in Optics*, pp. 1–109, North-Holland Publishing, 1980.

[40] D. Gloge and E. A. Marcatili, 'Multimode theory of graded-core fibers', *Bell Syst. Tech. J.*, **52**, pp. 1563–1578, 1973.

[41] J. E. Midwinter, *Optical Fibers for Transmission*, John Wiley, 1979.

[42] R. Olshansky and D. B. Keck, 'Pulse broadening in graded-index optical fibers', *Appl. Opt.*, **15**(12), pp. 483–491, 1976.

[43] R. E. Epworth, 'The phenomenon of modal noise in analogue and digital optical fibre systems', in *Proceedings of the 4th European Conference on Optical Communication*, Italy, pp. 492–501, 1978.

[44] A. R. Godwin, A. W. Davis, P. A. Kirkby, R. E. Epworth and R. G. Plumb, 'Narrow stripe semiconductor laser for improved performance of optical communication systems', *Proceedings of the 5th European Conference on Optical Communications*, The Netherlands, paper 4–3, 1979.

[45] K. Sato and K. Asatani, 'Analogue baseband TV transmission experiments using semiconductor laser diodes', *Electron. Lett.*, **15**(24), pp. 794–795, 1979.

[46] B. Culshaw, 'Minimisation of modal noise in optical-fibre connectors', *Electron. Lett.*, **15**(17), pp. 529–531, 1979.

[47] N. K. Cheung, A. Tomita and P. F. Glodis, 'Observation of modal noise in single-mode fiber transmission systems', *Electron. Lett.*, **21**, pp. 5–7, 1985.

[48] F. M. Sears, I. A. White, R. B. Kummer and F. T. Stone, 'Probability of modal noise in single-mode lightguide systems', *J. of Lightwave Technol.*, **LT-4**, pp. 652–655, 1986.

[49] D. Gloge, 'Dispersion in weakly guiding fibers', *Appl. Opt.*, **10**(11), pp. 2442–2445, 1971.

[50] W. A. Gambling, H. Matsumura and C. M. Ragdale, 'Mode dispersion, material dispersion and profile dispersion in graded index single-mode fibers', *IEEJ. Microwaves, Optics and Acoustics (GB)*, **3**(6), pp. 239–246, 1979.

[51] E. G. Neumann, *Single-Mode Fibers: Fundamentals*, Springer Verlag, 1988.

[52] J. W. Fleming, 'Material dispersion in lightguide glasses', *Electron. Lett.*, **14**(11), pp. 326–328, 1978.

[53] J. I. Yamada, M. Saruwatari, K. Asatani, H. Tsuchiya, A. Kawana, K. Sugiyama and T. Kimura, 'High speed optical pulse transmission at 1.29 μm wavelength using low-loss single-mode fibers', *IEEE J. Quantum Electron.*, **QE-14**, pp. 791–800, 1978.

[54] CCITT, COM.XV/TD 46-E, 'Revised version of Recommendation G.652 – Characteristics of a single-mode fiber cable', May 1984.

[55] F. P. Kapron, 'Chromatic dispersion format for single-mode and multimode fibers', *Conf. Dig. Opt. Fiber Commun., OFC'87* (USA), paper TUQ2, January 1987.

[56] F. P. Kapron, 'Dispersion-slope parameter for monomode fiber bandwidth', *Conf. Opt. Fiber Commun. OFC'84* (USA), pp. 90–92, January 1984.

[57] F. P. Kapron, 'Maximum information capacity of fibre-optic waveguides', *Electron. Lett.*, **13**(4), pp. 96–97, 1977.

[58] P. Kaiser and D. B. Keck, 'Fiber types and their status', in *Optical Fiber Telecommunications* II, S. E. Miller and I. P. Kaminow (Eds.), pp. 29–54, Academic Press Inc., 1988.

[59] V. A. Bhagavatula, J. C. Lapp, A. J. Morrow and J. E. Ritter, 'Segmented-core fiber for long-haul and local-area-network applications', *J. of Lightwave Technol.*, **6**(10), pp. 1466–1469, 1988.

[60] L. G. Cohen, 'Comparison of single mode fiber dispersion measurement techniques', *J. Lightwave Technol.*, **LT-3**, pp. 958–966, 1985.

[61] B. J. Ainslie and C. R. Day, 'A review of single-mode fibers with modified dispersion characteristics', *J. of Lightwave Technol.*, **LT-4**(8), pp. 967–979, 1986.

[62] B. J. Ainslie, K. J. Beales, C. R. Day and J. D. Rush, 'Interplay of design parameters and fabrication conditions on the performance of monomode fibers made by MCVD', *IEEE. J. Quantum Electron.*, **QE-17**, pp. 854–857, 1981.

[63] M. A. Saifi, 'Triangular index monomode fibres', *Proc. SPIE Int. Soc. Opt. Eng.* (USA), **374**, pp. 13–15, 1983.

[64] W. A. Gambling, H. Matsumura and C. M. Ragdale, 'Zero total dispersion in graded-index single mode fibres', *Electron. Lett.*, **15**, pp. 474–476, 1979.

[65] B. J. Ainslie, K. J. Beales, D. M. Cooper and C. R. Day, 'Monomode optical fibres with graded-index cores for low dispersion at 1.55 μm', *Br. Telecom Technol. J.*, **2**(2), pp. 25–34, 1984.

[66] H.-T. Shang, T. A. Lenahan, P. F. Glodis and D. Kalish, 'Design and fabrication of dispersion-shifted depressed-clad triangular-profile (DDT) single-mode fibre', *Electron. Lett.*, **21**, pp. 484–486, 1982.

[67] M. Miyamoto, T. Abiru, T. Ohashi, R. Yamauchi and O. Fukuda, 'Gaussian profile dispersion-shifted fibers made by VAD method', in *Proc. IOOC-ECOC'85* (Venice, Italy), pp. 193–196, 1985.

[68] D. M. Cooper, S. P. Craig, C. R. Day and B. J. Ainslie, 'Multiple index structures for dispersion shifted single mode fibers using multiple index structures', *Br. Telecom Technol. J.*, **3**, pp. 52–58, 1985.

[69] V. A. Bhagavatula and P. E. Blaszyk, 'Single mode fiber with segmented core', *Conf. Dig. Opt. Fiber Commun. OFC'83*, New Orleans, USA, Paper MF5, 1983.

[70] A. R. Hunwicks, P. A. Rosher, L. Bickers and D. Stanley, 'Installation of dispersion-shifted fibre in the British Telecom trunk network', *Electron. Lett.*, **24**(9), pp. 536–537, 1988.

[71] V. Bhagavatula, M. S. Spotz, W. F. Love and D. B. Keck, 'Segmented core single mode fibre with low loss and low dispersion', *Electron. Lett.*, **19**(9), pp. 317–318, 1983.

[72] N. Kuwaki, M. Ohashi, C. Tanaka, N. Uesugi, S. Seikai and Y. Negishi, 'Characteristics of dispersion-shifted dual shape core single-mode fiber', *J. Lightwave Technol.*, **LT-5**(6), pp. 792–797, 1987.

[73] K. Nishide, D. Tanaka, M. Miyamoto, R. Yamauchi and K. Inada, 'Long-length and high-strength dual-shaped core dispersion-shifted fibers made by a fully synthesized VAD method', in *Conf. Dig. Opt. Fiber Commun. OFC'88*, New Orleans, USA, Paper WI2, 1988.

[74] T. Miya, K. Okamoto, Y. Ohmori and Y. Sasaki, 'Fabrication of low dispersion single mode fibers over a wide spectral range', *IEEE J. Quantum Electron.*, **QE-17**, pp. 858–861, 1981.

[75] J. J. Bernard, C. Brehm, P. H. Dupont, G. M. Gabriagues, C. Le Sergeant, M. Liegois, P. L. Francois, M. Monerie and P. Sansonetti, 'Investigation of the properties of depressed inner cladding single-mode fibres', in *Proc. 8th Eur. Conf. Opt. Commun.* (Cannes, France), pp. 133–138, 1982.

[76] L. G. Cohen, W. L. Mammel and S. J. Jang, 'Low-loss quadruple-clad single-mode lightguides with dispersion below 2 ps/km nm over the 1.28 μm–1.65 μm wavelength range', *Electron. Lett.*, **18**, pp. 1023–1024, 1982.

[77] S. J. Jang, J. Sanchez, K. D. Pohl and L. D. L'Esperance, 'Graded-index single-mode fibers with multiple claddings', in *Proc. IOOC'83* (Tokyo, Japan), pp. 396–397, 1983.

[78] V. A. Bhagavatula, 'Dispersion-modified fibers', *Conf. Dig. Opt. Fiber Commun. OFC'88*, New Orleans, USA, Paper WI1, 1988.

[79] P. K. Backmann, D. Leers, H. Wehr, D. U. Wiechert, J. A. Steenwijk, D. L. A. Tjaden and E. R. Wehrhatim, 'Dispersion-flattened single-mode fibers prepared with PCVD: performance, limitations, design optimization', *J. Lightwave Technol.*, **LT-4**, pp. 858–863, 1986.

[80] M. Monerie, D. Moutonnet and L. Jeunhomme, 'Polarisation studies in long length single mode fibres', in *Proceedings of the 6th European Conference on Optical Communication*, UK, pp. 107–111, 1980.

[81] I. P. Kaminow, 'Polarization in fibers', *Laser Focus*, **16**(6), pp. 80–84, 1980.

[82] I. P. Kaminow, 'Polarization in optical fibers', *IEEE J. Quantum Electron.*, **QE-17**(1), pp. 15–22, 1981.

[83] S. C. Rashleigh and R. Ulrich, 'Polarization mode dispersion in single-mode fibers', *Opt. Lett.*, **3**, pp. 60–62, 1978.

[84] V. Ramaswamy, R. D. Standley, D. Sze and W. G. French, 'Polarisation effects in short length, single mode fibres', *Bell Syst. Tech. J.*, **57**, pp. 635–651, 1978.

[85] A. Papp and H. Harms, 'Polarization optics of index-gradient optical waveguide fibers', *Appl. Opt.*, **14**, pp. 2406–2411, 1975.

[86] K. Mochizuki, Y. Namihira and H. Wakabayashi, 'Polarization mode dispersion measurements in long single mode fibers', *Electron. Lett.*, **17**, pp. 153–154, 1981.

[87] Y. Sasaki, T. Hosaka and J. Noda, 'Fabrication of polarization-maintaining optical fibers with stress-induced birefringence', *Rev. Electr. Commun. Lab.*, NTT, Japan, **32**, pp. 452–460, 1984.

[88] D. N. Payne, A. J. Barlow and J. J. Ramskov Hansen, 'Development of low-and-high-birefringence optical fibers', *IEEE J. of Quantum Electron.*, **QE-18**(4) pp. 477–487, 1982.

[89] K. Kitayama, Y. Kato, S. Seikai and N. Uchida, 'Structural optimization for two-mode fiber: theory and experiment', *IEEE J. Quantum Electron*, **QE-17**, pp. 1057–1063, 1988.

[90] K. Suzuki, N. Shibata and Y. Ishida, 'Polarization-mode dispersion as a bandwidth-limiting factor in a long-haul single-mode optical-transmission system', *Electron. Lett.*, **19**, pp. 689–691, 1983.
[91] I. P. Kaminow, 'Polarization-maintaining fibers', *Appl. Scient. Research*, **41**, pp. 257–270, 1984.
[92] F. P. Kapron and P. D. Lazay, 'Monomode fiber measurement techniques and standards', *SPIE, Int. Soc. Opt. Eng.*, **425**, pp. 40–48, 1983.
[93] S. Heckmann, 'Modal noise in single-mode fibers', *Opt. Lett.*, **6**, pp. 201–203, 1981.
[94] R. H. Stolen and R. P. De Paula, 'Single-mode fiber components', *Proc. IEEE*, **75**(11), pp. 1498–1511, 1987.
[95] J. Noda, K. Okamoto and Y. Sasaki, 'Polarization-maintaining fibers and their applications', *J. of Lightwave Technol.*, **LT-4**(8), pp. 1071–1089, 1986.
[96] W. Eickhoff and E. Brinkmeyer, Scattering loss vs. polarization holding ability of single-mode fibers', *Appl. Opt.*, **23**, pp. 1131–1132, 1984.
[97] I. P. Kaminow and V. Ramaswamy, 'Single-polarization optical fibers: Slab model', *Appl. Phys. Lett.*, **34**, pp. 268–70, 1979.
[98] R. D. Birch, D. N. Payne and M. P. Varnham, 'Fabrication of polarization-maintaining fibers using gas-phase etching', *Electron. Lett.*, **18**, pp. 1036–1038, 1982.
[99] T. Hosaka, Y. Sasaki, J. Noda and M. Horiguchi, 'Low-loss and low-crosstalk polarization-maintaining optical fibers', *Electron. Lett.*, **21**, pp. 920–921, 1985.
[100] R. H. Stolen, W. Pleibel and J. R. Simpson, 'High-birefringence optical fibers by preform deformation', *J. Lightwave Technol.*, **LT-2**, pp. 639–641, 1985.
[101] H. Schneider, H. Harms, A. Rapp and H. Aulich, 'Low birefringence single-mode fibers: Preparation and polarization characteristics', *Appl. Opt.*, **17**(19), pp. 3035–3037, 1978.
[102] S. R. Norman, D. N. Payne, M. J. Adams and A. M. Smith, 'Fabrication of single-mode fibers exhibiting extremely low polarization birefringence', *Electron. Lett.*, **15**(11), pp. 309–311, 1979.
[103] D. N. Payne, A. J. Barlow and J. J. Ramskov Hansen, 'Development of low and high birefringent optical fibers', *IEEE J. Quantum Electron*, **QE-18**(4), pp. 477–487, 1982.
[104] W. A. Gambling and S. B. Poole, 'Optical fibers for sensors', in J. P. Dakin and B. Culshaw (Eds.), *Optical Fiber Sensors: Principles and components*, Artech House, pp. 249–276, 1988.
[105] R. D. Birch, 'Fabrication and characterisation of circularly-birefringent helical fibres', *Electron Lett.*, **23**, pp. 50–52, 1987.
[106] T. Hosaka, Y. Sasaki and K. Okamoto, '3-km long single-polarization single-mode fiber', *Electron. Lett.*, **21**(22), pp. 1023–1024, 1985.
[107] H. Kajioka, Y. Takuma, K. Yamada and T. Tokunaga, 'Low-loss polarization-maintaining single-mode fibers for 1.55 μm operation', in *Tech. Dig. Opt. Fiber. Commun. Conf. OFC'88* (New Orleans, USA), paper WA5, 1988.
[108] K. Tajima and Y. Sasaki, 'Transmission loss of a 125 μm diameter PANDA fiber with circular stress-applying parts', *J. Lightwave Technol.*, **7**(4), pp. 674–679, 1989.
[109] H. Winful, 'Nonlinear optical phenomena in single-mode fibers', in E. E. Basch (Ed.), *Optical-Fiber Transmission*, pp. 179–240, Howard W Sams & Co., 1987.
[110] L. B. Jeunhomme, *Single-Mode Fiber Optics*, Marcel Dekker Inc., 1983.
[111] W. J. Tomlinson and R. H. Stolen, 'Nonlinear phenomena in optical fibers', *IEEE Commun. Mag.*, **26**(4), pp. 36–44, 1988.

[112] R. H. Stolen, C. Lee and R. K. Jain, 'Development of the stimulated Raman spectrum in single-mode silica fibers', *J. Opt. Soc. Am.* **B**, **1**, pp. 652–657, 1984.
[113] M. Vampouille and J. Marty, 'Controlled phase modulation in single-mode optical fibres', *Opt. Quantum Electron.*, **13**, pp. 393–400, 1981.
[114] L. F. Mollenauer, R. H. Stolen, J. P. Gordon and W. J. Tomlinson, 'Extreme picosecond pulse narrowing by means of soliton effect in single-mode optical fibers', *Opt. Lett.*, **8**, pp. 289–291, 1983.
[115] E. E. Bordon and W. L. Andeson, 'Dispersion-adapted monomode fiber for propagation of nonlinear pulses', *J. Lightwave Technol.*, **7**(2), pp. 353–357, 1989.

4

Optical fibers and cables

4.1 Introduction

Optical fiber waveguides and their transmission characteristics have been considered in some detail in Chapters 2 and 3. However, we have yet to discuss the practical considerations and problems associated with the production, application and installation of optical fibers within a line transmission system. These factors are of paramount importance if optical fiber communication systems are to be considered as viable replacements for conventional metallic line communication systems. Optical fiber communication is of little use if the many advantages of optical fiber transmission lines outlined in the preceding chapters may not be applied in practice in the telecommunications network without severe degradation of their performance.

It is therefore essential that:

1. Optical fibers may be produced with good stable transmission characteristics in long lengths at a minimum cost and with maximum reproducibility.
2. A range of optical fiber types with regard to size, refractive indices and index profiles, operating wavelengths, materials, etc., be available in order to fulfil many different system applications.
3. The fibers may be converted into practical cables which can be handled in a similar manner to conventional electrical transmission cables without problems associated with the degradation of their characteristics or damage.
4. The fibers and fiber cables may be terminated and connected together (jointed) without excessive practical difficulties and in ways which limit the effect of this process on the fiber transmission characteristics to keep them within acceptable operating levels. It is important that these jointing techniques may be applied with ease in the field locations where cable connection takes place.

In this chapter, we therefore consider the first three of the above practical elements associated with optical fiber communications. The final element, however, concerned with fiber termination and jointing is discussed immediately following, in Chapter 5. The various methods of preparation for silica-based optical fibers (both liquid and vapour phase) with characteristics suitable for telecommunications applications are dealt with in Sections 4.2 to 4.4. Techniques employed for the preparation of heavy metal fluoride glass fibers designed for transmission in the mid-infrared wavelength range are then outlined in Section 4.5. This is followed in Section 4.6 by consideration of the major commercially available fibers describing in general terms both the types and their characteristics. In particular, consideration is given to the recent developments in the area of plastic or polymeric fibers for lower bandwidth, very short-haul applications. The requirements for optical fiber cabling in relation to fiber protection are then described in Section 4.7 prior to discussion in Section 4.8 of the factors which cause modification to the cabled fiber transmission characteristics in a practical operating environment (i.e. microbending, hydrogen absorption, nuclear radiation exposure). Finally, cable design strategies and their influence upon typical examples of optical fiber cable constructions are dealt with in Section 4.9.

4.2 Preparation of optical fibers

From the considerations of optical waveguiding of Chapter 2 it is clear that a variation of refractive index inside the optical fiber (i.e. between the core and the cladding) is a fundamental necessity in the fabrication of fibers for light transmission. Hence at least two different materials which are transparent to light over the current operating wavelength range (0.8 to 1.6 μm) are required. In practice these materials must exhibit relatively low optical attenuation and they must therefore have low intrinsic absorption and scattering losses. A number of organic and inorganic insulating substances meet these conditions in the visible and near-infrared regions of the spectrum.

However, in order to avoid scattering losses in excess of the fundamental intrinsic losses, scattering centres such as bubbles, strains and grain boundaries must be eradicated. This tends to limit the choice of suitable materials for the fabrication of optical fibers to either glasses (or glass-like materials) and monocrystalline structures (certain plastics).

It is also useful, and in the case of graded index fibers essential, that the refractive index of the material may be varied by suitable doping with another compatible material. Hence these two materials should have mutual solubility over a relatively wide range of concentrations. This is only achieved in glasses or glass-like materials, and therefore monocrystalline materials are unsuitable for the fabrication of graded index fibers, but may be used for step index fibers. However, it is apparent that glasses exhibit the best overall material characteristics for use in the fabrication of low loss optical fibers. They are therefore used almost exclusively in the preparation of fibers for telecommunications applications. Plastic clad [Ref. 1] and all plastic fibers find some use in short-haul, low bandwidth applications.

In this chapter the discussion will therefore be confined to the preparation of glass fibers. This is a two stage process in which initially the pure glass is produced and converted into a form (rod or preform) suitable for making the fiber. A drawing or pulling technique is then employed to acquire the end product. The methods of preparing the extremely pure optical glasses generally fall into two major categories which are:

(a) conventional glass refining techniques in which the glass is processed in the molten state (melting methods) producing a multicomponent glass structure;
(b) vapour-phase deposition methods producing silica-rich glasses which have melting temperatures that are too high to allow the conventional melt process.

These processes, with their respective drawing techniques, are described in the following sections.

4.3 Liquid-phase (melting) techniques

The first stage in this process is the preparation of ultra pure material powders which are usually oxides or carbonates of the required constituents. These include oxides such as SiO_2, GeO_2, B_2O_2 and A_2O_3, and carbonates such as Na_2CO_3, K_2CO_3, $CaCO_3$ and $BaCO_3$ which will decompose into oxides during the glass melting. Very high initial purity is essential and purification accounts for a large proportion of the material cost; nevertheless these compounds are commercially available with total transition metal contents below 20 parts in 10^9 and below 1 part in 10^9 for some specific impurities [Ref. 2]. The purification may therefore involve combined techniques of fine filtration and coprecipitation, followed by solvent extraction before recrystallization and final drying in a vacuum to remove any residual OH ions [Ref. 3].

The next stage is to melt these high purity, powdered, low melting point glass materials to form a homogeneous, bubble-free multicomponent glass. A refractive index variation may be achieved by either a change in the composition of the various constituents or by ion exchange when the materials are in the molten phase. The melting of these multicomponent glass systems occurs at relatively low temperatures between 900 and 1300 °C and may take place in a silica crucible as shown in Figure 4.1 [Ref. 4]. However, contamination can arise during melting from several sources including the furnace environment and the crucible. Both fused silica and platinum crucibles have been used with some success, although an increase in impurity content was observed when the melt was held in a platinum crucible at high temperatures over long periods [Ref. 5].

Silica crucibles can give dissolution into the melt which may introduce inhomogeneities into the glass, especially at high melting temperatures. A technique for avoiding this involves melting the glass directly into a radiofrequency (RF

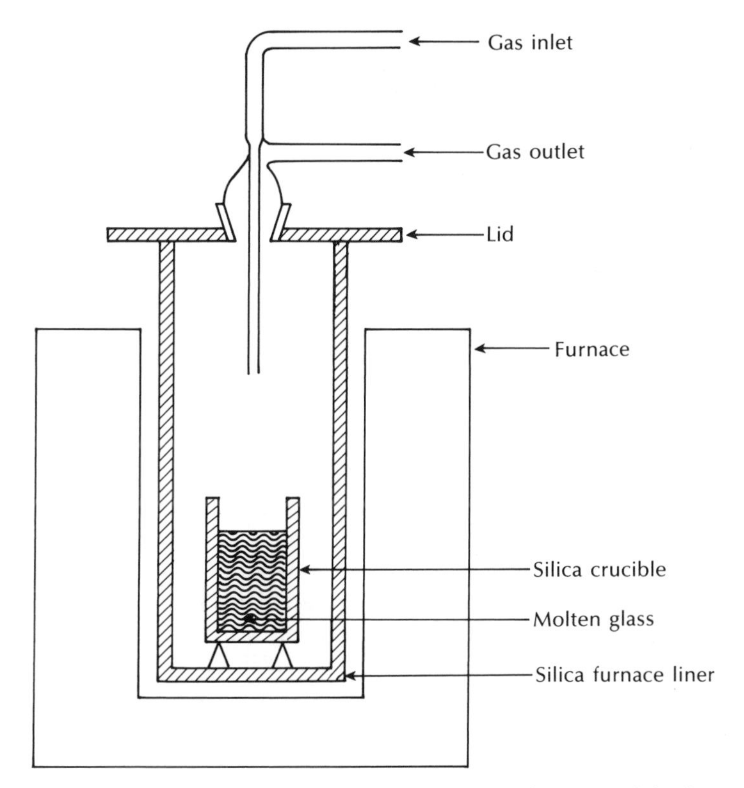

Figure 4.1 Glassmaking furnace for the production of high purity glasses [Ref. 4].

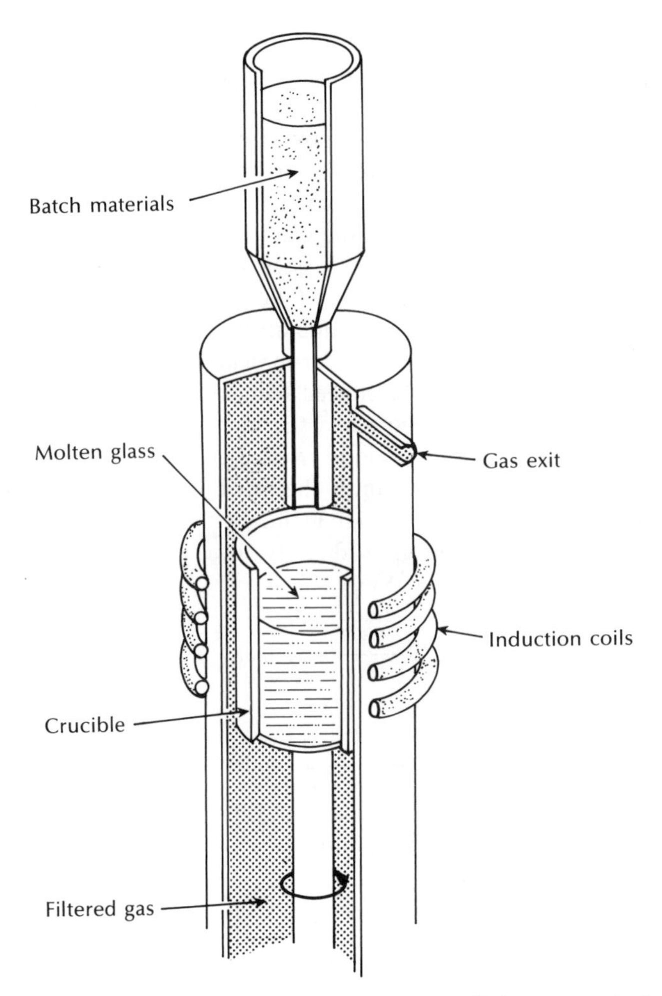

Figure 4.2 High-purity melting using a radiofrequency induction furnace [Refs. 6 to 8].

approximately 5 MHz) induction furnace while cooling the silica by gas or water flow, as shown in Figure 4.2 [Refs. 6 to 8]. The materials are preheated to around 1000 °C where they exhibit sufficient ionic conductivity to enable coupling between the melt and the RF field. The melt is also protected from any impurities in the crucible by a thin layer of solidified pure glass which forms due to the temperature difference between the melt and the cooled silica crucible.

In both techniques the glass is homogenized and dried by bubbling pure gases through the melt, whilst protecting against any airborne dust particles either originating in the melt furnace or present as atmospheric contamination. After the melt has been suitably processed, it is cooled and formed into long rods (cane) of multicomponent glass.

4.3.1 Fiber drawing

The traditional technique for producing fine optical fiber waveguides is to make a preform using the rod in tube process. A rod of core glass is inserted into a tube of cladding glass and the preform is drawn in a vertical muffle furnace, as illustrated in Figure 4.3 [Ref. 9]. This technique is useful for the production of step index fibers with large core and cladding diameters where the achievement of low attenuation is not critical as there is a danger of including bubbles and particulate matter at the core–cladding interface.

Another technique which is also suitable for the production of large core diameter step index fibers, and reduces the core–cladding interface problems, is called the stratified melt process. This process, developed by Pilkington Laboratories [Ref. 10], involves pouring a layer of cladding glass over the core glass in a platinum crucible as shown in Figure 4.4 [Ref. 11]. A bait glass rod is dipped into the molten combination and slowly withdrawn giving a composite core–clad preform which may then be drawn into a fiber.

Subsequent development in the drawing of optical fibers (especially graded index) produced by liquid-phase techniques has concentrated on the double crucible method. In this method the core and cladding glass in the form of separate rods is fed into two concentric platinum crucibles, as illustrated in Figure 4.5 [Ref. 4]. The assembly is usually located in a muffle furnace capable of heating the crucible contents to a temperature of between 800 and 1200 °C. The crucibles have nozzles in their bases from which the clad fiber is drawn directly from the melt, as shown in Figure 4.5. Index grading may be achieved through the diffusion of mobile ions across the core–cladding interface within the molten glass. It is possible to achieve

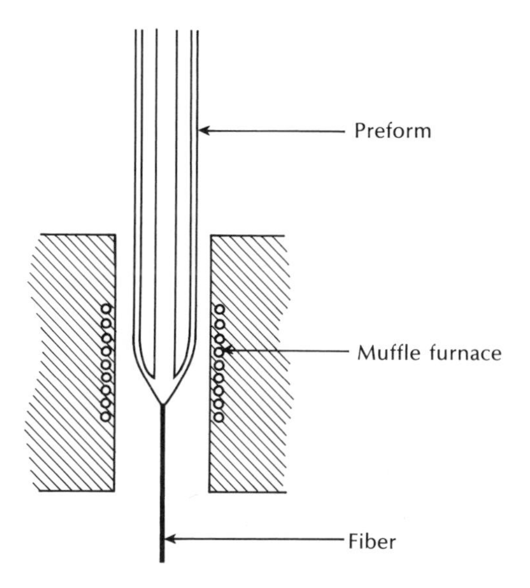

Figure 4.3 Optical fiber from a preform [Ref. 9].

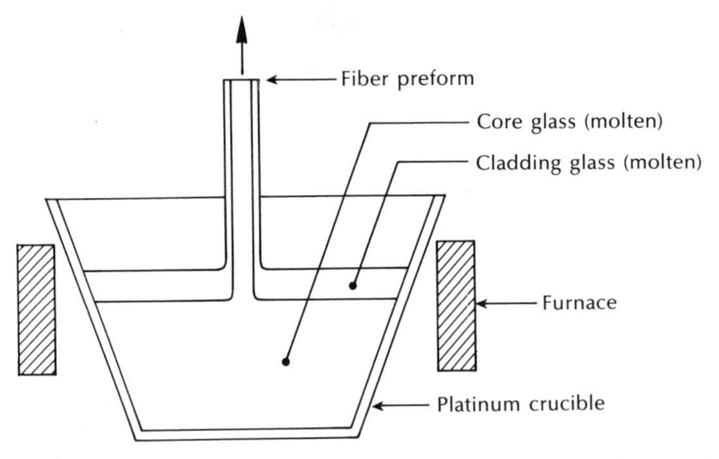

Figure 4.4 The stratified melt process (glass on glass technique) for producing glass clad rods or preforms [Ref. 11].

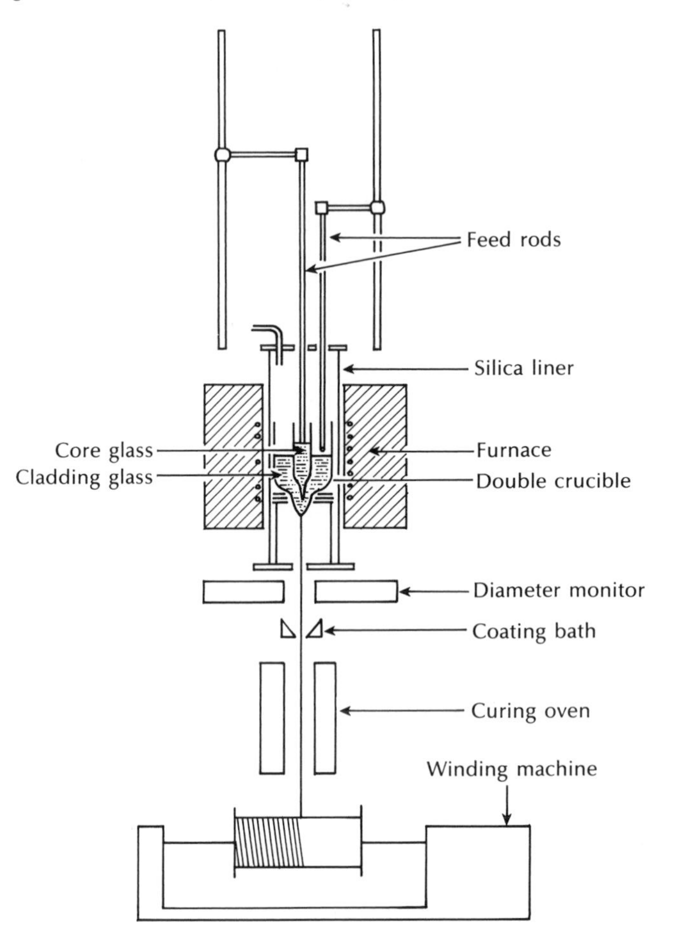

Figure 4.5 The double crucible method for fiber drawing [Ref. 4].

Table 4.1 Material systems used in the fabrication of multicomponent glass fibers by the double crucible technique

Step index	
Core glass	Cladding glass
Na_2–B_2O_3–SiO_2	Na_2O–B_2O_3–SiO_2
Na_2–LiO–CaO–SiO_2	Na_2O–Li_2O–CaO–SiO_2
Na_2–CaO–GeO_2	Na_2O–CaO–SiO_2
Tl_2O–Na_2O–B_2O_3–GeO_2–BaO–CaO–SiO_2	Na_2O–B_2O_3–SiO_2
Na_2O–BaO–GeO_2–B_2O_3–SiO_2	Na_2O–B_2O_5–SiO_2
P_2O_5–Ga_2O_3–GeO_2	P_2O_5–Ga_2O_3–SiO_2
Graded index	
Base glass	Diffusion mechanism
R_2O–GeO_2–CaO–SiO_2	$Na^+ \rightleftharpoons K^+$
R_2O–B_2O_3–SiO_2	$Tl^+ \rightleftharpoons Na^+$
Na_2O–B_2O_3–SiO_2	Na_2O diffusion
Na_2O–B_2O_3–SiO_2	CaO, BaO, diffusion

a reasonable refractive index profile via this diffusion process, although due to lack of precise control it is not possible to obtain the optimum near parabolic profile which yields the minimum pulse dispersion (see Section 3.10.2). Hence graded index fibers produced by this technique are subsequently less dispersive than step index fibers, but do not have the bandwidth–length products of optimum profile fibers. Pulse dispersion of 1 to 6 ns km^{-1} [Refs. 12, 13] is quite typical, depending on the material system used.

Some of the material systems used in the fabrication of multicomponent glass step index and graded index fibers are given in Table 4.1.

Using very high purity melting techniques and the double crucible drawing method, step index and graded index fibers with attenuations as low as 3.4 dB km^{-1} [Ref. 14] and 1.1 dB km^{-1} [Ref. 2], respectively, have been produced. However, such low losses cannot be consistently obtained using liquid-phase techniques and typical losses for multicomponent glass fibers prepared continuously by these methods are between 5 and 10 dB km^{-1}. Therefore, liquid-phase techniques have the inherent disadvantage of obtaining and maintaining extremely pure glass which limits their ability to produce low loss fibers. The advantage of these techniques is in the possibility of continuous production (both melting and drawing) of optical fibers.

4.4 Vapour-phase deposition techniques

Vapour-phase deposition techniques are used to produce silica-rich glasses of the highest transparency and with the optimal optical properties. The starting materials

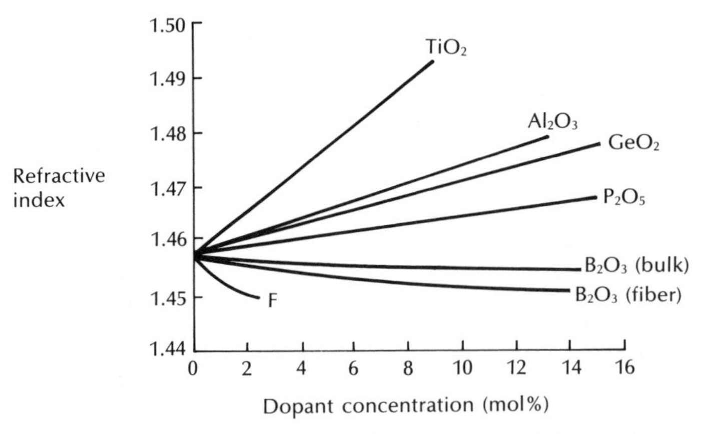

Figure 4.6 The variation in the refractive index of silica using various dopants. Reproduced with permission from the publishers, Society of Glass Technology, *Phys. Chem. Glasses*, **21**, p. 5, 1980.

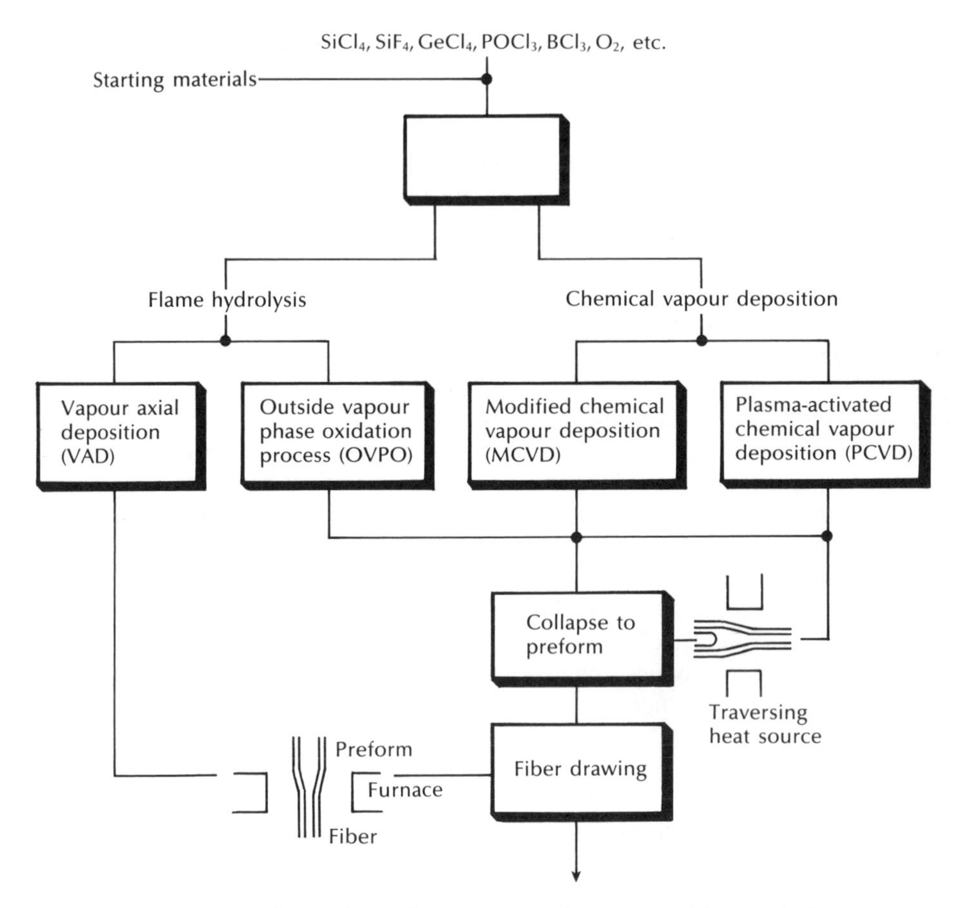

Figure 4.7 Schematic illustration of the vapour-phase deposition techniques used in the preparation of low loss optical fibers.

are volatile compounds such as $SiCl_4$, $GeCl_4$, SiF_4, BCl_3, O_2, BBr_3 and $POCl_3$ which may be distilled to reduce the concentration of most transition metal impurities to below one part in 10^9, giving negligible absorption losses from these elements. Refractive index modification is achieved through the formation of dopants from the nonsilica starting materials. These vapour-phase dopants include TiO_2, GeO_2, P_2O_5, Al_2O_3, B_2O_3 and F, the effects of which on the refractive index of silica are shown in Figure 4.6 [Ref. 2]. Gaseous mixtures of the silica-containing compound, the doping material and oxygen are combined in a vapour-phase oxidation reaction where the deposition of oxides occurs. The deposition is usually on to a substrate or within a hollow tube and is built up as a stack of successive layers. Hence the dopant concentration may be varied gradually to produce a graded index profile or maintained to give a step index profile. In the case of the substrate this directly results in a solid rod or preform whereas the hollow tube must be collapsed to give a solid preform from which the fiber may be drawn.

There are a number of variations of vapour-phase deposition which have been successfully utilized to produce low loss fibers. The major techniques are illustrated in Figure 4.7, which also indicates the plane (horizontal or vertical) in which, the deposition takes place as well as the formation of the preform. These vapour-phase deposition techniques fall into two broad categories: flame hydrolysis and chemical vapour deposition (CVD) methods. The individual techniques are considered in the following sections.

4.4.1 Outside vapour-phase oxidation (OVPO) process

This process which uses flame hydrolysis stems from work on 'soot' processes originally developed by Hyde [Ref. 16] which were used to produce the first fiber with losses of less than $20\ \text{dB km}^{-1}$ [Ref. 17]. The best known technique of this type is often referred to as the outside vapour-phase oxidation process. In this process the required glass composition is deposited laterally from a 'soot' generated by hydrolyzing the halide vapours in an oxygen–hydrogen flame. Oxygen is passed through the appropriate silicon compound (i.e. $SiCl_4$) which is vaporized, removing any impurities. Dopants such as $GeCl_4$ or $TiCl_4$ are added and the mixture is blown through the oxygen–hydrogen flame giving the following reactions:

$$\underset{\text{(vapour)}}{SiCl_4} + \underset{\text{(vapour)}}{2H_2O} \xrightarrow{\text{heat}} \underset{\text{(solid)}}{SiO_2} + \underset{\text{(gas)}}{4HCl} \tag{4.1}$$

and

$$\underset{\text{(vapour)}}{SiCL_4} + \underset{\text{(gas)}}{O_2} \xrightarrow{\text{heat}} \underset{\text{(solid)}}{SiO_2} + \underset{\text{(gas)}}{2Cl_2} \tag{4.2}$$

$$\underset{\text{(vapour)}}{GeCl_4} + \underset{\text{(gas)}}{O_2} \xrightarrow{\text{heat}} \underset{\text{(solid)}}{GeO_2} + \underset{\text{(gas)}}{2Cl_2} \tag{4.3}$$

or

$$\underset{\text{(vapour)}}{TiCl_4} + \underset{\text{(gas)}}{O_2} \xrightarrow{\text{heat}} \underset{\text{(solid)}}{TiO_2} + \underset{\text{(gas)}}{2Cl_2} \tag{4.4}$$

The silica is generated as a fine soot which is deposited on a cool rotating mandrel, as illustrated in Figure 4.8(a) [Ref. 18]. The flame of the burner is reversed back and forth over the length of the mandrel until a sufficient number of layers of silica (approximately 200) are deposited on it. When this process is completed the mandrel is removed and the porous mass of silica soot is sintered (to form a glass body), as illustrated in Figure 4.8(b). The preform may contain both core and cladding glasses by properly varying the dopant concentrations during the deposition process. Several kilometres (around 10 km of 120 μm core diameter fiber have been produced [Ref. 2]) can be drawn from the preform by collapsing and closing the central hole, as shown in Figure 4.8(c). Fine control of the index gradient for graded index fibers may be achieved using this process as the gas flows can be adjusted at the completion of each traverse of the burner. Hence fibers with bandwidth–length products as high as 3 GHz km have been achieved [Ref. 19] through accurate index grading with this process.

The purity of the glass fiber depends on the purity of the feeding materials and also upon the amount of OH impurity from the exposure of the silica to water vapour in the flame following the reactions given in Eqs. (4.1) to (4.4). Typically, the OH content is between 50 and 200 parts per million and this contributes to the fiber attenuation. It is possible to reduce the OH impurity content by employing gaseous chlorine as a drying agent during sintering. This has given losses as low as 1 dB km^{-1} and 1.8 dB km^{-1} at wavelengths of 1.2 and 1.55 μm respectively [Ref. 20] in fibers prepared using the OVPO process.

Other problems stem from the use of the mandrel which can create some difficulties in the formation of the fiber preform. Cracks may form due to stress

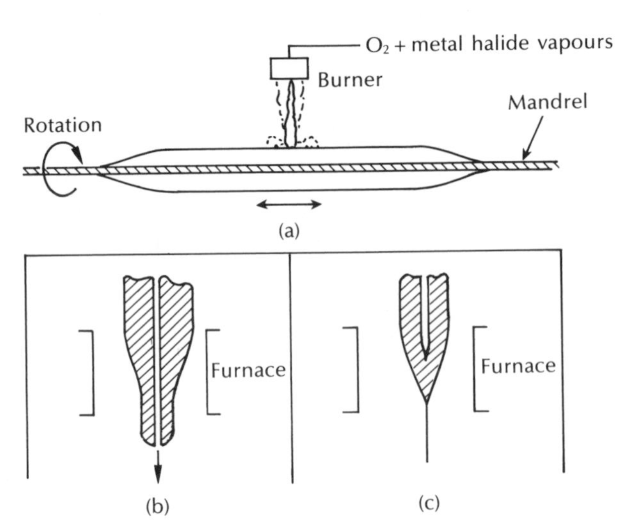

Figure 4.8 Schematic diagram of the OVPO process for the preparation of optical fibers; (a) soot deposition: (b) preform sintering; (c) fiber drawing [Ref. 19].

concentration on the surface of the inside wall when the mandrel is removed. Also the refractive index profile has a central depression due to the collapsed hole when the fiber is drawn. Therefore, although the OVPO process is a useful fiber preparation technique, it has several drawbacks. Furthermore, it is a batch process, which limits its use for the volume production of optical fibers. Nevertheless, a number of proprietary approaches to scaling-up the process have provided preforms capable of producing 250 km of fiber [Ref. 21].

4.4.2 Vapour axial deposition (VAD)

This process was developed by Izawa *et al.* [Ref. 22] in the search for a continuous (rather than batch) technique for the production of low loss optical fibers. The VAD technique uses an end-on deposition on to a rotating fused silica target, as illustrated in Figure 4.9 [Ref. 23]. The vaporized constituents are injected from burners and react to form silica soot by flame hydrolysis. This is deposited on the end of the starting target in the axial direction forming a solid porous glass preform in the shape of a boule. The preform which is growing in the axial direction is pulled upwards at a rate which corresponds to the growth rate. It is initially dehydrated by heating with $SOCl_2$ using the reaction:

$$\underset{\text{(vapour)}}{H_2O} + \underset{\text{(vapour)}}{SOCl_2} \xrightarrow{\text{heat}} \underset{\text{(gas)}}{2HCl} + \underset{\text{(gas)}}{SO_2} \qquad (4.5)$$

and is then sintered into a solid preform in a graphite resistance furnace at an elevated temperature of around 1500 °C. Therefore, in principle this process may be adapted to draw fiber continuously, although at present it tends to be operated as a batch process partly because the resultant preforms can yield more than 100 km of fiber [Ref. 21].

A spatial refractive index profile may be achieved using the deposition properties of SiO_2–GeO_2 particles within the oxygen–hydrogen flame. The concentration of these constituents deposited on the porous preform is controlled by the substrate temperature distribution which can be altered by changing the gas flow conditions. Fibers produced by the VAD process still suffer from some OH impurity content due to the flame hydrolysis and hence very low loss fibers have not been achieved using this method. Nevertheless, fibers with attenuation in the range 0.7 to 2.0 dB km^{-1} at a wavelength of 1.181 μm have been reported [Ref. 24].

4.4.3 Modified chemical vapour deposition (MCVD)

Chemical vapour deposition techniques are commonly used at very low deposition rates in the semiconductor industry to produce protective SiO_2 films on silicon semiconductor devices. Usually an easily oxidized reagent such as SiH_4 diluted by inert gases and mixed with oxygen is brought into contact with a heated silicon surface where it forms a glassy transparent silica film. This heterogeneous reaction (i.e. requires a surface to take place) was pioneered for the fabrication of optical

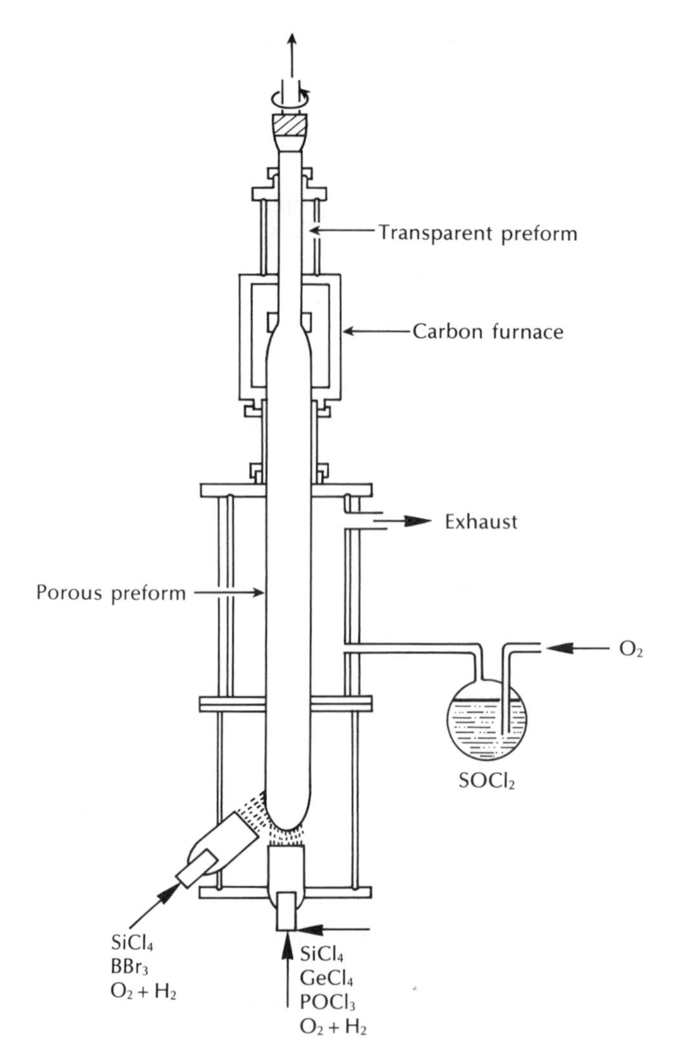

Figure 4.9 The VAD process [Ref. 23].

fibers using the inside surface of a fused quartz tube [Ref. 25]. However, these processes gave low deposition rates and were prone to OH contamination due to the use of hydride reactants. This led to the development of the modified chemical vapour deposition (MCVD) process by Bell Telephone Laboratories [Ref. 26] and Southampton University, UK [Ref. 27], which overcomes these problems and has found widespread application throughout the world.

The MCVD process is also an inside vapour-phase oxidation (IVPO) technique taking place inside a silica tube, as shown in Figure 4.10. However, the vapour-phase reactants (halide and oxygen) pass through a hot zone so that a substantial

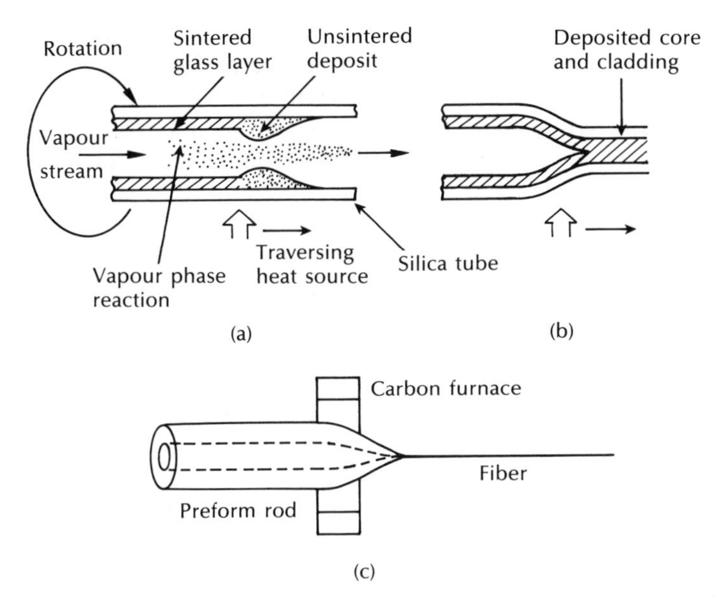

Figure 4.10 Schematic diagram showing the MCVD method for the preparation of optical fibers: (a) deposition; (b) collapse to produce a preform; (c) fiber drawing.

part of the reaction is homogeneous (i.e. involves only one phase; in this case the vapour phase). Glass particles formed during this reaction travel with the gas flow and are deposited on the walls of the silica tube. The tube may form the cladding material but usually it is merely a supporting structure which is heated on the outside by an oxygen–hydrogen flame to temperatures between 1400 °C and 1600 °C. Thus a hot zone is created which encourages high temperature oxidation reactions such as those given in Eqs. (4.2) and (4.3) or (4.4) (not Eq. (4.1)). These reactions reduce the OH impurity concentration to levels below those found in fibers prepared by hydride oxidation or flame hydrolysis.

The hot zone is moved back and forth along the tube allowing the particles to be deposited on a layer by layer basis giving a sintered transparent silica film on the walls of the tube. The film may be up to 10 μm in thickness and uniformity is maintained by rotating the tube. A graded refractive index profile can be created by changing the composition of the layers as the glass is deposited. Usually, when sufficient thickness has been formed by successive traverses of the burner for the cladding, vaporized chlorides of germanium ($GeCl_4$) or phosphorus ($POCl_3$) are added to the gas flow. The core glass is then formed by the deposition of successive layers of germanosilicate or phosphosilicate glass. The cladding layer is important as it acts as a barrier which suppresses OH absorption losses due to the diffusion of OH ions from the silica tube into the core glass as it is deposited. After the deposition is completed the temperature is increased to between 1700 and 1900 °C.

The tube is then collapsed to give a solid preform which may then be drawn into fiber at temperatures of 2000 to 2200 °C as illustrated in Figure 4.10.

This technique is the most widely used at present as it allows the fabrication of fiber with the lowest losses. Apart from the reduced OH impurity contamination the MCVD process has the advantage that deposition occurs within an enclosed reactor which ensures a very clean environment. Hence, gaseous and particulate impurities may be avoided during both the layer deposition and the preform collapse phases. The process also allows the use of a variety of materials and glass compositions. It has produced GeO_2 doped silica single-mode fiber with minimum losses of only 0.2 dB km^{-1} at a wavelength of 1.55 μm [Ref. 28]. More generally, the GeO_2–B_2O_3–SiO_2 system (B_2O_3 is added to reduce the viscosity and assist fining) has shown minimum losses of 0.34 dB km^{-1} with multimode fiber at a wavelength of 1.55 μm [Ref. 29]. Also, graded index germanium phosphosilicate fibers have exhibited losses near the intrinsic level for their composition of 2.8, 0.45 and 0.35 dB km^{-1} at wavelengths of 0.82, 1.3 and 1.5 μm respectively [Ref. 30].

The MCVD process has also demonstrated the capability of producing fibers with very high bandwidths, although still well below the theoretical values which may be achieved. Multimode graded index fibers with measured bandwidth–length products of 4.3 GHz km and 4.7 GHz km at wavelengths of 1.25 and 1.29 μm have been reported [Ref. 31]. Large-scale batch production (30 000 km) of 50 μm core graded index fiber has maintained bandwidth–length products of 825 MHz km and 735 MHz km at wavelengths of 0.825 and 1.3 μm respectively [Ref. 30]. The median attenuation obtained with this fiber was 3.4 dB km^{-1} at 0.825 μm and 1.20 dB km^{-1} at 1.3 μm. Hence, although it is not a continuous process, the MCVD technique has proved suitable for the mass production of high performance optical fiber and is the predominant technique used to prepare polarization maintaining fibers (see Section 3.13.2). Moreover it can be scaled-up to produce preforms which provide 100 to 200 km of fiber [Ref. 21].

4.4.4 Plasma-activated chemical vapour deposition (PCVD)

A variation on the MCVD technique is the use of various types of plasma to supply energy for the vapour-phase oxidation of halides. This method, first developed by Kuppers and Koenings [Ref. 32], involves plasma-induced chemical vapour deposition inside a silica tube, as shown in Figure 4.11. The essential difference between this technique and the MCVD process is the stimulation of oxide formation by means of a nonisothermal plasma maintained at low pressure in a microwave cavity (2.45 GHz) which surrounds the tube. Volatile reactants are introduced into the tube where they react heterogeneously within the microwave cavity, and no particulate matter is formed in the vapour phase.

The reaction zone is moved backwards and forwards along the tube by control of the microwave cavity and a circularly symmetric layer growth is formed. Rotation of the tube is unnecessary and the deposition is virtually 100% efficient. Film deposition can occur at temperatures as low as 500 °C but a high chlorine

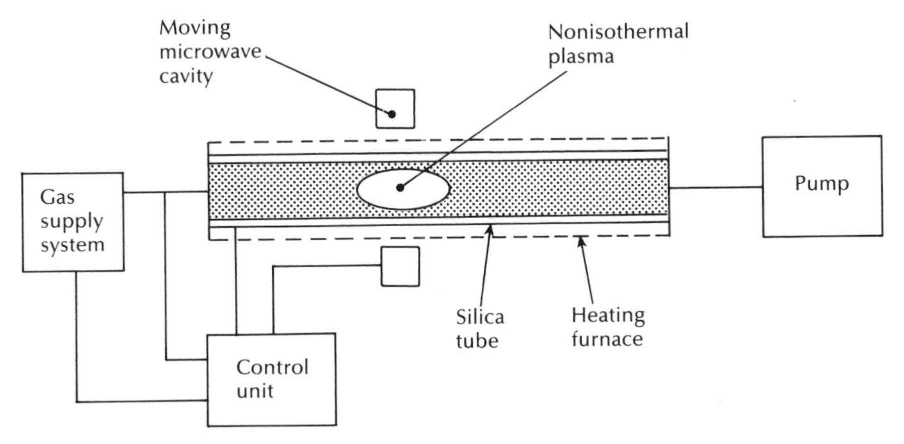

Figure 4.11 The apparatus utilized in the PCVD process.

content may cause expansivity and cracking of the film. Hence the tube is heated to around 1000 °C during deposition using a stationary furnace.

The high deposition efficiency allows the composition of the layers to be accurately varied by control of the vapour-phase reactants. Also, when the plasma zone is moved rapidly backwards and forwards along the tube, very thin layer deposition may be achieved giving the formation of up to 2000 individual layers. This enables very good graded index profiles to be realized which are a close approximation to the optimum near parabolic profile. Thus low pulse dispersion of less than 0.8 ns km^{-1}, for fibers with attenuations of between 3 and 4 dB km^{-1}, at a wavelength of 0.85 μm have been reported [Ref. 2].

A further PCVD technique uses an inductively coupled radiofrequency argon plasma which operates at a frequency of 3.4 MHz [Ref. 33]. The deposition takes place at 1 atmosphere pressure and is predominantly a homogeneous vapour-phase reaction which, via the high temperature discharge, causes the fusion of the deposited material into glass. This technique has proved to have a reaction rate five times faster than the conventional MCVD process. However, fiber attenuation was somewhat higher with losses of 6 dB km^{-1} at a wavelength of 1.06 μm. Variations on this theme operating at frequencies of 3 to 6 MHz and 27 MHz have produced GeO_2–P_2O_5–SiO_2 fibers with minimum losses of 4 to 5 dB km^{-1} at a wavelength of 0.85 μm [Ref. 34]. In addition, fiber attenuation as low as 0.3 dB km^{-1} at a wavelength of 1.55 μm has been obtained for dispersion flattened single-mode fibers (see Section 3.12.2) prepared using a low pressure PCVD process [Ref. 35]. The PCVD process also lends itself to the large scale production of optical fibers with preform sizes which would allow the preparation of over 200 km of fiber [Ref. 36].

4.4.5 Summary of vapour-phase deposition techniques

The salient features of the four major vapour-phase deposition techniques are summarized in Table 4.2 [Ref. 37]. These techniques have all demonstrated

Table 4.2 Summary of vapour-phase deposition techniques used in the preparation of low loss optical fibers

Reaction type	
Flame hydrolysis	OVPO, VAD
High temperature oxidation	MCVD
Low temperature oxidation	PCVD
Depositional direction	
Outside layer deposition	OVPO
Inside layer deposition	MCVD, PCVD
Axial layer deposition	VAD
Refractive index profile formation	
Layer approximation	OVPO, MCVD, PCVD
Simultaneous formation	VAD
Process	
Batch	OVPO, MCVD, PCVD
Continuous	VAD

relatively similar performance for the fabrication of both multimode and single-mode fiber of standard step and graded index designs [Ref. 21]. For the production of polarization maintaining fiber, however, the MCVD and VAD processes have been employed, together with a hybrid MCVD–VAD technique.

4.5 Fluoride glass fibers

Developments in heavy metal fluoride glass fibers for mid-infrared transmission (wavelength range 2 to 5 μm) were outlined in Section 3.7. It was indicated that the fluorozirconate (ZrF_4) and fluorohafnate (HfF_4) glass systems displayed the most promise and had received the greatest attention. In particular, a number of fluorozirconate glasses for fiber production have been investigated, the compositions of which are displayed in Table 4.3 [Ref. 40]. Such glasses have been synthesized from the melt by the same basic techniques as described in Section 4.3 for oxide (i.e. silica) glasses. Preforms for fiber drawing can then be cast from these melts. However, problems have occurred with contamination even using platinum or vitreous carbon crucibles. In addition, it has been found that ZrF_4 tends to be reduced to black ZrF_2 or ZrF_3 which results in melts containing dark particulate matter.

The use of inert glass glove boxes for raw material handling, glass melting and casting helps avoid major oxidation and OH pick-up which causes infrared absorption with the possibility of nucleation and growth of crystals. The latter conditions can also occur in reheating of the glass which is usually required for fiber

Table 4.3 Compositions of the major fluorozirconate glasses currently being employed for the fabrication of optical fibers

	Composition (mol %)							
Glass	ZrF_4	BaF_2	GdF_3	LaF_3	YF_3	AlF_3	LiF	NaF
ZBG	63	33	4	—	—	—	—	—
ZBGA	60	32	4	—	—	4	—	—
ZBLAL	52	20	—	5	—	3	20	—
ZBLYAL	49	22	—	3	3	3	20	—
ZBLAN	51	20	—	4.5	—	4.5	—	20
ZBLYAN	47.5	23.5	—	2.5	2	4.5	—	20

drawing. Typically, the glasses are very fluid at their liquid temperature and exhibit a very rapid decrease in viscosity above the glass transition temperature which results in sensitivity to crystallization and creates a very narrow temperature range for successful fiber drawing [Ref. 21]. Furthermore, oxidizing conditions are often found to be necessary to avoid fluoride reduction. This can be achieved by reactive atmosphere processing (RAP) where traces of CCl_4, SF_6, NF_3 or CS_2 entrained in the inert gaseous atmosphere are employed [Ref. 42].

An alternative approach to the purification of a number of the fluorides required for the preparation of heavy metal fluoride glasses is called chemical vapour purification (CVP) [Ref. 43]. This technique utilizes nonstoichiometric conditions in haloginating the metal. It is anticipated that the products will have transition metal contents close to undetectable levels, or less than 10^{-10} mole fraction of the fluoride. The process uses two sequential vapour-phase reactions to accomplish the preparation. Initially, a direct halogenation of the metal with Cl, Br or I is undertaken to form a volatile halide. Subsequently, this volatile halide is reacted with a fluorinating agent such as HF or SF to form the desired fluoride. This preparation method is considered to be appropriate for fluorides of Zr, Hf, Al, Ga and In [Ref. 44].

Using these purified raw materials and extreme care in handling the glass melting and preform casting it may be feasible to produce sufficiently pure glass to obtain fluorozirconate glass fibers with losses around 0.03 dB km^{-1} at a wavelength of 2.56 μm [Ref. 44]. This overall loss estimate takes into account extrinsic absorption losses together with the intrinsic losses associated with the infrared absorption edge and Rayleigh scattering which alone give the lower figure of 0.01 dB km^{-1} at a wavelength of 2.55 μm. Crucible drawing as well as preform techniques have been employed to prepare the fiber structures. Variations on the conventional double crucible method described in Section 4.3.1 have been proposed to overcome the crystallization and deformation problems which occur with these glasses [Ref.45].

Alternatively, preforms have been made by casting the core and cladding glasses in a mould. One such process is illustrated in Figure 4.12 where the cladding glass

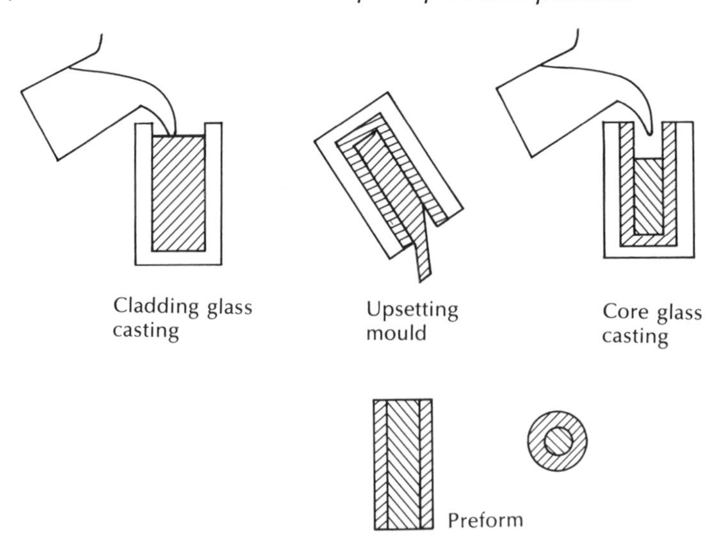

Figure 4.12 Schematic representation of the built-in casting technique for heavy metal fluoride glass preparation [Ref. 46].

is initially poured into the heated mould maintained near the glass transition temperature [Ref. 46]. The mould is then inverted allowing the unsolidified glass to flow out. Core glass is then poured into the hollow centre region and the entire structure is annealed to produce the core–cladding preform. Moreover, a rotational casting technique [Ref. 47] has provided improvements in preform uniformity and a reduction in the contamination obtained at the core–cladding interface. As indicated previously, fiber drawing from the preform requires stringent control of the hot zone and its temperature to avoid nucleation and growth of crystals. Drawing temperatures in the range 300 to 400 °C are generally used. Furthermore, since the glasses are hydroscopic and reactive with water and oxygen at these temperatures, then controlled atmospheres must be employed.

Following the achievements with low loss silica fiber preparation using chemical vapour deposition synthesis techniques (see Section 4.4), such approaches are also being investigated for the production of fluoride glass fibers [Ref. 44]. Unfortunately, whilst vapour-phase processing of silicate glasses is able to benefit from the large differential vapour pressure of the reagents used to prepare the glasses relative to the impurities contained within the raw materials, similar vapour-phase processing advantages do not exist at present for fluorozirconate glasses. Moreover, the complexity of the glass compositions creates a severe constraint to vapour-phase techniques and hence little progress has been reported [Ref. 21].

4.6 Optical fibers

In order to plan the use of optical fibers in a variety of line communication applications it is necessary to consider the various optical fibers currently available.

The following is a summary of the dominant optical fiber types with an indication of their general characteristics. The performance characteristics of the various fiber types discussed vary considerably depending upon the materials used in the fabrication process and the preparation technique involved. The values quoted are based upon both manufacturers' and suppliers' data, and practical descriptions [Refs. 48 to 52] for commercially available fibers, presented in a general form rather than for specific fibers. Hence in some cases the fibers may appear to have somewhat poorer performance characteristics than those stated for the equivalent fiber types produced by the best possible techniques and in the best possible conditions which were indicated in Chapter 3. It is interesting to note, however, that although the high performance values quoted in Chapter 3 were generally for fibers produced and tested in the laboratory, the performance characteristics of commercially available fibers in many cases are now quite close to these values. This factor is indicative of the improvements made over recent years in the fiber materials preparation and fabrication technologies.

This section therefore reflects the relative maturity of the technology associated with the production of both multicomponent and silica glass fibers. In particular, high performance silica-based fibers for operation in three major wavelength regions (0.8 to 0 9, 1.3 and 1.55 μm) are now widely commercially available. Moreover, complex refractive index profile single-mode fibers, including dispersion modified fibers (see Section 3.12) and polarization maintaining fibers (see Section 3.13.2), are also commercially available and in the former case are starting to find system application within communications. Nevertheless, in this section we concentrate on the conventional circularly symmetric step index design which remains at present the major single-mode fiber provision within telecommunications.

Another relatively new area of commercial fiber development is concerned with mid-infrared wavelength range (2 to 5 μm), often employing heavy metal fluoride glass technology. However, the fiber products that exist for this wavelength region tend to be multimode with relatively high losses* and hence at present are only appropriate for specialized applications. Such fibers are therefore considered no further in this section. Finally, it should be noted that the bandwidths quoted are specified over a 1 km length of fiber (i.e. $B_{opt} \times L$). These are generally obtained from manufacturers' data which does not always indicate whether the electrical or the optical bandwidth has been measured. It is likely that these are in fact optical bandwidths which are significantly greater than their electrical equivalents (see Section 7.4.3).

4.6.1 Multimode step index fibers

Multimode step index fibers may be fabricated from either multicomponent glass compounds or doped silica. These fibers can have reasonably large core diameters and large numerical apertures to facilitate efficient coupling to incoherent light

* For example, the loss obtained at a wavelength of 2.6 μm with a commercially available multimode step index zirconium fluoride glass fiber is 20 dB km^{-1}.

sources such as light emitting diodes (LEDs). The performance characteristics of this fiber type may vary considerably depending on the materials used and the method of preparation; the doped silica fibers exhibit the best performance. Multicomponent glass and doped silica fibers are often referred to as multicomponent glass/glass (glass-clad glass) and silica/silica (silica-clad silica), respectively, although the glass-clad glass terminology is sometimes used somewhat vaguely to denote both types. A typical structure for a glass multimode step index fiber is shown in Figure 4.13.

Structure

Core diameter:	50 to 400 μm
Cladding diameter:	125 to 500 μm
Buffer jacket diameter:	250 to 1000 μm
Numerical aperture:	0.16 to 0.5.

Performance characteristics

Attenuation: 2.6 to 50 dB km^{-1} at a wavelength of 0.85 μm, limited by absorption or scattering. The wide variation in attenuation is due to the large differences both within and between the two overall preparation methods (melting and deposition). To illustrate this point Figure 4.14 shows the attenuation spectra from suppliers' data [Ref. 48] for a multicomponent glass fiber (glass-clad glass) and a doped silica fiber (silica-clad silica). It may be observed that the multicomponent glass fiber has an attenuation of around 40 dB km^{-1} at a wavelength of 0.85 μm, whereas the doped silica fiber has an attenuation of less than 5 dB km^{-1} at a similar wavelength. Furthermore, at a wavelength of 1.3 μm losses reduced to around 0.4 dB km^{-1} can be obtained [Ref. 49].

Bandwidth: 6 to 50 MHz km.

Applications: These fibers are best suited for short-haul, limited bandwidth and relatively low cost applications.

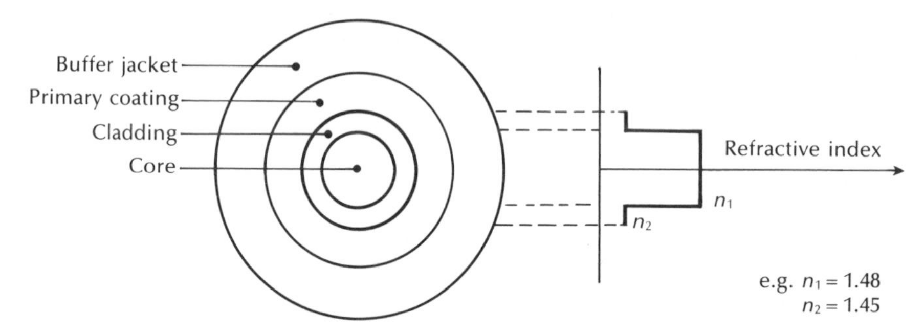

Figure 4.13 Typical structure for a glass multimode step index fiber.

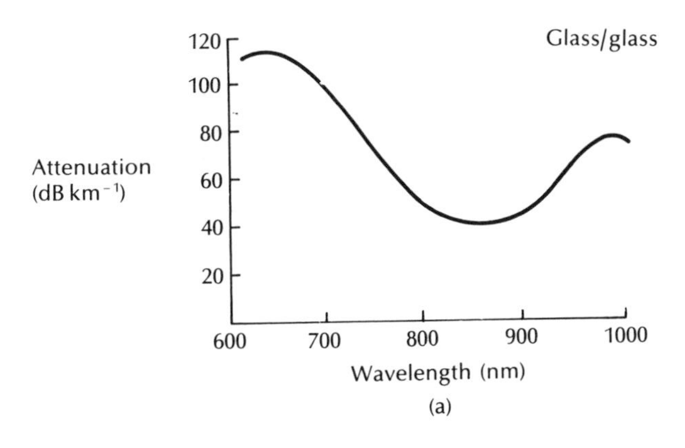

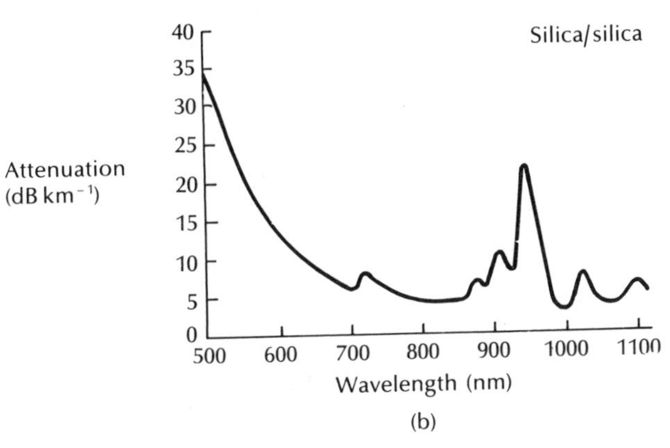

Figure 4.14 Attenuation spectra for multimode step index fibers: (a) multicomponent glass fiber; (b) doped silica fiber. Reproduced with the permission of Rayproof.

4.6.2 Multimode graded index fibers

These multimode fibers which have a graded index profile may also be fabricated using multicomponent glasses or doped silica. However, they tend to be manufactured from materials with higher purity than the majority of multimode step index fibers in order to reduce fiber losses. The performance characteristics of multimode graded index fibers are therefore generally better than those for multimode step index fibers due to the index grading and lower attenuation. Multimode graded index fibers tend to have smaller core diameters than multimode step index fibers, although the overall diameter including the buffer jacket is usually about the same. This gives the fiber greater rigidity to resist bending. A typical structure is illustrated in Figure 4.15.

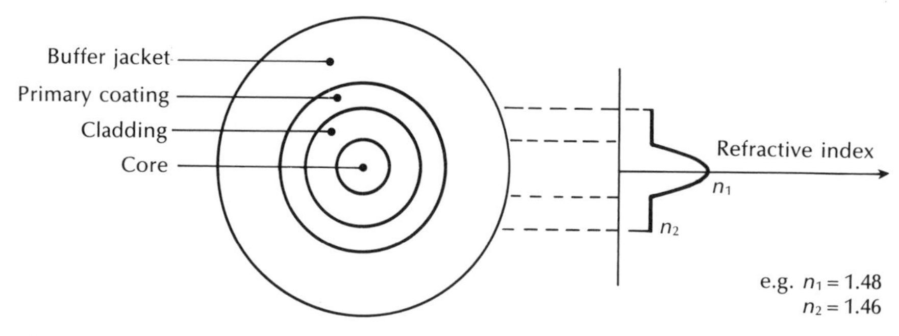

Figure 4.15 Typical structure for a glass multimode graded index fiber.

Structure

Core diameter:	30 to 100 μm.
Cladding diameter:	100 to 150 μm,
Buffer jacket diameter:	250 to 1000 μm.
Numerical aperture:	0.2 to 0.3.

Although the above general parameters encompass most of currently available multimode graded index fibers, in particular the following major groups have now emerged:

1. 50 μm/125 μm (core–cladding) diameter fibers with typical numerical apertures between 0.20 and 0.24. These fibers were originally developed and standardized by the CCITT (Recommendation G. 651) for telecommunication applications at wavelengths of 0.85 and 1.3 μm but now they are mainly utilized within data links and local area networks (LANs).
2. 62.5 μm/125 μm (core–cladding) diameter fibers with typical numerical apertures between 0.26 and 0.29. Although these fibers were developed for longer distance subscriber loop applications at operating wavelengths of 0.85 and 1.3 μm, they are now mainly used within LANs (see Section 14.7).
3. 85 μm/125 μm (core/cladding) diameter fibers with typical numerical apertures 0.26 and 0.30. These fibers were developed for operation at wavelengths of 0.85 and 1.3 μm in short-haul systems and LANs.
4. 100 μm/125 μm (core–cladding) diameter fibers with a numerical aperture of 0.29. These fibers were developed to provide high coupling efficiency to LEDs at a wavelength of 0.85 μm in low cost, short distance applications. They can, however, be utilized at the 1.3 μm operating wavelength and have therefore also found application within LANs.

Performance characteristics

Attenuation: 2 to 10 dB km^{-1} at a wavelength of 0.85 μm with generally a scattering limit. Average losses of around 0.4 and 0.25 dB km^{-1} can be obtained at wavelengths of 1.3 and 1.55 μm respectively [Ref. 49].

Bandwidth: 300 MHz km to 3 GHz km.
Applications: These fibers are best suited for medium-haul, medium to high bandwidth applications using incoherent and coherent multimode sources (i.e. LEDs and injection lasers respectively).

It is useful to note that there are a number of partially graded index fibers commercially available. These fibers generally exhibit slightly better performance characteristics than corresponding multimode step index fibers but are somewhat inferior to the fully graded index fibers described above.

4.6.3 Single-mode fibers

Single-mode fibers can have either a step index or graded index profile. The benefits of using a graded index profile are to provide dispersion modified single-mode fibers (see Section 3.12). The more sophisticated single-mode fiber structures used to produce polarization maintaining fibers (see Section 3.13.2) make these fibers quite expensive at present and thus they are not generally utilized within optical fiber communication systems. Therefore at present, commercially available single-mode fibers are still usually step index. They are high quality fibers for wideband, long-haul transmission and are generally fabricated from doped silica (silica-clad silica) in order to reduce attenuation.

Although single-mode fibers have small core diameters to allow single-mode propagation, the cladding diameter must be at least ten times the core diameter to avoid losses from the evanescent field. Hence with a buffer jacket to provide protection and strength, single-mode fibers have similar overall diameters to multimode fibers. A typical example of a single-mode step index fiber is shown in Figure 4.16.

Structure
Core diameter 5 to 10 μm, typically around 8.5 μm
Cladding diameter: generally 125 μm
Buffer jacket diameter: 250 to 1000 μm
Numerical aperture: 0.08 to 0.15, usually around 0.10.

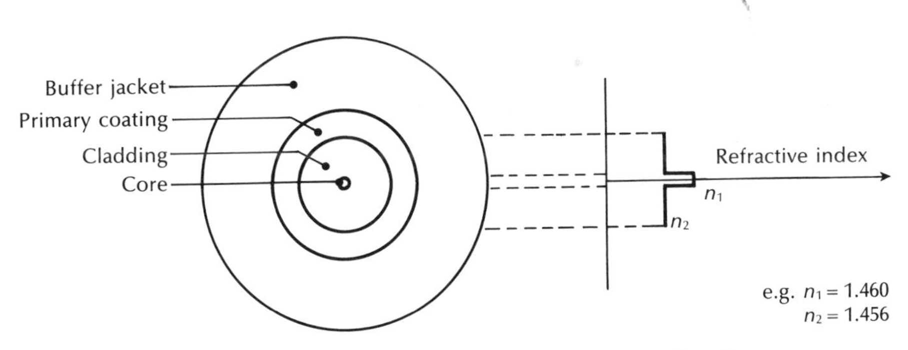

Figure 4.16 Typical structure for a silica single-mode step index fiber.

Performance characteristics

	2 to 5 dB km^{-1} with a scattering limit of around 1 dB km^{-1} at a wavelength of 0.85 μm. In addition, average losses of 0.35 and 0.21 dB km^{-1} at wavelengths of 1.3 and 1.55 μm can be obtained in a manufacturing environment.
Bandwidth:	Greater than 500 MHz km. In theory the bandwidth is limited by waveguide and material dispersion to approximately 40 GHz km at a wavelength of 0.85 μm. However, practical bandwidths in excess of 10 GHz km are obtained at a wavelength of 1.3 μm.
Applications:	These fibers are ideally suited for high bandwidth very long-haul applications using single-mode injection laser sources.

4.6.4 Plastic-clad fibers

Plastic-clad fibers are multimode and have either a step index or a graded index profile. They have a plastic cladding (often a silicone rubber) and a glass core which is frequently silica (i.e. plastic clad silica—PCS fibers). The PCS fibers exhibit lower radiation-induced losses than silica-clad silica fibers and, therefore, have an improved performance in certain environments. Plastic-clad fibers are generally slightly cheaper than the corresponding glass fibers, but usually have more limited performance characteristics. A typical structure for a step index plastic-clad fiber (which is more common) is shown in Figure 4.17.

Structure

Core diameter:	Step index	100 to 500 μm
	Graded index	50 to 100 μm
Cladding diameter:	Step index	300 to 800 μm
	Graded index	125 to 150 μm
Buffer jacket diameter:	Step index	500 to 1000 μm
	Graded index	250 to 1000 μm
Numerical aperture:	Step index	0.2 to 0.5
	Graded index	0.2 to 0.3.

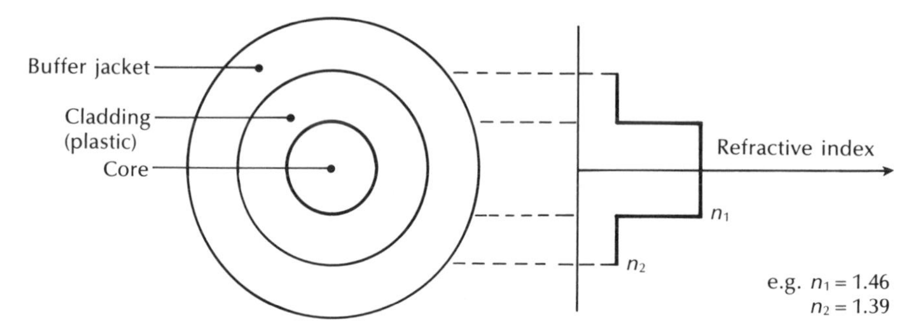

Figure 4.17 Typical structure for a plastic-clad silica multimode step index fiber.

Performance characteristics

Attenuation: Step index 5 to 50 dB km^{-1}
Graded index 4 to 15 dB km^{-1}.

4.6.5 All-plastic fibers

All-plastic or polymeric fibers are exclusively of the multimode step index type with large core and cladding diameters. Hence there is a reduced requirement for a buffer jacket for fiber protection and strengthening. These fibers are usually cheaper to produce and easier to handle than the corresponding silica-based glass variety. However, their performance (especially for optical transmission in the infrared) is restricted, giving them limited use in communication applications. All plastic fibers, however, generally have large numerical apertures which allow easier coupling of light into the fiber from a multimode source.

Early plastic fibers fabricated with a polymethyl methacrylate (PMMA) and a fluorinated acrylic cladding exhibited losses around 500 dB km^{-1}. Subsequently, a continuous casting process was developed for PMMA and losses as low as 110 dB km^{-1} were achieved in the visible wavelength region. The loss mechanisms in PMMA and polystyrene core fibers are similar to those in glass fibers. These fibers exhibit both intrinsic and extrinsic loss mechanisms including absorption and Rayleigh scattering which results from density fluctuations and the anisotropic structure of the polymers. Significant absorption occurs due to the long wavelength tail caused by the intrinsic effect, but this can be relatively small over the 0.5 to 0.9 μm wavelength range where these fibers tend to be utilized. Moreover, extrinsic absorption results from transition metal and organic contaminants as well as overtone bands from the OH ion.

Although substantial progress in the fabrication of higher performance plastic fibers has been made over recent years with losses as low as 20 dB km^{-1} being obtained in the laboratory for PMMA core fiber operating at a wavelength of 0.68 μm [Ref. 21], it is clearly the case that such fibers will not compete with doped silica fibers in any other than the very short distance applications. Current work, however, is focusing on improving their environmental performance, in particular at higher temperatures, as well as their chemical durability and mechanical properties to enhance their use as lower performance fibers for harsh conditions. A typical all-plastic fiber structure is illustrated in Figure 4.18.

Structure

Core diameter: 200 to 600 μm
Cladding diameter: 450 to 1000 μm
Numerical aperture: 0.5 to 0.6.

Performance characteristics

Attenuation: 50 to 1000 dB km^{-1} at a wavelength of 0.65 μm.
Bandwidth: This is not usually specified as transmission is generally limited to tens of metres.

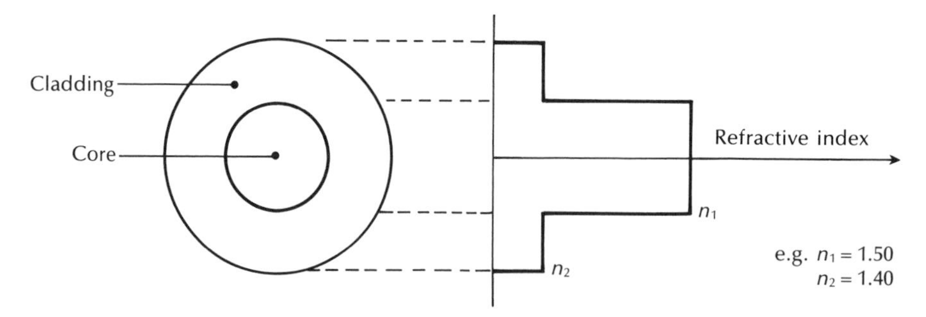

Figure 4.18 Typical structure for an all-plastic fiber.

Applications: These fibers can only be used for very short-haul (i.e. 'in-house') low cost links. However, fiber coupling and termination are relatively easy and do not require sophisticated techniques.

4.7 Optical fiber cables

It was indicated in Section 4.1 that if optical fibers are to be alternatives to electrical transmission lines it is imperative that they can be safely installed and maintained in all the environments (e.g. underground ducts) in which metallic conductors are normally placed. Therefore, when optical fibers are to be installed in a working environment their mechanical properties are of prime importance. In this respect the unprotected optical fiber has several disadvantages with regard to its strength and durability. Bare glass fibers are brittle and have small cross sectional areas which make them very susceptible to damage when employing normal transmission line handling procedures. It is therefore necessary to cover the fibers to improve their tensile strength and to protect them against external influences. This is usually achieved by surrounding the fiber with a series of protective layers, which is referred to as coating and cabling. The initial coating of plastic with high elastic modulus is applied directly to the fiber cladding, as illustrated in Section 4.6. It is then necessary to incorporate the coated and buffered fiber into an optical cable to increase its resistance to mechanical strain and stress as well as adverse environmental conditions.

The functions of the optical cable may be summarized into four main areas. These are as follows:

1. Fiber protection. The major function of the optical cable is to protect against fiber damage and breakage both during installation and throughout the life of the fiber.
2. Stability of the fiber transmission characteristics. The cabled fiber must have good stable transmission characteristics which are comparable with the uncabled

fiber. Increases in optical attenuation due to cabling are quite usual and must be minimized within the cable design.

3. Cable strength. Optical cables must have similar mechanical properties to electrical transmission cables in order that they may be handled in the same manner. These mechanical properties include tension, torsion, compression, bending, squeezing and vibration. Hence the cable strength may be improved by incorporating a suitable strength member and by giving the cable a properly designed thick outer sheath.
4. Identification and jointing of the fibers within the cable. This is especially important for cables including a large number of optical fibers. If the fibers are arranged in a suitable geometry it may be possible to use multiple jointing techniques rather than jointing each fiber individually.

In order to consider the cabling requirements for fibers with regard to (1) and (2), it is necessary to discuss the fiber strength and durability as well as any possible sources of degradation of the fiber transmission characteristics which are likely to occur due to cabling.

4.7.1 Fiber strength and durability

Optical fibers for telecommunications usage are almost exclusively fabricated from silica or a compound of glass (multicomponent glass). These materials are brittle and exhibit almost perfect elasticity until their breaking point is reached. The bulk material strength of flawless glass is quite high and may be estimated for individual materials using the relationship [Ref. 51]:

$$S_t = \left(\frac{\gamma_p E}{4 l_a}\right)^{\frac{1}{2}} \tag{4.6}$$

where S_t is the theoretical cohesive strength, γ_p is the surface energy of the material, E is Young's modulus for the material (stress/strain), and l_a is the atomic spacing or bond distance. However, the bulk material strength may be drastically reduced by the presence of surface flaws within the material.

In order to treat surface flaws in glass analytically, the Griffith theory [Ref. 56] is normally used. This theory assumes that the surface flaws are narrow cracks with small radii of curvature at their tips, as illustrated in Figure 4.19. It postulates that the stress is concentrated at the tip of the crack which leads to crack growth and eventually catastrophic failure. Figure 4.19 shows the concentration of stress lines at the crack tip which indicates that deeper cracks have higher stress at their tips. The Griffith theory gives a stress intensity factor K_I as:

$$K_I = SYC^{\frac{1}{2}} \tag{4.7}$$

where S is the macroscopic stress on the fiber, Y is a constant dictated by the shape of the crack (e.g. $Y = \pi^{\frac{1}{2}}$ for an elliptical crack, as illustrated in Figure 4.19) and C is the depth of the crack (this is the semimajor axis length for an elliptical crack).

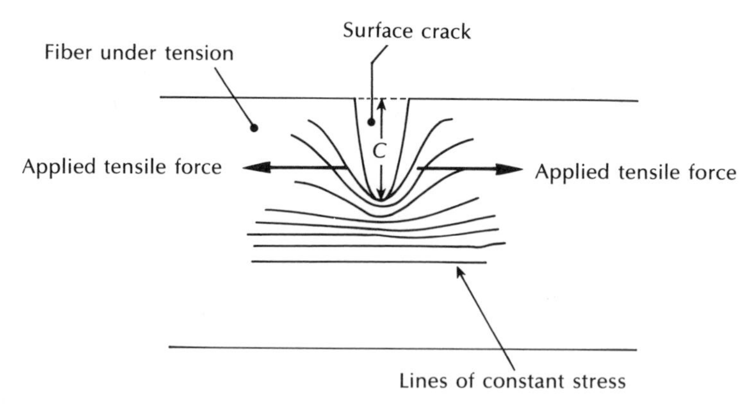

Figure 4.19 An elliptical surface crack in a tensioned optical fiber.

Further, the Griffith theory gives an expression for the critical stress intensity factor K_{IC} where fracture occurs as:

$$K_{IC} = (2E\gamma_p)^{\frac{1}{2}} \tag{4.8}$$

Combining Eqs. (4.7) and (4.8) gives the Griffith equation for fracture stress of a crack S_f as:

$$S_f = \left(\frac{2E\gamma_p}{Y^2C}\right)^{\frac{1}{2}} \tag{4.9}$$

It is interesting to note that S_f is proportional to $C^{-\frac{1}{2}}$. Therefore, S_f decreases by a factor of 2 for a fourfold increase in the crack depth C.

Example 4.1

The Si—O bond has a theoretical cohesive strength of 2.6×10^6 psi which corresponds to a bond distance of 0.16 nm. A silica optical fiber has an elliptical crack of depth 10 nm at a point along its length. Estimate

(a) the fracture stress in psi for the fiber if it is dependent upon this crack:
(b) the percentage strain at the break.

Young's modulus for silica is approximately 9×10^{10} N m^{-2} and 1 psi ≡ 6894.76 N m^{-2}.

Solution: (a) Using Eq. (4.6), the theoretical cohesive strength for the Si—O bond is:

$$S_t = \left(\frac{\gamma_p E}{4l_a}\right)^{\frac{1}{2}}$$

Hence

$$\gamma_p = \frac{4l_a S_t^2}{E} = \frac{4 \times 0.16 \times 10^{-9}(2.6 \times 10^6 \times 6894.76)^2}{9 \times 10^{10}}$$

$$= 2.29 \text{ J}$$

The fracture stress for the silica fiber may be obtained from Eq. (4.9) where:

$$S_f = \left(\frac{2E\gamma_p}{Y^2 C}\right)^{\frac{1}{2}}$$

For an elliptical crack:

$$S_f = \left(\frac{2E\gamma_p}{\pi C}\right)^{\frac{1}{2}} = \left(\frac{2 \times 9 \times 10^{10} \times 2.29}{\pi \times 10^{-8}}\right)^{\frac{1}{2}}$$

$$= 3.62 \times 10^9 \text{ N m}^{-1}$$
$$= 5.25 \times 10^5 \text{ psi}$$

It may be noted that the fracture stress is reduced from the theoretical value for flawless silica of 2.6×10^6 psi by a factor of approximately 5.

(b) Young's modulus is defined as:

$$E = \frac{stress}{strain}$$

Therefore

$$strain = \frac{stress}{E} = \frac{S_f}{E} = \frac{3.62 \times 10^9}{9 \times 10^{10}} = 0.04$$

Hence the strain at the break is 4%, which corresponds to the change in length over the original length for the fiber.

In Example 4.1 we considered only a single crack when predicting the fiber fracture. However, when a fiber surface is exposed to the environment and is handled, many flaws may develop. The fracture stress of a length of fiber is then dependent upon the dominant crack (i.e. the deepest) which will give a fiber fracture at the lowest strain. Hence, the fiber surface must be protected from abrasion in order to ensure high fiber strength. A primary protective plastic coating is usually applied to the fiber at the end of the initial production process so that mechanically induced flaws may be minimized. Flaws also occur due to chemical and structural causes. These flaws are generally smaller than the mechanically induced flaws and may be minimized within the fiber fabrication process.

There is another effect which reduces the fiber fracture stress below that predicted by the Griffith equation. It is due to the slow growth of flaws under the action of stress and water and is known as stress corrosion. Stress corrosion occurs because the molecular bonds at the tip of the crack are attacked by water when they are under stress. This causes the flaw to grow until breakage eventually occurs. Hence stress corrosion must be taken into account when designing and testing optical fiber cables. It is usual for optical fiber cables to have some form of water-protective barrier, as is the case for most electrical cable designs.

In order to predict the life of practical optical fibers under particular stresses it is necessary to use a technique which takes into account the many flaws a fiber may possess, rather than just the single surface flaw considered in Example 4.1. This is approached using statistical methods due to the nature of the problem which involves many flaws of varying depths over different lengths of fiber.

Calculations of strengths of optical fibers are usually conducted using Weibull statistics [Ref. 57] which describe the strength behaviour of a system that is dependent on the weakest link within the system. In the case of optical fibers this reflects fiber breakage due to the dominant or deepest crack. The empirical relationship established by Weibull and applied to optical fibers indicates that the probability of failure F at a stress S is given by:

$$F = 1 - \exp\left[-\left(\frac{S}{S_0}\right)^m\left(\frac{L}{L_0}\right)\right] \tag{4.10}$$

where m is the Weibull distribution parameter, S_0 is a scale parameter, L is the fiber length and L_0 is a constant with dimensions of length.

The expression given in Eq. (4.10) may be plotted for a fiber under test by breaking a large number of 10 to 20 m fiber lengths and measuring the strain at the break. The various strains are plotted against the cumulative probability of their occurrence to give the Weibull plot as illustrated in Figure 4.20 [Ref. 58]. It may be observed from Figure 4.20 that most of the fiber tested breaks at strain due to the prevalence of many shallow surface flaws. However, some of the fiber tested contains deeper flaws (possibly due to external damage) giving the failure at lower

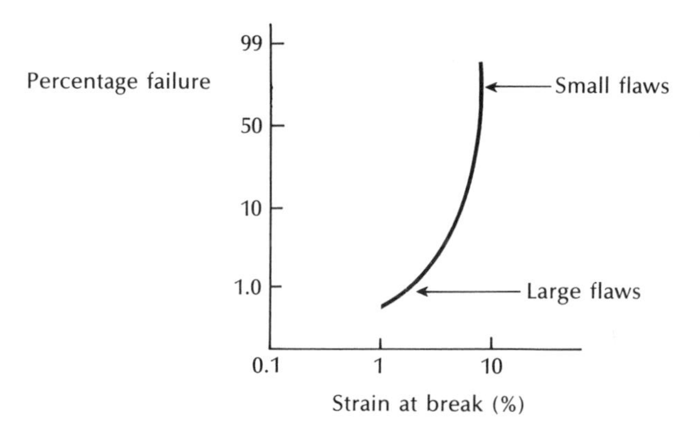

Figure 4.20 A schematic representation of a Weibull plot. Reproduced with permission from M. H. Reeve. *The Radio and Electron. Eng.*, **51**, p. 327.

strain depicted by the tail of the plot. This reduced strength region is of greatest interest when determining the fiber's lifetime under stress.

Finally, the additional problem of stress corrosion must be added to the information on the fiber under stress gained from the Weibull plot. The stress corrosion is usually predicted using an empirical relationship for the crack velocity v_c in terms of the applied stress intensity factor K_I, where [Ref. 58]:

$$v_c = AK_I^n \tag{4.11}$$

The constant n is called the stress corrosion susceptibility (typically in the range 15 to 50 for glass), and A is also a constant for the fiber material. Equation (4.11) allows estimation of the time to failure of a fiber under stress corrosion conditions. Therefore, from a combination of fiber testing (Weibull plot) and stress corrosion, information estimates of the maximum allowable fiber strain can be made available to the cable designer. These estimates may be confirmed by straining the fiber up to a specified level (proof testing) such as 1% strain. Fiber which survives this test can be accepted. However, proof testing presents further problems, as it may cause fiber damage. Also, it is necessary to derate the maximum allowable fiber strain from the proof test value to increase confidence in fiber survival under stress conditions. It is suggested [Ref. 58] that a reasonable derating for use by the cable designer for fiber which has survived a 1% strain proof test is around 0.3% in order that the fiber has a reasonable chance of surviving with a continuous strain for twenty years.

4.8 Stability of the fiber transmission characteristics

Optical fiber cables must be designed so that the transmission characteristics of the fiber are maintained after the cabling process and cable installation. Therefore, potential increases in the optical attenuation and reduction in the bandwidth of the cabled fiber should be avoided.

Certain problems can occur either within the cabling process or subsequently which can significantly affect the fiber transmission characteristics. In particular, a problem which often occurs in the cabling of optical fiber is the meandering of the fiber core axis on a microscopic scale within the cable form. This phenomenon, known as microbending, results from small lateral forces exerted on the fiber during the cabling process and it causes losses due to radiation in both multimode and single-mode fibers.

In addition to microbending losses caused by fiber stress and deformation on a micron scale, macrobending losses occur when the fiber cable is subjected to a significant amount of bending above a critical value of curvature. Such fiber bend losses are discussed in Section 3.6. However, additional optical losses can occur when fiber cables are *in situ*. These losses may result from hydrogen absorption by the fiber material or from exposure of the fiber cable to ionizing radiation. The above phenomena are discussed in this section in order to provide an insight into the problems associated with the stability of the cabled fiber transmission characteristics.

4.8.1 Microbending

Microscopic meandering of the fiber core axis, known as microbending, can be generated at any stage during the manufacturing process, the cable installation process or during service. This is due to environmental effects, particularly temperature variations causing differential expansion or contraction [Ref. 59]. Microbending introduces slight surface imperfections which can cause mode coupling between adjacent modes, which in turn creates a radiative loss which is dependent on the amount of applied fiber deformation, the length of fiber, and the exact distribution of power among the different modes.

It has become accepted to consider, in particular, two forms of modal power distribution. The first form occurs when a fiber is excited by a diffuse Lambertian source, launching all possible modes, and is referred to as a uniform, or fully filled mode distribution. The second form occurs when, due to a significant amount of mode coupling and mode attenuation, the distribution of optical power becomes essentially invariant with the distance of propagation along the fiber. This second distribution is generally referred to as a steady state, or equilibrium mode distribution, which typically occurs after transmission over approximately one kilometre of fiber (see Section 13.1).

Since microbending losses are mode dependent and from Eq. (2.69) the number of modes is an inverse function of the wavelength of the transmitted light within a particular fiber, it is to be expected that microbending losses will be wavelength dependent. This effect is demonstrated for multimode fiber in Figure 4.21 [Ref. 60], which illustrates the theoretical microbending loss for both the uniform and the steady state mode distributions as a function of applied linear pressure (i.e. simulated microbending), for a normalized frequency $V = 39$, corresponding to a wavelength of 0.82 μm and $V = 21$, corresponding to a wavelength of 1.55 μm. It may be observed from Figure 4.21 that the microbending loss decreases at longer wavelengths, and that it is also dependent on the modal power distribution present within the fiber; microbending losses corresponding to a uniform power distribution being approximately 1.75 times greater than those obtained with a steady state distribution. In addition it has been predicted [Ref. 61] that microbending losses for single-mode fiber follow an approximately exponential form, with increasing losses at longer wavelengths. Minimal losses were predicted at operating wavelengths below 1.3 μm, with a rapid rise in attenuation at wavelengths above 1.5 μm. Experimental measurements have confirmed these predictions.

It is clear that excessive microbending can create additional fiber losses to an unacceptable level. To avoid deterioration in the optical fiber transmission characteristics resulting from mode coupling induced microbending, it is important that the fiber is free from irregular external pressure within the cable. Carefully

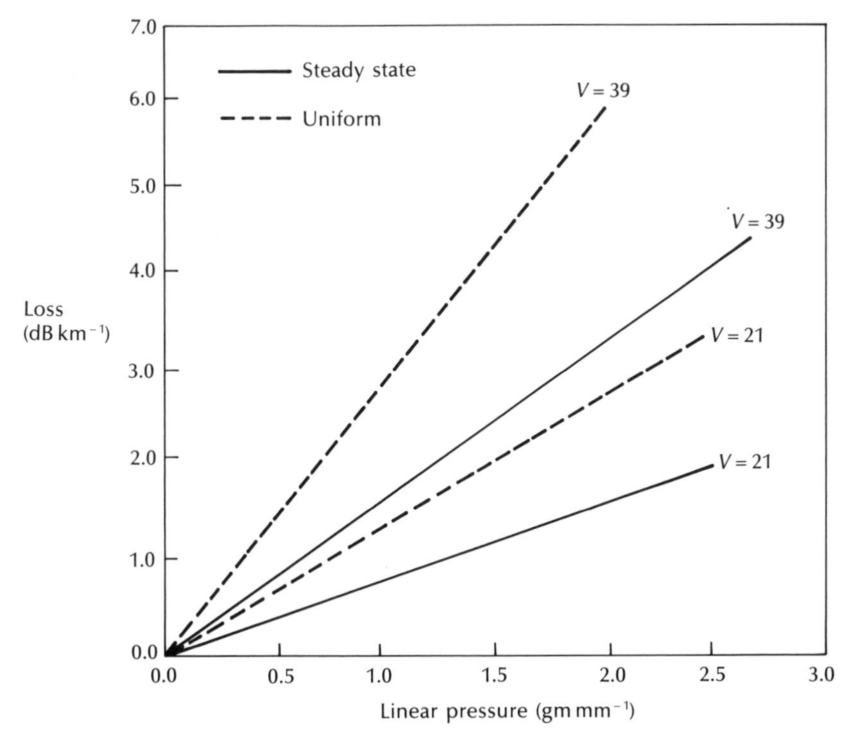

Figure 4.21 Theoretical microbending loss versus linear pressure for graded index multimode fibers ($NA = 0.2$, core diameter of 50 μm. Reproduced with permission from S. Das, C. G. Englefield and P. A. Goud, *Appl. Opt.*, **24**, p. 2323, 1985.

controlled coating and cabling of the fiber is therefore essential in order to minimize the cabled fiber attenuation. Furthermore, the fiber cabling must be capable of maintaining this situation under all the strain and environmental conditions envisaged in its lifetime.

4.8.2 Hydrogen absorption

The diffusion of hydrogen into optical fiber has been shown to affect the spectral attenuation characteristic [Ref. 62]. There are two fundamental mechanisms by which hydrogen absorption causes an increase in optical fiber losses [Ref. 59]. The first is where hydrogen diffuses into interstitial spaces in the glass, thereby altering the spectral loss characteristics through the formation of new absorption peaks. This phenomenon has been found to affect all silica-based glass fibers, both multimode and single-mode. However, the extra losses obtained can be reversed if the hydrogen source is removed. Typically, it causes losses in the range 0.2 to

0.3 dB km^{-1} Atm^{-1} at an optical wavelength of 1.3 μm and a temperature of 25 °C with 500 hours exposure [Ref. 62]. At higher temperatures these additional losses may be substantially increased, as can be observed from Figure 4.22, which displays the change in spectral attenuation obtained for a fiber with 68 hours' hydrogen exposure at a temperature of 150 °C [Ref. 63].

The second mechanism occurs when hydrogen reacts with the fiber deposits to give P–OH, Ge–OH or Si–OH absorption. These losses are permanent and can be greater than 25 dB km^{-1} [Ref. 59]. Studies suggest that hydrogen can be generated by either chemical decomposition of the fiber coating materials or through metal–electrolytic action (that is, moisture affecting the metal sheathing of the fiber cable). These effects can be minimized by careful selection of the cable, the prevention of immersion of the cable in water, or by pressurizing the cable to prevent water ingress. Alternatively, the fiber cable may be periodically purged using an inert gas.

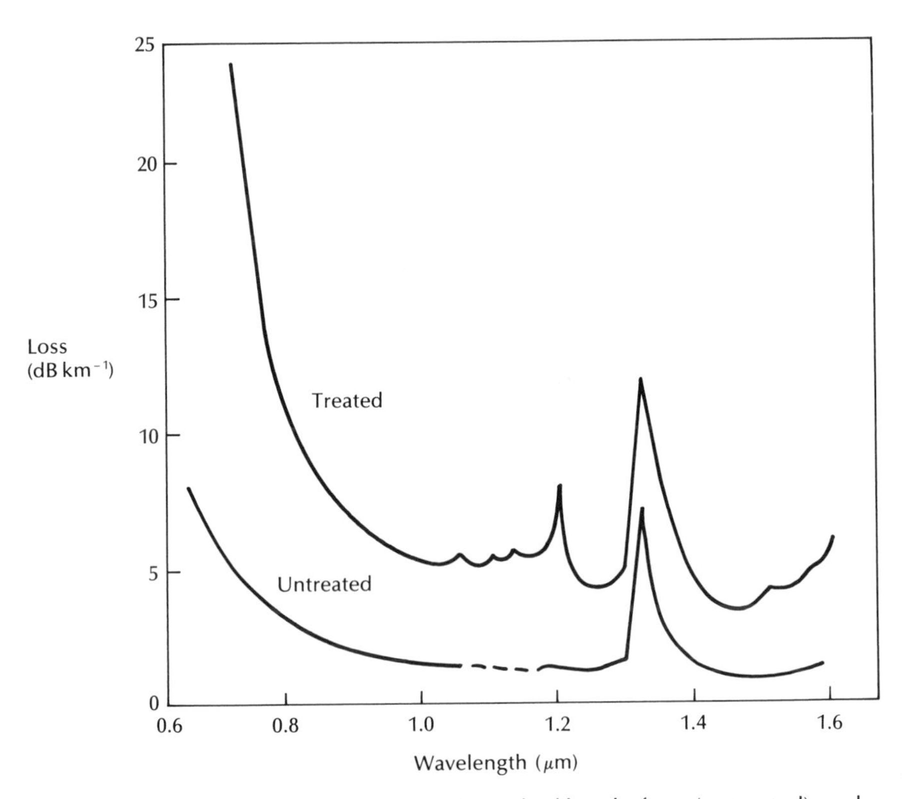

Figure 4.22 Attenuation spectra for multimode fiber before (untreated) and after (treated) hydrogen diffusion [Ref. 63].

4.8.3 Nuclear radiation exposure

The optical transmission characteristics of fiber cables can be seriously degraded by exposure to nuclear radiation. Such radiation forms colour centres in the fiber core which can cause spectral attenuation [Ref. 64]. The precise nature of this attenuation depends upon a number of factors including: fiber parameters such as structure; core and cladding material composition; system parameters such as optical intensity and wavelength as well as temperature; and radiation parameters such as total dose, dose rate and energy levels, together with the length of recovery time allowed. The radiation-induced attenuation comprises a permanent component which is irreversible, and a metastable component which is reversible and contains both a transient (with decay time less than 1 s) and steady state (with decay time less than 10 s) constituents. The nature of both the permanent and decaying components of the attenuation are dependent on the fiber composition.

Typical measured spectral loss characteristics for two single-mode fibers, following exposure to a 10 k rad dose of steady state radiation for 1 hour, are shown in Figure 4.23 [Ref. 64]. The Corning fiber under test had a Ge doped silica core and a pure silica cladding, whereas the Dainichi fiber had a pure silica core and

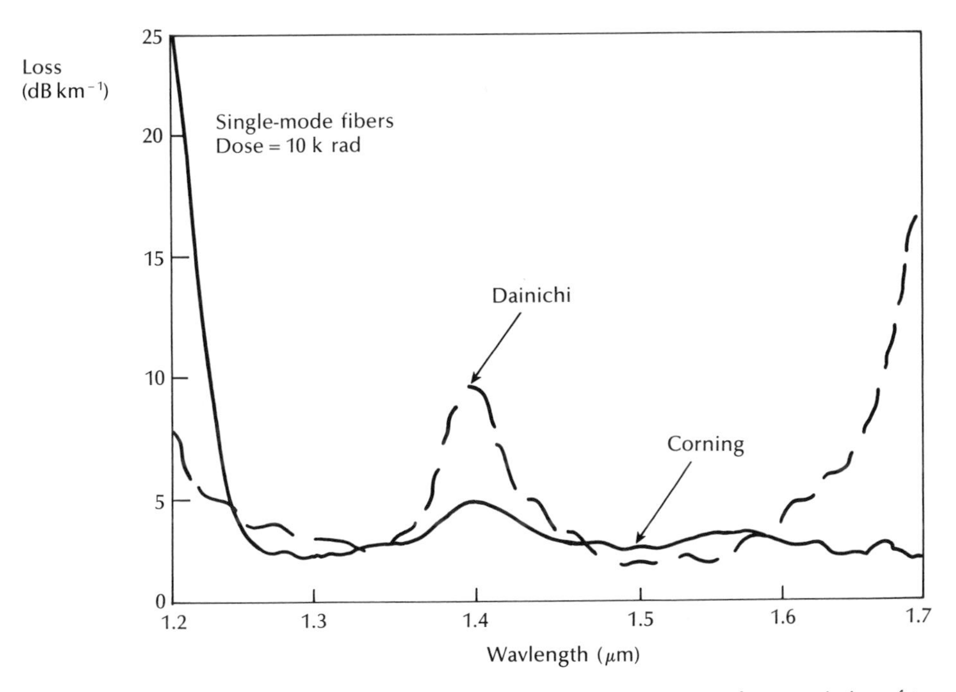

Figure 4.23 Effect of nuclear radiation on the spectral attenuation characteristics of two single-mode fibers. Reproduced with permission from E. J. Friebele, K. J. Long, C. G. Askins, M. E. Gingerich, M. J. Marrone and D. L. Griscom, 'Overview of radiation effects in fiber optics', *Proc. SPIE, Int. Soc. Opt. Eng., Radiation Effects in Optical Materials* **541**, p. 70, 1985.

an F–P doped silica cladding. Figure 4.23 displays the spectral loss characteristics over the wavelength range 1.2 to 1.7 μm, but it should also be noted that pure silica core fibers exhibit considerable radiation-induced losses at wavelengths around 0.85 μm. These losses initially increase linearly with increasing radiation dose and can become hundreds of dB km^{-1} [Ref. 64].

Radiation-resistant fibers have been developed which are less sensitive to the effects of nuclear radiation. For example, hydrogen treatment of a pure silica core fiber, or use of boron–fluoride doped silica cladding fiber has been found to reduce gamma-ray-induced attenuation in the visible wavelength region [Ref. 65]. It has also been reported that radiation-induced attenuation can be reduced through photobleaching [Ref. 66]. In general, however, the only fiber structures likely to have an acceptable performance over a wide wavelength range when exposed to ionizing radiation are those having pure undoped silica cores, or those with core dopants of germanium and germanium with small amounts of fluorine–phosphorus [Ref. 67].

Clearly, radiation exposure can induce a considerable amount of attenuation in optical fibers, although the number of possible variable parameters, relating both to fiber structure and the nature of the radiation, make it difficult to generalize on the precise spectral effects. Nevertheless, more specific details relating to these effects can be found in the literature [Refs. 64 to 68].

4.9 Cable design

The design of optical fiber cables must take account of the constraints discussed in Section 4.7. In particular, the cable must be designed so that the strain on the fiber in the cable does not exceed 0.2% [Ref. 49]. Alternatively, it is suggested that the permanent strain on the fiber should be less than 0.1% [Ref. 50]. In practice, these constraints may be overcome in various ways which are, to some extent, dependent upon the cable's application. Nevertheless, cable design may generally be separated into a number of major considerations. These can be summarized into the categories of fiber buffering, cable structural and strength members, and cable sheath and water barrier.

4.9.1 Fiber buffering

It was indicated in Section 4.7 that the fiber is given a primary coating during production in order to prevent abrasion of the glass surface and subsequent flaws in the material. The primary coated fiber is then given a secondary or buffer coating (jacket) to provide protection against external mechanical and environmental influences. This buffer jacket is designed to protect the fiber from microbending losses and may take several different forms. These generally fall into one of three distinct types which are illustrated in Figure 4.24 [Ref. 69]. A tight buffer jacket is shown in Figure 4.24(a) which usually consists of a hard plastic (e.g. nylon,

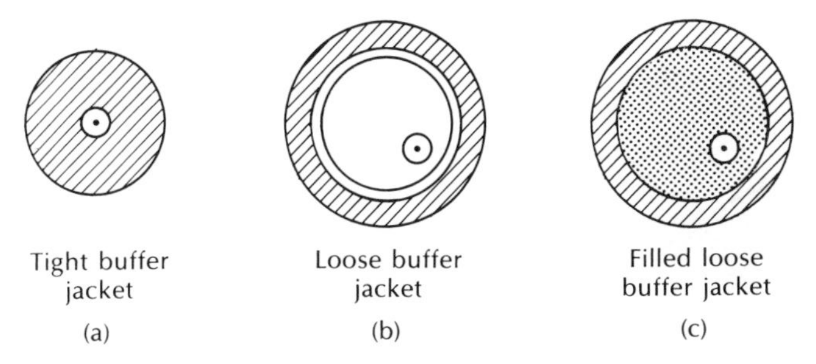

Figure 4.24 Techniques for buffering of optical fibers [Ref. 69]: (a) tight buffer jacket; (b) loose buffer jacket; (c) filled loose buffer jacket.

Hytrel, Tefzel) and is in direct contact with the primary coated fiber. This thick buffer coating (0.25 to 1 mm in diameter) provides stiffening for the fiber against outside microbending influences, but it must be applied in such a manner as not to cause microbending losses itself.

An alternative approach, which is shown in Figure 4.24(b), is the use of a loose buffer jacket. This produces an oversized cavity in which the fiber is placed and which mechanically isolates the fiber from external forces. Loose buffering is generally achieved by using a hard, smooth, flexible material in the form of an extruded tube, or sometimes a folded tape with a diameter between 1 and 2 mm.

Finally, Figure 4.24(c) shows a variation of the loose buffering in which the oversized cavity is filled with a moisture-resistant compound. This technique, which combines the advantages of the two previous methods, also provides a water barrier in the immediate vicinity of the fiber. The filling material must be soft, self-healing and stable over a wide range of temperatures and usually consists of specially blended petroleum or silicon-based compounds.

4.9.2 Cable structural and strength members

One or more structural members are usually included in the optical fiber cable to serve as a core foundation around which the buffered fibers may be wrapped, or into which they may be slotted, as illustrated in Figure 4.25 [Refs. 58 and 69]. The structural member may also be a strength member if it consists of suitable material (i.e. solid or stranded steel wire or Kevlar (DuPont Ltd) yarns). This situation is shown in Figure 4.25(a) where the central steel wire acts as both a structural and strength member. In this case the steel wire is the primary load-bearing element. Figure 4.25(b) shows an extruded plastic structural member around a central steel strength member commonly known as a slotted core. The primary function of the structural member in this case is not load-bearing, but to provide suitable accommodation for the buffered fibers within the cable.

Structural members may be nonmetallic with plastics, fiberglass and Kevlar often being used. However, for strength members the preferred features include a high

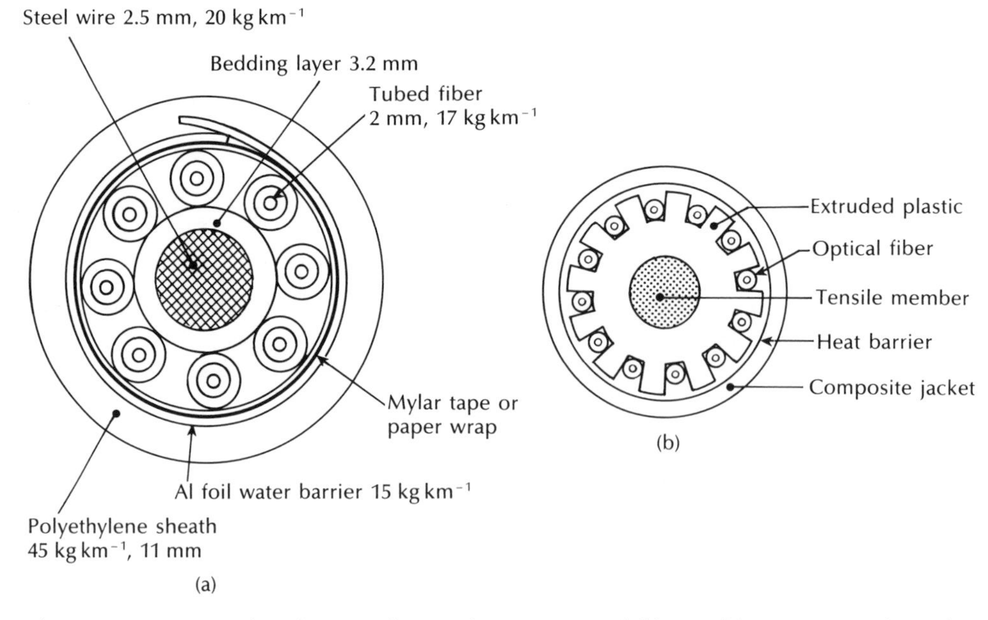

Figure 4.25 Structural and strength members in optical fiber cables: (a) central steel wire structural and strength member [Ref. 58]; (b) Northern Telecom unit core cable with central steel strength member and extruded plastic structural member [Ref. 69].

Young's modulus, high strain capability, flexibility and low weight per unit length. Therefore, although similar materials are frequently utilized for both strength and structural members, the requirement for additional tensile strength of the strength member must be considered within the cable design.

Flexibility in strength members formed of materials with high Young's modulii may be improved by using a stranded or bunched assembly of smaller units, as in the case of steel wire. Similar techniques are also employed with other materials used for strength members which include plastic monofilaments (i.e. specially processed polyester), textile fiber (nylon, Terylene, Dacron and the widely used Kevlar) and carbon and glass fibers. These materials provide a variety of tensile strengths for different cable applications. However, it is worth noting that Kevlar, an aromatic polyester, has a very high Young's modulus (up to 13×10^{10} N m^{-2}) which gives it a strength to weight ratio advantage four times that of steel.

It is usual when utilizing a stranded strength member to cover it with a coating of extruded plastic, or helically applied tape. This is to provide the strength member with a smooth (cushioned) surface which is especially important for the prevention of microbending losses when the member is in contact with the buffered optical fibers.

4.9.3 Cable sheath and water barrier

The cable is normally covered with a substantial outer plastic sheath in order to reduce abrasion and to provide the cable with extra protection against external mechanical effects such as crushing. The cable sheath is said to contain the cable core and may vary in complexity from a single extruded plastic jacket to a multilayer structure comprising two or more jackets with intermediate armouring. However, the plastic sheath material (e.g. polyethylene, polyurethane) tends to give very limited protection against the penetration of water into the cable. Hence, an additional water barrier is usually incorporated. This may take the form of an axially laid aluminium foil/polyethylene laminated film immediately inside the sheath as used by British Telecom [Ref. 70] and illustrated in Figure 4.25(a).

Alternatively, the ingress of water may be prevented by filling the spaces in the cable with moisture-resistant compounds. Specially formulated silicone rubber or petroleum-based compounds are often used which do not cause difficulties in identification and handling of individual optical fibers within the cable form. These filling compounds are also easily removed from the cable and provide protection from corrosion for any metallic strength members within the fiber. Also, the filling compounds must not cause degradation of the other materials within the cable and must remain stable under pressure and temperature variation.

4.9.4 Examples of fiber cables

A number of different cable designs have emerged and been adopted by various organizations throughout the world. Although no definite standards exist, due to the diversity of applications and the approaches chosen to achieve the objectives mentioned in Section 4.7, there is, however, a general consensus on the overall design requirements and on the various materials that can be employed for cable construction [Ref. 71]. In this section we therefore consider some of the more common designs used in optical fiber cable construction to provide the reader with an insight into the developments in this important field.

Leading cable designs in the late 1970s included the use of loose buffer tubes or, alternatively, fiber ribbons [Ref. 72]. In the former design the fibers are enclosed in tubes which are stranded around a central strength (see Figure 4.25(a)) prior to the application of a polymeric sheath, thus providing large fiber strain relief. In the latter case high fiber packing density as well as ease of connectorization can be obtained. More recently, other designs have also found widespread application. In particular the slotted core design, an example of which is shown in Figure 4.25(b), and the loose fiber-bundle design in which the fibers are packaged into bundles before being enclosed in a single, loose fitting tube [Ref. 69]. All of the aforementioned multifiber cable designs are available with both single-mode and multimode fibers.

Figure 4.26 [Ref. 52] shows two examples of cable construction for single fibers. In Figure 4.26(a) a tight buffer jacket of Hytrel is used surrounded by a layer of

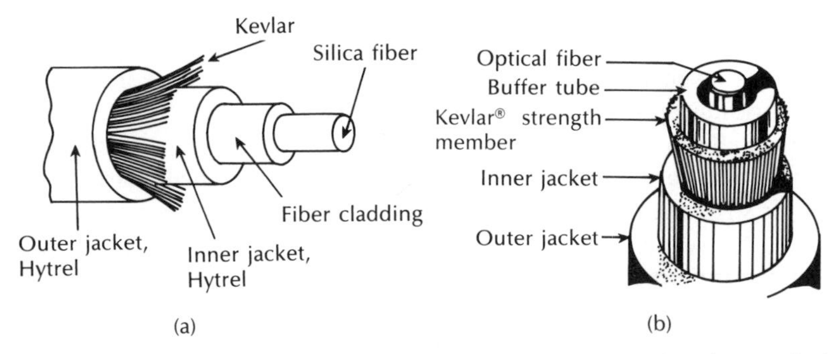

Figure 4.26 Single fiber cables [Ref. 52]: (a) tight buffer jacket design; (b) loose buffer jacket design.

Kevlar for strengthening. In this construction the optical fiber itself acts as a central strength member.

The cable construction illustrated in Figure 4.26(b) uses a loose tube buffer around the central optical fiber. This is surrounded by a Kevlar strength member which is protected by an inner sheath or jacket before the outer sheath layer. The strength members of single optical fiber cables are not usually incorporated at the centre of the cable (unless the fiber is acting as a strength member), but are placed in the surrounding cable form, as illustrated in Figure 4.26(b).

Although the single fiber cables shown in Figure 4.26 can be utilized for indoor applications, alternative simpler designs have been produced for such areas. A dual fiber indoor cable in which steel wire is employed for reinforcement is displayed in Figure 4.27(a) [Ref. 73]. In addition several flat-shaped cable types that can be laid under carpets have been developed [Ref. 74]. An example which is shown in Figure 4.27(b) can withstand pressures of more than 100 kg cm^{-1}, including that from high heeled shoes.

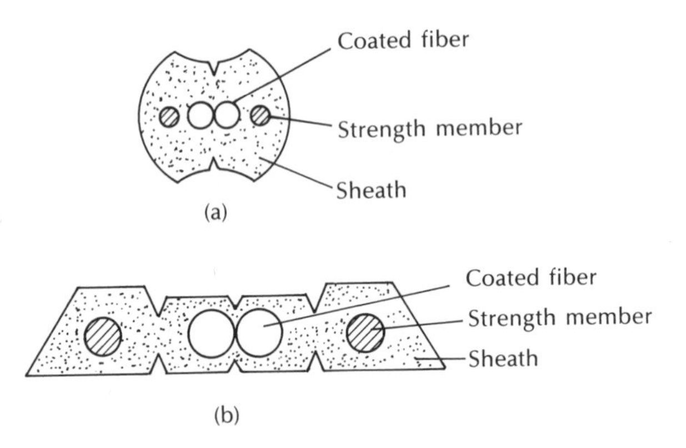

Figure 4.27 Dual fiber cables; (a) indoor cable; (b) flat cable.

Multifiber cables for outside plant applications, as mentioned previously, exhibit several different design methodologies. The use of a central strength member, as illustrated in Figure 4.25(a), is a common technique for the incorporation of either loose-buffer or tight buffer jacketed fibers. Such a structural member is utilized in the loose tube cable structure shown in Figure 4.28(a) in which filled loose tubes are extruded over fiber bundles with up to twelve fibers per tube [Ref. 72]. The tubes are then stranded around a central strength member, forming the cable core which is completed with a polyethylene sheath. A similar technique is used in the tight

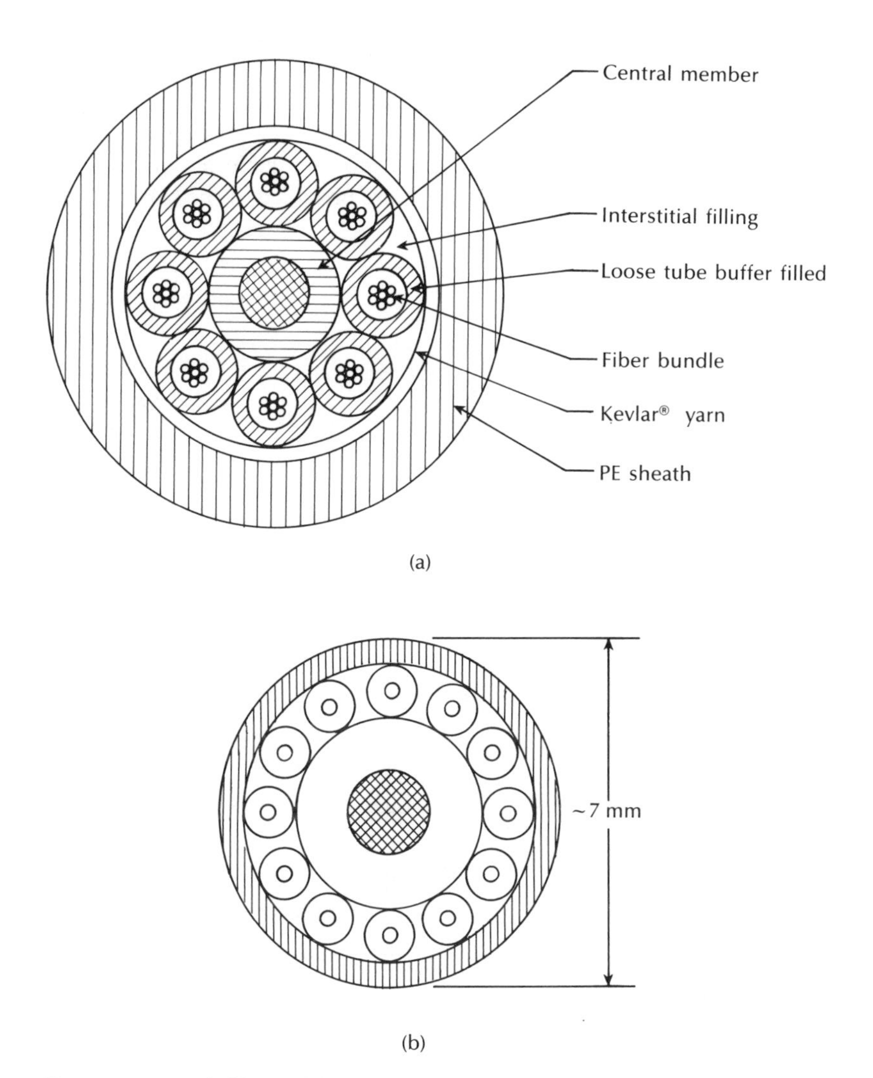

Figure 4.28 Multifiber cables for outside plant application: (a) loose tube cable; (b) layered cable construction [Ref. 72].

buffer cable design which often results in a layered construction for small cables of twelve fibers or less, as illustrated in Figure 4.28(b) [Ref. 72].

The slotted core structural member design, depicted in Figure 4.25(b) is also popular. A variation on this technique is illustrated in Figure 4.29(a) in which a number of fibers are placed into each of the slots [Ref. 71]. Within this design, radial movement of the fibers is facilitated in a similar manner to the loose tube approach, thus minimizing the possibility of residual fiber strain and the associated microbending losses.

Cable strength members may also be distributed or provided within an armoured cable design. An example of the former technique is shown in Figure 4.29(b) which comprises a multiunit design where each unit incorporates seven loose buffer jacketed fibers [Ref. 69]. In this case each unit contains a central strength member, which has the effect of dispersing the strength member throughout the cable core.

Structural members in the form of an armoured cable design are illustrated in Figure 4.30(a) in which stainless steel wires are provided in the cable sheath together

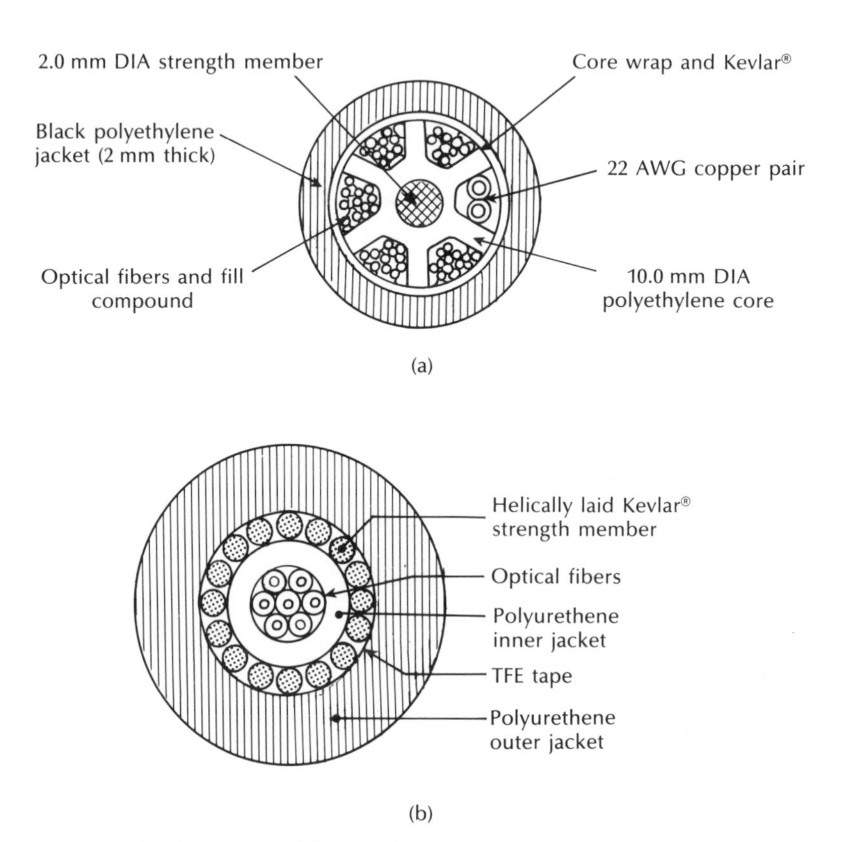

Figure 4.29 Multifiber cables: (a) slotted core cable [Ref. 71]; (b) ITT seven fiber external strength member cable. [Ref. 69].

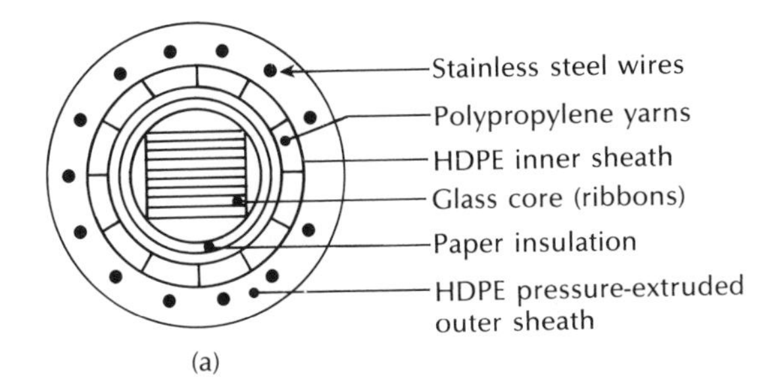

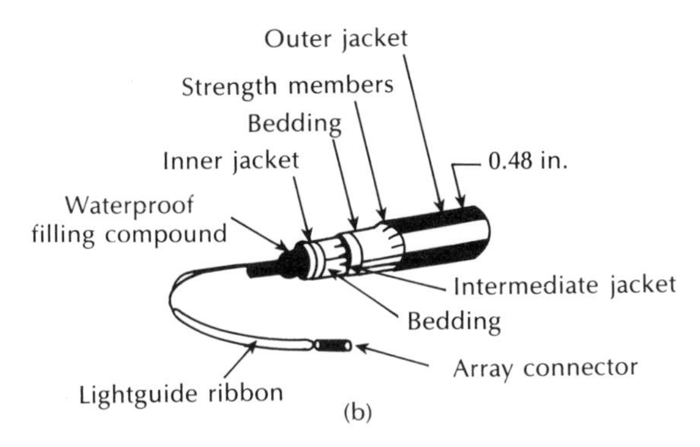

Figure 4.30 Ribbon fiber cables: (a) AT&T ribbon cable [Ref. 69]; (b) slotted core ribbon cable [Ref. 78].

with polypropylene yarns in the inner surrounding cable form. Furthermore, the cable depicted in Figure 4.30(a) is of a type that has found widespread application as it incorporates a ribbon fiber approach which at present provides the most efficient technique for simultaneously handling, splicing and connectorizing a large number of fibers. Between five and twelve fibers per ribbon have been placed into such linear ribbon arrays which may be stacked into twelve ribbon rectangular blocks [Ref. 71] to be placed in the cable core tube (Figure 4.30(a)). The ribbons may be formed with adhesive backed mylar tapes [Ref. 75], with extruded polymer substrates [Ref. 76] or by dipping the fiber array into a UV curable resin bath [Ref. 77]. Alternatively, the fiber ribbons can be placed into the voids of a slotted core structural member or several ribbon subunits may be stranded together to create very large fiber-count cables, as illustrated in Figure 4.30(b) [Ref. 78].

Finally, an approach commercialized by AT & T* is the loose fiber-bundle design shown in Figure 4.31 [Ref. 50]. In this cable up to twelve fibers are assembled into a bundle which is identified with a colour-coded binder. Several of the fiber bundles are then placed inside a core tube. This design does not employ a central strength member but relies upon an armoured design using a reinforced sheath with a

* It is called the Lightpack™ cable.

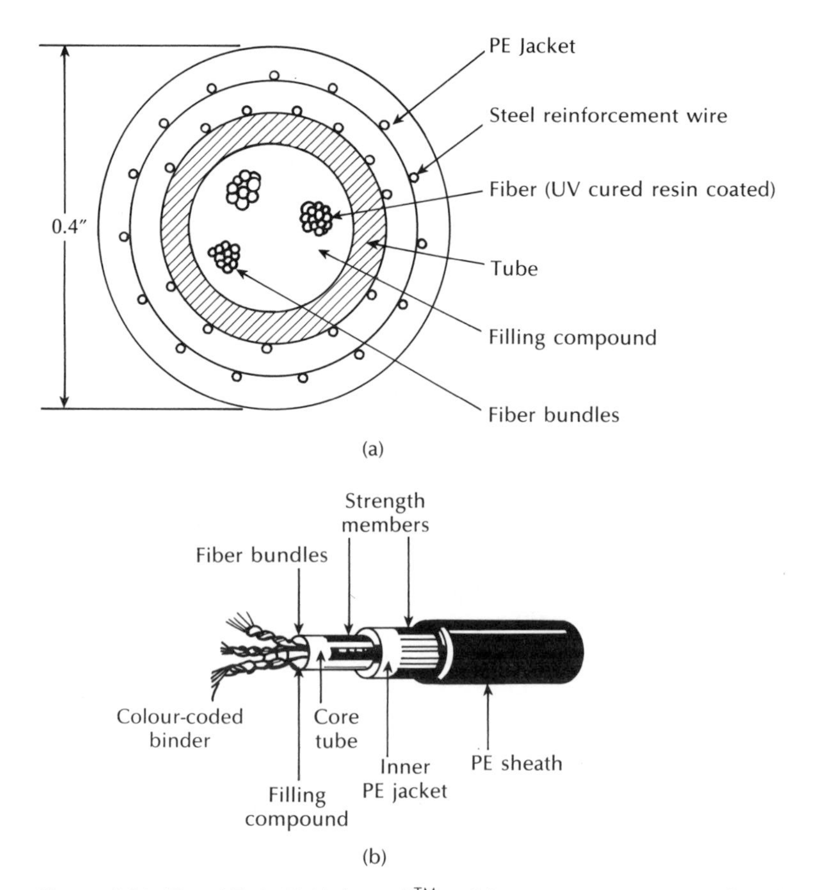

Figure 4.31 The AT & T Lightpack™ cable; (a) cross section showing steel reinforced sheath; (b) section illustrating the cable layers and the colour-coded fiber bundles [Ref. 50].

polyethylene inner jacket. Moreover, the core construction provides for compact, low weight cables with diameters of only 10 mm and 13 mm for fiber counts up to fifty and ninety-six fibers respectively [Ref. 72].

Problems

4.1 Describe in general terms liquid-phase techniques for the preparation of multicomponent glasses for optical fibers. Discuss with the aid of a suitable diagram one melting method for the preparation of multicomponent glass.

4.2 Indicate the major advantages of vapour-phase deposition in the preparation of glasses for optical fibers. Briefly describe the various vapour-phase techniques currently in use.

4.3 (a) Compare and contrast, using suitable diagrams, the outside vapour-phase oxidation (OVPO) process and the modified chemical vapour deposition (MCVD) technique for the preparation of low loss optical fibers.

(b) Briefly describe the salient features of vapour axial deposition (VAD) and the plasma-activated chemical vapour deposition (PCVD) when applied to the preparation of optical fibers.

4.4 Discuss the drawing of optical fibers from prepared glasses with regard to:

(a) multicomponent glass fibers;
(b) silica-rich fibers.

4.5 Indicate the primary advantage associated with the use of fluoride glass fibers and outline the possible methods of preparation for these fibers.

4.6 List the various optical fiber types currently on the market indicating their important features. Hence, briefly describe the general areas of application for each type.

4.7 Briefly describe the major reasons for the cabling of optical fibers which are to be placed in a field environment. Thus state the functions of the optical fiber cable.

4.8 Explain how the Griffith theory is developed in order to predict the fracture stress of an optical fiber with an elliptical crack.

Silica has a Young's modulus of 9×10^{10} N m^{-2} and a surface energy of 2.29 J. Estimate the fracture stress in psi for a silica optical fiber with a dominant elliptical crack of depth 0.5 μm. Also, determine the strain at the break for the fiber (1 psi $\equiv$ 6894.76 N m^{-2}).

4.9 Another length of the optical fiber described in Problem 4.7 is found to break at 1% strain. The failure is due to a single dominant elliptical crack. Estimate the depth of this crack.

4.10 Describe the effects of stress corrosion on optical fiber strength and durability.

It is found that a 20 m length of fused silica optical fiber may be extended to 24 m at liquid nitrogen temperatures (i.e. little stress corrosion) before failure occurs. Estimate the fracture stress in psi for the fiber under these conditions. Young's modulus for silica is 9×10^{10} N m^{-2} and 1 psi $\equiv$ 6894.76 N m^{-2}.

4.11 Outline the phenomena that can affect the stability of the transmission characteristics in optical fiber cables and describe any techniques by which these problems may be avoided.

4.12 Discuss optical fiber cable design with regard to:

(a) fiber buffering;
(b) cable strength and structural members;
(c) cable sheath and water barrier.

Further, compare and contrast possible cable designs for multifiber cables.

Answers to numerical problems

4.8 7.43×10^4 psi, 0.6%
4.9 0.2 μm
4.10 2.61×10^6 psi

References

[1] S. Tanaka, K. Inada, T. Akimoko and M. Kozima, 'Silicone-clad fused-silica-core fiber', *Electron. Lett.*, **11**(7), pp. 153–154, 1975.

[2] K. J. Beales and C. R. Day, 'A review of glass fibers for optical communications', *Phys. and Chem. of Glass*, **21**(1), pp. 5–21, 1980.

[3] T. Yamazuki and M. Yoshiyagawa, 'Fabrication of low-loss, multicomponent glass fibers with graded index and pseudo-step-index Borosilicate compound glass fibers', *Digest of International Conference on Integrated Optics and Optical Fiber Communication*, Osaka (Tokyo, IEEE, Japan), pp. 617–620, 1977.

[4] K. J. Beales, C. R. Day, W. J. Duncan, J. E. Midwinter and G. R. Newns, 'Preparation of sodium borosilicate glass fibers for optical communication', *Proc. IEE (London)*, **123**, pp. 591–595, 1976.

[5] G. R. Newns, P. Pantelis, J. L. Wilson, R. W. J. Uffen and R. Worthington, 'Absorption losses in glasses and glass fiber waveguides', *Opto-Electron*, **5**, pp. 289–296, 1973.

[6] B. Scott and H. Rawson, 'Techniques for producing low loss glasses for optical fibre communication systems', *Glass Technology*, **14**(5), pp. 115–124, 1973.

[7] C. E. E. Stewart, D. Tyldesley, B. Scott, H. Rawson and G. R. Newns, 'High-purity glasses for optical-fibre communication', *Electron. Lett.*, **9**(21), pp. 482–483, 1973.

[8] B. Scott and H. Rawson, 'Preparation of low loss glasses for optical fiber communication', *Opto-Electronics*, **5**(4), pp. 285–288, 1973.

[9] N. S. Kapany, *Fiber Optics*, Academic Press, 1967.

[10] A. M. Reid, W. W. Harper and A. Forbes, British Patent 50543, 1967.

[11] B. P. Pal, 'Optical communication, fiber waveguide fabrication: a review', *Fiber Int. Opt.*, **2**(2), pp. 195–252, 1979.

[12] G. R. Newns, 'Compound glass optical fibres', *2nd European Conference on Optical Fiber Communication* (Paris), pp. 21–26, 1976.

[13] K. J. Beales, C. R. Day, W. J. Duncan, A. G. Dunn, P. L. Dunn, G. R. Newns and J. V. Wright, 'Low loss graded index fiber by the double crucible technique', *5th European Conference on Optical Fiber Communication* (Amsterdam), paper 3.2, 1979.

[14] K. J. Beales, C. R. Day, W. J. Duncan and G. R. Newns, 'Low-loss compound-glass optical fibre', *Electron. Lett.*, **13**(24), pp. 755–756, 1977.

[15] H. Lydtin and F. Mayer, 'Review of techniques applied in optical fibre preparation', *Acta Electron.*, **22**(3), pp. 225–235, 1979.

[16] J. F. Hyde, US Patent 2 272 342, 1942.

[17] F. P. Kapron, D. B. Keck and R. D. Maurer, 'Radiation losses in optical waveguides', *Appl. Phys. Lett.*, **10**, pp. 423–425, 1970.

[18] B. Bendow and S. S. Mitra, *Fiber Optics*, Plenum Press, 1979.

[19] D. B. Keck and R. Bouillie, 'Measurements on high-bandwidth optical waveguides', *Optics Commun.*, **25**, pp. 43–48, 1978.

[20] B. S. Aronson, D. R. Powers and R. Sommer, 'Chloride drying of doped deposited silica preform simultaneous to consolidation', *Technical Digest of Topical Meeting on Optical Fiber Communication*, Washington, DC, p. 42, 1979.

[21] S. R. Nagel, 'Fiber materials and fabrication methods', in S. E. Miller and I. P. Kaminow (Eds.), *Optical Fiber Telecommunications II*, Academic Press, pp. 121–215, 1988.

[22] T. Izawa, T. Miyashita and F. Hanawa, US Patent 4 062 665, 1977.

[23] S. Sudo, M. Kawachi, M. Edahiro, T. Izawa, T. Shoida and H. Gotoh, 'Low-OH-content optical fiber fabricated by vapour-phase axial-deposition method', *Electron. Lett.*, **14**(17), pp. 534–535, 1978.

[24] T. Izawa, S. Sudo and F. Hanawa, 'Continuous fabrication process for high-silica fiber preforms (vapor phase axial deposition)', *Trans. Inst. Electron. Commun. Eng. Jpn. Section E* (Japan), **E62**(11), pp. 779–785, 1979.

[25] D. B. Keck and P. C. Schultz, US Patent 3 711 262, 1973.

[26] W. G. French, J. B. MacChesney, P. B. O'Conner and G. W. Tasker, 'Optical waveguides with very low losses', *Bell Syst. Tech. J.*, **53**, pp. 951–954, 1974.

[27] D. N. Payne and W. A. Gambling, 'New silica-based low-loss optical fibres', *Electron. Lett.*, **10**(15), pp. 289–90, 1974.

[28] T. Miya, Y. Terunuma, T. Mosaka and T. Miyashita, 'Ultimate low-loss single-mode fibre at 1.55 μm', *Electron. Lett.*, **15**(4), pp. 106–108, 1979.

[29] D. Gloge, 'The optical fibre as a transmission medium', *Rep. Prog. Phys.*, **42**, pp. 1778–1824, 1979.

[30] S. R. Nagel, J. B. MacChesney and K. L. Walker, 'An overview of the modified chemical vapour deposition (MCVD) process and performance', *IEEE J. Quantum Electron.*, **QE-18**(4), pp. 459–477, 1982.

[31] C. Lin, P. L. Lin, T. P. Lee, C. A. Burrus, F. T. Stone and A. J. Ritger, 'Measuring high bandwidth fibres in the 1.3 μm region with picosecond InGaAs injection lasers and ultrafast InGaAs detectors', *Electron. Lett.*, **17**(13), pp. 438–440, 1981.

[32] D. Kuppers and J. Koenings, 'Preform fabrication by deposition of thousands of layers with the aid of plasma activated CVD', *2nd European Conference on Optical Fiber Communication* (Paris), p. 49, 1976.

[33] R. E. Jaeger, J. B. MacChesney and T. J. Miller, 'The preparation of optical waveguide preforms by plasma deposition', *Bell. Syst. Tech. J.*, **57**, pp. 205–210, 1978.

[34] J. Irven and A. Robinson, 'Optical fibres prepared by plasma augmented vapour deposition', *Electron. Lett.*, **15**(9), pp. 252–4, 1979.

[35] P. K. Backman, D. Leers, H. Wehr, D. U. Wiechert, J. A. Van Steenwijk, D. L. A. Tjaden and E. R. Wehrhahn, 'Dispersion-flattened single-mode fibers prepared with PCVD: performance, limitations, design and optimization', *J. of Lightwave Technol.*, **LT-4**(7), pp. 858–863, 1986.

[36] H. Lydtin, 'PCVD: a technique suitable for large-scale fabrication of optical fibers', *J. of Lightwave Technol.*, **LT-4**(8), pp. 1034–1038, 1986.

[37] N. Nobukazu, 'Recent progress in glass fibers for optical communication', *Jap. J. Appl. Phys.*, **20**(8), pp. 1347–1360, 1981.

[38] P. W. Black, J. Irven and J. Titchmarsh, 'Fabrication of optical fibre waveguides', in C. P. Sandbank (Ed.), *Optical Fibre Communication Systems*, pp. 42–69, John Wiley, 1980.

[39] J. B. MacChesney, 'Materials and processes for preform fabrication – modified chemical vapour deposition and plasma chemical vapour deposition', *Proc. IEEE*, **68**(10), pp. 1181–1184, 1980.

[40] S. Sakaguchi and S. Takahashi, 'Low-loss fluoride optical fibers for midinfrared optical communication', *J. Lightwave Technol.*, **LT-5**(9), pp. 1219–1228, 1987.

[41] D. E. Quinn, 'Optical fibers', in F. C. Allard (ed.), *Fiber Optics Handbook: For engineers and scientists*, McGraw-Hill, pp. 1.1–1.50, 1990.

[42] T. Nakai, Y. Mimura, H. Tokiwa and O. Shinbori, 'Dehydration of fluoride glasses by NF_3 processing', *J. Lightwave Technol*, **LT-4**, pp. 87–89, 1986.

[43] R. C. Folweiler and D. E. Guenther, 'Chemical vapour purification of fluorides', *Mat. Sci. Forum*, **5**, pp. 43–48, 1985.

[44] J. A. Savage, 'Materials for infrared fibre optics', *Mat. Sci. Reports*, **2**, pp. 99–138, 1987.

[45] H. Iwasaki, 'Development of IR optical fibers in Japan', *Proc. SPIE,* **618**, *Infrared Optical Materials and Fibers IV*, pp. 2–9, 1986.

[46] S. Mitachi, T. Miyashita and T. Kanamori, 'Fluoride glass cladded optical fibers for mid infrared ray transmission', *Electron. Lett.*, **17**, pp. 591–592, 1981.

[47] D. C. Tran, C. F. Fischer and G. H. Sigel, 'Fluoride glass preforms prepared by a rotational casting process', *Electron. Lett.*, **18**, pp. 657–658, 1982.

[48] *Fibre Optical Components and Systems*, Belling Lee Limited, Electronic Components Group, UK (now Ray Proof).

[49] H. Murata, *Handbook of Optical Fibers and Cables*, Marcel Dekker, 1988.

[50] P. Kaiser and D. B. Keck, 'Fiber types and their status', in S. E. Miller and I. P. Kaminow (Eds.) *Optical Fiber Telecommunications II*, Academic Press, pp. 29–54, 1988.

[51] C. K. Koa, 'Optical fibres and cables', in M. J. Howes and D. V. Morgan (Eds.), *Optical Fibre Communications, Devices, Circuits and Systems*, pp. 189–249, John Wiley, 1980.

[52] J. McDermott, 'Fiber-optic-cable choices expand to fill design needs', *EDN*, pp. 95–99, May 1981.

[53] G. De Loane, 'Optical fibre cables', *Telecomm. J. (Eng. Ed.) Switzerland*, **48**(11), pp. 649–656, 1981.

[54] G. Galliano and F. Tosco, 'Optical fibre cables', *Optical Fibre Communications*, by Technical Staff of CSELT, pp. 501–540, McGraw-Hill, 1981.

[55] T. Nakahara and N. Uchida, 'Optical cable design and characterization in Japan', *Proc. IEEE*, **68**(10), pp. 1220–1226, 1980.

[56] A. A. Griffith, 'Phenomena of rupture and flow in solids', *Phil. Trans. R. Soc. Lond. Ser. A*, **221**, pp. 163–168, 1920.

[57] W. Weibull, 'A statistical theory of the strength of materials', *Proc. R. Swedish Inst. Res.*, No. 151, publication no. 4, 1939.

[58] M. H. Reeve, 'Optical fibre cables', *Ratio Electron. Eng. (IERE J.)*, **51**(7/8), pp. 327–332, 1981.

[59] B. Wiltshire and M. H. Reeve, 'A review of the environmental factors affecting optical cable design', *J. of Lightwave Technol.*, **LT-6**, pp. 179–185, 1988.

[60] S. Das, G. S. Englefield and P. A. Goud, 'Power loss, modal noise and distortion due to microbending of optical fibres', *Appl. Opt.*, **24**, pp. 2323–2333, 1985.

[61] P. Danielsen, 'Simple power spectrum of microbending in single mode fibers', *Electron. Lett.*, **19**, p. 318, 1983.

[62] R. S. Ashpole and R. J. W. Powell, 'Hydrogen in optical cables', *IEE Proc. Pt. J.*, **132**, pp. 162–168, 1985.

[63] P. J. Lemaire and A. Tomita, 'Hydrogen induced loss in MCVD fibres', *Optical Fiber Communications Conf., OFC '85* (USA) TUII, February 1985.

[64] E. J. Friebele, K. J. Long, C. G. Askins, M. E. Gingerich, M. J. Marrone and D. L. Griscom, 'Overview of radiation effects in fiber optics', *Proc. SPIE, Radiation Effects in Optical Materials*, **541**, pp. 70–88, 1985.

[65] A. Iino and J. Tamura, 'Radiation resistivity in silica optical fibers', *J. of Lightwave Technol.*, **LT-6**, pp. 145–149, 1988.

[66] E. J. Friebele and M. E. Gingerich, 'Photobleaching effects in optical fibre waveguides', *Appl. Opt.*, **20**, pp. 3448–3452, 1981.

[67] R. H. West, 'A local view of radiation effects in fiber optics', *J. of Lightwave Technol.*, **LT-6**, pp. 155–164, 1988.

[68] E. J. Friebele, E. W. Waylor, G. T. De Beauregard, J. A. Wall and C. E. Barnes, 'Interlaboratory comparison of radiation induced attenuation in optical fibers. Part 1: steady state exposure', *J. of Lightwave Technol.*, **LT-6**, pp. 165–178, 1988.

[69] P. R. Bank and D. O. Lawrence, 'Emerging standards in fiber optic telecommunications cable', *Proc. SPIE Int. Soc. Opt. Eng. (USA)*, **224**, pp. 149–158, 1980.

[70] J. C. Harrison, 'The metal foil/polyethylene cable sheath and its use in the Post Office', Institution of Post Office Engineers Paper No. 229, 1968.

[71] P. Kaiser and W. T. Anderson, 'Fiber cables for public communications: state-of-the-art technologies and the future', *J. Lightwave Technol.*, **LT-4**(8), pp. 1157–1165, 1986.

[72] C. H. Gartside III, P. D. Patel and M. R. Santana, 'Optical fiber cables', in S. E. Miller and I. P. Kaminow (Eds.), *Optical Fiber Telecommunications II*, Academic Press, 1988.

[73] F. Nihei, Y. Yamomoter and N. Kojima, 'Optical subscriber cable technologies in Japan', *J. Lightwave Technol.*, **LT-5**(6), pp. 809–821, 1987.

[74] S. Kukita, M. Kawase, H. Kobayashi and O. Ogasawara, 'Design and performance of optical cables and cabinets for local area networks', *ECL Tech. J.*, **34**(5) pp. 905–918, 1985.

[75] M. J. Buckler, M. R. Santana and M. J. Saunders, 'Lightwave cable manufacture and performance', *Bell Syst. Tech. J.*, **57**(6), p. 1745, 1978.

[76] M. Oda, M. Ogai, A. Ohtake, S. Tachigami, K. Ohkubo, F. Nihei and N. Kashima, 'Nylon extruded fiber ribbon and its connection', *Optical Fiber Communications Conf., OFC '82* (USA) paper THAA6, February 1982.

[77] Y. Katsuyama, S. Hatano, K. Hogari, T. Matsumoto and T. Kokubun, 'Single-mode optical-fiber ribbon cable', *Electron. Lett.*, **21**(4), p. 134, 1985.

[78] Y. Katsuyama, K. Hogari, S. Hatano and T. Kobubun, 'Optical loss characteristics of slotted rod cable composed of single-mode fiber ribbons', *Natl. Conv., IECEJ*, **2062**, pp. 9–91, March 1986.

5

Optical fiber connection: joints and couplers

5.1 Introduction

Optical fiber links, in common with any line communication system, have a requirement for both jointing and termination of the transmission medium. The number of intermediate fiber connections or joints is dependent upon the link length (between repeaters), the continuous length of fiber cable that may be produced by the preparation methods outlined in Sections 4.2 to 4.4, and the length of the fiber cable that may be practically or conveniently installed as a continuous section on the link. Although scaling up of the preparation processes now provides the capability to produce very large preforms allowing continuous single-mode fiber lengths of around 200 km, such fiber spans cannot be readily installed [Ref. 1]. However, continuous cable lengths of tens of kilometres have already been deployed, in particular within submarine systems where continuous cable laying presents fewer problems [Ref. 2].

Repeater spacing on optical fiber telecommunication links is a continuously increasing parameter with currently installed digital systems operating over spacings in the range 40 to 60 km at transmission rates of between 400 Mbit s^{-1} and 1.7 Gbit s^{-1} [Refs. 3 to 5]. Moreover, spacings in excess of 100 km are also readily achievable with practical systems at such transmission rates, and unrepeated distances of around 300 km have been obtained in the laboratory [Ref. 4]. For example, a fully operational 2.4 Gbit s^{-1} system operating with a 100 km repeater spacing using dispersion shifted single-mode fiber (see Section 3.12.1) has recently been reported [Ref. 6], together with the experimental transmission at 10 Gbit s^{-1} over a similar unrepeated distance [Ref. 7].

It is therefore apparent that fiber to fiber connection with low loss and minimum distortion (i.e. modal noise) remains an important aspect of optical fiber communication systems (fiber, sources, detectors, etc.) it is clear that, in recent systems. In addition, it also serves to increase the number of terminal connections permissible within the developing optical fiber communication networks (see Chapter 14). Although fiber jointing techniques appeared to lag behind the technologies associated with the other components required in optical fiber communication systems (fiber sources, detectors, etc.) it is clear that, in recent years, significant developments have been made. Therefore, in this and the sections immediately following we review the theoretical and practical aspects of fiber–fiber connection with regard to both multimode and single-mode systems. Fiber termination to sources and detectors is not considered since the important aspects of these topics are discussed in the chapters covering sources and detectors (Chapters 6, 7 and 8). Nevertheless, the discussion on fiber jointing is relevant to both source and detector coupling, as many manufacturers supply these electro-optical devices already terminated to a fiber optic pigtail in order to facilitate direct fiber–fiber connection to an optical fiber link.

Before we consider fiber–fiber connection in further detail it is necessary to indicate the two major categories of fiber joint currently both in use and development. These are as follows:

1. Fiber splices: these are semipermanent or permanent joints which find major use in most optical fiber telecommunication systems (analogous to electrical soldered joints).
2. Demountable fiber connectors or simple connectors: these are removable joints which allow easy, fast, manual coupling and uncoupling of fibers (analogous to electrical plugs and sockets).

The above fiber to fiber joints are designed ideally to couple all the light propagating in one fiber into the adjoining fiber. By contrast fiber couplers are branching devices that split all the light from a main fiber into two or more fibers or, alternatively, couple a proportion of the light propagating in the main fiber into a branch fiber. Moreover, these devices are often bidirectional, providing for the combining of light from one or more branch fibers into a main fiber. The importance and variety of these fiber couplers has increased substantially over recent years in order to facilitate the widespread deployment of optical fiber within communication networks. Although the requirement for such devices was less in earlier point-to-point fiber links, the growing demand for more sophisticated fiber network configurations (see Chapter 14) has made them essential components within optical fiber communications.

In this chapter we therefore consider the basic techniques and technology associated with both fiber joints and couplers. A crucial aspect of fiber jointing concerns the optical loss associated with the connection. This joint loss is critically dependent upon the alignment of the two fibers. Hence, in Section 5.2 the mechanisms which cause optical losses at fiber joints are outlined, with particular attention being paid to the fiber alignment. This discussion provides a grounding for consideration of the techniques employed for jointing optical fibers. Permanent fiber joints (i.e. splices) are then dealt with in Section 5.3 prior to discussion of the two generic types of demountable connector in Sections 5.4 and 5.5. Finally, in Section 5.6, the basic construction and performance characteristics of the various fiber coupler types are described.

5.2 Fiber alignment and joint loss

A major consideration with all types of fiber–fiber connection is the optical loss encountered at the interface. Even when the two jointed fiber ends are smooth and perpendicular to the fiber axes, and the two fiber axes are perfectly aligned, a small proportion of the light may be reflected back into the transmitting fiber causing attenuation at the joint. This phenomenon, known as Fresnel reflection, is associated with the step changes in refractive index at the jointed interface (i.e. glass–air–glass). The magnitude of this partial reflection of the light transmitted through the interface may be estimated using the classical Fresnel formula for light of normal incidence and is given by [Ref. 8]:

$$r = \left(\frac{n_1 - n}{n_1 + n}\right)^2 \tag{5.1}$$

where r is the fraction of the light reflected at a single interface, n_1 is the refractive index of the fiber core and n is the refractive index of the medium between the two jointed fibers (i.e. for air $n = 1$). However, in order to determine the amount of light reflected at a fiber joint, Fresnel reflection at both fiber interfaces must be taken into account. The loss in decibels due to Fresnel reflection at a single interface is given by:

$$Loss_{\text{Fres}} = -10 \log_{10}(1 - r) \tag{5.2}$$

Hence, using the relationships given in Eqs. (5.1) and (5.2) it is possible to determine the optical attenuation due to Fresnel reflection at a fiber–fiber joint.

It is apparent that Fresnel reflection may give a significant loss at a fiber joint even when all other aspects of the connection are ideal. However, the effect of

Fresnel reflection at a fiber–fiber connection can be reduced to a very low level through the use of an index matching fluid in the gap between the jointed fibers. When the index matching fluid has the same refractive index as the fiber core, losses due to Fresnel reflection are in theory eradicated.

Unfortunately, Fresnel reflection is only one possible source of optical loss at a fiber joint. A potentially greater source of loss at a fiber–fiber connection is caused by misalignment of the two jointed fibers. In order to appreciate the development and relative success of various connection techniques it is useful to discuss fiber alignment in greater detail.

Example 5.1

An optical fiber has a core refractive index of 1.5. Two lengths of the fiber with smooth and perpendicular (to the core axes) end faces are butted together. Assuming the fiber axes are perfectly aligned, calculate the optical loss in decibels at the joint (due to Fresnel reflection) when there is a small air gap between the fiber end faces.

Solution: The magnitude of the Fresnel reflection at the fiber–air interface is given by Eq. (5.1) where:

$$r = \left(\frac{n_1 - n}{n_1 + n}\right)^2 = \left(\frac{1.5 - 1.0}{1.5 + 1.0}\right)^2$$

$$= \left(\frac{0.5}{2.5}\right)^2$$

$$= 0.04$$

The value obtained for r corresponds to a reflection of 4% of the transmitted light at the single interface. Further, the optical loss in decibels at the single interface may be obtained using Eq. (5.2) where:

$$Loss_{\text{Fres}} = -10 \log_{10}(1 - r) = -10 \log_{10} 0.96$$
$$= 0.18 \text{ dB}$$

A similar calculation may be performed for the other interface (air–fiber). However, from considerations of symmetry it is clear that the optical loss at the second interface is also 0.18 dB.

Hence the total loss due to Fresnel reflection at the fiber joint is approximately 0.36 dB.

Any deviations in the geometrical and optical parameters of the two optical fibers which are jointed will affect the optical attenuation (insertion loss) through the connection. It is not possible within any particular connection technique to allow for all these variations. Hence, there are inherent connection problems when jointing fibers with, for instance:

(a) different core and/or cladding diameters;
(b) different numerical apertures and/or relative refractive index differences;
(c) different refractive index profiles;
(d) fiber faults (core ellipticity, core concentricity, etc.).

The losses caused by the above factors together with those of Fresnel reflection are usually referred to as intrinsic joint losses.

The best results are therefore achieved with compatible (same) fibers which are manufactured to the lowest tolerance. In this case there is still the problem of the quality of the fiber alignment provided by the jointing mechanism. Examples of possible misalignment between coupled compatible optical fibers are illustrated in Figure 5.1 [Ref. 9]. It is apparent that misalignment may occur in three dimensions, the separation between the fibers (longitudinal misalignment), the offset perpendicular to the fiber core axes (lateral/radial/axial misalignment) and the angle between the core axes (angular misalignment).

Optical losses resulting from these three types of misalignment depend upon the fiber type, core diameter and the distribution of the optical power between the propagating modes. Examples of the measured optical losses due to the various types of misalignment are shown in Figure 5.2. Figure 5.2(a) [Ref. 9] shows the attenuation characteristic for both longitudinal and lateral misalignment of a 50 μm core diameter graded index fiber. It may be observed that the lateral misalignment gives significantly greater losses per unit displacement than the longitudinal misalignment. For instance in this case a lateral displacement of 10 μm gives about 1 dB insertion loss whereas a similar longitudinal displacement gives an insertion loss of around 0.1 dB. Figure 5.2(b) [Ref. 10] shows the attenuation characteristic for the angular misalignment of two multimode step index fibers with numerical apertures of 0.22 and 0.3. An insertion loss of around 1 dB is obtained with angular

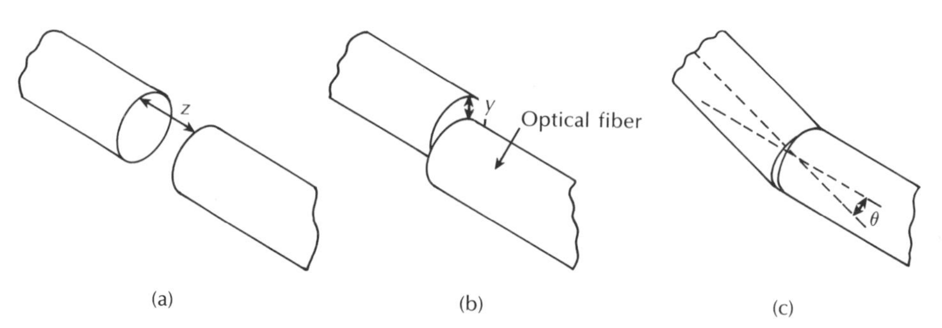

Figure 5.1 The three possible types of misalignment which may occur when jointing compatible optical fibers [Ref. 9]: (a) longitudinal misalignment; (b) lateral misalignment; (c) angular misalignment.

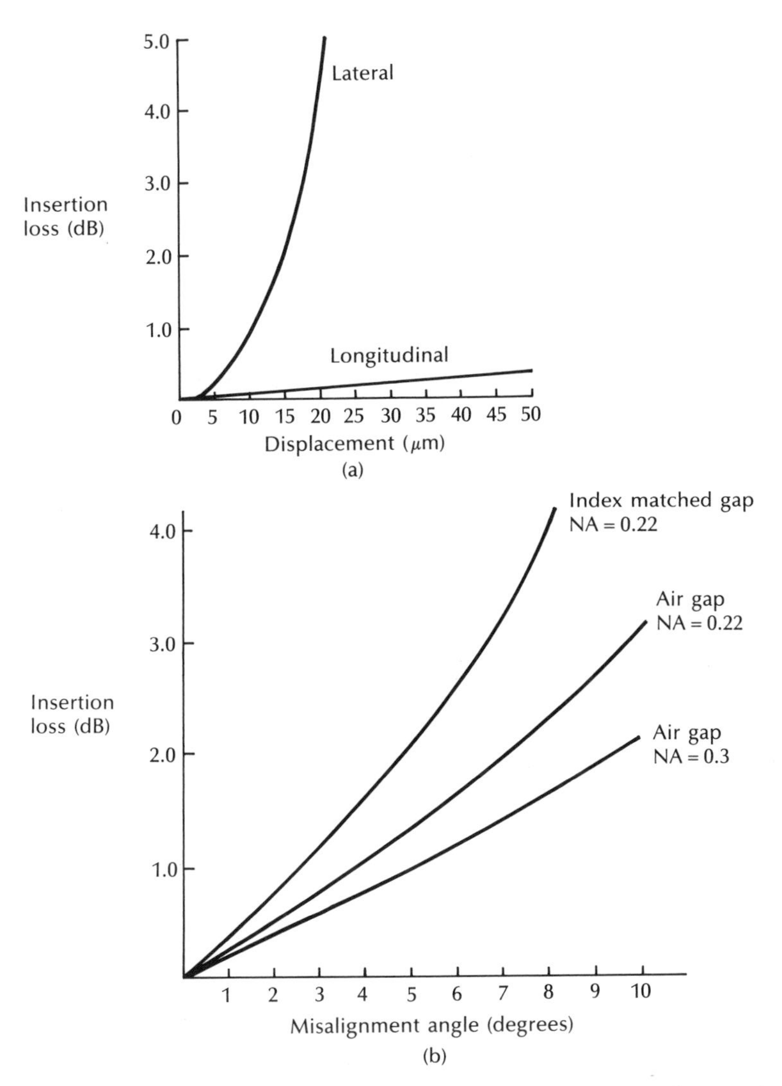

Figure 5.2 Insertion loss characteristics for jointed optical fibers with various types of misalignment: (a) insertion loss due to lateral and longitudinal misalignment for a 50 μm core diameter graded index fiber, reproduced with permission from P. Mossman, *The Radio and Electron. Eng.*, **51**, p. 333. 1981; (b) insertion loss due to angular misalignment for joints in two multimode step index fibers with numerical apertures of 0.22 and 0.3, reproduced with permission from C. P. Sandback (Ed.), *Optical Fiber Communication Systems*, John Wiley & Sons, 1980.

misalignment of 4° and 5° for the 0.22 NA and 0.3 NA fibers respectively. It may also be observed in Figure 5.2(b) that the effect of an index matching fluid in the fiber gap causes increased losses with angular misalignment. Therefore, it is clear that relatively small levels of lateral and/or angular misalignment can cause significant attenuation at a fiber joint. This is especially the case for small core diameter (less than 150 μm) fibers which are currently employed for most telecommunication purposes.

5.2.1 Multimode fiber joints

Theoretical and experimental studies of fiber misalignment in optical fiber connections [Refs. 11 to 19] allow approximate determination of the losses encountered with the various misalignments of different fiber types. We consider here some of the expressions used to calculate losses due to lateral and angular misalignment of optical fiber joints. Longitudinal misalignment is not discussed in detail as it tends to be the least important effect and may be largely avoided in fiber connection. Also there is some disagreement over the magnitude of the losses due to longitudinal misalignment when it is calculated theoretically between Miyazaki *et al.* [Ref. 12] and Tsuchiya *et al.* [Ref. 13]. Both groups of workers claim good agreement with experimental results, which is perhaps understandable when considering the number of variables involved in the measurement. However, it is worth noting that the lower losses predicted by Tsuchiya *et al.* agree more closely with a third group of researchers [Ref. 14]. Also, all groups predict higher losses for fibers with larger numerical apertures which is consistent with intuitive considerations (i.e. the larger the numerical aperture, the greater the spread of the output light and the higher the optical loss at a longitudinally misaligned joint).

Theoretical expressions for the determination of lateral and angular misalignment losses are by no means definitive, although in all cases they claim reasonable agreement with experimental results. However, experimental results from different sources tend to vary (especially for angular misalignment losses) due to difficulties of measurement. It is therefore not implied that the expressions given in the text are necessarily the most accurate, as at present the choice appears somewhat arbitrary.

Lateral misalignment reduces the overlap region between the two fiber cores. Assuming uniform excitation of all the optical modes in a multimode step index fiber the overlapped area between both fiber cores approximately gives the lateral coupling efficiency η_{lat}. Hence, the lateral coupling efficiency for two similar step index fibers may be written as [Ref. 13]:

$$\eta_{lat} \simeq \frac{16(n_1/n)^2}{(1+(n_1/n))^4}\,\frac{1}{\pi}\left\{2\cos^{-1}\left(\frac{y}{2a}\right) - \left(\frac{y}{a}\right)\left[1-\left(\frac{y}{2a}\right)^2\right]^{\frac{1}{2}}\right\} \tag{5.3}$$

where n_1 is the core refractive index, n is the refractive index of the medium between the fibers, y is the lateral offset of the fiber core axes, and a is the fiber core radius. The lateral misalignment loss in decibels may be determined using:

$$Loss_{\text{lat}} = -10 \log_{10} \eta_{\text{lat}} \text{ dB} \tag{5.4}$$

The predicted losses obtained using the formulae given in Eqs. (5.3) and (5.4) are generally slightly higher than the measured values due to the assumption that all modes are equally excited. This assumption is only correct for certain cases of optical fiber transmission. Also, certain authors [Refs. 12 and 18] assume index matching and hence no Fresnel reflection, which makes the first term in Eq. (5.3) equal to unity (as $n_1/n = 1$). This may be valid if the two fiber ends are assumed to be in close contact (i.e. no air gap in between) and gives lower predicted losses. Nevertheless, bearing in mind these possible inconsistencies, useful estimates for the attenuation due to lateral misalignment of multimode step index fibers may be obtained.

Lateral misalignment loss in multimode graded index fibers assuming a uniform distribution of optical power throughout all guided modes was calculated by Gloge [Ref. 16]. He estimated that the lateral misalignment loss was dependent on the refractive index gradient α for small lateral offset and may be obtained from:

$$L_t = \frac{2}{\pi}\left(\frac{y}{a}\right)\left(\frac{\alpha+2}{\alpha+1}\right) \qquad \text{for } 0 \leqslant y \leqslant 0.2a \tag{5.5}$$

where the lateral coupling efficiency was given by:

$$\eta_{\text{lat}} = 1 - L_t \tag{5.6}$$

Hence Eq. (5.6) may be utilized to obtain the lateral misalignment loss in decibels. With a parabolic refractive index profile where $\alpha = 2$, Eq. (5.5) gives:

$$L_t = \frac{8}{3\pi}\left(\frac{y}{a}\right) = 0.85\left(\frac{y}{a}\right) \tag{5.7}$$

A further estimate including the leaky modes, gave a revised expression for the lateral misalignment loss given in Eq. (5.6) of $0.75(y/a)$. This analysis was also extended to step index fibers (where $\alpha = \infty$) and gave lateral misalignment losses of $0.64(y/a)$ and $0.5(y/a)$ for the cases of guided modes only and both guided plus leaky modes respectively.

Example 5.2

A step index fiber has a core refractive index of 1.5 and a core diameter of 50 μm. The fiber is jointed with a lateral misalignment between the core axes of 5 μm. Estimate the insertion loss at the joint due to the lateral misalignment assuming a uniform distribution of power between all guided modes when:

(a) there is a small air gap at the joint;
(b) the joint is considered index matched.

Solution: (a) The coupling efficiency for a multimode step index fiber with uniform illumination of all propagating modes is given by Eq. (5.3) as:

$$\eta_{lat} \simeq \frac{16(n_1/n)^2}{(1+(n_1/n))^4}\frac{1}{\pi}\left\{2\cos^{-1}\left(\frac{y}{2a}\right)-\left(\frac{y}{a}\right)\left[1-\left(\frac{y}{2a}\right)^2\right]^{\frac{1}{2}}\right\}$$

$$= \frac{16(1.5)^2}{(1+1.5)^4}\frac{1}{\pi}\left\{2\cos^{-1}\left(\frac{5}{50}\right)-\left(\frac{5}{25}\right)\left[1-\left(\frac{5}{50}\right)^2\right]^{\frac{1}{2}}\right\}$$

$$= 0.293\{2(1.471)-0.2[0.99]^{\frac{1}{2}}\}$$

$$= 0.804$$

The insertion loss due to lateral misalignment is given by Eq. (5.4) where

$$Loss_{lat} = -10\log_{10}\eta_{lat} = -10\log_{10}0.804$$
$$= 0.95\ \text{dB}$$

Hence assuming a small air gap at the joint the insertion loss is approximately 1 dB when the lateral offset is 10% of the fiber diameter.

(b) When the joint is considered index matched (i.e. no air gap) the coupling efficiency may again be obtained from Eq. (5.3) where

$$\eta_{lat} \simeq \frac{1}{\pi}\left\{2\cos^{-1}\left(\frac{5}{50}\right)-\left(\frac{5}{25}\right)\left[1-\left(\frac{5}{50}\right)^2\right]^{\frac{1}{2}}\right]$$

$$= 0.318\{2(1.471)-0.2[0.99]^{\frac{1}{2}}\}$$

$$= 0.872$$

Therefore the insertion loss is:

$$Loss_{lat} = -10\log_{10}0.872 = 0.59\ \text{dB}$$

With index matching the insertion loss at the joint in Example 5.2 is reduced to approximately 0.36 dB. It may be noted that the difference between the losses obtained in parts (a) and (b) corresponds to the optical loss due to Fresnel reflection at the similar fiber–air–fiber interface determined in Example 5.1.

The result may be checked using the formulae derived by Gloge for a multimode step index fiber where the lateral misalignment loss assuming uniform illumination of all guided modes is obtained using:

$$L_t = 0.64\left(\frac{y}{a}\right) = 0.64\left(\frac{5}{25}\right) = 0.128$$

Hence the lateral coupling efficiency is given by Eq. (5.6) as:

$$\eta_{lat} = 1 - 0.128 = 0.872$$

Again using Eq. (5.4), the insertion loss due to the lateral misalignment assuming index matching is:

$$Loss_{lat} = -10 \log_{10} 0.872 = 0.59 \text{ dB}$$

Hence using the expression derived by Gloge we obtain the same value of approximately 0.6 dB for the insertion loss with the inherent assumption that there is no change in refractive index at the joint interface. Although this estimate of insertion loss may be shown to agree with certain experimental results [Ref. 12], a value of around 1 dB insertion loss for a 10% lateral displacement with regard to the core diameter (as estimated in Example 5.2(a)) is more usually found to be the case with multimode step index fibers [Refs. 8, 19 and 24]. Further, it is generally accepted that the lateral offset must be kept below 5% of the fiber core diameter in order to reduce insertion loss at a joint to below 0.5 dB [Ref. 19].

Example 5.3

A graded index fiber has a parabolic refractive index profile ($\alpha = 2$) and a core diameter of 50 μm. Estimate the insertion loss due to a 3 μm lateral misalignment at a fiber joint when there is index matching and assuming:

(a) there is uniform illumination of all guided modes only;
(b) there is uniform illumination of all guided and leaky modes.

Solution: (a) Assuming uniform illumination of guided modes only, the misalignment loss may be obtained using Eq. (5.7), where

$$L_t = 0.85\left(\frac{y}{a}\right) = 0.85\left(\frac{3}{25}\right) = 0.102$$

The coupling efficiency is given by Eq. (5.6) as:

$$\eta_{lat} = 1 - L_t = 1 - 0.102 = 0.898$$

Hence the insertion loss due to the lateral misalignment is given by Eq. (5.4), where:

$$Loss_{lat} = -10 \log_{10} 0.898 = 0.47 \text{ dB}$$

(b) When assuming the uniform illumination of both guided and leaky modes Gloge's formula becomes:

$$L_t = 0.75\left(\frac{y}{a}\right) = 0.75\left(\frac{3}{25}\right) = 0.090$$

Therefore the coupling efficiency is

$$\eta_{lat} = 1 - 0.090 = 0.910$$

and the insertion loss due to lateral misalignment is:

$$Loss_{lat} = -10 \log_{10} 0.910 = 0.41 \text{ dB}$$

It may be noted by observing Figure 5.2(a) which shows the measured lateral misalignment loss for a 50 μm diameter graded index fiber that the losses predicted above are very pessimistic (the loss for 3 μm offset shown in Figure 5.2(a) is less than 0.2 dB). A model which is found to predict insertion loss due to lateral misalignment in graded index fibers with greater accuracy was proposed by Miller and Mettler [Ref. 17]. In this model they assumed the power distribution at the fiber output to be of a Gaussian form. Unfortunately, the analysis is too detailed for this text as it involves integration using numerical techniques. We therefore limit estimates of insertion losses due to lateral misalignment in multimode graded index fibers to the use of Gloge's formula.

Angular misalignment losses at joints in multimode step index fibers may be predicted with reasonable accuracy using an expression for the angular coupling efficiency η_{ang} given by [Ref. 13]:

$$\eta_{ang} \simeq \frac{16(n_1/n)^2}{(1+(n_1/n))^4}\left[1 - \frac{n\theta}{\pi n_1(2\Delta)^{\frac{1}{2}}}\right] \tag{5.8}$$

where θ is the angular displacement in radians and Δ is the relative refractive index difference for the fiber. The insertion loss due to angular misalignment may be obtained from the angular coupling efficiency in the same manner as the lateral misalignment loss following:

$$Loss_{ang} = -10 \log_{10} \eta_{ang} \tag{5.9}$$

The formulae given in Eqs. (5.8) and (5.9) predict that the smaller the values of Δ the larger the insertion loss due to angular misalignment. This appears intuitively correct as small values of Δ imply small numerical aperture fibers, which will be more affected by angular misalignment. It is confirmed by the measurements shown in Figure 5.2(b) and demonstrated in Example 5.4.

Example 5.4

Two multimode step index fibers have numerical apertures of 0.2 and 0.4, respectively, and both have the same core refractive index which is 1.48. Estimate the insertion loss at a joint in each fiber caused by a 5° angular misalignment of the fiber core axes. It may be assumed that the medium between the fibers is air.

Solution: The angular coupling efficiency is given by Eq. (5.8) as

$$\eta_{ang} \simeq \frac{16(n_1/n)^2}{(1+(n_1/n))^4}\left[1 - \frac{n\theta}{\pi n_1(2\Delta)^{\frac{1}{2}}}\right]$$

The numerical aperture is related to the relative refractive index difference following Eq. (2.10) where:

$$NA \simeq n_1(2\Delta)^{\frac{1}{2}}$$

Hence

$$\eta_{\text{ang}} \simeq \frac{16(n_1/n)^2}{(1 + (n_1/n))^4} \left[1 - \frac{n\theta}{\pi NA}\right]$$

For the 0.2 NA fiber:

$$\eta_{\text{ang}} \simeq \frac{16(1.48)^2}{(1 + 1.48)^4} \left[1 - \frac{5\pi/180}{\pi 0.2}\right]$$

$$= 0.797$$

The insertion loss due to the angular misalignment may be obtained from Eq. (5.9), where:

$$Loss_{\text{ang}} = -10 \log_{10} \eta_{\text{ang}} = -10 \log_{10} 0.797$$
$$= 0.98 \text{ dB}$$

For the 0.4 NA fiber:

$$\eta_{\text{ang}} \simeq 0.926\left[1 - \frac{5\pi/180}{\pi\ 0.4}\right]$$

$$\simeq 0.862$$

The insertion loss due to the angular misalignment is therefore:

$$Loss_{\text{ang}} = -10 \log_{10} 0.862$$
$$= 0.64 \text{ dB}$$

Hence it may be noted from Example 5.4 that the insertion loss due to angular misalignment is reduced by using fibers with large numerical apertures. This is the opposite trend to the increasing insertion loss with numerical aperture for fiber longitudinal misalignment at a joint.

Factors causing fiber–fiber intrinsic losses were listed in Section 5.2; the major ones comprising a mismatch in the fiber core diameters, a mismatch in the fiber numerical apertures and differing fiber refractive index profiles are illustrated in Figure 5.3. Connections between multimode fibers with certain of these parameters being different can be quite common, particularly when a pigtailed optical source is used, the fiber pigtail of which has different characteristics from the main transmission fiber. Moreover, as indicated previously, diameter variations can occur with the same fiber type.

Assuming all the modes are equally excited in a multimode step or graded index fiber, and that the numerical apertures and index profiles are the same, then the loss resulting from a mismatch of core diameters (see Figure 5.3(a)) is given by [Refs. 13, 20]:

$$Loss_{\text{CD}} \begin{cases} = -10 \log_{10}\left(\dfrac{a_2}{a_1}\right)^2 \ (\text{dB}) & a_2 < a_1 \\ = 0 \qquad\qquad\qquad\quad (\text{dB}) & a_2 \geqslant a_1 \end{cases} \tag{5.10}$$

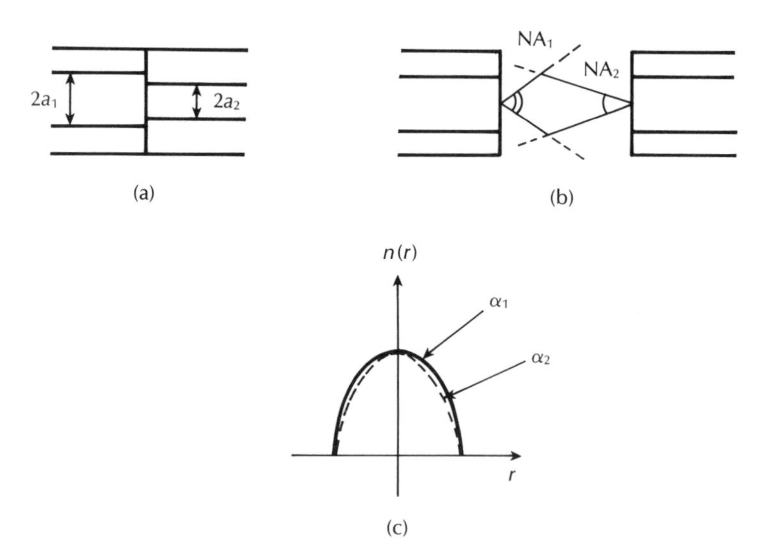

Figure 5.3 Some intrinsic coupling losses at fiber joints: (a) core diameter mismatch; (b) numerical aperture mismatch; (c) refractive index profile difference.

where a_1 and a_2 are the core radii of the transmitting and receiving fibers respectively. It may be observed from Eq. (5.10) that no loss is incurred if the receiving fiber has a larger core diameter than the transmitting one. In addition, only a relatively small loss (0.09 dB) is obtained when the receiving fiber core diameter is 1% smaller than that of the transmitting fiber.

When the transmitting fiber has a higher numerical aperture than the receiving fiber, then some of the emitted light rays will fall outside the acceptance angle of the receiving fiber and they will therefore not be coupled through the joint. Again assuming a uniform modal power distribution, and fibers with equivalent refractive index profiles and core diameters, then the loss caused by a mismatch of numerical apertures (see Figure 5.3(b)) can be obtained from [Refs. 18, 20]:

$$Loss_{NA} \begin{cases} = -10 \log_{10}\left(\dfrac{NA_2}{NA_1}\right)^2 \text{ (dB)} & NA_2 < NA_1 \\ = 0 \text{ (dB)} & NA_2 \geqslant NA_1 \end{cases} \tag{5.11}$$

where NA_1 and NA_2 are the numerical apertures for the transmitting and receiving fibers respectively. Equation (5.11) is valid for both step and graded index* fibers and in common with Eq. (5.10) it demonstrates that no losses occur when the receiving parameter (i.e. numerical aperture) is larger than the transmitting one.

* In the case of graded index fibers the numerical aperture on the fiber core axis must be used.

Finally, a mismatch in refractive index profiles (see Figure 5.3(a)) results in a loss which can be shown to be [Ref. 20]:

$$Loss_{RI}\begin{cases} = -10\log_{10}\dfrac{\alpha_2(\alpha_1+2)}{\alpha_1(\alpha_2+2)} \text{ (dB)} & \alpha_2 < \alpha_1 \\ = 0 \text{ (dB)} & \alpha_2 \geqslant \alpha_1 \end{cases} \tag{5.12}$$

where α_1 and α_2 are the profile parameters for the transmitting and receiving fibers respectively (see Section 2.4.4). When connecting from a step index fiber with $\alpha_1 = \infty$ to a parabolic profile graded index fiber with $\alpha_2 = 2$, both having the same core diameter and axial numerical aperture, then a loss of 3 dB is produced. The reverse connection, however, does not incur a loss due to refractive index profile mismatch.

The intrinsic losses obtained at multimode fiber–fiber joints provided by Eqs. (5.10) to (5.12) can be combined into a single expression as follows:

$$Loss_{\text{int}}\begin{cases} = -10\log_{10}\dfrac{(a_2NA_2)^2(\alpha_1+2)\alpha_2}{(a_1NA_1)^2(\alpha_2+2)\alpha_1} \text{ (dB)} & a_2 > a_1, NA_2 > NA_1, \alpha_2 > \alpha_1 \\ = 0 \text{ (dB)} & a_2 \leqslant a_1, NA_2 \leqslant NA_1, \alpha_2 \leqslant \alpha_1 \end{cases} \tag{5.13}$$

It should be noted that Eq. (5.13) assumes that the three mismatches occur together. Distributions of losses which are obtained when with particular distributions of parameters, various random combinations of mismatches occur in a long series of connections, are provided in Ref. 11.

5.2.2 Single-mode fiber joints

Misalignment losses at connections in single-mode fibers have been theoretically considered by Marcuse [Ref. 21] and Gambling *et al.* [Refs. 22 and 23]. The theoretical analysis which was instigated by Marcuse is based upon the Gaussian or near Gaussian shape of the modes propagating in single-mode fibers regardless of the fiber type (i.e. step index or graded index). Further development of this theory by Gambling *et al.* [Ref. 23] gave simplified formulae for both the lateral and angular misalignment losses at joints in single-mode fibers. In the absence of angular misalignment Gambling *et al.* calculated that the loss T_l due to lateral offset y was given by:

$$T_l = 2.17\left(\frac{y}{\omega}\right)^2 \text{ dB} \tag{5.14}$$

where ω is the normalized spot size of the fundamental mode.* However, the normalized spot size for the LP_{01} mode (which corresponds to HE mode) may be

* The spot size for single-mode fibers is discussed in Section 2.5.2. It should be noted, however, that the normalization factor for the spot size causes it to differ in Eq. (5.15) by a factor of $2^{\frac{1}{2}}$ from that provided in Eq. (2.125).

obtained from the empirical formula [Refs. 19 and 22]:

$$\omega = a \frac{(0.65 + 1.62V^{-\frac{3}{2}} + 2.88V^{-6})}{2^{\frac{1}{2}}} \tag{5.15}$$

where ω is the spot size in μm, a is the fiber core radius and V is the normalized frequency for the fiber. Alternatively, the insertion loss T_a caused by an angular misalignment θ (in radians) at a joint in a single-mode fiber may be given by:

$$T_a = 2.17\left(\frac{\theta\omega\ n_1 V}{a\ NA}\right)^2 \text{ dB} \tag{5.16}$$

where n_1 is the fiber core refractive index and NA is the numerical aperture of the fiber. It must be noted that the formulae given in Eqs. (5.15) and (5.16) assume that the spot sizes of the modes in the two coupled fibers are the same. Gambling *et al.* [Ref. 23] also derived a somewhat complicated formula which gave a good approximation for the combined losses due to both lateral and angular misalignment at a fiber joint. However, they indicate that for small total losses (less than 0.75 dB) a reasonable approximation is obtained by simply combining Eqs. (5.14) and (5.16).

Example 5.5

A single-mode fiber has the following parameters:

normalized frequency (V) = 2.40
core refractive index (n_1) = 1.46
core diameter ($2a$) = 8 μm
numerical aperture (NA) = 0.1

Estimate the total insertion loss of a fiber joint with a lateral misalignment of 1 μm and an angular misalignment of 1°.

Solution: Initially it is necessary to determine the normalized spot size in the fiber. This may be obtained from Eq. (5.15) where:

$$\omega = a \frac{(0.65 + 1.62V^{-\frac{3}{2}} + 2.88V^{-6})}{2^{\frac{1}{2}}}$$

$$= 4 \frac{(0.65 + 1.62(2.4)^{-1.5} + 2.88(2.4)^{-6})}{2^{\frac{1}{2}}}$$

$$= 3.12 \ \mu\text{m}$$

The loss due to the lateral offset is given by Eq. (5.14) as:

$$T_l = 2.17\left(\frac{y}{\omega}\right)^2 = 2.17\left(\frac{1}{3.12}\right)^2$$

$$= 0.22 \text{ dB}$$

The loss due to angular misalignment may be obtained from Eq. (5.16) where:

$$T_a = 2.17\left(\frac{\theta\omega\ n_1 V}{a\ NA}\right)^2$$

$$= 2.17\left(\frac{(\pi/180)\times 3.12\times 1.46\times 2.4}{4\times 0.1}\right)$$

$$= 0.49\ \text{dB}$$

Hence, the total insertion loss is

$$T_T \simeq T_l + T_a = 0.22 + 0.49$$
$$= 0.71\ \text{dB}$$

In this example the loss due to angular misalignment is significantly larger than that due to lateral misalignment. However, aside from the actual magnitudes of the respective misalignments, the insertion losses incurred are also strongly dependent upon the normalized frequency of the fiber. This is especially the case with angular misalignment at a single-mode fiber joint where insertion losses of less than 0.3 dB may be obtained when the angular misalignment is $1°$ with fibers of appropriate V value. Nevertheless, for low loss single-mode fiber joints it is important that angular alignment is better than $1°$.

The theoretical model developed by Marcuse [Ref. 21] has been utilized by Nemota and Makimoto [Ref. 26] in a derivation of a general equation for determining the coupling loss between single-mode fibers. Their full expression takes account of all the extrinsic factors (lateral, angular and longitudinal misalignments, and Fresnel reflection), as well as the intrinsic factor associated with the connection of fibers with unequal mode-field diameters. Moreover, good agreement with various experimental investigations has been obtained using this generalized equation [Ref. 27]. Although consideration of the full expression is beyond the scope of this text, a reduced equation, to allow calculation of the intrinsic factor which quite commonly occurs in the interconnection of single-mode fibers, may be employed. Hence, assuming that no losses are present due to the extrinsic factors, the intrinsic coupling loss is given by [Ref. 27]:

$$Loss_{\text{int}} = -10\log_{10}\left[4\left(\frac{\omega_{02}}{\omega_{01}} + \frac{\omega_{01}}{\omega_{02}}\right)^{-2}\right]\ (\text{dB}) \quad (5.17)$$

where ω_{01} and ω_{02} are the spot sizes of the transmitting and receiving fibers respectively. Equation (5.17) therefore enables the additional coupling loss resulting from mode-field diameter mismatch between two single-mode fibers to be calculated.

Example 5.6

Two single-mode fibers with mode-field diameters of 9.2 μm and 8.4 μm are to be connected together. Assuming no extrinsic losses, determine the loss at the connection due to the mode-field diameter mismatch.

Solution: The intrinsic loss is obtained using Eq. (5.17) where:

$$Loss_{\text{int}} = -10 \log_{10}\left[4\left(\frac{\omega_{02}}{\omega_{01}} + \frac{\omega_{01}}{\omega_{02}}\right)^{-2}\right]$$

$$= -10 \log_{10}\left[4\left(\frac{4.2}{5.6} + \frac{5.6}{4.2}\right)^{-2}\right]$$

$$= -10 \log_{10} 0.922$$

$$= 0.35 \text{ dB}$$

It should be noted from Example 5.6 that the same result is obtained irrespective of which fiber is transmitting or receiving through the connection. Hence, by contrast to the situation with multimode fibers (see Section 5.2.1), the intrinsic loss through a single-mode fiber joint is independent of the direction of propagation.

We have considered in some detail the optical attenuation at fiber–fiber connections. However, we have not yet discussed the possible distortion of the transmitted signal at a fiber joint. Although work in this area is in its infancy, increased interest has been generated with the use of highly coherent sources (injection lasers) and very low dispersion fibers. It is apparent that fiber connections strongly affect the signal transmission causing modal noise (see Section 3.10.3) and nonlinear distortion [Ref. 29] when a coherent light source is utilized with a multimode fiber. Also, it has been reported [Ref. 30] that the transmission loss of a connection in a coherent multimode system is extremely wavelength dependent, exhibiting a possible 10% change in the transmitted optical wavelength for a very small change (0.001 nm) in the laser emission wavelength. Although it has been found that these problems may be reduced by the use of single-mode optical fiber [Ref. 29], a theoretical model for the wavelength dependence of joint losses in single-mode fiber has been obtained [Ref. 31]. This model predicts that as the wavelength increases then the width of the fundamental mode field increases and hence for a given lateral offset or angular tilt the joint loss decreases. For example, the lateral offset loss at a wavelength of 1.5 μm was calculated to be only around 80% of the loss at a wavelength of 1.3 μm.

Furthermore, the above modal effects become negligible when an incoherent source (light emitting diode) is used with multimode fiber. However, in this instance there is often some mode conversion at the fiber joint which can make the connection effectively act as a mode mixer or filter [Ref. 32]. Indications are that this phenomenon, which has been investigated [Ref. 33] with regard to fiber splices, is more pronounced with fusion splices than with mechanical splices, both of which are described in Section 5.3.

5.3 Fiber splices

A permanent joint formed between two individual optical fibers in the field or factory is known as a fiber splice. Fiber splicing is frequently used to establish long-haul optical fiber links where smaller fiber lengths need to be joined, and there is no requirement for repeated connection and disconnection. Splices may be divided into two broad categories depending upon the splicing technique utilized. These are fusion splicing or welding and mechanical splicing.

Fusion splicing is accomplished by applying localized heating (e.g. by a flame or an electric arc) at the interface between two butted, prealigned fiber ends causing them to soften and fuse. Mechanical splicing, in which the fibers are held in alignment by some mechanical means, may be achieved by various methods including the use of tubes around the fiber ends (tube splices) or V-grooves into which the butted fibers are placed (groove splices). All these techniques seek to optimize the splice performance (i.e. reduce the insertion loss at the joint) through both fiber end preparation and alignment of the two joint fibers. Typical average splice insertion losses for multimode fibers are in the range 0.1 to 0.2 dB [Ref. 34] which is generally a better performance than that exhibited by demountable connections (see Sections 5.4 and 5.5). It may be noted that the insertion losses of fiber splices are generally much less than the possible Fresnel reflection loss at a butted fiber–fiber joint. This is because there is no large step change in refractive index with the fusion splice as it forms a continuous fiber connection, and some method of index matching (e.g. a fluid) tends to be utilized with mechanical splices. However, fiber splicing (especially fusion splicing) is at present a somewhat difficult process to perform in a field environment and suffers from practical problems in the development of field-usable tools.

A requirement with fibers intended for splicing is that they have smooth and square end faces. In general this end preparation may be achieved using a suitable tool which cleaves the fiber as illustrated in Figure 5.4 [Ref. 35]. This process is

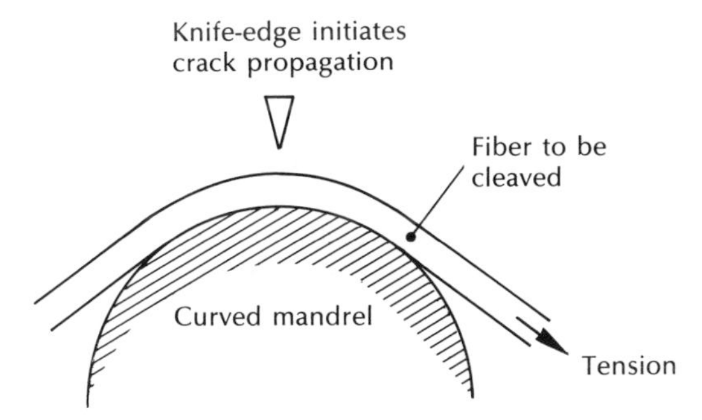

Figure 5.4 Optical fiber end preparation: the principle of scribe and break cutting [Ref. 35].

often referred to as scribe and break or score and break as it involves the scoring of the fiber surface under tension with a cutting tool (e.g. sapphire, diamond, tungsten carbide blade). The surface scoring creates failure as the fiber is tensioned and a clean, reasonably square fiber end can be produced. Figure 5.4 illustrates this process with the fiber tensioned around a curved mandrel. However, straight pull, scribe and break tools are also utilized, which arguably give better results [Ref. 36]. An alternative technique involves circumferential scoring which provides a controlled method of lightly scoring around the fiber circumference [Ref. 31]. In this case the score can be made smooth and uniform and large diameter fibers may be prepared by a simple straight pull with end angles less than 1°.

5.3.1 Fusion splices

The fusion splicing of single fibers involves the heating of the two prepared fiber ends to their fusing point with the application of sufficient axial pressure between the two optical fibers. It is therefore essential that the stripped (of cabling and buffer coating) fiber ends are adequately positioned and aligned in order to achieve good continuity of the transmission medium at the junction point. Hence the fibers are usually positioned and clamped with the aid of an inspection microscope.

Flame heating sources such as microplasma torches (argon and hydrogen) and oxhydric microburners (oxygen, hydrogen and alcohol vapour) have been utilized with some success [Ref. 37]. However, the most widely used heating source is an electric arc. This technique offers advantages of consistent, easily controlled heat with adaptability for use under field conditions. A schematic diagram of the basic arc fusion method is given in Figure 5.5(a) [Refs. 34 and 35] illustrating how the two fibers are welded together. Figure 5.5(b) [Ref. 24] shows a development of the basic arc fusion process which involves the rounding of the fiber ends with a low energy discharge before pressing the fibers together and fusing with a stronger arc. This technique, known as prefusion, removes the requirement for fiber end preparation which has a distinct advantage in the field environment. It has been utilized with multimode fibers giving average splice losses of 0.09 dB [Ref. 39].

Fusion splicing of single-mode fibers with typical core diameters between 5 and 10 μm presents problems of more critical fiber alignment (i.e. lateral offsets of less than 1 μm are required for low loss joints). However, splice insertion losses below 0.3 dB may be achieved due to a self alignment phenomenon which partially compensates for any lateral offset.

Self alignment, illustrated in Figure 5.6 [Refs. 38, 40 and 41], is caused by surface tension effects between the two fiber ends during fusing. An early field trial of single-mode fiber fusion splicing over a 31.6 km link gave mean splice insertion losses of 0.18 and 0.12 dB at wavelengths of 1.3 and 1.55 μm respectively [Ref. 42]. Mean splice losses of only 0.06 dB have also been obtained with a fully automatic single-mode fiber fusion splicing machine [Ref. 43].

A possible drawback with fusion splicing is that the heat necessary to fuse the fibers may weaken the fiber in the vicinity of the splice. It has been found that even

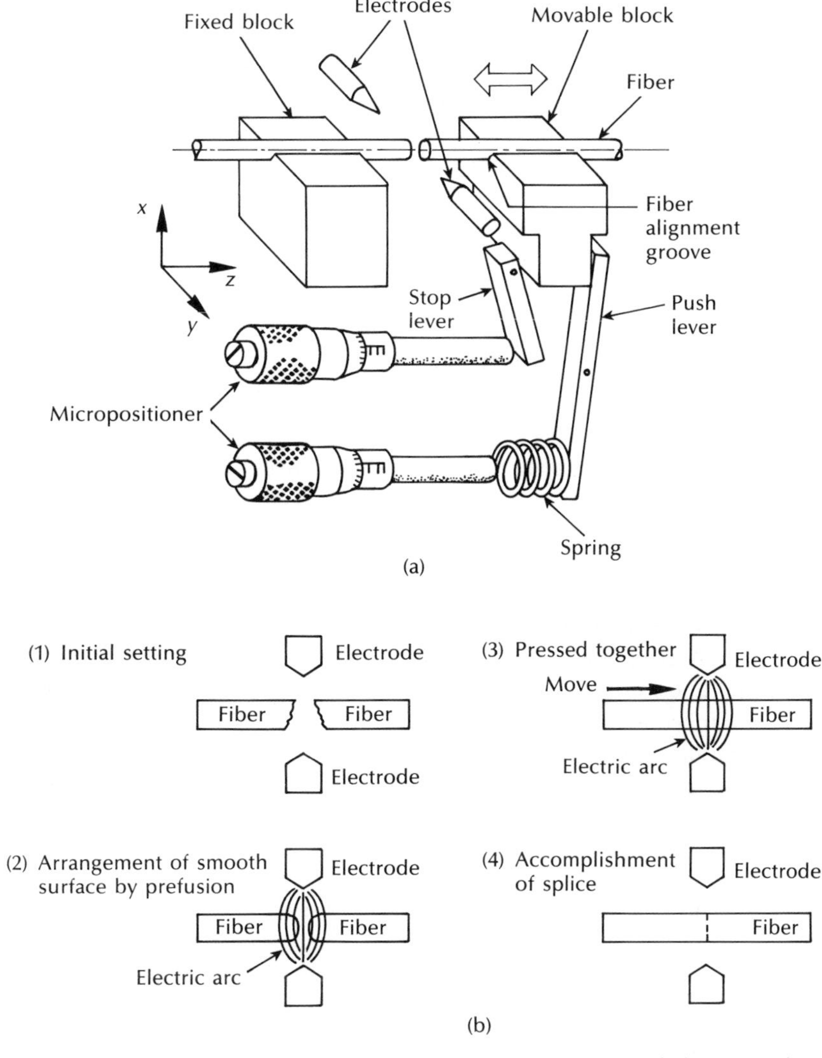

Figure 5.5 Electric arc fusion splicing: (a) an example of fusion splicing apparatus [Refs. 34 and 38]; (b) schematic illustration of the prefusion method for accurately splicing optical fibers [Ref. 24].

with careful handling, the tensile strength of the fused fiber may be as low as 30% of that of the uncoated fiber before fusion [Ref. 45]. The fiber fracture generally occurs in the heat-affected zone adjacent to the fused joint. The reduced tensile strength is attributed [Refs. 45 and 46] to the combined effects of surface damage caused by handling, surface defect growth during heating and induced residential stresses due to changes in chemical composition. It is therefore necessary that the

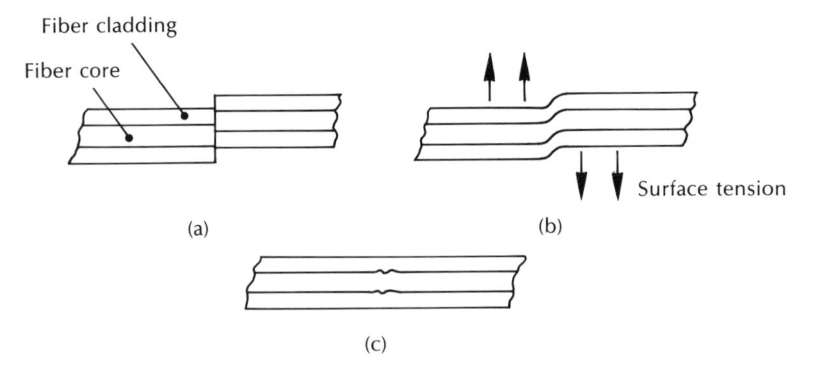

Figure 5.6 Self-alignment phenomenon which takes place during fusion splicing: (a) before fusion; (b) during fusion; (c) after fusion [Refs 38, 40 and 41].

completed splice is packaged so as to reduce tensile loading upon the fiber in the vicinity of the splice.

5.3.2 Mechanical splices

A number of mechanical techniques for splicing individual optical fibers have been developed. A common method involves the use of an accurately produced rigid alignment tube into which the prepared fiber ends are permanently bonded. This snug tube splice is illustrated in Figure 5.7(a) [Ref. 47] and may utilize a glass or ceramic capillary with an inner diameter just large enough to accept the optical fibers. Transparent adhesive (e.g. epoxy resin) is injected through a transverse bore in the capillary to give mechanical sealing and index matching of the splice. Average

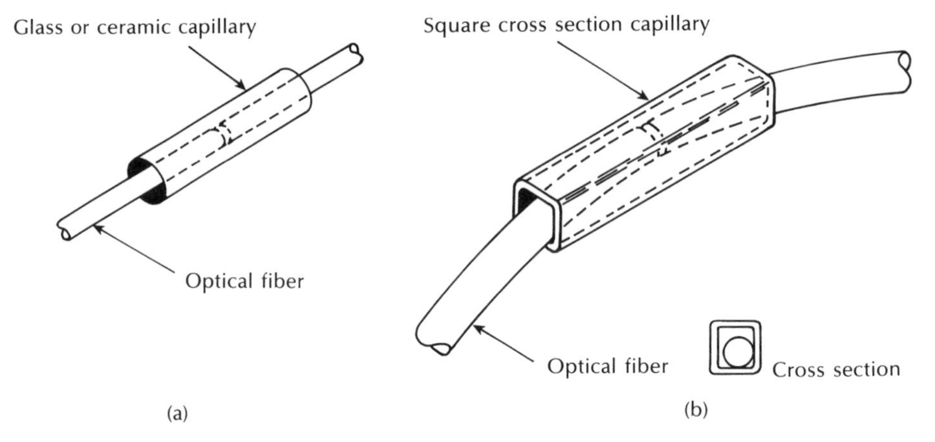

Figure 5.7 Techniques for tube splicing of optical fibers: (a) snug tube splice [Ref. 47]; (b) loose tube splice utilizing square cross section capillary [Ref. 50].

insertion losses as low as 0.1 dB have been obtained [Ref. 48] with multimode graded index and single-mode fibers using ceramic capillaries. However, in general, snug tube splices exhibit problems with capillary tolerance requirements. Hence as a commercial product they may exhibit losses of up to 0.5 dB [Ref. 49].

A mechanical splicing technique which avoids the critical tolerance requirements of the snug tube splice is shown in Figure 5.7(b) [Ref. 50]. This loose tube splice uses an oversized square section metal tube which easily accepts the prepared fiber ends. Transparent adhesive is first inserted into the tube followed by the fibers. The splice is self aligning when the fibers are curved in the same plane, forcing the fiber ends simultaneously into the same corner of the tube, as indicated in Figure 5.7(b). Mean splice insertion losses of 0.073 dB have been achieved [Refs. 41 and 51] using multimode graded index fibers with the loose tube approach.

Other common mechanical splicing techniques involve the use of grooves to secure the fibers to be jointed. A simple method utilizes a V-groove into which the two prepared fiber ends are pressed. The V-groove splice which is illustrated in Figure 5.8(a) [Ref. 52] gives alignment of the prepared fiber ends through insertion in the groove. The splice is made permanent by securing the fibers in the V-groove with epoxy resin. Jigs for producing V-groove splices have proved quite successful, giving joint insertion losses of around 0.1 dB [Ref. 35].

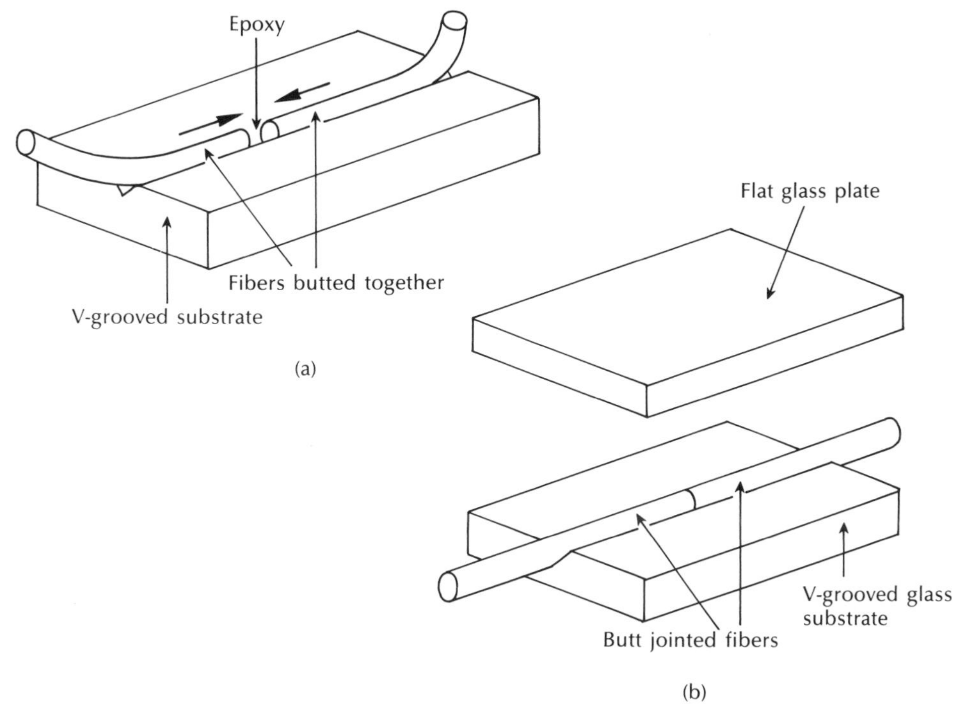

Figure 5.8 V-groove splices [Ref. 52].

V-groove splices formed by sandwiching the butted fiber ends between a V-groove glass substrate and a flat glass retainer plate, as shown in Figure 5.8(b) have also proved very successful in the laboratory. Splice insertion losses of less than 0.01 dB when coupling single-mode fibers have been reported [Ref. 53] using this technique. However, reservations are expressed regarding the field implementation of these splices with respect to manufactured fiber geometry, and housing of the splice in order to avoid additional losses due to local fiber bending.

A further variant on the V-groove technique is the elastic tube or elastomeric splice shown in Figure 5.9 [Ref. 54]. The device comprises two elastomeric internal parts, one of which contains a V-groove. An outer sleeve holds the two elastic parts in compression to ensure alignment of the fibers in the V-groove, and fibers with different diameters tend to be centred and hence may be successfully spliced. Although originally intended for multimode fiber connection, the device has become a widely used commercial product [Ref. 49] which is employed with single-mode fibres, albeit often as a temporary splice for laboratory investigations. The splice loss for the elastic tube device was originally reported as 0.12 dB or less [Ref. 54] but is generally specified as around 0.25 dB for the commercial product [Ref. 49]. In addition, index matching gel is normally employed within the device to improve its performance.

A slightly more complex groove splice known as the Springroove® splice utilizes a bracket containing two cylindrical pins which serve as an alignment guide for the two prepared fiber ends. The cylindrical pin diameter is chosen to allow the fibers to protrude above the cylinders, as shown in Figure 5.10(a) [Ref. 55]. An elastic element (a spring) is used to press the fibers into a groove and maintain the fiber end alignment, as illustrated in Figure 5.10(b). The complete assembly is secured using a drop of epoxy resin. Mean splice insertion losses of 0.05 dB [Ref. 41] have been obtained using multimode graded index fibers with the Springroove® splice. This device has found practical use in Italy.

The aforementioned mechanical splicing methods employ alignment of the bare fibers, whereas more recently alignment of secondary elements around the bare fibers is a technique which has gained favour [Ref. 31]. Secondary alignment generally gives increased ruggedness and provides a structure that can be ground and polished for fiber end preparation. Furthermore, with a good design the fiber coating can be terminated within the secondary element leaving only the fiber end

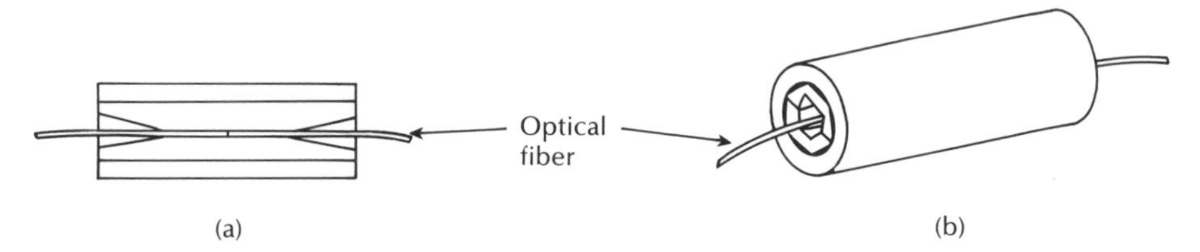

Figure 5.9 The elastomeric splice [Ref. 54]: (a) cross section; (b) assembly.

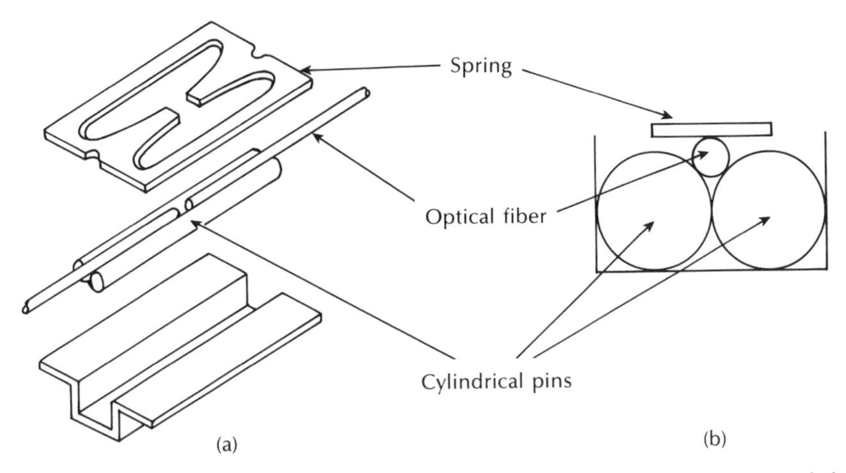

Figure 5.10 The Springroove® splice [Ref. 55]: (a) expanded overview of the splice; (b) schematic cross section of the splice.

face exposed. Hence when the fiber end face is polished flat to the secondary element, a very rugged termination is produced. This technique is particularly advantageous for use in fiber remountable connectors (see Section 5.4). However, possible drawbacks with this method include the time taken to make the termination and the often increased splice losses resulting from the tolerances on the secondary elements which tend to contribute to the fiber misalignment.

An example of a secondary aligned mechanical splice for multimode fiber is shown in Figure 5.11. This device uses precision glass capillary tubes called ferrules as the secondary elements with an alignment sleeve of metal or plastic into which the glass tubed fibers are inserted. Normal assembly of the splice using 50 μm core diameter fiber yields an average loss of around 0.2 dB [Ref. 56].

Finally, the secondary alignment technique has been employed in the realization of a low loss, single-mode fiber mechanical splice which has been used in several large installations in the United States. This device, known as a single-mode rotary splice, is shown in Figure 5.12 [Ref. 57]. The fibers to be spliced are initially terminated in precision glass capillary tubes which are designed to make use of the small eccentricity that is present, as illustrated in Figure 5.12(a). An ultraviolet curable adhesive is used to cement the fibers in the glass tubes and the fiber terminations are prepared with a simple grinding and polishing operation.

Alignment accuracies of the order of 0.05 μm are obtained using the three glass rod alignment sleeve shown in Figure 5.12(b). Such alignment accuracies are necessary to obtain low losses as the mode-field diameter for single-mode fiber is generally in the range 8 to 10 μm. The sleeve has a built-in offset such that when each ferrule is rotated within it, the two circular paths of the centre of each fiber core cross each other. Excellent alignment is obtained utilizing a simple algorithm, and strong metal springs provide positive alignment retention. Using index

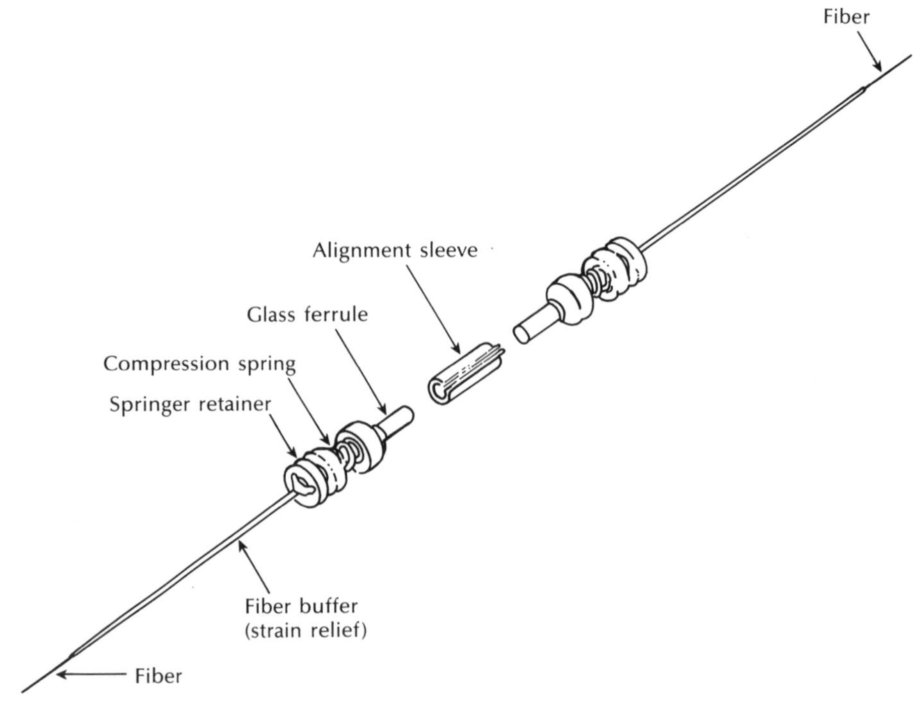

Figure 5.11 Multimode fiber mechanical splice using glass capillary tubes [Ref. 56].

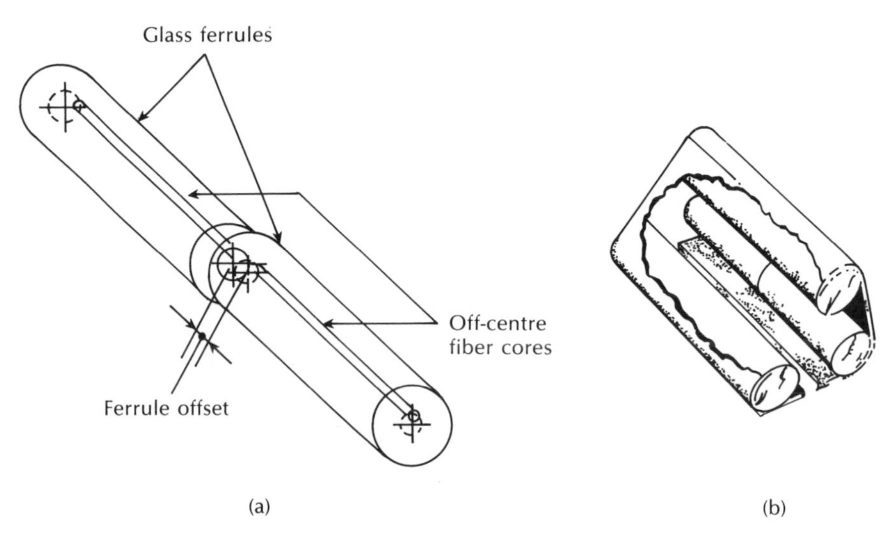

Figure 5.12 Rotary splice for single-mode fibers [Ref. 57]: (a) alignment technique using glass ferrules; (b) glass rod alignment sleeve.

matching gel such splices have demonstrated mean losses of 0.03 dB with a standard deviation of 0.018 dB [Ref. 31]. Moreover, these results were obtained in the field, suggesting that the rotary splicing technique was not affected by the skill level of the splicer in that harsh environment.

5.3.3 Multiple splices

Multiple simultaneous fusion splicing of an array of fibers in a ribbon cable has been demonstrated for both multimode [Ref. 58] and single-mode [Ref. 59] fibers. In both cases a five fiber ribbon was prepared by scoring and breaking prior to pressing the fiber ends on to a contact plate to avoid difficulties with varying gaps between the fibers to be fused. An electric arc fusing device was then employed to provide simultaneous fusion. Such a device is now commercially available to allow the splicing of five fibers simultaneously in a time of around 5 minutes, which compares favourably with the 15 minutes generally required for five single fusion splicings [Ref. 60]. Splice losses using this device with multimode graded index fiber range from an average of 0.04 dB to a maximum of 0.12 dB, whereas for single-mode fiber the average loss is 0.13 dB with a 0.4 dB maximum.

The most common technique employed for multiple simultaneous splicing to date involves mechanical splicing of an array of fibers, usually in a ribbon cable. A

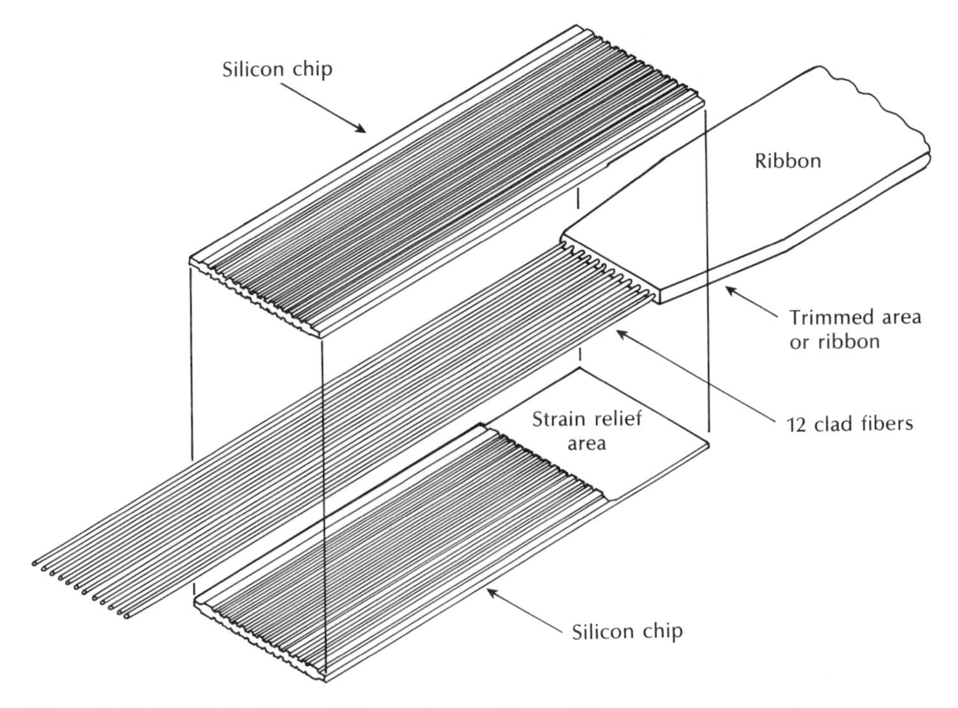

Figure 5.13 Multiple fiber splicing using a silicon chip array.

V-groove multiple splice secondary element comprising etched silicon chips has been used extensively in the United States [Ref. 31] for splicing multimode fibers. In this technique a twelve fiber splice is prepared by stripping the ribbon and coating material from the fibers. Then the twelve fibers are laid into the trapezoidal* grooves of a silicon chip using a comb structure, as shown in Figure 5.13. The top silicon chip is then applied prior to applying epoxy to the chip–ribbon interface. Finally, after curing, the front end face is ground and polished.

The process is normally carried out in the factory and the arrays are clipped together in the field, putting index matching silica gel between the fiber ends. The average splice loss obtained with this technique in the field is 0.12 dB, with the majority of the loss resulting from intrinsic fiber mismatch. Major advantages of this method are the substantial reduction in splicing time (by more than a factor of 10) per fiber and the increased robustness of the final connection. Although early array splicing investigations using silicon chips [Ref. 61] demonstrated the feasibility of connecting 12×12 fiber arrays, in practice only single twelve fiber

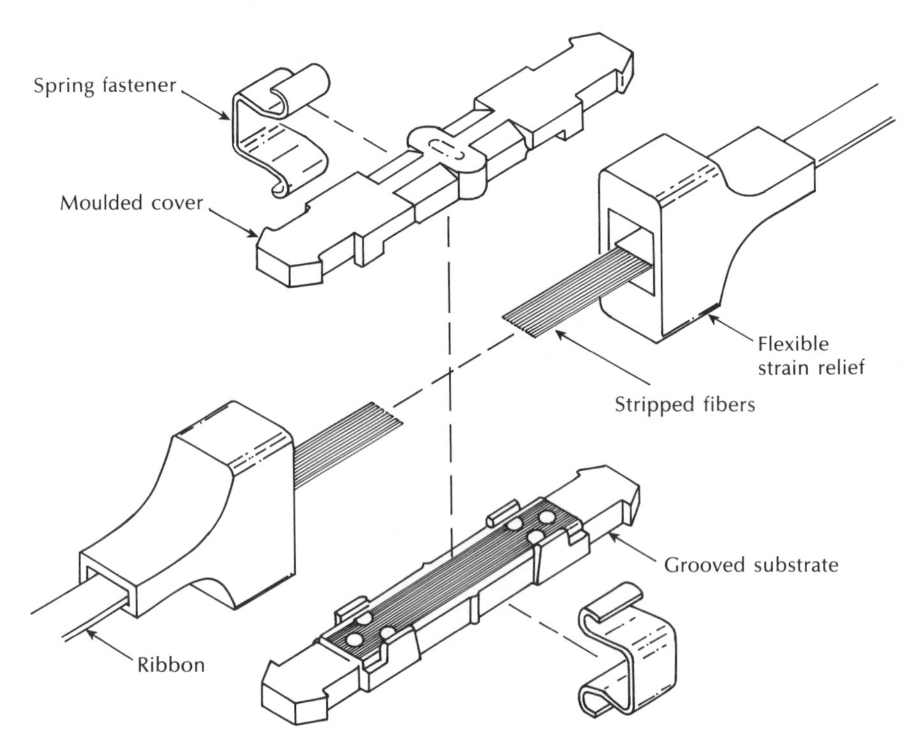

Figure 5.14 V-groove polymer resin ribbon fiber splice [Ref. 63].

* A natural consequence of etching.

ribbons have been spliced at one time due to concerns in relation to splice tolerance and the large number of telecommunication channels which would be present in the two dimensional array [Ref. 31].

An alternative V-groove flat chip moulded from a glass filled polymer resin has been employed in France [Ref. 62], in particular for the Biarritz project (see Section 14.2.3). Moreover, direct mass splicing of twelve fiber ribbons has been accomplished [Ref. 63]. In this technique simultaneous end preparation of all twenty-four fibers is achieved using a ribbon grinding and polishing procedure. The ribbons are then laid in guides and all twelve fibers are positioned in grooves in the glass filled plastic substrate shown in Figure 5.14. A vacuum technique is used to hold the fibers in position whilst the cover plate is applied, and spring clips hold the assembly together. Index matching gel is applied through a hole in the cover plate giving average splice losses of 0.18 dB with multimode fiber.

5.4 Fiber connectors

Demountable fiber connectors are more difficult to achieve than optical fiber splices. This is because they must maintain similar tolerance requirements to splices in order to couple light between fibers efficiently, but they must accomplish it in a removable fashion. Also, the connector design must allow for repeated connection and disconnection without problems of fiber alignment, which may lead to degradation in the performance of the transmission line at the joint. Hence to operate satisfactorily the demountable connector must provide reproducible accurate alignment of the optical fibers.

In order to maintain an optimum performance the connection must also protect the fiber ends from damage which may occur due to handling (connection and disconnection), must be insensitive to environmental factors (e.g. moisture and dust) and must cope with tensile load on the cable. Additionally, the connector should ideally be a low cost component which can be fitted with relative ease. Hence optical fiber connectors may be considered in three major areas, which are:

(a) the fiber termination, which protects and locates the fiber ends;
(b) the fiber end alignment to provide optimum optical coupling;
(c) the outer shell, which maintains the connection and the fiber alignment, protects the fiber ends from the environment and provides adequate strength at the joint.

The use of an index matching material in the connector between the two jointed fibers can assist the connector design in two ways. It increases the light transmission through the connection whilst keeping dust and dirt from between the fibers. However, this design aspect is not always practical with demountable connectors, especially where fluids are concerned. Apart from problems of sealing and replacement when the joint is disconnected and reconnected, liquids in this instance may have a detrimental effect, attracting dust and dirt to the connection.

There are a large number of demountable single fiber connectors, both commercially available and under development, which have insertion losses in the range 0.2 to 3 dB. Fiber connectors may be separated into two broad categories: butt jointed connectors and expanded beam connectors. Butt jointed connectors rely upon alignment of the two prepared fiber ends in close proximity (butted) to each other so that the fiber core axes coincide. Expanded beam connectors utilize interposed optics at the joint (i.e. lenses) in order to expand the beam from the transmitting fiber end before reducing it again to a size compatible with the receiving fiber end.

Butt jointed connectors are the most widely used connector type and a substantial number have been reported. In this section we review some of the more common butt jointed connector designs which have been developed for use with both multimode and single-mode fibers. In Section 5.5, following, expanded beam connectors are discussed.

5.4.1 Cylindrical ferrule connectors

The basic ferrule connector (sometimes referred to as a concentric sleeve connector), which is perhaps the simplest optical fiber connector design, is illustrated in Figure 5.15(a) [Ref. 9]. The two fibers to be connected are permanently bonded (with epoxy resin) in metal plugs known as ferrules which have an accurately drilled central hole in their end faces where the stripped (of buffer coating) fiber is located. Within the connector the two ferrules are placed in an alignment sleeve which, using accurately machined components, allows the fiber ends to be butt jointed. The ferrules are held in place via a retaining mechanism which, in the example shown in Figure 5.15(a), is a spring.

It is essential with this type of connector that the fiber end faces are smooth and square (i.e. perpendicular to the fiber axis). This may be achieved with varying success by either:

(a) cleaving the fiber before insertion into the ferrule;
(b) inserting and bonding before cleaving the fiber close to the ferrule end face;
(c) using either (a) or (b) and polishing the fiber end face until it is flush with the end of the ferrule.

Polishing the fiber end face after insertion and bonding provides the best results but it tends to be time consuming and inconvenient, especially in the field.

The fiber alignment accuracy of the basic ferrule connector is largely dependent upon the ferrule hole into which the fiber is inserted. Hence, some ferrule connectors have incorporated a watch jewel in the ferrule end face (jewelled ferrule connector), as illustrated in Figure 5.15(b) [Ref. 10]. In this case the fiber is centred with respect to the ferrule through the watch jewel hole. The use of the watch jewel allows the close diameter and tolerance requirements of the ferrule end face hole to be obtained more easily than simply through drilling of the metallic ferrule end face alone. Nevertheless, typical concentricity errors between the fiber core and the

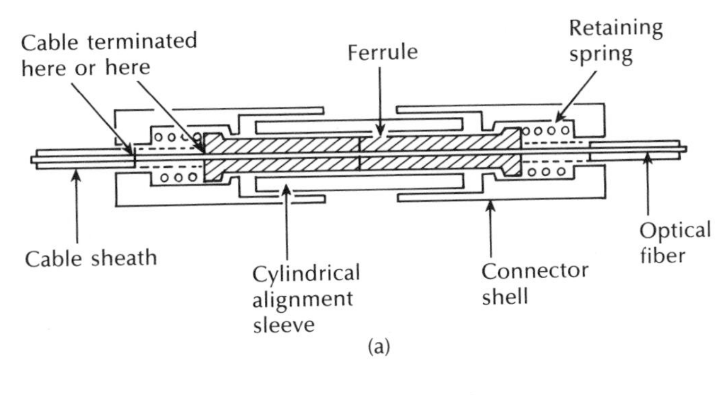

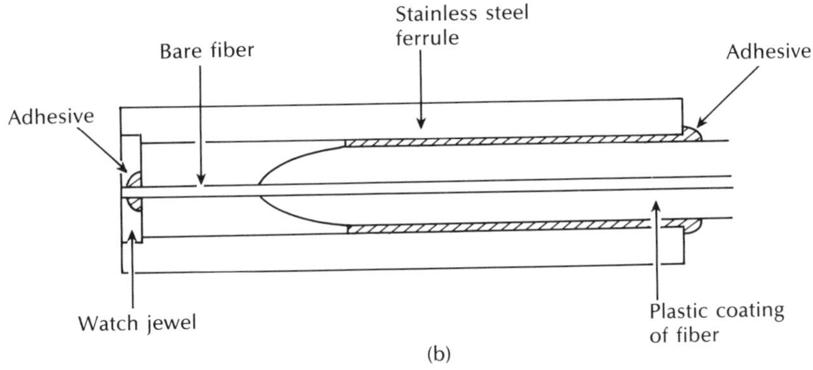

Figure 5.15 Ferrule connectors: (a) structure of a basic ferrule connector [Ref. 9]; (b) structure of a watch jewel connector ferrule [Ref. 10].

outside diameter of the jewelled ferrule are in the range 2 to 6 μm giving insertion losses in the range 1 to 2 dB with multimode step index fibers.

More recently capillary ferrules manufactured from ceramic materials (e.g. alumina porcelain) have found widespread application within precision ferrule connectors. Such capillary ferrules have a precision bore which is accurately centred in the ferrule. Final assembly of the connector includes the fixture of the fiber within the ferrule, using adhesive prior to the grinding and polishing for end preparation. The ceramic materials possess outstanding thermal, mechanical and chemical resistance characteristics in comparison to metals and plastics [Ref. 64]. In addition, unlike metal and plastic components, the ceramic ferrule material is harder than the optical fiber and is therefore unaffected by the grinding and polishing process, a factor which assists in the production of low loss fiber connectors. Typical average losses for multimode graded index fiber (i.e. core/cladding: 50/125 μm) and single-mode fiber (i.e. core/cladding: 9/125 μm) with the precision ceramic ferrule connector are 0.2 and 0.3 dB respectively [Ref. 60].

Numerous cylindrical sleeve ferrule connectors are commercially available for both multimode and single-mode fiber termination. The most common design types

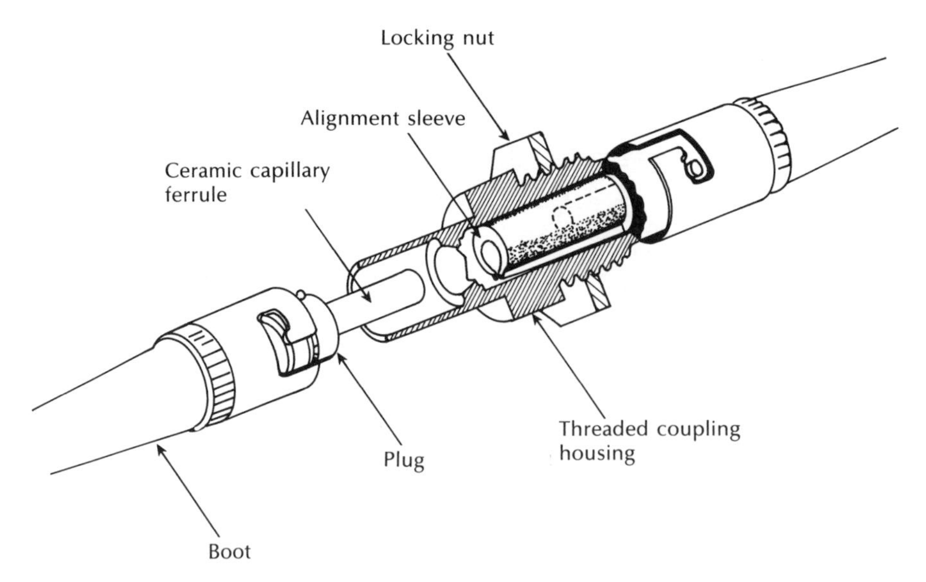

Figure 5.16 ST series multimode fiber connector using ceramic capillary ferrules.

are the straight tip (ST), the subminiature assembly (SMA), the fiber connector (FC), the physical contact (PC)*, the subscriber connector (SC) and the D3/D4 [Refs. 27, 31, 60, 65, 66]. An example of an ST series multimode fiber connector is shown in Figure 5.16, which exhibits an optimized cylindrical sleeve with a cross section designed to expand uniformly when the ferrules are inserted. Hence, the constant circumferential pressure provides accurate alignment, even when the ferrule diameters differ slightly. In addition, the straight ceramic ferrule may be observed in Figure 5.16 which contrasts with the stepped ferrule (i.e. a ferrule with a single step which reduces the diameter midway along its length) provided in the SMA connector design.

The average loss obtained using this connector with multimode graded index fiber (i.e. core/cladding: 62.5/125 μm) was 0.22 dB with less than 0.1 dB change in loss after 1000 reconnections [Ref. 65].

5.4.2 Biconical ferrule connectors

A ferrule type connector which is widely used as part of jumper cable in a variety of applications in the United States is the biconical plug connector† [Refs. 34 and 67]. The plugs are either transfer moulded directly on to the fiber or cast around the fiber using a silica-loaded epoxy resin ensuring concentricity to within 5 μm.

* It should be noted that combinations of the types exist including the ST–PC and the FC–PC designs.
† The device is also referred to as the biconic connector.

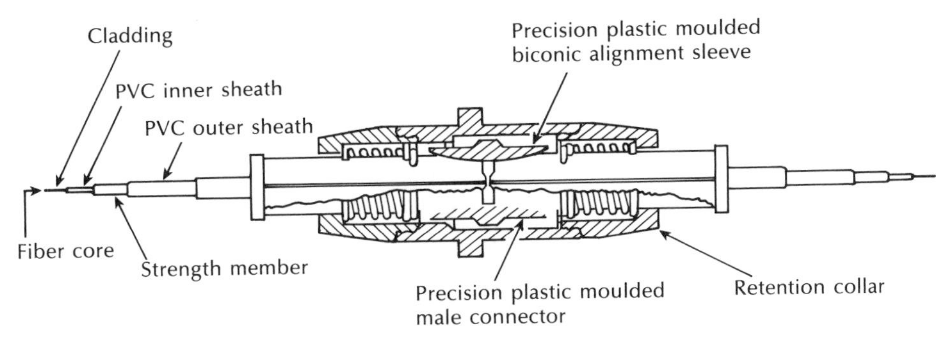

Figure 5.17 Cross section of the biconical connector [Refs. 34 and 67].

After plug attachment, the fiber end faces are polished before the plugs are inserted and aligned in the biconical moulded centre sleeve, as shown in Figure 5.17 [Ref. 34]. The conical ferrule geometry may also be observed in Figure 5.17. Mean insertion losses as low as 0.21 dB have been reported [Ref. 67] when using this connector with 50 μm core diameter graded index fibers. In the original design transparent silicon resin pads were placed over the fiber end faces to provide index matching. However, currently the polished fiber end faces are butted directly, the gap and parallelism of end faces being controlled to a degree that gives insertion losses better than the level normally exhibited by Fresnel reflection. This connector is also used with single-mode fibers by reducing the eccentricity of the tapered cone and also the fiber core eccentricity to 0.33 μm or less, whilst limiting the tilt angle of fibers to 0.35° or less. In this way an average connector loss of 0.28 dB can be obtained using single-mode fibers with a maximum loss of 0.7 dB [Ref. 60].

5.4.3 Double eccentric connector

The double eccentric connector does not rely on a concentric fixed sleeve approach but is an example of an active assembly which is adjustable, allowing close alignment of the fiber axes. The mechanism, which is shown in Figure 5.18 [Refs. 9 and 13], consists of two eccentric cylinders within the outer plug. It may be observed from Figure 5.18 that the optical fiber is mounted eccentrically within the inner cylinder. Therefore, when the two connector halves are mated it is always possible through rotation of the mechanism to make the fiber core axes coincide. This operation is performed on both plugs using either an inspection microscope or a peak optical adjustment. The mechanisms are then locked to give permanent alignment. This connector type has exhibited mean insertion losses of 0.48 dB with multimode graded index fibers: use of index matching fluid within the connector has reduced these losses to 0.2 dB. The double eccentric connector design has also been utilized with single-mode fibers where its adjustable nature has proved advantageous for alignment of the small core diameter fibers giving losses of 0.46 dB without index matching [Ref. 31].

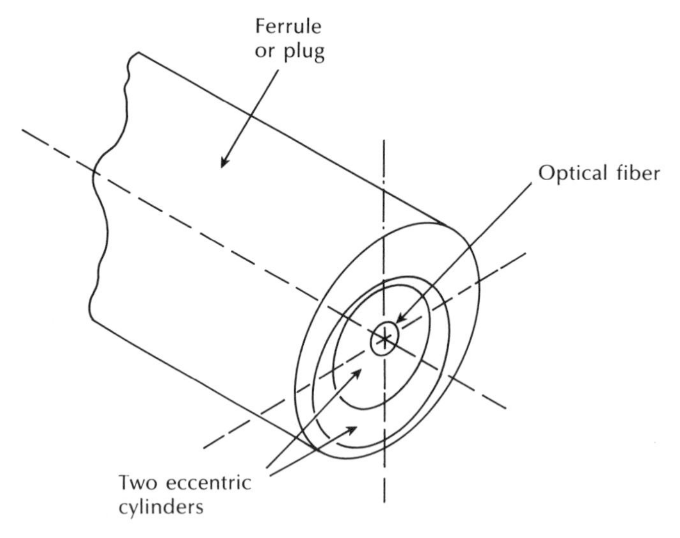

Figure 5.18 Structure of the double eccentric connector plug [Refs. 9 and 13].

5.4.4 Duplex and multiple fiber connectors

A number of duplex fiber connector designs have been developed in order to provide two way communication, but none have found widespread use to date [Ref. 27]. For example, AT & T have a duplex version of the ST single fiber connector (see Section 5.4.1). Moreover, the media interface connector plug shown in Figure 5.19 is part of a duplex fiber connector which has been developed to meet the American National Standards Institute (ANSI) specification for use within

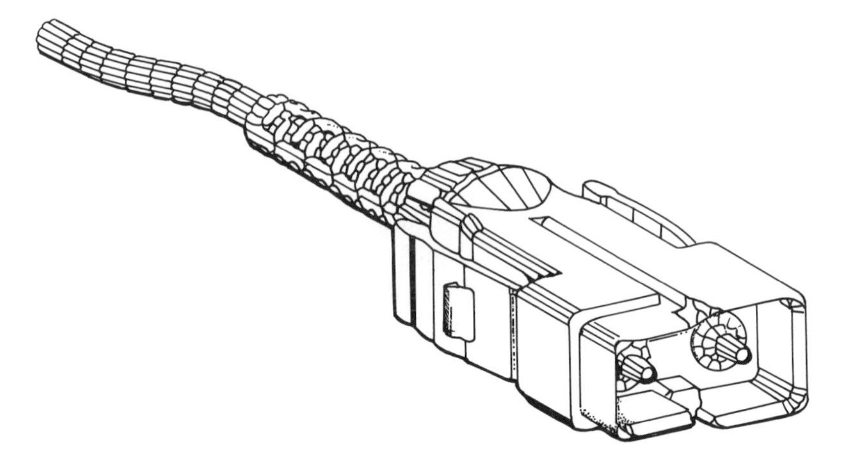

Figure 5.19 An example media interface plug for a duplex fiber connector.

optical fiber local area networks (LANs) [Ref. 68]. This connector plug will mate directly with connectorized optical LAN components (i.e. transmitters and receivers). A duplex fiber connector for use with the Fiber Distributed Data Interface (see Section 14.7.1) is now commercially available. It comprises two ST ferrules housed in a protective moulded shroud and exhibits a typical insertion loss of 0.6 dB. Hence, it is clear that such duplex connectors will be employed more extensively in the future.

Multiple fiber connection is obviously advantageous when interconnecting a large number of fibers. Both cylindrical and biconical ferrule connectors (see Sections 5.4.1 and 5.4.2) can be assembled in housings to form multiple fiber configurations

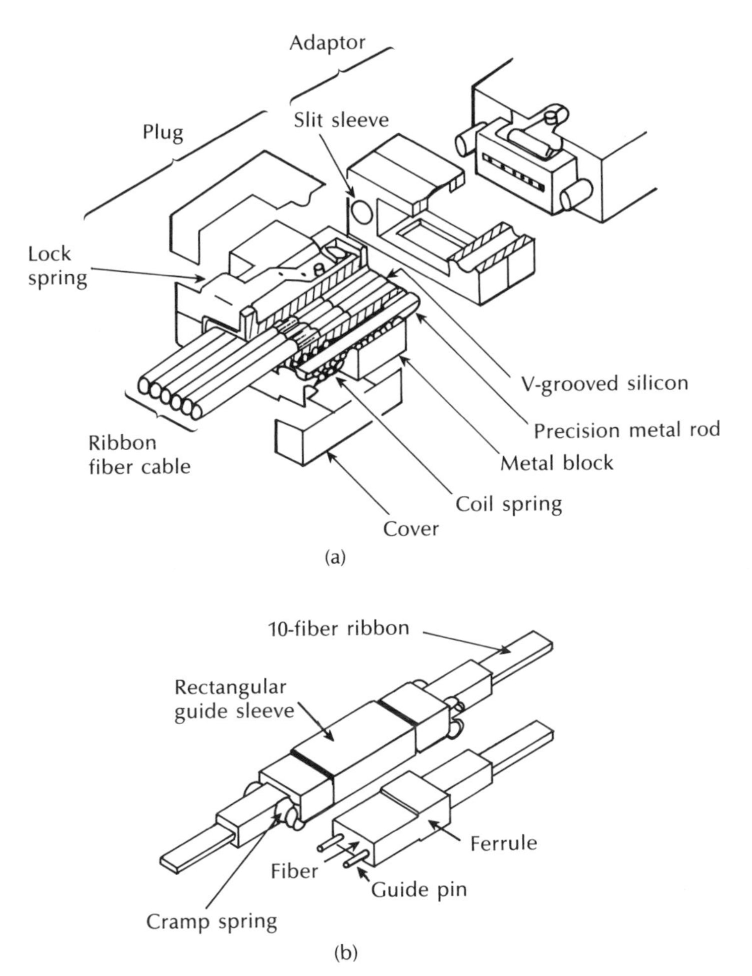

Figure 5.20 Multiple fiber connectors: (a) fiber ribbon connector using V-grooved silicon chips [Ref. 70]; (b) single-mode ten fiber connector [Ref. 73].

[Ref. 31]. Single ferrule connectors generally allow the alignment sleeve to float within the housing, thus removing any requirement for high tolerance on ferrule positioning within multiple ferrule versions. However, the force needed to insert multiple cylindrical ferrules can be large when many ferrules are involved. In this case multiple biconical ferrule connectors prove advantageous due to the low insertion force of the biconic configuration.

In addition to assembling a number of single fiber connectors to form a multiple fiber connector, other examples of multiple fiber connector exist in the literature. Silicon chip arrays have been suggested for the jointing of fiber ribbon cable for many years [Ref. 69]. However, difficulties were experienced in the design of an appropriate coupler for the two arrays. These problems have been largely overcome by the multiple connector design shown in Figure 5.20(a) which utilizes V-grooved silicon chips [Ref. 70]. In this connector, ribbon fibers are mounted and bonded into the V-grooves in order to form a plug together with precision metal guiding rods and coil springs. The fiber connections are then accomplished by butt jointing the two pairs of guiding rods in the slitted sleeves located in the adaptor, also illustrated in Figure 5.20(a). This multiple fiber connector has exhibited average insertion losses of 0.8 dB which were reduced to 0.4 dB by the use of index matching fluid. Improved loss characteristics were obtained with a more recent five fiber moulded connector, also used with fiber ribbons [Ref. 71]. In this case the mean loss and standard deviation without index matching were only 0.45 dB and 0.12 dB, respectively, when terminating 50 μm core multimode fibers [Ref. 72].

The structure of a small plastic moulded single-mode ten fiber connector is shown in Figure 5.20(b) [Ref. 73]. It comprises two moulded ferrules with ten fiber ribbon cables which are accurately aligned by guide pins, then held stable with a rectangular guide sleeve and a cramp spring. This compact multifiber connector which has dimensions of only 6×4 mm exhibited an average connection loss of 0.43 dB when used with single-mode fibers having a spot size (ω_0) of 5 μm.

5.5 Expanded beam connectors

An alternative to connection via direct butt joints between optical fibers is offered by the principle of the expanded beam. Fiber connection utilizing this principle is illustrated in Figure 5.21, which shows a connector consisting of two lenses for collimating and refocusing the light from one fiber into the other. The use of this interposed optics makes the achievement of lateral alignment much less critical than with a butt jointed fiber connector. Also, the longitudinal separation between the two mated halves of the connector ceases to be critical. However, this is achieved at the expense of more stringent angular alignment. Nevertheless, expanded beam connectors are useful for multifiber connection and edge connection for printed circuit boards where lateral and longitudinal alignment are frequently difficult to achieve.

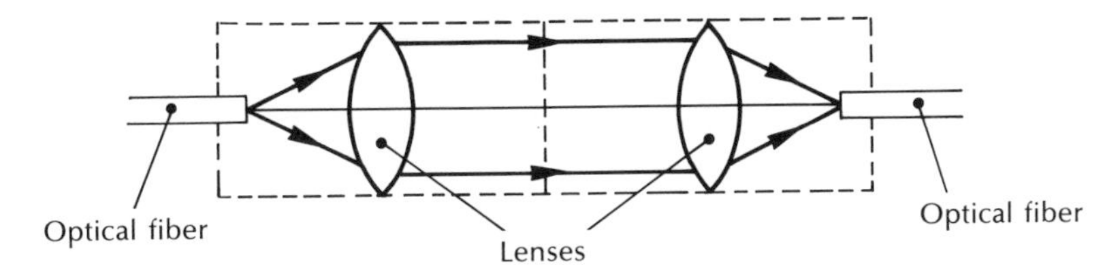

Figure 5.21 Schematic illustration of an expanded beam connector showing the principle of operation.

Two examples of lens coupled expanded beam connectors are illustrated in Figure 5.22. The connector shown in Figure 5.22(a) [Ref. 74] utilized spherical microlenses for beam expansion and reduction. It exhibited average losses of 1 dB which were reduced to 0.7 dB with the application of an antireflection coating on the lenses and the use of 50 μm core diameter graded index fiber.

A similar configuration has been used for single-mode fiber connection in which the lenses have a 2.5 mm diameter [Ref. 75]. Again with antireflection coated

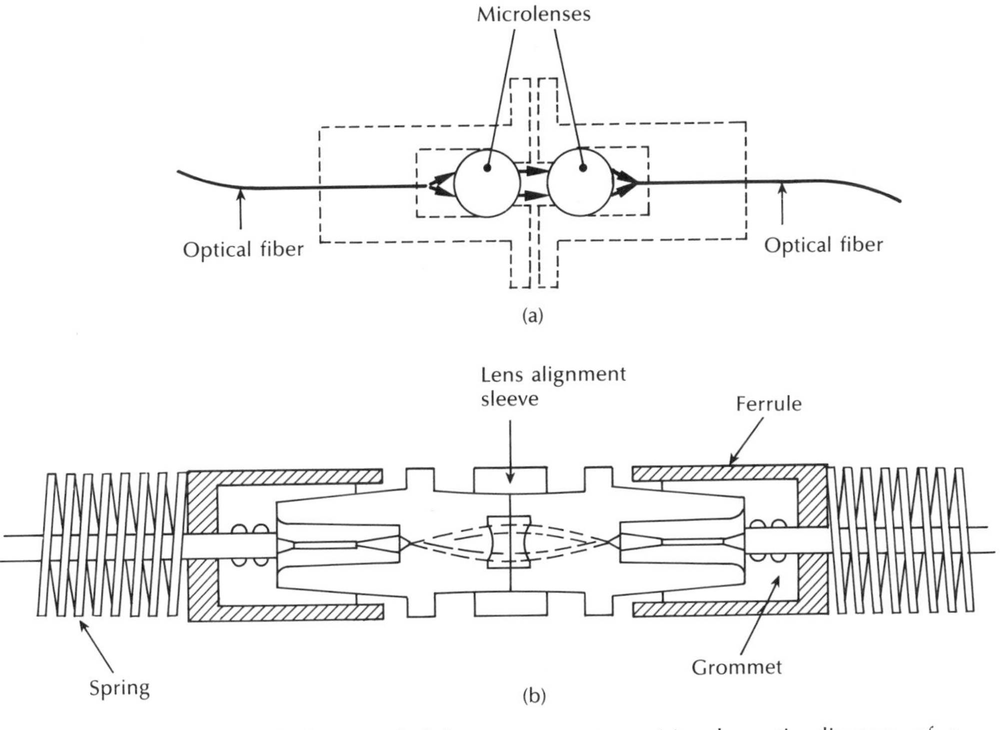

Figure 5.22 Lens coupled expanded beam connectors: (a) schematic diagram of a connector with two microlenses making a 1:1 image of the emitting fiber upon the receiving one [Ref. 74]; (b) moulded plastic lens connector assembly [Ref. 76].

lenses, average losses around 0.7 dB were obtained using 8 μm core diameter single-mode fibers. Furthermore, successful single-mode fiber connection has been achieved with a much smaller (250 μm diameter) sapphire ball lens expanded beam design [Ref. 31]. In this case losses in the range 0.4 to 0.7 dB were demonstrated over 1000 connections.

Figure 5.22(b) shows an expanded beam connector which employs a moulded spherical lens [Ref. 76]. The fiber is positioned approximately at the focal length of the lens in order to obtain a collimated beam and hence minimize lens to lens longitudinal misalignment effects. A lens alignment sleeve is used to minimize the effects of angular misalignment which, together with a ferrule, grommet, spring and external housing, provides the complete connector structure. The repeatability of this relatively straightforward lens design was found to be good, incurring losses of around 0.7 dB.

5.5.1 GRIN-rod lenses

An alternative lens geometry to facilitate efficient beam expansion and collimation within expanded beam connectors is that of the graded index (GRIN) rod lens [Refs. 77, 78]. In addition the focusing properties of such microlens devices have enabled them to find application within both fiber couplers (see Section 5.6) and source to fiber coupling (see Section 6.8).

The GRIN-rod lens which arose from developments on graded index fiber waveguides [Ref. 79] comprises a cylindrical glass rod typically 0.5 to 2 mm in diameter which exhibits a parabolic refractive index profile with a maximum at the axis similar to graded index fiber. Light propagation through the lens is determined by the lens dimensions and, because refractive index is a wavelength dependent parameter, by the wavelength of the light. The GRIN-rod lens can produce a collimated output beam with a divergent angle α of between 1° and 5° from a light source situated on, or near to, the opposite lens face, as illustrated in Figure 5.23. Conversely, it can focus an incoming light beam on to a small area located at the centre of the opposite lens face. Typically, light launched from a 50 μm diameter

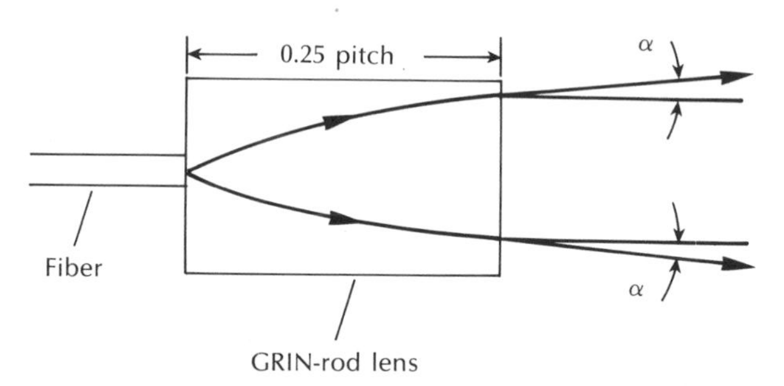

Figure 5.23 Formation of a collimated output beam from a GRIN-rod lens.

fiber core using a GRIN-rod lens results in a collimated output beam of between 0.5 and 1 mm.

Ray propagation through the GRIN-rod lens medium is approximately governed by the paraxial ray equation:

$$\frac{\mathrm{d}^2 r}{\mathrm{d}z^2} = \frac{1}{n}\frac{\mathrm{d}n}{\mathrm{d}r} \tag{5.18}$$

where r is the radial coordinate, z is the distance along the optical axis and n is the refractive index at a point.

Furthermore, the refractive index at r following Eq. (2.75) distance r from the optical axis in a gradient index medium may be expressed as [Ref. 80]:

$$n(r) = n_1\left(1 - \frac{Ar^2}{2}\right) \tag{5.19}$$

where n_1 is the refractive index on the optical axis and A is a positive constant.

Using Eqs. (5.18) and (5.19), the position r of the ray is given by:

$$\frac{\mathrm{d}^2 r}{\mathrm{d}z^2} = -Ar \tag{5.20}$$

Following Miller [Ref. 81], the general solution of Eq. (5.20) becomes:

$$r = K_1 \cos A^{\frac{1}{2}}r + K_2 \sin A^{\frac{1}{2}}r \tag{5.21}$$

where K_1 and K_2 are constants.

The refractive index variation with radius therefore causes all the input rays to follow a sinusoidal path through the lens medium. The traversion of one sinusoidal period is termed one full pitch and GRIN-rod lenses are manufactured with several pitch lengths. Three major pitch lengths are as follows:

1. The quarter pitch (0.25 pitch) lens, which produces a perfectly collimated output beam when the input light emanates from a point source on the opposite lens face. Conversely, the lens focuses an incoming light beam to a point at the centre of the opposite lens face (Figure 5.24(a)). Thus the focal point of the quarter pitch GRIN-rod lens is coincident with the lens faces, thus providing efficient direct butted connection to optical fiber.
2. The 0.23 pitch lens is designed such that its focal point lies outside the lens when a collimated beam is projected on the opposite lens face. It is often employed to convert the diverging beam from a fiber or laser diode into a collimated beam, as illustrated in Figure 5.24(b) [Ref. 82].
3. The 0.29 pitch lens is designed such that both focal points lie just outside the lens end faces. It is frequently used to convert a diverging beam from a laser diode into a converging beam. Hence, it proves useful for coupling the output from a laser diode into an optical fiber (Figure 5.24(c)), or alternatively for coupling the output from an optical fiber into a photodetector.

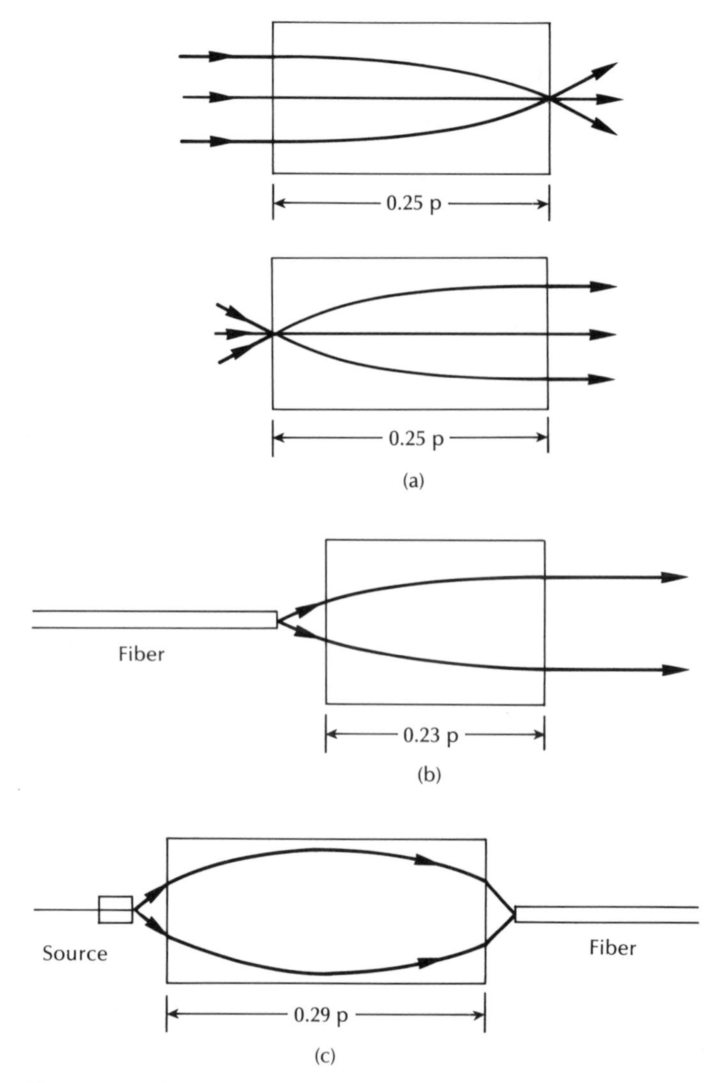

Figure 5.24 Operation of various GRIN-rod lenses: (a) the quarter pitch lens; (b) the 0.23 pitch lens; (c) the 0.29 pitch lens.

The majority of GRIN-rod lenses which have diameters in the range 0.5 and 2 mm may be employed with either single-mode or multimode (step or graded index) fiber. Various fractional pitch lenses, including those above as well as 0.5 p and 0.75 p, may be obtained from Nippon Sheet Glass Co. Ltd under the trade name SELFOC. They are available with numerical apertures of 0.37, 0.46 and 0.6.

A number of factors can cause divergence of the collimated beam from a GRIN-rod lens. These include errors in the lens cut length, the finite size of the fiber core and chromatic aberration. As indicated previously, divergence angles as small as 1° may be obtained which yield expanded beam connector losses of around 1 dB [Ref. 31]. Furthermore, in contrast to butt jointed multimode fiber connectors, GRIN-rod lens connectors have demonstrated loss characteristics which are independent of the modal power distribution in the fiber [Ref. 83].

5.6 Fiber couplers

An optical fiber coupler is a device that distributes light from a main fiber into one or more branch fibers.* The latter case is more normal and such devices are known as multiport fiber couplers. More recently, interest has grown in these devices to divide or combine optical signals for application within optical fiber information distribution systems including data buses, local area networks, computer networks and telecommunication access networks (see Chapter 14).

Optical fiber couplers are often passive devices in which the power transfer takes place either:

(a) through the fiber core cross section by butt jointing the fibers or by using some form of imaging optics between the fibers (core interaction-type); or
(b) through the fiber surface and normal to its axis by converting the guided core modes to both cladding and refracted modes which then enable the power sharing mechanism (surface interaction type).

The mechanisms associated with these two broad categories are illustrated in Figure 5.25.

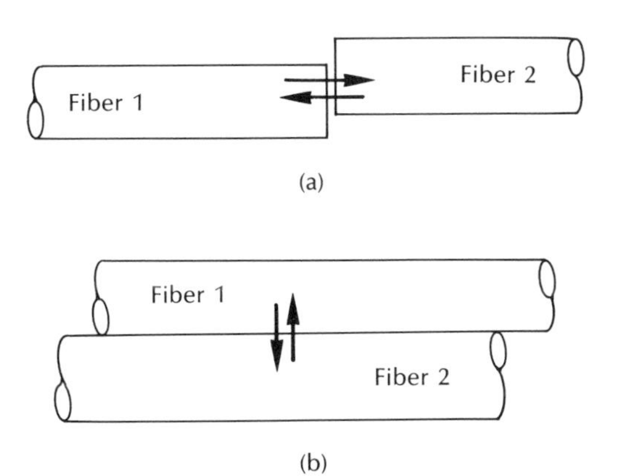

Figure 5.25 Classification of optical fiber couplers: (a) core interaction type; (b) surface interaction type.

* Devices of this type are also referred to as directional couplers.

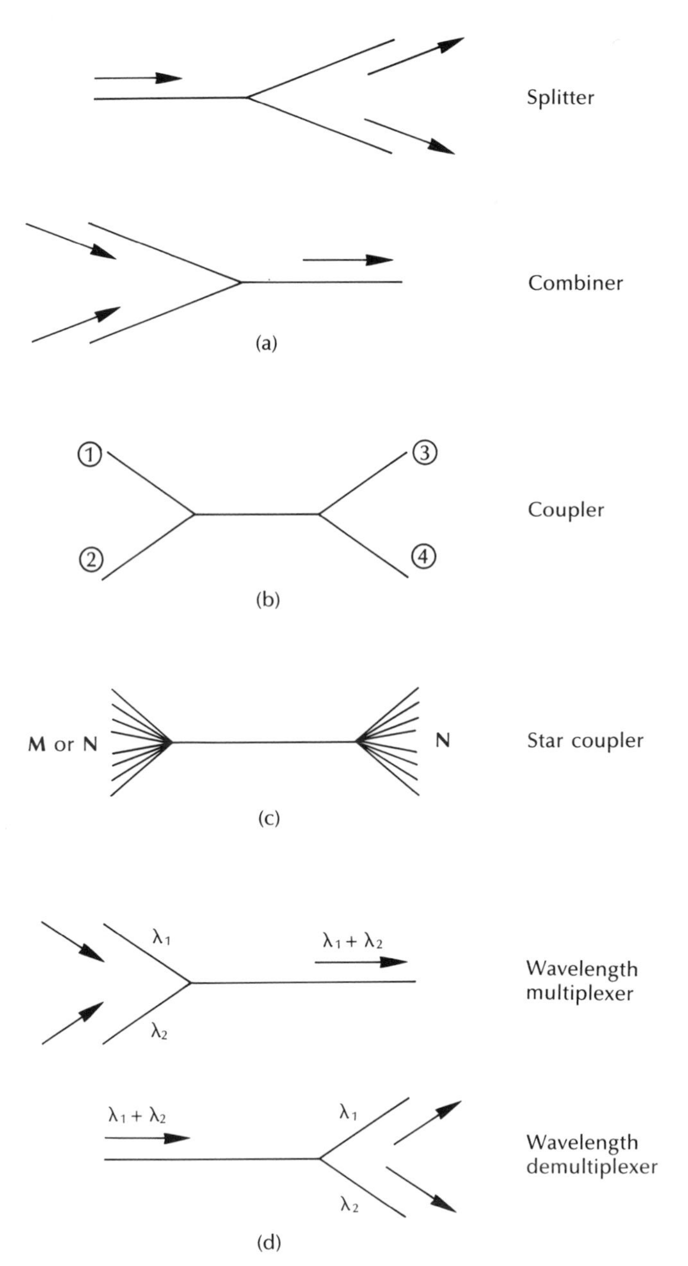

Figure 5.26 Optical fiber coupler types and functions: (a) three port couplers; (b) four port coupler; (c) star coupler; (d) wavelength division multiplexing and demultiplexing couplers.

Active waveguide directional couplers are also available which are realized using integrated optical fabrication techniques. Such device types, however, are dealt with in Section 10.6.1 and thus in this section the discussion is restricted to the above passive coupling strategies.

Multiport optical fiber couplers can also be subdivided into the following three main groups [Ref. 84], as illustrated in Figure 5.26:

1. Three and four port* couplers which are used for signal splitting, distribution and combining.
2. Star couplers which are generally used for distributing a single input signal to multiple outputs.
3. Wavelength division multiplexing (WDM) devices which are a specialized form of coupler designed to permit a number of different peak wavelength optical signals to be transmitted in parallel on a single fiber (see Section 11.9.3). In this context WDM couplers either combine the different wavelength optical signal on to the fiber (i.e. multiplex) or separate the different wavelength optical signals output from the fiber (i.e. demultiplex).

Ideal fiber couplers should distribute light among the branch fibers with no scattering loss† or the generation of noise, and they should function with complete insensitivity to factors including the distribution of light between the fiber modes, as well as the state of polarization of the light. Unfortunately, in practice passive fiber couplers do not display all of the above properties and hence the characteristics of the devices affect the performance of optical fiber networks. In particular, the finite scattering loss at the coupler limits the number of terminals that can be connected, or alternatively the span of the network, whereas the generation of noise and modal effects can cause problems in the specification of the network performance. Hence, couplers in a network cannot usually be treated as individual components with known parameters, a factor which necessitates certain compromises in their application. In this section, therefore, a selection of the more common fiber coupler types is described in relation to the coupling mechanisms, their performance and limitations.

5.6.1 Three and four port couplers

Several methods are employed to fabricate three and four port optical fiber couplers [Refs. 84 to 86]. The lateral offset method, illustrated in Figure 5.27(a) relies on the overlapping of the fiber end faces. Light from the input fiber is coupled to the output fibers according to the degree of overlap. Hence the input power can be distributed in a well defined proportion by appropriate control of the amount of lateral offset between the fibers. This technique, which can provide a bidirectional coupling capability, is well suited for use with multimode step index fibers but

* Four port couplers may also be referred to as 2×2 star couplers.

† The scattering loss through the coupler is often referred to as the excess loss.

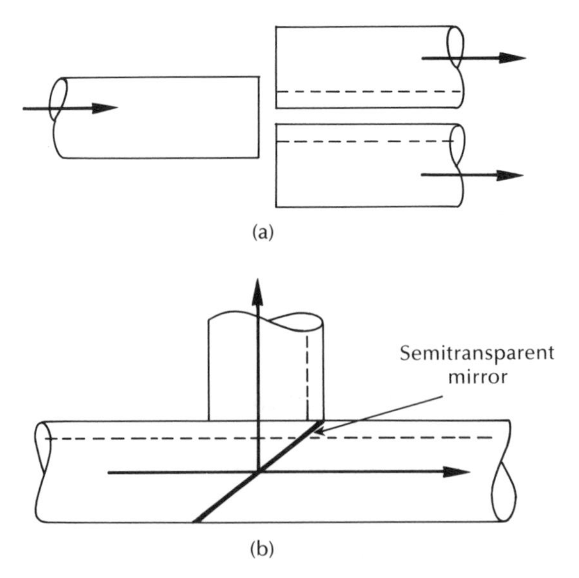

Figure 5.27 Fabrication techniques for three port fiber couplers: (a) the lateral offset method; (b) the semitransparent mirror method.

may incur higher excess losses than other methods as all the input light cannot be coupled into the output fibers.

Another coupling technique is to incorporate a beam splitter element between the fibers. The semitransparent mirror method provides an ingenious way to accomplish such a fiber coupler, as shown in Figure 5.27(b). A partially reflecting surface can be applied directly to the fiber end face cut at an angle of 45° to form a thin film beam splitter. The input power may be split in any desired ratio between the reflected and transmitted beams depending upon the properties of the intervening mirror, and typical excess losses for the device lie in the range 1 to 2 dB. Using this technology both three and four port couplers with both multimode and single-mode fibers have been fabricated [Ref. 86]. In addition, with suitable wavelength selective interference coatings this coupler type can form a wavelength division multiplexing device (see Section 5.6.3).

A fast growing category of optical fiber coupler is based on the use of microoptic components. In particular, a complete range of couplers has been developed which utilize the beam expansion and collimation properties of the graded index (GRIN) rod lens (see Section 5.5.1) combined with spherical retro-reflecting mirrors [Ref. 87]. These devices, two of which are displayed in Figure 5.28, are miniature optical assemblies of compact construction which generally exhibit low insertion loss (typically less than 1 dB) and are insensitive to modal power distribution.

Figure 5.28(a) shows the structure of a parallel surface type GRIN-rod lens three port coupler which comprises two quarter pitch lenses with a semitransparent mirror in between. Light rays from the input fiber F_1, collimate in the first lens

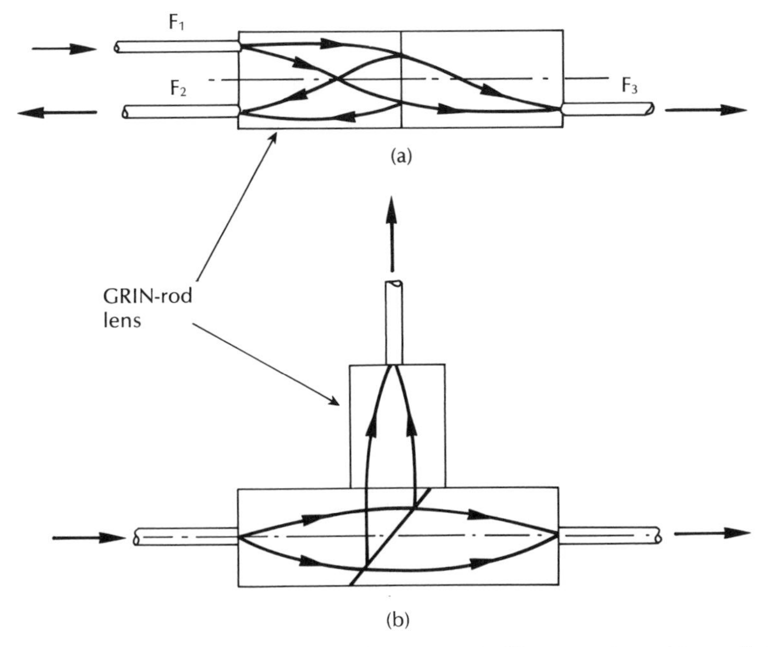

Figure 5.28 GRIN-rod lens based microoptic fiber couplers: (a) parallel surface type; (b) slant surface type.

before they are incident on the mirror. A portion of the incident beam is reflected back and is coupled to fiber F_2, whilst the transmitted light is focused in the second lens and then coupled to fiber F_3. The slant surface version of the similar coupler is shown in Figure 5.28(b). The parallel surface type, however, is the most attractive due to its ease of fabrication, compactness, simplicity and relatively low insertion loss. Finally, the substitution of the mirror by an interference filter* offers application of these devices to WDM (see Section 5.6.3).

Perhaps the most common method for manufacturing couplers is the fused biconical taper (FBT) technique, the basic structure and principle of operation of which is illustrated in Figure 5.29. In this method the fibers are generally twisted together and then spot fused under tension such that the fused section is elongated to form a biconical taper structure. A three port coupler is formed by removing one of the input fibers. Optical power launched into the input fiber propagates in the form of guided core modes. The higher order modes, however, leave the fiber core because of its reduced size in the tapered-down region and are therefore guided as cladding modes. These modes transfer back to guided core modes in the tapered-up region of the output fiber with an approximately even distribution between the two fibers.

* Such a dichroic device transmits only a certain wavelength band and reflects all other shorter or longer wavelengths.

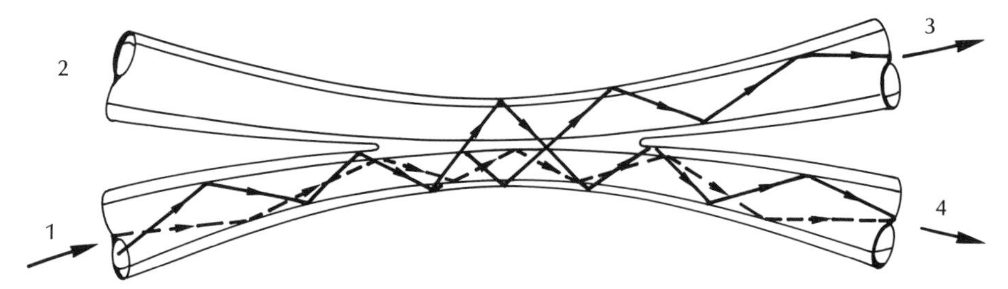

Figure 5.29 Structure and principle of operation for the fiber fused biconical taper coupler.

Often only a portion of the total power is coupled between the two fibers because only the higher order modes take part in the process, the lower order modes generally remaining within the main fiber. In this case a mode dependent (and therefore wavelength dependent) coupling ratio is obtained. However, when the waist of the taper is made sufficiently narrow, then the entire mode volume can be encouraged to participate in the coupling process and a larger proportion of input power can be shared between the output fibers. This strategy gives an improvement in both the power and modal uniformity of the coupler.

The various loss parameters associated with four port couplers may be written down with reference to Figure 5.29. Hence, the excess loss which is defined as the ratio of power input to power output is given by:

$$\textit{Excess loss} \text{ (four port coupler)} = 10 \log_{10} \frac{P_1}{(P_3 + P_4)} \text{ (dB)} \tag{5.22}$$

The insertion loss, however, is generally defined as the loss obtained for a particular port to port optical path.* Therefore, considering Figure 5.29:

$$\textit{Insertion loss} \text{ (ports 1 to 4)} = 10 \log_{10} \frac{P_1}{P_4} \text{ (dB)} \tag{5.23}$$

The crosstalk which provides a measure of the directional isolation† achieved by the device is the ratio of the backscattered power received at the second input port to the input power which may be written as:

$$\textit{Crosstalk} \text{ (four port coupler)} = 10 \log_{10} \frac{P_2}{P_1} \text{ (dB)} \tag{5.24}$$

* It should be noted that there is some confusion in the literature between coupler insertion loss and excess loss. Insertion loss is sometimes referred to when the value quoted is actually the excess loss. However, the author has not noticed the opposite where excess loss is used in place of insertion loss.

† The directional isolation and the crosstalk associated with a coupler are the same value in decibels but the former parameter is normally given as a positive value whereas the latter is negative value. Sometimes the directional isolation is referred to as the insertion loss between the two particular ports of the coupler which would be ports 1 to 2 in Figure 5.29.

Finally, the splitting or coupling ratio indicates the percentage division of optical power between the output ports. Again referring to Figure 5.29:

$$\textit{Split ratio} = \left[\frac{P_3}{(P_3 + P_4)}\right] \times 100\% \tag{5.25}$$

$$= \left[1 - \frac{P_4}{(P_3 + P_4)}\right] \times 100\% \tag{5.26}$$

Example 5.7

A four port multimode fiber FBT coupler has 60 μW optical power launched into port 1. The measured output powers at ports 2, 3 and 4 are 0.004, 26.0 and 27.5 μW respectively. Determine the excess loss, the insertion losses between the input and output ports, the crosstalk and the split ratio for the device.

Solution: The excess loss for the coupler may be obtained from Eq. (5.22) where:

$$\textit{Excess loss} = 10 \log_{10} \frac{P_1}{(P_3 + P_4)} = 10 \log_{10} \frac{60}{53.5}$$

$$= 0.5 \text{ dB}$$

The insertion loss is provided by Eq. (5.23) as

$$\textit{Insertion loss} \text{ (ports 1 to 3)} = 10 \log_{10} \frac{P_1}{P_3} = 10 \log_{10} \frac{60}{26}$$

$$= 3.63 \text{ dB}$$

$$\textit{Insertion loss} \text{ (ports 1 to 4)} = 10 \log_{10} \frac{60}{27.5} = 3.39 \text{ dB}$$

Crosstalk is given by Eq. (5.24) where:

$$\textit{Crosstalk} = 10 \log_{10} \frac{P_2}{P_1} = 10 \log_{10} \frac{0.004}{60}$$

$$= -41.8 \text{ dB}$$

Finally, the split ratio can be obtained from Eq. (5.25) as:

$$\textit{Split ratio} = \left[\frac{P_3}{P_3 + P_4} \times 100\right] = \frac{26}{53.5} \times 100$$

$$= 48.6\%$$

The split ratio for the FBT coupler is determined by the difference in the relative cross sections of the fibers, and the mode coupling mechanism is observed in both multimode and single-mode fibers [Ref. 85]. An advantage of the FBT structure is its relatively low excess loss which is typically less than 0.5 dB,* with low crosstalk being usually better than – 50 dB. A further advantage is the capability to fabricate FBT couplers with almost any fiber and geometry. Hence, they can be tailored to meet the specific requirements of a system or network. A major disadvantage, however, concerns the modal basis of the coupling action. The mode dependent splitting can result in differing losses through the coupler, a wavelength dependent performance, as well as the generation of modal noise when coherent light sources are employed [Ref. 88].

The precise spectral behaviour of FBT couplers is quite complex. It depends upon the dimensions and the geometry of the fused cross section, and on whether the fusing process produces a coupling region where the two cores are close (strongly fused) or relatively far apart (weakly fused) [Ref. 89]. It can also depend upon the refractive index of the surrounding medium [Ref. 90] and, in coherent systems, on the state of polarization of the optical field. Theoretical considerations [Ref. 89] show that for a single-mode FBT coupler, a minimum wavelength dependence on the splitting ratio is achieved for small cladding radii and strong fusing (i.e. the fiber cores placed close together). In order to obtain such performance it is necessary to taper the fibers down to a radius of around 15 μm or less, and to ensure that the rate of taper is such that the major proportion of the coupling occurs in the neck region. The wavelength dependent behaviour associated with single-mode FBT couplers follows an approximately sinusoidal pattern over the wavelength range 0.8 μm to 1.5 μm as a result of the single-mode coupling length between the two fibers [Ref. 91]. This mechanism has been used in the manufacture of WDM multiplexer/demultiplexer couplers (see Section 5.6.3).

More recently, single-mode fiber couplers have been fabricated from polarization maintaining fiber (so-called hi-birefringence couplers) which preserve the polarization of the input signals (see Section 3.13.2). Moreover, using polarization maintaining fiber, it is possible to fabricate polarization sensitive couplers, which effectively function as polarizing beamsplitters [Ref. 92].

An alternative technology to either fiber joint couplers, microoptic lensed devices or fused fiber couplers is the optical waveguide coupler. Corning have demonstrated [Ref. 93] the way in which such passive optical waveguide coupling components compatible with both multimode and single-mode fiber can be fabricated. Their production involves two basic processes. Firstly, a mask of the desired branching function is deposited on to a glass substrate using a photolithographic process. The substrate is then subjected to a two stage ion exchange [Ref. 94], which creates virtually circular waveguides embedded within the surface of the substrate on which the mask was deposited. An example of a three port integrated waveguide coupler fabricated using the above process is shown schematically in Figure 5.30. Multimode fibers are bonded to the structure using etched V-grooves. Excess

* Environmentally stable single-mode fused fiber couplers with excess losses less than 0.1 dB are commercially available.

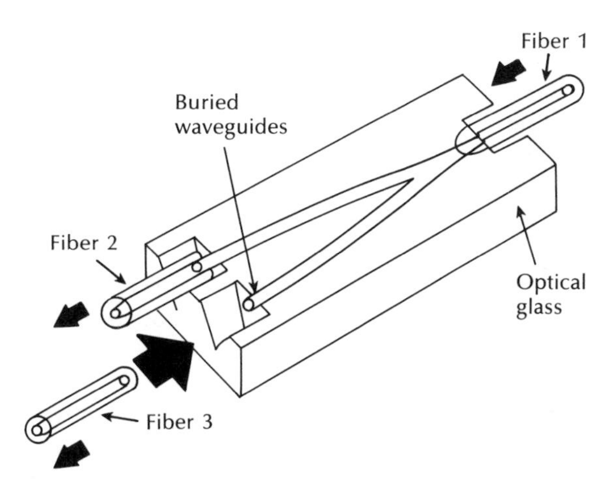

Figure 5.30 The Corning™ multimode fiber integrated waveguide three port coupler [Ref. 93].

losses were measured at 0.5 dB for the three port coupler and at 0.8 dB for the 1×8 star coupler [Ref. 93]. Clearly, this type of waveguide coupler is attractive because of the flexibility it allows at the masking stage. Furthermore, the same technique has been employed to fabricate WDM multiplexing and demultiplexing devices (see Section 5.6.3).

Finally, directional couplers have been produced which use the mode coupling that takes place between the guided and radiation modes when a periodic deformation is applied to the fiber. The principle of operation for this microbend* type coupler is illustrated in Figure 5.31 [Ref. 86]. Mode coupling between the guided and radiation modes may be obtained by pressing the fiber in close contact with a transparent mechanical grating. The radiated optical power can be collected by a lens or a shaped, curved glass plate. Interesting features of such devices are their variable coupling ratios which may be controlled over a wide range by altering the pressure on the fiber. In this context low light levels can be extracted from the fiber with very little excess loss (e.g. estimated at 0.05 dB [Ref. 95]).

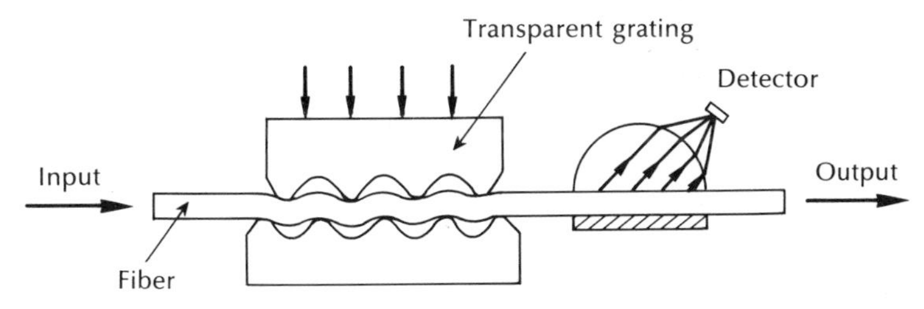

Figure 5.31 Schematic diagram of a microbend type coupler.

* This coupler operates in a similar manner to the microbend sensor described in Section 14.5.2.

5.6.2 Star couplers

Star couplers distribute an optical signal from a single input fiber to multiple output fibers, as may be observed in Figure 5.26. The two principal manufacturing techniques for producing multimode fiber star couplers are the mixer-rod and the fused biconical taper (FBT) methods. In the mixer-rod method illustrated in Figure 5.32 a thin platelet of glass is employed, which effectively mixes the light from one fiber, dividing it among the outgoing fibers. This method can be used to produce a transmissive star coupler or a reflective star coupler, as displayed in Figure 5.32. The typical insertion loss for an 8×8 mixer-rod transmissive star coupler with fiber pigtails is 12.5 dB with port to port uniformity of ± 0.7 dB [Ref. 84].

The manufacturing process for the fused biconical taper star coupler is similar to that discussed in Section 5.6.1 for the three and four port FBT coupler. Thus the fibers which constitute the star coupler are bundled, twisted, heated and pulled, to form the device illustrated in Figure 5.33. With multimode fiber this method

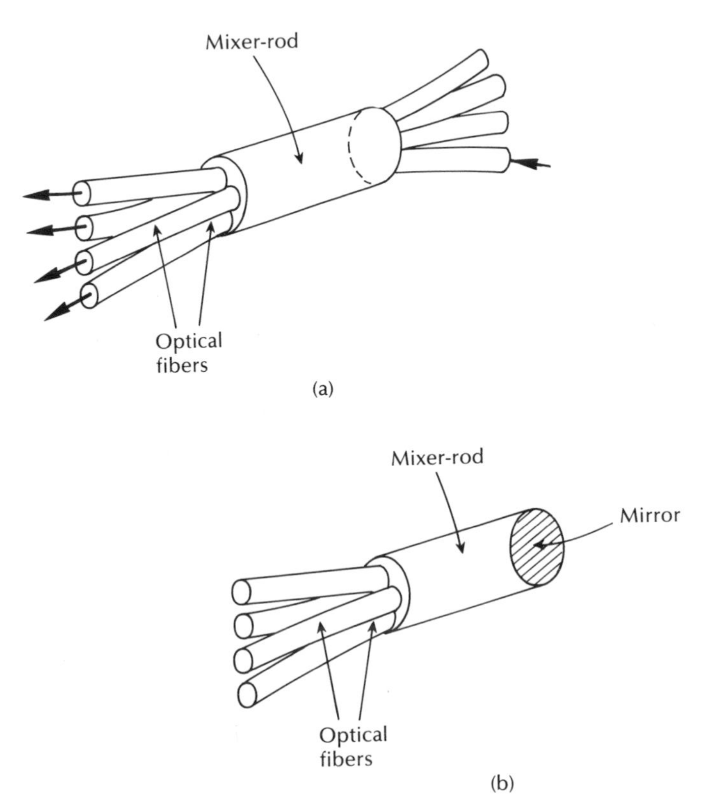

Figure 5.32 Fiber star couplers using the mixer-rod technique: (a) transmissive star coupler; (b) reflective star coupler.

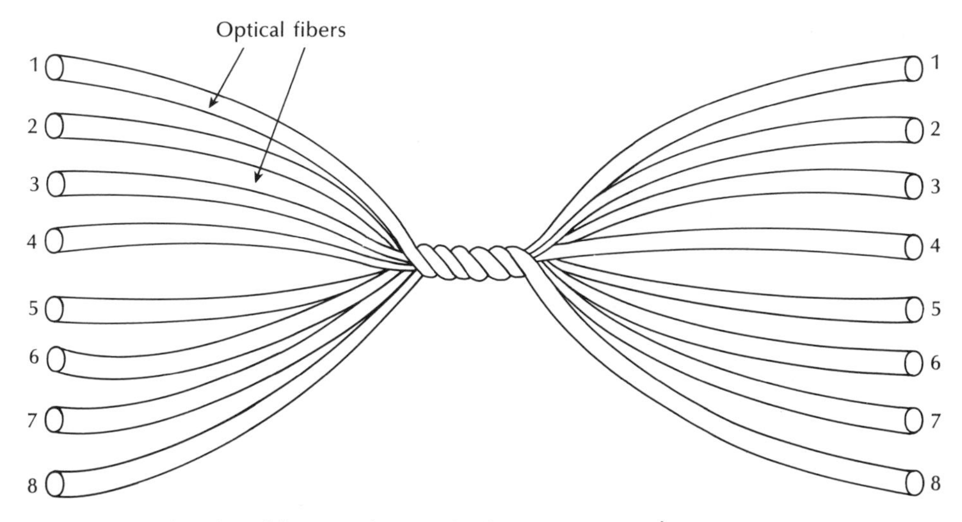

Figure 5.33 Fiber fused biconical taper 8 × 8 port star coupler.

relies upon the coupling of higher order modes between the different fibers. It is therefore highly mode dependent, which results in a relatively wide port to port output variation in comparison with star couplers based on the mixer-rod technique [Ref. 84].

In an ideal star coupler the optical power from any input fiber is evenly distributed among the output fibers. The total loss associated with the star coupler comprises its theoretical splitting loss together with the excess loss. The splitting loss is related to the number of output ports N following:

$$\textit{Splitting loss (star coupler)} = 10 \log_{10} N \text{ (dB)} \tag{5.27}$$

It should be noted that for a reflective star coupler N is equal to the total number of ports (both input and output combined).

For a single input port and multiple output ports where $j = 1, N$, then the excess loss is given by:

$$\textit{Excess loss (star coupler)} = 10 \log_{10}\left[P_i \Big/ \sum_{1}^{N} P_j\right] \text{ (dB)} \tag{5.28}$$

The insertion loss between any two ports on the star coupler may be obtained in a similar manner to the four port coupler using Eq. (5.23). Similarly, the crosstalk between any two input ports is given by Eq. (5.24).

Example 5.8

A 32 × 32 port multimode fiber transmissive star coupler has 1 mW of optical power launched into a single input port. The average measured optical power at each output port is 14 μW. Calculate the total loss incurred by the star coupler and the average insertion loss through the device.

Solution: The total loss incurred by the star coupler comprises the splitting loss and the excess loss through the device. The splitting loss is given by Eq. (5.27) as:

$$\begin{aligned}\textit{Splitting loss} &= 10 \log_{10} N = 10 \log_{10} 32 \\ &= 15.05 \text{ dB}\end{aligned}$$

The excess loss may be obtained from Eq. (5.28) where:

$$\textit{Excess loss} = 10 \log_{10}\left[P_i \Big/ \sum_{1}^{N} P_j\right] = 10 \log_{10}[10^3/32 \times 14] = 3.49 \text{ dB}$$

Hence the total loss for the star coupler:

$$\begin{aligned}\textit{Total loss} = \textit{splitting loss} + \textit{excess loss} &= 15.05 + 3.49 \\ &= 18.54 \text{ dB}\end{aligned}$$

The average insertion loss from the input port to an output port is provided by Eq. (5.23) as:

$$\textit{Insertion loss} = 10 \log_{10} \frac{10^3}{14} = 18.54 \text{ dB}$$

Therefore, as may have been anticipated, the total loss incurred by the star coupler is equivalent to the average insertion loss through the device. This result occurs because the total loss is the loss incurred on a single (average) optical path through the coupler which effectively defines the average insertion loss for the device.

An alternative strategy for the realization of a star coupler is to construct a ladder coupler, as illustrated in Figure 5.34. The ladder coupler generally comprises a number of cascaded stages, each incorporating three or four port FBT couplers in order to obtain a multiport output. Hence, the example shown in Figure 5.34 consists of three stages, which gives eight output ports. It must be noted, however, that when three port couplers are used such devices do not form symmetrical star couplers* in that they provide a $1 \times N$ rather than a $N \times N$ configuration. Nevertheless, the ladder coupler presents a useful device to achieve a multiport output with relatively low insertion loss. Furthermore, when four port couplers are employed, then a true $N \times N$ star coupler may be obtained. It may be deduced from Figure 5.34 that the number of output ports N obtained with an M stage ladder coupler is 2^M. These devices have found relatively widespread application for the production of single-mode fiber star couplers.

* Such devices are sometimes referred to as tree couplers.

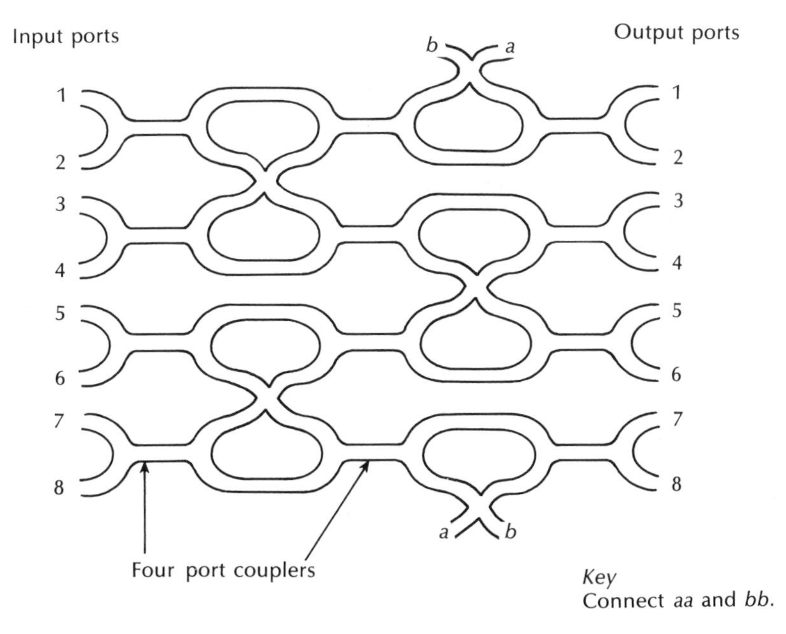

Figure 5.34 8 × 8 star coupler formed by cascading 12 four port couplers (ladder coupler). This strategy is often used to produce low loss single-mode fiber star or tree couplers.

Example 5.9

A number of three port single-mode fiber couplers are utilized in the fabrication of a tree (ladder) coupler with sixteen output ports. The three port couplers each have an excess loss of 0.2 dB with a split ratio of 50%. In addition, there is a splice loss of 0.1 dB at the interconnection of each stage. Determine the insertion loss associated with one optical path through the device.

Solution: The number of stages M within the ladder design is given by $2^M = 16$. Hence $M = 4$. Thus the excess loss through four stages of the coupler with three splices is:

$$\textit{Excess loss} = (4 \times 0.2) + (3 \times 0.1) = 1.1 \text{ dB}$$

Assuming a 50% split ratio at each stage, the splitting loss for the coupler may be obtained using Eq. (5.27) as:

$$\textit{Splitting loss} = 10 \log_{10} 16 = 12.04 \text{ dB}$$

Hence the insertion loss for the coupler which is equivalent to the total loss for one optical path though the device is:

$$\begin{aligned} \textit{Insertion loss} &= \textit{splitting loss} + \textit{excess loss} \text{ (four stages)} \\ &= 12.04 + 1.1 = 13.14 \text{ dB} \end{aligned}$$

Significantly lower excess losses than that indicated in Example 5.9 have been achieved with single-mode fiber ladder couplers. In particular, a mean excess loss of only 0.13 dB for an 8×8 star coupler constructed using this technique has been reported [Ref. 96]. Four port FBT couplers with mean excess losses of 0.05 dB were used in this device. Alternatively, 3×3 single-mode fiber FBT couplers have been employed as a basis for ladder couplers. For example, a 9×9 star coupler with an excess loss of 1.46 dB and output port power uniformity of ± 1.50 dB has been demonstrated [Ref. 97].

5.6.3 Wavelength division multiplexing couplers

It was indicated in Section 5.6 that wavelength division multiplexing (WDM) devices are a specialized coupler type which enable light from two or more optical sources of differing nominal peak optical wavelength to be launched in parallel into a single optical fiber. Hence such couplers perform as either wavelength multiplexers or wavelength demultiplexers (see Section 11.9.3). The spectral performance characteristic for a typical five channel WDM device is shown in Figure 5.35. The important optical parameters associated with the WDM coupler are the attenuation of the light over a particular wavelength band, the interband isolation and the wavelength band or channel separation. Ideally, the device should have a low loss transmission window for each wavelength band, giving a low

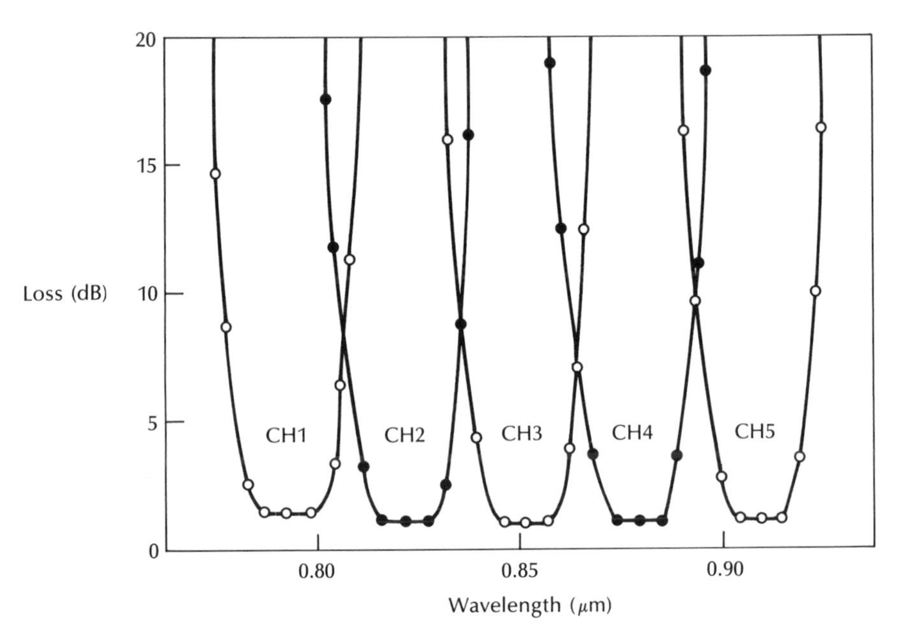

Figure 5.35 Typical flat passband spectral output characteristic for a WDM demultiplexer device (diffraction grating type) [Ref. 99].

insertion loss.* In addition, the device should exhibit high interband isolation, thus minimizing crosstalk. However, in practice, high interchannel isolation is only required at the receiver (demultiplexer) end of the link or at both ends in a bidirectional system. Finally, the channel separation should be as small as may be permitted by light source availability and stability together with crosstalk considerations.

Numerous techniques have been developed for the implementation of WDM couplers. Passive devices, however, may be classified into three major categories [Ref. 98], two of which are core interaction types: namely, angularly dispersive (usually diffraction grating) and filter, whilst the other is a surface interaction type which may be employed with single-mode fiber in the form of a directional coupler. Any other implementations tend to be hybrid combinations of the two core interaction types.

Although a glass prism may be utilized as an angularly dispersive element to facilitate wavelength multiplexing and demultiplexing, the principal angularly dispersive element used in this context is the diffraction grating. Any arrangement which is equivalent in its action to a number of parallel equidistant slits of the same width may be referred to as a diffraction grating. A common form of diffraction grating comprises an epoxy layer deposited on a glass substrate, on which lines are blazed. There are two main types of blazed grating. The first is produced by conventional mechanical techniques, whilst the other is fabricated by the anisotropic etching of single crystal silicon [Ref. 99] and hence is called a silicon grating. The silicon grating has been found to be superior to the conventional mechanically ruled device, since it provides greater design freedom in the choice of blazing angle θ_B (see Figure 5.36) and grating constant (number of lines per unit length). It is also highly efficient and produces a more environmentally stable surface.

A diffraction grating reflects light in particular directions according to the grating constant, the angle at which the light is incident on the grating and the optical wavelength. Two main structural types are used in the manufacture of WDM couplers: the Littrow device which employs a single lens and a separate plane grating, and the concave grating which does not utilize a lens since both focusing and diffraction functions are performed by the grating.

In a Littrow mounted grating, the blaze angle of the grating is such that the incident and reflected light beams follow virtually the same path, as illustrated in Figure 5.36, thereby maximizing the grating efficiency and minimizing lens astigmatism. For a given centre wavelength λ, the blaze angle is set such that [Ref. 100]

$$\theta_B = \sin^{-1}\left(\frac{\lambda}{2x}\right) \tag{5.29}$$

* In the case of the WDM coupler the device loss is specified by the insertion loss associated with a particular wavelength band. The use of excess loss as in the case of other fiber couplers is inappropriate because the optical signals are separated into different wavelength bands.

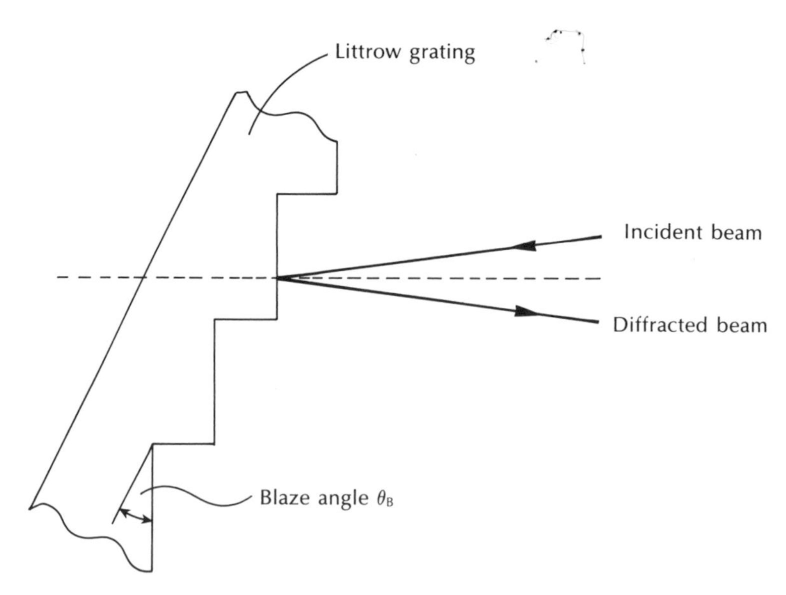

Figure 5.36 Littrow mounted diffraction grating.

where x is the line spacing on the grating. Schematic diagrams of Littrow type grating demultiplexers employing a conventional lens [Ref. 99] and a GRIN-rod lens [Ref. 101] are shown in Figure 5.37. The use of a spherical ball microlens has also been reported [Ref. 102]. Although all the lens type devices exhibit similar operating mechanisms and hence performance, the GRIN-rod lens configuration proves advantageous for its compactness and ease of alignment. Therefore the operation of a GRIN-rod lens type demultiplexer is considered in greater detail.

Referring to Figure 5.37(b), the single input fiber and multiple output fibers are arranged on the focal plane of the lens, which, for a quarter pitch GRIN-rod lens, is coincident with the fiber end face (see Section 5.5.1). The input wavelength multiplexed optical beam is collimated by the lens and hence transmitted to the diffraction grating, which is offset at the blaze angle so that the incoming light is incident virtually normal to the groove faces. The required offset angle can be produced by interposing a prism (glass wedge) between the lens and the grating, as illustrated in Figure 5.37(b) or, alternatively, by cutting and polishing the GRIN-rod lens and by mounting the grating on its end face. The former method gives superior performance since the optical properties of the GRIN-rod lens are not altered [Ref. 102]. On reflection from the grating, the diffraction process causes the light to be angularly dispersed according to the optical wavelength. Finally, the different optical wavelengths pass through the lens and are focused onto the different collecting output fibers. Devices of this type have demonstrated channel insertion losses of less than 2 dB and channel spacings of 18 nm with low crosstalk [Ref. 103].

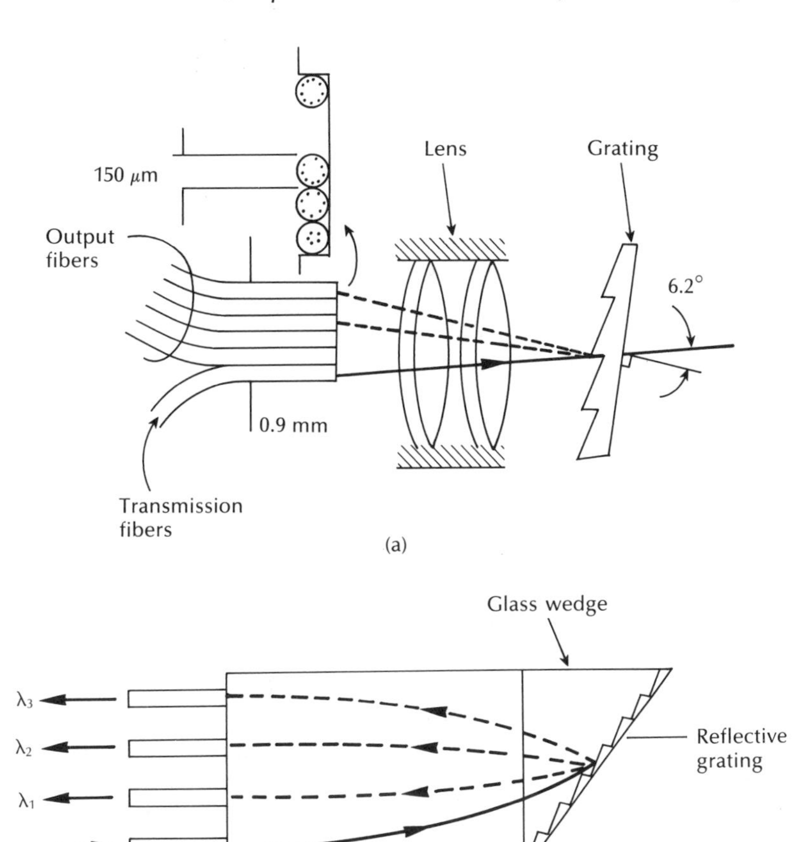

Figure 5.37 Littrow type grating demultiplexers: (a) using a conventional lens [Ref. 99]; (b) using a GRIN-rod lens [Ref. 101].

Finally, single-mode wavelength multiplexer and demultiplexer pairs based on a planar diffraction grating and a lithium niobate strip waveguide structure have also been reported [Ref. 104]. Six wavelength multiplexed channels were demonstrated, three over the wavelength region from 1275 nm to 1335 nm and three over the wavelength range from 1510 nm to 1570 nm. Crosstalk levels were less than −25 dB, with insertion losses for the multiplexer and demultiplexer of 5 to 8 dB and 1 to 2.2 dB respectively.

The other major core interaction type WDM devices employ optical filter technology. Optical spectral filters fall into two main categories: namely, interference filters and absorption filters. Dielectric thin film (DTF) interference filters can be constructed from alternate layers of high refractive index (e.g. zinc sulphide) and low refractive index (e.g. magnesium fluoride) materials, each of which is one quarter wavelength thick [Ref. 107]. In this structure, shown schematically in Figure 5.38, light which is reflected within the high index layers does not suffer any phase shift on reflection, while those optical beams reflected within the low index layers undergo a phase shift of 180°. Thus the successive reflected beams recombine constructively at the filter front face, producing a high reflectance over a limited wavelength region which is dependent upon the ratio between the high and low refractive indices. Outside this high reflectance region, the reflectance changes abruptly to a low value. Consequently, the quarter wave stack can be used either as a high pass filter, a low pass filter, or as a high reflectance coating.

Absorption filters comprise a thin film of material (e.g. germanium) which exhibits an absorption edge at a specific wavelength. Absorption filters usually display very high rejection in the cutoff region. However, as their operation is dependent upon the fundamental optical properties of the material structure, they tend to be inflexible because the edge positions are fixed. Nevertheless, by fabricating interference filters on to an absorption layer substrate, a filter can be obtained which combines the sharp rejection of the absorption filter together with the flexibility of the interference filter. Such combined structures can be used as high performance edge filters.

Specific filter WDM coupler designs are now considered in further detail. Firstly, edge filters are generally used in devices which require the separation of two wavelengths (generally reasonably widely separated by 10% or more of median wavelength). A configuration which has been adopted [Ref. 106] is one in which the fiber is cleaved at a specific angle and then an edge filter is interposed between

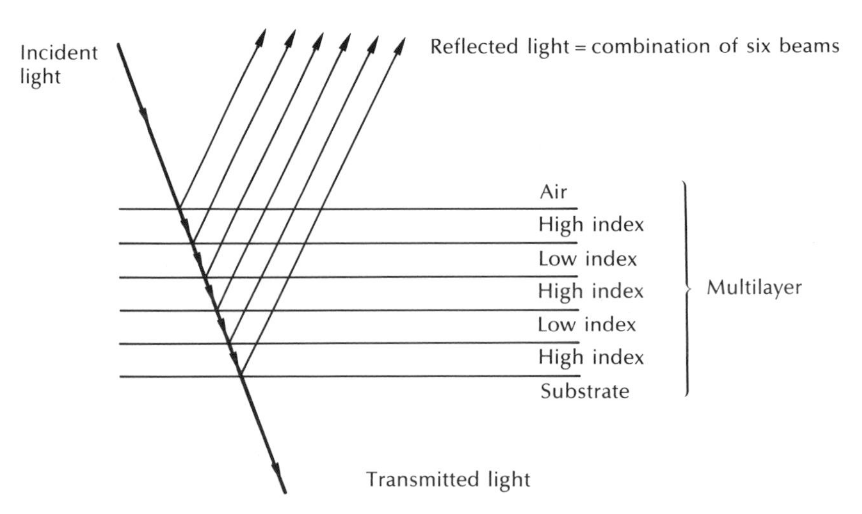

Figure 5.38 Multilayer interference filter structure.

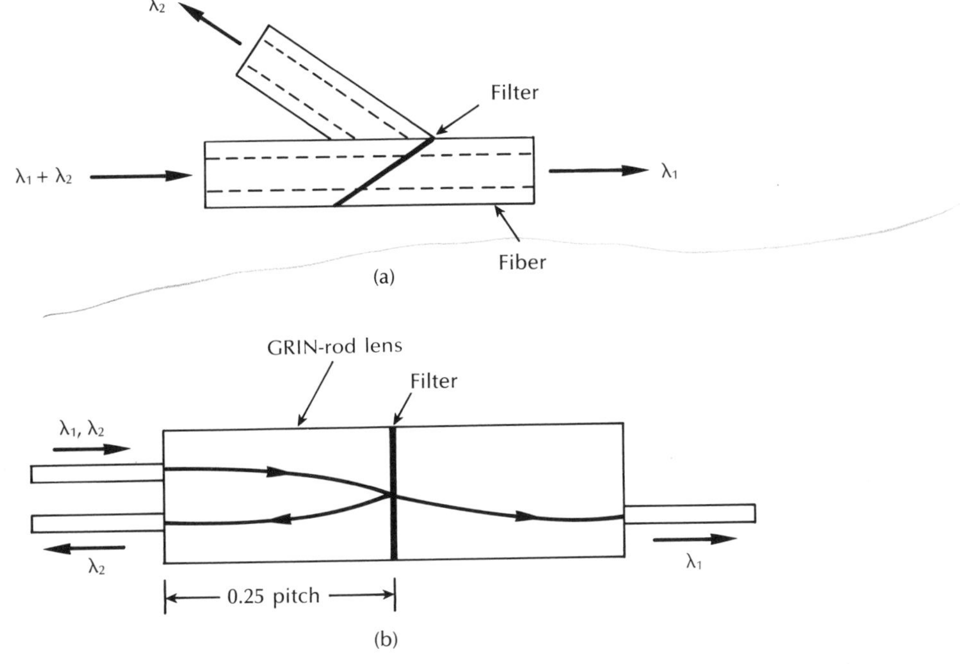

Figure 5.39 Two wavelength interference filter demultiplexers: (a) fiber-end device; (b) GRIN-rod lens device.

the two fiber ends, as illustrated in Figure 5.39(a). In a demultiplexing structure light at one wavelength is reflected by the filter and collected by a suitably positioned receive fiber, whilst the other optical wavelength is transmitted through the filter and then propagates down the cleaved fiber. Such a device, which has been tested with LED sources emitting at centre wavelengths of 755 nm and 825 nm, exhibited insertion losses of 2 to 3 dB with crosstalk levels less than −60 dB [Ref. 106]. An alternative two wavelength WDM device employing a cascaded BPF sandwiched between two GRIN-rod lenses is shown in Figure 5.39(b). A practical two channel (operating at wavelengths of 1.2 μm and 1.3 μm) multiplex/demultiplex system which is capable of operation in both directions using this WDM design has been reported [Ref. 107] to exhibit low insertion losses of around 1.5 dB with crosstalk levels less than −58 dB. This device also displayed acceptable environmental stability with insertion loss variations of less than 0.3 dB throughout a range of tests (i.e. vibration, temperature cycling and damp/heat tests).

Multiple wavelength multiplexer/demultiplexer devices employing DTF interference filters may be constructed from a suitably aligned series of bandpass filters with different passband wavelength regions, cascaded in such a way that each filter transmits a particular wavelength, but reflects all others. Such a multiple reflection demultiplexing device is illustrated in Figure 5.40. This structure has the

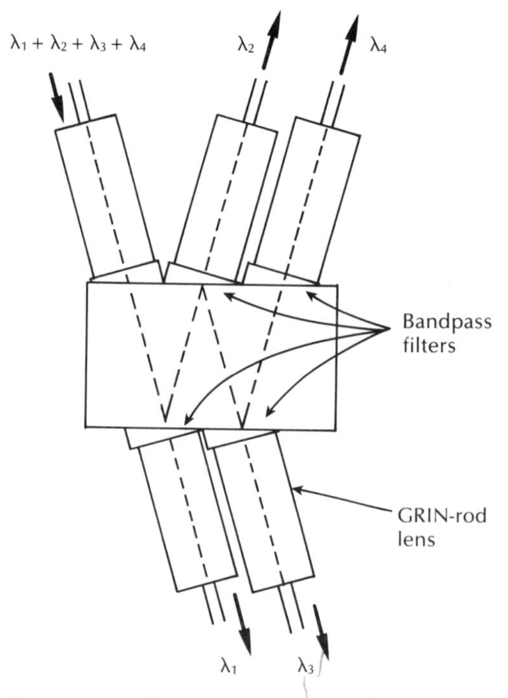

Figure 5.40 GRIN-rod lensed bandpass demultiplexer.

disadvantage that the insertion losses increase linearly with the number of multiplexed channels since losses are incurred at each successive reflection due to filter imperfections and the difficulties of maintaining good alignment [Ref. 98].

A two channel slab waveguide version of a filter WDM device has recently been introduced by Corning, which is based on the same technology as their optical waveguide coupler (see Section 5.6.1). The wavelength separation is accomplished within the waveguide using a diochroic filter which intersects the path of the incoming light beam. Longer wavelengths are transmitted and shorter wavelengths reflected. The multiplexer/demultiplexer device reported [Ref. 108] is compatible with both 50/125 μm and 85/125 μm graded index fibers. It combines/separates optical wavelength regions between 0.8 to 0.9 μm and 1.2 to 1.4 μm with an insertion loss lower than 1.5 dB and crosstalk levels less than −25 dB.

The wavelength dependent characteristics of single-mode fiber directional couplers were mentioned in Section 5.6.1. Both single-mode ground fiber and fused biconical taper fiber couplers can be fabricated to provide the complete transfer of optical power between the two fibers. However, since the optical power coupling characteristic of such single-mode fiber couplers is highly wavelength dependent they can be used to fabricate WDM devices.

Optical power transfer within multimode fiber couplers is a mode dependent phenomenon which, in general, takes place between the higher order modes propagating in the outer reaches of the fiber cores as well as in the cladding regions.

These higher order modes couple more freely when the f
Consequently, the spectral dependence of light tran
couplers is far less pronounced and predictable than
fiber structures. Therefore, multimode fiber WD
fabricated using the fused biconical taper or grour

Optical power is coupled between two single-m
cores close together over a region known as th
methods are generally used to perform this func
necessitates bending and fixing the two fibers in
(e.g. quartz) prior to grinding the two blocks down so
of the fiber cladding regions is worn away. Finally, the two
together, as shown in Figure 5.41.

Parallel single-mode fiber waveguides exchange energy with a spatial pe
(coupling length) $L = 2\pi/k$, where k is the coupling coefficient (units of inverse length) for the two interacting waveguide modes [Ref. 110]. This result can be extended to curved regions where spacing between the waveguides over the interaction length is no longer fixed [Ref. 111]. Thus, for a pair of fibers curved against each other, the coupling coefficient k is a nonlinear function of the interaction length (which in turn is proportional to the square root of the radius of curvature R), the minimum spacing between the fiber cores, the refractive index of the intervening material, the fiber parameters and the wavelength of the light. The wavelength dependent properties of a single-mode ground fiber coupler can therefore be altered by adjusting several different parameters.

An early demonstration of such a two channel ground fiber directional coupler was made from two identical 2 μm core diameter single-mode fibers [Ref. 112]. This device, with a radius of curvature $R_1 = R_2 = 70$ cm and a minimum core separation of 4.5 μm gave a measured coupling ratio which followed the typical sinusoidal pattern, with approximately two periods over the 0.45 μm to 0.9 μm

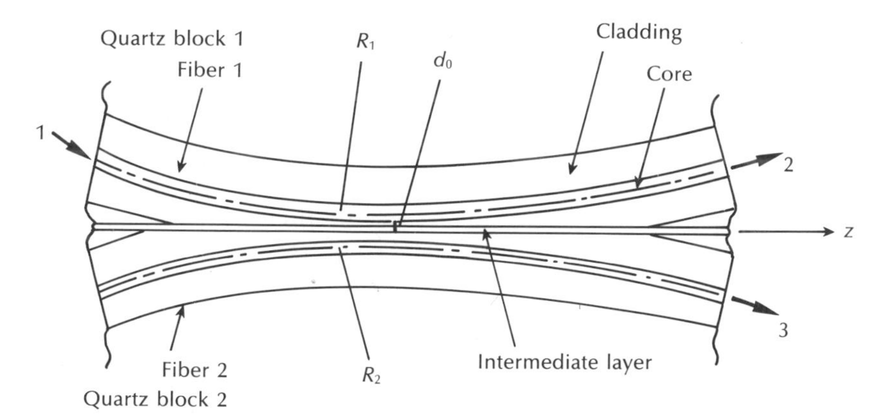

Figure 5.41 Schematic diagram showing ground (polished) single-mode fiber coupler.

region. By offsetting the cores laterally (i.e. in the direction z indicated 5.41), and effectively altering d_0, the spectral characteristics of the coupler ered. The sinusoidal response curve was shifted by around 400 nm with a offset of 5 μm. Interchannel wavelength spacings (wavelength separation een minimum and maximum on the sinusoidal pattern) were about 140 nm for is structure, which exhibited insertion losses as low as 0.1 dB using suitable index matching.

Wavelength selective ground fiber directional couplers constructed from single-mode fibers of different core diameters and refractive indices exhibit propagation constants which are matched at only one wavelength and hence can be used to produce true bandpass filters. The centre wavelength and spectral bandwidth of these couplers are essentially determined by the fiber parameters [Ref. 113]. Measured insertion losses for such devices fabricated for operation in the longer wavelength region were between 0.5 and 0.6 dB with crosstalk levels less than −22 dB [Ref. 109].

The second method of fabricating a single-mode fiber WDM coupler is the fused biconical taper (FBT) technique [Ref. 85]. Carefully fabricated fused couplers display very low insertion losses and provide a high degree of environmental stability. The manufacturing process requires the single-mode fibers to be fused together at around 1500 °C before being pulled whilst heat is still applied. The pulling process decreases the fiber core size causing the evanescent field of the transmitted optical signal to spread out further from the fiber core, which enables light to couple into the adjacent fiber. In practice this manufacturing process necessitates the monitoring of the optical power output from the two fibers, the process being halted when the required coupling ratio is reached [Ref. 114].

In common with the ground fiber coupler constructed by using similar fibers, the optical power transferred between the two fibers (or the coupling ratio) in a FBT coupler as a function of wavelength is sinusoidal with a period dependent on the dimensions and the geometry of the fused cross section, and on the refractive index of the surrounding medium [Ref. 90]. It can also depend upon whether the fusing process produces a coupling region where the two fiber cores are close (strongly fused) or relatively far apart (weakly fused) [Ref. 89]. The most popular method of varying the periodic coupling function in such fused WDM couplers is to extend the interaction length by continuing the stretching process during fusing. An increase in the interaction length has the effect of increasing the coupling ratio period. Such two channel devices have displayed insertion losses of 0.25 and 0.37 dB with crosstalk levels less than −22 dB [Ref. 115]. It should be noted, however, that a limitation with these WDM couplers is that they are not well suited for the provision of closely spaced or multiple channels.

In summary, the generalized performance characteristics of the various passive wavelength division multiplexing coupler generic types are provided in Table 5.1 [Ref. 98]. It should be stressed, however, that the entries in the table signify the characteristics of typical WDM devices of each type. In this context they are intended to act as a guide to the reader and are not a definitive statement of absolute device performance.

Table 5.1 Generalized characteristics of passive wavelength division multiplexer/demultiplexer device types [Ref. 98]

Device type	Mechanism	Implementation	Channel spectral bandwidth	Channel separation	No. of channels	Interchannel crosstalk levels	Insertion loss	Environmental stability	Drawbacks
Prism	Angularly dispersive	Bulk optic: difficult/ expensive	10–20 nm	Around 30 nm	Multi		Less than 10 dB	Acceptable	Optical material constraints produce low values of angular dispersion and bulkly devices
Diffraction grating	Angularly dispersive	Bulk optic: Straightforward compact	Dependent on fiber type/ dimensions and grating line constant but typically around 10 nm minimum	Dependent on fiber type/ dimensions and grating line constant but generally less than 40 nm	Multi, but generally in one wavelength region (e.g. 800–900 nm)	Better than −30 dB	Independent of no. of channels and typically less than 5 dB	Acceptable	High insertion loss for equal input fibers. Generally must be designed for use with particular fibers
Filter	Interference of absorption	Bulk optic: Straightforward compact	Minimum of 2.5 nm at 850 nm 4 nm at 1300 nm	May be used for widely separated wavelengths but also as low as 10 nm	Two for single filter. Multi requires cascaded filters	Better than −30 dB	Increases with number of channels generally greater than 1 dB	Generally acceptable apart from moisture absorption	Cascaded filters give increased insertion losses
Fiber directional coupler	Wavelength dependent power transfer between adjacent fibers	Fused biconical Ground fiber Straightforward compact	Generally periodic but large at around 40 nm	Generally widely separated around 100 nm	Two for single coupler. Multi requires cascaded couplers	Around −20 dB	Low (0.2 dB typical)	Good	Only single-mode fiber devices. Ground fiber type requires careful fabrication
Integrated waveguide	Slab, multimode Planar, single mode	Integrated. Very compact Do not require lens	100 nm available	Wide – 850/1300 nm available Grating types around 5 nm	Two Multi, Ten demonstrated	Better than −25 dB Better than −20 dB	Low – less than 1.5 dB Medium – high (2–5 dB)	Potentially good	Commerically available Commerical devices not generally available

Problems

5.1 State the two major categories of fiber–fiber joint, indicating the differences between them. Briefly discuss the problem of Fresnel reflection at all types of optical fiber joint, and indicate how it may be avoided.

A silica multimode step index fiber has a core refractive index of 1.46. Determine the optical loss in decibels due to Fresnel reflection at a fiber joint with:

(a) a small air gap;

(b) an index matching epoxy which has a refractive index of 1.40.

It may be assumed that the fiber axes and end faces are perfectly aligned at the joint.

5.2 The Fresnel reflection at a butt joint with an air gap in a multimode step index fiber is 0.46 dB. Determine the refractive index of the fiber core.

5.3 Describe the three types of fiber misalignment which may contribute to insertion loss at an optical fiber joint.

A step index fiber with a 200 μm core diameter is butt jointed. The joint which is index matched has a lateral offset of 10 μm but no longitudinal or angular misalignment. Using two methods, estimate the insertion loss at the joint assuming the uniform illumination of all guided modes.

5.4 A graded index fiber has a characteristic refractive index profile (α) of 1.85, and a core diameter of 60 μm. Estimate the insertion loss due to a 5 μm lateral offset at an index matched fiber joint assuming the uniform illumination of all guided modes.

5.5 A graded index fiber with a parabolic refractive index profile ($\alpha = 2$) has a core diameter of 40 μm. Determine the difference in the estimated insertion losses at an index matched fiber joint with a lateral offset of 1 μm (no longitudinal or angular misalignment). When performing the calculation assume (a) the uniform illumination of only the guided modes and (b) the uniform illumination of both guided and leaky modes.

5.6 A graded index fiber with a 50 μm core diameter has a characteristic refractive index profile (α) of 2.25. The fiber is jointed with index matching and the connection exhibits an optical loss of 0.62 dB. This is found to be solely due to a lateral offset of the fiber ends. Estimate the magnitude of the lateral offset assuming the uniform illumination of all guided modes in the fiber core.

5.7 A step index fiber has a core refractive index of 1.47, a relative refractive index difference of 2% and a core diameter of 80 μm. The fiber is jointed with a lateral offset of 2 μm, an angular misalignment of the core axes of 3° and a small air gap (no longitudinal misalignment). Estimate the total insertion loss at the joint which may be assumed to comprise the sum of the misalignment losses.

5.8 Briefly outline the factors which cause intrinsic losses of fiber–fiber joints.

(a) Plot the loss resulting from a mismatch in multimode fiber core diameters or numerical apertures over a mismatch range 0 to 50%.

(b) An optical source is packaged with a fiber pigtail comprising 62.5/125 μm graded index fiber with a numerical aperture of 0.28 and a profile parameter of 2.1. The fiber pigtail is spliced to a main transmission fiber which is 50/125 μm graded index fiber with a numerical aperture of 0.22 and a profile parameter of 1.9. When the fiber axes are aligned without either a gap, radial or angular misalignment, calculate the insertion loss at the splice.

5.9 Describe what is meant by the fusion splicing of optical fibers. Discuss the advantages and drawbacks of this jointing technique.

A multimode step index fiber with a core refractive index of 1.52 is fusion spliced. The splice exhibits an insertion loss of 0.8 dB. This insertion loss is found to be entirely due to the angular misalignment of the fiber core axes which is 7°. Determine the numerical aperture of the fiber.

5.10 Describe, with the aid of suitable diagrams, three common techniques used for the mechanical splicing of optical fibers.

A mechanical splice in a multimode step index fiber has a lateral offset of 16% of the fiber core radius. The fiber core has a refractive index of 1.49, and an index matching fluid with a refractive index of 1.45 is inserted in the splice between the butt jointed fiber ends. Assuming no longitudinal or angular misalignment, estimate the insertion loss of the splice.

5.11 Discuss the principles of operation of the two major categories of demountable optical fiber connector. Describe in detail a common technique for achieving a butt jointed fiber connector.

A butt jointed fiber connector used on a multimode step index fiber with a core refractive index of 1.42 and a relative refractive index difference of 1% has an angular misalignment of 9°. There is no longitudinal or lateral misalignment but there is a small air gap between the fibers in the connector. Estimate the insertion loss of the connector.

5.12 Briefly describe the types of demountable connector that may be used with single-mode fibers. Further, indicate the problems involved with the connection of single-mode fibers.

A single-mode fiber connector is used with a 6 μm core diameter silica (refractive index 1.46) step index fiber which has a normalized frequency of 2.2 and a numerical aperture of 0.9. The connector has a lateral offset of 0.7 μm and an angular misalignment of 0.8°. Estimate the total insertion loss of the connector assuming that the joint is index matched and that there is no longitudinal misalignment.

5.13 A 10 μm core diameter single-mode fiber has a normalized frequency of 2.0. A fusion splice at a point along its length exhibits an insertion loss of 0.15 dB. Assuming only lateral misalignment contributes to the splice insertion loss, estimate the magnitude of the lateral misalignment.

5.14 A 5 μm core diameter single-mode step index fiber has a normalized frequency of 1.7, a core refractive index of 1.48 and a numerical aperture of 0.14. The loss in decibels due to angular misalignment at a fusion splice with a lateral offset of 0.4 μm is twice that due to the lateral offset. Estimate the magnitude in degrees of the angular misalignment.

5.15 Given the following parameters for a single-mode step index fiber with a fusion splice estimate (a) the fiber core diameter: and (b) the numerical aperture for the fiber.

Fiber normalized frequency = 1.9
Fiber core refractive index = 1.46
Splice lateral offset = 0.5 μm
Splice lateral offset loss = 0.05 dB
Splice angular misalignment = 0.3°
Splice angular misalignment loss = 0.04 dB

5.16 Two single-mode fibers have mode-field diameters of 9 μm and 11 μm. Assuming that there are no extrinsic losses calculate the coupling loss between the fibers as a result of the mode-field diameter mismatch. Comment on the result in relation to the direction of transmission of the optical signal between the two fibers.

Determine the loss if the mode-field diameter mismatch between the fibers is increased to 30%.

5.17 With the aid of simple sketches outline the major categories of multiport optical fiber coupler.

Describe two common methods used in the fabrication of three and four port fiber couplers.

5.18 A four port FBT coupler is shown in Figure 5.29. In addition a section of a tapered multimode step index fiber from such a coupler may be observed in Figure 5.42. A meridional ray propagating along the taper (characterized by the taper angle γ) is shown to undergo an increase in its propagation angle (i.e. the angle formed with the fiber axis). However, as long as the angle of incidence remains larger than the critical angle, then the ray is still guided and it emerges from the taper region forming an angle θ_o with the fiber axis. When the taper is smooth and the number of reflections is high, then, in Figure 5.42, $\sin \theta_o = R_1/R_2 \sin \theta_i$ where R_1 and R_2 are the core radii before and after the taper respectively. Show that the numerical aperture for the tapered fiber NA_T is given by:

$$NA_T = \frac{R_2}{R_1}(n_1^2 - n_2^2)^{\frac{1}{2}}$$

where n_1 and n_2 are the refractive indices of the fiber core and cladding respectively. Comment on this result when considering the modes of the light launched into the coupler.

5.19 The measured optical output powers from ports 3 and 4 of a multimode fiber FBT coupler are 47.0 μW and 52.0 μW respectively. If the excess loss specified for the device is 0.7 dB, calculate the amount of optical power that is launched into port 1 in order to obtain these output power levels. Hence, determine the insertion losses between the input and two output ports, as well as the split ratio for the device.

When the specified crosstalk for the coupler is −45 dB, calculate the optical output power level that would be measured at port 2 when the above input power level is maintained.

5.20 Indicate the distinction between fiber star and tree couplers.

Discuss the major techniques used in the fabrication of multimode fiber star couplers and describe how this differs from the strategy that tends to be adopted to produce single-mode fiber star couplers.

5.21 A 64 × 64 port transmissive star coupler has 1.6 mW of optical power launched into a single input port. If the device exhibits an excess loss of 3.90 dB, determine the total loss through the device and the average optical power level that would be expected at each output port.

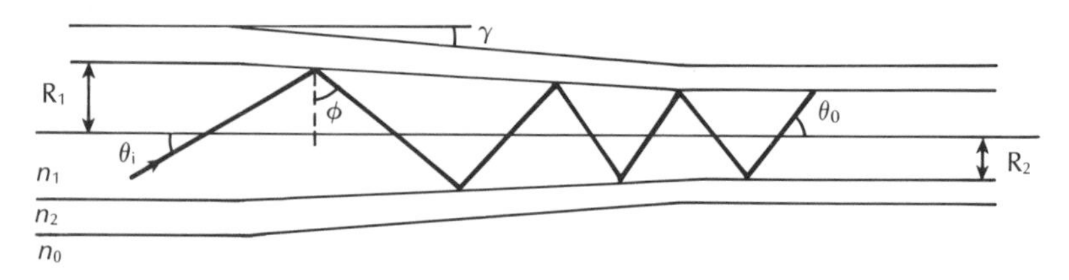

Figure 5.42 Section of a tapered multimode step index fiber for Problem 5.18.

5.22 An 8×8 port multimode fiber reflective star coupler has -8.0 dBm of optical power launched into a single port. The average measured optical power at each output port is -22.8 dBm. Obtain the excess loss for the device and hence the total loss experienced by an optical signal in transmission through the coupler. Check the result.

5.23 A number of four port single-mode fiber couplers are employed in the fabrication of a 32×32 port star coupler. Each four port coupler has a split ratio of 50% and when an optical input power level of -6 dBm is launched into port 1, the output power level from port 3 is found to be 122 μW. Furthermore, there is a splice loss of 0.06 dB at the interconnection of each stage within the ladder design. Calculate the optical power emitted from each of the output ports when the -6 dBm power level is launched into any one of the input ports. Check the result.

5.24 Outline the three major categories of passive wavelength division multiplexing coupler. Describe in detail one implementation of each category. Comment on the relative merits and drawbacks associated with each of the WDM devices you have described.

Answers to numerical problems

5.1 (a) 0.31 dB; (b) 3.8×10^{-4} dB
5.2 1.59
5.3 0.29 dB
5.4 0.67 dB
5.5 (a) 0.19 dB; (b) 0.17 dB; difference 0.02 dB
5.6 4.0 μm
5.7 0.71 dB
5.8 4.25 dB
5.9 0.35
5.10 0.47 dB
5.11 1.51 dB
5.12 0.54 dB
5.13 1.2 μm
5.14 0.65°
5.15 (a) 7.0 μm; (b) 0.10
5.16 0.17 dB, 0.54 dB
5.19 116.3 μm, 3.93 dB, 3.50 dB, 47.5%, 3.7 nW
5.21 21.96 dB, 10.18 μW
5.22 5.77 dB, 14.80 dB
5.23 6.40 μW

References

[1] S. R. Nagel, 'Fiber materials and fabrication methods', in S. E. Miller and I. P. Kaminov (Eds.), *Optical Fiber Telecommunications II*, Academic Press, pp. 121–215, 1988.

[2] T. R. Rowbotham, 'Submarine telecommunications', *Br. Telecom Technol. J.*, **5**(1), pp. 5–24, 1987.

[3] K. Nakagawa and K. Nosu, 'An overview of very high capacity transmission technology for NTT networks', *J. Lightwave Technol.*, **LT-5**(10), pp. 1498–1504, 1987.

[4] P. Cockrane and M. Brain, 'Future optical fiber transmission technology and networks', *IEEE Commun. Mag.*, pp. 45–60, November, 1988.

[5] D. C. Gloge and I. Jacobs, 'Terrestrial intercity transmission systems', in S. E. Miller and I. P. Kaminow (Eds.) *Optical Fiber Telecommunications II*, Academic Press, pp. 855–878, 1988.

[6] A. Stevenson, S. L. Arambepola, G. L. Blau, T. S. Brown, I. C. Catchpole and A. J. Flavin, 'A 2.4 Gbit/s long-reach optical transmission system', *Br. Telecom. Tech. J.*, **7**(1), pp. 92–99, 1989.

[7] S. Fujita, M. Fitamura, T. Torikai, N. Henmi, H. Yamada, T. Suzaki, I. Takano and M. Shikada, '10 Gbit/s, 100 km optical fiber transmission experiment using high-speed MQW DFB-LD and back-illuminated GaInAs APD', *Electron. Lett.*, **25**(11), pp. 702–703, 1989.

[8] M. Born and W. Wolf, *Principles of Optics*, (6th edn), Pergamon Press, 1980.

[9] P. Mossman, 'Connectors for optical fibre systems', *Radio Electron. Eng. (J. IERE)*, **51**(7/8), pp. 333–340, 1981.

[10] J. S. Leach, M. A. Matthews and E. Dalgoutte, 'Optical fibre cable connections', in C. P. Sandbank (Ed.), *Optical Fibre Communication Systems*, pp. 86–105, John Wiley, 1980.

[11] F. L. Thiel and R. M. Hawk, 'Optical waveguide cable connection', *Appl. Opt.*, **15**(11), pp. 2785–2791, 1976.

[12] K. Miyazaki *et al.*, 'Theoretical and experimental considerations of optical fiber connector', OSA Topical Meeting on Opt. Fiber Trans. Williamsburg, Va, paper WA 4-1, 1975.

[13] H. Tsuchiya, H. Nakagome, N. Shimizu and S. Ohara, 'Double eccentric connectors for optical fibers', *Appl. Opt.*, **16**(5), pp. 1323–1331, 1977.

[14] K. J. Fenton and R. L. McCartney, 'Connecting the thread of light', *Electronic Connector Study Group Symposium, 9th Annual Symposium Proc.*, p. 63, Cherry Hill, NJ, 1976.

[15] C. M. Miller, Transmission vs transverse offset for parabolic-profile fiber splices with unequal core diameters', *Bell Syst. Tech. J.*, **55**(7), pp. 917–927, 1976.

[16] D. Gloge, 'Offset and tilt loss in optical fiber splices', *Bell Syst. Tech. J.*, **55**(7), pp. 905–916, 1976.

[17] C. M. Miller and S. C. Mettler, 'A loss model for parabolic-profile fiber splices', *Bell Syst. Tech. J.*, **57**(9), pp. 3167–3180, 1978.

[18] J. J. Esposito, 'Optical connectors, couplers and switches', in H. F. Wolf (Ed.), *Handbook of Fiber Optics, Theory and Applications*, pp. 241–303, Granada, 1979.

[19] J. F. Dalgleish, 'Connections', *Electronics*, pp. 96–98, 5 Aug. 1976.

[20] W. van Etten and J. van Der Platts, *Fundamentals of Optical Fiber Communications*, Prentice Hall International, 1991.

[21] D. Marcuse, 'Loss analysis of single-mode fiber splices', *Bell Syst. Tech. J.*, **56**(5), pp. 703–718, 1977.

[22] W. A. Gambling, H. Matsumura and A. G. Cowley, 'Jointing loss in single-mode fibres', *Electron. Lett.*, **14**(3), pp. 54–55, 1978.

[23] W. A. Gambling, H. Matsumura and C. M. Ragdale, 'Joint loss in single-mode fibres', *Electron. Lett.*, **14**(15), pp. 491–493, 1978.

[24] D. Botez and G. J. Herskowitz, 'Components for optical communications systems: a review', *Proc. IEEE*, **68**(6), pp. 689–731, 1980.

[25] G. Coppa and P. Di Vita, 'Length dependence of joint losses in multimode optical fibres', *Electron. Lett.*, **18**(2), pp. 84–85, 1982.

[26] S. Nemoto and T. Makimoto, 'Analysis of splice loss in single-mode fibers using a Gaussian field approximation', *Opt. Quantum Electron.*, **11**, pp. 447–457, 1979.

[27] W. C. Young and D. R. Frey, 'Fiber connectors', in S. E. Miller and I. P. Kaminow (Eds.) *Optical Fiber Telecommunications II*, Academic Press, pp. 301–326, 1988.

[28] Y. Ushui, T. Ohshima, Y. Toda, Y. Kato and M. Tateda, 'Exact splice loss prediction for single-mode fiber', *IEEE J. Quantum Electron.*, **QE-18**(4), pp. 755–757, 1982.

[29] K. Petermann, 'Nonlinear distortions due to fibre connectors', *Proceedings of 6th European Conference on Optical Communication* (UK), pp. 80–83, 1980.
[30] K. Petermann, 'Wavelength-dependent transmission at fibre connectors', *Electron Lett.*, **15**(22), pp. 706–708, 1979.
[31] C. M. Miller, S. C. Mettler and I. A. White, *Optical Fiber Splices and Connectors: Theory and methods*, Marcel Dekker, 1986.
[32] M. Ikeda, Y. Murakami and K. Kitayama, 'Mode scrambler for optical fibers', *Appl. Opt.*, **16**(4), pp. 1045–1049, 1977.
[33] N. Nashima and N. Uchida, 'Relation between splice loss and mode conversion in a graded-index optical fibre', *Electron. Lett.*, **15**(12), pp. 336–338, 1979.
[34] A. H. Cherin and J. F. Dalgleish, 'Splices and connectors for optical fibre communications', *Telecommun. J. (Eng. Ed.) Switzerland*, **48**(11), pp. 657–665, 1981.
[35] J. E. Midwinter, *Optical Fibers for Transmission*, John Wiley, 1979.
[36] E. A. Lacy, *Fiber Optics*, Prentice Hall, 1982.
[37] R. Jocteur and A. Tardy, 'Optical fiber splicing with plasma torch and oxyhydric microburner', *2nd European Conference on Optical Fibre Communication* (Paris), 1976.
[38] I. Hatakeyama and H. Tsuchiya, 'Fusion splices for single-mode optical fibers', *IEEE J. Quantum Electron.*, **QE-14**(8), pp. 614–619, 1978.
[39] M. Hirai and N. Uchida, 'Melt splice of multimode optical fibre with an electric arc', *Electron. Lett.*, **13**(5), pp. 123–125, 1977.
[40] M. Tsuchiya and I. Hatakeyama, 'Fusion splices for single-mode optical fibres', *Optical Fiber Transmission II*, Williamsburg, pp. PD1, 1–4, Feb. 1977.
[41] F. Esposto and E. Vezzoni, 'Connecting and splicing techniques', *Optical Fibre Communication*, by Technical Staff of CSELT, pp. 541–643, McGraw-Hill, 1981.
[42] D. B. Payne, D. J. McCartney and P. Healey, 'Fusion splicing of a 31.6 km monomode optical fibre system', *Electron. Lett.*, **18**(2), pp. 82–84, 1982.
[43] O. Kawata, K. Hoshino, Y. Miyajima, M. Ohnishi and K. Ishihara, 'A splicing end inspection technique for single-mode fibers using direct core monitoring', *J. of Lightwave Technol.*, **LT-2**, pp. 185–190, 1984.
[44] D. R. Briggs and L. M. Jayne, 'Splice losses in fusion-spliced optical waveguide fibers with different core diameters and numerical apertures', *Proceedings of 27th International Wire and Cable Symposium*, pp. 356–361, 1978.
[45] I. Hatakeyama, M. Tachikura and H. Tsuchiya, 'Mechanical strength of fusion-spliced optical fibres', *Electron. Lett.*, **14**(19), pp. 613–614, 1978.
[46] C. K. Pacey and J. F. Dalgleish, 'Fusion splicing of optical fibres', *Electron. Lett.*, **15**(1), pp. 32–34, 1978.
[47] T. G. Giallorenzi, 'Optical communications research and technology', *Proc. IEEE*, **66**(7), pp. 744–780, 1978.
[48] K. Nawata, Y. Iwahara and N. Suzuki, 'Ceramic capillary splices for optical fibres', *Electron. Lett.*, **15**(15), pp. 470–472, 1979.
[49] J. G. Woods, 'Fiber optic splices', *Proc. SPIE, Fiber Optic Commun. Technol.*, **512**, pp. 44–56, 1984.
[50] C. M. Miller, 'Loose tube splice for optical fibres', *Bell Syst. Tech. J.*, **54**(7), pp. 1215–1225, 1975.
[51] D. Gloge, A. H. Cherin, C. M. Miller and P. W. Smith, 'Fiber splicing', in S. E. Miller (Ed.), *Optical Fiber Telecommunications*, pp. 455–482, Academic Press, 1979.

[52] P. Hensel, J. C. North and J. H. Stewart, 'Connecting optical fibers', *Electron. Power*, **23**(2), pp. 133–135, 1977.

[53] A. R. Tynes and R. M. Derosier, 'Low-loss splices for single-mode fibres', *Electron. Lett.*, **13**(22), pp. 673–674, 1977.

[54] D. N. Knecht, W. J. Carlsen and P. Melman, 'Fiber optic field splice', *Proc. SPIE Int. Soc. Opt. Eng. (USA)* pp. 44–50, 1982.

[55] G. Cocito, B. Costa, S. Longoni, L. Michetti, L. Silvestri, D. Tribone and F. Tosco, 'COS 2 experiment in Turin: field test on an optical cable in ducts', *IEEE Trans. on Commun.*, **COM-26**(7), pp. 1028–1036, 1978.

[56] J. A. Aberson and K. M. Yasinski, 'Multimode mechanical splices', *Proc. Tenth ECOC* (Germany), p. 182, 1984.

[57] C. M. Miller, G. F. DeVeau and M. Y. Smith, 'Simple high-performance mechanical splice for single mode fibers', *Proc. Opt. Fiber Commun. Conf., OFC '85* (USA) paper MI2, 1985.

[58] M. Kawase, M. Tachikura, F. Nihei and H. Murata, 'Mass fusion splices for high density optical fiber units', *Proc. Eighth ECOC* (France), paper AX-5, 1982.

[59] Y. Katsuyama, S. Hatano, K. Hogari, T. Matsumoto and T. Kokubun, 'Single mode optical fibre ribbon cable', *Electron. Lett.*, **21**, pp. 134–135, 1985.

[60] H. Murata, *Handbook of Optical Fibers and Cables*, Marcel Dekker, 1988.

[61] E. L. Chinnock, D. Gloge, D. L. Bisbee and P. W. Smith, 'Preparation of optical fiber ends for low-loss tape splices', *Bell Syst. Tech. J.*, **54**, pp. 471–477, 1975.

[62] R. Delebecque, E. Chazelas and D. Boscher, 'Flat mass splicing process for cylindrical V-groved cables', *Proc. IWCS '82*, Cherry Hill, NJ (USA), pp. 184–187, 1982.

[63] N. E. Hardwick and S. T. Davies, 'Rapid ribbon splice for multimode fiber splicing', *Proc. Opt. Fiber Commun. Conf., OFC '85* (USA) paper TUQ27, 1985.

[64] T. W. Tamulevich, 'Fiber optic ceramic capillary connectors', *Photonics Spectra*, pp. 65–70, October, 1984.

[65] M. D. Drake, 'A critical review of fiber optic connectors', *Proc. SPIE, Fiber Optic Commun. Technol.*, **512**, pp. 57–69, 1984.

[66] G. Kotelly, 'Special report: fiber optic connectors', *Lightwave, J. of Fiber Optics*, pp. 1, 32–39, April 1989.

[67] W. C. Young, P. Kaiser, N. K. Cheung, L. Curtis, R. E. Wagner and D. M. Folkes, 'A transfer molded biconic connector with insertion losses below 0.3 dB without index match', *Proceedings of 6th European Conference on Optical Communication*, pp. 310–313, 1980.

[68] T. King, 'Fibre optic components for the fibre distributed data interface (FDDI) 100 Mbit/s local area network', *Proc. SPIE, Fibre Optics '88*, **949**, pp. 2–13, 1988.

[69] P. W. Smith, D. L. Bisbee, D. Gloge and E. L. Chinnock, 'A moulded-plastic technique for connecting and splicing optical fiber tapes and cables', *Bell Syst. Tech. J.*, **54**(6), pp. 971–984, 1975.

[70] Y. Fujii, J. Minowa and N. Suzuki, 'Demountable multiple connector with precise V-grooved silicon', *Electron. Lett.*, **15**(14), pp. 424–425, 1979.

[71] M. Oda, M. Ogai, A. Ohtake, S. Tachigami, S. Ohkubo, F. Nihei and N. Kashima, 'Nylon extruded fiber ribbon and its connection', *Proc. Opt. Fiber Commun. Conf., OFC '82* (USA), p. 46, 1982.

[72] S. Tachigami, A. Ohtake, T. Hayashi, T. Iso and T. Shirasawa, 'Fabrication and evaluation of high density multi-fiber plastic connector', *Proc. IWCS*, Cherry Hill, N. J. (USA), pp. 70–75, 1983.

[73] T. Sakake, N. Kashima and M. Oki, 'Very small single-mode ten-fiber connector', *J. Lightwave Technol.*, **6**(2), pp. 269–272, 1988.
[74] A. Nicia, 'Practical low-loss lens connector for optical fibers', *Electron. Lett.*, **14**(16), pp. 511–512, 1978.
[75] A. Nicia and A. Tholen, 'High efficiency ball-lens connector and related functional devices for single-mode fibers', *Proc. Seventh ECOC*, (Denmark) paper 7.5, 1981.
[76] D. M. Knecht and W. J. Carlsen, 'Expanded beam fiber optic connectors', *Proc. SPIE* (USA), pp. 44–50, 1983.
[77] K. Kobayashi, R. Ishikawa, K. Minemura and S. Sugimoto, 'Micro-optic devices for fiber-optic communications', *Fiber and Integrated Optics*, **2**, pp. 1–17, 1979.
[78] W. J. Tomlinson, 'Applications of GRIN rod lenses in optical fiber communication systems', *Appl. Opt.*, **19**, pp. 1127–1138, 1980.
[79] T. Uchida, M. Furukawa, I. Kitano, K. Koizumi and H. Matsomura, 'Optical characteristics of a light focusing guide and its application', *IEEE J. of Quantum Electron.*, **QE-6**, pp. 606–612, 1970.
[80] D. Marcuse and S. E. Miller, 'Analysis of a tubular gas lens', *Bell Syst. Tech. J.*, **43**, pp. 1159–1782, 1965.
[81] S. E. Miller, 'Light Propagation in generalized lenslike media', *Bell Syst. Tech. J.*, **44**, pp. 2017–2064, 1965.
[82] K. Sono, 'Graded index rod lenses', *Laser Focus*, **17**, pp. 70–74, 1981.
[83] J. M. Senior, S. D. Cusworth, N. G. Burrow and A. D. Muirhead, 'Misalignment losses at multimode graded-index fiber splices and GRIN rod lens couplers', *Appl. Opt.*, **24**, pp. 977–982, 1985.
[84] S. van Dorn, 'Fiber optic couplers', *Proc. SPIE, 574, Fiber Optic Couplers, Connectors and Splice Technology II*, pp. 2–8, 1985.
[85] K. O. Hill, D. C. Johnson and R. G. Lamont, 'Optical fiber directional couplers: biconical taper technology and device applications', *Proc. SPIE, Fiber Optic Couplers, Connectors and Splice Technology II*, **574**, pp. 92–99, 1985.
[86] A. K. Agarwal, 'Review of optical fiber couplers', *Fiber Integr. Opt.*, **6**(1), pp. 27–53, 1987.
[87] K. Kobayashi, R. Ishikawa, K. Minemura and S. Sugimoto, 'Micro-optic devices for fiber-optic communications', *Fiber Integr. Opt.*, **2**(1), pp. 1–17, 1979.
[88] B. S. Kawasaki, K. O. Hill and Y. Tremblay, 'Modal-noise generation in biconical taper couplers', *Opt. Lett.*, **6**, p. 499, 1981.
[89] J. V. Wright, 'Wavelength dependence of fused couplers', *Electron. Lett.*, **22**, pp. 329–331, 1986.
[90] F. P. Payne, 'Dependence of fused taper couplers on external refractive index', *Electron. Lett.*, **22**, pp. 1207–1208, 1986.
[91] D. T. Cassidy, D. C. Johnson and K. O. Hill, 'Wavelength dependent transmission of monomode optical fiber tapers', *Appl. Opt.*, **24**, pp. 945–950, 1985.
[92] I. Yokohama, K. Okamoto and J. Noda, 'Polarization-independent optical circulator consisting of two fiber-optic polarizing beamsplitters and two YIG spherical lenses', *Electron. Lett.*, **22**, pp. 370–372, 1985.
[93] E. Paillard, 'Recent developments in integrated optics', *Proc. SPIE, Fiber Optics '87*, **734**, pp. 131–136, 1987.
[94] T. Findalky, 'Glass waveguides by ion exchange', *Opt. Eng.*, **24**, pp. 244–250, 1985.
[95] J. P. Dakin, M. G. Holliday and S. W. Hickling, 'Non invasive optical bus for video distribution', *Proc. SPIE, Fibre Optics '88*, **949**, pp. 36–40, 1988.
[96] G. D. Khoe and H. Lydtin, 'European optical fibers and passive components:

status and trends', *IEEE J. Selected Areas in Commun.*, **SAC-4**(4), pp. 457–471, 1986.

[97] C. C. Wang, W. K. Burns and C. A. Villaruel, '9×9 single-mode fiber optic star couplers', *Opt. Lett.*, **10**(1), pp. 49–51, 1985.

[98] J. M. Senior and S. D. Cusworth, 'Devices for wavelength multiplexing and demultiplexing', *IEE Proc., Pt J*, **136**(3), pp. 183–202, 1989.

[99] Y. Fujii, K. Aoyama and J. Minowa, 'Optical demultiplexer using a silicon echette grating', *IEEE J. Quantum Electron.*, **QE-16**, pp. 165–169, 1980.

[100] F. L. Pedrotti and L. S. Pedrotti, *Introduction to Optics*, Prentice Hall, 1987.

[101] R. Erdmann, 'Prism gratings for fiber optic multiplexing', *Proc. SPIE, Fiber Optics Multiplexing and Modulation*, **417**, pp. 12–17, 1983.

[102] A. Nicia, 'Wavelength multiplexing and demultiplexing systems for single mode and multimode fibers', *Seventh European Conf. on Opt. Commun. (ECOC 81)*, pp. 8.1–7, September 1981.

[103] J. Lipson, C. A. Young, P. D. Yeates, J. C. Masland, S. A. Wartonick, G. T. Harvey and P. H. Read, 'A four channel lightwave subsystem using wavelength multiplexing', *J. Lightwave Technol.*, **LT-3**, pp. 16–20, 1985.

[104] J. Lispon, W. J. Minford, E. J. Murphy, T. C. Rice, R. A. Linke and G. T. Harvey, 'A six-channel wavelength multiplexer and demultiplexer for single mode systems', *J. Lightwave Technol.*, **LT-3**, pp. 1159–1161, 1985.

[105] H. A. Macleod, *Thin Film Optical Filters* (2nd edn), Adam Hilger Ltd, 1986.

[106] G. Winzer, H. F. Mahlein and A. Reichelt, 'Single-mode and multimode all-fiber directional couplers for WDM', *Appl. Opt.*, **20**, pp. 3128–3135, 1981.

[107] Y. Fujii, J. Minowa and H. Tanada, 'Practical two-wavelength multiplexer and demultiplexer: design and performance', *Appl. Opt.*, **22**, pp. 3090–3097, 1983.

[108] M. McCourt and J. L. Malinge, 'Application of ion exchange techniques to the fabrication of multimode wavelength division multiplexers', *Proc. SPIE, Fiber Optics '88*, 949, pp. 131–137, 1988.

[109] R. Zengerle and O. G. Leminger, 'Wavelength-selective directional coupler made of non-identical single-mode fibers', *J. Lightwave Technol.*, **LT-4**, pp. 823–826, 1986.

[110] D. Marcuse, 'Coupling of degenerative modes in two parallel dielectric waveguides', *Bell Syst. Tech. J.*, **50**, pp. 1791–1816, 1971.

[111] B. S. Kawasaki and K. O. Hill, 'Low loss access coupler for multimode optical fiber distribution networks', *Appl. Opt.*, **16**, pp. 327–328, 1977.

[112] M. J. F. Digonnet and H. J. Shaw, 'Wavelength multiplexing in single mode fiber couplers', *Appl. Opt.*, **22**, pp. 484–492, 1983.

[113] O. Leminger and R. Zengerle, 'Bandwidth of directional-coupler wavelength filters made of dissimilar optical fibres', *Electron. Lett.*, **23**, pp. 241–242, 1987.

[114] R. Zengerle and O. Leminger, 'Narrow band wavelength selective directional-coupler made of dissimilar optical fibres', *J. of Lightwave Technol.*, **LT-5**, pp. 1196–1198, 1987.

[115] H. A. Roberts, 'Single-mode fused wavelength division multiplexer', *Proc. SPIE*, **574**, pp. 100–104, 1985.

[116] J. B. Straws and B. Kawasaki, 'Passive optical components', in E. E. Basch (Ed.), *Optical-Fiber Transmission*, H. W. Sams & Co., pp. 241–264, 1987.

[117] W. J. Tomlinson, 'Passive and low-speed active optical components for fiber systems', in S. E. Miller and I. P. Kaminow (Eds.), *Optical Fiber Telecommunications II*, Academic Press, pp. 369–419, 1988.

6

Optical sources 1: the laser

6.1 Introduction

The optical source is often considered to be the active component in an optical fiber communication system. Its fundamental function is to convert electrical energy in the form of a current into optical energy (light) in an efficient manner which allows the light output to be effectively launched or coupled into the optical fiber. Three main types of optical light source are available. These are:

(a) wideband 'continuous spectra' sources (incandescent lamps);
(b) monochromatic incoherent sources (light emitting diodes, LEDs);
(c) monochromatic coherent sources (lasers).

To aid consideration of the sources currently in major use the historical aspect must be mentioned. In the early stages of optical fiber communication the most powerful narrowband coherent light sources were necessary due to severe attenuation and dispersion in the fibers. Therefore, gas lasers (helium–neon) were utilized initially. However, the development of the semiconductor injection laser and the LED, together with the substantial improvement in the properties of optical fibers, has given prominence to these two specific sources.

To a large extent these two sources fulfil the major requirements for an optical fiber emitter which are outlined below:

1. A size and configuration compatible with launching light into an optical fiber. Ideally, the light output should be highly directional.
2. Must accurately track the electrical input signal to minimize distortion and noise. Ideally, the source should be linear.
3. Should emit light at wavelengths where the fiber has low losses and low dispersion and where the detectors are efficient.
4. Preferably capable of simple signal modulation (i.e. direct – see Section 7.5) over a wide bandwidth extending from audio frequencies to beyond the gigahertz range.
5. Must couple sufficient optical power to overcome attenuation in the fiber plus additional connector losses and leave adequate power to drive the detector.
6. Should have a very narrow spectral bandwidth (linewidth) in order to minimize dispersion in the fiber.
7. Must be capable of maintaining a stable optical output which is largely unaffected by changes in ambient conditions (e.g. temperature).
8. It is essential that the source is comparatively cheap and highly reliable in order to compete with conventional transmission techniques.

In order to form some comparison between these two types of light source the historical aspect must be enlarged upon. The first generation optical communication sources were designed to operate between 0.8 and 0.9 μm (ideally around 0.85 μm) because initially the properties of the semiconductor materials used lent themselves to emission at this wavelength. Also, as suggested in (3) this wavelength avoided the loss incurred in many fibers near 0.9 μm due to the OH ion (see Section 3.3.2). These early systems utilized multimode step index fibers which required the superior performance of semiconductor lasers for links of reasonable bandwidth (tens of megahertz) and distances (several kilometres). The LED (being a lower power source generally exhibiting little spatial or temporal coherence) was not suitable for long distance wideband transmission, although it found use in more moderate applications.

However, the role of the LED as a source for optical fiber communications was enhanced following the development of multimode graded index fiber. The substantial reduction in intermodal dispersion provided by this fiber type over multimode step index fiber allowed incoherent LEDs emitting in the 0.8 to 0.9 μm wavelength band to be utilized for applications requiring wider bandwidths. This position was further consolidated with the development of second generation optical fiber sources operating at wavelengths between 1.1 and 1.6 μm where both material losses and dispersion are greatly reduced. In this wavelength region, wideband graded index fiber systems utilizing LED sources may be operated over long distances without the need for intermediate repeaters. Furthermore, LEDs offer the advantages of relatively simple construction and operation with the inherent effects of these factors on cost and extended, trouble-free life.

In parallel with these later developments in multimode optical propagation came advances in single-mode fiber construction. This has stimulated the development of single-mode laser sources to take advantage of the extremely low dispersion offered by single-mode fibers. These systems are ideally suited to extra wideband, very long-haul applications and are currently under intensive investigation for long distance telecommunications. On the other hand, light is usually emitted from the LED in many spatial modes which cannot be as efficiently focused and coupled into single-mode fiber. Nevertheless, recently advanced LED sources have been developed that will allow moderate optical power levels to be launched into single-mode fiber (see Chapter 7). However, to date the LED has been utilized primarily as a multimode source giving acceptable coupling efficiencies into multimode fiber. Moreover, in this capacity the LED remains the major multimode source which is extensively used for increasingly wider bandwidth, longer-haul applications. Therefore at present the LED is chosen for many applications using multimode fibers and the injection laser diode (ILD) tends to find more use as a single-mode device in single-mode fiber systems. Although other laser types (e.g. Nd : YAG and glass fiber laser, Section 6.9), as well as the injection laser, may eventually find significant use in optical fiber communications, this chapter and the following one will deal primarily with major structures and configurations of semiconductor sources (ILD and LED), taking into account recent developments and possible future advances.

We begin by describing in Section 6.2 the basic principles of laser operation which may be applied to all laser types. Immediately following, in Section 6.3, is a discussion of optical emission from semiconductors in which we concentrate on the fundamental operating principles, the structure and the materials for the semiconductor laser. Aspects concerning practical semiconductor lasers are then considered in Section 6.4 prior to a more specific discussion of the structure and operation of some common injection laser types in Section 6.5. Following, in Section 6.6, the major single frequency injection laser structures which provide single-mode operation, primarily in the longer wavelength region (1.1 to 1.6 μm), are then described. In Section 6.7 we consider the operating characteristics which are common to all injection laser types, before a short discussion of injection laser to optical fiber coupling is presented in Section 6.8. Major nonsemiconductor laser devices which have found use in optical fiber communications (the neodymium-doped yttrium–aluminium–garnet (Nd : YAG) laser and the glass fiber laser) are then outlined in Section 6.9. This is followed in Section 6.10 with a discussion of advanced linewidth narrowed and wavelength tunable laser types. Finally, in Section 6.11, developments in laser sources for transmission in the mid-infrared wavelength region (2 to 5 μm) are briefly considered to give an insight into this potentially important future area.

6.2 Basic concepts

To gain an understanding of the light-generating mechanisms within the major optical sources used in optical fiber communications it is necessary to consider both the fundamental atomic concepts and the device structure. In this context the requirements for the laser source are far more stringent than those for the LED. Unlike the LED, strictly speaking, the laser is a device which amplifies light. Hence the derivation of the term LASER as an acronym for Light Amplification by Stimulated Emission of Radiation. Lasers, however, are seldom used as amplifiers since there are practical difficulties in relation to the achievement of high gain whilst avoiding oscillation from the required energy feedback. Thus the practical realization of the laser is as an optical oscillator. The operation of the device may be described by the formation of an electromagnetic standing wave within a cavity (or optical resonator) which provides an output of monochromatic highly coherent radiation. By contrast the LED provides optical emission without an inherent gain mechanism. This results in incoherent light output.

In this section we elaborate on the basic principles which govern the operation of both these optical sources. It is clear, however, that the operation of the laser must be discussed in some detail in order to provide an appreciation of the way it functions as an optical source. Hence we concentrate first on the general principles of laser action.

6.2.1 Absorption and emission of radiation

The interaction of light with matter takes place in discrete packets of energy or quanta, called photons. Furthermore, the quantum theory suggests that atoms exist only in certain discrete energy states such that absorption and emission of light causes them to make a transition from one discrete energy state to another. The frequency of the absorbed or emitted radiation f is related to the difference in energy E between the higher energy state E_2 and the lower energy state E_1 by the expression:

$$E = E_2 - E_1 = hf \tag{6.1}$$

where $h = 6.626 \times 10^{-34}$ J s is Planck's constant. These discrete energy states for the atom may be considered to correspond to electrons occurring in particular energy levels relative to the nucleus. Hence, different energy states for the atom correspond to different electron configurations, and a single electron transition between two energy levels within the atom will provide a change in energy suitable for the absorption or emission of a photon. It must be noted, however, that modern quantum theory [Ref. 1] gives a probabilistic description which specifies the energy levels in which electrons are most likely to be found. Nevertheless, the concept of stable atomic energy states and electron transitions between energy levels is still valid.

Figure 6.1(a) illustrates a two energy state or level atomic system where an atom is initially in the lower energy state E_1. When a photon with energy $(E_2 - E_1)$ is incident on the atom it may be excited into the higher energy state E_2 through absorption of the photon. This process is sometimes referred to as stimulated absorption. Alternatively, when the atom is initially in the higher energy state E_2 it can make a transition to the lower energy state E_1 providing the emission of a photon at a frequency corresponding to Eq. (6.1). This emission process can occur

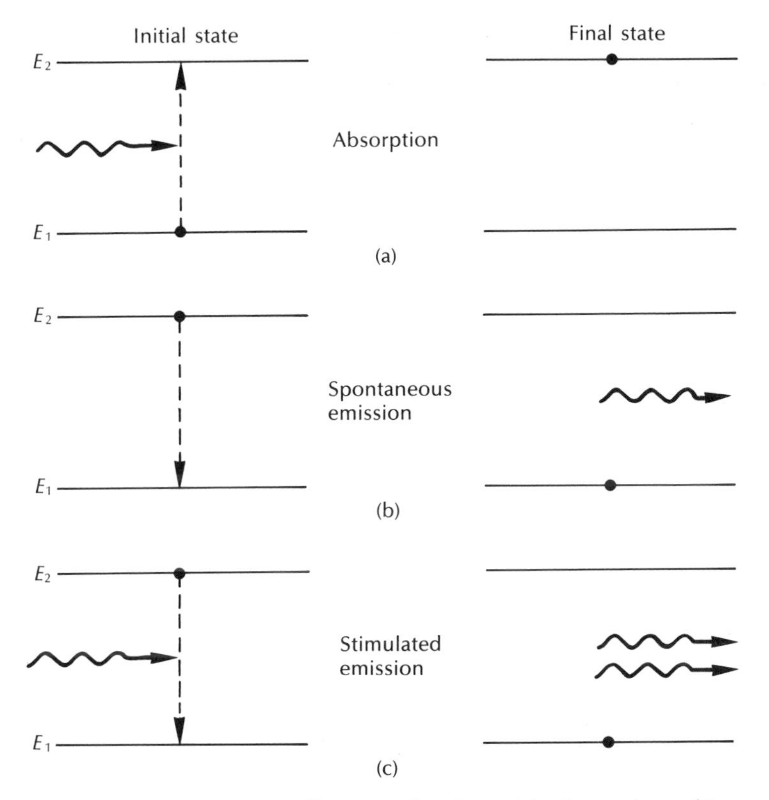

Figure 6.1 Energy state diagram showing: (a) absorption; (b) spontaneous emission; (c) stimulated emission. The black dot indicates the state of the atom before and after a transition takes place.

in two ways:

(a) by spontaneous emission in which the atom returns to the lower energy state in an entirely random manner;
(b) by stimulated emission when a photon having an energy equal to the energy difference between the two states ($E_2 - E_1$) interacts with the atom in the upper energy state causing it to return to the lower state with the creation of a second photon.

These two emission processes are illustrated in Figure 6.1(b) and (c) respectively. The random nature of the spontaneous emission process where light is emitted by electronic transitions from a large number of atoms gives incoherent radiation. A similar emission process in semiconductors provides the basic mechanism for light generation within the LED (see Section 6.3.2).

It is the stimulated emission process, however, which gives the laser its special properties as an optical source. Firstly, the photon produced by stimulated emission is generally* of an identical energy to the one which caused it and hence the light associated with them is of the same frequency. Secondly, the light associated with the stimulating and stimulated photon is in phase and has the same polarization. Therefore, in contrast to spontaneous emission, coherent radiation is obtained. Furthermore, this means that when an atom is stimulated to emit light energy by an incident wave, the liberated energy can add to the wave in a constructive manner, providing amplification.

6.2.2 The Einstein relations

Prior to discussion of laser action in semiconductors it is useful to consider optical amplification in the two level atomic system shown in Figure 6.1. In 1917 Einstein [Ref. 2] demonstrated that the rates of the three transition processes of absorption, spontaneous emission and stimulated emission were related mathematically. He achieved this by considering the atomic system to be in thermal equilibrium such that the rate of the upward transitions must equal the rate of the downward transitions. The population of the two energy levels of such a system are described by Boltzmann statistics which give:

$$\frac{N_1}{N_2} = \frac{g_1 \exp(-E_1/KT)}{g_2 \exp(-E_2/KT)} = \frac{g_1}{g_2} \exp(E_2 - E_1/KT)$$

$$= \frac{g_1}{g_2} \exp(hf/KT) \qquad (6.2)$$

where N_1 and N_2 represent the density of atoms in energy levels E_1 and E_2,

* A photon with energy hf will not necessarily always stimulate another photon with energy hf. Photons may be stimulated over a small range of energies around hf providing an emission which has a finite frequency or wavelength spread (linewidth).

respectively, with g_1 and g_2 being the corresponding degeneracies* of the levels, K is Boltzmann's constant and T is the absolute temperature.

As the density of atoms in the lower or ground energy state E_1 is N_1, the rate of upward transition or absorption is proportional to both N_1 and the spectral density ρ_f of the radiation energy at the transition frequency f. Hence, the upward transition rate R_{12} (indicating an electron transition from level 1 to level 2) may be written as:

$$R_{12} = N_1 \rho_f B_{12} \tag{6.3}$$

where the constant of proportionality B_{12} is known as the Einstein coefficient of absorption.

By contrast atoms in the higher or excited energy state can undergo electron transitions from level 2 to level 1 either spontaneously or through stimulation by the radiation field. For spontaneous emission the average time an electron exists in the excited state before a transition occurs is known as the spontaneous lifetime τ_{21}. If the density of atoms within the system with energy E_2, is N_2, then the spontaneous emission rate is given by the product of N_2 and $1/\tau_2$. This may be written as $N_2 A_{21}$ where A_{21}, the Einstein coefficient of spontaneous emission, is equal to the reciprocal of the spontaneous lifetime.

The rate of stimulated downward transition of an electron from level 2 to level 1 may be obtained in a similar manner to the rate of stimulated upward transition. Hence the rate of stimulated emission is given by $N_2 \rho_f B_{21}$, where B_{21} is the Einstein coefficient of stimulated emission. The total transition rates from level 2 to level 1, R_{21}, is the sum of the spontaneous and stimulated contributions. Hence:

$$R_{21} = N_2 A_{21} + N_2 \rho_f B_{21} \tag{6.4}$$

For a system in thermal equilibrium, the upward and downward transition rates must be equal and therefore $R_{12} = R_{21}$, or

$$N_1 \rho_f B_{12} = N_2 A_{21} + N_2 \rho_f B_{21} \tag{6.5}$$

It follows that:

$$\rho_f = \frac{N_2 A_{21}}{N_1 B_{12} - N_2 B_{21}}$$

and

$$\rho_f = \frac{A_{21}/B_{21}}{(B_{12} N_1 / B_{21} N_2) - 1} \tag{6.6}$$

* In many cases the atom has several sublevels of equal energy within an energy level which is then said to be degenerate. The degeneracy parameters g_1 and g_2 indicate the number of sublevels within the energy levels E_1 and E_2 respectively. If the system is not degenerate, then g_1 and g_2 may be set to unity [Ref. 1].

Substituting Eq. (6.2) into Eq. (6.6) gives

$$\rho_f = \frac{A_{21}/B_{21}}{[(g_1B_{12}/g_2B_{21})\exp(hf/KT)] - 1} \tag{6.7}$$

However, since the atomic system under consideration is in thermal equilibrium it produces a radiation density which is identical to black body radiation. Planck showed that the radiation spectral density for a black body radiating within a frequency range f to $f + df$ is given by [Ref.3]:

$$\rho_f = \frac{8\pi hf^3}{c^3}\left(\frac{1}{\exp(hf/KT) - 1}\right) \tag{6.8}$$

Comparing Eq. (6.8) with Eq. (6.7) we obtain the Einstein relations:

$$B_{12} = \left(\frac{g_2}{g_1}\right)B_{21} \tag{6.9}$$

and

$$\frac{A_{21}}{B_{21}} = \frac{8\pi hf^3}{c^3} \tag{6.10}$$

It may be observed from Eq. (6.9) that when the degeneracies of the two levels are equal ($g_1 = g_2$), then the probabilities of absorption and stimulated emission are equal. Furthermore, the ratio of the stimulated emission rate to the spontaneous emission rate is given by:

$$\frac{\textit{Stimulated emission rate}}{\textit{Spontaneous emission rate}} = \frac{B_{21}\rho_f}{A_{21}} = \frac{1}{\exp(hf/KT) - 1} \tag{6.11}$$

Example 6.1

Calculate the ratio of the stimulated emission rate to the spontaneous emission rate for an incandescent lamp operating at a temperature of 1000 K. It may be assumed that the average operating wavelength is 0.5 μm.

Solution: The average operating frequency is given by:

$$f = \frac{c}{\lambda} = \frac{2.998 \times 10^8}{0.5 \times 10^{-6}} \simeq 6.0 \times 10^{14} \text{ Hz}$$

Using Eq. (6.11) the ratio is:

$$\frac{\textit{Stimulated emission rate}}{\textit{Spontaneous emission rate}} = \frac{1}{\exp\left(\frac{6.626 \times 10^{-34} \times 6 \times 10^{14}}{1.381 \times 10^{-23} \times 1000}\right)}$$

$$= \exp(-28.8)$$

$$= 3.1 \times 10^{-13}$$

The result obtained in Example 6.1 indicates that for systems in thermal equilibrium spontaneous emission is by far the dominant mechanism. Furthermore, it illustrates that the radiation emitted from ordinary optical sources in the visible spectrum occurs in a random manner, proving these sources are incoherent.

It is apparent that in order to produce a coherent optical source and amplification of a light beam the rate of stimulated emission must be increased far above the level indicated by Example 6.1. From consideration of Eq. (6.5) it may be noted that for stimulated emission to dominate over absorption and spontaneous emission in a two level system, both the radiation density and the population density of the upper energy level N_2 must be increased in relation to the population density of the lower energy level N_1.

6.2.3 Population inversion

Under the conditions of thermal equilibrium given by the Boltzmann distribution (Eq. (6.2)) the lower energy level E_1 of the two level atomic system contains more atoms than the upper energy level E_2. This situation, which is normal for structures at room temperature, is illustrated in Figure 6.2(a). However, to achieve optical amplification it is necessary to create a nonequilibrium distribution of atoms such that the population of the upper energy level is greater than that of the lower energy level (i.e. $N_2 > N_1$). This condition, which is known as population inversion, is illustrated in Figure 6.2(b).

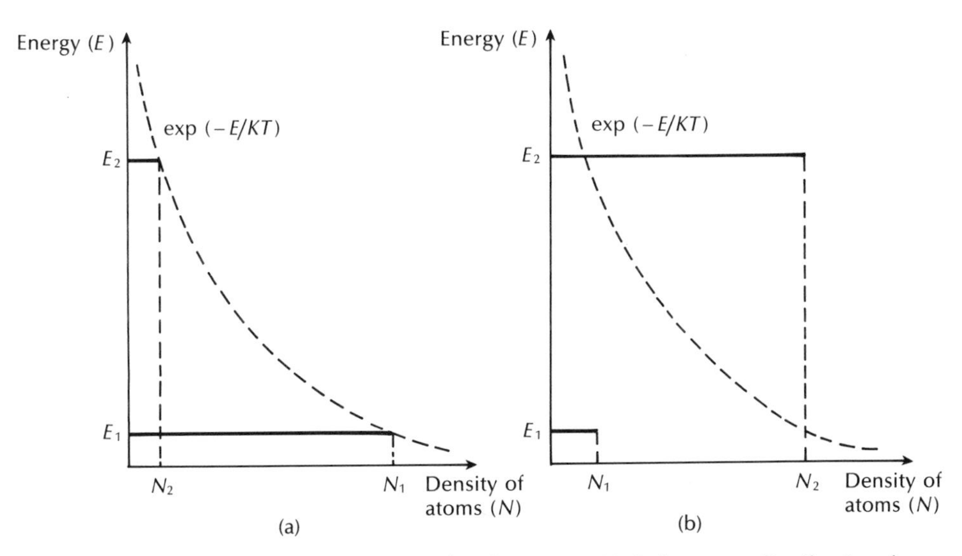

Figure 6.2 Populations in a two energy level system: (a) Boltzmann distribution for a system in thermal equilibrium; (b) a nonequilibrium distribution showing population inversion.

In order to achieve population inversion it is necessary to excite atoms into the upper energy level E_2 and hence obtain a nonequilibrium distribution. This process is achieved using an external energy source and is referred to as 'pumping'. A common method used for pumping involves the application of intense radiation (e.g. from an optical flash tube or high frequency radio field). In the former case atoms are excited into the higher energy state through stimulated absorption. However, the two level system discussed above does not lend itself to suitable population inversion. Referring to Eq. (6.9), when the two levels are equally degenerate (or not degenerate), then $B_{12} = B_{21}$. Thus the probabilities of absorption and stimulated emission are equal, providing at best equal populations in the two levels.

Population inversion, however, may be obtained in systems with three or four energy levels. The energy level diagrams for two such systems, which correspond to two nonsemiconductor lasers, are illustrated in Figure 6.3. To aid attainment of population inversion both systems display a central metastable state in which the atoms spend an unusually long time. It is from this metastable level that the stimulated emission or lasing takes place. The three level system (Figure 6.3(a)) consists of a ground level E_0, a metastable level E_1 and a third level above the metastable level E_2. Initially, the atomic distribution will follow Boltzmann's law. However, with suitable pumping the electrons in some of the atoms may be excited from the ground state into the higher level E_2. Since E_2 is a normal level the electrons will rapidly decay by nonradiative processes to either E_1 or directly to E_0. Hence empty states will always be provided in E_2. The metastable level E_1 exhibits

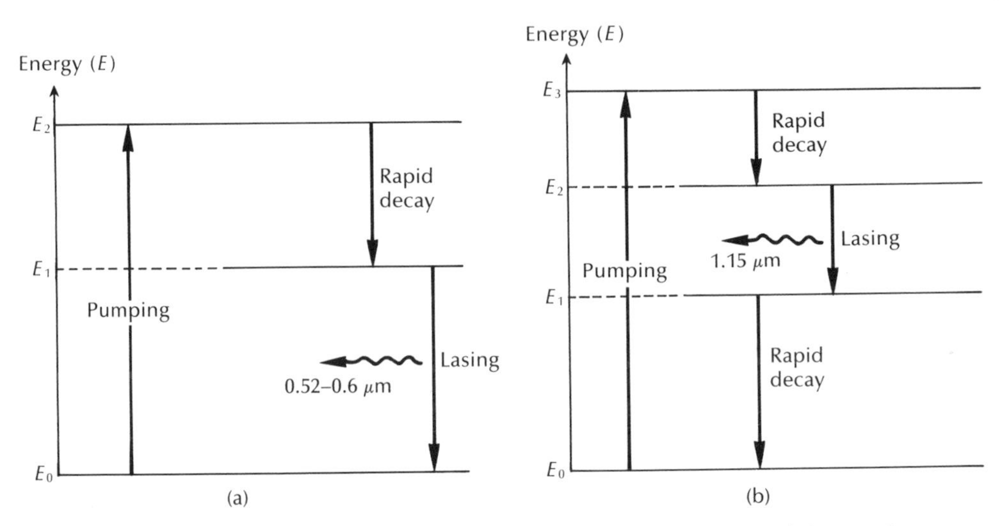

Figure 6.3 Energy level diagrams showing population inversion and lasing for two nonsemiconductor lasers: (a) three level system – ruby (crystal) laser; (b) four level system – He–Ne (gas) laser.

a much longer lifetime than E_2 which allows a large number of atoms to accumulate at E_1. Over a period the density of atoms in the metastable state N_1 increases above those in the ground state N_0 and a population inversion is obtained between these two levels. Stimulated emission and hence lasing can then occur creating radiative electron transitions between levels E_1 and E_0. A drawback with the three level system such as the ruby laser is that it generally requires very high pump powers because the terminal state of the laser transition is the ground state. Hence more than half the ground state atoms must be pumped into the metastable state to achieve population inversion.

By contrast, a four level system such as the He–Ne laser illustrated in Figure 6.3(b) is characterized by much lower pumping requirements. In this case the pumping excites the atoms from the ground state into energy level E_3 and they decay rapidly to the metastable level E_2. However, since the populations of E_3 and E_1 remain essentially unchanged a small increase in the number of atoms in energy level E_2 creates population inversion, and lasing takes place between this level and level E_1.

6.2.4 Optical feedback and laser oscillation

Light amplification in the laser occurs when a photon colliding with an atom in the excited energy state causes the stimulated emission of a second photon and then both these photons release two more. Continuation of this process effectively creates avalanche multiplication, and when the electromagnetic waves associated with these photons are in phase, amplified coherent emission is obtained. To achieve this laser action it is necessary to contain photons within the laser medium and maintain the conditions for coherence. This is accomplished by placing or forming mirrors (plane or curved) at either end of the amplifying medium, as illustrated in Figure 6.4. The optical cavity formed is more analogous to an oscillator than an amplifier as it provides positive feedback of the photons by reflection at the mirrors at either end of the cavity. Hence the optical signal is fed back many times whilst receiving amplification as it passes through the medium. The structure therefore acts as a Fabry–Perot resonator. Although the amplification of the signal from a single pass through the medium is quite small, after multiple passes the net gain can be large. Furthermore, if one mirror is made partially transmitting, useful radiation may escape from the cavity.

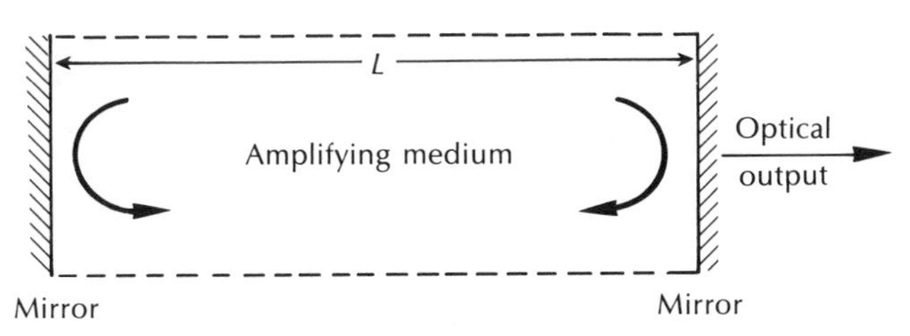

Figure 6.4 The basic laser structure incorporating plane mirrors.

A stable output is obtained at saturation when the optical gain is exactly matched by the losses incurred in the amplifying medium. The major losses result from factors such as absorption and scattering in the amplifying medium, absorption, scattering and diffraction at the mirrors and nonuseful transmission through the mirrors.

Oscillations occur in the laser cavity over a small range of frequencies where the cavity gain is sufficient to overcome the above losses. Hence the device is not a perfectly monochromatic source but emits over a narrow spectral band. The central frequency of this spectral band is determined by the mean energy level difference of the stimulated emission transition. Other oscillation frequencies within the spectral band result from frequency variations due to the thermal motion of atoms within the amplifying medium (known as Doppler broadening*) and by atomic collisions†. Hence the amplification within the laser medium results in a broadened laser transition or gain curve over a finite spectral width, as illustrated in Figure 6.5. The spectral emission from the device therefore lies within the frequency range dictated by this gain curve.

Since the structure forms a resonant cavity, when sufficient population inversion exists in the amplifying medium the radiation builds up and becomes established as standing waves between the mirrors. These standing waves exist only at frequencies for which the distance between the mirrors is an integral number of half wavelengths. Thus when the optical spacing between the mirrors is L the resonance condition along the axis of the cavity is given by [Ref. 4]:

$$L = \frac{\lambda q}{2n} \tag{6.12}$$

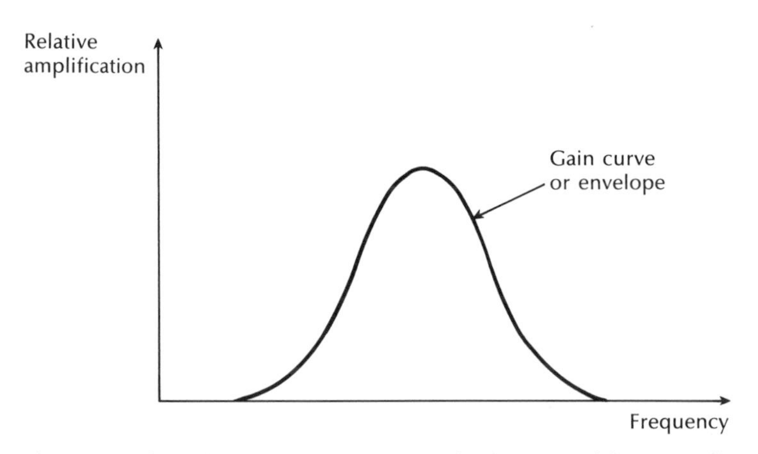

Figure 6.5 The relative amplification in the laser amplifying medium showing the broadened laser transition line or gain curve.

* Doppler broadening is referred to as an inhomogeneous broadening mechanism since individual groups of atoms in the collection have different apparent resonance frequencies.

† Atomic collisions provide homogeneous broadening as every atom in the collection has the same resonant frequency and spectral spread.

where λ is the emission wavelength, n is the refractive index of the amplifying medium and q is an integer. Alternatively, discrete emission frequencies f are defined by

$$f = \frac{qc}{2nL} \tag{6.13}$$

where c is the velocity of light. The different frequencies of oscillation within the laser cavity are determined by the various integer values of q and each constitutes a resonance or mode. Since Eqs. (6.12) and (6.13) apply for the case when L is along the longitudinal axis of the structure (Figure 6.4) the frequencies given by Eq. (6.13) are known as the longitudinal or axial modes. Furthermore, from Eq. (6.13) it may be observed that these modes are separated by a frequency interval δf where:

$$\delta f = \frac{c}{2nL} \tag{6.14}$$

The mode separation in terms of the free space wavelength, assuming $\delta f \ll f$ and as $f = c/\lambda$, is given by:

$$\delta\lambda = \frac{\lambda \delta f}{f} = \frac{\lambda^2}{c}\,\delta f \tag{6.15}$$

Hence substituting for δf from Eq. (6.14) gives:

$$\delta\lambda = \frac{\lambda^2}{2nL} \tag{6.16}$$

In addition it should be noted that Eq. (6.15) can be used to determine the device spectral linewidth as a function of wavelength when it is quoted in Hertz, or vice versa (see problem 6.4).

Example 6.2

A ruby laser contains a crystal length 4 cm with a refractive index of 1.78. The peak emission wavelength from the device is 0.55 μm. Determine the number of longitudinal modes and their frequency separation.

Solution: The number of longitudinal modes supported within the structure may be obtained from Eq. (6.12) where:

$$q = \frac{2nL}{\lambda} = \frac{2 \times 1.78 \times 0.04}{0.55 \times 10^{-6}} = 2.6 \times 10^5$$

Using Eq. (6.14) the frequency separation of the modes is:

$$\delta f = \frac{2.998 \times 10^8}{2 \times 1.78 \times 0.04} = 2.1 \text{ GHz}$$

Although the result of Example 6.2 indicates that a large number of modes may be generated within the laser cavity, the spectral output from the device is defined by the gain curve. Hence the laser emission will only include the longitudinal modes contained within the spectral width of the gain curve. This situation is illustrated in Figure 6.6 where several modes are shown to be present in the laser output. Such a device is said to be multimode.

Laser oscillation may also occur in a direction which is transverse to the axis of the cavity. This gives rise to resonant modes which are transverse to the direction of propagation. These transverse electromagnetic modes are designated in a similar manner to transverse modes in waveguides (Section 2.3.2) by TEM_{lm} where the integers l and m indicate the number of transverse modes (see Figure 6.7). Unlike the longitudinal modes which contribute only a single spot of light to the laser output, transverse modes may give rise to a pattern of spots at the output. This may be observed from the low order transverse mode patterns shown in Figure 6.7 on which the direction of the electric field is also indicated. In the case of the TEM_{00} mode all parts of the propagating wavefront are in phase. This is not so, however, with higher order modes (TEM_{10}, TEM_{11}, etc.) where phase reversals produce the

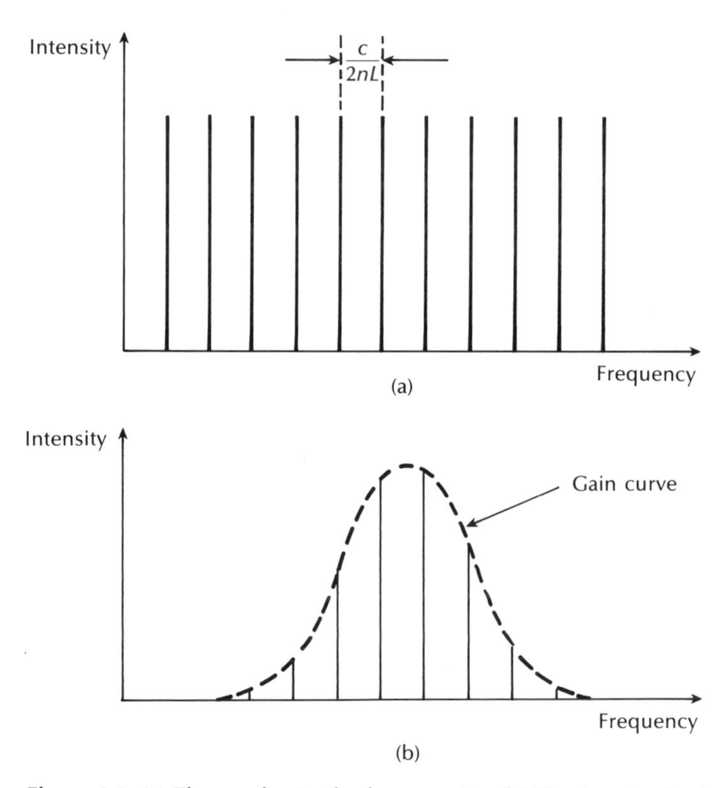

Figure 6.6 (a) The modes in the laser cavity. (b) The longitudinal modes in the laser output.

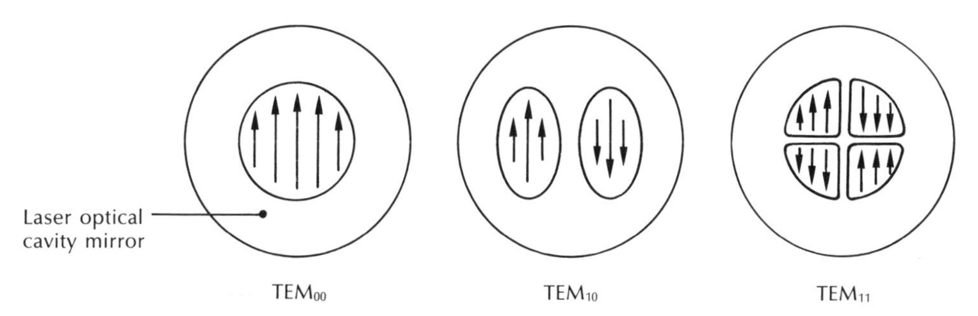

Figure 6.7 The lower order transverse modes of a laser.

various mode patterns. Thus the greatest degree of coherence, together with the highest level of spectral purity, may be obtained from a laser which operates in only the TEM_{00} mode. Higher order transverse modes only occur when the width of the cavity is sufficient for them to oscillate. Consequently, they may be eliminated by suitable narrowing of the laser cavity.

6.2.5 Threshold condition for laser oscillation

It has been indicated that steady state conditions for laser oscillation are achieved when the gain in the amplifying medium exactly balances the total losses.* Hence, although population inversion between the energy levels providing the laser transition is necessary for oscillation to be established, it is not alone sufficient for lasing to occur. In addition a minimum or threshold gain within the amplifying medium must be attained such that laser oscillations are initiated and sustained. This threshold gain may be determined by considering the change in energy of a light beam as it passes through the amplifying medium. For simplicity, all the losses except those due to transmission through the mirrors may be included in a single loss coefficient per unit length $\bar{\alpha}$ cm^{-1}. Again we assume the amplifying medium occupies a length L completely filling the region between the two mirrors which have reflectivities r_1 and r_2. On each round trip the beam passes through the medium twice. Hence the fractional loss incurred by the light beam is:

$$\textit{Fractional loss} = r_1 r_2 \exp(-2\bar{\alpha}L) \tag{6.17}$$

Furthermore, it is found that the increase in beam intensity resulting from stimulated emission is exponential [Ref. 4]. Therefore if the gain coefficient per unit length produced by stimulated emission is $\bar{g}$ cm^{-1}, the fractional round trip gain is given by

$$\textit{Fractional gain} = \exp(2\bar{g}L) \tag{6.18}$$

* This applies to a CW laser which gives a continuous output, rather than pulsed devices for which slightly different conditions exist. For oscillation to commence the fractional gain and loss must be matched.

Hence

$$\exp(2\bar{g}L) \times r_1 r_2 \exp(-2\bar{\alpha}L) = 1$$

and

$$r_1 r_2 \exp 2(\bar{g} - \bar{\alpha})L = 1 \tag{6.19}$$

The threshold gain per unit length may be obtained by rearranging the above expression to give:

$$\bar{g}_{th} = \bar{\alpha} + \frac{1}{2L} \ln \frac{1}{r_1 r_2} \tag{6.20}$$

The second term on the right hand side of Eq. (6.20) represents the transmission loss through the mirrors.*

For laser action to be easily achieved it is clear that a high threshold gain per unit length is required in order to balance the losses from the cavity. However, it must be noted that the parameters displayed in Eq. (6.20) are totally dependent on the laser type.

6.3 Optical emission from semiconductors

6.3.1 The *p–n* junction

To allow consideration of semiconductor optical sources it is necessary to review some of the properties of semiconductor materials, especially with regard to the *p–n* junction. A perfect semiconductor crystal containing no impurities or lattice defects is said to be intrinsic. The energy band structure [Ref. 1] of an intrinsic semiconductor is illustrated in Figure 6.8(a) which shows the valence and conduction bands separated by a forbidden energy gap or bandgap E_g, the width of which varies for different semiconductor materials.

Figure 6.8(a) shows the situation in the semiconductor at a temperature above absolute zero where thermal excitation raises some electrons from the valence band into the conduction band, leaving empty hole states in the valence band. These thermally excited electrons in the conduction band and the holes left in the valence band allow conduction through the material, and are called carriers.

For a semiconductor in thermal equilibrium the energy level occupation is described by the Fermi–Dirac distribution function (rather than the Boltzmann). Consequently, the probability $P(E)$ that an electron gains sufficient thermal energy at an absolute temperature T such that it will be found occupying a particular energy level E, is given by the Fermi–Dirac distribution [Ref. 1]:

$$P(E) = \frac{1}{1 + \exp(E - E_F)/KT} \tag{6.21}$$

* This term is sometimes expressed in the form $1/L \ln 1/r$, where r, the reflectivity of the mirrored ends, is equal to $\sqrt{(r_1 r_2)}$.

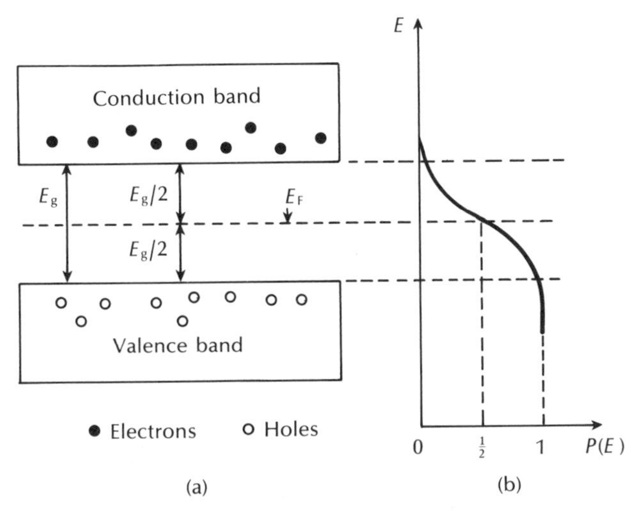

Figure 6.8 (a) The energy band structure of an intrinsic semiconductor at a temperature above absolute zero, showing an equal number of electrons and holes in the conduction band and the valence band respectively. (b) The Fermi–Dirac probability distribution corresponding to (a).

where K is Boltzmann's constant and E_F is known as the Fermi energy or Fermi level. The Fermi level is only a mathematical parameter but it gives an indication of the distribution of carriers within the material. This is shown in Figure 6.8(b) for the intrinsic semiconductor where the Fermi level is at the centre of the bandgap, indicating that there is a small probability of electrons occupying energy levels at the bottom of the conduction band and a corresponding number of holes occupying energy levels at the top of the valence band.

To create an extrinsic semiconductor the material is doped with impurity atoms which either create more free electrons (donor impurity) or holes (acceptor impurity). These two situations are shown in Figure 6.9 where the donor impurities form energy levels just below the conduction band whilst acceptor impurities form energy levels just above the valence band.

When donor impurities are added, thermally excited electrons from the donor levels are raised into the conduction band to create an excess of negative charge carriers and the semiconductor is said to be n type, with the majority carriers being electrons. The Fermi level corresponding to this carrier distribution is raised to a position above the centre of the bandgap, as illustrated in Figure 6.9(a). When acceptor impurities are added, as shown in Figure 6.9(b), thermally excited electrons are raised from the valence band to the acceptor impurity levels leaving an excess of positive charge carriers in the valence band and creating a p type semiconductor where the majority carriers are holes. In this case Fermi level is lowered below the centre of the bandgap.

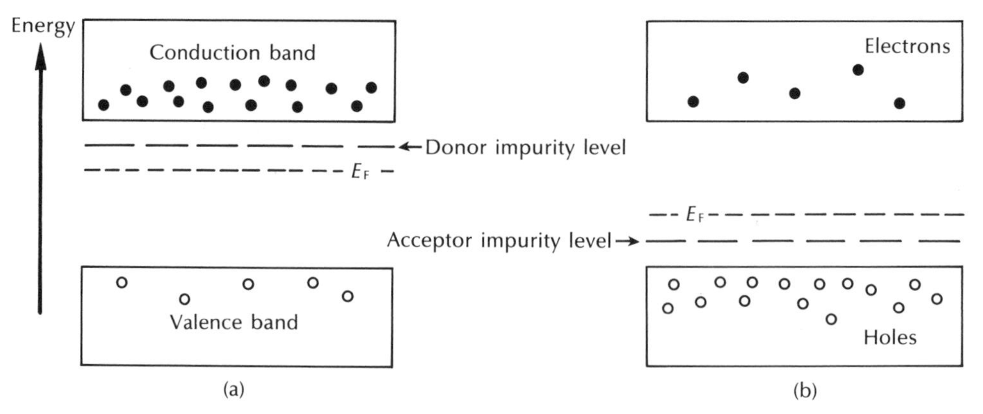

Figure 6.9 Energy band diagrams: (a) n type semiconductor; (b) p type semiconductor.

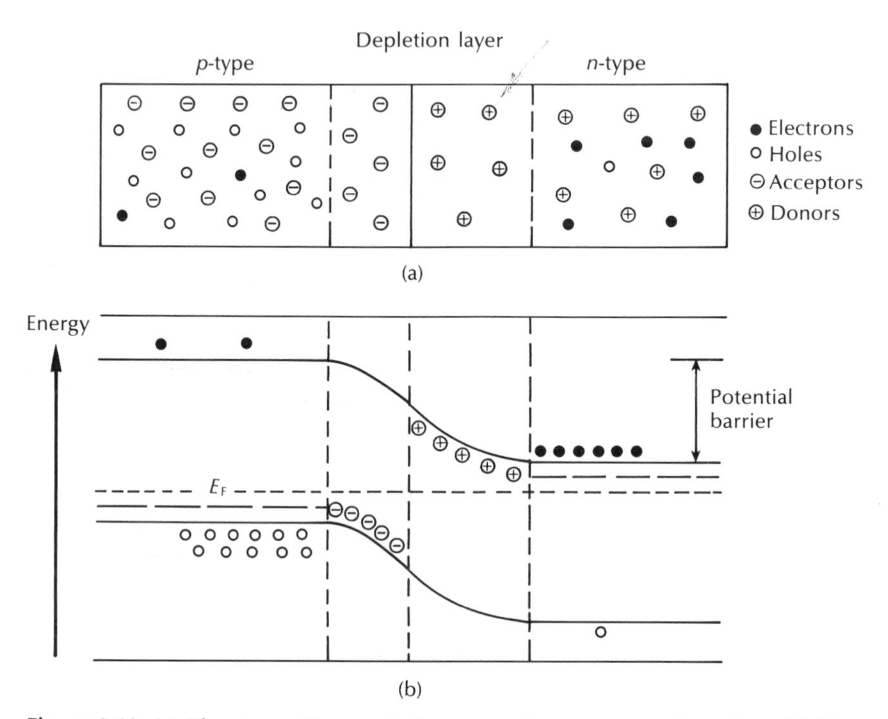

Figure 6.10 (a) The impurities and charge carriers at a p–n junction. (b) The energy band diagram corresponding to (a).

The p–n junction diode is formed by creating adjoining p and n type semiconductor layers in a single crystal, as shown in Figure 6.10(a). A thin depletion region or layer is formed at the junction through carrier recombination which effectively leaves it free of mobile charge carriers (both electrons and holes). This establishes a potential barrier between the p and n type regions which restricts the interdiffusion of majority carriers from their respective regions, as illustrated in Figure 6.10(b). In the absence of an externally applied voltage no current flows as the potential barrier prevents the net flow of carriers from one region to another. When the junction is in this equilibrium state the Fermi level for the p and n type semiconductor is the same as shown Figure 6.10(b).

The width of the depletion region and thus the magnitude of the potential barrier is dependent upon the carrier concentrations (doping) in the p and n type regions, and any external applied voltage. When an external positive voltage is applied to the p type region with respect to the n type, both the depletion region width and the resulting potential barrier are reduced and the diode is said to be forward biased. Electrons from the n type region and holes from the p type region can flow more readily across the junction into the opposite type region. These minority carriers are effectively injected across the junction by the application of the external voltage and form a current flow through the device as they continuously diffuse away from the interface. However, this situation in suitable semiconductor materials allows carrier recombination with the emission of light.

6.3.2 Spontaneous emission

The increased concentration of minority carriers in the opposite type region in the forward biased p–n diode leads to the recombination of carriers across the bandgap. This process is shown in Figure 6.11 for a direct bandgap (see Section

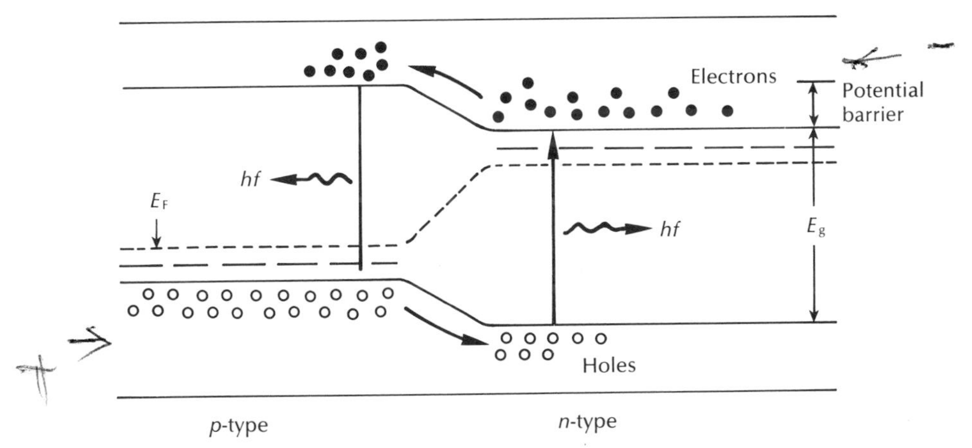

Figure 6.11 The p–n junction with forward bias giving spontaneous emission of photons.

6.3.3) semiconductor material where the normally empty electron states in the conduction band of the p type material and the normally empty hole states in the valence band of the n type material are populated by injected carriers which recombine across the bandgap. The energy released by this electron–hole recombination is approximately equal to the bandgap energy E_g. Excess carrier population is therefore decreased by recombination which may be radiative or nonradiative.

In nonradiative recombination the energy released is dissipated in the form of lattice vibrations and thus heat. However, in band to band radiative recombination the energy is released with the creation of a photon (see Figure 6.11) with a frequency following Eq. (6.1) where the energy is approximately equal to the bandgap energy E_g and therefore:

$$E_g = hf = \frac{hc}{\lambda} \tag{6.22}$$

where c is the velocity of light in a vacuum and λ is the optical wavelength. Substituting the appropriate values for h and c in Eq. (6.22) and rearranging gives:

$$\lambda = \frac{1.24}{E_g} \tag{6.23}$$

where λ is written in μm and E_g in eV.

This spontaneous emission of light from within the diode structure is known as electroluminescence.* The light is emitted at the site of carrier recombination which

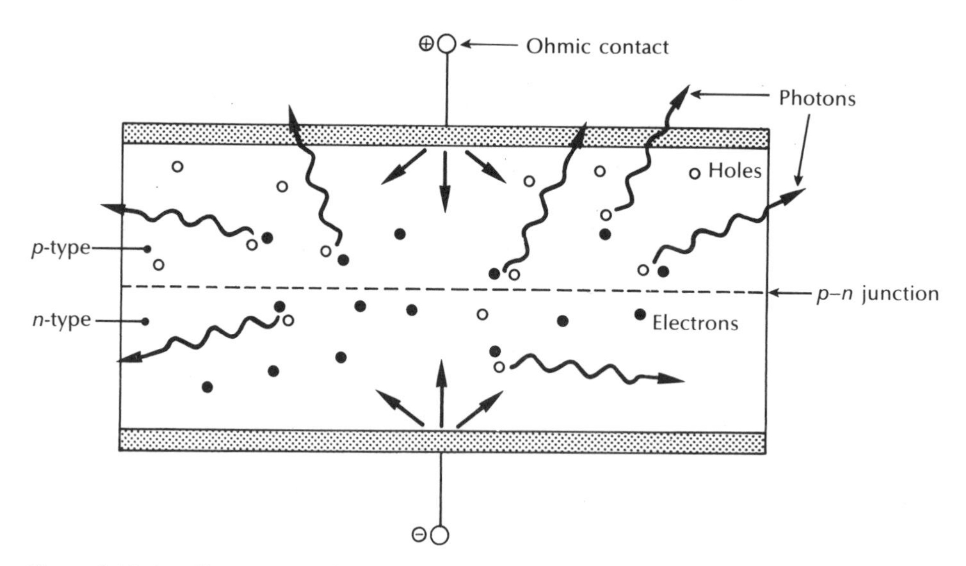

Figure 6.12 An illustration of carrier recombination giving spontaneous emission of light in a p–n junction diode.

* The term electroluminescence is used when the optical emission results from the application of an electric field.

is primarily close to the junction, although recombination may take place through hole diode structure as carriers diffuse away from the junction region (see Figure 6.12). However, the amount of radiative recombination and the emission area within the structure is dependent upon the semiconductor materials used and the fabrication of the device.

6.3.3 Carrier recombination

6.3.3.1 Direct and indirect bandgap semiconductors

In order to encourage electroluminescence it is necessary to select an appropriate semiconductor material. The most useful materials for this purpose are direct bandgap semiconductors in which electrons and holes on either side of the forbidden energy gap have the same value of crystal momentum and thus direct recombination is possible. This process is illustrated in Figure 6.13(a) with an energy-momentum diagram for a direct bandgap semiconductor. It may be observed that the energy maximum of the valence band occurs at the same (or very nearly the same) value of electron crystal momentum* as the energy minimum of the conduction band. Hence when electron–hole recombination occurs the

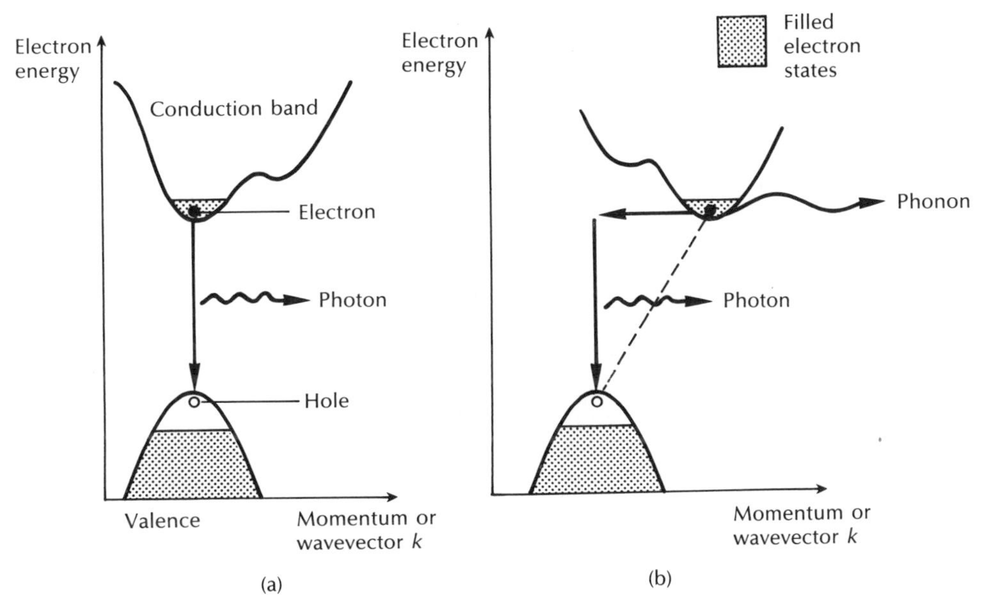

Figure 6.13 Energy–momentum diagrams showing the types of transition: (a) direct bandgap semiconductor; (b) indirect bandgap semiconductor.

* The crystal momentum p is related to the wavevector k for an electron in a crystal by $p = 2\pi hk$, where h is Planck's constant [Ref. 1]. Hence the abscissa of Figure 6.13 is often shown as the electron wavevector rather than momentum.

Table 6.1 Some direct and indirect bandgap semiconductors with calculated recombination coefficients

Semiconductor material	Energy bandgap (eV)	Recombination coefficient B_r ($cm^3 s^{-1}$)
GaAs	Direct: 1.43	7.21×10^{-10}
GaSb	Direct: 0.73	2.39×10^{-10}
InAs	Direct: 0.35	8.5×10^{-11}
InSb	Direct: 0.18	4.58×10^{-11}
Si	Indirect: 1.12	1.79×10^{-15}
Ge	Indirect: 0.67	5.25×10^{-14}
GaP	Indirect: 2.26	5.37×10^{-14}

momentum of the electron remains virtually constant and the energy released, which corresponds to the bandgap energy E_g, may be emitted as light. This direct transition of an electron across the energy gap provides an efficient mechanism for photon emission and the average time the minority carrier remains in a free state before recombination (the minority carrier lifetime) is short (10^{-8} to 10^{-10} s). Some commonly used direct bandgap semiconductor materials are shown in Table 6.1 [Ref. 3].

In indirect bandgap semiconductors, however, the maximum and minimum energies occur at different values of crystal momentum (Figure 6.13(b)). For electron–hole recombination to take place it is essential that the electron loses momentum such that it has a value of momentum corresponding to the maximum energy of the valence band. The conservation of momentum requires the emission or absorption of a third particle, a photon. This three particle recombination process is far less probable than the two particle process exhibited by direct bandgap semiconductors. Hence, the recombination in indirect bandgap semiconductors is relatively slow (10^{-2} to 10^{-4} s). This is reflected by a much longer minority carrier lifetime, together with a greater probability of nonradiative transitions. The competing nonradiative recombination processes which involve lattice defects and impurities (e.g. precipitates of commonly used dopants) become more likely as they allow carrier recombination in a relatively short time in most materials. Thus the indirect bandgap emitters such as silicon and germanium shown in Table 6.1 give insignificant levels of electroluminescence. This disparity is further illustrated in Table 6.1 by the values of the recombination coefficient B_r given for both the direct and indirect bandgap recombination semiconductors shown.

The recombination coefficient is obtained from the measured absorption coefficient of the semiconductor, and for low injected minority carrier density relative to the majority carriers it is related approximately to the radiative minority carrier lifetime* τ_r by [Ref. 4]:

$$\tau_r = [B_r(N + P)]^{-1} \tag{6.24}$$

* The radiative minority carrier lifetime is defined as the average time a minority carrier can exist in a free state before radiative recombination takes place.

where N and P are the respective majority carrier concentrations in the n and p type regions. The significant difference between the recombination coefficients for the direct and indirect bandgap semiconductors shown, underlines the importance of the use of direct bandgap materials for electroluminescent sources. Direct bandgap semiconductor devices in general have a much higher internal quantum efficiency. This is the ratio of the number of radiative recombinations (photons produced within the structure) to the number of injected carriers which is often expressed as a percentage.

Example 6.3

Compare the approximate radiative minority carrier lifetimes in gallium arsenide and silicon when the minority carriers are electrons injected into the p type region which has a hole concentration of 10^{18} cm^{-3}. The injected electron density is small compared with the majority carrier density.

Solution: Equation (6.24) gives the radiative minority carrier lifetime τ_r as

$$\tau_r \simeq [B_r(N + P)]^{-1}$$

In the p type region the hole concentration determines the radiative carrier lifetime as $P \gg N$. Hence,

$$\tau_r \simeq [B_r N]^{-1}$$

Thus for gallium arsenide:

$$\begin{aligned}\tau_r &\simeq [7.21 \times 10^{-10} \times 10^{18}]^{-1} \\ &= 1.39 \times 10^{-9} \\ &= 1.39 \text{ ns}\end{aligned}$$

For silicon

$$\begin{aligned}\tau_r &\simeq [1.79 \times 10^{-15} \times 10^{18}]^{-1} \\ &= 5.58 \times 10^{-4} \\ &= 0.56 \text{ ms}\end{aligned}$$

Thus the direct bandgap gallium arsenide has a radiative carrier lifetime factor of around 2.5×10^{-6} less than the indirect bandgap silicon.

6.3.3.2 Other radiative recombination processes

In the preceding sections, only full bandgap transitions have been considered to give radiative recombination. However, energy levels may be introduced into the bandgap by impurities or lattice defects within the material structure which may greatly increase the electron–hole recombination (effectively reduce the carrier lifetime). The recombination process through such impurity or defect centres may be either radiative or nonradiative. Major radiative recombination processes at 300 K other than band to band transitions are shown in Figure 6.14. These are band

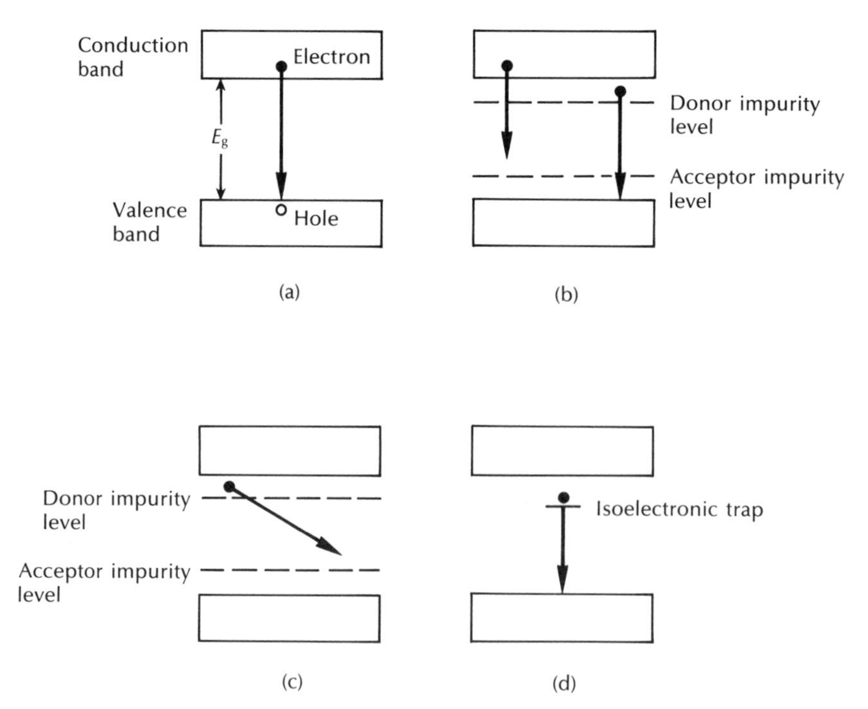

Figure 6.14 Major radiative recombination processes at 300 K (a) conduction to valence band (band to band) transition: (b) conduction band to acceptor impurity, and donor impurity to valence band transition; (c) donor impurity to acceptor impurity transition; (d) recombination from an isoelectronic impurity to the valence band.

to impurity centre or impurity centre to band, donor level to acceptor level and recombination involving isoelectronic impurities.

Hence, an indirect bandgap semiconductor may be made into a more useful electroluminescent material by the addition of impurity centres which will effectively convert it into a direct bandgap material. An example of this is the introduction of nitrogen as an impurity into gallium phosphide. In this case the nitrogen forms an isoelectronic impurity as it has the same number of valence (outer shell) electrons as phosphorus but with a different covalent radius and higher electronegativity [Ref. 1]. The nitrogen impurity centre thus captures an electron and acts as an isoelectronic trap which has a large spread of momentum. This trap then attracts the oppositely charged carrier (a hole) and a direct transition takes place between the impurity centre and the valence band. Hence gallium phosphide may become an efficient light emitter when nitrogen is incorporated. However, such conversion of indirect to direct bandgap transitions is only readily achieved in materials where the direct and indirect bandgaps have a small energy difference. This is the case with gallium phosphide but not with silicon or germanium.

6.3.4 Stimulated emission and lasing

The general concept of stimulated emission via population inversion was indicated in Section 6.2.3. Carrier population inversion is achieved in an intrinsic (undoped) semiconductor by the injection of electrons into the conduction band of the material. This is illustrated in Figure 6.15 where the electron energy and the corresponding filled states are shown. Figure 6.15(a) shows the situation at absolute zero when the conduction band contains no electrons. Electrons injected into the material fill the lower energy states in the conduction band up to the injection energy or the quasi-Fermi level for electrons. Since charge neutrality is conserved within the material an equal density of holes is created in the top of the valence band by the absence of electrons, as shown in Figure 6.15(b) [Ref. 5].

Incident photons with energy E_g but less than the separation energy of the quasi-Fermi levels $E_q = E_{Fc} - E_{Fv}$ cannot be absorbed because the necessary conduction band states are occupied. However, these photons can induce a downward transition of an electron from the filled conduction band states into the empty valence band states, thus stimulating the emission of another photon. The basic condition for stimulated emission is therefore dependent on the quasi-Fermi level separation energy as well as the bandgap energy and may be defined as:

$$E_{Fc} - E_{Fv} > hf > E_g \tag{6.25}$$

However, it must be noted that we have described an ideal situation whereas at normal operating temperatures the distribution of electrons and holes is less well defined but the condition for stimulated emission is largely maintained.

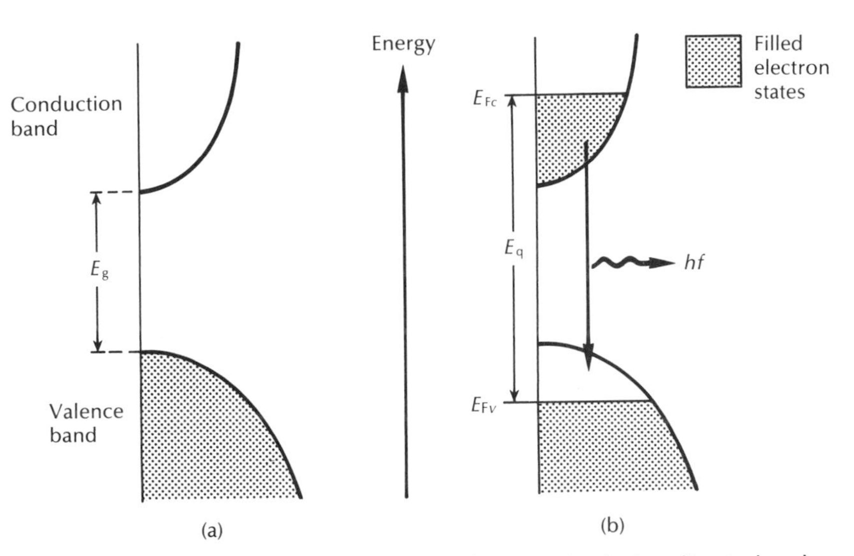

Figure 6.15 The filled electron states for an intrinsic direct bandgap semiconductor at absolute zero [Ref. 5]: (a) in equilibrium; (b) with high carrier injection.

Population inversion may be obtained at a *p–n* junction by heavy doping (degenerative doping) of both the *p* and *n* type material. Heavy *p* type doping with acceptor impurities causes a lowering of the Fermi level or boundary between the filled and empty states into the valence band. Similarly, degenerative *n* type doping causes the Fermi level to enter the conduction band of the material. Energy band diagrams of a degenerate *p–n* junction are shown in Figure 6.16. The position of the Fermi level and the electron occupation (shading) with no applied bias are shown in Figure 6.16(a). Since in this case the junction is in thermal equilibrium, the Fermi energy has the same value throughout the material. Figure 6.16(b) shows the *p–n* junction when a forward bias nearly equal to the bandgap voltage is applied and hence there is direct conduction. At high injection carrier density* in such a junction there exists an active region near the depletion layer that contains simultaneously degenerate populations of electrons and holes (sometimes termed doubly degenerate). For this region the condition for stimulated emission of Eq.

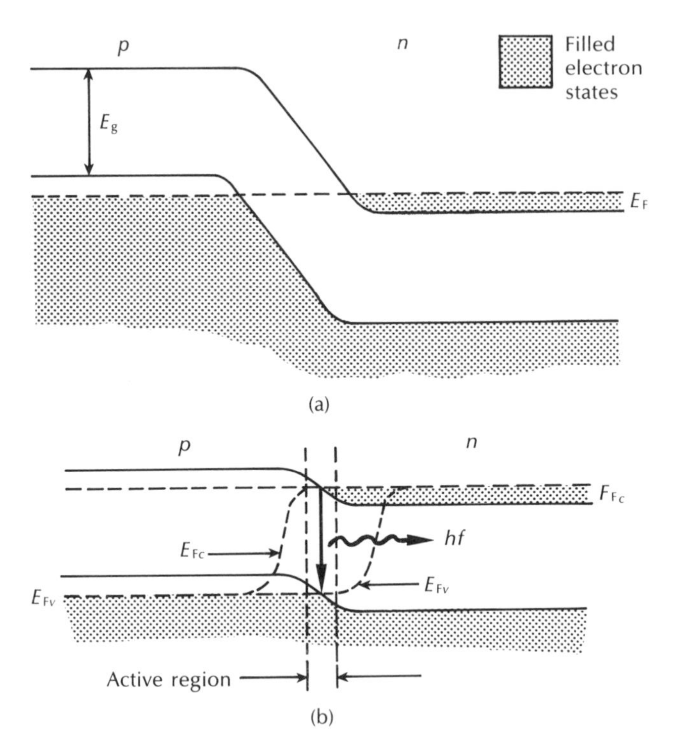

Figure 6.16 The degenerate *p–n* junction: (a) with no applied bias; (b) with strong forward bias such that the separation of the quasi-Fermi levels is higher than the electron–hole recombination energy *hf* in the narrow active region. Hence stimulated emission is obtained in this region.

* This may be largely considered to be electrons injected into the *p–n* region because of their greater mobility

(6.22) is satisfied for electromagnetic radiation of frequency $E_g/h < f < (E_{Fc} - E_{Fv})/h$. Therefore, any radiation of this frequency which is confined to the active region will be amplified. In general, the degenerative doping distinguishes a *p–n* junction which provides stimulated emission from one which gives only spontaneous emission as in the case of the LED.

Finally, it must be noted that high impurity concentration within a semiconductor causes differences in the energy bands in comparison with an intrinsic semiconductor. These differences are particularly apparent in the degeneratively doped *p–n* junctions used for semiconductor lasers. For instance at high donor level concentrations in gallium arsenide, the donor impurity levels form a band that merges with the conduction band. These energy states, sometimes referred to as 'bandtail' states [Ref. 7] extend into the forbidden energy gap. The laser transition may take place from one of these states. Furthermore the transitions may terminate on acceptor states which because of their high concentration also extend as a band into the energy gap. In this way the lasing transitions may occur at energies less than the bandgap energy E_g. When transitions of this type dominate, the lasing peak energy is less than the bandgap energy. Hence the effective lasing wavelength can be varied within the electroluminescent semiconductor used to fabricate the junction laser through variation of the impurity concentration. For example, the lasing wavelength of gallium arsenide may be varied between 0.85 and 0.95 μm, although the best performance is usually achieved in the 0.88 to 0.91 μm band (see Problem 6.6).

However, a further requirement of the junction diode is necessary to establish lasing. This involves the provision of optical feedback to give laser oscillation. It may be achieved by the formation of an optical cavity (Fabry–Perot cavity, see Section 6.2.4) within the structure by polishing the end faces of the junction diode to act as mirrors. Each end of the junction is polished or cleaved and the sides are roughened to prevent any unwanted light emission and hence wasted population inversion.

The behaviour of the semiconductor laser can be described by rate equations for electron and photon density in the active layer of the device. These equations assist in providing an understanding of the laser electrical and optical performance characteristics under direct current modulation as well as its potential limitations. The problems associated with high speed direct current modulation are unique to the semiconductor laser whose major application area is that of a source within optical fiber communications.

Although the rate equations may be approached with some rigour [Ref. 8] we adopt a simplified analysis which is valid within certain constraints [Ref. 9]. In particular, the equations represent an average behaviour for the active medium within the laser cavity and they are not applicable when the time period is short compared with the transit time of the optical wave in the laser cavity.* The two rate

* Thus performance characteristics derived from these rate equations become questionable when the time scale is less than 10 ps or the modulation bandwidth is greater than 100 GHz.

equations for electron density n, and photon density ϕ, are:

$$\frac{dn}{dt} = \frac{J}{ed} - \frac{n}{\tau_{sp}} - Cn\phi \qquad (m^{-3}\,s^{-1}) \tag{6.26}$$

and

$$\frac{d\phi}{dt} = Cn\phi + \delta\frac{n}{\tau_{sp}} - \frac{\phi}{\tau_{ph}} \qquad (m^{-3}\,s^{-1}) \tag{6.27}$$

where J is the current density, in amperes per square meter, e is the charge on an electron, d is the thickness of the recombination region, τ_{sp} is the spontaneous emission lifetime which is equivalent to τ_{21} in Section 6.2.2, C is a coefficient which incorporates the B coefficients in Section 6.2.2, δ is a small fractional value and τ_{ph} is the photon lifetime.

The rate equations given in Eqs. (6.26) and (6.27) may be balanced by taking into account all the factors which affect the numbers of electrons and holes in the laser structure. Hence, in Eq. (6.26), the first term indicates the increase in the electron concentration in the conduction band as the current flows into the junction diode. The electrons lost from the conduction band by spontaneous and stimulated transitions are provided by the second and third terms respectively. In Eq. (6.27) the first term depicts the stimulated emission as a source of photons. The fraction of photons produced by spontaneous emission which combine to the energy in the lasing mode is given by the second term. This term is often neglected, however, as δ is small. The final term represents the decay in the number of photons resulting from losses in the optical cavity.

Although these rate equations may be used to study both the transient and steady state behaviour of the semiconductor laser, we are particularly concerned with the steady state solutions. The steady state is characterized by the left hand side of Eqs. (6.26) and (6.27) being equal to zero, when n and ϕ have nonzero values. In addition, the fields in the optical cavity which are represented by ϕ must build up from small initial values, and hence $d\phi/dt$ must be positive when ϕ is small. Therefore, setting δ equal to zero in Eq. (6.27), it is clear that for any value of ϕ, $d\phi/dt$ will only be positive when:

$$Cn - \frac{1}{\tau_{ph}} \geqslant 0 \tag{6.28}$$

There is therefore a threshold value of n which satisfies the equality of Eq. (6.28). If n is larger than this threshold value, then ϕ can increase; however, when n is smaller it cannot. From Eq. (6.28) the threshold value for the electron density n_{th} is:

$$n_{th} = \frac{1}{C\tau_{ph}} \qquad (m^{-3}) \tag{6.29}$$

The threshold current written in terms of its current density J_{th}, required to maintain $n = n_{th}$ in the steady state when $\phi = 0$, may be obtained from Eq. (6.26) as:

$$\frac{J_{th}}{ed} = \frac{n_{th}}{\tau_{sp}} \qquad (m^{-3}\,s^{-1}) \tag{6.30}$$

Hence Eq. (6.30) defines the current required to sustain an excess electron density in the laser when spontaneous emission provides the only decay mechanism. The steady state photon density ϕ_s is provided by substituting Eq. (6.30) in Eq. (6.26) giving:

$$0 = \frac{(J - J_{th})}{ed} - Cn_{th}\phi_s.$$

Rearranging we obtain:

$$\phi_s = \frac{1}{Cn_{th}} \frac{(J - J_{th})}{ed} \qquad (\text{m}^{-3}) \tag{6.31}$$

Substituting for Cn_{th} from Eq. (6.29) we can write Eq. (6.31) in the form:

$$\phi_s = \frac{\tau_{ph}}{ed}(J - J_{th}) \qquad (\text{m}^{-3}) \tag{6.32}$$

The photon density ϕ_s cannot be a negative quantity as this is meaningless, and for ϕ_s to be greater than zero the current must exceed its threshold value. Moreover, ϕ_s is proportional to the amount by which J exceeds its threshold value. As each photon has energy hf it is possible to determine the optical power density in W m^{-2} by assuming that half the photons are travelling in each of two directions.

An idealized optical output power against current characteristic (also called light output against current characteristic) for a semiconductor laser is illustrated in Figure 6.17. The solid line represents the laser characteristic, whereas the dashed line is a plot of Eq. (6.32) showing the current threshold. It may be observed that the device gives little light output in the region below the threshold current which corresponds to spontaneous emission only within the structure. However, after the

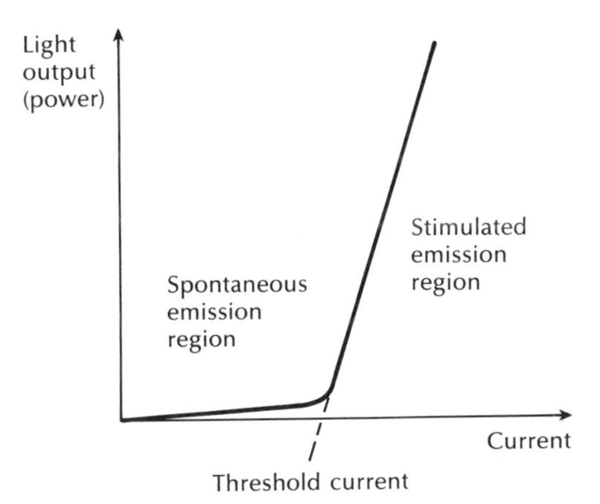

Figure 6.17 The ideal light output against current characteristic for an injection laser.

threshold current density is reached, the light output increases substantially for small increases in current through the device. This corresponds to the region of stimulated emission when the laser is acting as an amplifier of light.

In common with all other laser types a requirement for the initiation and maintenance of laser oscillation is that the optical gain matches the optical losses within the cavity (see Section 6.2.5). For the *p–n* junction or semiconductor laser this occurs at a particular photon energy within the spectrum of spontaneous emission (usually near the peak wavelength of spontaneous emission). Thus when extremely high currents are passed through the device (i.e. injection levels of around 10^{18} carriers cm^{-3}), spontaneous emission with a wide spectrum (linewidth) becomes lasing (when a current threshold is passed) and the linewidth subsequently narrows.

For strongly confined structures the threshold current density for stimulated emission J_{th} is to a fair approximation [Ref. 4] related to the threshold gain coefficient $\bar{g}_{th}$ for the laser cavity through:

$$\bar{g}_{th} = \bar{\beta} J_{th} \tag{6.33}$$

where the gain factor $\bar{\beta}$ is a constant appropriate to specific devices. Detailed discussion of the more exact relationship is given in Ref. 4.

Substituting for $\bar{g}_{th}$ from Eq. (6.18) and rearranging we obtain:

$$J_{th} = \frac{1}{\bar{\beta}}\left[\bar{\alpha} + \frac{1}{2L}\ln\frac{1}{r_1 r_2}\right] \tag{6.34}$$

Since for the semiconductor laser the mirrors are formed by a dielectric plane and are often uncoated, the mirror reflectivities r_1 and r_2 may be calculated using the Fresnel reflection relationship of Eq. (5.1).

Example 6.4

A GaAs injection laser has an optical cavity of length 250 μm and width 100 μm. At normal operating temperature the gain factor $\bar{\beta}$ is 21×10^{-3} A cm^{-3} and the loss coefficient $\bar{\alpha}$ per cm is 10. Determine the threshold current density and hence the threshold current for the device. It may be assumed that the cleaved mirrors are uncoated and that the current is restricted to the optical cavity. The refractive index of GaAs may be taken as 3.6.

Solution: The reflectivity for normal incidence of a plane wave on the GaAs–air interface may be obtained from Eq. (5.1) where:

$$r_1 = r_2 = r = \left(\frac{n-1}{n+1}\right)^2$$

$$= \left(\frac{3.6-1}{3.6+1}\right)^2 \simeq 0.32$$

The threshold current density may be obtained from Eq. (6.34) where:

$$J_{th} = \frac{1}{\bar{\beta}}\left[\bar{\alpha}\ \frac{1}{L}\ \ln \frac{1}{r}\right]$$

$$= \frac{1}{21 \times 10^{-3}}\left[10 + \frac{1}{250 \times 10^{-4}} \ln \frac{1}{0.32}\right]$$

$$= 2.65 \times 10^{3}\ \text{A cm}^{-2}$$

The threshold current I_{th} is given by:

$$I_{th} = J_{th} \times \text{area of the optical cavity}$$

$$= 2.65 \times 10^{3} \times 250 \times 100 \times 10^{-8}$$

$$\simeq 663\ \text{mA}$$

Therefore the threshold current for this device is 663 mA if the current flow is restricted to the optical cavity.

As the stimulated emission minority carrier lifetime is much shorter (typically 10^{-11} s) than that due to spontaneous emission, further increases in input current above the threshold will result almost entirely in stimulated emission, giving a high internal quantum efficiency (50 to 100%). Also, whereas incoherent spontaneous emission has a linewidth of tens of nanometres, stimulated coherent emission has a linewidth of a nanometre or less.

6.3.5 Heterojunctions

The preceding sections have considered the photoemissive properties of a single *p–n* junction fabricated from a single crystal semiconductor material. This is known as a homojunction. However, the radiative properties of a junction diode may be improved by the use of heterojunctions. A heterojunction is an interface between two adjoining single crystal semiconductors with different bandgap energies. Devices which are fabricated with heterojunctions are said to have heterostructure.

Heterojunctions are classified into either an isotype (*n–n* or *p–p*) or an anisotype (*p–n*). The isotype heterojunction provides a potential barrier within the structure which is useful for the confinement of minority carriers to a small active region (carrier confinement). It effectively reduces the carrier diffusion length and thus the volume within the structure where radiative recombination may take place. This technique is widely used for the fabrication of injection lasers and high radiance LEDs. Isotype heterojunctions are also extensively used in LEDs to provide a transparent layer close to the active region which substantially reduces the absorption of light emitted from the structure.

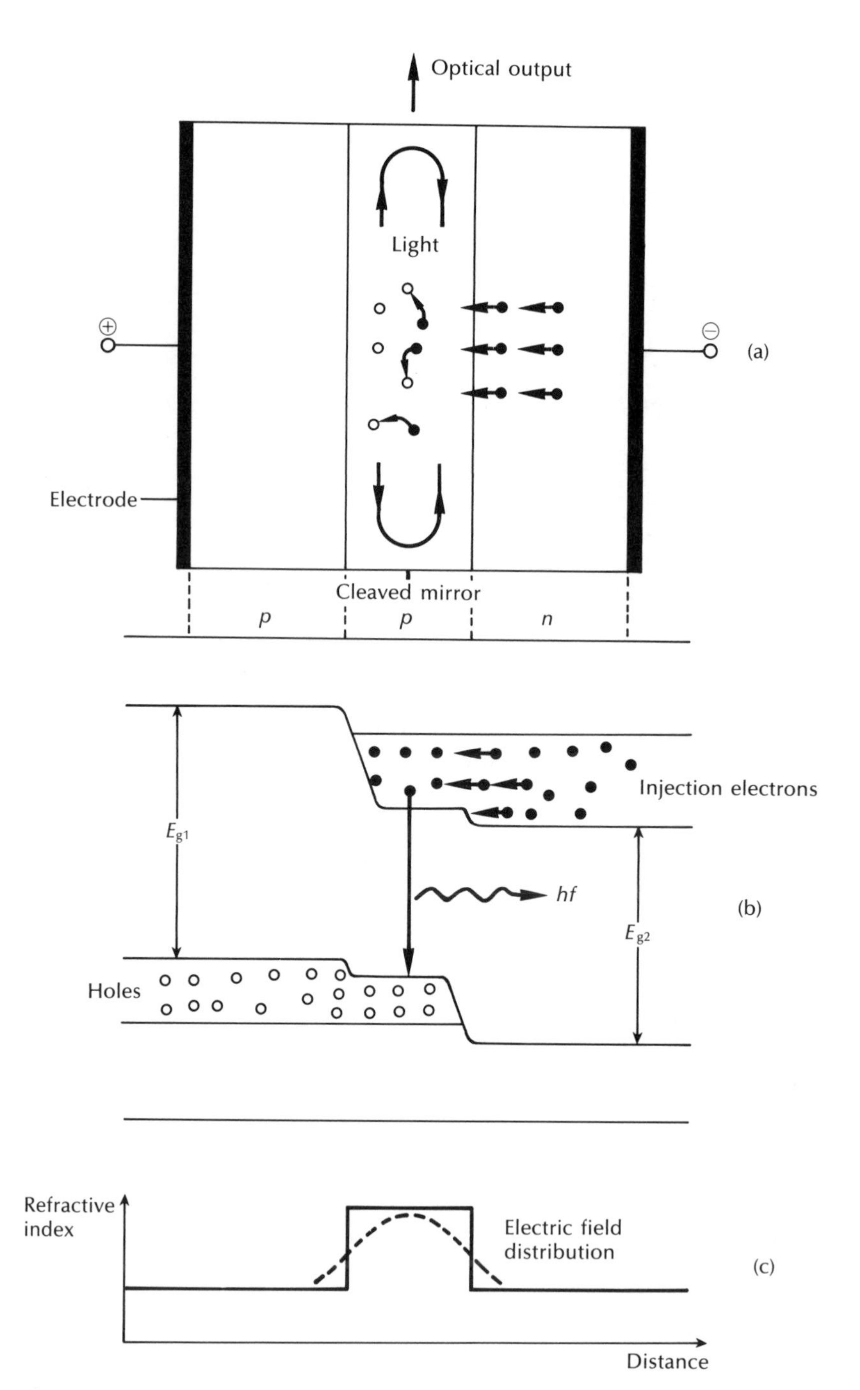

Figure 6.18 The double heterojunction injection laser: (a) the layer structure, shown with an applied forward bias; (b) energy band diagram indicating a *p–p* heterojunction on the left and a *p–n* heterojunction on the right; (c) the corresponding refractive index diagram and electric field distribution.

Alternatively, anisotype heterojunctions with sufficiently large bandgap differences improve the injection efficiency of either electrons or holes. Both types of heterojunction provide a dielectric step due to the different refractive indices at either side of the junction. This may be used to provide radiation confinement to the active region (i.e. the walls of an optical waveguide). The efficiency of the containment depends upon the magnitude of the step which is dictated by the difference in bandgap energies and the wavelength of the radiation.

It is useful to consider the application of heterojunctions in the fabrication of a particular device. They were first used to provide potential barriers in injection lasers. When a double heterojunction (DH) structure was implemented the resulting carrier and optical confinement reduced the threshold currents necessary for lasing by a factor of around 100. Thus stimulated emission was obtained with relatively small threshold currents (50 to 200 mA). The layer structure and an energy band diagram for a DH injection laser are illustrated in Figure 6.18. A heterojunction is shown either side of the active layer for laser oscillation. The forward bias is supplied by connecting a positive electrode of a supply to the *p* side of the structure and a negative electrode to the *n* side. When a voltage which corresponds to the bandgap energy of the active layer is applied, a large number of electrons (or holes) are injected into the active layer and laser oscillation commences. These carriers are confined to the active layer by the energy barriers provided by the heterojunctions which are placed within the diffusion length of the injected carriers. It may also be observed from Figure 6.18(c) that a refractive index step (usually a difference of 5 to 10%) at the heterojunctions provides radiation containment to the active layer. In effect the active layer forms the centre of a dielectric waveguide which strongly confines the electroluminescence within this region, as illustrated in Figure 6.18(c). The refractive index step shown is the same for each heterojunction which is desirable in order to prevent losses due to lack of waveguiding which can occur if the structure is not symmetrical.

Careful fabrication of the heterojunctions is also important in order to reduce defects at the interfaces such as misfit dislocations or inclusions which cause nonradiative recombination and thus reduce the internal quantum efficiency. Lattice matching is therefore an important criterion for the materials used to form the interface. Ideally, heterojunctions should have a very small lattice parameter mismatch of no greater than 0.1%. However, it is often not possible to obtain such good lattice parameter matching with the semiconductor materials required to give emission at the desired wavelength and therefore much higher lattice parameter mismatch is often tolerated ($\simeq 0.6\%$).

6.3.6 Semiconductor materials

The semiconductor materials used for optical sources must broadly fulfil several criteria. These are as follows:

1. *p–n* junction formation. The materials must lend themselves to the formation of *p–n* junctions with suitable characteristics for carrier injection.

2. Efficient electroluminescence. The devices fabricated must have a high probability of radiative transitions and therefore a high internal quantum efficiency. Hence the materials utilized must be either direct bandgap semiconductors or indirect bandgap semiconductors with appropriate impurity centres.
3. Useful emission wavelength. The materials must emit light at a suitable wavelength to be utilized with current optical fibers and detectors (0.8 to 1.7 μm). Ideally, they should allow bandgap variation with appropriate doping and fabrication in order that emission at a desired specific wavelength may be achieved.

Initial investigation of electroluminescent materials for LEDs in the early 1960s centred around the direct bandgap III–V alloy semiconductors including the binary compounds gallium arsenide (GaAs) and gallium phosphide (GaP) and the ternary gallium arsenide phosphide ($GaAs_xP_{1-x}$). Gallium arsenide gives efficient electroluminescence over an appropriate wavelength band (0.88 to 0.91 μm) and for the first generation optical fiber communication systems was the first material to be fabricated into homojunction semiconductor lasers operating at low temperature. It was quickly realized that improved devices could be fabricated with heterojunction structures which through carrier and radiation confinement would give enhanced light output for drastically reduced device currents. These heterostructure devices were first fabricated using liquid phase epitaxy (LPE) to produce $GaAs/Al_xGa_{1-x}As$ single heterojunction lasers. This process involves the precipitation of material from a cooling solution on to an underlying substrate. When the substrate consists of a single crystal and the lattice constant or parameter of the precipitating material is the same or very similar to that of the substrate (i.e. the unit cells within the two crystalline structures are of a similar dimension), the precipitating material forms an epitaxial layer on the substrate surface. Subsequently, the same technique was used to produce double heterojunctions consisting of $Al_xGa_{1-x}As/GaAs/Al_xGa_{1-x}As$ epitaxial layers, which gave continuous (CW) operation at room temperature [Refs. 10 and 11]. Some of the common material systems now utilized for double heterojunction device fabrication, together with their useful wavelength ranges, are shown in Table 6.2.

The GaAs/AlGaAs DH system is currently by far the best developed and is used for fabricating both lasers and LEDs for the shorter wavelength region. The bandgap in this material may be 'tailored' to span the entire 0.8 to 0.9 μm wavelength band by changing the AlGa composition. Also there is very little lattice mismatch (0.017%) between the AlGaAs epitaxial layer and the GaAs substrate which gives good internal quantum efficiency. In the longer wavelength region (1.1 to 1.6 μm) a number of III–V alloys have been investigated which are compatible with GaAs, InP and GaSb substrates. These include ternary alloys such as $GaAs_{1-x}Sb_x$ and $In_xGa_{1-x}As$ grown on GaAs.

However, although the ternary alloys allow bandgap tailoring they have a fixed lattice parameter. Therefore, quaternary alloys which allow both bandgap tailoring and control of the lattice parameter (i.e. a range of lattice parameters is available

Table 6.2 Some common material systems used in the fabrication of electroluminescent sources for optical fiber communications

Material systems active layer/confining layers	Useful wavelength range (μm)	Substrate
$GaAs/Al_xGa_{1-x}As$	0.8–0.9	GaAs
$GaAs/In_xGa_{1-x}P$	0.9	GaAs
$Al_yGa_{1-y}As/Al_xGa_{1-x}As$	0.65–0.9	GaAs
$In_yGa_{1-y}As/In_xGa_{1-x}P$	0.85–1.1	GaAs
$GaAs_{1-x}Sb_x/Ga_{1-y}Al_yAs_{1-x}Sb_x$	0.9–1.1	GaAs
$Ga_{1-y}Al_yAs_{1-x}Sb_x/GaSb$	1.0–1.7	GaSb
$In_{1-x}Ga_xAs_yP_{1-y}/InP$	0.92–1.7	InP

for each bandgap) appear to be of more use for the longer wavelength region. The most advanced are $In_{1-x}Ga_xAs_yP_{1-y}$ lattice matched to InP and $Ga_{1-y}Al_yAs_{1-x}Sb_x$ lattice matched to GaSb. Both these material systems allow emission over the entire 1.0 to 1.7 μm wavelength band. At present the InGaAsP/InP material system is the most favourable for both long wavelength light sources and detectors. This is due to the ease of fabrication with lattice matching on InP which is also a suitable material for the active region with a bandgap energy of 1.35 eV at 300 K. Hence, InP/InGaAsP (active/confining) devices may be fabricated. Conversely, GaSb is a low bandgap material (0.78 eV at 300 K) and the quaternary alloy must be used for the active region in the GaAlAsSb/GaSb system. Thus compositional control must be maintained for three layers in this system in order to minimize lattice mismatch in the active region, whereas it is only necessary for one layer in the InP/InGaAsP system.

6.4 The semiconductor injection laser

The electroluminescent properties of the forward biased *p–n* junction diode have been considered in the preceding sections. Stimulated emission by the recombination of the injected carriers is encouraged in the semiconductor injection laser (often called the injection laser diode (ILD) or simply the injection laser) by the provision of an optical cavity in the crystal structure in order to provide the feedback of photons. This gives the injection laser several major advantages over other semiconductor sources (e.g. LEDs) that may be used for optical communications. These are as follows:

1. High radiance due to the amplifying effect of stimulated emission. Injection lasers will generally supply milliwatts of optical output power.
2. Narrow linewidth of the order of 1 nm (10 Å) or less which is useful in minimizing the effects of material dispersion.

3. Modulation capabilities which at present extend up into the gigahertz range and will undoubtedly be improved upon.
4. Relative temporal coherence which is considered essential to allow heterodyne (coherent) detection in high capacity systems, but at present is primarily of use in single-mode systems.
5. Good spatial coherence which allows the output to be focused by a lens into a spot which has a greater intensity than the dispersed unfocused emission. This permits efficient coupling of the optical output power into the fiber even for fibers with low numerical aperture. The spatial fold matching to the optical fiber which may be obtained with the laser source is not possible with an incoherent emitter and, consequently, coupling efficiencies are much reduced.

These advantages, together with the compatibility of the injection laser with optical fibers (e.g. size) led to the early developments of the device in the 1960s. Early injection lasers had the form of a Fabry–Perot cavity often fabricated in gallium arsenide which was the major III–V compound semiconductor with electro-luminescent properties at the appropriate wavelength for first generation systems. The basic structure of this homojunction device is shown in Figure 6.19, where the cleaved ends of the crystal act as partial mirrors in order to encourage stimulated emission in the cavity when electrons are injected into the *p* type region. However, as mentioned previously these devices had a high threshold current density (greater than 10^4 A cm^{-2}) due to their lack of carrier containment and proved inefficient light sources. The high current densities required dictated that these devices when operated at 300 K were largely utilized in a pulsed mode in order to minimize the junction temperature and thus avert damage.

Improved carrier containment and thus lower threshold current densities (around 10^3 A cm^{-2}) were achieved using heterojunction structures (see Section 6.3.5). The double heterojunction injection laser fabricated from lattice matched III–V alloys

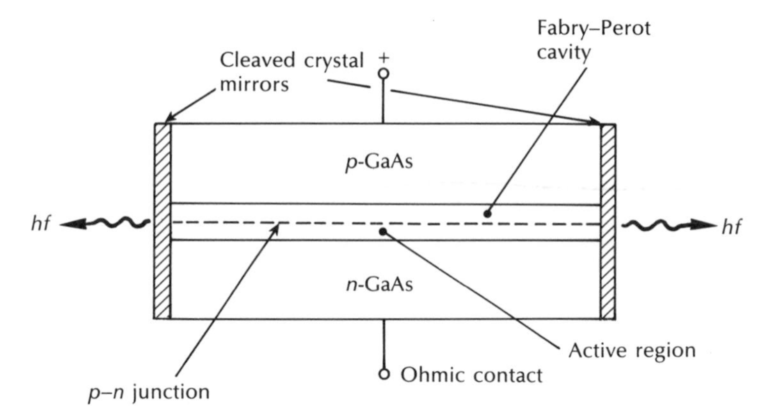

Figure 6.19 Schematic diagram of a GaAs homojunction injection laser with a Fabry–Perot cavity.

provided both carrier and optical confinement on both sides of the *p–n* junction, giving the injection laser a greatly enhanced performance. This enabled these devices with the appropriate heat sinking to be operated in a continuous wave (CW) mode at 300 K with obvious advantages for optical communications (e.g. analog transmission). However, in order to provide reliable CW operation of the DH injection laser it was necessary to provide further carrier and optical confinement which led to the introduction of stripe geometry DH laser configurations. Prior to discussion of this structure, however, it is useful to consider the efficiency of the semiconductor injection laser as an optical source.

6.4.1 Efficiency

There are a number of ways in which the operational efficiency of the semiconductor laser may be defined. A useful definition is that of the differential external quantum efficiency η_D which is the ratio of the increase in photon output rate for a given increase in the number of injected electrons. If P_e is the optical power emitted from the device, I is the current, e is the charge on an electron, and hf is the photon energy, then:

$$\eta_D = \frac{dP_e/hf}{dI/e} \simeq \frac{dP_e}{dI(E_g)} \qquad (6.35)$$

where E_g is the bandgap energy expressed in electronvolts. It may be noted that η_D gives a measure of the rate of change of the optical output power with current and hence defines the slope of the output characteristic (Figure 6.17) in the lasing region for a particular device. Hence η_D is sometimes referred to as the slope quantum efficiency. For a CW semiconductor laser it usually has values in the range 40 to 60%. Alternatively, the internal quantum efficiency of the semiconductor laser η_i, which was defined in Section 6.3.3.1 as:

$$\eta_i = \frac{\textit{number of photons produced in the laser cavity}}{\textit{number of injected electrons}} \qquad (6.36)$$

may be quite high with values usually in the range 50 to 100%. It is related to the differential external quantum efficiency by the expression [Ref. 4]:

$$\eta_D = \eta_i \left[\frac{1}{1 + (2\bar{\alpha}L/\ln(1/r_1 r_2))} \right] \qquad (6.37)$$

where $\bar{\alpha}$ is the loss coefficient of the laser cavity, L is the length of the laser cavity and r_1, r_2 are the cleaved mirror reflectivities.

Another parameter is the total efficiency (external quantum efficiency) η_T which is efficiency defined as:

$$\eta_T = \frac{\textit{total number of output photons}}{\textit{total number of injected electrons}} \qquad (6.38)$$

$$= \frac{P_e/hf}{I/e} \simeq \frac{P_e}{IE_g} \qquad (6.39)$$

As the power emitted P_e changes linearly when the injection current I is greater than the threshold current I_{th}, then:

$$\eta_T \simeq \eta_D\left(1 - \frac{I_{th}}{I}\right) \tag{6.40}$$

For high injection current (e.g. $I = 5I_{th}$) then $\eta_T \simeq \eta_D$, whereas for lower currents ($I \simeq 2I_{th}$) the total efficiency is lower and around 15 to 25%.

The external power efficiency of the device (or device efficiency) η_{ep} in converting electrical input to optical output is given by:

$$\eta_{ep} = \frac{P_e}{P} \times 100 = \frac{P_e}{IV} \times 100\% \tag{6.41}$$

where $P = IV$ is the d.c. electrical input power.

Using Eq. (6.39) for the total efficiency we find:

$$\eta_{ep} = \eta_T\left(\frac{E_g}{V}\right) \times 100\% \tag{6.42}$$

Example 6.5

The total efficiency of an injection laser with a GaAs active region is 18%. The voltage applied to the device is 2.5 V and the bandgap energy for GaAs is 1.43 eV. Calculate the external power efficiency of the device.

Solution: Using Eq. (6.42), the external power efficiency is given by:

$$\eta_{ep} = 0.18\left(\frac{1.43}{2.5}\right) \times 100 \simeq 10\%$$

This result indicates the possibility of achieving high overall power efficiencies from semiconductor lasers which are much larger than for other laser types.

6.4.2 Stripe geometry

The DH laser structure provides optical confinement in the vertical direction through the refractive index step at the heterojunction interfaces, but lasing takes place across the whole width of the device. This situation is illustrated in Figure 6.20 which shows the broad area DH laser where the sides of the cavity are simply formed by roughening the edges of the device in order to reduce unwanted emission in these directions and limit the number of horizontal transverse modes. However, the broad emission area creates several problems including difficult heat sinking, lasing from multiple filaments in the relatively wide active area and unsuitable light output geometry for efficient coupling to the cylindrical fibers.

To overcome these problems whilst also reducing the required threshold current, laser structures in which the active region does not extend to the edges of the device

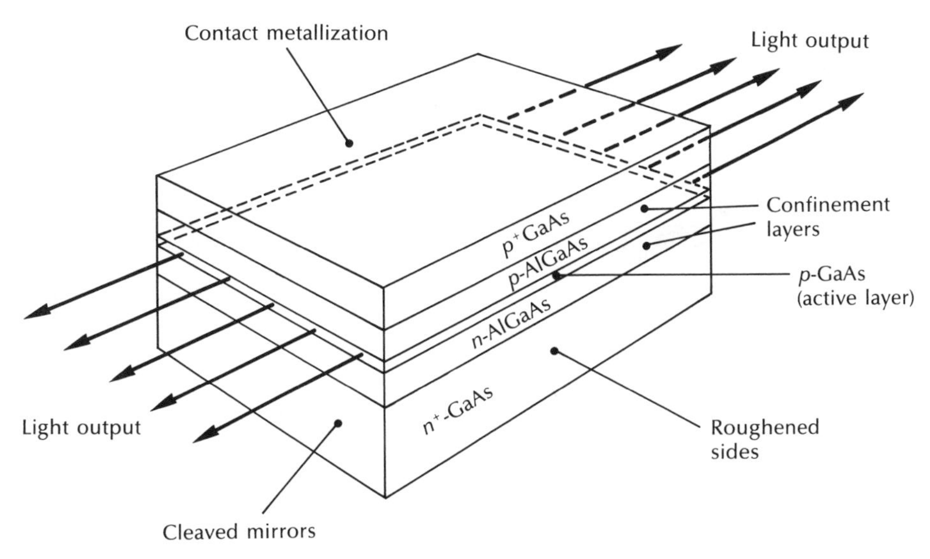

Figure 6.20 A broad area GaAs/AlGaAs DH injection laser.

were developed. A common technique involved the introduction of stripe geometry to the structure to provide optical containment in the horizontal plane. The structure of a DH stripe contact laser is shown in Figure 6.21 where the major current flow through the device and hence the active region is within the stripe. Generally, the stripe is formed by the creation of high resistance areas on either side by techniques such as proton bombardment [Ref. 10] or oxide isolation [Ref. 11]. The stripe therefore acts as a guiding mechanism which overcomes the major problems of the broad area device. However, although the active area width is reduced the light output is still not particularly well collimated due to isotropic emission from a small active region and diffraction within the structure. The optical output and far field emission pattern are also illustrated in Figure 6.21. The output beam divergence is typically 45° perpendicular to the plane of the junction and 9° parallel to it. Nevertheless, this is a substantial improvement on the broad area laser.

The stripe contact device also gives, with the correct balance of guiding, single transverse (in a direction parallel to the junction plane) mode operation, whereas the broad area device tends to allow multimode operation in this horizontal plane. Numerous stripe geometry laser structures have been investigated with stripe widths ranging from 2 to 65 μm, and the DH stripe geometry structure has been widely utilized for optical fiber communications. Such structures have active regions which are planar and continuous. Hence the stimulated emission characteristics of these injection lasers are determined by the carrier distribution (which provides optical gain) along the junction plane. The optical mode distribution along the junction plane is, however, decided by the optical gain and therefore these devices are said to be gain-guided laser structures (see Section 6.5.1).

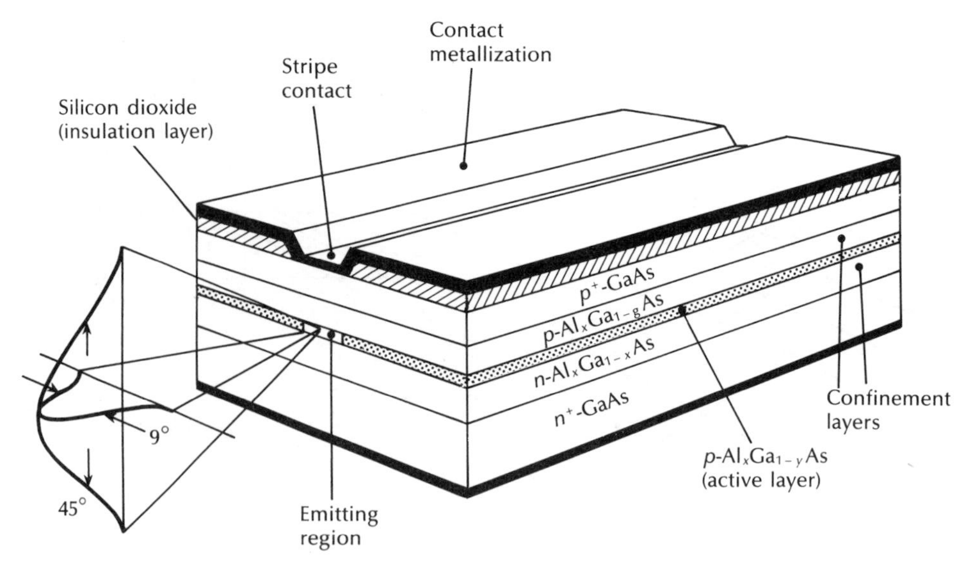

Figure 6.21 Schematic representation of an oxide stripe AlGaAs DH injection laser.

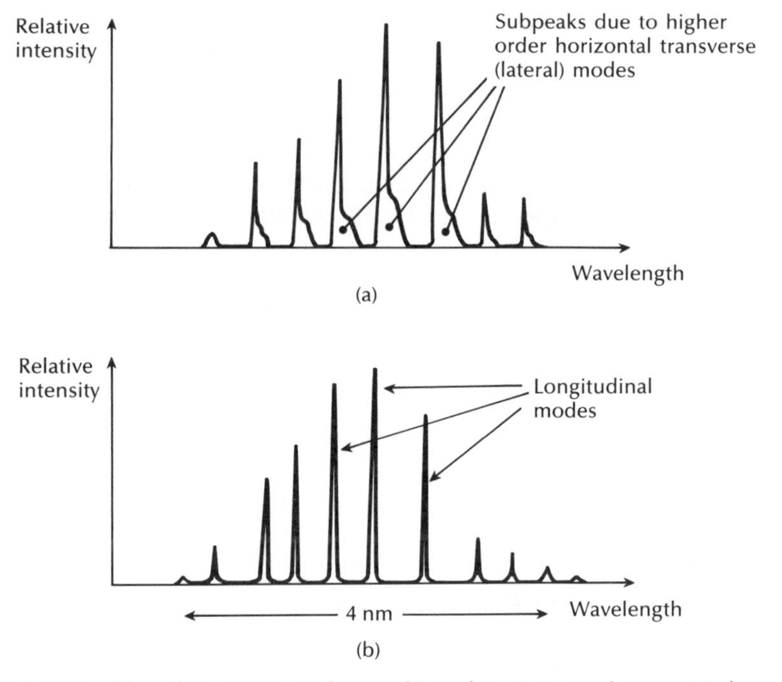

Figure 6.22 Output spectra for multimode injection lasers: (a) broad area device with multitransverse modes; (b) stripe geometry device with single transverse mode.

6.4.3 Laser modes

The typical output spectrum for a broad area injection laser is shown in Figure 6.22(a). It does not consist of a single wavelength output but a series of wavelength peaks corresponding to different longitudinal (in the plane of the junction, along the optical cavity) modes within the structure. As indicated in Section 6.2.4 the spacing of these modes is dependent on the optical cavity length as each one corresponds to an integral number of lengths. They are generally separated by a few tenths of a nanometre, and the laser is said to be a multimode device. However, Figure 6.22(a) also indicates some broadening of the longitudinal mode peaks due to subpeaks caused by higher order horizontal transverse modes.* These higher order lateral modes may exist in the broad area device due to the unrestricted width of the active region. The correct stripe geometry inhibits the occurrence of the higher order lateral modes by limiting the width of the optical cavity leaving only a single lateral mode which gives the output spectrum shown in Figure 6.22(b) where only the longitudinal modes may be observed. This represents the typical output spectrum for a good multimode injection laser.

6.4.4 Single-mode operation

For single-mode operation, the optical output from a laser must contain only a single longitudinal and single transverse mode. Hence the spectral width of the emission from the single-mode device is far smaller than the broadened transition linewidth discussed in Section 6.2.4. It was indicated that an inhomogeneously broadened laser can support a number of longitudinal and transverse modes simultaneously, giving a multimode output. Single transverse mode operation, however, may be obtained by reducing the aperture of the resonant cavity such that only the TEM_{00} mode is supported. To obtain single-mode operation it is then necessary to eliminate all but one of the longitudinal modes.

One method of achieving single longitudinal mode operation is to reduce the length L of the cavity until the frequency separation of the adjacent modes given by Eq. (6.14) as $\delta f = c/2nL$ is larger than the laser transition linewidth or gain curve. Then only the single mode which falls within the transition linewidth can oscillate within the laser cavity. However, it is clear that rigid control of the cavity parameters is essential to provide the mode stabilization necessary to achieve and maintain this single-mode operation.

The structures required to give mode stability are discussed with regard to the multimode injection laser in Section 6.5 and similar techniques can be employed to produce a laser emitting a single longitudinal and transverse mode. For example, the correct DH structure will restrict the vertical width of the waveguiding region to less than 0.4 μm allowing only the fundamental transverse mode to be supported

*Tranverse modes in the plane of the junction are often called lateral modes, transverse mode being reserved for modes perpendicular to the junction plane.

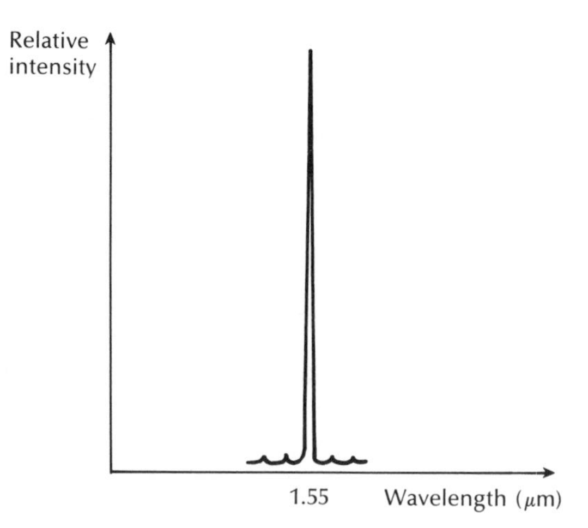

Figure 6.23 Typical single longitudinal mode output spectrum from a single-mode injection laser.

and removing any interference of the higher order transverse modes on the emitted longitudinal modes.

The lateral modes (in the plane of the junction) may be confined by the restrictions on the current flow provided by the stripe geometry. In general, only the lower order modes are excited which appear as satellites to each of the longitudinal modes. However, as will be discussed in Section 6.5.1, stripe contact devices often have instabilities and strong nonlinearities (e.g. kinks) in their light output against current characteristics. Tight current confinement as well as good waveguiding are therefore essential in order to achieve only the required longitudinal modes which form between the mirror facets in the plane of the junction. Finally, as indicated above, single-mode operation may be obtained through control of the optical cavity length such that only a single longitudinal mode falls within the gain bandwidth of the device. Figure 6.23 shows a typical output spectrum for a single-mode device.

However, injection lasers with short cavity lengths (around 50 μm) are difficult to handle and have not been particularly successful. Nevertheless, such devices, together with the major alternative structures which provide single-mode operation, are dealt with in Section 6.6 under the title of single frequency injection lasers.

6.5 Some injection laser structures

6.5.1 Gain-guided lasers

Fabrication of multimode injection lasers with a single or small number of lateral modes is achieved by the use of stripe geometry. These devices are often called gain-guided lasers as indicated in Section 6.4.2. The constriction of the current flow to the stripe is realized in the structure either by implanting the regions outside the stripe with protons (proton isolated stripe) to make them highly resistive, or by oxide or *p–n* junction isolation. The structure for an aluminium gallium arsenide oxide isolated stripe DH laser is shown in Figure 6.21. It has an active region of gallium arsenide bounded on both sides by aluminium gallium arsenide regions. This technique has been widely applied, especially for multimode laser structures used in the shorter wavelength region. The current is confined by etching a narrow stripe in a silicon dioxide film.

Two other basic techniques for the fabrication of gain-guided laser structures are illustrated in Figure 6.24(a) and (b), which show the proton isolated stripe and the *p–n* junction isolated stripe structures respectively. In Figure 6.24(a) the resistive region formed by the proton bombardment gives better current confinement than the simple oxide stripe and has superior thermal properties due to the absence of the silicon dioxide layer; *p–n* junction isolation involves a selective diffusion through the *n* type surface region in order to reach the *p* type layers, as illustrated in Figure 6.24(b). None of these structures confines all the radiation and current to the stripe region and spreading occurs on both sides of the stripe. With stripe widths of 10 μm or less, such planar stripe lasers provide highly efficient coupling into multimode fibers, but significantly lower coupling efficiency is achieved into small core diameter single-mode fibers.

The optical output power against current characteristic for the ideal semiconductor laser was illustrated in Figure 6.17. However, with certain practical laser diodes the characteristic is not linear in the simulated emission region, but exhibits

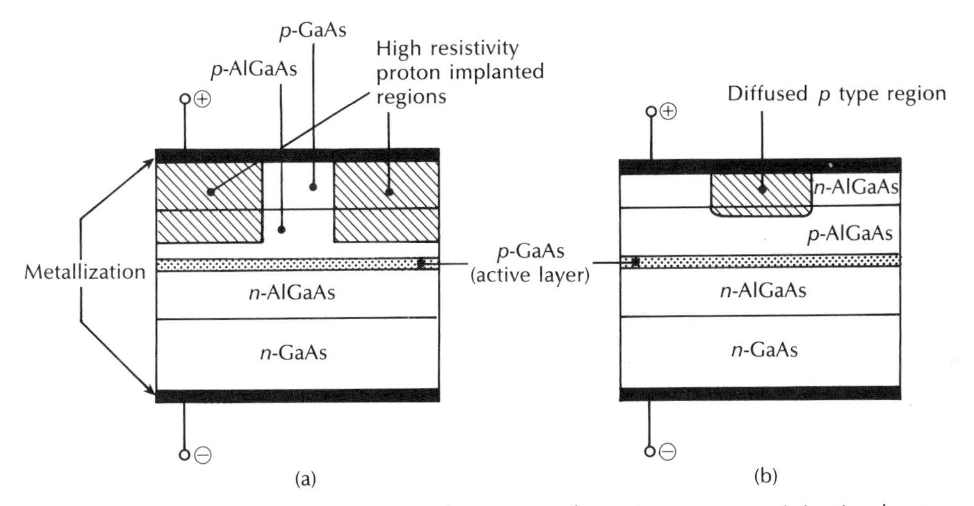

Figure 6.24 Schematic representation of structures for stripe geometry injection lasers: (a) proton isolated stripe GaAs/AlGaAs laser; (b) *p–n* junction isolated (diffused planar stripe) GaAs/AlGaAs laser.

kinks. This phenomenon is particularly prevalent with gain-guided injection laser devices. The kinks may be classified into two broad categories.

The first type of kink results from changes in the dominant lateral mode of the laser as the current is changed. The output characteristic for laser A in Figure 6.25(a) illustrates this type of kink where lasing from the device changes from the fundamental lateral mode to a higher order lateral mode (second order) in a current region corresponding to a change in slope. The second type of kink involves a 'spike', as observed for laser B of Figure 6.25(a). These spikes have been shown to be associated with filamentary behaviour within the active region of the device [Ref. 4]. The filaments result from defects within the crystal structure.

Both these mechanisms affect the near and far field intensity distributions (patterns) obtained from the laser. A typical near field intensity distribution corresponding to a single optical output power level in the plane of the junction is shown in Figure 6.25(b). As this distribution is in the lateral direction, it is determined by the nature of the lateral waveguide. The single intensity maximum shown indicates that the fundamental lateral mode is dominant. To maintain such a near field pattern the stripe geometry of the device is important. In general, relatively narrow stripe devices ($< 10\ \mu m$) formed by a planar process allow the fundamental lateral mode to dominate. This is especially the case at low power levels where near field patterns similar to Figure 6.25(b) may be obtained.

Although gain-guided lasers are commercially available for operation in both the shorter wavelength range (using GaAs active regions) and the longer wavelength

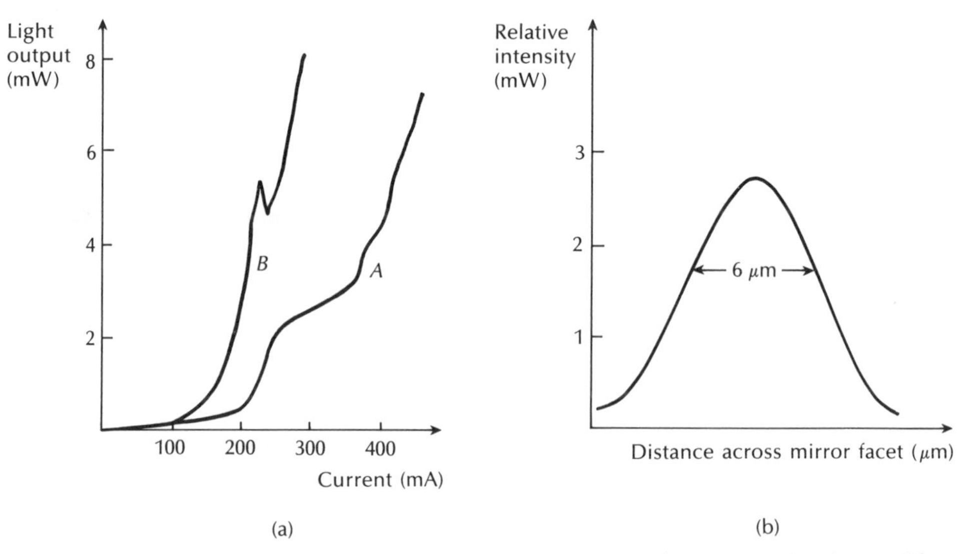

Figure 6.25 (a) The light output against current characteristic for an injection laser with nonlinearities or a kink in the stimulated emission region; (b) A typical near field intensity distribution (pattern) in the plane of the junction for an injection laser.

range (using InGaAsP active regions) they exhibit several undesirable characteristics. Apart from the nonlinearities in the light output versus current characteristics discussed above, gain-guided injection lasers have relatively high threshold currents (100 to 150 mA) as well as low differential quantum efficiency [Ref. 13]. These effects are primarily caused by the small carrier-induced refractive index reduction within the devices which results in the movement of the optical mode along the junction plane. The problems can be greatly reduced by introducing some real refractive index variation into the lateral structure of the laser such that the optical mode along the junction plane is essentially determined by the device structure.

6.5.2 Index-guided lasers

The drawbacks associated with the gain-guided laser structures were largely overcome through the development of index-guided injection lasers. In some such structures with weak index-guiding, the active region waveguide thickness is varied by growing it over a channel or ridge in the substrate. In the ridge waveguide laser shown in Figure 6.26(a), the ridge not only provides the loading for the weak index-guiding but also acts as a narrow current confining stripe [Ref. 14]. These devices have been fabricated to operate at various wavelengths with a single lateral mode, and room temperature CW threshold currents as low as 18 mA with output powers of 25 mW have been reported [Ref. 15]. More typically, the threshold currents for such weakly index-guided structures are in the range 40 to 60 mA, as illustrated in Figure 6.26(b) which compares a light output versus current characteristic for a ridge waveguide laser with that of an oxide stripe gain-guided device.

Alternatively, the application of a uniformly thick, planar active waveguide can be achieved through lateral variations in the confinement layer thickness or the refractive index. The inverted-rib waveguide device (sometimes called plano-convex waveguide) illustrated in Figure 6.26(c) is an example of this structure. However, room temperature CW threshold currents are between 70 and 90 mA with output powers of around 20 mW for InGaAsP devices operating at a wavelength of 1.3 μm [Ref. 16].

Strong index-guiding along the junction plane can provide improved transverse mode control in injection lasers. This can be achieved using a buried heterostructure (BH) device in which the active volume is completely buried in a material of wider bandgap and lower refractive index [Refs. 17, 18]. The structure of a BH laser is shown in Figure 6.27(a). The optical field is well confined both in the transverse and lateral directions within these lasers, providing strong index-guiding of the optical mode together with good carrier confinement. Confinement of the injected current to the active region is obtained through the reverse biased junctions of the higher bandgap material. It may be observed from Figure 6.27 that the higher bandgap, low refractive index confinement material is AlGaAs for GaAs lasers operating in the 0.8 to 0.9 μm wavelength range, whereas it is InP in InGaAsP devices operating in the 1.1 to 1.6 μm wavelength range.

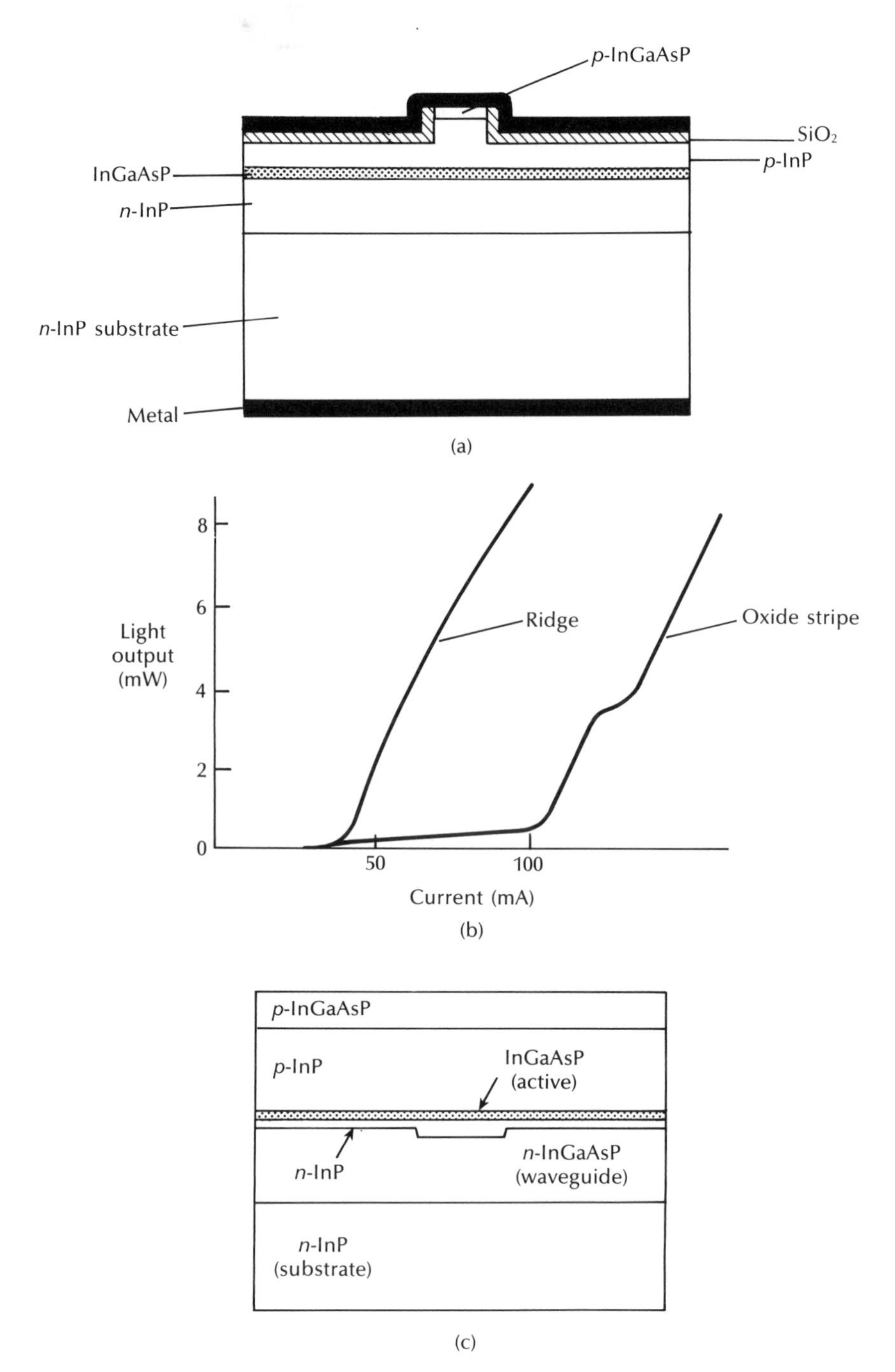

Figure 6.26 Index-guided lasers: (a) ridge waveguide injection laser structures; (b) light output versus current characteristic for (a) compared with that of an oxide stripe (gain-guided) laser; (c) rib (plano-convex) waveguide injection laser structure.

A wide variety of BH laser configurations are commercially available offering both multimode and single-mode operation. In general, the lateral current confinement provided by these devices leads to lower threshold currents (10 to 20 mA) than may be obtained with either weakly index-guided or gain-guided structures. A more complex structure called the double channel planar buried heterostructure (DCPBH) laser is illustrated in Figure 6.27(b) [Ref. 19]. This device which has a planar InGaAsP active region provides very high power operation with CW output powers up to 40 mW in the longer wavelength region. Room temperature threshold currents are in the range 15 to 20 mA for both 1.3 μm and 1.55 μm emitting devices [Ref. 20]. Lateral mode control may be achieved by reducing the dimension of the active region, with a cross sectional area of 0.3 μm^2 being required for fundamental mode operation [Ref. 13].

Parasitic capacitances resulting from the use of the reverse biased current confinement layers can reduce the high speed modulation capabilities of BH lasers. However, this problem has been overcome through either the regrowth of

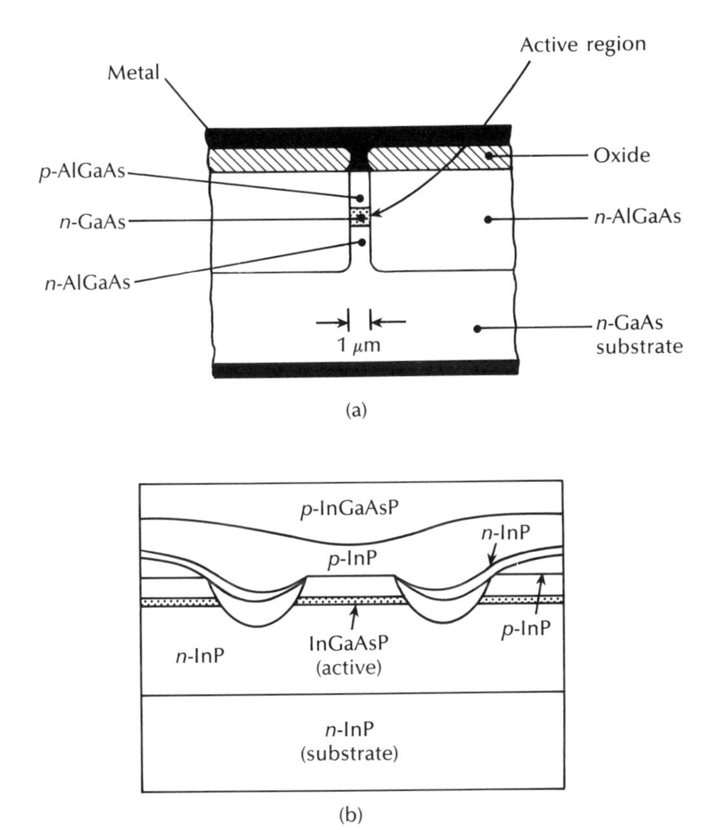

Figure 6.27 Buried heterostructure laser structures: (a) GaAs/AlGaAs BH device; (b) InGaAsP/InP double channel planar BH device.

semiinsulating material [Ref. 2] or the deposition of a dielectric material [Ref. 22]. Using these techniques, modulation speeds in excess of 20 GHz have been achieved which are limited by the active region rather than the parasitic capacitances [Ref. 18].

6.5.3 Quantum-well lasers

In recent years DH lasers have been fabricated with very thin active layer thicknesses of around 10 nm instead of the typical range for conventional DH

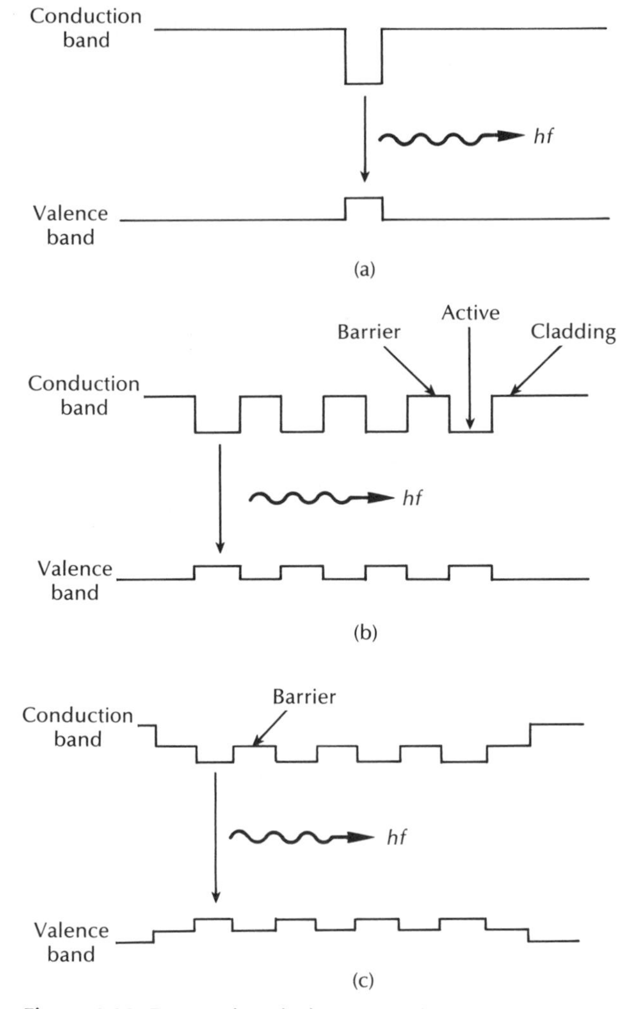

Figure 6.28 Energy band diagrams showing various types of quantum-well structure: (a) single quantum-well; (b) multiquantum-well; (c) modified multiquantum-well.

structures of 0.1 to 0.3 μm. The carrier motion normal to the active layer in these devices is restricted, resulting in a quantization of the kinetic energy into discrete energy levels for the carriers moving in that direction. This effect is similar to the well known quantum mechanical problem of a one dimensional potential well [Ref. 13] and therefore these devices are known as quantum-well lasers. In this structure the thin active layer causes drastic changes to the electronic and optical properties in comparison with a conventional DH laser. These changes are due to the quantized nature of the discrete energy levels with a step-like density of states which differs from the continuum normally obtained. Hence, quantum-well lasers exhibit an inherent advantage over conventional DH devices in that they allow high gain at low carrier density, thus providing the possibility of significantly lower threshold currents.

Both single quantum-well (SQW), corresponding to a single active region and multiquantum-well (MQW), corresponding to multiple active regions lasers have been fabricated [Ref. 13]. In the latter structure, the layers separating the active regions are called barrier layers. Energy band diagrams for the active regions of these structures are displayed in Figure 6.28. It may be observed in Figure 6.28(c) that when the bandgap energy of the barrier layer differs from the cladding layer in a MQW device it is usually referred to as a modified multiquantum-well laser [Ref. 23].

Better confinement of the optical mode is obtained in MQW lasers in comparison with SQW lasers, resulting in a lower threshold current density for these devices. A substantial amount of experimental work has been carried out on MQW lasers using the AlGaAs/GaAs material system. It has demonstrated the superior characteristics of MQW devices over conventional DH lasers in relation to lower threshold currents, narrower linewidths, higher modulation speeds, lower frequency chirp and less temperature dependence (see Section 6.7.1) [Ref. 18]. However, these potential performance advantages have not, as yet, been fully explored for longer wavelength MQW lasers using the InGaAsP/InP material system.

6.6 Single frequency injection lasers

Although the structures described in Section 6.5 provide control of the lateral modes of the laser, the Fabry–Perot cavity formed by the cleaved laser mirrors may allow several longitudinal modes to exist within the gain spectrum of the device (see Section 6.4.3). Nevertheless, such injection laser structures will provide single longitudinal mode operation, even though the mode discrimination obtained from the gain spectrum is often poor. However, improved longitudinal mode selectivity can be achieved using structures which give adequate loss discrimination between the desired mode and all of the unwanted modes of the laser resonator. It was indicated in Section 6.4.4 that such mode discrimination could be obtained by shortening the laser cavity. This technique has met with only limited success [Ref.

18] and therefore alternative structures have been developed to give the necessary electrical and optical containment to allow stable single longitudinal mode operation.

Longitudinal mode selectivity may be improved through the use of frequency-selective feedback so that the cavity loss is different for the various longitudinal modes. Devices which employ this technique to provide single longitudinal mode operation are often referred to as single frequency or dynamic single-mode (DSM) lasers [Refs. 24, 25]. Such lasers are of increasing interest not only to reduce fiber intramodal dispersion within high speed systems but also for the provision of suitable sources for coherent optical transmission (see Chapter 12). Strategies which have proved successful in relation to single frequency operation are the use of short cavity resonators, coupled cavity resonators and distributed feedback.

6.6.1 Short and coupled cavity lasers

It was suggested in Section 6.4.4 that a straightforward method for increasing the longitudinal mode discrimination of an injection laser is to shorten the cavity length; shortening from, say, 250 to 25 μm will have the effect of increasing the mode spacing from 1 to 10 nm. The peak of the gain curve can then be adjusted to provide the desired single-mode operation. Conventional cleaved mirror structures are, however, difficult to fabricate with cavity lengths below 50 μm and therefore configurations employing resonators, either microcleaved [Ref. 26] or etched [Ref. 27], have been utilized. Such resonators form a short cavity of length 10 to 20 μm in a direction normal to the active region providing stable single frequency operation.

Multiple element resonators or resonators with distributed reflectors also give a loss mechanism with a frequency dependence which is strong enough to provide single frequency oscillation under most operating conditions. Mode selectivity in such a coupled cavity laser is obtained when the longitudinal modes of each Fabry–Perot cavity coincide and therefore constitute the longitudinal modes of the coupled system for which both cavities are in resonance. One example of a three mirror resonator shown in Figure 6.29(a) uses a graded index (GRIN) rod lens (see Section 5.5.1) to enhance the coupling to an external mirror [Ref. 28].

An alternative approach is illustrated in Figure 6.29(b) in which two active laser sections are separated by a gap of approximately a single wavelength. When the gap is obtained by recleaving a finished laser chip into two partially attached segments it yields the cleaved-coupled-cavity (C^3) laser [Ref. 29]. This four mirror resonator device has provided dynamic single-mode operation with side mode suppression ratios of several thousand being achieved through control of the magnitudes and the relative phases of the two injection currents, as well as the temperature [Ref. 30]. Another attribute of the C^3 device is that its single frequency emission can be tuned discretely over a range of some 26 nm by varying the current through one section [Ref. 13]. This tunability, which occurs through mode jumps of around 2 nm each, is discussed further in Section 6.10.

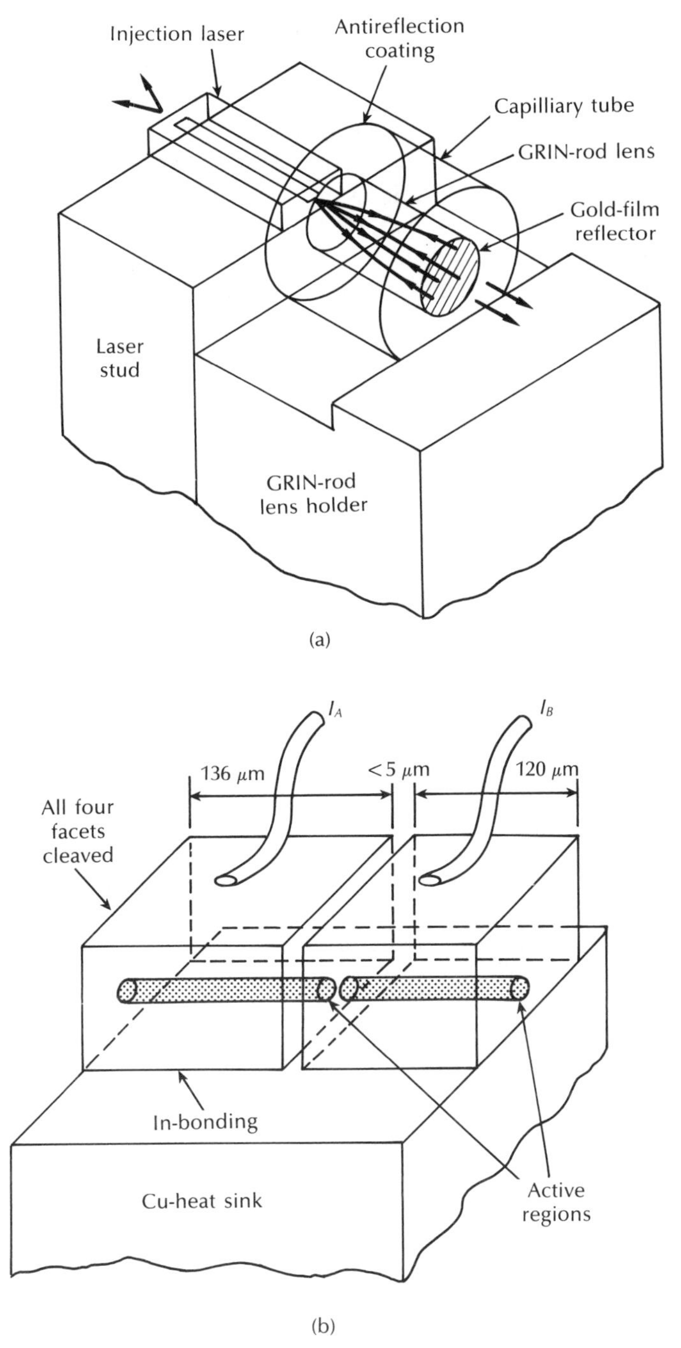

Figure 6.29 Coupled cavity lasers: (a) short external cavity laser using GRIN-rod lens; (b) cleaved-coupled-cavity laser.

6.6.2 Distributed feedback lasers

An elegant approach to single frequency operation which has recently found widespread application involves the use of distributed resonators, fabricated into the laser structure to give integrated wavelength selectivity. The structure which is employed is the distributed Bragg diffraction grating which provides periodic variation in refractive index in the laser heterostructure waveguide along the direction of wave propagation so that feedback of optical energy is obtained through Bragg reflection (see Section 10.6.3) rather than by the usual cleaved mirrors. Hence the corrugated grating structure shown in Figure 6.30(a) determines the wavelength of the longitudinal mode emission instead of the Fabry–Perot gain curve shown in Figure 6.30(b). When the period of the corrugation is equal to $l\lambda_B/2n_e$, where l is the integer order of the grating, λ_B is the Bragg wavelength and n_e is the effective refractive index of the waveguide; then only the mode near the Bragg wavelength λ_B is reflected constructively (i.e. Bragg reflection). Therefore, as may be observed in Figure 6.30(a), this particular mode will lase whilst the other modes exhibiting higher losses are suppressed from oscillation.

It should be noted that first order gratings (i.e. $l = 1$) provide the strongest coupling within the device. Nevertheless, second order gratings are sometimes used as their larger spatial period eases fabrication.

From the viewpoint of device operation, semiconductor lasers employing the distributed feedback mechanism can be classified into two broad categories, referred to as the distributed feedback (DFB) laser [Ref. 31] and the distributed Bragg reflector (DBR) laser [Ref. 32]. These two device structures are shown

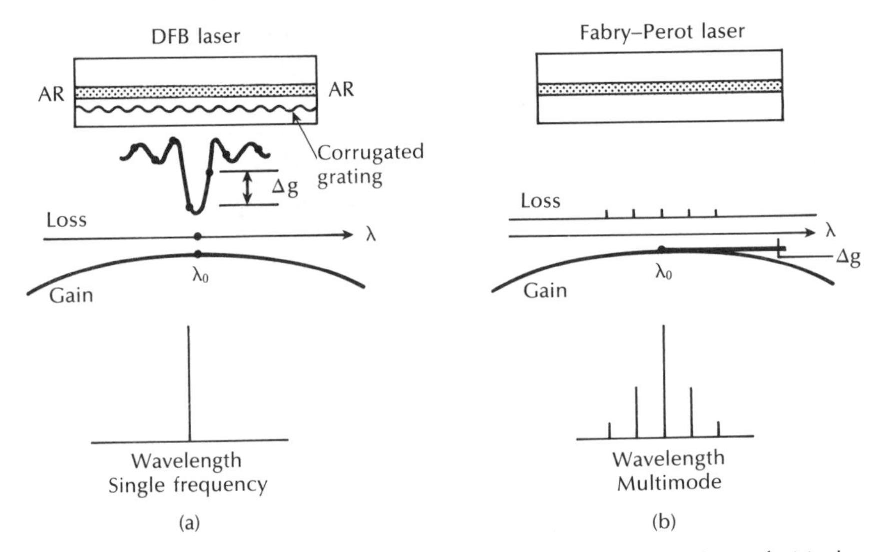

Figure 6.30 Illustration showing the single frequency operation of: (a) the distributed feedback (DFB) laser in comparison with; (b) the Fabry–Perot laser.

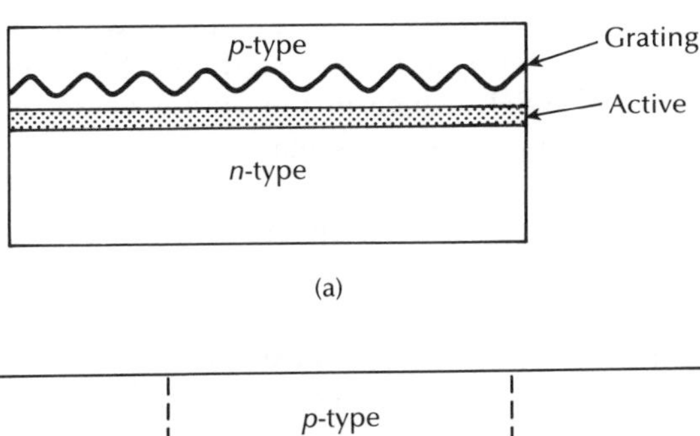

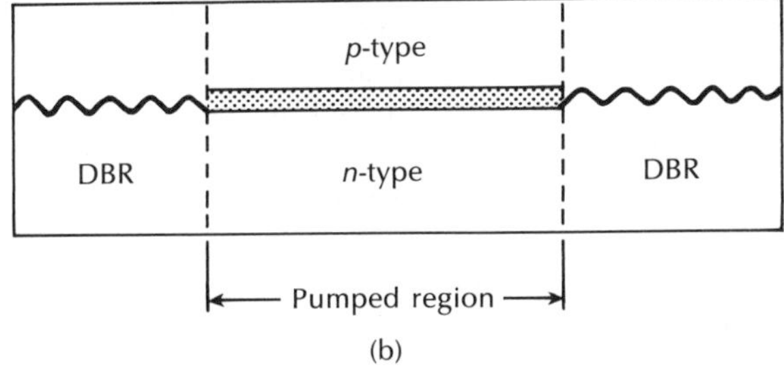

Figure 6.31 Schematic of distributed feedback lasers: (a) DFB laser; (b) DBR laser.

schematically in Figure 6.31. In the DFB laser the optical grating is usually applied over the entire active region which is pumped, whereas in the DBR laser the grating is etched only near the cavity ends and hence distributed feedback does not occur in the central active region. The unpumped corrugated end regions effectively act as mirrors whose reflectivity results from the distributed feedback mechanism which is therefore dependent on wavelength. In addition, this latter device displays the advantage of separating the perturbed regions from the active region but proves somewhat lossy due to optical absorption in the unpumped distributed reflectors. It should be noted that in Figure 6.31 the grating is shown in a passive waveguide layer adjacent to the active gain region for both device structures. This structure has evolved as a result of the performance deterioration with earlier devices (at temperatures above 80 K) in which the corrugations were applied directly to the active layer [Ref. 33].

At present, DBR lasers are less well developed than DFB lasers and it is the latter devices which are under intensive investigation for the provision of single frequency semiconductor optical sources. Any of the semiconductor laser structures discussed in Section 6.5 can be employed to fabricate a DFB laser after etching a grating into an appropriate cladding layer adjacent to the active layer. The grating period is determined by the desired emission frequency from the structure following the Bragg condition (see Section 10.6.3). In particular, buried heterostructure DFB lasers have been developed in many laboratories which exhibit low threshold currents (10 to 20 mA), high modulation speeds (several G bit s^{-1}) and output powers comparable with Fabry–Perot devices with similar BH geometries [Refs.

13, 18, 34]. The structure and the light output versus current characteristic for a low threshold current DFB–BH laser operating at a wavelength of 1.55 μm is displayed in Figure 6.32 [Ref. 35]. A substantial change in the output characteristic with increasing temperature may be observed for this separate confinement DH device (see Section 6.7.1).

In theory when considering a DFB laser with both end facets antireflection (AR) coated (see Figure 6.30(a)), then two modes located symmetrically on either side of the Bragg wavelength will experience the same lowest threshold gain within an ideal symmetrical structure and will therefore lase simultaneously. However, in practice, the randomness associated with the cleaving process creates different end phases, thus removing the degeneracy of the modal gain and providing only single-mode operation. Moreover, facet asymmetry can be increased by placing a high reflection coating on one end facet and a low reflection coating on the other (known as the hi–lo structure) in order to improve the power output for single frequency operation [Ref. 18].

Another technique to improve the performance of the DFB laser is to modify the grating at a central point to introduce an additional optical phase shift, typically a quarter wavelength or less [Refs. 36, 37]. Such a device is shown in Figure 6.33 which illustrates the structure of an InGaAsP/InP double channel planar buried

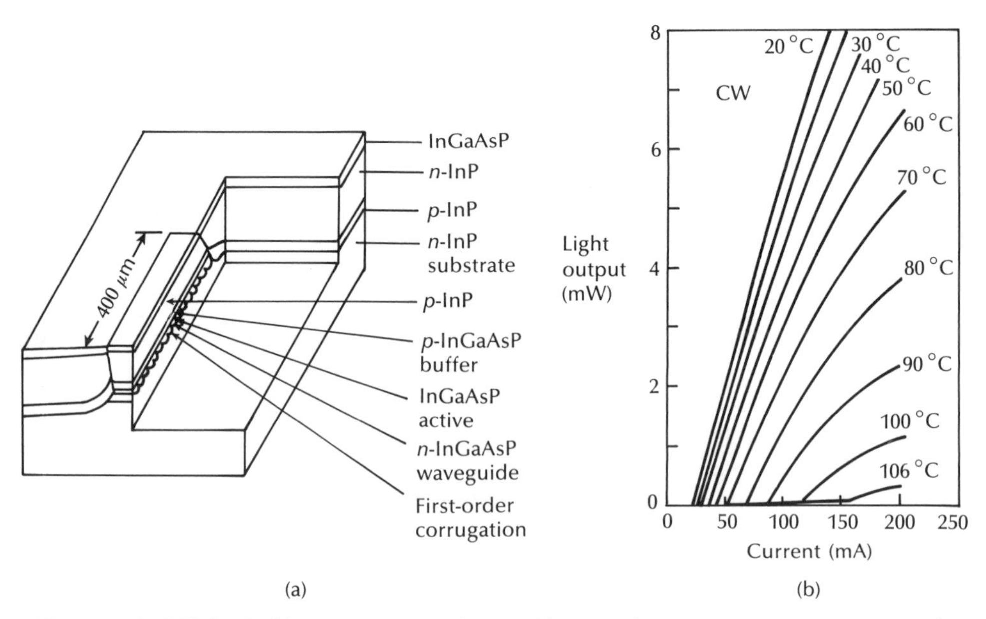

Figure 6.32 DFB buried heterostructure laser with a window structure: (a) structure; (b) light output versus current characteristics for various temperatures under CW operation. Reproduced with permission from S. Tsuji, A. Ohishi, H. Nakamura, M. Hirao, N. Chinone and H. Matsumura, 'Low threshold operation of 1.5 μm DFB laser diodes', *J. Lightwave Technol.*, **LT-5**, p. 822, 1987. Copyright © 1987 IEEE.

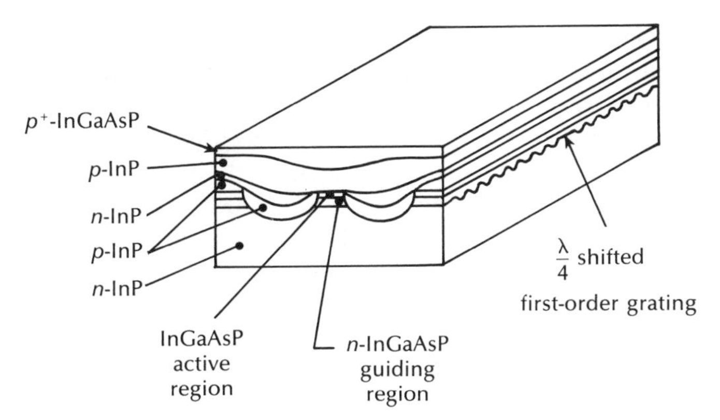

Figure 6.33 Quarter wavelength shifted double channel planar BH-DFB laser.

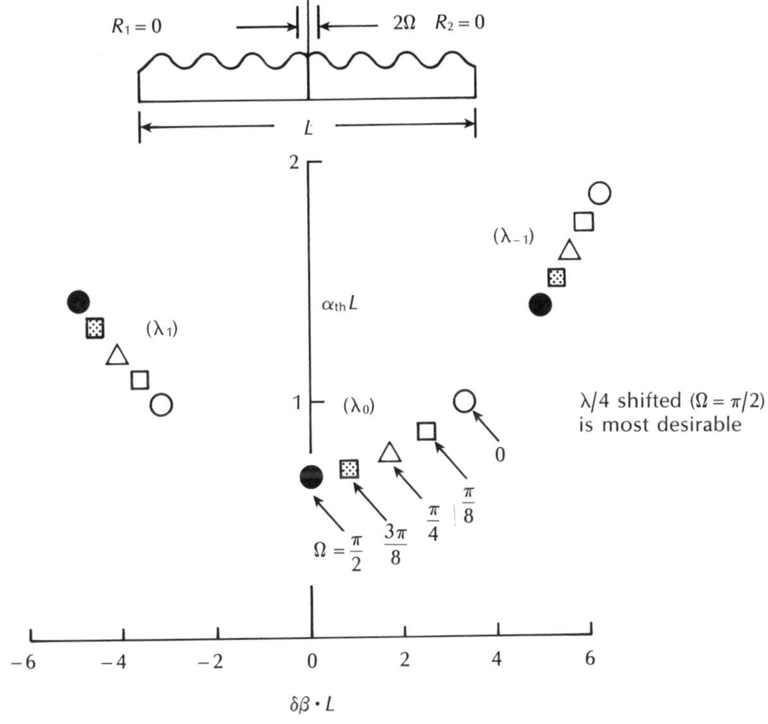

Figure 6.34 The threshold gain and mode frequency in a phase shifted DFB laser.

heterostructure DFB laser with a quarter wavelength shifted first order grating [Ref. 38]. This structure, which provides excellent, stable, single frequency operation, incorporates a $\pi/2$ phase shift (equivalent to one quarter wavelength) in the corrugation at the centre of the laser cavity with both end facets AR coated. The threshold gain and the mode frequency (relative to the Bragg wavelength) for the device as the phase shift is varied from 0 to $\pi/2$ is shown in Figure 6.34. It may be observed that the lowest threshold gain for the central mode (at λ_0) is obtained precisely at the Bragg wavelength when the phase shift is $\pi/2$. Furthermore, the gain difference between the central mode and the nearest side mode (at λ_1) has the largest value at this phase shift.

The performance of the quarter wavelength shifted DFB laser is superior to that of the conventional DFB structure because the large gain difference between the central mode and the side modes gives improved dynamic single-mode stability with negligible mode partition noise (see Section 6.7.4) at multigigabit s^{-1} modulation speeds [Ref. 38]. In addition, narrow linewidths of around 3 MHz ($\simeq 2 \times 10^{-5}$ nm) have been obtained under CW operation [Ref. 18], which is substantially less than the typical 100 MHz ($\simeq 6 \times 10^{-4}$ nm) linewidth associated with the Fabry–Perot injection laser. Linewidth narrowing is achieved within such DFB lasers by detuning the lasing wavelength towards the shorter wavelength side of the gain peak (i.e. towards λ_{-1} in Figure 6.34) in order to increase the differential gain between the central mode and the nearest side mode (λ_1 in Figure 6.34). This strategy is sometimes referred to as Bragg wavelength detuning.

6.7 Injection laser characteristics

When considering the use of the injection laser for optical fiber communications it is necessary to be aware of certain of its characteristics which may affect its efficient operation. The following sections outline the major operating characteristics of the device (the ones which have not been dealt with in detail previously) which generally apply to all the various materials and structures previously discussed, although there is substantial variation in behaviour between them.

6.7.1 Threshold current temperature dependence

Figure 6.35 shows the variation in threshold current with temperature for two gain-guided (oxide insulated stripe) injection lasers [Ref. 39]. Both devices had stripe widths of approximately 20 μm but were fabricated from different material systems for emission at wavelengths of 0.85 μm and 1.55 μm (AlGaAs and InGaAsP devices respectively).

In general terms the threshold current tends to increase with temperature, the temperature dependence of the threshold current density J_{th} being approximately exponential [Ref. 4] for most common structures. It is given by:

$$J_{th} \propto \exp \frac{T}{T_0} \tag{6.43}$$

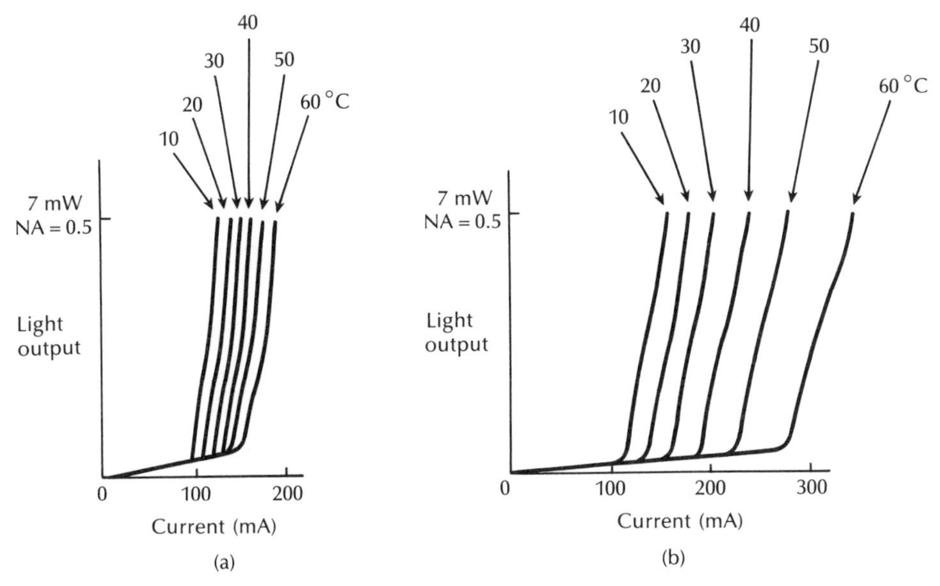

Figure 6.35 Variation in threshold current with temperature for gain-guided injection lasers: (a) AlGaAs device; (b) InGaAsP device. Reproduced with permission from P. A. Kirby, 'Semiconductor laser sources for optical communication', *The Radio and Electron. Eng.*, **51**, p. 362, 1981.

where T is the device absolute temperature and T_0 is the threshold temperature coefficient which is a characteristic temperature describing the quality of the material, but which is also affected by the structure of the device. For AlGaAs devices, T_0 is usually in the range 120 to 190 K, whereas for InGaAsP devices it is between 40 and 75 K [Ref. 40]. This emphasizes the stronger temperature dependence of InGaAsP structures which is illustrated in Figure 6.35 and Example 6.6. The increase in threshold current with temperature for AlGaAs devices can be accounted for with reasonable accuracy by consideration of the increasing energy spread of electrons and holes injected into the conduction and valence bands. It appears that the intrinsic physical properties of the InGaAsP material system may cause its higher temperature sensitivity; these include Auger recombination, intervalence band absorption and carrier leakage effects over the heterojunctions [Ref. 41]. The relative significance of these various mechanisms is, however, not clearly understood [Ref. 42].

Example 6.6

Compare the ratio of the threshold current densities at 20 °C and 80 °C for a AlGaAs injection laser with $T_0 = 160$ K and the similar ratio for an InGaAsP device with $T_0 = 55$ K.

Solution: From Eq. (6.43) the threshold current density:

$$J_{th} \propto \exp \frac{T}{T_0}$$

For the AlGaAs device:

$$J_{th}\ (20\ ^\circ\mathrm{C}) \propto \exp \frac{293}{160} = 6.24$$

$$J_{th}\ (80\ ^\circ\mathrm{C}) \propto \exp \frac{353}{160} = 9.08$$

Hence the ratio of the current densities:

$$\frac{J_{th}\ (80\ ^\circ\mathrm{C})}{J_{th}\ (20\ ^\circ\mathrm{C})} = \frac{9.08}{6.24} = 1.46$$

For the InGaAsP device:

$$J_{th}\ (20\ ^\circ\mathrm{C}) \propto \exp \frac{293}{55} = 205.88$$

$$J_{th}\ (80\ ^\circ\mathrm{C}) \propto \exp \frac{353}{55} = 612.89$$

Hence the ratio of the current densities:

$$\frac{J_{th}\ (80\ ^\circ\mathrm{C})}{J_{th}\ (20\ ^\circ\mathrm{C})} = \frac{612.89}{205.88} = 2.98$$

Thus in Example 6.6 the threshold current density for the AlGaAs device increases by a factor of 1.5 over the temperature range, whereas the threshold current density for the InGaAsP device increases by a factor of 3. Hence the stronger dependence of threshold current on temperature for InGaAsP structures is shown in this comparison of two average devices. It may also be noted that it is important to obtain high values of T_0 for the devices in order to minimize temperature dependence.

The increased temperature dependence for the InGaAsP/InP material system is also displayed by the more advanced, mode-stabilized device structures. Figure 6.36 provides the light output against current characteristic at various device temperatures for a strongly index-guided DCPBH injection laser (see Section 6.5.2) emitting at a wavelength of 1.55 μm [Ref. 13]. Moreover, the similar characteristic for a BH–DFB laser is shown in Figure 6.32(b). It is therefore necessary to pay substantial attention to thermal dissipation in order to provide efficient heat sinking arrangements (e.g. thermoelectric cooling, etc.) to achieve low operating currents. In addition, the need to minimize or eliminate the thermal resistance degradation

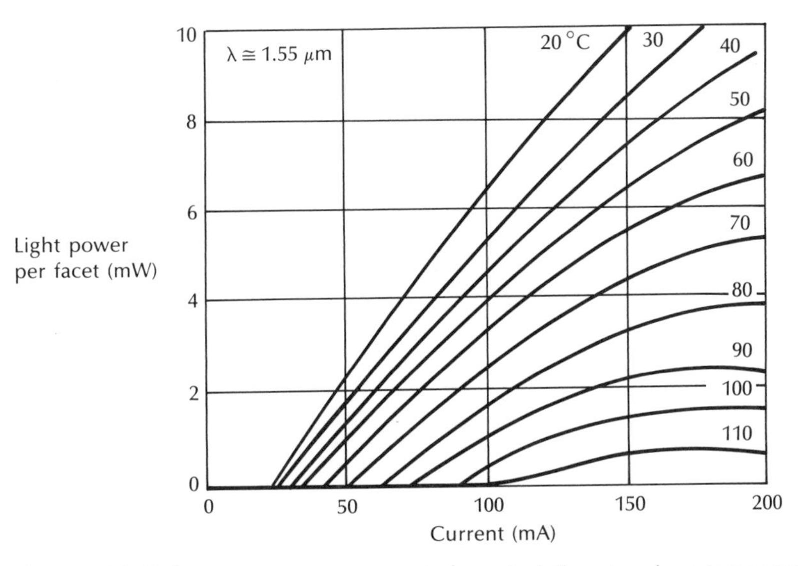

Figure 6.36 Light output versus current characteristics at various temperatures for a InGaAsP double channel planar BH laser emitting at a wavelength of 1.55 μm. Reproduced with permission from N. K. Dutta, 'Optical sources for lightwave system applications', in E. E. Basch (Ed.), *Optical-Fiber Transmission*, H. W. Sams & Co., p. 265, 1987.

associated with the solder bond on such devices (an effect which could, to a certain extent, be tolerated with GaAs injection lasers) has also become critically important [Ref. 42]. In all cases, however, adequate heat sinking along with consideration of the working environment are essential so that devices operate reliably over the anticipated current range.

6.7.2 Dynamic response

The dynamic behaviour of the injection laser is critical, especially when it is used in high bit rate (wideband) optical fiber communication systems. The application of a current step to the device results in a switch-on delay, often followed by high frequency (of the order of 10 GHz) damped oscillations known as relaxation oscillations (RO). These transient phenomena occur whilst the electron and photon populations within the structure come into equilibrium and are illustrated in Figure 6.37. In addition, when a current pulse reaches a laser which has significant parasitic capacitance after the initial delay time, the pulse will be broadened because the capacitance provides a source of current over the period that the photon density is high. Consequently, the injection laser output can comprise several pulses as the electron density is repetitively built up and quickly reduced, thus causing ROs. The switch-on delay t_d may last for 0.5 ns and the RO for perhaps twice that period. At data rates above 100 Mbit s^{-1} this behaviour can produce a serious deterioration

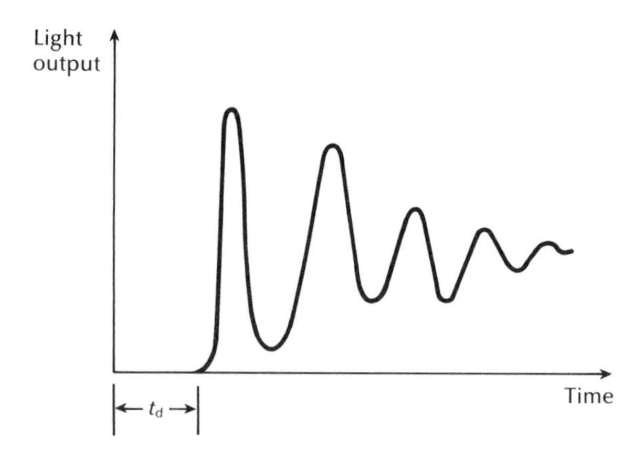

Figure 6.37 The possible dynamic behaviour of an injection laser showing the switch-on delay and relaxation oscillations.

in the pulse shape. Hence, reducing t_d and damping the relaxation oscillations is highly desirable.

The switch-on delay is caused by the initial build-up of photon density resulting from stimulated emission. It is related to the minority carrier lifetime and the current through the device [Ref. 7]. The current term, and hence the switch-on delay, may be reduced by biasing the laser near threshold (prebiasing). However, damping of the ROs is less straightforward. They are basic laser phenomena which vary with device structure and operating conditions; however, RO damping has been observed, and is believed to be due to several mechanisms including lateral carrier diffusion [Refs. 43 and 44], the feeding of the spontaneous emission into the lasing mode [Ref. 45] and gain nonlinearities [Ref. 46]. Narrow stripe geometry DH lasers and all the mode stabilized devices (see Sections 6.5 and 6.6) give RO damping, but it tends to coincide with a relatively slow increase in output power. This is thought to be the result of lateral carrier diffusion due to lack of lateral carrier confinement. However, it appears that RO damping and fast response may be obtained in BH structures with stripe widths less than the carrier diffusion length (i.e. less than 3 μm) [Ref. 47]. Moreover, this phenomenon has been employed within a digital transmission system by biasing the laser near threshold and then by using a single RO as a 'one' bit [Ref. 48].

6.7.3 Frequency chirp

The direct current modulation of a single longitudinal mode semiconductor laser can cause a dynamic shift of the peak wavelength emitted from the device [Ref. 49]. This phenomenon, which results in dynamic linewidth broadening under the direction modulation of the injection current, is referred to as frequency chirping. It arises from gain-induced variations in the laser refractive index due to the strong coupling between the free carrier density and the index of refraction which is present in any semiconductor structure. Hence, even small changes in carrier density, apart from producing relaxation oscillations in the device output, will also result in a phase shift of the optical field, giving an associated change in the resonance frequency within both Fabry–Perot and DFB laser structures.

The laser linewidth broadening or chirping combined with the chromatic dispersion characteristics of single-mode fibers (see Section 3.9) can cause a significant performance degradation within high transmission rate systems [Ref. 50]. In particular, it may result in a shift in operating wavelength from the zero-dispersion wavelength of the fiber, which can ultimately limit the achievable system performance. For example, theoretical predictions [Ref. 51] of the wavelength shift that may occur with an InGaAsP laser under modulation of a few gigabit s^{-1} are around 0.05 nm (6.4 GHz frequency shift).

A number of techniques can be employed to reduce frequency chirp. One approach is to bias the laser sufficiently above threshold so that the modulation current does not drive the device below the threshold where the rate of change of optical output power varies rapidly with time. Unfortunately, this strategy gives an extinction ratio penalty (see Section 11.2.1.6) of the order of several decibels at the receiver. Another method involves the damping of the relaxation oscillations that can occur at turn-on and turn-off which result in large power fluctuations. This has been achieved, for instance, by shaping the electrical drive pulses [Ref. 52].

Certain device structures also prove advantageous for chirp reduction. In particular, quantum-well lasers (see Section 6.5.3), Bragg wavelength detuned DFB lasers (see Section 6.6.2) and multielectrode DFB lasers (see Section 6.10.2) provide improved performance under direct current modulation in relation to frequency chirping. Such lasers, however, require complex fabricational processes. An alternative technique which has proved effective in minimizing the effects of chirp is to allow the laser to emit continuously and to impress the data on to the optical carrier using an external modulator [Ref. 51]. Such devices, which may be separate lithium niobate-based components or can be monolithically integrated with the laser [Ref. 53], are described in Chapter 10.

6.7.4 Noise

Another important characteristic of injection laser operation involves the noise behaviour of the device. This is especially the case when considering analog transmission. The sources of noise are:

(a) phase or frequency noise;
(b) instabilities in operation such as kinks in the light output against current characteristic (see Section 6.5.1) and self pulsation;
(c) reflection of light back into the device;
(d) mode partition noise.

It is possible to reduce, if not remove (b), (c) and (d) by using mode stabilized devices and optical isolators. Phase noise, however, is an intrinsic property of all laser types. It results from the discrete and random spontaneous or simulated transitions which cause intensity fluctuations in the optical emission and are an inevitable aspect of laser operation. Each event causes a sudden jump (of random magnitude and sign) in the phase of the electromagnetic field generated by the device. It has been observed that the spectral density of this phase or frequency noise has a characteristic represented by $1/f$ to $1/f^2$ up to a frequency (f) of around 1 MHz, as illustrated in Figure 6.38 [Ref. 55].

At frequencies above 1 MHz the noise spectrum is flat or white and is associated with quantum fluctuations (sometimes referred to as quantum noise, see Figure 6.38) which are a principal cause of linewidth broadening within semiconductor lasers [Ref. 56]. Although the low frequency components can easily be tracked and therefore are not a significant problem within optical fiber communications, this is not the case for the white noise component where as time elapses the phase executes a random walk away from the value it would have had in the absence of spontaneous emission.

For injection lasers operating at frequencies less than 100 MHz quantum noise levels are usually low (signal to noise ratios less than −80 dB) unless the device is biased within 10% of threshold. Over this region the noise spectrum is flat. However, for wideband systems when the laser is operating above threshold, quantum noise becomes more pronounced. This is especially the case with multimode devices (signal to noise ratios of around −60 dB). The higher noise level would appear to result from a peak in the noise spectrum due to a relaxation resonance which typically occurs between 200 MHz and 1 GHz [Ref. 7]. Single-mode lasers have demonstrated greater noise immunity by as much as 30 dB when the current is raised above threshold [Ref. 57]. Nevertheless, the wandering of the phase determines both the laser linewidth and the coherence time which are both major considerations, particularly within coherent optical fiber communications [Ref. 58].

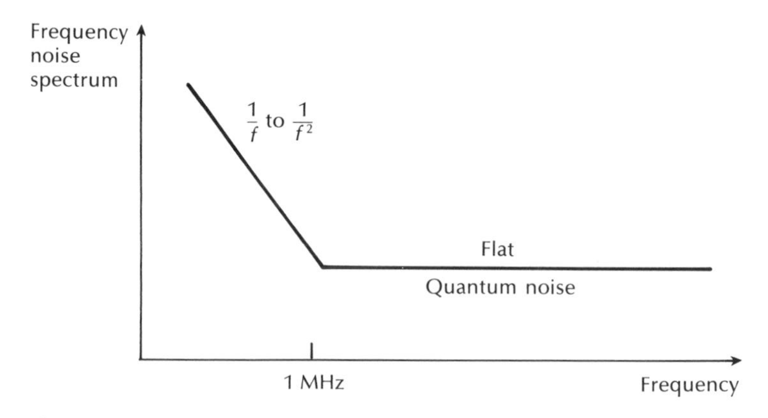

Figure 6.38 Spectral characteristic showing injection laser phase noise.

Fluctuations in the amplitude or intensity of the output from semiconductor injection lasers also leads to optical intensity noise. These fluctuations may be caused by temperature variations or, alternatively, they result from the spontaneous emission contained in the laser output, as mentioned previously. The random intensity fluctuations create a noise source referred to as relative intensity noise (RIN), which may be defined in terms of the mean square power fluctuation $\overline{\delta P_e^2}$ and the mean optical power squared $(\bar{P}_e)^2$ which is emitted from the device following:

$$RIN = \frac{\overline{\delta P_e^2}}{(\bar{P}_e)^2} \tag{6.44}$$

The above definition allows the RIN to be measured in dB Hz^{-1} where the power fluctuation is written as:

$$\overline{\delta P_e^2(t)} = \int_0^\infty S_{RIN}(f)\, \mathrm{d}f \tag{6.45}$$

where $S_{RIN}(f)$ is related to the power spectral density of the relative intensity noise $S_{RIN}(\omega)$ by:

$$S_{RIN}(f) = 2\pi S_{RIN}(\omega) \tag{6.46}$$

where $\omega = 2\pi f$.

Hence from Eq. (6.44), the RIN as a relative power fluctuation over a bandwidth B which is defined as 1 Hz:

$$RIN = \frac{S_{RIN}(f)\ B(=1\ \text{Hz})}{(\bar{P}_e)^2} \tag{6.47}$$

Typically, the RIN for a single-mode semiconductor laser would lie in the range 130 to 160 dB Hz^{-1}. However, it should be noted that the relative intensity noise decreases as the injection current level I increases following the relation:

$$RIN \propto \left(\frac{I}{I_{th}} - 1\right)^{-3} \tag{6.48}$$

where I_{th} is the laser threshold current.

From the discussion of optical detectors following in Section 8.6 it is clear that when an optical field at a frequency f is incident with power $P_o(t)$ on a photodetector whose quantum efficiency (electrons per photon) is η then the output photocurrent $I_p(t)$ is:

$$I_p(t) = \frac{\eta e P_o(t)}{hf} \tag{6.49}$$

where e is the charge on an electron and h is Planck's constant. Therefore an optical power fluctuation $\delta P_o(t)$ will cause a fluctuating current component

$\delta I_{\mathrm{p}}(t) = \eta e\ \delta P_{\mathrm{o}}(t)/hf$ which exhibits a mean square value:

$$\overline{i^2}(t) = \overline{\delta I_{\mathrm{p}}^2}(t) = \frac{\eta^2 e^2}{(hf)^2}\ \overline{\delta P_{\mathrm{o}}^2}(t) \tag{6.50}$$

Now, considering the fluctuation in the incident optical power at the detector to result from RIN in the laser emission, using Eqs. (6.44) and (6.47), and transposing P_{e} for P_{o}, then the mean square noise current in the output of the detector $\overline{i_{RIN}^2}$ due to these fluctuations is:

$$\overline{i_{RIN}^2} = \frac{\eta^2 e^2}{(hf)^2}\ (RIN)(\overline{P_{\mathrm{e}}})^2 B \tag{6.51}$$

Example 6.7

The output from a single-mode semiconductor laser with a RIN value of 10^{-15} dB Hz^{-1} is incident directly on an optical detector which has a bandwidth of 100 MHz. The device is emitting at a wavelength of 1.55 μm, at which the detector has a quantum efficiency of 60%. If the mean optical power incident on the detector is 2 mW, determine: (a) the rms value of the power fluctuation and (b) the rms noise current at the output of the detector.

Solution: (a) The relative mean square fluctuation in the detected current is equal to $\overline{\delta P_{\mathrm{e}}^2}/(\bar{P}_{\mathrm{e}})^2$, which, using Eqs. (6.44) and (6.47), can be written as:

$$\frac{\overline{\delta P_{\mathrm{e}}^2}}{(\bar{P}_{\mathrm{e}})^2} = \frac{S_{RIN}(f)}{(\bar{P}_{\mathrm{e}})^2}\ B = 10^{-15} \times 100 \times 10^6 = 10^{-7}$$

Hence the rms value of this power fluctuation is:

$$\frac{(\overline{\delta P_{\mathrm{e}}^2})^{\frac{1}{2}}}{\bar{P}_{\mathrm{e}}} = 3.16 \times 10^{-4}\ \mathrm{W}$$

(b) The rms noise current at the detector output may be obtained from Eq. (6.51) as:

$$(\overline{i_{RIN}^2})^{\frac{1}{2}} = \frac{e\eta}{hf}\ (RIN)^{\frac{1}{2}} \bar{P}_{\mathrm{e}} B^{\frac{1}{2}} = \frac{e\eta\lambda}{hc}\ (RIN)^{\frac{1}{2}} \bar{P}_{\mathrm{e}} B^{\frac{1}{2}}$$

$$= \frac{1.602 \times 10^{-19} \times 0.6 \times 1.55 \times 10^{-6} \times 3.16 \times 10^{-8} \times 2 \times 10^{-3} \times 10^4}{6.626 \times 10^{-34} \times 2.998 \times 10^8}$$

$$= 4.74 \times 10^{-7}\ \mathrm{A}$$

Optical feedback from unwanted external reflections can also affect the intensity and frequency stability of semiconductor lasers [Ref. 59]. With multimode lasers, however, this effect is reduced because the reflections are distributed among many fiber modes and therefore they are only weakly coupled back into the laser mode

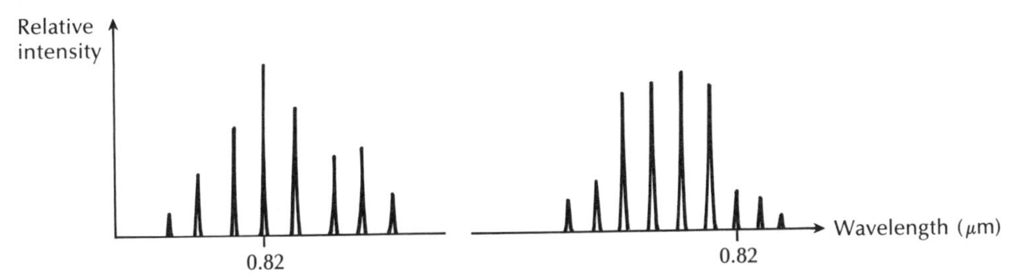

Figure 6.39 The effect of partition noise in a multimode injection laser. It is displayed as a variation in the distribution of the various longitudinal modes emitted from the device.

[Ref. 60]. The stronger fiber to laser coupling in single-mode systems, particularly those operating at 1.55 μm, can result in reflection-induced frequency hops and linewidth broadening [Ref. 51]. In these cases an optical isolator, which is a nonreciprocal device that allows light to pass in the forward direction but strongly attenuates it in the reverse direction, may be required to provide reliable single-mode operation.

Mode partition noise is a phenomenon which occurs in multimode semiconductor lasers when the modes are not well stabilized [Ref. 61]. Even when the total output power from a laser is maintained nearly constant, temperature changes can cause the relative intensities of the various longitudinal modes in the laser's output spectrum to vary considerably from one pulse to the next, as illustrated in Figure 6.39. These spectral fluctuations combined with the fiber dispersion produce random distortion of received pulses on a digital channel, causing an increase in bit error rate.

Mode partition noise can also occur in single-mode devices as a result of the residual side modes in the laser output spectrum. The effect varies between lasers emitting at 1.3 μm and those operating at 1.55 μm but, overall, a degree of side mode suppression is required in both cases in order to avoid additional errors at the receiver [Ref. 51].

6.7.5 Mode hopping

The single longitudinal mode output spectrum of a single-mode laser is illustrated in Figure 6.40(a). Mode hopping to a longer wavelength as the current is increased above threshold is demonstrated by comparison with the output spectrum shown in Figure 6.40(b). This behaviour occurs in all single-mode injection lasers and is a consequence of increases in temperature of the device junction. The transition (hopping) from one mode to another is not a continuous function of the drive current but occurs suddenly over only 1 to 2 mA. Mode hopping alters the light output against current characteristics of the laser, and is responsible for the kinks observed in the characteristics of many single-mode devices.

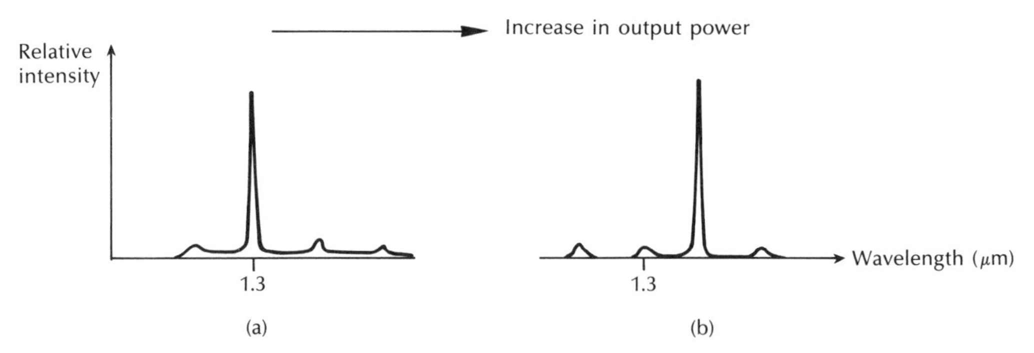

Figure 6.40 Mode hopping in a single-mode injection laser: (a) single longitudinal mode optical output: (b) mode hop to a longer peak emission wavelength at an increased optical output power.

Between hops the mode tends to shift slightly with temperature in the range 0.05 to 0.08 nm K^{-1}. Stabilization against mode hopping and mode shift may be obtained with adequate heat sinking or thermoelectric cooling. However, at constant heat sink temperature, shifts due to thermal increases can only be fully controlled by the use of feedback from external or internal grating structures (see Section 11.2.3).

6.7.6 Reliability

Device reliability has been a major problem with injection lasers and although it has been extensively studied, not all aspects of the failure mechanisms are fully understood [Ref. 13]. Nevertheless, much progress has been made since the early days when device lifetimes were very short (a few hours).

The degradation behaviour may be separated into two major processes known as 'catastrophic' and 'gradual' degradation. Catastrophic degradation is the result of mechanical damage of the mirror facets and leads to partial or complete laser failure. It is caused by the average optical flux density within the structure at the facet and therefore may be limited by using the device in a pulsed mode. However, its occurrence may severely restrict the operation (to low optical power levels) and lifetime of CW devices.

Gradual degradation mechanisms can be separated into two categories which are: (a) defect formation in the active region; and (b) degradation of the current confining junctions. These degradations are normally characterized by an increase in the threshold current for the laser which is often accompanied by a decrease in its external quantum efficiency [Ref. 62].

Defect formation in the active region can be promoted by the high density of recombining holes within the device [Ref. 63]. Internal damage may be caused by the energy released, resulting in the possible presence of strain and thermal gradients by these nonradiative carrier recombination processes. Hence if nonradiative electron–hole recombination occurs, for instance at the damaged surface of a laser where it has been roughened, this accelerates the diffusion of the point defects into the active region of the device. The emission characteristics of the active region therefore gradually deteriorate through the accumulation of point defects until the device is no longer useful. These defect structures are generally observed as dark spot defects (DSDs).

Mobile impurities formed by the precipitation process, such as oxygen, copper or interstitial beryllium or zinc atoms, may also be displaced into the active region of the laser. These atoms tend to cluster around existing dislocations encouraging high local absorption of photons. This causes dark lines in the output spectrum of the device which are a major problem associated with gradual degradation. Such defect structures are normally referred to as dark line defects (DLDs). Both DLDs and DSDs have been observed in ageing AlGaAs lasers as well as in InGaAsP lasers [Ref. 63].

Degradation of the current confining junctions occurs in many index-guided laser structures (see Section 6.5.2) which utilize current restriction layers so that most of the injected current will flow through the active region. For example, the current flowing outside the active region in buried heterostructure (BH) lasers is known as leakage current. Hence a mode of degradation that is associated with this laser structure is an increase in the leakage current which increases the device threshold and decreases the external differential quantum efficiency with ageing.

Over recent years techniques have evolved to reduce, if not eliminate, the introduction of defects, particularly into the injection laser active region. These include the use of substrates with low dislocation densities (i.e. less than 10^{-3} cm^{-2}), passivating the mirror facets to avoid surface-related effects and mounting with soft solders to avoid external strain. Together with improvements in crystal growth, device fabrication and material selection, this has led to CW injection lasers with reported mean lifetimes in excess of 10^6 hours, or more than 100 years. These projections have been reported [Ref. 64] for a variety of GaAs/AlGaAs laser structures. In the longer wavelength region where techniques were not as well advanced, earlier reported extrapolated lifetimes for CW InGaAsP/InP DH lasers were around 10^5 hours [Ref. 65]. More recently, however, InGaAsP/InP BH lasers emitting at 1.3 μm have been tested which display statistically estimated mean lifetimes in excess of 10^6 hours at operating temperatures of 50 °C [Ref. 66]. In addition DFB lasers emitting at 1.55 μm subject to accelerated ageing at a temperature of 60 °C have demonstrated stable ageing characteristics for more than 2000 hours of operating time.

6.8 Injection laser to fiber coupling

One of the major difficulties with using semiconductor lasers within optical fiber communication systems concerns the problems associated with the efficient coupling of light between the laser and the optical fiber (particularly single-mode fiber with its small core diameter and low numerical aperture). Although injection lasers are relatively directional they have diverging output fields which do not correspond to the narrow acceptance angles of single-mode fibers. Thus butt coupling (see Figure 6.41(a)) efficiency from the laser to the fiber is often low at around 10%, even with good alignment and the use of a fiber with a well cleaved end [Ref. 67]. In this case the optimum coupling efficiency is obtained by positioning the fiber end very close to the laser facet. Unfortunately, this technique allows back reflections from the fiber to couple strongly into the laser which produce noise at the device output that can cause performance degradations in high speed systems [Ref. 68].

The coupling efficiency can be substantially improved when the output field from the laser is matched to the output field of the fiber. Such matching is usually achieved using a lens (or lens system) positioned between the laser and the fiber. A simple and popular technique is to employ a hemispherical lens formed on the end of a tapered optical fiber,* as illustrated in Figure 6.41(b) [Refs. 69, 70]. The numbers of piece-parts are therefore minimized and only one alignment step is required. Measured coupling efficiencies up to 65% have been obtained using this

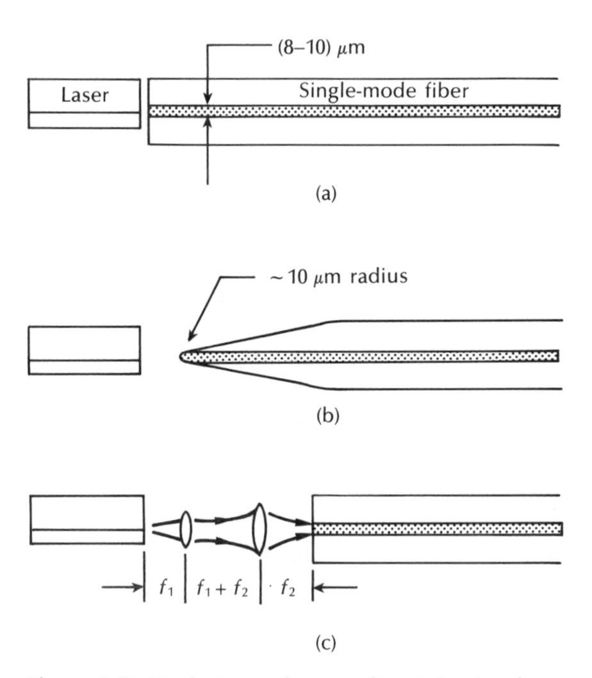

Figure 6.41 Techniques for coupling injection lasers to optical fiber, illustrated using single-mode fiber: (a) butt coupling; (b) tapered hemispherical fiber coupling; (c) confocal lens system.

* Such techniques are sometimes referred to as microlensed fibers [Ref. 71].

method [Ref. 71]. Alternative strategies for microlensed fiber coupling include the use of an etched fiber end with lens [Ref. 72] and a high index lens on the end of a fiber taper [Ref. 73]. Coupling efficiencies of 60% and 55%, respectively, have been achieved using these techniques.

Injection laser coupling using designs based on discrete lenses have also proved fruitful. In particular, such lens systems provide for a relaxation in the alignment tolerances normally required to achieve efficient microlensed fiber coupling. For example, the confocal lens system shown in Figure 6.41(c) allows a relaxation in the 1 dB tolerance by about a factor of 4 in comparison with an 8 μm radius microlensed fiber [Ref. 71]. The combination of the sphere lens and the GRIN-rod lens (see Section 5.5.1) is common within such systems because of the simplicity of the components. Coupling efficiencies of 40% have been obtained with the sphere and GRIN-rod lens in a confocal design. Furthermore, slightly higher efficiencies have been achieved using a GRIN-rod lens with one convex surface (49%) and with a silicon plano-convex lens (55%). Finally, the use of a silicon lens within a confocal system has provided coupling efficiencies of up to 70% [Ref.71].

6.9 Nonsemiconductor lasers

Although at present injection lasers are the major lasing source for optical fiber communications, certain nonsemiconductor sources are of increasing interest for application within this field. Both crystalline and glass waveguiding structures doped with rare earth ions (e.g. neodymium) show potential for use as optical communication sources. In particular, the latter devices in which the short waveguiding structures are glass optical fibers have formed an area of significant development only since 1985 [Ref. 74]. Prior to consideration of these rare earth doped fiber lasers, however, this section briefly discusses the most advanced of the crystalline solid state lasers which could find use within optical fiber communications: the Nd : YAG laser.

6.9.1 The Nd:YAG laser

The crystalline waveguiding material which forms the active medium for this laser is yttrium–aluminium–garnet ($Y_3Al_5O_{12}$) doped with the rare earth metal ion neodymium (Nd^{3+}) to form the Nd : YAG structure. The energy levels for both the lasing transitions and the pumping are provided by the neodymium ions which are randomly distributed as substitutional impurities on lattice sites normally occupied by yttrium ions within the crystal structure. However, the maximum possible doping level is around 1.5%. This laser, which is currently utilized in a variety of areas [Ref. 75], has the following several important properties that may enable its use as an optical fiber communication source:

1. Single-mode operation near 1.064 and 1.32 μm, making it a suitable source for single-mode systems.

2. A narrow linewidth (<0.01 nm) which is useful for reducing dispersion on optical links.
3. A potentially long lifetime, although comparatively few data are available.
4. The possibility that the dimensions of the laser may be reduced to match those of the single-mode fiber.

However, the Nd : YAG laser also has the following drawbacks which are common to all neodymium doped solid state devices:

1. The device must be optically pumped. However, long lifetime AlGaAs LEDs may be utilized which improve the overall lifetime of the laser.
2. A long fluorescence lifetime of the order of 10^{-4} seconds which only allows direct modulation (see Section 7.5) of the device at very low bandwidths. Thus an external optical modulator is necessary if the laser is to be usefully utilized in optical fiber communications.
3. The device cannot take advantage of the well developed technology associated with semiconductors and integrated circuits.
4. The above requirements (i.e. pumping and modulation) tend to give a cost disadvantage in comparison with semiconductor lasers.

An illustration of a typical end pumped Nd : YAG laser is shown in Figure 6.42. It comprises a Nd : YAG rod with its ends ground flat and then silvered. One mirror is made fully reflecting whilst the other is about 10% transmitting to give the output.

The Nd : YAG laser is a four level system (see Section 6.2.3) with a number of pumping bands and fluorescent transitions. The strongest pumping bands are at wavelengths of 0.75 and 0.81 μm, giving major useful lasing transitions at 1.064 and 1.32 μm. Single-mode emission is obtained at these wavelengths with devices which are usually only around 1 cm in length [Ref. 75]. Although the Nd : YAG laser has the specific advantages and drawbacks noted above, it also has a cost disadvantage in comparison with rare earth doped glass fiber lasers (see next section) in that it is far easier and less expensive to fabricate glass fiber than it is to grow YAG crystals.

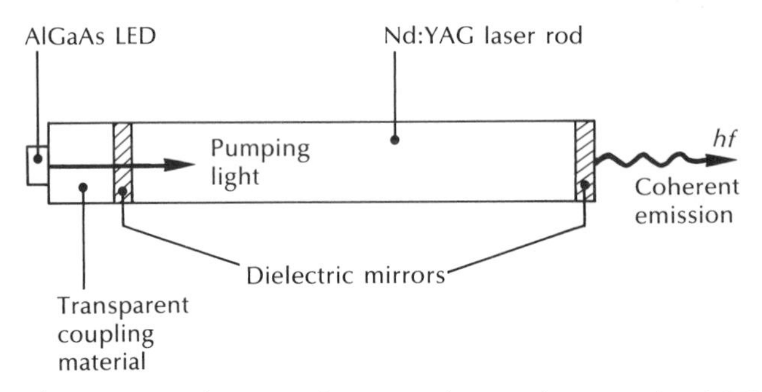

Figure 6.42 Schematic diagram of an end pumped Nd : YAG laser.

6.9.2 Glass fiber lasers

The basic structure of a glass fiber laser is shown in Figure 6.43. An optical fiber, the core of which is doped with rare earth ions, is positioned between two mirrors adjacent to its end faces which form the laser cavity. Light from a pumping laser source is launched through one mirror into the fiber core which is a waveguiding resonant structure forming a Fabry–Perot cavity. The optical output from the device is coupled through the mirror on the other fiber end face, as illustrated in Figure 6.43. Thus the fiber laser is effectively an optical wavelength converter in which the photons at the pumping wavelength are absorbed to produce the required population inversion and stimulated emission; this provides a lasing output at a wavelength which is characterized by the dopant in the fiber.

The rare earth elements, or lanthanides number fifteen and occupy the penultimate row of the periodic table. They range from lanthanum (La), with an atomic number of 57, to lutetium which has an atomic number of 71. Ionization of the rare earths normally takes place to form a trivalent state and the two major dopants currently employed for fiber lasers are neodymium (Nd^{3+}) and erbium (Er^{3+}). In common with the Nd : YAG laser (see Section 6.9.1) the former element provides a four level scheme with significant lasing outputs at wavelengths of 0.90, 1.06 and 1.32 μm. The latter element gives a three level scheme (see Section 6.2.3) with major useful lasing transitions at 0.80, 0.98 and 1.55 μm [Ref. 74]. One consequence of the number of levels involved in the laser action that is of particular significance to fiber lasers is the length dependence of the threshold power. Provided that the imperfection losses are low, then in a four level system the threshold power decreases inversely with the length of the fiber gain medium. In a three level system, however, there is an optimum length that gives the minimum threshold power which is independent of the value of the imperfection losses [Ref. 74].

The glasses which form the host materials for the rare earth doped fiber lasers mainly comprise covalently bonded molecules in the form of a disordered matrix with a wide range of bond lengths and bond angles [Ref. 76]. The rare earth ions which are impurities either act as network modifiers or are interstitially located within the glass network. To date, silica-based glasses have provided the major host material although fluorozirconate fibers (see Section 3.7) doped with both neodymium and erbium ions have produced lasers emitting at wavelengths of 1.05 and 1.35 μm, and 1.55 μm respectively. In addition, fluoride glasses with other dopants give lasing outputs in the mid-infrared wavelength range (see Section 6.11).

Both neodymium and erbium doped silica fiber lasers employ codopants such as

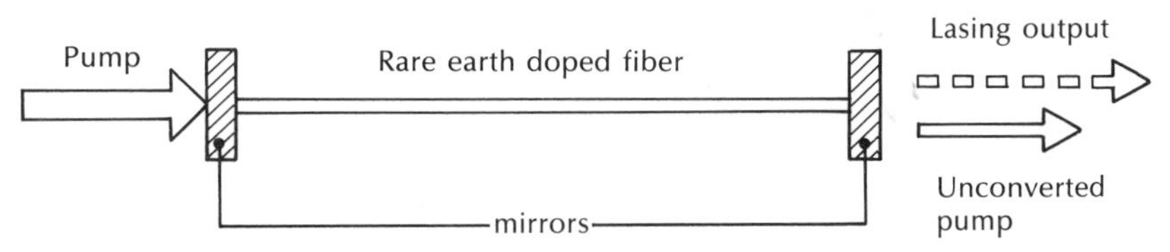

Figure 6.43 Schematic diagram showing the structure of a fiber laser.

phosphorous pentoxide (P_2O_5), germania (e.g. GeO_2, $GeCl_4$) or alumina (Al_2O_5). Dopant levels are generally low (at 400 parts per million) in order to avoid concentration quenching which causes a reduction in the population of the upper lasing levels as well as crystallization within the glass matrix [Ref. 74]. In addition, certain properties of the glass host materials lead to significant spectral broadening of the laser outputs through several mechanisms [Ref. 77] in contrast to what occurs with the Nd:YAG gain medium (see Section 6.9.1). For example, the different fluorescence spectra for an erbium doped silica fiber and a similarly doped fluorozirconate fiber (ZBLANP*) may be observed in Figure 6.44 [Ref. 78].

The light output versus absorbed pump power characteristics for two fiber lasers are displayed in Figure 6.45. The characteristic shown in Figure 6.45(a) corresponds to a neodymium doped silica fiber laser in which every effort was made to optimize the optical components in the cavity [Ref. 79]. This device in which the mirrors were dielectric coatings deposited directly on to the fiber end faces emitted at a wavelength of 1.06 μm. It may be observed from Figure 6.45(a) that the fiber laser provided a CW output power in excess of 4 mW with a threshold power of 1.51 mW. In addition, the characteristic is linear above threshold with a slope efficiency of 55%. Figure 6.45(b) corresponds to an erbium/ytterbium with alumina codoped silica fiber laser emitting at a wavelength of 1.56 μm [Ref. 80]. The device which could be injection laser pumped without the need for stringent pump-laser

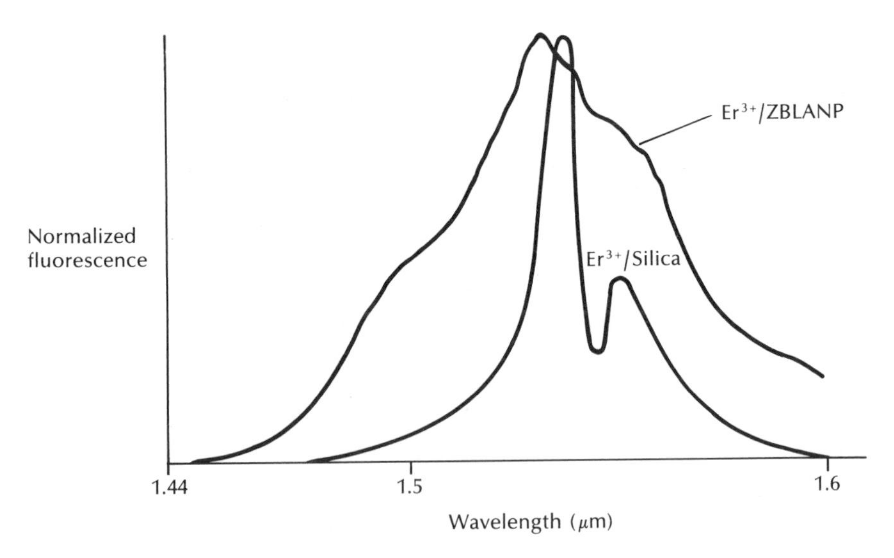

Figure 6.44 Normalized fluorescence from erbium doped silica and ZBLANP fibers. Reproduced with permission from C. A. Millar, M. C. Brierley and P. W. France, 'Optical amplification in an erbium-doped fluorozirconate fibre between 1480 nm and 1600 nm', *IEE Conf. Pub.*, **292**, Pt. 1, p. 66, 1988.

* ZBLANP fiber has lead fluoride added to the core glass to raise the relative refractive index.

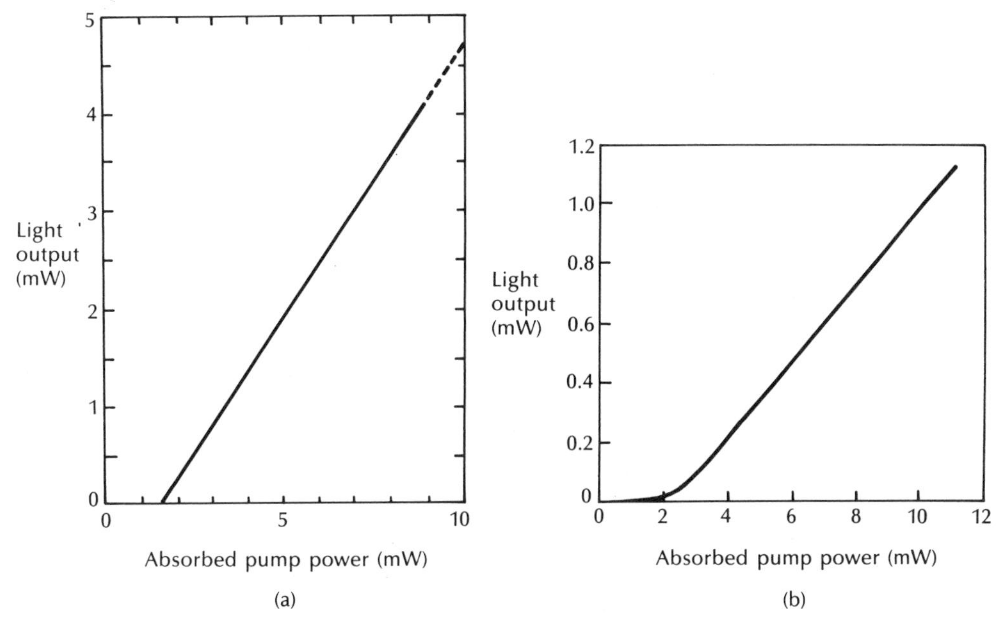

Figure 6.45 Light output against absorbed pump power characteristics for fiber lasers: (a) Neodymium doped silica fiber. Reproduced with permission from M. Shimitzu, H. Suda and M. Horiguchi, 'High efficiency Nd-doped fibre lasers using direct-coated dielectric mirrors', *Electron. Lett.*, **23**, p. 768, 1987 (IEE).
(b) Erbium/ytterbium with alumina codoped silica fiber. Reproduced with permission from D. N. Payne and L. Reekie, 'Rare-earth-doped fibre lasers and amplifers', *IEE Conf. Pub.*, **292**, Pt. 1, p. 49, 1988.

wavelength selection gave 1 mW of CW output power with a threshold power of 2 mW.

The basic Fabry–Perot cavity fiber laser shown in Figure 6.43 can be easily constructed from standard optical components but it has several limitations. In particular, the launching of light from the pump laser through one of the mirrored fiber ends can cause damage to the mirror coating as well as a substantial reduction in the launch efficiency. Furthermore, as mentioned previously, the gain spectrum of most rare earth ions extends over a wavelength range of some 50 nm. Unless the dielectric coatings on the mirrors are specially designed for broadband performance, however, the lasing output will be restricted to between 5 and 10 nm. Such a linewidth is too narrow for the provision of a broadband optical source but too wide to be used in single frequency laser applications such as coherent transmission. A number of alternative fiber laser structures have therefore been fabricated which do not require dielectric or metallic mirrors. Two of these structures, which are illustrated in Figure 6.46, are the fiber ring resonator [Ref. 81] and the fiber loop reflector made from a series concatenation of distributed reflectors using loops of fiber [Refs. 82, 83].

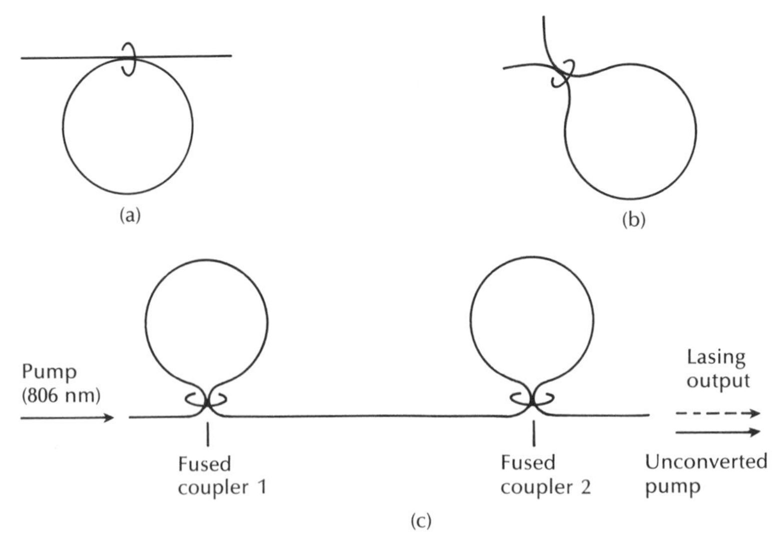

Figure 6.46 Fiber laser structures: (a) fiber ring resonator; (b) fiber loop reflector; (c) all-fiber laser made from two loops in series.

The fiber ring resonator may employ the coherent beam splitting properties of the single-mode fiber fused directional coupler (see Section 5.6.1). In this case two of the arms of the coupler are spliced together as shown in Figure 6.46(a) to form a circulating pathway in which light can travel. Hence an optical cavity without mirrors is formed, the finesse* of which is determined by the splitting ratio of the coupler. When the splitting ratio is low, the finesse is high and the energy storage on resonance is high, thus lowering the laser threshold. However, as with the Fabry–Perot laser, the lower threshold is obtained at the expense of a reduction in the slope efficiency. Alternatively, high performance fiber ring resonators can be fabricated by forming the two halves from a single fiber length. This technique has the effect of both reducing losses and of producing a higher finesse.

The structure of a fiber loop reflector which may also be based on a directional coupler is illustrated in Figure 6.46(b). However, in contrast to the fiber ring where there is energy storage within the resonant structure the fiber loop is a nonresonant interferometer (it constitutes a Sagnac interferometer, see Section 14.6.1). Light entering the loop through the input fiber end forms forward and backward (reflected) waves which are counter propagating, providing a coherent superposition of the clockwise and counterclockwise propagating fields. Hence the single fiber loop performs as a distributed all-fiber reflector which may be used to form a fiber laser [Ref. 82]. In addition, when two such loops are joined together in series a

* The finesse of the Fabry–Perot cavity provides a measure of its filtering properties and can be defined as the free spectral range divided by the full width half maximum permitted by the cavity.

resonator is obtained, as shown in Figure 6.46(c). This two loop structure provided the all-fiber laser which was fabricated from a single length of neodymium doped fiber without a splice [Ref. 83]. The excess loss of the couplers was only 0.04 dB, giving efficient laser action when the device was pumped with an AlGaAs injection laser at a wavelength of 0.806 μm and with a launch power of 470 μW. Lasing output from the device was obtained at a wavelength of 1.064 μm and was combined with the unconverted pump emission.

Narrow linewidth and frequency tunable rare earth doped fiber lasers are also under investigation and these devices are discussed in the following section.

6.10 Narrow linewidth and wavelength tunable lasers

The single frequency injection lasers described in Section 6.6 have been developed to minimize the transmission limitations resulting from fiber dispersion in high speed digital systems. For systems employing intensity modulation with direct detection of the optical signal, however, the laser linewidth and its absolute stability are of secondary importance. This is not the case with coherent optical fiber transmission where laser linewidth and stability are critical factors affecting the system performance (see Section 12.4.1). Laser linewidths in the range 1 MHz and below are required for such system applications which are around two orders of magnitude smaller than the 100 MHz linewidths obtained with 250 μm long Fabry–Perot or DFB devices which emit a few milliwatts without special linewidth control. In addition, wavelength or frequency tunable devices are considered to be key components for the provision of both the transmitter and local oscillator optical sources within coherent systems [Ref. 85].

Injection laser linewidth broadening occurs as a result of the change in lasing frequency with gain [Ref. 86]. It is a fundamental consequence of the spontaneous emission process which is directly related to fluctuations in the phase of the optical field. These phase fluctuations arise from the phase noise directly associated with the spontaneous emission process as well as the conversion of spontaneous emission amplitude noise to phase noise through a coupling mechanism between the photon and carrier densities. In the latter case, because the refractive index is strongly dependent on the carrier density which produces the gain, the fluctuations of gain due to spontaneous emission produce a substantial change in the refractive index which therefore increases the frequency/phase noise in the laser emission. The relationship for the linewidth Δf of an injection laser in terms of the emitted power P_e is given by [Ref.86]:

$$\Delta f = \frac{V_g^2 E n_{sp} \alpha_m}{8\pi P_e}(\alpha_i + \alpha_m)(1 + \alpha^2) \tag{6.52}$$

where V_g is the group velocity, E is the carrier (electron) energy, n_{sp} (in the range 2 to 3) is the spontaneous emission factor, α_i is the internal waveguide loss per unit length,* α_m is the mirror loss per unit length and α is called the linewidth enhancement factor. This latter parameter is defined as the ratio of the refractive index change with electron density to the differential gain change with electron density and is a measure of the amplitude to phase fluctuation conversion caused by the spontaneous emission. It can take up values between 2 and 16 depending upon the device material composition, structure and operating wavelength. The term $(1 + \alpha^2)$ in Eq. (6.52) results from the contributions to the linewidth of the two phase fluctuation effects.

It is clear that as the laser power increases, the spontaneous emission becomes relatively less important at the higher photon densities and hence the device linewidth decreases. However, as the output power of the laser cannot be made arbitrarily large, then a more effective method to reduce the linewidth is to make the cavity longer. The linewidth is decreased by increasing the laser length because the effective mirror loss α_m per unit length in Eq. (6.52) is decreased. Two techniques which can be utilized to increase the injection laser cavity length are to either use a long laser chip or to extend the cavity with a passive medium such as air, an optical fiber or an appropriate semiconductor integrated passive waveguide [Refs. 18, 38]. The latter external cavity devices also provide wavelength/frequency tunability.

6.10.1 Long external cavity lasers

Extension of the laser cavity length by the introduction of external feedback can be achieved by using an external cavity with a wavelength dispersive element as part of the cavity. Such devices are often referred to as long external cavity (LEC) lasers. A wavelength dispersive element is required because the long resonator structure has very closely spaced longitudinal modes which necessitates additional wavelength selectivity. A common technique for laboratory use is illustrated in Figure 6.47 where a diffraction grating is employed as an external mirror in order to filter the lasing emission from the wide gain spectrum of a laser chip giving a narrow linewidth at a desired wavelength [Refs. 87, 88]. Spectral linewidths as narrow as 10 kHz have been reported with such devices [Refs. 85, 87]. Furthermore, wavelength tuning of the output may be achieved by mechanical rotation of the grating such that the lasing wavelength moves with mode hops from one longitudinal mode to the next. In general, coarse spectral adjustment is obtained by rotation of the grating, whilst fine tuning can be achieved by lateral translation of the grating, as shown in Figure 6.47. Coarse tuning of a single-mode 1.5 μm laser over 90 nm through rotation of the external grating with fine tuning of the same device over approximately 1 GHz by lateral translation of the grating has been demonstrated [Ref. 89].

Another long external cavity method which has been proposed [Ref. 90] employs an external prism grating and graded-index (GRIN) rod lens (see Section 5.5.1)

* α_i is the injection laser equivalent of the laser loss coefficient per unit length $\bar{\alpha}$ defined in Section 6.2.5.

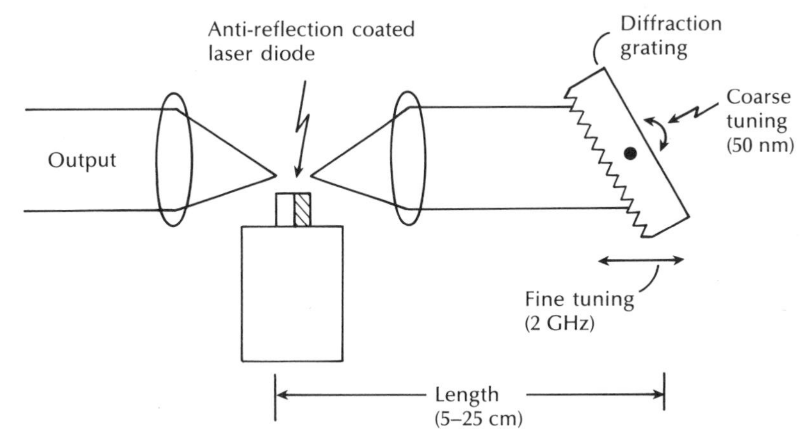

Figure 6.47 Wavelength tuning of an ILD using an external reflective diffraction grating (long external cavity technique).

combination. This technique enabled coarse wavelength adjustment of a buried heterostructure single-mode device over a range of 40 nm through the lateral displacement of the GRIN-rod lens relative to the laser chip. Fine tuning of around 6 GHz μm^{-1} could be achieved by slight variations in the separation between the laser chip and the GRIN-rod lens end face. The principal disadvantage with these mechanically tuned devices is their relatively low switching speeds. However, by using electro-optic [Ref. 91], acousto-optic [Ref. 92] devices to modulate the external cavity, much higher switching speeds can be achieved. Wavelength selection can then be produced by altering the electro-optic or acousto-optic drive frequency. For example, an acousto-optic filter and modulator pair has been used to select wavelengths over a range of 35 nm for a 0.85 μm laser, with switching speeds of 10 ns [Ref. 93].

6.10.2 Integrated external cavity lasers

An alternative technique for the provision of the external cavity is the integrated waveguide approach. Such monolithic integrated devices often utilize the distributed feedback (DFB) or the distributed Bragg reflector (DBR) structure. An example of an integrated external cavity DBR laser providing narrow linewidth dynamic single-mode (DSM) operation at a wavelength of 1.51 μm is shown in Figure 6.48 [Ref. 94]. This device which had a cavity length of 4.5 mm exhibited a spectral linewidth of 2 MHz with some 6 mW of optical output power.

Monolithic integrated DSM lasers also offer the potential for wavelength tuning. There are, in principle, two techniques which can be employed to tune these devices. One method is to use the mode selectivity of a coupled cavity structure such as a C^3 laser (see Section 6.6.1) [Ref. 29]. In this case the effective gain peak wavelength is controlled by the multicavity structure together with multisegment

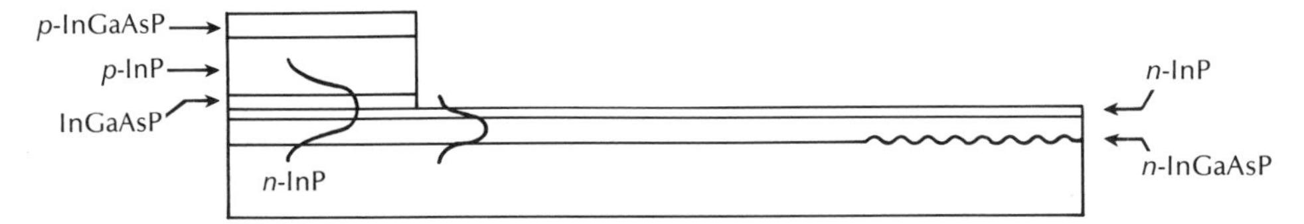

Figure 6.48 Structure of an integrated external cavity DBR laser.

electrodes, as illustrated in Figure 6.49(a). Hence the lasing wavelength can be varied within the effective gain width which is a range in excess of 15 nm for a 1.5 μm InGaAsP laser [Ref. 95]. The device wavelength changes, however, with mode jumps and thus this technique does not provide continuous wavelength tunability.

The other wavelength tuning method for monolithic integrated lasers is to use a refractive index change in the device cavity provided by current injection or the application of an electric field. Typically, this is achieved by employing a multiple electrode DFB or DBR structure [Ref. 38]. For example, with a single electrode DFB laser operated above threshold, the high injected carrier density (10^{18} cm^{-3}) reduces the effective refractive index in the corrugation region (Bragg region), thereby decreasing the lasing wavelength. Most of the injected carriers recombine, however, to produce photons which results in a very small increase in the carrier density leading to only a very small change in the lasing wavelength. The two electrode DFB laser shown in Figure 6.49(b) allows the wavelength tuning range to be improved by the application of a large current to one electrode and a small current to the other.

With the asymmetric DFB laser structure of Figure 6.49(b), the optical field is higher in the region near the output port where the facet is nonreflecting (antireflection (AR) coated, as shown in diagram), and the device operating wavelength is primarily determined by the effective refractive index in this region. When the aforementioned section is pumped at current densities at or slightly below the threshold density (under uniform pumping) simply to overcome the absorption losses, then it acts as a Bragg reflector. Furthermore, the injected carriers do not

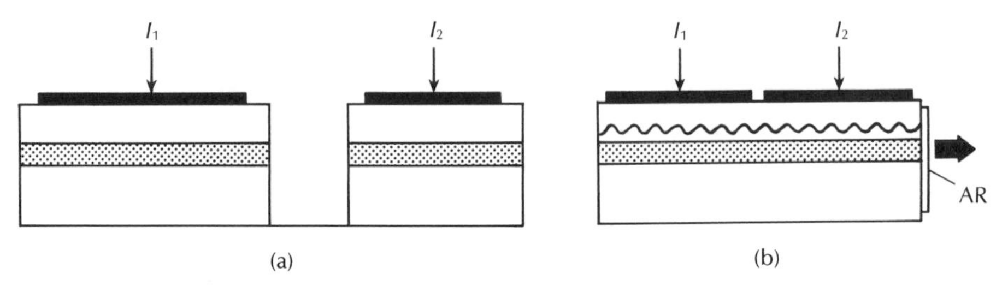

Figure 6.49 Monolithic integrated DSM lasers: (a) cleaved-coupled-cavity (C^3) laser; (b) double-sectioned DFB laser.

contribute significantly to the generation of photons because of the low pumping level. This factor results in a large change of refractive index which gives wavelength tuning. It should be noted that the gain is provided by the other section which is pumped well above threshold. A maximum continuous tuning range of 3.3 nm with 1 mW output power has been obtained with such a device [Ref. 96]. The spectral linewidth of this laser was 15 MHz and the tuning range reduced to 2 nm at an output power of 5 mW.

Three electrode DFB lasers have also demonstrated good tunability. A $\lambda/4$ shifted device (see Section 6.6.2) in which the two outer electrodes were electrically connected to a common current supply whilst the central electrode was supplied with a different current has given a continuous tuning range of 2 nm by varying the two currents [Ref. 97]. In addition, the device displayed a spectral linewidth of only 500 kHz. Although such tunable DFB lasers have a limited tuning range in comparison with coupled cavity devices, they exhibit advantages of ease of fabrication as well as providing continuous tuning rather than discrete jumps.

Multiple electrode DBR laser structures have also been developed to allow wavelength tuning [Ref. 95]. In particular, wider wavelength tuning ranges have been obtained by not only separating the Bragg region in the passive waveguide (a large bandgap material) from the active region (a small bandgap material) inside the laser cavity but also by introducing a phase region within the waveguide. The structure of such a three-sectioned DBR laser is illustrated in Figure 6.50(a). The wavelength of this device can simply be electronically tuned by current injection into the DBR section. This region exhibits a high reflectance within a certain wavelength band (the stop band) which is nominally between 2 and 4 nm wide. The mechanism which results from a refractive index change in the passive waveguide layer is known as Bragg wavelength control. A continuous tuning range, however, is limited to the resonant mode spacing which is defined from the effective cavity length of the laser. It is the mode which is nearest to the centre of the stop band and which simultaneously satisfies the 2π round trip phase condition that lases. Therefore, the introduction of the phase region in the waveguide (Figure 6.50(a)) which is independently controlled by the injection current allows the lasing wavelength to be tuned around each Bragg wavelength. Such a region provides phase control which again occurs through refractive index changes in the passive waveguide.

The combination of the two types of tuning (Bragg wavelength and phase tuning) provides a significantly larger tuning range because the lasing wavelength deviation from the Bragg wavelength can be compensated by phase control. With good design and the independent adjustment of the three currents in the active Bragg and phase regions, quasi-continuous tuning ranges between 8 and 10 nm have been obtained [Refs. 38, 98]. In addition, a continuous tuning range of 6.2 nm has been achieved with a similar device [Ref. 99]. Alternatively, for continuous wavelength tuning, one control current has been divided in a prescribed proportion into the Bragg and phase sections as illustrated in Figure 6.50(b). Continuous tuning ranges of between 2 and 4 nm have been reported using this method [Ref. 38].

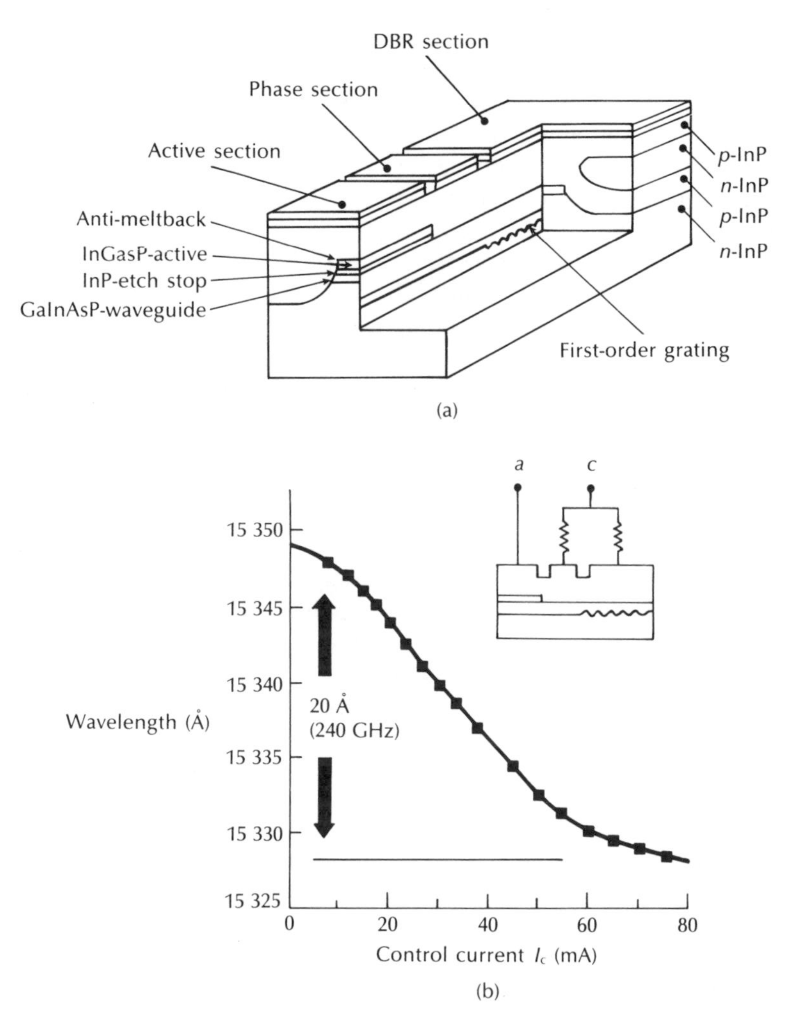

Figure 6.50 Three-sectioned DBR laser: (a) structure; (b) range of continuous wavelength tuning. Reproduced with permission from T. P. Lee and C. E. Zah, 'Wavelength-tunable and single-frequency semiconductor lasers for photonic communications networks', *IEEE Commun. Mag*, p. 42, Oct., 1989. Copyright © 1989 IEEE.

6.10.3 Fiber lasers

Techniques are also under investigation to obtain narrow linewidth output from glass fiber lasers [Ref. 74]. The rare earth doped fiber lasers described in Section 6.9.2 have spectral linewidths typically in the range 0.1 to 1 nm which are too broad for high speed transmission. One method to achieve narrower spectral linewidths employed polished silica blocks with surface gratings, as illustrated in Figure 6.51

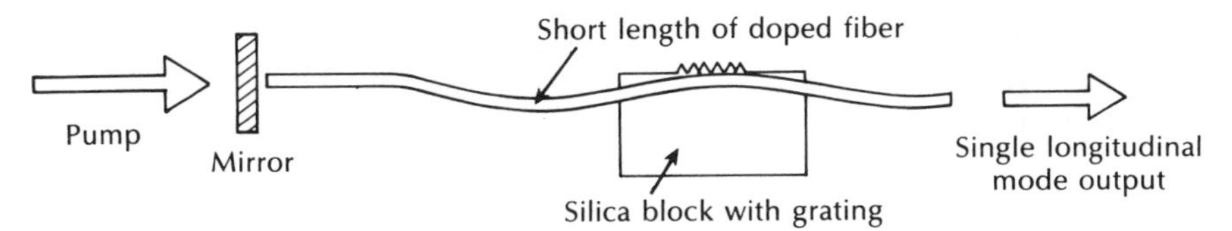

Figure 6.51 Fiber laser with a cavity incorporating a polished silica block and grating reflector [Ref. 100].

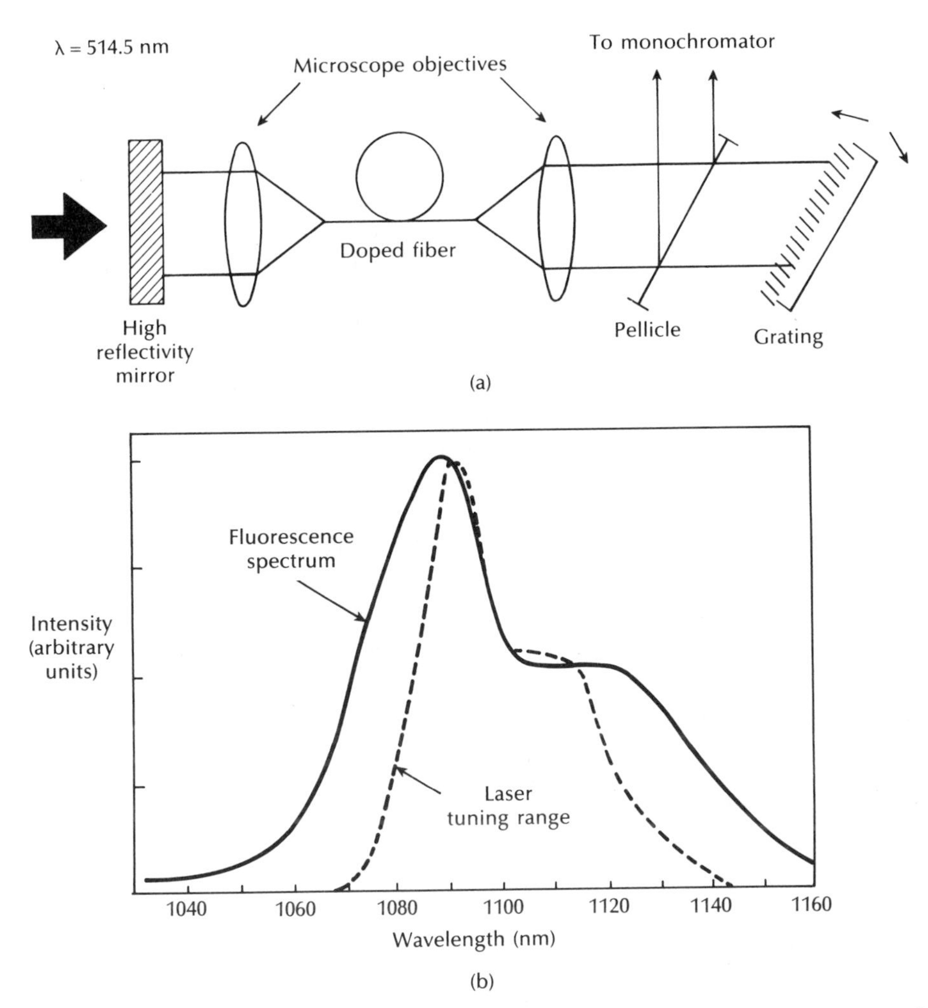

Figure 6.52 Tunable neodymium-doped single-mode fiber laser: (a) configuration; (b) fluorescence spectrum for doped fiber and the laser tuning range. Reproduced with permission from L. Reekie, R. J. Mears, S. B. Poole and D. N. Payne, 'Tunable single-mode fiber lasers', *J. Lightwave Technol.*, **LT-4**, p. 956, 1986. Copyright © 1986 IEEE.

[Ref. 100]. In this case the holographic gratings acted as distributed feedback reflectors which reflected only a narrow band of wavelengths. The reflector (Figure 6.51) through which the pump beam was launched was a dielectric mirror butted against the fiber end. Moreover, the fiber in the coupler block was undoped and one end was butt jointed to an erbium doped fiber. An output spectral linewidth of 0.04 nm (5 GHz) was obtained which is indicative of the relative state of development of fiber lasers in comparison with semiconductor devices.

Substantially narrower spectral linewidths have, however, been obtained with fiber lasers using a fiber Fox–Smith resonator design [Ref. 74]. This device which employs a fused coupler fabricated from erbium doped fiber has demonstrated a lasing linewidth of less than 8.5 MHz which compares favourably with the linewidths obtained from conventional semiconductor DFB lasers but not external cavity lasers.

Finally, wavelength tuning has also been obtained with fiber lasers. In particular, the use of silica as the laser medium provides good power handling characteristics and broadens the rare earth transitions, enabling tunable devices. An investigation of wavelength tuning in a neodymium (Nd^{3+}) doped single-mode fiber laser employed the experimental configuration shown in Figure 6.52(a) [Ref. 101]. Tuning was accomplished by changing the angle of the diffraction grating, which was mounted on a sine-bar-driven turntable. A tuning range of 80 nm was obtained, as may be observed from the characteristic (including the fluorescence spectrum of Nd^{3+} ions in silica) displayed in Figure 6.52(b). Furthermore, the wavelength tuning of an erbium (Er^{3+}) doped single-mode fiber laser was also reported [Ref. 101] to provide a tuning range of 25 nm around the 1.54 μm wavelength region using a similar experimental configuration.

An alternative method for wavelength tuning of fiber lasers employed the loop reflector discussed in Section 6.9.2. In this case a temperature shift was used to adjust the coupling ratio through the directional coupler which had a direct effect on the output optical wavelength. A 60 °C variation in temperature provided a tuning range of around 33 nm [Ref. 102].

6.11 Mid-infrared lasers

Laser sources for transmission at wavelengths beyond 2 μm, in particular gas and solid state lasers as well as low temperature injection lasers, have been utilized in nontelecommunications applications such as high resolution spectroscopy, materials processing and remote monitoring. More recently, however, progress in the potentially ultra-low-loss fibers for mid-infrared transmission (see Section 3.7) operating over the wavelength range 2 to 5 μm has encouraged greater activity in the pursuit of longer wavelength optical sources. For practical communication systems in the mid-infrared wavelength region the requirement is for semiconductor or fiber lasers which are capable of operating at, or close to, room temperature.

Semiconductor materials with direct bandgaps which encompass the mid-infrared wavelength range include many of the III–V, II–VI and IV–VI alloys. Injection lasers operating in this longer wavelength region, however, are subject to increased carrier losses over devices emitting at wavelengths up to 1.6 μm which result from nonradiative recombination via the Auger interaction [Ref. 103]. The recombination energy of the injected carriers is dissipated as thermal energy to the remaining free carriers by this process. Moreover, the probability of the occurrence of such a process increases as the bandgap of the semiconductor is reduced. In addition, optical losses due to free carrier absorption are also greater because of their dependence on the square of the wavelength. Both of these effects present more problems in the mid-infrared wavelength range and they exhibit increased importance at higher temperatures as a result of the higher concentration of free carriers. They therefore play a major role in the determination of the injection laser threshold current and efficiency, as well as providing a limit to the maximum operating temperature of the device.

The total current required to provide the injection laser threshold is greater than the amount attributable only to radiative recombination by the addition of an Auger current. Although the Auger current depends upon the precise electronic band structure of the material, and often consists of contributions from a number of different Auger transitions, it is generally large for materials with bandgaps which provide longer wavelength emission. In this context the results of calculations for threshold current and internal quantum efficiency for several long wavelength semiconductor alloys are displayed in Figure 6.53 [Ref. 104]. A comparison of the highest predicted oscillation temperatures of pulsed DH lasers fabricated from various compounds as a function of wavelength, based on estimates of the temperature at which the device internal quantum efficiency at current threshold falls to 2.5%, is shown. In addition, experimental observations are depicted as data points in the figure. It may be observed from Figure 6.53 that this data indicates an overall limit to room temperature laser action at wavelengths slightly above 2 μm for any of the semiconductor alloys investigated.

Room temperature operation of III–V alloy semiconductor lasers fabricated from InGaAsSb, and GaAlAsSb lattice matched to either GaSb or InAs has been obtained in the wavelength range 2.2 to 2.3 μm [Refs. 105, 106]. Low threshold current density of 1.7 kA cm^{-2} at room temperature has also been reported [Ref. 107] but although laser oscillation is predicted to occur up to a wavelength of 4.4 μm, it is at a temperature of only 77 K due to the presence of the Auger current [Ref. 18]. In addition, the InAsPSb lattice matched to InAs offers the potential for operation over the 2 to 3.5 μm wavelength region but calculations indicate a similar dependence of the maximum operating temperature on wavelength to GaInAsSb (see Figure 6.53).

An example of a II–VI alloy semiconductor is the HgCdTe material system, also shown in Figure 6.53, from which infrared detectors have been fabricated (see Section 8.10). Although LEDs and optically pumped lasers for operation over the wavelength range 2 to 4 μm have been demonstrated using this alloy, injection laser

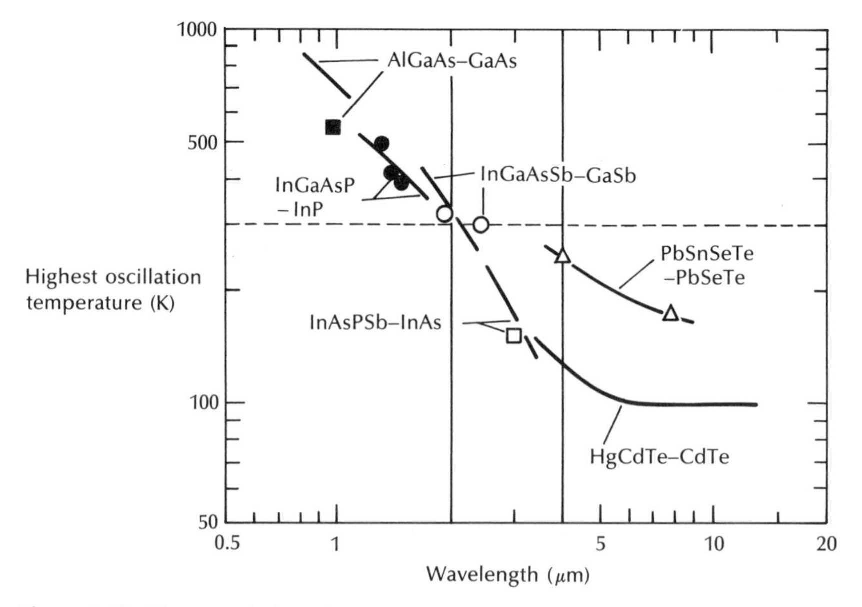

Figure 6.53 Characteristics showing maximum temperature of pulsed operation for DH lasers versus wavelength for several material systems [Ref. 104].

sources have as yet to be reported [Ref. 18]. Injection lasers, however, fabricated from IV–VI lead–salt alloys have been developed for high resolution spectroscopic as well as gas monitoring applications. Devices based on the quaternary PbSnSeTe and related ternary compounds generally emit at wavelengths longer than 4 μm. In this case the Auger effects have been calculated [Ref. 104] to be less in certain of these alloys than those obtained in III–V semiconductor materials, which could provide both lower current thresholds and higher maximum operating temperatures. The replacement of Sn with Eu, Cd or Ge increases the bandgap to provide shorter wavelength operation. For example, the structure of some recently reported ternary alloy PbEuTe/PbTe DH lasers [Ref. 108] is shown in Figure 6.54. These mesa-stripe devices which emitted over the 3.5 to 6.5 μm wavelength range provided in excess of 200 μW output power at temperatures up to 210 K in pulsed operation.

The investigation of rare earth doped fiber lasers for application in the mid-infrared wavelength region is also under way. In particular, fluorozirconate fiber lasers doped with erbium [Ref. 109], holium [Ref. 110] and thulium [Ref. 111] have been reported to provide emissions in the 2 to 3 μm wavelength range. The 2.702 μm transition in erbium which had only previously been obtained in bulk fluorozirconate glass samples [Ref. 74] was demonstrated in a CW fiber laser pumped at twice threshold [Ref. 109]. Lasing was obtained when 191 mW of pump light at a wavelength of 0.477 μm was launched into the doped fluorozirconate fiber.

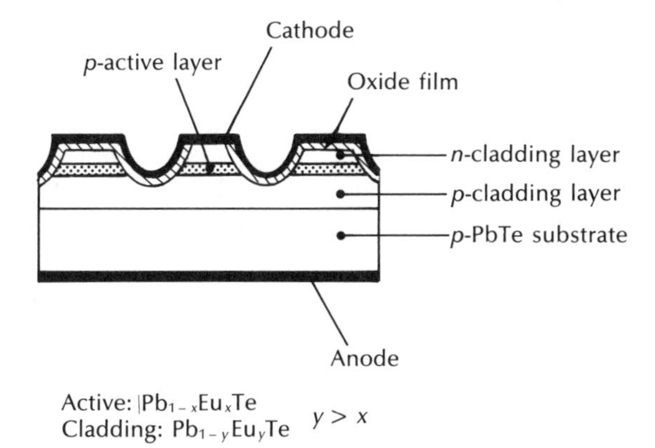

Figure 6.54 Structure of PbEuTe/PbTe DH laser [Ref. 108].

The holium doped fluorozirconate fiber was made to lase with a CW output at wavelengths of 1.38 μm and 2.08 μm [Ref. 110]. In both cases, pumping was obtained from an argon ion source at a wavelength of 0.488 μm and the 2.08 μm emission was the first report of the operation of a fiber laser at wavelengths beyond 1.55 μm [Ref. 74]. Finally, the thulium doped fiber laser emitted at a wavelength of 2.3 μm when pumped with the pulsed output from an alexandrite laser at 0.786 μm [Ref. 111]. Unlike the longer wavelength holium emission which originates from a three level system, the thulium system at 2.3 μm is four level in which the pump band is also the upper lasing level.

Problems

6.1 Briefly outline the general requirements for a source in optical fiber communications.
Discuss the areas in which the injection laser fulfils these requirements, and comment on any drawbacks of using this device as an optical fiber communication source.

6.2 Briefly describe the two processes by which light can be emitted from an atom. Discuss the requirement for population inversion in order that stimulated emission may dominate over spontaneous emission. Illustrate your answer with an energy level diagram of a common nonsemiconductor laser.

6.3 Discuss the mechanism of optical feedback to provide oscillation and hence amplification within the laser. Indicate how this provides a distinctive spectral output from the device.
The longitudinal modes of a gallium arsenide injection laser emitting at a wavelength of 0.87 μm are separated in frequency by 278 GHz. Determine the length of the optical cavity and the number of longitudinal modes emitted. The refractive index of gallium arsenide is 3.6.

6.4 An injection laser has a GaAs active region with a bandgap energy of 1.43 eV. Estimate the wavelength of optical emission from the device and determine its linewidth in Hertz when the measured spectral width is 0.1 nm.

6.5 When GaSb is used in the fabrication of an electroluminescent source, estimate the necessary hole concentration in the p type region in order that the radiative minority carrier lifetime is 1 ns.

6.6 The energy bandgap for lightly doped gallium arsenide at room temperature is 1.43 eV. When the material is heavily doped (degenerative) it is found that the lasing transitions involve 'bandtail' states which effectively reduce the bandgap transition by 8%. Determine the difference in the emission wavelength of the light between the lightly doped and this heavily doped case.

6.7 With the aid of suitable diagrams, discuss the principles of operation of the injection laser.

Outline the semiconductor materials used for emission over the wavelength range 0.8 to 1.7 μm and give reasons for their choice.

6.8 Determine the range of bandgap energies for:

(a) $Al_yGa_{1-y}As/Al_xGa_{1-x}As$;
(b) $In_{1-x}Ga_xAs_yP_{1-y}/InP$.

6.9 A DH injection laser has an optical cavity of length 50 μm and width 15 μm. At normal operating temperature the loss coefficient is 10 cm^{-1} and the current threshold is 50 mA. When the mirror reflectivity at each end of the optical cavity is 0.3, estimate the gain factor $\bar{\beta}$ for the device. It may be assumed that the current is confined to the optical cavity.

6.10 The coated mirror reflectivity at either end of the 350 μm long optical cavity of an injection laser is 0.5 and 0.65. At normal operating temperature the threshold current density for the device is 2×10^3 A cm^{-2} and the gain factor β is 22×10^{-3} cm A^{-1}. Estimate the loss coefficient in the optical cavity.

6.11 Describe the techniques used to give both electrical and optical confinement in multimode injection lasers. Contrast these techniques when used in gain-guided and index-guided lasers.

6.12 A gallium arsenide injection laser with a cavity of length 500 μm has a loss coefficient of 20 cm^{-1}. The measured differential external quantum efficiency of the device is 45%. Calculate the internal quantum efficiency of the laser. The refractive index of gallium arsenide is 3.6.

6.13 Compare the ideal light output against current characteristic for the injection laser with one from a typical gain-guided device. Describe the points of significance on the characteristics and suggest why the two differ.

6.14 Describe, with the aid of suitable diagrams, the major strategies and structures utilized in the fabrication of single frequency injection lasers. Indicate the reasons for the great interest in such devices.

6.15 The threshold current density for a stripe geometry AlGaAs laser is 3000 A cm^{-1} at a temperature of 15 °C. Estimate the required threshold current at a temperature of 60 °C when the threshold temperature coefficient T_0 for the device is 180 K, and the contact stripe is 20×100 μm.

6.16 Briefly describe what is meant by the following terms when they are used in relation to injection lasers:

(a) relaxation oscillations;
(b) frequency chirp;
(c) partition noise;
(d) mode hopping.

6.17 The rms value of the power fluctuation on the output from a single-mode semiconductor laser is 2×10^{-4} W when the relative intensity noise (RIN) is -160 dB Hz^{-1}. The emission, which is at a wavelength of 1.30 μm, is directly incident on an optical detector with a quantum efficiency of 70% at this wavelength. If the rms noise current at the detector output is 0.53 μA, and assuming that the RIN is the dominant noise source, calculate the mean optical power incident on the photodetector.

6.18 A single-mode injection laser launches light with a 3 dB linewidth Δf into a fiber link which has two connectors exhibiting reflectivities r_1 and r_2. It is known that the worst case relative intensity noise (RIN) occurs when the direct and doubly reflected optical fields interfere in quadrature [Ref. 112] following:

$$RIN(f) = \frac{4r_1r_2}{\pi}\frac{\Delta f}{f^2 + \Delta f^2}[1 + \exp(-4\pi\,\Delta f\,\tau) - 2\exp(-2\pi\,\Delta f\,\tau)\cos(2\pi f\tau)]$$

Demonstrate that the above expression reduces to:

$$RIN(f) = \frac{16r_1r_2}{\pi}\Delta f\tau^2 \qquad \text{for } \Delta f\cdot\tau \ll 1$$

and

$$RIN(f) = \frac{4r_1r_2}{\pi}\frac{\Delta f}{f^2 + \Delta f^2} \qquad \text{for } f\cdot\tau \gg 1$$

6.19 A DFB laser has a 3 dB linewidth Δf of 50 MHz. It is connected to a short optical jumper cable such that $\Delta f\cdot\tau$ is 0.1. Using the relationship given in Problem 6.18 when the frequency f is also 50 MHz, obtain the average reflectivity for each of the connectors so that the RIN is reduced below a level of -130 dB Hz^{-1}.

6.20 Discuss degradation mechanisms in injection lasers. Comment on these with regard to the CW lifetime of the devices.

6.21 Describe the structure and operation of a glass fiber laser. Comment on the glass compounds currently employed together with their fluorescence spectra.

6.22 Discuss linewidth narrowing and wavelength tunability associated with single frequency injection lasers. Outline the major techniques which are being adopted to facilitate these characteristics.

Answers to numerical problems

6.3 150 μm, 1241
6.4 0.87 μm, 39.6 GHz
6.5 4.2×10^{18} cm^{-3}
6.6 0.07 μm
6.8 (a) 1.38 to 1.91 eV
(b) 0.73 to 1.35 eV
6.9 3.76×10^{-2} cm A^{-1}
6.10 28 cm^{-1}
6.12 84.5%
6.15 77.0 mA
6.17 3.6 mW
6.19 -25.2 dB

References

[1] C. Kittel, *Introduction to Solid State Physics* (5th edn), John Wiley, 1976.
[2] E. S. Yang, *Fundamentals of Semiconductor Devices*, McGraw-Hill, 1978.

[3] Y. P. Varshni, 'Band to band radiative recombination in groups IV, VI and III–V semiconductor I', *Phys. Stat. Solidi (Germany)*, **19**(2), pp. 459–514, 1967.

[4] H. Kressel and J. K. Butler, *Semiconductor Lasers and Heterojunction LEDs*, Academic Press, 1977.

[5] H. Kressel, 'Electroluminescent sources for fiber systems', in M. K. Barnoski (Ed.), *Fundamentals of Optical Fiber communications*, pp. 109–141, Academic Press, 1976.

[6] S. M. Sze, *Physics of Semiconductor Devices* (2nd edn), John Wiley, 1981.

[7] H. C. Casey and M. B. Parish, *Heterostructure Lasers: Part A and B*, Academic Press, 1978.

[8] A. Yariv, *Optical Electronics* (4th edn), Holt, Rinehart and Winston, 1991.

[9] K. Y Lau and A. Yariv, 'High-frequency current modulation of semiconductor injection lasers', in *Semiconductors and Semimetals*, 22, Pt. B, pp. 70–152, Academic Press, 1985.

[10] J. C. Dyment, L. A. D'Asaro, J. C. Norht, B. I. Miller and J. E. Ripper, 'Proton-bombardment formation of stripe-geometry heterostructure lasers for 300 K CW operation', *Proc. IEEE*, **60**, pp. 726–728, 1982.

[11] H. Kressel and M. Ettenburg, 'Low-threshold double, heterojunction AlGaAs/GaAs laser diodes: theory and experiment', *J. Appl. Phys.*, **47**(8), pp. 3533–3537, 1976.

[12] P. R. Selway, A. R. Goodwin and P. A. Kirby, 'Semiconductor laser light sources for optical fiber communications', in C. P. Sandbank (Ed.), *Optical Fiber Communication Systems*, pp. 156–183, John Wiley, 1980.

[13] N. K. Dutta, 'Optical sources for lightwave systems applications', in E. E. Basch (Ed.) *Optical-Fiber Transmission*, H. W. Sams & Co., 1987.

[14] I. P. Kaminow, L. W. Stulz, J. S. Ko, A. G. Dentai, R. E. Nahory, J. C. DeWinter and R. L. Hartman, 'Low-Threshold InGaAsP ridge waveguide lasers at 1.3 μm', *IEEE J. Quantum Electron.*, **QE-19**, pp. 1312–1319, 1983.

[15] C. J. Armistead, S. A. Wheeler, R. G. Plumb and R. W. Musk, 'Low threshold ridge waveguide lasers at $\lambda = 1.5$ μm', *Electron. Lett.*, **22**, pp. 1145–1147, 1986.

[16] S. E. H. Turley, G. D. Henshall, P. D. Greene, V. P. Knight, D. M. Moule and S. A. Wheeler, 'Properties of inverted rib-waveguide lasers operating at 1.3 μm wavelength', *Electron. Lett.*, **17**, pp. 868–870, 1981.

[17] K. Saito and R. Ito, 'Buried-heterostructure AlGaAs lasers', *IEEE J. Quantum Electron.*, **QE-16**(2), pp. 205–215, 1980.

[18] J. E. Bowers and M. A. Pollack, 'Semiconductor lasers for telecommunications', in S. E. Miller and I. P. Kaminow (Eds.), *Optical Fiber Telecommunications II*, Academic Press, pp. 509–568, 1988.

[19] I. Mito, M. Kitamura, K. Kobayashi, S. Murata, M. Seki, Y. Odagiri, H. Nishimoto, M. Yamaguchi and K. Kobayashi, 'InGaAsP double-channel planar buried-heterostructure laser diode (DCPBH LD) with effective current confinement', *IEEE J. Lightwave Technol.*, **LT-1**, pp. 195–202, 1983.

[20] N. K. Dutta, R. B. Wilson, D. P. Wilt, P. Besomi, R. L. Brown, R. J. Nelson and R. W. Dixon, 'Performance comparison of InGaAsP lasers emitting at 1.3 and 1.55 μm for lightwave system applications', *AT&T Tech. J.*, **64**, pp. 1857–1884, 1985.

[21] S. E. Miller, 'Integrated low-noise lasers', *Electron., Lett.* **22**, pp. 256–257, 1986.

[22] J. E. Bowers, B. R. Hemenway, A. H. Gnauck, T. J. Bridges and E. G. Burkhardt, 'High-frequency constricted mesa lasers', *Appl. Phys. Lett.*, **47**, pp. 78–80, 1985.

[23] W. T. Tsang, 'Extremely low threshold AlGaAs modified multiquantum-well heterostructure lasers grown by MBE', *Appl. Phys. Lett.*, **39**, p. 786, 1981.

[24] T. E. Bell, 'Single-frequency semiconductor lasers', *IEEE Spectrum*, **20**, p. 38, 1983.

[25] T. Nakagami and T. Sakurai, 'Optical and optoelectronic devices for optical fiber transmission systems', *IEEE Commun. Mag.*, **26**(1), pp. 28–33, 1988.

[26] H. Blauvelt, N. Bar-Chaim, D. Fekete, S. Margalet and A. Yariv, 'AlGaAs lasers with micro-cleaved mirrors suitable for monolithic integration', *Appl. Phys. Lett.*, **40**, pp. 289–290, 1982.

[27] L. A. Coldren, K. Furuya, B. I. Miller and J. A. Rentschler, 'Etched mirror and groove-coupled GaInAsP/InP laser devices for integrated optics', *IEEE J. Quantum Electron.*, **QE-18**, pp. 1679–1688, 1982.

[28] K.-Y. Liou, C. A. Burrus, R. A. Linke, I. P. Kaminow, S. W. Granlund, C. B. Swan and P. Besomi, 'Single longitudinal-mode stabilized graded-index-rod external coupled-cavity laser', *Appl. Phys. Lett.*, **45**, p. 729, 1984.

[29] W. T. Tsang, N. A. Olsson and R. A. Logan, 'High-speed direct single-frequency modulation with large tuning rate in cleaved-coupled-cavity lasers', *Appl. Phys. Lett.*, **42**(8), pp. 650–651, 1983.

[30] L. A. Coldren, G. D. Boyd, J. E. Bowers and C. A. Burrus, 'Reduced dynamic linewidth in three-terminal two-section diode lasers', *Appl. Phys. Lett.*, **46**, pp. 125–127, 1985.

[31] D. R. Scifres, R. D. Burham and W. Streifer, 'Distributed feedback single heterojunction diode laser', *Appl. Phys. Lett.*, **25**, p. 203, 1974.

[32] W. T. Tsang and S. Wang, 'GaAs-AlGaAs double-heterostructure injection laser with distributed Bragg reflector', *Appl. Phys. Lett.*, **28**, p. 596, 1976.

[33] M. Nakumura, K. Aidi, J. Umeda and A. Yariv, 'CW operation of distributed-feedback GaAs-GaAlAs diode lasers at temperatures up to 300 K', *Appl. Phys. Lett.*, **27**(7), pp. 403–405, 1975.

[34] H. Ishiawa, H. Soda, K. Wakao, K. Kihara, K. Kamite, Y. Kotaki, M. Matsuda, H. Sudo, S. Yamakoshi, S. Isozumi and H. Imai, 'Distributed feedback laser emitting at 1.3 μm for gigabit communication systems', *J. Lightwave Technol.*, **LT-5**(6), pp. 848–855, 1987.

[35] S. Tsuji, A. Ohishi, H. Nakamura, M. Hirao, N. Chinone and H. Matsumura, 'Low threshold operation of 1.5 μm DFB laser diodes', *J. Lightwave Technol.*, **LT-5**(6), pp. 822–826, 1987.

[36] K. Utaka, S. Akiba, K. Sakai and Y. Matsushima, '$\lambda/4$ shifted InGaAsP/InP DFB lasers', *IEEE J. Quantum Electron*, **QE-22**, pp. 1042–1051, 1986.

[37] S. Akiba, M. Usami and K. Utaka, '1.5 μm $\lambda/4$ shifted InGaAsP/InP DFB lasers', *J. Lightwave Technol.*, **LT-5**(11), pp. 1564–1573, 1987.

[38] T-P. Lee and C. Zah, 'Wavelength-tunable and single-frequency semiconductor lasers for photonic communications networks', *IEEE Commun. Mag.*, pp. 42–51, October 1989.

[39] P. A. Kirby, 'Semiconductor laser sources for optical communications', *Radio Electron. Eng., J IERE*, **51**(7/8), pp. 363–376, 1981.

[40] D. Botez and G. J. Herskowitz, 'Components for optical communications systems: a review', *Proc., IEEE*, **68**(6), pp. 689–730, 1980.

[41] H. C. Casey, 'Temperature dependence of the threshold current density in InP-$Ga_{0.28}In_{0.72}As_{0.6}P_{0.4}$ ($\lambda = 1.3$ μm) double heterostructure lasers', *J. Appl. Phys.*, **56**, p. 1959, 1984.

[42] D. H. Newman and S. Ritchie, 'Sources and detectors for optical fibre communications applications: the first 20 years', *IEE Proc., Pt. J*, **133**(3), pp. 213–229, 1986.

[43] T. Ikegami, 'Spectrum broadening and tailing effect in direct-modulated injection lasers', *Proceedings of 1st European Conference on Optical Fiber Communication* (London, UK), p. 111, 1975.

[44] K. Furuya, Y. Suematsu and T. Hong, 'Reduction of resonance like peak in direct modulation due to carrier diffusion in injection laser', *Appl. Opt.*, **17**(12), pp. 1949–1952, 1978.

[45] P. M. Boers, M. T. Vlaardingerbroek and M. Danielson, 'Dynamic behaviour of semiconductor lasers', *Electron. Lett.*, **11**(10), pp. 206–208, 1975.

[46] D. J. Channin, 'Effect of gain saturation on injection laser switching', *J. Appl. Phys.*, **50**(6), pp. 3858–3860, 1979.

[47] N. Chinane, K. Aiki, M. Nakamura and R. Ito, 'Effects of lateral mode and carrier density profile on dynamic behaviour of semiconductor lasers', *IEEE J. Quantum Electron.*, **QE-14**, pp. 625–631, 1977.

[48] R. S. Tucker, A. H. Gnauck, J. M. Wiesenfield and J. E. Bowers, '8 Gb/s return to zero modulation of a semiconductor laser by gain switching', *Internet. Conf. on Integrated Optics and Optical Fiber Commun. Technical Digest Series 1987 (OSA)*, **3**, p. 178, 1987.

[49] R. A. Linke, 'Modulation induced transient chirping in single frequency lasers', *IEEE J. Quantum Electron.*, **QE-21**, pp. 593–597, 1985.

[50] J. C. Cartledge and G. S. Burley, 'The effect of laser chirping on lightwave system performance', *J. Lightwave Technol.*, **7**(3), pp. 568–573, 1989.

[51] P. S. Henry, R. A. Linke and A. H. Gnauck, 'Introduction to lightwave systems', in S. E. Miller and I. P. Kaminow (Eds.) *Optical Fiber Telecommunications II*, Academic Press, pp. 781–831, 1988.

[52] L. Bickers and L. P. Westbrook, 'Reduction in laser chirp in 1.5 μm DFB lasers by modulation pulse shaping', *Electron. Lett.*, **21**, pp. 103–104, 1985.

[53] M. Yamaguchi, K. Emura, M. Kitamura, I. Mito and K. Kobayashi, 'Frequency chirping suppression by a distributed-feedback laser diode and a monolithically integrated loss modulator', *Technical Digest of Optical Fiber Commun. Conf., OFC '85* (USA), paper WI3, 1985.

[54] G. Arnold, P. Russer and K. Peterman, 'Modulation of laser diodes', in H. Kressel (Ed.), *Semiconductor Devices for Optical Communication, Topics in Applied Physics*, **39**, pp. 213–242, 1985.

[55] J. Saltz, 'Modulation and detection for coherent lightwave communications', *IEEE Commun. Mag.*, **24**(6), pp. 38–49, 1986.

[56] F. G. Walther and J. E. Kaufmann, 'Characterization of GaAlAs laser diode frequency noise', *Sixth Top. Mtg. Opt. Fiber Commun.* (USA), paper TUJ5, 1983.

[57] Y. Suematsu and T. Hong, 'Suppression of relaxation oscillations in light output of injection lasers by electrical resonance circuit', *IEEE J. Quantum. Electron.*, **QE-13**(9), pp. 756–762, 1977.

[58] C. H. Henry, 'Phase noise in semiconductor lasers', *J. Lightwave Technol.*, **LT-4**(3), pp. 298–310, 1986.

[59] C. H. Henry and R. F. Kazarinov, 'Instability of semiconductor lasers due to optical feedback from distant reflectors', *IEEE J. Quantum. Electron.*, **QE-22**, pp. 294–301, 1986.

[60] S. D. Personick, *Fiber Optic Technology and Applications*, Plenum Press, 1985.
[61] K. Ogawa, 'Analysis of mode partition noise in laser transmission systems', *IEEE J. Quantum. Electron.*, **QE-18**, pp. 849–855, 1982.
[62] F. R. Nash, W. J. Sundberg, R. L. Hartman, J. R. Pawlik, D. A. Ackerman, N. K. Dutta and R. W. Dixon, 'Implementation of the proposed reliability assurance strategy for an InGaAsP/InP planar mesa B H laser for use in a submarine cable', *AT&T Tech. J.*, **64**, p. 809, 1985.
[63] N. K. Dutta and C. L. Zipfel, 'Reliability of lasers and LEDs', in S. E. Miller and I. P. Kaminow (Eds.) *Optical Fiber Telecommunications II*, Academic Press, pp. 671–687, 1988.
[64] H. Kogelnik, 'Devices for optical communications', *Solid State Devices Research Conf. (ESSDERC)* and *4th Symposium on Solid Device Technology* (Munich, W. Germany), **53**, pp. 1–19, 1980.
[65] T. Yamamoto, K. Sakai and S. Akiba, '10000-h continuous CW operation of $In_{1-x}Ga_xAs_yP_{1-y}$/InP DH lasers at room temperature', *IEEE J. Quantum Electron.*, **QE-15**(8), pp. 684–687, 1979.
[66] M. Hirao, K. Mizuishi and M. Nakamura, 'High-reliability semiconductor lasers for optical communications', *IEEE J. on Selected Areas in Commun.*, **SAC-4**(9), pp. 1494–1501, 1986.
[67] I. W. Marshall, 'Low loss coupling between semiconductor lasers and single-mode fibre using tapered lensed fibres', *Br. Telecom Technol. J.*, **4**(2), pp. 114–121, 1986.
[68] H. Kawahara, Y. Onada, M. Goto and T. Nakagami, 'Reflected light in the coupling of semiconductor lasers with tapered hemispherical end fibres', *Appl. Opt.,* **22**, pp. 2732–2738, 1983.
[69] H. Kawahara, M. Sasaki and N. Tokoyo, 'Efficient coupling from semiconductor lasers into single-mode fibres with tapered hemispherical ends', *Appl. Opt.*, **19**, pp. 2578–2583, 1980.
[70] T. Schwander, B. Schwaderer and H. Storm, 'Coupling of lasers to single-mode fibres with high efficiency and low optical feedback', *Electron. Lett.*, **21**, pp. 287–289, 1985.
[71] J. Lipson, R. T. Ku and R. E. Scotti, 'Opto-mechanical considerations for laser-fiber coupling and packaging', *Proc. SPIE Int. Soc. Opt. Eng. (USA)*, **5543**, pp. 308–312, 1985.
[72] R. T. Ku, 'Progress in efficient/reliable semiconductor laser-to-single mode fiber coupler development', *Conf. on Opt. Fiber Commun.* (Washington DC, USA), *Tech. Dig.* pp. 4–6, January 1984.
[73] G. D. Khoe, H. G. Kock, D. Kuppers, J. H. F. M. Poulissen and H. M. DeVrieze, 'Progress in monomode optical fiber interconnection devices', *J. of Lightwave Technol.*, **LT-2**(3), pp. 217–227, 1984.
[74] P. Urquhart, 'Review of rare earth doped fiber lasers and amplifiers', *IEE Proc., Pt. J. Optoelectronics*, **135**(6), pp. 385–407, 1988.
[75] J. Wilson and J. F. B. Hawkes, *Lasers: Principles and applications*, Prentice Hall, 1987.
[76] K. Patek, *Glass Lasers*, Butterworth, 1970.
[77] R. M. MacFarlane and R. M. Shelby, 'Coherent transient and holeburning spectroscopy of rare earth ions in solids', in R. M. MacFarlane and A. A. Kaplyanskii (Eds.), *Spectroscopy of Solids Containing Rare Earth Ions*, North-Holland, pp. 51–184, 1987.

[78] C. A. Millar, M. C. Brierley and P. W. France, 'Optical amplification in an erbium-doped fluorozirconate fibre between 1480 nm and 1600 nm', *Fourteenth European Conf. on Optical Commun., ECOC '88 (UK), IEE Conf. Pub.*, **292**, Pt. 1, pp. 66–69, 1988.

[79] M. Shimitzu, H. Suda and M. Horiguchi, 'High efficiency Nd-doped fibre lasers using direct-coated dielectric mirrors', *Electron. Lett.*, **23**, pp. 768–769, 1987.

[80] D. N. Payne and L. Reekie, 'Rare-earth-doped fibre lasers and amplifiers', *Fourteenth European Conf. on Optical Commun., ECOC '88, (UK), IEE Conf. Pub.*, **292**, Pt. 1, pp. 49–51, 1988.

[81] L. F. Stokes, M. Chodorow and H. J. Shaw, 'All single-mode fibre resonator', *Opt. Lett.*, **7**, p. 288, 1982.

[82] I. D. Miller, D. B. Mortimore, P. Urquhart, B. J. Ainslie, S. P. Craig, C. A. Millar and D. B. Payne, 'A Nd^{3+}-doped CW fibre laser using all-fibre reflectors', *Appl. Opt.*, **26**, pp. 2197–2201, 1987.

[83] I. D. Miller, D. B. Mortimore, B. J. Ainslie, P. Urquhart, S. P. Craig, C. A. Millar and D. B. Payne, 'New all-fiber laser', *Opt. Fiber Commun. Conf., OFC '87* (USA), January 1987.

[84] P. Urquhart, 'Transversely coupled fibre Fabry–Perot resonator: theory', *Appl. Opt.*, **26**, pp. 456–463, 1987.

[85] K. Kobayashi and I. Mito, 'Single frequency and tunable laser diodes', *J. Lightwave Technol.*, **6**(11), pp. 1623–1633, 1988.

[86] C. H. Henry, 'Theory of the linewidth of semiconductor lasers', *IEEE J. Quantum Electron.*, **QE-18**, pp. 259–264, 1982.

[87] R. Wyatt and W. J. Devlin, '10-kHz linewidth 1.5-μm InGaAsP external cavity laser with 55-nm tuning range', *Electron. Lett.*, **19**, pp. 110–112, 1983.

[88] N. A. Olsson and J. P. van der Ziel, 'Performance characteristics of 1.5 micron external cavity semiconductor lasers for coherent optical communication', *J. Lightwave Technol.*, **LT-5**, pp. 510–515, 1987.

[89] W. V. Severin and H. J. Shaw, 'A single-mode fiber evanescent grating reflector', *J. Lightwave Technol.*, **LT-3**, pp. 1041–1048, 1985.

[90] J. Wittmann and G. Gaukel, 'Narrow-linewidth laser with a prism grating: GRINrod lens combination serving an external cavity', *Electron. Lett.*, **23**, pp. 524–525, 1987.

[91] F. Heismann, R. C. Alferness, L. L. Buhl, G. Eisenstein, S. K. Korotky, J. J. Veselka, L. W. Stulz and C. A. Burrus, 'Narrow-linewidth, electro-optically tunable InGaAsP–Ti : $LiNbO_3$ extended cavity laser', *Appl. Phys. Lett.*, **51**, pp. 164–165, 1987.

[92] G. Coquin, K. W. Cheung and M. M. Choy, 'Single and multiple wavelength operation of acousto-optically tuned semiconductor lasers at 1.3 microns', *Proc. 11th IEEE Int. Semiconductor Laser Conf.* (USA), pp. 130–131, 1988.

[93] G. Coquin and K. W. Cheung, 'An electronically tunable external cavity semiconductor laser', *Electron. Lett.*, **24**, pp. 599–600, 1988.

[94] N. K. Dutta, T. Cella, A. B. Piccirilli, R. L. Brown, 'Integrated external cavity lasers', *Conf. on Lasers and Electrooptics, CLEO '87* (USA), MF2, April 1987.

[95] Y. Suematsu and S. Arai, 'Integrated optics approach for advanced semiconductor lasers', *Proc. IEEE*, **75**(11), pp. 1472–1487, 1987.

[96] M. Okai, S. Sakano and N. Chinone, 'Wide-range continuous tunable double-sectioned distributed feedback lasers', *Fifteenth European Conf. on Opt. Commun.* (Sweden) pp. 122–125, September, 1989.

[97] H. Imai, 'Tuning results of 3-sectioned DFB lasers', *Semiconductor Laser Workshop, Conf. on Lasers and Electrooptics, CLEO '89* (USA), 1989.

[98] S. Murata, I. Mito and K. Kobayashi, 'Tuning ranges for 1.5 μm wavelength tunable DBR lasers', *Electron. Lett.*, **24**, pp. 577–579, 1988.

[99] Y. Kotaki, M. Matsuda, H. Ishikawa and H. Imni, 'Tunable DBR laser with wide tuning range', *Electron. Lett.*, **24**, pp. 503–505, 1988.

[100] I. M. Jauncey, L. Reekie, J. E. Townsend and D. N. Payne, 'Single longitudinal-mode operation of an Nd^{3+}-doped fibre laser', *Electron. Lett.*, **24**, pp. 24–26, 1988.

[101] L. Reekie, R. J. Mears, S. B. Poole and D. N. Payne, 'Tunable single-mode fibre lasers', *J. Lightwave Technol.*, **LT-4**(7), pp. 956–960, 1986.

[102] C. A. Millar, I. D. Miller, D. B. Mortimore, B. J. Ainslie and P. Urquhart, 'Fibre laser with adjustable fibre reflector for wavelength tuning and variable output coupling', *IEE Proc., Part J*, **135**, pp. 303–304, 1988.

[103] G. H. B. Thompson, *Physics of Semiconductor Laser Devices*, John Wiley, 1980.

[104] Y. Horikoshi, 'Semiconductor lasers with wavelengths exceeding 2 μm', in W. T. Tsang (Vol. Ed.) *Semiconductors and Semimetals: Lightwave communication technology*, Academic Press, 22C, pp. 93–151, 1985.

[105] C. Caneau, A. K. Srivastava, A. G. Dentai, J. L. Zyskind and M. A. Pollack, 'Room temperature GaInAsSb/AlGaAsSb DH injection lasers at 2.2 μm', *Electron. Lett.*, **21**, pp. 815–817, 1985.

[106] A. E. Bockarev, L. M. Dolginov, A. E. Drakin, L. V. Druzhinina, P. G. Eliseev and B. N. Sverdlov, 'Injection InGaAsSb lasers emitting radiation of wavelengths 1.2–2.3 μm at room temperature', *Sov. J. Quantum Electron.*, **15**, pp. 869–870, 1985.

[107] C. Caneau, J. L. Zyskind, J. W. Sulhoff, T. E. Glover, J. Centanni, C. A. Burrus, A. G. Dentai and M. A. Pollack, '2.2 μm GaInAsSb/AlGaAsSb injection lasers with low threshold current density', *Appl. Phys. Lett.*, **51**, pp. 764–766, 1987.

[108] H. Ebe, Y. Nishijima and K. Shinohara, 'PbEuTe lasers with 4–6 μm wavelength mode with hot-well epitaxy', *IEEE J. Quantum Electron.*, **25**(6), pp. 1381–1384, 1989.

[109] M. C. Brierley and P. W. France, 'Continuous wave lasing at 2.7 μm in an erbium-doped fluorozirconate fibre', *Electron. Lett.*, **24**, pp. 935–937, 1988.

[110] M. C. Brierley, P. W. France and C. A. Millar, 'Lasing at 2.08 μm in a holmium-doped fluorozirconate fibre laser', *Electron. Lett.*, **24**, pp. 539–540, 1988.

[111] L. Esterowitz, R. Allen and I. Aggarwal, 'Pulsed laser emission at 2.3 μm in a thulium-doped fluorozirconate fibre', *Electron. Lett.*, **24**, p. 1104, 1988.

[112] R. W. Tkach and A. R. Chaplyry, 'Phase noise and linewidth in an InGaAsP laser', *J. Lightwave Technol.*, **LT-4**, pp. 1711–1716, 1986.

[113] T.-P. Lee, 'Recent advances in long-wavelength semiconductor lasers for optical fiber communication', *Proc. IEEE*, **79**(3), pp. 253–276, 1991.

7

Optical sources 2: The light emitting diode

7.1 Introduction

Spontaneous emission of radiation in the visible and infrared regions of the spectrum from a forward biased *p–n* junction was discussed in Section 6.3.2. The normally empty conduction band of the semiconductor is populated by electrons injected into it by the forward current through the junction, and light is generated when these electrons recombine with holes in the valence band to emit a photon. This is the mechanism by which light is emitted from an LED, but stimulated emission is not encouraged, as it is in the injection laser, by the addition of an optical cavity and mirror facets to provide feedback of photons.

The LED can therefore operate at lower current densities than the injection laser, but the emitted photons have random phases and the device is an incoherent optical source. Also, the energy of the emitted photons is only roughly equal to the bandgap energy of the semiconductor material, which gives a much wider spectral linewidth (possibly by a factor of 100) than the injection laser. The linewidth for an LED corresponds to a range of photon energy between 1 and 3.5 KT, where K is Boltzmann's constant and T is the absolute temperature. This gives linewidths of 30 to 40 nm for GaAs-based devices operating at room temperature. Thus the LED supports many optical modes within its structure and is therefore often used as a multimode source, although more recently the coupling of LEDs to single-mode fibers has been pursued with success, particularly when advanced structures have been employed. At present, LEDs have several further drawbacks in comparison with injection lasers. These include:

(a) generally lower optical power coupled into a fiber (microwatts);
(b) usually lower modulation bandwidth;
(c) harmonic distortion.

However, although these problems may initially appear to make the LED a less attractive optical source than the injection laser, the device has a number of distinct advantages which have given it a prominent place in optical fiber communications:

1. *Simpler fabrication*. There are no mirror facets and in some structures no striped geometry.
2. *Cost*. The simpler construction of the LED leads to much reduced cost which is always likely to be maintained.
3. *Reliability*. The LED does not exhibit catastrophic degradation and has proved far less sensitive to gradual degradation than the injection laser. It is also immune to self pulsation and modal noise problems.
4. *Generally less temperature dependence*. The light output against current characteristic is less affected by temperature than the corresponding characteristic for the injection laser. Furthermore, the LED is not a threshold device and therefore raising the temperature does not increase the threshold current above the operating point and hence halt operation.
5. *Simpler drive circuitry*. This is due to the generally lower drive currents and reduced temperature dependence which makes temperature compensation circuits unnecessary.
6. *Linearity*. Ideally, the LED has a linear light output against current characteristic (see Section 7.4.1), unlike the injection laser. This can prove advantageous where analog modulation is concerned.

These advantages combined with the development of high radiance, relatively high bandwidth devices have ensured that the LED remains an extensively used source for optical fiber communications.

Structures fabricated using the GaAs/AlGaAs material system are well tried for operation in the shorter wavelength region. In addition, more recently there have been substantial advances in devices based on the InGaAsP/InP material structure for use in the longer wavelength region especially around 1.3 μm. At this wavelength, the material dispersion in silica glass fibers goes through zero and hence the wider linewidth of the LED imposes a far slighter limitation on link length than does intermodal dispersion within multimode fiber. Furthermore, the reduced fiber attenuation at this operating wavelength can allow longer-haul LED systems.

Although longer wavelength LED systems using multimode graded index fiber are continuing to be developed, particularly for nontelecommunication applications (see Sections 14.5 to 14.7), much recent activity has been concerned with both high speed operation and with the coupling of these InGaAsP LEDs to single-mode fiber. A major impetus for these strategies is the potential deployment of such single-mode LED systems in the telecommunication access network or subscriber loop (see Section 14.2.3). In this context, theoretical studies of both LED coupling [Ref. 1] and transmission [Ref. 2] with single-mode fiber have been undertaken, as well as numerous practical investigations, some of which are outlined in the following sections. It is therefore apparent that LEDs are likely to remain a prominent optical fiber communication source for many system applications including operation over significant distances with single-mode fiber at transmission rates that may exceed a Gbit s^{-1}.

Structures fabricated using the GaAs/AlGaAs material system are well advanced for the shorter wavelength region. There has also been much interest in LEDs for the longer wavelength region, especially around 1.3 μm where material dispersion in silica-based fibers goes through zero and where the wide linewidth of the LED imposes far less limitation on link length than intermodal dispersion within the fiber. Furthermore, the reduced attenuation allows longer-haul LED systems. As with injection lasers InGaAsP/InP is the material structure currently favoured in this region for the high radiance devices. These longer wavelength systems utilizing graded index fibers have led to the development of wider bandwidth devices providing transmission rates of hundreds of Mbit s^{-1}.

LEDs therefore remain the primary optical source for nontelecommunication applications (i.e. shorter-haul) whilst injection lasers find major use as single-mode devices within single-mode fiber systems for long-haul, wideband applications. In addition, LEDs have been shown to launch acceptable, albeit often modest (5 to 10 μW) levels of optical power into single-mode fiber and therefore may well find use in short-haul single-mode fiber telecommunication systems (i.e. the subscriber loop) in the future.

Having dealt with the basic operating principles for the LED in Section 6.3.2, we continue in Section 7.2 with a discussion of LED power and efficiency in relation to the launching of light into optical fibers. Moreover, at the end of this section we include a brief account of the operation of an efficient LED which employs a double heterostructure. This leads into a discussion in Section 7.3 of the major practical LED structures where again we have regard to their light coupling efficiency. Also included in this section are the more advanced device structures such as the edge-emitting and the superluminescent LED. The various operating characteristics and limitations on LED performance are then described in Section 7.4. Finally in Section 7.5, we include a brief discussion on the possible modulation techniques for semiconductor optical sources.

7.2 LED power and efficiency

The absence of optical amplification through stimulated emission in the LED tends to limit the internal quantum efficiency (ratio of photons generated to injected electrons) of the device. Reliance on spontaneous emission allows nonradiative recombination to take place within the structure due to crystalline imperfections and impurities giving, at best, an internal quantum efficiency of 50% for simple homojunction devices. However, as with injection lasers double heterojunction (DH) structures have been implemented which recombination lifetime measurements suggest [Ref. 3] give internal quantum efficiencies of 60 to 80%.

The power generated internally by an LED may be determined by consideration of the excess electrons and holes in the p and n type material respectively (i.e. the minority carriers) when it is forward biased and carrier injection takes place at the device contacts (see Section 6.3.2). The excess density of electrons Δn and holes Δp is equal since the injected carriers are created and recombined in pairs such that charge neutrality is maintained within the structure. In extrinsic materials one carrier type will have a much higher concentration than the other and hence in the p type region, for example, the hole concentration will be much greater than the electron concentration. Generally, the excess minority carrier density decays exponentially with time t [Ref. 4] according to the relation:

$$\Delta n = \Delta n(0) \exp(-t/\tau) \tag{7.1}$$

where $\Delta n(0)$ is the initial injected excess electron density and τ represents the total carrier recombination lifetime. In most cases, however, Δn is only a small fraction of the majority carriers and comprises all of the minority carriers. Therefore, in these cases, the carrier recombination lifetime becomes the minority or injected carrier lifetime τ_i.

When there is a constant current flow into the junction diode, an equilibrium condition is established. In this case, the total rate at which carriers are generated will be the sum of the externally supplied and the thermal generation rates. The current density J in amperes per square metre may be written as J/ed in electrons per cubic metre per second, where e is the charge on an electron and d is the thickness of the recombination region. Hence a rate equation for carrier recombination in the LED can be expressed in the form [Ref. 4]:

$$\frac{d(\Delta n)}{dt} = \frac{J}{ed} - \frac{\Delta n}{\tau} \quad (\text{m}^{-3}\,\text{s}^{-1}) \tag{7.2}$$

The condition for equilibrium is obtained by setting the derivative in Eq. (7.2) to zero. Hence

$$\Delta n = \frac{J\tau}{ed} \quad (\text{m}^{-3}) \tag{7.3}$$

Equation (7.3) therefore gives the steady state electron density when a constant current is flowing into the junction region.

It is also apparent from Eq. (7.2) that in the steady state the total number of carrier recombinations per second or the recombination rate r_t will be:

$$r_t = \frac{J}{ed} \quad (\mathrm{m}^{-3}) \tag{7.4}$$

$$= r_r + r_{nr} \quad (\mathrm{m}^{-3}) \tag{7.5}$$

where r_r is the radiative recombination rate per unit volume and r_{nr} is the nonradiative recombination rate per unit volume. Moreover, when the forward biased current into the device is i, then from Eq. (7.4) the total number of recombinations per second R_t becomes:

$$R_t = \frac{i}{e} \tag{7.6}$$

It was indicated in Section 6.3.3.1 that excess carriers can recombine either radiatively or nonradiatively. Whilst in the former case a photon is generated, in the latter case the energy is released in the form of heat (i.e. lattice vibrations). Moreover, for a DH device with a thin active region (a few μm), then the nonradiative recombination tends to be dominated by surface recombination at the heterojunction interfaces.

The LED internal quantum efficiency* η_{int}, which can be defined as the ratio of the radiative recombination rate to the total recombination rate, following Eq. (7.5) may be written as [Ref. 5]:

$$\eta_{int} = \frac{r_r}{r_t} = \frac{r_r}{r_r + r_{nr}} \tag{7.7}$$

$$= \frac{R_r}{R_t} \tag{7.8}$$

where R_r is the total number of *radiative* recombinations per second. Rearranging Eq. (7.8) and substituting from Eq. (7.6) gives:

$$R_r = \eta_{int} \frac{i}{e} \tag{7.9}$$

Since R_r is also equivalent to the total number of photons generated per second and from Eq. (6.1) each photon has an energy equal to hf joules, then the optical power

* The internal quantum efficiency for the LED is obtained only from the spontaneous radiation and hence is written as η_{int}. By contrast, the internal quantum efficiency for the injection laser combined the internal quantum efficiencies for both spontaneous and simulated radiation. It was therefore denoted as η_i (see Section 6.4.1).

generated internally by the LED, P_{int} is:

$$P_{int} = \eta_{int} \frac{i}{e} hf \qquad \text{(W)} \tag{7.10}$$

Using Eq. (6.22) to express the internally generated power in terms of wavelength rather than frequency gives:

$$P_{int} = \eta_{int} \frac{hci}{e\lambda} \qquad \text{(W)} \tag{7.11}$$

It is interesting to note that Eqs. (7.10) and (7.11) display a linear relationship between the optical power generated in the LED and the drive current into the device (see Section 7.4.1). Similar relationships may be obtained for the optical power emitted from an LED but in this case the constant of proportionality η_{int} must be multiplied by a factor representing the external quantum efficiency* η_{ext} to provide an overall quantum efficiency for the device.

For the exponential decay of excess carriers depicted by Eq. (7.1) the radiative minority carrier lifetime is $\tau_r = \Delta n/r_r$ and the nonradiative minority carrier lifetime is $\tau_{nr} = \Delta n/r_{nr}$. Therefore, from Eq. (7.7) the internal quantum efficiency is:

$$\eta_{int} = \frac{1}{1 + (r_{nr}/r_r)} = \frac{1}{1 + (\tau_r/\tau_{nr})} \tag{7.12}$$

Furthermore, the total recombination lifetime τ can be written as $\tau = \Delta n/r_t$ which, using Eq. (7.5), gives:

$$\frac{1}{\tau} = \frac{1}{\tau_r} + \frac{1}{\tau_{nr}} \tag{7.13}$$

Hence Eq. (7.12) becomes

$$\eta_{int} = \frac{\tau}{\tau_r} \tag{7.14}$$

It should be noted that the same expression for the internal quantum efficiency could be obtained from Eq. (7.7).

Example 7.1

The radiative and nonradiative recombination lifetimes of the minority carriers in the active region of a double heterojunction LED are 60 ns and 100 ns respectively. Determine the total carrier recombination lifetime and the power internally generated within the device when the peak emission wavelength is 0.87 μm at a drive current of 40 mA.

* The external quantum efficiency may be defined as the ratio of the photons emitted from the device to the photons internally generated. However, it is sometimes defined as the ratio of the number of photons emitted to the total number of carrier recombinations (radiative and nonradiative).

Solution: The total carrier recombination lifetime is given by Eq. (7.13) as:

$$\tau = \frac{\tau_r \tau_{nr}}{\tau_r + \tau_{nr}} = \frac{60 \times 100 \text{ ns}}{(60 + 100)} = 37.5 \text{ ns}$$

To calculate the power internally generated it is necessary to obtain the internal quantum efficiency of the device. Hence using Eq. (7.14):

$$\eta_{\text{int}} = \frac{\tau}{\tau_r} = \frac{37.5}{60} = 0.625$$

Thus from Eq (7.11):

$$P_{\text{int}} = \eta_{\text{int}} \frac{\text{hci}}{e\lambda} = \frac{0.625 \times 6.626 \times 10^{-34} \times 2.998 \times 10^{8} \times 40 \times 10^{-3}}{1.602 \times 10^{-19} \times 0.87 \times 10^{-6}}$$

$$= 35.6 \text{ mW}$$

The LED which has an internal quantum efficiency of 62.5% generates 35.6 mW of optical power, internally. It should be noted, however, that this power level will not be readily emitted from the device.

Although the possible internal quantum efficiency can be relatively high the radiation geometry for an LED which emits through a planar surface is essentially Lambertian in that the surface radiance (the power radiated from a unit area into a unit solid angle, given in $\text{W sr}^{-1} \text{m}^{-2}$) is constant in all directions. The Lambertian intensity distribution is illustrated in Figure 7.1 where the maximum intensity I_0 is perpendicular to the planar surface but is reduced on the sides in proportion to the cosine of the viewing angle θ as the apparent area varies with this angle. This reduces the external power efficiency to a few per cent as most of the light generated within the device is trapped by total internal reflection (see Section 2.2.1) when it is radiated at greater than the critical angle for the crystal–air

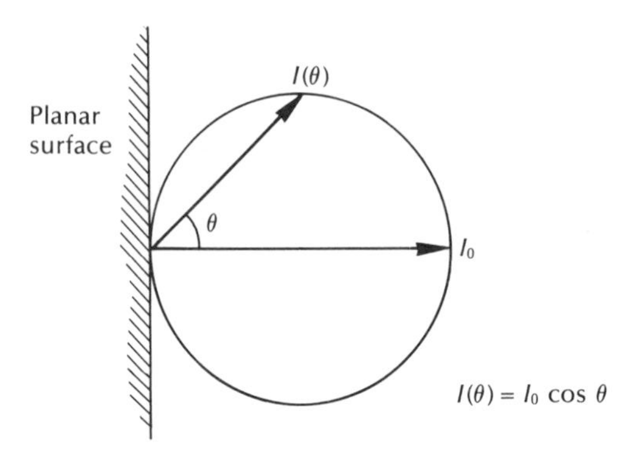

Figure 7.1 The Lambertian intensity distribution typical of a planar LED.

interface. As with the injection laser (see Section 6.4.1) the external power efficiency η_{ep} is defined as the ratio of the optical power emitted externally P_e to the electrical power provided to the device P or:

$$\eta_{ep} \simeq \frac{P_e}{P} \times 100\% \tag{7.15}$$

Also, the optical power emitted P_e into a medium of low refractive index n from the face of a planar LED fabricated from a material of refractive index n_x is given approximately by [Ref. 6]:

$$P_e = \frac{P_{int} F n^2}{4 n_x^2} \tag{7.16}$$

where P_{int} is the power generated internally and F is the transmission factor of the semiconductor–external interface. Hence it is possible to estimate the percentage of optical power emitted.

Example 7.2

A planar LED is fabricated from gallium arsenide which has a refractive index of 3.6.

(a) Calculate the optical power emitted into air as a percentage of the internal optical power for the device when the transmission factor at the crystal–air interface is 0.68.

(b) When the optical power generated internally is 50% of the electrical power supplied, determine the external power efficiency.

Solution: (a) The optical power emitted is given by Eq. (7.16), in which the refractive index n for air is 1.

$$P_e \simeq \frac{P_{int} F n^2}{4 n_x^2} = \frac{P_{int} 0.68 \times 1}{4(3.6)^2} = 0.013 P_{int}$$

Hence the power emitted is only 1.3% of the optical power generated internally.

(b) The external power efficiency is given by Eq. (7.15), where

$$\eta_{ep} = \frac{P_e}{P} \times 100 = 0.013 \frac{P_{int}}{P} \times 100$$

Also, the optical power generated internally $P_{int} = 0.5P$.

Hence

$$\eta_{ep} = \frac{0.013 P_{int}}{2 P_{int}} \times 100 = 0.65\%$$

A further loss is encountered when coupling the light output into a fiber. Considerations of this coupling efficiency are very complex; however, it is possible to use an approximate simplified approach [Ref. 7]. If it is assumed for step index fibers that all the light incident on the exposed end of the core within the acceptance angle θ_a is coupled, then for a fiber in air, using Eq. (2.8):

$$\theta_a = \sin^{-1}(n_1^2 - n_2^2)^{\frac{1}{2}} = \sin^{-1}(NA) \tag{7.17}$$

Also, incident light at angles greater than θ_a will not be coupled. For a Lambertian source, the radiant intensity at an angle θ, $I(\theta)$ is given by (see Figure 7.1):

$$I(\theta) = I_0 \cos \theta \tag{7.18}$$

where I_0 is the radiant intensity along the line $\theta = 0$. Considering a source which is smaller than, and in close proximity to, the fiber core, and assuming cylindrical symmetry, the coupling efficiency η_c is given by:

$$\eta_c = \frac{\int_0^{\theta_a} I(\theta) \sin \theta \, d\theta}{\int_0^{\pi/2} I(\theta) \sin \theta \, d\theta} \tag{7.19}$$

Hence substituting from Eq. (7.18):

$$\eta_c = \frac{\int_0^{\theta_a} I_0 \cos \theta \sin \theta \, d\theta}{\int_0^{\pi/2} I_0 \cos \theta \sin \theta \, d\theta}$$

$$= \frac{\int_0^{\theta_a} I_0 \sin 2\theta \, d\theta}{\int_0^{\pi/2} I_0 \sin 2\theta \, d\theta}$$

$$\eta_c = \frac{[-I_0 \cos 2\theta/2]_0^{\theta_a}}{[-I_0 \cos 2\theta/2]_0^{\pi/2}}$$

$$= \sin^2 \theta_a \tag{7.20}$$

Furthermore, from Eq. (7.17):

$$\eta_c = \sin^2 \theta_a = (NA)^2 \tag{7.21}$$

Equation (7.21) for the coupling efficiency allows estimates for the percentage of optical power coupled into the step index fiber relative to the amount of optical power emitted from the LED.

Example 7.3

The light output from the GaAs LED of Example 7.2 is coupled into a step index fiber with a numerical aperture of 0.2, a core refractive index of 1.4 and a diameter larger than the diameter of the device. Estimate:

(a) The coupling efficiency into the fiber when the LED is in close proximity to the fiber core.
(b) The optical loss in decibels, relative to the power emitted from the LED, when coupling the light output into the fiber.
(c) The loss relative to the internally generated optical power in the device when coupling the light output into the fiber when there is a small air gap between the LED and the fiber core.

Solution: (a) From Eq. (7.21), the coupling efficiency is given by:

$$\eta_c = (NA)^2 = (0.2)^2 = 0.04$$

Thus about 4% of the externally emitted optical power is coupled into the fiber.

(b) Let the optical power coupled into the fiber be P_c. Then the optical loss in decibels relative to P_e when coupling the light output into the fiber is

$$Loss = -10 \log_{10} \frac{P_c}{P_e}$$

$$= -10 \log_{10} \eta_c$$

Hence,

$$Loss = -10 \log_{10} 0.04$$
$$= 14.0 \text{ dB}$$

(c) When the LED is emitting into air, from Example 7.2:

$$P_e = 0.013 P_{int}$$

Assuming a very small air gap (i.e. cylindrical symmetry unaffected); then from (a) the power coupled into the fiber is:

$$P_c = 0.04 P_e = 0.04 \times 0.013 P_{int}$$
$$= 5.2 \times 10^{-4} \; P_{int}$$

Hence in this case only about 0.05% of the internal optical power is coupled into the fiber.

The loss in decibels relative to P_{int} is:

$$Loss = -10 \log_{10} \frac{P_c}{P_{int}} = -10 \log_{10} 5.2 \times 10^{-4} = 32.8 \text{ dB}$$

If significant optical power is to be coupled from an incoherent LED into a low NA fiber the device must exhibit very high radiance. This is especially the case when considering graded index fibers where the Lambertian coupling efficiency with the same NA (same refractive index difference) and $\alpha \simeq 2$ (see Section 2.4.4) is about half that into step index fibers [Ref. 8]. To obtain the necessary high radiance, direct bandgap semiconductors (see Section 6.3.3.1) must be used fabricated with DH structures which may be driven at high current densities. The principle of operation of such a device will now be considered prior to discussion of various LED structures.

7.2.1 The double heterojunction LED

The principle of operation of the DH LED is illustrated in Figure 7.2. The device shown consists of a *p* type GaAs layer sandwiched between a *p* type AlGaAs

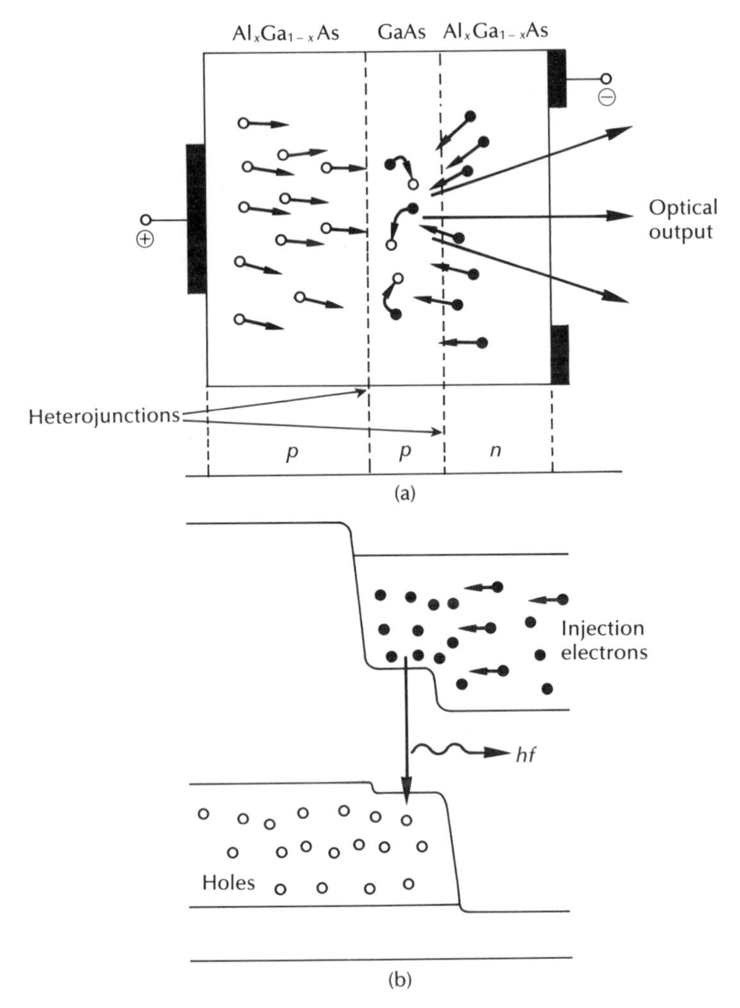

Figure 7.2 The double heterojunction LED: (a) the layer structure, shown with an applied forward bias; (b) the corresponding energy band diagram.

and an *n* type AlGaAs layer. When a forward bias is applied (as indicated in Figure 7.2(a)) electrons from the *n* type layer are injected through the *p–n* junction into the *p* type GaAs layer where they become minority carriers. These minority carriers diffuse away from the junction [Ref. 9], recombining with majority carriers (holes) as they do so. Photons are therefore produced with energy corresponding to the bandgap energy of the *p* type GaAs layer. The injected electrons are inhibited from diffusing into the *p* type AlGaAs layer because of the potential barrier presented by the *p–p* heterojunction (see Figure 7.2(b)). Hence, electroluminescence only occurs in the GaAs junction layer, providing both good internal quantum efficiency and high radiance emission. Furthermore, light is emitted from the device without reabsorption because the bandgap energy in the AlGaAs layer is large in comparison with that in GaAs. The DH structure is therefore used to provide the most efficient incoherent sources for application within optical fiber communications. Nevertheless, these devices generally exhibit the previously discussed constraints in relation to coupling efficiency to optical fibers. This and other LED structures are considered in greater detail in the following section.

7.3 LED structures

There are five major types of LED structure and although only two have found extensive use in optical fiber communications a third is becoming of increasing interest. These are the surface emitter, the edge emitter and the superluminescent LED respectively. The other two structures, the planar and dome LEDs, find more application as cheap plastic encapsulated visible devices for use in such areas as intruder alarms, TV channel changes and industrial counting. However, infrared versions of these devices have been used in optical communications mainly with fiber bundles and it is therefore useful to consider them briefly before progressing on to the high radiance LED structures.

7.3.1 Planar LED

The planar LED is the simplest of the structures that are available and is fabricated by either liquid- or vapour-phase epitaxial processes over the whole surface of a GaAs substrate. This involves a *p* type diffusion into the *n* type substrate in order to create the junction illustrated in Figure 7.3. Forward current flow through the junction gives Lambertian spontaneous emission and the device emits light from all surfaces. However, only a limited amount of light escapes the structure due to total internal reflection, as discussed in Section 7.2, and therefore the radiance is low.

7.3.2 Dome LED

The structure of a typical dome LED is shown in Figure 7.4. A hemisphere of *n* type GaAs is formed around a diffused *p* type region. The diameter of the dome is chosen to maximize the amount of internal emission reaching the surface within the

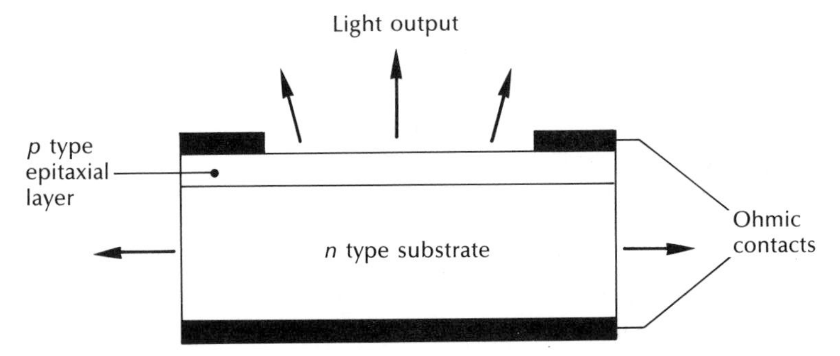

Figure 7.3 The structure of a planar LED showing the emission of light from all surfaces.

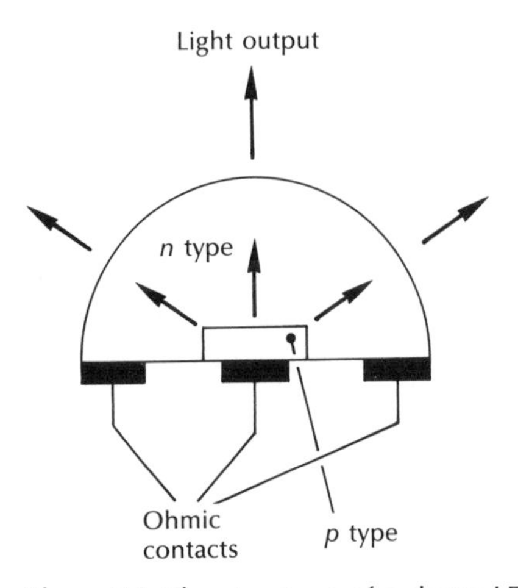

Figure 7.4 The structure of a dome LED.

critical angle of the GaAs–air interface. Hence this device has a higher external power efficiency than the planar LED. However, the geometry of the structure is such that the dome must be far larger than the active recombination area, which gives a greater effective emission area and thus reduces the radiance.

7.3.3 Surface emitter LEDs

A method for obtaining high radiance is to restrict the emission to a small active region within the device. The technique pioneered by Burrus and Dawson [Ref. 10] with homostructure devices was to use an etched well in a GaAs substrate in order to prevent heavy absorption of the emitted radiation, and physically to accommodate the fiber. These structures have a low thermal impedance in the active region allowing high current densities and giving high radiance emission into the optical fiber. Furthermore, considerable advantage may be obtained by employing

DH structures giving increased efficiency from electrical and optical confinement as well as less absorption of the emitted radiation. This type of surface emitter LED (SLED) is now widely employed within optical fiber communications.

The structure of a high radiance etched well DH surface emitter* for the 0.8 to 0.9 μm wavelength band is shown in Figure 7.5 [Ref. 11]. The internal absorption in this device is very low due to the larger bandgap confining layers, and the reflection coefficient at the back crystal face is high giving good forward radiance. The emission from the active layer is essentially isotropic, although the external emission distribution may be considered Lambertian with a beam width of 120° due to refraction from a high to a low refractive index at the GaAs–fiber interface. The power coupled P_c into a multimode step index fiber may be estimated from the relationship [Ref. 12]:

$$P_c = \pi(1 - r)AR_D(NA)^2 \tag{7.22}$$

where r is the Fresnel reflection coefficient at the fiber surface, A is the smaller of the fiber core cross section or the emission area of the source and R_D is the radiance of the source. However, the power coupled into the fiber is also dependent on many other factors including the distance and alignment between the emission area and the fiber, the SLED emission pattern and the medium between the emitting area and the fiber. For instance the addition of epoxy resin in the etched well tends to reduce the refractive index mismatch and increase the external power efficiency of the device. Hence, DH surface emitters often give more coupled optical power than

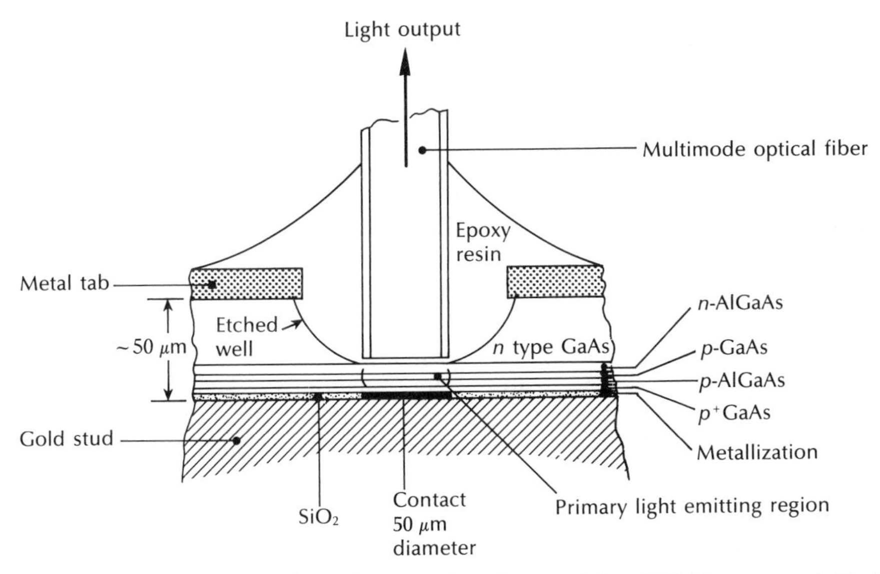

Figure 7.5 The structure of an AlGaAs DH surface-emitting LED (Burrus type) [Ref. 11].

* These devices are also known as Burrus type LEDs

predicted by Eq. (7.22). Nevertheless Eq. (7.22) may be used to gain an estimate of the power coupled, although accurate results may only be obtained through measurement.

Example 7.4

A DH surface emitter which has an emission area diameter of 50 μm is butt jointed to an 80 μm core step index fiber with a numerical aperture of 0.15. The device has a radiance of 30 W sr^{-1} cm^{-2} at a constant operating drive current. Estimate the optical power coupled into the fiber if it is assumed that the Fresnel reflection coefficient at the index matched fiber surface is 0.01.

Solution: Using Eq. (7.22), the optical power coupled into the fiber P_c is given by:

$$P_c = \pi(1 - r)AR_D(NA)^2$$

In this case A represents the emission area of the source.

Hence:

$$A = \pi(25 \times 10^{-4})^2 = 1.96 \times 10^{-5} \text{ cm}^2$$

Thus,

$$\begin{aligned} P_c &= \pi(1 - 0.01)1.96 \times 10^{-5} \times 30 \times (0.15)^2 \\ &= 41.1 \ \mu\text{W} \end{aligned}$$

In this example around 41 μW of optical power is coupled into the step index fiber.

However, for graded index fiber optimum direct coupling requires that the source diameter be about one half the fiber core diameter. In both cases lens coupling may give increased levels of optical power coupled into the fiber but at the cost of additional complexity. Other factors which complicate the LED fiber coupling are the transmission characteristics of the leaky modes or large angle skew rays (see Section 2.4.1). Much of the optical power from an incoherent source is initially coupled into these large angle rays, which fall within the acceptance angle of the fiber but have much higher energy than meridional rays. Energy from these rays goes into the cladding and may be lost. Hence much of the light coupled into a multimode fiber from an LED is lost within a few hundred metres. It must therefore be noted that the effective optical power coupled into a short length of fiber significantly exceeds that coupled into a longer length.

The planar structure of the Burrus type LED and other nonetched well SLEDs [Ref. 13] allows significant lateral current spreading, particularly for contact diameters less than 25 μm. This current spreading results in a reduced current density as well as an effective emission area substantially greater than the contact area. A technique which has been used to reduce the current spreading in very small

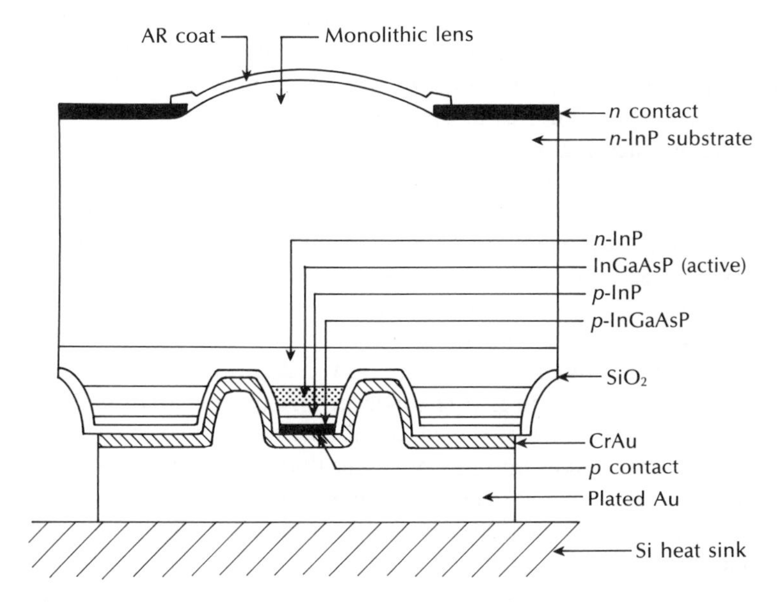

Figure 7.6 Small area InGaAsP mesa-etched surface-emitting LED structure [Ref. 14]

devices is to fabricate a mesa structure SLED, as illustrated in Figure 7.6 [Ref. 14]. In this case mesas with diameters in the range 20 to 25 μm at the active layer were formed by chemical etching.

These InGaAsP/InP devices which emitted at a wavelength of 1.3 μm had an integral lens formed at the exit face of the InP substrate in order to improve the coupling efficiency, particularly to single-mode fiber. Such monolithic lens structures provide a common strategy for improving the power coupled into fiber from LEDs, and alternative lens coupling techniques are discussed in Section 7.3.6. Moreover, there is increasing interest in coupling LEDs to single-mode fiber for shorter-haul applications which, in the case of SLEDs, necessitates efficient lens coupling to obtain acceptable launch powers. For example, the LED illustrated in Figure 7.6 with a drive current of 50 mA was found to couple only around 2 μW of optical power into single-mode fiber [Ref. 14].

7.3.4 Edge emitter LEDs

Another basic high radiance structure currently used in optical communications is the stripe geometry DH edge emitter LED (ELED). This device has a similar geometry to a conventional contact stripe injection laser, as shown in Figure 7.7. It takes advantage of transparent guiding layers with a very thin active layer (50 to 100 μm) in order that the light produced in the active layer spreads into the transparent guiding layers, reducing self absorption in the active layer. The consequent waveguiding narrows the beam divergence to a half power width of around 30° in the plane perpendicular to the junction. However, the lack of waveguiding in the plane of the junction gives a Lambertian output with a half power width of around 120°, as illustrated in Figure 7.7.

Most of the propagating light is emitted at one end face only due to a reflector on the other end face and an antireflection coating on the emitting end face. The effective radiance at the emitting end face can be very high giving an increased coupling efficiency into small NA fiber compared with the surface emitter. However, surface emitters generally radiate more power into air (2.5 to 3 times) than edge emitters since the emitted light is less affected by reabsorption and interfacial recombination. Comparisons [Refs. 15 to 17] have shown that edge emitters couple more optical power into low NA (less than 0.3) than surface emitters, whereas the opposite is true for large NA (greater than 0.3).

The enhanced waveguiding of the edge emitter enables it in theory [Ref. 16] to couple 7.5 times more power into low NA fiber than a comparable surface emitter. However, in practice the increased coupling efficiency has been found to be slightly less than this (3.5 to 6 times) [Refs. 16 and 17]. Similar coupling efficiencies may be achieved into low NA fiber with surface emitters by the use of a lens.

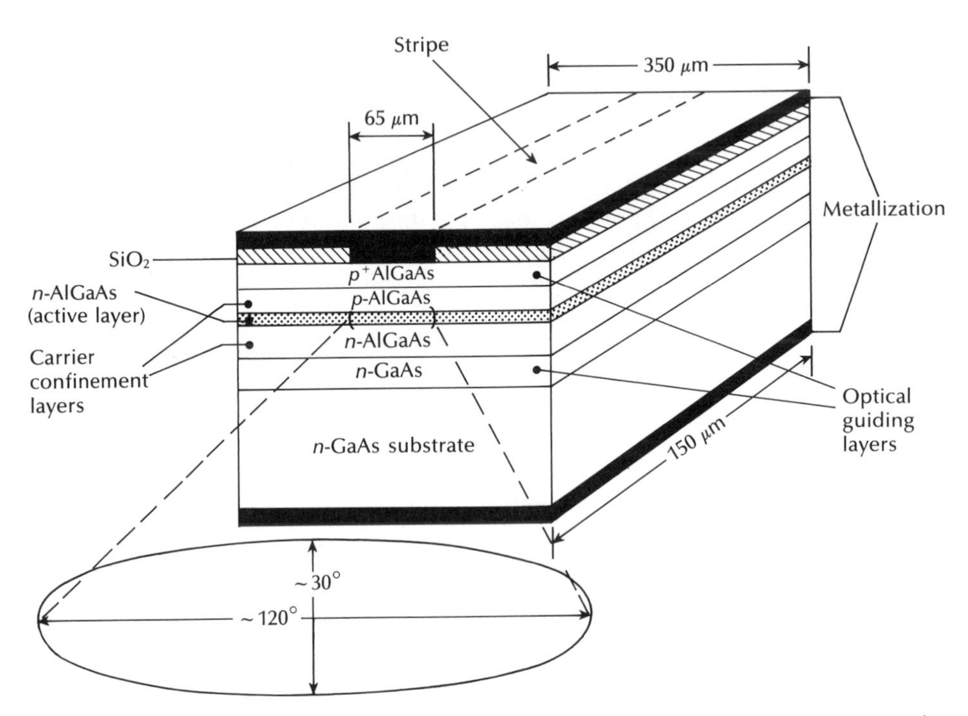

Figure 7.7 Schematic illustration of the structure of a strip geometry DH AlGaAs edge-emitting LED.

Furthermore, it has been found that lens coupling with edge emitters may increase the coupling efficiencies by comparable factors (around five times).

The stripe geometry of the edge emitter allows very high carrier injection densities for given drive currents. Thus it is possible to couple approaching a milliwatt of optical power into low NA (0.14) multimode step index fiber with edge-emitting LEDs operating at high drive currents (500 mA) [Ref. 18].

Edge emitters have also been found to have a substantially better modulation bandwidth of the order of hundreds of megahertz than comparable surface-emitting structures with the same drive level [Ref. 17]. In general it is possible to construct edge-emitting LEDs with a narrower linewidth than surface emitters, but there are manufacturing problems with the more complicated structure (including difficult heat sinking geometry) which moderate the benefits of these devices.

Nevertheless, a number of ELED structures have been developed using the InGaAsP/InP material system for operation at a wavelength of 1.3 μm. A common device geometry which has also been utilized for AlGaAs/GaAs ELEDs [Ref. 19] is shown in Figure 7.8 [Ref. 20]. This DH edge-emitting device is realized in the form of a restricted length, stripe geometry *p*-contact arrangement. Such devices are also referred to as truncated-stripe ELEDs. The short stripe structure (around 100 μm long) improves the external efficiency of the ELED by reducing its internal absorption of carriers.

It was mentioned in Section 7.1 that a particular impetus for the development of high performance LEDs operating at a wavelength of 1.3 μm was their potential application in the future optical fiber subscriber loop. In this context the capacity to provide both high speed transmission and significant launch powers into single-mode fiber are of prime concern. Aspects of these attributes are displayed by the two device structures shown in Figure 7.9.

The ELED illustrated in Figure 7.9(a) [Ref. 21] comprises a mesa structure with a width of 8 μm and a length of 150 μm for current confinement. The tilted back facet of the device was formed by chemical etching in order to suppress laser

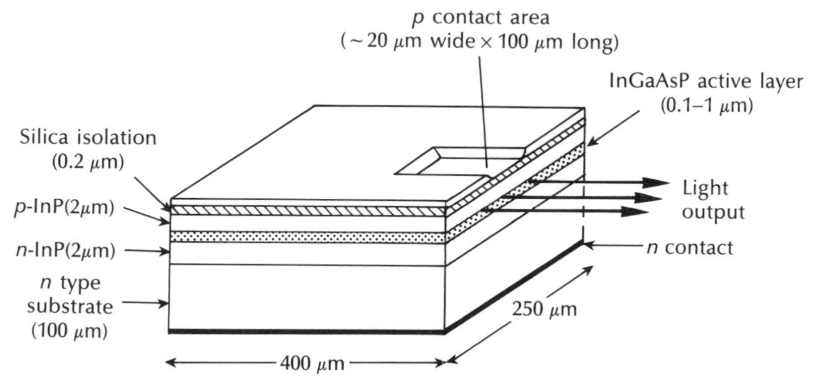

Figure 7.8 Truncated stripe InGaAsP edge-emitting LED [Ref. 20].

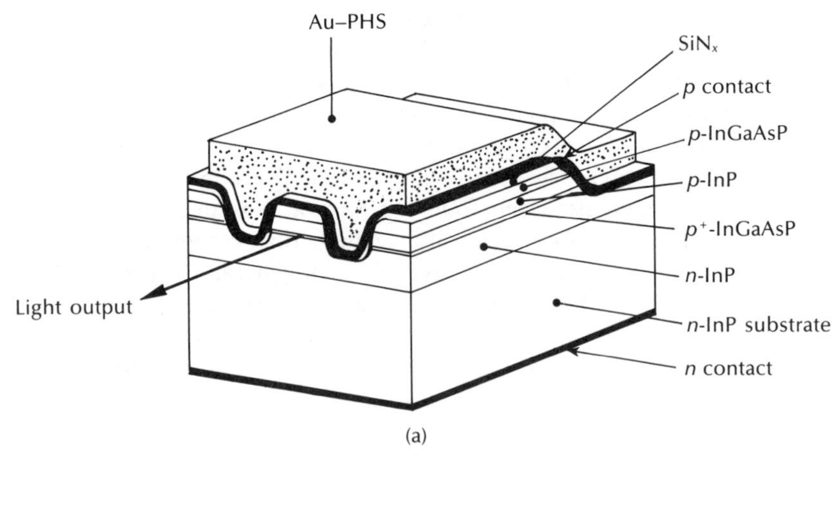

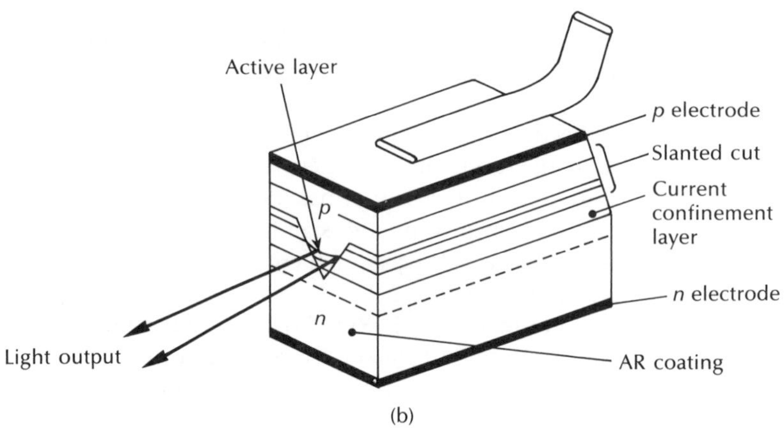

Figure 7.9 High speed InGaAsP edge-emitting LEDs: (a) mesa structure ELED [Ref. 21]. (b) V-grooved substrate BH ELED [Ref. 23].

oscillation. It should be noted that such ELED device structures, being very similar to injection structures, could lase unless this mechanism is specifically avoided by removing the potential Fabry–Perot cavity. This point is discussed in further detail in Section 7.3.5.

The ELED active layer was heavily doped with Zn to reduce the minority carrier lifetime and thus improve the device modulation bandwidth. In this way a 3 dB modulation bandwidth of 600 MHz was obtained [Ref. 21]. When operating at a speed of 600 Mbit s^{-1} the device, with lens coupling (see Section 7.3.6), launched an average optical power of approximately 4 μW into single-mode fiber at a peak drive current of 100 mA. An increase in the peak drive current to 240 mA provided an improvement in the coupled power to slightly over 6 μW. By contrast a buried heterostructure ELED has been reported which couples 7 μW of optical power into single-mode fiber with a drive current of only 20 mA [Ref. 22]. This short cavity device (100 μm) had a spectral width (FWHP) of 70 nm in comparison with a linewidth of 90 nm for the high speed ELED shown in Figure 7.9(a).

Figure 7.9(b) displays another advanced InGaAsP ELED which was fabricated as a V-grooved substrate buried heterostructure device [Ref. 23]. In this case the front facet was antireflection coated and the rear facet was also etched at a slat to prevent laser action. This device, which again emitted at a centre wavelength of 1.3 μm, was reported to have a 3 dB modulation bandwidth around 350 MHz, with the possibility of launching 30 μW of optical power into single-mode fiber [Ref. 23].

Very high coupled optical power levels into single-mode fiber in excess of 100 μW have been obtained with InGaAsP ELEDs at drive currents as low as 50 mA [Ref. 24]. This device structure was based on the configuration of the *p*-substrate buried crescent injection laser [Ref. 25] with the rear facet bevelled by chemical etching to suppress laser oscillation. Butt coupling to 10 μm core diameter single-mode fiber provided launch powers of only 12 μW which were increased to over 200 μW using lens coupling (see Section 7.3.6) and drive currents of 100 mA. Moreover, the spectral widths of the ELEDs were as narrow as 50 nm which gave device characteristics approaching those of the superluminescent LEDs dealt with in the following section.

7.3.5 Superluminescent LEDs

A third device geometry which is already providing significant benefits over both SLEDs and ELEDs for communication applications is the superluminescent diode or SLD. This device type offers advantages of: (a) a high output power; (b) a directional output beam; and (c) a narrow spectral linewidth; all of which prove useful for coupling significant optical power levels into optical fiber (in particular to single-mode fiber [Ref. 22]). Furthermore, the superradiant emission process within the SLD tends to increase the device modulation bandwidth over that of more conventional LEDs.

Figure 7.10 shows two forms of construction for the SLD. It may be observed that the structures in both cases are very similar to those of ELEDs or, for that matter, injection lasers. In effect, the SLD has optical properties that are bounded by the ELED and the injection laser. Similar to this latter device the SLD structure requires a *p–n* junction either in the form of a long rectangular stripe (Figure 7.10(a) [Ref. 26]), a ridge waveguide [Ref. 27] or a buried heterostructure (Figure 7.10(b) [Refs. 22, 28]). However, one end of the device is made optically lossy to prevent reflections and thus suppress lasing, the output being from the opposite end.

For operation the injected current is increased until stimulated emission, and hence amplification, occurs (i.e. the initial step towards laser action), but because there is high loss at one end of the device, no optical feedback takes place.

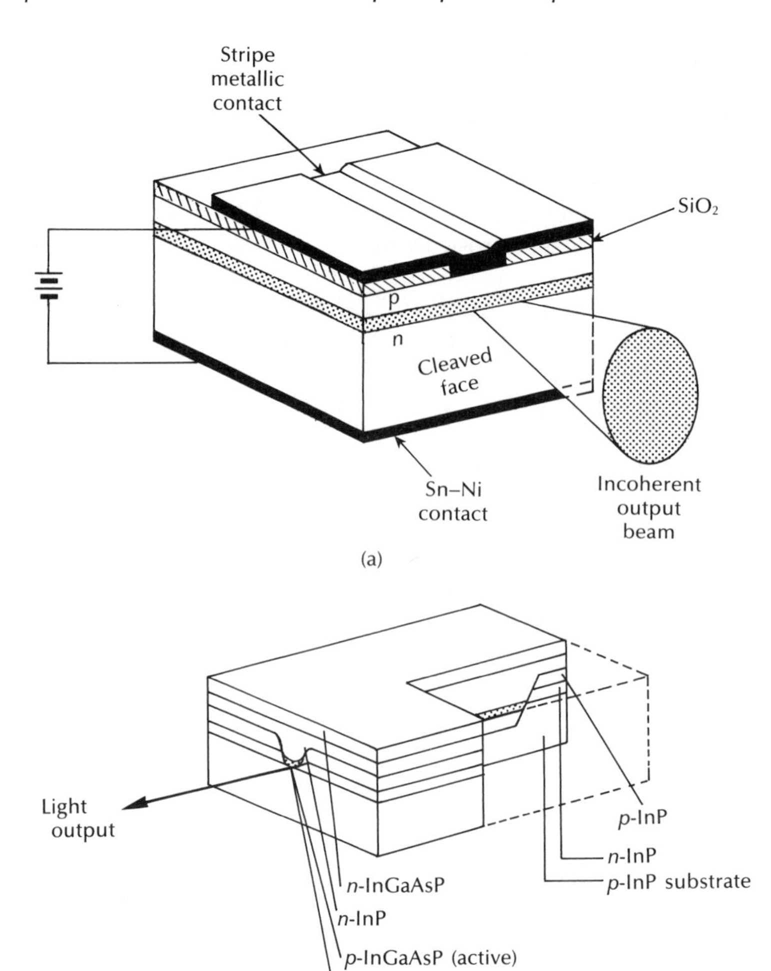

Figure 7.10 Superluminescent LED structures: (a) AlGaAs contact stripe SLD [Ref. 26]; (b) high output power InGaAsP SLD [Ref. 28].

Therefore, although there is amplification of the spontaneous emission, no laser oscillation builds up. However, operation in the current region for stimulated emission provides gain causing the device output to increase rapidly with increases in drive current due to what is effectively single pass amplification. High optical output power can therefore be obtained, together with a narrowing of the spectral width which also results from the stimulated emission.

An early SLD is shown in Figure 7.10(a) which employs a contact stripe together with an absorbing region at one end to suppress laser action. Such devices have provided peak output power of 60 mW at a wavelength of 0.87 μm in pulsed mode [Ref. 26]. More recently, antireflection (AR) coatings have been applied to the cleaved facets of SLDs in order to suppress Fabry–Perot resonance [Refs. 22, 27, 29]. Such devices have launched 550 μW of optical power in 50 μm diameter multimode graded index fiber at drive currents of 250 mA [Ref. 22] and 250 μW into single-mode fiber using drive currents of 100 mA [Ref. 29]. In both cases the device linewidths were in the range 30 to 40 nm rather than the 60 to 90 nm spectral widths associated with conventional ELEDs.

The structure of a recently developed InGaAsP/InP SLD is illustrated in Figure 7.10(b) [Ref. 28]. The device which emits at 1.3 μm comprises a buried active layer within a V-shaped groove on the *p* type InP substrate. This technique provides an appropriate structure for high power operation because of its low leakage current. Unlike the aforementioned SLD structures which incorporate AR coatings on both end facets to prevent feedback, a light diffusion surface is placed within this device. The surface, which is applied diagonally on the active layer of length 350 μm, serves to scatter the backward light emitted from the active layer and thus decreases feedback into this layer. In addition, an AR coating is provided on the output facet. As it is not possible to achieve a perfect antireflection coating, the above structure is therefore not left totally dependent on this feedback suppression mechanism. The coupling of 1 mW of optical power into the spherically lensed end of a single-mode fiber (10 μm core diameter) has been demonstrated with this device operating at a drive current of 150 mA [Ref. 28]. Moreover, the spectral distribution from the SLD was observed to be a smooth envelope with a FWHP of 30 nm, whilst the device modulation bandwidth reached 350 MHz at the −1.5 dB point (see Section 7.4.3).

Although the incoherent optical power output from SLDs can approach that of the coherent output from injection lasers, the required current density is substantially higher (by around a factor of three times), necessitating high drive currents due to the long device active lengths (i.e. large areas). Recent improvements, however, in injection laser structures (see Sections 6.5 and 6.6) providing lower threshold currents for specific output powers have also made the SLD a more practical proposition. Nevertheless, other potential drawbacks associated with the SLD in comparison with conventional LEDs are the nonlinear output characteristic and the increased temperature dependence of the output power (see Section 7.4.1).

7.3.6 Lens coupling to fiber

It is apparent that much of the light emitted from LEDs is not coupled into the generally narrow acceptance angle of the fiber. Even with the etched well surface emitter where the low NA fiber is butted directly into the emitting aperture of the device, coupling efficiencies are poor (of the order of 1 to 2%). However, it has been found that greater coupling efficiency may be obtained if lenses are used to collimate the emission from the LED, particularly when the fiber core diameter is significantly larger than the width of the emission region. There are several lens

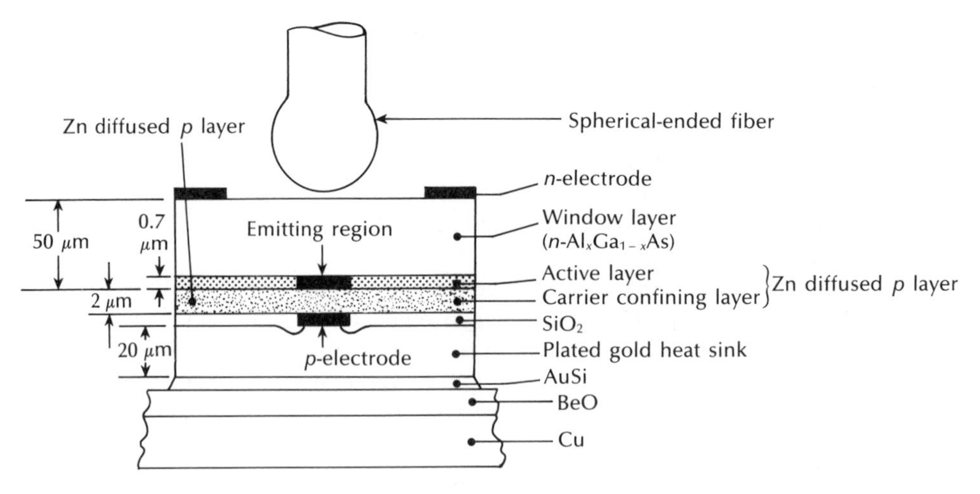

Figure 7.11 Schematic illustration of the structure of a spherical-ended fiber coupled AlGaAs LED [Ref. 30].

coupling configurations which include spherically polished structures not unlike the dome LED, spherical-ended or tapered fiber coupling, truncated spherical microlenses, GRIN-rod lenses and integral lens structures.

A GaAs/AlGaAs spherical-ended fiber coupled LED is illustrated in Figure 7.11 [Ref. 30]. It consists of a planar surface emitting structure with the spherical-ended fiber attached to the cap by epoxy resin. An emitting diameter of 35 μm was fabricated into the device and the light was coupled into fibers with core diameters of 75 and 110 μm. The geometry of the situation is such that it is essential that the active diameter of the device be substantially less (factor of 2) than the fiber core diameter if increased coupling efficiency is to be obtained. In this case good performance was obtained with coupling efficiencies around 6%. This is in agreement with theoretical [Ref. 31] and other experimental [Ref. 32] results which suggest an increased coupling efficiency of 2 to 5 times through the spherical fiber lens.

Another common lens coupling technique employs a truncated spherical microlens. This configuration is shown in Figure 7.12 for an etched well InGaAsP/InP DH surface emitter [Ref. 33] operating at a wavelength of 1.3 μm. Again, a requirement for efficient coupling is that the emission region diameter is much smaller than the core diameter of the fiber. In this case the best results were obtained with a 14 μm active diameter and an 85 μm core diameter step index fiber with a numerical aperture of 0.16. The coupling efficiency was increased by a factor of 13, again supported by theory [Ref. 31] which suggests possible increases of up to thirty times.

However, the overall power conversion efficiency η_{pc} which is defined as the ratio of the optical power coupled into the fiber P_c to the electrical power applied at the

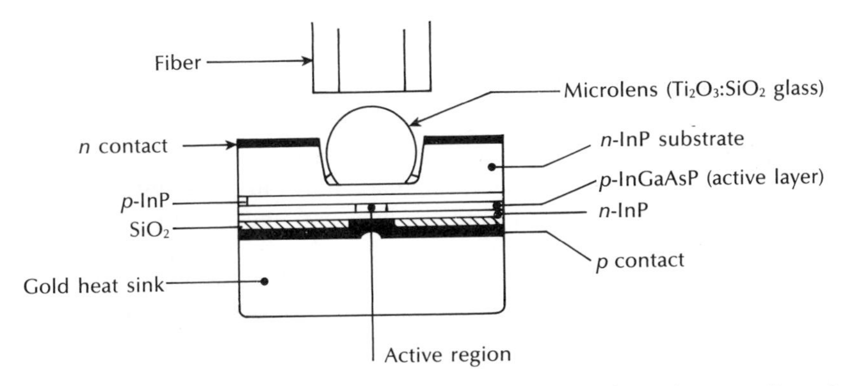

Figure 7.12 The use of a truncated spherical microlens for coupling the emission from an InGaAsP surface-emitting LED to the fiber [Ref. 33].

terminals of the device P and is therefore given by:

$$\eta_{pc} = \frac{P_c}{P} \tag{7.23}$$

is still quite low. Even with the increased coupling efficiency η_{pc} was found to be around 0.4%.

Example 7.5

A lens coupled surface-emitting LED launches 190 μW of optical power into a multimode step index fiber when a forward current of 25 mA is flowing through the device. Determine the overall power conversion efficiency when the corresponding forward voltage across the diode is 1.5 V.

Solution: The overall power conversion efficiency may be obtained from Eq. (7.23) where:

$$\eta_{pc} = \frac{P_c}{P} = \frac{190 \times 10^{-6}}{25 \times 10^{-3} \times 1.5} = 5.1 \times 10^{-3}$$

Hence the overall power conversion efficiency is 0.5%.

The integral lens structure shown in Figure 7.6 has become a favoured power coupling strategy for use with surface emitters. In this technique a low absorption lens is formed at the exit face of the substrate material instead of it being fabricated in glass and attached to a planar SLED with epoxy. The method benefits from the elimination of the semiconductor-epoxy-lens interface which can limit the maximum lens gains of the SLEDs discussed above. An early example gave an improved coupling efficiency of around three times that of a planar SLED [Ref.

32], but for optimized devices it is predicted that coupling efficiencies should exceed 15% [Ref. 33].

It was mentioned in Section 7.3.4 that lens coupling can also be usefully employed with edge-emitting devices. In practice, lenses attached to the fiber ends or tapered fiber lenses are widely utilized to increase the coupling efficiency [Refs. 13, 19]. An example of the former technique is illustrated in Figure 7.13(a) in which a hemispherical lens is epoxied onto the fiber end and positioned adjacent to the ELED emission region. The coupling efficiency has been increased by a factor of three to four times using this strategy [Refs. 13, 19]. Alternatively, a truncated spherical lens glued onto the emitting facet of a superradiant ELED has given a coupling gain of a factor of five or 7 dB [Ref. 34].

Tapered fiber lenses have been extensively used to couple power from ELEDs into single-mode fiber. Butt coupling of optical power from LEDs into single-mode fiber is substantially reduced in comparison with that obtained into multimode fiber. It ranges from between 0.5 and 2 μW for a standard SLED up to around 10 to 12 μW for an ELED. The small core diameter of single-mode fiber does not allow significant lens coupling gain to be achieved with SLEDs. For edge emitters, however, a coupling gain of around 5 dB may be realized using tapered fiber [Ref. 35].

An alternative strategy to improve the coupling efficiency from an ELED into single-mode fiber is depicted in Figure 7.13(b) [Ref. 24]. In this case a tapered GRIN-rod lens (see Section 5.5.1) was positioned between the high power ELED and the fiber. A coupling efficiency defined as the ratio of the coupled power to the total emitted power of around 15% was obtained [Ref. 24].

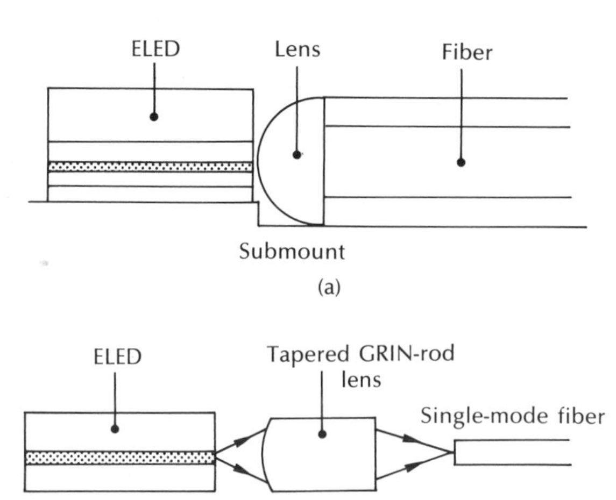

Figure 7.13 Lens coupling with edge-emitting LEDs: (a) lens-ended fiber coupling; (b) tapered (plano-convex) GRIN-rod lens coupling to single-mode fiber.

7.4 LED characteristics

7.4.1 Optical output power

The ideal light output power against current characteristic for an LED is shown in Figure 7.14. It is linear corresponding to the linear part of the injection laser optical power output characteristic before lasing occurs. Intrinsically the LED is a very linear device in comparison with the majority of injection lasers and hence it tends to be more suitable for analog transmission where severe constraints are put on the linearity of the optical source. However, in practice LEDs do exhibit significant nonlinearities which depend upon the configuration utilized. It is therefore often necessary to use some form of linearizing circuit technique (e.g. predistortion linearization or negative feedback) in order to ensure the linear performance of the device to allow its use in high quality analog transmission systems [Ref. 37]. Figure 7.15(a) and (b) show the light output against current characteristics for

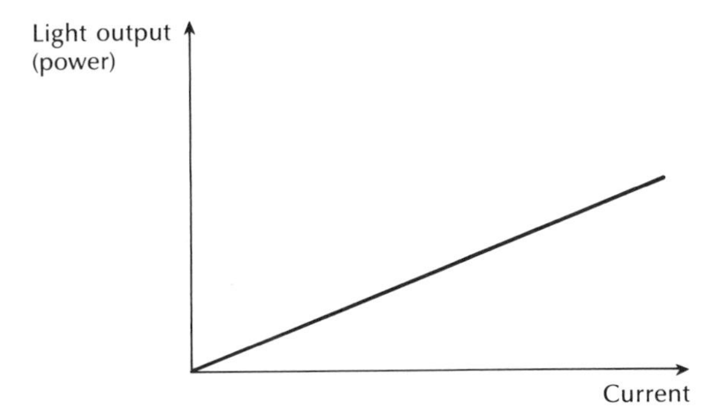

Figure 7.14 An ideal light output against current characteristic for an LED.

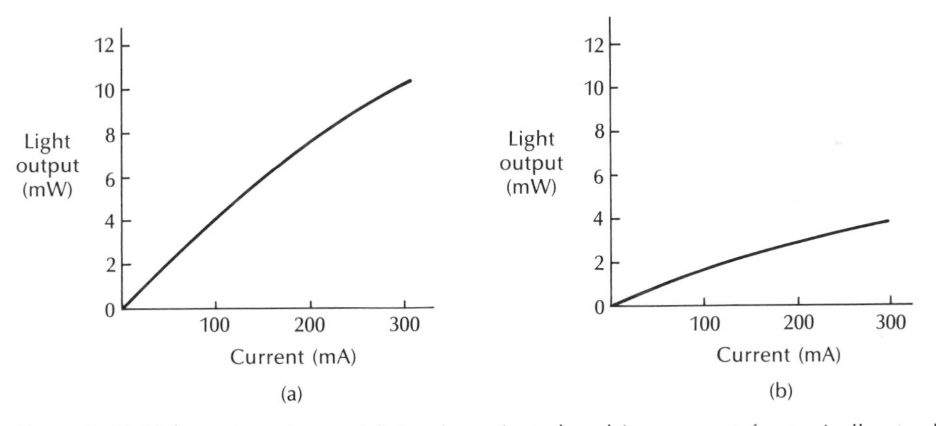

Figure 7.15 Light output (power) into air against d.c. drive current for typically good LEDs [Ref. 17]: (a) an AlGaAs surface emitter with a 50 μm diameter dot contact; (b) an AlGaAs edge emitter with a 65 μm wide stripe and 100 μm length.

typically good surface and edge emitters respectively [Ref. 17]. It may be noted that the surface emitter radiates significantly more optical power into air than the edge emitter, and that both devices are reasonably linear at moderate drive currents.

In a similar manner to the injection laser, the internal quantum efficiency of LEDs decreases exponentially with increasing temperature (see Section 6.7.1) Hence the light emitted from these devices decreases as the p–n junction temperature increases. The light output power against temperature characteristics for the three major LED structures operating at a wavelength of 1.3 μm are shown, for comparison, in Figure 7.16 [Ref. 13]. It may be observed that the edge-emitting device exhibits a greater temperature dependence than the surface emitter and that the output of the SLD with its stimulated emission is *strongly* dependent on the junction temperature. This latter factor is further emphasized in the light output against current characteristics for a superluminescent LED displayed in Figure 7.17 [Ref. 27]. These characteristics show the variation in output power at a specific drive current over the temperature range 0 to 40 °C for a ridge waveguide device providing lateral current confinement. The nonlinear nature of the output characteristic typical of SLDs can also be observed with a knee becoming apparent at an operating temperature around 20 °C. Hence to utilize the high power potential of such devices at elevated temperatures, the use of thermoelectric coolers may be necessary [Ref. 13].

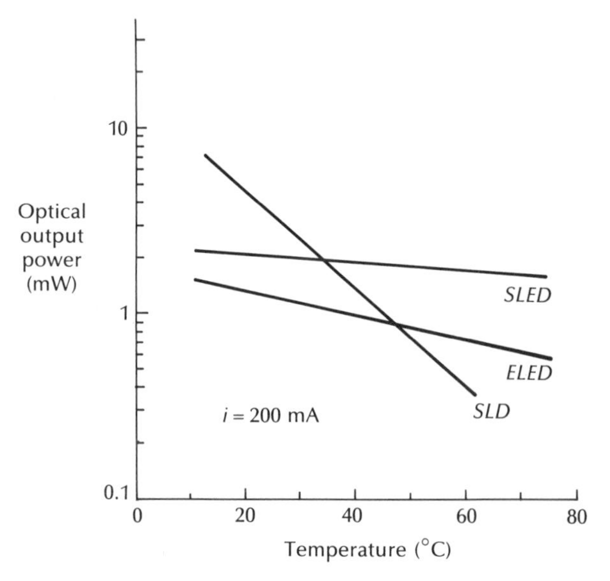

Figure 7.16 Light output temperature dependence for the three major LED structures emitting at a wavelength of 1.3 μm [Ref. 13].

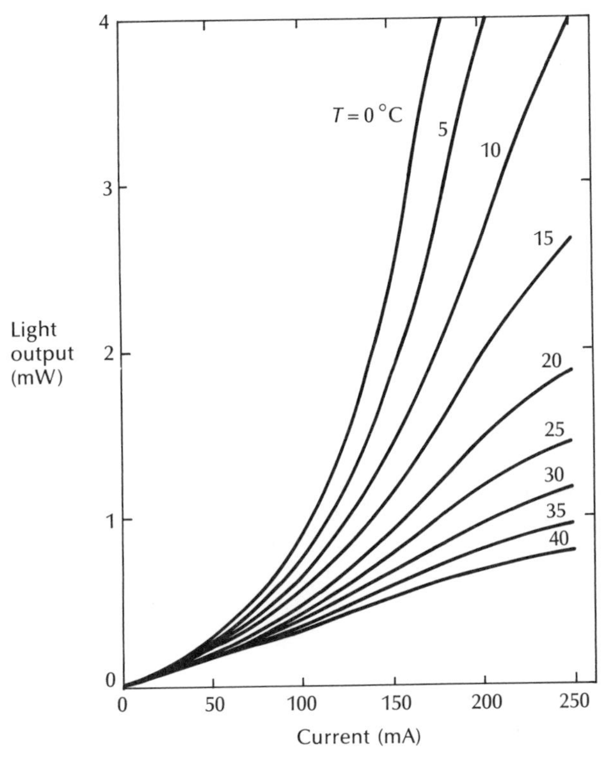

Figure 7.17 Light output against current characteristic at various ambient temperatures for an InGaAsP ridge waveguide SLD. Reproduced with permission from I. P. Kaminow, G. Eisenstein, L. W. Stulz and A. G. Dentai 'Lateral confinement InGaAsP superluminescent diode at 1.3 μm', *IEEE J. Quantum Electron.*, **QE19**, p. 78, 1983. Copyright © 1983, IEEE.

7.4.2 Output spectrum

The spectral linewidth of an LED operating at room temperature in the 0.8 to 0.9 μm wavelength band is usually between 25 and 40 nm at the half maximum intensity points (full width at half power (FWHP) points). For materials with smaller bandgap energies operating in the 1.1 to 1.7 μm wavelength region the linewidth tends to increase to around 50 to 160 nm. Examples of these two output spectra are shown in Figure 7.18 [Refs. 7 and 38]. Also illustrated in Figure 7.18(b) are the increases in linewidth due to increased doping levels and the formation of bandtail states (see Section 6.3.4). This becomes apparent in the differences in the output spectra between surface- and edge-emitting LEDs where the devices have generally heavily doped and lightly doped (or undoped) active layers respectively. It may also be noted that there is a shift to lower peak emission wavelength (i.e. higher energy) through reduction in doping in Figure 7.18(b), and hence the

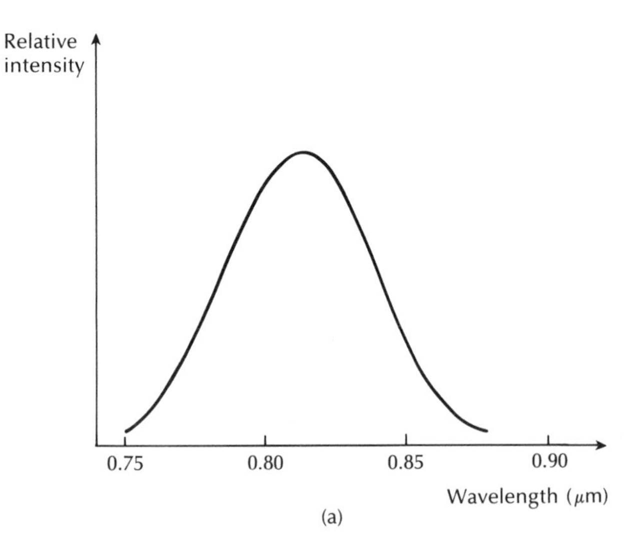

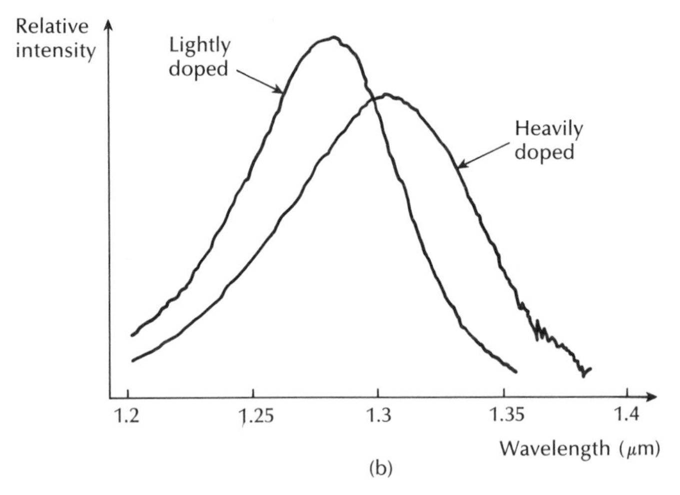

Figure 7.18 LED output spectra: (a) output spectrum for an AlGaAs surface emitter with doped active region [Ref. 7]; (b) output spectra for an InGaAsP surface emitter showing both the lightly doped and heavily doped cases. Reproduced with permission from A. C. Carter. *The Radio and Electron. Eng.*, **51**, p. 41, 1981.

active layer composition must be adjusted if the same centre wavelength is to be maintained.

The differences in the output spectra between InGaAsP SLEDs and ELEDs caused by self absorption along the active layer of the devices are displayed in Figure 7.19. It may be observed that the FWHP points are around 1.6 times smaller for the ELED than the SLED [Ref. 13]. In addition, the spectra of the ELED

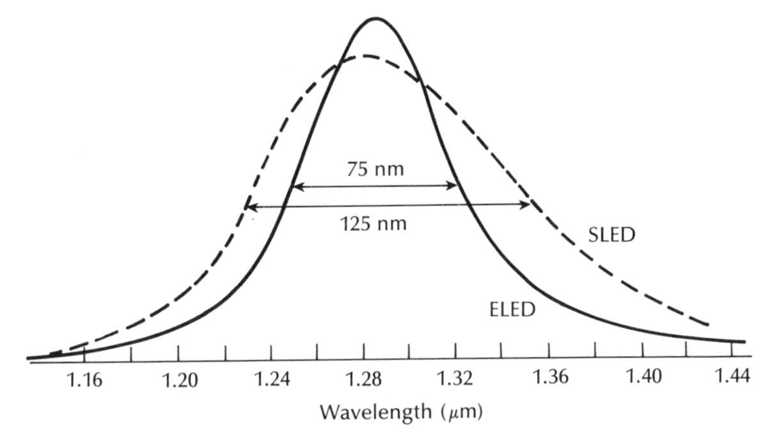

Figure 7.19 Typical spectral output characteristics for InGaAsP surface and edge-emitting LEDs operating in the 1.3 μm wavelength region [Ref. 13].

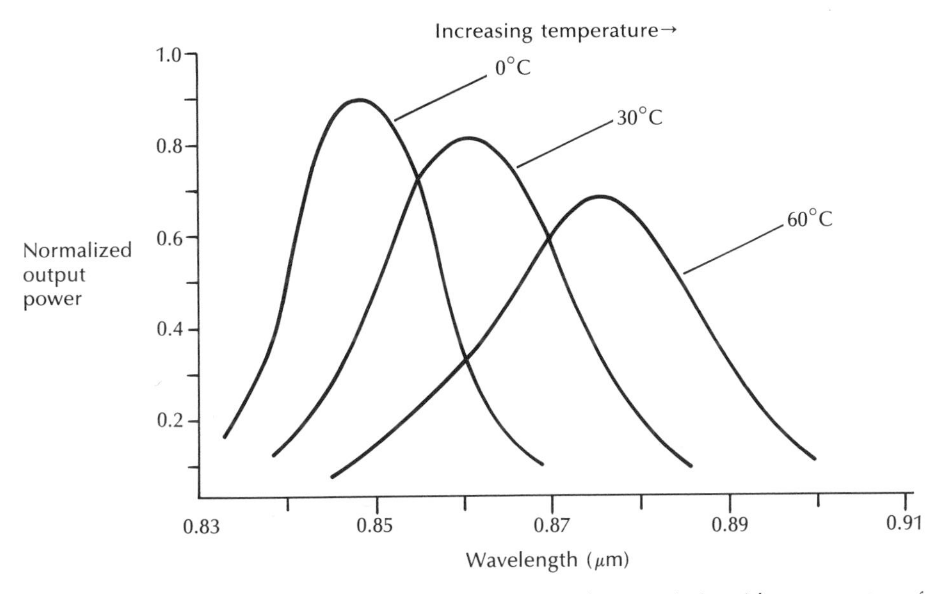

Figure 7.20 Typical spectral variation of the output characteristic with temperature for an AlGaAs surface-emitting LED.

may be further narrowed by the superluminescent operation due to the onset of stimulated gain and in this case the linewidth can be far smaller (e.g. 30 nm) than that obtained with the SLED.

The output spectra also tend to broaden at a rate of between 0.1 and 0.3 nm $^{\circ}C^{-1}$ with increase in temperature due to the greater energy spread in carrier distributions at higher temperatures. Increases in temperature of the junction affect the peak emission wavelength as well, and it is shifted by +0.3 to 0.4 nm $^{\circ}C^{-1}$ for AlGaAs devices [Ref. 11] and by +0.6 nm $^{\circ}C^{-1}$ for InGaAsP devices [Ref. 39]. The combined effects on the output spectrum from a typical AlGaAs surface emitter are illustrated in Figure 7.20. It is clear that it may therefore be necessary to utilize heat sinks with LEDs for certain optical fiber communication applications, although this is far less critical (normally insignificant compared with the device linewidth) than the cooling requirements for injection lasers.

7.4.3 Modulation bandwidth

The modulation bandwidth in optical communications may be defined in either electrical or optical terms. However, it is often more useful when considering the associated electrical circuitry in an optical fiber communication system to use the electrical definition where the electrical signal power has dropped to half its constant value due to the modulated portion of the optical signal. This corresponds to the electrical 3 dB point or the frequency at which the output electrical power is reduced by 3 dB with respect to the input electrical power. As optical sources operate down to d.c. we only consider the high frequency 3 dB point, the modulation bandwidth being the frequency range between zero and this high frequency 3 dB point.

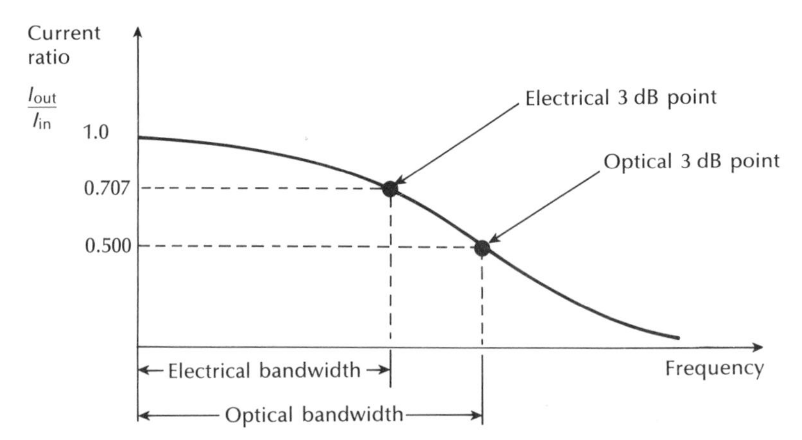

Figure 7.21 The frequency response for an optical fiber system showing the electrical and optical bandwidths.

Alternatively, if the 3 dB bandwidth of the modulated optical carrier (optical bandwidth) is considered, we obtain an increased value for the modulation bandwidth. The reason for this inflated modulation bandwidth is illustrated in Example 7.6 and Figure 7.21. In considerations of bandwidth within the text the electrical modulation bandwidth will be assumed unless otherwise stated, following current practice.

Example 7.6

Compare the electrical and optical bandwidths for an optical fiber communication system and develop a relationship between them.

Solution: In order to obtain a simple relationship between the two bandwidths it is necessary to compare the electrical current through the system. Current rather than voltage (which is generally used in electrical systems) is compared as both the optical source and optical detector (see Section 8.6) may be considered to have a linear relationship between light and current.

Electrical bandwidth: The ratio of the electrical output power to the electrical input power in decibels RE_{dB} is given by:

$$RE_{dB} = 10 \log_{10} \frac{\text{electrical power out (at the detector)}}{\text{electrical power in (at the source)}}$$

$$= 10 \log_{10} \frac{I_{out}^2/R_{out}}{I_{in}^2/R_{in}}$$

$$\propto 10 \log_{10} \left[\frac{I_{out}}{I_{in}}\right]^2$$

The electrical 3 dB points occur when the ratio of electrical powers shown above is $\frac{1}{2}$. Hence it follows that this must occur when:

$$\left[\frac{I_{out}}{I_{in}}\right]^2 = \frac{1}{2}, \quad \text{or} \quad \frac{I_{out}}{I_{in}} = \frac{1}{\sqrt{2}}$$

Thus in the electrical regime the bandwidth may be defined by the frequency when the output current has dropped to $1/\sqrt{2}$ or 0.707 of the input current to the system.

Optical bandwidth: The ratio of the optical output power to the optical input power in decibels RO_{dB} is given by,

$$RO_{dB} = 10 \log_{10} \frac{\text{optical power out (received at detector)}}{\text{optical power in (transmitted at source)}}$$

$$\propto 10 \log_{10} \frac{I_{out}}{I_{in}}$$

(due to the linear light/current relationships of the source and detector). Hence the optical 3 dB points occur when the ratio of the currents is equal to $\frac{1}{2}$, and:

$$\frac{I_{out}}{I_{in}} = \frac{1}{2}$$

Therefore in the optical regime the bandwidth is defined by the frequencies at which the output current has dropped to $\frac{1}{2}$ or 0.5 of the input current to the system. This corresponds to an electrical power attenuation of 6 dB.

The comparison between the two bandwidths is illustrated in Figure 7.21 where it may be noted that the optical bandwidth is significantly greater than the electrical bandwidth. The difference between them (in frequency terms) depends on the shape of the frequency response for the system. However, if the system response is assumed to be Gaussian, then the optical bandwidth is a factor of $\sqrt{2}$ greater than the electrical bandwidth [Ref. 40].

The modulation bandwidth of LEDs is generally determined by three mechanisms. These are:

(a) the doping level in the active layer;
(b) the reduction in radiative lifetime due to the injected carriers;
(c) the parasitic capacitance of the device.

Assuming negligible parasitic capacitance, the speed at which an LED can be directly current modulated is fundamentally limited by the recombination lifetime of the carriers, where the optical output power $P_e(\omega)$ of the device (with constant peak current) and angular modulation frequency ω is given by [Ref. 41],

$$\frac{P_e(\omega)}{P_{dc}} = \frac{1}{[1 + (\omega\tau_i)^2]^{\frac{1}{2}}} \tag{7.24}$$

where τ_i is the injected (minority) carrier lifetime in the recombination region and P_{dc} is the d.c. optical output power for the same drive current.

Example 7.7

The minority carrier recombination lifetime for an LED is 5 ns. When a constant d.c. drive current is applied to the device the optical output power is 300 μW. Determine the optical output power when the device is modulated with an rms drive current corresponding to the d.c drive current at frequencies of (a) 20 MHz; (b) 100 MHz.

It may be assumed that parasitic capacitance is negligible. Further, determine the 3 dB optical bandwidth for the device and estimate the 3 dB electrical bandwidth assuming a Gaussian response.

Solution: (a) From Eq. (7.24), the optical output power at 20 MHz is:

$$P_e(20\ \text{MHz}) = \frac{P_{dc}}{[1 + (\omega\tau_i)^2]^{\frac{1}{2}}}$$

$$= \frac{300 \times 10^{-6}}{[1 + (2\pi \times 20 \times 10^6 \times 5 \times 10^{-9})^2]^{\frac{1}{2}}}$$

$$= \frac{300 \times 10^{-6}}{[1.39]^{\frac{1}{2}}}$$

$$= 254.2\ \mu\text{W}$$

(b) Again using Eq. (7.24):

$$P_e(100\ \text{MHz}) = \frac{300 \times 10^{-6}}{[1 + (2\pi \times 100 \times 10^6 \times 5 \times 10^{-9})^2]^{\frac{1}{2}}}$$

$$= \frac{300 \times 10^{-6}}{[10.87]^{\frac{1}{2}}}$$

$$= 90.9\ \mu\text{W}$$

This example illustrates the reduction in the LED optical output power as the device is driven at higher modulating frequencies. It is therefore apparent that there is a somewhat limited bandwidth over which the device may be usefully utilized.

To determine the optical 3 dB bandwidth: the high frequency 3 dB point occurs when $P_e(\omega)/P_{dc} = \frac{1}{2}$. Hence, using Eq. (7.24):

$$\frac{1}{[1 + (\omega\tau_i)^2]^{\frac{1}{2}}} = \frac{1}{2}$$

and $1 + (\omega\tau_i)^2 = 4$. Therefore $\omega\tau_i = \sqrt{3}$, and

$$f = \frac{\sqrt{3}}{2\pi\tau} = \frac{\sqrt{3}}{\pi \times 10^{-8}} = 55.1\ \text{MHz}$$

Thus the 3 dB optical bandwidth B_{opt} is 55.1 MHz as the device similar to all LEDs operates down to d.c.

Assuming a Gaussian frequency response, the 3 dB electrical bandwidth B will be:

$$B = \frac{55.1}{\sqrt{2}} = 39.0\ \text{MHz}$$

Thus the corresponding electrical bandwidth is 39 MHz. However, it must be remembered that parasitic capacitance may reduce the modulation bandwidth below this value.

The carrier lifetime is dependent on the doping concentration, the number of injected carriers into the active region, the surface recombination velocity and the thickness of the active layer. All these parameters tend to be interdependent and are adjustable within limits in present-day technology. In general, the carrier lifetime may be shortened by either increasing the active layer doping or by decreasing the thickness of the active layer. However, in surface emitters this can reduce the external power efficiency of the device due to the creation of an increased number of nonradiative recombination centres.

Edge-emitting LEDs have a very thin, virtually undoped active layer and the carrier lifetime is controlled only by the injected carrier density. At high current densities the carrier lifetime decreases with injection level because of a bimolecular recombination process [Ref. 41]. This bimolecular recombination process allows edge-emitting LEDs with narrow recombination regions to have short recombination times, and therefore relatively high modulation capabilities at reasonable operating current densities. For instance, edge-emitting devices with electrical modulation bandwidths of 145 MHz have been achieved with moderate doping and extremely thin (approximately 50 nm) active layers [Ref. 42].

However, LEDs tend to be slower devices with significantly lower output powers than injection lasers because of the longer lifetime of electrons in their donor regions resulting from spontaneous recombination rather than stimulated emission,* coupled with the increased numbers of nonradiative centres at higher doping levels. Thus at high modulation bandwidths the optical output power from conventional LED structures decreases as illustrated in Example 7.7 and also as shown in Figure 7.22.

The reciprocal relationship between modulation bandwidth and output power may be observed in Figure 7.22 which illustrates experimental results obtained with both AlGaAs and InGaAsP LEDs [Refs. 43, 44]. The solid line gives an indication of the best results for AlGaAs LEDs, whereas, for comparison, the dotted line represents these AlGaAs results shifted by the ratio of the photon energy at 0.85 μm to that at 1.3 μm. Finally, the dashed line provides a contour of the best reported results for InGaAsP LEDs. It may be observed that the output power from AlGaAs LEDs is a factor of two higher than that of InGaAsP devices at all bandwidths which partly results from the photon energy at the 1.3 μm wavelength being smaller (by a factor of 1.53) than that at 0.85 μm. Hence the centre dotted line displays the adjustment of the AlGaAs LEDs for this factor showing that the best performance of InGaAsP LEDs is not far below that of AlGaAs LEDs. Moreover the difference is probably due to the more advanced technology which is available for the latter devices combined with the enhanced wavelength saturation in the longer wavelength material [Ref. 13].

For surface-emitting AlGaAs LEDs high output power of 15 mW has been obtained at modest bandwidths (17 MHz) [Ref. 45], whereas the very large

* The superluminescent LED is an exception in this respect and is therefore capable of high output power at relatively high modulation bandwidths (see Section 7.3.5).

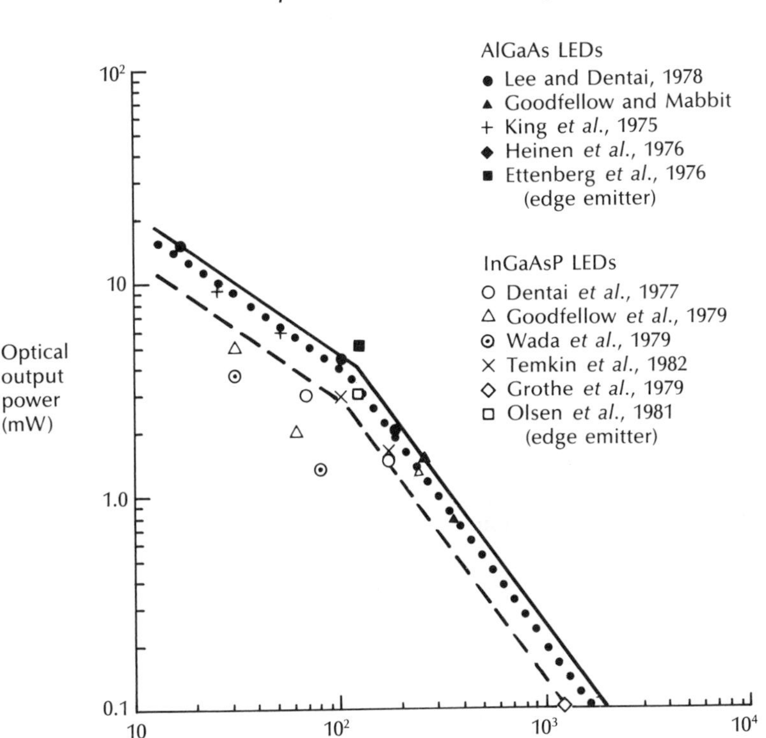

Figure 7.22 Reported optical output power against bandwidth for both AlGaAs and InGaAsP LEDs [Refs. 43, 44]. Best results for AlGaAs devices (solid line). These AlGaAs LED results shifted to a wavelength of 1.3 μm (dotted line). Best results for InGaAsP devices (dashed line).

bandwidth of 1.1 GHz was only achieved at the far lower output power of 0.2 mW [Ref. 46]. In general terms, to maximize the output power from SLEDs exhibiting low modulation bandwidths in the range 20 to 50 MHz, a thick active layer (2 to 2.5 μm) with low doping levels (less than 5×10^{17} cm^{-3}) can be employed. Thinner active layers (1 to 1.5 μm) and higher doping levels (0.5 to 1.0×10^{-18} cm^{-3}) are required for devices operating in the 50 to 100 MHz bandwidth region. In order to increase the modulation bandwidth into and beyond the 100 to 200 MHz range, however, very high doping levels in excess of 5×10^{18} cm^{-3} are necessary in combination with thin active layers.

Longer wavelength LEDs fabricated from the InGaAsP/InP material system for operation at a wavelength around 1.3 μm have become widely commercially available. Such devices with undoped (i.e. with a residual n type concentration between 1×10^{-17} and 5×10^{17} cm^{-3}) active layers provide modulation bandwidths in the range 50 to 100 MHz [Ref. 39]. Moreover, with higher doping densities

(5×10^{18} cm^{-3}) and relatively thin active layers (400 nm), bandwidths of 690 MHz have been obtained [Ref. 47]. Modulation rates in the range 600 Mbit s^{-1} to 1.2 Gbit s^{-1} have also been achieved using high levels of Zn doping (1×10^{-19} to 1.3×10^{-19} cm^{-3}) in InGaAsP devices [Refs. 21, 48].

7.4.4 Reliability

LEDs are not generally affected by the catastrophic degradation mechanisms which can severely affect injection lasers (see Section 6.7.6). Early or infant failures do, however, occur as a result of random and not always preventable fabricational defects. Such failures can usually be removed from the LED batch population over an initial burn-in operational period [Ref. 49]. In addition, LEDs do exhibit gradual degradation which may take the form of a rapid degradation mode* or a slow degradation mode.

Rapid degradation in LEDs is similar to that in injection lasers, and is due to both the growth of dislocations and precipitate-type defects in the active region giving rise to dark line defects (DLDs) and dark spot defects (DSDs), respectively, under device ageing [Ref. 43]. DLDs tend to be the dominant cause of rapid degradation in GaAs-based LEDs. The growth of these defects does not depend upon substrate orientation but on the injection current density, the temperature, and the impurity concentration in the active layer.

Good GaAs substrates have dislocation densities around 5×10^4 cm^{-2}. Hence, there is less probability of dislocations in devices with small active regions. DSDs, and the glide of existing misfit dislocations, however, predominate as the cause of rapid degradation in InP-based LEDs.

LEDs may be fabricated which are largely free from these defects and are therefore subject to a slower long term degradation process. This homogeneous degradation is thought to be due to recombination enhanced point defect generation (i.e. vacancies and interstitials), or the migration of impurities into the active region [Ref. 50]. The optical output power $P_e(t)$ may be expressed as a function of the operating time t, and is given by [Ref. 50]:

$$P_e(t) = P_{out} \exp(-\beta_r t) \tag{7.25}$$

where P_{out} is the initial output power and β_r is the degradation rate. The degradation rate is characterized by the activation energy of homogeneous degradation E_a and is a function of temperature. It is given by:

$$\beta_r = \beta_0 \exp[-E_a/KT] \tag{7.26}$$

where β_0 is a proportionality constant, K is Boltzmann's constant and T is the absolute temperature of the emitting region. The activation energy E_a is a variable which is dependent on the material system and the structure of the device. The value

* LEDs which display rapid degradation are sometimes referred to as freak failures [Ref. 49] because they pass the burn-in period but fail earlier in operational life than the main device population.

of E_a is in the range 0.56 to 0.65 eV, and 0.9 to 1.0 eV for surface-emitting GaAs/AlGaAs and InGaAsP/InP LEDs respectively [Ref. 9]. These values suggest 10^6 to 10^7 hours (100 to 1000 years) CW operation at room temperature for AlGaAs devices, and in excess of 10^9 hours for surface-emitting InGaAsP LEDs.

Example 7.8

An InGaAsP surface emitter has an activation energy of 1 eV with a constant of proportionality (β_0) of 1.84×10^7 h^{-1}. Estimate the CW operating lifetime for the LED with a constant junction temperature of 17 °C, if it is assumed that the device is no longer useful when its optical output power has diminished to 0.67 of its original value.

Solution: Initially, it is necessary to obtain the degradation rate β_r, thus from Eq. (7.26):

$$\begin{aligned}\beta_r &= \beta_0 \exp[-E_a/KT] \\ &= 1.84 \times 10^7 \exp\left[\frac{-1 \times 1.602 \times 10^{-19}}{1.38 \times 10^{-23} \times 290}\right] \\ &= 1.84 \times 10^7 \exp[-40] \\ &= 7.82 \times 10^{-11}\ \text{h}^{-1}\end{aligned}$$

Now using Eq. (7.25):

$$\frac{P_e(t)}{P_{out}} = \exp(-\beta_r t) = 0.67$$

Therefore

$$\beta_r t = -\ln 0.67$$

and

$$\begin{aligned}t &= \frac{\ln 0.67}{7.82 \times 10^{-11}} = \frac{0.40}{7.82 \times 10^{-11}} \\ &= 5.1 \times 10^9\ \text{h}\end{aligned}$$

Hence the estimated lifetime of the device under the specified conditions in Example 7.8 is 5.1×10^9 hours. It must be noted that the junction temperature, even for a device operating at room temperature, is likely to be well in excess of room temperature when substantial drive currents are passed. Also the diminished level of optical output in the example is purely arbitrary and for many applications this reduced level may be unacceptable.

Nevertheless it is quite common for the device lifetime or median life to be determined for a 50% drop in light output power from the device [Ref. 49]. It is clear, however, that with the long term LED degradation process there is no absolute end-of-life power level and therefore to a large extent it is system dependent such that a trade-off can be made between the required system end-of-life power margin and the device reliability [Ref. 44]. Hence the allocated drop to end-of-life power can be substantially reduced to, say, 20% which will provide for an enhanced system power margin (e.g. increased repeater spacing) at the expense of the device median life. Overall, even with these more rigorous conditions the anticipated median life for such LEDs is excellent and it is unlikely to cause problems in most optical fiber communication system applications.

Extrapolated accelerated lifetime tests are also in broad agreement with the theoretical estimates [Refs. 47, 49 to 53] for the less sophisticated device structures. For example, a planar GaAs/AlGaAs DH LED exhibited a median life for a 50% output power reduction of 9×10^7 hours at a temperature of 25 °C [Ref. 52]. By comparison, extrapolated half-power lifetimes in excess of 10^8 hours at a temperature of 60 °C have been obtained with high speed (greater than 200 Mbit s^{-1}) InGaAsP/InP LEDs [Ref. 53].

7.5 Modulation

In order to transmit information via an optical fiber communication system it is necessary to modulate a property of the light with the information signal. This property may be intensity, frequency, phase or polarization (direction) with either digital or analog signals. The choices are indicated by the characteristics of the optical fiber, the available optical sources and detectors, and considerations of the overall system.

However, at present in optical fiber communications considerations of the above for practical systems tend to dictate some form of intensity modulation of the source. Although much effort has been expended and considerable success has been achieved in the area of coherent optical communications (see Chapter 12) the widescale deployment of such systems will take some further time. Therefore intensity modulation (IM) of the optical source and envelope or direct detection (DD) at the optical receiver is likely to remain the major modulation strategy* in the immediate future.

Intensity modulation is easy to implement with the electroluminescent sources available at present (LEDs and injection lasers). These devices can be directly modulated simply by variation of their drive currents at rates up to several gigahertz. Thus direct modulation of the optical source is satisfactory for the modulation bandwidths currently under investigation. However, considering the recent interest in integrated optical devices (see Chapter 10) it is likely that external

* This strategy is often referred to as intensity modulation/direct detection, or IM/DD.

optical modulators [Ref. 54] may be utilized more in the future in order to achieve greater bandwidths and to allow the use of nonsemiconductor sources (e.g. Nd : YAG laser) which cannot be directly modulated at high frequency (see Section 6.9.1). External optical modulators are active devices which tend to be used primarily to modulate the frequency or phase of the light, but may also be used for time division multiplexing and switching of optical signals. However, modulation considerations within this text (excepting Chapter 12) will mainly be concerned with the direct modulation of the intensity of the optical source.

Intensity modulation may be utilized with both digital and analog signals. Analog intensity modulation is usually easier to apply but requires comparatively large signal to noise ratios (see Section 9.2.5) and therefore it tends to be limited to relatively narrow bandwidth, short distance applications. Alternatively, digital intensity modulation gives improved noise immunity but requires wider bandwidths, although these may be small in comparison with the available bandwidth. It is therefore ideally suited to optical fiber transmission where the available bandwidth is large. Hence at present most fiber systems in the medium to long distance range use digital intensity modulation.

Problems

7.1 Describe with the aid of suitable diagrams the mechanism giving the emission of light from an LED. Discuss the effects of this mechanism on the properties of the LED in relation to its use as an optical source for communications.

7.2 Briefly outline the advantages and drawbacks of the LED in comparison with the injection laser for use as a source in optical fiber communications.

7.3 The power generated internally within a double heterojunction LED is 28.4 mW at a drive current of 60 mA. Determine the peak emission wavelength from the device when the radiative and nonradiative recombination lifetimes of the minority carriers in the active region are equal.

7.4 The diffusion length L_D or the average distance moved by charge carriers before recombination in the active region of an LED is given by:

$$L_D = (D\tau)^{\frac{1}{2}}$$

where D is the diffusion coefficient and τ is the total carrier recombination lifetime. Calculate the diffusion coefficient in gallium arsenide when the diffusion length is 21 μm and the radiative and nonradiative carrier recombination lifetimes are equal at 90 ns.

7.5 Estimate the external power efficiency of a GaAs planar LED when the transmission factor of the GaAs–air interface is 0.68 and the internally generated optical power is 30% of the electrical power supplied. The refractive index of GaAs may be taken as 3.6.

7.6 The external power efficiency of an InGaAsP/InP planar LED is 0.75% when the internally generated optical power is 30 mW. Determine the transmission factor for the InP–air interface if the drive current is 37 mA and the potential difference across the device is 1.6 V. The refractive index of InP may be taken as 3.46.

7.7 A GaAs planar LED emitting at a wavelength of 0.85 μm has an internal quantum efficiency of 60% when passing a forward current of 20 mA s^{-1}. Estimate the optical power emitted by the device into air, and hence determine the external power efficiency if the potential difference across the device is 1 V. It may be assumed that the transmission factor at the GaAs–air interface is 0.68 and that the refractive index of GaAs is 3.6. Comment on any assumptions made.

7.8 The external power efficiency of a planar GaAs LED is 1.5% when the forward current is 50 mA and the potential difference across its terminals is 2 V. Estimate the optical power generated within the device if the transmission factor at the coated GaAs–air interface is 0.8.

7.9 Outline the common LED structures for optical fiber communications discussing their relative merits and drawbacks. In particular, compare surface- and edge-emitting devices. Comment on the distinction between multimode and single-mode devices.

7.10 Derive an expression for the coupling efficiency of a surface-emitting LED into a step index fiber, assuming the device to have a Lambertian output. Determine the optical loss in decibels when coupling the optical power emitted from the device into a step index fiber with an acceptance angle of $14°$. It may be assumed that the LED is smaller than the fiber core and that the two are in close proximity.

7.11 Considering the LED of problem 7.5, calculate:

(a) the coupling efficiency and optical loss in decibels of coupling the emitted light into a step index fiber with an NA of 0.15, when the device is in close proximity to the fiber and is smaller than the fiber core;

(b) the optical loss relative to the optical power generated internally if the device emits into a thin air gap before light is coupled into the fiber.

7.12 Estimate the optical power coupled into a 50 μm diameter core step index fiber with an NA of 0.18 from a DH surface emitter with an emission area diameter of 75 μm and a radiance of 60 W sr^{-1} cm^{-2}. The Fresnel reflection at index matched semiconductor–fiber interface may be considered negligible.

Further, determine the optical loss when coupling light into the fiber relative to the power emitted by the device into air if the Fresnel reflection at the semiconductor–air interface is 30%.

7.13 Comment on the differences in the performance characteristics between the conventional LEDs used for optical fiber communications and superluminescent LEDs.

Describe, with the aid of a diagram the structure of an SLD used for operation in the longer wavelength region and suggest potential application areas for such devices.

7.14 The Fresnel reflection coefficient at a fiber core of refractive index n_1 is given approximately from the classical Fresnel formulae by

$$r = \left[\frac{n_1 - n}{n_1 + n}\right]^2$$

where n is the refractive index of the surrounding medium.

(a) Estimate the optical loss due to Fresnel reflection at a fiber core from GaAs each of which have refractive indices of 1.5 and 3.6 respectively.

(b) Calculate the optical power coupled into a 200 μm diameter core step index fiber with an NA of 0.3 from a GaAs surface-emitting LED with an emission diameter of 90 μm and a radiance of 40 W sr^{-1} cm^{-2}. Comment on the result.

(c) Estimate the optical power emitted into air for the device in (b).

7.15 Determine the overall power conversion efficiency for the LED in Problem 7.14 if it is operating with a drive current of 100 mA and a forward voltage of 1.9 V.

7.16 Discuss lens coupling of LEDs to optical fibers and outline the various techniques employed.

7.17 Discuss the relationship between the electrical and optical modulation bandwidths for an optical fiber communication system. Estimate the 3 dB optical bandwidth corresponding to a 3 dB electrical bandwidth of 50 MHz. A Gaussian frequency response may be assumed.

7.18 Determine the optical modulation bandwidth for the LED of Problem 7.14 if the device emits 840 μW of optical power into air when modulated at a frequency of 150 MHz.

7.19 Estimate the electrical modulation bandwidth for an LED with a carrier recombination lifetime of 8 ns. The frequency response of the device may be assumed to be Gaussian.

7.20 Discuss the reliability of LEDs in comparison with injection lasers.

Estimate the CW operating lifetime for an AlGaAs LED with an activation energy of 0.6 eV and a constant of proportionality (β_0) of 2.3×10^3 h^{-1} when the junction temperature of the device is constant at 50 °C. It may be assumed that the LED is no longer useful when its optical output power is 0.8 of its original value.

7.21 What is meant by the intensity modulation of an optical source? Give reasons for the major present use of direct intensity modulation of semiconductor optical sources and comment on possible alternatives.

Answers to numerical problems

7.3 1.31 μm
7.4 9.8×10^{-3} m s^{-1}
7.5 0.4%
7.6 0.70
7.7 230 μW, 1.15%
7.8 97.2 mW
7.10 12.3 dB
7.11 (a) 16.7 dB
(b) 35.2 dB
7.12 0.12 mW, 16.9 dB
7.14 (a) 0.81 dB
(b) 600 μW
(c) 5.44 mW
7.15 0.32%
7.17 70.7 MHz
7.18 40.6 MHz
7.19 24.4 MHz
7.20 2.21×10^5 hours

References

[1] D. N. Christodoulides, L. A. Reith and M. A. Saifi, 'Theory of LED coupling to single-mode fibers', *J. Lightwave Technol.*, **LT-5**(11), pp. 1623–1629, 1987.

[2] L. Hafskjaer and A. S. V. Sudbo, 'Attenuation and bit-rate limitations in LED/single-mode fiber transmission systems', *J. Lightwave Technol.*, **6**(12), pp. 1793–1797, 1988.

[3] T. P. Lee and A. G. Dentai, 'Power and modulation bandwidth of GaAs–AlGaAs high radiance LEDs for optical communication systems', *IEEE J. Quantum Electron.*, **QE-14**(3), pp. 150–156, 1978.

[4] G. Keiser, *Optical Fiber Communications*, McGraw-Hill, 1983.

[5] H. Kressel, 'Electroluminescent sources for fiber systems', in M. K. Barnoski (Ed.), *Fundamentals of Optical Fiber Communications*, pp. 109–141, Academic Press, 1976.

[6] R. C. Goodfellow and R. Davis, 'Optical source devices', in M. J. Howes and D. V. Morgan (Eds.), *Optical Fibre Communications*, pp. 27–106, John Wiley, 1980.

[7] J. P. Wittke, M. Ettenburg and H. Kressel, 'High radiance LED for single fiber optical links', *RCA Rev.*, **37**(2), pp. 160–183, 1976.
[8] T. G. Giallorenzi, 'Optical communications research and technology: fiber optics', *Proc. IEEE*, **66**, pp. 744–780, 1978.
[9] A. A. Bergh and P. J. Dean, *Light-Emitting Diodes*, Oxford University Press, 1976.
[10] C. A. Burrus and R. W. Dawson, 'Small area high-current density GaAs electroluminescent diodes and a method of operation for improved degradation characteristics', *Appl. Phys. Lett.*, **17**(3), pp. 97–99, 1970.
[11] C. A. Burrus and B. I. Miller, 'Small-area double heterostructure aluminum-gallium arsenide electroluminescent diode sources for optical fiber transmission lines', *Opt. Commun.*, **4**, pp. 307–369, 1971.
[12] T. P. Lee, 'Recent developments in light emitting diodes for optical fiber communication systems', *Proc. SPIE Int. Soc. Opt. Eng.* (USA), **224**, pp. 92–101, 1980.
[13] T. P. Lee, C. A. Burrus Jr and R. H. Saul, 'Light-emitting diodes for telecommunications', in S. E. Miller and I. P. Kaminow (Eds.) *Optical Fiber Telecommunications II*, Academic Press, pp. 467–507, 1988.
[14] T. Uji and J. Hayashi, 'High-power single-mode optical-fiber coupling to InGaAsP 1.3 μm mesa-structure surface emitting LEDs', *Electron. Lett.*, **21**(10), pp. 418–419, 1985.
[15] D. Gloge, 'LED design for fibre system', *Electron. Lett.*, **13**(4), pp. 399–400, 1977.
[16] D. Marcuse, 'LED fundamentals: Comparison of front and edge emitting diodes', *IEEE J. Quantum Electron.*, **QE-13**(10), pp. 819–827, 1977.
[17] D. Botez and M. Ettenburg, 'Comparison of surface and edge emitting LEDs for use in fiber-optical communications', *IEEE Trans. Electron. Devices*, **ED-26**(3), pp. 1230–1238, 1979.
[18] M. Ettenburg, H. Kressel and J. P. Wittke, 'Very high radiance edge-emitting LED', *IEEE J. Quantum Electron.*, **QE-12**(6), pp. 360–364, 1979.
[19] D. H. Newman, M. R. Matthews and I. Garrett, 'Sources for optical fiber communications', *Telecommun. J. (Eng. Ed.) Switzerland*, **48**(2), pp. 673–680, 1981.
[20] D. H. Newman and S. Ritchie, 'Sources and detectors for optical fiber communications applications: the first 20 years', *IEE Proc. Pt. J*, **133**(3), pp. 213–228, 1986.
[21] S. Fujita, J. Hayashi, Y. Isoda, T. Uji and M. Shikada, '2 Gbit/s and 600 Mbit/s single-mode fibre transmission experiments using a high speed Zn-doped 1.3 μm edge-emitting LED', *Electron. Lett.*, **13**(12), pp. 636–637, 1987.
[22] D. M. Fye, 'Low-current 1.3 μm edge-emitting LED for single-mode fiber subscriber loop applications', *J. Lightwave Technol.*, **LT-4**(10), pp. 1546–1551, 1986.
[23] T. Ohtsuka, N. Fujimoto, K. Yamaguchi, A. Taniguchi, N. Naitou and Y. Nabeshima, 'Gigabit single-mode fiber transmission using 1.3 μm edge-emitting LEDs for broadband subscriber loops', *J. Lightwave Technol.*, **LT-5**(10), pp. 1534–1541, 1987.
[24] S. Takahashi, K. Goto, T. Shiba, K. Yoshida, E. Omura, H. Namizaki and W. Susaki, 'High-coupled-power high-speed 1.3 μm edge-emitting LED with buried crescent structure on p-InP substrate', *Tech. Dig. Optical Fiber Communications Conf., OFC 88* (USA), paper WB5, January 1988.
[25] Y. Sakakibara, H. Higuchi, E. Oomura, Y. Nakajima, Y. Yamamoto, K. Goto, H. Namizaki, K. Ikeda and W. Susaki, 'High-power 1.3 μm InGaAsP p-substrate buried cresent lasers', *J. Lightwave Technol.*, **LT-3**(5), pp. 978–984, 1985.
[26] T. P. Lee, C. A. Burrus and B. I. Miller, 'A Stripe-geometry double-heterostructure

amplified-spontaneous-emission (superluminescent) diode', *IEEE J. Quantum Electron.*, **QE-9**, p. 820, 1973.

[27] I. P. Kaminow, G. E. Eisenstein, L. W. Stulz and A. G. Dentai, 'Lateral confinement InGaAsP superluminescent diode at 1.3 μm', *IEEE J. Quantum Electron.*, **QE-19**(1), pp. 78–82, 1983.

[28] Y. Kashima, M. Kobayashi and T. Takano, 'High output power GaInAsP/InP superluminescent diode at 1.3 μm', *Electron., Lett.*, **24**(24), pp. 1507–1508, 1988.

[29] G. Arnold, H. Gottsman, O. Krumpholz, E. Schlosser and E. A. Schurr, '1.3 μm edge-emitting diodes launching 250 μW into single-mode fiber at 100 mA', *Electron. Lett.*, **21**(21), pp. 993–994, 1985.

[30] M. Abe, I. Umebu, O. Hasegawa, S. Yamakoshi, T. Yamaoka, T. Kotani, H. Okada and H. Takamashi, 'Highly efficient long lived GaAlAs LEDs for fiber-optical communications', *IEEE Trans. Electron. Devices*, **ED-24**(7), pp. 990–994, 1977.

[31] R. A. Abram, R. W. Allen and R. C. Goodfellow, 'The coupling of light emitting diodes to optical fibres using sphere lenses', *J. Appl. Phys.*, **46**(8), pp. 3468–3474, 1975.

[32] O. Wada, S. Yamakoshi, A. Masayuki, Y. Nishitani and T. Sakurai, 'High radiance InGaAsP/InP lensed LEDs for optical communication systems at 1.2–1.3 μm, *IEEE J. Quantum Electron.*, **QE-17**(2), pp. 174–178, 1981.

[33] R. C. Goodfellow, A. C. Carter, I. Griffith and R. R. Bradley, 'GaInAsP/InP fast, high radiance, 1.05–1.3 μm wavelength LEDs with efficient lens coupling to small numerical aperture silica optical fibers', *IEEE Trans. Electron. Devices*, **ED-26**(8), pp. 1215–1220, 1979.

[34] J. Ure, A. C. Carter, R. C. Goodfellow and M. Harding, 'High power lens coupled 1.3 μm edge-emitting LED for long haul 140 Mb/s fiber optics systems', *IEEE Specialist Conf. on Light Emitting Diodes and Photodetectors* (Canada), paper 20, p. 204, 1982.

[35] R. H. Saul, W. C. King, N. A. Olsson, C. L. Zipfel, B. H. Chin, A. K. Chin, I. Camlibel and G. Minneci, '180 Mbit/s, 35 km transmission over single-mode fiber using 1.3 μm edge-emitting LEDs', *Electron. Lett.*, **21**(17), pp. 773–775, 1985.

[36] J. Straus, 'The nonlinearity of high-radiance light-emitting diodes', *IEEE J. Quantum Electron.*, **QE-14**(11), pp. 813–819, 1979.

[37] J. Straus, 'Linearized transmitters for analog fiber links', *Laser Focus (USA)*, **14**(10), pp. 54–61, 1978.

[38] A. C. Carter, 'Light-emitting diodes for optical fibre systems', *Radio Electron. Eng. J. IERE*, **51**(7/8), pp. 341–348, 1981.

[39] H. Tempkin, C. L. Zipfel, M. A. DiGiuseppe, A. K. Chin, V. G. Keramides and R. H. Saul, 'InGaAsP LEDs for 1.3 μm optical transmission', *Bell Syst. Tech. J.*, **62** (1) pp. 1–24, 1983.

[40] I. Garrett and J. E. Midwinter, 'Optical communication systems', in M. J. Howes and D. V. Morgan (Eds.), *Optical Fibre Communications*, pp. 251–300, John Wiley, 1980.

[41] H. Kressel and J. K. Butler, *Semiconductor Lasers and Heterojunction LEDs*, Academic Press, 1977.

[42] H. F. Lockwood, J. P. Wittke and M. Ettenburg, 'LED for high data rate, optical communications', *Opt. Commun.*, **16**, p. 193, 1976.

[43] T. P. Lee, 'Recent development in light emitting diodes (LEDs) for optical fiber communications systems', *Proc. SPIE Int. Soc. Opt. Eng. (USA)*, **340**, pp. 22–31, 1982.

[44] R. H. Saul, 'Recent advances in the performance and reliability of InGaAsP LEDs for lightwave communication systems', *IEEE Trans. Electron. Devices*, **ED-30**(4), pp. 285–295, 1983.

[45] T. P. Lee and A. G. Dentai, 'Power and modulation bandwidth of GaAs–AlGaAs high radiance LEDs for optical communication systems', *IEEE J. Quantum Electron.*, **QE-14**, pp. 150–159, 1978.

[46] J. Heinen, W. Huber and W. Harth, 'Light-emitting diodes with a modulation bandwidth of more than 1 GHz', *Electron. Lett.*, **12**, p. 533, 1976.

[47] W. C. King, B. H. Chin, I. Camlibel and E. L. Zipfel, 'High-speed high-power 1.3 μm InGaAsP/InP surface emitting LEDs for short-haul wide-bandwidth optical fiber communications', *IEEE Electron. Device Lett.*, **EDL-6**, p. 335, 1985.

[48] A. Suzuki, Y. Inomoto, J. Hayashi, Y. Isoda, T. Uji and H. Nomura, 'Gbit/s modulation of heavily Zn-doped surface-emitting InGaAsP/InP DH LED', *Electron. Lett.*, **20**, p. 274, 1984.

[49] N. K. Dutta and C. L. Zipfel, 'Reliability of lasers and LEDs', in S. E. Miller and I. P. Kaminow (Eds.), *Optical Fiber Telecommunications II*, Academic Press, pp. 671–687, 1988.

[50] S. Yamakoshi, A. Masayuki, O. Wada, S. Komiya and T. Sakurai, 'Reliability of high radiance InGaAsP/InP LEDs operating in the 1.2–1.3 μm wavelength', *IEEE J. Quantum Electron.*, **QE-17**(2), pp. 167–173, 1981.

[51] S. Yamakoshi, T. Sugahara, O. Hasegawa, Y. Toyama and H. Takanashi, 'Growth mechanism of ⟨100⟩ dark-line defects in high radiance GaAlAs LEDs', *International Electronic Devices Meeting*, pp. 642–645, 1978.

[52] C. L. Zipfel, A. K. Chin, V. G. Keramidas and R. H. Saul, 'Reliability of DH $Ga_{1-x}Al_xAs$ LEDs for lightwave communications', *Proc. 19th Ann., IEEE Int. Reliab. Phys. Symp.*, pp. 124–129, 1981.

[53] A. Suzuki, T. Uji, Y. Inomoto, J. Hayashi, Y. Isoda and H. Nomura, 'InGaAsP/InP 1.3 μm wavelength surface-emitting LED's for high-speed short-haul optical communication systems', *J. of Lightwave Technol.*, **LT-3**(6), pp. 1217–1222, 1985.

[54] S. M. Stone, 'Modulation of optical sources', in E. E. Basch (Ed.) *Optical-Fiber Transmission*, H. W. Sams & Co., pp. 303–334, 1987.

8

Optical detectors

8.1 Introduction

We are concerned in this chapter with photodetectors currently in use and under investigation for optical fiber communications.

The detector is an essential component of an optical fiber communication system and is one of the crucial elements which dictate the overall system performance. Its function is to convert the received optical signal into an electrical signal, which is then amplified before further processing. Therefore when considering signal attenuation along the link, the system performance is determined at the detector. Improvement of detector characteristics and performance thus allows the installation of fewer repeater stations and lowers both the capital investment and maintenance costs.

The role the detector plays demands that it must satisfy very stringent requirements for performance and compatibility. The following criteria define the important performance and compatibility requirements for detectors which are generally similar to the requirements for sources.

1. *High sensitivity at the operating wavelengths.* The first generation systems have wavelengths between 0.8 and 0.9 μm (compatible with AlGaAs laser and LED emission lines). However, considerable advantage may be gained at the detector from second generation sources with operating wavelengths above 1.1 μm as both fiber attenuation and dispersion are reduced. There is much research activity at present in this longer wavelength region, especially concerning wavelengths around 1.3 μm where attenuation and material dispersion can be minimized. In this case semiconductor materials are currently under investigation (see Section 8.4.3) in order to achieve good sensitivity at normal operating temperatures (i.e. 300 K).
2. *High fidelity.* To reproduce the received signal waveform with fidelity, for analogy transmission the response of the photodetector must be linear with regard to the optical signal over a wide range.
3. *Large electrical response to the received optical signal.* The photodetector should produce a maximum electrical signal for a given amount of optical power, i.e. the quantum efficiency should be high.
4. *Short response time to obtain a suitable bandwidth.* Present systems extend into the hundreds of megahertz. However, it is apparent that future systems (single-mode fiber) will operate in the gigahertz range, and possibly above.
5. *A minimum noise introduced by the detector.* Dark currents, leakage currents and shunt conductance must be low. Also the gain mechanism within either the detector or associated circuitry must be of low noise.
6. *Stability of performance characteristics.* Ideally, the performance characteristics of the detector should be independent of changes in ambient conditions. However, the detectors currently favoured (photodiodes) have characteristics (sensitivity, noise, internal gain) which vary with temperature, and therefore compensation for temperature effects is often necessary.
7. *Small size.* The physical size of the detector must be small for efficient coupling to the fiber and to allow easy packaging with the following electronics.
8. *Low bias voltages.* Ideally the detector should not require excessive bias voltages or currents.
9. *High reliability.* The detector must be capable of continuous stable operation at room temperature for many years.

10. *Low cost*. Economic considerations are often of prime importance in any large scale communication system application.

We continue the discussion in Section 8.2 by briefly indicating the various types of device which could be employed for optical detection. From this discussion it is clear that semiconductor photodiodes currently provide the best solution for detection in optical fiber communications. Therefore, in Sections 8.3 and 8.4 we consider the principles of operation of these devices, together with the characteristics of the semiconductor materials employed in their construction. Sections 8.5–8.7 then briefly outline the major operating parameters (quantum efficiency, responsivity, long wavelength cutoff) of such photodiodes. Following, in Sections 8.8 and 8.9, we discuss the structure, operation and performance characteristics of the major device types (*p–n*, *p–i–n* and avalanche photodiodes) for optical detection over the wavelength range 0.8 to 1.6 μm. Then in Section 8.10 recent developments associated with photodiodes for mid-infrared detection (particularly up to 2.6 μm) are considered prior to discussion in Sections 8.11 and 8.12 of other semiconductor devices (heterojunction phototransistors and photoconductive detectors) which may eventually find wider use as detectors for optical fiber communications.

8.2 Device types

To detect optical radiation (photons) in the near-infrared region of the spectrum, both external and internal photoemission of electrons may be utilized. External photoemission devices typified by photomultiplier tubes and vacuum photodiodes meet some of the performance criteria but are too bulky, and require high voltages for operation. However, internal photoemission devices especially semiconductor photodiodes with or without internal (avalanche) gain provide good performance and compatibility with relatively low cost. These photodiodes are made from semiconductors such as silicon, germanium and an increasing number of III–V alloys, all of which satisfy in various ways most of the detector requirements. They are therefore used in all major current optical fiber communication systems.

The internal photoemission process may take place in both intrinsic and extrinsic semiconductors. With intrinsic absorption, the received photons excite electrons from the valence to the conduction bands in the semiconductor, whereas extrinsic absorption involves impurity centres created within the material. However, for fast response coupled with efficient absorption of photons, the intrinsic absorption process is preferred and at present all detectors for optical fiber communications use intrinsic photodetection.

Silicon photodiodes [Ref. 1] have high sensitivity over the 0.8–0.9 μm wavelength band with adequate speed (hundreds of megahertz), negligible shunt conductance, low dark current and long term stability. They are therefore widely used in first generation systems and are currently commercially available. Their usefulness is limited to the first generation wavelength region as silicon has an indirect bandgap energy (see Section 8.4.1) of 1.14 eV giving a loss in response above 1.09 μm. Thus for second generation systems in the longer wavelength range 1.1–1.6 μm research is devoted to the investigation of semiconductor materials which have narrower bandgaps. Interest has focused on germanium and III–V alloys which give a good response at the longer wavelengths. Again, the performance characteristics of such devices has improved considerably over recent years and a wide selection of III–V alloy photodiodes as well as germanium photodiodes are now commercially available.

In addition to the development of advanced photodiode structures fabricated from III–V semiconductor alloys for operation at wavelengths of 1.3 and 1.55 μm, similar material systems are under investigation for use at the even longer wavelengths required for mid-infrared transmission (2 to 5 μm). Interest has also been maintained in other semiconductor detector types, namely, the heterojunction phototransistor and the photoconductive detector, both of which can be usefully fabricated from III–V alloy material systems. In particular, the latter device type has more recently found favour as a potential detector over the 1.1 to 1.6 μm wavelength range. Nevertheless, at present the primary operating wavelength regions remain 0.8 to 0.9 μm, 1.3 μm and 1.55 μm, with the major device types being the *p–i–n* and avalanche photodiodes. We shall therefore consider these devices in greater detail before discussing mid-infrared photodiodes, phototransistors and photoconductive detectors.

8.3 Optical detection principles

The basic detection process in an intrinsic absorber is illustrated in Figure 8.1 which shows a *p–n* photodiode. This device is reverse biased and the electric field developed across the *p–n* junction sweeps mobile carriers (holes and electrons) to their respective majority sides (*p* and *n* type material). A depletion region or layer is therefore created on either side of the junction. This barrier has the effect of stopping the majority carriers crossing the junction in the opposite direction to the field. However, the field accelerates minority carriers from both sides to the opposite side of the junction, forming the reverse leakage current of the diode. Thus intrinsic conditions are created in the depletion region.

A photon incident in or near the depletion region of this device which has an energy greater than or equal to the bandgap energy E_g of the fabricating material (i.e. $hf \geqslant E_g$) will excite an electron from the valence band into the conduction band. This process leaves an empty hole in the valence band and is known as the photogeneration of an electron–hole (carrier) pair, as shown in Figure 8.1(a). Carrier pairs so generated near the junction are separated and swept (drift) under the influence of the electric field to produce a displacement by current in the external circuit in excess of any reverse leakage current (Figure 8.1(b)). Photo-

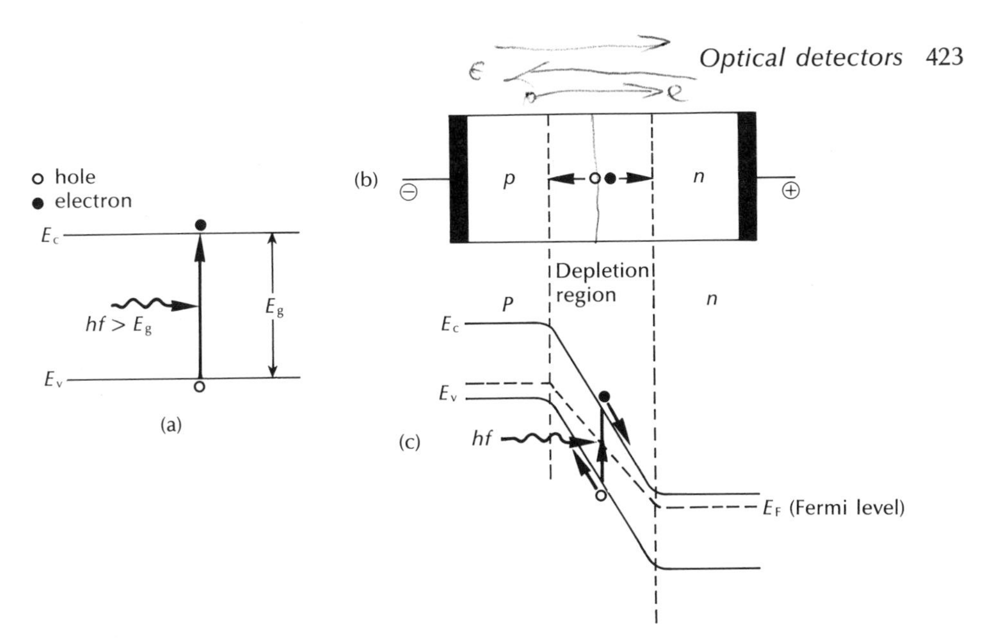

Figure 8.1 Operation of the *p–n* photodiode: (a) photogeneration of an electron–hole pair in an intrinsic semiconductor; (b) the structure of the reverse biased *p–n* junction illustrating carrier drift in the depletion region; (c) the energy band diagram of the reverse biased *p–n* junction showing photo-generation and the subsequent separation of an electron–hole pair.

generation and the separation of a carrier pair in the depletion region of this reverse biased *p–n* junction is illustrated in Figure 8.1(c).

The depletion region must be sufficiently thick to allow a large fraction of the incident light to be absorbed in order to achieve maximum carrier-pair generation. However, since long carrier drift times in the depletion region restrict the speed of operation of the photodiode it is necessary to limit its width. Thus there is a trade-off between the number of photons absorbed (sensitivity) and the speed of response.

8.4 Absorption

8.4.1 Absorption coefficient

The absorption of photons in a photodiode to produce carrier pairs and thus a photocurrent, is dependent on the absorption coefficient α_0 of the light in the semiconductor used to fabricate the device. At a specific wavelength and assuming only bandgap transitions (i.e. intrinsic absorber) the photocurrent I_p produced by incident light of optical power P_o is given by [Ref. 4]:

$$I_p = \frac{P_o e(1-r)}{hf}\,[1 - \exp(-\alpha_0 d)] \tag{8.1}$$

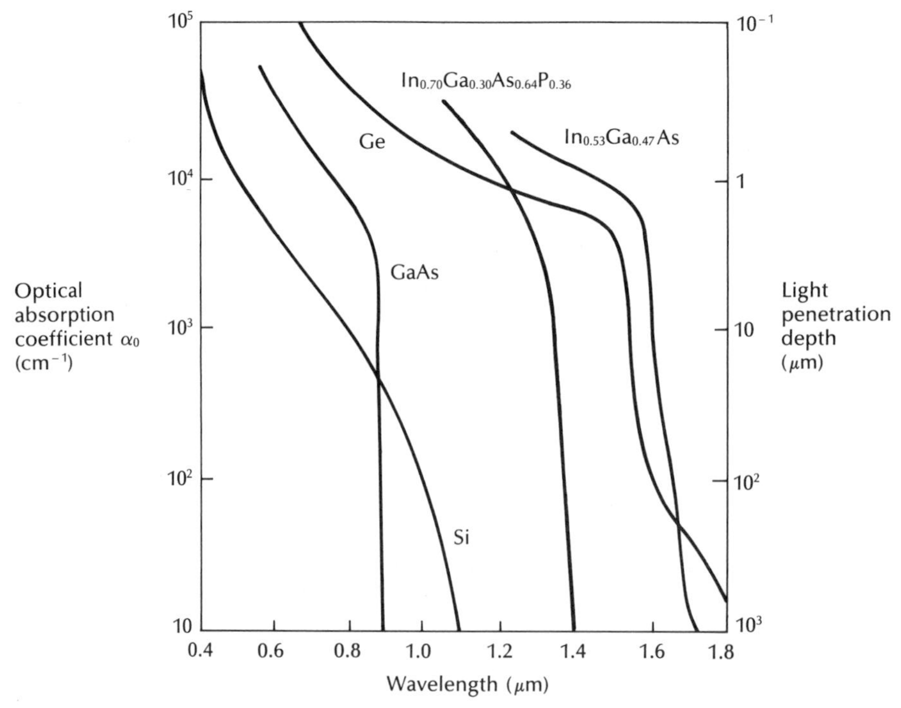

Figure 8.2 Optical absorption curves for some common semiconductor photodiode materials (silicon, germanium, gallium arsenide, indium gallium arsenide and indium gallium arsenide phosphide).

Table 8.1 Bandgaps for some semiconductor photodiode materials at 300 K

	Bandgap (eV) at 300 K	
	Indirect	Direct
Si	1.14	4.10
Ge	0.67	0.81
GaAs	—	1.43
InAs	—	0.35
InP	—	1.35
GaSb	—	0.73
$In_{0.53}Ga_{0.47}As$	—	0.75
$In_{0.14}Ga_{0.86}As$	—	1.15
$GaAs_{0.88}Sb_{0.12}$	—	1.15

where e is the charge on an electron, r is the Fresnel reflection coefficient at the semiconductor–air interface and d is the width of the absorption region.

The absorption coefficients of semiconductor materials are strongly dependent on wavelength. This is illustrated for some common semiconductors [Ref. 4] in Figure 8.2. It may be observed that there is a variation between the absorption curves for the materials shown and that they are each suitable for different wavelength applications. This results from their differing bandgaps energies, as show in Table 8.1. However, it must be noted that the curves depicted in Figure 8.2 also vary with temperature.

8.4.2 Direct and indirect absorption: silicon and germanium

Table 8.1 indicates that silicon and germanium absorb light by both direct and indirect optical transitions. Indirect absorption requires the assistance of a photon so that momentum as well as energy are conserved. This makes the transition probability less likely for indirect absorption than for direct absorption where no photon is involved. In this context direct and indirect absorption may be contrasted with direct and indirect emission discussed in Section 6.3.3.1. Therefore as may be seen from Figure 8.2 silicon is only weakly absorbing over the wavelength band of interest in optical fiber communications (i.e. first generation 0.8 to 0.9 μm). This is because transitions over this wavelength band in silicon are due only to the indirect absorption mechanism. As mentioned previously (Section 8.2) the threshold for indirect absorption occurs at 1.09 μm. The bandgap for direct absorption in silicon is 4.10 eV, corresponding to a threshold of 0.30 μm in the ultraviolet, and thus is well outside the wavelength range of interest.

Germanium is another semiconductor material for which the lowest energy absorption takes place by indirect optical transitions. However, the threshold for direct absorption occurs at 1.53 μm, below which germanium becomes strongly absorbing, corresponding to the kink in the characteristic shown in Figure 8.2. Thus germanium may be used in the fabrication of detectors over the whole of the wavelength range of interest (i.e. first and second generation 0.8 to 1.6 μm), especially considering that indirect absorption will occur up to a threshold of 1.85 μm.

Ideally, a photodiode material should be chosen with a bandgap energy slightly less than the photon energy corresponding to the longest operating wavelength of the system. This gives a sufficiently high absorption coefficient to ensure a good response, and yet limits the number of thermally generated carriers in order to achieve a low dark current (i.e. displacement current generated with no incident light (see Figure 8.5)). Germanium photodiodes have relatively large dark currents due to their narrow bandgaps in comparison to other semiconductor materials. This is a major disadvantage with the use of germanium photodiodes, especially at shorter wavelengths (below 1.1 μm).

8.4.3 III–V alloys

The drawback with germanium as a fabricating material for semiconductor photodiodes has led to increased investigation of direct bandgap III–V alloys for the longer wavelength region. These materials are potentially superior to germanium because their bandgaps can be tailored to the desired wavelength by changing the relative concentrations of their constituents, resulting in lower dark currents. They may also be fabricated in heterojunction structures (see Section 6.3.5) which enhances their high speed operations.

Ternary alloys such as InGaAs and GaAlSb deposited on InP and GaSb substrates, respectively, have been used to fabricate photodiodes for the longer wavelength band. Although difficulties were experienced in the growth of these alloys, with lattice matching causing increased dark currents, these problems have now been reduced. In particular the alloy $In_{0.53}Ga_{0.47}As$ lattice matched to InP, which responds to wavelengths up to around 1.7 μm (see Figure 8.2), has been extensively utilized in the fabrication of photodiodes for operation at both 1.3 and 1.55 μm. Quaternary alloys are also under investigation for detection at these wavelengths. Both InGaAsP grown on InP and GaAlAsSb grown on GaSb have been studied, with the former material system finding significant application within advanced photodiode structures.

8.5 Quantum efficiency

The quantum efficiency η is defined as the fraction of incident photons which are absorbed by the photodetector and generate electrons which are collected at the detector terminals:

$$\eta = \frac{\text{number of electrons collected}}{\text{number of incident photons}} \tag{8.2}$$

Hence,

$$\eta = \frac{r_e}{r_p} \tag{8.3}$$

where r_p is the incident photon rate (photons per second) and r_e is the corresponding electron rate (electrons per second).

One of the major factors which determines the quantum efficiency is the absorption coefficient (see Section 8.4.1) of the semiconductor material used within the photodetector.The quantum efficiency is generally less than unity as not all of the incident photons are absorbed to create electron–hole pairs. Furthermore, it should be noted that it is often quoted as a percentage (e.g. a quantum efficiency of 75% is equivalent to 75 electrons collected per 100 incident photons). Finally, in common with the absorption coefficient, the quantum efficiency is a function of the photon wavelength and must therefore only be quoted for a specific wavelength.

8.6 Responsivity

The expression for quantum efficiency does not involve photon energy and therefore the responsivity R is often of more use when characterizing the performance of a photodetector. It is defined as:

$$R = \frac{I_p}{P_o} \ (AW^{-1}) \tag{8.4}$$

where I_p is the output photocurrent in amperes and P_o is the incident optical power in watts. The responsivity is a useful parameter as it gives the transfer characteristic of the detector (i.e. photocurrent per unit incident optical power).

The relationship for responsivity (Eq. (8.4)) may be developed to include quantum efficiency as follows. Considering Eq. (6.1) the energy of a photon $E = hf$. Thus the incident photon rate r_p may be written in terms of incident optical power and the photon energy as:

$$r_p = \frac{P_o}{hf} \tag{8.5}$$

In Eq. (8.3) the electron rate is given by:

$$r_e = \eta r_p \tag{8.6}$$

Substituting from Eq. (8.5) we obtain

$$r_e = \frac{\eta P_o}{hf} \tag{8.7}$$

Therefore, the output photocurrent is:

$$I_p = \frac{\eta P_o e}{hf} \tag{8.8}$$

where e is the charge on an electron. Thus from Eq. (8.4) the responsivity may be written as:

$$R = \frac{\eta e}{hf} \tag{8.9}$$

Equation (8.9) is a useful relationship for responsivity which may be developed a stage further to include the wavelength of the incident light.

The frequency f of the incident photons is related to their wavelength λ and the velocity of light in air c, by:

$$f = \frac{c}{\lambda} \tag{8.10}$$

Substituting into Eq. (8.9) a final expression for the responsivity is given by:

$$R = \frac{\eta e \lambda}{hc} \tag{8.11}$$

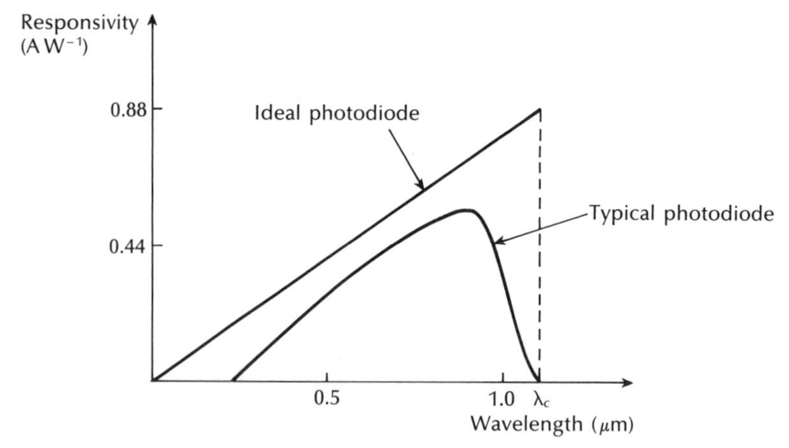

Figure 8.3 Responsivity against wavelength characteristic for an ideal silicon photodiode. The responsivity of a typical device is also shown.

It may be noted that the responsivity is directly proportional to the quantum efficiency at a particular wavelength.

The ideal responsivity against wavelength characteristic for a silicon photodiode with unit quantum efficiency is illustrated in Figure 8.3. Also shown is the typical responsivity of a practical silicon device.

Example 8.1

When 3×10^{11} photons each with a wavelength of 0.85 μm are incident on a photodiode, on average 1.2×10^{11} electrons are collected at the terminals of the device. Determine the quantum efficiency and the responsivity of the photodiode at 0.85 μm.

Solution: From Eq. (8.2),

$$\textit{Quantum efficiency} = \frac{\textit{number of electrons collected}}{\textit{number of incident photons}}$$

$$= \frac{1.2 \times 10^{11}}{3 \times 10^{11}}$$

$$= 0.4$$

The quantum efficiency of the photodiode at 0.85 μm is 40%.

From Eq. (8.11),

$$\textit{Responsivity } R = \frac{\eta e \lambda}{hc}$$

$$= \frac{0.4 \times 1.602 \times 10^{-19} \times 0.85 \times 10^{-6}}{6.626 \times 10^{-34} \times 2.998 \times 10^{8}}$$

$$= 0.274 \text{ A W}^{-1}$$

The responsivity of the photodiode at 0.85 μm is 0.27 A W^{-1}.

Example 8.2

A photodiode has a quantum efficiency of 65% when photons of energy 1.5×10^{-19} J are incident upon it.

(a) At what wavelength is the photodiode operating?
(b) Calculate the incident optical power required to obtain a photocurrent of 2.5 μA when the photodiode is operating as described above.

Solution: (a) From Eq. (6.1), the photon energy $E = hf = hc/\lambda$. Therefore

$$\lambda = \frac{hc}{E} = \frac{6.626 \times 10^{-34} \times 2.998 \times 10^{8}}{1.5 \times 10^{-19}}$$

$$= 1.32 \ \mu\text{m}$$

The photodiode is operating at a wavelength of 1.32 μm.

(b) From Eq. (8.9),

$$\textit{Responsivity } R = \frac{\eta e}{hf} = \frac{0.65 \times 1.602 \times 10^{-19}}{1.5 \times 10^{-19}}$$

$$= 0.694 \text{ A W}^{-1}$$

Also from Eq. (8.4).

$$R = \frac{I_p}{P_o}$$

Therefore

$$P_o = \frac{2.5 \times 10^{-6}}{0.694} = 3.60 \ \mu\text{W}$$

The incident optical power required is 3.60 μW.

8.7 Long wavelength cutoff

It is essential when considering the intrinsic absorption process that the energy of incident photons be greater than or equal to the bandgap energy E_g of the material used to fabricate the photodetector. Therefore, the photon energy

$$\frac{hc}{\lambda} \geqslant E_g \tag{8.12}$$

giving

$$\lambda \leqslant \frac{hc}{E_g} \tag{8.13}$$

Thus the threshold for detection, commonly known as the long wavelength cutoff point λ_c, is:

$$\lambda_c = \frac{hc}{E_g} \tag{8.14}$$

The expression given in Eq. (8.14) allows the calculation of the longest wavelength of light to give photodetection for the various semiconductor materials used in the fabrication of detectors.

It is important to note that the above criterion is only applicable to intrinsic photodetectors. Extrinsic photodetectors violate the expression given in Eq. (8.12), but are not currently used in optical fiber communications.

Example 8.3

GaAs has a bandgap energy of 1.43 eV at 300 K. Determine the wavelength above which an intrinsic photodetector fabricated from this material will cease to operate.

Solution: From Eq. (8.14), the long wavelength cutoff:

$$\lambda_c = \frac{hc}{E_g} = \frac{6.626 \times 10^{-34} \times 2.998 \times 10^{8}}{1.43 \times 1.602 \times 10^{-19}}$$

$$= 0.867\ \mu\text{m}$$

The GaAs photodetector will cease to operate above 0.87 μm.

8.8 Semiconductor photodiodes without internal gain

Semiconductor photodiodes without internal gain generate a single electron-hole pair per absorbed photon. This mechanism was outlined in Section 8.3, and in order to understand the development of this type of photodiode it is now necessary to elaborate upon it.

8.8.1 *p–n* Photodiode

Figure 8.4 shows a reverse biased *p–n* photodiode with both the depletion and diffusion regions. The depletion region is formed by immobile positively charged donor atoms in the *n* type semiconductor material and immobile negatively charged acceptor atoms in the *p* type material, when the mobile carriers are swept to their majority sides under the influence of the electric field. The width of the depletion region is therefore dependent upon the doping concentrations for a given applied reverse bias (i.e. the lower the doping, the wider the depletion region). For the interested reader expressions for the depletion layer width are given in Ref. 5.

Photons may be absorbed in both the depletion and diffusion regions, as indicated by the absorption region in Figure 8.4. The absorption region's position and width depends upon the energy of the incident photons and on the material from which the photodiode is fabricated. Thus in the case of the weak absorption of photons, the absorption region may extend completely throughout the device. Electron–hole pairs are therefore generated in both the depletion and diffusion regions. In the depletion region the carrier pairs separate and drift under the influence of the electric field, whereas outside this region the hole diffuses towards the depletion region in order to be collected. The diffusion process is very slow compared to drift and thus limits the response of the photodiode (see Section 8.8.3).

It is therefore important that the photons are absorbed in the depletion region. Thus it is made as long as possible by decreasing the doping in the *n* type material. The depletion region width in a *p–n* photodiode is normally 1 to 3 μm and is optimized for the efficient detection of light at a given wavelength. For silicon

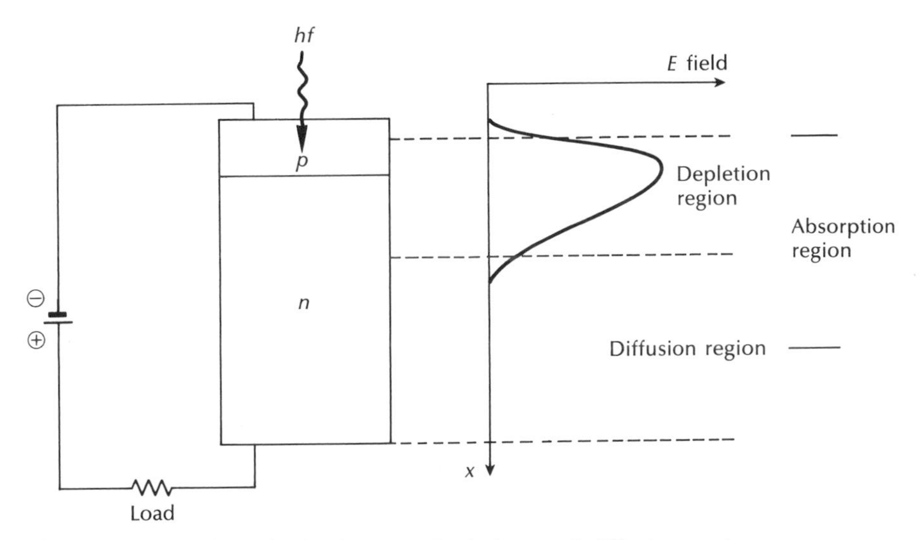

Figure 8.4 *p–n* photodiode showing depletion and diffusion regions.

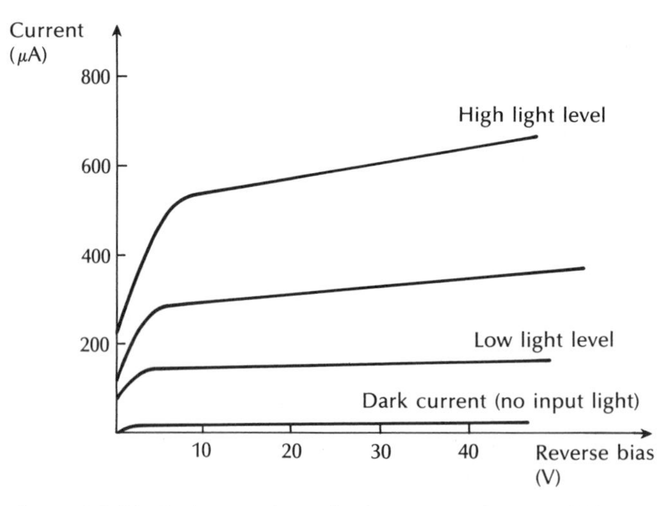

Figure 8.5 Typical *p–n* photodiode output characteristics.

devices this is in the visible spectrum (0.4 to 0.7 μm) and for germanium in the near infrared (0.7 to 0.9 μm).

Typical output characteristics for the reverse-biased *p–n* photodiode are illustrated in Figure 8.5. The different operating conditions may be noted moving from no light input to a high light level.

8.8.2 *p–i–n* Photodiode

In order to allow operation at longer wavelengths where the light penetrates more deeply into the semiconductor material a wider depletion region is necessary. To achieve this the *n* type material is doped so lightly that it can be considered intrinsic, and to make a low resistance contact a highly doped *n* type (n^+) layer is added. This creates a *p–i–n* (or PIN) structure, as may be seen in Figure 8.6 where all the absorption takes place in the depletion region.

Figure 8.7 shows the structures of two types of silicon *p–i–n* photodiode for operation in the shorter wavelength band below 1.09 μm. The front illuminated photodiode, when operating in the 0.8 to 0.9 μm band (Figure 8.7(a)), requires a depletion region of between 20 and 50 μm in order to attain high quantum efficiency (typically 85%) together with fast response (less than 1 ns) and low dark current (1 nA). Dark current arises from surface leakage currents as well as generation–recombination currents in the depletion region in the absence of illumination. The side illuminated structure (Figure 8.7(b)), where light is injected parallel to the junction plane, exhibits a large absorption width ($\simeq 500$ μm) and hence is particularly sensitive at wavelengths close to the bandgap limit (1.09 μm) where the absorption coefficient is relatively small.

Germanium *p–i–n* photodiodes which span the entire wavelength range of interest are also commercially available, but as mentioned previously (Section 8.4.2)

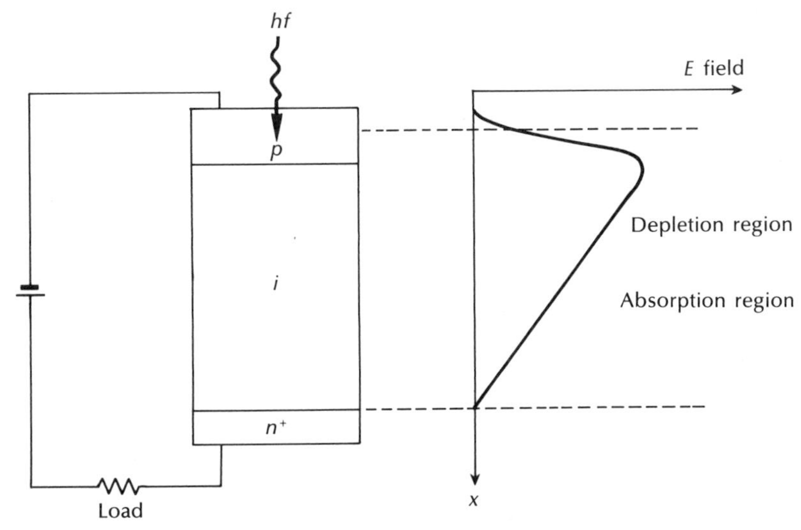

Figure 8.6 *p–i–n* photodiode showing combined absorption and depletion region.

the relatively high dark currents are a problem (typically 100 nA at 20 °C increasing to 1 μA at 40 °C). However, as outlined in Section 8.4.3, III–V semiconductor alloys have been employed in the fabrication of longer wavelength region detectors. At present, the favoured material is the lattice matched $In_{0.53}Ga_{0.47}As/InP$ system [Ref. 6] which can detect at wavelengths up to 1.67 μm. A typical planar device structure is shown in Figure 8.8(a) [Ref. 7] which requires epitaxial growth of several layers on an *n* type InP substrate. The incident light is absorbed in the low doped *n* type InGaAs layer generating carriers, as illustrated in the energy band diagram Figure 8.8(b) [Ref. 8]. The discontinuity due to the homojunction between the n^+–InP substrate and the *n*–InGaAs absorption region may be noted. This can be reduced by the incorporation of an *n* type InP buffer layer.

The top entry* device shown in Figure 8.8(a) is the simplest structure, with the light being introduced through the upper p^+ layer. However, a drawback with this structure is a quantum efficiency penalty which results from optical absorption in the undepleted p^+ region. In addition, there is a limit to how small such a device can be fabricated as both light access and metallic contact are required on the top. To enable smaller devices with lower capacitances to be made a substrate entry technique is employed. In this case light enters through a transparent InP substrate and the device area can be fabricated as small as may be practical for bonding.

Conventional growth techniques for III–V semiconductors can be employed to fabricate these devices, although liquid phase epitaxy (LPE) tends to be preferred because of the relative ease in obtaining the low doping levels needed (around

* Top entry is also referred to as front illumination.

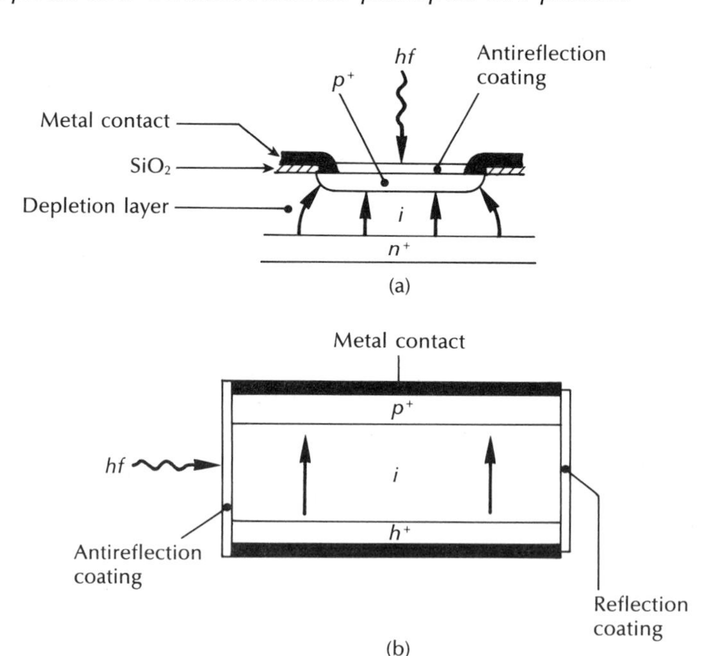

Figure 8.7 (a) Structure of a front illuminated silicon *p–i–n* photodiode. (b) Structure of a side illuminated (parallel to junction) *p–i–n* photodiode.

10^5 cm^{-3}) to obtain low capacitance (less than 0.2 pF). However, LPE does not easily allow low impurity level concentrations and it is necessary to use long baking procedures over several days to purify the source material. High quality devices have been produced using metal oxide vapour-phase epitaxy (MOVPE) [Ref. 9], a technique which appears much more appropriate for large scale production of such devices.

A substrate entry* *p–i–n* photodiode is shown in Figure 8.9(a). This device incorporates a p^+–InGaAsP layer to provide a heterojunction structure (Schottky barrier) which improves quantum efficiency. Moreover, it is fabricated as a mesa structure which reduces parasitic capacitances [Ref. 10]. Unfortunately, charge trapping can occur at the n^-p^+–InGaAs/InGaAsP interface which may be observed in the energy band diagram of Figure 8.9(b). This may cause limitations in the response time of the device [Ref. 8]. However, small area substrate entry devices can be produced with extremely low capacitance (less than 0.1 pF), quantum efficiency between 75% and 100%, and dark currents less than 1 nA.

In both device types a depleted InGaAs layer of around 3 μm is used which provides high quantum-efficiency and bandwidth. Furthermore, low doping permits full depletion of the InGaAs layer at low voltage (5 V). The short transit times in

* Substrate entry is also referred to as back illumination.

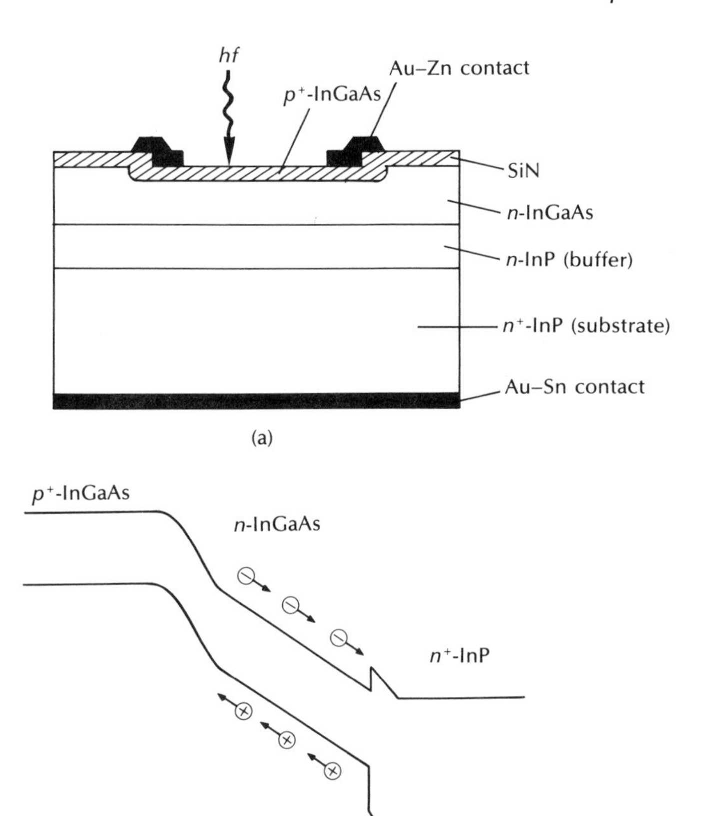

Figure 8.8 Planar InGaAs *p–i–n* photodiode: (a) structure; (b) energy band diagram showing homojunction associated with the conventional *p–i–n* structure.

the relatively narrow depletion layers give a theoretical bandwidth of approximately 15 GHz. However, the bandwidth of commercially available packaged detectors is usually between 1 and 2 GHz due to limitations of the packaging.

8.8.3 Speed of response

Three main factors limit the speed of response of a photodiode. These are [Ref. 11]:

(a) *Drift time of carriers through the depletion region*
The speed of response of a photodiode is fundamentally limited by the time it takes photogenerated carriers to drift across the depletion region. When the field in the depletion region exceeds a saturation value then the carriers may be assumed to travel at a constant (maximum) drift velocity v_d. The longest transit

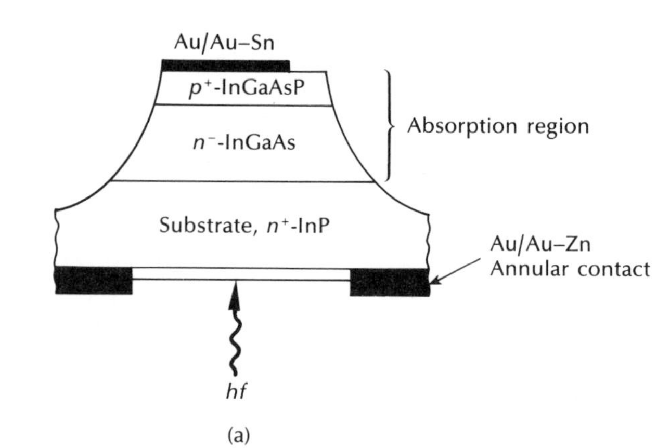

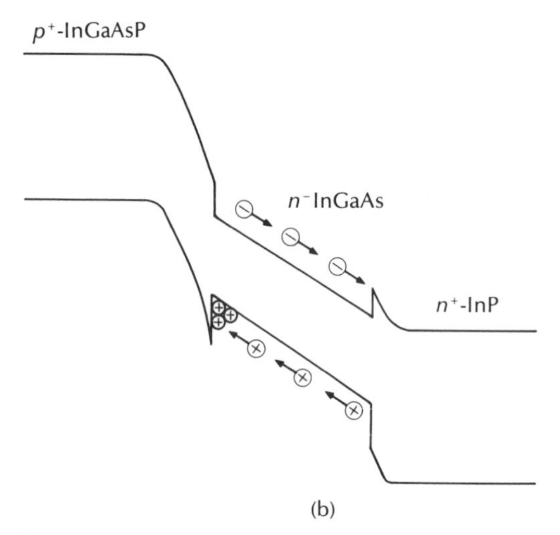

Figure 8.9 Substrate entry InGaAs *p–i–n* photodiode: (a) structure; (b) energy band diagram illustrating the heterojunction and charge trapping.

time, t_{drift}, is for carriers which must traverse the full depletion layer width w and is given by

$$t_{\text{drift}} = \frac{w}{v_{\text{d}}} \tag{8.15}$$

A field strength above 2×10^4 V cm^{-1} in silicon gives maximum (saturated) carrier velocities of approximately 10^7 cm s^{-1}. Thus the transit time through a depletion layer width of 10 μm is around 0.1 ns.

(b) *Diffusion time of carriers generated outside the depletion region*
Carrier diffusion is a comparatively slow process where the time taken, t_{diff}, for

carriers to diffuse a distance d may be written as

$$t_{\mathrm{diff}} = \frac{d^2}{2D_{\mathrm{c}}} \tag{8.16}$$

where D_{c} is the minority carrier diffusion coefficient. For example, the hole diffusion time through 10 μm of silicon is 40 ns whereas the electron diffusion time over a similar distance is around 8 ns.

(c) *Time constant incurred by the capacitance of the photodiode with its load*
A reversed biased photodiode exhibits a voltage dependent capacitance caused by the variation in the stored charge at the junction. The junction capacitance C_{j} is given by

$$C_{\mathrm{j}} = \frac{\varepsilon_{\mathrm{s}} A}{w} \tag{8.17}$$

where ε_{s} is the permittivity of the semiconductor material and A is the diode junction area. Hence, a small depletion layer width w increases the junction capacitance. The capacitance of the photodiode C_{d} is that of the junction together with the capacitance of the leads and packaging. This capacitance must be minimized in order to reduce the RC time constant which also limits the detector response time (see Section 9.3.2).

Although all the above factors affect the response time of the photodiode, the ultimate bandwidth of the device is limited by the drift time of carriers through the depletion region t_{drift}. In this case when assuming no carriers are generated outside the depletion region and that there is negligible junction capacitance, then the maximum photodiode 3 dB bandwidth B_{m} is given by [Ref.12]:

$$B_{\mathrm{m}} = \frac{1}{2\pi t_{\mathrm{drift}}} = \frac{v_{\mathrm{d}}}{2\pi w} \tag{8.18}$$

Moreover, when there is no gain mechanism present within the device structure, the maximum possible quantum efficiency is 100%. Hence the value for the bandwidth given by Eq. (8.18) is also equivalent to the ultimate gain–bandwidth product for the photodiode.

Example 8.4

The carrier velocity in a silicon p–i–n photodiode with a 25 μm depletion layer width is 3×10^4 m s^{-1}. Determine the maximum response time for the device.

Solution: The maximum 3 dB bandwidth for the photodiode may be obtained from Eq. (8.18) where:

$$B_{\mathrm{m}} = \frac{v_{\mathrm{d}}}{2\pi_{\mathrm{w}}} = \frac{3 \times 10^4}{2\pi \times 25 \times 10^{-6}} = 1.91 \times 10^8 \text{ Hz}$$

The maximum response time for the device is therefore:

$$\text{Max. response time} = \frac{1}{B_m} = 5.2 \text{ ns}$$

It must be noted, however, that the above response time takes no account of the diffusion of carriers in the photodiode or the capacitance associated with the device junction and the external connections.

The response of a photodiode to a rectangular optical input pulse for various device parameters is illustrated in Figure 8.10. Ideally, to obtain a high quantum efficiency for the photodiode the width of the depletion layer must be far greater than the reciprocal of the absorption coefficient (i.e. $1/\alpha_o$) for the material used to fabricate the detector so that most of the incident light will be absorbed. Hence the response to a rectangular input pulse of a low capacitance photodiode meeting this condition, and exhibiting negligible diffusion outside the depletion region, is shown in Figure 8.10(a). It may be observed in this case that the rising and falling edges of the photodiode output follow the input pulse quite well. When the detector capacitance is larger, however, the speed of response becomes limited by the RC time constant of this capacitance and the load resistor associated with the receiver circuit (see Section 9.3.2), and thus the output pulse appears as illustrated in Figure 8.10(b).

Furthermore, when there is significant diffusion of carriers outside the depletion region, as is the case when the depletion layer is too narrow ($w \leqslant 1/\alpha_o$) and carriers are therefore created by absorption outside this region, then the output pulse displays a long tail caused by the diffusion component to the input optical pulse, as shown in Figure 8.10(c). Thus devices with very thin depletion layers have a tendency to exhibit distinctive fast response and slow response components to their output pulses, as may be observed in Figure 8.10(c). The former response resulting from absorption in the thin depletion layer.

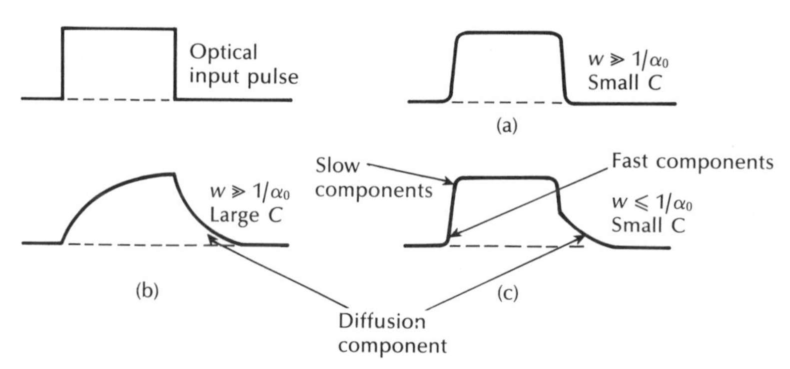

Figure 8.10 Photodiode responses to rectangular optical input pulses for various detector parameters.

8.8.4 Noise

The overall sensitivity of a photodiode results from the random current and voltage fluctuations which occur at the device output terminals in both the presence and absence of an incident optical signal. Although the factors that determine the sensitivity of the optical receiver are dealt with in Chapter 9, it is appropriate at this stage to consider the sources of noise that arise within photodiodes, which do not have an internal gain mechanism. The photodiode dark current mentioned in Section 8.8.2 corresponds to the level of the output photocurrent when there is no intended optical signal present. However, there may be some photogenerated current present due to background radiation entering the device.

The inherent dark current can be minimized through the use of high quality, defect-free material which reduces the number of carriers generated in the depletion region as well as those which diffuse into this layer from the p^+ and n^+ regions. Moreover, the surface currents can be minimized by careful fabrication and surface passivation such that the surface state and impurity ion concentrations are reduced. Nevertheless, it is the case that the detector average current $\bar{I}$ always exhibits a random fluctuation about its mean value as a result of the statistical nature of the quantum detection process (see Section 9.2.3). This fluctuation is exhibited as shot noise [Ref. 13] where the mean square current variation $\overline{i_s^2}$ is proportional to $\bar{I}$ and the photodiode received bandwidth B. Thus the rms value of this shot noise current is:

$$(\overline{i_s^2})^{\frac{1}{2}} = (2eB\bar{I})^{\frac{1}{2}} \tag{8.19}$$

Various figures of merit have traditionally been employed to assess the noise performance of optical detectors. Although these parameters are not always appropriate for the evaluation of the high speed photodiodes used in optical fiber communications, it is instructive to define those most commonly utilized. These are: the noise equivalent power (NEP); the detectivity (D); and the specific detectivity (D^*).

The NEP is defined as the incident optical power, at a particular wavelength or with a specified spectral content required to produce a photodetector current equal to the rms noise current within a unit bandwidth (i.e. $B = 1$ Hz). To obtain an expression for the NEP at a specific wavelength, Eq. (8.8) must be rearranged as follows to give:

$$P_o = \frac{I_p hf}{\eta e} = \frac{I_p hc}{\eta e \lambda} \tag{8.20}$$

Then putting the photocurrent I_p equal to the rms shot noise current in Eq. (8.19) gives:

$$I_p = (2e\bar{I}B)^{\frac{1}{2}} \tag{8.21}$$

Moreover, the photodiode average current $\bar{I}$ may be represented by $(I_p + I_d)$ where I_d is the dark current within the device. Hence:

$$I_p = [2e(I_p + I_d)B]^{\frac{1}{2}} \tag{8.22}$$

When $I_p \gg I_d$, then:

$$I_p \simeq 2eB \tag{8.23}$$

Substituting Eq. (8.23) into Eq. (8.20) and putting $B = 1$ Hz gives the noise equivalent power as:

$$NEP = P_o \simeq \frac{2hc}{\eta\lambda} \tag{8.24}$$

It should be noted that the *NEP* for an ideal photodetector is given by Eq. (8.24) when the quantum efficiency $\eta = 1$.

When $I_p \ll I_d$, then from Eq. (8.22) the photocurrent becomes:

$$I_p \simeq [2eI_dB]^{\frac{1}{2}} \tag{8.25}$$

Hence for a photodiode in which the dark current noise is dominant, the use of Eq. (8.20) with $B = 1$ Hz gives an expression for the noise equivalent power of:

$$NEP = P_o \simeq \frac{hc(2eI_d)^{\frac{1}{2}}}{\eta e\lambda} \tag{8.26}$$

The detectivity D is defined as the inverse of the *NEP*; thus:

$$D = \frac{1}{NEP} \tag{8.27}$$

Considering a photodiode receiving monochromatic radiation with the dark current as its dominant noise source, then from Eqs. (8.26) and (8.27):

$$D = D_\lambda = \frac{\eta e\lambda}{hc(2eI_d)^{\frac{1}{2}}} \tag{8.28}$$

The specific detectivity D^* is a parameter which incorporates the area of the photodetector A in order to take account of the effect of this factor on the amplitude of the device dark current. This proves necessary when background radiation and thermal generation rather than surface conduction are the major causes of dark current. Therefore the specific detectivity is given by:

$$D^* = DA^{\frac{1}{2}} = \frac{\eta e\lambda}{hc(2eI_d/A)^{\frac{1}{2}}} \tag{8.29}$$

It should be noted, however, that the above definition for D^* assumes a bandwidth of 1 Hz. Hence the specific detectivity over a bandwidth B would be equal to $D(AB)^{\frac{1}{2}}$.

Example 8.5

A germanium *p–i–n* photodiode with active dimensions of $100 \times 50\ \mu$m has a quantum efficiency of 55% when operating at a wavelength of 1.3 μm. The measured dark current at this wavelength is 8 nA. Calculate the noise equivalent power and specific detectivity for the device. It may be assumed that dark current is the dominant noise source.

Solution: The noise equivalent power is given by Eq. (8.26) as:

$$NEP \simeq \frac{hc(2eI_d)^{\frac{1}{2}}}{\eta e \lambda}$$

$$= \frac{6.626 \times 10^{-34} \times 2.998 \times 10^{8}(2 \times 1.602 \times 10^{-19} \times 8 \times 10^{-9})^{\frac{1}{2}}}{0.55 \times 1.602 \times 10^{-19} \times 1.3 \times 10^{-6}}$$

$$= 8.78 \times 10^{-14} \text{ W}$$

Substituting for the detectivity D in Eq. (8.29) from Eq. (8.27) allows the specific detectivity to be written as:

$$D^* = \frac{A^{\frac{1}{2}}}{NEP} = \frac{(100 \times 10^{-6} \times 50 \times 10^{-6})^{\frac{1}{2}}}{8.78 \times 10^{-14}}$$

$$= 8.1 \times 10^{8} \text{ m Hz}^{\frac{1}{2}} \text{W}^{-1}$$

The above parameters are solely concerned with the noise performance of the photodiodes used within optical fiber communications. However, it is the noise associated with the optical receiver which also incorporates a load resistance and a preamplifier that is the major concern. This more general issue is dealt with in Chapter 9.

8.9 Semiconductor photodiodes with internal gain

8.9.1 Avalanche photodiodes

The second major type of optical communications detector is the avalanche photodiode (APD). This has a more sophisticated structure than the *p–i–n* photodiode in order to create an extremely high electric field region (approximately 3×10^5 V cm^{-1}), as may be seen in Figure 8.11(a). Therefore, as well as the depletion region where most of the photons are absorbed and the primary carrier pairs generated there is a high field region in which holes and electrons can acquire sufficient energy to excite new electron–hole pairs. This process is known as impact ionization and is the phenomenon that leads to avalanche breakdown in ordinary reverse biased diodes. It often requires high reverse bias voltages (50 to 400 V) in order that the new carriers created by impact ionization can themselves produce additional carriers by the same mechanism as shown in Figure 8.11(b). More recently, however, it should be noted that devices which will operate at much lower bias voltages (15 to 25 V) have become available.

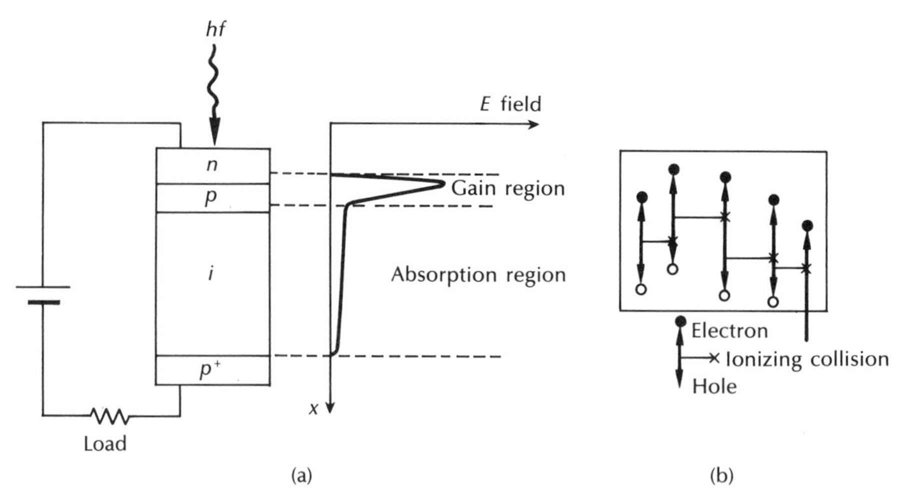

Figure 8.11 (a) Avalanche photodiode showing high electric field (gain) region. (b) Carrier pair multiplication in the gain region.

Carrier multiplication factors as great as 10^4 may be obtained using defect-free materials to ensure uniformity of carrier multiplication over the entire photosensitive area. However, other factors affect the achievement of high gain within the device. Microplasmas, which are small areas with lower breakdown voltages than the remainder of the junction, must be reduced through the selection of defect-free materials together with careful device processing and fabrication [Ref. 14]. In addition, excessive leakage at the junction edges can be eliminated by the use of a guard ring structure as shown in Figure 8.12. At present both silicon and germanium APDs are available.

Operation of these devices at high speed requires full depletion in the absorption region. As indicated in Section 8.8.1, when carriers are generated in undepleted

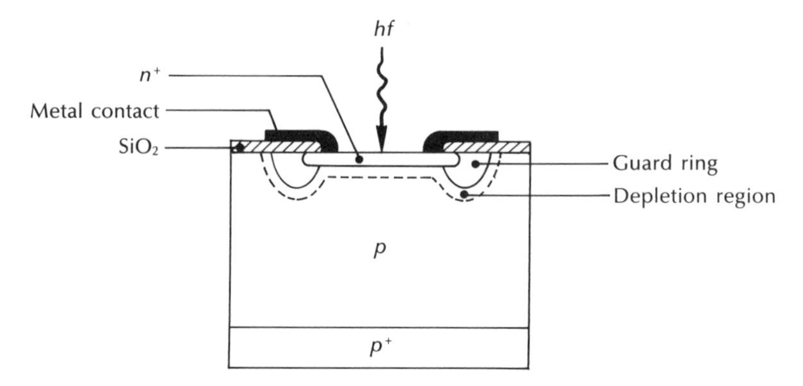

Figure 8.12 Structure of a silicon avalanche photodiode with guard ring.

material, they are collected somewhat slowly by the diffusion process. This has the effect of producing a long 'diffusion tail' on a short optical pulse. When the APD is fully depleted by employing electric fields in excess of 10^4 V m^{-1}, all the carriers drift at saturation-limited velocities. In this case the response time for the device is limited by three factors. These are:

(a) the transit time of the carriers across the absorption region (i.e. the depletion width);
(b) the time taken by the carriers to perform the avalanche multiplication process; and
(c) the RC time constant incurred by the junction capacitance of the diode and its load.

At low gain the transit time and RC effects dominate giving a definitive response time and hence constant bandwidth for the device. However, at high gain the avalanche build-up time dominates and therefore the device bandwidth decreases proportionately with increasing gain. Such APD operation is distinguished by a constant gain–bandwidth product.

Often an asymmetric pulse shape is obtained from the APD which results from a relatively fast rise time as the electrons are collected and a fall time dictated by the transit time of the holes travelling at a slower speed. Hence, although the use of suitable materials and structures may give rise times between 150 and 200 ps, fall times of 1 ns or more are quite common which limit the overall response of the device.

8.9.2 Silicon reach through avalanche photodiodes

To ensure carrier multiplication without excess noise for a specific thickness of multiplication region within the APD it is necessary to reduce the ratio of the ionization coefficients for electrons and holes k (see Section 9.3.4). In silicon this ratio is a strong function of the electric field varying from around 0.1 at 3×10^5 V m^{-1} to 0.5 at 6×10^5 V m^{-1}. Hence for minimum noise, the electric field at avalanche breakdown must be as low as possible and the impact ionization should be initiated by electrons. To this end a 'reach through' structure has been implemented with the silicon avalanche photodiode. The silicon 'reach through' APD(RAPD) consists of $p^+–\pi–p–n^+$ layers as shown in Figure 8.13(a). As may be seen from the corresponding field plot in Figure 8.13(b), the high field region where the avalanche multiplication takes place is relatively narrow and centred on the $p–n^+$ junction. Thus under low reverse bias most of the voltage is dropped across the $p–n^+$ junction.

When the reverse bias voltage is increased the depletion layer widens across the p region until it 'reaches through' to the nearly intrinsic (lightly doped) π region. Since the π region is much wider than the p region the field in the π region is much lower than that at the $p–n^+$ junction (see Figure 8.13(b)). This has the effect of removing some of the excess applied voltage from the multiplication region to the

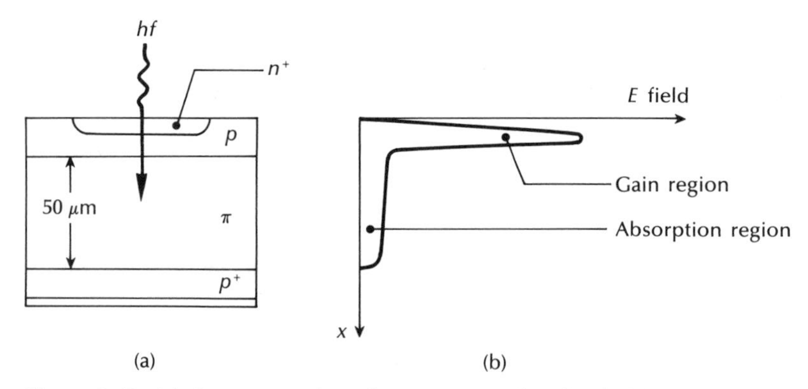

Figure 8.13 (a) Structure of a silicon RAPD. (b) The field distribution in the RAPD showing the gain region across the p–n^+ junction.

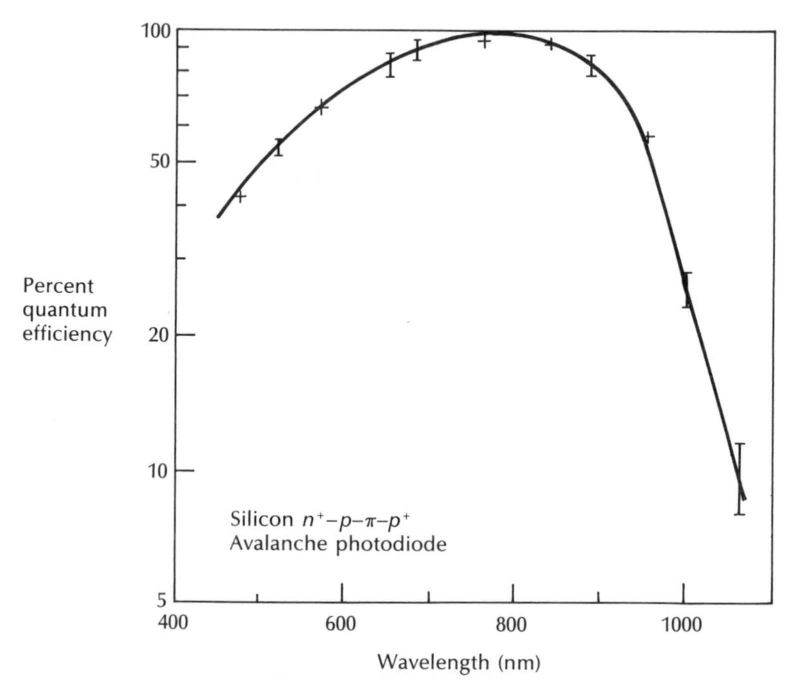

Figure 8.14 Measurement of quantum efficiency against wavelength for a silicon RAPD. After Ref. 16. Reprinted with permission from *The Bell System Technical Journal*. © 1978, AT&T.

π region giving a relatively slow increase in multiplication factor with applied voltage. Although the field in the π region is lower than in the multiplication region it is high enough (2×10^4 V cm^{-1}) when the photodiode is operating to sweep the carriers through to the multiplication region at their scattering limited velocity (10^7 cm s^{-1}). This limits the transit time and ensures a fast response (as short as 0.5 ns).

Measurements [Ref. 16] for a silicon RAPD for optical fiber communication applications at a wavelength of 0.825 μm have shown a quantum efficiency (without avalanche gain) of nearly 100% in the working region, as may be seen in Figure 8.14. The dark currents for this photodiode are also low and depend only slightly on bias voltage.

8.9.3 Germanium avalanche photodiodes

The elemental semiconductor germanium has been used to fabricate relatively sensitive and fast APDs that may be used over almost the entire wavelength range of primary interest at present (0.8–1.6 μm). However, it was clear from an early stage that higher dark currents together with larger excess noise factors (see Section 9.3.3) than those in silicon APDs were a problem with these devices. The large dark currents were associated with edge and surface effects resulting from difficulties in passivating germanium, and were also a direct consequence of the small energy bandgap as mentioned earlier in Section 8.4.2.

In the late 1970s when interest increased in the fabrication of detectors for longer wavelength operation (1.1–1.6 μm), germanium APDs using a conventional n^+p structure similar to the silicon APD shown in Figure 8.12 were produced [Ref. 18]. However, such devices exhibited dark currents near breakdown of between 100 nA and 300 nA which were very sensitive to temperature variations [Ref. 6]. Furthermore, unlike the situation with silicon APDs, these dark currents had significant components of both bulk (multiplied) and surface (unmultiplied) current. It was the multiplied component (typically 100 nA for the n^+p structure) which needed to be reduced (to around 1 nA) in order to provide low noise operation. In addition, large excess noise factors associated with the avalanche multiplication process were obtained as a result of electrons rather than holes (which have a higher impact ionization coefficient in germanium) initiating the multiplication process. One advantage, however, of such germanium APDs over their silicon counterparts is that because of the relatively high absorption coefficient exhibited by germanium at 1.3 μm, avalanche breakdown voltages are quite low (typically 25 V).

Germanium APD structures have been fabricated to provide multiplication initiated by holes thus to reduce the excess noise factor in the longer wavelength region. For example, a n^+np structure has been demonstrated [Ref. 19] which goes some way to achieving this performance by reducing the factor by some 30% on that obtained in n^+p devices. However, multiplied dark current around 1 μA was obtained when operating at a wavelength of 1.3 μm and a multiplication factor of

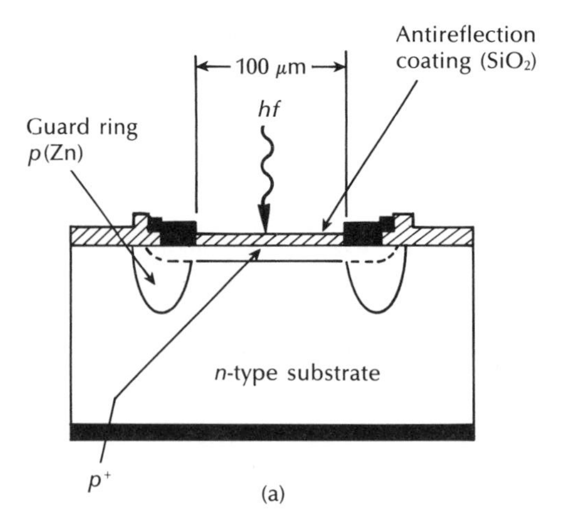

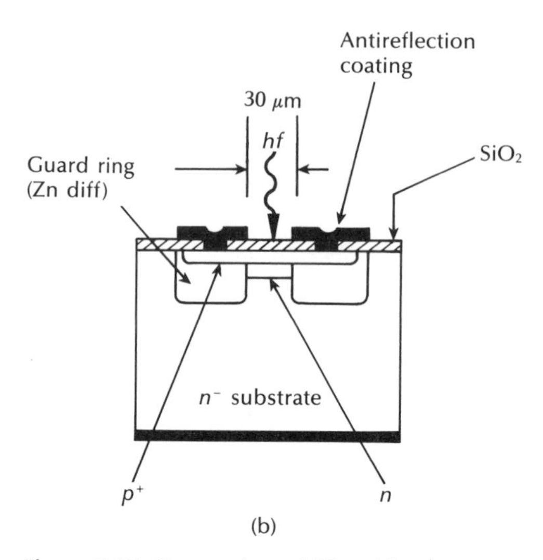

Figure 8.15 Germanium APDs: (a) p^+n structure [Ref. 20]; (b) Hi-Lo (p^+nn^-) structure [Ref. 23].

ten. An alternative device providing similar results utilizes the p^+n structure [Ref. 20] shown in Figure 8.15(a). In this case dark currents were reduced to between 150 and 250 nA by using an ion implanted technology [Ref. 21] and subsequently to around 5 nA by reducing the device sensitive area from 100 μm to 30 μm [Ref. 22].

Unfortunately, the speed of the p^+n structure at a wavelength of 1.5 μm is poor because most of the absorption in germanium at this wavelength takes place outside the depletion region.* This has led to the development of the p^+nn^- structure

* The absorption length in germanium at a wavelength of 1.5 μm is 10 μm.

shown in Figure 8.15(b) [Ref. 23] which resembles the reach through structure used for silicon APDs (see Section 8.9.2). It is known as a Hi–Lo structure as it combines high bandwidth (700 MHz) with low multiplied dark current (33 nA) and good excess noise performance. However, the breakdown voltage is higher at +85 V and the unmultiplied dark currents are around 1 μA. Nevertheless, these Hi–Lo devices appear to be among the highest performance germanium APDs for longer wavelength operation and are only eclipsed by the emerging III–V alloy APDs which do not exhibit quite the same fundamental material limitations.

8.9.4 III–V alloy avalanche photodiodes

Due to the drawbacks with germanium APDs for longer wavelength operation much effort has been expended in the study of III–V semiconductor alloys for the fabrication of APDs. In particular, the ternary InGaAs/InP and quaternary InGaAsP/InP material systems have been successfully employed. In common with the silicon reach through APD (see Section 8.9.2) separate absorption and multiplication regions are provided, as illustrated in Figure 8.16. This defines the so-called SAM (separate absorption and multiplication) APD which is a heterostructure device designed so that the multiplication takes place in the InP *p–n* junction [Ref. 24]. The performance of such long wavelength APDs is limited, however, by the fundamental properties of the material systems.

A first limitation is related to the large tunnelling currents associated with the narrow bandgap required for longer wavelength optical absorption. The band to band or defect assisted tunnelling currents become large before the electric field is high enough to obtain significant avalanche gain. This problem is substantially reduced using a separate absorption and multiplication region with the gain

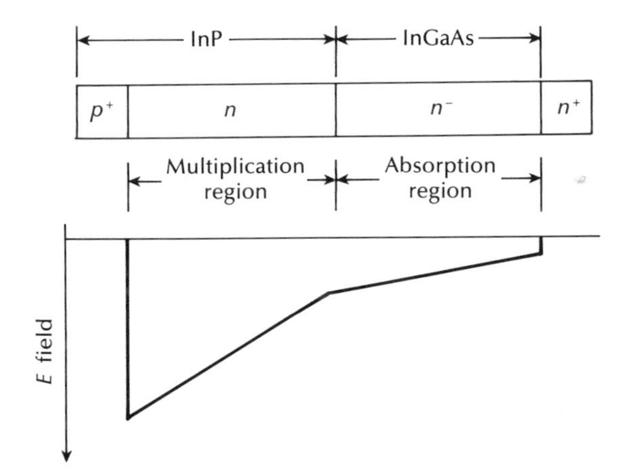

Figure 8.16 Separate absorption and multiplication (SAM) APD layer composition and electric field profile.

occurring at the InP p–n junction where the tunnelling is much less [Ref. 6]. However, control over the doping and thickness of the n type InP layer is critical in order to avoid excessive leakage current. Nevertheless, it is possible to obtain low dark currents of less than 10 nA (unmultiplied) together with quantum efficiencies of 80%, capacitance of approximately 0.5 pF and an operating voltage of around 100 V.

A second limitation associated with SAM APDs concerns the trapping of holes in the valence band discontinuity at the InGaAs/InP heterointerface, as illustrated in Figure 8.17(a) [Ref. 25]. This factor results in a slow component of the photoresponse which causes a speed limitation. However, the problem can be alleviated by incorporating a thin grading layer of InGaAsP (whose bandgap is intermediate between InGaAs and InP) between these two layers (Figure 8.17(b)) to smooth out the discontinuity and thus provide improved speed performance [Ref. 26]. Nevertheless, the gain–bandwidth products for such devices are still only between 10 and 20 GHz, not quite sufficient for high bit rate systems in the gigabit s^{-1} region [Ref. 6]. For example, with a gain around ten such devices will only provide operation to between 1 and 2 GHz.

An improved technique for increasing the speed of response of the device is to provide several (two or three) InGaAsP buffer layers to create compositional grading at the heterojunction interface [Refs. 12, 24]. This may be achieved by interposing a thin multiquantum-well (MQW) structure between the narrow and wideband gap layer. The configuration of a recent back illuminated mesa-structure separate absorption, grading and multiplication (SAGM) InGaAs APD is shown in Figure 8.18. This device type has displayed a gain–bandwidth product of up to 70 GHz [Ref. 27] thus allowing operation at bandwidths of 5 GHz or higher.

Overall, advanced photodiode developments are targeted at devices with improved sensitivities for operation at very high bandwidths, together with the

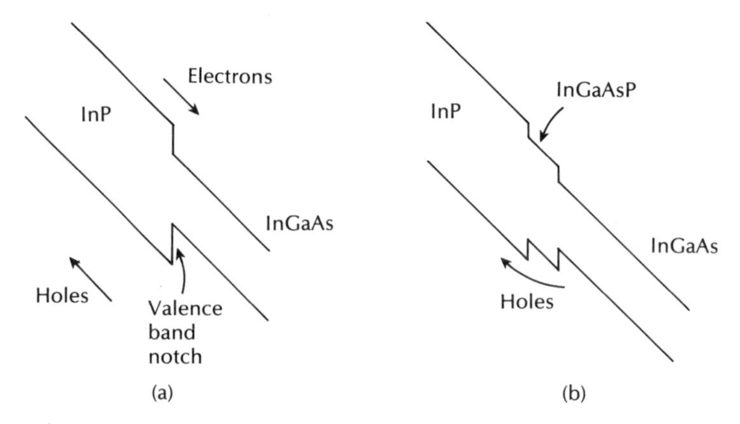

Figure 8.17 Energy band diagrams for SAM APDs: (a) InGaAs/InP heterojunction illustrating the notch in which holes may be trapped; (b) similar heterojunction to (a) with InGaAsP layer to reduce the effect of the notch.

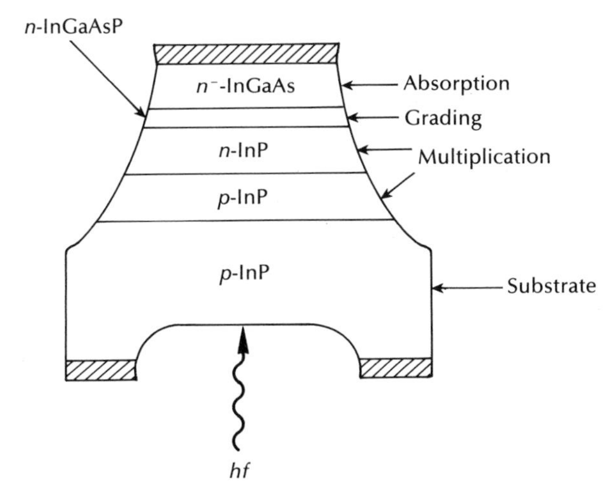

Figure 8.18 Separate absorption, grading and multiplication (SAGM) APD structure.

fabrication of structures with improved functionality at low cost [Ref. 12]. More recently, efforts to improve sensitivity have focused upon semiconductor superlattices in the form of MQW structures [Ref. 29] and staircase APDs [Ref. 30]. Both of these APD structures have gain regions comprising multiquantum wells formed by alternately growing thin layers of wide and narrow bandgap materials such as AlGaAs and GaAs respectively. By using materials exhibiting these properties the conduction and valence band discontinuities differ significantly, resulting in different ionization coefficients for electrons and holes. This factor should therefore give improvements in the noise performance of such III–V alloy APDs by reducing the ratio of the ionization coefficients for electrons and holes (k value, see Section 9.3.4) because the ionization coefficients of the two carrier types are normally approximately equal. In addition the other major advantage in using MQW APDs results from their improved bandwidth capabilities caused by the reduction in avalanche build-up time provided by the multilayer structures.

The step-like MQW energy band structure where the discontinuity in the conduction band is greater than that in the valence band is shown in Figure 8.19. The structure, which is illustrated both unbiased and biased, could comprise about 100 layers of alternate wide and narrow bandgap semiconductors. Although such devices have been fabricated using the AlGaAs/GaAs material system by molecular beam epitaxy (MBE), the structure does not provide the same favourable k value reduction when using InP-based alloys for longer wavelength operation. In this case problems with tunnelling in the high field regions containing the narrow bandgap layers tend to destroy the sensitivity improvement provided by the multiquantum wells [Ref. 12].

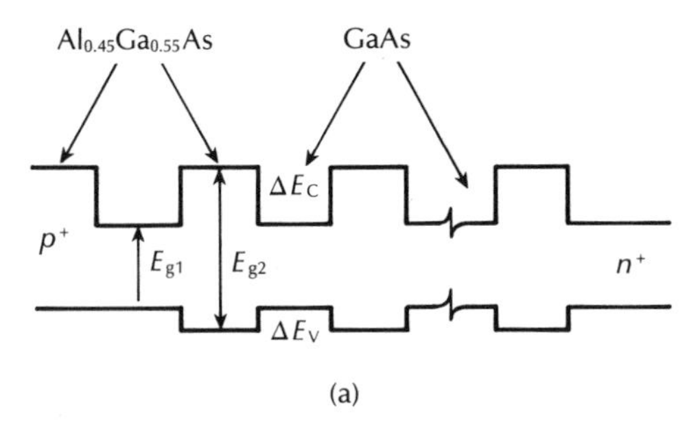

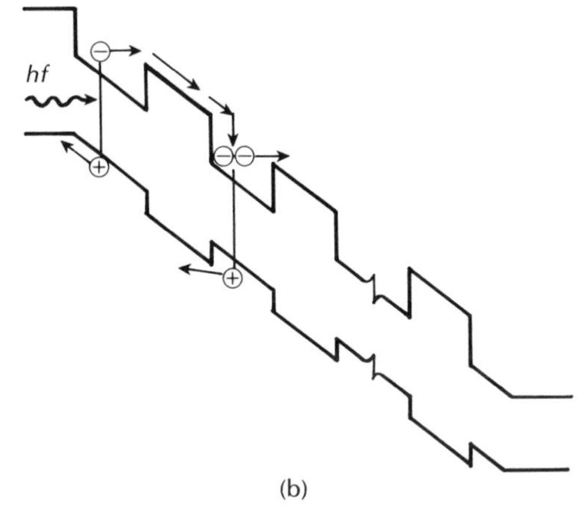

Figure 8.19 Energy band diagrams for MQW superlattice APD structure: (a) unbiased showing alternate layers of wide and narrow bandgap semiconductors; (b) the biased device [Ref. 29].

The energy band structure of a more complex scheme known as a staircase APD is shown in Figure 8.20. In this technique a narrow bandgap region is compositionally graded over a distance of 10 to 20 nm into a material with a minimum of twice the bandgap at the narrow end of the step. Again, the composition is abruptly changed to obtain the narrow bandgap as a second step is formed. The primary advantage of this staircase structure is that carrier multiplication caused by carrier transitions from the wide to the narrow bandgap material can occur at much lower electric field densities than that required with MQW devices.

In principle this APD could operate with a very low bias voltage, thereby removing the possibility of tunnelling in the narrow bandgap material. However, the structure and grading presents substantial fabricational problems and therefore has not, as yet, been realized. Moreover, it is suggested that the only suitable candidate material with the ability to grade continuously in alloy composition is GaAlAsSb [Ref. 31].

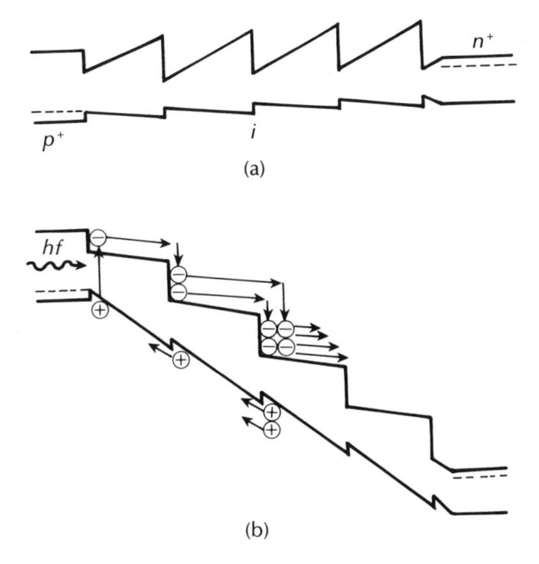

Figure 8.20 Energy band diagrams for the staircase APD: (a) the unbiased device: (b) the biased device under normal operation [Ref. 30].

8.9.5 Benefits and drawbacks with the avalanche photodiode

APDs have a distinct advantage over photodiodes without internal gain for the detection of the very low light levels often encountered in optical fiber communications. They generally provide an increase in sensitivity of between 5 and 15 dB over *p–i–n* photodiodes whilst often giving a wider dynamic range as a result of their gain variation with response time and reverse bias.

The optimum sensitivity improvement of APD receivers over *p–i–n* photodiode devices is illustrated in the characteristics shown in Figure 8.21. The characteristics display the minimum detectable optical power for direct detection (see Section 7.5) versus the transmitted bit rate in order to maintain a bit error rate (BER) of 10^{-9} (see Section 11.6.3) in the shorter and longer wavelength regions. Figure 8.21(a) compares silicon photodiodes operating at a wavelength of 0.82 μm where the APD is able to approach within 10 to 13 dB of the quantum limit. In addition, it may be observed that the *p–i–n* photodiode receiver has a sensitivity around 15 dB below this level. InGaAs photodiodes operating at a wavelength of 1.55 μm are compared in Figure 8.21(b). In this case the APD requires around 20 dB more power than the quantum limit, whereas the *p–i–n* photodiode receiver is some 10 to 12 dB less sensitive than the APD.

APDs, however, also have several drawbacks which include:

(a) fabrication difficulties due to their more complex structure and hence increased cost;

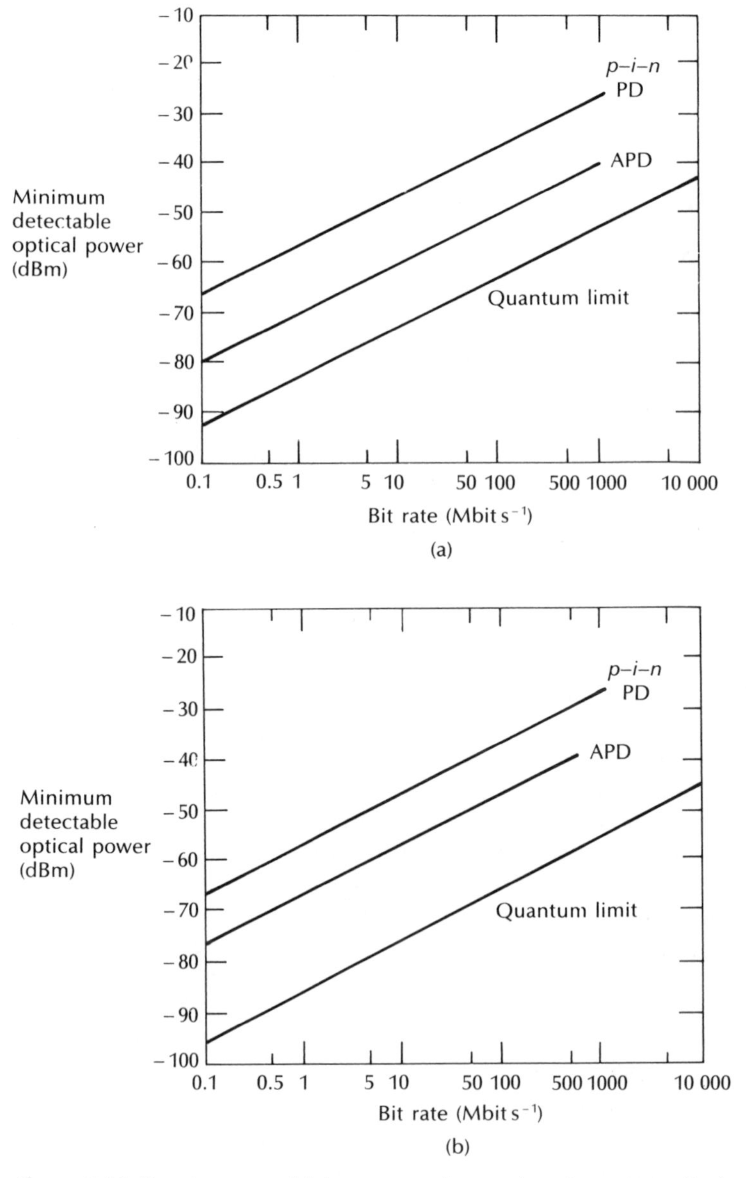

Figure 8.21 Receiver sensitivity comparison of *p–i–n* photodiode and APD devices at BER of 10^{-9}: (a) using silicon detectors operating at a wavelength of 0.82 μm; (b) using InGaAs detectors operating at a wavelength of 1.55 μm.

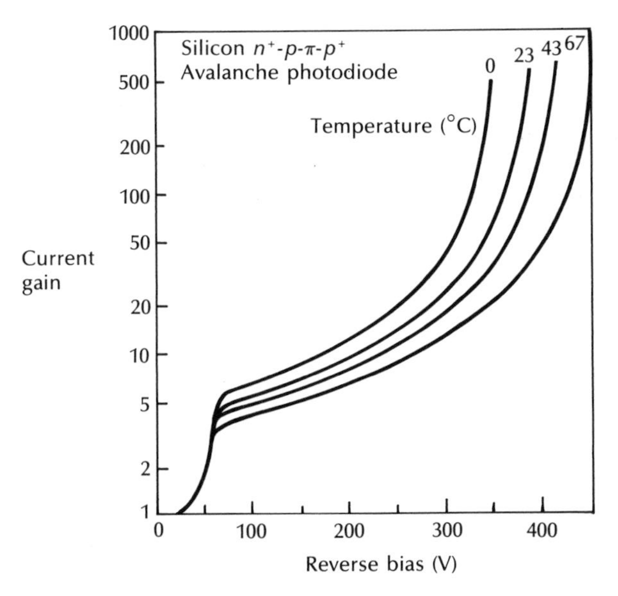

Figure 8.22 Current gain against reverse bias for a silicon RAPD operating at a wavelength of 0.825 μm. After Ref. 16. Reprinted with permission from *The Bell System Technical Journal*. © 1978, AT&T.

(b) the random nature of the gain mechanism which gives an additional noise contribution (see Section 9.3.3);
(c) the often high bias voltages required (50 to 400 V) which are wavelength dependent;
(d) the variation of the gain (multiplication factor) with temperature as shown in Figure 8.22 for a silicon RAPD [Ref. 16]; thus temperature compensation is necessary to stabilize the operation of the device.

8.9.6 Multiplication factor

The multiplication factor M is a measure of the internal gain provided by the APD. It is defined as:

$$M = \frac{I}{I_p} \tag{8.30}$$

where I is the total output current at the operating voltage (i.e. where carrier multiplication occurs) and I_p is the initial or primary photocurrent (i.e. before carrier multiplication occurs).

Example 8.6

The quantum efficiency of a particular silicon RAPD is 80% for the detection of radiation at a wavelength of 0.9 μm. When the incident optical power is 0.5 μW, the output current from the device (after avalanche gain) is 11 μA. Determine the multiplication factor of the photodiode under these conditions.

Solution: From Eq. (8.11), the responsivity

$$R = \frac{\eta e \lambda}{hc} = \frac{0.8 \times 1.602 \times 10^{-19} \times 0.9 \times 10^{-6}}{6.626 \times 10^{-34} \times 2.998 \times 10^{8}}$$

$$= 0.581 \text{ A W}^{-1}$$

Also from Eq. (8.4), the photocurrent

$$I_{\text{p}} = P_{\text{o}}R$$

$$= 0.5 \times 10^{-6} \times 0.581$$

$$= 0.291 \ \mu\text{A}$$

Finally, using Eq. (8.30):

$$M = \frac{I}{I_{\text{p}}} = \frac{11 \times 10^{-6}}{0.291 \times 10^{-6}}$$

$$= 37.8$$

The multiplication factor of the photodiode is approximately 38.

8.10 Mid-infrared photodiodes

Developments of photodiodes for mid-infrared transmission systems are at a relatively early stage; however, several potential devices have been demonstrated over recent years. Obtaining suitable lattice matching for III–V alloy materials is a problem when operating at wavelengths greater than 2.0 μm. A lattice matched InGaAsSb/GaSb material system has been utilized in a *p–i–n* photodiode for high speed operation at wavelengths up to 2.3 μm [Ref. 32].

An alternative approach which has achieved some success is the use of indium alloys that, due to the high indium content for operation above 2 μm, are mismatched with respect to the InP substrate causing inherent problems of dislocation-induced junction leakage and low quantum efficiency [Ref. 33]. However, these problems have been reduced by utilizing a compositionally graded buffer layer to accommodate the lattice mismatch. One technique has involved the replacement of the conventional *p–i–n* homojunction with an InGaAs/AlInAs heterojunction in which a wider bandgap *p*-type AlInAs layer acted as a transparent window at long wavelengths to ensure that optical absorption occurred in a lightly doped *n* type region of the device. This device, which exhibited a useful response out to a wavelength of 2.4 μm, displayed a quantum efficiency as high as 95% over the wavelength range 1.3 to 2.25 μm with dark currents as low as 35 nA [Ref. 33]. A similar approach has been demonstrated with the ternary alloys $In_xGa_{1-x}As$/$InAs_yP_{1-y}$ to produce mesa structure photodiodes which operate at a wavelength of 2.55 μm [Ref. 34]. A compositionally graded region of InGaAs or InAsP accommodates the lattice mismatch between the ternary layers and the InP substrate. These devices which have $In_{0.85}Ga_{0.15}As$ absorbing layers and lattice matched $InAs_{0.68}P_{0.32}$ capping layers have displayed a quantum efficiency of 52% at a wavelength of 2.55 μm but with dark currents of between 10 and 20 μA. It is suggested [Ref. 34] that these high dark currents are a result of electrically active defects associated with misfit dislocations in the active regions of the diode which can be reduced by improved materials growth.

In common with injection lasers, the HgCdTe material system has been utilized to fabricate long wavelength photodiodes (see Section 6.11) $Hg_{1-x}Cd_xTe$ ternary alloys form a continuous family of semiconductors whose bandgap energy variation with x enables optical detection from 0.8 μm to the far-infrared. Furthermore, with this material system the hole ionization coefficient exhibits a resonant characteristic which is a function of the alloy composition [Ref. 31]. This phenomenon is a band structure effect which is also displayed by the $Ga_{1-x}Al_xSb$ alloys [Ref. 34]. Resonant impact ionization processes in such materials yield a high ionization coefficient ratio* providing enhanced sensitivity at particular operating wavelengths. However, for the HgCdTe material system this occurs for a composition in the vicinity of $x = 0.7$, which is suitable for detection at a wavelength of 1.3 μm. Hence both HgCdTe *p–i–n* photodiodes and APDs that exhibit low sensitivity and high speed response (500 MHz) have been produced for operation at this wavelength [Ref. 36]. In addition, $Hg_{0.4}Cd_{0.6}Te$ APDs have been fabricated for detection at a wavelength of 1.55 μm [Ref. 37]. With this material system, the potential remains, however, to provide even longer wavelength operation into the mid-infrared region, although this would be greatly assisted by improvements in the material quality to facilitate higher performance devices [Ref. 31].

8.11 Phototransistors

The problems encountered with APDs for use in the longer wavelength region stimulated a renewed interest in bipolar phototransistors in the late 1970s. Hence, although these devices have been investigated for a number of years, they have yet to find use in major optical fiber communication systems. In common with the APD the phototransistor provides internal gain of the photocurrent. This is achieved through transistor action rather than avalanche multiplication. A symbolic representation of the *n–p–n* bipolar phototransistor is shown in Figure 8.23(a). It differs from the conventional bipolar transistor in that the base is unconnected, the

* In this case the ratio of hole ionization coefficient to electron ionization coefficient.

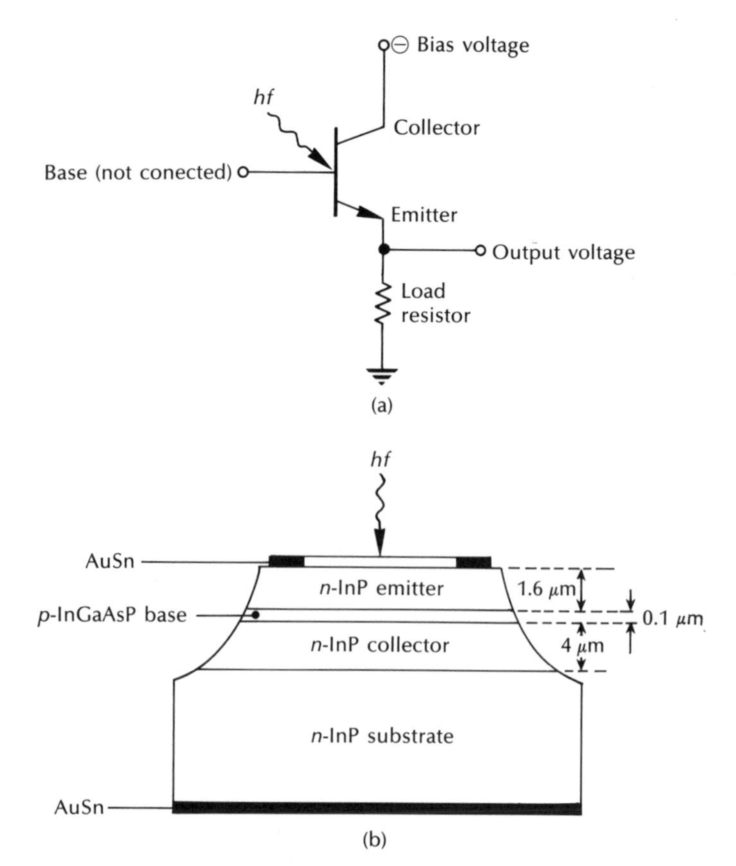

Figure 8.23 (a) Symbolic representation of the *n–p–n* phototransistor showing the external connections. (b) Cross section of an *n–p–n* InGaAsP/InP heterojunction phototransistor [Ref. 38].

base–collector junction being photosensitive to act as a light-gathering element. Thus absorbed light affects the base current giving multiplication of primary photocurrent through the device.

The structure of a *n–p–n* InGaAsP/InP heterojunction phototransistor is shown in Figure 8.23(b) [Ref. 38]. The three layer heterostructure (see Section 6.3.5) is grown on an InP substrate using liquid-phase epitaxy (LPE). It consists of an *n* type InP collector layer followed by a thin (0.1 μm)*p* type InGaAsP base layer. The third layer is a wide bandgap *n* type InP emitter layer. Radiation incident on the device passes unattenuated through the wide bandgap emitter and is absorbed in the base, base–collector depletion region and the collector. A large secondary photocurrent between the emitter and collector is obtained as the photogenerated holes are swept into the base, increasing the forward bias on the device. The use of the heterostructure permits low emitter–base and collector–base junction capacitances together with low base resistance. This is achieved through low emitter and collector doping levels coupled with heavy doping of the base, and allows large current gain. In addition the potential barrier created by the heterojunction at the emitter–base junction effectively eliminates hole injection from the base when the junction is forward biased. This gives good emitter base injection efficiency. The optical gain G_o of the device is given approximately by [Ref. 38]:

$$G_o \simeq \eta h_{FE} = \frac{hf}{e}\frac{I_c}{P_o} \tag{8.31}$$

where η is the quantum efficiency of the base–collector photodiode, h_{FE} is the common emitter current gain, I_s is the collector current, P_o is the incident optical power, e is the electronic charge and hf is the photon energy.

The phototransistor shown in Figure 8.23(b) is capable of operating over the 0.9 to 1.3 μm wavelength band giving optical gains in excess of one hundred, as demonstrated in Example 8.7. Moreover, the InGaAs/InP heterojunction phototransistor has also been the subject of significant development work but it is not anticipated that these devices will replace photodiodes within high sensitivity, high transmission rate applications [Ref. 31].

Example 8.7

The phototransistor of Figure 8.23(b) has a collector current of 15 mA when the incident optical power at a wavelength of 1.26 μm is 125 μW. Estimate:

(a) the optical gain of the device under the above operating conditions;
(b) the common emitter current gain if the quantum efficiency of the base–collector photodiode at a wavelength of 1.26 μm is 40%.

Solution: (a) Using Eq. (8.31), the optical gain is given by:

$$G_o \simeq \frac{hf}{e}\frac{I_c}{P_o} = \frac{hc}{\lambda e}\frac{I_c}{P_o}$$

$$= \frac{6.626 \times 10^{-34} \times 2.998 \times 10^{8} \times 15 \times 10^{-3}}{1.26 \times 10^{-6} \times 1.602 \times 10^{-19} \times 125 \times 10^{-6}}$$

$$= 118.1$$

(b) The common emitter current gain is:

$$h_{FE} = \frac{G_o}{\eta} = \frac{118.1}{0.4} = 295.3$$

In this example a common emitter current gain of 295 gives an optical gain of 118. It is therefore possible that this type of device will become an alternative to the APD for optical detection at wavelengths above 1.1 μm [Refs. 39–41].

8.12 Photoconductive detectors

The photoconductive detector or photoconductor, which provides what is conceptually the simplest form of semiconductor optical detection, has not until recently been considered as a serious contender for photodetection within optical fiber communications. Lately, however, there has been renewed interest in such devices, particularly for use in the longer wavelength region because of the suitability of III–V semiconductors for photoconductive detection applications [Ref. 12]. The basic detection process in a semiconductor discussed in Section 8.3 indicated that an electron may be raised from the valence band to the conduction band by the absorption of a photon provided that the photon energy was greater than the bandgap energy (i.e. $hf \geqslant E_g$). In this case, as long as the electron remains in the conduction band it will cause an increase in the conductivity of the semiconductor, a phenomenon referred to as photoconductivity. This forms the basic mechanism for the operation of photoconductive detectors.

A typical photoconductor device structure designed for operation in the longer wavelength region is shown in Figure 8.24 [Ref. 12]. The conducting channel comprises a thin layer (1 to 2 μm) of n type InGaAs which can absorb a significant amount of the incident light over the wavelength range 1.1 to 1.6 μm. In particular, good sensitivities at reasonably high bandwidths have been obtained with lightly doped (less than 5×10^{14} cm^{-3}) n type $In_{0.53}Ga_{0.47}As$ channel layers. Moreover, the composition of the InGaAs is arranged to be lattice-matched to the semi-insulating InP substrate to avoid the formation of dislocations and other crystalline imperfections in the epitaxial layer. Low resistance contacts are made to the conducting layer through the use of interdigital anodes and cathodes, as illustrated in Figure 8.24. In addition, these contacts are designed to maximize the coupling of light into the absorbing region by minimizing their obstruction of the active area whilst reducing the distance that photogenerated carriers have to travel prior to being collected at one of the electrodes. The optical coupling efficiency can also be improved by the application of an antireflection coating to the surface of the photoconductor facing the optical input.

In operation the incident light on the channel region is absorbed, thereby generating additional electron–hole pairs. These photogenerated carriers increase the channel conductivity which results in an increased current in the external circuit. The optical receiver must therefore be sensitive to very small changes in resistance induced by the incident light. Furthermore, once the carriers have been generated the electrons will be swept by the applied electric field towards the anode whilst the holes move towards the cathode. In general, however, the mobility of the holes is considerably smaller than that of the electrons in III–V alloys such as GaAs and InGaAs. Thus the electrons, being the fastest charge carriers, provide the minimum time for detection and hence the limitation on the speed of response of the photoconductive detector.

In addition, whilst the fast electrons are collected at the anode, the corresponding holes are still proceeding across the channel. This creates an absence of electrons

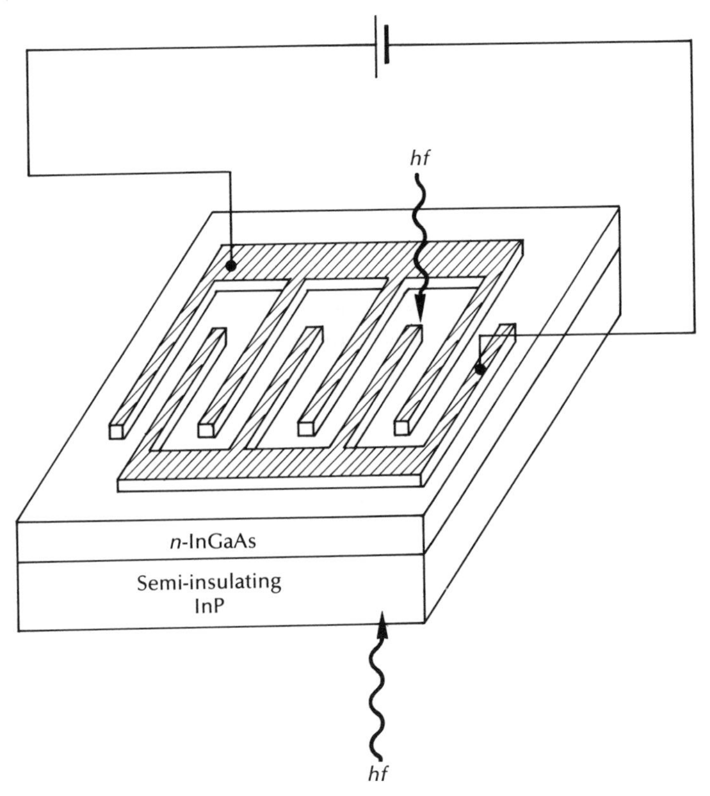

Figure 8.24 Photoconductive detector structure for operation in the 1.1 to 1.6 μm wavelength range.

and hence a net positive charge in the channel region. However, the excess charge is immediately compensated by the injection of further electrons from the cathode into the channel. Thus further electrons may be generated from the absorption of a single photon. This factor creates what is known as the photoconductive gain G which may be defined as the ratio of the slow carrier transit time (or lifetime) t_s to the fast carrier transit time t_f. Hence:

$$G = \frac{t_s}{t_f} \tag{8.32}$$

Moreover, the photocurrent I_p produced by the photoconductor following Eq. (8.8) can be written as [Ref. 42]:

$$I_p = \frac{\eta P_o e}{hf} G \tag{8.33}$$

where η is the device quantum efficiency which in a similar manner to the absorption process (see Eq. (8.1)) follows an exponential distribution, P_o is the incident optical power and e is the charge on an electron.

Since the current response in the photoconductor remains for a time t_s after the end of an incident optical pulse and exhibits an exponential decay with a time constant equal to this slow carrier transit time (or lifetime), then the maximum 3 dB bandwidth B_m for the device is given by:

$$B_m = \frac{1}{2\pi t_s} \tag{8.34}$$

Equations (8.32) and (8.34) can be combined to provide an expression for the gain–bandwidth product of a photoconductor as:

$$G \cdot B_m = \frac{1}{2\pi t_f} \tag{8.35}$$

It may be noted that an implication of Eq. (8.35) is that when t_f is fixed, photoconductive gain can only be obtained at the expense of the maximum bandwidth permitted by the device. There is therefore a trade-off between gain (and hence sensitivity) and speed of response.

Example 8.8

The electron transit time in an InGaAs photoconductive detector is 5 ps. Determine the maximum 3 dB bandwidth permitted by the device when its photoconductive gain is 70.

Solution: Using Eq. (8.35) the maximum 3 dB bandwidth provided by the photoconductor may be written as:

$$B_m = \frac{1}{2\pi t_f G} = \frac{1}{2\pi \times 5 \times 10^{-12} \times 70}$$

$$= 454.7 \text{ MHz}$$

The result obtained in Example 8.8 typifies the bandwidth (i.e. less than 500 MHz) currently achievable with InGaAs photoconductors with gains in the range 50 to 100. By contrast, silicon photoconductors can provide very large gains of around 1000 but this is accompanied by a reduction in the 3 dB bandwidth obtained from these devices (response times typically in the range 1 μs to 1 ms).

The sensitivity of a photoconductive detector is limited by the noise generated within the device, even though the quantum efficiency may be quite high. For example, in an InGaAs photoconductor with a 2 μm channel the quantum efficiency is around 88%. However, there are numerous sources contributing to the noise component within photoconductors [Ref. 43]. In particular, noise arises from two sources: Johnson noise associated with the thermal noise from the bulk resistance of the photoconductor slab, and generation–recombination noise caused by fluctuations in the generation and recombination rates of the photogenerated carrier pairs. The former noise source, which is often dominant, results in a finite dark conductivity for the device which generates a randomly varying background dark current. It can be shown [Ref. 12] that the signal to noise ratio of a photoconductor receiver increases with increasing channel resistance and gain. Therefore, a method for increasing the sensitivity of the photoconductor is to increase its photoconductive gain. Unfortunately, as indicated previously, this reduces the device response time.

Alternatively the photoconductor can be designed such that the dark current is reduced to a level such that the generation–recombination noise, which is fundamental, tends to dominate. Some success has been achieved in this direction using *p* type substrates [Ref. 44]. In this case, however, the speed of response for the device was limited by the electrode spacing required to improve the sensitivity. Improved performance has been obtained with a recent photoconductive-like detector which effectively combines a photoconductor and a *p–i–n* photodiode [Ref. 45]. This hybrid device type also illustrates another useful factor concerning photoconductive detectors in that they are very suitable for monolithic integration to provide optoelectronic integrated circuits (see Section 10.7).

Problems

8.1 Outline the reasons for the adoption of the materials and devices used for photodetection in optical fiber communications. Discuss in detail the *p–i–n* photodiode with regard to performance and compatibility requirements in photodetectors.

8.2 A *p–i–n* photodiode on average generates one electron–hole pair per three incident photons at a wavelength of 0.8 μm. Assuming all the electrons are collected calculate:

(a) the quantum efficiency of the device;
(b) its maximum possible bandgap energy;
(c) the mean output photocurrent when the received optical power is 10^{-7} W.

8.3 Explain detection process in the *p–n* photodiode. Compare this device with the *p–i–n* photodiode.

8.4 Define the quantum efficiency and the responsivity of a photodetector.

Derive an expression for the responsivity of an intrinsic photodetector in terms of the quantum efficiency of the device and the wavelength of the incident radiation.

Determine the wavelength at which the quantum efficiency and the responsivity are equal.

8.5 A *p–n* photodiode has a quantum efficiency of 50% at a wavelength of 0.9 μm. Calculate:

(a) its responsivity at 0.9 μm;
(b) the received optical power if the mean photocurrent is 10^{-6} A;
(c) the corresponding number of received photons at this wavelength.

8.6 When 800 photons per second are incident on a *p–i–n* photodiode operating at a wavelength of 1.3 μm they generate on average 550 electrons per second which are collected. Calculate the responsivity of the device.

8.7 Explain what is meant by the long wavelength cutoff point for an intrinsic photodetector, deriving any relevant expressions.

Considering the bandgap energies given in Table 8.1, calculate the long wavelength cutoff points for both direct and indirect optical transitions in silicon and germanium.

8.8 A *p–i–n* photodiode ceases to operate when photons with energy greater than 0.886 eV are incident upon it; of which material is it fabricated?

8.9 (a) The time taken for electrons to diffuse through a layer of *p* type silicon is 28.8 ns. If the minority carrier diffusion coefficient is 3.4×10^{-3} m^2 s^{-1}, determine the thickness of the silicon layer.

(b) Assuming the depletion layer width in a silicon photodiode corresponds to the layer thickness obtained in part (a) and that the maximum response time of the photodiode is 877 ps, estimate the carrier (hole) drift velocity.

8.10 A silicon *p–i–n* photodiode with an area of 1.5 mm^2 is to be used in conjunction with a load resistor of 100 Ω. If the requirement for the device is a fast response time, estimate the thickness of the intrinsic region that should be provided. It may be assumed that the permittivity for silicon is 1.04×10^{-10} F m^{-1} and that the electron saturation velocity is 10^7 m s^{-1}.

8.11 Define the noise equivalent power (NEP) for a photodetector. Commencing with Eq. (8.8) obtain an expression for the NEP of a photodiode in which the dark current noise dominates.

A silicon *p–i–n* photodiode with active dimensions 10 μm has a specific detectivity of 7×10^{10} m Hz$^{\frac{1}{2}}$ W^{-1} when operating at a wavelength of 0.85 μm. The device quantum efficiency at this wavelength is 64%. Assuming that it is the dominant noise source, calculate the dark current over a 1 Hz bandwidth in the device.

8.12 The specific detectivity of a wide area silicon photodiode at its operating wavelength is 10^{11} m Hz$^{\frac{1}{2}}$ W^{-1}. Estimate the smallest detectable signal power at this wavelength when the sensitive area of the device is 25 mm^2 and the signal bandwidth is 1 kHz.

8.13 Discuss the operation of the silicon RAPD, describing how it differs from the *p–i–n* photodiode.

Outline the advantages and drawbacks with the use of the RAPD as a detector for optical fiber communications.

8.14 Compare and contrast the structure and performance characteristics of germanium and III–V semiconductor alloy APDs for operation in the wavelength range 1.1 to 1.6 μm.

8.15 An APD with a multiplication factor of 20 operates at a wavelength of 1.5 μm. Calculate the quantum efficiency and the output photocurrent from the device if its responsivity at this wavelength is 0.6 A W^{-1} and 10^{10} photons of wavelength 1.5 μm are incident upon it per second.

8.16 Given that the following measurements were taken for an APD, calculate the multiplication factor for the device.

Received optical power at 1.35 μm = 0.2 μW
Corresponding output photocurrent = 4.9 μA
(after avalanche gain)
Quantum efficiency at 1.35 μm = 40%

8.17 An APD has a quantum efficiency of 45% at 0.85 μm. When illuminated with radiation of this wavelength it produces an output photocurrent of 10 μA after avalanche gain with a multiplication factor of 250. Calculate the received optical power to the device. How many photons per second does this correspond to?

8.18 When 10^{11} photons per second each with an energy of 1.28×10^{-19} J are incident on an ideal photodiode, calculate:

(a) the wavelength of the incident radiation;
(b) the output photocurrent;
(c) the output photocurrent if the device is an APD with a multiplication factor of 18.

8.19 A silicon RAPD has a multiplication factor of 10^3 when operating at a wavelength of 0.82 μm. At this operating point the quantum efficiency of the device is 90% and the dark current is 1 nA.

Determine the number of photons per second of wavelength 0.82 μm required in order to register a light input to the device corresponding to an output current (after avalanche gain) which is greater than the level of the dark current (i.e. $I > 1$ nA).

8.20 Indicate the material systems under investigation and discuss their application in the fabrication of photodiodes for use in the mid-infrared wavelength region.

8.21 An InGaAsP heterojunction phototransistor has a common emitter current gain of 170 when operating at a wavelength of 1.3 μm with an incident optical power of 80 μW. The base collector quantum efficiency at this wavelength is 65%. Estimate the collector current in the device.

8.22 Describe the basic detection process in a photoconductive detector.

The maximum 3 dB bandwidth allowed by an InGaAs photoconductive detector is 380 MHz when the electron transit time through the device is 7.6 ps. Calculate the photocurrent obtained from the device when 10 μW of optical power at a wavelength of 1.32 μm is incident upon it, and the device quantum efficiency is 75%.

Answers to numerical problems

8.2 (a) 33%; (b) 24.8×10^{-20} J; (c) 21.3 nW
8.4 1.24 μm
8.5 (a) 0.36 A W^{-1}; (b) 2.78 μW; (c) 1.26×10^{13} photon s^{-1}
8.6 0.72 A W^{-1}
8.7 0.3 μm, 1.09 μm, 1.53 μm, 1.85 μm
8.8 $In_{0.7}Ga_{0.3}As_{0.64}P_{0.36}$
8.9 (a) 14 μm; (b) 10^5 ms^{-1}
8.10 395 μm
8.11 1.23×10^{-14} A
8.12 1.58 pW
8.15 50%, 15.9 nA
8.16 24.1
8.17 77.8 nW, 3.33×10^{11} photon s^{-1}
8.18 (a) 1.55 μm; (b) 1.6 μA; (c) 28.8 μA
8.19 6.94×10^6 photon s^{-1}
8.21 9.3 mA
8.22 0.44 mA

References

[1] H. Melchior, M. B. Fisher and F. R. Arams, 'Photodetectors for optical communication systems', *Proc. IEEE*, **58**, pp. 1446–1486, 1970.

[2] H. Melchior, 'Detectors for lightwave communications', *Phys. Today*, **30**, pp. 32–39, 1977.

[3] S. D. Personick, 'Photodetectors for fiber systems', in M. K. Barnoski (Ed.), *Fundamentals of Optical Fiber Communications* (2nd edn), pp. 257–293, Academic Press, 1981.

[4] T. P. Lee and T. Li, 'Photodetectors', in S. E. Miller and A. G. Chynoweth (Eds.), *Optical Fiber Telecommunications*, pp. 593–626, Academic Press, 1979.

[5] S. M. Sze, *Physics of Semiconductor Devices* (2nd edn), John Wiley, 1981.

[6] D. H. Newman and S. Ritchie, 'Sources and detectors for optical fibre communications applications: the first 20 years', *IEE Proc., Pt. J*, **133**(3), pp. 213–229, 1986.

[7] R. W. Dixon and N. K. Dutta, 'Lightwave device technology', *AT&T Tech. J.*, **66**(1), pp.73–83, 1987.

[8] J. E. Bowers and C. A. Burrus, 'Ultrawide-band long-wave *p–i–n* photodetectors', *J. of Lightwave Technol.*, **LT – 5**(10), pp. 1339–1350, 1987.

[9] A. W. Nelson, S. Wong, S. Ritchie and S. K. Sargood, 'GaInAs PIN photodiodes grown by atmospheric-pressure MOVPE', *Electron. Lett.*, **21**(19), pp. 838–840, 1985.

[10] S. Miura, H. Kuwatsuka, T. Mikawa and O. Wada, 'Planar embedded InP/GaInAs *p–i–n* photodiode for very high speed operation', *J. of Lightwave Technol.*, **LT-5** (10), pp. 1371–1376, 1987.

[11] G. Keiser, *Optical Fiber Communications* (2nd edn), McGraw-Hill, 1991.

[12] S. R. Forrest, 'Optical detectors for lightwave communication', in S. E. Miller and I. P. Kaminow (Eds.), *Optical Fiber Telecommunications II*, Academic Press, pp. 569–599, 1988.

[13] M. Schwartz, *Information Transmission, Modulation and Noise* (4th edn), McGraw-Hill, 1990.

[14] T. P. Lee, C. A. Burrus Jr and A. G. Dentai, 'InGaAsP/InP photodiodes microplasma-limited avalanche multiplication at 1–1.3 μm wavelength', *IEEE J. Quantum Electron.*, **QE-15**, pp. 30–35, 1979.

[15] P. P. Webb, R. J. McIntyre and J. Conradi, 'Properties of avalanche photodiodes', *RCA Rev.*, **35**, pp. 235–277, 1974.

[16] A. R. Hartman, H. Melchior, D. P. Schinke and T. E. Seidel, 'Planar epitaxial silicon avalanche photodiode', *Bell Sys. Tech. J.*, **57**, pp. 1791–1807, 1978.

[17] T. Pearsall, 'Photodetectors for communication by optical fibres', in M. J. Howes and D. V Morgan (Eds.), *Optical Fibre Communications*, pp. 107–165, John Wiley, 1980.

[18] H. Ando, H. Kanbe, T. Kimura, T. Yamaoka and T. Kaneda, 'Characteristics of germanium avalanche photodiodes in the wavelength region 1–1.6 μm, *IEEE J. Quantum Electron.*, **QE-14**(11), pp. 804–809, 1978.

[19] T. Mikawa, S. Kagawa, T. Kaneda, T. Sakwai, H. Ando and O. Mikami, 'A low-noise n^+np germanium avalanche photodiode', *IEEE J. Quantum Electron.* **QE-17**(2), pp. 210–216, 1981.

[20] O. Mikami, H. Ando, H. Kanbe, T. Mikawa, T. Kaneda and Y. Toyama, 'Improved germanium avalanche photodiodes', *IEEE J. Quantum Electron.*, **QE-16**(9), pp. 1002–1007, 1980.

[21] S. Kagawa, T. Kaneda, T. Mikawa, Y. Banba, Y. Toyama and O. Mikami, 'Fully ion-implanted p^+-n germanium avalanche photodiodes', *Appl. Phys. Lett.*, **38**(6), pp. 429–431, 1981.

[22] T. Mikawa, T. Kaneda, H. Nishimoto, M. Motegi and H. Okushima, 'Small-active-area germanium avalanche photodiode for single-mode fibre at 1.3 μm wavelength', *Electron. Lett.*, **19**(12), pp. 452–453, 1983.

[23] M. Niwa, Y. Tashiro, K. Minemura and H. Iwasaki, 'High-sensitivity Hi–Lo

germanium avalanche photodiode for 1.5 μm wavelength optical communication', *Electron. Lett.*, **20**(13), pp. 552–553, 1984.

[24] G. E. Stillman, 'Detectors for optical-waveguide communications', in E. E. Basch (Ed.), *Optical-Fiber Transmission*, H. W. Sams & Co., pp. 335–374, 1987.

[25] S. R. Forrest, O. K. Kim and R. G. Smith, 'Optical response time in $In_{0.53}Ga_{0.47}As/InP$ avalanche photodiodes', *Appl. Phys. Lett.*, **41**, pp. 95–98, 1982.

[26] J. C. Campbell, A. G. Dentai, W. S. Holder and B. L. Kasper, 'High performance avalanche photodiode with separate absorption "grading" and multiplication regions', *Electron. Lett.*, **19**, pp. 818–820, 1983.

[27] B. L. Kasper and J. C. Campbell, 'Multigigabit-per-second avalanche photodiode lightwave receivers', *J. of Lightwave Technol.*, **LT-5**(10), pp. 1351–1364, 1987.

[28] C. Fujihashi, 'Dark-current multiplication noises in avalanche photodiodes and optimum gains', *J. of Lightwave Technol.*, **LT-5**(6), pp. 798–808, 1987.

[29] R. Chin, N. Holonyak, G. E. Stillman, J. Y. Tang and K. Hess, 'Impact ionisation in multi-layer heterojuction structures', *Electron. Lett.*, **16**(12), pp. 467–468, 1980.

[30] F. Capasso, 'Band-gap engineering via graded gap, superlattice and periodic doping structures: applications and novel photodetectors and other devices', *J. Vac. Sci. Technol. B*, **1**(2), pp. 457–461, 1983.

[31] B. T. Debney and A. C. Carter, 'Optical detectors and receivers', in J. Dakin and B. Culshaw (Eds.) *Optical Fiber Sensors: Principles and components*, Artech House, pp. 107–149, 1988.

[32] J. E. Bowers, A. K. Srivastrava, C. A. Burrus, J. C. Dewinter, M. A. Pollack, J. L. Zyskind, 'High-speed GaInAsSb/GaSb PIN photodetectors for wavelengths to 2.3 μm', *Electron. Lett.*, **22**(3), pp. 137–138, 1986.

[33] A. J. Moseley, M. D. Scott, A. H. Moore and R. H. Wallis, 'High-efficiency, low-leakage MOCVD-grown GaInAs/AlInAs heterojunction photodiodes for detection to 2.4 μm, *Electron Lett.*, **22**(22), pp. 1206–1207, 1986.

[34] R. U. Martinelli, T. J. Zamerowski and P. A. Longeway, '$In_xG_{1-x}As/InAs_yP_{1-y}$ lasers and photodiodes for 2.55 μm optical fiber communications', *Opt. Fiber. Commun. Conf., OFC '88* (USA), paper TUC6, January 1988.

[35] O. Hildebrand, W. Kuebart, K. W. Benz and M. H. Pilkuhn, 'GaAlSb avalanche photodiodes: resonant impact ionisation with very high ratio of ionisation coefficients', *IEEE J. Quantum Electron.*, **17**(2), pp. 284–288, 1981.

[36] G. Pichard, J. Meslage, T. Nguyen Duy and F. Raymond, '1.3 μm CdHgTe avalanche photodiodes for fibre-optic applications', in *Proc. 9th ECOC 83* (Switzerland) pp. 479–482, 1983.

[37] H. Haupt and O. Hildebrand, 'Lasers and photodetectors in Europe', *IEEE J. on Selected Areas in Commun.*, **SAC-4**(4), pp. 444–456, 1986.

[38] P. D. Wright, R. J. Nelson and T. Cella, 'High gain InGaAsP–InP heterojunction phototransistors'. *Appl. Phys. Lett.*, **37**(2), pp. 192–194, 1980.

[39] R. A. Milano, P. D. Dapkus and G. E. Stillman, 'Heterojunction phototransistors for fiber-optic communications', *Proc. SPIE Int. Soc. Opt. Eng.*, **272**, pp. 43–50, 1981.

[40] K. Tubatabaie-Alavi and C. G. Fonstad, 'Recent advances in InGaAs/InP phototransistors', *Proc. SPIE Int. Soc. Opt. Eng.*, **272**, pp. 38–42, 1981.

[41] G. E. Stillman, L. W. Cook, G. E. Bulman, N. Tabatabaie, R. Chin and P. D. Dapkus, 'Long-wavelength (1.3 to 1.6 μm) detectors for fiber-optical communications', *IEEE Trans. Electron. Dev.*, **ED-29**(9), pp. 1355–1371, 1982.

[42] S. M. Sze, *Semiconductor Devices: Physics and technology*, John Wiley 1985.
[43] S. R. Forrest, 'The sensitivity of photoconductor receivers for long-wavelength optical communications', *J. of Lightwave Technol.*, **LT-3**(2), pp. 347–360, 1985.
[44] C. Y. Chen, A. G. Dentai, B. L. Kasper and P. A. Garbinski, 'High-speed junction-depleted $Ga_{0.47}In_{0.53}As$ photoconductive detectors', *Appl. Phys. Lett*, **46**, pp. 1164–1166, 1985.
[45] T. Morita, M. Murata, K. Koike and K. Ono, 'High speed GaInAs photoconductive-like detectors for long wavelength optical communication', *Proc. 14th ECOC '88* (UK) pp. 424–427, September 1988.

9

Direct detection receiver performance considerations

9.1 Introduction

The receiver in an intensity modulated/direct detection (IM/DD) optical fiber communication system (see Section 7.5) essentially consists of the photodetector plus an amplifier with possibly additional signal processing circuits. Therefore the receiver initially converts the optical signal incident on the detector into an electrical signal, which is then amplified before further processing to extract the information originally carried by the optical signal.

The importance of the detector in the overall system performance was stressed in Chapter 8. However, it is necessary to consider the properties of this device in the context of the associated circuitry combined in the receiver. It is essential that the detector performs efficiently with the following amplifying and signal processing circuits. Inherent to this process is the separation of the information originally contained in the optical signal from the noise generated within the rest of the system and in the receiver itself, as well as any limitations on the detector response imposed by the associated circuits. These factors play a crucial role in determining the performance of the system.

In order to consider receiver design it is useful to regard the limit on the performance of the system set by the signal to noise ratio (SNR) at the receiver. It is therefore necessary to outline noise sources within optical fiber systems. The noise in these systems has different origins from that of copper-based systems. Both types of system have thermal noise generated in the receiver. However, although optical fiber systems exhibit little crosstalk the noise generated within the detector must be considered, as well as the noise properties associated with the electromagnetic carrier.

In Section 9.2 we therefore briefly review the major noise mechanisms which are present in direct detection optical fiber communication receivers prior to more detailed discussion of the limitations imposed by photon (or quantum) noise in both digital and analog transmission. This is followed in Section 9.3 with a more specific discussion of the noise associated with the two major receiver types (i.e. employing *p–i–n* and avalanche photodiode detectors). Expressions for the SNRs of these two receiver types are also developed in this section. Section 9.4 considers the noise and bandwidth performance of common preamplifier structures utilized in the design of optical fiber receivers. In Section 9.5 we present a brief account of low noise field effect transistor (FET) preamplifiers which find wide use within optical fiber communication receivers. This discussion also includes consideration of *p–i–n* photodiode/FET (PIN–FET) hybrid receiver circuits which have been developed for optical fiber communications. Finally, major high performance receiver design strategies to provide low noise and high bandwidth operation as well as wide dynamic range are described in Section 9.6.

9.2 Noise

Noise is a term generally used to refer to any spurious or undesired disturbances that mask the received signal in a communication system. In optical fiber communication systems we are generally concerned with noise due to spontaneous fluctuations rather than erratic disturbances which may be a feature of copper-based systems (due to electromagnetic interference, etc.).

There are three main types of noise due to spontaneous fluctuations in optical fiber communication systems: thermal noise, dark current noise and quantum noise.

9.2.1 Thermal noise

This is the spontaneous fluctuation due to thermal interaction between, say, the free electrons and the vibrating ions in a conducting medium, and it is especially prevalent in resistors at room temperature.

The thermal noise current i_t in a resistor R may be expressed by its mean square value [Ref. 1] and is given by:

$$\overline{i_t^2} = \frac{4KTB}{R} \tag{9.1}$$

where K is Boltzmann's constant, T is the absolute temperature and B is the post-detection (electrical) bandwidth of the system (assuming the resistor is in the optical receiver).

9.2.2 Dark current noise

When there is no optical power incident on the photodetector a small reverse leakage current still flows from the device terminals. This dark current (see Section 8.4.2) contributes to the total system noise and gives random fluctuations about the average particle flow of the photocurrent. It therefore manifests itself as shot noise [Ref. 1] on the photocurrent. Thus the dark current noise $\overline{i_d^2}$ is given by:

$$\overline{i_d^2} = 2eBI_d \tag{9.2}$$

where e is the charge on an electron and I_d is the dark current. It may be reduced by careful design and fabrication of the detector.

9.2.3 Quantum noise

The quantum nature of light was discussed in Section 6.2.1 and the equation for the energy of this quantum or photon was stated as $E = hf$. The quantum behaviour of electromagnetic radiation must be taken into account at optical frequencies since $hf > KT$ and quantum fluctuations dominate over thermal fluctuations.

The detection of light by a photodiode is a discrete process since the creation of an electron–hole pair results from the absorption of a photon, and the signal emerging from the detector is dictated by the statistics of photon arrivals. Hence the statistics for monochromatic coherent radiation arriving at a detector follows a discrete probability distribution which is independent of the number of photons previously detected.

It is found that the probability $P(z)$ of detecting z photons in time period τ when it is expected on average to detect z_m photons obeys the Poisson distribution [Ref. 2]:

$$P(z) = \frac{z_m^z \exp(-z_m)}{z!} \tag{9.3}$$

where z_m is equal to the variance of the probability distribution. This equality of the mean and the variance is typical of the Poisson distribution. From Eq. (8.7) the electron rate r_e generated by incident photons is $r_e = \eta P_o/hf$. The number of electrons generated in time τ is equal to the average number of photons detected

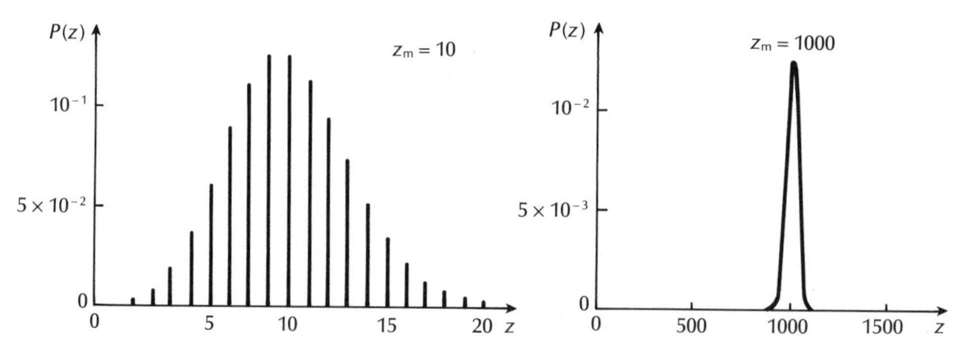

Figure 9.1 Poisson distributions for $z_m = 10$ and $z_m = 1000$.

over this time period z_m. Therefore:

$$z_m = \frac{\eta P_o \tau}{hf} \tag{9.4}$$

The Poisson distributions for $z_m = 10$ and $z_m = 1000$ are illustrated in Figure 9.1 and represent the detection process for monochromatic coherent light.

Incoherent light is emitted by independent atoms and therefore there is no phase relationship between the emitted photons. This property dictates exponential intensity distribution for incoherent light which if averaged over the Poisson distribution [Ref. 2] gives

$$P(z) = \frac{z_m^z}{(1 + z_m)^{z+1}} \tag{9.5}$$

Equation (9.5) is identical to the Bose–Einstein distribution [Ref. 3] which is used to describe the random statistics of light emitted in black body radiation (thermal light). The statistical fluctuations for incoherent light are illustrated by the probability distributions shown in Figure 9.2.

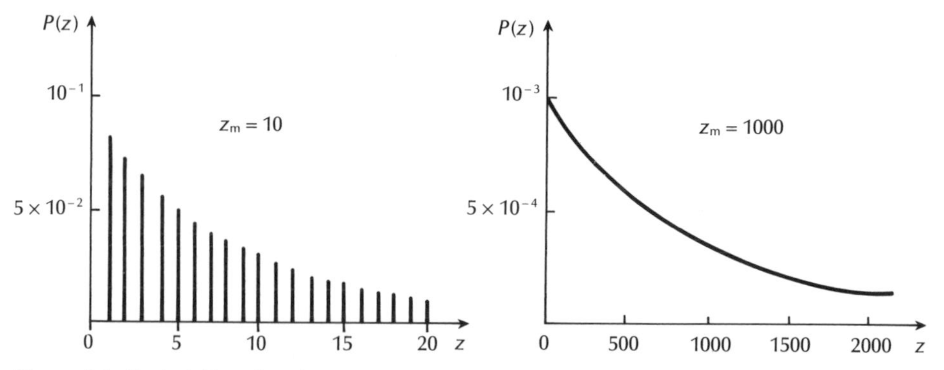

Figure 9.2 Probability distributions indicating the statistical fluctuations of incoherent light for $z_m = 10$ and $z_m = 1000$.

9.2.4 Digital signalling quantum noise

For digital optical fiber systems it is possible to calculate a fundamental lower limit to the energy that a pulse of light must contain in order to be detected with a given probability of error. The premise on which this analysis is based is that the ideal receiver has a sufficiently low amplifier noise to detect the displacement current of a single electron–hole pair generated within the detector (i.e. an individual photon may be detected). Thus in the absence of light, and neglecting dark current, no current will flow. Therefore the only way an error can occur is if a light pulse is present and no electron–hole pairs are generated. The probability of no pairs being generated when a light pulse is present may be obtained from Eq. (9.3) and is given by:

$$P(0/1) = \exp(-z_m) \tag{9.6}$$

Thus in the receiver described $P(0/1)$ represents the system error probability $P(e)$ and therefore:

$$P(e) = \exp(-z_m) \tag{9.7}$$

However, it must be noted that the above analysis assumes that the photodetector emits no electron–hole pairs in the absence of illumination. In this sense it is considered perfect. Equation (9.7) therefore represents an absolute receiver sensitivity and allows the determination of a fundamental limit in digital optical communications. This is the minimum pulse energy E_{min} required to maintain a given bit error rate (BER) which any practical receiver must satisfy and is known as the quantum limit.

Example 9.1

A digital optical fiber communication system operating at a wavelength of 1 μm requires a maximum bit error rate of 10^{-9}. Determine:

(a) the theoretical quantum limit at the receiver in terms of the quantum efficiency of the detector and the energy of an incident photon;
(b) the minimum incident optical power required at the detector in order to achieve the above bit error rate when the system is employing ideal binary signalling at 10 Mbits s^{-1}, and assuming the detector is ideal.

Solution: (a) From Eq. (9.7) the probability of error.

$$P(e) = \exp(-z_m) = 10^{-9}$$

and thus $z_m = 20.7$.

z_m corresponds to an average number of photons detected in a time period τ for a BER of 10^{-9}.

From Eq. (9.4):

$$z_{\mathrm{m}} = \frac{\eta P_{\mathrm{o}} \tau}{hf} = 20.7$$

Hence the minimum pulse energy or quantum limit

$$E_{\min} = P_{\mathrm{o}} \tau = \frac{20.7\, hf}{\eta}$$

Thus the quantum limit at the receiver to maintain a maximum BER of 10^{-9} is

$$\frac{20.7\, hf}{\eta}$$

(b) From part (a) the minimum pulse energy:

$$P_{\mathrm{o}} \tau = \frac{20.7\, hf}{\eta}$$

Therefore the average received optical power required to provide the minimum pulse energy is:

$$P_{\mathrm{o}} = \frac{20.7\, hf}{\tau \eta}$$

However, for ideal binary signalling there are an equal number of ones and zeros (50% in the on state and 50% in the off state). Thus the average received optical power may be considered to arrive over two bit periods, and

$$P_{\mathrm{o}}(\text{binary}) = \frac{20.7\, hf}{2\tau\eta} = \frac{20.7\, hf\, B_{\mathrm{T}}}{2\eta}$$

where B_{T} is the bit rate. At a wavelength of 1 μm, $f = 2.998 \times 10^{14}$ Hz, and assuming an ideal detector, $\eta = 1$.

Hence

$$P_{\mathrm{o}}\ (\text{binary}) = \frac{20.7 \times 6.626 \times 10^{-34} \times 2.998 \times 10^{14} \times 10^{7}}{2}$$

$$= 20.6 \text{ pW}$$

In decibels (dB)

$$P_{\mathrm{o}} \text{ in dB} = 10 \log_{10} \frac{P_{\mathrm{o}}}{P_{\mathrm{r}}}$$

where P_{r} is a reference power level.

When the reference power level is one watt:

$$P_{\mathrm{o}} = 10 \log_{10} P_{\mathrm{o}} \qquad \text{where } P_{\mathrm{o}} \text{ is expressed in watts}$$

$$= 10 \log_{10} 2.06 \times 10^{-11}$$

$$= 3.14 - 110$$

$$= -106.9 \text{ dBW}$$

When the reference power level is one milliwatt

$$P_o = 10 \log_{10} 2.06 \times 10^{-8}$$
$$= 3.14 - 80$$
$$= -76.9 \text{ dBm}$$

Therefore the minimum incident optical power required at the receiver to achieve an error rate of 10^{-9} with ideal binary signalling is 20.6 pW or -76.9 dBm.

The result of Example 9.1 is a theoretical limit and in practice receivers are generally found to be at least 10 dB less sensitive. Furthermore, although some 20.7 photons are required in order to detect a binary one with a bit error rate of 10^{-9}, it is clear that these photons can arrive at the receiver over two bit periods if an equal number of transmitted ones and zeros are assumed (i.e. there are no photons transmitted in the zero bit periods). Hence the 20.7 photon per pulse requirement can be considered as an average of around 10.4 photons per bit at the quantum limit.

9.2.5 Analog transmission quantum noise

In analog optical fiber systems quantum noise manifests itself as shot noise which also has Poisson statistics [Ref. 1]. The shot noise current i_s on the photocurrent I_p is given by:

$$\overline{i_s^2} = 2eBI_p \tag{9.8}$$

Neglecting other sources of noise the SNR at the receiver may be written as:

$$\frac{S}{N} = \frac{I_p^2}{\overline{i_s^2}} \tag{9.9}$$

Substituting for $\overline{i_s^2}$ from Eq. (9.8) gives:

$$\frac{S}{N} = \frac{I_p}{2eB} \tag{9.10}$$

The expression for the photocurrent I_p given in Eq. (8.8) allows the SNR to be obtained in terms of the incident optical power P_o.

$$\frac{S}{N} = \frac{\eta P_o e}{hf 2eB} = \frac{\eta P_o}{2hfB} \tag{9.11}$$

Equation (9.11) allows calculation of the incident optical power required at the receiver in order to obtain a specified SNR when considering quantum noise in analog optical fiber systems.

Example 9.2

An analog optical fiber system operating at a wavelength of 1 μm has a post detection bandwidth of 5 MHz. Assuming an ideal detector and considering only quantum noise on the signal, calculate the incident optical power necessary to achieve an SNR of 50 dB at the receiver.

Solution: From Eq. (9.11), the SNR is

$$\frac{S}{N} = \frac{\eta P_o}{2hfB}$$

Hence

$$P_o = \left(\frac{S}{N}\right)\frac{2hfB}{\eta}$$

For $S/N = 50$ dB, when considering signal and noise powers:

$$10 \log_{10} \frac{S}{N} = 50$$

and therefore $S/N = 10^5$

At 1 μm, $f = 2.998 \times 10^{14}$ Hz. For an ideal detector $\eta = 1$ and, thus the incident optical power:

$$P_o = \frac{10^5 \times 2 \times 6.626 \times 10^{-34} \times 2.998 \times 10^{14} \times 5 \times 10^6}{1}$$

$$= 198.6 \text{ nW}$$

In dBm

$$P_o = 10 \log_{10} 198.6 \times 10^{-6}$$
$$= -40 + 2.98$$
$$= -37.0 \text{ dBm}$$

Therefore the incident optical power required to achieve an SNR of 50 dB at the receiver is 198.6 nW which is equivalent to -37.0 dBm.

In practice receivers are less sensitive than the Example 9.2 suggests and thus in terms of the absolute optical power requirements analog transmission compares unfavourably with digital signalling.

However, it should be noted that there is a substantial difference in information transmission capacity between the digital and analog cases (over similar bandwidths) considered in Examples 9.1 and 9.2. For example, a 10 Mbit s^{-1} digital optical fiber communication system would provide only about 150 speech channels using standard baseband digital transmission techniques (see Section 11.5). In contrast a 5 MHz analog system, again operating in the baseband, could provide as many as 1250 similar bandwidth ($\simeq 3.4$ kHz) speech channels. A comparison of signal to quantum noise ratios between the two transmission methods, taking account of this information capacity aspect, yields less disparity although digital signalling still proves far superior. For instance, applying the figures quoted above within Examples 9.1 and 9.2, in order to compare two systems capable of transmitting the same number of speech channels (e.g. digital bandwidth of 10 Mbit s^{-1} and analog bandwidth of 600 kHz) gives a difference in absolute sensitivity in favour of digital transmission of approximately 31 dB. This indicates a reduction of around 9 dB on the 40 dB difference obtained by simply comparing the results over similar bandwidths. Nevertheless, it is clear that digital signalling techniques still provide a significant benefit in relation to quantum noise when employed within optical fiber communications.

9.3 Receiver noise

In order to investigate the optical receiver in greater detail it is necessary to consider the relative importance and interplay of the various types of noise mentioned in the preceding section. This is dependent on both the method of demodulation and the type of device used for detection.

The conditions for coherent detection are not met in IM/DD optical fiber systems for the reasons outlined in Section 7.5. Thus heterodyne and homodyne detection, which are very sensitive techniques and provide excellent rejection of adjacent channels, are not used, as the optical signal arriving at the receiver tends to be incoherent. In practice the vast majority of installed optical fiber communication systems use incoherent or direct detection in which the variation of the optical power level is monitored and no information is carried in the phase or frequency content of the signal. Therefore, the noise considerations in this section are based on a receiver employing direct detection of the modulated optical carrier which gives the same signal to noise ratio as an unmodulated optical carrier. The substantial developments in coherent optical fiber transmission, however, which have taken place over recent years are described in Chapter 12. Nevertheless, the major performance parameters associated with direct detection receivers which are discussed in this section and the following ones also apply to coherent optical receivers.

Figure 9.3 shows a block schematic of the front end of an optical receiver and the various noise sources associated with it. The majority of the noise sources shown apply to both main types of optical detector (*p–i–n* and avalanche photodiode). The noise generated from background radiation, which is important in atmospheric propagation and some copper-based systems, is negligible in both types of optical fiber receiver, and thus is often ignored. Also the beat noise generated from the various spectral components of the incoherent optical carrier can be shown to be insignificant [Ref. 4] with multimode propagation and hence

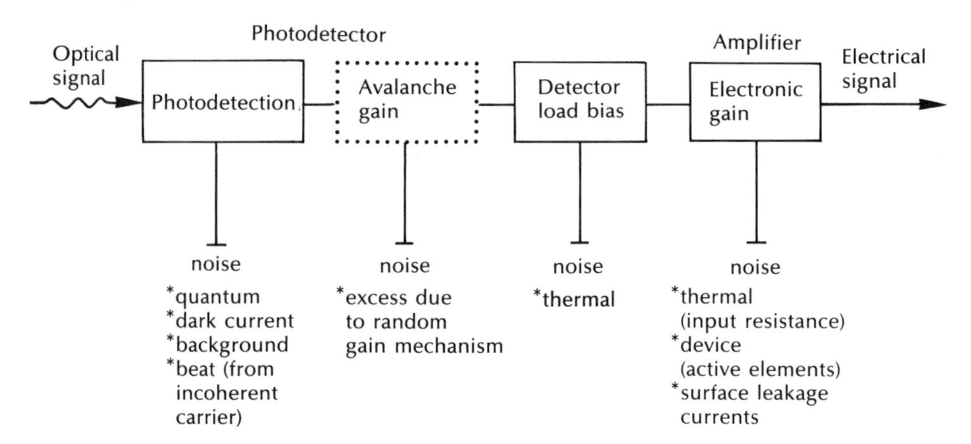

Figure 9.3 Block schematic of the front end of an optical receiver showing the various sources of noise.

will not be considered. It is necessary, however, to take into account the other sources of noise shown in Figure 9.3.

The avalanche photodiode receiver is the most complex case as it includes noise resulting from the random nature of the internal gain mechanism (dotted in Figure 9.3). It is therefore useful to consider noise in optical fiber receivers employing photodiodes without internal gain, before avalanche photodiode receivers are discussed.

9.3.1 *p–n* and *p–i–n* photodiode receiver

The two main sources of noise in photodiodes without internal gain are dark current noise and quantum noise, both of which may be regarded as shot noise on the photocurrent (i.e. effectively, analog quantum noise). When the expressions for these noise sources given in Eqs. (9.2) and (9.4) are combined the total shot noise $\overline{i_{\mathrm{TS}}^2}$ is given by:

$$\overline{i_{\mathrm{TS}}^2} = 2eB(I_{\mathrm{p}} + I_{\mathrm{d}}) \tag{9.12}$$

If it is necessary to take the noise due to the background radiation into account then the expression given in Eq. (9.12) may be expanded to include the background radiation induced photocurrent I_{b} giving

$$\overline{i_{\mathrm{TS}}^2} = 2eB(I_{\mathrm{p}} + I_{\mathrm{d}} + I_{\mathrm{b}}) \tag{9.13}$$

However, as I_{b} is usually negligible the expression given in Eq. (9.12) will be used in the further analysis.

When the photodiode is without internal avalanche gain, thermal noise from the detector load resistor and from active elements in the amplifier tends to dominate. This is especially the case for wideband systems operating in the 0.8 to 0.9 μm wavelength band because the dark currents in well-designed silicon photodiodes can be made very small. The thermal noise $\overline{i_t^2}$ due to the load resistance R_L may be obtained from Eq. (9.1) and is given by:

$$\overline{i_t^2} = \frac{4KTB}{R_L} \tag{9.14}$$

The dominating effect of this thermal noise over the shot noise in photodiodes without internal gain may be observed in Example 9.3.

Example 9.3

A silicon *p–i–n* photodiode incorporated into an optical receiver has a quantum efficiency of 60% when operating at a wavelength of 0.9 μm. The dark current in the device at this operating point is 3 nA and the load resistance is 4 kΩ.

The incident optical power at this wavelength is 200 nW and the post detection bandwidth of the receiver is 5 MHz. Compare the shot noise generated in the photodiode with the thermal noise in the load resistor at a temperature of 20 °C.

Solution: From Eq. (8.8) the photocurrent is given by:

$$I_p = \frac{\eta P_o e}{hf} = \frac{\eta P_o e \lambda}{hc}$$

Therefore

$$I_p = \frac{0.6 \times 200 \times 10^{-9} \times 1.602 \times 10^{-19} \times 0.9 \times 10^{-6}}{6.626 \times 10^{-34} \times 2.998 \times 10^{8}}$$

$$= 87.1 \text{ nA}$$

From Eq. (9.12) the total shot noise is:

$$\overline{i_{TS}^2} = 2eB(I_d + I_p)$$

$$= 2 \times 1.602 \times 10^{-19} \times 5 \times 10^{6}[(3 + 87.1) \times 10^{-9})]$$

$$= 1.44 \times 10^{-19} \text{ A}^2$$

and the root mean square (rms) shot noise current is

$$(\overline{i_{TS}^2})^{\frac{1}{2}} = 3.79 \times 10^{-10} \text{ A}$$

The thermal noise in the load resistor is given by Eq. (9.14):

$$\overline{i_t^2} = \frac{4KTB}{R_L}$$

$$= \frac{4 \times 1.381 \times 10^{-23} \times 293 \times 5 \times 10^{6}}{4 \times 10^{3}}$$

$$= 2.02 \times 10^{-17} \text{ A}^2$$

$(T = 20\,^\circ\text{C} = 293 \text{ K})$

Therefore the rms thermal noise current is

$$(\overline{i_t^2})^{\frac{1}{2}} = 4.49 \times 10^{-9} \text{ A}$$

In this example the rms thermal noise current is a factor of 12 greater than the total rms shot noise current.

Example 9.3 does not include the noise sources within the amplifier, shown in Figure 9.3. These noise sources, associated with both the active and passive elements of the amplifier, can be represented by a series voltage noise source $\overline{v_a^2}$ and a shunt current noise source $\overline{i_a^2}$.

Thus the total noise associated with the amplifier $\overline{i_{amp}^2}$ is given by:

$$\overline{i_{amp}^2} = \int_0^B (\overline{i_a^2} + \overline{v_a^2}|Y|^2)\, df \tag{9.15}$$

where Y is the shunt admittance (combines the shunt capacitances and resistances) and f is the frequency. An equivalent circuit for the front end of the receiver, including the effective input capacitance C_a and resistance R_a of the amplifier is shown in Figure 9.4. The capacitance of the detector C_d is also shown and the noise resulting from C_d is usually included in the expression for $\overline{i_{amp}^2}$ given in Eq. (9.15).

The SNR for the p–n or p–i–n photodiode receiver may be obtained by summing the noise contributions from Eqs. (9.12), (9.14) and (9.15). It is given by:

$$\frac{S}{N} = \frac{I_p^2}{2eB(I_p + I_d) + \dfrac{4KTB}{R_L} + \overline{i_{amp}^2}} \tag{9.16}$$

The thermal noise contribution may be reduced by increasing the value of the load resistor R_L, although this reduction may be limited by bandwidth considerations which are discussed later. Also, the noise associated with the amplifier $\overline{i_{amp}^2}$ may be reduced with low detector and amplifier capacitance.

However, when the noise associated with the amplifier $\overline{i_{amp}^2}$ is referred to the load resistor R_L, the noise figure F_n [Ref. 1] for the amplifier may be obtained. This

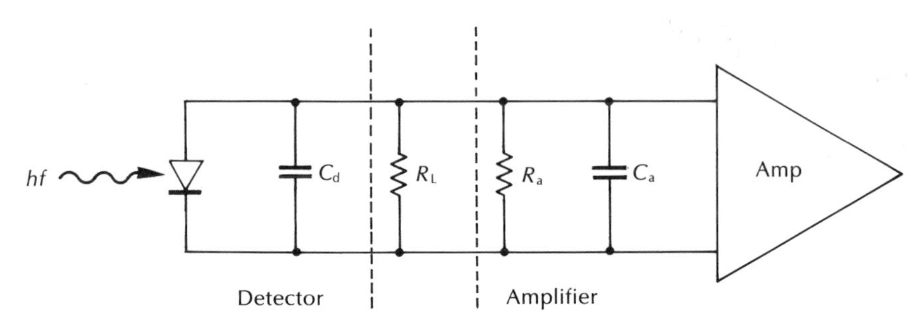

Figure 9.4 The equivalent circuit for the front end of an optical fiber receiver.

allows $\overline{i_{amp}^2}$ to be combined with the thermal noise from the load resistor $\overline{i_t^2}$ to give:

$$\overline{i_t^2} + \overline{i_{amp}^2} = \frac{4KTBF_n}{R_L} \qquad (9.17)$$

The expression for the SNR given in Eq. (9.16) can now be written in the form:

$$\frac{S}{N} = \frac{I_p^2}{2eB(I_p + I_d) + \dfrac{4KTBF_n}{R_L}} \qquad (9.18)$$

Thus if the noise figure F_n for the amplifier is known, Eq. (9.18) allows the SNR to be determined.

Example 9.4

The receiver in Example 9.3 has an amplifier with a noise figure of 3 dB. Determine the SNR at the output of the receiver under the same conditions as Example 9.3.

Solution: From Example 9.3:

$$I_p = 87.1 \times 10^{-9} \text{ A}$$
$$\overline{i_{TS}^2} = 1.44 \times 10^{-19} \text{ A}^2$$
$$\overline{i_t^2} = 2.02 \times 10^{-17} \text{ A}^2$$

The amplifier noise figure

$$F_n = 3 \text{ dB}$$
$$= 10 \log_{10} 2$$

Thus F_n may be considered as $\times 2$.

In Eq. (9.18) the SNR is given by:

$$\frac{S}{N} = \frac{I_p^2}{2eB(I_p + I_d) + \dfrac{4KTBF_n}{R_L}}$$
$$= \frac{I_p^2}{\overline{i_{TS}^2} + (\overline{i_t^2} \times F_n)}$$
$$= \frac{(87.1 \times 10^{-9})^2}{(1.44 \times 10^{-19}) + (2.02 \times 10^{-17} \times 2)}$$
$$= 1.87 \times 10^2$$

SNR in dB $= 10 \log_{10} 1.87 \times 10^2 = 22.72$ dB.

Alternatively it is possible to conduct the calculation in dB if we neglect the shot noise (say, $\overline{i_{TS}^2} = 0$).

In dB:

$$I_p = 9.40 - 80 = -70.60$$

Hence

$$I_p^2 = -141.20 \text{ dB}$$

and

$$\overline{i_t^2} = 3.05 - 170 = -166.95 \text{ dB.}$$

The amplifier noise figure $F_n = 3$ dB.

Therefore the

$$\begin{aligned} SNR &= -141.20 + 166.95 - 3 \\ &= 22.75 \text{ dB} \end{aligned}$$

A slight difference in the final answer may be noted. This is due to the neglected shot noise term.

A quantity discussed in Section 8.8.3 which is often used in the specification of optical detectors (or detector–amplifier combinations) is the noise equivalent power (NEP). It is defined as the amount of incident optical power P_o per unit bandwidth required to produce an output power equal to the detector (or detector–amplifier combination) output noise power. The NEP is therefore the value of P_o which gives an output SNR of unity. Thus the lower the NEP for a particular detector (or detector–amplifier combination), the less optical power is needed to obtain a particular SNR.

9.3.2 Receiver capacitance and bandwidth

Considering the equivalent circuit shown in Figure 9.4, the total capacitance for the front end of an optical receiver C_T is given by:

$$C_T = C_d + C_a \tag{9.19}$$

where C_d is the detector capacitance and C_a is the amplifier input capacitance. It is important that this total capacitance is minimized not only from the noise considerations discussed previously but also from the bandwidth penalty which is incurred due to the time constant of C_T and the load resistance R_L. We assume here that R_L is the total loading on the detector and therefore have neglected the amplifier input resistance R_a. However, in practical receiver configurations R_a may have to be taken into account (see Section 9.4.1). The reciprocal of the time constant $2\pi R_L C_T$ must be greater than, or equal to, the post detection bandwidth B;

$$\frac{1}{2\pi R_L C_T} \geqslant B \tag{9.20}$$

When the equality exists in Eq. (9.20) it defines the maximum possible value of B for the straightforward termination indicated in Figure 9.4.

Assuming that the total capacitance may be minimized, then the other parameter which affects B is the load resistance R_L. To increase B it is necessary to reduce R_L. However, this introduces a thermal noise penalty as may be seen from Eq. (9.14) where both the increase in B and decrease in R_L contribute to an increase in the thermal noise. A trade-off therefore exists between the maximum bandwidth and the level of thermal noise which may be tolerated. This is especially important in receivers which are dominated by thermal noise.

Example 9.5

A photodiode has a capacitance of 6 pF. Calculate the maximum load resistance which allows an 8 MHz post detection bandwidth.

Determine the bandwidth penalty with the same load resistance when the following amplifier also has an input capacitance of 6 pF.

Solution: From Eq. (9.20) the maximum bandwidth is given by:

$$B = \frac{1}{2\pi R_L C_d}$$

Therefore the maximum load resistance

$$R_L(\max) = \frac{1}{2\pi C_d B} = \frac{1}{2\pi \times 6 \times 10^{-12} \times 8 \times 10^6}$$

$$= 3.32 \text{ k}\Omega$$

Thus for an 8 MHz bandwidth the maximum load resistance is 3.32 kΩ.

Also, considering the amplifier capacitance, the maximum bandwidth

$$B = \frac{1}{2\pi R_L (C_d + C_a)} = \frac{1}{2\pi \times 3.32 \times 10^3 \times 12 \times 10^{-12}}$$

$$= 4 \text{ MHz}$$

As would be expected the maximum post detection bandwidth is halved.

9.3.3 Avalanche photodiode (APD) receiver

The internal gain mechanism in an APD increases the signal current into the amplifier and so improves the SNR because the load resistance and amplifier noise remain unaffected (i.e. the thermal noise and amplifier noise figure are unchanged). However, the dark current and quantum noise are increased by the multiplication process and may become a limiting factor. This is because the random gain mechanism introduces excess noise into the receiver in terms of increased shot noise above the level that would result from amplifying only the primary shot noise. Thus if the photocurrent is increased by a factor M (mean avalanche multiplication factor), then the shot noise is also increased by an excess noise factor M^x, such that the total shot noise $\overline{i_{SA}^2}$ is now given by:

$$\overline{i_{SA}^2} = 2eB(I_p + I_d)M^{2+x} \tag{9.21}$$

where x is between 0.3 and 0.5 for silicon APDs and between 0.7 and 1.0 for germanium or III–V alloy APDs.

Equation (9.21) is often used as the total shot noise term in order to compute the SNR, although there is a small amount of shot noise current which is not multiplied through impact ionization. The shot noise current in the detector which is not multiplied is a device parameter and may be considered as an extra shot noise term. However, it tends to be insignificant in comparison with the multiplied shot noise and is therefore neglected in the further analysis (i.e. all shot noise is assumed to be multiplied).

The SNR for the avalanche photodiode may be obtained by summing the combined noise contribution from the load resistor and the amplifier given in Eq. (9.17), which remains unchanged, with the modified noise term given in Eq. (9.21). Hence the SNR for the APD is:

$$\frac{S}{N} = \frac{M^2 I_p^2}{2eB(I_p + I_d)M^{2+x} + \dfrac{4KTBF_n}{R_L}} \tag{9.22}$$

It is apparent from Eq. (9.22) that the relative significance of the combined thermal and amplifier noise term is reduced due to the avalanche multiplication of the shot noise term. When Eq. (9.22) is written in the form:

$$\frac{S}{N} = \frac{I_p^2}{2eB(I_p + I_d)M^x + \dfrac{4KTBF_n}{R_L} M^{-2}} \tag{9.23}$$

it may be seen that the first term in the denominator increases with increasing M whereas the second term decreases. For low M the combined thermal and amplifier noise term dominates and the total noise power is virtually unaffected when the signal level is increased, giving an improved SNR. However, when M is large, the thermal and amplifier noise term becomes insignificant and the SNR decreases with increasing M at the rate of M^x. An optimum value of the multiplication factor M_{op} therefore exists which maximizes the SNR. It is given by:

$$\frac{2eB(I_p + I_d)M_{op}^x}{(4KTBF_n/R_L)M_{op}^{-2}} = \frac{2}{x} \tag{9.24}$$

and therefore

$$M_{op}^{2+x} = \frac{4KTF_n}{xeR_L(I_p + I_d)} \tag{9.25}$$

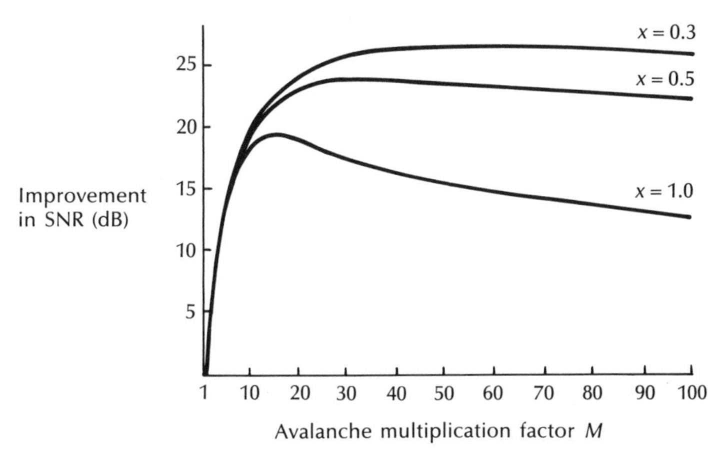

Figure 9.5 The improvement in SNR as a function of avalanche multiplication factor M for different excess noise factors M^x. Reproduced with permission from I. Garrett, *The Radio and Electron. Eng.*, **51**, p. 349, 1981.

The variation in M_{op} for both silicon and germanium APDs is illustrated in Figure 9.5 [Ref. 5]. This shows a plot of Eq. (9.22) with F_n equal to unity and neglecting the dark current. For good silicon APDs where x is 0.3, the optimum multiplication factor covers a wide range. In the case illustrated in Figure 9.5 M_{op} commences at about 40 where the possible improvement in SNR above a photodiode without internal gain is in excess of 25 dB. However, for germanium and III–V alloy APDs where x may be equal to unity it can be seen that less SNR improvement is possible (less than 19 dB). Moreover, the maximum is far sharper, occurring at a multiplication factor of about 12. Also it must be noted that Figure 9.5 demonstrates the variation of M_{op} with x for a specific case, and therefore only represents a general trend. It may be observed from Eq. (9.25) that M_{op} is dependent on a number of other variables apart from x.

Example 9.6

A good silicon APD ($x = 0.3$) has a capacitance of 5 pF, negligible dark current and is operating with a post detection bandwidth of 50 MHz. When the photocurrent before gain is 10^{-7} A and the temperature is 18 °C; determine the maximum SNR improvement between $M = 1$ and $M = M_{op}$ assuming all operating conditions are maintained.

Solution: Determine the maximum value of the load resistor from Eq. (9.20):

$$R_L = \frac{1}{2\pi C_d B} = \frac{1}{2\pi \times 5 \times 10^{-12} \times 50 \times 10^{6}}$$

$$= 635.5\ \Omega$$

When $M = 1$, the SNR is given by Eq. (9.22),

$$\frac{S}{N} = \frac{I_p^2}{2eBI_p + \dfrac{4KTB}{R_L}}$$

where $I_d = 0$ and $F_n = 1$

The shot noise is:

$$2eBI_p = 2 \times 1.602 \times 10^{-19} \times 50 \times 10^6 \times 10^{-7}$$
$$= 1.602 \times 10^{-18}\ \text{A}^2$$

and the thermal noise is:

$$\frac{4KTB}{R_L} = \frac{4 \times 1.381 \times 10^{-23} \times 291 \times 50 \times 10^6}{636.5}$$
$$= 1.263 \times 10^{-15}\ \text{A}^2$$

It may be noted that the thermal noise is dominating.

Therefore

$$\frac{S}{N} = \frac{10^{-14}}{1.602 \times 10^{-18} \times 1.263 \times 10^{-15}} = 7.91$$

and the SNR in dBs is:

$$\frac{S}{N} = 10 \log_{10} 7.91 = 8.98\ \text{dB}$$

Thus the SNR when $M = 1$ is 9.0 dB.

When $M = M_{op}$ and $x = 0.3$, from Eq. (9.25):

$$M_{op}^{2+x} = \frac{4KT}{xeR_L I_p}$$

where $I_d = 0$ and $F_n = 1$. Hence:

$$M_{op}^{2.3} = \frac{4 \times 1.381 \times 10^{-23} \times 291}{0.3 \times 1.602 \times 10^{-19} \times 636.5 \times 10^{-7}}$$

and

$$M_{op} = (5.255 \times 10^3)^{0.435}$$
$$= 41.54$$

The SNR at M_{op} may be obtained from Eq. (9.22):

$$\frac{S}{N} = \frac{M^2 I_p^2}{2eBI_p M^{2.3} + \dfrac{4KTB}{R_L}}$$

$$= \frac{(41.54)^2 \times 10^{-14}}{\{1.602 \times 10^{-18} \times (41.54)^{2.3}\} + 1.263 \times 10^{-15}}$$

$$= 1.78 \times 10^3$$

and the SNR in dBs is

$$\frac{S}{N} = 10 \log_{10} 1.78 \times 10^3 = 32.50 \, \text{dB}$$

Therefore the SNR when $M = M_{op}$ is 32.5 dB and the SNR improvement over $M = 1$ is 23.5 dB.

Example 9.7

A germanium APD (with $x = 1$) is incorporated into an optical fiber receiver with a 10 kΩ load resistance. When operated at a temperature of 120 K, the minimum photocurrent required to give a SNR of 35 dB at the output of the receiver is found to be a factor of 10 greater than the dark current. If the noise figure of the following amplifier at this temperature is 1 dB and the post detection bandwidth is 10 MHz, determine the optimum avalanche multiplication factor.

Solution: From Eq. (9.22) with $x = 1$ and $M = M_{op}$ (i.e. minimum photocurrent specifies that $M = M_{op}$) the SNR is:

$$\frac{S}{N} = \frac{M_{op}^2 I_p^2}{2eB(I_p + I_d)M_{op}^3 + \dfrac{4KTB}{R_L}}$$

Also from Eq. (9.25)

$$M_{op}^3 = \frac{4KTF_n}{eR_L(I_p + I_d)}$$

Therefore

$$M_{op} = \left\{\frac{4KTF_n}{eR_L(I_p + I_d)}\right\}^{\frac{1}{3}}$$

Substituting into Eq. (9.22), this gives:

$$\frac{S}{N} = \frac{\left\{\dfrac{4KTF_n}{eR_L(I_p + I_d)}\right\}^{\frac{2}{3}} I_p^2}{\dfrac{8KTBF_n}{R_L} + \dfrac{4KTBF_n}{R_L}}$$

and as $I_d = 0.1 \; I_p$ the SNR is:

$$\frac{S}{N} = \frac{\left(\dfrac{4KTF_n}{1.1eR_L}\right)^{\frac{2}{3}} I_p^{\frac{4}{3}}}{\dfrac{12KTBF_n}{R_L}}$$

Therefore the minimum photocurrent I_p:

$$I_p^{\frac{4}{3}} = \left(\frac{S}{N}\right) \frac{\dfrac{12KTBF_n}{R_L}}{\left(\dfrac{4KTF_n}{1.1 \; eR_L}\right)^{\frac{2}{3}}}$$

where the SNR is:

$$\frac{S}{N} = 35 \text{ dB} = 3.16 \times 10^{3}$$

and as $F_n = 1$ dB which is equivalent to 1.26:

$$\frac{12KTBF_n}{R_L} = \frac{12 \times 1.381 \times 10^{-23} \times 120 \times 10^{7} \times 1.26}{10^{4}}$$

$$= 2.51 \times 10^{-17}$$

Also

$$\left(\frac{4KTF_n}{1.1\ eR_L}\right)^{\frac{2}{3}} = \left(\frac{4 \times 1.381 \times 10^{-23} \times 120 \times 1.26}{1.1 \times 1.602 \times 10^{-19} \times 10^{4}}\right)^{\frac{2}{3}}$$

$$= 2.82 \times 10^{-4}$$

Therefore

$$I_p = \left(\frac{3.16 \times 10^{3} \times 2.51 \times 10^{-17}}{2.82 \times 10^{-4}}\right)^{\frac{3}{4}}$$

$$= 6.87 \times 10^{-8} \text{ A}$$

To obtain the optimum avalanche multiplication factor we substitute back into Eq. (9.25), where:

$$M_{op} = \left(\frac{4 \times 1.381 \times 10^{-23} \times 120 \times 1.26}{1.602 \times 10^{-19} \times 10^{3} \times 1.1 \times 6.87 \times 10^{-8}}\right)^{\frac{1}{3}}$$

$$= 8.84$$

In Example 9.7 the optimum multiplication factor for the germanium APD is found to be approximately 9. It shows the dependence of the optimum multiplication factor on the variables in Eq. (9.25), and although the example does not necessarily represent a practical receiver (some practical germanium APD receivers are cooled to reduce dark current), the optimum multiplication factor is influenced by device and system parameters as well as operating conditions.

9.3.4 Excess avalanche noise factor

The value of the excess avalanche noise factor is dependent upon the detector material, the shape of the electric field profile within the device and whether the avalanche is initiated by holes or electrons. It is often represented as $F(M)$ and in the preceding section we have considered one of the approximations for the excess noise factor, where:

$$F(M) = M^x \tag{9.26}$$

and the resulting noise is assumed to be white with a Gaussian distribution.

However, a second and more exact relationship is given by [Ref. 6]:

$$F(M) = M\left[1 - (1 - k)\left(\frac{M-1}{M}\right)^2\right] \tag{9.27}$$

where the only carriers are injected electrons and k is the ratio of the ionization coefficients of holes and electrons. If the only carriers are injected holes:

$$F(M) = M\left[1 + \left(\frac{1-k}{k}\right)\left(\frac{M-1}{M}\right)^2\right] \tag{9.28}$$

The best performance is achieved when k is small, and for silicon APDs k is between 0.02 and 0.10, whereas for germanium and III–V alloy APDs k is between 0.3 and 1.0.

With electron injection in silicon photodiodes, the smaller values of k obtained correspond to a larger ionization rate for the electrons than for the holes. As k departs from unity, only the carrier with the larger ionization rate contributes to the impact ionization and the excess avalanche noise factor is reduced. When the impact ionization is initiated by electrons this corresponds to fewer ionizing collisions involving the hole current which is flowing in the opposite direction (i.e. less feedback). In this case the amplified signal contains less excess noise. The carrier ionization rates in germanium photodiodes are often nearly equal and hence k approaches unity, giving a high level of excess noise.

9.4 Receiver structures

A full equivalent circuit for the digital optical fiber receiver, in which the optical detector is represented as a current source i_{det}, is shown in Figure 9.6. The noise sources (i_t, i_{TS} and i_{amp}) and the immediately following amplifier and equalizer are also shown. Equalization [Ref. 7] compensates for distortion of the signal due to the combined transmitter, medium and receiver characteristics. The equalizer is

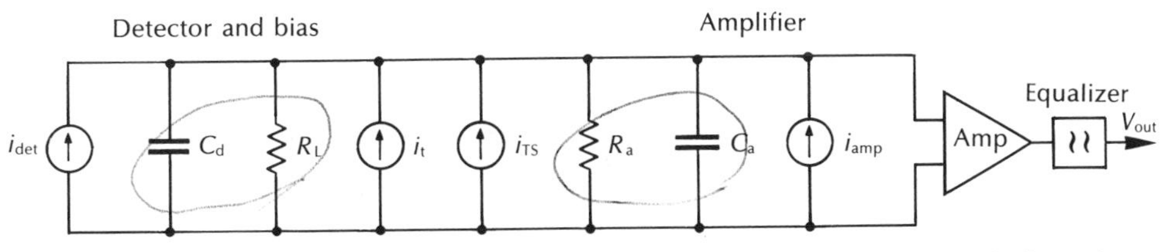

Figure 9.6 A full equivalent circuit for a digital optical fiber receiver including the various noise sources.

often a frequency shaping filter which has a frequency response that is the inverse of the overall system frequency response. In wideband systems this will normally boost the high frequency components to correct the overall amplitude of the frequency response. To acquire the desired spectral shape for digital systems (e.g. raised cosine, see Figure 11.37), in order to minimize intersymbol interference, it is important that the phase frequency response of the system is linear. Thus the equalizer may also apply selective phase shifts to particular frequency components.

However, the receiver structure immediately preceding the equalizer is the major concern of this section. In both digital and analog systems it is important to minimize the noise contributions from the sources shown in Figure 9.6 so as to maximize the receiver sensitivity whilst maintaining a suitable bandwidth. It is therefore useful to discuss various possible receiver structures with regard to these factors.

9.4.1 Low impedance front end

Three basic amplifier configurations are frequently used in optical fiber communication receivers. The simplest, and perhaps the most common, is the voltage amplifier with an effective input resistance R_a as shown in Figure 9.7.

In order to make suitable design choices, it is necessary to consider both bandwidth and noise. The bandwidth considerations in Section 9.3.2 are treated solely with regard to a detector load resistance R_L. However, in most practical receivers the detector is loaded with a bias resistor R_b and an amplifier (see Figure 9.7). The bandwidth is determined by the passive impedance which appears across the detector terminals which is taken as R_L in the bandwidth relationship given in Eq. (9.20).

However, R_L may be modified to incorporate the parallel resistance of the detector bias resistor R_b and the amplifier input resistance R_a. The modified total load resistance R_{TL} is therefore given by:

$$R_{TL} = \frac{R_b R_a}{R_b + R_a} \tag{9.29}$$

Considering the expressions given in Eqs. (9.20) and (9.29), to achieve an optimum bandwidth both R_b and R_a must be minimized. This leads to a low

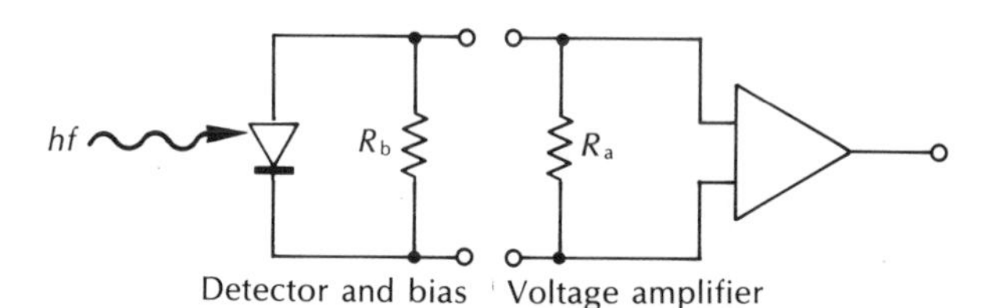

Figure 9.7 Low impedance front end optical fiber receiver with voltage amplifier.

impedance front end design for the receiver amplifier. Unfortunately this design allows thermal noise to dominate within the receiver (following Eq. (9.14)), which may severely limit its sensitivity. Therefore this structure demands a trade-off between bandwidth and sensitivity which tends to make it impractical for long-haul, wideband optical fiber communication systems.

9.4.2 High impedance (integrating) front end

The second configuration consists of a high input impedance amplifier together with a large detector bias resistor in order to reduce the effect of thermal noise. However, this structure tends to give a degraded frequency response as the bandwidth relationship given in Eq. (9.20) is not maintained for wideband operation. The detector output is effectively integrated over a large time constant and must be restored by differentiation. This may be performed by the correct equalization at a later stage [Ref. 8] as illustrated in Figure 9.8. Therefore the high impedance (integrating) front end structure gives a significant improvement in sensitivity over the low impedance front end design, but it creates a heavy demand for equalization and has problems of limited dynamic range (the ratio of maximum to minimum input signals).

The limitations on dynamic range result from the attenuation of the low frequency signal components by the equalization process which causes the amplifier to saturate at high signal levels. When the amplifier saturates before equalization has occurred the signal is heavily distorted. Thus the reduction in dynamic range is dependent upon the amount of integration and subsequent equalization employed.

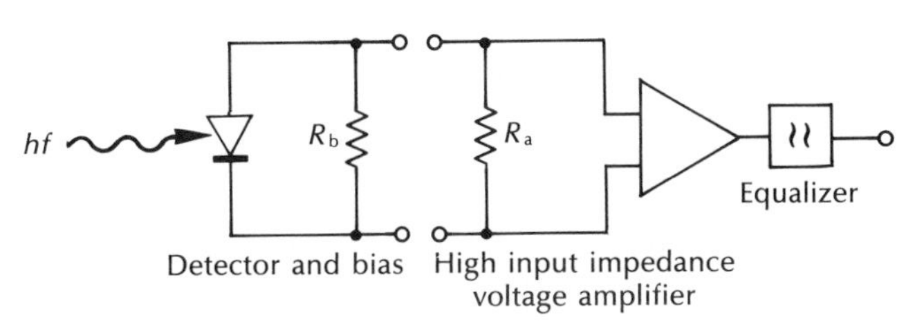

Figure 9.8 High impedance integrating front end optical fiber receiver with equalized voltage amplifier.

9.4.3 The transimpedance front end

This configuration largely overcomes the drawbacks of the high impedance front end by utilizing a low noise, high input impedance amplifier with negative feedback. The device therefore operates as a current mode amplifier where the high input impedance is reduced by negative feedback. An equivalent circuit for an optical fiber receiver incorporating a transimpedance front end structure is shown in Figure 9.9. In this equivalent circuit the parallel resistances and capacitances are combined

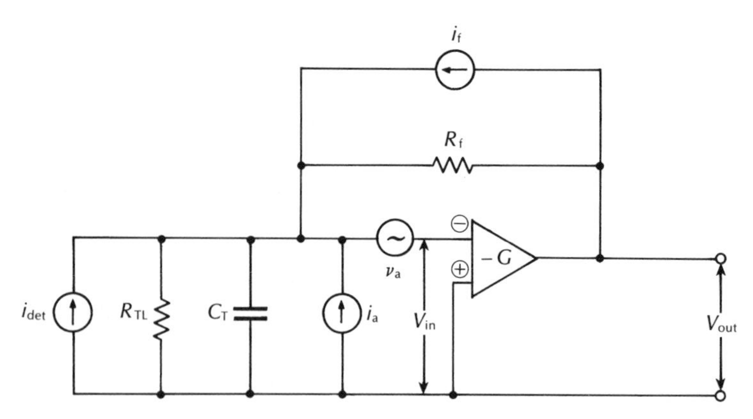

Figure 9.9 An equivalent circuit for the optical fiber receiver incorporating a transimpedance (current mode) preamplifier.

into R_{TL} and C_T respectively. The open loop current to voltage transfer function $H_{OL}(\omega)$ for this transimpedance configuration corresponds to the transfer function for the two structures described previously which do not employ feedback (i.e. the low and high impedance front ends). It may be written as:

$$H_{OL}(\omega) = -G\frac{V_{in}}{i_{det}} = -G\frac{R_{TL}\dfrac{1}{j\omega C_T}}{R_{TL}+\dfrac{1}{j\omega C_T}} = \frac{-GR_{TL}}{1+j\omega R_{TL}C_T}\ (\text{V A}^{-1}) \tag{9.30}$$

where G is the open loop voltage gain of the amplifier and ω is the angular frequency of the input signal.

In this case the bandwidth (without equalization) is constrained by the time constant given in Eq. (9.20).*

When the feedback is applied, the closed loop current to voltage transfer function $H_{CL}(\omega)$ for the transimpedance configuration is given by (see Appendix E)

$$H_{CL}(\omega) \simeq \frac{-R_f}{1+(j\omega R_f C_T/G)}\ (\text{V A}^{-1}) \tag{9.31}$$

where R_f is the value of the feedback resistor. In this case the permitted electric bandwidth B (without equalization) may be written as:

$$B \leqslant \frac{G}{2\pi R_f C_T} \tag{9.32}$$

Hence, comparing Eq. (9.32) with Eq. (9.20) it may be noted that the trans-

* The time constant can be obtained directly from Eq. (9.30) where the maximum bandwidth is defined by $\omega = 2\pi B = 1/R_{TL}C_T$.

impedance (or feedback) amplifier provides a much greater bandwidth than do the amplifiers without feedback. This is particularly pronounced when G is large.

Moreover, it is interesting to consider the thermal noise generated by the transimpedance front end. Using a referred impedance noise analysis it can be shown [Ref. 12] that to a good approximation the feedback resistance (or impedance) may be referred to the amplifier input in order to establish the noise performance of the configuration. Thus when $R_f \ll R_{TL}$, the major noise contribution is from thermal noise generated in R_f. The noise performance of this configuration is therefore improved when R_f is large, and it approaches the noise performance of the high impedance front end when $R_f = R_{TL}$. Unfortunately, the value of R_f cannot be increased indefinitely due to problems of stability with the closed loop design. Furthermore, it may be observed from Eq. (9.32) that increasing R_f reduces the bandwidth of the transimpedance configuration. This problem may be alleviated by making G as large as the stability of the closed loop will allow. Nevertheless, it is clear that the noise in the transimpedance amplifier will always exceed that incurred by the high impedance front end structure.

Example 9.8

A high input impedance amplifier which is employed in an optical fiber receiver has an effective input resistance of 4 MΩ which is matched to a detector bias resistor of the same value. Determine:

(a) The maximum bandwidth that may be obtained without equalisation if the total capacitance C_T is 6 pF.
(b) The mean square thermal noise current per unit bandwidth generated by this high input impedance amplifier configuration when it is operating at a temperature of 300 K.
(c) Compare the values calculated in (a) and (b) with those obtained when the high input impedance amplifier is replaced by a transimpedance amplifier with a 100 kΩ feedback resistor and an open loop gain of 400. It may be assumed that $R_f \ll R_{TL}$, and that the total capacitance remains 6 pF.

Solution: (a) Using Eq. (9.29), the total effective load resistance:

$$R_{TL} = \frac{(4 \times 10^6)^2}{8 \times 10^6} = 2 \text{ M}\Omega$$

Hence from Eq. (9.20) the maximum bandwidth is given by:

$$B = \frac{1}{2\pi R_{TL} C_T} = \frac{1}{2\pi \times 2 \times 10^6 \times 6 \times 10^{-12}}$$
$$= 1.33 \times 10^4 \text{ Hz}$$

The maximum bandwidth that may be obtained without equalization is 13.3 kHz.

(b) The mean square thermal noise current per unit bandwidth for the high impedance configuration following Eq. (9.14) is:

$$\overline{i_t^2} = \frac{4KT}{R_{TL}} = \frac{4 \times 1.381 \times 10^{-23} \times 300}{2 \times 10^6}$$

$$= 8.29 \times 10^{-27}\ \text{A}^2\,\text{Hz}^{-1}$$

(c) The maximum bandwidth (without equalization) for the transimpedance configuration may be obtained using Eq. (9.32), where

$$B = \frac{G}{2\pi R_f C_T} = \frac{400}{2\pi \times 10^5 \times 6 \times 10^{-12}}$$

$$= 1.06 \times 10^8\ \text{Hz}$$

Hence a bandwidth of 106 MHz is permitted by the transimpedance design.

Assuming $R_f \ll R_{TL}$, the mean square thermal noise current per unit bandwidth for the transimpedance configuration is given by:

$$\overline{i_t^2} = \frac{4KT}{R_f} = \frac{4 \times 1.381 \times 10^{-23} \times 300}{10^5}$$

$$= 1.66 \times 10^{-25}\ \text{A}^2\,\text{Hz}^{-1}$$

The mean square thermal noise current in the transimpedance configuration is therefore a factor of 20 greater than that obtained with the high input impedance configuration.

The equivalent value in decibels of the ratio of these noise powers is:

$$\frac{\textit{Noise power in the transimpedance configuration}}{\textit{Noise power in the high input impedance configuration}} = 10 \log_{10} 20$$

$$= 13\ \text{dB}$$

Thus the transimpedance front end in Example 9.8 provides a far greater bandwidth without equalization than the high impedance front end. However, this advantage is somewhat offset by the 13 dB noise penalty incurred with the transimpedance amplifier over that of the high input impedance configuration. Nevertheless it is apparent, even from this simple analysis, that transimpedance amplifiers may be optimized for noise performance, although this is usually obtained at the expense of bandwidth. This topic is pursued further in Ref. 13. However, wideband transimpedance designs generally give a significant improvement in noise performance over the low impedance front end structures using simple voltage amplifiers (see Problem 9.18). Finally it must be emphasized that the approach adopted in Example 9.8 is by no means rigorous and includes two important simplifications: firstly, that the thermal noise in the high impedance amplifier is assumed to be totally generated by the effective input resistance of the device; and secondly, that the thermal noise in the transimpedance configuration is assumed to be totally generated by the feedback resistor when it is referred to the amplifier input. Both these assumptions are approximations, the accuracy of which is largely dependent on the parameters of the particular amplifier. For example, another factor which tends to reduce the bandwidth of the transimpedance amplifier is the stray capacitance C_f generally associated with the feedback resistor R_f. When C_f is taken into account the closed loop response of Eq. (9.31) becomes:

$$H_{CL}(\Omega) \simeq \frac{-R_f}{1 + j\omega R_f(C_T/G + C_f)} \tag{9.33}$$

However, the effects of C_f may be cancelled by employing a suitable compensating network [Ref. 14].

The other major advantage which the transimpedance configuration has over the high impedance front end is a greater dynamic range. This improvement in dynamic range obtained using the transimpedance amplifier is a result of the different attenuation mechanism for the low frequency components of the signal. The attenuation is accomplished in the transimpedance amplifier through the negative feedback and therefore the low frequency components are amplified by the closed loop rather than the open loop gain of the device. Hence for a particular amplifier the improvement in dynamic range is approximately equal to the ratio of the open loop to the closed loop gains. The transimpedance structure therefore overcomes some of the problems encountered with the other configurations and is often preferred for use in wideband optical fiber communication receivers [Ref. 15].

9.5 FET Preamplifiers

The lowest noise amplifier device which is widely available is the silicon field effect transistor (FET). Unlike the bipolar transistor, the FET operates by controlling the current flow with an electric field produced by an applied voltage on the gate of the device (see Figure 9.10) rather than with a base current. Thus the gate draws virtually no current, except for leakage, giving the device an extremely high input impedance (can be greater than 10^{14} ohms). This, coupled with its low noise and capacitance (no greater than a few picofarads), makes the silicon FET appear an ideal choice for the front end of the optical fiber receiver amplifier. However, the superior properties of the FET over the bipolar transistor are limited by its comparatively low transconductance g_m (no better than 5 millisiemens in comparison with at least 40 millisiemens for the bipolar). It can be shown [Ref. 13] that a figure of merit with regard to the noise performance of the FET amplifier is g_m/C_T^2. Hence the advantage of high transconductance together with low total capacitance C_T is apparent. Moreover, as $C_T = C_d + C_a$, it should be noted that the figure of merit is optimized when $C_a = C_d$. This requires FETs to be specifically matched to particular detectors, a procedure which device availability does not generally permit in current optical fiber receiver design. As indicated above, the gain of the FET is restricted.

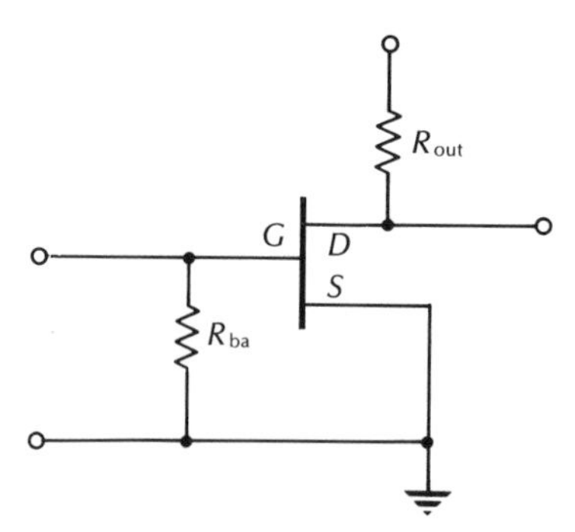

Figure 9.10 Grounded source FET configuration for the front end of an optical fiber receiver amplifier.

This is especially the case for silicon FETs at frequencies above 25 MHz where the current gain drops to values near unity as the transconductance is fixed with a decreasing input impedance. Therefore at frequencies above 25 MHz, the bipolar transistor is a more useful amplifying device.*

Figure 9.10 shows the grounded source FET configuration which increases the device input impedance especially if the amplifier bias resistor R_{ba} is large. A large bias resistor has the effect of reducing the thermal noise but it will also increase the low frequency impedance of the detector load which tends to integrate the signal (i.e. high impedance integrating front end). Thus compensation through equalization at a later stage is generally required.

9.5.1 Gallium arsenide MESFETs

Although silicon FETs have a limited useful bandwidth, much effort has been devoted to the development of high performance microwave FETs since the mid-1970s. These FETs are fabricated from gallium arsenide and, being Schottky barrier devices [Refs. 16 to 19], are called GaAs metal Schottky field effect transistors (MESFETs). They overcome the major disadvantage of silicon FETs in that they will operate with both low noise and high gain at microwave frequencies (GHz). Thus in optical fiber communication receiver design they present an alternative to bipolar transistors for wideband operation. These devices have therefore been incorporated into high performance receiver designs using both *p–i–n* and avalanche photodiode detectors [Refs. 21 to 32]. In particular, there has been much interest in hybrid integrated receiver circuits utilizing *p–i–n* photodiodes with GaAs MESFET amplifier front ends. The hybrid integration of a photodetector with a GaAs MESFET preamplifier having low leakage current,

* The figure of merit in relation to noise performance for the bipolar transistor amplifier may be shown [Ref. 13] to be $(h_{FE})^{\frac{1}{2}}/C_T$ where h_{FE} is the common emitter current gain of the device. Hence the noise performance of the bipolar amplifier may be optimized in a similar manner to that of the FET amplifier.

low capacitance (less than 0.5 pF) and high transconductance (greater than 30 millisiemens) provides a strategy for low noise optical receiver design [Ref. 33].

9.5.2 PIN–FET hybrid receivers

The *p–i–n*/FET, or PIN–FET, hybrid receiver utilizes a high performance *p–i–n* photodiode followed by a low noise preamplifier often based on a GaAs MESFET, the whole of which is fabricated using thick film integrated circuit technology. This hybrid integration on a thick film substrate reduces the stray capacitance to negligible levels giving a total input capacitance which is very low (e.g. 0.4 pF). The MESFETs employed have a transconductance of approximately 15 millisiemens at the bandwidths required (e.g. 140 Mbit s^{-1}). Early work [Refs. 22 and 23] in the 0.8 to 0.9 μm wavelength band utilizing a silicon *p–i–n* detector showed the PIN–FET hybrid receiver to have a sensitivity of -45.8 dBm for a 10^{-9} bit error rate which is only 4 dB worse than current silicon RAPD receivers (see Section 8.9.2).

The work was subsequently extended into the longer wavelength band (1.1 to 1.6 μm) utilizing III–V alloy *p–i–n* photodiode detectors. An example of a PIN–FET hybrid high impedance (integrating) front end receiver for operation at a wavelength of 1.3 μm using an InGaAs *p–i–n* photodiode is shown in Figure 9.11 [Refs. 24 to 27]. This design, used by British Telecom, consists of a preamplifier with a GaAs MESFET and microwave bipolar transistor cascode followed by an emitter follower output buffer. The cascode circuit is chosen to ensure that sufficient gain is obtained from the first stage to give an overall gain of 18 dB. As

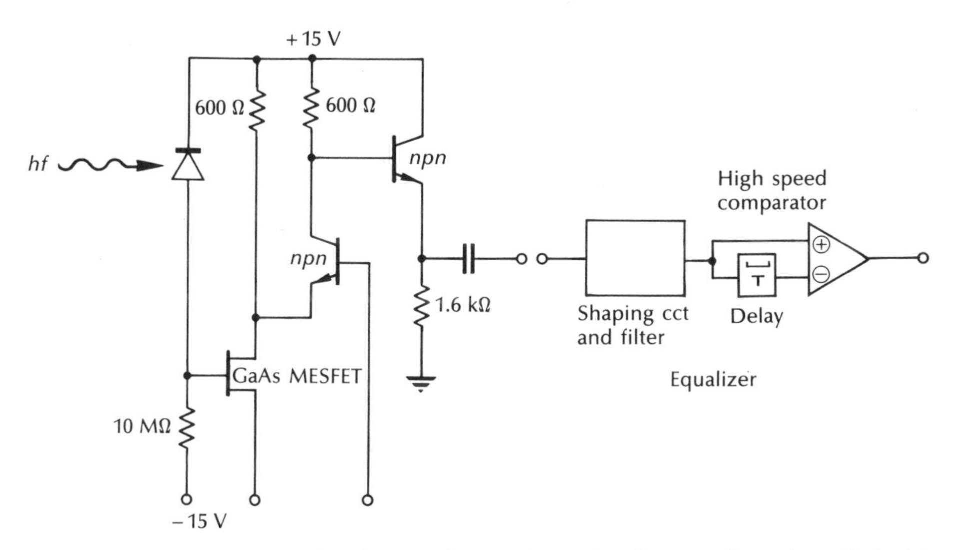

Figure 9.11 PIN–FET hybrid high impedance integrating front end receiver [Refs. 24 to 27].

the high impedance front end effectively integrates the signal, the following digital equalizer is necessary. The pulse shaping and noise filtering circuits comprise two passive filter sections to ensure that the pulse waveform shape is optimized and the noise is minimized. Equalization for the integration (i.e. differentiation) is performed by monitoring the change in the integrated waveform over one period with a subminiature coaxial delay line followed by a high speed low level comparator. The receiver is designed for use at a transmission rate of 140 Mbit s^{-1} where its performance is found to be comparable to germanium and III–V alloy APD receivers. For example, the receiver sensitivity at a bit error rate of 10^{-9} is -44.2 dBm.

When compared with the APD receiver the PIN–FET hybrid has both cost and operational advantages especially in the longer wavelength region. The low voltage operation (e.g. +15 and −15 V supply rails) coupled with good sensitivity and ease of fabrication makes the incorporation of this receiver into wideband optical fiber communication systems commercially attractive. A major drawback with the PIN–FET receiver is the possible lack of dynamic range. However, the configuration shown in Figure 9.11 gave adequate dynamic range via a control circuit which maintained the mean voltage at the gate at 0 V by applying a negative voltage proportional to the mean photocurrent to the MESFET bias resistor. With a −15 V supply rail an optical dynamic range of some 20 dB was obtained. This was increased to 27 dB by reducing the value of the MESFET bias resistor from 10 to 2 MΩ which gave a slight noise penalty of 0.5 dB. These figures compare favourably with practical APD receivers.

Transimpedance front end receivers have also been fabricated using the PIN–FET hybrid approach. An example of this type of circuit [Ref. 29] is shown

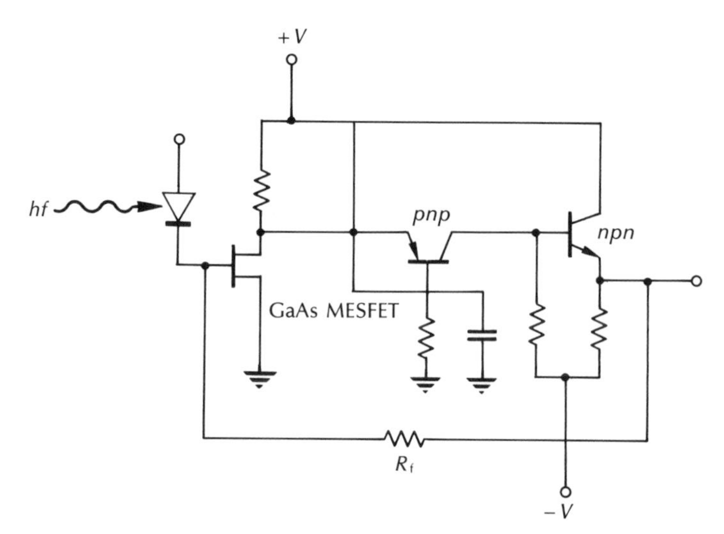

Figure 9.12 PIN–FET hybrid transimpedance front end receiver [Ref. 29].

in Figure 9.12. The amplifier consists of a GaAs MESFET followed by two complementary bipolar microwave transistors. A silicon *p–i–n* photodiode was utilized with the amplifier and the receiver was designed to accept data at a rate of 274 Mbits s^{-1}. In this case the effective input capacitance of the receiver was 4.5 pF giving a sensitivity around -35 dBm for a bit error rate of 10^{-9}.

These figures are somewhat worse than the high impedance front end design discussed previously. However, this design has the distinct advantage of a flat frequency response to a wider bandwidth which requires little, if any, equalization.

9.6 High performance receivers

It is clear from the discussions in Sections 9.3 to 9.5 that noise performance is a major design consideration providing a limitation to the sensitivity which may be obtained with a particular receiver structure and component mix. However, two other important receiver performance criteria were also outlined in the aforementioned sections, namely, bandwidth and dynamic range. Moreover, distinct tradeoffs exist between these three performance attributes such that an optimized design for one criterion may display a degradation in relation to one or both of the other criteria. Nevertheless, although high performance receiver design may seek to provide optimization for one particular attribute, attempts are generally made to minimize the degradations associated with the other performance parameters. In this section we describe further the strategies that have been adopted to produce high performance receivers for optical fiber communications, together with some of the performance results which have been obtained over the last few years.

As mentioned in Section 9.5.2, low noise performance combined with potential high speed operation has been a major pursuit in the hybrid integration of *p–i–n* photodiodes with GaAs MESFETs. In this context it is useful to compare the noise performance of various transistor preamplifiers over a range of bandwidths. A theoretical state-of-the-art performance comparison for the silicon junction FET and (JFET), the silicon metal oxide semiconductor FET (MOSFET) and the silicon bipolar transistor preamplifier with a GaAs MESFET device for transmission rates from 1 Mbit s^{-1} to 10 Gbit s^{-1} is shown in Figure 9.13 [Ref. 34]. It may be observed that at low speeds the three FET preamplifiers provide higher sensitivity than the Si bipolar device. In addition it is apparent that below 10 Mbit s^{-1} the Si MOSFET preamplifier provides a lower noise performance than the GaAs MESFET. Above 20 Mbit s^{-1}, however, the highest sensitivity is obtained with the GaAs MESFET device, even though at very high speeds the Si MOSFET and Si bipolar transistor preamplifiers exhibit a noise performance that is only slightly worse than the aforementioned device. Furthermore, it is clear that, as indicated in Section 9.5, the Si bipolar transistor preamplifier displays a noise improvement over the Si JFET, in this case at speeds above 50 Mbit s^{-1}.

The optimization of PIN–FET receiver designs for sensitivity and high speed operation have been investigated [Refs. 35, 36]. Also a wideband (10 GHz) low

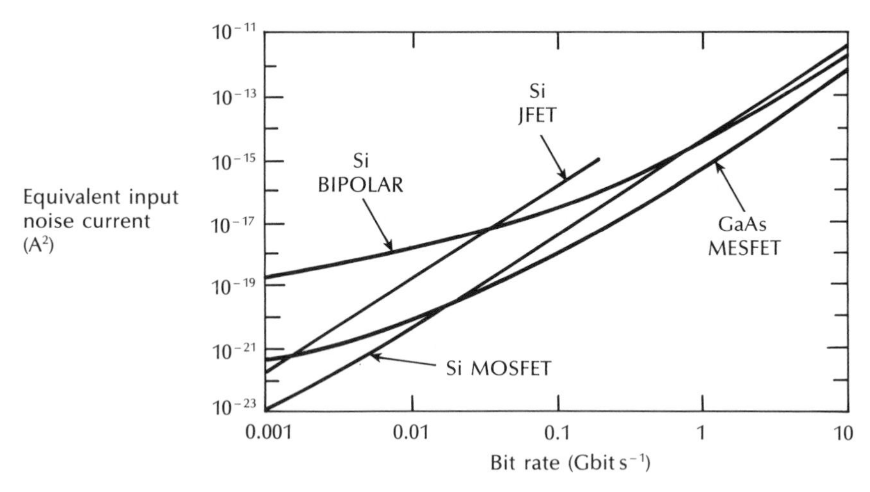

Figure 9.13 Noise characteristics for various optical receiver transistor preamplifiers. Reproduced with permission from B. L. Kasper, 'Receiver design', in S. E. Miller and I. P. Kaminow (Eds.), *Optical Fiber Telecommunications II*, Academic Press Inc., p. 689, 1988.

noise device using discrete commercial components has recently been reported [Ref. 37]. In addition, new high speed, low noise transistor types are under investigation for optical receiver preamplifiers. These devices include the heterojunction bipolar transistor (HBT) [Ref. 38] and high electron mobility transistor (HEMT) [Refs. 39, 40]. The latter device type comprises a selectively doped heterojunction FET which has displayed 3 dB bandwidths up to 20 GHz within the three stage optical preamplifier illustrated in Figure 9.14(a) [Ref. 39]. Each stage comprised a shunt feedback configuration containing a single HEMT with mutual conductance of 70 millisiemens and a gate to source capacitance of 0.36 pF (see Figure 9.14(b)). When operated with an InGaAs *p–i–n* photodiode the preamplifier exhibited a 21.5 dB gain with an averaged input equivalent noise current density of 7.6 pA $Hz^{-\frac{1}{2}}$ over the range 100 MHz to 18 GHz.

Although the above discussion has centred on *p–i–n* receiver preamplifier designs, high speed APD optical receivers are also under investigation [Refs. 41 to 43]. In particular, a high sensitivity APD–FET receiver designed to operate at speeds up to 8 Gbit s^{-1} and at wavelengths in the range 1.3 to 1.5 μm is shown in Figure 9.15 [Ref. 42]. The receiver employed a 60 GHz gain–bandwidth product InGaAs/InGaAsP/InP APD followed by a hybrid GaAs MESFET high impedance front end. Moreover, a receiver sensitivity of −25.8 dBm was obtained for a bit error rate of 10^{-9}.

An additional strategy for the provision of wideband, low noise receivers, especially using the *p–i–n* photodiode detector, involves the monolithic integration of this device type with III–V semiconductor alloy FETs or HBTs [Refs. 44 to 48].

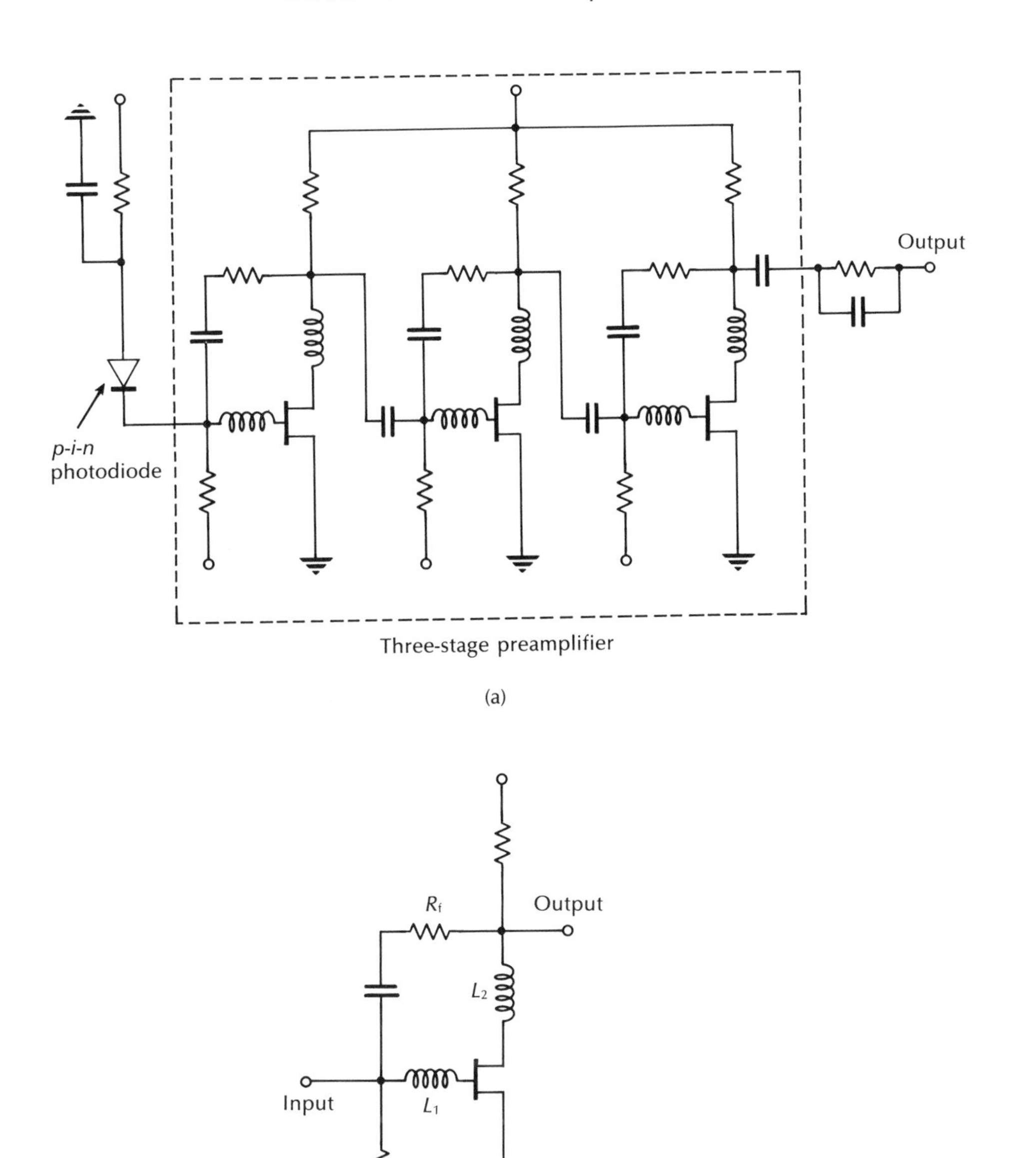

Figure 9.14 Circuit configuration for a high speed optical receiver using a HEMT preamplifier [Ref. 39]: (a) *p–i–n*–HEMT optical receiver; (b) single shunt feedback stage.

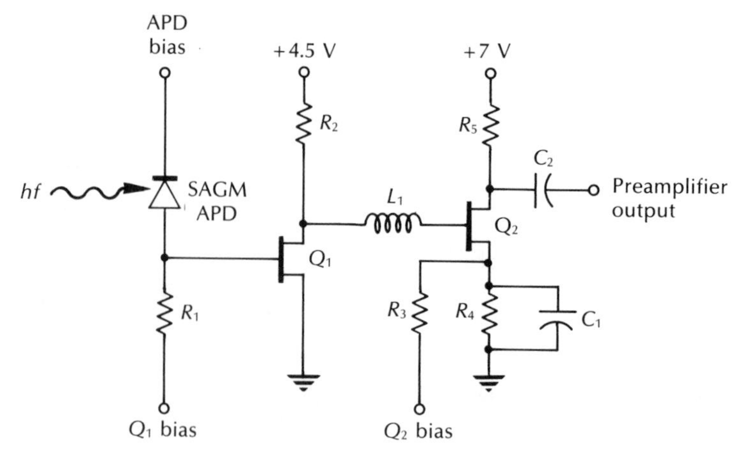

Figure 9.15 Circuit configuration for a high sensitivity APD–FET optical receiver [Ref. 42].

Such monolithic integrated receivers or optoelectronic integrated circuits (OEICs) are discussed further in Section 10.7. However, it should be noted that the major recent activities in this area have concerned devices for operation in the 1.1 to 1.6 μm wavelength range. An example of the circuit configuration of a monolithic PIN–FET receiver is illustrated in Figure 9.16 [Ref. 48]. The design comprises a voltage variable FET feedback resistor which produces active feedback as an input shunt automatic gain control (AGC) circuit which extends the dynamic range by diverting excess photocurrent away from the input of the basic receiver.

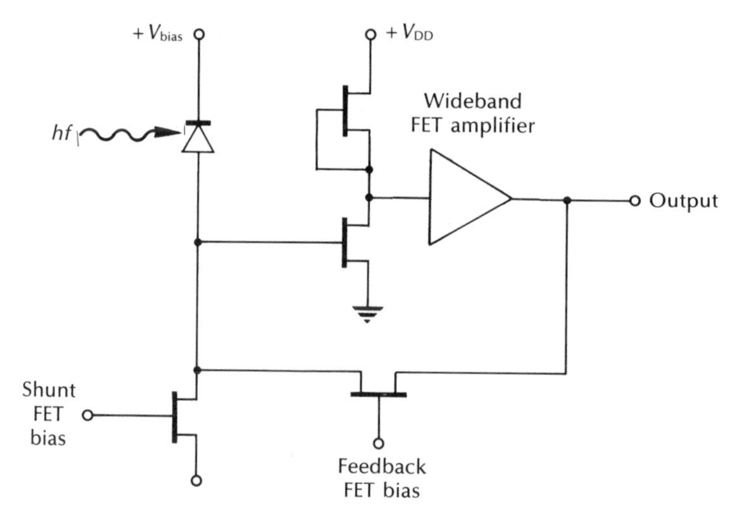

Figure 9.16 Monolithic PIN–FET optical receiver circuit configuration.

Furthermore, the shunt FET gives additional dynamic range extension through the mechanism of active receiver bias compensation, which is discussed further in relation to Figure 9.18.

The receiver dynamic range is an important performance parameter as it provides a measure of the difference between the device sensitivity and its saturation or overload level. A receiver saturation or overload level is largely determined by the value of the photodiode bias resistor or, alternatively, the feedback resistor in the transimpedance configuration. Because the photodiode bias resistor has a small value in the low impedence front end design, then the saturation level is high.* Similarly, the relatively low value of feedback resistor in the transimpedance configuration gives a high saturation level which combined with a high sensitivity, provides a wide dynamic range, as indicated in Section 9.4.3. By contrast the high value of photodiode bias resistor in the high impedance front end causes a low saturation level which, even taking account of the high sensitivity of the configuration, gives a relatively narrow dynamic range. The difference between the two latter receiver structures may be observed in the dynamic range and sensitivity characteristics shown in Figure 9.17. Although the sensitivity decreases in moving from the high impedance design (left hand side) to the transimpedance configuration (right hand side) as the value of the feedback resistor R_f is reduced, the saturation level increases at a faster rate, producing a significantly wider dynamic range for the transimpedance front end receiver.

The significance of the receiver dynamic range becomes apparent when the reader considers the ideal multipurpose use of such a device for operation with a variety of optical source powers over different fiber lengths. Moreover, when a high impedance receiver with a 1 MΩ bias resistor is utilized, the saturation level occurs at an input optical power of 0.5 μW or -33 dBm. Therefore, this device can only be employed in long-haul communication applications where the input power level is low. Corresponding figures for the transimpedance configuration (1 kΩ feedback resistor) and the low impedance front end (200 Ω bias resistor) are 0.5 mW (-3 dBm) and 2.5 mW ($+4$ dBm). In all cases the saturation level can be substantially improved by using active receiver bias compensation, as illustrated schematically in Figure 9.18. Hence, as the d.c. voltage at the input to the amplifier increases with the incident optical power, the control loop applies an equal but opposite shift in the voltage to the other side of the bias resistor. In this way the voltage at the input to the preamplifier becomes independent of the detected power level. However, in practice the feedback voltage in the control loop cannot be unbounded and therefore the technique has limitations. Nevertheless, saturation levels for high impedance front end receiver designs may be improved to around 20 μW, or 17 dBm using this technique.

Even when using bias compensation with a high impedance front end receiver to improve the saturation level, the overall dynamic range tends to be poor. For such

* Unfortunately, the sensitivity of the low impedance configuration is poor and hence the dynamic range is generally not large.

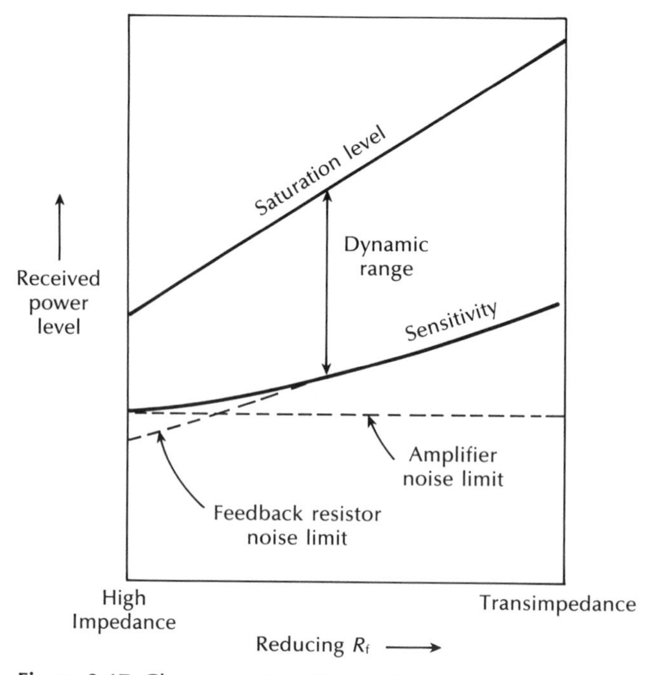

Figure 9.17 Characteristics illustrating the variation in received power level against the value of the feedback resistor R_f in the transimpedance front end receiver structure. The high impedance front end receiver corresponds to $R_f = \infty$.

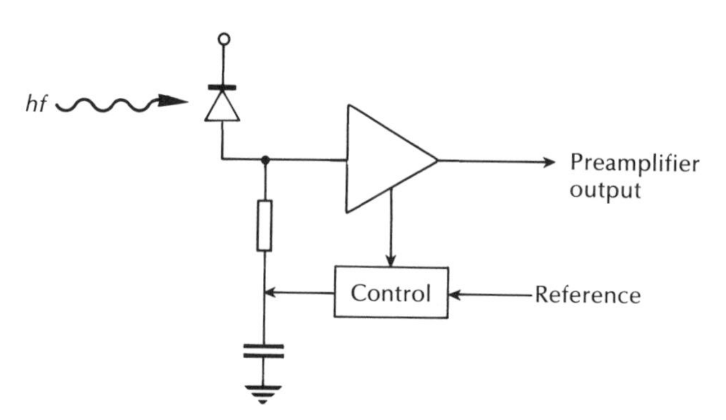

Figure 9.18 Active receiver bias compensation.

a receiver operating at a speed of 1 Gbit s^{-1} it is usually in the range 20 to 27 dB, whereas for a corresponding transimpedance receiver configuration without bias compensation, the dynamic range can be 30 to 39 dB.* Furthermore, in the latter case alternative design strategies have proved successful in increasing the receiver dynamic range. In particular the use of optically coupled feedback has demonstrated dynamic ranges of around 40 dB for *p–i–n* receivers operating at modest bit rates [Refs. 49, 50].

The optical feedback technique, which is shown schematically in Figure 9.19, eliminates the thermal noise associated with the feedback resistor in the transimpedance front end design. This strategy proves most useful at low transmission rates because in this case the feedback resistors employed are normally far smaller than the optimum value for low noise performance so as to maintain the resistor at a practical size (e.g. 1 MΩ). Moreover, large values of feedback resistor limit the dynamic range of the conventional transimpedance receiver structure, whilst also introducing parasitic shunt capacitance which can cause signal integration and hence restrict the bandwidth of the preamplifier. It may be observed from Figure 9.19 that the optical feedback signal is provided by an LED that is driven from the preamplifier output through a small resistor. This resistor acts as a load to generate an output voltage for the following amplifiers. The current feedback to the signal *p–i–n* photodetector is obtained from a second *p–i–n* photodiode which detects the optical feedback signal

The removal of the feedback resistor in the optical feedback technique allows low noise performance and hence high receiver sensitivity of the order of −64 dBm at transmission rates of 2 Mbit s^{-1} [Ref. 50]. In addition, as the feedback LED is a low impedance device that can be driven with a low output voltage, the problem

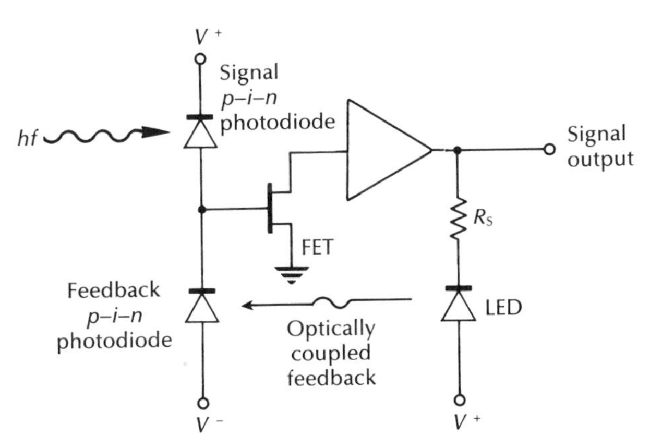

Figure 9.19 Optical feedback transimpedence receiver schematic.

* It should be noted that in both cases the bottom end of the range refers to *p–i–n* photodiode receivers whilst the top end of the range is only obtained with APD receivers.

associated with amplifier saturation is much reduced. Therefore this factor, combined with the high sensitivity produced by the strategy, enables wide dynamic range. It should be noted, however, that some penalties occur when employing this technique in that there is an increase in receiver input capacitance and also an increase in detector dark current noise (resulting from the use of two photodiodes) in comparison with the conventional transimpedance preamplifier structure. Nevertheless, it is suggested that the optical feedback receiver component costs can be comparable to resistive feedback designs [Ref. 50] whilst providing a significant performance improvement.

An alternative strategy for the realization of high sensitivity receivers, in this case for high speed operation, is to employ preamplification using an optical amplifier prior to the receiver [Refs. 51 to 55]. The two basic optical amplifier technological types, namely the semiconductor laser amplifier (SLA) and the fiber amplifier, are discussed in Sections 10.3 and 10.4 respectively. It is clear, however, that both device types may be utilized in this preamplification role which is illustrated schematically for an SLA device in Figure 9.20.* The SLA shown in Figure 9.20 operates as a near-travelling wave amplifier and therefore the output emissions are predominantly spontaneous creating a spectral bandwidth which is determined by the gain profile of the device. Because the typical spectral bandwidth is in the range 30 to 40 nm, a bandpass optical filter† is employed to reduce the intensity of the spontaneous emission reaching the optical detector. This has the effect of reducing the spontaneous noise products and thus improving the overall receiver sensitivity.

Although the sensitivity improvement introduced by the laser preamplifier is a function of the device internal gain, the coupling losses between the various elements and the bandwidth of the bandpass filter, it is typically in the range 10 to 15 dB when using an SLA. Moreover, it is interesting to observe from the sensitivity versus transmission rate characteristics shown in Figure 9.21 [Ref. 53] that the SLA preamplifier *p–i–n* photodiode configuration illustrated in Figure 9.20 displays a

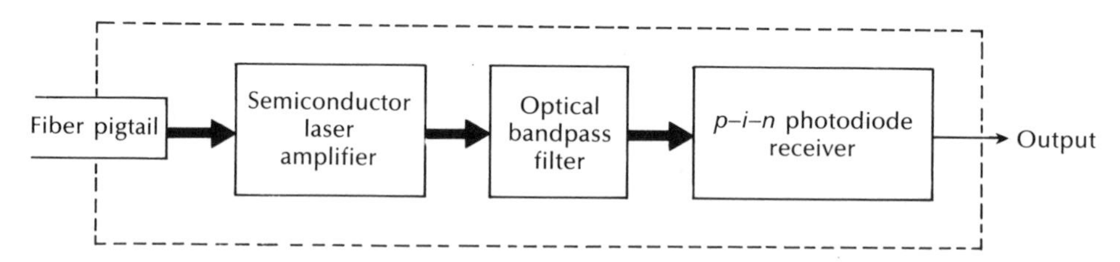

Figure 9.20 Block schematic of a SLA preamplified *p–i–n* photodiode receiver.

* It should be noted that the corresponding schematic showing a fiber amplifier fulfilling this role is provided in Figure 10.9(c).

† The optimum filter bandwidth is determined by a number of factors including the detector noise, the transmission rate, the transmitter chirp characteristics and the filter insertion loss but is typically in the range 0.5 to 3 nm.

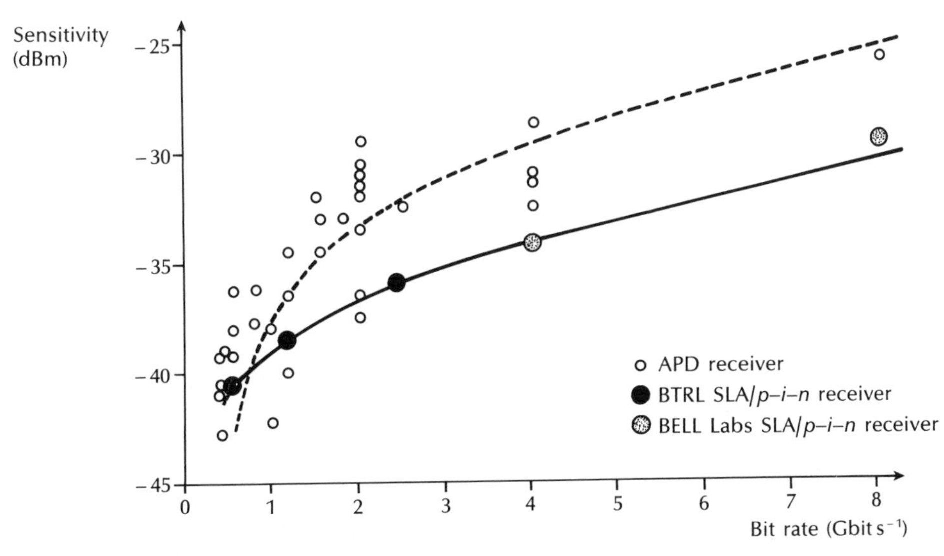

Figure 9.21 Characteristics showing receiver sensitivity against transmission bit rate for SLA–preamplified *p-i-n* photodiode receivers and APD receivers. Reproduced with permission from C. A. Hunter, L. N. Barker, D. J. T. Heatley, K. H. Cameron and R. L. Calton, 'The design and performance of semiconductor laser preamplified optical receivers', *IEE Colloq. Dig.* No. 1989/119, paper 9, 1989.

significant improvement over high performance APD receivers, particularly at speeds of 2.4 Gbit s^{-1} and above.

Improvements in overall receiver sensitivity have also been demonstrated using erbium doped fiber preamplifiers [Refs. 53, 54]. Such fiber amplifiers can provide much lower input and output coupling losses together with smaller amounts of spontaneous emission because of their narrower spectral bandwidths (4 to 10 nm) in comparison with SLAs. This latter attribute means that fiber preamplifiers can be used without the requirement for an optical bandpass filter between the device and the receiver. In this context the spontaneous noise bandwidth is simply determined by the gain characteristic of the fiber amplifier. A recent demonstration of an erbium doped fiber preamplified *p–i–n* photodiode receiver displayed an improvement in receiver sensitivity of 10.5 dB [Ref. 55].

Although in this chapter we have focused on receiver performance and design techniques for intensity modulated/direct detection optical fiber communication systems, many of the strategies discussed are also utilized within the generally more complex receiver structures required to enable coherent transmission. The various coherent demodulation schemes are discussed in some detail in Section 12.6 and the coherent receiver sensitivities are compared both with each other and with direct detection in Section 12.7. However, the specific preamplifier noise and technological considerations are not repeated as they apply equally to both detection techniques.

Problems

9.1 Briefly discuss the possible sources of noise in optical fiber receivers. Describe in detail what is meant by quantum noise. Consider this phenomenon with regard to:

(a) digital signalling;
(b) analog transmission,

giving any relevant mathematical formulae.

9.2 A silicon photodiode has a responsivity of $0.5\ \mathrm{A\,W^{-1}}$ at a wavelength of $0.85\ \mu\mathrm{m}$. Determine the minimum incident optical power required at the photodiode at this wavelength in order to maintain a bit error rate of 10^{-7}, when utilizing ideal binary signalling at a rate of $35\ \mathrm{Mbit\,s^{-1}}$.

9.3 An analog optical fiber communication system requires an SNR of 40 dB at the detector with a post detection bandwidth of 30 MHz. Calculate the minimum optical power required at the detector if it is operating at a wavelength of $0.9\ \mu\mathrm{m}$ with a quantum efficiency of 70%. State any assumptions made.

9.4 A digital optical fiber link employing ideal binary signalling at a rate of $50\ \mathrm{Mbit\,s^{-1}}$ operates at a wavelength of $1.3\ \mu\mathrm{m}$. The detector is a germanium photodiode which has a quantum efficiency of 45% at this wavelength. An alarm is activated at the receiver when the bit error rate drops below 10^{-5}. Calculate the theoretical minimum optical power required at the photodiode in order to keep the alarm inactivated. Comment briefly on the reasons why in practice the minimum incident optical power would need to be significantly greater than this value.

9.5 Discuss the implications of the load resistance on both thermal noise and post detection bandwidth in optical fiber communication receivers.

9.6 A silicon *p–i–n* photodiode has a quantum efficiency of 65% at a wavelength of $0.8\ \mu\mathrm{m}$. Determine:

(a) the mean photocurrent when the detector is illuminated at a wavelength of $0.8\ \mu\mathrm{m}$ with $5\ \mu\mathrm{W}$ of optical power;
(b) the rms quantum noise current in a post detection bandwidth of 20 MHz;
(c) the SNR in dB, when the mean photocurrent is the signal.

9.7 The photodiode in Problem 9.6 has a capacitance of 8 pF. Calculate:

(a) the minimum load resistance corresponding to a post detection bandwidth of 20 MHz;
(b) the rms thermal noise current in the above resistor at a temperature of 25 °C;
(c) the SNR in dB resulting from the illumination in Problem 9.6 when the dark current in the device is 1 nA.

9.8 The photodiode in Problems 9.6 and 9.7 is used in a receiver where it drives an amplifier with a noise figure of 2 dB and an input capacitance of 7 pF. Determine:

(a) the maximum amplifier input resistance to maintain a post detection bandwidth of 20 MHz without equalization;
(b) the minimum incident optical power required to give an SNR of 50 dB.

9.9 A germanium photodiode incorporated into an optical fiber receiver working at a wavelength of $1.55\ \mu\mathrm{m}$ has a dark current of 500 nA at the operating temperature. When the incident optical power at this wavelength is 10^{-6} W and the responsivity of the device is $0.6\ \mathrm{A\,W^{-1}}$, shot noise dominates in the receiver. Determine the SNR in dB at the receiver when the post detection bandwidth is 100 MHz.

9.10 Discuss the expression for the SNR in an APD receiver given by:

$$\frac{S}{N} = \frac{M^2 I_p^2}{2eB(I_p + I_d)M^{2+x} + \dfrac{4KTBF_n}{R_L}}$$

with regard to the various sources of noise present in the receiver. How may this expression be modified to give the optimum avalanche multiplication factor?

9.11 A silicon RAPD has a quantum efficiency of 95% at a wavelength of 0.9 μm, has an excess avalanche noise factor of $M^{0.3}$ and a capacitance of 2 pF. It may be assumed that the post detection bandwidth (without equalization) is 25 MHz, and that the dark current in the device is negligible at the operating temperature of 290 K. Determine the minimum incident optical power which can yield an SNR of 23 dB.

9.12 With the device and conditions given in Problem 9.11, calculate:

(a) the SNR obtained when the avalanche multiplication factor for the RAPD falls to half the optimum value calculated;

(b) the increased optical power necessary to restore the SNR to 23 dB with $M = 0.5M_{op}$.

9.13 What is meant by the excess avalanche noise factor $F(M)$? Give two possible ways of expressing this factor in analytical terms. Comment briefly on their relative merits.

9.14 A germanium APD (with $x = 1.0$) operates at a wavelength of 1.35 μm where its responsivity is 0.45 A W^{-1}. The dark current is 200 nA at the operating temperature of 250 K and the device capacitance is 3 pF. Determine the maximum possible SNR when the incident optical power is 8×10^{-7} W and the post detection bandwidth without equalization is 560 MHz.

9.15 The photodiode in Problem 9.14 drives an amplifier with a noise figure of 3 dB and an input capacitance of 3 pF. Determine the new maximum SNR when they are operated under the same conditions.

9.16 Discuss the three main amplifier configurations currently adopted for optical fiber communications. Comment on their relative merits and drawbacks.

A high impedance integrating front end amplifier is used in an optical fiber receiver in parallel with a detector bias resistor of 10 MΩ. The effective input resistance of the amplifier is 6 MΩ and the total capacitance (detector and amplifier) is 2 pF.

It is found that the detector bias resistor may be omitted when a transimpedance front end amplifier design is used with a 270 kΩ feedback resistor and an open loop gain of 100.

Compare the bandwidth and thermal noise implications of these two cases, assuming an operating temperature of 290 K.

9.17 A *p–i–n* photodiode operating at a wavelength of 0.83 μm has a quantum efficiency of 50% and a dark current of 0.5 nA at a temperature of 295 K. The device is unbiased but loaded with a current mode amplifier with a 50 kΩ feedback resistor and an open loop gain of 32. The capacitance of the photodiode is 1 pF and the input capacitance of the amplifier is 6 pF.

Determine the incident optical power required to maintain a SNR of 55 dB when the post detection bandwidth is 10 MHz. Is equalization necessary?

9.18 A voltage amplifier for an optical fiber receiver is designed with an effective input resistance of 200 Ω which is matched to the detector bias resistor of the same value. Determine:

(a) The maximum bandwidth that may be obtained without equalization if the total capacitance (C_T) is 10 pF.

(b) The rms thermal noise current generated in this configuration when it is operating over the bandwidth obtained in (a) and at a temperature of 290 K. The thermal noise generated by the voltage amplifier may be assumed to be from the effective input resistance to the device.

(c) Compare the values calculated in (a) and (b) with those obtained when the voltage amplifier is replaced by a transimpedance amplifier with a 10 kΩ feedback resistor and an open loop gain of 50. It may be assumed that the feedback resistor is also used to bias the detector, and the total capacitance remains 10 pF.

9.19 What is a PIN–FET hybrid receiver? Discuss in detail its merits and possible drawbacks in comparison with the APD receiver.

9.20 Identify the characteristics which are of greatest interest in the pursuit of high performance receivers.

Discuss the major techniques which have been adopted in order to produce such high performance receivers for use in long-haul optical fiber communications.

Answers to numerical problems

9.2 −70.4 dBm
9.3 −37.2 dBm
9.4 −70.1 dBm
9.6 (a) 2.01 μA; (b) 3.59 nA; (c) 55.0 dB
9.7 (a) 994.7 Ω; (b) 18.19 nA; (c) 39.3 dB
9.8 (a) 1.137 kΩ; (b) 19.58 μW
9.9 40.1 dB
9.11 −50.3 dBm
9.12 (a) 14.2 dB; (b) −49.6 dBm
9.14 23.9 dB
9.15 21.9 dB
9.16 High impedance front end: 21.22 kHz, 4.27×10^{-27} A^2 Hz^{-1}; Transimpedance front end: 29.47 MHz, 5.93×10^{-26} A^2 Hz^{-1}
9.17 −23.1 dBm, equalization is unnecessary
9.18 (a) 159.13 MHz; (b) 160 nA; (c) 79.56 MHz, 11.3 nA, noise power 23 dB down

References

[1] (a) M. Schwartz, *Information Transmission, Modulation and Noise* (4th edn), McGraw-Hill, 1990. (b) F. R. Conner, *Noise*, (2nd edn), Edward Arnold, 1982.
[2] P. Russer, 'Introduction to optical communications', in M. J. Howes and D. V. Morgan (Eds.), *Optical Fibre Communications*, pp. 1–26, John Wiley, 1980.
[3] M. Garbuny, *Optical Physics*, Academic Press, 1965.
[4] W. M. Hubbard, 'Efficient utilization of optical frequency carriers for low and moderate bit rate channels', *Bell Syst. Tech. J.*, **50**, pp. 713–718, 1973.
[5] I. Garrett, 'Receivers for optical fibre communications', *Electron. and Radio Eng.*, **51**(7/8), pp. 349–361, 1981.
[6] P. P. Webb, R. J. McIntyre and J. Conradi, 'Properties of avalanche photodiodes', *RCA Rev.*, **35**, pp. 234–278, 1974.
[7] W. R. Bennett and J. R. Davey, *Data Transmission*, McGraw-Hill, 1965.
[8] S. D. Personick, 'Receiver design for digital fiber optic communication systems (Part I and II)', *Bell Syst. Tech. J.*, **52**, pp. 843–886, 1973.

[9] T. P. Lee and T. Li. 'Photodetectors', in. S. E. Miller and A. G. Chynoweth (Eds.), *Optical Fiber Telecommunications*, pp. 593–623. Academic Press, 1979.

[10] S. D. Personick, 'Receiver design', in S. E. Miller and A. G. Chynoweth (Eds.), *Optical Fiber Telecommunications*, pp. 627–651, Academic Press, 1979.

[11] J. E. Goell, 'Input amplifiers for optical PCM receivers', *Bell Syst. Tech. J.*, **54**, pp. 1771–1793. 1974.

[12] J. L. Hullett and T. V. Muoi, 'Referred imepedance noise analysis for feedback amplifiers', *Electron. Lett.*, **13**(13), pp. 387–389, 1977.

[13] R. G. Smith and S. D. Personick, 'Receiver design for optical fiber communication systems', in H. Kressel (Ed.), *Semiconductor Devices for Optical Communication* (2nd edn), Springer-Verlag, 1982.

[14] J. L. Hullett, 'Optical communication receivers', *Proc. IREE Australia*, **40**(4), pp. 127–136, 1979.

[15] J. L. Hullett and T. V. Muoi, 'A feedback receiver amplifier for optical transmission systems', *Trans. IEEE*, **COM 24**, pp. 1180–1185, 1976.

[16] J. S. Barrera, 'Microwave transistor review, Part 1. GaAs field-effect transistors'. *Microwave J.* (USA), **19**(2), pp. 28–31, 1976.

[17] B. S. Hewitt, H. M. Cox, H. Fukui, J. V. Dilorenzo, W. O. Scholesser and D. E. Iglesias, 'Low noise GaAs MESFETs', *Electron. Lett.*, **12**(12), pp. 309–310, 1976.

[18] D. V. Morgan, F. H. Eisen and A. Ezis, 'Prospects for ion bombardment and ion implantation in GaAs and InP device fabrication', *IEE Proc.*, **128**(1–4), pp. 109–129, 1981.

[19] J. Mun, J. A. Phillips and B. E. Barry, 'High-yield process for GaAs enhancement-mode MESFET integrated circuits', *IEE Proc.*, **128**(1–4), pp. 144–147, 1981.

[20] S. D. Personick, P. Balaban, J. H. Bobsin and P. R. Kumar, 'A detailed comparison of four approaches to the calculation of the sensitivity of optical fiber system receivers', *IEEE Trans. Commun.*, **COM-25**, pp. 541–549, 1977.

[21] S. D. Personick, 'Design of receivers and transmitters for fiber systems', in M. K. Barnoski (Ed.), *Fundamentals of Optical Fiber Communications* (2nd edn), Academic Press, 1981.

[22] D. R. Smith, R. C. Hooper and I. Garrett, 'Receivers for optical communications: A comparison of avalanche photodiodes with PIN–FET hybrids', *Opt. Quant. Electron.*, **10**, pp. 293–300, 1978.

[23] R. C. Hooper and D. R. Smith, 'Hybrid optical receivers using PIN photodiodes, *IEE (London) Colloquium on Broadband High Frequency Amplifiers*, pp. 9/1–9/5, 1979.

[24] K. Ahmad and A. W. Mabbitt, '$Ga_{1-x}In_xAs$ photodetectors for 1.3 micron PIN–FET receiver', *IEEE NY (USA) International Electronic Devices Meeting* (Washington, DC), pp. 646–649, 1978.

[25] D. R. Smith, R. C. Hooper and R. P. Webb, 'High performance digital optical receivers with PIN photodiodes', *IEEE (NY) Proceedings of the International Symposium on Circuits and Systems* (Tokyo), pp. 511–514, 1979.

[26] D. R. Smith, R. C. Hooper, K. Ahmad, D. Jenkins, A. W. Mabbitt and R. Nicklin, '*p–i–n*/FET hybrid optical receiver for longer wavelength optical communication systems', *Electron. Lett.*, **16**(2), pp. 69–71, 1980.

[27] R. C. Hooper, D. R. Smith and B. R. White, 'PIN–FET Hybrids for digital optical receivers', *IEEE NY (USA) 30th Electronic Components Conference*, San Francisco, pp. 258–260, 1980.

[28] S. Hata, Y. Sugeta, Y. Mizushima, K. Asatani and K. Nawata, 'Silicon *p–i–n* photodetectors with integrated transistor amplifiers', *IEEE Trans. Electron. Devices*, **ED-26**(6), pp. 989–991, 1979.

[29] K. Ogawa and E. L. Chinnock, 'GaAs FET transimpedance front-end design for a wideband optical receiver', *Electron. Lett*, **15**(20), pp. 650–652, 1979.

[30] S. M. Abbott and W. M. Muska, 'Low noise optical detection of a 1.1 Gb/s optical data stream, *Electron. Lett.*, **15**(9), pp. 250–251, 1979.

[31] L. A. Godfrey, 'Designing for the fastest response ever – ultra high speed photodetection', *Opt. Spectra (USA)*, **13**(10), pp. 43, 46, 1979.

[32] R. I. MacDonald, 'High gain optical detection with GaAs field effect transistors', *Appl. Opt. (USA)*, **20**(4), pp. 591–594, 1981.

[33] M. Brain and T. P. Lee, 'Optical receivers for lightwave communication systems', *J. of Lightwave Technol.*, **LT-3**(6), pp. 1281–1300, 1985.

[34] B. L. Kasper, 'Receiver design', in S. E. Miller and I. P. Kaminow (Eds.), *Optical Fiber Telecommunications II*, Academic Press, pp. 689–722, 1988.

[35] G. P. Vella-Coleiro, 'Optimization of optical sensitivity of *p–i–n* FET receivers', *IEEE Electron. Device Lett.*, **9**(6), pp. 269–271, 1988.

[36] R. A. Minasian, 'Optimum design of 4-Gbit/s GaAs MESFET optical preamplifier', *J. of Lightwave Technol.*, **LT-5**(3), pp. 373–379, 1987.

[37] M.A.R. Violas, D. J. T. Heatley, A. M. O. Duarte and D. M. Beddow, '10 GHz bandwidth low noise optical receiver using discrete commercial devices', *Electron. Lett.*, **26**(1), pp. 35–36, 1990.

[38] C. W. Farley, M. F. Chang, P. M. Asbeck, N. H. Sheng, R. Pierson, G. J. Sullivan, K. C. Wang and R. B. Nubling, 'High-speed ($f_t = 78$ GHz) AlInAs/GaInAs single heterojunction HBT', *Electron. Lett.*, **25**(13), pp. 846–847, 1989.

[39] N. Ohkawa, '20 GHz bandwidth low-noise HEMT preamplifier for optical receivers', *Electron. Lett.*, **24**(7), pp. 1061–1062, 1988.

[40] S. D. Walker, L. C. Blank, R. A. Garnham and J. M. Boggis, 'High electron mobility transistor lightwave receiver for broadband optical transmission system applications', *J. of Lightwave Technol.*, **7**(3), pp. 454–458, 1989.

[41] B. L. Kasper and J. C. Campbell, 'Multigigabit-per-second avalanche photodiode lightwave receivers', *J. of Lightwave Technol.*, **LT-5**(10), pp. 1351–1364, 1987.

[42] B. L. Kasper, J. C. Campbell, J. R. Talman, A. H. Gnauck, J. E. Bowers and W. S. Holden, 'An APD/FET optical receiver operating at 8 Gbit/s', *J. of Lightwave Technol.*, **LT-5**(3), pp. 344–347, 1987.

[43] J. J. O'Reilly and R. S. Fyath, 'Performance of optical receivers employing ultralow noise avalanche photodiodes', *J. of Optical Communications*, **9**(3), pp. 82–84, 1988.

[44] M. J. N. Sibley, R. T. Unwin, D. R. Smith, B. A. Boxall and R. J. Hawkins, 'A monolithic common collector front-end optical preamplifier', *J. of Lightwave Technol.*, **LT-3**(1), pp. 13–15, 1985.

[45] Y. Archambault, D. Pavlidis and J. P. Guet, 'GaAs monolithic integrated optical preamplifier', *J. of Lightwave Technol.*, **LT-5**(3), pp. 355–366, 1987.

[46] W. T. Colleran and A. A. Abidi, 'Wideband monolithic GaAs amplifier using cascodes', *Electron. Lett.*, **23**(18), pp. 951–952, 1987.

[47] S. Miura, T. Mikawa, T. Fujii and O. Wada, 'High-speed monolithic GaInAs pinFET', Electron. Lett., **24**(7), pp. 394–395, 1988.

[48] G. F. Williams and H. P. Leblanc, 'Active feedback lightwave receivers', *J. of Lightwave Technol.*, **LT-4**(10), pp. 1502–1508, 1986.

[49] B. L. Kasper, A. R. McCormick, C. A. Burrus Jr and J. R. Talman, 'An optical-feedback transimpedance receiver for high sensitivity and wide dynamic range at low bit rates', *J. of Lightwave Technol.*, **6**(2), pp. 329–338, 1988.

[50] S. G. Methley, 'An optical feedback receiver, with high sensitivity', *Proc. SPIE Int. Soc. Opt. Eng., Fibre Optics' 88*, **949**, pp. 51–55, April 1988.

[51] I. W. Marshall and M. J. O'Mahony, '10 GHz optical receiver using a travelling wave semiconductor laser preamplifier', *Electron. Lett.*, **23**(20), p. 1052, 1987.

[52] N. A. Olsson and M. G. Oberg, 'Ultra low reflectivity 1.5 μm semiconductor laser preamplifier', *Electron. Lett.*, **24**(9), pp. 569–570, 1988.

[53] C. A. Hunter, L. N. Barker, D. J. T. Heatley, K. H. Cameron and R. L. Calton, 'The design and performance of semiconductor laser preamplifier optical receivers', *IEE Colloquium on Optical Amplifiers for Communication*, Dig.No. 1989/119, paper 9, 27 October 1989.

[54] C. R. Giles, J. L. Desurvire, J. L. Zyskind and J. R. Simpson, 'Near quantum limited erbium doped fiber preamplifier with 215 photons/bit sensitivity at 1.8 Gbit/s', *IOOC 89*, Kobe, Japan, paper 20PDA-5, 1989.

[55] M. J. Pettitt, R. A. Baker and A. Hadifotiou, 'System performance of optical fibre preamplifier', *Electron. Lett.*, **25**(4), pp. 273–275, 1989.

10

Optical amplification and integrated optics

10.1 Introduction

The preceding four chapters have been concerned with the devices employed to provide the electrical–optical interfaces within optical fiber communications. Optical sources were dealt with in Chapters 6 and 7 followed by optical detectors in Chapter 8 prior to consideration of optical receiver noise and its effect on receiver design in Chapter 9. These electrical–optical and optical–electrical conversion devices are crucial components for the realization of optical fiber communications, as may be observed in the following two chapters which discuss optical fiber systems. However, these devices are also, in a number of respects, a limiting factor within the implementation of optical fiber systems. The conversion of the information signal from the electrical domain to the optical domain and vice-versa often provides a bottleneck within optical fiber communications which may restrict both the operating bandwidth and the quality of the transmitted signal. Performing operations on signals in the optical domain combined with the pursuit of more efficient mechanisms to provide the electrical–optical interfaces has therefore assumed increasing significance within optical fiber communications and its associated application areas.

The above considerations have stimulated a growing activity in the area of active devices and components which allow optical signals to be manipulated without returning them back to the electrical regime where such operations have normally been carried out in the past. Potentially, such devices alleviate the possible bottleneck associated with the interfaces as well as providing more efficient, and hence cost effective, methods for processing the optical signals. Moreover, in some cases the use of these devices and components may represent the only realistic solution for the implementation of particular optical fiber transmission techniques and systems.

This chapter therefore discusses the major developments in the area of active optical devices which may be utilized for a variety of functions within optical fiber communications. In addition it deals with the technologies associated with the integration of such optical and optoelectronic devices into circuits and subsystems (i.e. integrated optics and optoelectronic integration) which are becoming important areas within the design strategies for advanced optical fiber communication systems. A major development which was stimulated by the massive effort to produce laser sources for optical fiber communications is that of optical amplification. The technology associated with such active optical devices is now well established and their use within optical fiber communications is assured.

Section 10.2 introduces the concept of an optical amplifier and outlines the various generic types that are under investigation. Semiconductor laser amplifiers are then considered in some detail in Section 10.3. This is followed in Section 10.4 with a discussion of the various types of fiber amplifier which have evolved more recently, assisted by the activities which have led to the realization of fiber lasers (see Section 6.9.2).

The concepts and technology associated with integrated optics are then introduced in Section 10.5 prior to consideration of some common integrated optical devices in Section 10.6. In particular the latter section concentrates on directional couplers and switches, together with modulator devices which provide an alternative to direct current modulation for optical sources. Developments in the area of optoelectronic integration are then dealt with in Section 10.7, particular emphasis in this technology being concerned with the implementation of transmitters/receivers and switches on a single substrate.

Optical bistability and digital optics are then discussed in Section 10.8 to provide the reader with an insight into this important area, together with an understanding of how these phenomena may be utilized within optical fiber communications.

Finally, in Section 10.9 developments in the field of optical computation are outlined. Although it is clear that these latter developments do not, at present, significantly influence optical fiber communications, it is likely that in the future there will be a requirement for the combination of optical communication, optical switching and optical computational technologies, within what is fast becoming a predominantly optical fiber telecommunication network.

10.2 Optical amplifiers

Optical amplifiers, as their name implies, operate solely in the optical domain with no interconversion of photons to electrons. Therefore, instead of using regenerative repeaters which, as currently implemented, require optoelectronic devices for source and detector, together with substantial electronic circuitry for pulse slicing, retiming and shaping (see Section 11.6.1), optical amplifiers can be placed at intervals along a fiber link to provide linear amplification of the transmitted optical signal. The optical amplifier, in principle, provides a much simpler solution in that it is a single in-line component which can be used for any kind of modulation at virtually any transmission rate. Moreover, such a device can be bidirectional and if it is sufficiently linear it may allow multiplex operation of several signals at different optical wavelengths (i.e. wavelength division multiplexing). In particular with single-mode fiber systems, the effects of signal dispersion can be small and hence the major limitation on repeater spacing becomes attenuation due to fiber losses. Such systems do not require full regeneration of the transmitted digital signal at each repeater, and optical amplification of the signal proves sufficient. Hence over recent years optical amplifiers have emerged as promising network elements not just for use as linear repeaters but as optical gain blocks, optical receiver preamplifiers and, when used in a nonlinear mode, as optical gates, pulse shapers and routeing switches [Ref. 1].

The two main approaches to optical amplification to date have concentrated on semiconductor laser amplifiers which utilize stimulated emission from injected carriers and fiber amplifiers in which gain is provided by either stimulated Raman or Brillouin scattering* (see Sections 3.5 and 3.14), or by rare earth dopants (see Section 6.9.2). Both amplifier types (i.e. semiconductor and fiber; specifically rare earth and Raman) have the ability to provide high gain over wide spectral bandwidths, making them eminently suitable for future optical fiber system applications.

The typical gain profiles for various optical amplifier types based around the 1.5 μm wavelength region are illustrated in Figure 10.1 [Ref. 2]. It may be observed that the InGaAsP travelling wave semiconductor laser amplifier (TWSLA), the erbium doped fiber amplifier and the Raman fiber amplifier all provide wide

* Amplification from both stimulated Raman or Brillouin scattering can occur in undoped relatively long fiber lengths ($\simeq$ 10 km) or doped short lengths ($\simeq$ 10 m) of fiber [Ref. 2].

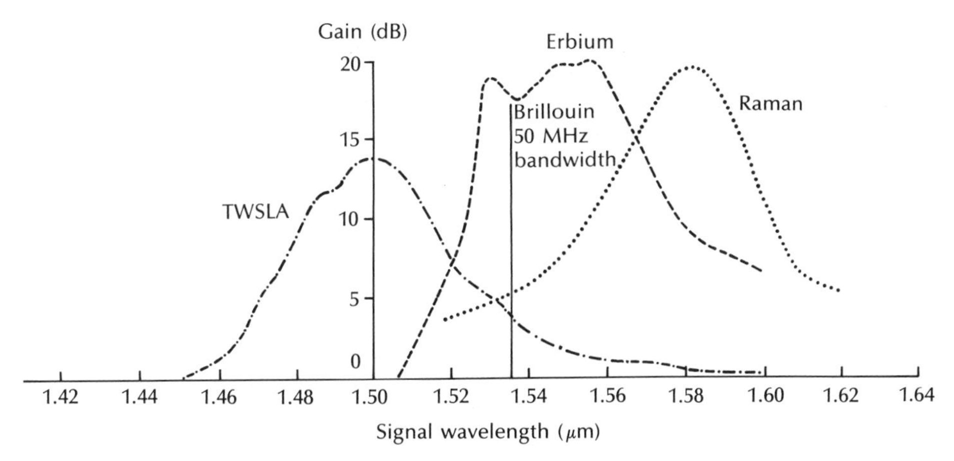

Figure 10.1 Optical amplifier gain characteristics. Reproduced with permission from P. Cochrane, *Br. Telecom. Technol. J.*, **8** (2), p. 5, 1990.

spectral bandwidths. Hence these optical amplifier types lend themselves to applications involving wavelength division multiplexing [Ref. 3]. By contrast, the Brillouin fiber amplifier has a very narrow spectral bandwidth, possibly around 50 MHz and therefore cannot be employed for wideband amplification. It could, however, be used for channel selection within a WDM system by allowing amplification of a particular channel without boosting other nearby channels.

Whereas semiconductor laser amplifiers exhibit low power consumption and their single-mode waveguide structures make them particularly appropriate for use with single-mode fiber, it is fiber amplifiers which present fewer problems of compatibility for in-line interconnection within optical fiber links [Ref. 4]. At present, semiconductor laser amplifiers are the most developed optical amplifier generic type but research into fiber amplifiers has also made rapid progress towards commercial products over the last few years.

10.3 Semiconductor laser amplifiers

The semiconductor laser amplifier (SLA) is based on the conventional semiconductor laser structure where the output facet reflectivities are between 30 and 35% [Refs. 4, 5]. SLAs can be used in both nonlinear and linear modes of operation [Ref. 1]. Various types of SLA may be distinguished including the resonant or Fabry–Perot amplifier which is an oscillator biased below oscillation threshold [Ref. 6], the travelling wave (TW) and the near travelling wave (NTW) amplifiers which are effectively single pass devices [Refs. 1, 7] and the injection locked laser amplifier, which is a laser oscillator designed to oscillate at the incident signal frequency [Ref. 8]. Such devices are capable of providing high internal gain

(15 to 35 dB) with low power consumption and their single-mode waveguide structure makes them particularly suitable for use with single-mode fiber. SLAs can, however, be classified into two main groups which are Fabry–Perot amplifiers (FPAs) and travelling wave amplifiers (TWAs)* [Refs. 1, 9], the difference between these groups being the facet reflectivities. A schematic diagram of an SLA is shown in Figure 10.2. It is based on the conventional semiconductor laser structure (gain- or index-guided) with an active region width w, thickness d and length L. When the input and output laser facet reflectivities denoted by R_1 and R_2 are each around 0.3, which depicts a normal semiconductor laser, then an FPA is obtained.† In this case, as the facet reflectivity is large, a highly resonant amplifier is formed and the transmission characteristic comprises very narrow passbands, as displayed in Figure 10.3. The mode zero corresponds to the peak gain wavelength and the mode

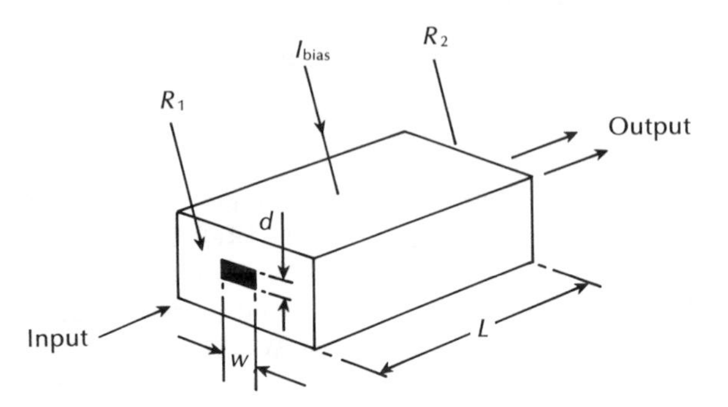

Figure 10.2 Schematic structure of the semiconductor laser amplifier.

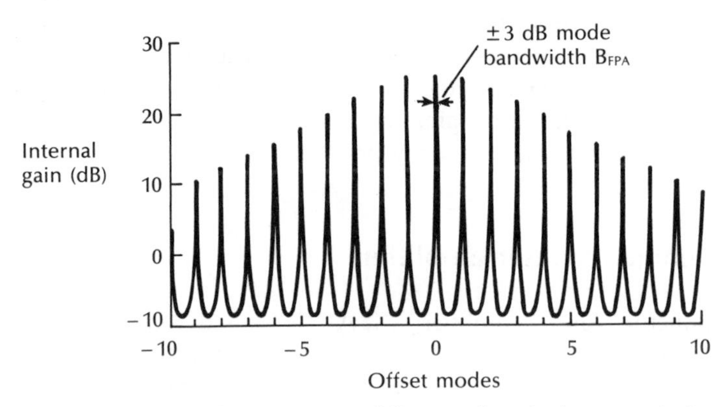

Figure 10.3 The Fabry–Perot amplifier passband where mode 0 corresponds to the peak gain wavelength [Ref. 1].

* The neumonics for the travelling wave semiconductor laser amplifiers are sometimes given as TWSLA [e.g. Refs. 2, 10].
† A FPA may be defined as an amplifier with facet reflectivities in the order 0.01 to 0.3 [Ref. 1].

spacing $\delta\lambda$ can be obtained from Eq. (6.16). For operation, the FPA is biased below the normal lasing threshold current, and light entering one facet appears amplified at the other facet together with inherent noise. In practice, the amplifier chip is bonded into a package with single-mode fiber pigtails which are used to guide light into and out of the amplifier. The inherent filtering of the FPA, although useful in certain applications, means the device is very sensitive to fluctuations in bias current, temperature and signal polarization. However, because of the resonant nature of FPAs, combined with their high internal fields, they are used within nonlinear applications: for example, to provide pulse shaping and bistable elements (see Section 10.8).

To form a travelling wave SLA antireflection coatings may be applied to the laser facets to reduce or eliminate the end reflectivities. This can be achieved by depositing a thin layer of silicon oxide or silicon nitride on the end facets such that the reflectivities are reduced to 1×10^{-3} or less. Such a device becomes a TWA operating in the single-pass amplification mode in which the Fabry–Perot resonance is suppressed by the reduction in facet reflectivity.* This has the effect of substantially increasing the amplifier spectral bandwidth and it makes the transmission characteristics less dependent upon fluctuations in bias current, temperature and input signal polarization. Hence the TWA proves superior to the FPA (particularly for linear applications) and also provides advantages in relation to both signal gain saturation and noise characteristics [Ref. 9]. Moreover, antireflection facet coatings have the effect of increasing the lasing current threshold, as illustrated in Figure 10.4 and so in practice such SLAs are operated at currents far beyond the normal lasing threshold current.

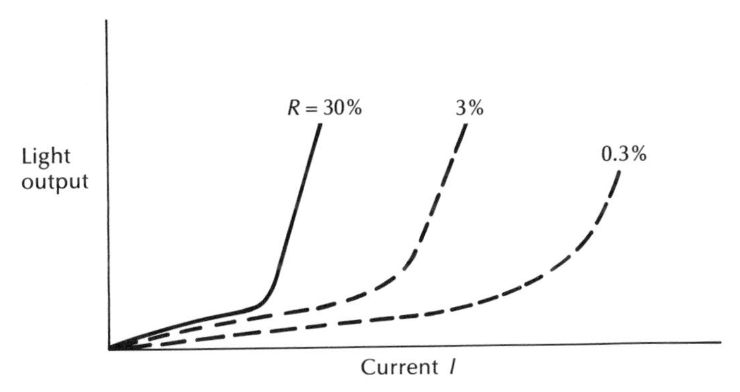

Figure 10.4 Light output against current characteristic for the semiconductor laser amplifier with different values of facet reflectivity R.

* In theory, a true TWA is the limiting case of a device with facets exhibiting zero reflectivity. However, in practice, even with the best antireflection coatings, some residual facet reflectivity remains (e.g. a low reflectivity of 1×10^{-4} has been obtained at a wavelength of 1.5 μm [Ref. 11]). Hence such devices are also referred to as near travelling wave amplifiers [Refs. 1, 12].

10.3.1 Theory

The general equation for the cavity gain G of a SLA as a function of signal frequency f takes the form [Refs. 1, 9, 10]:

$$G(f) = \frac{(1 - R_1)(1 - R_2)G_s}{(1 - \sqrt{R_1 R_2} G_s)^2 + 4\sqrt{R_1 R_2} G_s \sin^2 \phi} \tag{10.1}$$

where R_1 and R_2 are the input and output facet reflectivities respectively, G_s is the single pass gain and ϕ is the single pass phase shift through the amplifier. It should be noted that Eq. (10.1) does not include coupling losses to and from the amplifier and that the phase shift ϕ may be written as [Ref. 9]:

$$\phi = \frac{\pi(f - f_o)}{\delta f} \tag{10.2}$$

where f_o is the Fabry–Perot resonant frequency and δf is the free spectral range of the SLA.

The 3 dB spectral bandwidth of an FPA, or essentially the ± 3 dB single longitudinal mode bandwidth defined by the FWHP points B_{FPA}, is shown in Figure 10.3. It may be observed that using Eqs. (10.1) and (10.2), B_{FPA} may be expressed as:

$$B_{FPA} = 2(f - f_o) = \frac{2\delta f}{\pi} \sin^{-1}\left[\frac{1 - \sqrt{R_1 R_2} G_s}{2(\sqrt{R_1 R_2} G_s)^{\frac{1}{2}}}\right]$$

$$= \frac{c}{\pi n L} \sin^{-1}\left[\frac{1 - \sqrt{R_1 R_2} G_s}{2(\sqrt{R_1 R_2} G_s)^{\frac{1}{2}}}\right] \tag{10.3}$$

where the mode separation frequency interval δf given by Eq. (6.14) combines the velocity of light c and the refractive index of the amplifier medium n with its length L. Alternatively the 3 dB spectral or optical bandwidth may be expressed as a function of the FPA cavity gain G following [Ref. 1]:

$$B_{FPA} = \frac{c}{\pi n L} \sin^{-1}\left[\frac{1}{2}\left(\frac{(1 - R_1)(1 - R_2)}{\sqrt{R_1 R_2 G}}\right)\right] \tag{10.4}$$

Example 10.1

An uncoated FPA has facet reflectivities of 30% and a single pass gain of 4.8 dB. The amplifier has a 300 μm long active region, a mode spacing of 1 nm and a peak gain wavelength of 1.5 μm. Determine the refractive index of the active medium and the 3 dB spectral bandwidth of the device.

Solution: The refractive index of the active medium at the peak gain wavelength may be obtained by rearranging Eq. (6.16) such that:

$$n = \frac{\lambda^2}{2\delta\lambda L} = \frac{(1.5 \times 10^{-6})^2}{2 \times 1 \times 10^{-9} \times 300 \times 10^{-6}}$$

$$= 3.75$$

Using Eq. (10.3) the 3 dB spectral bandwidth is given by:

$$B_{\mathrm{FPA}} = \frac{c}{\pi n L} \sin^{-1}\left[\frac{1 - \sqrt{R_1 R_2 G_s}}{2(\sqrt{R_1 R_2 G_s})^{\frac{1}{2}}}\right]$$

$$= \frac{2.998 \times 10^8}{\pi \times 3.75 \times 300 \times 10^{-6}} \sin^{-1}\left[\frac{1 - \sqrt{0.09} \times 3.020}{2(\sqrt{0.09} \times 3.020)^{\frac{1}{2}}}\right]$$

$$= 8.482 \times 10^{10} \sin^{-1}\left[\frac{0.040}{1.904}\right]$$

$$= 8.482 \times 10^{10} \times 0.494 = 4.2 \text{ GHz}.$$

The above result demonstrates the narrow spectral bandwidth obtained with an uncoated FPA.

The single pass gain G_s defined in terms of the device parameters and the applied bias current following Eq. (6.18) for the semiconductor laser* is generally written in the form:

$$G_s = \exp[\bar{g}L] \tag{10.5}$$

where $\bar{g}$ is the nett gain coefficient per unit length and L is the amplifier active length. However, the nett gain per unit length $\bar{g}$ may be defined in terms of the material gain coefficient g_m, the optical confinement factor Γ, and the effective loss coefficient per unit length $\bar{\alpha}$ as [Ref. 1]:

$$\bar{g} = \Gamma g_m - \bar{\alpha} \tag{10.6}$$

Furthermore, the material gain coefficient g_m is related to the signal intensity I following [Ref. 1]:

$$g_m = \frac{g_o}{1 + I/I_s} \tag{10.7}$$

where g_o is the unsaturated material gain coefficient in the absence of the input signal and I_s is the saturation intensity. Hence substitution of Eqs. (10.6) and (10.7)

* The round trip gain of Eq. (6.18) includes a factor of 2 which is omitted for the single pass gain.

in Eq. (10.5) for the single pass gain gives:

$$G_s = \exp[(\Gamma g_m - \bar{\alpha})L] \tag{10.8}$$

$$= \exp\left[\left(\frac{\Gamma g_o}{1 + I/I_s} - \bar{\alpha}\right)L\right] \tag{10.9}$$

It may be observed from Eqs. (10.8) and (10.9) that the single pass gain decreases with increasing intensity and that the material gain coefficient is reduced by a factor of 2 when the internal signal intensity I is equal to the saturation intensity I_s.

The phase shift ϕ_s associated with the single pass amplifier includes the nominal phase shift ϕ_o and an additional component resulting from the change in carrier density from the nominal density in the absence of a signal. Hence the total phase shift is given by [Ref. 13]:

$$\phi_s = \phi_o + \frac{g_o bL}{2}\left(\frac{I}{I + I_s}\right) \tag{10.10}$$

where b is the linewidth broadening factor, and the nominal phase shift is:

$$\phi_o = \frac{2\pi nL}{\lambda} \tag{10.11}$$

where n is the material refractive index. Thus Eqs. (10.9) and (10.10) indicate that both single pass gain and phase are functions of optical intensity. It is clear that for a constant signal intensity (i.e. with frequency modulation) there is no inherent signal distortion; however, with a time-varying intensity the gain and phase may also change with time, causing signal distortion. Furthermore, as G_s and ϕ_s are functions of the input signal intensity, then the SLA will exhibit nonlinear and bistable characteristics at high input powers.

It may be noted that Figure 10.3 shows the general form of the gain versus wavelength for an SLA obtained from Eq. (10.1). Furthermore, Eqs. (10.3) and (10.4) give the 3 dB spectral bandwidth for an FPA as the bandwidth of one longitudinal mode. The 3 dB spectral bandwidth, however, of a TWA is determined by the full gain width of the amplifier medium itself, as illustrated in Figure 10.5(a) rather than the Fabry–Perot gain profile. Hence the 3 dB bandwidth of a TWA is three orders of magnitude larger than that of an FPA [Ref. 9]. Nevertheless, the passband comprises peaks and troughs whose relative amplitudes are determined by the facet reflectivities, the single pass gain (and hence the applied bias current) and the input intensity. This gain undulation or peak–trough ratio of the passband ripple ΔG, which is defined as the difference between the resonant and nonresonant signal gain, may be observed in Figure 10.5(b). It is given by [Refs. 1, 9]:

$$\Delta G = \left(\frac{1 + \sqrt{R_1 R_2 G_s}}{1 - \sqrt{R_1 R_2 G_s}}\right)^2 \tag{10.12}$$

For wideband operation the peak–trough ratio must be small and for convenience is normally considered to be less than 3 dB for TWAs over their signal-

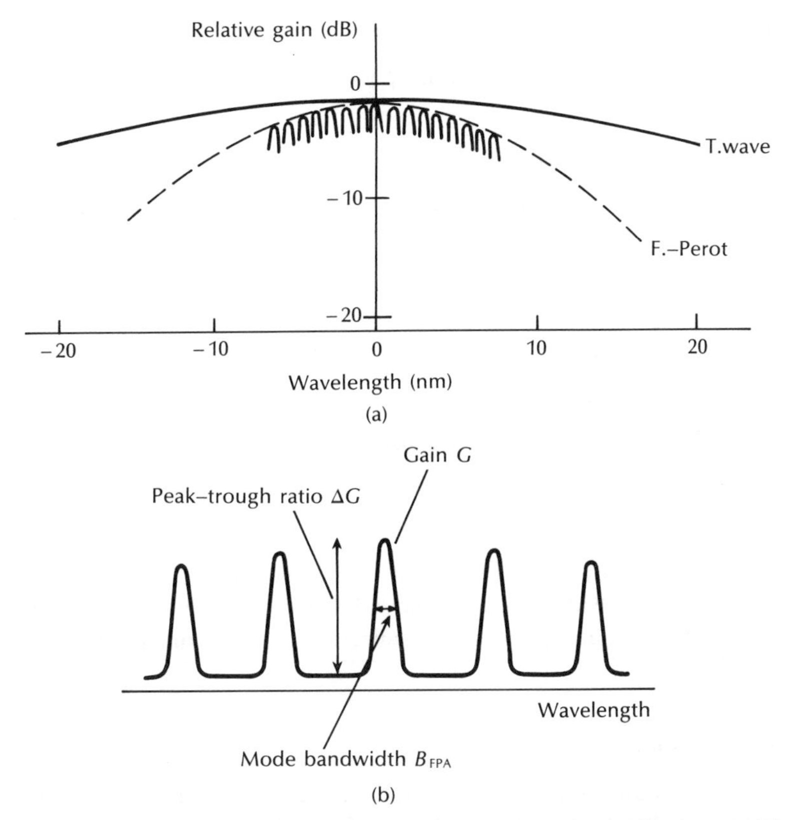

Figure 10.5 Passband characteristics for semiconductor laser amplifiers: (a) overall characteristics for both the travelling wave and Fabry–Perot amplifiers showing the large passband ripple in the latter case; (b) illustration of the peak–trough ratio of the passband ripple given in Eq. (10.12).

gain spectrum* [Refs. 1, 9]. Hence an amplifier whose gain ripple significantly exceeds 3 dB is usually categorized as an FPA.

Example 10.2

Derive an approximate expression for the cavity gain of a TWA in the limiting case of a 3 dB peak–trough ratio.

Solution: For a 3 dB peak–trough ratio Eq. (10.12) becomes:

$$\left(\frac{1+\sqrt{R_1R_2G_s}}{1-\sqrt{R_1R_2G_s}}\right)^2 = 0.5$$

* Sometimes this definition is said to apply to a near travelling wave amplifier as, in theory, a gain ripple of zero would correspond to a pure TWA.

Therefore,

$$1 + \sqrt{R_1R_2}G_s = 0.707\ (1 - \sqrt{R_1R_2}G_s)$$

$$\sqrt{R_1R_2}G_s = \frac{0.293}{1.707} = 0.172$$

For a TWA $R_1, R_2 \ll 1$ and, assuming a zero single pass phase shift, Eq. (10.1) becomes

$$G \simeq \frac{G_s}{(1 - \sqrt{R_1R_2}G_s)^2}$$

Substituting for $\sqrt{R_1R_2}G_s$ gives:

$$G \simeq \frac{G_s}{(1 - 0.172)^2} = \frac{0.172}{(1 - 0.172)^2\sqrt{R_1R_2}}$$

$$= \frac{0.25}{\sqrt{R_1R_2}}$$

The approximate expression for the cavity gain for a TWA in the limiting case is $0.25/\sqrt{R_1R_2}$. Thus for wide spectral bandwidth operation the available cavity gain is determined by the quality of the antireflection coatings on the device.

10.3.2 Performance characteristics

The wide spectral bandwidths that may be achieved using high quality antireflection facet coatings on TWAs are in the region of 50 to 70 nm. However, in comparison with FPAs, such devices require significantly higher bias currents for operation as may be observed from Figure 10.4. In addition, whereas the narrow spectral bandwidth of FPAs provides inherent noise filtering, it is not obtained with TWAs and therefore they are subject to increased levels of noise.

The residual facet reflectivity in TWAs introduces a further problem when considering the use of such amplifiers within optical fiber communication systems. This problem results from the effect of backward gain within the devices. The gain of the backward travelling signal G_b is defined as the ratio of the power in the backward travelling signal P_b to the input signal power P_{in} into the amplifier. Hence the gain of the backward travelling signal is given by [Ref. 14]:

$$G_b = \frac{P_b}{P_{in}} = \frac{(\sqrt{R_1} - \sqrt{R_2}G_s)^2 + 4\sqrt{R_1R_2}G_s \sin^2\phi}{(1 - \sqrt{R_1R_2}G_s)^2 + 4\sqrt{R_1R_2}G_s \sin^2\phi} \quad (10.13)$$

A graphical representation of Eq. (10.13) in which $R_1 = R_2$ for $G_s = 25$ dB is shown in Figure 10.6. It may be observed that the backward gain is approaching the potential forward gain at high facet reflectivity. Moreover, even at low facet reflectivity (0.01%) the backward gain is still very significant (10 dB). In systems with cascaded amplifiers, optical isolators may therefore be required to avoid the

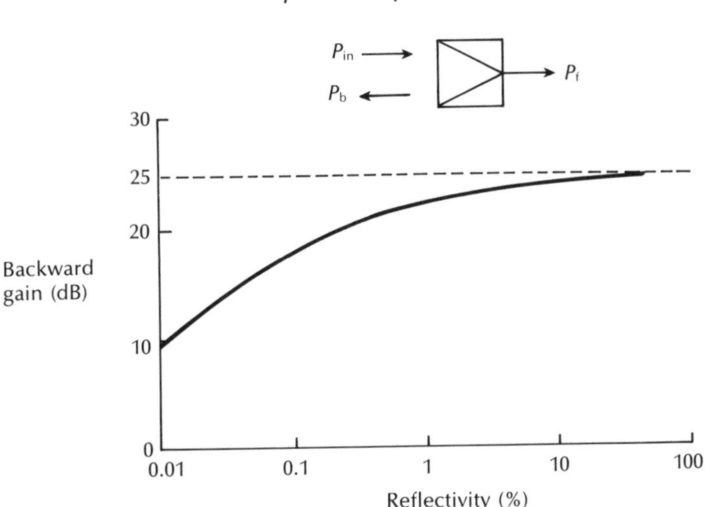

Figure 10.6 Backward gain against facet reflectivity for the travelling wave amplifier determined from Eq. (10.13). P_f is the power in the forward travelling signal.

interaction of backward signals between the devices, unless the backward wave amplitude can be made sufficiently small [Ref. 1].

Another very important characteristic of SLAs is noise since it largely determines the maximum number of devices which can be cascaded as linear repeaters within an optical fiber communication system [Ref. 9]. The overall noise generated by an SLA comprises signal spontaneous beat noise, spontaneous–spontaneous beat noise, spontaneous emission shot noise and amplified signal shot noise [Ref. 15]. The beat noise components occur between the signal and the spontaneously emitted photons as well as between the spontaneously emitted photons themselves. In addition the former noise component tends to dominate in the higher power operating region required for repeater applications [Ref. 9]. The theoretical treatment of the SLA noise characteristics is beyond the scope of this text but interested readers are directed to Ref. 15. It is clear, however, that unlike an electronic amplifier, the noise output from an SLA is a function of the signal intensity. The theoretical mean noise power as a function of facet reflectivity for an SLA, of length 500 μm with a fixed gain of 25 dB, is shown in Figure 10.7 [Ref. 1]. It may be observed that in the travelling wave region the noise power tends towards a fixed value of -3 dBm. This value is in good agreement with the results obtained from measurement [Ref. 1].

It was mentioned previously that the gain of the FPA was very sensitive to changes in temperature and signal polarization. A dependence on these two parameters is also observed in TWAs. For example, at an operating wavelength of 1.5 μm the gain decreases by around 3 dB when the temperature of a TWA of length 500 μm is increased by 5 °C [Ref. 1]. Moreover, although a decrease in temperature increases the device gain, it also increases the passband ripple when

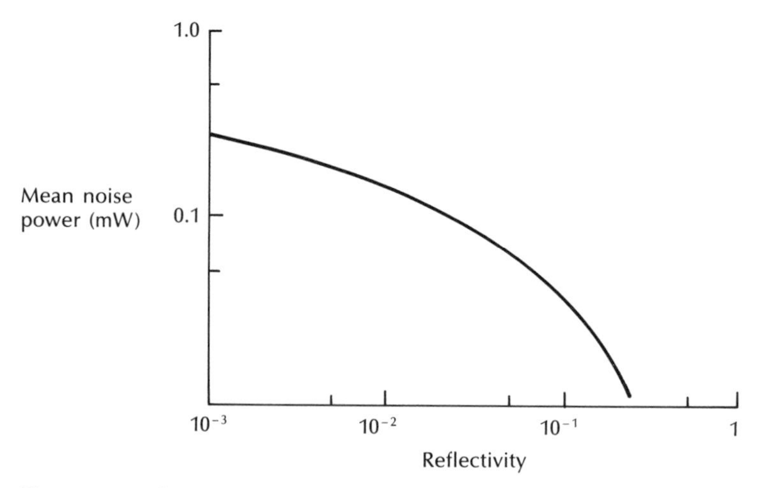

Figure 10.7 Theoretical mean noise power against facet reflectivity characteristic for a semiconductor laser amplifier of length 500 μm with a gain of 25 dB [Ref. 1].

there is residual reflectivity. With the FPA these effects are compounded by a shift in mode wavelength of approximately 10 GHz $^{\circ}C^{-1}$ caused by the variation in refractive index with temperature. Hence it is suggested that for high gain FPAs the temperature must be controlled to within 0.1 °C [Ref. 1].

The dependence of the gain on the polarization of the input signal results from the difference in the single pass gain for the TE and TM polarization modes. It is caused by a difference in their optical confinement factors (i.e. $\Gamma_{TE} \neq \Gamma_{TM}$). Furthermore, this effect is magnified in a resonant cavity because the mode propagation constants and the modal facet reflectivities are also polarization dependent. Hence the use of polarization controllers may be necessary when employing FPAs. The gain difference is minimized, however, with TWAs. For example, the gain difference for an FPA and a TWA both operating at a wavelength of 1.5 μm was found to be 10 dB and 2.5 dB respectively [Ref. 1].

10.4 Fiber amplifiers

System studies employing fiber based optical amplifiers have at present not progressed as far as those using semiconductor laser amplifiers [Ref. 2]. Nevertheless, investigations and demonstrations have indicated that such fiber devices offer significant potential within optical fiber communications [Refs. 17 to 19]. Hence it appears certain that fiber amplifiers will complement the growing device technology associated with SLAs. Although the various fiber amplifier types have significantly different performance characteristics, some of which do not match those obtained with SLAs, in all fiber amplifier devices the spectral bandwidths and centre

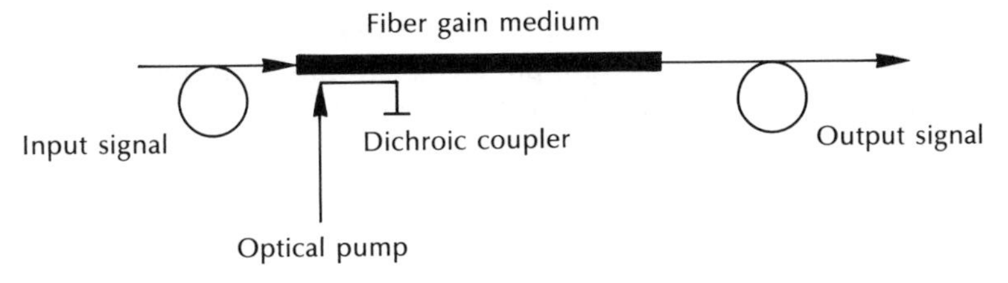

Figure 10.8 Schematic of a fiber amplifier.

wavelengths are largely defined by the atomic structure and not the mechanical geometry. Variations resulting from temperature changes, ageing and pump power are therefore less significant in fiber amplifiers than in SLAs.

A general representation of a fiber amplifier is shown in Figure 10.8. The gain medium normally comprises a length of single-mode fiber connected to a dichroic coupler (i.e. a wavelength division multiplexing coupler; see Section 5.6.3) which provides low insertion loss at both signal and pump wavelengths. Excitation occurs

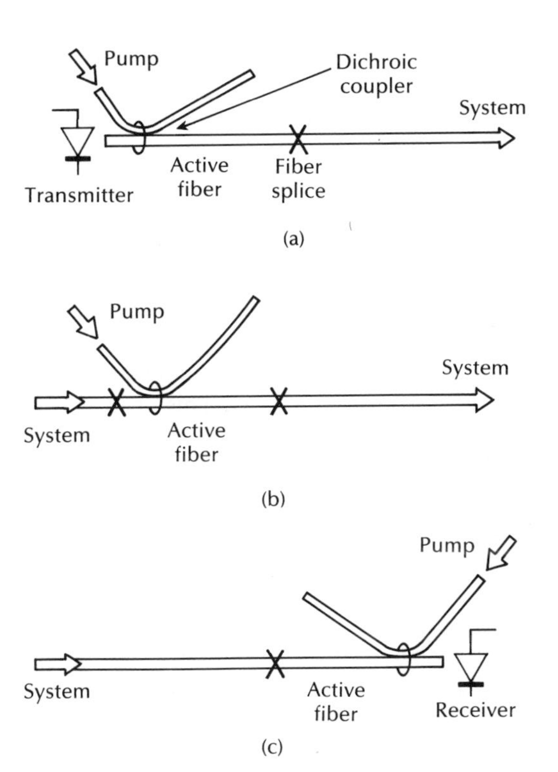

Figure 10.9 Some potential system applications for the fiber amplifier: (a) a power amplifier at the transmitter; (b) an optical repeater; (c) a preamplifier at the receiver.

through optical pumping from a high power solid state or semiconductor laser which is combined with the optical input signal within the coupler. The amplified optical signal is therefore emitted from the other end of the gain medium.

The major options for implementing fiber amplifiers were mentioned in Section 10.2; namely, rare earth doped fiber amplifiers, Raman fiber amplifiers and Brillouin fiber amplifiers.* In particular, the former two device types, in common with travelling wave SLAs, are expected to be used across a broad range of system applications, some of which are illustrated in Figure 10.9. These include use as: a power amplifier at the transmitter, an in-line optical repeater amplifier and an optical preamplifier at the receiver.

10.4.1 Rare earth doped fiber amplifiers

Both neodymium and erbium doped fiber lasers were discused in Section 6.9.2. To date work on rare earth doped fiber amplifiers has concentrated on the erbium dopant, particularly in silica based single-mode fibers. High gains of between 30 and 40 dB with low noise have been demonstrated [Refs. 3, 20, 21] with optical pump powers in the range 50 to 100 mW. Such devices can be made to lase over the longer wavelength region (1.5 to 1.6 μm) of interest within optical fiber communications. Practical pump bands exist at wavelengths of 532 nm, 670 nm, 807 nm, 980 nm and 1480 nm [Refs. 2, 22]. However, the latter three wavelengths comprise the most important pump bands. The amplification is dependent on the material gain of a relatively short section (1 to 100 m) of the fiber. Aluminium codoping can be used to broaden the spectral bandwidth to around 40 nm (see Figure 10.1). It should be noted, however, that the spectral dependence on gain is not always as constant as that illustrated in Figure 10.1 (for example, see Figure 6.44) and hence the spectral bandwidth for erbium doped silica fibers may be restricted to around 300 GHz (2.4 nm) [Ref. 20].

A factor which limits the gain available from an erbium doped fiber amplifier is a phenomenon known as excited state absorption (ESA). This process is illustrated in the energy level diagrams for an erbium doped fiber system shown in Figure 10.10. Erbium provides a three level lasing scheme which is illustrated in Figure 10.10(a). However, in the erbium fiber amplifier photons at the pump wavelength tend to promote the electrons in the upper lasing level into a still higher state of excitation, as shown in Figure 10.10(b). These electrons then decay nonradiatively to intermediate levels, such as the pump bands, and then eventually back to the upper lasing level. Hence ESA reduces the pumping efficiency of the device and as a result it is necessary to pump at a higher power to obtain a specific gain.

The reduction of ESA in erbium doped fiber amplifiers is therefore being pursued [Refs. 19, 22]. This may be achieved by changing the location of the energy levels through codoping of the erbium–silica fiber amplifier with other compounds such as phosphorus pentoxide. Another technique is to pump the fiber amplifier at a

* In these devices the gain medium is often a standard single-mode fiber.

wavelength which does not cause the population of an excited state. Unfortunately, significant ESA is present at the favoured 807 nm pump band. Nevertheless, improved efficiency is obtained with the 980 and 1480 nm pump wavelengths. In particular the 980 nm wavelength displays high efficiency (twice the dB W^{-1} gain figure of the 1480 nm wavelength) but pump sources are not readily available, whereas operation at 1480 nm can be facilitated by both semiconductor and solid state laser sources.

An alternative solution to avoid ESA, however, is to change to another glass technology in place of silica. In this context the success that has been achieved in lasing with a fluorozirconate host glass (see Section 6.9.2) may provide a way forward. Signal amplification has already been obtained in an erbium doped

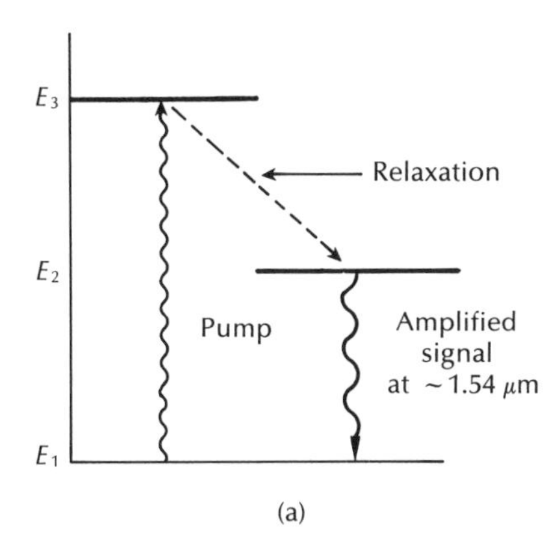

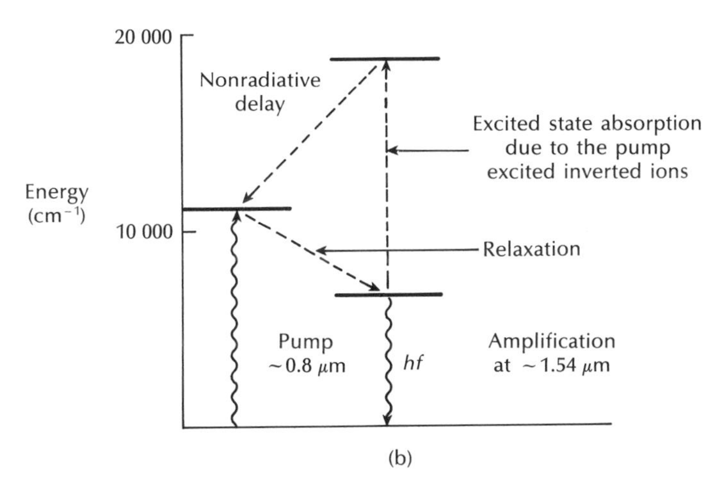

Figure 10.10 Energy level diagrams for erbium doped silica fiber laser: (a) the three level lasing scheme provided by Er^{3+} doping; (b) an illustration of excited state absorption, which is a major limitation.

multimode fluorozirconate fiber using a 488 nm pump wavelength to provide gain at a wavelength of 1.525 μm [Ref. 27].

10.4.2 Raman and Brillouin fiber amplifiers

Nonlinear effects within optical fiber may also be employed to provide optical amplification. Such amplification can be achieved by using stimulated Raman scattering, stimulated Brillouin scattering or stimulated four photon mixing, giving parametric gain (see Sections 3.5 and 3.14) by injecting a high power laser beam into undoped (or doped) optical fiber. Among these Raman amplification exhibits advantages of self phase matching between the pump and signal together with a broad gain–bandwidth or high speed response in comparison with the other nonlinear processes. In particular the broad gain–bandwidth associated with Raman amplification is attractive for application to possible future wavelength division multiplexed (WDM) systems [Ref. 18].

The pump signal optical wavelengths in Raman fiber amplifiers are typically 500 cm^{-1} higher in frequency than the signal to be amplified, and the pumping signal can propagate in either direction along the fiber. A schematic representation of both the forward and backward pumping capability of Raman fiber amplifiers is shown in Figure 10.11. Moreover, continuous-wave Raman gains exceeding 20 dB have been demonstrated experimentally in silica fiber [Ref. 24] which in principle exhibits a broad spectral bandwidth of up to 40 nm with suitable doping of the fiber [Ref. 2]. In addition, Raman gain in excess of 40 dB has been obtained using fluoride glass fiber in which the Raman shift is 590 cm^{-1} [Ref. 25]. More recently, Raman fiber amplifiers have been investigated for WDM system applications. For example, the simultaneous amplification using 60 mW pump power of three DFB

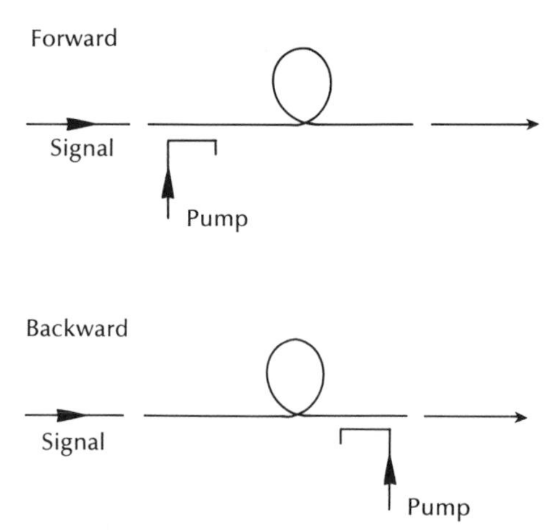

Figure 10.11 Illustrations of the forward and backward pumping capability associated with the fiber Raman amplifier.

laser diodes operating at wavelengths between 1570 nm and 1580 nm provided each channel with 5 dB gain [Ref. 26]. Furthermore, the gain–bandwidth of this fiber Raman amplifier was estimated to be in the range 20 to 30 nm.

The Raman gain G_R is dependent on a number of factors including the fiber length, the fiber attenuation and the fiber core diameter.* It may be expressed as a function of the optical pump power P_p as [Ref. 23]:

$$G_R = \exp\left(\frac{g_R P_p L_{eff}}{A_{eff} k}\right) \tag{10.14}$$

where g_R is the power Raman gain coefficient, and A_{eff} and L_{eff} are the effective fiber core area and length, respectively, and k is a numerical factor that accounts for polarization scrambling between the optical pump and signal [Ref. 28]. It should be noted that for complete polarization scrambling, as in conventional single-mode fiber, $k = 2$. The effective fiber core area and length are given by:

$$A_{eff} = \pi r_{eff}^2 \tag{10.15}$$

$$L_{eff} = \frac{1 - \exp(-\alpha_p L)}{\alpha_p} \tag{10.16}$$

where r_{eff} is the effective core radius, α_p is the fiber transmission loss at the pump wavelength and L is the actual fiber length.

The theoretical Raman gain characteristics as a function of fiber length for standard 10 μm core single-mode fibers with a pump input power of 1.6 W are shown in Figure 10.12 [Ref. 18]. It may be observed that the Raman gain becomes larger as the fiber lengths increase up to around 50 km where asymptotically it reaches a constant value. Moreover, it is clear that higher Raman gains can be obtained with lower loss fibers. Although not apparent from Figure 10.12, it is also the case that the Raman gain is increased as the fiber core diameter is decreased (see Eq. (10.14)). Nevertheless, in general the optical pump power required for Raman amplification tends to be high.

By contrast, stimulated Brillouin scattering is a very efficient nonlinear amplification mechanism that can provide high gains at modest optical pump powers of around 1 mW [Ref. 19]. However, it results from the scattering process in which the pump wavelength is often only around 20 GHz distance from the frequency of the optical signal to be amplified. Moreover, it is a narrow band process and the gain–bandwidth may only be in the range 15 to 20 MHz in silica fiber at a wavelength of 1.5 μm [Ref. 29]. The limitation on the spectral bandwidth in a pure silica fiber is around 50 MHz [Ref. 2] which fundamentally restricts the use of Brillouin amplifiers to relatively low speed communications. Although it is possible to extend the spectral bandwidth to 100 to 200 MHz with germanium doping of the fiber core, it does not significantly alleviate this problem.

* In standard single-mode fibers there are relatively low concentrations of germanium in the core which increases the peak Raman gain in comparison with pure silica core fiber.

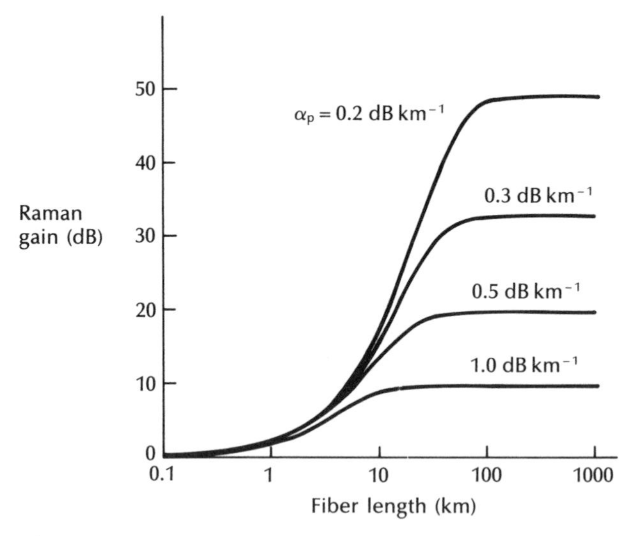

Figure 10.12 Raman gain dependence on fiber length and pump loss (α_p) for a pump input power of 1.6 W and a fiber core diameter of 10 μm. Reproduced with permission from Y. Aoki, 'Properties of fiber Raman amplifiers and their applicability to digital optical communication systems', *J. Lightwave Technol.*, **6**, p. 1225, 1988. Copyright © 1988 IEEE.

Nevertheless, when the fiber is pumped with a CW laser at a power in the range 5 to 10 mW, gains in excess of 15 dB can be obtained [Ref. 2]. A very precise frequency difference of around 11 GHz, however, must be maintained between the optical pump and the signal to ensure that the Brillouin scattering phenomenon continues unabated. This fiber amplifier type is therefore perceived to have a rather restricted range of application. However, the narrowband process could be useful in the provision of tunable filters within WDM systems. It can provide channel selection by allowing amplification of a particular channel without boosting other nearby channels. For example, Brillouin amplification has been investigated for channel selection in densely packed, single-mode fiber systems [Ref. 30]. In this case data transmitted at a rate of 45 Mbit s^{-1} were detected without errors with an interfering channel spaced only 140 MHz away at an operating wavelength of 1.5 μm.

A possible limitation of Brillouin amplification, however, for bidirectional WDM applications results from crosstalk due to Brillouin gain if the frequency difference between the counterpropagating waves coincides with the Brillouin shift of around 20 GHz [Ref. 31]. The power level at which significant crosstalk can occur is only of the order of 100 μW [Ref. 32]. Fortunately, since the Brillouin gain–bandwidth is particularly narrow such crosstalk can generally be avoided by a correct choice of signal wavelengths, without restricting the channel packing density.

10.5 Integrated optics

The multitude of potential application areas for optical fiber communications coupled with the tremendous advances in the field have over recent years stimulated a resurgence of interest in the area of integrated optics (IO). The concept of IO involves the realization of optical and electro-optical elements which may be integrated in large numbers on to a single substrate. Hence, IO seeks to provide an alternative to the conversion of an optical signal back into the electrical regime prior to signal processing by allowing such processing to be performed on the optical signal. Thin transparent dielectric layers on planar substrates which act as optical waveguides are used in IO to produce miniature optical components and circuits.

The birth of IO may be traced back to basic ideas outlined by Anderson in 1966 [Ref. 33]. He suggested that a microfabrication technology could be developed for single-mode optical devices with semiconductor and dielectric materials in a similar manner to that which had taken place with electronic circuits. It was in 1969, however, after Miller [Ref. 34] had introduced the term 'integrated optics' whilst discussing the long term outlook in the area, that research began to gain momentum.

Developments in IO have now reached the stage where simple signal processing and logic junctions may be physically realized. Furthermore, such devices may form the building blocks for future digital optical computers. Nevertheless, at present, these advances are closely linked with developments in lightwave communication employing optical fibers.

A major factor in the development of integrated optics is that it is essentially based on single-mode optical waveguides and therefore tends to be incompatible with multimode fiber systems. Hence IO did not make a significant contribution to first and second generation optical fiber systems (see Section 14.1). The advent, however, of single-mode transmission technology has further stimulated work in IO in order to provide devices* and circuits for these more advanced third generation systems. It is apparent that the continued expansion of single-mode optical fiber communications will create a growing market for such IO components. Furthermore, it is predicted that the next generation of optical fiber communication systems employing coherent transmission will lean heavily on IO techniques for their implementation (see Chapter 12).

The proposals for IO devices and circuits which in many cases involve reinventions of electronic devices and circuits exhibits major advantages other than solely a compatibility with optical fiber communications. Electronic circuits have a practical limitation on speed of operation at a frequency of around 10^{10} Hz resulting from their use of metallic conductors to transport electronic charges and build-up signals. The large transmission bandwidths (over 1 GHz) currently under

* This is especially the case in relation to the fabrication of single-mode injection lasers (see Section 6.6).

investigation for optical fiber communications are already causing difficulties for electronic signal processing within the terminal equipment. The use of light with its property as an electromagnetic wave of extremely high frequency (10^{14} to 10^{15} Hz) offers the possibility of high speed operation around 10^4 times faster than that conceivable employing electronic circuits. Interaction of light with materials such as semiconductors or transparent dielectrics occurs at speeds in the range 10^{12} (pico) to approaching 10^{-15} (femto) seconds, thus providing a basis for subpicosecond optical switching.

The other major attribute provided by optical signals interacting within a responsive medium is the ability to utilize lightwaves of different frequencies (or wavelengths) within the same guided wave channel or device. Such frequency division multiplexing allows an information transfer capacity far superior to anything offered by electronics. Moreover, in signal processing terms it facilitates parallel access to information points within an optical system. This possibility for powerful parallel signal processing coupled with ultrahigh speed operation offers tremendous potential for applications within both communications and computing.

The devices of interest in IO are often the counterparts of microwave or bulk optical devices. These include junctions and directional couplers, switches and modulators, filters and wavelength multiplexers, lasers and amplifiers, detectors and bistable elements. It is envisaged that developments in this technology will provide the basis for the fourth generation systems mentioned in Section 14.1 where full monolithic integration may be achieved.

10.5.1 Planar waveguides

The use of circular dielectric waveguide structures for confining light is universally utilized within optical fiber communications. IO involves an extension of this guided wave optical technology through the use of planar optical waveguides to confine and guide the light in guided wave devices and circuits. The mechanism of optical confinement in symmetrical planar waveguides was discussed in Section 2.3 prior to investigation of circular structures. In fact the simplest dielectric waveguide structure is the planar slab guide shown in Figure 10.13. It comprises a planar film of refractive index n_1 sandwiched between a substrate of refractive index n_2 and a cover layer of refractive index n_3 where $n_1 > n_2 \geqslant n_3$. Often the cover layer consists of air where $n_3 = n_0 = 1$, and it exhibits a substantially lower refractive index than the other two layers. In this case the film has layers of different refractive index above and below the guiding layer and hence performs as an asymmetric waveguide.

In the discussions of optical waveguides given in Chapter 2 we were solely concerned with symmetrical structures. When the dimensions of the guide are reduced so are the number of propagating modes. Eventually the waveguide dimensions are such that only a single-mode propagates, and if the dimensions are reduced further this single-mode still continues to propagate. Hence there is no cutoff for the fundamental mode in a symmetric guide. This is not the case for an

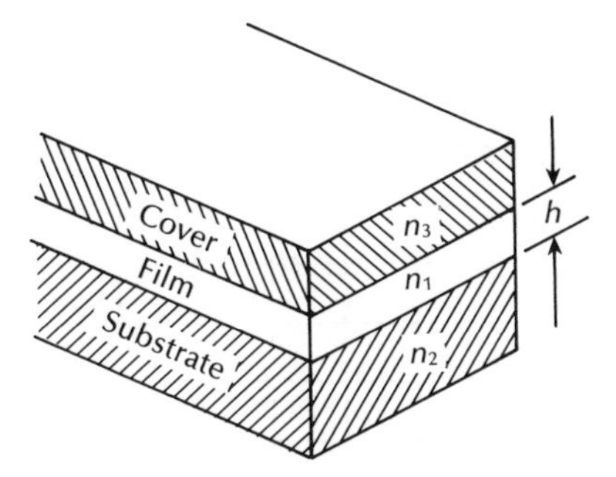

Figure 10.13 A planar slab waveguide. The film with high refractive index n_1 acts as the guiding layer and the cover layer is usually air where $n_3 = n_0 = 1$.

asymmetric guide where the dimensions may be reduced until the structure cannot support any modes and even the fundamental is cutoff. If the thickness or height of the guide layer of a planar asymmetric guide is h (see Figure 10.13), then the guide can support a mode of order m with a wavelength λ when [Ref. 35]:

$$h \geqslant \frac{(m + \frac{1}{2})\lambda}{2(n_1^2 - n_2^2)^{\frac{1}{2}}} \tag{10.17}$$

Equation (10.17) which assumes, $n_2 > n_3$ defines the limits of the single-mode region for h between values when $m = 0$ and $m = 1$. Hence for a typical thin film glass guide with $n_1 = 1.6$ and $n_2 = 1.5$, single-mode operation is maintained only when the guide has a thickness in the range $0.45\lambda \leqslant h \leqslant 1.35\lambda$.

An additional consideration of equal importance is the degree of confinement of the light to the guiding layer. The light is not exclusively confined to the guiding region and evanescent fields penetrate into the substrate and cover. An effective guide layer thickness h_{eff} may be expressed as:

$$h_{\text{eff}} = h + x_2 + x_3 \tag{10.18}$$

where x_2 and x_3 are the evanescent field penetration depths for the substrate and cover regions respectively. Furthermore, we can define a normalized effective thickness H for an asymmetric slab guide as:

$$H = kh_{\text{eff}}(n_1^2 - n_2^2)^{\frac{1}{2}} \tag{10.19}$$

where k is the free space propagation constant equal to $2\pi/\lambda$. The normalized frequency (sometimes called the normalized film thickness) for the planar slab guide following Eq. (2.68) is given by:

$$V = kh(n_1^2 - n_2^2)^{\frac{1}{2}} \tag{10.20}$$

An indication of the degree of confinement for the asymmetric slab waveguide may be observed by plotting the normalized effective thickness against the normalized frequency for the TE modes. A series of such plots is shown in Figure

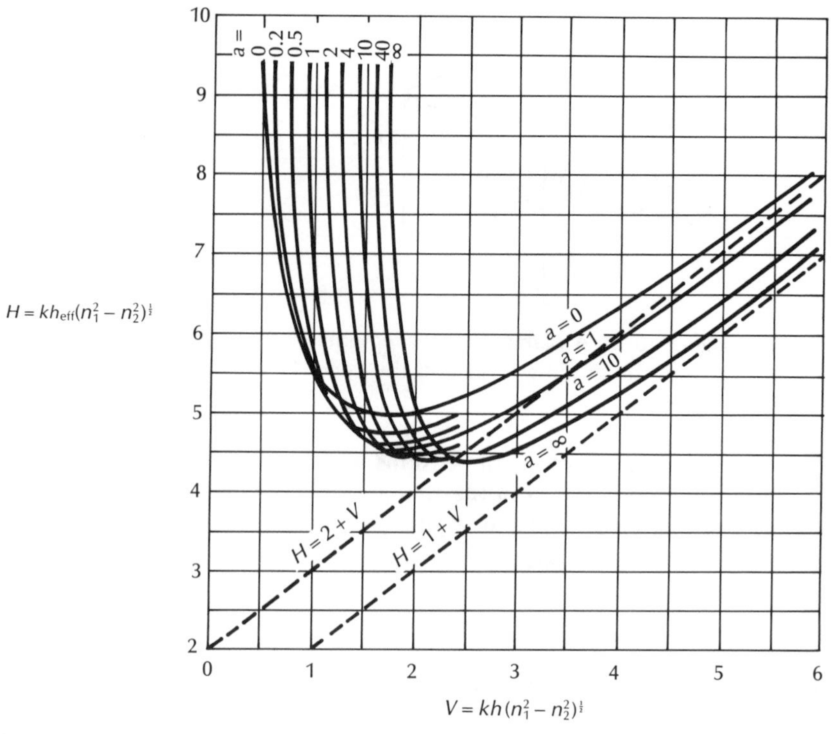

Figure 10.14 The normalized effective thickness H as a function of the normalized frequency V for a slab waveguide with various degrees of asymmetry. Reproduced with permission from H. Kogelnik and V. Ramaswamy, *Appl. Opt.*, **13**, p. 1857, 1974.

10.14 [Ref. 36] for various values of the parameter a which indicates the asymmetry of the guide, and is defined as:

$$a = \frac{n_2^2 - n_3^2}{n_1^2 - n_2^2} \tag{10.21}$$

It may be observed in Figure 10.14 that the confinement improves with decreasing film thickness only up to a point where $V \simeq 2.5$. For example, the minimum effective thickness for a highly asymmetric guide ($a = \infty$) occurs when $H_{\min} = 4.4$ at $V = 2.55$. Using Eq. (10.19) this gives a minimum effective thickness of:

$$(h_{\text{eff}})_{\min} = \frac{4.4}{k}(n_1^2 - n_2^2)^{-\frac{1}{2}}$$

$$= 0.7\,\lambda(n_1^2 - n_2^2)^{-\frac{1}{2}} \tag{10.22}$$

Therefore considering a typical glass waveguide ($n_1 = 1.6$ and $n_2 = 1.5$), we obtain a minimum effective thickness of:

$$(h_{\text{eff}})_{\min} = 1.26\ \lambda \tag{10.23}$$

Assuming a minimum operating wavelength to be 0.8 μm limits the effective thickness of the guide, and hence the confinement to around 1 μm. Therefore it appears there is a limit to possible fabrication with IO which is not present in other technologies* [Ref. 38]. At present there is still ample scope but confinement must be considered along with packing density and the avoidance of crosstalk.

The planar waveguides for IO may be fabricated from glasses and other isotropic materials such as silicon dioxide and polymers. Although these materials are used to produce the simplest integrated optical components, their properties cannot be controlled by external energy sources and hence they are of limited interest. In order to provide external control of the entrapped light to cause deflection, focusing, switching and modulation, active devices employing alternative materials must be utilized. A requirement for these materials is that they have the correct crystal symmetry to allow the local refractive index to be varied by the application of either electrical, magnetic or acoustic energy.†

To date, interest has centred on the exploitation of the electro-optic effect due to the ease of controlling electric fields through the use of electrodes together with the generally superior performance of electro-optic devices. Acousto-optic devices have, however, found a lesser role, primarily in the area of beam deflection. Magneto-optic devices [Ref. 39] utilizing the Faraday effect are not widely used, as in general, electric fields are easier to generate than magnetic fields.

A variety of electro-optic and acousto-optic materials have been employed in the fabrication of individual devices. Two basic groups can be distinguished by their refractive indices. These are materials with a refractive index near 2 ($LiNbO_3$, $LiTaO_3$, NbO_5, ZnS and ZnO) and materials with a refractive index greater than 3 (GaAs, InP and compounds of Ga and In with elements of Al, As and Sb).

Planar waveguide structures are produced using several different techniques which have in large part been derived from the microelectronics industry. For example, passive devices may be fabricated by radiofrequency sputtering to deposit thin films of glass onto glass substrates. Alternatively, active devices are often produced by titanium (Ti) diffusion into lithium niobate ($LiNbO_3$) or by ion implantation into gallium arsenide [Ref. 41].

The planar slab waveguide shown in Figure 10.13 confines light in only one direction, allowing it to spread across the guiding layer. In many instances it is useful to confine the light in two dimensions to a particular path on the surface of the substrate. This is achieved by defining the high index guiding region as a thin strip (strip guide) where total internal reflection will prevent the spread of the light beam across the substrate. In addition the strips can be curved or branched as

* The 1 μm barrier to confinement applies with all suitable waveguide materials. However, metal clad waveguides are not so limited but are plagued by high losses [Ref. 37].

† Using the electro-optic, magneto-optic or acousto-optic effects [Ref. 40].

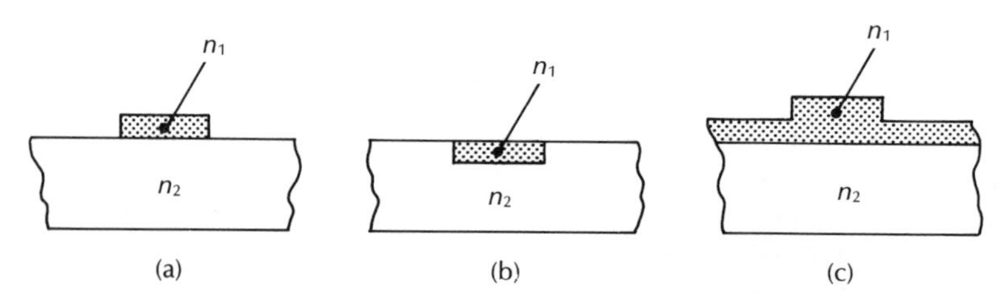

Figure 10.15 Cross section of some strip waveguide structures: (a) ridge guide; (b) diffused channel (embedded strip) guide; (c) rib guide.

required. Examples of such strip waveguide structures are shown in Figure 10.15. They may be formed as either a ridge on the surface of the substrate or by diffusion to provide a region of higher refractive index below the substrate, or as a rib of increased thickness within a thin planar slab. Techniques employed to obtain the strip pattern include electron and laser beam lithography as well as photolithography. The rectangular waveguide configurations illustrated in Figure 10.15 prove very suitable for use with electro-optic deflectors and modulators giving a reduction in the voltage required to achieve a particular field strength. In addition they allow a number of optical paths to be provided on a given substrate.

A trade-off also exists between the minimum radius of curvature which is required for high density integration and the ease of fabrication. It is clear from Eq. (10.22) that the waveguide dimensions are dependent upon the refractive index change. When the change is large, the dimensions of the waveguide may be reduced, even though the scattering losses become larger. As the maximum confinement of the single-mode guide occurs when it is operated near to the cutoff of the second order mode, then when the refractive index change is large, the radius of curvature of the waveguide can also be made very small. It is therefore necessary to find a compromise for the waveguide material used.

Titanium in-diffusion of $LiNbO_3$ gives rise to refractive index increases in the order of 0.01 to 0.02 which dictates a bend radius of the order of a few centimetres for negligible losses. It is, however, possible to use a proton exchange technique to increase the refractive index change up to 0.15 [Ref. 42]. By contrast, semiconductor III–V alloy waveguides based on compositional modification of the crystal give an index change of around 0.1 or more [Ref. 43]. Therefore, bend radii of the order of 1 mm or less may be obtained using these compounds. Moreover, although the effects of interest in IO are usually exhibited over short distances of around one wavelength, efficient devices require relatively long interaction lengths, the effects being cumulative. Hence, typical device lengths range from 0.5 to 10 mm.

Optical connections to and from waveguide devices are normally made by optical fibers. The overall insertion loss for such devices therefore comprises a waveguide-fiber coupling loss as well as the waveguide optical propagation loss. Careful fabrication of $Ti:LiNbO_3$ waveguides with mode spot sizes well matched to that of typical single-mode fibers has yielded coupling losses in the range 0.5 to 1.0 dB per connection [Ref. 44]. In general, however, semiconductor waveguide devices exhibit larger fiber coupling losses because they operate with smaller spot sizes.

Propagation losses within both slab and strip waveguides are generally much greater than those obtained in single-mode optical fibers. However, more recently, propagation losses for $Ti:LiNbO_3$ waveguides have gone below 0.2 $dB\,cm^{-1}$, with excess bend losses being maintained below 0.1 dB per bend [Ref. 45]. By contrast propagation losses in semiconductor waveguides around 1 $dB\,cm^{-1}$ are obtained when operating at wavelengths corresponding to the bandgap energy. Much lower losses of approximately 0.2 $dB\,cm^{-1}$, however, have to be achieved at operating wavelengths far below the bandgap energy [Ref. 46].

10.6 Some integrated optical devices

In this section some examples of various types of integrated optical devices together with their salient features are considered. However, the numerous developments in this field exclude any attempt to provide other than general examples in the major areas of investigation which are pertinent to optical fiber communications. The requirement for multichannel communication within the various systems considered in Chapters 11 and 12 demands the combination of information from separate channels, transmission of the combined signals over a single optical fiber link, and separation of the individual channels at the receiver prior to routeing to their individual destinations. Hence the application of IO in this area is to provide optical methods for multiplexing, modulation and routeing. These various functions may be performed with a combination of optical beam splitters, switches, modulators, filters, sources and detectors.

10.6.1 Beam splitters, directional couplers and switches

Beam splitters are a basic element of many optical fiber communication systems often providing a Y-junction by which signals from separate sources can be combined, or the received power divided between two or more channels. A passive Y-junction beam splitter fabricated from $LiNbO_3$ is shown in Figure 10.16. Unfortunately, the power transmission through such a splitter decreases sharply with increasing half angle γ, the power being radiated into the substrate. Hence the total power transmission depends critically upon γ which, for the example chosen, must not exceed 0.5° if an acceptable insertion loss is to be achieved [Ref. 47]. In order to provide effective separation of the output arms so that access to each is possible, the junction must be many times the width of the guide. For example, around 3000 wavelengths are required to give a separation of about 30 μm between the output arms. Therefore, for practical reasons, the device is relatively long.

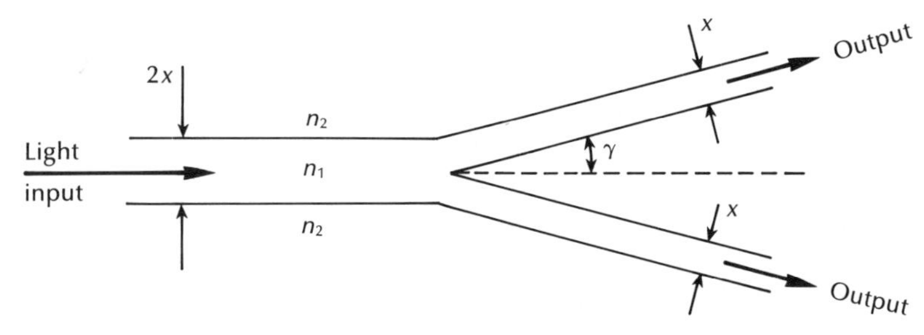

Figure 10.16 A passive Y-junction beam splitter.

The passive Y-junction beam splitter finds application where equal power division of the incident beam is required. However, the Y-junction is of wider interest when it is fabricated from an electro-optic material, in which case it may be used as a switch. Such materials exhibit a change in refractive index δn which is directly proportional to an applied electric field* E following,

$$\delta n = \pm \tfrac{1}{2} n_1^3 r E \tag{10.24}$$

where n_1 is the original refractive index, and r is the electro-optic coefficient. Hence an active Y-junction may be fabricated from a single crystal electro-optic material as illustrated in Figure 10.17. Lithium niobate is often utilized as it combines relatively low loss with large values of electro-optic coefficients† (as high as 30.8×10^{-12} m V^{-1}). Metal electrodes are attached so that when biasing is applied, one side of the waveguide structure exhibits an increased refractive index whilst the value of refractive index on the other side is reduced. The light beam is therefore

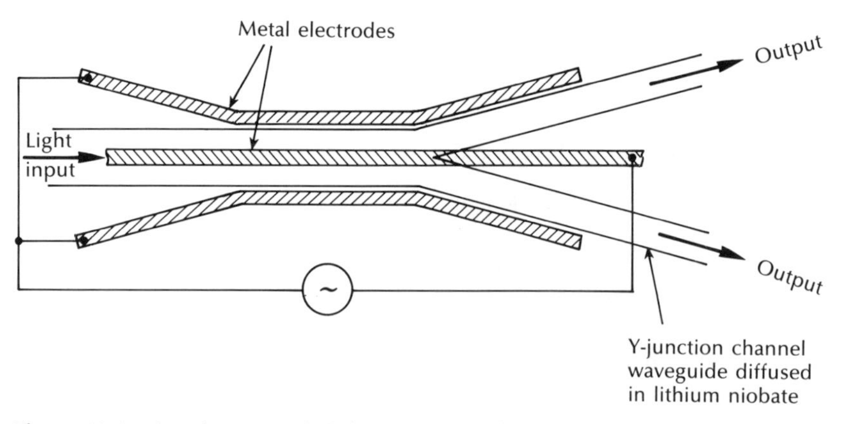

Figure 10.17 An electro-optic Y-junction switch.

* The linear variation of refractive index with the electric field is known as the Pockels effect [Ref. 40].
† The change in refractive index is related by the applied field via the linear and quadratic electro-optic coefficients [Ref. 39].

deflected towards the region of higher refractive index causing it to follow the corresponding output arm. Furthermore, the field is maintained in the electrodes which extend beyond the junction ensuring continuation of the process. With switching voltages around 30 V, these devices prove to be quite efficient allowing for larger junction angles to be tolerated than those of the passive Y-junction beam splitter. However, a physical length of several hundred wavelengths is still required for the switch. These devices therefore serve the function of optical signal routeing. In addition, high speed switches can be used to provide time division multiplexing of several lower bit rate channels onto a single-mode fiber link.

Switches may also be fabricated by placing two parallel strip waveguides in close proximity to each other as illustrated in Figure 10.18. The evanescent fields generated outside the guiding region allow transverse coupling between the guides. When the two waveguide modes have equal propagation constants β with amplitudes A and B (Figure 10.18), then the coupled mode equations may be written as [Ref. 48]:

$$\frac{\mathrm{d}A}{\mathrm{d}z} = \mathrm{j}\beta A + \mathrm{jCB}$$

$$\frac{\mathrm{d}B}{\mathrm{d}z} = \mathrm{j}\beta B + \mathrm{j}CA \tag{10.25}$$

where C is the coupling coefficient per unit length. In this case, assuming no losses, all the energy from waveguide X will be transferred to waveguide Y over a coupling length l_0. Furthermore it can be shown [Ref. 49] that for this complete energy transfer l_0 is given by $\pi/2C$. If the waveguide modes have different propagation constants, however, only part of the energy from guide X will be coupled into guide Y, and this energy will be subsequently recoupled back into X.

It is also noted that when the propagation constants differ the coupling length l is reduced from the matched value l_0 and although less energy is transferred, the

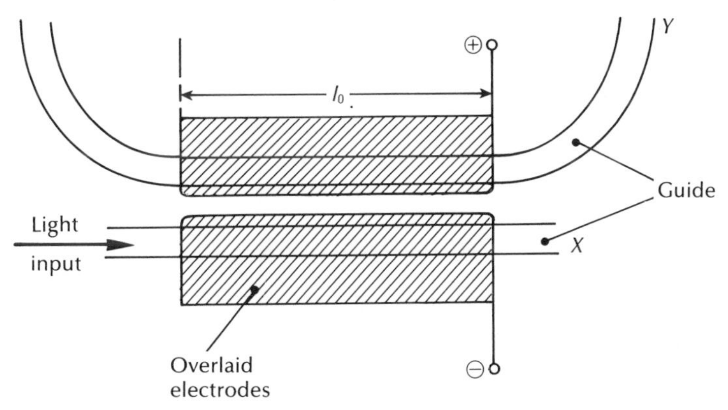

Figure 10.18 Electro-optically switched directional coupler. The COBRA configuration using two electrodes [Ref. 50].

exchange occurs more rapidly. This property may be utilized to good effect in the formation of an optical switch. The mismatch in propagation constants can be adjusted such that the coupling length l is reduced to $l_0/2$. In this case, energy coupled from one guide into the other over a distance $l_0/2$ will be recoupled into the original guide over a similar distance. Hence two distinct cases exist for a switch of length l_0, namely the matched case whereby all the energy is transferred from one guide to the other and the mismatched case when $l = l_0/2$ where over a distance l_0 the energy is recoupled into the original guide.

Optical switches of the above type use electrodes placed on the top of each matched waveguide (Figure 10.18) so that the refractive indices of the guides are differentially altered to produce the differing propagation constants for the mismatched case. A widely used switch utilizing this technique is called the COBRA (*Commutateur Optique Binaire Rapide*) [Ref. 50] and is normally formed from titanium diffused lithium niobate. Fabrication of the device, however, is critical in order to provide a coupling length which is exactly l_0 or an odd multiple of l_0. An electrode structure which avoids this problem by dividing the electrodes into halves with opposite polarities on each half is shown in Figure 10.19. With this device, which is called the stepped $\Delta\beta$ reversal coupler, it is always possible to obtain both the matched and mismatched cases described previously by applying suitable values of the reversed voltage. Hence the fabricated coupling length is no longer critical as the effective coupling length of the device may be adjusted electrically to achieve l_0.

The increasing deployment of optical fiber, particularly in the telecommunications network, has stimulated a great interest in optical or photonic switching in order to provide routeing in what is, at present, a circuit switched network [Refs. 51 to 56]. The technology discussed in Sections 10.6 to 10.8 provides the basic building blocks for such optical switching systems. Such switching systems can be classified in terms of their switching mechanism into space division switches, time division switches and wavelength or frequency* division switches [Ref. 51].

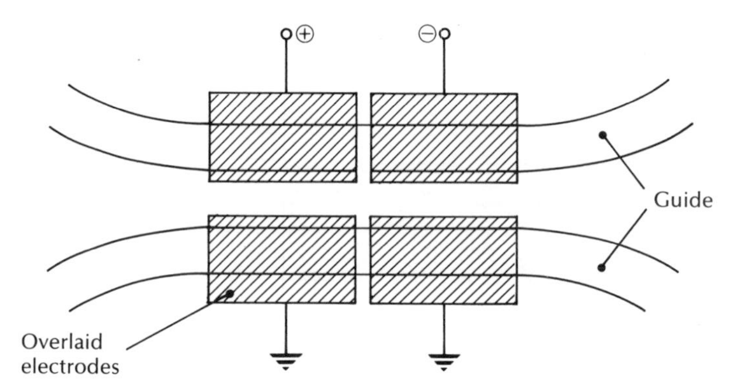

Figure 10.19 The stepped $\Delta\beta$ reversal coupler switch.

* In the optical domain these two terms are often used to indicate the same principle.

Although at a relatively early stage of development, optical switching matrices have been realized using IO technology. Optical space division switches incorporating electro-optically controlled directional couplers have been demonstrated. An example is illustrated in Figure 10.20 [Ref. 53]. The device is an 8×8 space switch comprising sixty-four directional couplers on a single lithium niobate substrate. It exhibited insertion losses in the range 6 to 8 dB and required a modulation voltage of 40 V. In addition a four channel time division switch using optical fiber delay lines combined with 4×4 lithium niobate optical switches has also been reported [Ref. 57]. Optical time division switching at 32 Mbit s^{-1} was obtained with this device.

Finally, an optical wavelength division switch for two channels using an acousto-optic deflector (see Section 10.6.2) with a photodiode and injection laser array has also been demonstrated [Ref. 58]. The WDM input signal is deflected by the acousto-optic deflector according to the frequency of an electrical signal applied to control transducer. Each electrical signal frequency corresponds to a deflection to

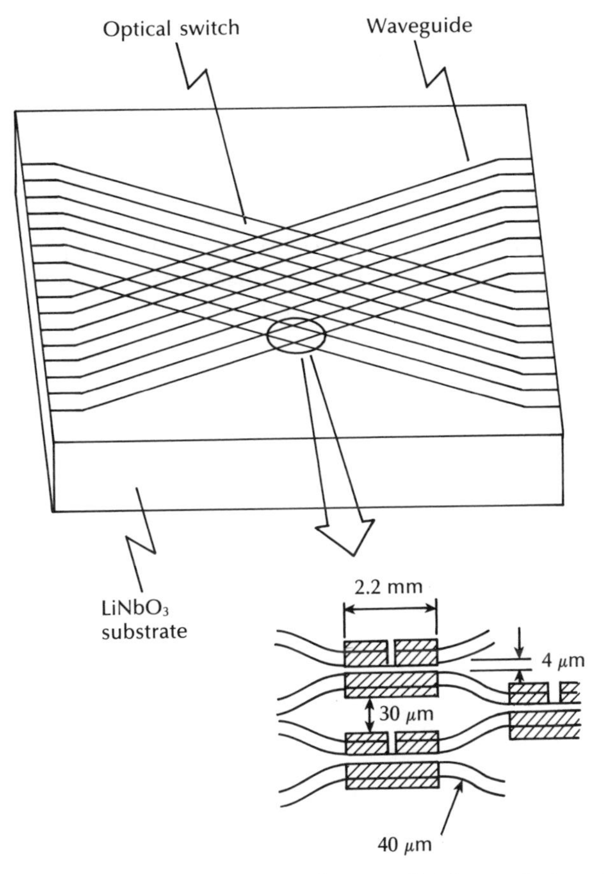

Figure 10.20 An 8×8 optical space division switch matrix.

a particular photodetector. Hence when the electrical signal frequency is altered, the optical signal with a wavelength corresponding to the desired frequency is deflected to the appropriate detector. Moreover, as each optical detector is connected to an individual injection laser emitting at a different wavelength to the one received, then the optical signal wavelength is converted to another wavelength. The device demonstrated the ability to wavelength switch two 400 Mbit s^{-1} optical signals.

10.6.2 Modulators

The limitations imposed by direct current modulation of semiconductor injection lasers currently restricts the maximum achievable modulation frequencies to a few gigahertz. Furthermore, with most injection lasers high speed current modulation also creates undesirable wavelength modulation which imposes problems for systems employing wavelength division multiplexing. Thus to extend the bandwidth capability of single-mode fiber systems there is a requirement for high speed modulation which can be provided by integrated optical waveguide intensity modulators. Simple on/off modulators may be based on the techniques utilized for the active beam splitters and switches described in Section 10.6.1. In addition a large variety of predominantly electro-optic modulators have been reported [Ref. 59] which exhibit good characteristics. For example, an important waveguide modulator is based upon a Y-branch interferometer which employs optical phase shifting produced by the electro-optic effect.

The change in refractive index exhibited by an electro-optic material with the application of an electric field given by Eq. (10.24) also provides a phase change for light propagating in the material. This phase change $\delta\phi$ is accumulative over a distance L within the material and is given by [Ref. 60]:

$$\delta\phi = \frac{2\pi}{\lambda}\,\delta n L \tag{10.26}$$

When the electric field is applied transversely to the direction of optical propagation we may substitute for δn from Eq. (10.24) giving:

$$\delta\phi = \frac{\pi}{\lambda}\,n_1^3 r E L \tag{10.27}$$

Furthermore taking E equal to V/d, where V is the applied voltage and d is the distance between electrodes gives:

$$\delta\phi = \frac{\pi}{\lambda}\,n_1^3 r\,\frac{VL}{d} \tag{10.28}$$

It may be noted from Eq. (10.28) that in order to reduce the applied voltage V required to provide a particular phase change, the ratio L/d must be made as large as possible.

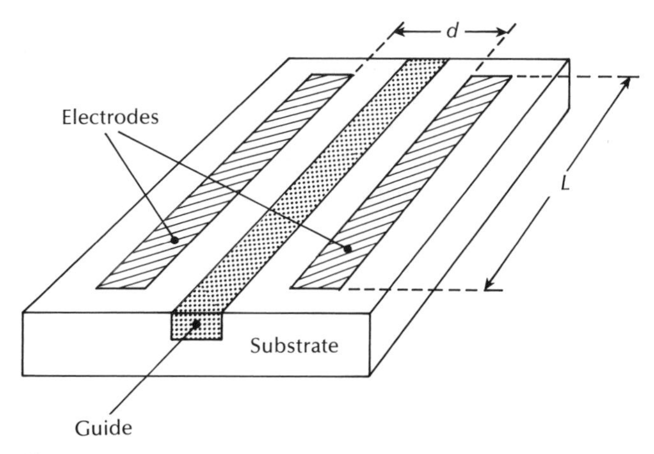

Figure 10.21 A simple strip waveguide phase modulator.

A simple phase modulator may therefore be realized on a strip waveguide in which the ratio L/d is large as shown in Figure 10.21. These devices when, for example, fabricated by diffusion of Nb into $LiTaO_3$ provide a phase change of π radians with an applied voltage in the range 5 to 10 V.

Example 10.3

A lithium niobate strip waveguide phase modulator designed for operation at a wavelength of 1.3 μm is 2 cm long with a distance between the electrodes of 25 μm. Determine the voltage required to provide a phase change of π radians given that the electro-optic coefficient for lithium niobate is $30.8 \times 10^{-12}\,\text{mV}^{-1}$ and its refractive index is 2.1 at 1.3 μm.

Solution: When the phase change is π radians, using Eq. (10.28) we can write:

$$\delta\phi = \pi = \frac{\pi}{\lambda} n_1^3 r \frac{V_\pi L}{d}$$

Hence the voltage required to provide a π radian phase change is:

$$V_\pi = \frac{\lambda}{n_1^3 r} \frac{d}{L}$$

$$= \frac{1.3 \times 10^{-6} \times 25 \times 10^{-6}}{(2.1)^3 \times 30.8 \times 10^{-12} \times 2 \times 10^{-2}}$$

$$= 5.7\ \text{V}$$

The result obtained in Example 10.3 has assumed the spatially uniform electric field of an ideal parallel plate capacitor. However, because the electro-optic refractive index change is small this is rarely the case and its effect on the optical phase velocity is dependent on the overlap integral of the electrical and optical fields. The consequence of these nonuniform fields can be incorporated into an overlap integral α, having a value between 0 and 1 which gives a measure of the overlap between the electrical and optical fields [Refs. 42, 45]. The electro-optic refractive index change of Eq. (10.24) therefore becomes:

$$\delta n = \frac{\pm \alpha n_1^3 r}{2} \frac{V}{d} \tag{10.29}$$

where the factor α represents the efficiency of the electro-optic interaction relative to an idealized parallel plate capacitor with the same distance between the electrodes.

As mentioned previously the electro-optic property can be employed in an interferometric intensity modulator. Such a Mach–Zehnder type interferometer is shown in Figure 10.22. The device comprises two Y-junctions which give an equal division of the input optical power. With no potential applied to the electrodes, the input optical power is split into the two arms at the first Y-junction and arrives at the second Y-junction in phase giving an intensity maximum at the waveguide output. This condition corresponds to the 'on' state. Alternatively when a potential is applied to the electrodes, which operate in a push-pull mode on the two arms of the interferometer, a differential phase change is created between the signals in the two arms. The subsequent recombination of the signals gives rise to constructive or destructive interference in the output waveguide. Hence the process has the effect of converting the phase modulation into intensity modulation. A phase shift of π between the two arms gives the 'off' state for the device.

High speed interferometric modulators have been demonstrated with titanium doped lithium niobate waveguides. A 1.1 GHz modulation bandwidth has been reported [Ref. 61] for a 6 mm interferometer employing a 3.8 V on/off voltage across a 0.9 μm gap. Similar devices incorporating electrodes on one arm only

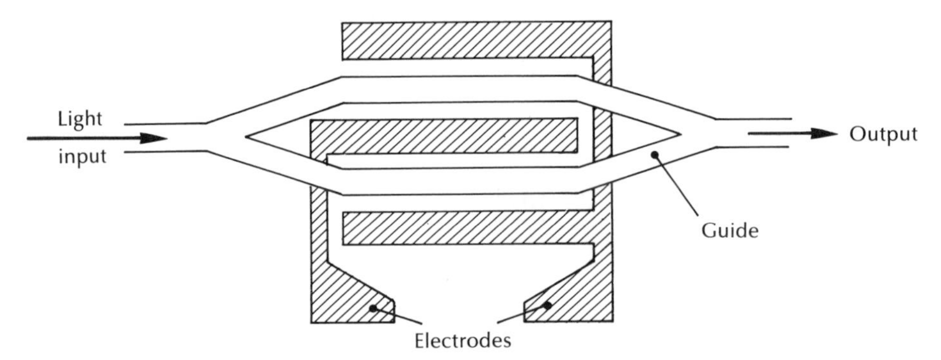

Figure 10.22 A Y-junction interferometric modulator based on the Mach–Zehnder interferometer.

may be utilized as switches and are generally referred to as balanced bridge interferometric switches [Ref. 59].

Useful modulators may also be obtained employing the acousto-optic effect. These devices which deflect a light beam are based on the diffraction of light produced by an acoustic wave travelling through a transparent medium. The acoustic wave produces a periodic variation in density (i.e. mechanical strain) along its path which, in turn, gives rise to corresponding changes in refractive index within the medium due to the photoelastic effect. Therefore, a moving optical phase-diffraction grating is produced in the medium. Any light beam passing through the medium and crossing the path of the acoustic wave is diffracted by this phase grating from the zero order into higher order modes.

Two regimes of operation are of interest: the Bragg regime and the Raman–Nath regime. The interaction, however, is of greatest magnitude in the Bragg regime where the zero order mode is partially deflected into only one higher order (i.e. first order) mode, rather than the multiplicity of higher order modes obtained in the Raman–Nath regime. Hence most acousto-optic modulators operate in the Bragg regime providing the highest modulation depth for a given acoustic power.

The Bragg regime is obtained by effecting a suitably long interaction length for the device so that it performs as a 'thick' diffraction grating. An IO acousto-optic Bragg deflection modulator is shown in Figure 10.23. It consists of a piezoelectric substrate (e.g. lithium niobate) on to the surface of which a thin film optical waveguide is formed by, for example, titanium indiffusion or lithium outdiffusion. An acoustic wave is launched parallel to the surface of the waveguide forming a surface acoustic wave (SAW) in which most of the wave energy is concentrated within a depth of one acoustic wavelength. The wave is generated from an interdigital electrode system comprising parallel electrodes deposited on the substrate.

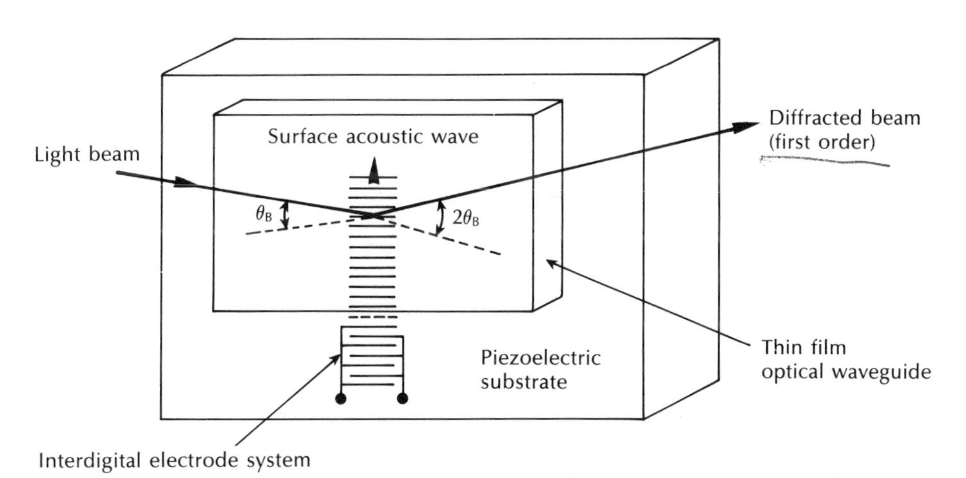

Figure 10.23 An acousto-optic waveguide modulator. The device gives deflection of a light beam due to Bragg diffraction by surface acoustic waves.

A light beam guided by the thin film waveguide interacts with the SAW giving beam deflection since both the light and the acoustic energy are confined to the same surface layer. The conditions for Bragg diffraction between the zero and first order mode are met when [Ref. 41]:

$$\sin \theta_B = \frac{\lambda_1}{2\Lambda} \tag{10.30}$$

where θ_B is the angle between the light beam and the acoustic beam wavefronts, λ_1 is the wavelength of light in the thin film waveguide and Λ is the acoustic wavelength. In this case the light is deflected by $2\theta_B$ from its original path as illustrated in Figure 10.23.

The fraction of the light beam deflected depends upon the generation efficiency and the width of the SAW, the latter also defining the interaction length for the device. Although diffraction efficiencies are usually low (no more than 20%), the diffracted on/off ratio can be very high. Hence these devices provide effective switches as well as amplitude or frequency modulators.

10.6.3 Periodic structures for filters and injection lasers

Periodic structures may be incorporated into planar waveguides to form integrated optical filters and resonators. Light is scattered in such a guide in a similar manner to light scattered by a diffraction grating. A common example of a periodic waveguide structure is the corrugated slab waveguide shown in Figure 10.24. When light propagating in the guide impinges on the corrugation, some of the energy will be diffracted out of the guide into either the cover or the substrate. The device, however, acts as a one-dimensional Bragg diffraction grating, and light which satisfies the Bragg condition is reflected back along the guide at 180° to the original direction of propagation (Figure 10.24).

The Bragg condition for the case of 180° reflection can be obtained from Eq. (10.30) if we let the corrugation period D (Figure 10.24) equal the acoustic

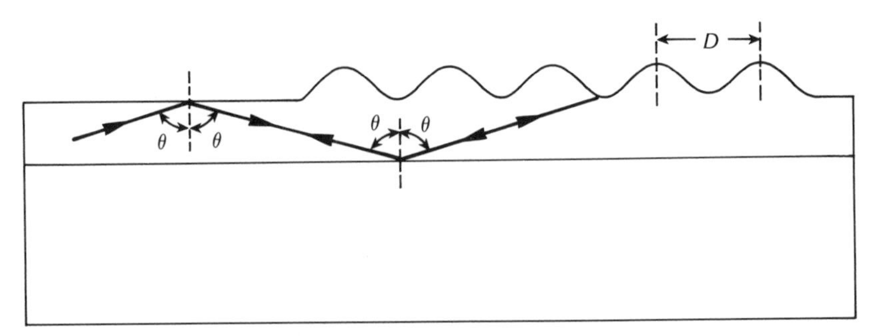

Figure 10.24 A slab waveguide with surface corrugation giving reflection back along the guide when the Bragg condition is met. Hence the structure performs as a one dimensional Bragg diffraction grating.

wavelength Λ and let λ_1 equal λ_B/n_e, where λ_B (the Bragg wavelength) is the optical wavelength in a vacuum and n_e is the effective refractive index of the guide. If we also assume that λ_B is equal to 90°, then Eq. (10.30) becomes

$$D = \frac{l\lambda_B}{2n_e} \tag{10.31}$$

where $l = 1, 2, 3, \ldots, m$ is the order of the grating which was unity in Eq. (10.30) because diffraction took place between the zero and first order mode. The vacuum wavelength of light that will be reflected through 180° by such a grating is therefore:

$$\lambda_B = \frac{2n_e D}{l} \tag{10.32}$$

When the reflected light is incident at an angle (Figure 10.24) then [Ref. 39]:

$$n_e = n_1 \sin 2\theta \tag{10.33}$$

where n_1 is the refractive index of the guide. Hence depending on the corrugation period of the structure all the incident power at a particular wavelength will be reflected. Devices of this type therefore behave as frequency selective rejection filters or mirrors. An example of such a reflection filter is shown in Figure 10.25. It comprises a InGaAsP/InP grating waveguide device in which the surface corrugation is typically written as a photoresist mask using two interfering ultraviolet beams before chemical or physical etching. The filter bandwidth can be quite small (6Å) with modest interaction lengths using this technique [Ref. 62]. Moreover in glass waveguides filter bandwidths as narrow as 0.1 Å have been obtained with a 1 cm long grating filter [Ref. 63]. The low substrate–waveguide refractive index difference using lithium niobate devices, however, combined with the inherent etching difficulties have limited the development of Ti : $LiNbO_3$ waveguide reflection filters.

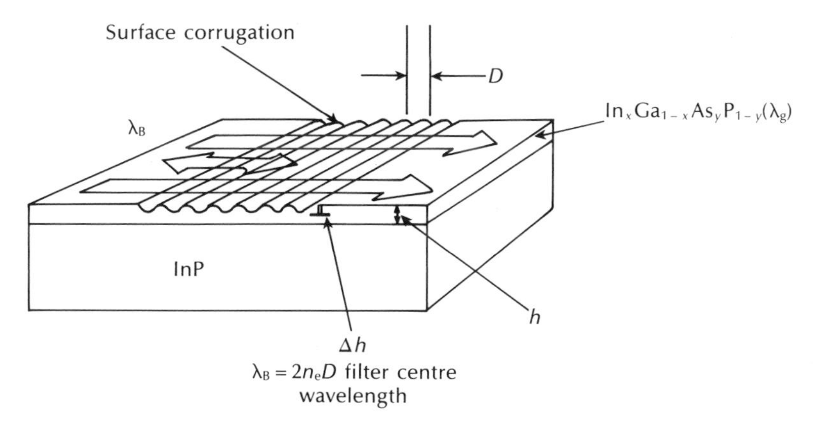

Figure 10.25 An InGaAsP/InP waveguide grating filter.

For a waveguide grating filter which exhibits a large change in effective refractive index with a fine grating period, the 3 dB fractional bandwidth is given approximately by [Ref. 45]:

$$\frac{\delta\lambda}{\lambda} \simeq \frac{D}{L} \tag{10.34}$$

where L is the grating length. Hence Eq. (10.34) allows estimates of the filter 3 dB bandwidth, $\delta\lambda$, to be obtained.

Example 10.4

A 1 cm long InGaAsP/InP first order grating filter is designed to operate at a centre wavelength of 1.52 μm. The reflected light is incident at an angle of 1° and the refractive index of InGaAsP is 3.1. Determine the corrugation period and estimate the filter 3 dB bandwidth. A large change in effective refractive index may be assumed.

Solution: The effective refractive index of the waveguide is given by Eq. (10.33) as:

$$\begin{aligned} n_e = n_1 \sin 2\theta &= 3.1 \sin 2^\circ \\ &= 0.11 \end{aligned}$$

The corrugation period for the first order grating may be obtained from Eq. (10.31) as:

$$D = \frac{\lambda_B}{2n_e} = \frac{1.52 \times 10^{-6}}{2 \times 0.11} = 6.9 \ \mu\text{m}$$

Finally, the filter 3 dB bandwidth can be estimated from Eq. (10.34) where

$$\begin{aligned} \delta\lambda \simeq \frac{D\lambda}{L} &= \frac{6.9 \times 10^{-6} \times 1.52 \times 10^{-6}}{10^{-2}} \\ &= 10.5 \ \text{Å} \ (\simeq 1 \ \text{nm}) \end{aligned}$$

It may be observed that a relatively narrow filter bandwidth is obtained in Example 10.4. Such devices could find use for wavelength demultiplexing of a larger number of channels. Alternatively, wide bandwidth filters may be realized by forming gratings which exhibit a gradual change in the corrugation period. Such grating devices are said to have a chirped structure [Ref. 64].

Finally, it should be noted that the corrugated gratings discussed above are also incorporated into advanced single-mode injection laser structures; namely, the distributed feedback and the distributed Bragg reflector lasers (see Section 6.6.2).

10.6.4 Polarization transformers and frequency translators

The electro-optic effect typically in lithium niobate waveguide devices can be used to facilitate TE–TM mode conversion. However, to allow the transformation of an arbitrary input polarization, not just TE or TM, it is necessary to control the relative phase between the TE and TM components. Such polarization transformers which operate as TE–TM mode converters can be employed as elements within intensity modulators (when combined with a polarizer), optical filters or polarization controllers. A basic example of the latter device is shown in Figure 10.26 [Ref. 65]. It comprises two phase modulators and a single TE–TM mode converter on X-cut* lithium niobate.

The first phase modulator is required to adjust the phase difference between the incoming TE and TM modes to be $\pi/2$ so that the polarization controller can operate with all incoming polarization states. When this condition is satisfied the central phase matched mode converter is operated as a linear polarization rotator. Although a linear output polarization of either TE or TM is sufficient in some applications, for full polarization control a second phase shifter is required to adjust the output phase to a desired value of elliptical output polarization.

A number of electro-optic waveguide devices can be used to provide frequency translation of an optical signal [Refs. 42, 66]. A common technique is to employ a phase modulator in a serrodyne configuration to alter the optical frequency by a linearly increasing voltage applied to the device electrodes [Ref. 67]. In practice a continuously increasing ramp signal voltage cannot readily be produced and hence a sawtooth voltage waveform is used. However, sawtooth waveforms with instantaneous fall-times can be generated and hence additional frequency components

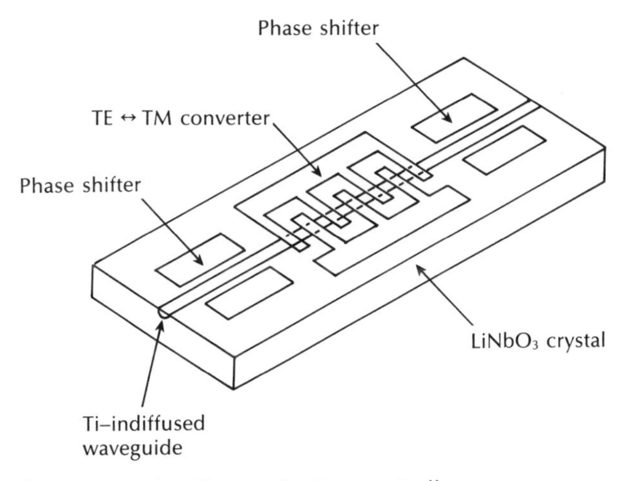

Figure 10.26 An IO polarization controller.

* Conventional Y-cut lithium niobate is not normally used as the electro-optic coefficient is smaller, necessitating higher operating voltages.

tend to be produced. This factor, combined with the need to vary the rate of change of the applied voltage to alter the extent of the frequency shift, limits the use of this device to applications where a small constant frequency shift is required. In applications where large frequency translations are necessary and where the device is used as a control element in a feedback loop (e.g. coherent optical receivers), alternative frequency translators are utilized [Ref. 42].

Devices based on TE–TM mode conversion are also capable of generating frequency translated optical signals [Ref. 68]. When the region where the mode conversion takes place is made to move relative to the direction of the optical wave, then the source of the converted signal appears to be moving to a stationary observer and the light is therefore Doppler shifted. To generate the effect of a moving coupling grating in practice, a mode converter is divided into several sections and each is driven with a correctly phase shifted sinusoidal signal which has a frequency equal to the desired up or down frequency translation. In principle, this technique should be highly efficient and generate no unwanted optical signals. However, significant unwanted sidebands have been observed with such devices which appear to arise from parasitic electrical fields [Ref. 42]. Careful device design is therefore necessary to maintain these signals at an acceptable level.

Mach–Zehnder interferometric Y-junction modulators (see Figure 10.22) can also be used to generate double sideband frequency translated optical signals when they are modulated with a sinusoidal voltage waveform. In this case the optical frequency shift is proportional to the frequency of the electrical modulating signal

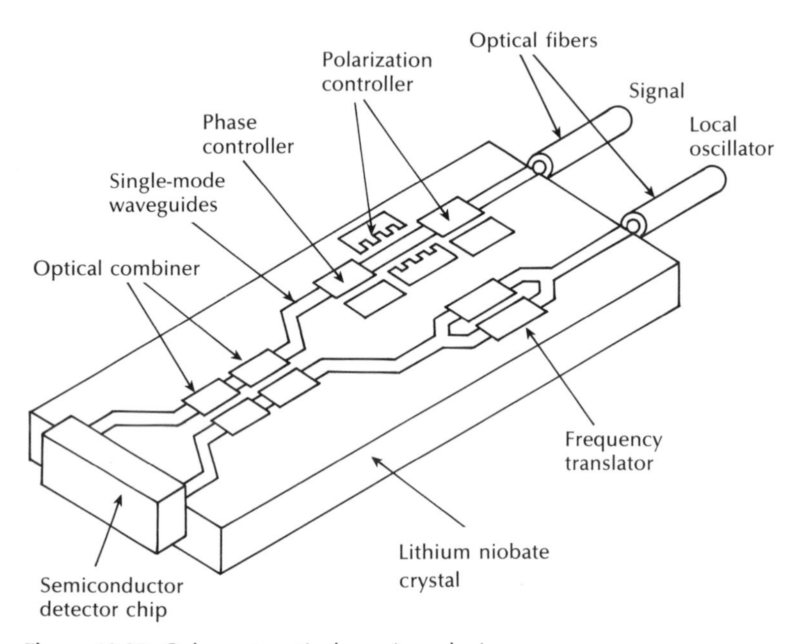

Figure 10.27 Coherent optical receiver device.

[Ref. 69]. However, in simple device structures the charging of the electrode capacitance limits the maximum modulation frequency and thus the magnitude of the frequency translation that can be obtained. To overcome this problem Mach–Zehnder interferometers with travelling wave electrode structures have been designed which provide multigigahertz bandwidths. Frequency translations of 3 GHz and 6.5 GHz at wavelengths of 0.85 and 1.52 μm, respectively, have been reported [Refs. 70, 71] using devices of this type.

Integration of the aforementioned electro-optic devices into a single lithium niobate substrate, particularly for use in coherent optical fiber communications systems (see Chapter 12) has become an increasing area of interest both to reduce losses between individual devices as well as system cost. For example, the configuration of a potential coherent optical receiver device is illustrated in Figure 10.27 [Refs. 42, 73]. It was fabricated on Z-cut lithium niobate and comprises a polarization controller with output phase controller, and a frequency translator together with a directional coupler for mixing the two optical signals. Successful operation of this integrated device was demonstrated and a similar X-cut lithium niobate device requiring a lower operating voltage has been proposed [Ref. 73].

10.7 Optoelectronic integration

The integration of interconnected optical and electronic devices is an important area of investigation for applications within optical fiber systems [Ref. 73]. Monolithic optoelectronic integrated circuits (OEICs) incorporating both optical sources and detectors have been successfully realized for a number of years. Monolithic integration for optical sources has been generally confined to the use of group III–V semiconductor compounds. These materials prove useful as they possess both optical and electronic properties which can be exploited to produce high performance devices. Circuits are often fabricated from GaAs/AlGaAs for operation in the shorter wavelength region between 0.8 and 0.9 μm. Such a circuit is shown in Figure 10.28(a) where an injection laser is fabricated on a GaAs substrate with a MESFET (metal-Schottky FET, see Section 9.5.1) which is used to bias and modulate the laser. Alternatively, Figure 10.28(b) demonstrates the integration of a longer wavelength (1.1 to 1.6 μm) injection laser fabricated from InGaAsP/InP together with a MISFET (metal integrated-semiconductor FET) where the conventional *n* type substrate is replaced by a semi-insulating InP substrate.

The realization of OEICs has, however, lagged behind other developments in IO using dielectric materials such as lithium niobate. This situation has been caused by the inherent difficulties in the fabrication of OEICs even when III–V compound semiconductors are employed [Ref. 74]. Compositional and structural differences between photonic devices and electronic circuits create problems in epitaxial crystal growth, planarization for lithography, electrical interconnections, thermal and chemical stability of materials, electrical matching between photonic and electrical

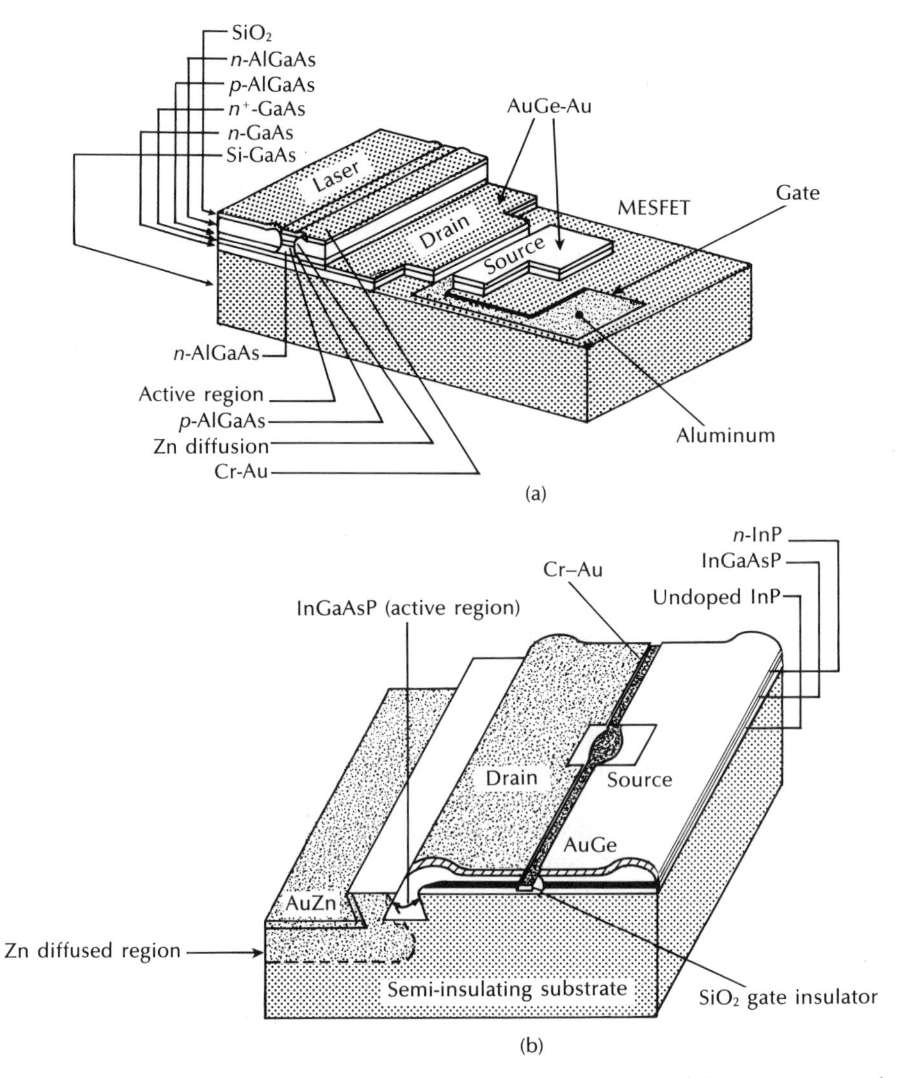

Figure 10.28 Monolithic integrated transmitter circuits: (a) GaAs/AlGaAs injection laser fabricated with a MESFET on a GaAs substrate; (b) InGaAsP/InP injection laser fabricated with a MISFET on a semi-insulating InP substrate.

devices together with heat dissipation. Nevertheless, the maturing of gallium arsenide technology for integrated circuits (as opposed to OEICs) [Ref. 75] has helped stimulate the more recent research activities into high speed OEICs. For example, a 2 Gbit s^{-1} optical transmitter incorporating five active devices was reported in 1986 [Ref. 76]. Moreover, the structure of a monolithically integrated DFB laser with an optical intensity modulator is shown in Figure 10.29 [Ref. 77].

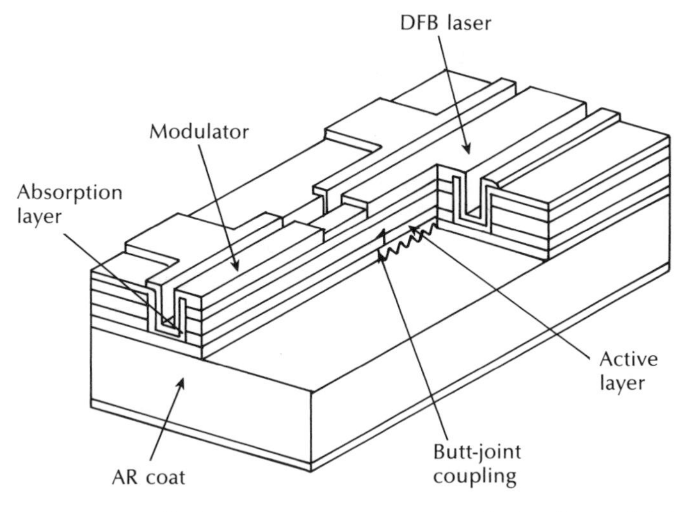

Figure 10.29 Device structure for an optical intensity modulator monolithically integrated with a DFB laser [Ref. 77].

This InGaAsP/InP device, which was designed to avoid the large chirp associated with directly modulated semiconductor lasers, displayed good dynamic characteristics at a modulation rate of 5 Gbit s^{-1} when operating at a wavelength of 1.55 μm. Hence OEIC transmitters for operation in both the short wavelength [Ref. 78] and longer wavelength regions [Refs. 79, 80] have received significant attention in recent years.

The monolithic integration of optical detectors with other active components has also been achieved using the group III–V semiconductor alloys. The structure of an OEIC photoreceiver fabricated for operation in the 0.8 to 0.85 μm wavelength range is illustrated in Figure 10.30 [Ref. 81]. This device incorporates a Schottky

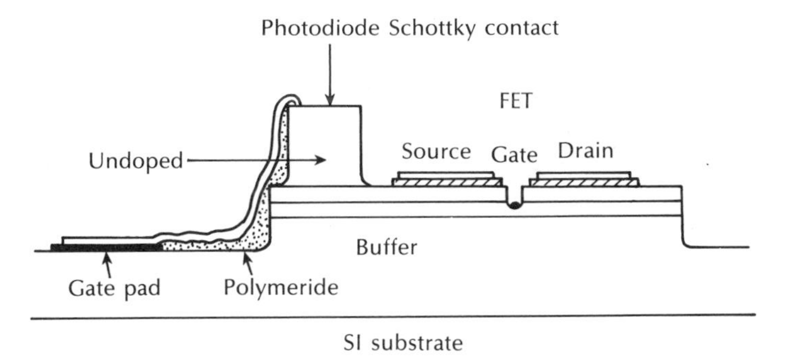

Figure 10.30 A planar monolithic integrated photoreceiver incorporating a Schottky photodiode and field effect transistor [Ref. 81].

photodiode and an FET on a GaAs semi-insulating substrate. It displayed a sensitivity of -30 dBm at a transmission rate of 250 Mbit s^{-1} for a bit error rate of 10^{-9}.

Receiver OEICs for the longer wavelength region have also been given attention. A design incorporating a *p–i–n* photodiode with three junction FETs forming a preamplifier was fabricated in InGaAsP on a semi-insulating InP substrate [Ref. 82]. Photoreceiver sensitivies of -25.5 dBm at 565 Mbit s^{-1} and -14.2 dBm at 1.2 Gbit s^{-1} were obtained for a bit error rate of 10^{-9} with this device. Furthermore, an improved process for fabricating the junction FETs has enabled a longer wavelength receiver OEIC to be fabricated which demonstrates higher receiver sensitivity of -22.8 dBm at a transmission rate of 1.2 Gbit s^{-1} [Ref. 80]. Nevertheless, this improved receiver sensitivity value is still around 10 dB lower than those achieved with a well designed PIN–FET hybrid receiver (see Section 9.5.2).

Optoelectronic integrated transmitter and receiver arrays have also been fabricated for applications such as wavelength division multiplexing and optical interconnection. For the latter application a four channel OEIC transmitter array comprising single quantum-well lasers integrated with FET drive circuits, as well as monitor photodiodes, has been demonstrated [Ref. 83]. This device which was fabricated using the GaAs/AlGaAs material system operated at a wavelength of 0.83 μm with a transmission rate in excess of 1.5 Gbit s^{-1}. GaAs-based photoreceiver arrays have also been fabricated. An example, also a four channel device, integrated a metal-semiconductor–metal photodiode with six MESFETs for each channel [Ref. 85]. A single element sensitivity for the receiver circuit of -26 dBm at 1 Gbit s^{-1} was obtained with low crosstalk up to transmission rates of 1.5 Gbit s^{-1}.

More complex optoelectronic integration is shown in Figure 10.31 [Ref. 87] where two possible designs of monolithic integrated circuits which serve as receive terminals in a wavelength division multiplex system are illustrated. These wavelength demultiplexers utilized micrograting filters (either transmission or reflection

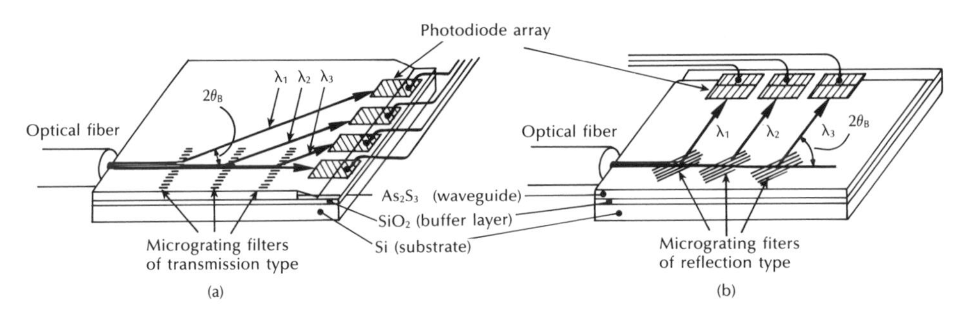

Figure 10.31 Monolithic integrated optical wavelength demultiplexers fabricated with micrograting filters and a Schottky barrier photodiode array on a silicon substrate: (a) using transmission gratings; (b) using reflection gratings [Ref. 87].

type) together with an array of Schottky barrier photodiodes fabricated on a silicon substrate. In each case the filters picked out individual transmission wavelengths directing them to the appropriate photodiode for detection.

A more recent monolithic integrated wavelength demultiplexer device is displayed in Figure 10.32 [Ref. 88]. This structure utilized a novel single-mode optical waveguide which is fabricated on a semiconductor substrate. The antiresonant reflecting optical waveguide (ARROW) shown in Figure 10.32(a) comprised an interference cladding inserted between the core and the substrate which consisted of two different films having a large refractive index difference. Therefore, the first cladding layer had a high refractive index whilst the second cladding layer exhibited a low refractive index. Using this mechanism the waveguide maintained single-mode propagation through loss discrimination of the higher order modes. Furthermore, the propagation loss of the fundamental mode displayed a wavelength dependence resulting from any changes in the resonant condition inside the interference cladding.

The structure of the ARROW-type wavelength demultiplexer and photodetector device is illustrated in Figure 10.32(b). It may be observed that the wavelength division multiplexed optical signal propagates through the low loss silicon dioxide

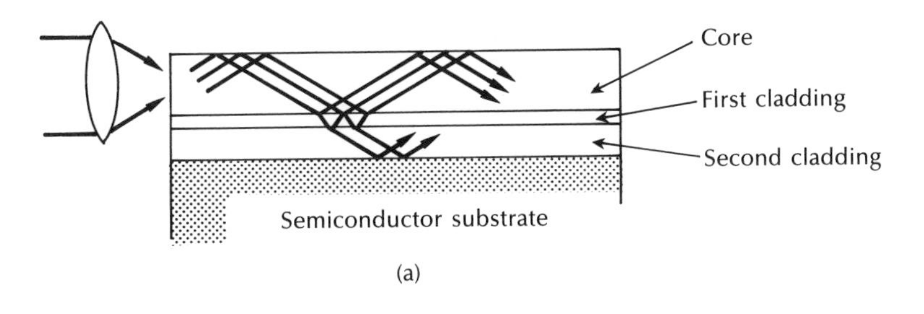

(a)

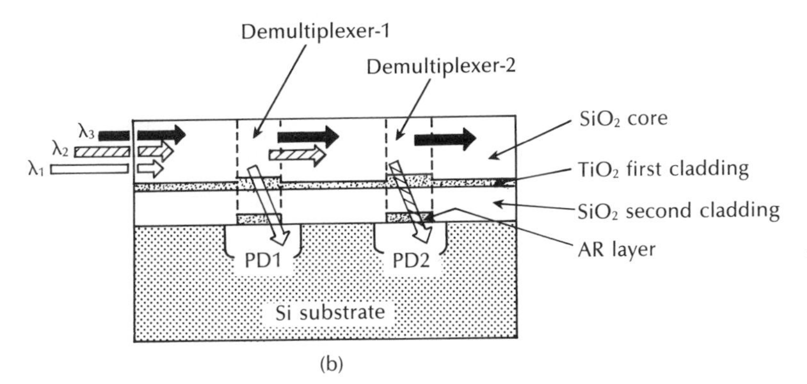

(b)

Figure 10.32 Wavelength demultiplexing using the antiresonant reflecting optical waveguide (ARROW): (a) structure of the ARROW; (b) an ARROW-type wavelength demultiplexer and photodetector integrated device [Ref. 88].

core layer. Wavelength demultiplexing was obtained in a region where the thickness of the first cladding layer had been adjusted to the resonant condition for which a particular wavelength band was radiated to a photodetector fabricated in the substrate. An antireflection (AR) layer was provided adjacent to the photodetector in order to enhance the coupling efficiency for the demultiplexed optical wavelength band. However, the remaining optical signal, being at other wavelengths, passed through the demultiplexing region with low loss. This device has been successfully demonstrated for demultiplexing of two wavelengths at 0.78 and 0.88 μm with a crosstalk isolation of -21.6 dB [Ref. 88]. It is therefore likely that IO circuits based on the above types will find application within WDM systems in the near future.

The monolithic integration of both sources and detectors on the same substrate has been achieved using group III–V semiconductor compounds. For example, a rudimentary AlGaAs OEIC optical repeater was implemented in 1984 [Ref. 89]. This monolithic integrated device, which is illustrated in Figure 10.33, incorporated a BH laser, *p–i–n* photodiode and a MESFET together to provide amplification on a semi-insulating GaAs substrate. Moreover, a longer wavelength OEIC repeater has also been demonstrated [Ref. 90]. The device fabricated on a semi-insulating InP substrate also incorporated a BH with a *p–i–n* photodiode and two FETs. In this case an overall repeater gain of 5.5 dB was obtained.

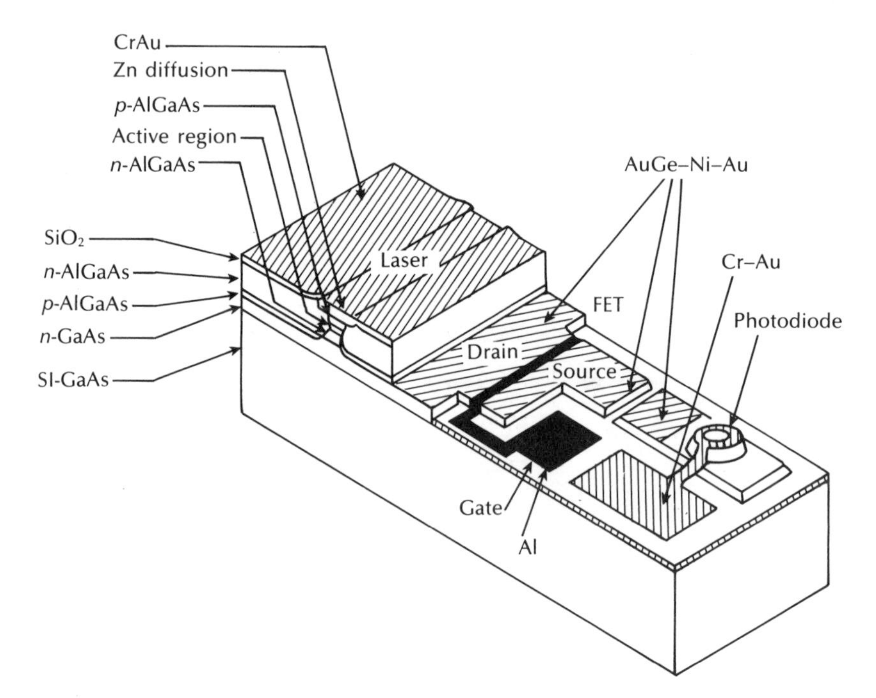

Figure 10.33 Monolithic integrated optoelectronic repeater chip [Ref. 89].

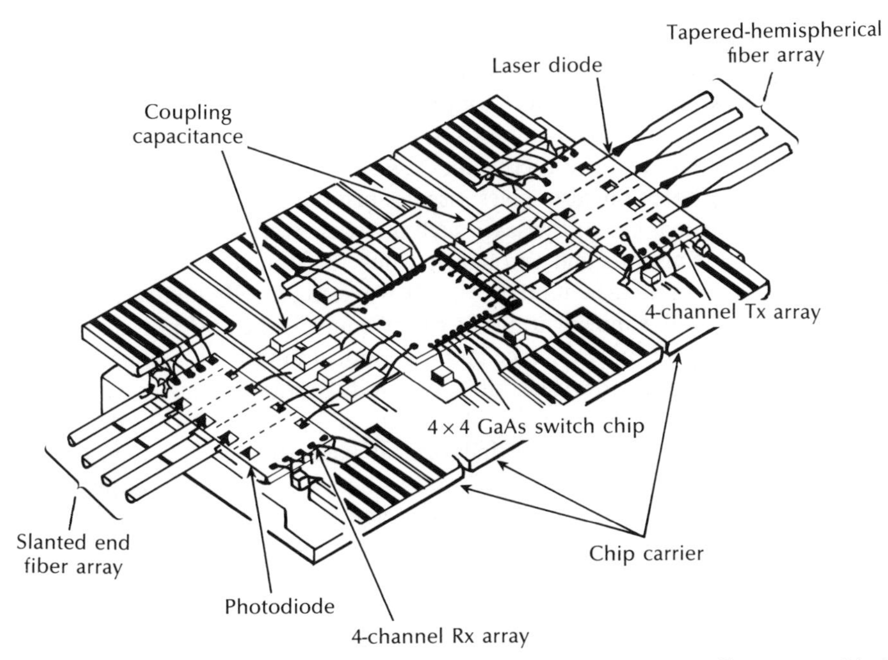

Figure 10.34 A 4 × 4 hybrid crossbar switch fabricated using gallium arsenide-based OEIC technology [Ref. 92].

There is also significant interest in the use of OEICs to provide optical switching matrices within future optical communication networks [Ref. 91]. Such optoelectronic switches form a hybrid optical switching technology which can facilitate space, time, and frequency division switching. An example of a 4 × 4 space division crossbar switch module is shown in Figure 10.34 [Ref. 92]. This hybrid device, which was fabricated using GaAs OEIC technology, comprises a laser array and the receiver array outlined above combined with a GaAs switch circuit. The all-GaAs device has been successfully operated at transmission rates up to 560 Mbit s^{-1}.

Although the present state of the art for OEICs centres on high speed transmitters, receivers and optoelectronic switching matrices for optical communications, it is perceived that OEICs will in the future perform a variety of functions. These include intrachip optical connection and parallel processing as well as interchip optical connection.

10.8 Optical bistability and digital optics

Bistable optical devices have been under investigation for a number of years to provide a series of optical processing functions. These include optical logic and memory elements, power limiters and pulse shapers, differential amplifiers, and

A–D converters. Moreover, the bistable optical device (BOD) in providing for digital optical logic – namely, a family of logic gates whose response to light is nonlinear – gives the basis for optical computation.

In its simplest form the BOD comprises a Fabry–Perot cavity containing a material in which variations in refractive index with optical intensity are nonlinear (nonlinear optical absorption also gives rise to bistability), as shown in Figure 10.35(a). In a similar manner to the laser such a cavity exhibits a sharp resonance to optical power passing into and through it when the optical path length in the nonlinear medium is an integer number of half wavelengths. By contrast with the laser the value of refractive index within the cavity controls the optical transmission giving high optical output on resonance and low optical output off resonance. The transfer characteristic for the device exhibits two-state hysteresis which results from tuning into and out of resonance, as illustrated in Figure 10.35(b). BODs are therefore able to latch between two distinct optical states (0 or 1) in response to an external signal to act as a memory or flip flop. Furthermore, by careful adjustments of the device bias and input levels, the BOD can act as an AND-gate, an OR-gate, or a NOT-gate, hence providing logic functions [Ref. 94].

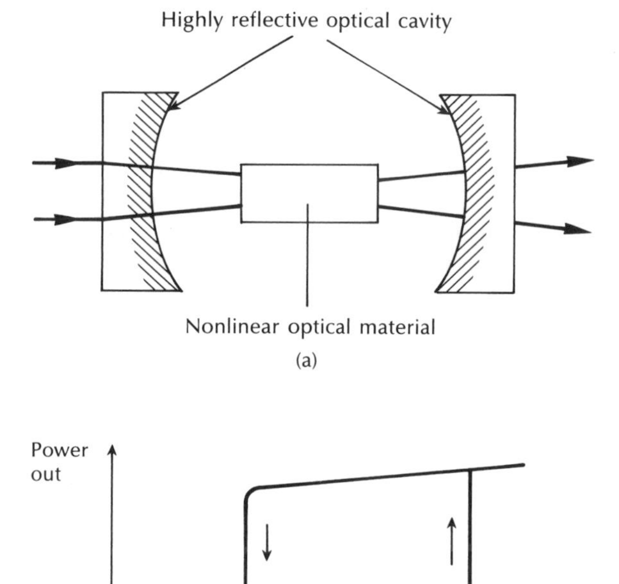

Figure 10.35 A generalized bistable optical device: (a) schematic structure; (b) typical transfer characteristic.

Although the switching speed of BODs is dependent on drive power, they offer the potential for very fast switching at low power levels. Investigations are therefore directed towards the possibility of picosecond switching using only picojoules of energy. A BOD exhibiting these properties would prove far superior to an electronic device which performs the same function. However, suitable nonlinear materials and device structures to give this performance are still under investigation. BODs may be separated into two basic classes: all-optical or intrinsic devices which utilize a nonlinear optical medium between a pair of partially reflecting mirrors forming a nonlinear etalon in which the feedback is provided optically; and hybrid devices where the feedback is provided electrically.

In some cases hybrid devices employ an artificial nonlinearity such as an electro-optic medium within the cavity to produce variations in refractive index via the electro-optic effect. In materials such as lithium niobate and gallium arsenide this produces strong artificial nonlinearity which can be combined with an electronic feedback loop. Such hybrid BODs have been fabricated in integrated optical form. A typical device is shown in Figure 10.36. [Ref. 95]. It consists of a titanium diffused optical waveguide on a lithium niobate substrate with cleaved and silvered end faces to form the resonant optical cavity. The light emitted from the cavity is detected and amplified by an avalanche photodiode (APD). The electrical signal thus obtained is then fed back to the electrodes deposited on either side of the cavity in order to produce refractive index variations. Such a device therefore exhibits hysteresis and bistability. Although these hybrid BODs provide flexibility for experimental study their switching speeds are ultimately limited by the use of electrical feedback. Nevertheless, it is possible that several such devices could be interconnected to provide a more complex logic circuit.

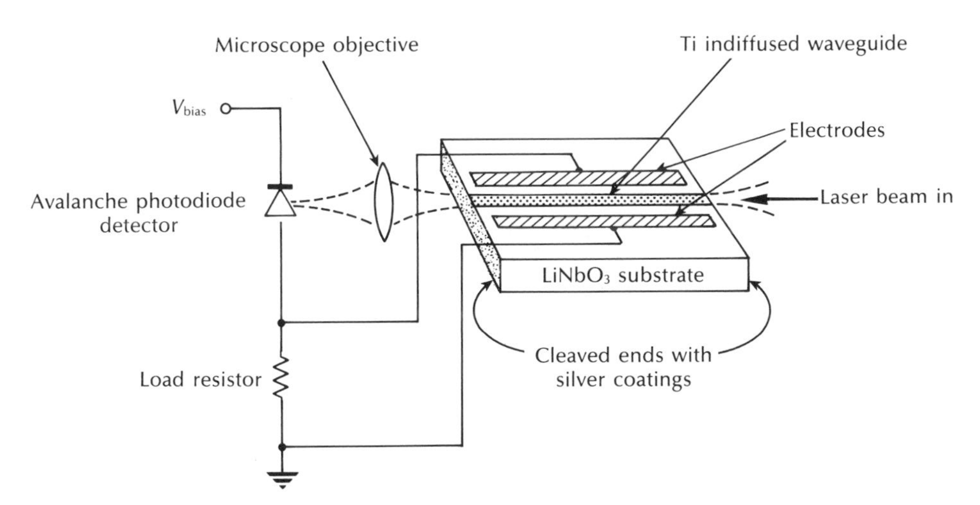

Figure 10.36 A hybrid integrated bistable optical device [Ref. 95].

An alternative hybrid approach based on the use of inorganic superlattices has been pursued at AT & T Bell Laboratories [Ref. 96], and elsewhere [Ref. 97]. These materials are constructed by alternating thin films of two different semiconductor materials which exhibit nonlinear properties. Combinations used include gallium arsenide and gallium aluminium arsenide, mercury telluride and cadmium telluride, silicon, and indium phosphide. This work has resulted in the development of the, so-called, self-electro-optic effect device (SEED) which exhibits hysteresis and bistate transmission. The device, a schematic of which is shown in Figure 10.37(a), comprises a single chip of alternating layers. Although the device is activated by light an electric field is required to 'prime' the material for switching. The switching results from wavelength sensitive absorption within the superlattice structure which causes current flow, thus decreasing the bias voltage which in turn

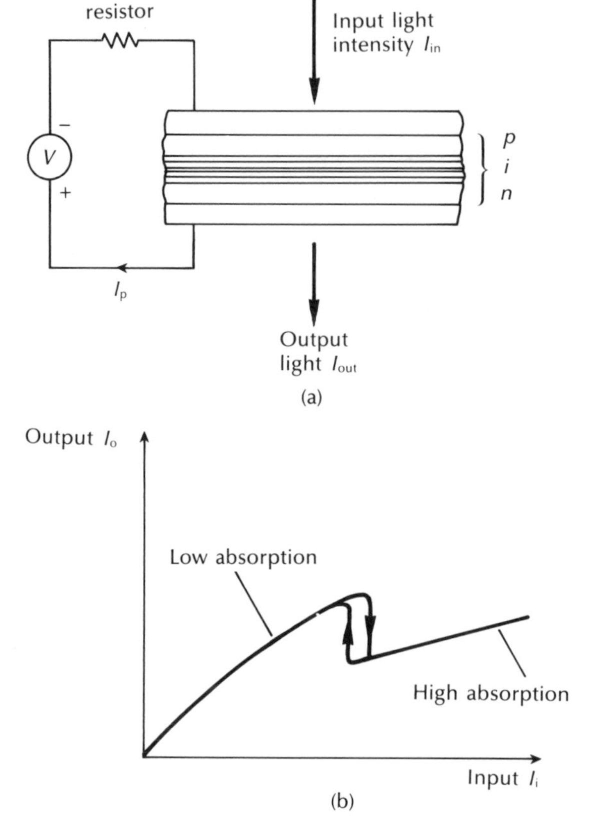

Figure 10.37 The self-electro-optic effect device (SEED): (a) schematic structure; (b) input/output response characteristic.

increases the absorption. Eventually a point is reached at which switching occurs and the device switches to high absorption as indicated in Figure 10.37(b).

Alternatively, all-optical or intrinsic BODs may employ an appropriate nonlinear optical medium. Investigations at present are centred around materials such as indium antimonide, zinc selenide, cadmium sulphide, gallium arsenide, indium arsenide, gallium aluminium arsenide and indium gallium arsenide phosphide in which optical absorption gives a change in refractive index. Unfortunately, these effects are generally weak and often require low temperatures to display themselves adequately. However, the possibility of low power, low energy, fast switching integratable devices for use in real time optical processing and digital optical computing is a proposition which has encouraged a concentrated activity in this area. Studies have involved the use of indium antimonide (InSb) in the near-infrared [Ref. 98] and more recently in the visible region, although operation is only achieved at low temperature (77 K). A more recent success, however, is the observation of bistability at room temperatures using thermal nonlinearities in zinc selenide (ZnSe) interference filter configurations [Refs. 99, 100]. This work was undertaken with visible light which provides the advantage that the switching and hysteresis effects can actually be seen.

Intrinsic optical bistability may also be obtained from large resonant nonlinearities available near the bandgaps of other semiconductor materials [Ref. 13]. Such bistability can be further distinguished as an active system which incorporates its own optical source, or a passive system which does not. The input to a passive bistable device is always optical while the input to an active device depends upon the method by which the source is to be excited. For example, in the former case room temperature bistability in bulk gallium arsenide at switching speeds of 30 ps has been observed [Ref. 101]. In addition, nonlinear channel waveguide structures in GaAs/GaAlAs multiple quantum-well material have shown optical bistability at relatively low power levels, but with slow switching speeds [Ref. 102].

The source of excitation for active bistability in semiconductors is normally provided by an injection current giving the configuration of a bistable laser diode [Ref. 103]. Semiconductor lasers exhibit optical bistability due to nonlinearities in absorption, gain, dispersion, waveguiding and the selection of the output polarization. One approach to laser diode bistability through nonlinear absorption is illustrated in Figure 10.38(a) [Ref. 104]. In this case the device is fabricated with a tandem electrode which provides two gain sections, with a loss region between them. The loss region acts as a saturable absorber creating the hysteresis characteristic displayed in Figure 10.38(b). Such devices fabricated in GaAs/GaAlAs and InP/InGaAsP have demonstrated nanosecond switching times with milliwatt power levels at room temperature [Ref. 13].

The above BODs have been discussed primarily in relation to the provision of optical logic and memory elements. However, investigations of optical bistability have also included the other functions mentioned previously. Optical pulse shaping can be achieved using a BOD with a very narrow bistable loop. Such a device can

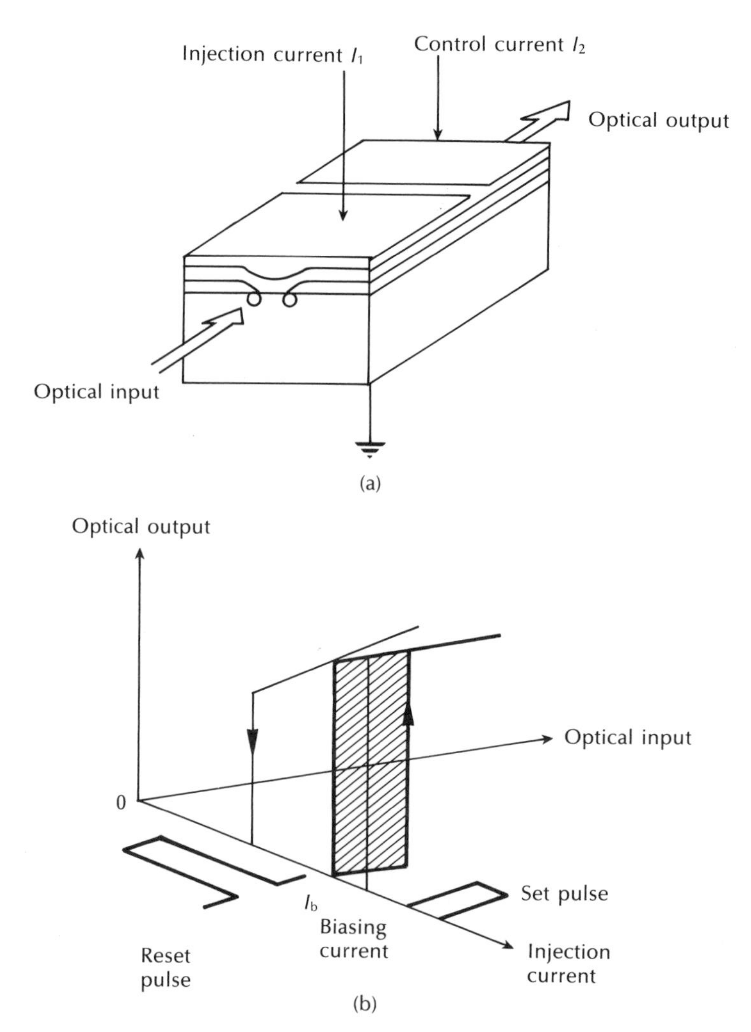

Figure 10.38 Bistable laser diode: (a) structure; (b) response characteristic.

be used to shape, clean-up and amplify a noisy input pulse, as illustrated in Figure 10.39(a).

Nonlinear optical amplification can also be obtained with certain BODs. In particular, the Fabry–Perot semiconductor laser amplifier can display dispersive bistability [Ref. 103] which, unlike its linear counterpart (see Section 10.3), provides a nonlinear gain characteristic, as shown in Figure 10.39(b). The optical amplification mechanism in this case can involve the interaction of at least two optical fields through the field dependent dielectric constant of the nonlinear material. The operation of such a BOD differential amplifier is also illustrated in Figure 10.39(b). The introduction of a weak second beam into the nonlinear optical

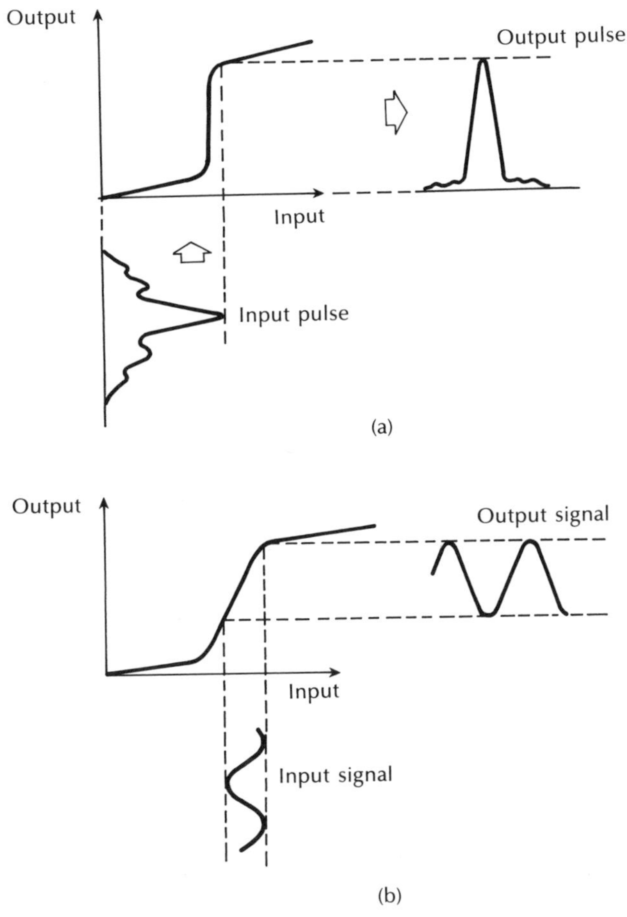

Figure 10.39 Illustration of two functions provided by the nonlinear characteristic of BODs: (a) optical pulse shaping; (b) optical amplification.

cavity is used to control the resonance and transmission of the main beam through the additive effects of its own stored energy. Hence differential optical gain is provided by the device. With this configuration a weak beam can control an intense main beam producing the optical, equivalent* of the electronic transistor.

Linear optical amplifiers exhibit the drawback of amplifying low level noise signals together with the desired signal. Bistable amplifiers, however, are useful because of their signal regeneration capability [Ref. 105]. In the ideal case no amplification is provided for signals below a particular intensity level. Once an intensity threshold has been surpassed, the large gain can be determined by the

* This two beam optical transistor has been dubbed 'the transphasor' by the authors of Ref. 98.

slope of the curve in Figure 10.39(b). Moreover, a saturated or maximum value of output intensity is also provided, displaying the power limiter function of the device.

10.9 Optical computation

Although the maximum potential switching speeds of individual IO logic devices have as yet to be accomplished, the use of parallel processing with optical signals mentioned in Section 10.5 can provide a net benefit over a similar serial electronic system, even at much slower optical device speeds. Conventional digital computers suffer from a bottleneck resulting from the limited number of interconnections which can be practically supported by an electronic-based communications technology. This restriction led to the classical von Neumann architecture for computing systems shown in Figure 10.40(a) in which the memory is addressed sequentially from the central processing unit (CPU). The CPU accesses the memory through a binary addressing unit and the memory contents are returned to the CPU via a small number of lines. This serial addressing of memory reduces the communications requirements and minimizes the number of lines, but this is achieved at the expense of overall computing speed. The problem, which is referred to as the von Neumann bottleneck, eventually limits the speed of the computer system.

With optical systems the situation is changed as they are capable of communicating many high bandwidth channels in parallel without interference. Thus parallel communication can easily be provided within an optical computer system at relatively low cost. In theory this lends itself to the use of non-von Neumann architecture (see Figure 10.40(b)) in which all memory elements are accessible in parallel, thus removing the speed limitation caused by the bottleneck. The potential

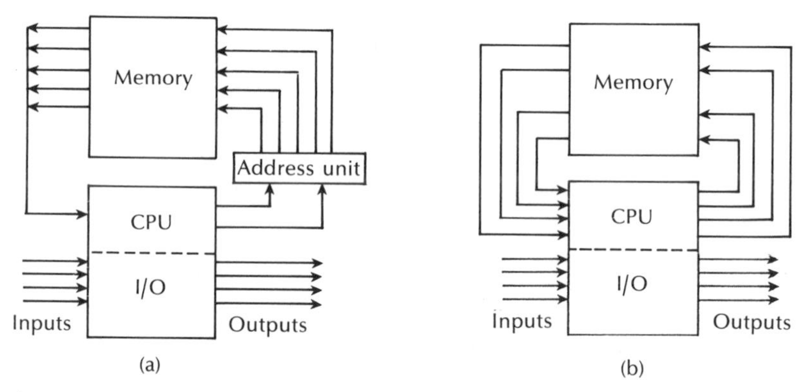

Figure 10.40 Computer architecture: (a) von Neuman; (b) non-von Neuman.

advantages offered by the digital optical computer are not therefore solely dependent upon the realization of subpicosecond optical switching devices.

For some time work in optical computation [Ref. 100] has been directed towards particular requirements which are necessary to provide a practical optical computing system. These include:

(a) High contrast. Logic devices must exhibit a large change between logic 0 and logic 1 levels.

(b) Steady state bias. To provide various different logic gates it is necessary that optical bias levels may be altered. For a BOD this implies that the device can be held indefinitely at any point on the characteristic with a CW laser beam. However, this holding beam necessitates a degree of thermal stability. Such stability has been demonstrated with devices based on InSb at 77 K and on ZnSe at 300 K (Ref. 99).

(c) External addressing. The function of external addressing is to provide for separate external optical signals which can be combined with the holding beam to switch the device, thus giving logic functions. The switching energy can be derived from the holding beam which is then switched and propagates in transmission or reflection as the output beam to further devices in the optical circuit.

(d) Cascadability. The output from a particular device must be sufficient to switch at least one following device. This condition may be fulfilled by setting a holding beam near the switch point since the extra increment is then small in comparison with the change in output.

(e) Fan-out and fan-in. The advantage of parallel processing requires that a particular device has the ability to drive a large number of following devices. This could be achieved using free space propagation for addressing purposes. Furthermore, the summed effect of several elements could be focused onto one device to achieve fan-in.

(f) Gain. In order to maintain (d) and (e) above, there is a requirement for differential gain. This could possibly be achieved by the use of optical amplifier devices.

(g) Arrays. The easy construction of two dimensional (2-D) arrays within the technology must ideally be available.

(h) Speed and power. For 1-D circuits, subnanosecond or picosecond switching times are desirable, although this may be relaxed to microseconds for parallel arrays. Speed and power tend to be interchangeable but a low power device is a necessity. The power requirements for a device should be in the milliwatt region or less.

Certain, although not all, of the above requirements are met by specific nonlinear devices described in Section 10.8. For example, it is suggested [Ref. 100] that the ZnSe interference devices, used as separate elements activated by external addresses, exhibit the possibilities of projection and display. Proposals for logic subsystems of this type are shown in Figure 10.41. In addition, Refs. 100 and 106 indicate the possible arrangements for more complex optical logic subsystems,

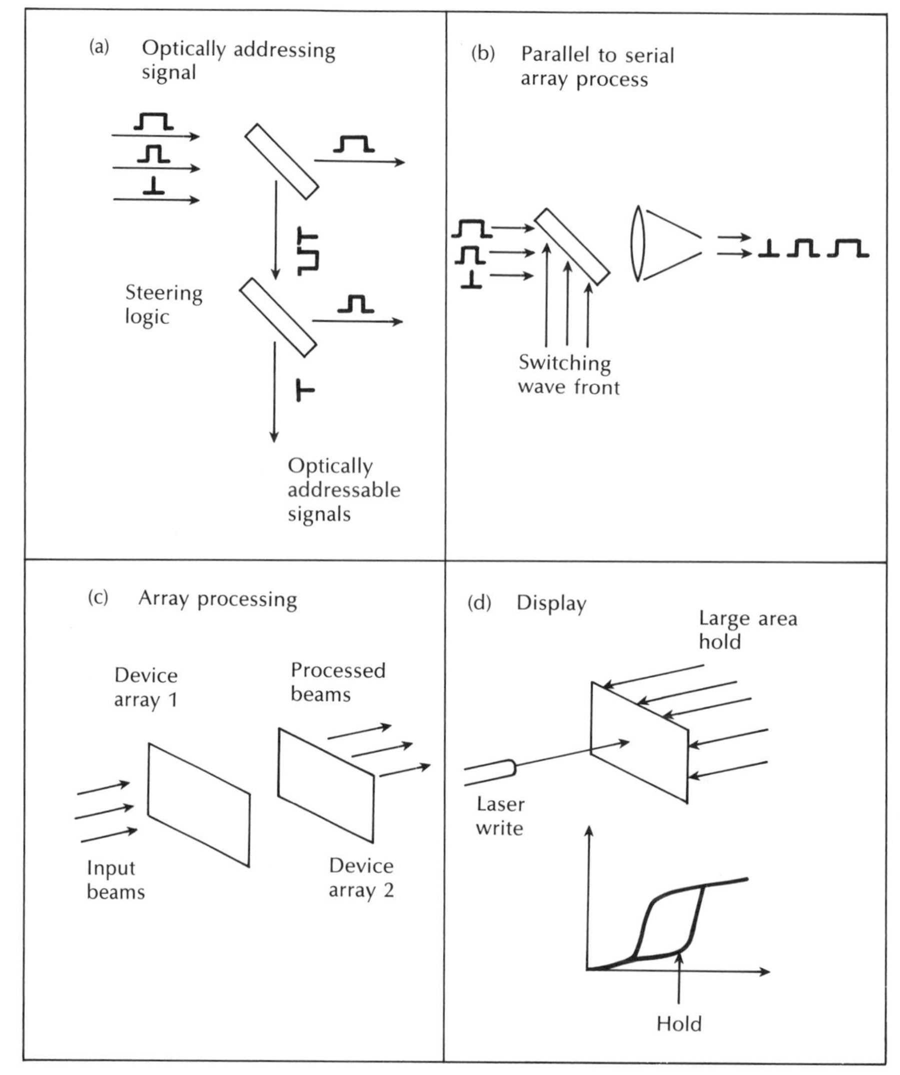

Figure 10.41 Proposals for logic subsystems based on the zinc selenide interference device, or similar [Ref. 100].

including a simple parallel processor, a serial to parallel convertor, a shift register and a packet switch. At present such proposals necessitate a solution which involves bulk optic, discrete elements together with possible monolithic IO devices. Moreover, a hybrid approach for the implementation of high speed switching matrices has been suggested [Ref. 107] which incorporates electronic logic elements with optical interconnections in order to exploit the best features of each technology.

In order to implement more complex optical logic subsystems, alternative nonlinear materials are already under investigation including inorganic insulators and organic nonlinear compounds [Refs. 108, 109]. Apart from lithium niobate, the three leading inorganic insulator materials are strontium barium niobate, bismuth silicon oxide and barium titanate. Unfortunately, these materials exhibit drawbacks in relation to poor thermal and mechanical properties, as well as slow response times (milliseconds). However, the organic nonlinear materials listed in Table 10.1 have displayed the potential for greater degrees of nonlinearity and much shorter response times. At present the major disadvantage with these materials is their relative environmental instability compared to inorganic materials (e.g. oxidation). However, work is still at a preliminary stage and more favourable results may be anticipated in the future.

Success with such materials, together with further developments of the other optical devices mentioned in Section 10.8, could lead to the implementation of an all-optical computer. However, it is more likely that initially hybrid optical/electronic computational machines will evolve. For example, a multiprocessing system under development, called the Connection Machine [Ref. 110], comprises a large array of printed circuit boards, each containing 512 processing elements divided equally between 32 electronic chips. This particular concept is illustrated in Figure 10.42 where, for simplicity, only four chips per level are shown. It may be observed that each electronic board contains a frequency selective filter (hologram) in addition to optoelectronic chips. These OEICs contain semiconductor lasers and

Table 10.1 Some organic nonlinear materials

Substituted and disubstituted acetylenes and diacetylenes
Anthracines and derivatives
Dyes
Macrocyclics
Polybenzimidazole
Polybenzimidazole and polybenzobisoxazole
Polyester and polyesteramids
Polyetherketone
Polyquinoxalines
Porphyrins and metal-porphyrin complexes
Metal complexes of TCNQ or TNAP
Urea

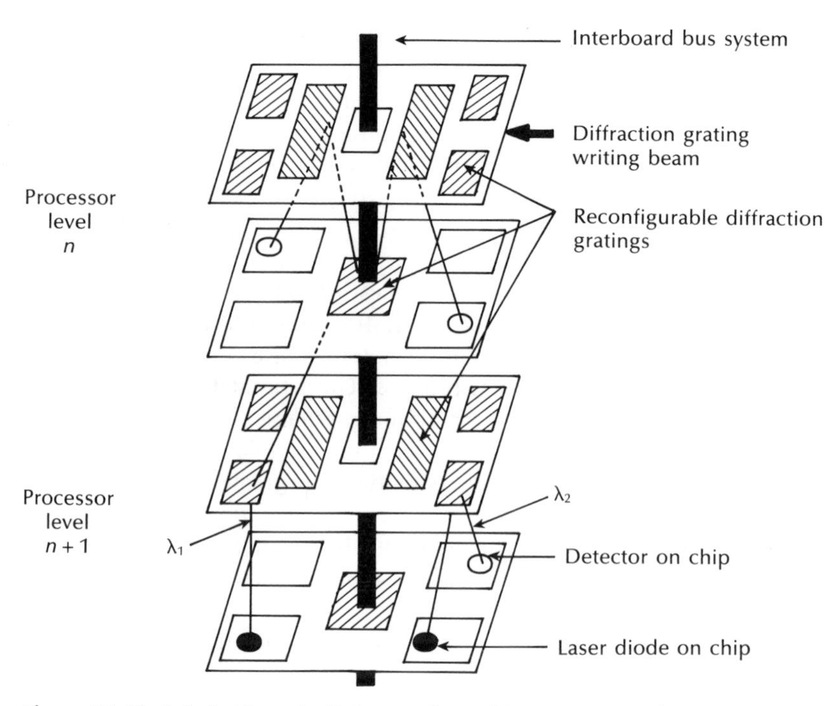

Figure 10.42 A hybrid optical/electronic multiprocessor architecture.

photodiodes which enable communications to be established between the chips. The switching operations for the interconnection process are performed by a planar array of reconfigurable diffraction gratings located on each board. Moreover, the architecture employs WDM to direct bit streams to the appropriate board as illustrated by λ_1 and λ_2 in Figure 10.42.

A possible all-optical digital multiprocessor architecture is shown in Figure 10.43 [Refs. 109, 111]. The input to the machine is via either an array of independently addressable semiconductor lasers, or alternatively via a 2-D spatial light modulator (SLM). A laser array is capable of higher modulation speeds but necessitates more complex circuitry, especially when uniformity is required over the complete array. The gate array illustrated in Figure 10.43 comprises either another 2-D SLM with a nonlinear response or an array of BODs. In theory the BODs would provide much greater switching speed but currently exhibit the drawbacks mentioned in Section 10.8. The beam controller employs reconfigurable diffraction gratings in order to provide switching and interconnection. However, as a result of the large number of channels required in the all-optical computer it is likely that multiple planes of real time hologram arrays would be utilized.

It may be observed in Figure 10.43 that all three computer interconnect systems (CPU–CPU, CPU–memory and CPU–I/O) are combined in the beam controller, although they could be implemented by three different components. Nevertheless,

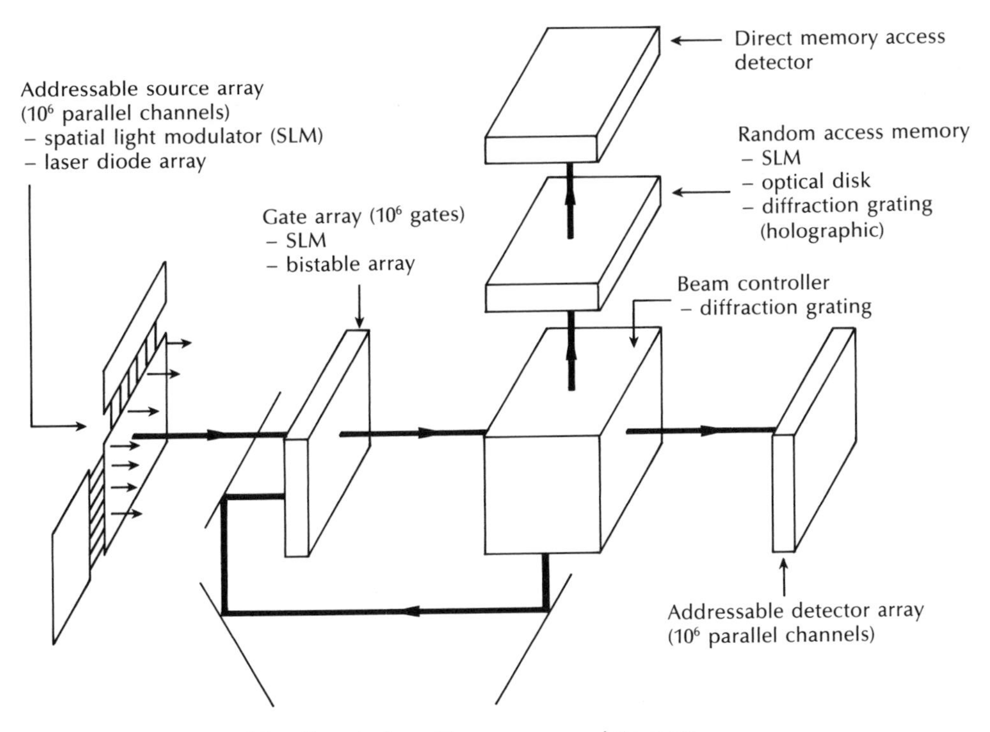

Figure 10.43 A possible all-optical multiprocessor architecture.

the ideal solution involves the beam controller directing any beam emerging from the gate array to any particular location on either the detector array, the memory, or the input of the gate array. Moreover, several logic elements can be interconnected via the beam controller to form a processing element. An example of a possible structure for a processing element, or node, in which individual elements in the gate array are designated to typical functions such as logic unit, clock, cache memory, etc., is illustrated in Figure 10.44. The example shown depicts a 5×5 rectangular array of logic elements, or gates, which gives 25 per processor. A practical multiprocessor might require 4×10^4 nodes giving 10^6 switching elements in the gate array.

A variant on the above architecture has been developed by AT & T Laboratories who recently announced the first demonstration of a digital optical processor [Ref. 112]. This processor, which operates at 10^6 cycles per second, is shown in schematic in Figure 10.45. The hybrid bistable switching element is a GaAs/AlGaAs symmetric self-electro-optic effect device (S-SEED) which is claimed to offer a potential of 10^9 operations per second with a switching energy of 1 pJ (see Section 10.8). The S-SEEDs which are 5 μm square and contain two mirrors with controllable reflectivity are formed into thirty-two device arrays. Each array also

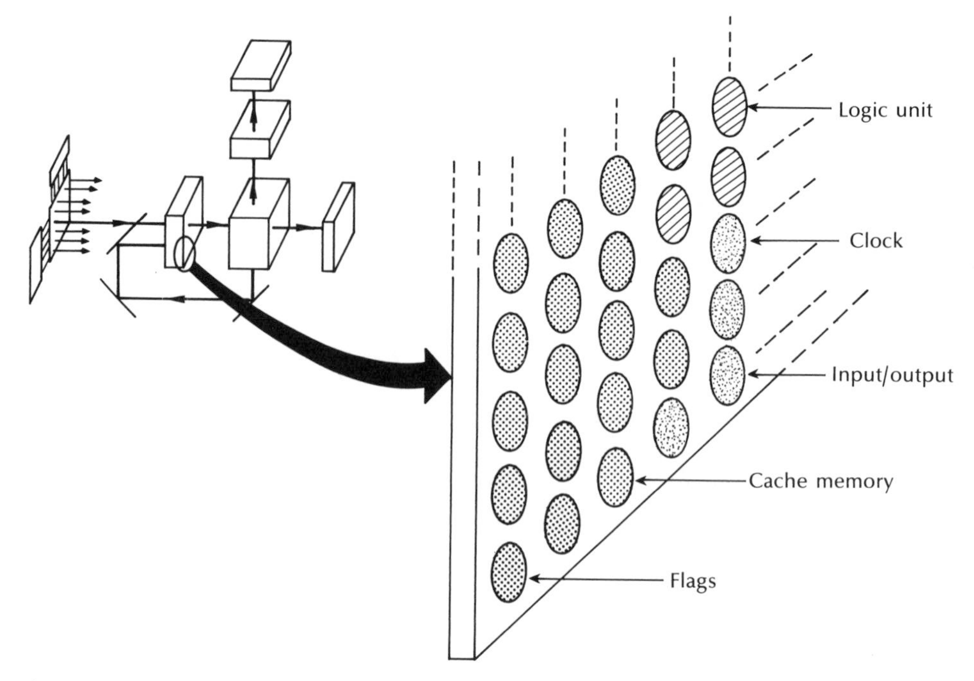

Figure 10.44 A processing element within the all-optical multiprocessor of Figure 10.43.

contains two 10 mW modulated injection laser diodes emitting at a wavelength of 0.85 μm as illustrated in Figure 10.45(a).

In the demonstration four S-SEED arrays were located within the processor, as may be observed in Figure 10.45(b). The injection lasers emitted many separate beams to provide communication between the arrays whilst each S-SEED drove two inputs. Interconnection between the four arrays was controlled by the lenses and masks, also shown in Figure 10.45. The masks comprised glass slides with patterns of transparent and opaque spots that allowed or impeded the transmission of light. Hence these patterns defined the connectivity within the processor.

The processor logic was accomplished by each S-SEED operating as a NOR-gate. Thus the output from each device array served as an input for the next array where the logic state of the S-SEEDs in the second array were determined by the state of the devices in the first array. Changing the on–off status of the switches in successive arrays allowed calculations to be performed. The memory resided in each S-SEED which did not change its state until the information represented by that state (i.e. a 0 or 1) was processed. In this way extensive pipelining of information was utilized within the processor (i.e. the output from one part of the machine served as the input for another part). Finally, the I/O was accomplished using both optical fibers and laser beams transmitted through free space.

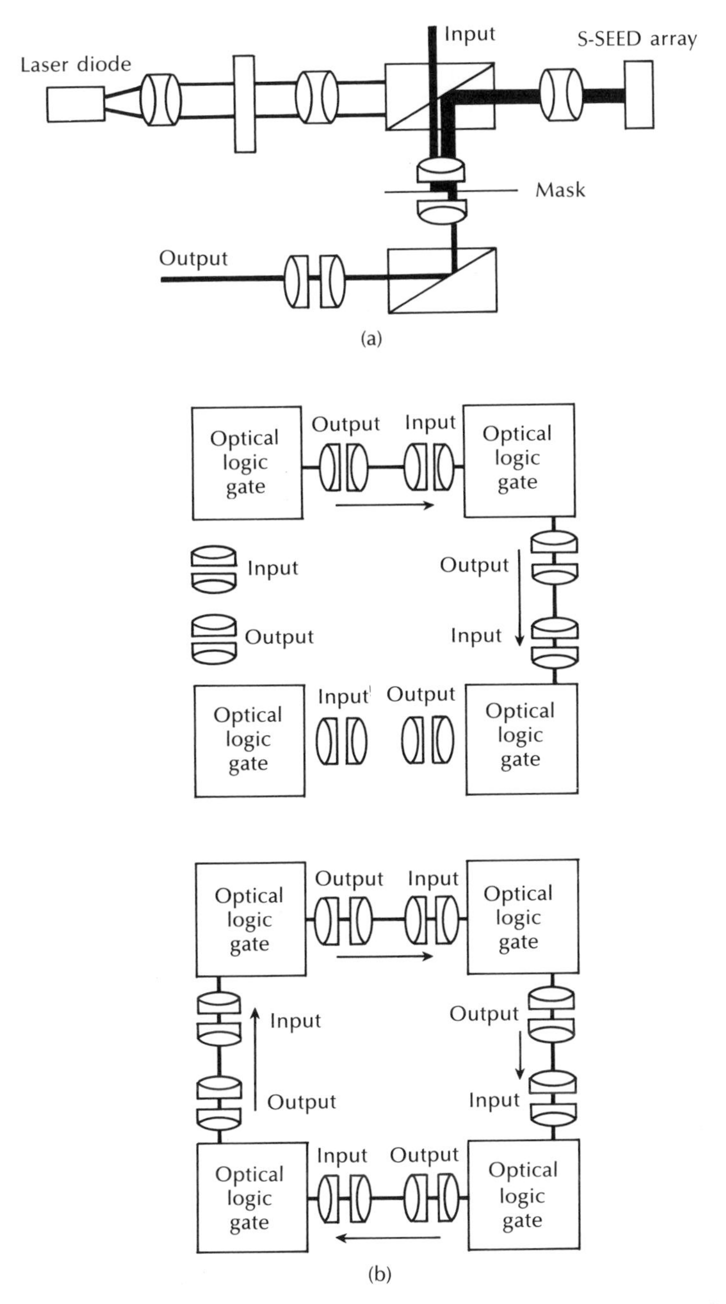

Figure 10.45 The A T & T digital optical processor: (a) structure of the processor S-SEED array module; (b) demonstration system using four array modules.

It is suggested that larger device arrays operating at faster switching speeds combined with the interconnection of greater numbers of modules will be demonstrated in the near future [Ref. 112]. Hence it would appear that the optical computer dimension to the future telecommunication network is firmly on the agenda.

Problems

10.1 Give the major reasons which have led to the development of optical amplifiers, outlining the attributes and application areas for these devices.

Describe the two main SLA types and indicate their distinguishing features.

10.2 A Fabry–Perot SLA has facet reflectivities of 23% and a single pass gain of 6 dB. The device has an active region with a refractive index of 3.6, a peak gain wavelength of 1.55 μm with a spectral bandwidth of 5 GHz. Determine the length of the active region for the FPA and also its mode spacing.

10.3 The following parameter values apply to a semiconductor TWA operating at a wavelength of 1.3 μm:

material gain coefficient	1000 cm^{-1}
effective loss coefficient	22 cm^{-1}
active region length	200 μm
facet reflectivities	0.1%
optical confinement factor	0.3

Calculate in decibels both the minimum and the maximum optical gain that could be obtained from the device.

10.4 Determine the peak–trough ratio of the passband ripple for the TWA of Problem 10.3. Compare this value with that obtained using a device with the same specification as Problem 10.3 excepting that the facet reflectivities are reduced to 0.03%.

Estimate the cavity gain for the latter semiconductor TWA.

10.5 Describe the phenomenon of backward gain in a semiconductor TWA and suggest a way in which it might be limited in systems that employ cascaded amplifiers.

A semiconductor TWA has a maximum cavity gain of 17 dB with a peak–trough ratio of 3 dB. Estimate the backward gain exhibited by the device under maximum gain operation.

10.6 (a) Sketch the major elements of a fiber amplifier and describe the operation of the device. Indicate the benefits of fiber amplifier technology in comparison with that associated with SLAs.

(b) Using an energy band diagram, briefly discuss the mechanism for the provision of stimulated emission in the erbium doped silica fiber amplifier. Name and describe a phenomenon occurring in this material system which creates a limitation to the optical gain that may be obtained from the device.

10.7 Explain the gain process in a Raman fiber amplifier and comment upon the flexibility associated with the pumping process in this fiber amplifier type.

The Raman gain coefficient for a 10 μm core diameter silica based fiber at a pump wavelength of 1.2 μm is 6.3×10^{-14} m W^{-1}. Determine the Raman gain obtained in a

25 km length of the fiber when it is pumped at this wavelength with an input power of 1.4 W and when the transmission loss is 0.8 dB km^{-1}. It may be assumed that the effective core radius is 1.15 times as large as the actual core radius and that complete polarization scrambling occurs.

10.8 Briefly describe the waveguide structures employed for IO.

The normalized effective thickness H for an asymmetric slab guide is given by Eq. (10.19), obtain an expression for the minimum effective thickness in order to provide optical confinement.

Determine the minimum effective thickness for a lithium niobate IO waveguide structure which has film and substrate refractive indices of 2.1 and 2.0, respectively, when it is operated at a wavelength of 1.3 μm.

10.9 Calculate the minimum effective thickness to provide optical confinement in a III–V semiconductor compound IO waveguide operating at a wavelength of 1.3 μm which has film and substrate refractive indices of 3.5 and 2.7, respectively, at this wavelength.

Comment on the value obtained in comparison to that determined in Problem 10.8.

10.10 Outline the techniques that can be employed to provide directional coupling between waveguides with IO.

Commencing with Eq. (10.25), show that the power coupled from one waveguide to another when their propagation constants are equal is proportional to the factor $\sin^2(Cz)$. The boundary conditions $A(z = 0) = A(0)$ and $B(z = 0) = 0$ may be assumed.

10.11 Compare the voltages required to operate lithium niobate and III–V semiconductor compound strip waveguide phase modulators in order to produce a phase change of π radians when using 3 cm long devices with a distance between the electrodes of 30 μm. The electro-optic coefficients for lithium niobate and the III–V semiconductor compound may be taken as 30.8×10^{-12} m V^{-1} and 1.3×10^{-12} m V^{-1}, respectively, whilst the refractive indices are 2.1 and 3.1, respectively, at the operating wavelength of 1.3 μm.

Comment on the values for the voltages obtained.

10.12 Assuming that the waveguide size for the III–V semiconductor waveguide modulator of problem 10.11 can be reduced by a factor of five, determine the reduced voltage needed to obtain a phase shift of π radians.

Comment on the value calculated in comparison with that obtained for the lithium niobate phase modulator of Problem 10.11.

10.13 A lithium niobate phase modulator has $V_\pi L$ of 45 V mm. Determine the interaction length required so that an applied voltage of 10 V will produce a phase shift of 2π radians.

10.14 A first order InGaAsP/InP waveguide grating filter is required for operation at a wavelength of 1.56 μm. The reflected light is expected to be incident over an angle of 3°. Estimate the filter length required to provide a 3 dB bandwidth of 2.5 Å for the device.

10.15 Discuss the function and operation of polarization transformers and frequency translators with specific reference to coherent optical transmission.

10.16 Outline the importance of optoelectronic integration in relation to the future developments in optical fiber communications. Illustrate your answer with descriptions of OEICs which have been fabricated to provide optical transmitter, receiver and multiplexing functions.

10.17 Describe the generalized bistable optical device and mention the applications in which it is finding use. Indicate the primary reasons for the evolution of optical computational devices and discuss recent developments in this process which augur well for their future implementation.

Answers to numerical problems

10.2 234 μm, 1.43 nm
10.3 22.1 dB, 26.8 dB
10.4 4.6 dB, 0.7 dB
29.2 dB
10.5 12.1 dB
10.7 31.8 dB
10.8 1.42 μm
10.9 0.41 μm
10.11 4.6 V, 33.6 V
10.12 6.7 V
10.13 9.0 mm
10.14 1.5 cm

References

[1] M. J. O'Mahony, 'Semiconductor laser optical amplifiers for use in future fiber systems', *J. of Lightwave Technol.*, **6**(4), pp. 531–544, 1988.

[2] P. Cockrane, 'Future directions in long haul fibre optic systems', *Br. Telecom Technol. J.*, **8**(2), pp. 5–17, 1990.

[3] J. M. Senior and S. D. Cusworth, 'Devices for wavelength multiplexing and demultiplexing', *IEE Proc., Pt. J.*, **136**(3), pp.183–202, 1989.

[4] M. J. O'Mahony, 'Optical amplification techniques using semiconductors and fibres', *Proc. SPIE Int. Soc. Opt. Eng. USA, Fibre Optics'89*, **1120**, pp. 43–44, 1989.

[5] R. Baker, 'Optical amplification', *Physics World*, pp. 41–44, March 1990.

[6] J. Buus and R. Plastow, 'Theoretical and experimental investigations of 1.3 μm Fabry–Perot amplifiers, *IEEE J. Quantum Electron.*, **QE-21**(6), pp. 614–618, 1985.

[7] G. Eisenstein, B. L. Johnson and G. Raybon, 'Travelling-wave optical amplifier at 1.3 μm', *Electron. Lett.*, **23**(19), pp. 1020–1022, 1987.

[8] G. N. Brown, 'A study of the static locking properties of injection locked laser amplifiers', *Br. Telecom. Technol. J.*, **4**(1), pp. 71–80, 1986.

[9] T. Saitoh and T. Mukai, 'Recent progress in semiconductor laser amplifiers', *J. of Lightwave Technol.*, **6**(11), pp. 1656–1664, 1988.

[10] M. J. O'Mahony, I. W. Marshall and H. J. Westlake, 'Semiconductor laser amplifiers for optical communication systems', *Br. Telecom Technol J.*, **5**(3), pp. 9–18, 1987.

[11] T. Saitoh, T. Mukai and O. Mikami, 'Theoretical analysis of antireflection coatings on laser diode facets', *J. of Lightwave Technol.*, **LT-3**(2), pp. 288–293, 1985.

[12] G. Eisentein and R. M. Jopson, 'Measurements of the gain spectrum of near-travelling-wave and Fabry–Perot semiconductor optical amplifiers at 1.5 μm', *Int. J. Electron.*, **60**(1), pp. 113–121, 1986.

[13] M. J. Adams, H. J. Westlake, M. J. O'Mahony and I. D. Henning, 'A comparison of active and passive bistability in semiconductors', *IEEE J. Quantum Electron.*, **QE-21**(9), pp. 1498–1504, 1985.

[14] I.D. Henning, M. J. Adams and J. V. Collins, 'Performance predictions from a new optical amplifier model', *IEEE J. Quantum Electron.*, **QE-21**, pp. 609–613, 1985.

[15] T. Mukai, Y. Yamamoto and T. Kimwa, 'Optical amplification by semiconductor lasers', in *Semiconductor and Semimetals*, **22-E**, R. K. Willardson and A. C. Beer (Eds.), Academic Press, pp. 265–319, 1985.

[16] N. A. Olsson, 'Lightwave systems with optical amplifiers', *J. Lightwave Technol.*, **7**(7), pp. 1071–1082, 1989.

[17] P. Urquhart, 'Review of rare earth doped fibre lasers and amplifiers', *IEE Proc., Pt. J.*, **135**(6), pp. 385–407, 1988.

[18] Y. Aoki, 'Properties of fiber Raman amplifiers and their applicability to digital optical communication systems', *J. of Lightwave Technol.*, **6**(7), pp. 1225–1239, 1988.

[19] G. N. Brown and D. M. Spirit, 'Gain saturation and laser linewidth effects in a Brillouin fibre amplifier', *15th European Conf. on Opt. Commun.* (*ECOC-89*), Gottenburg, pp. 70–73, September 1989.

[20] R. J. Mears, L. Reekie, I. M. Jauncey and D. N. Payne, 'Low-noise erbium-doped fibre amplifier at 1.54 μm', *Electron Lett.*, **23**, pp. 1026–1028, 1987.

[21] E. Desurvire, J. R. Simpson and P. C. Parker, 'High-gain erbium-doped travelling-wave fibre amplifier', *Opt. Lett.*, **12**, pp. 888–890, 1987.

[22] D. N. Payne and L. Reekie, 'Rare-earth-doped fibre lasers and amplifiers', *14th European Conf. on Opt. Commun.* (*ECOC'88*), pp. 49–53, September 1988.

[23] C. A. Millar, M. C. Brierley and P.W. France, 'Optical amplification in an erbium-doped fluorozirconate fibre between 1480 nm and 1600 nm', *IEE Conf. Publ.*, **292**, *Pt. 1.*, pp. 66–69, 1988.

[24] Y. Aoki, S. Kishida, H. Honomon, K. Washio and M. Sugimoto, 'Efficient backward and forward pumping CW Raman amplification for InGaAsP laser light in silica fibres', *Electron. Lett.*, **19**, pp 620–622, 1983.

[25] Y. Durteste, M. Monerie and P. Lamouler, 'Raman amplification in fluoride glass fibres', *Electron. Lett.*, **21**, p. 723, 1985.

[26] N. Edegawa, K. Mochizuki and Y. Imamoto, 'Simultaneous amplification of wavelength division multiplexed signals by highly efficient amplifier pumped by higher power semiconductor lasers', *Electron. Lett.*, **23**, pp. 556–557, 1987.

[27] M. L. Dakss and P. Melman, 'Amplified stimulated Raman scattering and gain in fiber Raman amplifiers', *J. Lightwave Technol.*, **LT-3**, pp. 806–813, 1985.

[28] R. J. Stolen, 'Polarization effects in fiber Raman and Brillouin lasers', *IEEE J. Quantum Electron.*, **QE-15**, pp. 1157–1160, 1979.

[29] C. G. Atkins, D. Cotter, D. W. Smith and R. Wyatt, 'Application of Brillouin amplification in coherent optical transmission', *Electron. Lett.*, **22**, pp. 556–557, 1986.

[30] A. R. Charplyvy and R. W. Tkach, 'Narrow-band tunable optical filter for channel selection in densely packed WDM systems', *Electron. Lett.*, **22**, pp. 1084–1085, 1986.

[31] R. G. Waarts and R. P. Braun, 'Crosstalk due to stimulated Brillouin scattering in monomode fiber', *Electron. Lett.*, **21**, p. 1114, 1985.

[32] E. J. Bachus, R. P. Braun, W. Eutin, E. Grossman, H. Foisel, K. Heims and B. Strebel, 'Coherent optical fibre subscriber line', *Electron. Lett.*, **21**, p. 1203, 1985.

[33] D. B. Anderson, *Optical and Electrooptical Information Processing*, pp. 221–234, MII Press, 1965.

[34] S. E. Miller, 'Integrated optics: an introduction', *Bell Syst. Tech. J.*, **48**(7), pp. 2059–2069, 1969.

[35] L. Levi, *Applied Optics*, Vol. 2, Chapter 13, John Wiley, 1980.

[36] H. Kogelnik and V. Ramaswamy, 'Scaling rules for thin-film optical waveguides', *Appl. Opt.*, **13**(8), pp. 1857–1862, 1974.

[37] A. Reisinger, 'Attenuation properties of optical waveguides with a metal boundary', *Appl. Phys. Lett.*, **23**(5), pp. 237–239, 1973.

[38] H. Kogelnik, 'Limits in integrated optics', *Proc. IEEE*, **69**(2), 232–238, 1981.

[39] T. Tamir (Ed.), *Integrated Optics* (2nd edn), Springer-Verlag, New York, 1979.

[40] J. Wilson and J. F. B. Hawkes, *Optoelectronics: An introduction* (2nd edn), Chapter 3, Prentice Hall, 1989.

[41] P. J. R. Laybourne and J. Lamb, 'Integrated optics: a tutorial review', *Radio Electron. Eng.* (*IERE J.*), **51**(7/8), pp. 397–413, 1981.

[42] B. K. Nayar and R. C. Booth, 'An introduction to integrated optics', *Br. Telecom Technol. J.*, **4**(4), pp. 5–15, 1986.

[43] R. Th. Kersten, 'Integrated optics for sensors', in J. Dakin and B. Culshaw (Eds.), *Optical Fiber Sensors: Principles and components*, Artech House, 1988.

[44] J. J. Veselka and S. K. Korotky, 'Optimization of Ti : $LiNbO_3$ optical waveguides and directional coupler switches for 1.5 μm wavelength, *IEEE J. Quantum Electron.*, **QE-22**, pp. 933–938, 1986.

[45] S. K. Korotky and R. C. Alferness, 'Waveguide electrooptic devices for optical fiber communication', in S. E. Miller and I. P. Kaminow (Eds.), *Optical Fiber Telecommunications II*, Academic Press, pp. 421–465, 1988.

[46] E. Kapron and R. Bhat, 'Low-loss GaAs/AlGaAs ridge waveguides grown by organmetallic vapour-phase epitaxy', *Tech. Dig. of Conf. on Lasers and Electrooptics*, Baltimore, USA, paper WQ3, 1987.

[47] H. Sasaki and I. Anderson, 'Theoretical and experimental studies on active Y-junctions in optical waveguides', *IEEE J. Quantum Electron*, **QE-14**, pp. 883–892, 1978.

[48] D. Marcuse, 'The coupling of degenerate modes in two parallel dielectric waveguides', *Bell Syst. Tech. J.*, **50**(6), pp. 1791–1816, 1971.

[49] A. Yariv, 'Coupled mode theory for guided wave optics', *IEEE J. Quantum Electron.*, **QE-9**, pp. 919–933, 1973.

[50] M. Papuchon, Y. Combemale, X. Mathieu, D. B. Ostrowsky, L. Reiber, A. M. Roy, B. Sejourne and M. Werner, 'Electrically switched optical directional coupler: COBRA', *Appl. Phys. Lett.*, **27**(5), pp. 289–291, 1975.

[51] S. F. Su, L. Jou and J. Lenart, 'A review on classification of optical switching systems', *IEEE Commun. Mag.*, **24**(5), pp. 50–55, 1986.

[52] S. D. Personick, 'Photonic switching: technology and applications', *IEEE Commun. Mag.*, **25**(5), pp. 5–8, 1987.

[53] T. Yasui and H. Goto, 'Overview of optical switching technologies in Japan', *IEEE Commun. Mag.*, pp. 10–15, 1987.

[54] M. Sakaguchi and K. Kaede, 'Optical switching device technologies', *IEEE Commun. Mag.*, **25**(5), pp. 27–32, 1987.

[55] W. A. Payne and H. S. Hinton, 'Design of lithium niobate based photonic switching systems', *IEEE Commun. Mag.*, pp. 37–41, 1987.

[56] S. D. Personick and W. O. Fleckenstein, 'Communications switching – from operators to photonics', *Proc. IEEE*, **75**(10), pp. 1380–1403, 1987.

[57] H. Goto, K. Nagashima, S. Suzuki, M. Kondo and Y. Ohta, 'Optical time-division digital switching: an experiment', *Topical Meeting on Opt. Fiber Commun.* (USA), MJ6, pp. 22–23, 1983.

[58] Y. Shimazu and S. Nishi, 'Wavelength-division optical switch using acousto-optic device', *Record of National Convention of IECEJ*, **510**, p. 2563, 1986.

[59] R. C. Alferness, 'Guided-wave devices for optical communication', *IEEE J. Quantum Electron.*, **QE-17**(6), pp. 946–959, 1981.

[60] D. B. Ostrowsky, 'Optical waveguide components' in M. J. Howes and D. V. Morgan (Eds.), *Optical Fibre Communications*, pp. 165–188, John Wiley, 1980.

[61] F. Auracher and R. Keil, 'Method for measuring the rf modulation characteristics of Mach-Zehnder-type modulators', *Appl. Phys. Lett.*, **36**, pp. 626–628, 1980.

[62] R. C. Alferness, C. H. Joyner, M. D. Divino and L. L. Buhl, 'InGaAsP/InP, waveguide grating filters for $\lambda = 1.5\ \mu$m', *Appl. Phys. Lett.*, **45**, pp. 1278–1280, 1984.

[63] R. V. Schmidt, D. C. Flanders, C. V. Shank and R. D. Standby, 'Narrow-band grating filters for thin-film optical waveguides', *Appl. Phys. Lett.*, **25**, pp. 651–652, 1974.

[64] A. Katzir, A. C. Livanos, J. B. Shellan and A. Yariv, 'Chirped gratings in integrated optics', *IEEE J. Quantum Electron.*, **QE-13**(4), pp. 296–304, 1977.

[65] R. V. Alferness and L. L. Buhl, 'Waveguide electro-optic polarization transformer', *Appl. Phys. Lett.*, **38**(9), pp. 655–657, 1981.

[66] W. A. Stallard, D. J. T. Heatley, R. A. Lobbett, A. R. Beaumont, D. J. Hunkin, B. E. Daymond-John, R. C. Booth and G. R. Hill, 'Electro-optic frequency translators and their application in coherent optical fibre systems', *Br. Telecom Technol. J.*, **4**(4), pp. 16–22, 1986.

[67] K. K. Wong, R. De La Rue and S. Wright, 'Electro-optic waveguide frequency translator in $LiNbO_3$ fabricated by proton exchange', *Opt. Lett.*, **7**(11), pp. 546–548, 1982.

[68] F. Heismann and R. Ulrich, 'Integrated optical frequency translator with stripe waveguide', *Appl. Phys. Lett.*, **45**(5), pp. 490–492, 1984.

[69] F. Auracher and R. Keil, 'Method for measuring the RF modulating characteristics of Mach–Zehnder-type modulators', *Appl. Phys. Lett.*, **36**(8), pp. 626–629, 1980.

[70] C. M. Gee and G. D. Thurmond, 'High speed integrated optic travelling wave modulator', *Proc. 2nd European Conf. on Integrated Optics* (Italy), pp. 118–121, October 1983.

[71] B. E. Daymond-John, A. R. Beaumont, W. A. Stallard and R. C. Booth, 'Lithium niobate electro-optic frequency translators for coherent optical systems', *Proc. Integrated Optical Circuit Engineering II* (USA), pp. 214–219, September 1985.

[72] W.A. Stallard, A. R. Beaumont and R. C. Booth, 'Integrated optic devices for coherent transmission', *J. Lightwave Technol.*, **LT-4**(7), pp. 852–857, 1986.

[73] M. Nakamura and T. Ozeki 'Optoelectronic integration and its impact on system application', *IEEE J. on Selected Areas in Commun.*, **SAC-4**(9), 1509–1514, 1986.

[74] H. Matsueda, S. Sasaki and M. Nakamura, 'GaAs optoelectronic integrated light sources', *J. of Lightwave Technol.*, **LT-1**(1), pp. 261–269, 1983.

[75] P. J. T. Mellor, 'Gallium arsenide integrated circuits for telecommunication systems', *Br. Telecom Technol. J.*, **5**(4), pp. 5–18, 1987.

[76] H. Nobuhara, T. Sanada, M. Kuno, M. Makinchi, T. Funjii and O. Wada, 'OEIC transmitter fabricated by planar integration process', *18th Conf. Solid State Devices and Materials*, Tokyo, Japan, pp. 185–188, August 1986.

[77] H. Soda, M. Furutsu, K. Sato, M. Matsuda and H. Ishikawa, '5 Git/s modulation characteristics of optical intensity modulator monolithically integrated with DFB laser', *Electron. Lett.*, **25**(5), pp. 334–355, 1989.

[78] H. Matsueda, 'AlGaAs OEIC transmitters', *J. of Lightwave Technol.*, **LT-5**(10), pp. 1382–1390, 1987.

[79] K. Kasahara, T. Terakado, A. Suzuki and S. Murata, 'Monolithically integrated high-speed light source using 1.3 μm wavelength DFB-DC-PBH laser', *J. of Lightwave Technol.*, **LT-4**(7), pp. 908–912, 1986.

[80] A. Suzuki, K. Kasahara and M. Shikada, 'InGaAsP/InP long wavelength optoelectronic integrated circuits (OEIC's) for high-speed optical fibre communication systems', *J. of Lightwave Technol.*, **LT-5**(10), pp. 1479–1487, 1987.

[81] H. Verriele, J. L. Lorriaux, P. Legry, J. P. Gouy, J. P. Vilcot and D. Decoster, 'GaAs monolithic integrated photoreceiver for 0.8 μm wavelength: association of Schottky photodiode and FET', *IEE Proc., Pt. J*, **135**(2), 1988.

[82] K. Kasahara, A. Suzuki, S. Fujita, Y. Inomoto, T. Terakado and M. Shikada, 'InGaAsP/InP long wavelength transmitter and receiver OEIC's for high speed optical transmission systems', in *12th European Conf. Optical Fiber Commun.*, (Spain), pp. 119–122, 1986.

[83] O. Wada, H. Nobuhara, T. Sanada, M. Kuno, M. Makiuchi, T. Fujii and T. Sakurai, 'Optoelectronic integrated four-channel transmitter array incorporating AlGaAs/GaAs quantum well lasers', *J. of Lightwave Technol.*, **7**(1), pp. 186–197, 1989.

[84] S. R. Forrest, 'Optoelectronic integrated circuits', *Proc. IEEE*, **75**(11), pp. 1488–1497, 1987.

[85] O. Wada, H. Hamaguchi, M. Makiuchi, T. Kumai, M. Ito, K. Nakai, T. Horimatsu and T. Sakurai, 'Monolithic four-channel photodiode/amplifier receiver array integrated on a GaAs substrate', *J. of Lightwave Technol.*, **LT-4**, pp. 1694–1703, 1986.

[86] K. Kobayashi, 'Integrated optical and electronic devices', in S. E. Miller and I. P. Kaminow (Eds.), *Optical Fiber Telecommunications II*, Academic Press, pp. 601–630, 1988.

[87] T. Suhara, Y. Hunda, H. Nishihara and J. Koyama, 'Monolithic integrated micrograting and photodiodes for wavelength demultiplexing', *Appl. Phys. Lett.*, **40**(2), pp. 120–122, 1982.

[88] T. Baba, Y. Kokubun and H. Watanabe, 'Monolithic integration of an ARROW-type demultiplexer and photodetector in the shorter wavelength region', *J. of Lightwave Technol.*, **8**(1), pp. 99–104, 1990.

[89] N. Bar-Chain, K. Y. Lau, I. Ury and A. Yariv, 'Gallium aluminium arsenide integrated optical repeater', *Proc. SPIE Int. Soc. Eng. (USA)*, **466**, pp. 65–68, 1984.

[90] S. Hata, M. Ikeda, S. Kondo and Y. Noguchi, 'PIN-FET-LD, integrated device in long wavelength region', *Tech. Dig. Domestic Conf., IECE*, Japan, p. 58, 1985.

[91] R. I. MacDonald, 'Optolectronic switching', *IEEE Commun. Mag.*, **25**(5), pp. 33–36, 1987.

[92] T. Iwama, Y. Oikawa, K. Yamaguchi, T. Horimatsu, M. Makiuchi and H. Hamaguchi, 'A 4 × 4 GaAs OEIC switch module', *Optical Fiber Commun. Conf.* (USA), **3**, paper WG3, 1987.

[93] W. J. Tomlinson and C. A. Brackett, 'Telecommunications applications of integrated optics and optoelectronics', *Proc. IEEE*, **75**(11), pp. 1512–1523, 1987.

[94] R. Cush and M. Goodwin, 'Optical logic devices', *National Electronic Review*, UK, pp. 51–55, 1987.

[95] P. W. Smith, I. P. Kaminow, P. J. Maloney and L. W. Stulz, 'Integrated bistable optical devices', *Appl. Phys. Lett.*, **33**(1), pp. 24–26, 1978.

[96] D. A. B. Miller, 'Multiple quantum well optical non-linearities'; bistability from increasing absorption and the self-electro-optic effect', *Phil. Trans. R. Soc. Lond.*, **A313**, pp. 239–248, 1984.

[97] D. Jager, F. Forsman and H. C. Zhai, 'Hybrid optical bistability based on increasing absorption in depletion layer of an Si Schottky SEED device', *Electron. Lett.*, **23**(10), 490–491, 1987.

[98] A. C. Walker, F. A. P. Tooley, M. E. Prise, J. G. H. Mathew, A. K. Kar, M. R. Taghizadeh and S. D. Smith, 'InSb devices: transphasors with high gain, bistable switches and sequential logic gates', *Phil. Trans. R. Soc. Lond.*, **A313**, pp. 249–256, 1984.

[99] S. D. Smith, J. G. H. Mathew, M. R. Taghizadeh, A. C. Walker, B. S. Wherret and A. Henry, 'Room temperature, visible wavelength optical bistability in ZnSe interference filters', *Optics Communications*, **51**, pp. 357–362, 1984.

[100] S. D. Smith, 'Optical bistability, phononic logic and optical computation', *Appl. Opt.*, **25**, pp.1550–1564, 1986.

[101] N. Peyghambarian, 'Recent advances in optical bistability', *Fiber and Integ. Optics*, **6**(2), pp. 117–123, 1987.

[102] P. Li Kam Wa, J. W. Sitch, N. J. Mason, J. S. Roberts and P. N. Robson, 'All-optical multiple-quantum well waveguide switch', *Electron. Lett.*, **21**, pp. 26–27, 1985.

[103] J. G. McInerney and D. M. Heffernan, 'Optical bistability in semiconductor injection lasers', *IEE Proc., Pt. J.*, **134**(1), pp. 41–50, 1987.

[104] Y. Odagiri, K. Komastu and S. Suzuki, 'Bistable laser diode memory for optical time-division switching applications', *Internat. Conf. on Lasers and Electro-optics*, Anaheim, California, USA, paper ThJ3, 1984.

[105] Y. Silberberg, 'All-optical repeater', *Opt. Lett.*, **11**(6), pp. 392–394, 1986.

[106] J. E. Midwinter, 'Light electronics, myth or reality', *IEE Proc., Pt. J.*, **132**, pp. 371–383, 1985.

[107] J. E. Midwinter, 'Novel approach to the design of optically activated wide band switching matrices', *IEE Proc., Pt. J.*, **134**(5), pp. 261–268, 1987.

[108] 'Research on nonlinear optical materials: an assessment', *Appl. Opt.*, **26**(2), pp. 211–234, 1987.

[109] J. A. Neff, 'Major initiatives for optical computing', *Opt. Eng.*, **26**(1), pp. 2–9, 1987.

[110] W. D. Hillis, *The Connection Machine*. The MIT Press, Cambridge, USA, 1985.

[111] B. K. Jenkins, P. Chavel, R. Forchheimer, A. A. Sawchuk and T. C. Strand', Architectural implications of a digital optical processor', *Appl. Opt.*, **23**(19), pp. 3465–3474, 1984.

[112] 'Optical computer: is concept becoming reality', *OE Reports, SPIE Int. Soc. Opt. Eng. (USA)*, **75**, pp. 1–2, March 1990.

11

Optical fiber systems 1: intensity modulation/direct detection

11.1 Introduction

The transfer of information in the form of light propagating within an optical fiber requires the successful implementation of an optical fiber communication system. This system, in common with all systems, is composed of a number of discrete components which are connected together in a manner that enables them to perform a desired task. Hence, to achieve reliable and secure communication using optical fibers it is essential that all the components within the transmission system are compatible so that their individual performances, as far as possible, enhance rather than degrade the overall system performance.

The principal components of a general optical fiber communication system for either digital or analog transmission are shown in the system block schematic of Figure 11.1. The transmit terminal equipment consists of an information encoder or signal shaping circuit preceding a modulation or electronic driver stage which operates the optical source. Light emitted from the source is launched into an optical fiber incorporated within a cable which constitutes the transmission medium. The light emerging from the far end of the transmission medium is converted back into an electrical signal by an optical detector positioned at the input of the receive terminal equipment. This electrical signal is then amplified prior to decoding or demodulation in order to obtain the information originally transmitted.

The operation and characteristics of the optical components of this general system have been discussed in some detail within the preceding chapters. However, to enable the successful incorporation of these components into an optical fiber communication system it is necessary to consider the interaction of one component with another, and then to evaluate the overall performance of the system. Furthermore, to optimize the system performance for a given application it is often helpful to offset a particular component characteristic by trading it off against the performance of another component, in order to provide a net gain within the overall system. The electronic components play an important role in this context, allowing the system designer further choices which, depending on the optical components utilized, can improve the system performance.

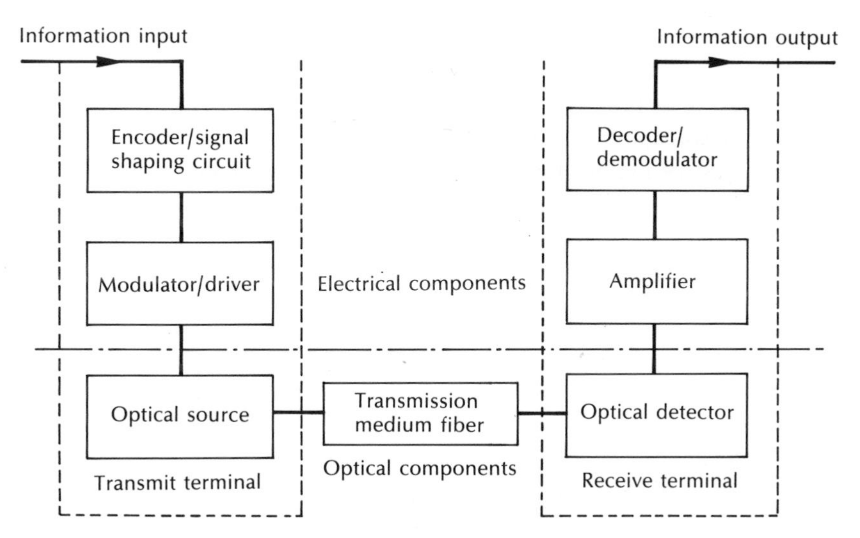

Figure 11.1 The principal components of an optical fiber communication system.

The purpose of this chapter is to bring together the important performance characteristics of the individual system elements, and to consider their interaction within optical fiber communication systems. In particular we concentrate on the major current implementations of optical fiber communication systems which employ some form of intensity modulation (IM) of the optical source, together with simple direct detection (DD) of the modulated optical signal at the receiver (see Chapter 9). Such IM/DD optical fiber systems are in widespread use within many application areas and do not employ the more sophisticated coherent detection techniques which are discussed in Chapter 12.

It is intended that this chapter provide guidance in relation to the various possible component configurations which may be utilized for different IM/DD system applications, whilst also giving an insight into system design and optimization. Hence, the optical components and the associated electronic circuits will be discussed prior to consideration of general system design procedures. Although the treatment is by no means exhaustive, it will indicate the various problems involved in system design and provide a description of the basic techniques and practices which may be adopted to enable successful system implementation.

We commence in Section 11.2 with a discussion of the optical transmitter circuit. This includes consideration of the source limitations prior to description of various LED and laser drive circuits for both digital and analog transmission. In Section 11.3 we present a similar discussion for the optical receiver including examples of preamplifier and main amplifier circuits. General IM/DD system design considerations are then dealt with in Section 11.4. This is followed by a detailed discussion of digital systems, commencing with an outline of the operating principles of pulse code modulated (PCM) systems in Section 11.5, before continuing to consider the various aspects of digital IM/DD optical fiber systems in Section 11.6.

Analog IM/DD optical fiber systems are dealt with in Section 11.7 where the various possible analog modulation techniques are described and analysed. Following, in Section 11.8, consideration is given to IM/DD optical fiber systems configured not simply as point-to-point links but as distribution systems. Then, in Section 11.9 the major multiplexing strategies (both digital and analog) which can be employed in IM/DD optical fiber systems are discussed in further detail to provide a greater understanding of the techniques which are being adopted to increase the information transmission capacity of both currently deployed and potential future systems. Finally, the deployment of optical amplifiers (see Chapter 10), particularly on long-haul IM/DD optical fiber links, is dealt with in Section 11.10 to complement the earlier consideration of multiplexing strategies and to enable the reader to appreciate the interaction of these advanced optical fiber communication techniques.

11.2 The optical transmitter circuit

The unique properties and characteristics of the injection laser and the light emitting diode (LED) which make them attractive sources for optical fiber communications were discussed in Chapters 6 and 7.

Although both device types exhibit a number of similarities in terms of their general performance and compatibility with optical fibers, striking differences exist between them in relation to both system application and transmitter design. It is useful to consider these differences, as well as the limitations of the two source types, prior to discussion of transmitter circuits for various applications.

11.2.1 Source limitations

11.2.1.1 Power

The electrical power required to operate both injection lasers and LEDs is generally similar, with typical current levels of between 20 and 300 mA (certain laser thresholds may be substantially higher than this – of the order of 1 to 2 A), and voltage drops across the terminals of 1.5 to 2.5 V. However, the optical output power against current characteristic for the two devices varies considerably, as indicated in Figure 11.2. The injection laser is a threshold device which must be operated in the region of stimulated emission (i.e. above the threshold) where continuous optical output power levels are typically in the range 1 to 10 mW.

Much of this light output may be coupled into an optical fiber because the isotropic distribution of the narrow linewidth, coherent radiation is relatively directional. In addition, the spatial coherence of the laser emission allows it to be readily focused by appropriate lenses within the numerical aperture of the fiber. Coupling efficiencies near 30% may be obtained by placing a fiber close to a laser mirror, and these can approach 80% with a suitable lens arrangement [Refs. 1 and

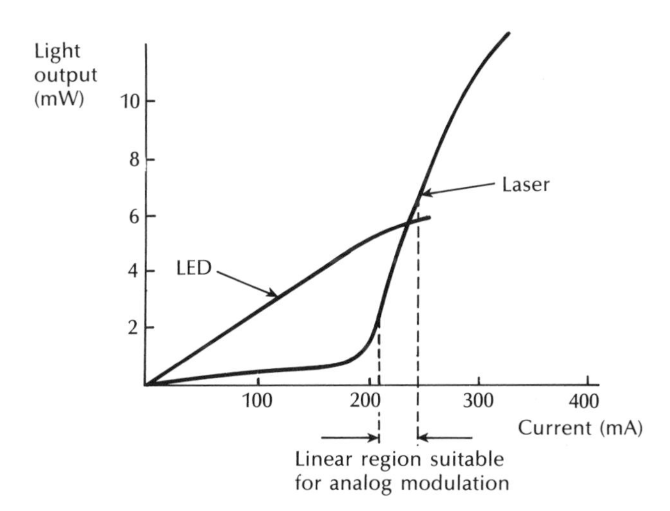

Figure 11.2 Light output (power) emitted into air as a function of d.c. drive current for a typical high radiance LED and for a typical injection laser. The curves exhibit nonlinearity at high currents due to junction heating.

2]. Therefore injection lasers are capable of launching between 0.5 and several milliwatts of optical power into a fiber.

LEDs are capable of similar optical output power levels to injection lasers depending on their structure and quantum efficiency, as indicated by the typical characteristic for a surface emitter shown in Figure 11.2. However, the spontaneous emission of radiation over a wide linewidth from the LED generally exhibits a Lambertian intensity distribution which gives poor coupling into optical fibers. Consequently only between 1% and perhaps 10% (using a good edge emitter) of the emitted optical power from an LED may be launched into a multimode fiber, even with appropriate lens coupling (see Section 7.3.6). These considerations translate into optical power levels from a few to several hundred microwatts launched into individual multimode fibers. Thus the optical power coupled into a fiber from an LED can be 10 to 20 dB below that obtained with a typical injection laser. The power advantage gained with the injection laser is a major factor in the choice of source, especially when considering a long-haul optical fiber link.

11.2.1.2 Linearity

Linearity of the optical output power against current characteristic is an important consideration with both the injection laser and LED. It is especially pertinent to the design of analog optical fiber communication systems where source nonlinearities may cause severe distortion of the transmitted signal. At first sight the LED may appear to be ideally suited to analog transmission as its output is approximately proportional to the drive current. However, most LEDs display some degree of nonlinearity in their optical output against current characteristic because of junction heating effects which may either prohibit their use or necessitate the incorporation of a linearizing circuit within the optical transmitter. Certain LEDs (e.g. etched-well surface emitters) do display good linearity, with distortion products (harmonic and intermodulation) between 35 and 45 dB below the signal level [Refs. 3 and 4].

An alternative approach to obtaining a linear source characteristic is to operate an injection laser in the light-generating region above its threshold, as indicated in Figure 11.2. This may prove more suitable for analog transmission than would the use of certain LEDs. However, gross nonlinearities due to mode instabilities may occur in this region. These are exhibited as kinks in the laser output characteristic (see Section 6.5.1). Therefore, many of the multimode injection lasers have a limited use for analog transmission without additional linearizing circuits within the transmitter, although some of the single-mode structures have demonstrated linearity suitable for most analog applications. Alternatively, digital transmission, especially that utilizing a binary (2 level) format, is far less sensitive to source nonlinearities and is therefore often preferred when using both injection lasers and LEDs.

11.2.1.3 Thermal

The thermal behaviour of both injection lasers and LEDs can limit their operation within the optical transmitter. However, as indicated in Section 6.7.1 the variation of injection laser threshold current with the device junction temperature can cause a major operating problem. Threshold currents of typical AlGaAs devices increase by approximately 1% per degree centigrade increase in junction temperature. Hence any significant increase in the junction temperature may cause loss of lasing and a subsequent dramatic reduction in the optical output power. This limitation cannot usually be overcome by simply cooling the device on a heat sink, but must be taken into account within the transmitter design, through the incorporation of optical feedback, in order to obtain a constant optical output power level from the device.

The optical output from an LED is also dependent on the device junction temperature, as indicated in Section 7.4.2. Most LEDs exhibit a decrease in optical output power following an increase in junction temperature, which is typically around -1% per degree centigrade. This thermal behaviour, however, although significant is not critical to the operation of the device due to its lack of threshold. Nevertheless, this temperature dependence can result in a variation in optical output power of several decibels over the temperature range 0 to 70 °C. It is therefore a factor within system design considerations which, if not tolerated, may be overcome by providing a circuit within the transmitter which adjusts the LED drive current with temperature.

11.2.1.4 Response

The speed of response of the two types of optical source is largely dictated by their respective radiative emission mechanisms. Spontaneous emission from the LED is dependent on the effective minority carrier lifetime in the semiconductor material (see Section 7.4.3). In heavily doped (10^{18} to 10^{19} cm^{-3}) gallium arsenide this is typically between 1 and 10 ns. However, the response of an optical fiber source to a current step input is often specified in terms of the 10 to 90% rise time, a parameter which is reciprocally related to the device frequency response (see Section 11.6.5). The rise time of the LED is at least twice the effective minority carrier lifetime, and often much longer because of junction and stray capacitance. Hence, the rise times for many available LEDs lie between 2 and 50 ns and give 3 dB bandwidths of around 7 to at best 175 MHz. Therefore, LEDs have tended to be restricted to lower bandwidth applications, although suitable drive circuits can maximize their bandwidth capabilities (i.e. reduce rise times).

Stimulated emission from injection lasers occurs over a much shorter period giving rise times of the order of 0.1 to 1 ns, thus allowing 3 dB bandwidths above 1 GHz. However injection laser performance is limited by the device switch-on delay (see Section 6.7.2). To achieve the highest speeds it is therefore necessary to minimize the switch-on delay. Transmitter circuits, which prebias the laser to just below or just above threshold in conjunction with high speed drive currents which take the device well above threshold, prove useful in the reduction of this limitation.

11.2.1.5 Spectral width

The finite spectral width of the optical source causes pulse broadening due to material dispersion on an optical fiber communication link. This results in a limitation on the bandwidth–length product which may be obtained using a particular source and fiber. The incoherent emission from an LED usually displays a spectral linewidth of between 20 and 50 nm (full width at half power (FWHP) points) when operating in the 0.8 to 0.9 μm wavelength range. This limits the bandwidth–length product with a silica fiber to around 100 and 160 MHz km at wavelengths of 0.8 and 0.9 μm respectively. Hence the overall system bandwidth for an optical fiber link over several kilometres may be restricted by material dispersion rather than the response time of the source.

The problem may be alleviated by working at a longer wavelength where the material dispersion in high silica fibers approaches zero (i.e. near 1.3 μm, see Section 3.9.1). In this region the source spectral width is far less critical and bandwidth–length products approaching 1 GHz km are feasible using LEDs.

Alternatively, an optical source with a narrow spectral linewidth may be utilized in place of the LED. The coherent emission from an injection laser generally has a linewidth of 1 nm or less (FWHP). Use of the injection laser greatly reduces the effect of material dispersion within the fiber, giving bandwidth–length products of 1 GHz km at 0.8 μm, and far higher at longer wavelengths. Hence, the requirement for a system operating at a particular bandwidth over a specific distance will influence both the choice of source and operating wavelength.

11.2.1.6 Nonzero extinction ratio

When the optical source is either intentionally prebiased-on during a 0 bit period, as indicated in Section 11.2.1.4, or simply not turned fully off, then some optical power will be emitted during the 0 pulse. This is particularly important with injection lasers when they are biased just below threshold and hence they launch spontaneous emissions into the fiber. In the case when optical power is incident on the photodetector during the 0 bit period, then the system is said to exhibit a nonzero extinction ratio.

The extinction ratio ε is usually defined as the ratio of the optical energy emitted in the 0 bit period to that emitted during the 1 bit period. For an ideal system $\varepsilon = 0$, and the extinction ratio therefore varies between this value and unity. It should be noted, however, that in some cases the extinction ratio is defined as the reciprocal of the above, which implies that it takes up values between 1 and ∞.

Typical values for the extinction ratio are between 0.05 and 0.10 and such nonzero ratios give rise to a noise penalty (often called an extinction ratio penalty) within the optical fiber communication system. The extinction ratio penalty can be evaluated and in practice it is often found to be in the range 1 to 2 dB. Any dark current present in the photodetector will also appear to increase the extinction ratio as it adds to the signal current in both the 0 and 1 bit periods. The greatest penalty occurs, however, in the case of quantum noise limited detection.

11.2.2 LED drive circuits

Although the LED is somewhat restricted in its range of possible applications in comparison with the more powerful, higher speed injection laser, it is generally far easier to operate. Therefore in this section we consider some of the circuit configurations that may be used to convert the information voltage signal at the transmitter into a modulation current suitable for an LED source. In this context it is useful to discuss circuits for digital and analog transmission independently.

11.2.2.1 Digital transmission

The operation of the LED for binary digital transmission requires the switching on and off of a current in the range of several tens to several hundreds of milliamperes. This must be performed at high speed in response to logic voltage levels at the driving circuit input. A common method of achieving this current switching operation for an LED is shown in Figure 11.3. The circuit illustrated uses a bipolar transistor switch operated in the common emitter mode. This single stage circuit provides current gain as well as giving only a small voltage drop across the switch when the transmitter is in saturation (i.e. when the collector–base junction is forward biased, the emitter to collector voltage V_{CE} (sat) is around 0.3 V).

The maximum current flow through the LED is limited by the resistor R_2 whilst independent bias to the device may be provided by the incorporation of resistor R_3. However, the switching speed of the common emitter configuration is limited by space charge and diffusion capacitance; thus bandwidth is traded for current gain. This may, to a certain extent, be compensated by overdriving (pre-emphasizing) the base current during the switch-on period. In the circuit shown in Figure 11.3 pre-emphasis is accomplished by use of the speed-up capacitor C.

Increased switching speed may be obtained from an LED without a pulse shaping or speed-up element by use of a low impedance driving circuit, whereby charging of the space charge and diffusion capacitance occurs as rapidly as possible. This may be achieved with the emitter follower drive circuit shown in Figure 11.4 [Ref.

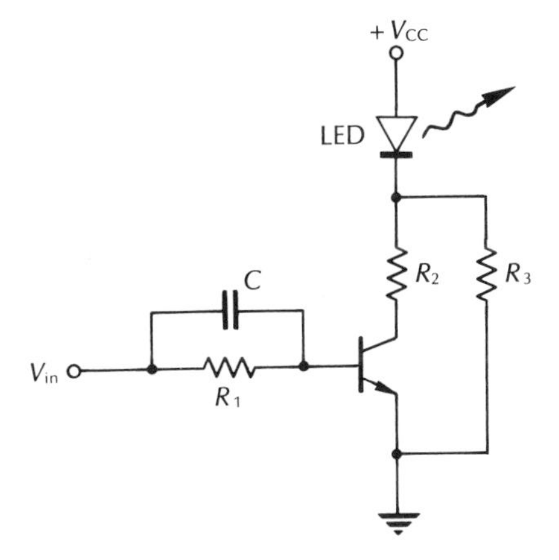

Figure 11.3 A simple drive circuit for binary digital transmission consisting of a common emitter saturating switch.

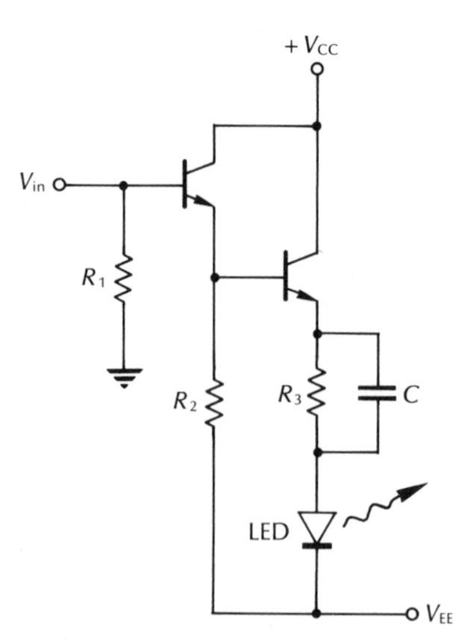

Figure 11.4 Low impedance drive circuit consisting of an emitter follower with compensating matching network [Ref. 5].

5]. The use of this configuration with a compensating matching network (R_3C) provides fast direct modulation of LEDs with relatively low drive power. A circuit, with optimum values for the matching network, is capable of giving optical rise times of 2.5 ns for LEDs with capacitance of 180 pF, thus allowing 100 Mbit s^{-1} operation [Ref. 6].

Another type of low impedance driver is the shunt configuration shown in Figure 11.5. The switching transistor in this circuit is placed in parallel with the LED, providing a low impedance path for switching off the LED by shunting current around it. The switch-on performance of the circuit is determined by the combination of resistor R and the LED capacitance. Stored space charge may be removed by slightly reverse biasing the LED when the device is switched off. This may be achieved by placing the transistor emitter potential V_{EE} below ground. In this case a Schottky clamp (shown dotted) may be incorporated to limit the extent of the reverse bias without introducing any extra minority carrier stored charge into the circuit.

A frequent requirement for digital transmission is the interfacing of the LED by drive circuit with a common logic family, as illustrated in the block schematic of Figure 11.6(a). In this case the logic interface must be considered along with possible drive circuits. Compatibility with TTL may be achieved by use of commercial integrated circuits, as shown in Figure 11.6(b) and (c). The configuration shown in Figure 11.6(b) uses a Texas Instruments' 74S140 line driver which provides a drive current of around 60 mA to the LED when R_1 is 50 Ω. Moreover, the package contains two sections which may be connected in parallel

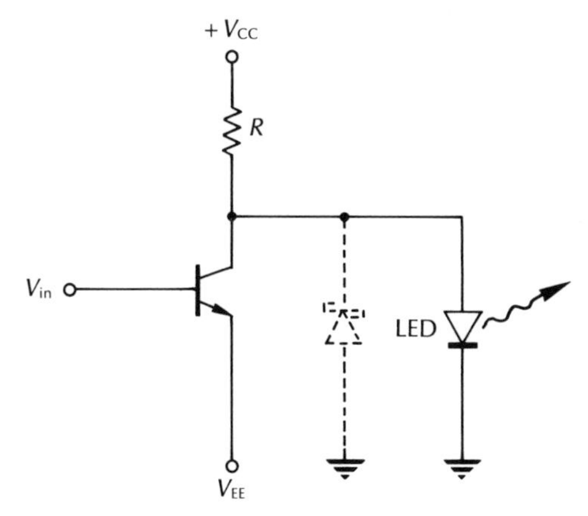

Figure 11.5 Low impedance drive circuit consisting of a simple shunt configuration.

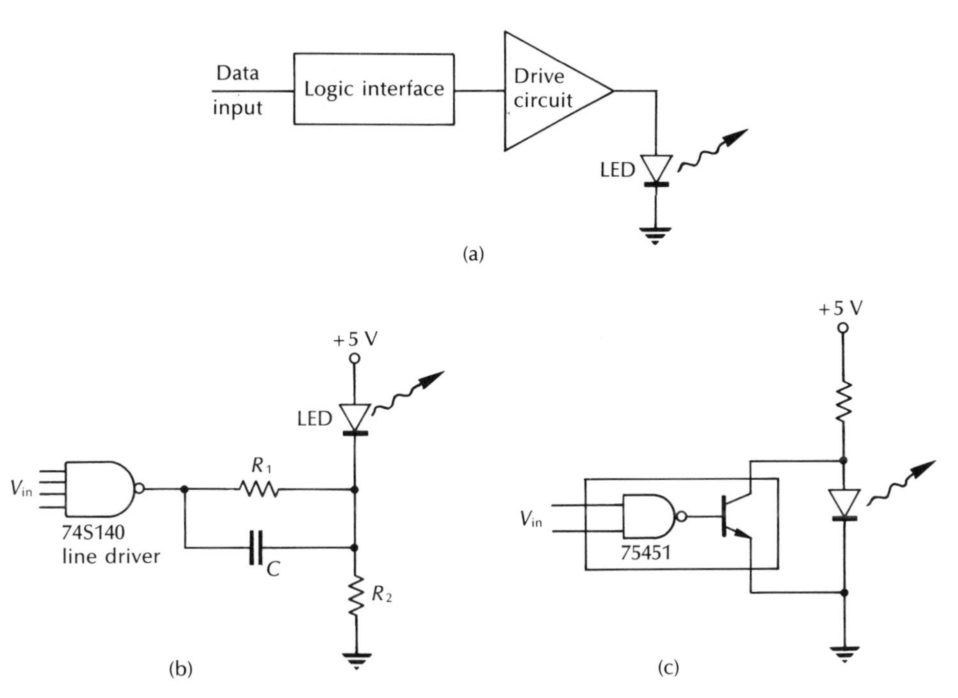

Figure 11.6 Logic interfacing for digital transmission: (a) block schematic showing the interfacing of the LED drive circuit with logic input levels; (b) a simple TTL compatible LED drive circuit employing a Texas Instruments' 74S140 line driver [Ref. 7]; (c) a TTL shunt drive circuit using a commercially available integrated circuit [Ref. 7].

in order to obtain a drive current of 120 mA. The incorporation of a suitable speed-up capacitor (e.g. $C = 47$ pF) gives optical rise times of around 5 ns when using LEDs with between 150 and 200 pF capacitance [Ref. 7]. Figure 11.6(c) illustrates the shunt configuration using a standard TTL 75451 integrated circuit. The rise time of this shunt circuit may be improved through maintenance of charge on the LED capacitance by placing a resistor between the shunt switch collector and the LED [Ref. 7].

An alternative important drive circuit configuration is the emitter coupled circuit shown in Figure 11.7 [Ref. 7]. The LED acts as a load in one collector so that the circuit provides current gain and hence a drive current for the device. Thus the circuit resembles a linear differential amplifier, but it is operated outside the linear range and in the switching mode. Fast switching speeds may be obtained due to the configuration's nonsaturating characteristic which avoids switch-off time degradations caused by stored charge accumulation on the transistor base region. The lack of saturation also minimizes the base drive requirements for the transistors, thus preserving their small signal current gain. The emitter coupled driver configuration shown in Figure 11.7 is compatible with commercial emitter coupled logic (ECL). However, to achieve this compatibility the circuit includes two level shifting transistors which give ECL levels (high -0.8 V, low -1.8 V) when the positive terminal of the LED is at earth potential. The response of this circuit is specified [Ref. 7] at up to 50 Mbit s^{-1}, with a possible extension to 300 Mbit s^{-1} when using a faster ECL logic family and high speed transistors. The emitter coupled drive circuit configuration may also be interfaced with other logic families, and a TTL compatible design is discussed in Ref. 8.

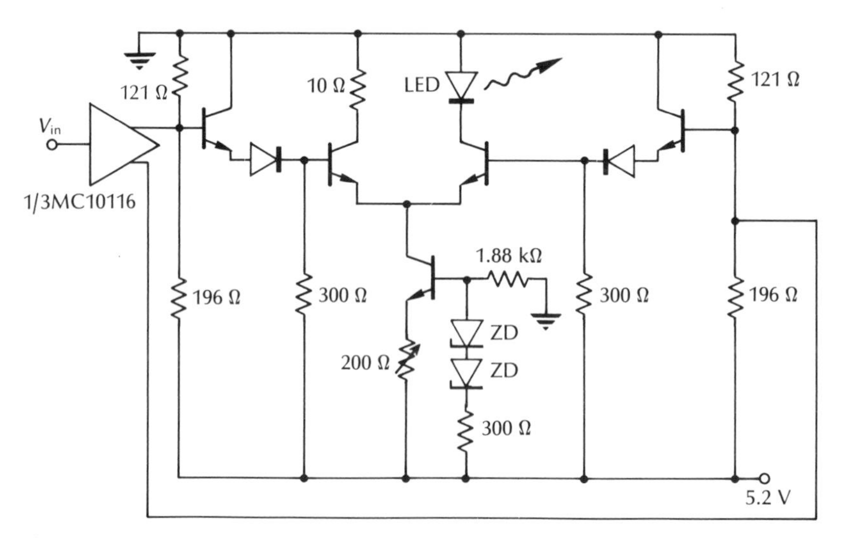

Figure 11.7 An emitter coupled drive circuit which is compatible with ECL [Ref. 7].

11.2.2.2 Analog transmission

For analog transmission the drive circuit must cause the light output from an LED source to follow accurately a time-varying input voltage waveform in both amplitude and phase. Therefore, as indicated previously, it is important that the LED output power responds linearly to the input voltage or current. Unfortunately, this is not always the case because of inherent nonlinearities within LEDs which create distortion products on the signal. Thus the LED itself tends to limit the performance of analog transmission systems unless suitable compensation is incorporated into the drive circuit. However, unless extremely low distortion levels are required, simple transistor drive circuits may be utilized.

Two possible high speed drive circuit configurations are illustrated in Figure 11.8. Figure 11.8(a) shows a driver consisting of a common emitter transconductance amplifier which converts an input base voltage into a collector current. The circuit is biased for a class A mode of operation with the quiescent collector current about half the peak value. A similar transconductance configuration which utilizes a Darlington transistor pair in order to reduce the impedance of the source is shown in Figure 11.8(b). A circuit of this type has been used to drive high radiance LEDs at frequencies of 70 MHz [Ref. 9].

Another simple drive circuit configuration is shown in Figure 11.9. It consists of a differential amplifier operated over its linear region which directly modulates the LED. The LED operating point is controlled by a reference voltage V_{ref} whilst the current generator provided by the transistor T_3 feeding the differential stage (T_1 and T_2) limits the maximum current through the device. The transimpedance of the driver is reduced through current series feedback provided by the two resistors R_1 and R_2 which are normally assigned equal values. Furthermore, variation between these feedback resistors can be used to compensate for the transfer function of both the drive circuit and the LED.

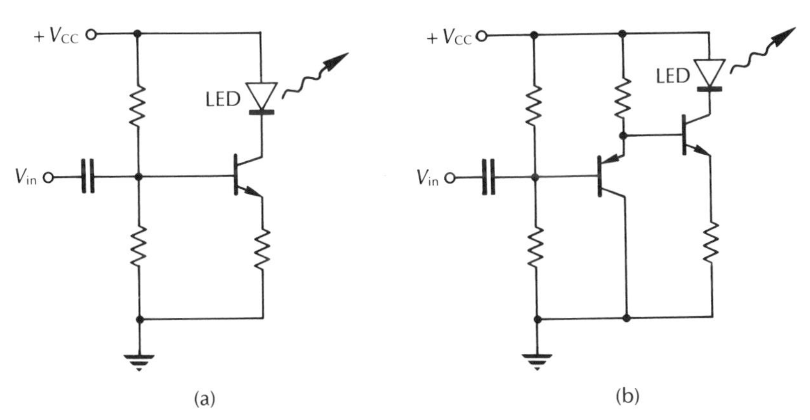

Figure 11.8 Transconductance drive circuits for analog transmission: (a) common emitter configuration; (b) Darlington transistor pair.

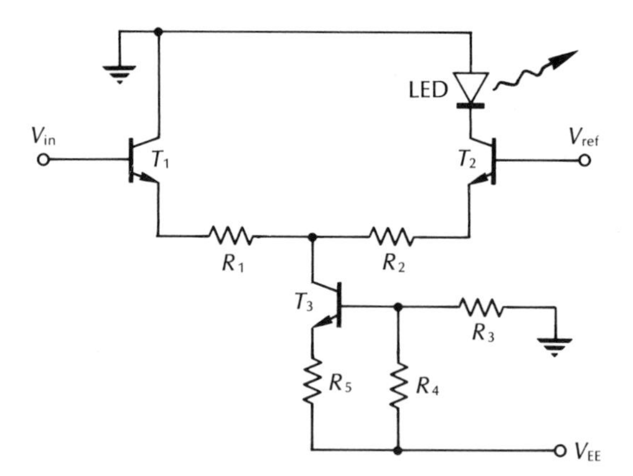

Figure 11.9 A differential amplifier drive circuit.

Although in many communication applications where a single analog signal is transmitted certain levels of amplitude and phase distortion can be tolerated, this is not the case in frequency multiplexed systems (see Section 11.4.2) where a high degree of linearity is required in order to minimize interference between individual channels caused by the generation of intermodulation products. Also, baseband video transmission of TV signals requires the maintenance of extremely low levels of amplitude and phase distortion. For such applications the simple drive circuits described previously are inadequate without some form of linearization to compensate for both LED and drive circuit nonlinearities. A number of techniques have been reported [Ref. 10], some of which are illustrated in Figure 11.10. Figure 11.10(a) shows the complementary distortion technique [Ref. 11] where additional nonlinear devices are included in the system. It may take the form of predistortion compensation (before the source drive circuit) or postdistortion compensation (after the receiver). This approach has been shown [Ref. 12] to reduce harmonic distortion by up to 20 dB over a limited range of modulation amplitudes.

In the negative feedback compensation technique shown in Figure 11.10(b), the LED is included in the linearization scheme. The optical output is detected and compared with the input waveform, the amount of compensation being dependent on the gain of the feedback loop. Although the technique is straightforward, large bandwidth requirements (i.e. video) can cause problems at high frequencies [Ref. 13].

The technique shown in Figure 11.10(c) employs phase shift modulation for selective harmonic compensation using a pair of LEDs with similar characteristics [Ref. 14]. The input signal is divided into equal parts which are phase shifted with respect to each other. These signals then modulate the two LEDs giving a cancellation of the second and third harmonic with a 90° and 60° phase shift respectively. However, although there is a high degree of distortion cancellation, both harmonics cannot be reduced simultaneously.

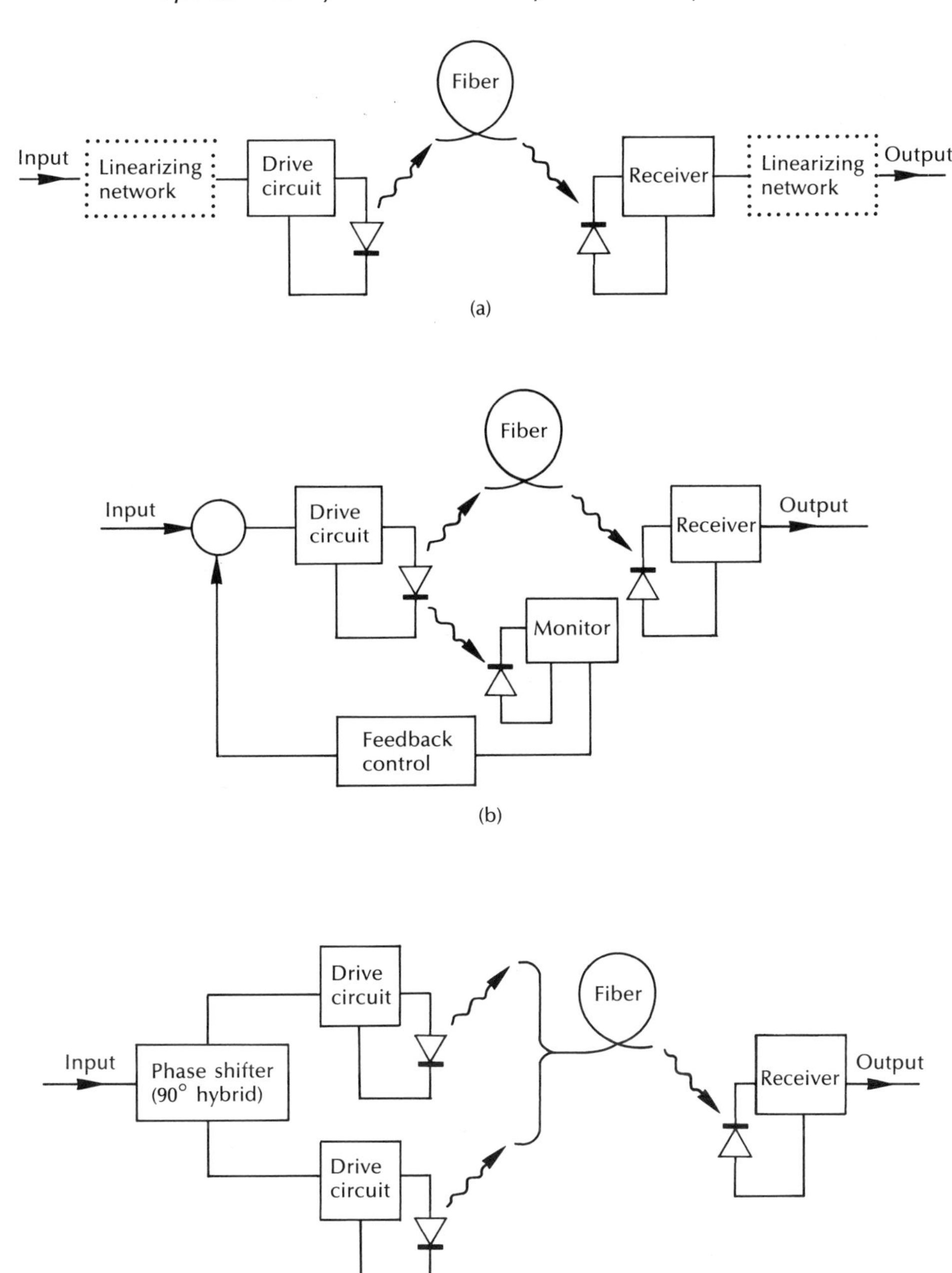

Figure 11.10 Block schematics of some linearization methods for LED drive circuits: (a) complementary distortion technique; (b) negative feedback compensation technique; (c) selective harmonic compensation technique.

Other linearization techniques include cascade compensation [Ref. 15], feed-forward compensation [Ref. 16] and quasi-feedforward compensation [Refs. 17 and 18].

11.2.3 Laser drive circuits

A number of configurations described for use as LED drive circuits for both digital and analog transmission may be adapted for injection laser applications with only minor changes. The laser, being a threshold device, has somewhat different drive current requirements from the LED. For instance, when digital transmission is considered, the laser is usually given a substantial applied bias, often referred to as prebias, in the off state. Reasons for biasing the laser near but below threshold in the off state are as follows:

1. It reduces the switch-on delay and minimizes any relaxation oscillations.
2. It allows easy compensation for changes in ambient temperature and device ageing.
3. It reduces the junction heating caused by the digital drive current since the on and off currents are not widely different for most lasers.

Although biasing near threshold causes spontaneous emission of light in the off state, this is not normally a problem for digital transmission because the stimulated emission in the on state is generally greater by, at least, a factor of 10.

A simple laser drive circuit for digital transmission is shown in Figure 11.11. This circuit is a shunt driver utilizing a field effect transistor (FET) to provide high speed laser operation. Sufficient voltage is maintained in series with the laser using the resistor R_2 and the compensating capacitor C such that the FET is biased into its

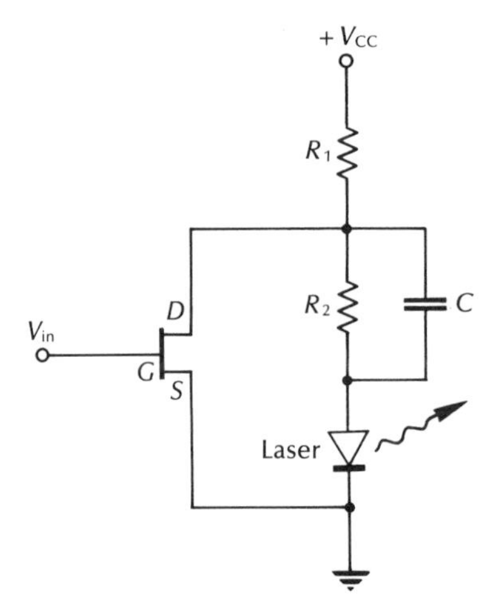

Figure 11.11 A shunt drive circuit for use with an injection laser.

active or pinch-off region. Hence for a particular input voltage V_{in} (i.e. V_{GS}) a specific amount of the total current flowing through R_1 is diverted around the laser leaving the balance of the current to flow through R_2 and provide the off state for the device. Using stable gallium arsenide MESFETs (see Section 9.5.1) the circuit shown in Figure 11.11 has modulated lasers at rates in excess of 1 Gbit s^{-1} [Ref. 19].

An alternative high speed laser drive circuit employing bipolar transistors is shown in Figure 11.12 [Ref. 20]. This circuit configuration, again for digital transmission, consists of two differential amplifiers connected in parallel. The input stage, which is ECL compatible, exhibits a 50 Ω input impedance by use of an emitter follower T_1 and a 50 Ω resistor in parallel with the input. The transistor T_2 acts as a current source with the zener diode ZD adjusting the signal level for ECL operation. The two differential amplifiers provide sufficient modulation current amplitude for the laser under the control of a d.c. control current I_E through the two emitter resistors R_{E1} and R_{E2}; I_E is provided by an optical feedback control circuit, to be discussed shortly. Finally, a prebias current is applied to the laser from a separate current source. This circuit when utilizing microwave transistors was operated with a return to zero digital format (see Section 3.8) at 1 Gbit s^{-1} [Ref. 20].

A major difference between the drive circuits of Figures 11.11 and 11.12 is the absence and use, respectively, of feedback control for adjustment of the laser output level. For this reason it is unlikely that the shunt drive circuit of Figure 11.11 would be used for a system application. Some form of feedback control is generally

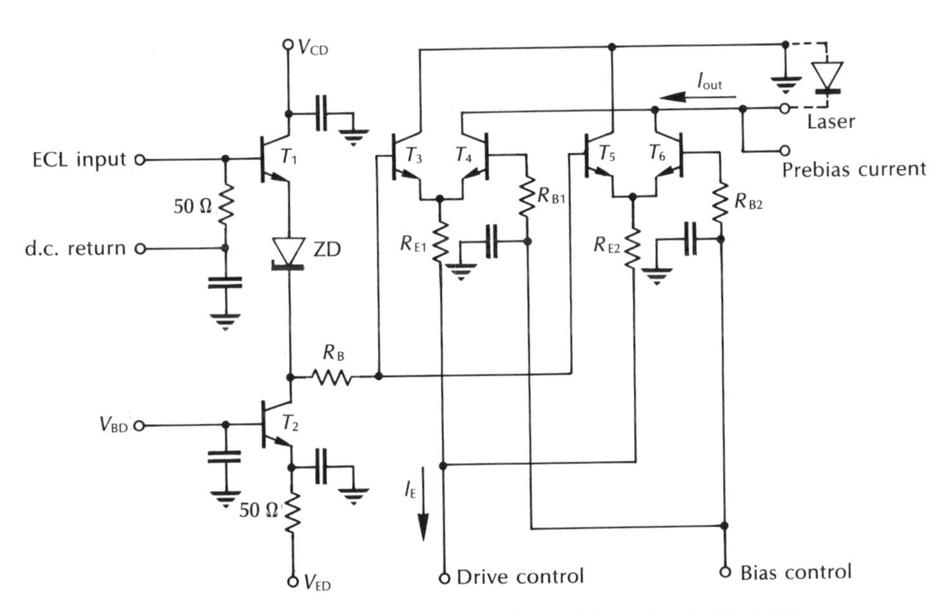

Figure 11.12 An ECL compatible high speed laser drive circuit [Ref. 20].

required to ensure continuous laser operation because the device lasing threshold is a sensitive function of temperature. Also, the threshold level tends to increase as the laser ages following an increase in internal device losses. Although lasers may be cooled to compensate for temperature variations, ageing is not so easily accommodated by the same process. However, both problems may be overcome through control of the laser bias using a feedback technique. This may be achieved using low speed feedback circuits which adjust the generally static bias current when necessary. For this purpose it is usually found necessary to monitor the light output from the laser in order to keep some aspect constant.

Several strategies of varying complexity are available to provide automatic output level control for the laser. The simplest and perhaps most common form of laser drive circuit incorporating optical feedback is the mean power control circuit shown in Figure 11.13. Often the monitor detector consists of a cheap, slow photodiode positioned next to the rear face of the laser package, as indicated in Figure 11.13. Alternatively, an optical coupler at the fiber input can be used to direct some of the radiation emitted from the laser into the monitor photodiode. The detected signal is integrated and compared with a reference by an operational amplifier which is used to servo-control the d.c. bias applied to the laser. Thus the mean optical power is maintained constant by varying the threshold current level. This technique is suitable for both digital and analog transmission.

An alternative control method for digital systems which offers accurate threshold tracking and very little device dependence is the switch-on delay technique illustrated in Figure 11.14 [Ref. 24]. This circuit monitors the switch-on delay of an optical pulse in order to control the laser bias current. The switch-on delay is measured for a zero level set below threshold and the feedback is set to a constant fixed delay to control it. Hence, the circuit provides a reference signal proportional to the delay period. This signal is used to control the bias level. The technique

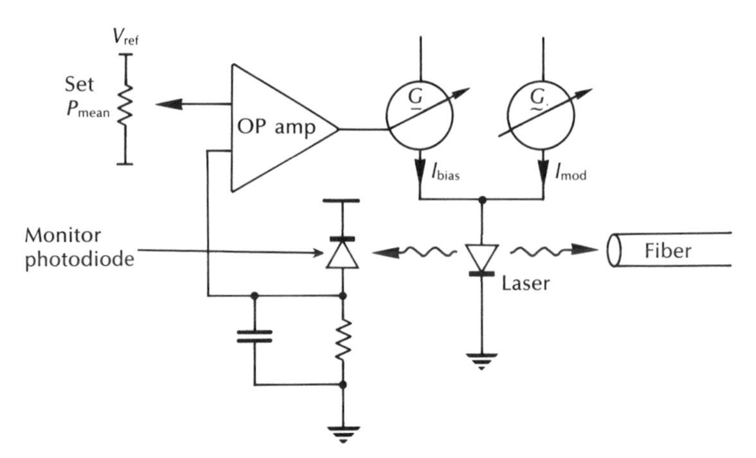

Figure 11.13 Mean power feedback circuit for control of the laser bias current.

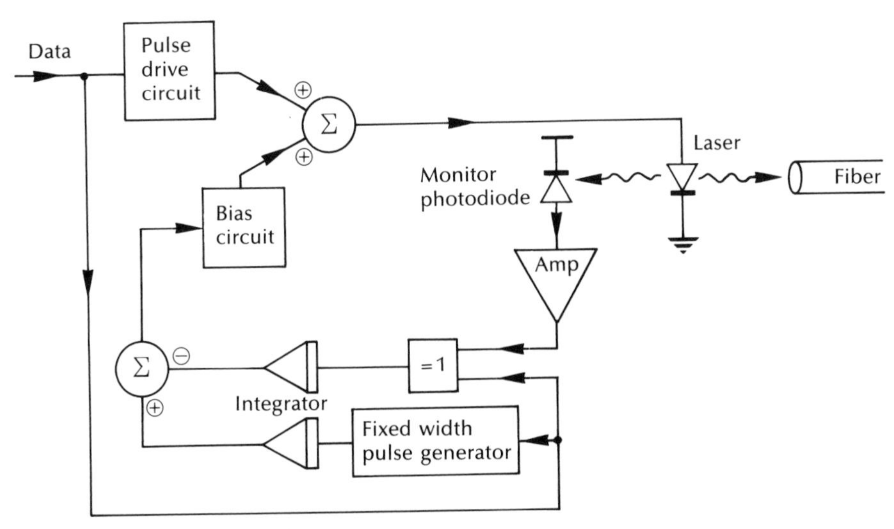

Figure 11.14 Switch-on delay feedback laser control circuit [Ref. 24].

requires a fast monitor photodiode as well as a wideband amplifier to allow measurement of the small delay periods. It is also essential that the zero level is set below the lasing threshold because the feedback loop will only stabilize for a finite delay (i.e. the delay falls to zero at the threshold).

A major disadvantage, however, with just controlling the laser bias current is that it does not compensate for variations in the laser slope efficiency. The modulation current for the device is preset and does not take into account any slope changes with temperature and ageing. In order to compensate for such changes, the a.c. and d.c. components of the monitored light output must be processed independently. This is especially important in the case of high bit rate digital systems where control of the on and off levels as well as the light level is required. A circuit which utilizes both a.c. and d.c. information in the laser output control the device drive current and bias independently is shown in Figure 11.15 [Ref. 20]. The electrical output from the monitor photodiode is fed into a low drift d.c. amplifier $A1$ and into a wideband amplifier $A2$. Therefore the mean value of the laser output power P_e(ave) is proportional to the output from $A1$ whilst the a.c. content of the monitoring signal is peak detected after the amplifier $A2$. The peak signals correspond to the maximum P_e(max) and the minimum P_e(min) laser output powers within a certain time interval. The difference signal proportional to (P_e(max) – P_e(min)) is acquired in $A3$ and compared with a drive reference voltage in order to control the current output from $A4$ and, consequently, the laser drive current. In this way the modulation amplitude of the laser is controlled. Control of the laser bias current is achieved from the difference between the output signal of $A1$ (P_e(ave)) and P_e(min) which is acquired in $A5$. The output voltage of $A5$, which is proportional to P_e(min), is compared with a bias reference voltage in $A6$

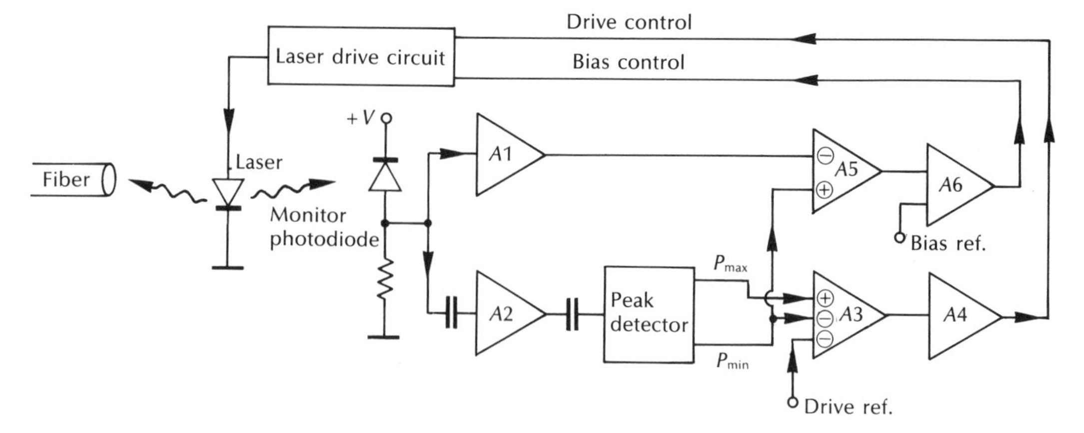

Figure 11.15 A laser feedback control circuit which uses a.c. and d.c. information in the monitored light output to control the laser drive and bias currents independently [Ref. 20].

which supplies a current output to control the laser d.c. bias. This feedback control circuit was designed for use with the laser drive circuit shown in Figure 11.12 to give digital operation at bit rates in the gigahertz range.

11.3 The optical receiver circuit

The noise performance for optical fiber receivers incorporating both major detector types (the p–i–n and avalanche photodiode) was discussed in Chapter 9. Receiver noise is of great importance within optical fiber communications as it is the factor which limits receiver sensitivity and therefore can dictate the overall system design. It was necessary within the analysis given in Chapter 9 to consider noise generated by electronic amplification (i.e. within the preamplifier) of the low level signal as well as the noise sources associated with the optical detector. Also, the possible strategies for the configuration of the preamplifier were considered (see Section 9.4) as a guide to optimization of the receiver noise performance for a particular application. In this section we extend the discussion to consider different possible circuit arrangements which may be implemented to achieve low noise preamplification, as well as further amplification (main amplification) and processing of the detected optical signal.

A block schematic of an optical fiber receiver is shown in Figure 11.16. Following the linear conversion of the received optical signal into an electrical current at the detector, it is amplified to obtain a suitable signal level. Initial amplification is performed in the preamplifier circuit where it is essential that additional noise is kept to a minimum in order to avoid corruption of the received signal. As noise sources within the preamplifier may be dominant, its configuration and design are

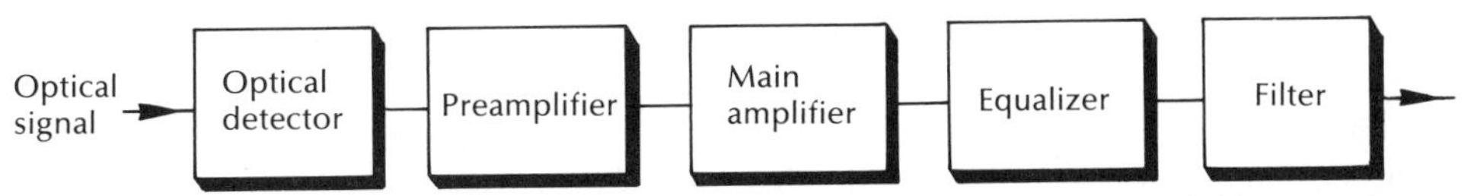

Figure 11.16 Block schematic showing the major elements of an optical fiber receiver.

major factors in determining the receiver sensitivity. The main amplifier provides additional low noise amplification of the signal to give an increased signal level for the following circuits.

Although optical detectors are very linear devices and do not themselves introduce significant distortion on to the signal, other components within the optical fiber communication system may exhibit nonlinear behaviour. For instance, the received optical signal may be distorted due to the dispersive mechanisms, within the optical fiber. Alternatively, the transfer function of the preamplifier–main amplifier combination may be such that the input signal becomes distorted (especially the case with the high impedance front end preamplifier). Hence, to compensate for this distortion and to provide a suitable signal shape for the filter, an equalizer is often included in the receiver. It may precede or follow the main amplifier, or may be incorporated in the functions of the amplifier and filter. In Figure 11.16 the equalizer is shown as a separate element following the amplifier and preceding the filter.

The function of the final element in the receiver, the filter, is to maximize the received signal to noise ratio whilst preserving the essential features of the signal. In digital systems the function of the filter is primarily to reduce intersymbol interference, whereas in analog systems it is generally required to hold the amplitude and phase response of the received signal within certain limits. The filter is also designed to reduce the noise bandwidth as well as inband noise levels.

Finally, the general receiver consisting of the elements depicted in Figure 11.16 is often referred to as a linear channel because all operations on the received optical signal may be considered to be mathematically linear.

11.3.1 The preamplifier

The choice of circuit configuration for the preamplifier is largely dependent upon the system application. Bipolar or field effect transistors (FETs) can be operated in three useful connections. These are the common emitter or source, the common base or gate, and the emitter or source follower for the bipolar and field effect transistors respectively. Each connection has characteristics which will contribute to a particular preamplifier configuration. It is therefore useful to discuss the three basic preamplifier structures (low impedance, high impedance and transimpedance front end) and indicate possible choices of transistor connection. In this context the discussion is independent of the type of optical detector utilized. However, it must be noted that there are a number of significant differences in the performance characteristics between the *p–i–n* and avalanche photodiode (see Chapter 8) which must be considered within the overall design of the receiver.

The simplest preamplifier structure is the low input impedance voltage amplifier. This design is usually implemented using a bipolar transistor configuration because of the high input impedance of FETs. The common emitter and the grounded emitter (without an emitter resistor) amplifier shown in Figure 11.17 are favoured connections, as they may be designed with reasonably low input impedance and therefore give operation over a moderate bandwidth without the need for equalization. However, this is achieved at the expense of increased thermal noise due to the low effective load resistance presented to the detector. Nevertheless, it is possible to reduce the thermal noise contribution of this preamplifier by choosing a transistor with characteristics which give a high current gain at a low emitter current in order to maintain the bandwidth of the stage. Also, an inductance may be inserted at the collector to provide partial equalization for any integration performed by the stage. The alternative connection giving very low input impedance is the common base circuit. Unfortunately, this configuration has an input impedance which gives insufficient power gain when connected to the high impedance of the optical detector.

The preferred preamplifier configurations for low noise operation use either a high impedance integrating front end or a transimpedance amplifier (see Sections 9.4.2 and 9.4.3). Careful design employing these circuit structures can facilitate high gain coupled with low noise performance and therefore enhanced receiver sensitivity. Although the bipolar transistor incorporated in the emitter follower circuit may be used to realize a high impedance front end amplifier, the FET is generally employed for this purpose because of its low noise operation. It was indicated in Section 9.5 that the grounded source FET connection was a useful circuit to provide a high impedance front end amplifier. The same configuration with a source resistor (common source connection) shown in Figure 11.18 provides

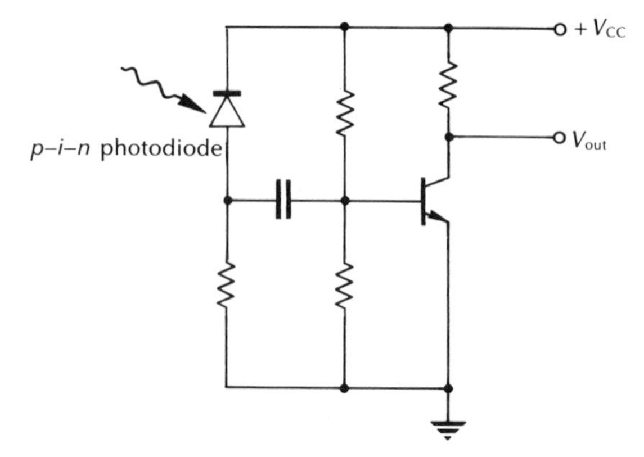

Figure 11.17 A *p–i–n* photodiode with a grounded emitter, low input impedance voltage preamplifier.

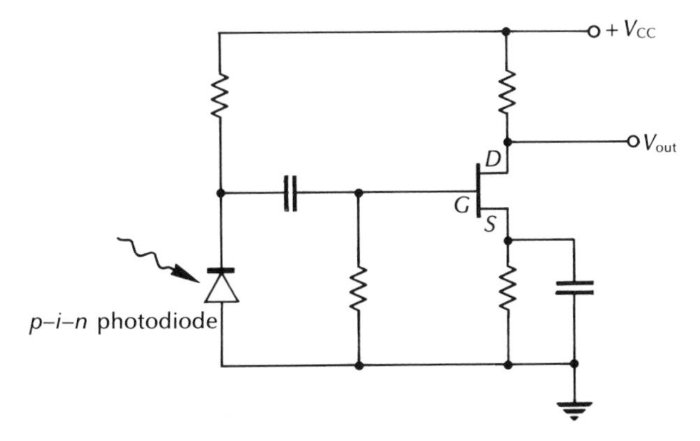

Figure 11.18 An FET common source preamplifier configuration which provides high input impedance for the *p–i–n* photodiode.

a similar high input impedance and may also be used (often both configurations are referred to as the common source connection). When operating in this mode the FET power gain and output impedance are both high, which tends to minimize any noise contributions from the following stages. It is especially the case when the voltage gain of the common source stage is minimized in order to reduce the Miller capacitance [Ref. 27] associated with the gate to drain capacitance of the FET. This may be achieved by following the common source stage with a stage having a low input impedance.

Two configurations which provide a low input impedance stage are shown in Figure 11.19. Figure 11.19(a) shows the grounded source FET followed by a bipolar transistor in the common emitter connection with shunt feedback over the stage. Another favoured configuration to reduce Miller capacitance in the first stage FET is shown in Figure 11.19(b). In this case the second stage consists of a bipolar transistor in the common base configuration which, with the initial grounded source FET, forms the cascode configuration.

The high impedance front end structure provides a very low noise preamplifier design but suffers from two major drawbacks. The first is with regard to equalization, which must generally be tailored to the amplifier in order to compensate for distortion introduced on to the signal. Secondly, the high input–impedance approach suffers from a lack of dynamic range which occurs because the charge on the input capacitance from the low frequency components in the signal builds up over a period of time, causing premature saturation of the amplifier at high input signal levels. Therefore, although the circuits shown in Figure 11.19 are examples of possible high impedance integrating front end amplifier configurations, similar connections may be employed with overall feedback (to the first stage) to obtain a transimpedance preamplifier.

The transimpedance or shunt feedback amplifier finds wide application in preamplifier design for optical fiber communications. This front end structure

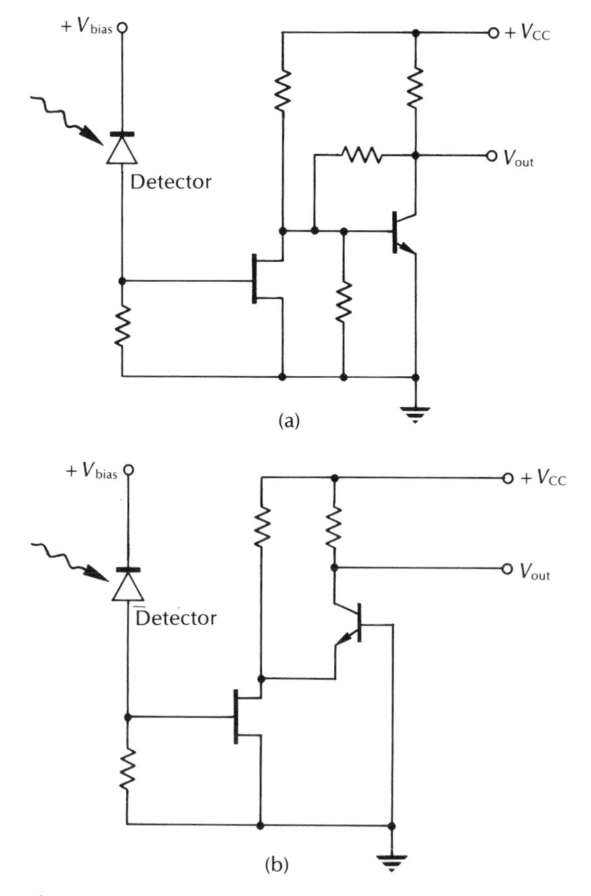

Figure 11.19 High input impedance preamplifier configurations: (a) grounded source FET followed by common emitter connection with shunt feedback; (b) cascode connection. The separate bias voltage indicates the use of either *p–i–n* or avalanche photodiode.

which acts as a current–voltage converter gives low noise performance without the severe limitations on bandwidth imposed by the high input impedance front end design. It also provides greater dynamic range than the high input impedance structure. However, in practice the noise performance of the transimpedance amplifier is not quite as good as that achieved with the high impedance structure due to the noise contribution from the feedback resistor (see Section 9.4.3). Nevertheless, the transimpedance design incorporating a large value of feedback resistor can achieve a noise performance which approaches that of the high impedance front end.

Two example of transimpedance front end configurations are shown in Figure 11.20. Figure 11.20(a) illustrates a bipolar transistor structure consisting of a

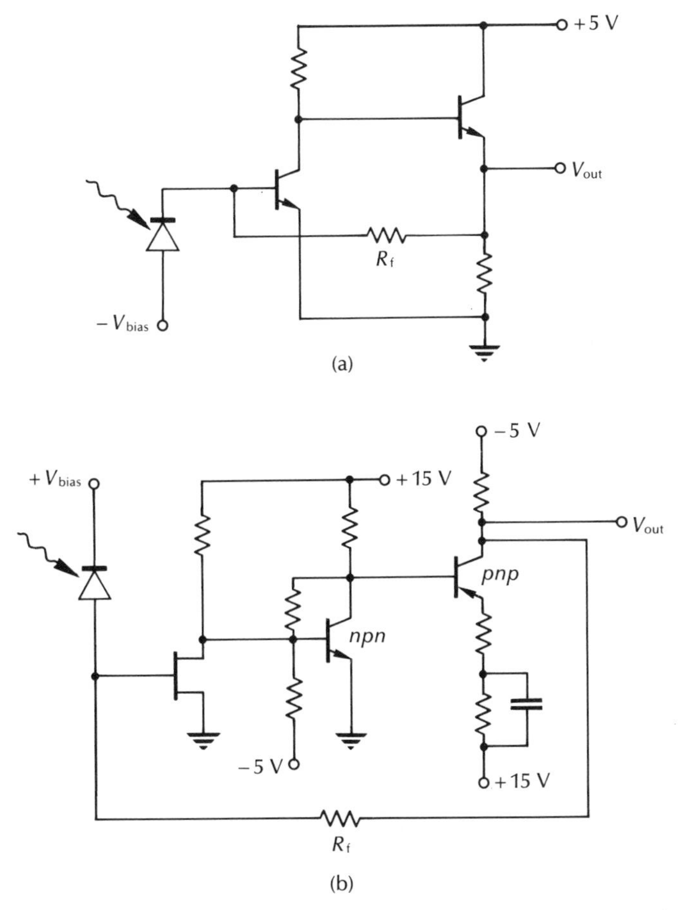

Figure 11.20 Transimpedance front end configuration: (a) bipolar transistor design [Refs. 28 and 29]; (b) FET front end and bipolar transistor cascade structure [Ref. 32].

common emitter stage followed by an emitter follower [Refs. 28 and 29] with overall feedback through resistor R_f. The output signal level from this transimpedance pair may be increased by the addition of a second common emitter stage [Ref. 30] after the emitter follower. This stage is not usually included in the feedback loop. An FET front end transimpedance design is shown in Figure 11.20(b) [Ref. 32].

The circuit consists of a grounded source configuration followed by a bipolar transistor cascade with feedback over the three stages. In this configuration the bias currents for the bipolar stages and the feedback resistance may be chosen to give good open loop bandwidth whilst making the noise contribution from these stages negligible.

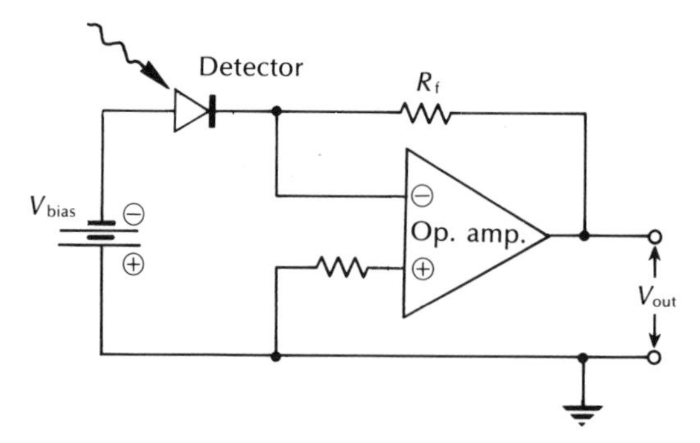

Figure 11.21 A typical circuit for an operational amplifier transimpedance front end [Ref. 33].

Finally, for lower bandwidth, shorter-haul applications an FET operational amplifier front end is often adequate [Ref. 33]. Such a transimpedance preamplifier circuit, which is generally used with a *p–i–n* photodiode, is shown in Figure 11.21. The choice of the operational amplifier is dependent on the gain versus bandwidth product for the device. In a simple digital receiver design all that may be required in addition to the circuit shown in Figure 11.21 is a logic (e.g. TTL) interface stage following the amplifier.

11.3.2 Automatic gain control (AGC)

It may be noted from the preceding section that the receiver circuit must provide a steady reverse bias voltage for the optical detector. With a *p–i–n* photodiode this is not critical and a voltage of between 5 and 80 V supplying an extremely low current is sufficient. The avalanche photodiode requires a much larger bias voltage of between 100 and 400 V which defines the multiplication factor for the device. An optimum multiplication factor is usually chosen so that the receiver signal to noise ratio is maximized (see Section 9.3.3). The multiplication factor for the APD varies with the device temperature (see Section 8.9.5) making provision of fine control for the bias voltage necessary in order to maintain the optimum multiplication factor. However, the multiplication factor can be held constant by some form of automatic gain control (AGC). An additional advantage in the use of AGC is that it reduces the dynamic range of the signals applied to the preamplifier giving increased optical dynamic range at the receiver input.

One method of providing AGC is simply to bias the APD with a constant d.c. current source I_{bias}, as illustrated in Figure 11.22. The constant current source is decoupled by a capacitor C at all signal frequencies to prevent gain modulation. When the mean optical input power is known, the mean current to the APD is defined by the bias which gives a constant multiplication factor (gain) at all temperatures. Any variation in the multiplication factor will produce a variation in the charge on C, thus adjusting the biasing of the APD back to the required

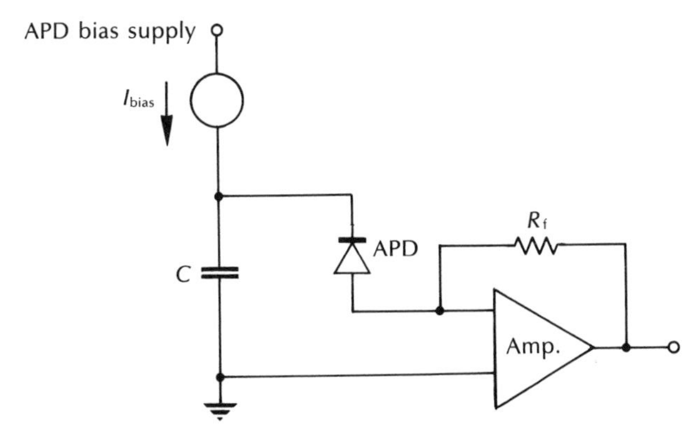

Figure 11.22 Bias of an APD with a constant current source to provide simple AGC.

multiplication factor. Therefore, the output current from the photodetector is only defined by the input current from the constant current source, giving full automatic gain control. However, this simple AGC technique is dependent on a constant, mean optical input power level, and takes no account of dark current generated within the detector.

A more widely used method which allows for the effect of variations in the detector dark current whilst providing critical AGC is to peak detect the a.c. coupled signal after suitable low noise amplification, as shown in Figure 11.23. The signal from the final stage of the main amplifier is compared with a preset reference level and fed back to adjust the high voltage bias supply in order to maintain a constant signal level. This effectively creates a constant current source with the dark current subtracted.

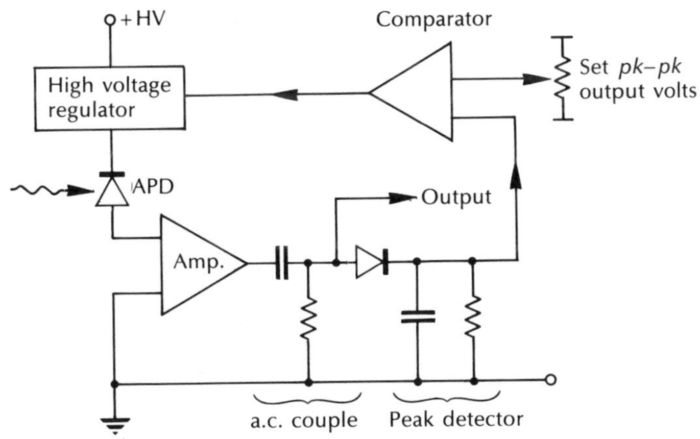

Figure 11.23 Bias of an APD by peak detection and feedback to provide AGC.

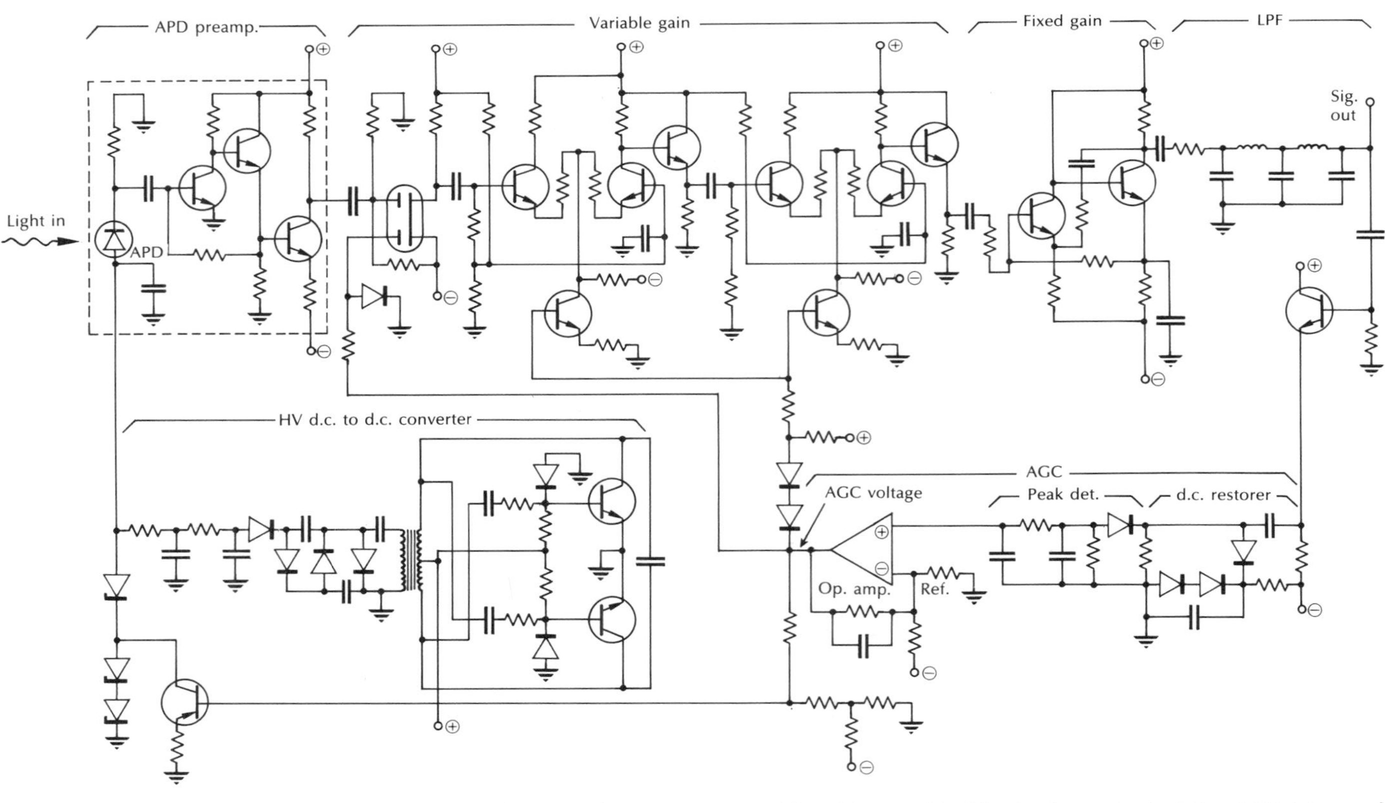

Figure 11.24 An optical fiber receiver circuit for digital transmission with AGC provided by both control of the APD bias and the main amplifier gain [Ref. 34].

A further advantage of this technique is that it may also be used to provide AGC for the main amplifier giving full control of the receiver gain.

A digital receiver circuit for an APD employing full AGC is shown in Figure 11.24 [Ref. 34]. The APD is followed by a transimpedance preamplifier employing bipolar transistors, the output of which is connected into a main amplifier consisting of a variable gain amplifier followed by a fixed gain amplifier. The first stage of the main amplifier is provided by a dual gate FET which gives a variable gain over a range of 20 dB. This variable gain amplifier also incorporates two stages, each of which consists of an emitter coupled pair with a gain variation of 14 dB. The following fixed gain amplifier gives a 2 V peak to peak signal to the low pass filter, the output of which is maintained at 1 V peak to peak by the AGC. Peak detection is provided in the AGC where the signal level is compared with a preset reference prior to control of the gain for both the APD and the main amplifier. The gain of the APD is controlled via a simple d.c. to d.c. converter which supplies the bias from a low voltage input, whereas the gain of the main amplifier is controlled by an input on the dual gate FET front end. This circuit allows a gain variation of 26 and 47 dB for the APD and the main amplifier respectively. The APD bias circuit is designed to protect the device against possible excess power dissipation at very high optical input power levels, as well as excess power dissipation when there is no optical input.

11.3.3 Equalization

The linear channel provided by the optical fiber receiver is often required to perform equalization as well as amplification of the detected optical signal. In order to discuss the function of the equalizer it is useful to assume the light falling on the detector to consist of a series of pulses given by:

$$P_o(t) = \sum_{k=-\infty}^{+\infty} a_k h_p(t - k\tau) \tag{11.1}$$

where $h_p(t)$ is the received pulse shape, $a_k = 0$ or 1 corresponding to the binary information transmitted and τ is the pulse repetition time or pulse spacing. In digital transmission τ corresponds to the bit period, although the pulse length does not necessarily fill the entire time period τ. For a typical optical fiber link, the received pulse shape is dictated by the transmitted pulse shape $h_t(t)$ and the fiber impulse response $h_f(t)$ following:

$$h_p(t) = h_t(t) * h_f(t) \tag{11.2}$$

where * denotes convolution. Hence determination of the received pulse shape requires knowledge of the fiber impulse response which is generally difficult to characterize. However, it can be shown [Ref. 37] for fiber which exhibits mode coupling that the impulse response is close to a Gaussian shape in both the time and frequency domain.

It is likely that the pulses given by Eq. (11.1) will overlap due to pulse broadening caused by dispersion on the link giving intersymbol interference (ISI). Following detection and amplification Eq. (11.1) may be written in terms of a voltage $v_A(t)$ as:

$$v_A(t) = \sum_{k=-\infty}^{+\infty} a_k h_A(t - k\tau) \tag{11.3}$$

where the response $h_A(t)$ includes any equalization required to compensate for distortion (e.g. integration) introduced by the amplifier. Therefore, although there is equalization for degradations caused by the amplifier, distortion caused by the channel and the resulting intersymbol interference is still included in $h_A(t - k\tau)$. The pulse overlap causing this intersymbol interference, may be reduced through the incorporation of a suitable equalizer with a frequency response $H_{eq}(\omega)$ such that:

$$H_{eq}(\omega) = \frac{\mathcal{F}\{h_{out}(t)\}}{\{h_A(t)\}} = \frac{H_{out}(\omega)}{H_A(\omega)} \tag{11.4}$$

where $h_{out}(t)$ is the desired output pulse shape and $\mathcal{F}$ indicates Fourier transformation. A block diagram indicating the pulse shapes in the time and frequency domains at the various points in an optical fiber system is shown in Figure 11.25.

An equalizer characterized by Eq. (11.4) will provide high frequency enhancement in the linear channel to compensate for high frequency roll off in the received pulses, thus giving the desired pulse shape. However, in order to construct such an equalizer we require knowledge of $h_A(t)$ and therefore $h_p(t)$. In turn, this

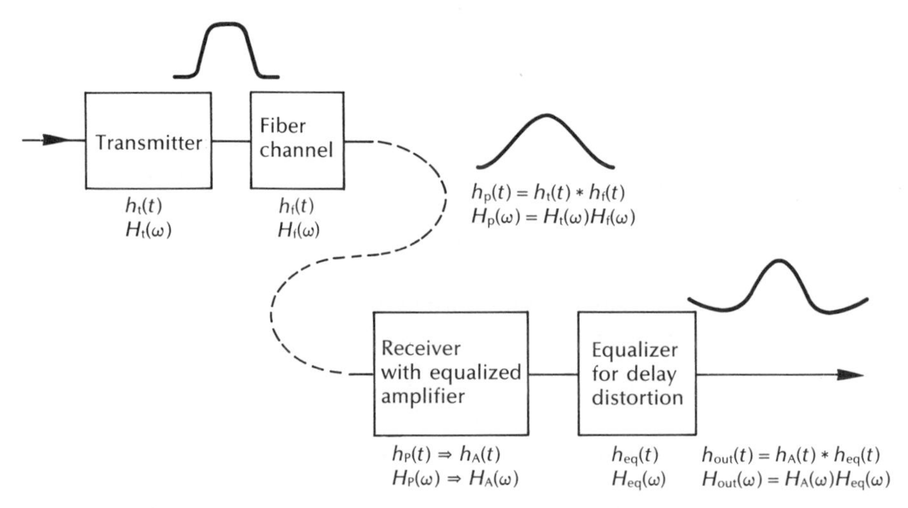

Figure 11.25 Block schematic of an optical fiber system illustrating the transmitted and received optical pulse shapes, together with electrical pulse shape, at the linear channel output.

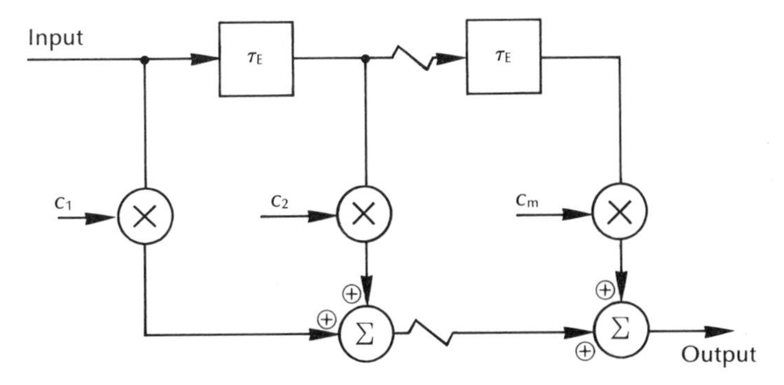

Figure 11.26 The transversal equalizer employing a tapped delay line.

necessitates information on the fiber impulse response $h_f(t)$ which may not be easily obtained.

Nevertheless, the conventional transversal equalizer shown in Figure 11.26 may be incorporated into the linear channel to keep ISI at tolerable levels, even if it is difficult to design a circuit which gives the optimum system response indicated in Eq. (11.4).

The transversal equalizer consists of a delay line tapped at τ_E second intervals. Each tap is connected through a variable gain device with tap coefficients c_i to a summing amplifier. Intersymbol interference is reduced by filtering the input signal and by the computing values for the tap coefficients which minimize the peak ISI. It is likely that further reduction in ISI will be accomplished using adaptive equalization which has yet to be rigorously applied to optical fiber communications. This is discussed further in Ref. 40.

11.4 System design considerations

Many of the problems associated with the design of optical fiber communication systems occur as a result of the unique properties of the glass fiber as a transmission medium. However, in common with metallic line transmission systems, the dominant design criteria for a specific application using either digital or analog transmission techniques are the required transmission distance and the rate of information transfer.

Within optical fiber communications these criteria are directly related to the major transmission characteristics of the fiber, namely optical attenuation and dispersion. Unlike metallic conductors where the attenuation (which tends to be the dominant mechanism) can be adjusted by simply changing the conductor size, entirely different factors limit the information transfer capability of optical fibers (see Chapter 3). Nevertheless, it is mainly these factors, together with the associated constraints within the terminal equipment, which finally limit the maximum distance that may be tolerated between the optical fiber transmitter and receiver. Where the terminal equipment is more widely spaced than this maximum distance, as in long-haul telecommunication applications, it is necessary to insert repeaters at regular intervals, as shown in Figure 11.27. The repeater incorporates a line receiver in order to convert the optical signal back into the electrical regime where, in the case of analog transmission, it is amplified and equalized (see Section 11.3.3) before it is retransmitted as an optical signal via a line transmitter. When digital transmission techniques are used the repeater also regenerates the original digital signal in the electrical regime (a regenerative repeater which is often simply called a regenerator) before it is retransmitted as a digital optical signal. In this case the repeater may additionally provide alarm, supervision and engineering order wire facilities.

The installation of repeaters substantially increases the cost and complexity of any line communication system. Hence a major design consideration for long-haul telecommunication systems is the maximum distance of unrepeatered transmission so that the number of intermediate repeaters may be reduced to a minimum. In this respect optical fiber systems display a marked improvement over alternative line transmission systems using metallic conductors. However, this major advantage of optical fiber communications is somewhat reduced due to the present requirement for electrical signal processing at the repeater. This necessitates the supply of electrical power to the intermediate repeaters via metallic conductors, as may be observed in Figure 11.27.

Before any system design procedures can be initiated it is essential that certain basic system requirements are specified. These specifications include:

(a) transmission type: digital or analog;
(b) acceptable system fidelity, generally specified in terms of the received BER for digital systems or the received SNR and signal distortion for analog systems;
(c) required transmission bandwidth;
(d) acceptable spacing between the terminal equipment or intermediate repeaters;
(e) cost;
(f) reliability.

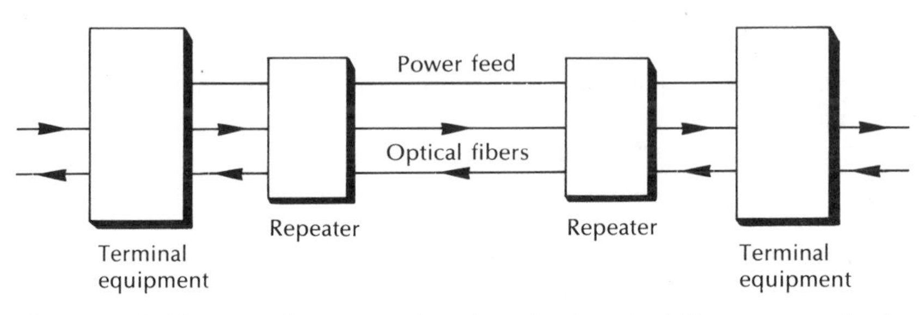

Figure 11.27 The use of repeaters in a long-haul optical fiber communication system.

However, the exclusive use of the above specifications inherently assumes that system components are available which will allow any system, once specified, to be designed and implemented. Unfortunately, this is not always the case, especially when the desired result is a wideband, long-haul system. In this instance it may be necessary to make choices by considering factors such as availability, reliability, cost and ease of installation and operation, before specifications (a) to (d) can be fully determined. A similar approach must be adopted in lower bandwidth, shorter-haul applications where there is a requirement for the use of specific components which may restrict the system performance. Hence it is likely that the system designer will find it necessary to consider the possible component choices in conjunction with the basic system requirements.

11.4.1 Component choice

The system designer has many choices when selecting components for an optical fiber communication system. In order to exclude certain components at the outset it is useful if the operating wavelength of the system is established (i.e. shorter wavelength region 0.8 to 0.9 μm or longer wavelength region 1.1 to 1.6 μm). This decision will largely be dictated by the overall requirements for the system performance, the ready availability of suitable reliable components, and cost. Hence the major component choices are:

1. Optical fiber type and parameters. Multimode or single-mode; size, refractive index profile, attenuation, dispersion, mode coupling, strength, cabling, jointing, etc.
2. Source type and characteristics. Laser or LED; optical power launched into the fiber, rise and fall time, stability, etc.
3. Transmitter configuration. Design for digital or analog transmission; input impedance, supply voltage, dynamic range, optical feedback, etc.
4. Detector type and characteristics, *p–n*, *p–i–n*, or avalanche photodiode; responsivity, response time, active diameter, bias voltage, dark current, etc.
6. Receiver configuration. Preamplifier design (low impedance, high impedance or transimpedance front end), BER or SNR, dynamic range, etc.
7. Modulation and coding. Source intensity modulation; using pulse modulation techniques for either digital (e.g. pulse code modulation, adaptive delta modulation) or analog (pulse amplitude modulation, pulse frequency modulation, pulse width modulation, pulse position modulation) transmission. Also, encoding schemes for digital transmission such as biphase (Manchester) and delay modulation (Miller) codes [Ref. 7]. Alternatively analog transmission using direct intensity modulation or frequency modulation of the electrical subcarrier (subcarrier FM). In the latter technique the frequency of an electrical subcarrier is modulated rather than the frequency of the optical source, as would be the case with direct frequency modulation. The electrical subcarrier, in turn, intensity modulates the optical source (see Section 11.7.5).

Digital and analog modulation techniques which require coherent detection are under investigation but system components which will permit these modulation methods to be utilized are not yet widely available (see Chapter 12).

Decisions in the above areas are interdependent and may be directly related to the basic system requirements. The potential choices provide a wide variety of economic optical fiber communication systems. However, it is necessary that the choices are made in order to optimize the system performance for a particular application.

11.4.2 Multiplexing

In order to maximize the information transfer over an optical fiber communication link it is usual to multiplex several signals on to a single fiber. It is possible to convey these multichannel signals by multiplexing in the electrical time or frequency domain, as with conventional electrical line or radio communication, prior to intensity modulation of the optical source. Hence, digital pulse modulation schemes may be extended to multichannel operation by time division multiplexing (TDM) narrow pulses from multiple modulators under the control of a common clock. Pulses from the individual channels are interleaved and transmitted sequentially, thus enhancing the bandwidth utilization of a single fiber link.

Alternatively, a number of baseband channels may be combined by frequency division multiplexing (FDM). In FDM the optical channel bandwidth is divided into a number of nonoverlapping frequency bands and each signal is assigned one of these bands of frequencies. The individual signals can be extracted from the combined FDM signal by appropriate electrical filtering at the receive terminal. Hence, frequency division multiplexing in an IM/DD system is generally performed electrically at the transmit terminal prior to intensity modulation of a single optical source. However, it is possible to utilize a number of optical sources, each operating at a different wavelength on the single fiber link. In this technique, often referred to as wavelength division multiplexing (WDM), the separation and extraction of the multiplexed signals (i.e. wavelength separation) is performed with optical filters (e.g. interference filters, diffraction grating filters, or prism filters) [Ref. 41].

Finally, a multiplexing technique which does not involve the application of several message signals on to a single fiber is known as space division multiplexing (SDM). In SDM, each signal channel is carried on a separate fiber within a fiber bundle or multifiber cable form. The good optical isolation offered by fibers means that cross coupling between channels can be made negligible. However, this technique necessitates an increase in the number of optical components required (e.g. fiber, connectors, sources, detectors) within a particular system and therefore has not been widely used to date.

11.5 Digital systems

Most of the future expansion of the telecommunication network is being planned around digital telephone exchanges linked by digital transmission systems. The shift towards digitizing the network followed the introduction of digital circuit techniques and, especially, integrated circuit technology which made the transmission of discrete time signals both advantageous and economic. Digital transmission systems generally give superior performance over their analog counterparts, as well as providing an ideal channel for data communications and compatibility with digital computing techniques.

Optical fiber communication is well suited to baseband digital transmission in several important ways. For instance, it offers a tremendous advantage with regard to the acceptable signal to noise ratio (SNR) at the optical fiber receiver over analog transmission by some 20 to 30 dB (for practical systems), as indicated in the noise considerations of Section 9.2. Also, the use of baseband digital signalling reduces problems involved with optical source (and sometimes detector) nonlinearities and temperature dependence which may severely affect analog transmission. Therefore, most high capacity optical fiber communication systems convey digital information in the baseband using intensity modulation of the optical source.

In common with electrical transmission systems, analog signals (e.g. speech) may be digitized for transmission utilizing pulse code modulation (PCM). Encoding the analog signal into a digital bit pattern is performed by initially sampling the analog signal at a frequency in excess of the Nyquist rate (i.e. greater than twice the maximum signal frequency). Within the European telecommunication network where the 3 dB telephone bandwidth is defined as 3.4 kHz, the sampling rate is 8 kHz. Hence, the amplitude of the constant width sampling pulses varies in proportion to the sample values of the analog signal giving a discrete signal known as pulse amplitude modulation (PAM), as indicated in Figure 11.28. The sampled analog signal is then quantized into a number of discrete levels, each of which is designated by a binary code which provides the PCM signal. This process is also illustrated in Figure 11.28 using a linear quantizer with eight levels (or seven steps) so that each PAM sample is encoded into three binary bits. The analog signal is thus digitized and may be transmitted as a baseband signal or, alternatively, be modulated by amplitude, frequency or phase shift keying [Ref. 43]. However, in practical PCM systems for speech transmission, nonlinear encoding (A law in Europe and μ law in North America) is generally employed over 256 levels (2^8), giving eight binary bits per sample (seven bits for code levels plus one polarity bit). Hence, the bandwidth requirement for PCM transmission is substantially greater (in this case by a factor of approximately 16) than the corresponding baseband analog transmission. This is not generally a problem with optical fiber communications because of the wideband nature of the optical channel.

Nonlinear encoding may be implemented via a mechanism known as companding where the input signal is compressed before transmission to give a nonlinear

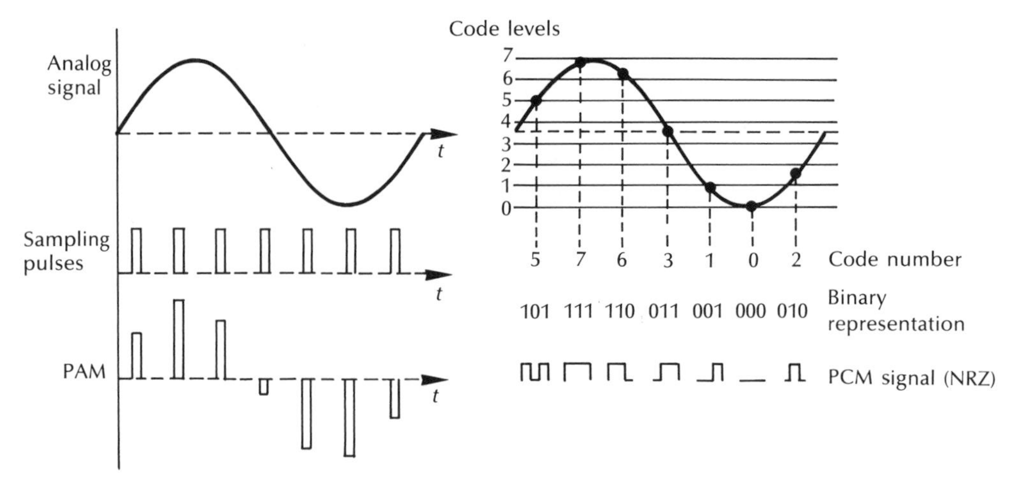

Figure 11.28 The quantization and encoding of an analog signal into PCM using a linear quantizer with eight levels.

encoding characteristic and expanded again at the receive terminal after decoding. A typical nonlinear input–output characteristic giving compression is shown in Figure 11.29. Companding is used to reduce the quantization error on small amplitude analog signal levels when they are encoded from PAM to PCM. The quantization error (i.e. the rounding off to the nearest discrete level) is exhibited as distortion or noise on the signal (often called quantization noise). Companding tapers the step size, thus reducing the distance between levels for small amplitude signals whilst increasing the distance between levels for higher amplitude signals. This substantially reduces the quantization noise on small amplitude signals at the expense of slightly increased quantization noise, in terms of signal amplitude, for

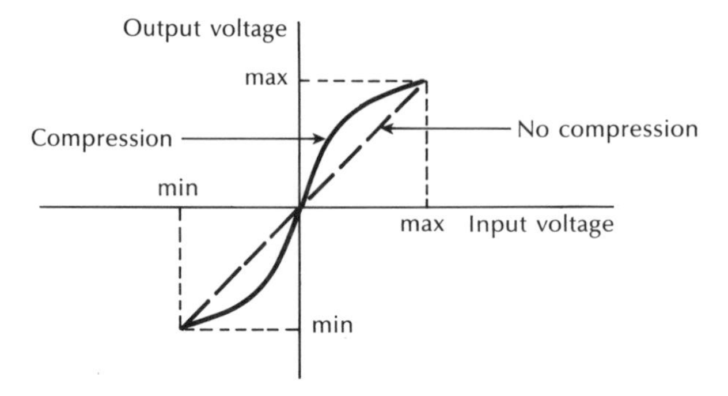

Figure 11.29 A typical nonlinear input–output characteristic which provides compression.

the larger signal levels. The corresponding SNR improvement for small amplitude signals significantly reduces the overall signal degradation of the system due to the quantization process.

A block schematic of a simplex (one direction only) baseband PCM system is shown in Figure 11.30(a). The optical interface is not shown but reference may be made to Figure 11.1 which illustrates the general optical fiber communication system. It may be noted from Figure 11.30(a) that the received PCM waveform is decoded back to PAM via the reverse process to encoding, and then simply passed through a low pass filter to recover the original analog signal.

The conversion of a continuous analog waveform into a discrete PCM signal allows a number of analog channels to be time division multiplexed (TDM) for simultaneous transmission down one optical fiber link, as illustrated in Figure 11.30(b). The encoded samples from the different channels are interleaved within the multiplexer to give a single composite signal consisting of all the interleaved pulses. This signal is then transmitted over the optical channel. At the receive terminal the interleaved samples are separated by a synchronous switch or demultiplexer before each analog signal is reconstructed from the appropriate set of samples. Time division multiplexing a number of channels on to a single link can be used with any form of digital transmission and is frequently employed in the transmission of data as well as with the transmission of digitised analog signals. However, the telecommunication network is primarily designed for the transmission of analog speech signals, although the compatibility of PCM with data signals has encouraged the adoption of digital transmission systems.

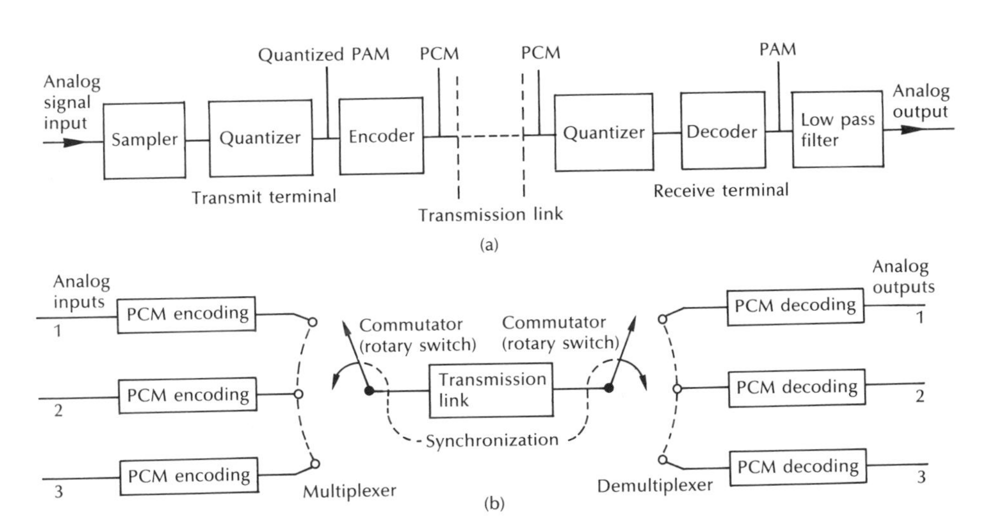

Figure 11.30 PCM transmission: (a) block schematic of a baseband PCM transmission system for single channel transmission; (b) time division multiplexing of three PCM channels on to a single transmission link and subsequent demultiplexing at the link output.

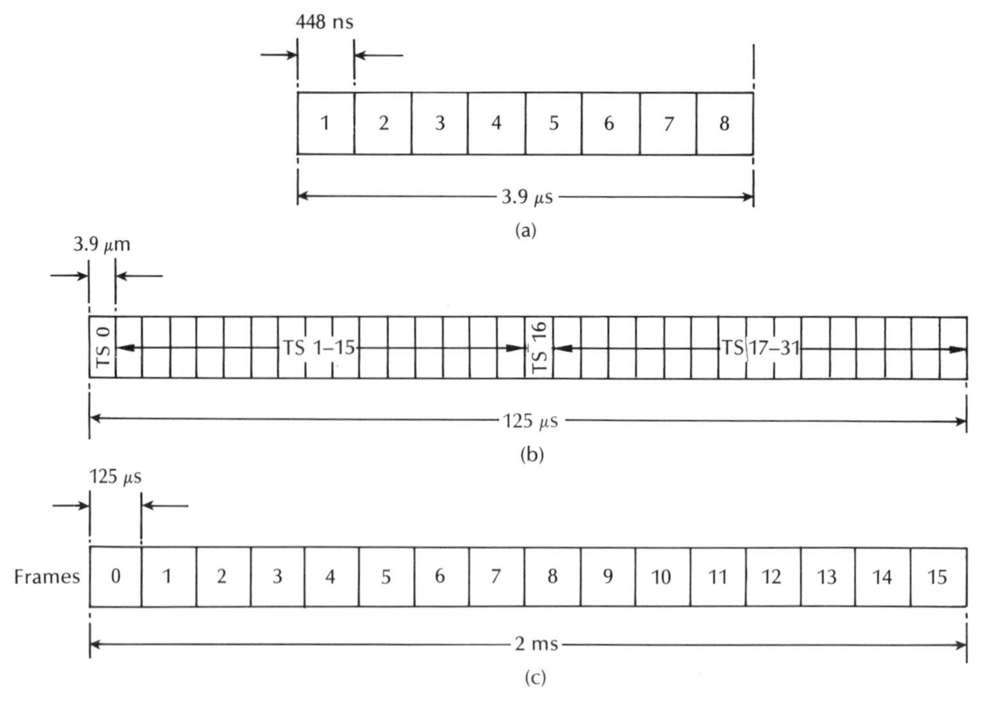

Figure 11.31 The timing for the line signalling structure of the European standard thirty channel PCM system: (a) bits per time slot; (b) time slots per frame; (c) frames per multiframe.

A current European standard for speech transmisson using PCM on metallic conductors (i.e. coaxial line) is the thirty channel system. In this system the PAM samples from each channel are encoded into eight binary bits which are incorporated into a single time slot. Time slots from respective channels are interleaved (multiplexed) into a frame consisting of thirty-two time slots. The two additional time slots do not carry encoded speech but signalling and synchronization information. Finally, sixteen frames are incorporated into a multiframe which is a self contained timing unit. The timing for this line signalling structure is shown in Figure 11.31 and calculated in Example 11.1.

Example 11.1

The sampling rate for each speech channel on the 30 channel PCM system is 8 kHz and each sample is encoded into eight bits. Determine:

(a) the transmission or bit rate for the system:
(b) the duration of a time slot;
(c) the duration of a frame and multiframe.

Solution: (a) The 30 channel PCM system has 32 time slots each eight bits wide which make up a frame. Therefore,

$$\textit{Number of bits in a frame} = 32 \times 8 = 256 \text{ bits}$$

This frame must be transmitted within the sampling period and thus 8×10^3 frames are transmitted per second. Hence, the transmission rate for the system is:

$$8 \times 10^3 \times 256 = 2.048 \text{ Mbit s}^{-1}$$

(b) The bit duration is simply:

$$\frac{1}{2.048 \times 10^6} = 488 \text{ ns}$$

Therefore, the duration of a time slot is:

$$8 \times 488 \text{ ns} = 3.9\ \mu\text{s}$$

(c) The duration of a frame is thus:

$$32 \times 3.9\ \mu\text{s} = 125\ \mu\text{s}$$

and the duration of a multiframe is:

$$16 \times 125\ \mu\text{s} = 2 \text{ ms}$$

The signalling structure shown in Figure 11.31 applies to thirty channel PCM systems which were originally designed to transmit over metallic conductors using a high density bipolar line code (HDB 3). The increased bandwidth with optical fiber communications allows transmission rates far in excess of 2.048 Mbit s^{-1}. Therefore an increased number of telephone channels may be sampled, encoded, multiplexed and transmitted on an optical fiber link. In Europe the increased bit rates were chosen as multiples of the thirty channel system, whereas in North America they tend to be multiples of a twenty-four channel system. These bit rates and the corresponding number of transmitted telephone channels are specified in Table 11.1.

It must be noted that a bipolar code with a zero mean level (i.e. with positive and negative going pulses in the electrical regime) such as HDB 3 cannot be transmitted directly over an optical fiber link unless the mean level is raised to allow both positive and negative going pulses to be transmitted by the intensity modulated optical source. The resultant ternary (three level) optical transmission is not always suitable for telecommunication applications and therefore binary coding after appropriate scrambling, biphase (Manchester encoding), delay modulation (Miller encoding), etc., is often employed. This involves additional complexity at the transmit and receive terminals as well as necessitating extra redundancy (i.e. bits which do not contain the transmitted information, thus giving a reduction in the information per transmitted symbol) in the line code. This topic is considered in greater detail in Section 11.6.7.

Table 11.1 Digital bit rates for multichannel PCM transmission in Europe and North America

Europe		North America	
Telephone channels	Bit rates Mbit s^{-1}	Telephone channels	Bit rates Mbit s^{-1}
30	2.048	24	1.544
120	8.448	48	3.152
480	34.368	96	6.312
1920	139.264	672	44.736
7680	565.148	4032	274.176

11.6 Digital system planning considerations

The majority of digital optical fiber communication systems for the telecommunication network or local data applications utilize binary intensity modulation of the optical source. Therefore, we choose to illustrate the planning considerations for digital transmission based on this modulation technique. Baseband PCM transmission using source intensity modulation is usually designated as PCM–IM.

11.6.1 The regenerative repeater

In the case of the long-haul, high capacity digital systems, the most important overall system performance parameter is the spacing of the regenerative repeaters. It is therefore useful to consider the performance of the digital repeater, especially as it is usually designed with the same optical components as the terminal equipment. Figure 11.32 shows the functional parts of a typical regenerative repeater for optical fiber communications. The attenuated and dispersed optical pulse train is detected and amplified in the receiver unit. This consists of a photodiode followed by a low noise preamplifier. The electrical signal thus acquired is given a further increase in power level in a main amplifier prior to reshaping in order to compensate for the transfer characteristic of the optical fiber (and the amplifier) using an equalizer. Depending on the photodiode utilized, automatic gain control may be provided at this stage for both the photodiode bias current and the main amplifier (see Section 11.3.2).

Accurate timing (clock) information is then obtained from the amplified and equalized waveform using a timing extraction circuit such as a ringing circuit or phase locked loop. This enables precise operation of the following regenerator circuit within the bit intervals of the original pulse train. The function of the regenerator circuit is to reconstitute the originally transmitted pulse train, ideally without error. This can be achieved by setting a threshold above which a binary one is registered, and below which a binary zero is recorded, as indicated in Figure

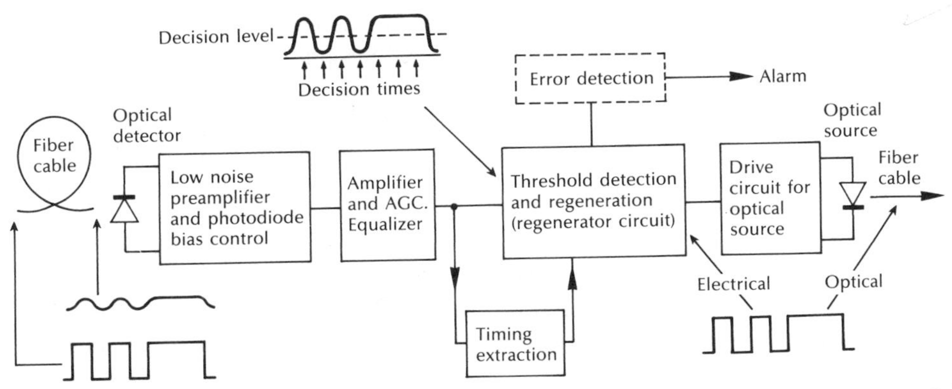

Figure 11.32 Block schematic showing a typical regenerative repeater for digital optical fiber communications.

11.32. The regenerator circuit makes these decisions at times corresponding to the centre of the bit intervals based on the clock information provided by the timing circuit.

Hence the decision times are usually set at the mid-points between the decision level crossings of the pulse train. The pulse train is sampled at a regular frequency equal to the bit rate, and at each sample instant a decision is made of the most probable symbol being transmitted. The symbols are then regenerated in their original form (either a binary one or zero) before retransmission as an optical signal using a source operated by an electronic drive circuit. Hence the possible regeneration of an exact replica of the originally transmitted waveform is a major advantage of digital transmission over corresponding analog systems. Repeaters in analog systems filter, equalize and amplify the received waveform, but are unable to reconstitute the originally transmitted waveform entirely free from distortion and noise. Signal degradation in long-haul analog systems is therefore accumulative, being a direct function of the number of repeater stages. In contrast the signal degradation encountered in PCM systems is purely a function of the quantization process and the system bit error rate.

Errors may occur in the regeneration process in the following situations:

1. The signal to noise ratio at the decision instant is insufficient for an accurate decision to be made. For instance, with high noise levels, the binary zero may occur above the threshold and hence be registered as a binary one.
2. There is intersymbol interference due to dispersion on the optical fiber link. This may be reduced by equalization which forces the transmitted binary one to pass through zero at all neighbouring decision times.
3. There is a variation in the clock rate and phase degradations (jitter) such as distortion of the zero crossings and static decision time misalignment.

A method which is often used to obtain a qualitative indication of the performance of a regenerative repeater or a PCM system is the examination of the received waveform on an oscilloscope using a sweep rate which is a fraction of the bit rate. The display obtained over two bit intervals' duration, which is the result of superimposing all possible pulse sequences, is called an eye pattern or diagram. An illustration of an eye pattern for a binary system with little distortion and no additive noise is shown in Figure 11.33(a).It may be observed that the pattern has the shape of a human eye which is open and that the decision time corresponds to the centre of the opening. To regenerate the pulse sequence without error the eye must be open thereby indicating that a decision area exists, and the decision crosshair (provided by the decision time and the decision threshold) must be within this open area. The effect of practical degradations on the pulses (i.e. intersymbol interference and noise) is to reduce the size of, or close, the eye, as shown in Figure 11.33(b). Hence for reliable transmission it is essential that the eye is kept open, the margin against an error occurring being the minimum distance between the decision crosshair and the edge of the eye.

In practice, a low bit error rate (BER) in the region 10^{-7} to 10^{-10} may be tolerated with PCM transmission. However, with data transmission (e.g. computer communications) any error can cause severe problems, and it is necessary to incorporate error detecting and possibly correcting circuits into the regenerator. This invariably requires the insertion of a small amount of redundancy into the transmitted pulse train (see Section 11.6.7).

Calculation of the possible repeater spacing must take account of the following system component performances:

(a) the average optical power launched into the fiber based on the end of life transmitter performance;

(b) the receiver input power required to achieve an acceptably low BER (e.g. 10^{-9}), taking into account component deterioration during the system's lifetime;

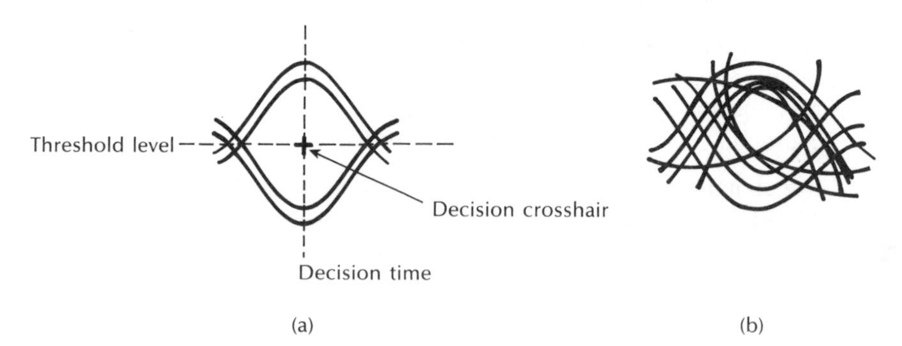

Figure 11.33 Eye patterns in binary digital transmission: (a) the pattern obtained with a bandwidth limitation but no additive noise (open eye); (b) the pattern obtained with a bandwidth limitation and additive noise (partially closed eye).

(c) the installed fiber cable loss including jointing and coupling (to source and detector) losses as well as the effects of ageing and from anticipated environmental changes;
(d) the temporal response of the system including the effects of pulse dispersion on the channel; this becomes an important consideration with high bit rate multimode fiber systems which may be dispersion limited.

These considerations are discussed in detail in the following sections.

11.6.2 The optical transmitter

The average optical power launched into the fiber from the transmitter depends upon the type of source used and the required system bit rate, as indicated in Section 11.2.1. These factors may be observed in Figure 11.34 [Ref. 45] which compares the optical power available from an injection laser and an LED for transmission over a multimode fiber with a core diameter of 50 μm and a numerical aperture of 0.2. Typically, the laser launches around 1 mW, whereas usually the LED is limited to about 100 μW. It may also be noted that both device types emit less optical power at higher bit rates. However, the LED gives reduced output at modulation bandwidths in excess of 50 MHz, whereas laser output is unaffected below 200 MHz. Also, the fact that generally the optical power which may be

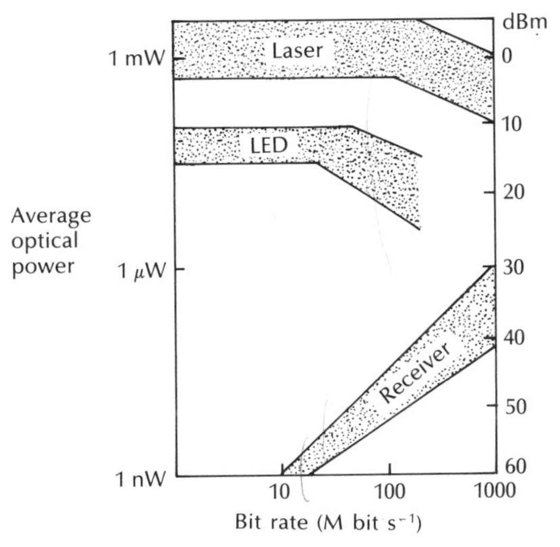

Figure 11.34 The average power launched into multimode optical fiber from typical injection lasers and LEDs as a function of digital bit rate (upper bands). Also included in the lower band is the received optical power required for binary NRZ pulses transmitted with a BER of 10^{-9}. Reproduced with permission from D. C. Gloge and T. Li, 'Multimode-fiber technology for digital transmission', *Proc. IEEE*, **68**, p. 1269, 1980. Copyright © 1980 IEEE.

launched into a fiber from an LED even at low bit rates is 10 to 15 dB down on that available from a laser is an important consideration, especially when receiver noise is a limiting factor within the system.

11.6.3 The optical receiver

The input optical power required at the receiver is a function of the detector combined with the electrical components within the receiver structure. It is strongly dependent upon the noise (i.e. quantum, dark current and thermal) associated with the optical fiber receiver. The theoretical minimum pulse energy or quantum limit required to maintain a given BER was discussed in Section 9.2.4.

It was predicted that approximately twenty-one incident photons were necessary at an ideal photodetector in order to register a binary one with a BER of 10^{-9}. However, this is a fundamental limit which cannot be achieved in practice and therefore it is essential that estimates of the minimum required optical input power are made in relation to practical devices and components.

Although the statistics of quantum noise follow a Poisson distribution, other important sources of noise within practical receivers (e.g. thermal) are characterized by a Gaussian probability distribution. Hence estimates of the required SNR to maintain particular bit error rates may be obtained using the procedure adopted for error performance of electrical digital systems where the noise distribution is considered to be white Gaussian. This Gaussian approximation [Ref. 46] is

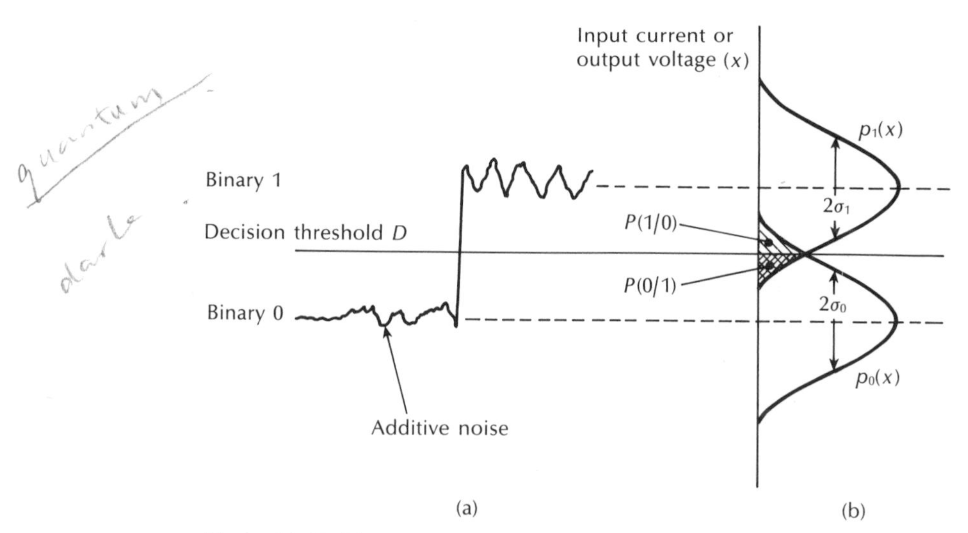

Figure 11.35 Binary transmission: (a) the binary signal with additive noise; (b) probability density functions for the binary signal showing the decision case. P(0/1) is the probability of falsely identifying a binary one and P(1/0) is the probability of falsely identifying a binary zero.

sufficiently accurate for design purposes and is far easier to evaluate than the more exact probability distribution within the receiver [Ref. 47]. The receiver sensitivities calculated by using the Gaussian approximation are generally within 1 dB of those calculated by other methods [Ref. 29].

Although the transmitted signal consists of two well defined light levels, in the presence of noise the signal at the receiver is not as well defined. This situation is shown in Figure 11.35(a) which illustrates a binary signal in the presence of noise. The signal plus the additive noise at the detector may be defined in terms of the probability density functions (PDFs) shown in Figure 11.35(b). These PDFs describe the probability that the input current (or output voltage) has a value i (or v) within the incremental range di (or dv). The expected values of the signal in the two transmitted states, namely 0 and 1, are indicated by $p_0(x)$ and $p_1(x)$ respectively. When the additive noise is assumed to have a Gaussian distribution, the PDFs of the two states will also be Gaussian. The Gaussian PDF which is continuous is defined by:

$$p(x) = \frac{1}{\sigma\sqrt{(2\pi)}} \exp - [(x-m)^2/2\sigma^2] \tag{11.5}$$

where m is the mean value and σ the standard deviation of the distribution. When $p(x)$ describes the probability of detecting a noise current or voltage, σ corresponds to the rms value of that current or voltage.

If a decision threshold D is set between the two signal states, as indicated in Figure 11.35, signals greater than D are registered as a one and those less than D as a zero. However when the noise current (or voltage) is sufficiently large it can either decrease a binary one to a zero or increase a binary zero to a one. These error probabilities are given by the integral of the signal probabilities outside the decision region. Hence the probability that a signal transmitted as a 1 is received as a 0, $P(0/1)$, is proportional to the shaded area indicated in Figure 11.35(b). The probability that a signal transmitted as a 0 is received as a 1, $P(1/0)$, is similarly proportional to the other shaded area shown in the diagram. If $P(1)$ and $P(0)$ are the probabilities of transmission for binary ones and zeros, respectively, then the total probability of error $P(\mathrm{e})$ may be defined as

$$P(\mathrm{e}) = P(1)P(0/1) + P(0)P(1/0) \tag{11.6}$$

Now let us consider a signal current i_{sig} together with an additive noise current i_{N} and a decision threshold set at $D = i_{\mathrm{D}}$. If at any time when a binary 1 is transmitted the noise current is negative such that:

$$i_{\mathrm{N}} < -(i_{\mathrm{sig}} - i_{\mathrm{D}}) \tag{11.7}$$

then the resulting current $i_{\mathrm{sig}} + i_{\mathrm{N}}$ will be less than i_{D} and an error will occur. The corresponding probability of the transmitted 1 being received as a 0 may be written as:

$$P(0/1) = \int_{-\infty}^{i_{\mathrm{D}}} p(i, i_{\mathrm{sig}})\, \mathrm{d}i \tag{11.8}$$

and following Eq. (11.5)

$$p_1(x) = p(i, i_{sig}) = \frac{1}{(\overline{i_N^2})^{\frac{1}{2}}\sqrt{(2\pi)}} \exp - \left[\frac{(i - i_{sig})^2}{2(\overline{i_N^2})}\right] \tag{11.9}$$

$$= \text{Gsn}\,[i, i_{sig}, (\overline{i_N^2})^{\frac{1}{2}}] \tag{11.10}$$

where i is the actual current, i_{sig} is the peak signal current during a binary 1 (this corresponds to the peak photocurrent I_p when only a signal component is present), and $\overline{i_N^2}$ is the mean square noise current. Substituting Eq. (11.10) into Eq. (11.8) gives:

$$P(0/1) = \int_{-\infty}^{i_D} \text{Gsn}\,[i, i_{sig}, (\overline{i_N^2})^{\frac{1}{2}}]\ \text{d}i \tag{11.11}$$

Similarly, the probability that a binary 1 will be received when a 0 is transmitted is the probability that the received current will be greater than i_D at some time during the zero bit interval. It is given by:

$$P(1/0) = \int_{i_D}^{\infty} p(i, 0) \tag{11.12}$$

Assuming the mean square noise current in the zero state is equal to the mean square noise current in the one state $(\overline{i_N^2})$ (this is an approximation if shot noise is dominant), and that for a zero bit $i_{sig} = 0$, then following Eq. (11.5):

$$p_0(x) = p(i, 0) = \frac{1}{(\overline{i_N^2})^{\frac{1}{2}}\sqrt{(2\pi)}} \exp - \left[\frac{(i - 0)^2}{2(\overline{i_N^2})}\right] \tag{11.13}$$

$$= \text{Gsn}\,[i, 0, (\overline{i_N^2})^{\frac{1}{2}}] \tag{11.14}$$

Hence substituting Eq. (11.14) into Eq. (11.12) gives:

$$P(1/0) = \int_{i_D}^{\infty} \text{Gsn}\,[i, 0, (\overline{i_N^2})^{\frac{1}{2}}]\ \text{d}i \tag{11.15}$$

The integrals of Eqs. (11.11) and (11.15) are not readily evaluated but may be written in terms of the error function (erf)* where:

$$\text{erf}(u) = \frac{2}{\sqrt{\pi}} \int_0^u \exp(-z^2)\ \text{d}z \tag{11.16}$$

and the complementary error function is:

$$\text{erfc}(u) = 1 - \text{erf}(u) = \frac{2}{\sqrt{\pi}} \int_u^{\infty} \exp(-z^2)\ \text{d}z \tag{11.17}$$

* Another form of the error function denoted by erf is defined in Problem 11.10.

Hence

$$P(0/1) = \frac{1}{2}\left[1 - \text{erf}\left(\frac{|i_{\text{sig}} - i_{\text{D}}|}{(\overline{i_{\text{N}}^2})^{\frac{1}{2}}\sqrt{2}}\right)\right]$$

$$= \tfrac{1}{2}\,\text{erfc}\left(\frac{|i_{\text{sig}} - i_{\text{D}}|}{(\overline{i_{\text{N}}^2})^{\frac{1}{2}}\sqrt{2}}\right) \tag{11.18}$$

and

$$P(1/0) = \tfrac{1}{2}\,\text{erfc}\left(\frac{|0 - i_{\text{D}}|}{(\overline{i_{\text{N}}^2})^{\frac{1}{2}}\sqrt{2}}\right) = \tfrac{1}{2}\,\text{erfc}\left(\frac{|-i_{\text{D}}|}{(\overline{i_{\text{N}}^2})^{\frac{1}{2}}\sqrt{2}}\right) \tag{11.19}$$

If we assume that a binary code is chosen such that the number of transmitted ones and zeros are equal, then $P(0) = P(1) = \frac{1}{2}$, and the net probability of error is one half the sum of the shaded areas in Figure 11.35(b). Therefore Eq. (11.6) becomes:

$$P(\text{e}) = \tfrac{1}{2}[P(0/1) + P(1/0)] \tag{11.20}$$

and substituting for $P(0/1)$ and $P(1/0)$ from Eqs. (11.18) and (11.19) gives:

$$P(\text{e}) = \frac{1}{2}\left[\tfrac{1}{2}\,\text{erfc}\left(\frac{|i_{\text{sig}} - i_{\text{D}}|}{(\overline{i_{\text{N}}^2})^{\frac{1}{2}}\sqrt{2}}\right) + \tfrac{1}{2}\,\text{erfc}\left(\frac{|-i_{\text{D}}|}{(\overline{i_{\text{N}}^2})^{\frac{1}{2}}\sqrt{2}}\right)\right] \tag{11.21}$$

Equation (11.21) may be simplified by setting the threshold decision level at the mid-point between zero current and the peak signal current such that $i_{\text{D}} = i_{\text{sig}}/2$. In electrical systems this situation corresponds to an equal minimum probability of error in both states due to the symmetrical nature of the PDFs. It must be noted that for optical fiber systems this is not generally the case since the noise in each signal state contains shot noise contributions proportional to the signal level. Nevertheless, assuming a Gaussian distribution for the noise and substituting $i_{\text{D}} = i_{\text{sig}}/2$ into Eq. (11.21) we obtain:

$$P(\text{e}) = \frac{1}{2}\left[\tfrac{1}{2}\,\text{erfc}\left(\frac{|i_{\text{sig}}/2|}{(\overline{i_{\text{N}}^2})^{\frac{1}{2}}\sqrt{2}}\right) + \tfrac{1}{2}\,\text{erfc}\left(\frac{|-i_{\text{sig}}/2|}{(\overline{i_{\text{N}}^2})^{\frac{1}{2}}\sqrt{2}}\right)\right]$$

$$= \tfrac{1}{2}\,\text{erfc}\left(\frac{i_{\text{sig}}}{2(\overline{i_{\text{N}}^2})^{\frac{1}{2}}\sqrt{2}}\right) \tag{11.22}$$

The electrical SNR at the detector may be written in terms of the peak signal power to rms noise power (mean square noise current) as:

$$\frac{S}{N} = \frac{i_{\text{sig}}^2}{\overline{i_{\text{N}}^2}} \tag{11.23}$$

Comparison of Eq. (11.23) with Eq. (11.22) allows the probability of error to be expressed in terms of the analog SNR as:

$$P(\text{e}) = \tfrac{1}{2}\,\text{erfc}\left(\frac{(S/N)^{\frac{1}{2}}}{2\sqrt{2}}\right) \tag{11.24}$$

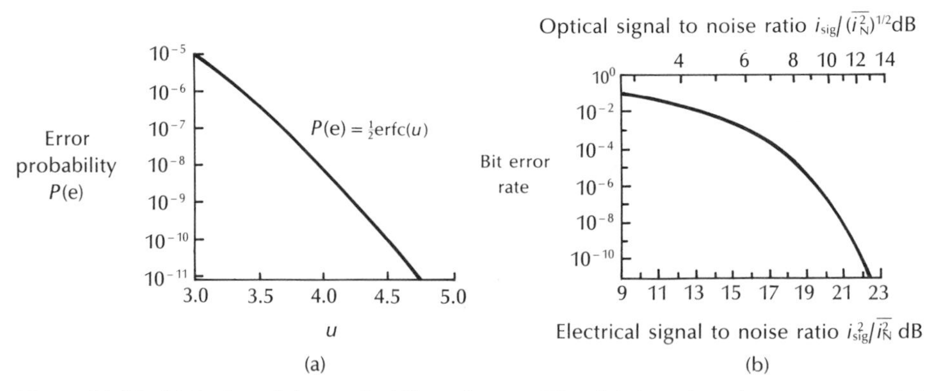

Figure 11.36 (a) A plot of the probability of error 1/2 erfc(u) against the argument of the error function u. (b) The bit error rate as a function of both the ratio of peak signal power to rms noise power (electrical SNR) and the ratio of peak signal current to rms noise current (optical SNR) for binary transmission.

Estimates of the required SNR to maintain a given error rate may be obtained using the standard table for the complementary error function. A plot of P(e) against $\frac{1}{2}$ erfc(u) is shown in Figure 11.36(a). This may be transposed into the characteristic illustrated in Figure 11.36(b) where the bit error rate which is equivalent to the error probability P(e) is shown as a function of the SNR following Eq. (11.24).

Example 11.2

Using the Gaussian approximation determine the required signal to noise ratios (optical and electrical) to maintain a BER of 10^{-9} on a baseband binary digital optical fiber link. It may be assumed that the decision threshold is set midway between the one and the zero level and that $2 \times 10^{-9} \simeq \text{erfc } 4.24$.

Solution: Under the above conditions, the probability of error is given by Eq. (11.24) where,

$$P(\text{e}) = \tfrac{1}{2}\,\text{erfc}\left(\frac{(S/N)^{\frac{1}{2}}}{2\sqrt{2}}\right) = 10^{-9}$$

Hence

$$\text{erfc}\left(\frac{(S/N)^{\frac{1}{2}}}{2\sqrt{2}}\right) = 2 \times 10^{-9}$$

and

$$\frac{(S/N)^{\frac{1}{2}}}{2\sqrt{2}} = 4.24$$

giving

$$(S/N)^{\frac{1}{2}} = 4.24 \times 2\sqrt{2} \simeq 12$$

The optical SNR may be defined in terms of the peak signal current and rms noise current as $i_{sig}/(\overline{i_N^2})^{\frac{1}{2}}$. Therefore using Eq. (11.23):

$$\frac{i_{sig}}{(\overline{i_N^2})^{\frac{1}{2}}} = \left(\frac{S}{N}\right)^{\frac{1}{2}} = 12 \text{ or } 10.8 \text{ dB}$$

The electrical SNR is defined by Eq. (11.23) as:

$$\frac{i_{sig}^2}{\overline{i_N^2}} = \frac{S}{N} = 144 \text{ or } 21.6 \text{ dB}$$

These results for the SNRs may be seen to correspond to a bit error rate of 10^{-9} on the curve shown in Figure 11.36(b).

However, the plot shown in Figure 11.36(b) does not reflect the best possible results, or those which may be obtained with an optimized receiver design. In this case, if the system is to be designed with a particular BER, the appropriate value of the error function is established prior to adjustment of the parameter values (signal levels, decision threshold level, avalanche gain, component values, etc.) in order to obtain this BER [Ref. 48]. It is therefore necessary to use the generalized forms of Eqs. (11.18) and (11.19) where:

$$P(0/1) = \tfrac{1}{2}\,\text{erfc}\left(\frac{|i_{sig\,1} - i_D|}{(\overline{i_{N1}^2})^{\frac{1}{2}}\sqrt{2}}\right) \quad (11.25)$$

$$P(0/1) = \tfrac{1}{2}\,\text{erfc}\left(\frac{|i_D - i_{sig\,0}|}{(\overline{i_{N0}^2})^{\frac{1}{2}}\sqrt{2}}\right) \quad (11.26)$$

where $i_{sig\,1}$ and $i_{sig\,0}$ are the signal currents, in the 1 and 0 states, respectively, and $\overline{i_{N1}^2}$ and $\overline{i_{N0}^2}$ are the corresponding mean square noise currents which may include both shot and thermal noise terms. Equations (11.25) and (11.26) allow a more exact evaluation of the error performance of the digital optical fiber system under the Gaussian approximation [Refs. 48 and 49]. Unfortunately, this approach does not give as simple a direct relationship between the BER and the analog SNR as the one shown in Eq. (11.24). Thus for estimates of SNR within this text we will make use of the slightly poorer approximation given by Eq. (11.24). Although this approximation does not give the correct decision threshold level or optimum avalanche gain it is reasonably successful at predicting bit error rate as a function of signal power and hence provides realistic estimates of the number of photons required at a practical detector in order to maintain given bit error rates.

For instance, let us consider a good avalanche photodiode receiver which we assume to be quantum noise limited. Hence we ignore the shot noise contribution from the dark current within the APD, as well as the thermal noise generated by the electronic amplifier. In practice, this assumption holds when the multiplication factor M is chosen to be sufficiently high to ensure that the SNR is determined by photon noise rather than by electronic amplifier noise, and the APD used has a low dark current. To determine the SNR for this ideal APD receiver it is useful to define the quantum noise on the primary photocurrent I_p within the device in terms of shot noise following Eq. (9.8). Therefore, the mean square shot noise current is given by:

$$\overline{i_s^2} = 2eBI_pM^2 \tag{11.27}$$

where e is the electronic charge and B is the post detection or effective noise bandwidth. It may be observed that the mean square shot noise current $\overline{i_s^2}$ given in Eq. (11.27) is increased by a factor M^2 due to avalanche gain in the APD. However, Eq. (11.27) does not give the total noise current at the output of the APD as there is an additional noise contribution from the random gain mechanism. The excess avalanche noise factor $F(M)$ incurred was discussed in Section 9.3.4 and defined by Eqs. (9.27) and (9.28). Equation (9.27) may be simplified [Ref. 50] to give an expression for electron injection in the low frequency limit of:

$$F(M) = kM + \left(2 - \frac{1}{M}\right)(1 - k) \tag{11.28}$$

where k is the ratio of the carrier ionization rates. Hence, the excess avalanche noise factor may be combined into Eq. (11.27) to give a total mean square shot noise current $\overline{i_n^2}$ as:

$$\overline{i_n^2} = 2eBI_pM^2F(M) \tag{11.29}$$

Furthermore, the avalanche multiplication mechanism raises the signal current to MI_p and therefore the SNR in terms of the peak signal power to rms noise power may be written as:

$$\frac{S}{N} = \frac{(MI_p)^2}{2eBI_pM^2F(M)} = \frac{I_p}{2eBF(M)} \tag{11.30}$$

Now, if we let z_{md} correspond to the average number of photons detected in a time period of duration τ, then

$$I_p = \frac{z_{md}e}{\tau} = \frac{z_m e\eta}{\tau} \tag{11.31}$$

where z_m is the average number of photons incident on the APD and η is the quantum efficiency of the device. Substituting for I_p in Eq. (11.30) we have:

$$\frac{S}{N} = \frac{z_m\eta}{2B\tau F(M)} \tag{11.32}$$

Rearranging Eq. (11.32) gives an expression for the average number of photons required within the signalling interval τ to detect a binary one in terms of the

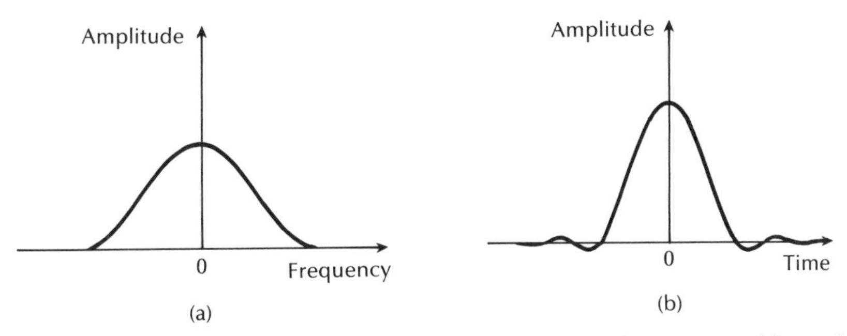

Figure 11.37 (a) Raised cosine spectrum. (b) Output of a system with a raised cosine output spectrum for a single input pulse.

received SNR for the good APD receiver as:

$$z_{\mathrm{m}} = \frac{2B_T \mathrm{F}(M)}{\eta}\left(\frac{S}{N}\right) \tag{11.33}$$

A reasonable pulse shape obtained at the receiver in order to reduce intersymbol interference has the raised cosine spectrum shown in Figure 11.37. The raised cosine spectrum for the received pulse gives a pulse response resulting in a binary pulse train passing through either full or zero amplitude at the centres of the pulse intervals and with transitions passing through half amplitude at points which are midway in time between pulse centres. For raised cosine pulse shaping and full τ signalling B_T is around 0.6. Hence the average number of photons required to detect a binary one using a good APD receiver at a specified BER may be estimated using Eq. (11.33) in conjunction with Eq. (11.24).

Example 11.3

A good APD is used as a detector in an optical fiber PCM receiver designed for baseband binary transmission with a decision threshold set midway between the zero and one signal levels. The APD has a quantum efficiency of 80%, a ratio of carrier ionization rates of 0.02 and is operated with a multiplication factor of 100. Assuming a raised cosine signal spectrum at the receiver, estimate the average number of photons which must be incident on the APD to register a binary one with a BER of 10^{-9}.

Solution: The electrical SNR required to obtain a BER of 10^{-9} at the receiver is given by the curve shown in Figure 11.36(b), or the solution to example 11.2 as 21.6 dB or 144. Also, the excess avalanche noise factor $F(M)$ may be determined using Eq. (11.28) where:

$$F(M) = kM + \left(2 - \frac{1}{M}\right)(1 - k)$$

$$= 2 + (2 - 0.01)(1 - 0.02)$$

$$= 3.95 \simeq 4$$

The average number of photons which must be incident at the receiver in order to maintain the BER can be estimated using Eq. (11.33) (assuming $B\mathcal{T} = 0.6$ for the raised cosine pulse spectrum) as:

$$z_m = \frac{2B\mathcal{T}F(M)}{\eta}\left(\frac{S}{N}\right)$$

$$= \frac{2 \times 0.6 \times 4 \times 144}{0.8}$$

$$= 864 \text{ photons}$$

The estimate in Example 10.3 gives a more realistic value for the average number of incident photons required at a good APD receiver in order to register a binary one with a BER of 10^{-9} than the quantum limit of twenty-one photons determined for an ideal photodetector in Example 9.1. However, it must be emphasized that the estimate in Example 11.3 applies to a good silicon APD receiver (with high sensitivity and low dark current) which is quantum noise limited, and that no account has been taken of the effects of either dark current within the APD or thermal noise generated within the preamplifier. It is therefore likely that at least 1000 incident photons are required at a good APD receiver to register a binary one and provide a BER of 10^{-9} [Ref. 51]. Nevertheless somewhat lower values may be achieved by setting the decision threshold below the half amplitude level because the shot noise on the zero level is lower than the shot noise on the one level.

The optical power required at the receiver P_o is simply the optical energy divided by the time interval over which it is incident. The optical energy E_o may be obtained directly from the average number of photons required at the receiver in order to maintain a particular BER following:

$$E_o = z_m hf \tag{11.34}$$

where hf is the energy associated with a single photon which is given by Eq. (6.1). In order that a binary one is registered at the receiver, the optical energy E_o must be incident over the bit interval τ. For system calculations we can assume a zero disparity code which has an equal density of ones and zeros. In this case the optical power required to register a binary one may be considered to be incident over two bit intervals giving:

$$P_o = \frac{E_o}{2\tau} \tag{11.35}$$

Substituting for E_o from Eq. (10.34) we obtain:

$$P_o = \frac{z_m hf}{2\tau} \tag{11.36}$$

Also as the bit rate B_T for the channel is the reciprocal of the bit interval τ, Eq. (11.36) may be written as:

$$P_o = \frac{z_m hf B_T}{2} \tag{11.37}$$

Equation (11.37) allows estimates of the incident optical power required at a good APD receiver in order to maintain a particular BER, based on the average number of incident photons. In system calculations these optical power levels are usually expressed in dBm. It may also be observed that the required incident optical power is directly proportional to the bit rate B_T which typifies a shot noise limited receiver.

Example 11.4

The receiver of example 11.3 operates at a wavelength of 1 μm. Assuming a zero disparity binary code, estimate the incident optical power required at the receiver to register a binary one with a BER of 10^{-9} at bit rates of 10 Mbit s^{-1} and 140 Mbit s^{-1}.

Solution: Under the above conditions, the required incident optical power may be obtained using Eq. (11.37) where,

$$P_o = \frac{z_m hf B_T}{2} = \frac{z_m hc B_T}{2\lambda}$$

At 10 Mbit s^{-1}:

$$P_o = \frac{864 \times 6.626 \times 10^{-34} \times 2.998 \times 10^8 \times 10^7}{2 \times 1 \times 10^{-6}}$$

$$= 858.2 \text{ pW}$$

$$= -60.7 \text{ dBm}$$

At 140 Mbit s^{-1}:

$$P_o = \frac{864 \times 6.626 \times 10^{-34} \times 2.998 \times 10^8 \times 14 \times 10^7}{2 \times 1 \times 10^{-14}}$$

$$= 12.015 \text{ nW}$$

$$= -49.2 \text{ dBm}$$

Example 11.4 illustrates the effect of direct proportionality between the optical power required at the receiver and the system bit rate. In the case considered, the required incident optical power at the receiver to give a BER of 10^{-9} must be increased by around 11.5 dB (factor of 14) when the bit rate is increased from 10 to 140 Mbit s^{-1} Also, comparison with Example 9.1 where a similar calculation was performed for an ideal photodetector operating at 10 Mbit s^{-1} emphasizes the necessity of performing the estimate for a practical photodiode. The good APD receiver considered in Example 11.4 exhibits around 16 dB less sensitivity than the ideal photodetector (i.e. quantum limit).

The assumptions made in the evaluation of Examples 11.3 and 11.4 are not generally valid when considering *p–i–n* photodiode receivers because these devices are seldom quantum noise limited due to the absence of internal gain within the photodetector. In this case thermal noise generated within the electronic amplifier is usually the dominating noise contribution and is typically 1×10^3 to 3×10^3 times larger than the peak response produced by the displacement current of a single electron–hole pair liberated in the detector. Hence, for reliable performance with a BER of 10^{-9}, between 1 and 3×10^4 photons must be detected when a binary one is incident on the receiver [Ref. 53].

This translates into sensitivities which are about 30 dB or more, less than the quantum limit. Finally, for a thermal noise limited receiver the input optical power is proportional to the square root of both the post detection or effective noise bandwidth and the SNR (i.e. $P_o \propto |(S/N)B|^{\frac{1}{2}}$). However, this result is best obtained from purely analog SNR considerations and therefore is dealt with in Section 11.7.1.

11.6.4 Channel losses

Another important factor when estimating the permissible separation between regenerative repeaters or the overall link length is the total loss encountered between the transmitter(s) and receiver(s) within the system. Assuming there are no dispersion penalties on the link, the total channel loss may be obtained by simply summing in decibels the installed fiber cable loss, the fiber–fiber jointing losses and the coupling losses of the optical source and detector. The fiber cable loss in decibels per kilometre α_{fc} is normally specified by the manufacturer, or alternatively it may be obtained by measurement (see Sections 13.2 and 13.10). It must be noted that the cabled fiber loss is likely to be greater than the uncabled fiber loss usually measured in the laboratory due to possible microbending of the fiber within the cabling process (see Section 4.8.1).

Loss due to joints (generally splices) on the link may also, for simplicity, be specified in terms of an equivalent loss in decibels per kilometre α_j. In fact, it is more realistic to regard α_j as a distributed loss since the optical attenuation resulting from the disturbed mode distribution at a joint does not only occur in the vicinity of the joint. Finally, the loss contribution attributed to the connectors α_{cr} (in decibels) used for coupling the optical source and detector to the fiber must be included in the overall channel loss. Hence the total channel loss C_L (in decibels) may be written as:

$$C_L = (\alpha_{fc} + \alpha_j)L + \alpha_{cr} \tag{11.38}$$

where L is the length in kilometres of the fiber cable either between regenerative repeaters or between the transmit and receive terminals for a link without repeaters.

Example 11.5

An optical fiber link of length 4 km comprises a fiber cable with an attenuation of 5 dB km^{-1}. The splice losses for the link are estimated at 2 dB km^{-1}, and the connector losses at the source and detector are 3.5 and 2.5 dB respectively. Ignoring the effects of dispersion on the link determine the total channel loss.

Solution: The total channel loss may be simply obtained using Eq. (11.38) where:

$$\begin{aligned} C_L &= (\alpha_{fc} + \alpha_j)L + \alpha_{cr} \\ &= (5 + 2)4 + 3.5 + 2.5 \\ &= 34 \text{ dB} \end{aligned}$$

11.6.5 Temporal response

The system design considerations must also take into account the temporal response of the system components. This is especially the case with regard to pulse dispersion on the optical fiber channel. The formula given in Eq. (11.38) allows determination of the overall channel loss in the absence of any pulse broadening due to the dispersion mechanisms within the transmission medium. However, the finite bandwidth of the optical system may result in overlapping of the received pulses or intersymbol interference, giving a reduction in sensitivity at the optical receiver. Therefore, either a worse BER must be tolerated, or the ISI must be compensated by equalization within the receiver (see Section 11.3.3). The latter necessitates an increase in optical power at the receiver which may be considered as an additional loss penalty. This additional loss contribution is usually called the dispersion–equalization or ISI penalty. The dispersion–equalization penalty D_L becomes especially significant in high bit rate multimode fiber systems and has been determined analytically for Gaussian shaped pulses [Ref. 48]. In this case it is given by:

$$D_L = \left(\frac{\tau_e}{\tau}\right)^4 \text{ dB} \tag{11.39}$$

where τ_e is the $1/e$ full width pulse broadening due to dispersion on the link and τ is the bit interval or period. For Gaussian shaped pulses, τ_e may be written in terms of the rms pulse width σ as (see Appendix B):

$$\tau_e = 2\sigma\sqrt{2} \tag{11.40}$$

Hence, substituting into Eq. (11.39) for τ_e and writing the bit rate B_T as the reciprocal of the bit interval τ gives:

$$D_L = 2(2\sigma B_T\sqrt{2})^4 \text{ dB} \tag{11.41}$$

Since the dispersion–equalization penalty as defined by Eq. (11.41) is measured in decibels, is may be included in the formula for the overall channel loss given by Eq. (11.38). Therefore, the total channel loss including the dispersion–equalization penalty C_{LD} is given by:

$$C_{LD} = (\alpha_{fc} + \alpha_j)L + \alpha_{cr} + D_L \text{ dB} \tag{11.42}$$

The dispersion–equalization penalty is usually only significant in wideband multimode fiber systems which exhibit intermodal as well as intramodal dispersion. Single-mode fiber systems which are increasingly being utilized for wideband long-haul applications are not generally limited by pulse broadening on the channel because of the absence of intermodal dispersion. However, it is often the case that intermodal dispersion is the dominant mechanism within multimode fibers. In Section 3.10.1 intermodal pulse broadening was considered to be a linear function of the fiber length L. Furthermore, it was indicated that the presence of mode coupling within the fiber made the pulse broadening increase at a slower rate proportional to $L^{\frac{1}{2}}$. Hence it is useful to consider the dispersion–equalization penalty in relation to fibers without and with mode coupling operating at various bit rates.

Example 11.6

The rms pulse broadening resulting from intermodal dispersion within a multimode optical fiber is 0.6 ns km^{-1}. Assuming this to be the dominant dispersion mechanism, estimate the dispersion–equalization penalty over an unrepeatered fiber link of length 8 km at bit rates of (a) 25 Mbit s^{-1} and (b) 150 Mbit s^{-1}. In both cases evaluate the penalty without and with mode coupling. The pulses may be assumed to have a Gaussian shape.

Solution: (a) *Without mode coupling.* The total rms pulse broadening over 8 km is given by:

$$\sigma_T = \sigma \times L = 0.6 \times 8 = 4.8 \text{ ns}$$

The dispersion–equalization penalty is given by Eq. (11.41) where:

$$\begin{aligned} D_L = 2(2\sigma_T B_T\sqrt{2})^4 &= 2(2 \times 4.8 \times 10^{-9} \times 25 \times 10^6\sqrt{2})^4 \\ &= 0.03 \text{ dB} \end{aligned}$$

With mode coupling. The total rms pulse broadening is:

$$\sigma_T \simeq \sigma\sqrt{L} = 0.6 \times \sqrt{8} = 1.7 \text{ ns}$$

Hence the dispersion–equalization penalty is:

$$D_L = 2(2 \times 1.7 \times 10^{-9} \times 25 \times 10^6\sqrt{2})^4$$
$$= 4.2 \times 10^{-4} \text{ dB (i.e. negligible)}$$

(b) *Without mode coupling.*

$$\sigma_T = 4.8 \text{ ns}$$
$$D_L = 2(2 \times 4.8 \times 10^{-9} \times 150 \times 10^6\sqrt{2})^4 = 34.38 \text{ dB}$$

With mode coupling.

$$\sigma_T = 1.7 \text{ ns}$$
$$D_L = 2(2 \times 1.7 = 10^{-9} \times 150 \times 10^6\sqrt{2})^4 = 0.54 \text{ dB}$$

Example 11.6(a) demonstrates that at low bit rates the dispersion–equalization penalty is very small if not negligible. In this case the slight advantage of the effect of mode coupling on the penalty is generally outweighed by increased attenuation on the link because of the mode coupling, which may be of the order of 1 dB km^{-1}. Example 11.6(b) indicates that at higher bit rates with no mode coupling the dispersion–equalization penalty dominates to the extent that it would be necessary to reduce the repeater spacing to between 4 and 5 km. However, it may be observed that encouragement of mode coupling on the link greatly reduces this penalty and outweighs any additional attenuation incurred through mode coupling within the fiber. In summary, it is clear that the dispersion equalization penalty need only be applied when considering wideband systems. Moreover, it is frequently the case that lower bit rate systems may be up-graded at a later date to a higher capacity without incurring a penalty which might necessitate a reduction in repeater spacing.

An alternative approach involving the calculation of the system rise time can be employed to determine the possible limitation on the system bandwidth resulting from the temporal response of the system components. Therefore, if there is not a pressing need to obtain the maximum possible bit rate over the maximum possible distance, it is sufficient within the system design to establish that the total temporal response of the system is adequate for the desired system bandwidth. Nevertheless this approach does allow for a certain amount of optimization of the system components, but at the exclusion of considerations regarding equalization and the associated penalty.

The total system rise time may be determined from the rise times of the individual system components which include the source (or transmitter), the fiber cable, and the detector (receiver). These times are defined in terms of a Gaussian response as the 10–90% rise (or fall) times of the individual components. The fiber cable 10–90% rise time may be separated into rise times arising from intermodal T_n

and intramodal or chromatic dispersion T_c. The total system rise time is given by [Ref. 56]:

$$T_{syst} = 1.1(T_S^2 + T_n^2 + T_c^2 + T_D^2)^{\frac{1}{2}} \quad (11.43)$$

where T_S and T_D are the source and detector 10 to 90% rise times, respectively, and all the rise times are measured in nanoseconds. Comparison of the rise time edge with the overall pulse dispersion results in the weighting factor of 1.1.

The maximum system bit rate $B_T(\max)$ is usually defined in terms of T_{syst} by consideration of the rise time of the simple RC filter circuit shown in Figure 11.38(a). For a voltage step input of amplitude V, the output voltage waveform $v_{out}(t)$ as a function of time t is:

$$v_{out}(t) = V[1 - \exp(-t/RC)] \quad (11.44)$$

Hence the 10 to 90% rise time t_r for the circuit is given by:

$$t_r = 2.2\ RC \quad (11.45)$$

The transfer function for this circuit is shown in Figure 11.38 (b) and is given by:

$$|H(\omega)| = \frac{1}{(1 + \omega^2 C^2 R^2)^{\frac{1}{2}}} \quad (11.46)$$

Therefore the 3 dB bandwidth for the circuit is

$$B = \frac{1}{2\pi RC} \quad (11.47)$$

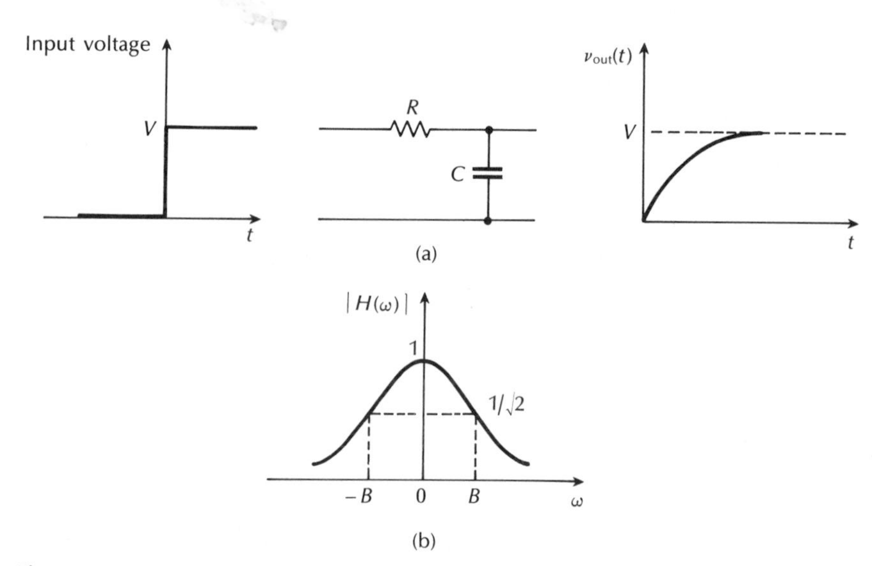

Figure 11.38 (a) The response of a low pass RC filter circuit to a voltage step input. (b) The transfer function $H(\omega)$ for the circuit in (a).

Combining Eqs. (11.45) and (11.47) gives,

$$t_r = \frac{2.2}{2\pi B} = \frac{0.35}{B} \tag{11.48}$$

The result for the 10 to 90% rise time indicated in Eq. (11.48) is of general validity, but a different constant term may be obtained with different filter circuits. However, for rise time calculations involving optical fiber systems the constant 0.35 is often utilized and hence in Eq. (11.48), $t_r = T_{syst}$. Alternatively, if an ideal (unrealizable) filter with an arbitrarily sharp cutoff is considered, the constant in Eq. (11.48) becomes 0.44. However, although this value for the constant is frequently employed when calculating the bandwidth of fiber from pulse dispersion measurements (see Section 13.3.1), the more conservative estimate obtained using a constant term of 0.35 is generally favoured for use in system rise time calculations [Refs. 56 and 57]. Also, in both cases it is usually accepted [Ref. 43] that to conserve the shape of a pulse with a reasonable fidelity through the RC circuit then the 3 dB bandwidth must be at least large enough to satisfy the condition $B\tau = 1$, where τ is the pulse duration. Combining this relation with Eq. (11.48) gives:

$$T_{syst} = t_r = 0.35\tau \tag{11.49}$$

For an RZ pulse format, the bit rate $B_T = B = 1/\tau$ (see Section 3.8) and hence substituting into Eq. (11.49) gives:

$$B_T(\max) = \frac{0.35}{T_{syst}} \tag{11.50}$$

Alternatively for an NRZ pulse format $B_T = B/2 = 1/2\tau$ and therefore the maximum bit rate is given by:

$$B_T(\max) = \frac{0.7}{T_{syst}} \tag{11.51}$$

Thus the upper limit on T_{syst} should be less than 35% of the bit interval for an RZ pulse format and less than 70% of the bit interval for an NRZ pulse format.

The effects of mode coupling are usually neglected in calculations involving system rise time, and hence the pulse dispersion is assumed to be a linear function of the fiber length. This results in a pessimistic estimate for the system rise time and therefore provides a conservative value for the maximum possible bit rate.

Example 11.7

An optical fiber system is to be designed to operate over an 8 km length without repeaters. The rise times of the chosen components are:

Source (LED)	8 ns
Fiber: intermodal	5 ns km^{-1}
(pulse broadening) intramodal	1 ns km^{-1}
Detector (*p–i–n* photodiode)	6 ns

From system rise time considerations, estimate the maximum bit rate that may be achieved on the link when using an NRZ format.

Solution: The total system rise time is given by Eq. (11.43) as:

$$\begin{aligned} T_{\text{syst}} &= 1.1(T_S^2 + T_n^2 + T_c^2 + T_D^2)^{\frac{1}{2}} \\ &= 1.1(8^2 + (8 \times 5)^2 + (8 \times 1)^2 + 6^2)^{\frac{1}{2}} \\ &= 46.2 \text{ ns} \end{aligned}$$

Hence the maximum bit rate for the link using an NRZ format is given by Eq. (11.51) where:

$$B_T(\max) = \frac{0.7}{T_{\text{syst}}} = \frac{0.7}{46.2 \times 10^{-9}} \simeq 15.2 \text{ Mbit s}^{-1}$$

The rise time calculations indicate that this will support a maximum bit rate of 15.2 Mbit s^{-1} which for an NRZ format is equivalent to a 3 dB optical bandwidth of 7.6 MHz (i.e. the NRZ format has two bit intervals per wavelength).

Once it is established that pulse dispersion is not a limiting factor, the major design exercise is the optical power budget for the system.

11.6.6 Optical power budgeting

Power budgeting for a digital optical fiber communication system is performed in a similar way to power budgeting within any communication system. When the transmitter characteristics, fiber cable losses and receiver sensitivity are known, the relatively simple process of power budgeting allows the repeater spacing or the maximum transmission distance for the system to be evaluated. However, it is necessary to incorporate a system margin into the optical power budget so that small variations in the system operating parameters do not lead to an unacceptable decrease in system performance. The operating margin is often included in a safety margin M_a which also takes into account possible source and modal noise, together with receiver impairments such as equalization error, noise degradations and eye opening impairments. The safety margin depends to a large extent on the system components as well as the system design procedure and is typically in the range 5 to 10 dB. Systems using an injection laser transmitter generally require a larger safety margin (e.g. 8 dB) than those using an LED source (e.g. 6 dB) because the temperature variation and ageing of the LED are less pronounced.

The optical power budget for a system is given by the following expression:

$$P_i = P_o + C_L + M_a \text{ dB} \tag{11.52}$$

where P_i is the mean input optional power launched into the fiber, P_o is the mean incident optical power required at the receiver and C_L (or C_{LD} when there is a dispersion–equalization penalty) is the total channel loss given by Eq. (11.38) (or

Eq. (11.42)). Therefore the expression given in Eq. (11.52) may be written as:

$$P_{\mathrm{i}} = P_{\mathrm{o}} + (\alpha_{\mathrm{fc}} + \alpha_{\mathrm{j}})L + \alpha_{\mathrm{cr}} + M_{\mathrm{a}} \text{ dB} \tag{11.53}$$

Alternatively, when a dispersion–equalization penalty is included Eq. (11.52) becomes:

$$P_{\mathrm{i}} = P_{\mathrm{o}} + (\alpha_{\mathrm{fc}} + \alpha_{\mathrm{j}})L + \alpha_{\mathrm{cr}} + D_{\mathrm{L}} + M_{\mathrm{a}} \text{ dB} \tag{11.54}$$

Equations (11.53) and (11.54) allow the maximum link length without repeaters to be determined, as demonstrated in Example 11.8.

Example 11.8

The following parameters are established for a long-haul single-mode optical fiber system operating at a wavelength of 1.3 μm.

Mean power launched from the laser transmitter	-3 dBm
Cabled fiber loss	0.4 dB km^{-1}
Splice loss	0.1 dB km^{-1}
Connector losses at the transmitter and receiver	1 dB each
Mean power required at the APD receiver:	
when operating at 35 Mbit s^{-1} (BER 10^{-9})	-55 dBm
when operating at 400 Mbit s^{-1} (BER 10^{-9})	-44 dBm
Required safety margin	7 dB

Estimate:

(a) the maximum possible link length without repeaters when operating at 35 Mbit s^{-1} (BER 10^{-9}). It may be assumed that there is no dispersion–equalization penalty at this bit rate.

(b) the maximum possible link length without repeaters when operating at 400 Mbit s^{-9} (BER 10^{-9}) and assuming no dispersion–equalization penalty.

(c) the reduction in the maximum possible link length without repeaters of (b) when there is a dispersion–equalization penalty of 1.5 dB. It may be assumed for the purposes of this estimate that the reduced link length has the 1.5 dB penalty.

Solution: (a) When the system is operating at 35 Mbit s^{-1} an optical power budget may be performed using Eq. (11.53), where

$$P_{\mathrm{i}} - P_{\mathrm{o}} = (\alpha_{\mathrm{fc}} + \alpha_{\mathrm{j}})L + \alpha_{\mathrm{cr}} + M_{\mathrm{a}} \text{ dB}$$

$$-3 \text{ dBm} - (-55 \text{ dBm}) = (\alpha_{\mathrm{fc}} + \alpha_{\mathrm{j}})L + \alpha_{\mathrm{cr}} + M_{\mathrm{a}}$$

Hence,

$$(\alpha_{\mathrm{fc}} + \alpha_{\mathrm{j}})L = 52 - \alpha_{\mathrm{cr}} - M_{\mathrm{a}}$$

$$0.5L = 52 - 2 - 7$$

$$L = \frac{43}{0.5} = 86 \text{ km}$$

(b) Again using Eq. (11.53) when the system is operating at 400 Mbit s^{-1}:

$$-3 \text{ dBm} - (-44 \text{ dBm}) = (\alpha_{fc} + \alpha_j)L + \alpha_{cr} + M_a$$

$$(a_{fc} + \alpha_j)L = 41 - 2 - 7$$

$$L = \frac{32}{0.5} = 64 \text{ km}$$

(c) Performing the optical power budget using Eq. (11.54) gives:

$$P_i - P_o = (\alpha_{fc} + \alpha_j)L + \alpha_{cr} + D_L + M_a$$

Hence,

$$0.5L = 41 - 2 - 1.5 - 7$$

and

$$L = \frac{30.5}{0.5} = 61 \text{ km}$$

Thus there is a reduction of 3 km in the maximum possible link length without repeaters.

Although in Example 11.8 we have demonstrated the use of the optical power budget to determine the maximum link length without repeaters, it is also frequently used to aid decisions in relation to the combination of components required for a particular optical fiber communication system. In this case the maximum transmission distance and the required bandwidth may already be known. Therefore, the optical power budget is used to provide a basis for optimization in the choice of the system components, whilst also establishing that a particular component configuration meets the system requirements.

Example 11.9

Components are chosen for a digital optical fiber link of overall length 7 km and operating at a 20 Mbit s^{-1} using an RZ code. It is decided that an LED emitting at 0.85 μm with graded index fiber to a *p–i–n* photodiode is a suitable choice for the system components, giving no dispersion–equalization penalty. An LED which is capable of launching an average of 100 μW of optical power (including the connector loss into a 50 μm core diameter graded index fiber is chosen. The proposed fiber cable has an attenuation of 2.6 dB km^{-1} and requires splicing every kilometre with a loss of 0.5 dB per splice. There is also a connector loss at the receiver of 1.5 dB. The receiver requires mean incident optical power of -41 dBm in order to give the necessary BER of 10^{-10}, and it is predicted that a safety margin of 6 dB will be required.

Write down the optical power budget for the system and hence determine its viability.

Solution:

Mean optical power launched into the fiber from the transmitter (100 μm)	-10 dBm
Receiver sensitivity at 20 Mbit s^{-1} (BER 10^{-10})	-41 dBm
Total system margin	31 dB
Cabled fiber loss (7×2.6 dB km^{-1})	18.2 dB
Splice losses (6×0.5 dB)	3.0 dB
Connector loss (1×1.5 dB)	1.5 dB
Safety margin	6.0 dB
Total system loss	28.7 dB
Excess power margin	2.3 dB

Based on the figures given the system is viable and provides a 2.3 dB excess power margin. This could give an extra safety margin to allow for possible future splices if these were not taken into account within the original safety margin.

11.6.7 Line coding

The preceding discussions of digital system design have assumed that only information bits are transmitted, and that the 0 and 1 symbols are equally likely. However, within digital line transmission there is a requirement for redundancy in the line coding to provide efficient timing recovery and synchronization (frame alignment) as well as possible error detection and correction at the receiver. Line coding also provides suitable shaping of the transmitted signal power spectral density. Hence the choice of line code is an important consideration within digital optical fiber system design.

Binary line codes are generally preferred because of the large bandwidth available in optical fiber communications. In addition, these codes are less susceptible to any temperature dependence of optical sources and detectors. Under these conditions two level codes are more suitable than codes which utilize an increased number of levels (multilevel codes). Nevertheless, these factors do not entirely exclude the use of multilevel codes, and it is likely that ternary codes (three levels 0, $\frac{1}{2}$, 1) which give increased information transmission per symbol over binary codes will be considered for some system applications. The corresponding symbol transmission rate (i.e. bit rate) for a ternary code may be reduced by a factor of 1.58 ($\log_2 3$), whilst still providing the same information transmission rate as a similar system using a binary code. It must be noted that this gain in information capacity for a particular bit rate is obtained at the expense of the dynamic range between adjacent levels as there are three levels inserted in place of two. This is exhibited as a 3 dB SNR penalty at the receiver when compared with a binary system at a given BER. Therefore ternary codes (and higher multilevel codes) are not attractive for long-haul systems.

For the reasons described above most digital optical fiber communication systems currently in use employ binary codes. In practice, binary codes are designed which insert extra symbols into the information data stream on a regular and logical basis to minimize the number of consecutive identical received symbols, and to facilitate efficient timing extraction at the receiver by producing a high density of decision level crossings. The reduction in consecutive identical symbols also helps to minimize the variation in the mean signal level which provides a reduction in the low frequency response requirement of the receiver. This shapes the transmitted signal spectrum by reducing the d.c. component. However, this factor is less important for optical fiber systems where a.c. coupling is performed with capacitors, unlike metallic cable systems where transformers are often used, and the avoidance of d.c. components is critical. A further advantage is apparent within the optical receiver with a line code which is free from long identical symbol sequences, and where the continuous presence of 0 and 1 levels aids decision level control and avoids gain instability effects.

Two level block codes of the *nBmB* type fulfil the above requirements through the addition of a limited amount of redundancy. These codes convert blocks of n bits into blocks of m bits where $m > n$ so that the difference between the number of transmitted ones and zeros is on average zero. A simple code of this type is the 1*B*2*B* code in which a 0 may be transmitted as 01, and a 1 as 10. This encoding format is shown in Figure 11.39(b) and is commonly referred to as biphase or Manchester encoding. It may be observed that with this code there are never more than two consecutive identical symbols, and that two symbols must be transmitted

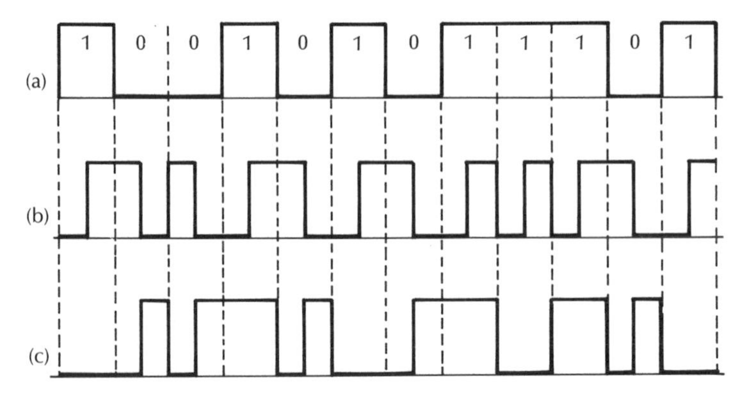

Figure 11.39 Examples of binary 1B2B codes used in optical fiber communications: (a) unencoded NRZ data; (b) biphase or Manchester encoding; (c) coded mark inversion (CMI) encoding.

for one information bit, giving 50% redundancy. Thus twice the transmission bandwidth is required for the 1*B*2*B* code which restricts its use to systems where pulse dispersion is not a limiting factor. Another example of a 1*B*2*B* code which is illustrated in Figure 11.39(c) is the coded mark inversion (CMI) code. In this code a digit 0 is transmitted as 01 and the digit 1 alternately as 00 or 11.

Timing information is obtained from the frequent positive to negative transitions, but, once again, the code is highly redundant requiring twice as many transmitted bits as input information bits.

More efficient codes of this type requiring less redundancy exist such as the 3*B*4*B*, 5*B*6*B* and the 7*B*8*B* codes. There is a trade-off within this class of code between the complexity of balancing the number of zeros and ones, and the added redundancy. The increase in line symbol rate (bit rate) and the corresponding power penalty over encoded binary transmission is given by the ratio $m:n$. Hence, considering the widely favoured 5*B*6*B* code, the symbol rate is increased by a factor of 1.2 whilst the power penalty is also equal to 1.2 or about 0.8 dB. It is therefore necessary to take into account the increased bandwidth requirement and the power penalty resulting from coding within the optical fiber system design.

Simple error monitoring may be provided with block codes, at the expense of a small amount of additional redundancy, by parity checking. Each block of N bits can be made to have an even (even parity) or odd (odd parity) number of ones so that any single error in a block can be identified. More extensive error detection and error correction may be provided with increased redundancy and equipment complexity. This is generally not considered worth while unless it is essential that the digital transmission system is totally secure (e.g. data transmission applications). Alternatively, error monitoring when using block codes may be performed by measuring the variation in disparity between the numbers of ones and zeros within the received bit pattern. Any variation in the accumulated disparity above an upper limit or below a lower limit allowed by a particular code is indicated as an error. Further discussion of error correction with relation to disparity may be found in Ref. 65. Moreover, variations on the above block codes to provide efficient high speed digital transmission have been devised (e.g. Ref. 66). Such line coding schemes possess a good balance of ones and zeros together with jitter suppression and the capability to provide a simple error monitoring function.

11.7 Analog systems

In Section 11.5 we indicated that the vast majority of optical fiber communication systems are designed to convey digital information (e.g. analog speech encoded as PCM). However, in certain areas of the telecommunication network or for particular applications, information transfer in analog form is still likely to remain for some time to come, or be advantageous. Therefore, analog optical fiber transmission will undoubtedly have a part to play in future communication networks, especially in situations where the optical fiber link is part of a larger analog network (e.g. microwave relay network). Use of analog transmission in these areas avoids the cost and complexity of digital terminal equipment, as well as degradation due to quantization noise. This is especially the case with the transmission of video signals over short distances where the cost of high speed analog to digital (A–D) and D–A converters is not generally justified. Hence, there are many applications such as direct cable television and common antenna television (CATV) where analog optical fiber systems may be utilized.

There are limitations, however, inherent to analog optical fiber transmission, some of which have been mentioned previously. For instance, the unique requirements of analog transmission over digital are for high signal to noise ratios at the receiver output which necessitates high optical input power (see Section 9.2.5), and high end to end linearity to avoid distortion and prevent crosstalk between different channels of a multiplexed signal (see Section 11.4.2). Furthermore, it is instructive to compare the SNR constraints for typical analog optical fiber and coaxial cable systems.

In a coaxial cable system the fundamental limiting noise is $4KTB$, where K is Boltzmann's constant, T is the absolute temperature, and B is the effective noise bandwidth for the channel. If we assume for simplicity that the coaxial cable loss is constant and independent of frequency, the SNR for a coaxial system is

$$\left(\frac{S}{N}\right)_{\text{coax}} = \frac{V^2 \exp(-\alpha_N)}{Z_0 4KTB} \tag{11.55}$$

where α_N is the attenuation in nepers between the transmitter and receiver, V is the peak output voltage, and Z_0 is the impedance of the coaxial cable.

The SNR for an analog optical fiber system may be obtained by referring to Eq. (9.11) where

$$\left(\frac{S}{N}\right)_{\text{fiber}} = \frac{\eta P_0}{2hfB} \tag{11.56}$$

The expression given in Eq. (11.56) includes the fundamental limiting noise for optical fiber systems which is $2hfB$. Although Eq. (11.56) is sufficiently accurate for the purpose of comparison it applies to an unmodulated optical carrier. A more accurate expression would take into account the depth of modulation for the analog optical fiber system which cannot be unity [Ref. 53].* The average received optical power P_o may be expressed in terms of the average input (transmitted) optical power P_i as

$$P_o = P_i \exp(-\alpha_N) \tag{11.57}$$

Substituting for P_o into Eq. (11.56) gives

$$\left(\frac{S}{N}\right)_{\text{fiber}} = \frac{P_i \exp(-\alpha_N)}{2hfB} \tag{11.58}$$

* Strictly speaking, Eq. (11.56) depicts the optical carrier to noise ratio (CNR).

Equations (11.55) and (11.58) allow a simple comparison to be made of available SNR (or CNR) between analog coaxial and optical fiber systems, as demonstrated in Example 11.10.

Example 11.10

A coaxial cable system operating at a temperature of 17 °C has a transmitter peak output voltage of 5 V with a cable impedance of 100 Ω. An analog optical fiber system uses an injection laser source emitting at 0.85 μm and launches an average of 1 mW of optical power into the fiber cable. The optical receiver comprises a photodiode with a quantum efficiency of 70%. Assuming the effective noise bandwidth and the attenuation between the transmitter and receiver for the two systems is identical, estimate in decibels the ratio of the SNR of the coaxial system to the SNR of the fiber system.

Solution: Using Eqs. (11.55) and (11.58) for the SNRs of the coaxial and fiber systems respectively:

$$\text{Ratio} = \frac{\left(\dfrac{S}{N}\right)_{\text{coax}}}{\left(\dfrac{S}{N}\right)_{\text{fiber}}} = \frac{\dfrac{V^2 \exp(-\alpha_{\text{N}})}{Z_0 4KTB}}{\dfrac{\eta P_{\text{i}} \exp(-\alpha_{\text{N}})}{2hfB}} = \frac{V^2 hf}{2KTZ_0 \eta P_{\text{i}}}$$

$$= \frac{V^2 hc}{2KTZ_0 \eta P_{\text{i}} \lambda}$$

Hence

$$\text{Ratio} = \frac{25 \times 6.626 \times 10^{-34} \times 2.998 \times 10^{8}}{2 \times 1.385 \times 10^{-23} \times 290 \times 100 \times 0.7 \times 1 \times 10^{-3} \times 0.85 \times 10^{-6}}$$

$$= 1.04 \times 10^{4} \simeq 40 \text{ dB}$$

The optical fiber channel in Example 11.10 has around 40 dB less SNR available than the alternative coaxial channel exhibiting similar channel losses. This results both from $2hfB$ being larger than $4KTB$ and from the far smaller transmitted power within the optical system. Furthermore, it must be noted that the comparison was made using an injection laser transmitter. If an LED transmitter with 10 to 20 dB less optical output power was compared, the coaxial system would display an advantage in the region 50 to 60 dB. For this reason it is difficult to match with fiber systems the SNR requirements of some analog coaxial links, even though the fiber cable attenuation may be substantially lower than that of the coaxial cable.

The analog signal can be transmitted within an optical fiber communication system using one of several modulation techniques. The simplest form of analog modulation for optical fiber communications is direct intensity modulation (D–IM) of the optical source. In this technique the optical output from the source is modulated simply by varying the current flowing in the device around a suitable bias or mean level in proportion to the message. Hence the information signal is transmitted directly in the baseband.

Alternatively, the baseband signal can be translated on to an electrical subcarrier by means of amplitude, phase or frequency modulation using standard techniques, prior to intensity modulation of the optical source. Pulse analog techniques where a sequence of pulses are used for the carrier may also be utilized. In this case a suitable parameter such as the pulse amplitude, pulse width, pulse position or pulse frequency is electrically modulated by the baseband signal. Again, the modulated electrical carrier is transmitted optically by intensity modulation of the optical source.

Direct modulation of the optical source in frequency, phase or polarization rather than by intensity requires these parameters to be well defined throughout the optical fiber system. There is much interest in this area and optical component technology has been developed which will allow practical system implementation. These techniques concerned with coherent optical transmission are discussed in Chapter 12.

11.7.1 Direct intensity modulation (D–IM)

A block schematic for an analog optical fiber system which uses direct modulation of the optical source intensity with the baseband signal is shown in Figure 11.40(a). Obviously, no electrical modulation or demodulation is required with this technique, making it both inexpensive and easy to implement.

The transmitted optical power waveform as a function of time $P_{opt}(t)$, an example of which is illustrated in Figure 11.40(b), may be written as:

$$P_{opt}(t) = P_i(1 + m(t)) \tag{11.59}$$

where P_i is the average transmitted optical power (i.e. the unmodulated carrier power) and $m(t)$ is the intensity modulating signal which is proportional to the source message $a(t)$. For a cosinusoidal modulating signal:

$$m(t) = m_a \cos \omega_m t \tag{11.60}$$

where m_a is the modulation index or the ratio of the peak excursion from the average to the average power as shown in Figure 11.40(b), and ω_m is the angular frequency of the modulating signal. Combining Eqs. (11.59) and (11.60) we get:

$$P_{opt}(t) = P_i(1 + m_a \cos \omega_m t) \tag{11.61}$$

Furthermore, assuming transmission medium has zero dispersion, the received optical power will be of the same form as Eq. (11.61), but with an average received optical power P_o. Hence the secondary photocurrent $I(t)$ generated at an APD receiver with a multiplication factor M is given by:

$$I(t) = I_p M(1 + m_a \cos \omega_m t) \tag{11.62}$$

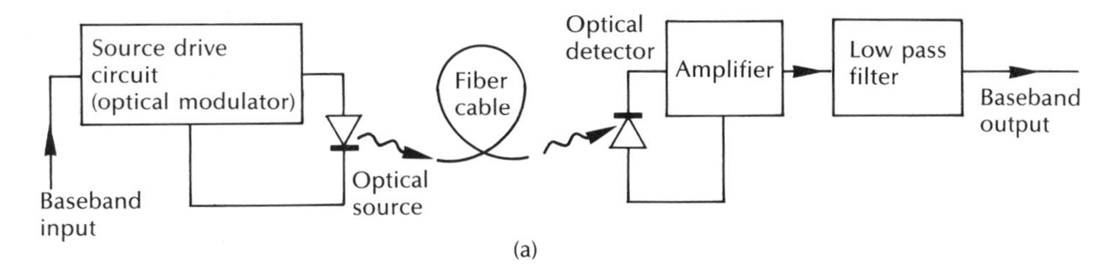

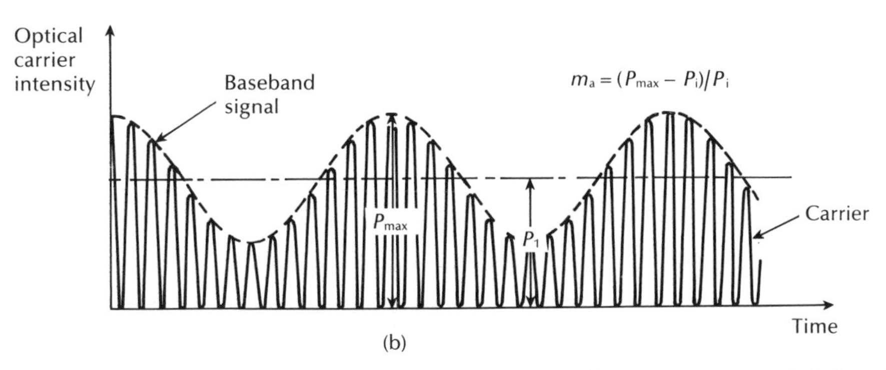

Figure 11.40 (a) Analog optical fiber system employing direct intensity modulation. (b) Time domain representation showing direct intensity modulation of the optical carrier with a baseband analog signal.

where the primary photocurrent obtained with an unmodulated carrier I_p is given by Eq. (8.8) as,

$$I_p = \frac{\eta e}{hf} P_o \tag{11.63}$$

The mean square signal current $\overline{i_{sig}^2}$ which is obtained from Eq. (11.62) is given by:

$$\overline{i_{sig}^2} = \tfrac{1}{2}(m_a M I_p)^2 \tag{11.64}$$

The total average noise in the system is composed of quantum, dark current, and thermal (circuit) noise components. The noise contribution from quantum effects and detector dark current may be expressed as the mean square total shot noise current for the APD receiver $\overline{i_{SA}^2}$ given by Eq. (9.21) where the excess avalanche noise factor is written following Eq. (9.26) as $F(M)$ such that:

$$\overline{i_{SA}^2} = 2eB(I_p + I_d)M^2F(M) \tag{11.65}$$

where B is the effective noise or post detection bandwidth.

Thermal noise generated by the load resistance R_L and the electronic amplifier noise can be expressed in terms of the amplifier noise figure F_n referred to R_L as given by Eq. (9.17). Thus the total mean square noise current $\overline{i_N^2}$ may be written as:

$$\overline{i_N^2} = 2eB(I_p + I_d)M^2F(M) + \frac{4KTBF_n}{R_L} \tag{11.66}$$

The SNR defined in terms of the ratio of the mean square signal current to the mean square noise current (rms signal power to rms noise power) for the APD receiver is therefore given by:

$$\left(\frac{S}{N}\right)_{\text{rms}} = \frac{\overline{i_{\text{sig}}^2}}{\overline{i_N^2}} = \frac{\frac{1}{2}(m_a M I_p)^2}{2eB(I_p + I_d)M^2F(M) + (4KTBF_n/R_L)} \quad \text{(APD)} \tag{11.67}$$

It must be emphasized that the SNR given in Eq. (11.67) is defined in terms of rms signal power rather than peak signal power used previously. When a unity gain photodetector is utilized in the receiver (i.e. *p–i–n* photodiode) Eq. (11.67) reduces to:

$$\left(\frac{S}{N}\right)_{\text{rms}} = \frac{\frac{1}{2}(m_a I_p)^2}{2eB(I_p + I_d) + (4KTBF_n/R_L)} \quad (p\text{–}i\text{–}n) \tag{11.68}$$

Moreover, the SNR for video transmission is often defined in terms of the peak to peak picture signal power to the rms noise power and may include the ratio of luminance to composite video b. Using this definition in the case of the unity gain detector gives:

$$\left(\frac{S}{N}\right)_{p-p} = \frac{(2m_a I_p b)^2}{2eB(I_p + I_d) + (4KTBF_n/R_L)} \quad (p\text{–}i\text{–}n) \tag{11.69}$$

It may be observed that excluding b, the SNR defined in terms of the peak to peak signal power given in Eq. (11.69) is a factor of 8 (or 9 dB) greater than that defined in Eq. (11.68).

Example 11.11

A single TV channel is transmitted over an analog optical fiber link using direct intensity modulation. The video signal which has a bandwidth of 5 MHz and a ratio of luminance to composite video of 0.7 is transmitted with a modulation index of 0.8. The receiver contains a *p–i–n* photodiode with a responsivity of 0.5 A W^{-1} and a preamplifier with an effective input impedance of 1 MΩ together with a noise figure of 1.5 dB. Assuming the receiver is operating at a temperature of 20 °C and neglecting the dark current in the photodiode, determine the average incident optical power required at the receiver (i.e. receiver sensitivity) in order to maintain a peak to peak signal power to rms noise power ratio of 55 dB.

Solution: Neglecting the photodiode dark current, the peak to peak signal rms noise power ratio is given following Eq. (11.69) as:

$$\left(\frac{S}{N}\right)_{p-p} = \frac{(2m_a I_p b)^2}{2eBI_p + (4KTBF_n/R_L)}$$

The photocurrent I_p may be expressed in terms of the average incident optical power at the receiver P_o using Eq. (8.4) as:

$$I_p = RP_o$$

where R is the responsivity of the photodiode. Hence

$$\left(\frac{S}{N}\right)_{p-p} = \frac{(2m_a RP_o b)^2}{2eBRP_o + (4KTBF_n/R_L)}$$

and

$$\left(\frac{S}{N}\right)_{p-p}\left(2eBRP_o = \frac{4KTBF_n}{R_L}\right) = (2m_a RP_o b)^2$$

Rearranging

$$(2m_a Rb)^2 P_o^2 - \left(\frac{S}{N}\right)_{p-p} 2eBRP_o - \left(\frac{S}{N}\right)_{p-p} \frac{4KTBF_n}{R_L} = 0$$

where

$$\begin{aligned}(2m_a Rb)^2 &= 4 \times 0.64 \times 0.25 \times 0.49\\ &= 0.314\end{aligned}$$

$$\begin{aligned}\left(\frac{S}{N}\right)_{p-p} 2eBR &= 3.162 \times 10^5 \times 2 \times 1.602 \times 10^{-19} \times 5 \times 10^6 \times 0.5\\ &= 2.533 \times 10^{-7}\end{aligned}$$

$$\begin{aligned}\left(\frac{S}{N}\right)_{p-p} \frac{4KTBF_n}{R_L} &= \frac{3.162 \times 10^5 \times 4 \times 1.381 \times 10^{-23} \times 293 \times 5 \times 10^6 \times 1.413}{10^6}\\ &= 3.616 \times 10^{-14}\end{aligned}$$

Therefore,

$$0.314P_o^2 - 2.533 \times 10^{-7} P_o - 3.616 \times 10^{-14} = 0$$

and

$$P_o = \frac{2.533 \times 10^{-7} \pm \sqrt{[(2.533 \times 10^{-7})^2 - (-4 \times 0.314 \times 3.616 \times 10^{-14})]}}{0.628}$$

$$\begin{aligned}&= 0.93\ \mu\text{W}\\ &= -30.3\ \text{dBm}\end{aligned}$$

It must be noted that the low noise preamplification depicted in Example 11.11 may not always be obtained, and that higher thermal noise levels will adversely affect the receiver sensitivity for a given SNR. This is especially the case with lower SNRs, as illustrated in the peak to peak signal power to rms noise power ratio against average received optical power characteristics for a video system shown in Figure 11.41. The performance of the system for various values of mean square thermal noise current $\overline{i_t^2} = 4KTBF_n/R_L$, where $\overline{i_t^2}$ is expressed as a spectral density in $A^2\ Hz^{-1}$, as indicated. The value for the receiver sensitivity obtained in Example 11.11 is approaching the quantum limit, also illustrated in Figure 11.41, which is the best that could possibly be achieved with a noiseless amplifier.

The quantum or shot noise (when ignoring the photodetector dark current) limit occurs with large values of signal current (i.e. primary photocurrent) at the receiver. Considering a *p–i–n* photodiode receiver, this limiting case which corresponds to large SNR is given by Eq. (11.68) when neglecting the device dark current as:

$$\left(\frac{S}{N}\right)_{\text{rms}} \simeq \frac{m_a^2 I_p}{4eB} \text{ (quantum noise limit)} \tag{11.70}$$

Using the relationship between the average received optical power P_o and the primary photocurrent given in Eq. (11.63) allows Eq. (11.70) to be expressed as:

$$P_o \simeq \frac{4hf}{m_a^2 \eta}\left(\frac{S}{N}\right)_{\text{rms}} B \tag{11.71}$$

Equation (11.71) indicates that for a quantum noise limited analog receiver, the optical input power is directly proportional to the effective noise or post detection bandwidth B. A similar result was obtained in Eq. (11.37) for the digital receiver.

Alternatively at low SNRs thermal noise is dominant, and the thermal noise limit when I_p is small, which may also be obtained from Eq. (11.68), is given by:

$$\left(\frac{S}{N}\right)_{\text{rms}} \simeq \frac{(m_a I_p)^2 R_L}{8KTBF_n} \text{ (thermal noise limit)} \tag{11.72}$$

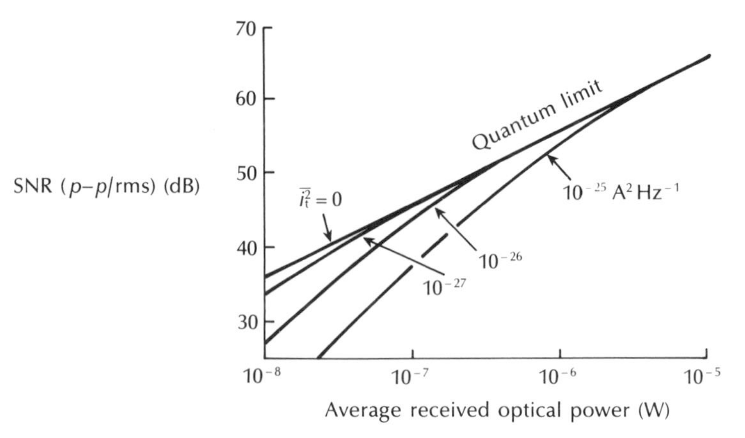

Figure 11.41 Peak to peak signal power to rms noise power ratio against the average received optical power for a direct intensity modulated video system and various levels of thermal noise given by $\overline{i_t^2}$. Reproduced with permission from G. G. Windus, *Marconi Rev*, **XLI**, p. 77. 1981.

Again substituting for I_p from Eq. (11.63) gives:

$$P_o \simeq \frac{hf}{e\eta m_a^2}\left(\frac{8KTF_n}{R_L}\right)^{\frac{1}{2}}\left(\frac{S}{N}\right)_{rms}^{\frac{1}{2}} B^{\frac{1}{2}} \tag{11.73}$$

Therefore it may be observed from Eq. (11.73) that in the thermal noise limit the average incident optical power is directly proportional to $B^{\frac{1}{2}}$ instead of the direct dependence on B shown in Eq. (11.71) for the quantum noise limit. The dependence expressed in Eq. (11.73) is typical of the *p–i–n* photodiode receiver operating at low optical input power levels. Thus Eq. (11.73) may be used to estimate the required input optical power to achieve a particular SNR for a *p–i–n* photodiode receiver which is dominated by thermal noise.

Example 11.12

An analog optical fiber link employing D–IM has a *p–i–n* photodiode receiver in which thermal noise is dominant. The system components have the following characteristics and operating conditions.

p–i–n photodiode quantum efficiency	60%
effective load impedance for the photodiode	50 kΩ
preamplifier noise figure	6 dB
operating wavelength	1 μm
operating temperature	300 K
receiver post detection bandwidth	10 MHz
modulation index	0.5

Estimate the required average incident optical power at the receiver in order to maintain an SNR, defined in terms of the mean square signal current to mean square noise current, of 45 dB.

Solution: The average incident optical power for a thermal noise limited *p–i–n* photodiode receiver may be estimated using Eq. (11.73) where:

$$P_o \simeq \frac{hf}{e\eta m_a^2}\left(\frac{8KTF_n}{R_L}\right)^{\frac{1}{2}}\left(\frac{S}{N}\right)_{rms}^{\frac{1}{2}} B^{\frac{1}{2}}$$

and

$$\frac{hf}{e\eta m_a^2} = \frac{hc}{e\eta m_a^2\lambda} = \frac{6.626 \times 10^{-34} \times 2.998 \times 10^{8}}{1.602 \times 10^{-19} \times 0.6 \times 0.25 \times 1 \times 10^{-6}}$$

$$= 8.267$$

$$\left(\frac{8KTF_n}{R_L}\right)^{\frac{1}{2}} = \left(\frac{8 \times 1.381 \times 10^{-23} \times 300 \times 4}{50 \times 10^{3}}\right)^{\frac{1}{2}}$$

$$= 1.628 \times 10^{-12}$$

$$\left(\frac{S}{N}\right)_{rms}^{\frac{1}{2}} B^{\frac{1}{2}} = (3.162 \times 10^{4} \times 10^{7})^{\frac{1}{2}}$$

$$= 5.623 \times 10^{5}$$

Hence,

$$P_o \simeq 8.267 \times 1.628 \times 10^{-12} \times 5.623 \times 10^5$$
$$= 7.57\ \mu W$$
$$= -21.2\ dBm$$

Therefore, as anticipated, the receiver sensitivity in the thermal noise limit is low.

11.7.2 System planning

Many of the general planning considerations for optical fiber systems outlined in Section 11.4 may be applied to analog transmission. However, extra care must be taken to ensure that the optical source and, to a lesser extent, the detector have linear input–output characteristics, in order to avoid distortion of the transmitted optical signal. Furthermore, careful optical power budgeting is often necessary with analog systems because of the generally high SNRs required at the optical receiver (40 to 60 dB) in comparison with digital systems (20 to 25 dB), to obtain a similar fidelity. Therefore, although analog system optical power budgeting may be carried out in a similar manner to digital systems (see Section 11.6.6), it is common for the system margin, or the difference between the optical power launched into the fiber and the required optical power at the receiver, for analog systems to be quite small (perhaps only 10 to 20 dB when using an LED source to *p–i–n* photodiode receiver). Consequently, analog systems employing direct intensity modulation of the optical source tend to have a limited transmission distance without repeaters which generally prohibits their use for long-haul applications.

Example 11.13

A D–IM analog optical fiber link of length 2 km employs an LED which launches mean optical power of -10 dBm into a multimode optical fiber. The fiber cable exhibits a loss of 3.5 dB km^{-1} with splice losses calculated at 0.7 dB km^{-1}. In addition there is a connector loss at the receiver of 1.6 dB. The *p–i–n* photodiode receiver has a sensitivity of -25 dBm for an SNR ($\overline{i_{sig}^2}/\overline{i_N^2}$) of 50 dB and with a modulation index of 0.5. It is estimated that a safety margin of 4 dB is required. Assuming there is no dispersion–equalization penalty:

(a) Perform an optical power budget for the system operating under the above conditions and ascertain its viability.
(b) Estimate any possible increase in link length which may be achieved using an injection laser source which launches mean optical power of 0 dBm into the fiber cable. In this case the safety margin must be increased to 7 dB.

Solution: (a) Optical power budget:

Mean power launched into the fiber cable from the LED transmitter	−10 dBm
Mean optical power required at the *p–i–n* photodiode receiver for SNR of 50 dB and a modulation index of 0.5	−25 dBm
Total system margin	15 dB
Fiber cable loss (2 × 3.5)	7.0 dB
Splice losses (2 × 0.7)	1.4 dB
Connector loss at the receiver	1.6 dB
Safety margin	4.0 dB
Total system loss	14.0 dB
Excess power margin	1.0 dB

Hence the system is viable, providing a small excess power margin.

(b) In order to calculate any possible increase in link length when using the injection laser source we refer to Eq. (11.53), where

$$P_{\mathrm{i}} - P_{\mathrm{o}} = (\alpha_{\mathrm{fc}} + \alpha_{\mathrm{j}})L + \alpha_{\mathrm{cr}} + M_{\mathrm{a}} \text{ dB}$$

Therefore,

$$0 \text{ dBm} - (-25 \text{ dBm}) = (3.5 + 0.7)L + 1.6 + 7.0$$

and

$$4.2L = 25 - 8.6 = 16.4 \text{ dB}$$

giving

$$L = \frac{16.4}{4.2} = 3.9 \text{ km}$$

Hence the use of the injection laser gives a possible increase in the link length of 1.9 km or almost a factor of 2. It must be noted that in this case the excess power margin has been reduced to zero.

The transmission distance without repeaters for the analog link of Example 11.13 could be extended further by utilizing an APD receiver which has increased sensitivity. This could facilitate an increase in the maximum link length to around 7 km, assuming no additional power penalties or excess power margin. Although this is quite a reasonable transmission distance, it must be noted that a comparable digital system could give in the region of 13 km transmission without repeaters.

The temporal response of analog systems may be determined from system rise time calculations in a similar manner to digital systems (see Section 11.6.5). The maximum permitted 3 dB optical bandwidth for analog systems in order to avoid dispersion penalties follows from Eq. (11.49) and is given by:

$$B_{\text{opt}}(\text{max}) = \frac{0.35}{T_{\text{syst}}} \tag{11.74}$$

Hence calculation of the total system 10 to 90% rise time T_{syst} allows the maximum system bandwidth to be estimated. Often this calculation is performed in order to establish that the desired system bandwidth may be achieved using a particular combination of system components.

Example 11.14

The 10 to 90% rise times for possible components to be used in a D–IM analog optical fiber link are specified below:

Source (LED)	10 ns
Fiber cable: intermodal	9 ns km^{-1}
intramodal	2 ns km^{-1}
Detector (APD)	3 ns

The desired link length without repeaters is 5 km and the required optical bandwidth is 6 MHz. Determine whether the above combination of components give an adequate temporal response.

Solution: Equation (11.74) may be used to calculate the maximum permitted system rise time which gives the desired bandwidth where:

$$T_{\text{syst}}(\text{max}) = \frac{0.35}{B_{\text{opt}}} = \frac{0.35}{6 \times 10^6} = 58.3 \text{ ns}$$

The total system rise time using the specified components can be estimated using Eq. (11.43) as:

$$\begin{aligned} T_{\text{syst}} &= 1.1(T_S^2 + T_n^2 + T_c^2 + T_D^2)^{\frac{1}{2}} \\ &= 1.1(10^2 + (9 \times 5)^2 + (2 \times 5)^2 + 3^2)^{\frac{1}{2}} \\ &\simeq 52 \text{ ns} \end{aligned}$$

Therefore the specified components give a system rise time which is adequate for the bandwidth and distance requirements of the optical fiber link. However, there is little leeway for upgrading the system in terms of bandwidth or distance without replacing one or more of the system components.

11.7.3 Subcarrier intensity modulation

Direct intensity modulation of the optical source is suitable for the transmission of a baseband analog signal. However, if the wideband nature of the optical fiber medium is to be fully utilized it is essential that a number of baseband channels are multiplexed on to a single fiber link. This may be achieved with analog transmission through frequency division multiplexing of the individual baseband channels. Initially, the baseband channels must be translated on to carriers of different frequency by amplitude modulation (AM), frequency modulation (FM) or phase modulation (PM) prior to being simultaneously transmitted as an FDM signal. The frequency translation may be performed in the electrical regime where the baseband analog signals modulate electrical subcarriers and are then frequency division multiplexed to form a composite electrical signal prior to intensity modulation of the optical source.

A block schematic of an analog system employing this technique, which is known as subcarrier intensity modulation, is shown in Figure 11.42. The baseband signals are modulated onto radiofrequency (RF) subcarriers by either AM, FM or PM and multiplexed before being applied to the optical source drive circuit.* Hence an intensity modulated (IM) optical signal is obtained which may be either AM–IM, FM–IM or PM–IM. In practice, however, system output SNR considerations dictate that generally only the latter two modulation formats are used. Nevertheless, systems may incorporate two levels of electrical modulation whereby the baseband channels are initially amplitude modulated prior to frequency or phase modulation [Ref. 68]. The FM or PM signal thus obtained is then used to intensity modulate the optical source. At the receive terminal the transmitted optical signal is detected prior to electrical demodulation and demultiplexing (filtering) to obtain the originally transmitted baseband signals.

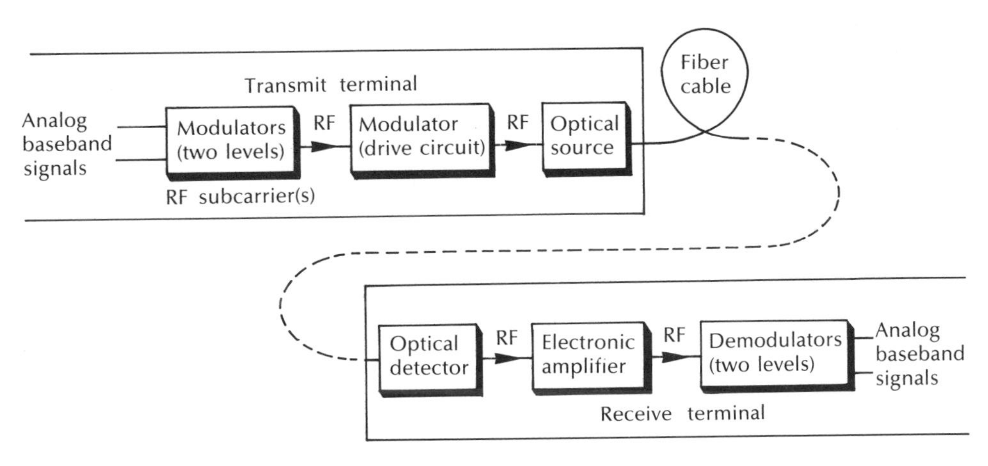

Figure 11.42 Subcarrier intensity modulation system for analog optical fiber transmission.

* When microwave frequency rather than radiofrequency subcarriers are employed the strategy is usually referred to as subcarrier multiplexing or SCM (see Section 11.9.2).

A further major advantage of subcarrier intensity modulation is the possible improvement in SNR that may be obtained during subcarrier demodulation. In order to investigate this process it is necessary to obtain a general expression for the SNR of the intensity modulated optical carrier which may then be applied to the subcarrier intensity modulation formats. Therefore, as with D–IM, considered in the preceding section, an electrical signal $m(t)$ modulates the source intensity. The transmitted optical power waveform is of the same form as Eq. (11.59), where:

$$P_{\mathrm{opt}}(t) = P_{\mathrm{i}}(1 + m(t)) \tag{11.75}$$

Also the secondary photon $I(t)$ generated at an APD receiver following Eq. (11.62) is given by:

$$I(t) = I_{\mathrm{p}}M(1 + m(t)) \tag{11.76}$$

The mean square signal current $\overline{i_{\mathrm{sig}}^2}$ may be written as [Ref. 65]:

$$\overline{i_{\mathrm{sig}}^2} = (I_{\mathrm{p}}M)^2 P_{\mathrm{m}} \tag{11.77}$$

where P_{m} is the total power of $m(t)$, which can be defined in terms of the spectral density $S_{\mathrm{m}}(\omega)$ of $m(t)$ occupying a one-sided bandwidth B_{m} Hz as:

$$P_{\mathrm{m}} = \frac{1}{2\pi}\int_{-2\pi B_{\mathrm{m}}}^{2\pi B_{\mathrm{m}}} S_{\mathrm{m}}(\omega)\,\mathrm{d}\omega \tag{11.78}$$

Hence the SNR defined in terms of the mean square signal current to mean square noise current (i.e. rms signal power to rms noise power) using Eqs. (11.77) and (11.66) can now be written as:

$$\begin{aligned}\left(\frac{S}{N}\right)_{\mathrm{rms}} &= \frac{\overline{i_{\mathrm{sig}}^2}}{\overline{i_{\mathrm{N}}^2}} = \frac{(I_{\mathrm{p}}M)^2 P_{\mathrm{m}}}{2eB_{\mathrm{m}}(I_{\mathrm{p}} + I_{\mathrm{d}})M^2F(M) + (4KTBF_{\mathrm{n}}/R_{\mathrm{L}})} \\ &= \frac{I_{\mathrm{p}}^2 P_{\mathrm{m}}}{2B_{\mathrm{m}}e(I_{\mathrm{p}} + I_{\mathrm{d}})F(M) + (4KTBF_{\mathrm{n}}/M^2R_{\mathrm{L}})} \\ &= \frac{(RP_{\mathrm{o}})^2 P_{\mathrm{m}}}{2B_{\mathrm{m}}N_{\mathrm{o}}} \qquad \text{(D–IM)}\end{aligned} \tag{11.79}$$

where we substitute for I_{p} from Eq. (8.4) and for notational simplicity write:

$$N_{\mathrm{o}} = e(I_{\mathrm{p}} + I_{\mathrm{d}})F(M) + \frac{4KTBF_{\mathrm{n}}}{M^2R_{\mathrm{L}}} \tag{11.80}$$

The result obtained in Eq. (11.79) gives the SNR for a direct intensity modulated optical source where the total modulating signal power is P_{m}. In this context Eq. (11.79) is simply a more general form of Eq. (11.67). However, we are now in a position to examine the signal to noise performance of various subcarrier intensity modulation formats.

11.7.4 Subcarrier double sideband modulation (DSB–IM)

A simple way to translate the spectrum of the baseband message signal $a(t)$ is by direct multiplication with the subcarrier waveform $A_c \cos \omega_c t$ giving the modulated waveform $m(t)$ as:

$$m(t) = A_c a(t) \cos \omega_c t \qquad (11.81)$$

where A_c is the amplitude, and ω_c the angular frequency of the subcarrier waveform. For a cosinusoidal modulating signal ($\cos \omega_m t$) the subcarrier electric field $E_m(t)$ becomes:

$$E_m(t) = \frac{A_c}{2} \cos (\omega_c + \omega_m)t + \cos (\omega_c - \omega_m)t \qquad (11.82)$$

giving the upper and lower sidebands. The time and frequency domain representations of the modulated waveform are shown in Figure 11.43. It may be observed from the frequency domain representation that only the two sideband components are present as indicated in Eq. (11.82). This modulation technique is known as double sideband modulation (DSB) or double sideband suppressed carrier (DSBSC) amplitude modulation. It provides a more efficient method of translating the spectrum of the baseband message signal than conventional full amplitude modulation where a large carrier component is also present in the modulated waveform.

The DSB signal shown in Figure 11.43 intensity modulates the optical source. Therefore the transmitted optical power waveform is obtained by combining Eqs. (11.75) and (11.81) where for simplicity we set the carrier amplitude A_c to unity,

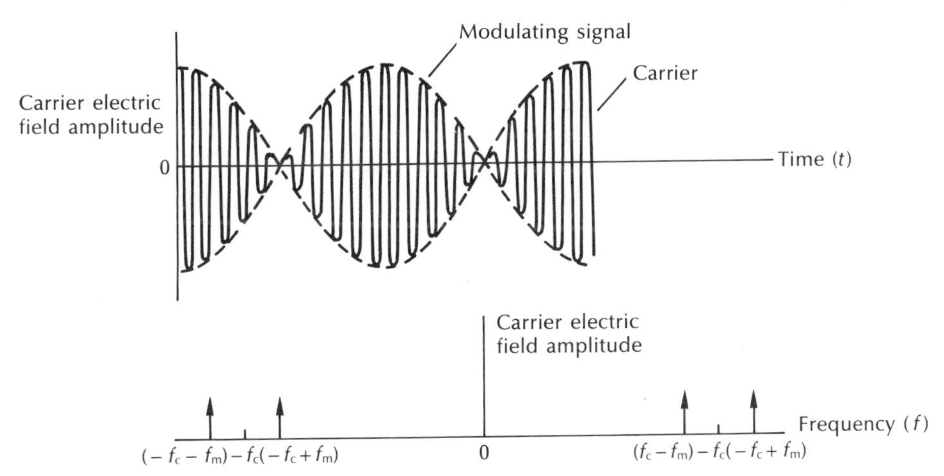

Figure 11.43 Time and frequency domain representations of double sideband modulation.

giving:

$$P_{opt}(t) = P_i(1 + a(t) \cos \omega_c t) \tag{11.83}$$

Furthermore, in order to prevent overmodulation, the value of the message signal is normalized such that $|a(t) \leqslant 1|$ with power $P_a \leqslant 1$. The DSB modulated electrical subcarrier occupies a bandwidth $B_m = 2B_a$, and with a carrier amplitude of unity, $P_m = P_a/2$. Hence, the ratio of rms signal power to rms noise power obtained within the subcarrier bandwidth at the input to the double sideband demodulator is given by Eq. (11.79), where:

$$\left(\frac{S}{N}\right)_{rms} \textit{input DSB} = \frac{(RP_o)^2 P_a/2}{2 \times 2B_a N_o} = \frac{(RP_o)^2 P_a}{8B_a N_o} \tag{11.84}$$

However, an ideal DSB demodulator gives a detection gain of 2 or 3 dB improvement in SNR [Ref. 68]. This yields an output SNR of:

$$\left(\frac{S}{N}\right)_{rms} \textit{output DSB} = 2\left(\frac{S}{N}\right)_{rms} \textit{input DSB} = \frac{(RP_o)^2 P_a}{4B_a N_o} \tag{11.85}$$

Comparison of the result obtained in Eq. (11.85) with that using direct intensity modulation of the baseband signal given by Eq. (11.79) shows a 3 dB degradation in SNR when employing DSB–IM under the same conditions of bandwidth (i.e. $B_m = B_a$), modulating signal power (i.e. $P_m = P_a$), detector photocurrent and noise. For this reason DSB–IM systems (and also AM–IM systems in general) are usually not considered efficient for optical fiber communications. Therefore far more attention is devoted to both FM–IM and PM–IM systems.

11.7.5 Subcarrier frequency modulation (FM–IM)

In this modulation format, the subcarrier is frequency modulated by the message signal. The conventional form for representing the baseband signal which intensity modulates the optical source is [Ref. 68]:

$$m(t) = A_c \cos\left[\omega_c t + k_f \int_0^t a(\tau)\, d\tau\right] \tag{11.86}$$

where k_f is the angular frequency deviation in radians per second per unit of $a(t)$. To prevent intensity over modulation, the carrier amplitude, $A_c \leqslant 1$. The generally accepted expression for the bandwidth which is referred to as Carson's rule is given by:

$$B_m \simeq 2(D_f + 1)B_a \tag{11.87}$$

where D_f is the frequency deviation ratio defined by:

$$D_f = \frac{\textit{peak frequency deviation}}{\textit{bandwidth of } a(t)} = \frac{f_d}{B_a} \tag{11.88}$$

The peak frequency deviation in the subcarrier FM signal f_d is given by:

$$f_d = k_f \max |a(t)| \tag{11.89}$$

Hence the SNR at the input to the subcarrier FM demodulator is:

$$\left(\frac{S}{N}\right)_{rms} \textit{input FM} = \frac{(RP_o)^2(A_c^2/2)}{2B_m N_o} \tag{11.90}$$

The subcarrier demodulator operating above threshold yields an output SNR [Ref. 65]:

$$\left(\frac{S}{N}\right)_{rms} \textit{output FM} = 6D_f^2(D_f^2 + 1)\,\frac{P_a(RP_o)^2(A_c^2/2)}{2B_m N_o} \tag{11.91}$$

Substituting for B_m from Eq. (11.87) gives:

$$\left(\frac{S}{N}\right)_{rms} \textit{output FM} = \frac{3D_f^2 P_a(RP_o)^2(A_c^2/2)}{2B_a N_o} \tag{11.92}$$

The result obtained in Eq. (11.92) indicates that a significant improvement in the postdetection SNR may be achieved by using wideband FM–IM as demonstrated in the following example.

Example 11.15

(a) A D–IM and an FM–IM optical fiber communication system are operated under the same conditions of modulating signal power and bandwidth, detector photocurrent and noise. Furthermore, in order to maximize the SNR in the FM–IM system, the amplitude of the subcarrier is set to unity. Derive an expression for the improvement in post detection SNR of the FM–IM system over the D–IM system. It may be assumed that the SNR is defined in terms of the rms signal power to rms noise power.

(b) The FM–IM system described in (a) has an 80 MHz subcarrier which is modulated by a baseband signal with a bandwidth of 4 kHz such that the peak frequency deviation is 400 kHz. Use the result obtained in (a) to determine the improvement in post detection SNR (in decibels) over the D–IM system operating under the same conditions. Also estimate the bandwidth of the FM signal.

Solution: (a) The output SNR for the D–IM system is given by Eq. (11.79) where we can write $P_m = P_a$ and $B_m = B_a$. Hence:

$$\left(\frac{S}{N}\right)_{rms} \textit{output D–IM} = \frac{(RP_o)^2 P_a}{2B_a N_o}$$

The corresponding output SNR for the FM–IM system is given by Eq. (11.92) where setting A_c to unity gives:

$$\left(\frac{S}{N}\right)_{rms} \textit{output FM} = \frac{3D_f^2 P_a (RP_o)^2}{4B_a N_o}$$

Therefore the improvement in SNR of the FM–IM system over the D–IM system is given by:

$$\textit{SNR improvement} = \frac{[3D_f^2 P_a (RP_o)^2]/(4B_a N_o)}{[(RP_o)^2 P_a]/(2B_a N_o)}$$

$$= \frac{3D_f^2}{2}$$

and,

$$\textit{SNR improvement in decibels} = 10 \log_{10} \frac{3}{2} D_f^2$$

$$= 1.76 + 20 \log_{10} D_f$$

(b) The frequency deviation ratio is given by Eq. (11.88) where:

$$D_f = \frac{f_d}{B_a} = \frac{400 \times 10^3}{4 \times 10^3} = 100$$

Therefore the SNR improvement is:

$$\textit{SNR improvement} = 1.76 + 20 \log_{10} 100$$
$$= 41.76 \text{ dB}$$

The bandwidth of the FM–IM signal may be estimated using Eq. (11.87) where:

$$B_m \simeq 2(D_f + 1)B_a = 2(100 + 1)4 \times 10^3$$
$$= 808 \text{ kHz}$$

This result indicates that the system is operating as a wideband FM–IM system.

Example 11.15 illustrates that a substantial improvement in the post detection SNR over D–IM may be obtained using FM–IM. However, it must be noted that this is at the expense of a tremendous increase in the bandwidth required (808 kHz) for transmission of the 4 kHz baseband channel.

11.7.6 Subcarrier phase modulation (PM–IM)

With this modulation technique the instantaneous phase of the subcarrier is set proportional to the modulating signal. Hence in a PM–IM system the modulating signal $m(t)$ may be written as [Ref. 68]:

$$m(t) = A_c \cos(\omega_c t + k_p a(t)) \tag{11.93}$$

where k_p is the phase deviation constant in radians per unit of $a(t)$. Again the carrier amplitude $A_c \leqslant 1$ to prevent intensity overmodulation. Moreover, the bandwidth of the PM–IM signal is given by Carson's rule as:

$$B_m \simeq 2(D_p + 1)B_a \tag{11.94}$$

where D_p is the frequency deviation ratio for the PM–IM system. In common with subcarrier frequency modulation the frequency deviation ratio is defined as:

$$D_p = \frac{f_d}{B_a} \tag{11.95}$$

where f_d is the peak frequency deviation of the subcarrier PM signal, which is given by:

$$f_d = k_p \max \left| \frac{da(t)}{dt} \right| \tag{11.96}$$

The SNR at the input to the subcarrier PM modulator is:

$$\left(\frac{S}{N}\right)_{rms} \textit{input PM} = \frac{(RP_o)^2 A_c^2/2}{2B_m N_o} \tag{11.97}$$

The output SNR from an ideal subcarrier PM demodulator operating above threshold is [Ref. 65]:

$$\left(\frac{S}{N}\right)_{rms} \textit{output PM} = \frac{D_p^2 P_a (RP_o)^2 A_c^2/2}{2B_a N_o} \tag{11.98}$$

The result given in Eq. (11.98) suggests that an improvement in SNR over D–IM may be obtained using PM–IM, especially when the SNR is maximized with $A_c = 1$. However, comparison of PM–IM with FM–IM indicates that the latter modulation format gives the greatest improvement.

Example 11.16

A PM–IM and an FM–IM optical fiber communication system are operated under the same conditions of bandwidth, baseband signal power, subcarrier amplitude, frequency deviation, detector photocurrent and noise. Assuming the demodulators for both systems are ideal, determine the ratio (in decibels) of the output SNR from the FM–IM system.

Solution: The output SNR from the FM–IM system is given by Eq. (11.92) where:

$$\left(\frac{S}{N}\right)_{rms} \textit{output FM} = \frac{3D_f^2 P_a (RP_o)^2 A_c^2/2}{2B_a N_o}$$

Substituting for D_f from Eq. (11.88) gives:

$$\left(\frac{S}{N}\right)_{rms} output\ FM = \frac{3f_d^2 P_a (RP_o)^2 A_c^2/2}{2B_a^3 N_o}$$

The output SNR for the PM–IM system is given by Eq. (11.98) where:

$$\left(\frac{S}{N}\right)_{rms} output\ PM = \frac{D_p^2 P_a (RP_o)^2 A_c^2/2}{2B_a N_o}$$

Substituting for D_p from Eq. (11.95) gives:

$$\left(\frac{S}{N}\right)_{rms} output\ PM = \frac{f_d^2 P_a (RP_o)^2 A_c^2/2}{2B_a^3 N_o}$$

The ratio of the output SNRs from the FM–IM and the PM–IM system is:

$$Ratio = \frac{[3f_d^2 P_a (RP_o)^2 A_c^2/2]/(2B_a^3 N_o)}{[f_d^2 P_a (RP_o)^2 A_c^2/2]/(2B_a^3 N_o)}$$

$$= 3$$

$$= 4.77\ \text{dB}$$

Example 11.16 shows that the FM–IM system has a superior output SNR by some 4.77 dB over the corresponding PM–IM system. Nevertheless, this does not prohibit the use of PM–IM systems for analog optical fiber communications as they still exhibit a substantial improvement in output SNR over D–IM systems, as well as allowing frequency division multiplexing. It should be noted, however, that a similar bandwidth penalty to FM–IM is incurred using this modulation format.

11.7.7 Pulse analog techniques

Pulse modulation techniques for analog transmission, rather than encoding the analog waveform into PCM, were mentioned within the system design considerations of Section 11.4. The most common techniques are pulse amplitude modulation (PAM), pulse width modulation (PWM), pulse position modulation (PPM) and pulse frequency modulation (PFM). All the pulse analog techniques employ pulse modulation in the electrical regime prior to intensity modulation of the optical source. However, PAM–IM is affected by source nonlinearities and is less efficient than D–IM, and therefore is usually discounted. PWM–IM is also inefficient since a large part of the transmitted energy conveys no information as only variations of the pulse width about a nominal value are of interest [Ref. 69]. Alternatively, PPM–IM and PFM–IM offer distinct advantages since the modulation affects the timing of the pulses, thus allowing the transmission of very narrow pulses. Hence, PPM–IM and PFM–IM provide similar signal to noise

may be obtained using Eq. (11.99), where:

$$\left(\frac{S}{N}\right)_{p-p} = \frac{3(T_0 f_D MRP_{po})^2}{(2\pi T_R B)^2 \overline{i_N^2}}$$

$$= \frac{3(5 \times 10^{-8} \times 5 \times 10^6 \times 60 \times 0.7 \times 10^{-7})^2}{(2\pi \times 12 \times 10^{-9} \times 6 \times 10^6)^2 \times 10^{-17}}$$

$$= 1.62 \times 10^6$$

$$= 62.1 \text{ dB}$$

The result of Example 11.17(b) illustrates the possibility of acquiring high SNRs at the output to a PFM–IM system using a regenerative receiver with achievable receiver noise levels and with moderate input optical signal power to the receiver.

11.8 Distribution systems

Thus far, the considerations in this chapter have effectively concerned only point-to-point and primarily unidirectional optical fiber communication systems. A strategy for obtaining bidirectional optical transmission on the same fiber link is described in Section 11.9.3, whilst in this section we discuss the implementation aspects of a growing area of activity within optical fiber communications, namely that of multiterminal distribution systems. For example, two major areas of application for such multiterminal distribution systems or networks which are dealt with in Chapter 14 are the telecommunication local access network (Section 14.2.3) and local area networks (Section 14.7).

Although many variants or hybrid topologies have been explored, the three basic multiterminal system architectures comprise the ring, bus and star configurations. The former topology, which has largely found implementation as a closed path or loop where consecutive nodes or terminals are connected by a series of point-to-point fiber links, is discussed in relation to the fiber distributed data interface covered in Section 14.7.1. With the latter topologies, however, substantial progress has been made into the realization of multiterminal distribution systems and networks which do not simply comprise a series of point-to-point fiber links. In particular, they make use of the basic passive coupling devices described in Section 5.6

It is instructive to form a comparison between the topological implementations of the bus and star distribution systems when each employ passive optical couplers to direct the signals to particular nodes. Block schematics for these two configurations are shown in Figure 11.45 where, for the purposes of the comparison, the linear nature of the bus is replicated in the star network through the positioning of the nodes in a linear manner. It is clear, however, that the star network configurations shown in Figures 14.4(c) and 14.28(d) are more representative of the use of the star topology to provide a widely distributed multiterminal network. Moreover the star–bus network implementation displayed in Figure 11.45(b) is not very economic in its use of fiber cable in comparison with the bus topology (Figure 11.45(a)) for a linear ordering of the network nodes.

The bus configuration illustrated in Figure 11.45(a) utilizes three port fiber couplers (see Section 5.6.1) to act as both beam splitter/combiner devices for the transmit and receive paths at each node, as well as passive fiber access couplers or taps along the bus link. However, whereas in the former case the split ratio is around 50%, in the latter tapping application the split ratio is often reduced to between 5 and 10% for the tap fiber so that the throughput optical power is a factor of 9 to 18 times greater than the optical power tapped off. Such an arrangement enables a larger optical power level to be transmitted down the bus and thus ensures adequate power at nodes distant from the transmit terminal.

Let us consider the total loss between node 1 and node $N-1$. It should be noted in the configuration shown in Figure 11.45(a) that the path between nodes 1 and $N-1$ exhibits the maximum loss because the final fiber tap couples only 10% of the incident optical power into the beam splitter of node $N-1$. By contrast, the path to node N obtains a factor of nine times this power level. Clearly, this situation could be modified by using a fiber beam splitter in place of the fiber tap in order to connect these two final nodes on to the bus.

Notwithstanding the above point we now consider the optical power budget for the worst case node interconnection (nodes 1 to $N-1$) for the multiterminal bus system of Figure 11.45(a). It is apparent that to obtain the total channel loss $C_L(1, N-1)$ between these two nodes the losses through each of the components must be summed. Let us commence at the transmit terminal, node 1. Then designating the connector losses in decibels as α_{cr} and assuming no excess loss in combining the transmitted signal on to the bus, a loss of $2\alpha_{cr}$ is obtained after the first beam splitter. The loss per kilometre exhibited by the fiber cable α_{fc} enables the total fiber cable loss between the two terminals to be written as $(N-1)\alpha_{fc}L_{bu}$ where L_{bu} is equal to the fiber length between each of the access couplers. Furthermore, the total loss incurred by the signal in passing through the access couplers or taps between nodes 1 and $N-1$ (excepting the final access coupler at which the signal to node $N-1$ is tapped off) is given by $(2\alpha_{cr} + L_{ac})$ $(N-3)$ where L_{ac} is the insertion loss of the access coupler. At the final access coupler before node $N-1$ the loss obtained is $(2\alpha_{cr} + L_{tr})$ where L_{tr} is the loss due to the tap ratio of the device. Finally, a splitting loss L_{sp} occurs at the beam splitter together with a further connector loss α_{cr} at the optical receiver of node $N-1$. The total channel

Figure 11.45 (overleaf) Distribution system implementations: (a) linear bus system/network; (b) star system/network configured as a bus for comparative purposes.

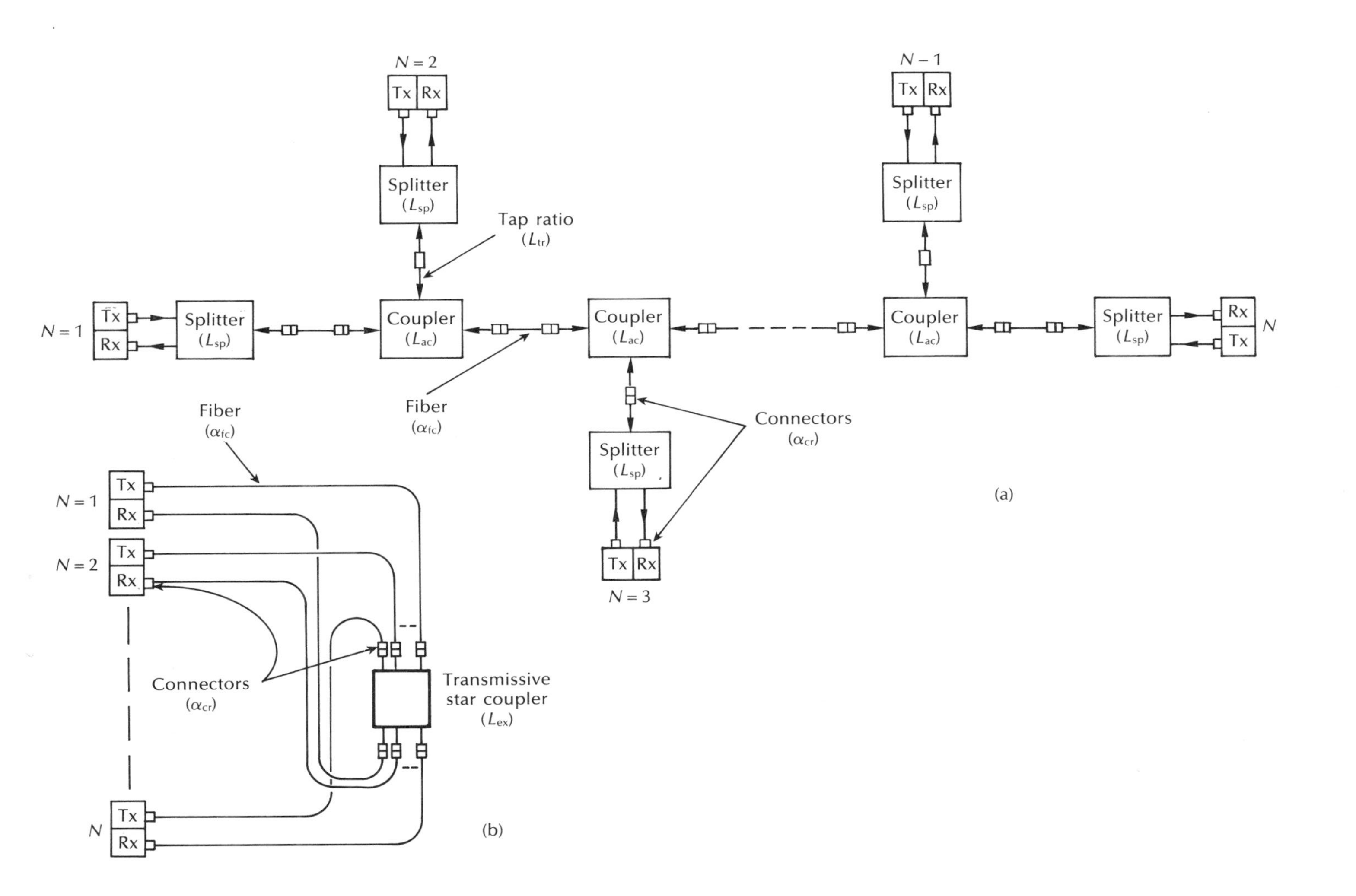

N = 2
Tx
Rx
Splitter
(L_{sp})
Tap ratio
(L_{tr})
N − 1
N = 1
Coupler
(L_{ac})
Fiber
(α_{fc})
Connectors
(α_{cr})
N = 3
N
(a)
Transmissive
star coupler
(L_{ex})
(b)

loss between nodes 1 and $N-1$ can therefore be written as:

$$C_L(1, N-1) = 2\alpha_{cr} + (N-1)\alpha_{fc}L_{bu} + (2\alpha_{cr} + L_{ac})(N-3) + (2\alpha_{cr} + L_{tr}) + L_{sp} + \alpha_{cr} \quad (11.101)$$

To incorporate the overall channel losses into an optical power budget for the multiterminal bus distribution system, the mean power obtained at the optical transmitter P_t at node 1, together with the mean incident optical power at the receiver, P_o of node $N-1$ must be included. Hence the optical power budget may be written as:

$$P_t = P_o + 2\alpha_{cr} + (N-1)\alpha_{fc}L_{bu} + (2\alpha_{cr} + L_{ac})(N-3) + (2\alpha_{cr} + L_{cr}) + L_{sp} + \alpha_{cr} + M_a \text{ dB} \quad (11.102)$$

where M_a is the system safety margin (see Section 11.6.6).

The star distribution system configuration displayed in Figure 11.45(b) employs a passive transmissive star coupler which provides two fibers to each node terminal (see Section 5.6.2). Hence an $N \times N$ star coupler allows the interconnection of N terminals. Assuming that the fiber cable lengths to each node are equal, then the same system loss is incurred for transmission between any two nodes. In this case the total system loss comprises the four connector losses at the transmitter, the receiver and the input and output ports of the star coupler $4\alpha_{cr}$; the total fiber cable loss $\alpha_{fc}L_{st}$ where L_{st} is the total fiber length in both arms of the star; the star splitting loss given by Eq. (5.18) as $10 \log_{10} N$ and the star excess loss L_{ex} provided by Eq. (5.19). In the case of equal fiber lengths the total channel loss between any two nodes is given by:

$$C_L(\text{star}) = 4\alpha_{cr} + \alpha_{fc}L_{st} + 10 \log_{10} N + L_{ex} \quad (11.103)$$

Again to incorporate the overall channel losses into an optical power budget for the multiterminal star distribution system, we designate the mean power obtained at the output of the optical transmitter P_t and the mean optical power incident at the receiver P_o so that:

$$P_t = P_o + 4\alpha_{cr} + \alpha_{fc}L_{st} + 10 \log_{10} N + L_{ex} + M_a \text{ dB} \quad (11.104)$$

where M_a is the system safety margin. A comparison of the optical power efficiencies of the two distribution systems is illustrated in the following example.

Example 11.18

Form a graphical comparison showing total channel loss against number of nodes for the bus and star distribution systems which incorporate components with the following performance parameters.

Connector loss:	1 dB
Access coupler insertion loss:	1 dB

Fiber cable loss:	5 dB km^{-1}
Access coupler tap ratio:	10 dB
Splitter loss:	3 dB
Star coupler excess loss:	0 dB

The distance between nodes on the bus system should be taken as 100 m and the worst case channel loss should be considered. It can be assumed that the total fiber cable length between all nodes on the star system is equal to 100 m.

Solution: Bus distribution system: using Eq. (11.101) the total channel loss is

$$\begin{aligned} C_L(1, N-1) &= 2 \times 1 + (N-1)5 \times 0.1 + (2 \times 1 + 1)(N-3) \\ &\quad + (2 \times 1 + 10) + 3 + 1 \text{ dB} \\ &= 0.5(N-1) + 3(N-3) + 18 \text{ dB} \\ &= 3.5N + 8.5 \text{ dB} \end{aligned}$$

Star distribution system: the total loss is given by Eq. (11.103) as

$$\begin{aligned} C_L(\text{star}) &= 4 \times 1 + 5 \times 0.1 + 10 \log_{10} N + 0 \text{ dB} \\ &= 4.5 + 10 \log_{10} N \text{ dB} \end{aligned}$$

The two expressions above for the bus and star distribution systems are plotted in Figure 11.46. It may be observed that the star configuration provides substantially greater efficiency in the utilization of optical power than the bus topology,

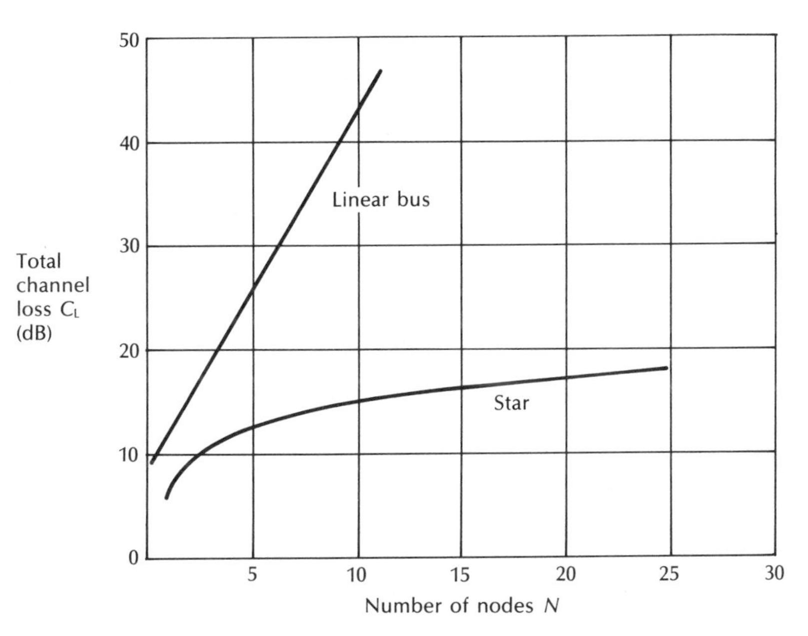

Figure 11.46 Characteristics showing the total channel loss against the number of nodes for the two distribution systems specified in Example 11.18.

particularly when the number of nodes becomes larger. It must be noted, however, that no excess loss for the star coupler has been included in the calculation and therefore it is anticipated that the total losses for the two distribution systems would be a little closer. Nevertheless, this factor would only make the optical power budgetary performance of the two configurations become similar when less than five terminals are interconnected.

11.9 Advanced multiplexing strategies

The basic multiplexing techniques which can be employed with IM/DD optical fiber systems were outlined in Section 11.4.2. Furthermore, the major baseband digital strategy, namely time division multiplexing was discussed in some detail in Section 11.5. In this section, however, the most significant of these multiplexing techniques are discussed in greater detail with particular emphasis on those strategies which allow greater exploitation of the available fiber bandwidth. We commence by further consideration of the multiplexing of digital signals prior to the more detailed description of techniques that may be employed for multiplexing either digital or analog intensity modulated signals, or a combination of both signal types.

11.9.1 Optical time division multiplexing

It was indicated in Section 10.5 that electronic circuits meet practical limitations on their speed of operation at frequencies around 10 GHz. Therefore, although more recently the feasibility of 10 Gbit s^{-1} direct intensity modulation and transmission over substantial distances (100 km) has been demonstrated (e.g. Ref. 80), electronic multiplexing at such speeds remains difficult and presents a restriction on the bandwidth utilization of a single-mode fiber link. An alternative strategy for increasing the bit rate of digital optical fiber systems beyond the bandwidth capabilities of the drive electronics is known as optical time division multiplexing (OTDM) [Ref. 81]. A block schematic of an OTDM system which has demonstrated 16 Gbit s^{-1} transmission over 8 km is shown in Figure 11.45 [Ref. 82]. The principle of this technique is to extend time division multiplexing by optically combining a number of lower speed electronic baseband digital channels. In the case illustrated in Figure 11.47, the optical multiplexing and demultiplexing ratio is 1 : 4, with a baseband channel rate of 4 Gbit s^{-1}. Hence the system can be referred to as a four channel OTDM system.

The four optical transmitters in Figure 11.47 were driven by a common 4 GHz clock using quarter bit period time delays. Mode-locked semiconductor laser sources which produced short optical pulses (around 15 ps long) were utilized at the transmitters to provide low duty cycle pulse streams for subsequent time multiplexing. Data was encoded onto these pulse streams using integrated optical

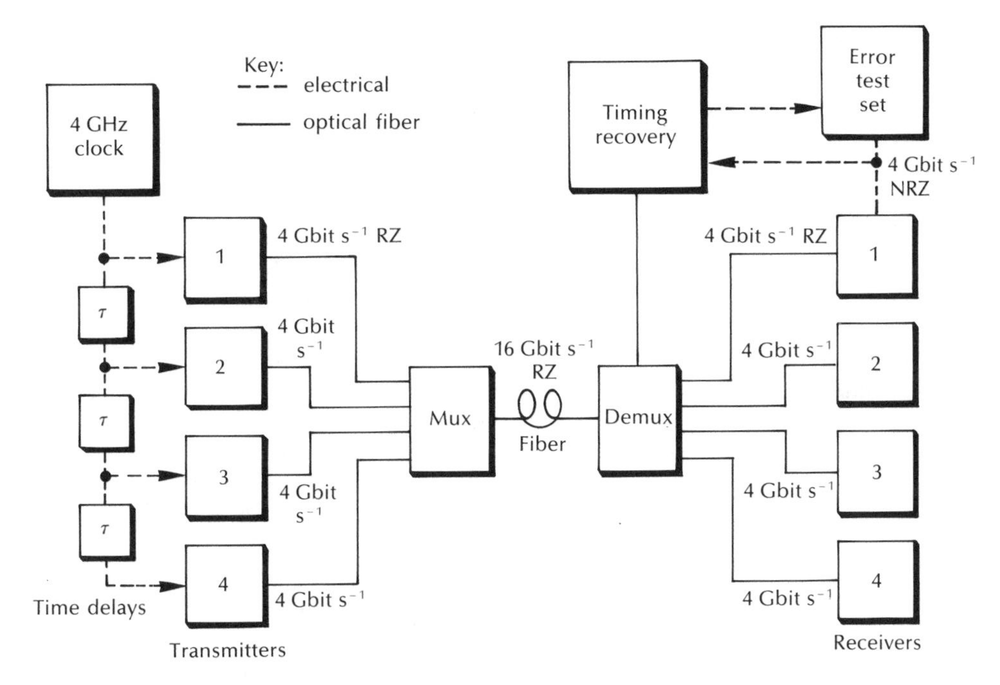

Figure 11.47 Four channel OTDM fiber system.

intensity modulators (see Section 10.6.2) which gave return to zero transmitter outputs at 4 Gbit s^{-1}. These IO devices were employed to eliminate the laser chirp (see Section 6.7.3) which would result in dispersion of the transmitted pulses as they propagated within the single-mode fiber, thus limiting the achievable transmission distance.

The four 4 Gbit s^{-1} data signals were combined in a passive optical power combiner but, in principle, an active switching element could be utilized. Although four optical sources were employed, they all emitted at the same optical wavelength within a tolerance of ± 0.1 nm and hence the 4 Gbit s^{-1} data streams were bit interleaved to produce the 16 Gbit s^{-1} signal. At the receive terminal the incoming signal was decomposed into the 4 Gbit s^{-1} baseband components in a demultiplexer which comprised two levels. Again, IO waveguide devices were used to provide a switching function at each level (see Section 10.6.1). At the first level the IO switch was driven by a sinusoid at 8 GHz to demultiplex the incoming 16 Gbit s^{-1} stream into two 8 Gbit s^{-1} signals. At the second level two similar switches, each operating at 4 GHz, demultiplexed each of the 8 Gbit s^{-1} data streams into two 4 Gbit s^{-1} signals. Hence single wavelength 16 Gbit s^{-1} optical transmission was obtained with electronics which only required a maximum bandwidth of about 2.5 GHz, as return to zero pulses were employed.

11.9.2 Subcarrier multiplexing

The use of radiofrequency subcarriers modulated by analog signals prior to intensity modulation of an optical source was discussed in Section 11.7.3. More recently, however, the utilization of substantially higher frequency microwave subcarriers multiplexed in the frequency domain before being applied to intensity modulate a high speed injection laser source has generated significant interest [Refs. 83 to 87]. Such microwave subcarrier multiplexing (SCM) enables multiple broadband signals to be transmitted over single-mode fiber and appears particularly attractive for video distribution systems [Refs. 86 to 88]. In addition, with SCM, conventional microwave techniques can be employed to subdivide the available intensity modulation bandwidth in a convenient way. The result is a useful multiplexing technique which does not require sophisticated optics or source wavelength specification (see Section 11.9.3). Either digital or analog modulation of the subcarriers can be utilized by upconverting to a narrowband channel at high frequency employing either amplitude, frequency or phase shift keying (i.e. ASK, FSK or PSK), and either amplitude, frequency or phase modulation (i.e. AM, FM or PM) respectively. For digital signals, FSK has the advantage of being simple to implement, both at the modulator and demodulator, whereas for analog video signals the modulation of the high frequency carrier (upconversion) is often carried out using either AM–VSB (vestigial sideband) or FM techniques. In both cases, the multicarrier signal is formed by frequency division multiplexing (FDM) of the modulated microwave subcarriers in the electrical domain prior to conversion to an intensity modulated optical signal.

A block schematic of a basic subcarrier multiplexed system is shown in Figure 11.48 [Ref. 86]. The modulated microwave subcarrier signals are obtained by frequency upconversion from the baseband using voltage controlled oscillator (VCOs). These subcarrier signals f_i are then summed in a microwave power combiner prior to the application of the composite signal to an injection laser which is d.c. biased at around 5 mW in order to produce the desired intensity modulation. The IM optical signal is then transmitted over single-mode fiber and directly detected using a wideband photodiode before demultiplexing and demodulation using a conventional microwave receiver.

Although relatively straightforward to implement using available components, SCM does exhibit some disadvantages, the most important of which is the problem associated with source nonlinearity [Ref. 87]. Distortion caused by this phenomenon can be particularly noticeable when several subcarriers are transmitted from a single optical source. Moreover, despite the fact that the receivers require narrow bandwidth, SCM systems, with the exception of those employing AM–VSB modulation, must operate at high frequency, often in the gigahertz range. In addition, for digital systems SCM requires more bandwidth per channel than a time division multiplexed system. The upconversion results in the bandwidth expansion so that a 50 Mbit s^{-1} channel may require some 80 MHz of bandwidth [Ref. 83]. Any reduction of this bandwidth overhead necessitates the adoption of more

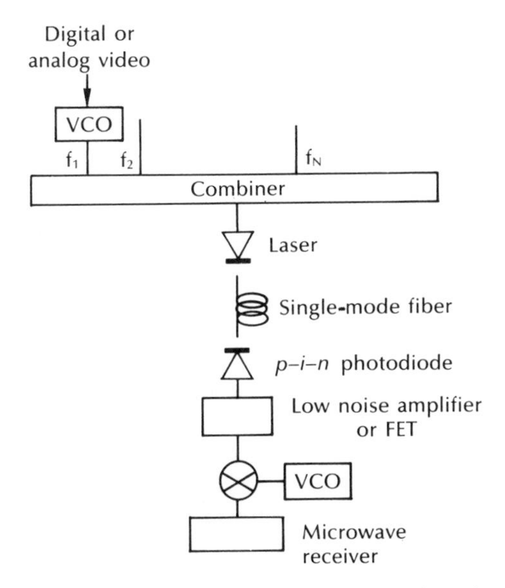

Figure 11.48 Basic subcarrier multiplexed (SCM) fiber system.

complex and less robust modulation techniques. For example, AM–VSB systems transmitting a standard cable television (CATV) multichannel spectrum tend to minimize the required bandwidth, but the signal must be received with a carrier to noise ratio of between 45 and 55 dB to avoid degradation of picture quality [Ref. 87].

The transmission of multiple CATV channels over substantial unrepeatered distances with good quality reception has, however, been demonstrated with SCM using frequency modulation (i.e. FM–FDM). For example, thirty-four multiple sub-nyquist-sampling encoding (MUSE) high definition television (HDTV) channels, each requiring an FM bandwidth of 27 MHz, have been transmitted over an unrepeatered distance of 42 km [Ref. 88]. This transmission system, which operated at a wavelength of 1.3 μm, provided a carrier to noise ratio of 17.5 dB at the receive terminal. Furthermore, an unrepeatered transmission distance in excess of 100 km has also been demonstrated with SCM when operating at a wavelength of 1.54 μm [Ref. 89]. In this case some eight baseband video channels, each of which was frequency modulated to occupy around 30 MHz of bandwidth, were then frequency multiplexed over a range 840 to 1160 MHz before directly modulating a distributed feedback laser.

Apart from the possibility of combining digital and analog subcarrier multiplexed signals into a composite signal, an alternative attractive strategy is the so-called hybrid SCM system which combines a baseband digital signal with a high frequency composite microwave signal [Ref. 86]. In this case the receiver shown in Figure 11.48 cannot be narrowband but must have a bandwidth from d.c. to beyond the highest microwave signal frequency employed. As mentioned previously, SCM is under investigation for application in video distribution systems and networks. In such systems only a single channel needs to be selected for demodulation. Hence a tunable local oscillator, mixer and narrow band filter can be utilized at the receive terminals (Figure 11.48) to simultaneously select the desired SCM channel and downconvert it to a more convenient intermediate frequency (IF) signal. Finally, the IF signal can be input to an appropriate demodulator to recover the baseband video signal.

11.9.3 Wavelength division multiplexing

Wavelength division multiplexing (WDM) involves the transmission of a number of different peak wavelength optical signals in parallel on a single optical fiber. Although in spectral terms optical WDM is analogous to electrical frequency division multiplexing, it has the distinction that each WDM channel effectively has access to the entire intensity modulation fiber bandwidth which with current technology is of the order of several gigahertz. The technique is illustrated in Figure 11.49, where a conventional (i.e. single nominal wavelength) optical fiber

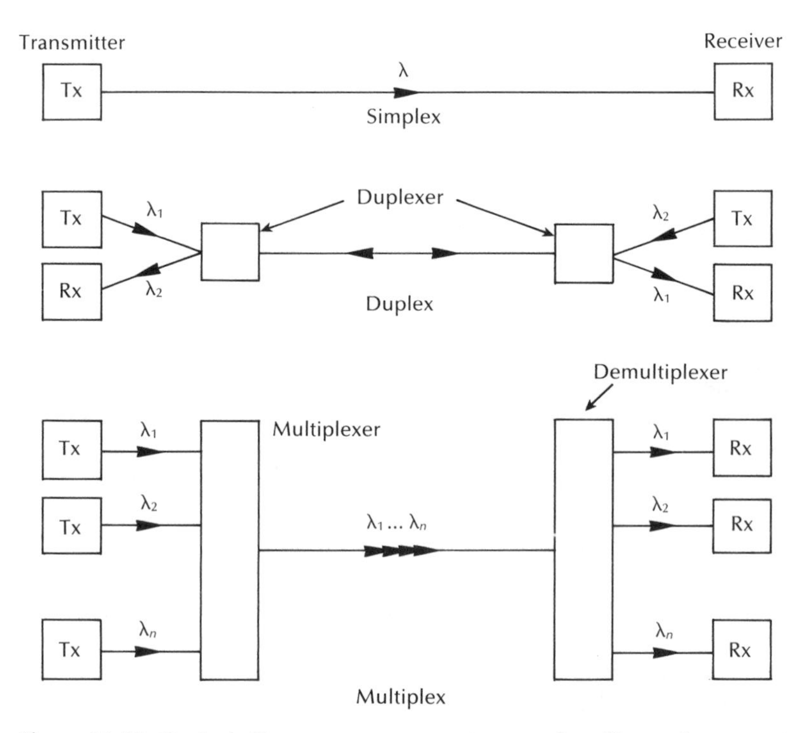

Figure 11.49 Optical fiber system operating modes illustrating wavelength division multiplexing (WDM).

communication system is shown together with a duplex (i.e. two different nominal wavelength optical signals travelling in opposite directions providing bidirectional transmission), and also a multiplex (i.e. two or more different nominal wavelength optical signals transmitted in the same direction) fiber communication system. It is the latter wavelength division multiplex operation which has generated particular interest within telecommunications. For example, two channel WDM is very attractive for a simple system enhancement such as piggybacking a 565 Mbit s^{-1} system onto an installed 140 Mbit s^{-1} link, or for doubling the capacity of a 565 Mbit s^{-1} link [Ref. 90]. Moreover, this multiplexing strategy overcomes certain power budgetary restrictions associated with electrical time division multiplexing. When the transmission rate over a particular optical link is doubled using TDM, a further 3 to 6 dB of optical power is generally required at the receiver (see Section 11.6.3). In the case of WDM, however, additional losses are also incurred from the incorporation of wavelength multiplexers and demultiplexers (see Section 5.6.3).

Wavelength division multiplexing in IM/DD optical fiber systems can be implemented using either LED or injection laser sources with either multimode or single-mode fiber. More recently, however, the wide scale deployment of single-mode fiber has encouraged the investigation of WDM on this transmission medium. In particular, the potential utilization of the separate wavelength channels to provide dedicated communication services to individual subscriber terminals is an attractive concept within telecommunications. For example, a multiwavelength, single-mode optical star network called LAMBDANET has been developed using commercial components [Ref. 91]. This network, which is internally nonblocking, has been configured to allow the integration of point-to-point and point-to-multipoint wideband services, including video distribution applications.

A block schematic of the LAMBDANET star network is shown in Figure 11.50. The network incorporated a sixteen port passive transmission star coupler (see Section 5.6.2). Each node was equipped with a single distributed feedback laser selected with centre wavelengths spaced at 2 nm intervals over the range 1527 to 1561 nm. Hence, each node transmitted at a unique wavelength, providing a contention-free broadcast capability to all other nodes. At the receive terminals every node could detect transmissions from all other nodes using a wavelength demultiplexer and sixteen optical receivers. Moreover, the wavelength channels on the network were demonstrated to operate at a transmission rate of 2 Gbit s^{-1} over a distance of 40 km [Ref. 92].

Another WDM strategy which has been investigated for both telecommunication and nontelecommunication applications is illustrated in Figure 11.51. [Ref. 93]. In this case, in place of narrow linewidth injection laser sources, wide spectral width (63 nm) edge-emitting LEDs were utilized to provide the multiwavelength optical carrier signals which were transmitted on single-mode optical fiber. The full spectral output from each ELED was not, however, transmitted for each wavelength channel. Instead, a relatively narrow spectral slice (3.65 nm) for each separate channel was obtained using the diffraction grating WDM multiplexer device, as shown in Figure 11.51, prior to transmission down the optical link. This technique,

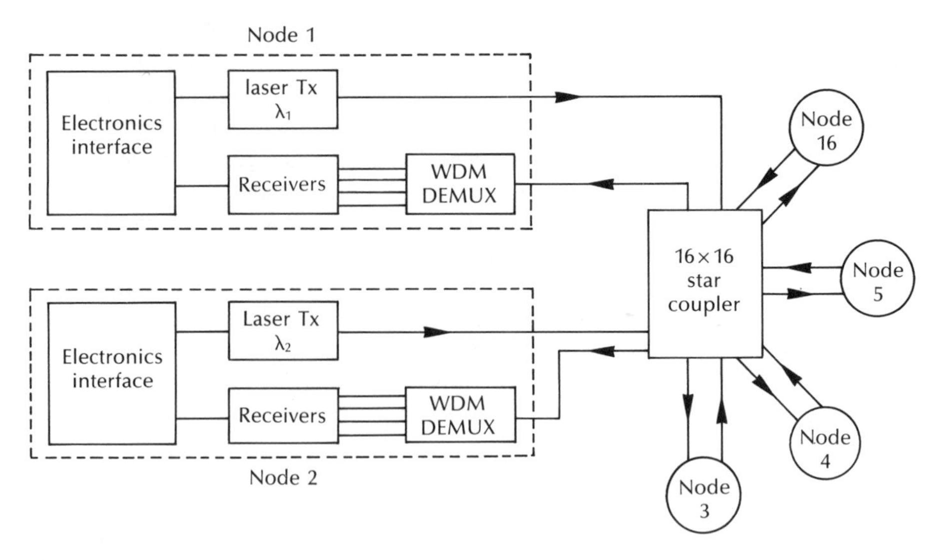

Figure 11.50 The LAMBDANET star network [Ref. 91].

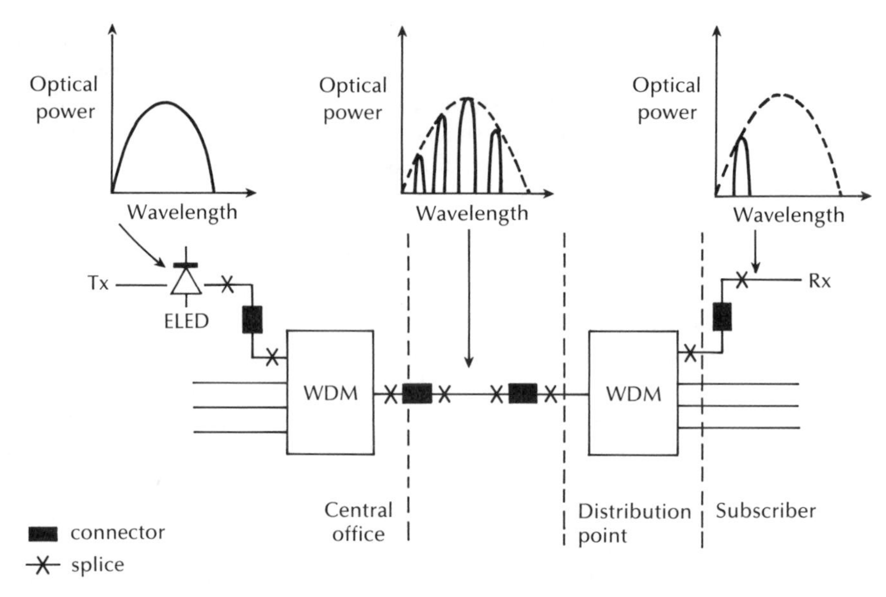

Figure 11.51 Spectral slicing of LED outputs to form several WDM channels.

which is known as spectral slicing, could enable LEDs with the same overall spectral output to be employed whilst still providing the distinctive wavelength channels for transmission between each subscriber terminal. In this case a WDM demultiplexer device is located at a distribution point in order to separate and distribute the different wavelength optical channels to the appropriate subscriber receive terminals. A similar strategy has been demonstrated for sixteen channels using superluminescent LEDs, again transmitting on single-mode fiber [Ref. 94]. Moreover, the technique has also been employed in nontelecommunication areas to provide multiple wavelength channels from single LED sources, usually on multimode fiber, in order to, for example, service a multiple optical sensor system (see Section 14.5) in which each wavelength channel supplies a signal to a different optical sensor device [Ref. 95].

11.10 Application of optical amplifiers

The use of electronics-based regenerative repeaters in long-haul optical fiber communications was discussed in Sections 11.4 and 11.6.1. It is clear, however, that such devices not only increase the cost and complexity of the optical communication system, but they may also act as a bottle-neck by restricting the system operational bandwidth. Hence the recent developments in optical amplifier technology described in Sections 10.2 to 10.4 have started to provide an additional, welcome flexibility in the design and implementation of IM/DD optical fiber systems.

The above flexibility stems from the ability for the transmitted optical signal to remain in the optical domain over the entire length of a long-haul link. Optical amplifiers (both semiconductor and fiber devices) therefore exhibit interesting features which assist in the system design, as illustrated in Figure 11.52. It may be observed from Figure 11.52(a) that, in a similar manner to electronic repeaters (see Figure 11.27), optical amplifiers may be employed in a simplex mode where each transmitted optical signal is carried on a separate fiber link. However, optical amplifiers have the ability to operate simultaneously in both directions at the same carrier wavelength,* as shown in Figure 11.52(b). Moreover, in this bidirectional mode they offer an added degree of reliability in that a single fiber break would only disable one half of communication capacity per fiber pair rather than causing a complete system failure, as is the case with the present unidirectional systems (i.e. one transmission path between all user pairs is disabled).

A further range of flexibility associated with optical amplifiers concerns the ability of particular devices to simultaneously amplify multiple wavelength division multiplexed (WDM) optical signals (see Section 11.9.3). Both SLAs and fiber amplifiers with spectral bandwidths in the range 20 to 50 nm can be realized (see

* It is obviously necessary to intensity modulate the optical carriers at different speeds to avoid signal interference.

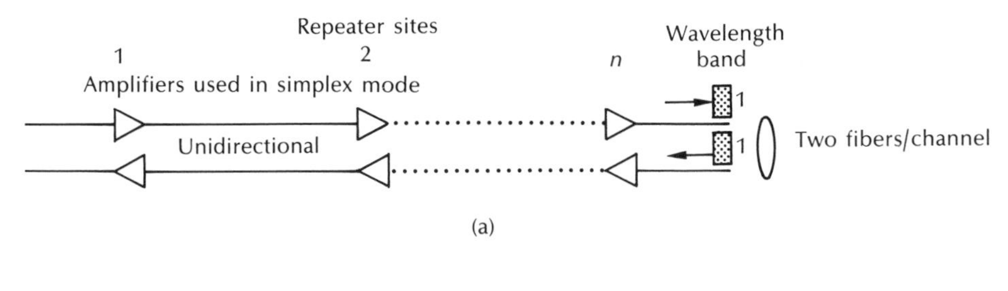

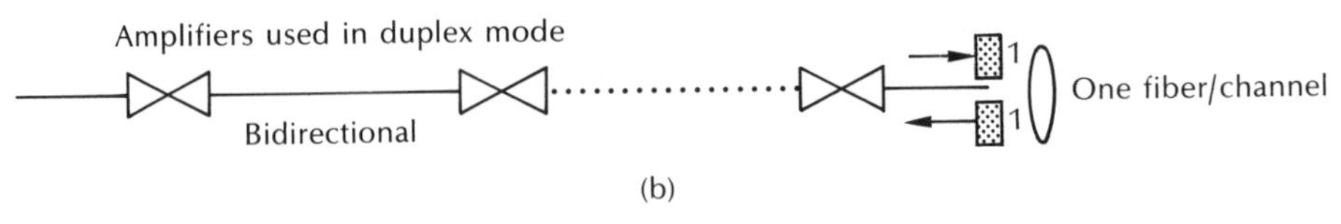

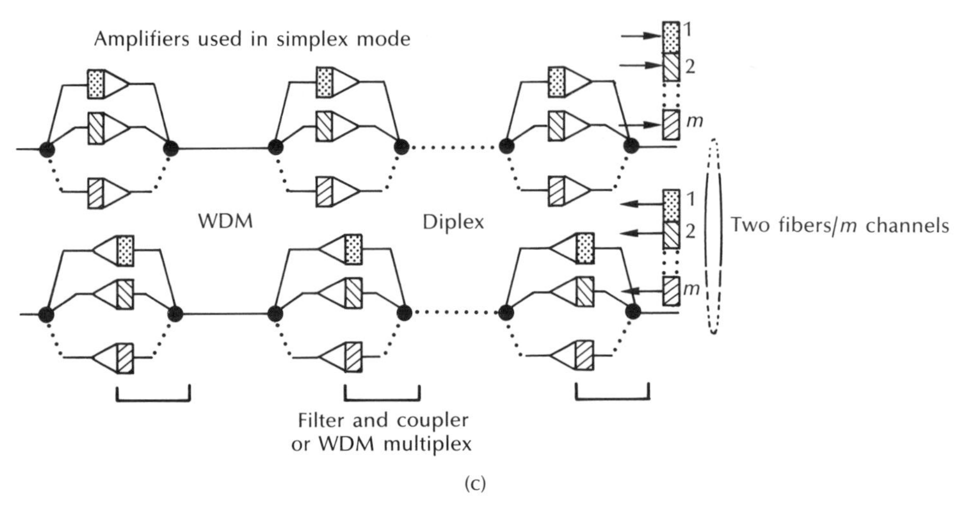

Figure 11.52 Potential point-to-point system applications for optical amplifiers: (a) simplex mode; (b) duplex mode; (c) multiamplifier configuration for wavelength division multiplex (WDM) operation.

Sections 10.3 and 10.4) which will allow single amplifiers to support in excess of ten intensity modulated WDM channels. Moreover, the parallel multiamplifier configuration illustrated in Figure 11.52(c) could be envisaged which would enable contiguously spectrally aligned amplifiers to span a complete wavelength window (say around 1.55 μm). Such configurations could increase system reliability in the event of an individual amplifier failure, whilst also relaxing the linearity and overload characteristics for amplifiers operating with densely packed WDM hierarchies [Ref. 96].

It was mentioned in Section 10.4 that optical amplifiers could be used in a broad range of system applications: namely, as power amplifiers at the optical transmitter; as in-line repeater amplifiers; and as preamplifiers at optical receivers. Moreover, the latter system application was dealt with in Section 9.6. Therefore, in this section we concentrate on the utilization of optical amplifiers as in-line repeaters within IM/DD optical fiber systems. It must be remembered, however, that in contrast to regenerative repeaters, optical amplifiers simply act as gain blocks on an optical fiber link and hence they do not reconstitute a transmitted digital optical signal. A drawback with this operation is that both noise and signal distortions are continuously amplified as the optical signal passes down a link which uses cascaded amplifiers. A benefit, however, is that optical amplifiers are transparent to any type of signal modulation (i.e. digital or analog) and to any modulation bandwidth.

It was mentioned in Section 10.3.2 that noise in travelling wave SLAs may determine the number of devices that can be cascaded as linear repeaters. Since the mean noise power resulting from spontaneous emission in optical amplifiers accumulates in proportion to the number of repeaters, gain saturation occurs when the total noise power becomes equal to the signal power. A simple model for the noise behaviour of such devices which is valid under conditions of high signal to noise ratio and high gain is shown in Figure 11.53. It comprises an ideal noiseless amplifier with gain G preceded by an additive noise source of spectral density $S(f)$ given by [Ref. 97]:

$$S(f) = Khf \tag{11.105}$$

where K, which is dependent on the population inversion and the cavity loss, provides a measure of the amplifier quality and hf constitutes photon energy. A minimum theoretical value for K is unity, which would only occur for the ideal case of complete inversion and no cavity loss. However, in this case $S = hf$ which indicates that it is theoretically impossible for the TWA to be noiseless. Nevertheless, in practice $K < 2$ has been obtained, demonstrating that amplifiers with less than 3 dB more noise than the theoretical limit can be achieved [Ref. 98].

The cascading of optical amplifiers in a long-haul communication system is illustrated in Figure 11.54(a). Following each section of fiber cable length L there is an optical amplifier with gain G which just compensates for the fiber cable loss

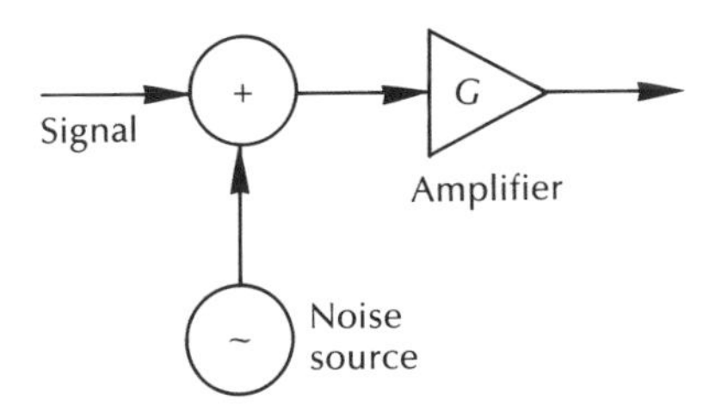

Figure 11.53 Noise model for travelling wave optical amplifier.

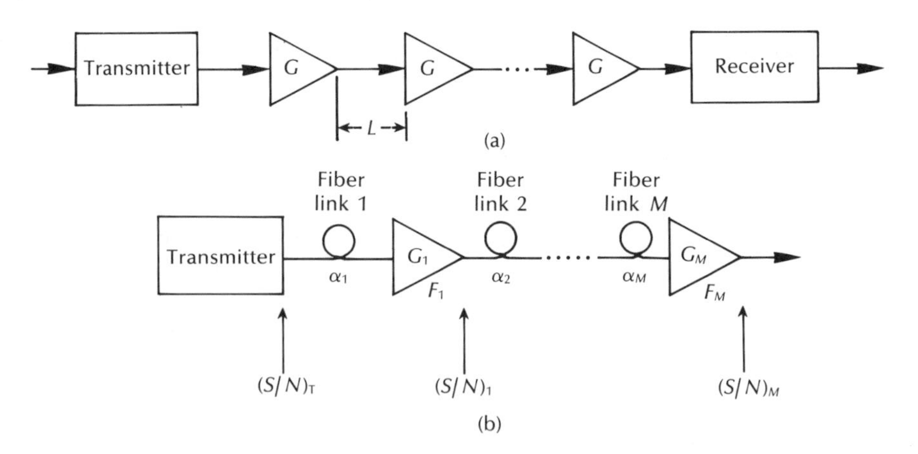

Figure 11.54 Cascaded optical amplifiers: (a) fiber system with cascaded optical amplifiers; (b) signal to noise ratios in a cascaded amplifier chain.

such that:

$$G = 10^{-(\alpha_{fc} + \alpha_j)L/10} \tag{11.106}$$

where α_{fc} and α_j are the fiber cable losses and joint losses respectively, both in dB km^{-1}. Furthermore, as the optical signal travels through the amplifier cascade the noise levels increase because the additive noise from each device is cumulative. Hence, using Eqs. (11.105) and (11.106) the signal to noise ratio at the end of a cascaded link may be written as:

$$\frac{S}{N} \simeq \frac{P_i 10^{-(\alpha_{fc} + \alpha_j)L/10}}{(L_{to}/L)KhfB} \tag{11.107}$$

where P_i is the power launched into the link at the transmitter, L_{to} the total system length and hence (L_{to}/L) is approximately equal to the total number of amplifier repeaters,* and the noise bandwidth equals the system bandwidth B. Equation (11.107) therefore enables the maximum transmission distance for a system using cascaded travelling wave SLAs to be deduced.

Example 11.19

A long-haul digital single-mode fiber system operating at a wavelength of 1.55 μm is envisaged employing travelling wave SLAs spaced at intervals of 100 km. The power launched into the link at the transmitter is 0 dBm and the fiber cable attenuation is 0.22 dB km^{-1}. In addition, there are splice losses which average out

* The total number of amplifier repeaters is actually $(L_{to}/L - 1)$; however for a long link and a large number of repeaters this approximates to (L_{to}/L).

at 0.03 dB km^{-1}. A signal to noise ratio of 17 dB is required at the system receive terminal to provide an acceptable BER at the operating transmission rate of 1.2 Gbit s^{-1}. Assuming that the system bandwidth is equal to the transmission bit rate and that K for the amplifiers is equal to 4, estimate the maximum system length such that satisfactory performance is maintained.

Solution: Using Eq. (11.107), then

$$(L_{to}/L) \simeq \frac{P_i \lambda 10^{-(\alpha_{fc}+\alpha_j)L/10}}{KhcB}\left(\frac{S}{N}\right)^{-1}$$

Hence for a link with a large number of cascaded amplifiers:

$$L_{to} \simeq \left(\frac{P_i \lambda 10^{-(\alpha_{fc}+\alpha_j)L/10}}{KhcB}\right)\left(\frac{S}{N}\right)^{-1} L$$

$$= \frac{(10^{-3} \times 1.55 \times 10^{-6} \times 10^{-2.5})100 \times 10^3}{4 \times 6.626 \times 10^{-34} \times 2.998 \times 10^8 \times 1.2 \times 10^9 \times 50}$$

$$\simeq 1 \times 10^4 \text{ km}$$

Thus the maximum system length obtained in Example 11.19 is very large and would allow the interconnection of most points on the earth using a chain of optical amplifiers. However, the calculation does not take account of the nonregenerative nature of the amplifier repeaters in which pulse spreading as well as noise down the link is accumulated. Fiber dispersion therefore imposes serious limitations on the system performance, as discussed in Section 11.6.5, and it will restrict both the maximum system span as well as the maximum transmission rate.

Another parameter which is often specified in relation to the noise performance of optical amplifiers is the noise figure of the devices. The noise figure F for an optical amplifier is defined in a similar manner to an electrical amplifier as the signal to noise degradation between the device input and the device output:

$$F = \frac{(S/N)_{in}}{(S/N)_{out}} \tag{11.108}$$

Again, it is governed by factors including the population inversion, the number of transverse modes in the amplifier cavity, the number of incident photons on the amplifier and the optical bandwidth of the amplified spontaneous emissions. Typical noise figures range from 4 to 8 dB, with SLAs generally towards the bottom end of the range and fiber amplifiers towards the top end.

We now consider a system with M cascaded optical amplifiers, as illustrated in Figure 11.54(b). In this case the link attenuation (both fiber cable and joint losses) in front of the kth amplifier is denoted by α_k, whilst the amplifier has a signal gain of G_k and a noise figure F_k. The input and output SNRs for such a cascaded link can be defined at the transmitter output T and the Mth amplifier output, respectively, so that in a similar manner to electrical amplifiers the total noise figure for the system F_{to} is:

$$F_{\mathrm{to}} = \frac{(S/N)_T}{(S/N)_M}$$

$$= \frac{F_1}{\alpha_1} + \frac{F_2}{\alpha_1 G_1 \alpha_2} + \frac{F_3}{\alpha_1 G_1 \alpha_2 G_2 \alpha_3} + \cdots + \frac{F_M}{\alpha_M \sum_{k=1}^{M-1} (\alpha_k G_k)} \quad (11.109)$$

The above expression can therefore enable determination of the total noise figure for the amplifier cascade.

Example 11.20

An optical fiber system is configured with a series of M optical amplifiers in cascade. The fiber cable and joint losses on each span between amplifiers on the link are compensated by the following amplifier gain. Obtain an expression for the total noise figure for the system and determine its value when all the amplifiers are identical.

Solution: As the amplifier gain compensates for the losses, then $\alpha_k G_k = 1$. Hence using Eq. (11.109) the total noise figure is given by:

$$F_{\mathrm{to}} = \frac{F_1 G_1}{\alpha_1 G_1} + \frac{F_2 G_2}{\alpha_1 G_1 \alpha_2 G_2} + \cdots + \frac{F_M G_M}{\alpha_m \sum_{k=1}^{M-1} (\alpha_k G_k)}$$

$$= F_1 G_1 + F_2 G_2 + \cdots + F_M G_M$$

$$= \sum_{k=1}^{M} F_k G_k$$

When all the repeaters are identical then $F_1 G_1$, $F_2 G_2, \ldots, F_M G_M$ are equal to FG. Therefore, the total noise figure becomes:

$$F_{\mathrm{to}} = MFG$$

At the output from the first amplifier repeater a degradation in signal to noise ratio of FG occurs followed by a decrease $1/M$.

A typical experimental system configuration employing five travelling wave SLAs and operating over a distance of some 500 km at a transmission rate of 565 Mbit s^{-1} is shown in Figure 11.55 [Refs. 99, 100]. The system was designed to give a bit error rate better than 10^{-9}. In addition, as discussed above, the noise in TWSLAs can largely determine the number of devices that may be cascaded as linear repeaters. However, the spontaneous emission noise profile is relatively broadband, typically

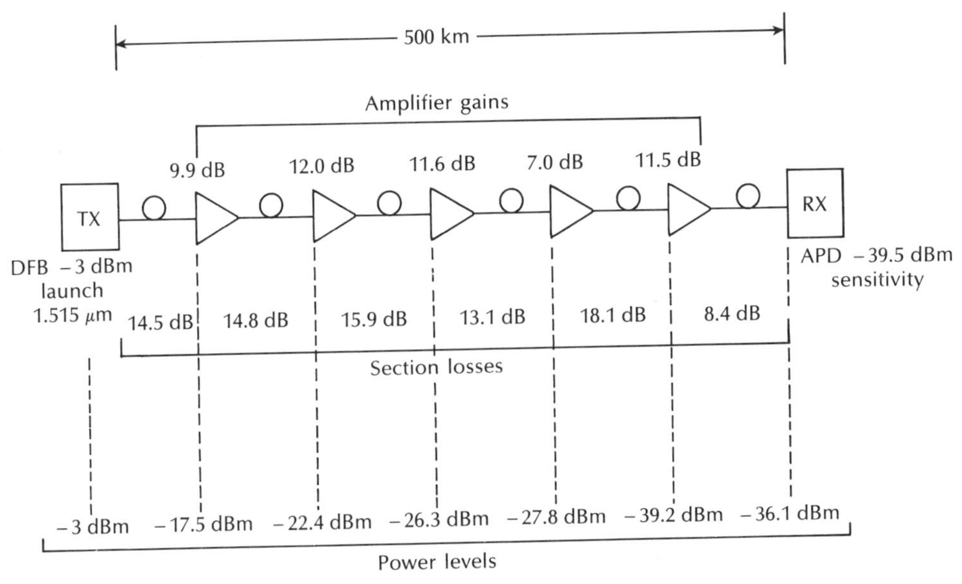

Figure 11.55 Experimental system incorporating five cascaded travelling wave semiconductor laser amplifiers [Refs. 99, 100].

occupying around 30 nm, and therefore optical filtering can be used as a method of reducing overall noise levels. It is suggested that filters with bandwidths in the range 5 to 10 nm would be required for systems spanning greater than 500 km [Ref. 100]. Nevertheless, a higher speed IM/DD optical fiber system operating at 2.4 Gbit s^{-1} over a distance of 516 km has been demonstrated by cascading ten TWSLAs [Ref. 101].

Although the technology associated with SLAs is at present more established than that of fiber amplifiers, there appear to be significant drawbacks with these former devices in relation to their active nature, mechanical structure, reliability and yield performance which may inhibit their application in future systems. By contrast, in fiber amplifiers these parameters are largely defined by the atomic structure and hence they exhibit greater stability. Furthermore, although the signal gain–bandwidth of TWSLAs is generally more appropriate to WDM applications, they are also prone to crosstalk problems which tends to limit their suitability in this area [Ref. 102].

When considering fiber amplifiers for use in WDM systems the Raman device (see Section 10.4.2) offers a gain bandwidth similar to the TWSLA (i.e. 20 to 50 nm). However, in practice there are a number of difficulties associated with the use of these fiber amplifiers. In particular, very high pump source powers of the order of 300 mW into the fiber are required to provide gains of around 15 dB [Ref. 103]. Also, the fiber length needed for the Raman fiber amplifier is generally greater than 1 km. For example, recent predictions obtained using a 10 km length device indicated a Raman gain of 13.5 dB (net gain of 8.5 dB) when the pump power was 200 mW [Ref. 104]. Raman amplification has, however, been demonstrated in a 310 km span remotely pumped 1.8 Gbit s^{-1} IM/DD system experiment [Ref. 105]. In this case Raman amplification in the dispersion shifted fiber after the transmitter as well as before the receiver (an in-line erbium fiber amplifier was also employed) provided gains of 4.3 dB and 5.5 dB respectively. By contrast the gain provided by the remotely pumped erbium doped fiber device was 13.2 dB.

There is increasing interest in the investigation of erbium doped fiber amplifiers to provide optical gain within IM/DD optical fiber systems. Both long span and high gain operation of these devices around 1.53 μm has been demonstrated [Refs. 106–108]. Moreover, typical good performance characteristics display a signal gain of 25 dB with a spectral bandwidth of 35 nm for a pump power of 50 mW at 1.49 μm [Ref. 107]. In particular, the realization of a relatively broad 3 dB spectral bandwidth for these devices has provided an impetus for their application within WDM optical fiber systems.

Recent system demonstrations have underlined the use of single erbium doped fiber amplifiers within WDM systems for small numbers of wavelength channels [Refs. 110, 111]. Furthermore, four wavelength channels, each modulated at 2.4 Gbit s^{-1} have been transmitted over 459 km using six cascaded erbium fiber amplifiers [Ref. 112]. A block schematic of this experimental system is shown in Figure 11.56. Distributed feedback (DFB) laser transmitters emitting over a range 1548.8 to 1554.7 nm were externally modulated using Mach–Zehnder IO devices (see Section 10.6.2). It may be observed from Figure 11.56 that a booster power amplifier was employed immediately after the four port coupler followed by five in-line repeater amplifiers. Moreover, the 3 dB spectral bandwidths for the amplifiers utilized were only around 10 nm.

Increased numbers of WDM channels have been demonstrated for broadcast

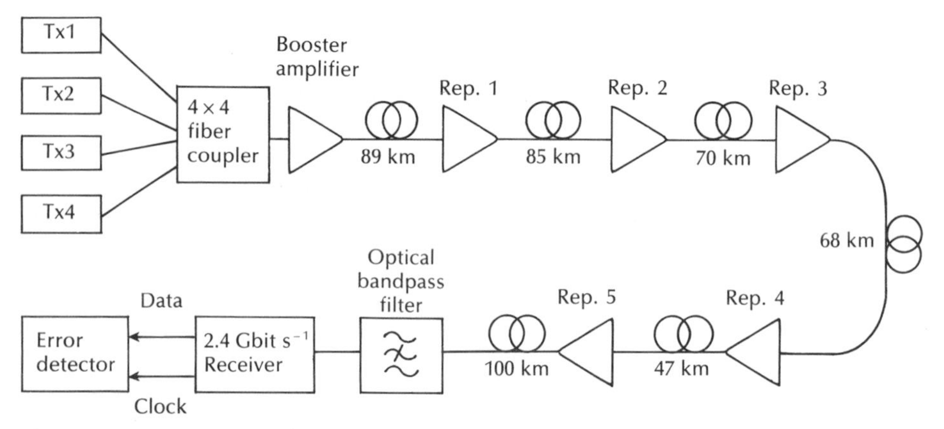

Figure 11.56 Experimental system demonstrating the transmission of four WDM channels through six cascaded erbium doped fiber amplifiers covering a distance of 459 km [Ref. 112].

network applications targeted at the future local access network and cable TV applications (see Sections 14.2.3 and 14.4.1). In particular, twelve closely spaced wavelength channels in the range 1530 to 1534 nm obtained from twelve separate DFB lasers were successfully transmitted through two erbium doped fiber amplifier stages [Ref. 113]. The spectrum provided by the twelve DFB lasers after two stages of amplification is shown in Figure 11.57. Furthermore, the use of twelve wavelength channels and two stages of erbium doped fiber amplification demonstrated the feasibility of broadcasting up to 384 digital video channels over a passive optical network with a range of 27.7 km to some 39.5 million customer terminals [Ref. 113].

The amplification of sixteen wavelength channels separated at 2 nm intervals over a range from 1527 to 1561 nm has also been demonstrated through a single erbium doped fiber amplifier stage [Ref. 114]. Again, DFB laser transmitters were employed for the transmission of either ten analog FM video signals subcarrier multiplexed (see Section 11.8.2) between 300 and 700 MHz or data at 622 Mbit s^{-1}. In this case the amplifier provided between 17 and 24 dB of gain per channel.

An advantage provided by erbium doped fiber amplifiers when used in WDM optical fiber systems is that the lifetimes associated with the lasing process in these devices are such that crosstalk in the presence of a number of wavelengths is significantly reduced. Moreover, recent a.c. crosstalk measurements suggest that the crosstalk is heavily dependent upon the modulating frequency and is negligible above a transmission rate of 140 Mbit s^{-1} [Ref. 115]. In addition, spectral broadening resulting from amplifier phase noise does not appear to be a significant

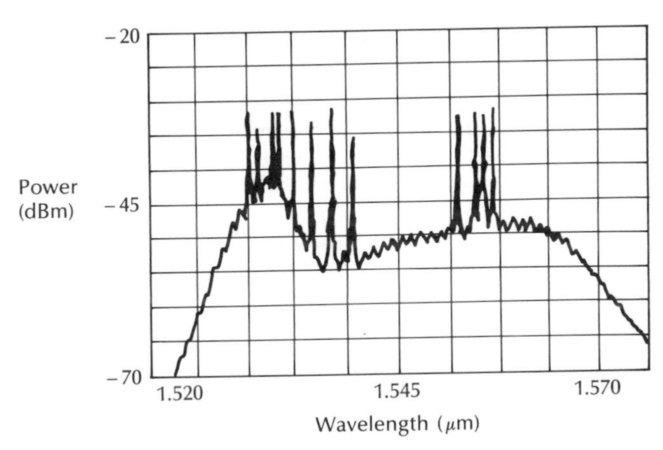

Figure 11.57 Spectrum showing twelve WDM channels after two stages of amplification using erbium doped fiber amplifiers. Reproduced with permission from A. M. Hill, R. Wyatt, J. F. Massicott, K. J. Blyth, D. S. Forrester, R. A. Lobbett, P. J. Smith and D. B. Payne, '39.5 million-way WDM broadcast network employing two stages of erbium-doped fibre amplifiers', *Electron. Lett.*, **26**, p. 1882, 1990.

problem for the operation of intensity modulated WDM long-haul or distribution systems which utilize erbium doped fiber devices [Ref. 116].

Problems

11.1 Discuss the major considerations in the design of digital drive circuits for:

(a) an LED source;
(b) an injection laser source.

Illustrate your answer with an example of a drive circuit for each source.

11.2 Outline, with the aid of suitable diagrams, possible techniques for:

(a) the linearization of LED transmitters;
(b) the maintenance of constant optical output power from an injection laser transmitter.

11.3 Discuss, with the aid of a block diagram, the function of the major elements of an optical fiber receiver. In addition, describe possible techniques for automatic gain control in APD receivers.

11.4 Equalization within an optical receiver may be provided using the simple frequency 'rollup' circuit shown in Figure 11.58(a). The normalized frequency response for this circuit is illustrated in Figure 11.58(b).

The amplifier indicated in Figure 11.58(a) presents a load of 5 kΩ to the photodetector and together with the photodetector gives a total capacitance of 5 pF. However, the desired response from the amplifier–equalizer configuration has an upper 3 dB point or corner frequency at 30 MHz. Assuming R_2 is fixed at 100 Ω determine the required values for C_1 and R_1 in order to obtain such a response.

11.5 Describe the conversion of an analog signal into a pulse code modulated waveform for transmission on a digital optical fiber link. Furthermore, indicate how several signals may be multiplexed on to a single fiber link.

A speech signal is sampled at 8 kHz and encoded using a 256 level binary code. What is the minimum transmission rate for this single pulse code modulated speech signal? Comment on the result.

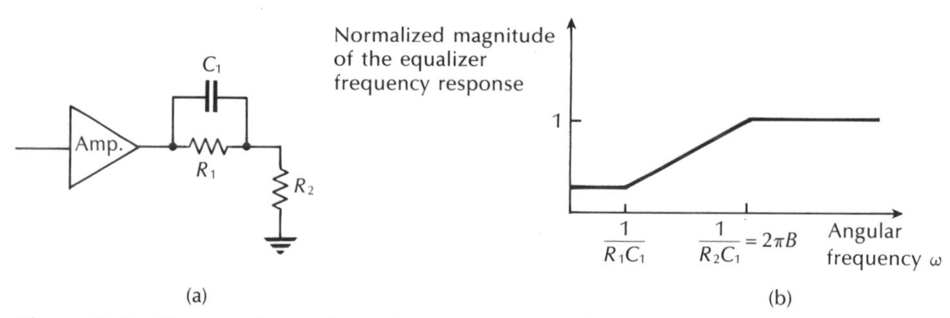

Figure 11.58 The equalizer of problem 11.4: (a) the frequency 'rollup' circuit; (b) the spectral transfer characteristic for the circuit.

11.6 A 1.5 MHz information signal with a dynamic range of 64 mV is sampled, quantized and encoded using a direct binary code. The quantization is linear with 512 levels. Determine:

(a) the maximum possible bit duration;
(b) the amplitude of one quantization level.

11.7 Describe, with the aid of a suitable block diagram, the operation of an optical fiber regenerative repeater. Indicate reasons for the occurrence of bit errors in the regeneration process and outline a technique for establishing the quality of the channel.

11.8 Twenty-four 4 kHz speech channels are sampled, quantized, encoded and then time division multiplexed for transmission as binary PCM on a digital optical fiber link. The quantizer is linear with 0.5 mV steps over a dynamic range of 2.048 V.

Calculate:

(a) the frame length of the PCM transmission, assuming an additional channel time slot is used for signalling and synchronization;
(b) the required channel bandwidth assuming NRZ pulses.

11.9 Develop a relationship between the error probability and the received SNR (peak signal power to rms noise power ratio) for a baseband binary optical fiber system. It may be assumed that the number of ones and zeros are equiprobable and that the decision threshold is set midway between the one and zero level.

The electrical SNR (defined as above) at the digital optical receiver is 20.4 dB. Determine:

(a) the optical SNR;
(b) the BER.

It may be assumed that erfc $(3.71) \simeq 1.7 \times 10^{-7}$.

11.10 The error function (erf) is defined in the text by Eq. (11.16). However, an error function also used in communications is defined as:

$$\text{Erf}(u) = \frac{1}{\sqrt{(2\pi)}} \int_{-\infty}^{u} \exp\,(-x^2/2)\,\mathrm{d}x$$

where a capital E is used to denote this form of the error function. The corresponding complementary error function is:

$$\text{Erfc}(u) = 1 - \text{Erf}(u) = \frac{1}{\sqrt{(2\pi)}} \int_{u}^{\infty} \exp\,(-x^2/2)\,\mathrm{d}x$$

This complementary error function is also designated as $Q(u)$ in certain texts. Use of Erfc(u) or $Q(u)$ is sometimes considered more convenient within communication systems.

Develop a relationship for erfc(u) in terms of Erfc(u). Hence, obtain an expression for the error probability $P(\mathrm{e})$ as a function of the Erfc for a binary digital optical fiber system where the decision threshold is set midway between the one and zero levels and the number of transmitted ones and zeros are equiprobable. In addition, given that Erfc $(4.75) \simeq 1 \times 10^{-6}$ estimate the required peak signal power to rms noise power ratios (both optical and electrical) at the receiver of such a system in order to maintain a BER of 10^{-6}.

11.11 Show that Eq. (9.27) reduces to Eq. (11.28). Hence determine $F(M)$ when $k = 0.3$ and $M = 20$.

11.12 A silicon APD detector is utilized in a baseband binary PCM receiver where the decision threshold is set midway between the one and zero signal level. The device has a quantum efficiency of 70% and a ratio of carrier ionization rates of 0.05. In operation the APD has a multiplication factor of 65. Assuming a raised cosine signal spectrum and a zero disparity code, and given that erfc $(4.47) \simeq 2 \times 10^{-10}$:

(a) estimate the number of photons required at the receiver to register a binary one with a BER of 10^{-10};

(b) calculate the required incident optical power at the receiver when the system is operating at a wavelength of 0.9 μm and a transmission rate of 34 Mbit s^{-1};

(c) indicate how the value obtained in (b) should be modified to compensate for a 3*B*4*B* line code.

11.13 A *p–i–n* photodiode receiver requires 2×10^4 incident photons in order to register a binary one with a BER of 10^{-9}. The device has a quantum efficiency of 65%. Estimate in decibels the additional signal level required in excess of the quantum limit for this photodiode to maintain a BER of 10^{-9}.

11.14 An optical fiber system employs an LED transmitter which launches an average of 300 μW of optical power at a wavelength of 0.8 μm into the optical fiber cable. The cable has an overall attenuation (including joints) of 4 dB km^{-1}. The APD receiver requires 1200 incident photons in order to register a binary one with a BER of 10^{-10}. Determine the maximum transmission distance (without repeaters) provided by the system when the transmission rate is 1 Mbit s^{-1} and 1 Gbit s^{-1} such that a BER of 10^{-10} is maintained.

Hence sketch a graph showing the attenuation limit on transmission distance against the transmission rate for the system.

11.15 An optical fiber system uses fiber cable which exhibits a loss of 7 dB km^{-1}. Average splice losses for the system are 1.5 dB km^{-1}, and connector losses at the source and detector are 4 dB each. After safety margins have been allowed, the total permitted channel loss is 37 dB. Assuming the link to be attenuation limited, determine the maximum possible transmission distance without a repeater.

11.16 Assuming a linear increase in pulse broadening with fiber length, show that the transmission rate $B_T(DL)$ at which a digital optical fiber system becomes dispersion limited is given by:

$$B_T(DL) \simeq \frac{(\alpha_{dB} + \alpha_j)}{5\sigma_T(\text{km})} \frac{1}{10 \log_{10}(P_i/P_o)}$$

where $\sigma_T(\text{km})$ is the total rms pulse broadening per kilometre on the link (hint: refer to Eqs. (3.3) and (3.11)).

(a) A digital optical fiber system using an injection laser source displays rms pulse broadening of 1 ns km^{-1}. The fiber cable has an attenuation of 3.5 dB km^{-1} and joint losses average out to 1 dB km^{-1}. Estimate the transmission rate at the dispersion limit when the difference in optical power levels between the input and output is 40 dB.

(b) Calculate the dispersion limited transmission distance for the system described in (a) when the transmission rates are 1 Mbit s^{-1} and 1 Gbit s^{-1}. Hence sketch a graph showing the dispersion limit on transmission distance against the transmission rate for the system.

11.17 The digital optical fiber system described in problem 11.16(a) has a transmission rate of 50 Mbit s^{-1} and operates over a distance of 12 km without repeaters. Assuming Gaussian shaped pulses, calculate the dispersion–equalization penalty exhibited by the system for the cases when:

(a) there is no mode coupling; and
(b) there is mode coupling.

11.18 A digital optical fiber system uses an RZ pulse format. Show that the maximum bit rate for the system $B_T(\text{max})$ may be estimated using the expression:

$$B_T(\text{max}) = \frac{0.35}{T_{\text{syst}}}$$

where T_{syst} is the total system rise time. Comment on the possible use of the factor 0.44 in place of 0.35 in the above relationship.

An optical fiber link is required to operate over a distance of 10 km without repeaters. The fiber available exhibits a rise time due to intermodal dispersion of 0.7 ns km^{-1}, and a rise time due to intramodal dispersion of 0.2 ns km^{-1}. In addition the APD detector has a rise time of 1 ns. Estimate the maximum rise time allowable for the source in order for the link to be successfully operated at a transmission rate of 40 Mbit s^{-1} using an RZ pulse format.

11.19 An edge-emitting LED operating at a wavelength of 1.3 μm launches -22 dBm of optical power into a single-mode fiber pigtail. The pigtail is connected to a single-mode fiber link which exhibits an attenuation of 0.4 dB km^{-1} at this wavelength. In addition, the splice losses on the link provide an average loss of 0.05 dB km^{-1}. The transmission rate of the system is 280 Mbit s^{-1} so that the sensitivity of the *p–i–n* photodiode receiver is -35 dBm. Penalties on the link require an allowance of 1.5 dB and a safety margin of 6 dB is also specified. If the connector losses at the LED transmitter and *p–i–n* photodiode receiver are each 1 dB, calculate the unrepeatered distance over which the link will operate.

11.20 A digital single-mode optical fiber system is designed for operation at a wavelength of 1.5 μm and a transmission rate of 560 Mbit s^{-1} over a distance of 50 km without repeaters. The single-mode injection laser is capable of launching a mean optical power of -13 dBm into the fiber cable which exhibits a loss of 0.25 dB km^{-1}. In addition, average splice losses are 0.1 dB at 1 km intervals. The connector loss at the receiver is 0.5 dB and the receiver sensitivity is -39 dBm. Finally, an extinction ratio penalty of 1 dB is predicted for the system. Perform an optical power budget for the system and determine the safety margin.

11.21 Briefly discuss the reasons for the use of block codes in digital optical fiber transmission. Indicate the advantages and drawbacks when a *5B6B* code is employed.

11.22 A D–IM analog optical fiber system utilizes a *p–i–n* photodiode receiver. Derive an expression for the rms signal power to rms noise power ratio in the quantum limit for this system.

The *p–i–n* photodiode in the above system has a responsivity of 0.5 at the operating wavelength of 0.85 μm. Furthermore the system has a modulation index of 0.4 and transmits over a bandwidth of 5 MHz. Sketch a graph of the quantum limited receiver sensitivity against the received SNR (rms signal power to rms noise power) for the system over the range 30 to 60 dB. It may be assumed that the photodiode dark current is negligible.

11.23 In practice, the analog optical fiber receiver of Problem 11.22 is found to be thermal noise limited. The mean square thermal noise current for the receiver is

2×10^{-23} $A^2 Hz^{-1}$. Determine the peak to peak signal power to rms noise power ratio at the receiver when the average incident optical power is -17.5 dBm.

11.24 An analog optical fiber system has a modulation bandwidth of 40 MHz and a modulation index of 0.6. The system utilizes an APD receiver with a responsivity of 0.7 and is quantum noise limited. An SNR (rms signal power to rms noise power) of 35 dB is obtained when the incident optical power at the receiver is -30 dBm. Assuming the detector dark current may be neglected, determine the excess avalanche noise factor at the receiver.

11.25 A simple analog optical fiber link operates over a distance of 15 m. The transmitter comprises an LED source which emits an average of 1 mW of optical power into air when the drive current is 40 mA. Plastic fiber cable with an attenuation of 500 dB km^{-1} at the transmission wavelength is utilized. The minimum optical power level required at the receiver for satisfactory operation of the system is 5 μW. The coupling losses at the transmitter and receiver are 8 and 2 dB respectively. In addition, a safety margin of 4 dB is necessary. Calculate the minimum LED drive current required to maintain satisfactory system operation.

11.26 An analog optical fiber system employs an LED which emits 3 dBm mean optical power into air. However, a coupling loss of 17.5 dB is encountered when launching into a fiber cable. The fiber cable which extends for 6 km without repeaters exhibits a loss of 5 dB km^{-1}. It is spliced every 1.5 km with an average loss of 1.1 dB per splice. In addition, there is a connector loss at the receiver of 0.8 dB. The PIN–FET receiver has a sensitivity of -54 dBm at the operating bandwidth of the system. Assuming there is no dispersion–equalization penalty, perform an optical power budget for the system and establish a safety margin.

11.27 Indicate the techniques which may be used for analog optical fiber transmission where an electrical subcarrier is employed. Illustrate your answer with a system block diagram showing the multiplexing of several signals onto a single analog optical fiber link.

11.28 Subcarrier amplitude modulation (AM–IM) is employed on the RF carriers of an analog optical fiber system. When a large number of the RF subcarriers, each with random phases, are frequency division multiplexed, then they add on a power basis so that the optical modulation index m is related to the per-channel modulation index m_k by:

$$m = \left(\sum_{k=1}^{N} m_k^2 \right)^{\frac{1}{2}}$$

where N equals the number of channels. An FDM signal incorporates eighty AM subcarriers. When forty of these signals have a per-channel modulation index of 2%, twenty signals have a 3% and the other twenty signals a 4% per channel modulation index, calculate the optical modulation index of the transmitter.

11.29 A narrowband FM–IM optical signal has a maximum frequency deviation of 120 kHz when the frequency deviation ratio is 0.2. Compare the post detection SNR of this signal with that of a DSB–IM optical signal having the same modulating signal power, bandwidth, detector photocurrent and noise. Also estimate the bandwidth of the FM–IM signal. Comment on both results.

11.30 A frequency division multiplexed optical fiber system uses FM–IM. It has fifty equal amplitude voice channels each bandlimited to 3.5 kHz. A 1 kHz guard band is provided between the channels and below the first channel. The peak frequency deviation for the system is 1.35 MHz. Determine the transmission bandwidth for this FDM system.

11.31 An FM–IM system utilizes pre-emphasis and de-emphasis to enhance its performance in noise [Ref. 65]. The de-emphasis filter is a first order RC low pass filter placed at the demodulator to reduce the total noise power. This filter may be assumed to have an amplitude response $H_{de}(\omega)$ given by:

$$|H_{de}(\omega)| = \frac{1}{1 + (\omega/\omega_c)^2}$$

where $\omega_c = 2\pi f_c = 1/RC$.

The SNR improvement over FM–IM without pre-emphasis and de-emphasis is given by:

$$SNR_{de}\ improvement = \frac{1}{3}\left(\frac{B_a}{f_c}\right)^2$$

where B_a is the bandwidth of the baseband signal and $f_c \ll B_a$.

(a) Write down an expression for the amplitude response of the pre-emphasis filter so that there is no overall signal distortion.

(b) Deduce an expression for the post detection SNR improvement in decibels for the FM–IM system with pre-emphasis and de-emphasis over a D–IM system operating under the same conditions of modulating signal power and bandwidth, photocurrent and noise. It may be assumed that $f_c \ll B_a$.

(c) A baseband signal with a bandwidth of 300 kHz is transmitted using the FM–IM system with pre-emphasis and de-emphasis. The maximum frequency deviation for the system is 4 MHz. In addition the de-emphasis filter comprises a 500 Ω resistor and a 0.1 μF capacitor. Determine the post detection SNR improvement for this system over a D–IM system operating under the same conditions of modulating signal power and bandwidth, photocurrent and noise.

11.32 A PM–IM optical fiber system operating above threshold has a frequency deviation ratio of 15 and a transmission bandwidth of 640 kHz.

(a) Estimate the bandwidth of the baseband message signal.

(b) Compute the post detection SNR improvement for the system over a D–IM system operating with the same modulating signal power and bandwidth, detector photocurrent and noise.

11.33 Discuss the advantages and drawbacks of the various pulse analog techniques for optical fiber transmission. Describe the operation of a PFM–IM optical fiber system employing regenerative baseband recovery.

11.34 An optical fiber PFM–IM system uses regenerative baseband recovery. The optical receiver which incorporates a *p–i–n* photodiode has an optimized bandwidth of 125 MHz. The other system parameters are:

Nominal pulse rate	35 MHz
Peak to peak frequency deviation	8 MHz
p–i–n photodiode responsivity	$0.6\ \mathrm{A\,W^{-1}}$
Baseband noise bandwidth	10 MHz
Receiver mean square noise current	$3 \times 10^{-25}\ \mathrm{A^2\,Hz^{-1}}$

(a) Calculate the peak level of incident optical power necessary at the receiver to maintain a peak to peak signal power to rms noise power ratio of 60 dB.

(b) The source and detector have rise times of 3.5 and 5.0 ns respectively. Estimate the maximum permissible total rise time for the fiber cable utilized in the system such that satisfactory operation is maintained. Comment on the value obtained.

11.35 Considering the bus and star distribution systems of Example 11.18, compare the losses associated with the addition of an extra node when the original number of nodes is twelve. How does this alter if the connector losses increase to 1.5 dB and the distance between the nodes on the bus/combined length of two arms from the star hub increases to 400 m? Comment on the results.

11.36 An optical fiber data bus is to be implemented to interconnect nine stations, each separated by 50 m. Multimode fiber cable with an attenuation of 3 dB km^{-1} is to be used along with LED transmitters which launch 200 μW of optical power into their fiber pigtails. The PIN–FET hybrid receivers have a sensitivity of -50 dBm at the desired BER, whilst the connector losses are 1.1 dB each. The access couplers to be used have a tap ratio or power tap-off factor of 8% together with an insertion loss of 0.9 dB. Finally, the beam splitter can be assumed to have a loss of 3 dB. Determine the safety margin for the system when considering the highest loss terminal interconnection path.

11.37 Repeat Problem 11.36 when considering a thirty terminal star distribution network where the combined distance in the two fiber cable arms is 200 m and the excess loss of the star coupler hub is 3.4 dB. Comment on the result.

11.38 Compare and contrast the merits and drawbacks associated with the advanced multiplexing techniques discussed in Section 11.9. Comment on the possibilities for combining two of the techniques in order to provide increased information transfer.

11.39 Describe, with the aid of simple sketches, the ways in which optical amplifiers may be configured on long-haul telecommunication links in order to provide both bidirectional and multichannel optical transmission.

Suggest how optical amplifiers might be incorporated into optical fiber distribution systems to facilitate the interconnection of a larger number of nodes.

11.40 It is desired to obtain the maximum signal to noise ratio on a single-mode fiber communication link incorporating cascaded travelling wave SLAs. If only noise considerations are taken into account, determine the optimum length between the amplifier repeaters when the system is operating at:

(a) a wavelength of 1.3 μm where the fiber cable and joint losses are 0.41 dB km^{-1} and 0.04 dB km^{-1} respectively; and

(b) a wavelength of 1.55 μm where the fiber cable and joint losses are 0.20 dB km^{-1} and 0.02 dB km^{-1} respectively.

Comment on the values obtained.

11.41 The following specifications are envisaged for a single-mode fiber communication system employing cascaded travelling wave SLAs:

operating wavelength:	1.30 μm
power launched at transmitter:	2 dBm
fiber cable loss:	0.40 dB km^{-1}
fiber splice losses:	0.02 dB km^{-1}
K (amplifiers):	6
amplifier separation:	50 km
received SNR:	20 dB

Assuming amplifier noise to be the limiting factor estimate the maximum transmission rate allowed for the system when the total system length is 8000 km.

11.42 The SNR at the transmitter output for a single-mode fiber system employing cascaded travelling wave SLAs is 48.6 dB. The final SLA is positioned as a preamplifier to the optical receiver on the link and the output SNR from this device is 15 dB. In addition the amplifiers each have a noise figure of 4.5 dB and a fiber to fiber gain of 9.7 dB, and the attenuation of the link prior to each amplifier is 5.2 dB. Assuming the SNR position to remain constant, how many amplifiers can be cascaded on the link?

Answers to numerical problems

11.4 53.0 pF, 472 Ω
11.5 64 kbit s^{-1}
11.6 (a) 37 ns; (b) 125 μV
11.8 (a) 300 bits; (b) 1.2 MHz
11.9 (a) 10.2 dB; (b) 3.4×10^{-7}
11.10 $P(e) = \text{Erfc}\,|(S/N)^{\frac{1}{2}}/2|$, 9.8 dB, 19.6 dB
11.11 7.4
11.12 (a) 1400; (b) −52.8 dBm; (c) −51.8 dBm
11.13 27.9 dB
11.14 15.76 km, 8.26 km
11.15 3.41 km
11.16 (a) 22.5 Mbit s^{-1}; (b) 200 km, 0.2 km
11.17 (a) 16.6 dB; (b) 0.1 dB.
11.18 3.04 ns
11.19 7.8 km
11.20 2.1 dB
11.22 40 nW, 40 μW
11.23 52.1 dB
11.24 3.1
11.25 35.5 mA
11.26 5.4 dB
11.28 25.7%
11.29 Ratio of output SNRs (rms signal power to rms noise power) for FM–IM to DSB–IM is −9.21 dB, 1200 kHz
11.30 3.15 MHz
11.31 (a) $1 + (\omega/\omega_c)^2$; (b) $-3 + 20 \log_{10}(D_f B_a/f_c)$; (c) 59.0 dB
11.32 (a) 20 kHz; (b) 20.5 dB
11.34 (a) −24.4 dBm; (b) 4.6 ns
11.35 3.5 dB, 0.35 dB, 6.0 dB, 0.35 dB
11.36 3.7 dB
11.37 19.8 dB
11.40 (a) 9.6 km; (b) 19.7 km
11.41 858 Mbit s^{-1} (RZ)
11.42 87

References

[1] C. C. Timmermann, 'Highly efficient light coupling from GaAlAs lasers into optical fibers', *Appl. Opt.*, **15**(10), pp. 2432–2433, 1976.

[2] M. Maeda, I. Ikushima, K. Nagano, M. Tanaka, H. Naskshima, R. Itoh, 'Hybrid laser to fiber coupler with a cylindrical lens', *Appl. Opt.*, **16**(7), pp. 1966–1970, 1977.

[3] R. W. Dawson, 'Frequency and bias dependence of video distortion in Burrus-type homostructure and heterostructure LED's', *IEEE Trans Electron. Devices*, **ED-25**(5), pp. 550–553, 1978.

[4] J. Strauss, 'The nonlinearity of high-radiance light-emitting diodes', *IEEE J. Quantum Electron.*, **QE-14**(11), pp. 813–819, 1978.

[5] K. Asatani and T. Kimura, 'Non-linear phase distortion and its compensation in LED direct modulation', *Electron. Lett.*, **13**(6), pp. 162–163, 1977.

[6] G. White and C. A. Burrus, 'Efficient 100 Mb/s driver for electroluminescent diodes', *Int. J. Electron.*, **35**(6), pp. 751–754, 1973.

[7] P. W. Shumate Jr and M. DiDomenico Jr, 'Lightwave transmitters', in H. Kressel (Ed.), *Semiconductor Devices for Optical Communications, Topics in Applied Physics*, Volume 39, pp. 161–200, Springer-Verlag, 1982.

[8] L. Foltzer, 'Low-cost transmitters, receivers serve well in fibre-optic links', *EDN*, pp. 141–146, 20 October 1980.

[9] A. Albanese and H. F. Lenzing, 'Video transmission tests, performed on intermediate-frequency light wave entrance links', *J. SMPTE* (*USA*), **87**(12), pp. 821–824, 1978.

[10] J. Strauss, 'Linearized transmitters for analog fiber links', *Laser Focus* (*USA*), **14**(10), pp. 54–61, 1978.

[11] A. Prochazka, P. Lancaster and R. Neumann, 'Amplifier linearization by complementary pre or post distortion', *IEEE Trans. Cable Telev.*, **CATV-1**(1), pp. 31–39, 1976.

[12] K. Asatani and T. Kimura, 'Nonlinear distortions and their compensations in light emitting diodes', *Proceeding of International Conference on Integrated Optics and Optical Fiber Communications*, p. 105, 1977.

[13] K. Asatani and T. Kimura, 'Linearization of LED nonlinearity by predistortions', *IEEE J. Solid State Circuits*, **SC-13**(1), pp. 133–138, 1978.

[14] J. Strauss, A. J. Springthorpe and O. I. Szentesi, 'Phase shift modulation technique for the linearisation of analogue optical transmitters', *Electron. Lett.*, **13**(5), pp. 149–151, 1977.

[15] J. Strauss and D. Frank, 'Linearisation of a cascaded system of analogue optical links', *Electron. Lett.*, **14**(14), 436–437, 1978.

[16] H. S. Black, US Patent 1686792, issued Oct 9, 1929.

[17] B. S. Kawasaki and K. O. Hill, 'Low-loss access coupler for multimode optical fiber distribution network', *Appl. Opt.*, **16**(7), p. 1794, 1977.

[18] J. Strauss and O. I. Szentesi, 'Linearisation of optical transmitters by a quasifeedforward compensation technique', *Electron. Lett.*, **13**(6), pp. 158–159, 1977.

[19] S. M. Abbott, W. M. Muska, T. P. Lee, A. G. Dentai and C. A. Burrus, '1.1 Gb/s pseudorandom pulse-code modulation of 1.27μm wavelength CW InGaAsP/InP DH lasers', *Electron. Lett.*, **14**(11), pp. 349–350, 1978.

[20] J. Gruber, P. Marten, R. Petschacher and P. Russer, 'Electronic circuits for high bit rate digital fiber optic communication systems', *IEEE Trans. Commun.*, **COM-26**(7), pp. 1088–1098, 1978.

[21] P. K. Runge, 'An experimental 50 Mb/s fiber optic PCM repeater', *IEEE Trans. Commun.*, **COM-24**(4), pp. 413–418, 1976.

[22] U. Wellens, 'High-bit-rate pulse regeneration and injection laser modulation using a diode circuit', *Electron. Lett.*, **13**(18), pp. 529–530, 1977.

[23] A. Chappell (Ed.), *Optoelectronics: Theory and Practice*, McGraw-Hill, 1978.

[24] S. R. Salter, D. R. Smith, B. R. White and R. P. Webb, 'Laser automatic level control for optical communications systems', *Third European Conf. on Optical Communications*, Munich, September 1977, VDE-Verlag GmbH, Berlin, 1977.

[25] A. Fausone, 'Circuit considerations', *Optical Fibre Communication*, by Tech. Staff of CSELT, pp. 777–800, McGraw-Hill, 1981.

[26] A. Moncalvo and R. Pietroiusti, 'Transmission systems using optical fibres', *Telecommunication J. (Switzerland)*, **49**, pp. 84–92, 1982.
[27] S. D. Personick, 'Design of receivers and transmitters for fibre systems', in M. K. Barnoski (Ed.), *Fundamentals of Optical Fiber Communications* (2nd edn), pp. 295–328, Academic Press, 1981.
[28] T. L. Maione and D. D. Sell, 'Experimental fiber-optic trransmission system for interoffice trunks', *IEEE Trans. Commun.*, **COM-25**(5), pp. 517–522, 1977.
[29] R. G. Smith and S. D. Personick, 'Receiver design for optical fibre communication systems', in H. Kressel (Ed.), *Semiconductor Devices for Optical Communications*, Topics in Advanced Physics, Vol. 39, pp. 88–160, Springer-Verlag, 1982.
[30] R. G. Smith, C. A. Brackett and H. W. Reinbold, 'Atlanta fiber system experiment, optical detector package', *Bell Syst. Tech. J.*, **57**(6), pp. 1809–1822, 1978.
[31] J. L. Hullett and T. V. Muoi, 'A feedback amplifier for optical transmission systems', *IEEE Trans. Commun.*, **COM-24**, pp. 1180–1185, 1976.
[32] J. L. Hullett, 'Optical communication receivers', *Proc. IREE Australia*, pp. 127–134, September 1979.
[33] N. J. Bradley, 'Fibre optic systems design', *Electronic. Eng.*, pp. 98–101, mid April 1980.
[34] T. L. Maione, D. D. Sell and D. H. Wolaver, 'Practical 45 Mb/s regenerator for lightwave transmission', *Bell Syst. Tech. J.*, **57**(6), pp. 1837–1856, 1978.
[35] S. D. Personick, 'Receiver design for optical systems', *Proc. IEEE*, **65**(12), pp. 1670–1678, 1977.
[36] J. E. Goell, 'Input amplifiers for optical PCM receivers', *Bell Syst. Tech. J.*, **53**(9), pp. 1771–1793, 1974.
[37] S. D. Personick, 'Time dispersion in dielectric waveguides', *Bell Syst. Tech. J.*, **50**(3), pp. 843–859, 1971.
[38] I. Garrett 'Receivers for optical fibre communications', *Radio Electron. Eng. J. IERE*, **51**(7/8). pp. 349–361, 1981.
[39] S. D. Personick, 'Receiver design', in S. E. Miller and A. G. Chynoweth (Eds.), *Optical Fiber Telecommunications*, pp. 627–651, Academic Press Inc., 1979.
[40] A. Moncalvo and L. Sacchi, 'System considerations', *Optical Fibre Communication*, by Tech. Staff of CSELT, pp. 723–776, McGraw-Hill, 1981.
[41] M. Rocks and R. Kerstein, 'Increase in fiber bandwidth for digital systems by means of multiplexing', *ICC 80 1980 International Conf. on Commun.*, Seattle, WA, USA, Part 28, 5/1–5, June 1980.
[42] W. Koester and F. Mohr, 'Bidirectional optical link', *Electrical Commun.*, **55**(4), pp. 342–349, 1980.
[43] H. Taub and D. L. Schilling, *Principles of Communication System* (2nd edn), McGraw-Hill, 1986.
[44] P. Hensel and R. C. Hooper, 'The development of high performance optical fibre data links', *IERE Conference Proceedings Fibre Optics*, 1–2 March 1982 (London) pp. 91–98, 1982.
[45] D. C. Gloge and T. Li, 'Multimode-fiber technology for digital transmission', *Proc. IEEE*, **68**(10), pp. 1269–1275, 1980.
[46] G. E. Stillman,: 'Design consideration, for fibre optic detectors', *Proc. SPIE Int. Soc. Opt. Eng. (USA)*, **239**, pp. 42–52, 1980.
[47] P. P. Webb, R. J. McIntyre and J. Conradi, 'Properties of avalanche photodiodes', *RCA Rev.*, **35**, pp. 234–278, 1974.

[48] J. E. Midwinter, *Optical Fibers for Transmission*, John Wiley, 1979.
[49] I. Garrett and J. E. Midwinter, 'Optical communication systems', in M. J. Howes and D. V. Morgan (Eds.), *Optical Fibre Communications: Devices, Circuits, and Systems*, pp. 251–300, John Wiley, 1980.
[50] R. J. McIntyre and J. Conradi, 'The distribution of gains in uniformly multiplying avalanche photodiodes', *IEEE Trans. Electron. Devices*, **ED-19**, pp. 713–718, 1972.
[51] K. Mouthaan, 'Teleconmunications via glass–fibre cables', *Philips Telecommun. Rev.*, **37**(4), pp. 201–214, 1979.
[52] H. F. Wolf, 'System aspects', in H. F. Wolf (Ed.), *Handbook of Fiber Optics: Theory and applications*, pp. 377–427, Granada, 1979.
[53] S. D. Personick, N. L. Rhodes, D. C. Hanson and K. H. Chan, 'Contrasting fiber-optic-component-design requirements in telecommunications, analog, and local data communications applications', *Proc. IEEE*, **68**(10), pp. 1254–1262, 1980.
[54] S. E. Miller, 'Transmission system design', in S. E. Miller and A. G. Chynoweth (Eds.), *Optical Fiber Telecommunications*, pp. 653–683, Academic Press, 1979.
[55] C. K. Koa, *Optical Fiber Systems: Technology, design and applications*, McGraw-Hill, 1982.
[56] C. Kleekamp and B. Metcalf, *Designer's Guide to Fiber Optics*, Cahners Publishing Company, 1978.
[57] G. R. Elion and H. A. Elion, *Fiber Optics in Communications Systems*, Marcel Dekker, 1978.
[58] S. Shimada, 'Systems engineering for long-haul optical-fiber transmission', *Proc. IEEE*, **68**(10), pp. 1304–1309, 1980.
[59] J. H. C. van Heuven, 'Techniques for optical transmission', in *Proceedings of 11th European Microwave Conference*, Amsterdam, Netherlands, pp. 3–10, 1981.
[60] R. Tell and S. T. Eng, 'Optical fiber communication at 5 Gbit/sec', *Appl. Opt.*, **20**(22), pp. 3853–3858, 1981.
[61] P. Wells, 'Optical-fibre systems for telecommunications', *GEC J. Sci. Tech.*, **46**(2), pp. 51–60, 1980.
[62] M. Chown and K. C. Koa, 'Some broadband fiber system design considerations', in *Proceedings of IEEE International Conference on Communications*, Philadelphia PA, 1972, pp. 12/1–5, IEEE, 1972.
[63] I. Garrett and C. J. Todd, 'Optical fiber transmission systems at 1.3 and 1.5 μm wavelength', in *Proceedings of IEEE 1981 International Conference on Communications*, New York, Vol. 1, Pt. 16,2/1–5, IEEE, 1981.
[64] J. L. Hullett and T. V. Muoi, 'Optical fiber systems analysis', *Proc. IREE Australia*, **38**(1–2), pp. 390–397, 1977.
[65] A. Luvison, Topics in optical fibre communication theory', *Optical Fibre Communications*, by Technical Staff of CSELT, pp. 647–721, McGraw-Hill, 1981.
[66] S. Kawanishi, N. Yoshikai, J.-I. Yamada and K. Nakagawa, 'DmBIM code and its performance, in a very high-speed optical transmission system', *IEEE Trans. on Commun.*, **36**(8), pp. 951–956, 1988.
[67] M. Chown, A. W. Davis, R. E. Epworth and J. G. Farrington, 'System design', in C. P. Sandbank (Ed.), *Optical Fibre Communication Systems*, pp. 206–283, John Wiley, 1980.
[68] K. Sam Shanmugan, *Digital and Analog Communication Systems*, John Wiley, 1979.
[69] S. Y. Suh, 'Pulse width modulation for analog fiber-optic communications', *J. of Lightwave Technol.*, **LT-5**(1), pp. 102–112, 1987.

[70] G. G. Windus, 'Fibre optic systems for analogue transmission', *Marconi Rev.*, **XLIV**(221), pp. 78–100, 1981.

[71] W. Horak, 'Analog TV signal transmission over multimode optical waveguides', *Siemens Research and Development Reports*, **5**(4), pp. 192–202, 1976.

[72] K. Sato and K. Asatani, 'Analogue baseband TV transmission experiments using semiconductor laser diodes', *Electron. Lett.*, **15**(24), pp. 794–795, 1979.

[73] R. M. Gagliadi and S. Karp, *Optical Communications*, John Wiley, 1976.

[74] C. C. Timmerman, 'Signal to noise ratio of a video signal transmitted by a fiber-optic system using pulse-frequency modulation', *IEEE Trans. Broadcasting*, **BC-23**(1), pp. 12–16, 1976.

[75] C. C. Timmerman, 'A fiber optical system using pulse frequency modulation', *NTZ*, **30**(6), pp. 507–508, 1977.

[76] D. J. Brace and D. J. Heatley, 'The application of pulse modulation schemes for wideband distribution to customers (integrated optical fibre systems)', in *Sixth European Conference on Optical Communication*, York, UK, 16–19 Sept. 1980, pp. 446–449, 1980.

[77] E. Yoneda, T. Kanada and K. Hakoda, 'Design and performance of optical fibre transmission systems for color television signals', *Rev. Elect. Commun. Lab.*, **29**(11–12), pp. 1107–1117, 1981.

[78] T. Kanada, K. Hakoda and E. Yoneda, 'SNR fluctuation and nonlinear distortion in PFM optical NTSC video transmission systems', *IEEE Trans. Commun.*, **COM-30**(8), pp. 1868–1875, 1982.

[79] S. F. Heker, G. J. Herskowitz, H. Grebel and H. Wichansky, 'Video transmission in optical fiber communication systems using pulse frequency modulation', *IEEE Trans. on Commun.*, **36**(2), pp. 191–194, 1988.

[80] S. Fujita, M. Kitamura, T. Torikai, N. Henmi, H. Yamada, T. Suzaki, I. Takano and M. Shikada, '10 Gbit/s, 100 km optical fibre transmission experiment using high-speed MQW DFB-LD and back illuminated GaInAs APD', *Electron. Lett.*, **25**(11), pp. 702–703, 1989.

[81] R. S. Tucker, G. Eisenstein, S. K. Korotky, U. Koren, G. Raybon, J. J. Veselka, L. L. Buhl, B. L. Kasper and R. C. Alferness, 'Optical time-division multiplexing in a multigigabit/second fibre transmission system', *Electron. Lett.*, **23**(5), pp. 208–209, 1987.

[82] R. C. Alferness, 'Multigigabit fibre optics', *Communications International*, **15**(4), pp. 42–51, 1988.

[83] T. E. Darcie, M. E. Dixon, B. L. Kasper and C. A. Burrus, 'Lightwave system using microwave subcarrier multiplexing', *Electron. Lett.*, **22**(15), pp. 774–775, 1986.

[84] R. Olshansky and V. A. Lanziera, '60 channel FM video subcarrier multiplexed optical communications system', *Electron. Lett.*, **23**, pp. 1196–1197, 1987.

[85] T. E. Darcie, P. P. Iannone, B. L. Kasper, J. R. Talman, C. A. Burrus and T. A. Baker, 'Wide-band lightwave distribution system using subcarrier multiplexing', *J. of Lightwave Technol.*, **7**(6), pp. 997–1005, 1989.

[86] R. Olshansky and V. A. Lanziera, 'Subcarrier multiplexed lightwave systems for broad-band distribution', *J. of Lightwave Technol.*, **7**(9), pp. 1329–1342, 1989.

[87] T. E. Darcie, 'Subcarrier multiplexing for lightwave networks and video distribution systems', *IEEE J. on Selected Areas in Commun.*, **8**(7), pp. 1240–1248, 1990.

[88] M. Maeda and M. Yamamoto, 'FM–FDM optical CATV transmission experiment

and system design for MUSE HDTV signals', *IEEE J. on Selected Areas in Commun.*, **8**(7), pp. 1257–1267, 1990.

[89] P. A. Rosher and S. C. Fenning, 'Multichannel video transmission over 100 km of step index single mode fibre using a directly modulated distributed feedback laser', *Electron. Lett.*, **26**(8), pp. 534–536, 1990.

[90] A. C. Carter, 'Wavelength multiplexing for enhanced fibre-optic performance', *Telecommunications*, pp. 30–36, October 1986.

[91] H. Kobrinski, R. M. Bulley, M. S. Goodman, M. P. Vecchi and C. A. Brackett, 'Demonstration of high capacity in the LAMBDANET architecture: a multiwavelength optical network', *Electron. Lett.*, **23**(16), pp. 824–826, 1987.

[92] M. P. Vecchi, R. M. Bulley, M. S. Goodman, H. Kobrinski and C. A. Brackett, 'High-bit-rate measurements in the LAMBDANET multiwavelength optical star network', *Opt. Fiber Commun. Conf. OFC'88* (USA), paper WO2, January 1988.

[93] A. Hunwicks, L. Bickers, M. H. Reeve and S. Hornung, 'An optical transmission system for single-mode local-loop applications using a sliced spectrum technique', *Twelfth Internat. Fiber Optic Commun. and Local Area Networks Exposition, FOC/LAN'88* (USA), pp. 237–240, September 1988.

[94] S. S. Wagner and T. E. Chapuran, 'Broadband high-density WDM transmission using superluminescent diodes', *Electron. Lett.*, **26**(11), pp. 696–697, 1990.

[95] J. M. Senior and S. D. Cusworth, 'Wavelength division multiplexing in optical sensor systems and networks: a review', *Opt. and Laser Technol.*, **22**(2), pp. 113–126, 1990.

[96] P. Cockrane, 'Future directions in undersea fibre optic system technology', *Proc. IOOC'89*, Kobe, Japan, paper 21B1–2, July 1989.

[97] P. S. Henry, R. A. Linke and A. H. Gnauck, 'Introduction to Lightwave Systems', in S. E. Miller and I. P. Kaminow (Eds.), *Optical Fiber Telecommunications II*, Academic Press, pp. 781–831, 1988.

[98] C. H. Henry, 'Theory of spontaneous emission noise in optical resonators and its application to lasers and optical amplifiers', *J. Lightwave Technol.*, **LT-4**, pp. 288–297, 1986.

[99] W. A. Stallard, J. D. Cox, D. J. Malyon, A. E. Ellis and K. H. Cameron, 'Long span high capacity optical transmission system employing laser amplifier repeaters', *Proc. Internat. Conf. on Integrated Optics and Fibre Commun., IOOC'89*, Kobe (Japan), paper 21B2–2, July 1989.

[100] P. Cockrane, 'Future directions in long haul fibre optic systems, *Br. Telecom Technol. J.*, **8**(2), pp. 5–17, 1990.

[101] S. Yamamoto, H. Taga, N. Edagawa, K. Mochizuki anbd H. Wakabayashi, '516 km, 2.4 Gbit/s optical fiber transmission experiment using 10 semiconductor laser amplifiers and measurement of jitter accumulation', *Proc. Internat. Conf. on Integrated Optics and Fibre Commun., IOOC'89*, Kobe (Japan), paper 20PDA-9. July 1989.

[102] N. H. Taylor and A. Hadjitotiou, 'Optical amplification and its applications', *Proc. SPIE, Int. Soc. Opt. Eng*, **1314**, pp. 64–67, 1990.

[103] M. J. O'Mahony, 'Progress in optical amplifiers', *Internat. Workshop on Digital Commun.*, Session 6, paper 3, Tirrenia, Italy, September 1989.

[104] D. M. Spirit, L. C. Blank, S. T. Davey and D. L. Williams, 'Systems aspects of Raman fibre amplifiers, *IEE Proc., Pt. J.*, **137**(4), pp. 221–224, 1990.

[105] K. Aida, S. Nishi, Y. Sato, K. Hagimoto and K. Nakagawa, 'Design and performance of a long-span IM/DD optical transmission system using remotely pumped optical amplifiers', *IEE Proc., Pt. J.*, **137**(4), pp. 225–229, 1990.

[106] N. Edagawa, K. Mochizuki and H. Wakabayashi, '1.2 Gbit/s, 218 km transmission using inline Er-doped optical fibre amplifier', *Electron. Lett.*, **25**(5), pp. 363–365, 1989.

[107] K. Hagimoto, K. Iwatsuki, A. Takada, M. Nakazawa, M. Saruwatari, K. Aida and K. Nakagawa, '250 km nonrepeated transmission experiment at 1.8 Gb/s using LD pumped E^{3+}-doped fibre amplifiers in IM/direct detection system', *Electron. Lett.*, **25**(10), pp. 662–664, 1989.

[109] J. F. Massicott, R. Wyatt, B. J. Ainslie and S. P. Craig-Ryan, 'Efficient, high power, high gain doped silica fibre amplifier', *Electron. Lett.*, **26**(14), pp. 1038–1039, 1990.

[109] C. G. Atkins, J. F. Massicott, J. R. Armitage, R. Wyatt, B. J. Ainslie and S. P. Craig-Ryan, 'A high gain, broad spectral bandwidth erbium doped fibre amplifier pumped near 1.5 μm', *Electron. Lett.*, **25**(14), pp. 910–911, 1989.

[110] P. D. D. Kilkelly, P. J. Chidgey and G. Hill, 'Experimental demonstration of a three channel WDM system over 110 km using superluminescent diodes', *Electron. Lett.*, **26**(20), pp. 1671–1673, 1990.

[111] P. J. Chidgey and G. R. Hill, 'Diverse routing in wavelength selective networks', *Electron. Lett.*, **26**(20), pp. 1709–1711, 1990.

[112] H. Taga, Y. Yoshida, N. Edagawa, S. Yamamoto and H. Wakabayashi, '459 km, 2.4 Gbit/s four wavelength multiplexing optical fibre transmission experiment using six Er-doped fibre amplifiers', *Electron. Lett.*, **26**(8), pp. 500–501, 1990.

[113] A. M. Hill, R. Wyatt, J. F. Massicott, K. J. Blyth, D. S. Forrester, R. A. Lobbett, P. J. Smith and D. B. Payne, '39.5 million-way WDM broadcast network employing two stages of erbium-doped fibre amplifiers', *Electron. Lett.*, **26**(22), pp. 1882–1884, 1990.

[114] W. I. Way, S. S. Wagner, M. M. Choy, C. Lin, R. C. Menendez, H. Tohme, A. Yi-Yan, A. C. von Lehman, R. E. Spicer, M. Andrejco, M. A. Saifi and H. Lemberg, 'Distribution of 100 FM-TV channels and six 622 Mb/s channels to 4096 terminals using high-density WDM and a broadband in-line erbium-doped fiber amplifier', *Optical Fiber Communications Conf.*, *OFC'90* (USA), Post Deadline Paper PD21, January 1990.

[115] M. J. Pettitt, A. Hadjifotiou and R. A. Baker, 'Crosstalk in erbium doped fibre amplifiers', *Electron. Lett.*, **25**(6), pp. 416–417, 1989.

[116] G. J. Cowle, P. R. Morkel, R. I. Laming and D. N. Payne, 'Spectral broadening due to fibre amplifier phase noise', *Electron. Lett.*, **26**(7), pp. 424–425, 1990.

12

Optical fiber systems 2: coherent

12.1 Introduction

The direct detection of an intensity modulated optical carrier is basically a photon counting process where each detected photon is converted into an electron–hole pair (or, in the case of the APD, a number of pairs due to avalanche gain). It was indicated in Section 7.5 that this process which ignores the phase and polarization of the electromagnetic carrier may be readily implemented with currently available optical components. Thus all the preceding discussion in Chapter 11 involving both digital and analog systems concerned only an intensity modulated optical carrier being transmitted to a direct detection optical receiver or IM/DD optical fiber systems.

Conventional direct detection receivers, however, are generally limited by noise generated in the detector and preamplifier (see Chapter 9) except at very high signal to noise ratios (SNRs). The sensitivity of such square law detection systems is therefore reduced below the fundamental quantum noise limit by at least 10 to 20 dB [Ref. 1]. This is particularly the case at longer wavelengths (i.e. 1.3 to 1.6 μm) and at higher transmission rates since the electronic preamplifier usually has a rising input optical power with frequency requirement* (see Eq. (11.37)). For a good APD receiver operating in the wavelength range 1.3 to 1.6 μm this corresponds to between 700 and 1000 photons per bit required to maintain a bit error rate (BER) of 10^{-9} [Ref. 2].

Improvements in receiver sensitivity, together with wavelength selectivity, may be obtained using the well-known coherent detection techniques (i.e. heterodyne and homodyne detection) for the optical signal [Ref. 3]. Unlike direct detection in which the optical signal is converted directly into a demodulated electrical output, such coherent optical receivers first add to the incoming optical signal from a locally generated optical wave prior to detecting the sum.† The resulting photocurrent is a replica of the original signal which is translated down in frequency from the optical domain (around 10^5 GHz) to the radio domain (up to several GHz) and where conventional electronic techniques can be used for further signal processing and demodulation. Hence an ideal coherent receiver operating in the 1.3 to 1.6 μm wavelength region requires a signal energy of only 10 to 20 photons per bit to achieve a BER of 10^{-9}. It should therefore be noted that coherent detection provides the greatest benefit for high speed systems operating at longer wavelengths.

A potential improvement in receiver sensitivity using coherent detection of up to 20 dB can be obtained over direct detection [Ref. 4]. Furthermore, such enhanced receiver sensitivity could translate into increases in repeater spacings of the order of 100 km when using low loss fiber at a wavelength of 1.55 μm. Hence, the improved sensitivity of 5 to 20 dB which results from the photomixing gain in the coherent receiver could provide:

(a) increased repeater spacings for both inland and undersea transmission systems;
(b) higher transmission rates over existing routes without reducing repeater spacings;
(c) increased power budgets to compensate for losses associated with couplers and optical multiplexer/demultiplexer devices (see Section 5.6.3) in distribution networks;
(d) improved sensitivity to optical test equipment such as optical time domain reflectometers (see Section 13.10.1).

* This requirement corresponds to a rising noise versus frequency characteristic.

† Current usage in optical fiber communications is that the term 'coherent' refers to any system or technique which employs nonlinear mixing between two optical waves. Typically, one of these is the information carrying signal and the other a locally generated signal (by a local oscillator) at the receiver. The result of this process is a new signal (for heterodyne detection, the intermediate frequency) which appears at a microwave frequency given by the difference between the frequencies of the incoming signal and the local oscillator.

Although possible increases in transmission distance between repeaters initially created the impetus for the pursuit of coherent transmission within optical fiber communications, such techniques also allow a further massive step to be taken in the exploitation of the transmission capacity of optical fiber systems. The inherent wavelength selectivity afforded by the coherent receiver could be used to access efficiently the vast optical bandwidth available in single-mode fiber. For example, the optical bandwidth provided by the low loss window between 1.3 and 1.6 μm is over 50 000 GHz [Ref. 5]. Therefore, as a result of its improved selectivity over direct detection, a coherent receiver could permit wavelength division multiplexing* with channel spacings of only a few hundred MHz [Ref. 6] instead of the minimum of around 100 GHz (i.e. 0.5 nm) provided by conventional optical multiplexing technology. It is this factor in particular which has more recently focused interest on coherent optical fiber transmission because wavelength/frequency selectivity may well be more important than improved receiver sensitivity in the provision of future wideband distribution networks within telecommunications [Refs. 7, 8]. Finally, a further potentially important attribute of coherent reception is that it allows the use of electronic equalization to compensate for the effects of optical pulse dispersion in the fiber.

The modulation formats that may be employed within coherent optical fiber communications are essentially the same as those used in coherent electrical line and radio communications. Modulation formats of this type were discussed in Sections 11.7.3 to 11.7.6 in a slightly different context, namely the generation of subcarriers for electrical frequency division multiplexing prior to intensity modulation of the optical source. In these cases direct detection of the optical signal is carried out at the receiver with subsequent electrical demodulation for the subcarriers. Such systems only provide improvements in the SNR over baseband IM/DD systems at the expense of a substantial bandwidth penalty. When a narrow linewidth injection laser (less than 1 MHz) is used in an optical fiber communication system, however, it is possible to directly modulate the coherent optical carrier in amplitude (direct AM), frequency (direct FM) and phase (direct PM) prior to demodulation using a coherent optical receiver. In the case of digital transmission this implies amplitude, frequency or phase shift keying (i.e. ASK, FSK or PSK) modulation techniques [Ref. 3].

The discussion is continued in Section 12.2 through a brief historical review of the development of coherent optical transmission prior to the description of the basic coherent optical fiber communication system, together with its important features. This leads into the consideration of the fundamental detection principles associated with the coherent optical receiver (i.e. heterodyne and homodyne detection) in Section 12.3. There are, however, a number of practical constraints which have in the past inhibited the development of coherent optical fiber systems, and even now they create certain limitations on the choice of system components. These issues are therefore dealt with in Section 12.4. This is followed in Section 12.5

* In this case it is often referred to as frequency division multiplexing.

by discussion of the various modulation formats that may be employed for coherent optical transmission prior to the description of the numerous demodulation schemes which have been applied within the coherent detection process in Section 12.6. A comparison of the various major modulation and demodulation techniques in relation to receiver sensitivity is then provided in Section 12.7. Finally, in Section 12.8 we describe the major features and performance characteristics of some advanced coherent optical fiber transmission systems. In particular, the field trial of a single carrier system is discussed prior to the consideration of recent demonstrations of multicarrier coherent systems using optical frequency division multiplexing techniques.

12.2 Basic system

Since the invention of the laser in 1960, research efforts have focused on techniques by which the coherent properties of laser light could be utilized for coherent optical communications. Improved SNRs over direct detection were demonstrated in free space optical communication systems using gas lasers in the late 1960s [Refs. 9, 10]. In addition, the concept of optical frequency division multiplexing using coherent detection schemes was proposed in 1970 [Ref. 11]. However, it was only in the latter half of the 1970s, when single-mode transmission from a narrow linewidth AlGaAs semiconductor laser was demonstrated [Ref. 12], that the proposals for coherent optical fiber transmission began taking shape. Nevertheless, it was appreciated that the polarization stability of the transmission medium was crucial for successful coherent detection. Ideally, for coherent transmission the fiber would be required to maintain a single linear polarization state throughout its length. This factor, in part, led to the investigations on polarization maintaining (PM) fibers in the early 1980s (see Section 3.13.2). Moreover, for a brief period the use of PM fibers was a favoured potential approach to the problem [Ref. 13]. For example, the use of optical adaptors (e.g. birefringent plates) at the transmit and receive terminals was suggested in order to allow only a single polarization state of the fundamental mode to be launched into, and received from, the PM fiber [Ref. 14].

It was clear, however, that if conventional circularly symmetric single-mode fiber, which did not maintain a single polarization state over its length (see Section 13.3.1), were to be employed then some form of polarization matching of the incoming optical signal with the locally generated optical signal would be necessary. Although the first successful demonstration of optical frequency shift keyed heterodyne detection using a semiconductor laser source and local oscillator was reported in 1989 [Ref. 15], a period elapsed before the polarization stability measurements on installed conventional single-mode fiber indicated the real possibility of its use within coherent optical transmission systems [Ref. 16]. It is therefore only comparatively recently that coherent transmission techniques have proved feasible within optical fiber communications [Ref. 5].

A block schematic of a generalized coherent optical fiber communication system is illustrated in Figure 12.1. The broken lines enclose the main elements which distinguish the coherent system from its direct detection equivalent. At the transmitter a CW narrow linewidth semiconductor laser is shown which acts as an optical frequency oscillator. An external optical modulator usually provides amplitude, frequency or phase shift keying of the optical carrier by the information signal. At present external modulators are generally waveguide devices fabricated from lithium niobate or the group III–V compound semiconductors (see Section 10.6.2). Internal modulation of the injection laser drive current may, however, also be utilized to produce either ASK or FSK [Refs. 17, 18].

Modulated carrier waveforms for the three standard modulation techniques with binary data are illustrated in Figure 12.2. It may be observed from Figure 12.2(a) where binary ASK is often referred to as on-off keying (OOK). Figure 12.2(b) shows FSK in which the binary 1 is transmitted at a higher optical frequency than the binary 0 bit. The 180° phase shift between the binary 1 and 0 bits displayed in Figure 12.2(c) depicts PSK. Furthermore, it should be noted that whereas with ASK the amplitude of the carrier waveform is effectively switched on and off, the amplitude of the optical carrier remains constant in the other two modulation schemes shown in Figure 12.2. Variants on these standard modulation techniques exist, such as continuous phase FSK and differential PSK, which has also been applied in experimental coherent optical fiber transmission systems. Moreover, an alternative digital scheme based on the modulation of the polarization properties of the optical signal has more recently come under investigation. This strategy, known

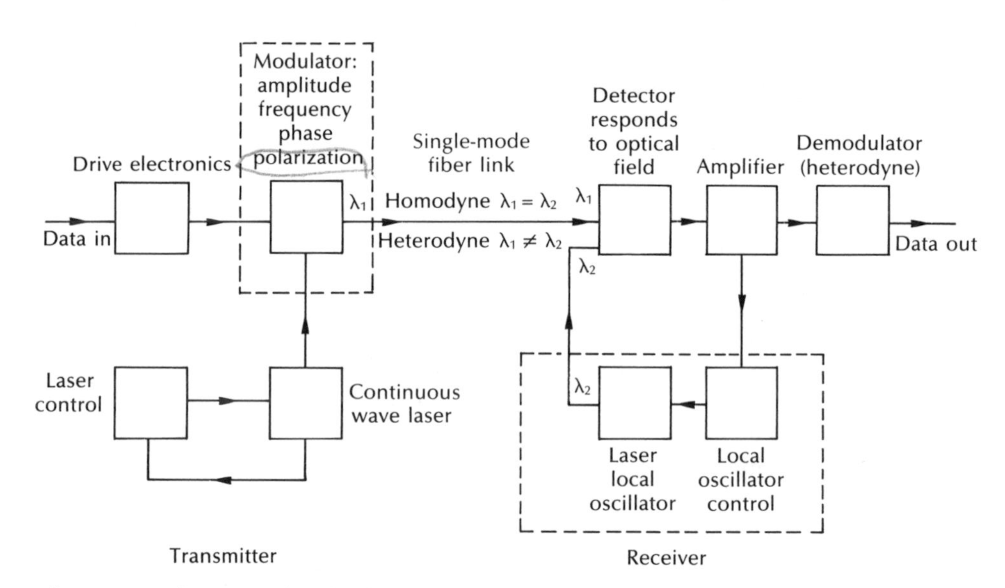

Figure 12.1 The generalized coherent optical fiber system.

homodyne
heterodyne

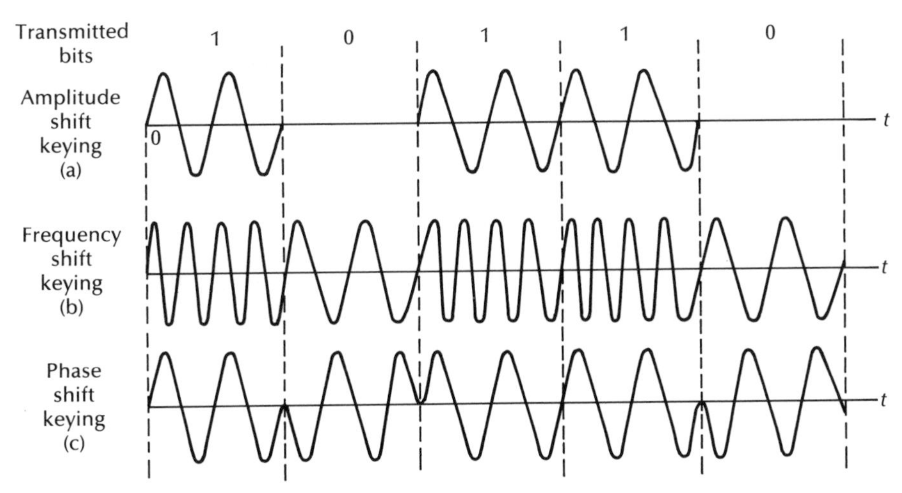

Figure 12.2 Modulated carrier waveforms used for binary data transmission: (a) amplitude shift keying (ASK); (b) frequency shift keying (FSK); (c) phase shift keying (PSK).

as polarization shift keying (PolSK), is discussed further along with the other modulation formats in Section 12.5.

Referring now to the receiver shown in Figure 12.1, then the incoming signal is combined (or mixed) with the optical output from a semiconductor laser local oscillator. This function can be provided by a single-mode fiber fused biconical coupler (see Section 5.6.1), a device which gives excellent wavefront matching of the two optical signals. However, integrated optical waveguide couplers (see Section 10.6.1) may also be utilized. The combined signal is then fed to a photodetector for direct detection in the conventional square law device. Nevertheless, to permit satisfactory optical coherent detection the optical coupler device must combine the polarized optical information-bearing signal field with the similarly polarized local oscillator signal field in the most efficient manner.

When the optical frequencies (or wavelengths) of the incoming signal and the local oscillator laser output are identical, then the receiver operates in a homodyne mode and the electrical signal is recovered directly in the baseband. For heterodyne detection, however, the local oscillator frequency is offset from the incoming signal frequency and therefore the electrical spectrum from the output of the detector is centred on an intermediate frequency (IF) which is dependent on the offset and is chosen according to the information transmission rate and the modulation characteristics. This IF, which is a difference signal (or difference frequency), contains the information signal and can be demodulated using standard electrical techniques [Ref. 3].

The electrical demodulator block shown in Figure 12.1 is required in particular for an optical heterodyne detection system which can utilize either synchronous or nonsynchronous/asynchronous electrical detection. Synchronous or coherent* demodulation implies an estimation of phase of the IF signal in transferring it to the baseband. Such an approach requires the use of phase-locking techniques in order to follow phase fluctuations in the incoming and local oscillator signals. Alternatively, nonsynchronous or noncoherent (envelope) IF demodulation schemes may be employed which are less demanding but generally produce a lower performance than synchronous detection techniques [Ref. 5]. Optical homodyne detection is by definition, however, a synchronous demodulation scheme and as the detected signal is brought directly into the baseband, then optical phase estimation is required. These issues are discussed further in Section 12.6.

12.3 Detection principles

A simple coherent receiver model for ASK is displayed in Figure 12.3. The low level incoming signal field e_S is combined with a second much larger signal field e_L derived from the local oscillator laser. It is assumed that the electromagnetic fields obtained from the two lasers (i.e. the incoming signal and local oscillator devices) can be represented by cosine functions and that the angle $\phi = \phi_S - \phi_L$ represents the phase relationship between the incoming signal phase ϕ_S and the local oscillator signal phase ϕ_L defined at some arbitrary point in time. Hence as depicted in Figure 12.3 the two fields may be written as [Ref. 19]:

$$e_S = E_S \cos(\omega_S t + \phi) \tag{12.1}$$

and

$$e_L = E_L \cos(\omega_L t) \tag{12.2}$$

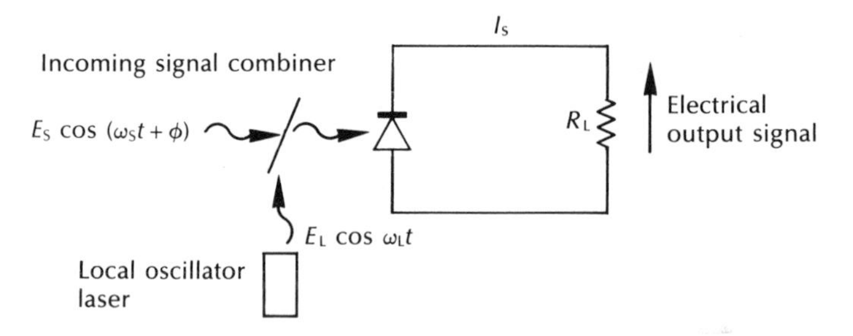

Figure 12.3 Basic coherent receiver model.

* It is a little confusing but reference is made in the literature to (a) heterodyne, coherent detection and (b) to heterodyne, noncoherent or incoherent detection, both of which are coherent optical detection schemes. The former terminology means an optical heterodyne receiver using a synchronous electrical demodulation technique, whereas the latter corresponds to a heterodyne receiver with a nonsynchronous demodulation scheme.

where E_S is the peak incoming signal field and ω_S is its angular frequency, and E_L is the peak local oscillator field and ω_L is its angular frequency. The angle $\phi(t)$ representing the phase relationship between the two fields contains the transmitted information in the case of FSK or PSK. However, with ASK $\phi(t)$ is constant and hence it is simply written as ϕ in Eq. (12.1), the information being contained in the variation of E_S for ASK as may be observed in Figure 12.2(a)

For heterodyne detection, the local oscillator frequency ω_L is offset from the incoming signal frequency ω_S by an intermediate frequency (IF) such that:

$$\omega_S = \omega_L + \omega_{IF} \tag{12.3}$$

where ω_{IF} is the angular frequency of the IF. As mentioned in Section 12.1 the IF is usually in the radiofrequency region and may be a few tens or hundreds of megahertz. By contrast, within homodyne detection there is no offset between ω_S and ω_L and hence $\omega_{IF} = 0$. In this case the combined signal is therefore recovered in the baseband.

The two wavefronts from the incoming signal and the local oscillator laser must be perfectly matched at the surface of the photodetector for ideal coherent detection. This factor creates the normal requirement for polarization control of the incoming optical signal which is discussed further in Section 12.4.

In the case of both heterodyne and homodyne detection the optical detector produces a signal photocurrent I_p which is proportional to the optical intensity (i.e. the square of the total field for the square law photodetection process) so that:

$$I_p \propto (e_S + e_L)^2 \tag{12.4}$$

Substitution in the expression Eq. (12.4) from Eqs. (12.1) and (12.2) gives:

$$I_p \propto [E_S(\cos \omega_S t + \phi) + E_L \cos \omega_L t]^2 \tag{12.5}$$

Assuming perfect optical mixing expansion of the right hand side of the expression shown in Eq. (12.5) gives:

$$\begin{aligned}[E_S^2 \cos^2(\omega_S t + \phi) + E_L^2 \cos^2 \omega_L t + 2E_S E_L \cos(\omega_S t + \phi)\cos \omega_L t] \\ = [\tfrac{1}{2}E_S^2 + \tfrac{1}{2}E_S \cos(2\omega_S t + \phi) + \tfrac{1}{2}E_L^2 + \tfrac{1}{2}\cos 2\omega_L t \\ + E_S E_L(\cos \omega_S t + \phi - \omega_L t) + E_S E_L \cos(\omega_S t + \phi + \omega_L t)]\end{aligned}$$

Removing the higher frequency terms oscillating near the frequencies of $2\omega_S$ and $2\omega_L$ which are beyond the response of the detector and therefore do not appear in its output, we have:

$$I_p \propto \tfrac{1}{2}E_S^2 + \tfrac{1}{2}E_L^2 + 2E_S E_L \cos(\omega_S t - \omega_L t + \phi) \tag{12.6}$$

Then recalling that the optical power contained within a signal is proportional to the square of its electrical field strength, Eq.(12.6) may be written as:

$$I_p \propto P_S + P_L + 2\sqrt{P_S P_L} \cos(\omega_S t - \omega_L t + \phi) \tag{12.7}$$

where P_S and P_L are the optical powers in the incoming signal and local oscillator signal respectively.

Furthermore, a relationship was obtained between the output photocurrent from an optical detector and the incident optical power P_o in Eq. (8.8) of the form $I_p = \eta e P_o / hf$. Hence the expression in Eq. (12.7) becomes:

$$I_P = \frac{\eta e}{hf} [P_S + P_L + 2\sqrt{P_S P_L} \cos(\omega_S t - \omega_L t + \phi)] \tag{12.8}$$

where η is the quantum efficiency of the photodetector, e is the charge on an electron, h is Planck's constant and f is optical frequency. When the local oscillator signal is much larger than the incoming signal, then the third a.c. term in Eq. (12.8) may be distinguished from the first two d.c. terms and I_p can be replaced by the approximation I_S where [Ref. 5]:

$$I_S = \frac{\eta e}{hf} [2\sqrt{P_S P_L} \cos(\omega_S t - \omega_L t + \phi)] \tag{12.9}$$

Equation (12.9) allows the two coherent detection strategies to be considered. For heterodyne detection $\omega_S \neq \omega_L$ and substituting from Eq. (12.3) gives:

$$I_S = \frac{2\eta e}{hf} \sqrt{P_S P_L} \cos(\omega_{IF} t + \phi) \tag{12.10}$$

indicating that the output from the photodetector is centred on an intermediate frequency. This IF is stabilized by incorporating the local oscillator laser in a frequency control loop. Temperature stability for the signal and local oscillator lasers is also a factor which must be considered (see Example 12.1). The stabilized IF current is usually separated from the d.c. current by filtering prior to electrical amplification and demodulation.

For the special case of homodyne detection, however, $\omega_S = \omega_L$ and therefore Eq. (12.9) reduces to:

$$I_S = \frac{2\eta e}{hf} \sqrt{P_S P_L} \cos \phi \tag{12.11}$$

or

$$I_S = 2R\sqrt{P_S P_L} \cos \phi \tag{12.12}$$

where R is the responsivity of the optical detector. In this case the output from the photodiode is in the baseband and the local oscillator laser needs to be phase locked to the incoming optical signal.

Example 12.1

A semiconductor laser used to provide the local oscillator signal in an ASK optical heterodyne receiver exhibits an output frequency change of 19 GHz $^{\circ}C^{-1}$. If the receiver has a nominal IF of 1.5 GHz and assuming that there is no other form of laser frequency control, estimate the maximum temperature change that could be allowed for the local oscillator laser in order that satisfactory detection could take place.

Solution: Initially it is necessary to estimate the maximum frequency excursion allowed for the IF signal such that detection can still be facilitated. This must be no greater than 10% of the frequency of the IF. Hence the maximum allowed frequency change to the local oscillator laser output is around 150 MHz.

The maximum temperature change allowed for the local oscillator laser is therefore:

$$\textit{Max. temp. change} = \frac{150 \times 10^6}{19 \times 10^9}$$

$$\simeq 8 \times 10^{-3}\ ^\circ\mathrm{C}\ (0.008\ ^\circ\mathrm{C})$$

Very small temperature changes can therefore adversely affect the detection process if the IF is not otherwise stabilized.

It may be observed from the expressions given in Eqs. (12.10) and (12.11) that the signal photocurrent is proportional to $\sqrt{P_S}$, rather than P_S as in the case of direct detection (Eq. (8.8)). Moreover, the signal photocurrent is effectively amplified by a factor $\sqrt{P_L}$ proportional to the local oscillator field. This local oscillator gain factor has the effect of increasing the optical signal level without affecting the receiver preamplifier thermal noise or the photodetector dark current noise (see Sections 9.2 and 9.3); hence the reason why coherent detection provides improved receiver sensitivities over direct detection.

The requirement for coherence between the incoming and local oscillator signals in order to obtain coherent detection was mentioned in Section 12.2 and is discussed further in Section 12.4.2. Hence for successful mixing to occur, then some correlation must exist between the two signals shown in Figure 12.3. Care must therefore be taken to ensure that this is the case when two separate laser sources are employed to provide the signal and local oscillator beams. It may be noted that this problem is reduced when a single laser source is used with an appropriate path length difference as, for example, when taking measurements by interferometric techniques (see Section 14.5.1).

When the local oscillator signal power is much greater than the incoming signal power then the dominant noise source in coherent detection schemes becomes the local oscillator quantum noise. In this limit the quantum noise may be expressed as shot noise following Eq. (9.8) where the mean square shot noise current from the local oscillator is given by:

$$\overline{i_{SL}^2} = 2eBI_{pL} \tag{12.13}$$

Substituting for I_{pL} from Eq. (8.8), where the photocurrent generated by the local oscillator signal is assumed to be by far the major contribution to the photocurrent, gives:

$$\overline{i_{SL}^2} = \frac{2e^2\eta P_L B}{hf} \tag{12.14}$$

The detected signal power S being the square of the average signal photocurrent* is given by Eq. (12.9) as:

$$S = \left(\frac{\eta e}{hf}\right)^2 P_S P_L \tag{12.15}$$

Hence the SNR for the ideal heterodyne detection receiver when the local oscillator power is large (ignoring the electronic preamplifier thermal noise and photodetector dark current noise terms) may be obtained from Eqs. (12.14) and (12.15) as:

$$\left(\frac{S}{N}\right)_{\text{het-lim}} = \left(\frac{\eta e}{hf}\right)^2 P_S P_L \Big/ \frac{2e^2\eta P_L B}{hf}$$

$$= \frac{\eta P_S}{hf\,2B} = \frac{\eta P_S}{hf\,B_{IF}} \tag{12.16}$$

Equation (12.16) provides the so-called shot noise limit for optical heterodyne detection in which the IF amplifier bandwidth B_{IF} is assumed to be equal to $2B$† (i.e. $B_{IF} = 2B$) [Ref. 20]. It is also interesting to note that this heterodyne shot noise limit corresponds to the quantum noise limit for analog direct detection derived in Eq. (9.11). However, it is clear that optical heterodyne detection allows a much closer approach to this limit than does direct detection.

The shot noise SNR limit for optical homodyne detection can be deduced from Eq. (12.16) by reducing the receiver bandwidth requirement from B_{IF} to B as the output signal from the photodetector appears in the baseband when using the homodyne scheme. Hence the SNR limit for optical homodyne detection is:

$$\left(\frac{S}{N}\right)_{\text{hom-lim}} = \frac{\eta P_S}{hf\,B} \tag{12.17}$$

It should be remembered that the expressions given in Eqs. (12.16) and (12.17) are based on simple on-off keying (i.e. OOK) and have effectively been derived in terms of carrier to noise ratio. Nevertheless, they display the potential 3 dB improvement in SNR when using optical homodyne detection over heterodyne detection. The improvement occurs as a direct result of the reduction in the receiver bandwidth provided by the former technique. Therefore, homodyne detection displays the twin advantages over heterodyne detection of increased sensitivity coupled with a reduced receiver bandwidth requirement. The latter factor implies that a higher

* It is implicit from Eq. (12.9) that the photodetector is a unity gain device (e.g. *p–i–n* photodiode) and not an APD. In the latter case the effect of the multiplication factor M on both the signal and noise powers must be taken into account (see Sections 9.3.3 and 9.3.4).

† This constitutes the minimum bandwidth requirement for optical heterodyne detection.

maximum transmission rate should be facilitated by coherent optical fiber systems employing homodyne detection as they will be less restricted by the speed of response of the photodetector.

Example 12.2

The incoming signal power to an optical homodyne receiver operating at a wavelength of 1.54 μm, and at its shot noise limit, is −55.45 dBm. When the photodetector in the receiver has a quantum efficiency of 86% at this wavelength and the received SNR is 12 dB, determine the operating bandwidth of the receiver.

Solution: The incoming signal power P_S is given by:

$$-85.45 = 10 \log_{10} P_S$$

Hence

$$P_S = 10^{0.455} \times 10^{-9} = 2.851 \text{ nW}$$

The operating bandwidth B of the homodyne receiver may be obtained from Eq. (12.17) as:

$$B = \frac{\eta P_S}{hf}\left(\frac{S}{N}\right)^{-1}_{\text{hom-lim}} = \frac{\eta P_S \lambda}{hc}\left(\frac{S}{N}\right)^{-1}_{\text{hom-lim}}$$

$$= \frac{0.86 \times 2.851 \times 10^{-9} \times 1.54 \times 10^{-6} \times 10^{-1.2}}{6.626 \times 10^{-34} \times 2.998 \times 10^{8}}$$

$$= 1.2 \text{ GHz}$$

The above analysis of SNR for optical heterodyne and homodyne detection applies only to ASK and is not appropriate for FSK and PSK. The final SNR at the signal decision point is therefore dependent on the modulation scheme utilized, and in the case of heterodyne detection on the type of IF demodulator and baseband filter employed. More detailed considerations of the SNR for different modulation schemes are dealt with in Section 12.7.

12.4 Practical constraints

It was indicated in Section 12.1 that until quite recently various practical constraints had inhibited the development of coherent optical fiber communications. These constraints are largely derived from factors associated with the elements of the coherent optical fiber communication system shown in Figure 12.1, and they are exacerbated by the stringent demands of coherent transmission. Substantial developments, however, in the component technology associated with optical fiber communications particularly over the last few years, have allowed the earlier difficulties experienced with coherent optical fiber transmission to be largely overcome. Nevertheless, the practical constraints still exist and they still dictate the performance characteristics required from components and devices which are to be utilized in coherent optical fiber systems. It is therefore important to consider the major constraints and their effect on the choice of system elements. We start by discussing the aspects which determine specific requirements for the achievement of coherent optical transmission at both the transmit and receive terminals prior to outlining certain limitations of the fiber transmission medium which may affect the performance of future coherent optical communication systems.

12.4.1 Injection laser linewidth

For coherent transmission the several hundred gigahertz wide linewidths of earlier multimode semiconductor injection lasers required a substantial reduction to very narrow linewidth (ideally, less than a megahertz) single-mode spectra. The major reasons for the use of a narrow linewidth injection laser within the coherent detection process are the phase locking requirement (for synchronous detection) as well as the minimum frequency locking requirement for nonsynchronous detection. In addition, both the amplitude and phase of semiconductor laser emissions are noisy, causing a reduction in the SNR performance of the coherent system. Laser linewidth reduction therefore improves the spectral purity of the device output and thus reduces its noise content. However, it was not until the latter half of the 1970s that semiconductor device technology evolved to a point where injection lasers, with good reproducibility which could operate in a single longitudinal mode, could be fabricated. However, the spectral linewidths associated with the most sophisticated of these devices such as the distributed feedback laser (see Section 6.6.2) were of the order of 5 to 50 MHz which was too broad for most of the coherent techniques [Ref. 18].

Several approaches to the solution of this laser linewidth problem have recently evolved. They include the narrowing of injection laser linewidths through the use of an external resonator cavity in the long external cavity (LEC) laser (see Section 6.10.1), the use of integrated external cavity lasers in the form of advanced DFB and DBR structures (see Section 6.10.2) and the potential provided by injection locked semiconductor lasers [Refs. 21, 22]. This latter technique is dependent upon the injection of a sufficiently strong optical signal from an external source such that the frequency and phase of the output from the primary semiconductor laser will follow those of the injected signal. Hence in a coherent optical system this technique potentially allows for direct control of the local oscillator laser by the incoming signal. Unfortunately, injection locking in this way requires excessive power at the receiver input which far exceeds the target sensitivity for most homodyne* detection systems [Ref. 23].

* The injection locking technique described can only be employed with homodyne detection as the incoming signal and local oscillator will have the same frequency.

In particular, injection laser phase or frequency noise (see Section 6.7.4) can affect the coherent system performance as it is the principal cause of linewidth broadening in such devices [Ref. 24]. Randomly occurring spontaneous emission events, which are an inevitable aspect of injection laser operation, lead to sudden shifts (of random magnitude and sign) in the phase of the electromagnetic field generated by the laser causing the broadening effect. Hence, phase noise together with other linewidth broadening factors [Ref. 25] must be minimized in devices which are to be employed for coherent optical transmission. Nevertheless, phase locking techniques are often employed within the coherent receiver (see Section 12.6.1).

Overall, the injection laser linewidth requirements are critically dependent on the modulation format employed (i.e. ASK, FSK or PSK), the coherent detection mechanism (i.e. heterodyne or homodyne) and the electrical demodulation technique (i.e. synchronous, nonsynchronous or other). Although these issues are discussed in greater detail in Section 12.6, it is useful to comment on the narrowed linewidths achieved by the extended cavity lasers currently favoured for coherent optical fiber communications. At present the most exacting coherent optical transmission techniques necessitate the use of long external cavity (LEC) lasers which can provide linewidths of around 10 kHz or less [Ref. 17]. Whereas typical lengths for these devices are between 10 and 20 cm, developments in more rugged prototype units with cavity lengths in the range 2 to 3 cm have already taken place [Ref. 26]. This commercially available device provides linewidths of less than 100 kHz. Although submegahertz linewidths have been achieved with integrated external cavity lasers (see Section 6.10.2), these devices at present generally display significantly larger linewidths than LEC lasers. Hence, their use can only be contemplated with the less demanding coherent optical fiber transmission schemes.

Another important factor concerning the favoured narrow linewidth injection lasers for coherent optical transmission is their inherent tunability. This aspect, which is discussed in more detail in Sections 6.10.1 and 6.10.2, provides the ability to tune the frequency of the local oscillator laser to that of the incoming optical signal for homodyne detection, or alternatively to tune the appropriate frequency difference to maintain the correct IF signal for heterodyne detection.

12.4.2 State of polarization

To enable either heterodyne or homodyne detection the polarization states of the incoming optical signal and the local oscillator laser output must be well matched in order to provide efficient mixing of the two signals within the coupling element shown in Figure 12.3. Conventional circularly symmetric single-mode fiber allows two orthogonally polarized fundamental modes to propagate. Within a perfectly formed fiber both modes would travel together but in practice the fiber contains random manufacturing irregularities which produce geometric and strain related anisotropic effects. This results in a progressive spatial separation between the two polarization modes as they propagate along the fiber, an effect which is usually referred to as modal birefringence (see Section 3.13.1). Hence at any particular point along the fiber the state of polarization (SOP) can be linear, elliptical or circular. To date, several counter-measures have been investigated to overcome this fluctuation in the SOP with coherent transmission. They are:

(a) the use of polarization maintaining (PM) single-mode fiber;
(b) the use of an SOP control device at the coherent optical receiver;
(c) the use of a polarization diversity receiver, or a polarization scrambling transmitter.

As mentioned in Section 12.2 early studies focused on the polarization stability of the transmission medium which was considered sufficiently important to necessitate the use of specially fabricated PM fiber (see Section 3.13.2). Such fibers, however, generally exhibit higher losses and are more expensive to fabricate than conventional single-mode fiber. Furthermore, much circularly symmetric single-mode fiber has already been installed and therefore coherent transmission techniques which utilize this medium are desirable.

Measurements of polarization stability for light propagating in conventional single-mode fiber over a ninety-six hour period are shown in Figure 12.4 [Ref. 16]. A single frequency linearly polarized 1523 nm emitting helium–neon gas laser together with a receiver which contained a polarization dependent beam splitter to

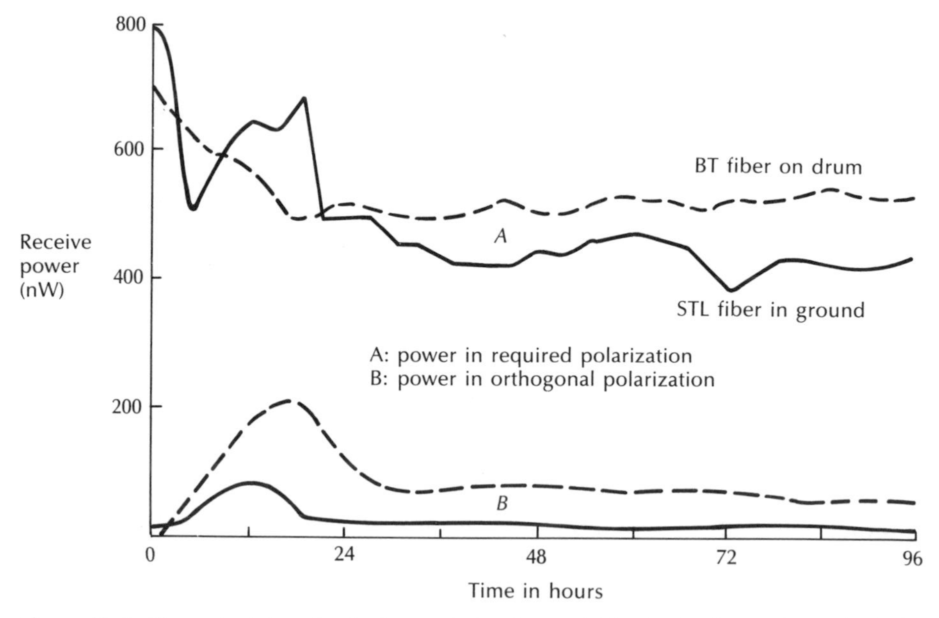

Figure 12.4 Characteristics displaying polarization stability in single-mode optical fiber. Reproduced with permission from D. W. Smith, R. A. Harmon and T. G. Hodgkinson, *Br. Telecom Technol. J.*, **1** (2), p. 12, 1983.

isolate the components were used in these measurements. It may be observed from Figure 12.4 that, as expected, the polarization state of the optical signal was not temporarily constant. Nevertheless, although polarization changes did occur, it was over periods of minutes or hours. These observations of the relatively slow changes in the polarization state of the transmitted signal, which were also verified on long cable links [Ref. 27], provided the potential for polarization matching at the coherent optical receiver. Polarization control devices with achievable response times could therefore be located at the receiver to provide polarization correction and hence matching of the SOP of the incoming signal and the local oscillator signal.

12.4.2.1 Polarization-state control

Active polarization-state control can be accomplished using mechanical, electro-optical or magneto-optical techniques. A polarization error signal may be generated and fed back to the polarization control device in all cases. In general, the SOP is described by the amplitude ratio of the x and y components of the electric field vector and their relative phase difference (see Section 3.13.1). As the two parameters which vary randomly at the coherent receiver are the ellipticity of the SOP and its orientation, the error signal must correct for both of these factors such that the incoming signal and the local oscillator output have identical states of polarization.

A number of mechanisms have been developed for polarization-state control within coherent optical fiber communications. Initial implementations of such devices which were based on birefringent elements (i.e. elements which induce modal birefringence in order to correct the SOP) included fiber squeezers [Ref. 28], electro-optic crystals [Ref. 29], rotatable fiber coils [Ref. 30], rotatable phase plates [Ref. 31] and rotatable fiber cranks [Ref. 32]. In addition, the possibility of using the Faraday effect to rotate the SOP of the incoming optical signal has also been demonstrated [Ref. 33]. Developments have also encompassed integrated optical electro-optic polarization control devices. Such controllers have been incorporated into integrated optical coherent receiver devices (see Section 10.6.4). At least two compensator devices are required to provide full polarization-state control. They can be placed in either the incoming signal path or the local oscillator output path; however the latter position is preferable if the device introduces significant signal attenuation.

Until recently a major concern was the insufficient range exhibited by these polarization control schemes in tracking the continuously varying SOP which can change unpredictably over virtually unlimited range. Polarization-state control schemes with infinite ranges of adjustments have, however, now been demonstrated [Refs. 34 to 37]. In particular, a control technique using four fiber squeezers, which is illustrated in Figure 12.5(a), was found to provide an infinite range of adjustment or so-called endless polarization control [Ref. 35]. Stress was applied to the fiber in the local oscillator path using the squeezers which were angled at 45° to each other. The SOP could therefore be manipulated to the appropriate matching point.

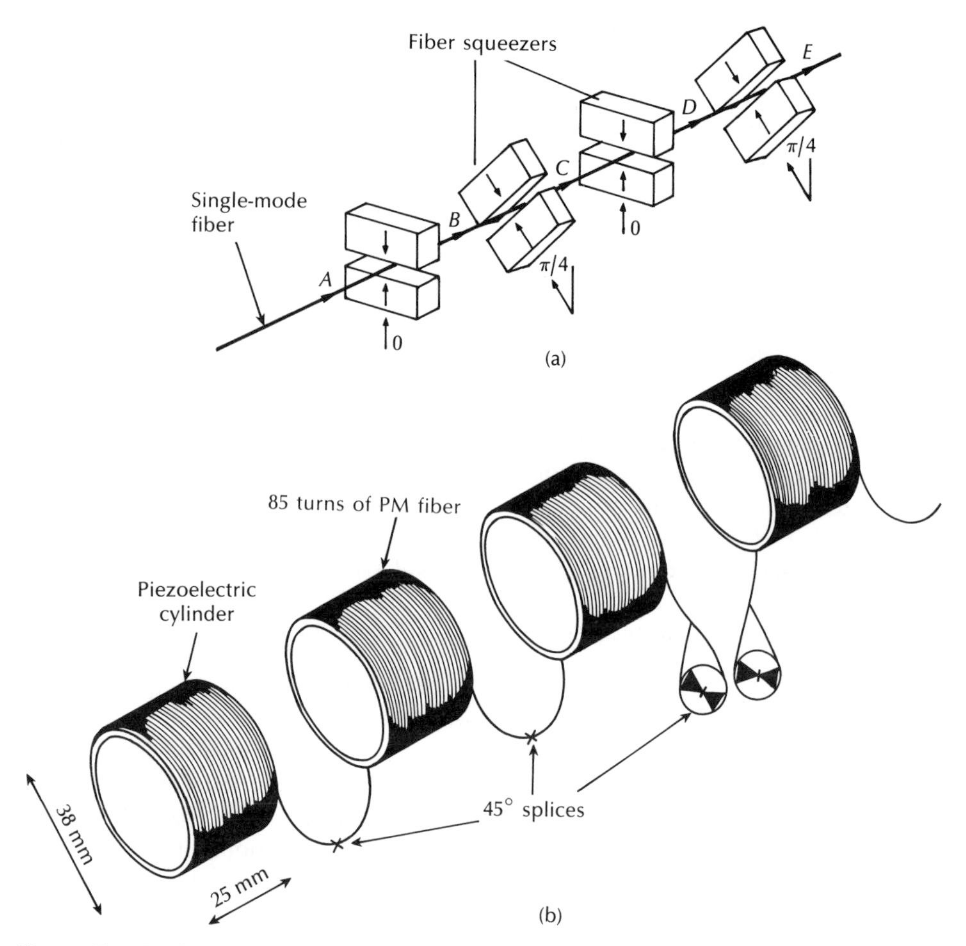

Figure 12.5 Techniques for endless polarization control: (a) four fiber squeezers; (b) polarization maintaining (PM) fiber controller.

Moreover, the polarization control system was automated using a control algorithm whereby a dither signal was applied to the bias on each squeezer in a defined order, and the variation of the received demodulated signal was used to identify the optimum operating point. This strategy removed the need to optically sense the SOP.

Although the squeezers are simple to configure, a drawback with the technique is that they tend to damage the fiber. Moreover, it is questionable as to whether they could be engineered into reliable transducers for practical systems. A more rugged and reliable controller is shown in Figure 12.5(b) [Ref. 36]. It comprised four piezoelectric cylinders each wound around with 85 turns of polarization maintaining (PM) fiber. The PM fibers on each cylinder were spliced together with the principal axis of the fibers mutually aligned at 45°, as illustrated in Figure 12.5(b). As the PM fiber used was highly birefringent with a beat length of only a few millimetres, then the SOP changed many times along each element. The application of a voltage caused the piezoelectric cylinders to expand slightly and stretch the fiber, thus modifying the fiber birefringence. Hence, the overall effect was a variable retardation which gave polarization control. Again a control algorithm was devised to provide automatic operation.

An analogous control technique has also been demonstrated with an integrated optical, electro-optic polarization control device [Refs. 36, 37]. This lithium niobate waveguide structure which comprised two elements is shown in Figure 12.6. Each element consisted of three longitudinal electrodes placed symmetrically over the Z-propagating waveguide diffused into an X-cut substrate. Voltages applied to the electrodes produced an electric field that could be orientated in any direction transverse to the waveguide to provide a virtually infinite range of polarization-state control. This technique appears to offer a robust mechanically stable method of polarization control which has been demonstrated in both a laboratory-based and a field-installed coherent optical fiber systems, with no measurable sensitivity penalties [Refs. 36, 37].

Although the above are encouraging developments in polarization-state control techniques, problems with such methods may exist in advanced network applications using optical FDM and passive routeing (see Section 12.8.2). For example, a significant spectral variation in the SOP has been observed over relatively large spectral bandwidths (100 nm) for short interaction lengths in polarization couplers [Ref. 23]. Such spectral variation could cause difficulties for coherent optical FDM systems with active polarization correction. Furthermore, within advanced networking applications the time to acquire polarization matching may be an unacceptable overhead.

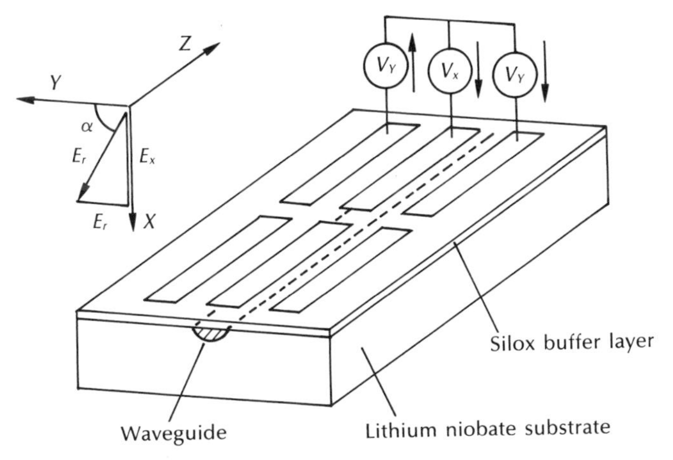

Figure 12.6 Lithium niobate waveguide polarization controller.

12.4.2.2 Polarization diversity reception and polarization scrambling transmission

Alternative approaches which avoid the requirement for polarization-state control devices, but which also allow the use of conventional circularly symmetric single-mode fiber, include polarization diversity reception [Refs. 38 to 42] and polarization scrambling transmission [Refs. 43, 44] techniques. A block schematic of a polarization diversity receiver is shown in Figure 12.7. This scheme, which is essentially polarization insensitive, employs separate heterodyne or homodyne detection for the two orthogonal polarization states of the incoming optical signal.

The incoming signal is therefore combined with a circularly polarized local oscillator signal and the composite signal is passed through a polarizing beam splitter. The two orthogonally polarized outputs from the beam splitter are then detected on separate photodiodes and, in the heterodyne case, demodulated down to baseband prior to recombination. Such electrical recombination provides a polarization independent signal as the IF is produced in one or other of the receivers regardless of the SOP of the incoming signal. The basic configuration illustrated in Figure 12.7 can incur a SNR performance degradation of 3 dB [Ref. 41] for particular input polarization states in which phase matching is difficult to maintain through both receivers. However, this penalty can be reduced to less than 1 dB with appropriate post demodulation processing.

Polarization insensitive heterodyne detection has also been demonstrated using polarization scrambling transmission. In this technique the polarization state of the

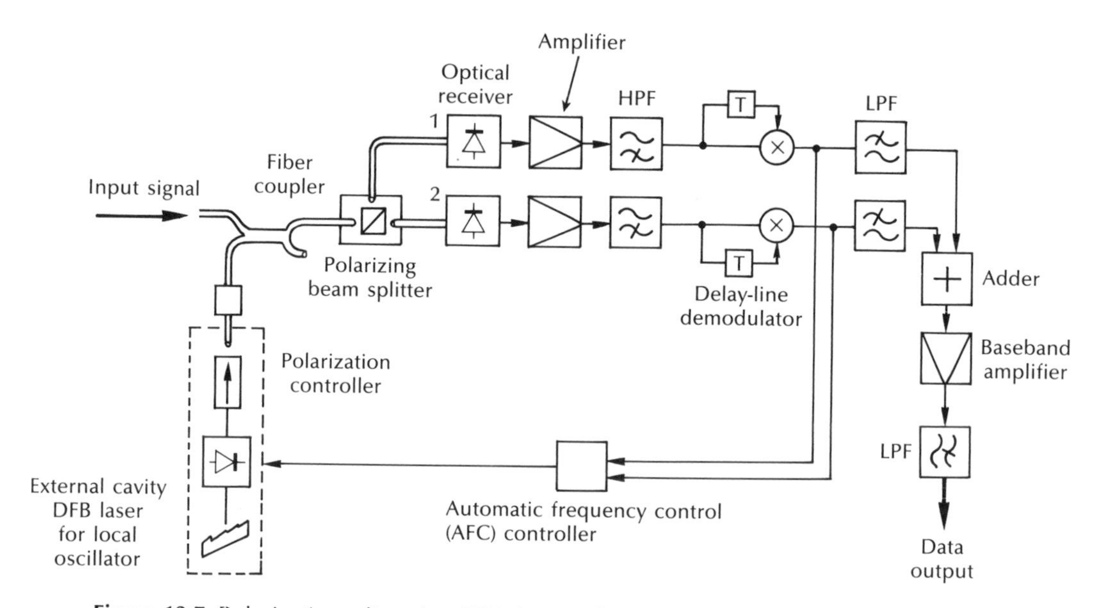

Figure 12.7 Polarization diversity FSK heterodyne receiver using one bit delayed demodulation [Ref. 39].

optical signal is deliberately changed at the coherent transmitter so that all possible polarization states are propagated during a single bit period. Although this method significantly reduces the receiver complexity in comparison with polarization diversity reception, it incurs a reduction in sensitivity at the coherent optical receiver. Moreover, the polarization scrambler is a relatively sophisticated arrangement which incorporates a polarization beam splitter. Hence, an earlier difficulty with this technique, in common with polarization diversity reception, concerned the realization of the optical devices in a rugged form. More recently, however, polarization maintaining fiber directional couplers have become commercially available.

12.4.3 Local oscillator power

In a practical coherent receiver the theoretical performance may not be attained for the reason already outlined in Sections 12.4.1 and 12.4.2. In addition, there may be insufficient local oscillator power to achieve the shot noise detection limits discussed in Section 12.3. This factor, which highlights the need to ensure a low loss signal path, can be facilitated by an appropriate choice of an incoming signal/local oscillator combiner which has high coupling efficiency. In particular, it is clear that when the basic coherent detector shown in Figure 12.3 is considered, the optical combiner or coupler has only one output port utilized, whereas in reality such devices have two output ports (see Section 5.6.1). There is, therefore, an optical loss associated with the power which is coupled to the other output port. Although the combiner or coupler can be designed so that the majority of the incoming signal power is coupled into the optical detector, a consequence of this process is that there will be reduction in the power from the local oscillator laser coupled into the detector. It therefore becomes more difficult to maintain a high local oscillator signal power and thus to obtain shot noise limited receiver performance.

The dramatic effect of the local oscillator power on an optical homodyne receiver sensitivity is illustrated by the theoretical characteristic shown in Figure 12.8 which corresponds to a PIN–FET receiver operating at 140 Mbit s^{-1} [Ref. 45]. One method to overcome limited local oscillator power is by the use of a low noise photodiode/preamplifier combination such as the PIN–FET hybrid configuration (see Section 9.5.2) at the front end of the coherent receiver. Near shot noise limited detection has been obtained with just 1 μW of local oscillator power at a transmission rate of 140 Mbit s^{-1} when employing this strategy [Ref. 46]. In addition a heterodyne PIN–FET receiver with 8 GHz bandwidth and 10 pA $Hz^{-\frac{1}{2}}$ equivalent circuit noise current at an IF of 4.6 GHz has been demonstrated [Ref. 47].

An alternative approach which compensates for the losses due to coupling optics and also suppresses excess noise in the local oscillator signal is the use of a balanced receiver.* This scheme, which has often been employed for heterodyne detection in

* It is also referred to as the balanced-mixer receiver [Refs. 23, 48].

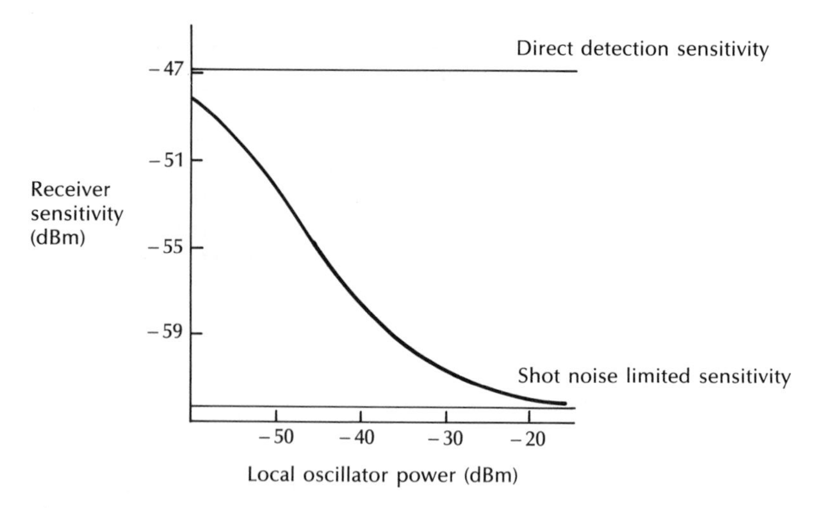

Figure 12.8 Theoretical characteristic showing the effect of local oscillator power on homodyne receiver sensitivity for a PIN–FET receiver operating at 140 Mbit s^{-1}. Reproduced with permission from D. W. Smith, 'Techniques for multigigabit coherent optical transmission', *J. Lightwave Technol.*, **LT-5**, p. 1466, 1987. Copyright © 1987 IEEE.

microwave communications to suppress local oscillator fluctuations [Refs. 49, 50], is shown in Figure 12.9. In this technique the local oscillator output and the incoming optical signal are usually combined using a four port (i.e. 3 dB) single-mode fiber directional coupler.* The signal in one fiber in this device suffers a $\pi/2$ phase shift upon transfer to the other fiber. In effect complete coupling is only possible because this phase shift in the coupled signal is $\pi/2$ out of phase with the throughput signal.† Hence the throughput and coupled signals can be represented

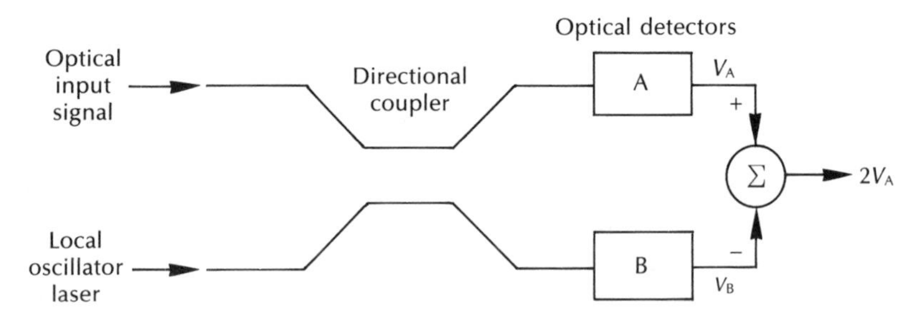

Figure 12.9 Schematic of a balanced optical receiver using two matched optical detectors.

* Other devices which may be utilized are bulk optic or waveguide beam splitters [Ref. 22].
† This signal becomes out of phase if it is coupled back to the throughput fiber where it would interfere destructively with the throughput signal.

as a sine wave and cosine wave respectively. Considering Figure 12.9 the inputs to the optical detectors A and B can therefore be written as $E_S \sin \omega_S t + E_L \cos \omega_L t$ and $E_S \cos \omega_S t + E_L \sin \omega_L t$ respectively.

The two detector output voltages are thus given by:

$$V_A = E_S E_L \sin (\omega_S - \omega_L)t \tag{12.18}$$

$$V_B = E_S E_L \sin (\omega_L - \omega_S)t \tag{12.19}$$

It may be observed that these output voltages are similar but of opposite sign in that $V_A = -V_B$.

The two output voltages are operated upon by the combiner function (Σ) depicted in Figure 12.9 and as one is a positive input and the other is a negative input, then the output from the combiner function V_o will form the difference between the two inputs such that:

$$V_o = V_A - V_B = 2V_A \tag{12.20}$$

Equation (12.20) indicates that twice the voltage, or four times the power, is provided in comparison with the single optical detector scheme (Figure 12.3). This technique therefore gives a 6 dB improvement over the single optical detector. Furthermore, as the two photocurrents are effectively subtracted, the process results in both a cancelling out of the large d.c. term produced by the local oscillator signal, together with any local oscillator excess noise. It is particularly useful in reducing the excess AM noise generated by the local oscillator [Ref. 18]. Moreover, as a result of the efficient use of the local oscillator and incoming signal powers, the constraints imposed by the earlier requirement for a widespread ultra-low noise preamplifier are relaxed. Close matching of the two arms of the balanced receiver is essential, however, if good excess noise cancellation is to be obtained. For example, a balanced receiver with an estimated equivalent circuit noise of 5.4 pA Hz$^{-\frac{1}{2}}$ at 1 GHz and 10.7 pA Hz$^{-\frac{1}{2}}$ at 2 GHz has been reported [Ref. 51]. Furthermore, the similar receiver design was operated with an IF of 3 GHz within a 2 Gbit s^{-1} system experiment [Ref. 52].

12.4.4 Transmission medium limitations

Although, in common with IM/DD systems, the fiber loss is the major limitation on the performance of single carrier coherent optical systems that can be ascribed to the transmission medium, there are nevertheless other factors that may well affect the operation of future coherent systems. These include intramodal or chromatic dispersion, polarization dispersion and the nonlinear scattering effects [Ref. 44].

The intramodal dispersion in conventional single-mode fiber which has a dispersion zero at a wavelength of 1.3 μm is around 15 ps km^{-1} nm^{-1} when the fiber is operated at a wavelength of 1.55 μm. This factor can lead to significant dispersion penalties (i.e. receiver sensitivity degradations) for IM/DD systems even at modest transmission rates and distances. It results from the transmitted spectrum usually being far wider than the information spectrum due to laser frequency chirp (see Section 6.7.3) caused by the direct amplitude modulation of the semiconductor laser. By contrast coherent optical transmission systems have the advantage of a compact spectrum even if the injection laser is directly modulated but particularly when an external modulator is employed. Receiver sensitivity degradation due to intramodal dispersion has, however, been observed in FSK transmission experiments at transmission rates greater than 4 Gbit s^{-1} [Ref. 44]. Furthermore, as both the transmission rates and distances for coherent transmission are increased, then greater chromatic dispersion penalties will be incurred. For example, calculations have indicated the maximum transmission rates to incur a 2 dB penalty after a 100 km distance on a conventional 1.3 μm single-mode fiber when operating at a wavelength of 1.55 μm to be in the range 5 to 9 Gbit s^{-1} depending on the modulation format used [Ref. 53].

As in the case of IM/DD the intermodal dispersion problem can be reduced through the use of dispersion shifted fiber which exhibits a dispersion zero in the 1.55 μm wavelength window (see Section 3.12.1). However, this solution is only partially satisfactory as there is a massive base of installed conventional single-mode fiber. An alternative strategy is to compensate for the chromatic dispersion in either the optical domain [Ref. 44] or the electrical domain [Ref. 54]. The latter method in particular has shown some limited success using stripline delay equalizers which have demonstrated the potential to compensate for dispersion-induced distortion up to transmission rates of 10 Gbit s^{-1}.

Polarization mode dispersion results from birefringence in the single-mode fiber and it corresponds to the difference in the propagation time associated with the two principal orthogonal polarization states (see Section 3.13.1). Whereas in IM/DD systems polarization mode dispersion simply results in pulse broadening due to the different spectral components arriving at different times, in coherent systems these components can also arrive with different polarizations. Moreover, both of these effects may be detrimental to coherent system performance. However, the differential propagation time is dependent upon the amount of mode mixing which takes place in the fiber, an effect which results from internal and external fiber perturbations. Assuming some mode mixing, the effects of polarization mode dispersion are therefore not expected to become important in a single carrier system with a fiber distance of 100 km until transmission rates exceed 10 Gbit s^{-1} [Ref. 44]. Although in multicarrier systems (see Section 12.8.2) the problem associated with receiving optical carriers at different wavelengths in different polarization states will generally be at lower transmission rates, it may be avoided by using either PM fiber, polarization diversity reception or polarization scrambling transmission (see Section 12.4.2.2).

The nonlinear phenomena which may be of importance within coherent optical transmission include stimulated Raman scattering (SRS), stimulated Brillouin scattering (SBS), cross-phase modulation and four wave mixing [Ref. 44]. ASK systems prove particularly susceptible to such nonlinear effects which result from optical power level changes. Moreover, a potential advantage of the FSK and PSK modulation formats over ASK is that they produce a constant amplitude signal which provides some immunity to certain nonlinear effects. Although SRS should not be a consideration in single carrier coherent optical systems operating at power levels below 1 W, Raman induced crosstalk may be a concern in multicarrier systems [Ref. 55].

Stimulated Brillouin scattering may be a problem at lower light levels than SRS as its threshold power level is significantly smaller (see Section 3.5). However, unlike SRS, SBS is a narrowband process with a bandwidth of only around 20 MHz at a wavelength of 1.55 μm (see Section 3.14). The maximum SBS gain which also is maximized in the reverse direction will therefore occur for lasers with linewidths less than 20 MHz. Moreover, as a result of information broadening of the linewidth, SBS will be greatly reduced when using modulation formats which do not contain a residual carrier component (i.e. PSK and narrow deviation FSK, see Section 12.5).

Self-phase modulation is a phenomenon which occurs in single carrier systems due to small refractive index changes induced by optical power fluctuations which affect the phase of the transmitted signal (see Section 3.14). For digitally modulated coherent optical systems this effect is perceived to be negligibly small at launched power levels up to a few hundred milliwatts [Ref. 44]. With multicarrier systems, however, a cross-phase modulation phenomenon occurs which can cause high levels of phase noise in long fiber lengths. In this case it has been shown that the power of each carrier should be restricted in order to limit the degradation caused by this phase crosstalk [Ref. 55]. Nevertheless, this limitation on transmitter power in a multicarrier system may not be as severe as the one imposed by the four wave mixing nonlinear phenomenon. It is suggested that this latter process will be present in all frequency division multiplexed systems with channel separations less than 10 GHz and that the crosstalk will restrict the maximum power per carrier to around 0.1 mW when the fiber lengths exceed 10 km [Refs. 44, 55].

12.5 Modulation formats

12.5.1 Amplitude shift keying

Several techniques may be employed to amplitude modulate an optical signal. Digital intensity modulation used in direct detection systems is essentially a crude form of amplitude shift keying* (ASK) in which the received signal is simply detected using the photodetector as a square law device (see Figure 12.2(a)). It is apparent, therefore, that the simplest approach to ASK is by direct modulation of the laser drive current. A problem exists, however, with this approach because of the inability of semiconductor lasers to maintain a stable output frequency with

* It should be noted that ASK is also referred to as on–off keying (OOK).

changing drive current. The resulting frequency deviation, which can be of the order of 200 MHz mA^{-1} [Ref. 23], broadens the linewidth of the modulated laser which creates difficulties for coherent optical detection (see Section 12.4.1).

Although direct modulation of the semiconductor laser in ASK coherent optical fiber systems has been demonstrated [Ref. 4], external modulation using active integrated optical devices, such as the directional coupler or the Mach–Zehnder interferometer (see Section 10.6.2), present attractive alternatives [Refs. 23, 45]. It should be noted, however, that all external ASK modulators suffer the drawback that around half of the transmitter power is wasted. Nonsynchronous detection can also be employed with the ASK format which puts the least demands on the injection laser phase stability. In principle this modulation scheme can be used with laser sources exhibiting linewidths comparable with the bit transmission rate. In practice the linewidth in the range 10 to 50% of bit rate is normally specified for ASK heterodyne detection [Refs. 45, 56], although some authors indicate 10 to 20% [Refs. 5, 44]. Nevertheless, nonsynchronous heterodyne and phase diversity (see Section 12.6) detection receivers which use ASK can tolerate the linewidths of currently available DFB lasers.

12.5.2 Frequency shift keying

The frequency deviation property of a directly modulated semiconductor laser can be usefully employed with wideband frequency shift keying (FSK) coherent optical fiber systems. Hence optical FSK (see Figure 12.2(b)) in common with ASK has the advantage that it does not necessarily require an external modulator, thus allowing higher launch powers and a more compact transmitter configuration, as illustrated in Figure 12.10(a). The direct frequency modulation characteristics of the laser are determined by changes in the device carrier density in the high modulation frequency region, and by the temperature modulation effect in the low frequency region. Thus at frequencies above 1 MHz where the carrier modulation effect occurs, the frequency deviation is typically 100 to 500 MHz mA^{-1}, whereas below 1 MHz it is around 1 GHz mA^{-1} due to the predominant temperature effect [Ref. 45]. Although the response under frequency modulation of the semiconductor laser is therefore not uniform, a frequency shift of between 100 MHz to 1 GHz is readily obtained from the device without serious intensity modulation effects [Ref. 18].

The laser linewidth requirements for wide frequency deviation FSK heterodyne nonsynchronous detection are similar to those of ASK with nonsynchronous heterodyne detection and are in the range 10 to 50% of the transmission bit rate [Refs. 45, 55]. Therefore, the use of FSK with broad linewidth injection lasers has been relatively successful, particularly at low bit rates. For example, the direct modulation of the laser drive current causing variations in the lasing wavelength has provided FSK transmission over 300 km at a rate of 34 Mbit s^{-1} with a receiver sensitivity of 165 photons per bit [Ref. 57]. However, due to the presence of large thermal effects at linewidths less than 10 MHz the FM response of semiconductor lasers is typically nonuniform and hence electronic equalization may be required

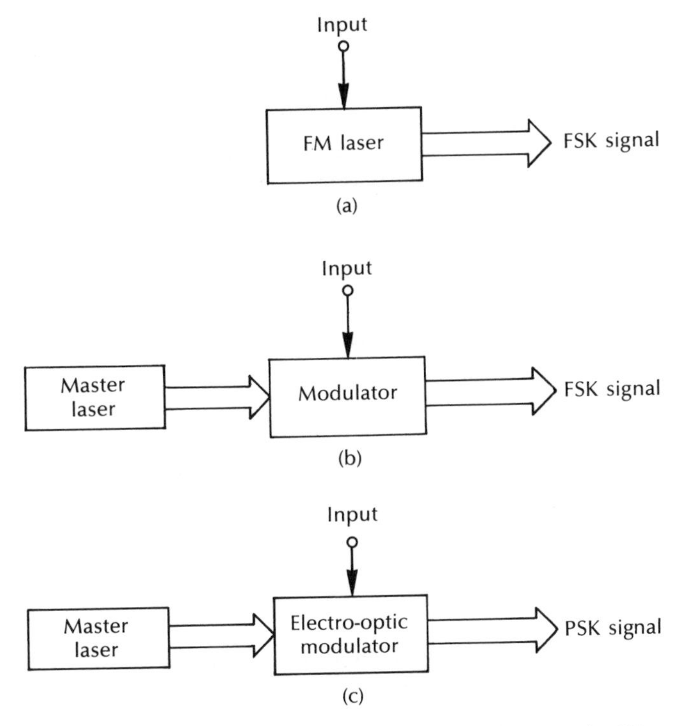

Figure 12.10 Transmitter configurations for FSK and PSK modulation: (a) FSK by direct modulation of an FM semiconductor laser injection current; (b) FSK using an external modulator; (c) PSK using an external electro-optic modulator.

[Ref. 18]. Although it has been observed that the use of a split electrode on the injection laser largely eliminates this effect [Ref. 58], alternative strategies have also been utilized. These include the use of bipolar optical FSK transmission [Ref. 59] and alternate mark inversion encoding [Ref. 60]. The former technique provided a transmission rate of 1 Gbit s^{-1} over 121 km whilst the latter operated at a rate of 565 Mbit s^{-1} using commercial DFB lasers.

When a single oscillator is switched between two frequencies, as is often the case with a semiconductor laser source, the phase of the signal is a continuous function of time and the modulation is known as continuous phase frequency shift keying (CPFSK) [Ref. 44]. This modulation scheme has been successfully demonstrated [Ref. 61] using integrated external cavity lasers (see Section 6.10.2). In this experiment transmission at 2 Gbit s^{-1} over 197 km of single-mode fiber was achieved. CPFSK is attractive because it allows direct current modulation of the injection laser whilst also providing high receiver sensitivity. Furthermore, it is suitable for high speed transmission since it creates no laser chirping degradation (see Section 6.7.3), as experienced in IM/DD systems. More recently the multichannel properties of CPFSK with a small frequency deviation have been reported [Ref. 62] as a potential technique for optical frequency division multiplexing (OFDM).

External modulation techniques for FSK, shown schematically in Figure 12.10(b), include both acousto-optic and electro-optic approaches. Using bulk optic devices, FSK may be accomplished using a Bragg cell which employs travelling acoustic waves in a crystal to simultaneously diffract and frequency shift the optical signal. Alternatively, an equivalent effect can be obtained by using surface acoustic waves on an integrated optical waveguide device (see Section 10.6.2). FSK modulation can also be provided by a Mach–Zehnder interferometer with sinusoidal modulation applied to one of its branches. Such devices have been operated at modulation frequencies in excess of 1 GHz [Ref. 23].

Finally, multilevel frequency shift keying (MFSK) offers the potential for improving the coherent optical receiver sensitivity by increasing the choice of signalling frequencies [Ref. 63]. In principle this M'ary scheme provides the best receiver performance in the limit of large channel spacing [Ref. 23]. Thus eight-level FSK yields an equivalent sensitivity to binary PSK but at the expense of a greater receiver bandwidth requirement.

12.5.3 Phase shift keying

Although rarely employed, optical phase modulation can be achieved by direct current modulation of a semiconductor laser into which external coherent laser light is injected [Ref. 64]. When the injected laser frequency is exactly tuned to the modulating signal frequency, the output signal phase relative to the modulating signal phase is zero. A relative phase change of $\pi/2$ is obtained when the injected laser frequency is detuned away from the modulated light frequency to the injection locking limit. Hence the cutoff modulation frequency is determined by the injection locking bandwidth. Furthermore, this technique has the effect of reducing the linewidth of the injection locked laser to that of the injected signal device [Ref. 18].

External modulation for phase shift keying (PSK) is relatively straightforward and therefore normally utilized to provide the modulation format (see Figure 12.10(c)) which allows the most sensitive coherent detection mechanism within the binary modulation schemes (see Section 12.7). Simple integrated optical phase modulators fabricated from electro-optic materials such as lithium niobate or III–V compound semiconductors may be employed to give the appropriate shift with the application of an electric field (see Section 10.6.2). Such devices which exhibit a fiber to fiber insertion loss of 2 to 5 dB require around 5 V drive to produce a phase shift of π radians. Moreover, modulation bandwidths in excess of 10 GHz have been obtained from travelling-wave structures.

The phase detection process for PSK, however, necessitates synchronous detection with the requirement for corresponding narrow laser linewidths. These very narrow linewidth requirements for both PSK heterodyne and homodyne detection may be observed in Table 12.1 which presents the laser linewidths as a percentage of the transmission bit rate for the major modulation formats

Table 12.1 Laser linewidth requirements for various modulation formats as a percentage of the bit rate.

Modulation format	Homodyne	Heterodyne Synchronous	Heterodyne Nonsynchronous
ASK	0.005–0.1%	0.05–0.1%	10–50%
FSK (wide deviation)	No	0.05–0.1%	10–50%
FSK (narrow deviation)	No		0.3–2.0%
PSK	0.005–0.01%	0.1–0.5%	No
DPSK	No	0.3–0.5%	No

considered in Section 12.5. Moreover, it may be noted that the most stringent laser linewidth requirement is for homodyne detection with binary PSK where for efficient detection linewidths of the order of 0.01% of the transmission rate are required [Ref. 65].

By contrast differential phase shift keying (DPSK) also indicated in Table 12.1 is a less demanding form of PSK since information is encoded as a change (or the absence of a change) in the optical phase on a bit by bit basis. The relationship between DPSK and PSK is illustrated in Figure 12.11 where it may be observed that with DPSK the incoming bit is delayed in order that its phase can be compared to the next received bit. Hence the technique does not require phase comparisons over more than two bit intervals. Moreover, the SNR performance of DPSK is only a fraction of a decibel less than that of heterodyne (synchronous) PSK [Ref. 2]. As

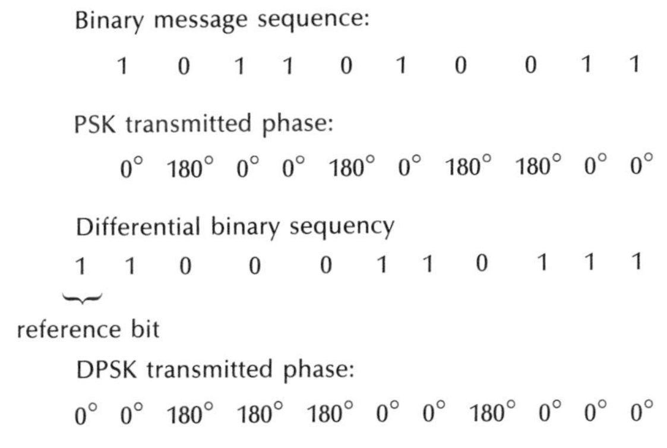

Figure 12.11 Comparison of a one-bit-at-a-time DPSK scheme with binary PSK. The differential binary sequence is obtained by repeating the preceding bit in the sequence if the message bit is a 1 or by changing to the opposite bit if the message bit is a 0.

laser linewidths of the order of 0.3 to 0.5% of the transmission rate can be tolerated with DPSK, experimental systems operating at 1.2 Gbit s^{-1} have been demonstrated using both an integrated external cavity DFB laser [Ref. 66] and an external fiber cavity DFB laser [Ref. 67]. In addition DPSK is technically straightforward to implement at high transmission rates because the phase fluctuation between the two signal bits is reduced [Ref. 20].

Unlike multilevel FSK which can provide improved receiver sensitivity by spectral expansion, M'ary PSK (and also, for that matter, M'ary ASK) could potentially provide spectral conservation through the use of multilevel signalling. Alternatively, M'ary PSK, M'ary ASK and their combinations such as quadrature amplitude modulation (QAM) [Ref. 3] can avoid noise degradation in the electronic preamplifier by increasing the utilization of the IF band frequency within optical heterodyne detection [Ref. 68]. The receiver sensitivities associated with these multilevel transmission techniques are discussed further in Section 12.7.

12.5.4 Polarization shift keying

An additional modulation format which has been investigated within coherent optical fiber communications involves use of the polarization characteristics of the transmitted optical signal. The digital transmission implementation of such polarization modulation is known as polarization shift keying (PolSK). A realization of coherent optical transmission using heterodyne detection with PolSK was obtained through external modulation by a lithium niobate phase modulator [Ref. 69]. This device produced a phase shift of π radians between the TE and TM modes, which rotated the signal polarization by 90°. These orthogonal polarization states were then maintained during transmission within a single-mode fiber. In this context a prerequisite for the fiber was that no coupling occurred between the two orthogonal polarization modes. The system was, however, successfully operated at a transmission rate of 560 Mbit s^{-1} and proved between 2 and 3 dB more sensitive than ASK modulation with heterodyne detection.

The differential variant of polarization shift keying (DPolSK) has also been demonstrated [Refs. 70, 71]. This modulation format eliminates the ambiguity involved in deciding whether a particular polarization represents a binary zero or a one. Furthermore, the DPolSK scheme can lead to the removal of the phase noise associated with both the laser source and the fluctuations from the transmission medium [Ref. 71]. This factor provides an improvement over PolSK where only the phase jitter of the laser may be cancelled.

The above binary schemes have been concerned with the two polarization modes of a single-mode fiber. Multilevel PolSK is also possible in which the transmitted symbols are each associated with different polarization states within the fiber [Refs. 72, 73]. Moreover, such a modulation format can provide a performance improvement over the more traditional multilevel systems outlined in Sections 12.5.2 and 12.5.3 [Ref. 73].

12.6 Demodulation schemes

Basic receiver configurations for optical heterodyne and homodyne detection are shown in Figure 12.12. In both cases it has been assumed that some form of polarization control is required to match the incoming signal SOP to that of the local oscillator signal (see Section 12.4.2). This factor therefore implies the use of conventional circularly symmetric single-mode fiber. For heterodyne detection (Figure 12.12(a)), a beat-note signal between the incoming optical signal and the local oscillator signal produces the IF signal which is obtained using the square law optical detector (see Section 12.3). The IF signal, which generally has a frequency of between three and four times the transmission rate, is then demodulated into the baseband using either a synchronous or nonsynchronous detection technique.* An

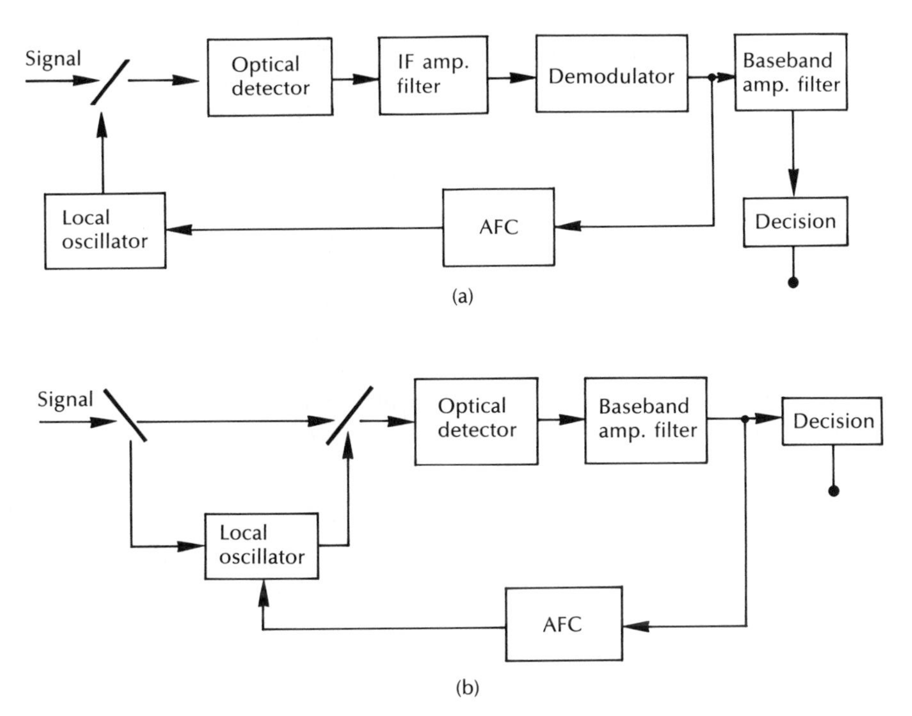

Figure 12.12 Basic coherent receiver configurations: (a) optical heterodyne receiver; (b) optical homodyne receiver illustrating the phase locking between the local oscillator and incoming signals.

* A brief explanation of the terminology was provided in Section 12.2. It should be noted that nonsynchronous heterodyne detection does not strictly require phase matching between the incoming signal and the local oscillator. Spatial coherence between the two signals is, however, required when they are combined so that nonsynchronous detection schemes do fall within the broad heading of coherent optical fiber systems.

optical receiver bandwidth several times greater than that of a direct detection receiver is therefore required for a specific transmission rate. Moreover, as IF frequency fluctuation degrades the heterodyne receiver performance, then frequency stabilization may be achieved by feeding back from the demodulator through an automatic frequency control (AFC) circuit to the local oscillator drive circuit.

In the case of homodyne detection in which the phase of the local oscillator signal is locked to the incoming signal, then, by definition, a synchronous detection scheme must be employed. Moreover, the result of the mixing process in the optical detector produces an information signal which is in the baseband (see Section 12.3) and thus requires no further demodulation. An AFC loop is also shown within the homodyne receiver configuration of Figure 12.12(b) to provide the necessary frequency stabilization between the two signals. Hence any variant detection schemes based on homodyne detection, but in which the local oscillator laser is not phase locked to the incoming signal such as phase diversity or multiport detection, could be considered as a form of heterodyne rather than homodyne detection [Ref. 5]. However, this technique is dealt with separately in Section 12.6.4.

In both optical heterodyne and homodyne detection, where the incoming signal is demodulated using a local oscillator laser, FM noise in this device together with that resulting from the source laser causes SNR degradation in the receivers through FM to AM, or PM to AM conversion which generally determines the lower limit of bit error rate performance [Ref. 74]. FM noise which basically results from the spontaneous emission coupled to the lasing mode, is, in the semiconductor laser, enhanced by AM noise caused by photon number fluctuation which is generated through the same mechanism [Ref. 75]. Moreover, excess AM noise within the local oscillator laser due to its resonance characteristics also deteriorates the SNR performance and hence degrades the receiver sensitivity. To reduce the effect of local oscillator FM noise a semiconductor laser with a narrowed or suppressed spectral linewidth must be used [see Section 12.4.1]. The excess AM noise in the semiconductor laser decreases with an increase in the bias level so that high bias operation is effective in suppressing this mechanism [Ref. 75]. Furthermore, excess AM noise associated with the local oscillator can be suppressed by employing the balanced receiver configuration described in Section 12.4.3.

12.6.1 Heterodyne synchronous detection

Optical heterodyne synchronous detection necessitates an estimation of the phase of the IF signal in translating it to the baseband. Such an approach generally requires the use of phase locking techniques at the receiver in order to track phase fluctuations in the incoming and local oscillator signals. Since the information signal is to be processed on an IF carrier, then electrical phase estimation may be employed. Hence the phase locked loop (PLL) techniques and configurations appropriate to radiofrequency and microwave communications can be utilized

[Refs. 2, 23]. Such techniques have been investigated primarily for PSK demodulation where an estimation of the phase of the signal is required [Ref. 5]. Furthermore, synchronous PSK demodulation is the most sensitive of the heterodyne detection techniques (see Section 12.7). In order to achieve a measurement of the phase of a fully modulated PSK signal, it is necessary to obtain a phase reference from the phase of the average incoming optical signal within a particular time interval. Therefore the purpose of the PLL is to provide that reference where, in general, the time average is defined by the bandwidth of the loop.

An examination of the spectrum of a PSK signal reveals that no signal energy is present at the carrier frequency when the phase shift from the binary one to zero states is a full 180°.* The introduction of a nonlinear element within the PLL is therefore necessary to enable efficient carrier recovery. A squaring loop technique illustrated in Figure 12.13(a) is particularly applicable to binary PSK, phase noise-sensitive coherent optical fiber systems [Ref. 23]. By squaring the PSK signal frequency this method produces a carrier at twice the original frequency which can be filtered out and then used for phase estimation. A similar result may be obtained with the statistically equivalent Costas loop shown in Figure 12.13(b) [Ref. 76].

An alternative approach to carrier recovery is to reduce slightly the depth of the phase modulation so that a small component of the transmitted energy lies at the carrier frequency. This pilot carrier signal can then be amplified and recovered as a phase reference at the receiver. In this case of what is essentially a weak carrier, a much reduced loop bandwidth (i.e. reduced integration time) is required in order adequately to recover the carrier. Furthermore, whilst providing a stable reference, long integration times increase the sensitivity of the receiver to carrier phase noise. However, to detect the PSK signal adequately using a pilot carrier, a significant amount of signal power may be sacrificed [Ref. 23].

A variant on the pilot carrier technique for PSK synchronous demodulation is shown in Figure 12.14 in which carrier recovery takes place at the IF stage [Ref. 77]. In this case the detected IF signal is divided into two routes, one being the signal route and the other being the carrier recovery route. Following the carrier recovery route the signal is doubled by a frequency doubler (FD) to remove the $(0, \pi)$ phase modulation component. Twice the IF frequency of the resultant signal is then divided by a frequency halver (FH) to recover the reference carrier signal. Finally, the recovered carrier and signal are mixed to give the demodulator output. This technique which provides for suppression of phase noise has been demonstrated in an optical PSK system operating at 560 Mbit s^{-1} [Refs. 77, 78].

The above synchronous demodulation schemes can also be used within ASK and FSK heterodyne optical fiber systems but they do not provide the same potential receiver sensitivity performance as that of PSK. Moreover, alternative nonsynchronous techniques for ASK and FSK are often more reliable in the presence of phase noise and provide receiver sensitivities only slightly less than the

* This situation typifies a suppressed carrier modulation type [Ref. 3].

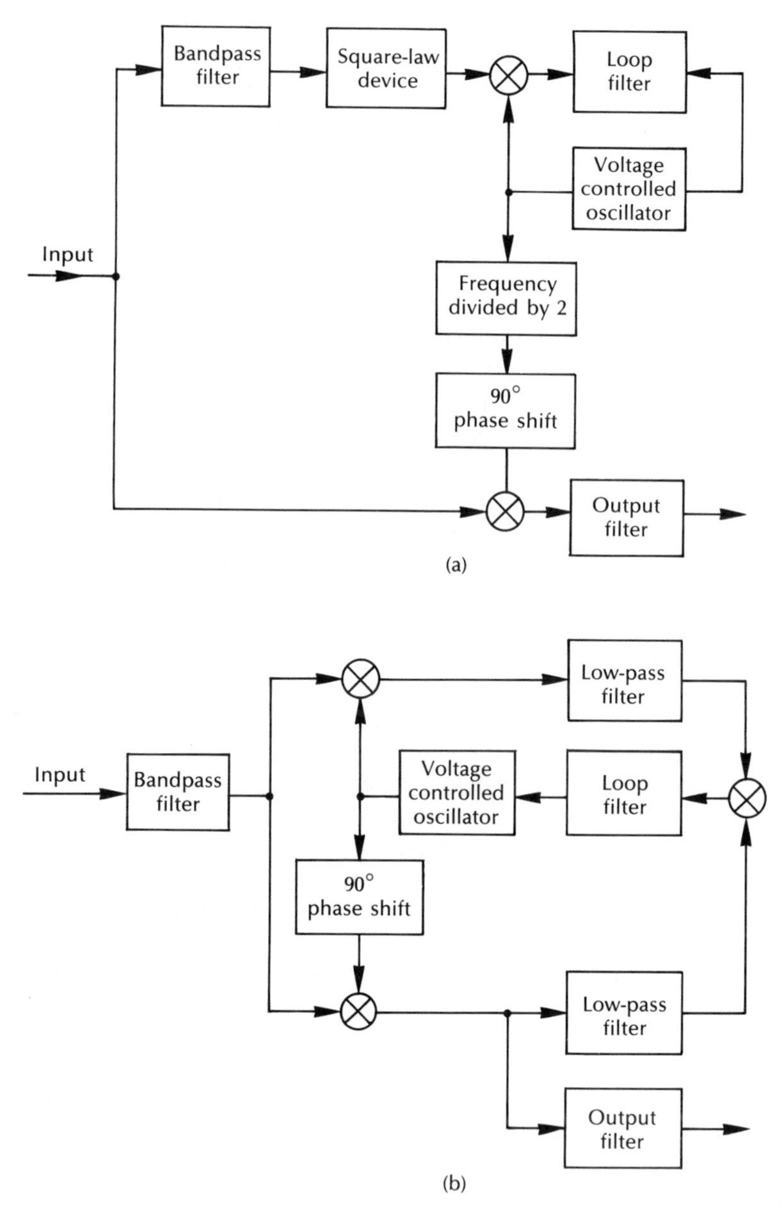

Figure 12.13 Techniques for carrier recovery used in coherent optical PSK receivers: (a) squaring loop; (b) Costas loop.

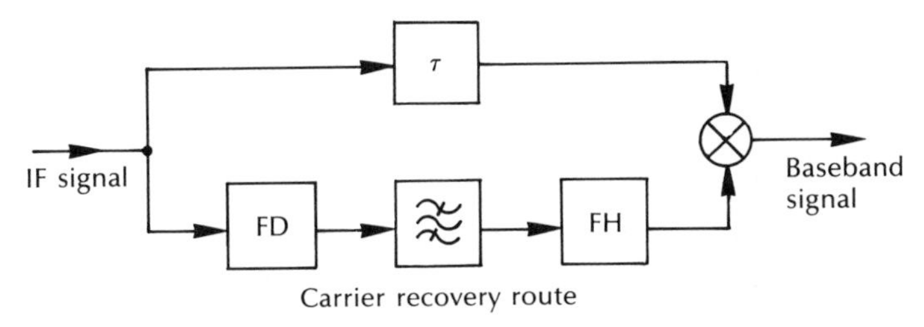

Figure 12.14 A carrier recovery synchronous demodulator [Ref. 77].

corresponding synchronous methods. These nonsynchronous demodulation schemes are therefore discussed in Section 12.6.2.

Although nonsynchronous demodulation of a PSK signal is strictly not possible because the message information resides in the phase of the carrier signal, the phase comparison detection associated with differential phase shift keying (DPSK) reduces the synchronization problems associated with PSK (see Figure 12.11). Therefore, as indicated in Section 12.5.3, a broader linewidth laser source and local oscillator laser may be utilized with DPSK and heterodyne detection than that necessary for binary PSK. In addition, optical DPSK reception is only a fraction of a decibel less sensitive than synchronous PSK.

In the optical heterodyne receiver the DPSK signal is first translated down to a suitable IF frequency prior to demodulation by a conventional DPSK, or phase comparison detector [Ref. 2]. Such a detection system is illustrated in Figure 12.15 where the incoming signal from the heterodyne process is band limited to reduce the phase noise. In the phase comparison system, the bandlimited version of the DPSK signal is delayed by a bit period T, and the product with the undelayed version is formed. This product is then integrated to eliminate residual noise, a positive voltage being obtained if the phases of the combined signals are the same. Alternatively, when the two phases differ by π radians, a negative voltage occurs at the output.

The use of a balanced heterodyne receiver (see Section 12.4.3) in an experimental DPSK system operating at 1.4 Gbit s^{-1} resulted in a receiver sensitivity of

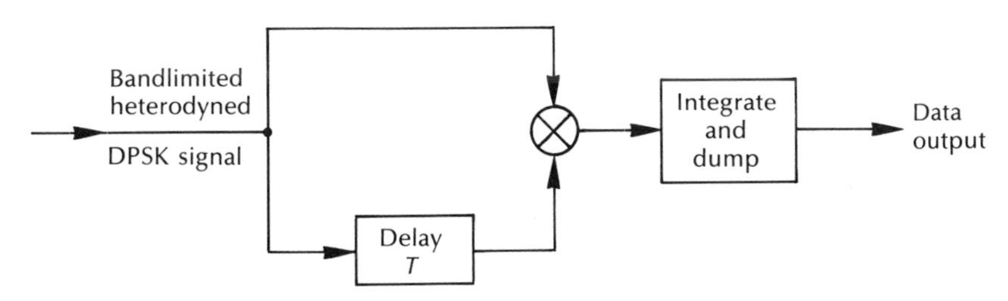

Figure 12.15 Heterodyne differential phase shifting keying demodulator.

-42.8 dBm (or 285 photons per bit) at a bit error rate of less than 10^{-9} [Ref. 79]. Furthermore, a local oscillator laser output power of only 0.5 mW was required.

12.6.2 Heterodyne nonsynchronous detection

It was indicated in Section 12.5 that both ASK and FSK may be demodulated using nonsynchronous detection techniques which puts the least demands on laser linewidth and phase stability. Such demodulation schemes, which include ASK envelope detection as well as FSK single and dual filter detection, do not therefore require the extremely narrow laser linewidths associated with synchronous binary PSK demodulation or even DPSK (see Section 12.5.3). Heterodyne envelope detection of an ASK signal may be achieved using an intermediate frequency bandpass filter followed by a peak detector to recover the baseband signal as shown in Figure 12.16(a). Such a scheme, however, incurs a receiver sensitivity penalty as a result of nonlinear filtering of the Gaussian distributed noise in the peak detection process combined with the phase noise broadening of the signal spectrum such that a significant proportion of the signal energy can be translated outside the IF signal band. An optimum receiver bandwidth balances these factors and, for combined

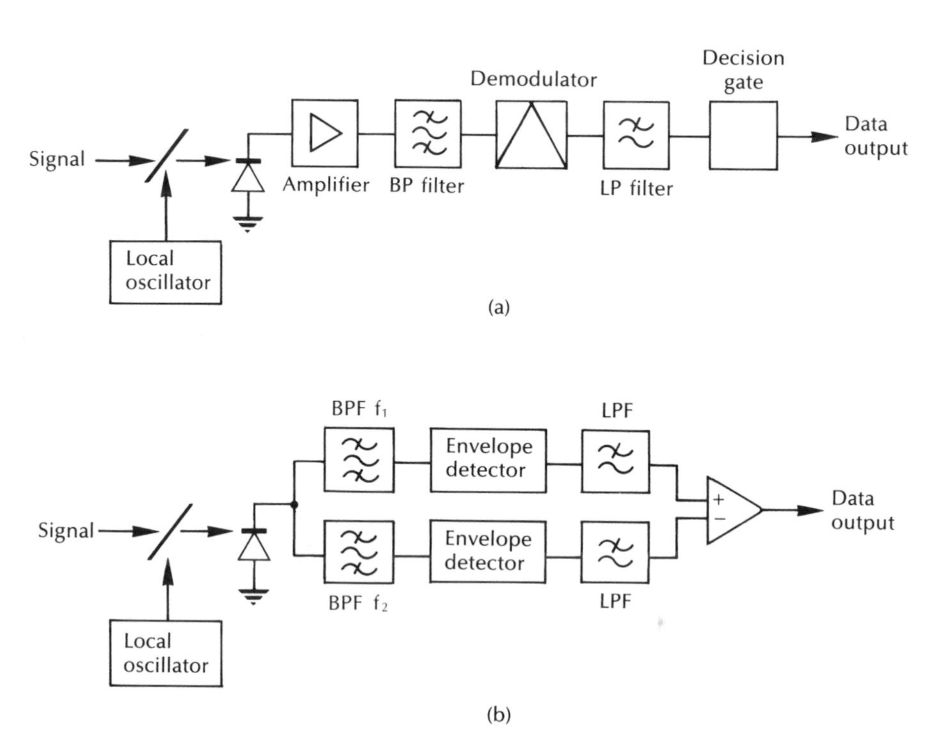

Figure 12.16 Nonsynchronous heterodyne detection: (a) ASK single envelope detector receiver; (b) FSK dual filter receiver.

source and local oscillator linewidths of 10% of the transmitted data rate, the receiver penalty is around 3 dB [Ref. 42].

By employing parallel filters with channels centred on the two transmitted frequencies it is possible to use envelope detection on each channel of a binary FSK signal. The configuration for this dual filter demodulation technique is provided in Figure 12.16(b) [Ref. 44]. At the output it produces a differential ASK signal with a receiver sensitivity which is slightly better than nonsynchronous ASK demodulation in the presence of phase noise (i.e. a 2 dB penalty when source and local oscillator laser linewidths are 10% of the transmitted data rate [Ref. 42]). This improvement results from the complementary behaviour of the dual filter approach where, for large spacing between the two FSK signal channels, it is possible to have a significant spectral broadening of the signal but with insufficient energy in the complementary channel to register an error [Ref. 23].

12.6.3 Homodyne detection

The attraction of optical homodyne detection is not just the potential 3 dB improvement in receiver sensitivity (see Section 12.3) but also that it can ease the receiver bandwidth requirement considerably. This factor is illustrated in Figure 12.17 which compares the spectra at the output of the detector for PSK homodyne and PSK heterodyne detection. It may be observed that homodyne detection requires only the normal direct detection receiver bandwidth whereas heterodyne detection requires at least twice this bandwidth and often a factor of three or four times it. Unfortunately, optical homodyne detection using independent source and local oscillator lasers (i.e. not self homodyne) has proved extremely difficult to achieve because of the problems associated with remotely optical locking a local oscillator laser to a low level modulated signal [Ref. 45]. Such phase locking is essential because the phase difference ϕ in Eq. (12.12) must be held near zero for high sensitivity reception. Furthermore, if ϕ drifts to $\pi/2$, then the output signal current I_S will become zero and the detection process will cease.

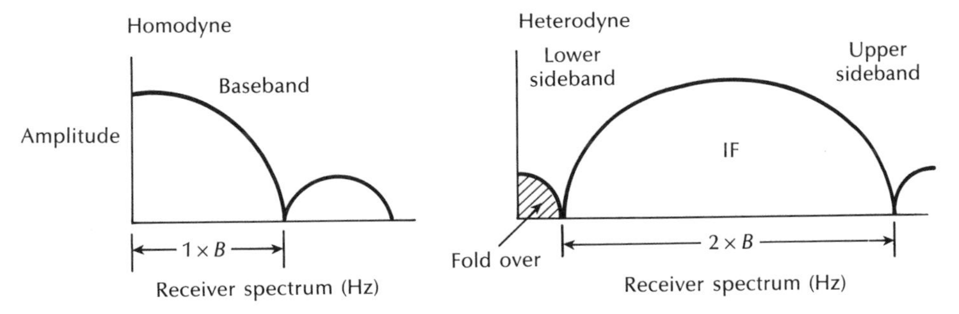

Figure 12.17 Comparison of the electrical spectra at the optical detector output for homodyne and heterodyne detection of a PSK signal.

Two homodyne demodulation strategies have, however, proved successful in demonstration at optical frequencies. They are the use of an optical phase lock loop and the selective amplification of the carrier [Refs. 2, 18, 23, 45]. Unlike the electrical PLL techniques described in Section 12.6.1 which comprised an electrical phase detector, loop filter and voltage controlled oscillator (VCO), in the optical PLL, the photodetector and laser local oscillator act as the phase detector and VCO respectively. Hence the optical phase difference between the incoming signal and the local oscillator signal, or the phase error signal, is detected by the photodetector prior to being fed back to correct the local oscillator frequency and phase. Although homodyne detection using an optical PLL has been demonstrated [Ref. 80], it is somewhat difficult to realize and puts stringent demands on the laser linewidths [Ref. 65].

The optical PPL configuration shown in Figure 12.18 [Ref. 80] employs a pilot carrier strategy for PSK homodyne detection. In common with other pilot carrier techniques (see Section 12.6.1) this carrier is generated by using incomplete (less than 180°) phase modulation. The pilot carrier signal, together with the incoming signal, are combined in a 3 dB fiber directional coupler and then detected using a balanced receiver. The output signal from difference amplifier (Figure 12.18) is therefore a function of the phase error which may be used for phase locking through the loop filter to the optical local oscillator which performs as the VCO. It should be noted, however, that any carrier power used in this phase locking process directly reduces the receiver sensitivity by an equivalent amount. Furthermore, the signal power required to track the phase of the incoming carrier to a specified accuracy (i.e. the tracking error relates directly to a degradation in the bit error rate performance) is dependent upon the combined phase noise of the source and local oscillator lasers as well as the PLL bandwidth. Hence, extending the optical PLL bandwidth improves the tracking performance until a point is reached where the increased shot noise significantly degrades the loop SNR. There is, therefore, an optimum loop bandwidth to provide a minimum phase error and

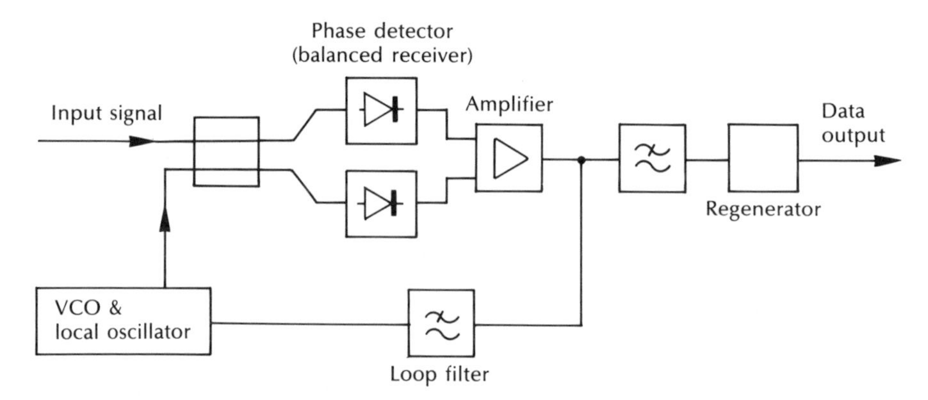

Figure 12.18 Pilot carrier optical phase locked loop receiver [Ref. 80].

it is possible to improve the performance of the optical homodyne receiver when the local oscillator laser has substantial phase noise by simply increasing the PLL bandwidth.

The basic principle of an optical Costas PLL has also been demonstrated at a wavelength of 10.6 μm using carbon dioxide lasers to provide the source and local oscillator signals [Ref. 81]. A hypothetical optical Costas PLL homodyne receiver for antipodal PSK modulation is illustrated in Figure 12.19 [Ref. 45]. In this case the incoming and local oscillator signals are combined in an optical hybrid which ensures a 90° phase difference between the output signals from the two detectors.

The 90° hybrid device can be realized using the phase shift properties of polarized light [Ref. 82]. A 90° phase shift can be obtained by combining a circularly polarized local oscillator signal with a linearly polarized incoming signal and then by resolving the combined signal into two orthogonal components with a polarization beam splitter. The linearly polarized incoming signal must, however, be aligned at 45° to the beam splitter plane.

The two outputs from the optical hybrid are detected, amplified prior to multiplication in the mixer shown in Figure 12.19. The phase of the suppressed carrier is then determined by the low pass filtering of this product. Moreover, the control signal is also filtered and then used to adjust the local oscillator frequency in a similar manner to that employed in the pilot carrier optical PLL. Using the optical Costas PLL, however, provides the advantage that all the low level circuits, prior to the mixer, can be a.c. coupled and there is no wastage of transmitted power in a pilot carrier component.

It is possible to achieve optical homodyne detection without the use of an optical PLL if a residual carrier component of the incoming signal is selectively amplified (i.e. the modulation sidebands are not amplified) and then recombined with the sidebands before photodetection. In this way the amplified carrier functions as the

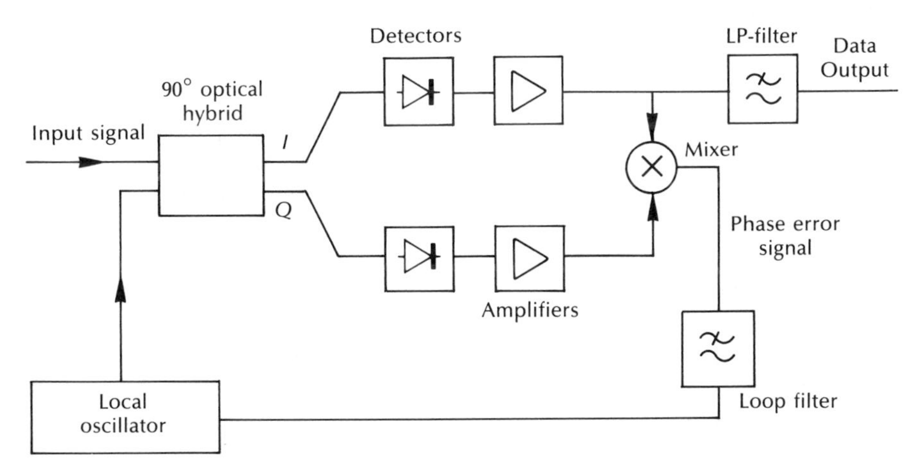

Figure 12.19 Hypothetical optical Costas loop receiver [Ref. 45].

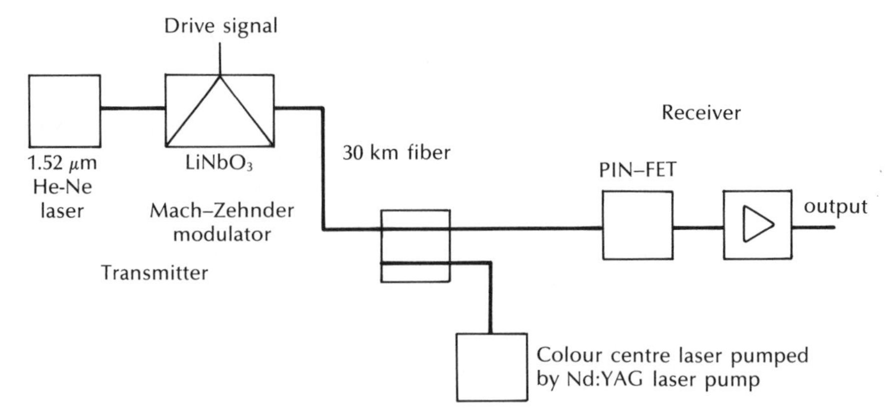

Figure 12.20 Brillouin amplification of the carrier signal providing homodyne detection [Ref. 84].

local oscillator signal. The principle was first demonstrated with a resonant gas laser amplifier [Ref. 83] but a more recent strategy for such self homodyne detection utilized the Brillouin gain process (see Section 3.14) which has a bandwidth of about 20 MHz in silica fiber. The experimental arrangement for this homodyne receiver is shown in Figure 12.20 [Ref. 84]. The fiber was pumped with a colour centre laser* in a direction back towards the transmission fiber at a frequency of 11 GHz (the Stokes shift) higher than the incoming carrier signal. Pump powers of between 3 and 5 mW were typically required to obtain carrier gains of around 50 dB. The homodyne gain, however, was equal to one half of the carrier gain and hence a homodyne gain of 26 dB was reported [Ref. 83]. This Brillouin homodyne technique can be used directly for ASK demodulation, but for PSK demodulation it would be necessary to form a pilot carrier at the optical transmitter.

12.6.4 Phase diversity reception

An additional demodulation scheme employed in microwave systems which has been applied to coherent optical systems is phase diversity reception, or multiport detection [Refs. 85, 86]. In these techniques the local oscillator laser is operated at a frequency comparable to the frequency of the incoming signal but the two signals are not phase locked. The phase diversity receiver does, however, convert the incoming signal directly to the baseband and therefore has the bandwidth advantage of homodyne detection. As the optical mixing is not phase synchronized, the demodulation strategy avoids this major problem associated with homodyne detection, but at best the receiver sensitivity is equivalent to that of heterodyne detection. Hence, from the viewpoint of receiver bandwidth requirements, phase

* The colour centre laser was itself pumped by an Nd : YAG laser.

diversity reception behaves like an optical homodyne receiver but it is essentially nonsynchronous with the sensitivity performance of, or worse than, a heterodyne receiver. It cannot therefore be regarded strictly as a homodyne technique* and certain authors suggest it is more appropriate to classify such multiport detection schemes with heterodyne receivers [Ref. 5].

A number of optical phase diversity reception schemes have been investigated operating with two or more ports and two or more matched receivers. The technique utilizes a fixed phase relationship between the ports of a multiport coupler to provide the direct demodulation to the baseband without the requirement for an optical PLL. One variant of the optical phase diversity receiver known as the in-phase and quadrature (I & Q) receiver is shown in Figure 12.21. In this two phase [Refs. 86 to 90] scheme the incoming and local oscillator signals are combined in a 90° optical hybrid similar to the one described in Section 12.6.3. The optical hybrid is connected to two detectors, the outputs of which are amplified and then passed through square law devices prior to electrical recombination. The output signals from each receiver path prior to recombination can be written in terms of their voltages V_1 and V_2 as:

$$V_1^2 = k_1^2 m^2(t) \sin^2 \delta\phi \tag{12.21}$$

$$V_2^2 = k_2^2 m^2(t) \cos^2 \delta\phi \tag{12.22}$$

where k_1 and k_2 are constants, $m(t)$ is the modulation and $\delta\phi$ is the phase error.

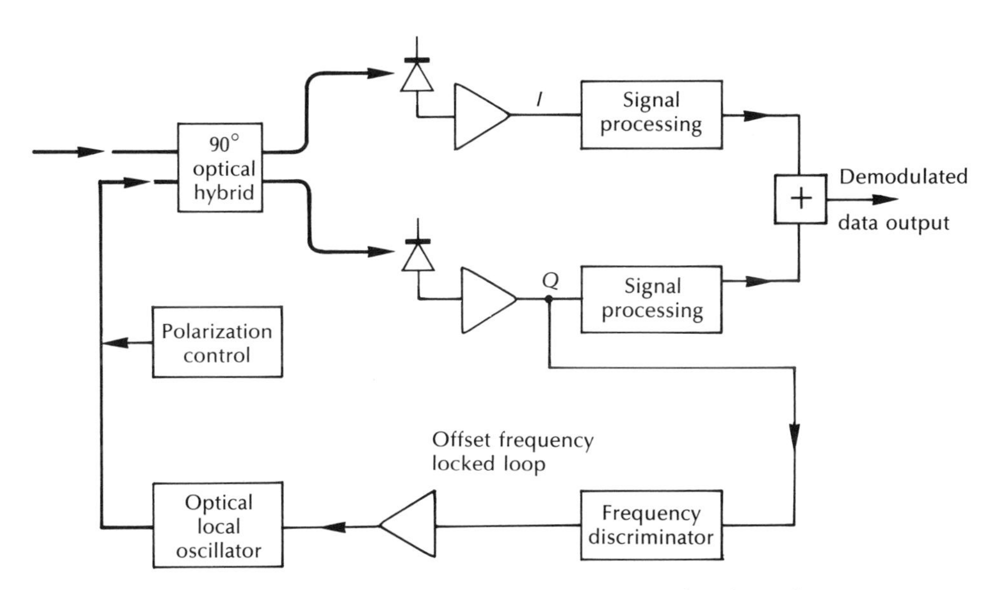

Figure 12.21 In-phase and quadrature phase diversity (two phase) receiver.

* Many authors do, however, refer to it as a homodyne strategy [Refs. 87 – 89].

Hence the output signal from the receiver is:

$$V_1^2 + V_2^2 = k^2[m(t)]^2 \tag{12.23}$$

where we assume $k = k_1 = k_2$. It may be observed from Eq. (12.23) that for ASK the demodulated signal is constant irrespective of the relative phase between the incoming and local oscillator signals. Therefore, with ASK modulation the laser linewidth requirements are comparable to heterodyne detection with nonsynchronous IF demodulation (see Section 12.5.1). For example, the experimental demonstration of such two phase ASK demodulation has been achieved at a transmission rate of 150 Mbit s^{-1} [Ref. 89]. The system utilized commercial 1.5 μm DFB laser with linewidths of 38 MHz and achieved a sensitivity of -55 dBm.

For PSK modulation, however, the signal may be differentially demodulated by the inclusion of a one bit delay in one of the inputs to the square law mixer. Hence the change in phase during a single bit period only is of concern and any longer term phase drift is removed. This DPSK I & Q demodulation is expected to have similar laser linewidth requirements to DPSK heterodyne detection (see Section 12.5.3) [Ref. 45]. The I & Q phase diversity receiver is more sensitive to fluctuations of the incoming SOP than a conventional optical heterodyne receiver [Ref. 90]. Polarization control should, however, be more straightforward in the I & Q case because it is possible to obtain electrical signals directly from the two receiver arms which provide exact information on the received polarization state. Nevertheless, problems do exist with two port phase diversity reception, as in practice the electrical square law demodulation is imperfect and produces additional terms which tend to appear in the baseband along with the demodulated signal. Moreover, the two detected currents must be 90° out of phase which may only be achieved at the expense of additional signal processing [Ref. 82], and also the two arms of the receiver must be well matched.

Other phase diversity techniques can be used as alternatives to I & Q detection. In particular, the phase diversity receiver using three phase reception has proved successful [Refs. 86, 87, 91, 92]. A schematic diagram of the generalised three phase receiver is shown in Figure 12.22. It may be observed that in this phase diversity scheme a 120° hybrid is required which can be conveniently realized from the intrinsic symmetry associated with the construction of a three fiber fused biconical coupler [Ref. 93]. Furthermore, this strategy avoids the polarization sensitivity of the two phase arrangement. However, the receiver sensitivity of this approach is poorer than the conventional optical heterodyne strategy as the additional port for the third detector introduces an extra 1.8 dB degradation [Ref. 85].

Demonstrations of three phase schemes have taken place using helium–neon lasers operating at 650 Mbit s^{-1} [Ref. 91] and DFB lasers at 140 Mbit s^{-1} [Ref. 92]. Furthermore, demodulation of direct FSK signals generated from a DFB laser operating at 5 Gbit s^{-1} has recently been demonstrated using a three port single filter phase diversity receiver [Ref. 88]. The single filter detection was achieved by setting the local oscillator frequency equal to the centre frequency of the stronger

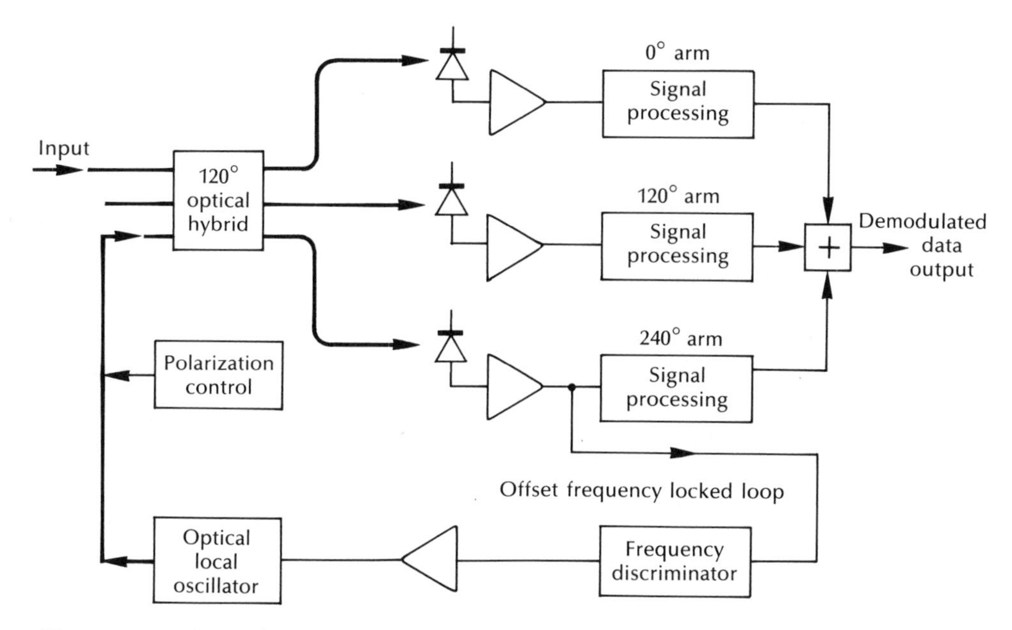

Figure 12.22 Phase diversity receiver using three phase detection.

FSK sideband [Ref. 93]. Receiver sensitivities for a bit error rate of 10^{-9} of -30.5 dBm and -27.0 dBm were obtained at transmission rates of 4 and 5 Gbit s^{-1} respectively [Ref. 88].

In the longer term, it is suggested [Ref. 45] that multiport detection will be extended to provide both phase and polarization diversity reception (see Section 12.4.2.2). Furthermore, greater numbers of ports than three for phase diversity receivers are envisaged (i.e. four, six or eight) combined with a balanced receiver approach (see Section 12.4.3) in order to reduce excess noise from the local oscillator laser [Refs. 23, 45, 85, 94].

12.7 Receiver sensitivities

The basic detection principles for the ASK coherent optical receiver were discussed in Section 12.3. In addition, the 3 dB SNR improvement for the ASK homodyne receiver in comparison with the corresponding heterodyne receiver in the shot noise limit was demonstrated (Eqs. (12.16) and (12.17)). Although a synchronous detection process was assumed in Section 12.3 for ASK with heterodyne detection, ASK with nonsynchronous detection can achieve approximately the same SNR limit [Ref. 95]. It is now important to consider the receiver sensitivities for the other major modulation schemes and detection processes so that the choices regarding the implementation of coherent optical fiber systems can be understood.

In Section 12.3 comparison between ASK heterodyne and homodyne reception was undertaken from determination of their respective shot or quantum noise limited SNRs. As in this section we propose to extend this comparison to receiver sensitivities of other digital modulation schemes, it is useful to transfer from considerations of SNR to those of bit error rate (BER). In addition we will continue to consider minimum receiver sensitivities in the presence of quantum noise only, neglecting thermal and other noise sources in the electronic preamplifier discussed in Section 9.3, as well as excess noise sources in the local oscillator laser. Although these other noise sources are normally present, comparison of the modulation formats under quantum noise limited detection assists in the deliberations regarding the desirability of specific schemes. Furthermore, near-quantum noise limited reception is more readily achieved using heterodyne or homodyne detection than by employing direct detection. We concentrate on synchronous demodulation for the major modulation formats (i.e. ASK, FSK, PSK) prior to consideration of homodyne detection for the ASK and PSK modulation schemes.

12.7.1 ASK heterodyne detection

The ASK or OOK modulation format has similarities to digital transmission in a IM/DD optical fiber system, a BER analysis for the latter system being provided in Section 11.6.3. In a heterodyne receiver, however, the analyses of signal and noise phenomena are more complicated than in the IM–DD case because the optical detector output appears as an IF signal and not as a baseband signal. Hence the IF output current from the photodetector $I_S(t)$ which corresponds to the input current to the preamplifier from Eq. (12.9) can be written as:

$$I_S(t) = \begin{cases} I_{SH} \cos(\omega_{IF}t + \phi), & \text{for a 1 bit} \\ 0, & \text{for a 0 bit} \end{cases} \tag{12.24}$$

where

$$I_{SH} = \frac{2\eta e}{hf} \sqrt{P_S P_L} \tag{12.25}$$

To obtain the IF noise current, two assumptions can be made. Firstly, it is assumed that the local oscillator signal power is much larger than the incoming signal power so that the total noise current is approximately equal to $\overline{i_{SL}^2}$, given by Eq. (12.14) and this applies for both the 1 and the 0 bit. Secondly, it is assumed that this IF noise current $N(t)$ can be considered as narrowband noise which can be expressed as [Ref. 3]:

$$N(t) = x(t) \cos \omega_{IF}t + y(t) \sin \omega_{IF}t \tag{12.26}$$

where $x(t)$ and $y(t)$ are functions of time which vary at a much slower rate than the IF signal. It should be noted that the first and second terms in Eq. (12.26) represent the in-phase and quadrature components respectively. Hence the mean square values of $x(t)$ and $y(t)$ may be written as:

$$\overline{x^2}(t) = \overline{y^2}(t) = \overline{i_{SL}^2} \tag{12.27}$$

For heterodyne synchronous detection, the IF amplifier is followed by a demodulation circuit which has a phase synchronised reference signal proportional to cos $\omega_{IF}t$. Therefore, the detector output $V_d(t)$ is given by:

$$V_d(t) = k[I_S(t) + x(t)] \tag{12.28}$$

The probability density functions of $V_d(t)$ for the ASK signal $I_S(t)$ represented by Eq. (12.24) are illustrated in Figure 12.23. Moreover, it may be observed that these probability density functions are similar to those for digital IM–DD reception shown in Figure 11.35. Assuming this to be the case, the optimum decision threshold level D is set midway between the zero current (0 state) and the peak signal current (1 state) such that:

$$I_D \simeq \frac{I_{SH}}{2} = \frac{\eta e}{hf}\sqrt{P_S P_L} \tag{12.29}$$

The optical detector output given by Eq. (12.28) can now be considered as a baseband signal and noise contribution. Hence the analysis for BER can follow the method utilized for IM/DD in Section 11.6.3. We are therefore in a position to move straight to Eq. (11.21) and to substitute in the appropriate values from this

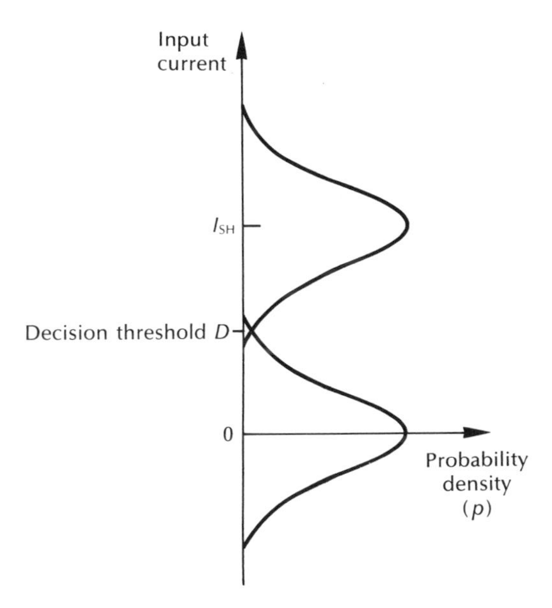

Figure 12.23 Probability density functions for ASK heterodyne synchronous detection.

derivation. Thus the probability of error $P(\mathrm{e})$ for ASK heterodyne synchronous detection can be written as:

$$P(\mathrm{e}) = \tfrac{1}{2}\left[\tfrac{1}{2}\,\mathrm{erfc}\left(\frac{|I_{SH} - I_D|}{(\overline{i_{SL}^2})^{\frac{1}{2}}\sqrt{2}}\right) + \tfrac{1}{2}\,\mathrm{erfc}\left(\frac{|-I_D|}{(\overline{i_{SL}^2})^{\frac{1}{2}}\sqrt{2}}\right)\right] \quad (12.30)$$

Substituting for I_D from Eq. (12.29) gives:

$$P(\mathrm{e}) = \tfrac{1}{2}\left[\tfrac{1}{2}\,\mathrm{erfc}\left(\frac{|I_{SH}/2|}{(\overline{i_{SL}^2})^{\frac{1}{2}}\sqrt{2}}\right) + \tfrac{1}{2}\,\mathrm{erfc}\left(\frac{|-I_{SH}/2|}{(\overline{i_{SL}^2})^{\frac{1}{2}}\sqrt{2}}\right)\right]$$

$$= \tfrac{1}{2}\,\mathrm{erfc}\left(\frac{I_{SH}}{2(\overline{i_{SL}^2})^{\frac{1}{2}}\sqrt{2}}\right) \quad (12.31)$$

Finally substituting for I_{SH} using Eq. (12.25) and for $\overline{i_{SL}^2}$ from Eq. (12.14), where we replace B with the IF signal bandwidth B_{IF} because $I_S(t)$ was originally an IF signal exhibiting this bandwidth, then Eq. (12.31) can be written as:

$$P(\mathrm{e}) = \tfrac{1}{2}\mathrm{erfc}\ \frac{2\eta e\sqrt{P_S P_L}}{hf}\Bigg/ 2\sqrt{2}\left(\frac{2e^2\eta P_L B_{IF}}{hf}\right)^{\frac{1}{2}}$$

$$= \tfrac{1}{2}\,\mathrm{erfc}\left(\frac{\eta P_S}{4hf\,B_{IF}}\right)^{\frac{1}{2}} \quad (12.32)$$

It is more appropriate to specify the probability of error in terms of the transmission bit rate B_T rather than the receiver bandwidth and, therefore, assuming transmission at a rate equivalent to twice the baseband bandwidth (see Eq. (3.12)), then $B_T = 2B \simeq B_{IF}$. Hence Eq. (12.32) becomes:

$$P(\mathrm{e}) = \tfrac{1}{2}\,\mathrm{erfc}\left(\frac{\eta P_S}{4hf\,B_T}\right)^{\frac{1}{2}} \quad (12.33)$$

This expression allows the shot noise limited performance of ASK heterodyne synchronous detection to be compared with alternative detection schemes.

For ASK heterodyne nonsynchronous or envelope detection it can be shown that the probability of error in the shot noise limit, under similar assumptions to those used above for synchronous detection, is given by the approximate expression [Refs. 95, 96]:

$$P(\mathrm{e}) \simeq \tfrac{1}{2}\exp - \left(\frac{I_{SH}^2}{8(\overline{i_{SL}^2})}\right) \quad (12.34)$$

Again substituting for I_{SH} and $\overline{i_{SL}^2}$ from Eqs. (12.25) and (12.14), respectively, gives:

$$P(\mathrm{e}) \simeq \tfrac{1}{2}\exp - \left(\frac{\eta P_S}{4hf\,B_T}\right) \quad (12.35)$$

Moreover, as indicated in Section 12.7.5, these formulae give approximately equivalent BER results to those provided by Eqs. (12.31) and (12.33) for ASK heterodyne synchronous detection. The reason for this situation is that the approximation erfc $(u) \simeq \exp - (u)^2$ holds for large u and hence for a low BER [Ref. 96].

12.7.2 FSK heterodyne detection

We commence by considering FSK heterodyne synchronous detection in the shot or quantum noise limit. The two angular frequencies for the transmitted 1 and 0 bits are assumed to be ω_1 and ω_2 so that:

$$I_S(t) = \begin{cases} I_{SH} \cos(\omega_1 + \phi), & \text{for a 1 bit} \\ I_{SH} \cos(\omega_2 + \phi), & \text{for a 0 bit} \end{cases} \tag{12.36}$$

where I_{SH} is defined in Eq. (12.25) and ϕ, which is a function of time, represents the phase noise associated with the semiconductor laser. This is neglected, as before in Section 12.7.1, because we are concerned with shot noise limited detection.

It is assumed that the signal $I_S(t)$ is received using two receivers tuned to ω_1 and ω_2 and that the output voltages from receivers 1 and 2 are V_1 and V_2 respectively. Furthermore it is assumed that the two receivers exhibit ideal frequency selectivity such that there is no crosstalk between ω_1 and ω_2 and therefore any additional voltages are generated by shot noise effects only. It is possible to just consider the time slot when a 1 bit (ω_1) is transmitted without losing generality. Hence the PDF of the output from receiver 1 is given by [Ref. 95]:

$$p_1(V) = \frac{1}{(\overline{i_{SL}^2})^{\frac{1}{2}}\sqrt{2\pi}} \exp - \left[\frac{(I_{SH} - V_1)^2}{2(\overline{i_{SL}^2})}\right] \tag{12.37}$$

Again we assume the local oscillator output power to be much higher than that of the incoming signal so that the total noise current is approximately equal to $\overline{i_{SL}^2}$ provided by Eq. (12.14). The noise output from receiver 2 can therefore be written as:

$$p_2(V) = \frac{1}{(\overline{i_{SL}^2})^{\frac{1}{2}}\sqrt{2\pi}} \exp - \left[\frac{V_2^2}{2(\overline{i_{SL}^2})}\right] \tag{12.38}$$

As an error occurs when $V_2 > V_1$, then the probability of error $P(e)$ is equivalent to the probability that $V_1 - V_2 < 0$. Hence:

$$P(e) = \int_{-\infty}^{0} \frac{1}{[2\pi 2(\overline{i_{SL}^2})]^{\frac{1}{2}}} \exp - \left[\frac{(w - I_{SH})^2}{4(\overline{i_{SL}^2})}\right] \mathrm{d}w \tag{12.39}$$

Changing the limits of the integration gives:

$$P(e) = \int_{I_{SH}/(\overline{i_{SL}^2})}^{\infty} \frac{1}{\sqrt{\pi}} \exp(-z^2)\, \mathrm{d}z \tag{12.40}$$

Comparison with the definition for the complementary error function given in

Eq. (11.17) allows Eq. (12.40) to be written as:

$$P(\mathrm{e}) = \tfrac{1}{2}\,\mathrm{erfc}\left(\frac{I_{\mathrm{SH}}}{2(\overline{i_{\mathrm{SL}}^2})^{\frac{1}{2}}}\right) \tag{12.41}$$

Finally, substituting from Eq. (12.25) for I_{SH} and from Eq. (12.13) for $\overline{i_{\mathrm{SL}}^2}$ where the bandwidth, which in this heterodyne detection case is B_{IF}, is again (see Section 12.7.1) written in terms of the transmission bit rate B_{T}; then the probability of error becomes:

$$P(\mathrm{e}) = \tfrac{1}{2}\,\mathrm{erfc}\left(\frac{\eta P_{\mathrm{S}}}{2hf\,B_{\mathrm{T}}}\right)^{\frac{1}{2}} \tag{12.42}$$

The comparison of Eq. (12.42) with Eq. (12.33) indicates that FSK heterodyne synchronous detection has a receiver sensitivity which in the shot noise limit is 3 dB higher than that of ASK heterodyne synchronous detection. This improvement in sensitivity for FSK modulation may be attributed to the use of two frequencies (and hence dimensions) rather than only the one in the case of ASK. It should be noted, however, that a similar BER is obtained with the two modulation schemes when the same average power is transmitted. Whereas with ASK, zero signal power is transmitted for a binary 0 bit, in FSK a similar signal power is continuously transmitted [Ref. 97]. Nevertheless, there are advantages associated with the use of FSK over ASK, even when two systems with the same average signal power are considered. In particular, the optimization of the decision level proves easier and the spectrum broadening as a result of switching between a one and a zero state in practice is much reduced on that obtained with ASK.

Considering FSK heterodyne nonsynchronous or envelope detection, it can be shown that the probability of error in the shot noise limit under similar assumptions to those above for synchronous detection is given by the expression [Refs. 95, 96]:

$$P(\mathrm{e}) = \tfrac{1}{2}\exp - \left(\frac{I_{\mathrm{SH}}^2}{4(\overline{i_{\mathrm{SL}}^2})}\right) \tag{12.43}$$

Substituting for I_{SH} and $\overline{i_{\mathrm{SL}}^2}$ from Eqs. (12.25) and (12.14), respectively, gives:

$$P(\mathrm{e}) = \tfrac{1}{2}\exp - \left(\frac{\eta P_{\mathrm{S}}}{2hf\,B_{\mathrm{T}}}\right) \tag{12.44}$$

This result for FSK nonsynchronous detection is approximately equivalent to the one obtained for synchronous detection (Eq. (12.42)) and shows a 3 dB improvement over ASK nonsynchronous detection (Eq. (12.35)).

12.7.3 PSK heterodyne detection

In this modulation format the information is transmitted by a carrier of one phase for a binary 1 and a different phase for a binary 0. The phase shift employed is normally π radians so that:

$$I_S(t) = \begin{cases} I_{SH}\cos(\omega_{IF}t + \phi), & \text{for a 1 bit} \\ I_{SH}\cos(\omega_{IF}t + \pi + \phi), & \\ \text{or} & \text{for a 0 bit} \\ -I_{SH}\cos(\omega_{IF}t + \phi), & \end{cases} \tag{12.45}$$

Therefore the synchronously detected signal $I_S(t)$ is positive for the 1 bits and negative for the 0 bits. In this case the optimum decision level current is given as $I_D = 0$ instead of that obtained in Eq. (12.29) for ASK synchronous detection. Nevertheless, a similar method to obtain the probability of error $P(\mathrm{e})$ for the PSK heterodyne synchronous detection to that used in Section 12.7.1 for ASK detection may be employed. Hence assuming that the output voltage from the receiver for a binary one is V_1 and for a binary zero it is V_2 then:

$$\begin{aligned} P(\mathrm{e}) &= \frac{1}{2}\int_{-\infty}^{0} \frac{1}{(\overline{i_{SL}^2})^{\frac{1}{2}}\sqrt{2\pi}} \exp -\left(\frac{(I_{SH} - V_1)^2}{2(\overline{i_{SL}^2})}\right) \mathrm{d}V_1 \\ &\quad + \frac{1}{2}\int_{0}^{\infty} \frac{1}{(\overline{i_{SL}^2})^{\frac{1}{2}}\sqrt{2\pi}} \exp -\left(\frac{(-I_{SH} - V_2)^2}{2(\overline{i_{SL}^2})}\right) \mathrm{d}V_2 \\ &= \frac{1}{2}\operatorname{erfc}\left(\frac{I_{SH}}{(\overline{i_{SL}^2})^{\frac{1}{2}}\sqrt{2}}\right) \end{aligned} \tag{12.46}$$

Substituting from Eq. (12.25) for I_{SH} and from Eq. (12.13) for $\overline{i_{SL}^2}$ where the bandwidth, which in this heterodyne detection case is B_{IF}, and following Section 12.7.1, it is written in terms of the transmission bit rate B_T; then the probability of error becomes:

$$P(\mathrm{e}) = \tfrac{1}{2}\operatorname{erfc}\left(\frac{\eta P_S}{hf B_T}\right)^{\frac{1}{2}} \tag{12.47}$$

It may be noted that in the shot noise limit PSK heterodyne synchronous detection exhibits 3 dB and 6 dB more sensitivity than the FSK and ASK heterodyne synchronous detection schemes respectively (Eqs. (12.42) and (12.33)). In practice, however, very small levels of phase fluctuation at the transmitter can significantly deteriorate the potential low BER of the PSK system.

Although nonsynchronous PSK detection is not strictly realizable, a more relaxed synchronous detection process is afforded by differential PSK (DPSK) in which the transmitted information is contained by the phase difference between two consecutive bit periods (see Sections 12.5.3 and 12.7.1). The probability of error in the detection of this modulation format at the shot noise limit and under the previous assumption is given by [Refs. 95, 96]:

$$P(\mathrm{e}) = \tfrac{1}{2}\exp -\left(\frac{I_{SH}^2}{2(\overline{i_{SL}^2})}\right) \tag{12.48}$$

Furthermore, substituting for I_{SH} and $\overline{i_{SL}^2}$ from Eqs. (12.25) and (12.14) gives:

$$P(e) = \tfrac{1}{2} \exp - \left(\frac{\eta P_S}{hf B_T}\right) \tag{12.49}$$

This expression for DPSK indicates a roughly equivalent probability of error for this scheme to that obtained with PSK synchronous detection (Eq. (12.47)). Moreover, it demonstrates a potential 3 dB and 6 dB improvement over the FSK and ASK nonsynchronous detection schemes respectively.

12.7.4 ASK and PSK homodyne detection

From the three basic modulation formats, ASK and PSK signals can be demodulated using a homodyne detection scheme, provided that both the frequency and the phase of the local oscillator output signal can be synchronized to the incoming carrier signal (see Section 12.6.3). It should be noted that FSK modulation can only be detected using a homodyne type receiver when the device has two phase controlled local oscillators. The only exception to this occurs when phase diversity reception (see Section 12.6.4) is employed, but these techniques cannot be regarded as true homodyne detection schemes.

It was shown in Section 12.3 that the reduction in the bandwidth requirement for homodyne detection produced a sensitivity improvement of 3 dB over the corresponding ASK heterodyne detection scheme. The probability of error for ASK homodyne detection can be derived from a slight modification to the result obtained in Eq. (12.31) for ASK heterodyne synchronous detection. This modification involves the noise power term ($\overline{i_{SL}^2}$) which in the homodyne case is reduced by a half because of the factor of two bandwidth reduction. Hence the probability of error for ASK homodyne detection is given by [Ref. 96]:

$$P(e) = \tfrac{1}{2} \operatorname{erfc}\left(\frac{I_{SH}}{2(\overline{i_{SL}^2}/2)^{\frac{1}{2}}\sqrt{2}}\right)$$

$$= \tfrac{1}{2} \operatorname{erfc}\left(\frac{I_{SH}}{2(\overline{i_{SL}^2})^{\frac{1}{2}}}\right) \tag{12.50}$$

It should be noted, however, that the signal power in Eq. (12.50) remains the same as in the heterodyne case which may be observed to be correct by comparing Eqs. (12.10) and (12.11). Substitution from Eq. (12.25) for I_{SH} and from Eq. (12.14) for $\overline{i_{SL}^2}$ where in this case the bit rate B_T is set equal to the baseband bandwidth B gives:

$$P(e) = \tfrac{1}{2} \operatorname{erfc}\left(\frac{\eta P_S}{2hf B_T}\right)^{\frac{1}{2}} \tag{12.51}$$

Considering now the PSK homodyne detection scheme, Eq. (12.46) for PSK synchronous detection can be modified in a similar manner to the above so that the probability of error is given by:

$$P(\mathrm{e}) = \tfrac{1}{2}\ \mathrm{erfc}\left(\frac{I_{\mathrm{SH}}}{(\overline{i_{\mathrm{SL}}^2})^{\frac{1}{2}}}\right) \tag{12.52}$$

Again substituting from Eqs. (12.25) and (12.14) gives the probability of error for PSK homodyne detection in the shot noise limit as:

$$P(\mathrm{e}) = \tfrac{1}{2}\ \mathrm{erfc}\left(\frac{2\eta P_{\mathrm{S}}}{hf\,B_{\mathrm{T}}}\right)^{\frac{1}{2}} \tag{12.53}$$

The result obtained in Eq. (12.53) represents the lowest error probability and hence the highest receiver sensitivity of all the coherent detection schemes. As anticipated, it displays a 3 dB improvement over PSK heterodyne synchronous detection.

12.7.5 Comparison of sensitivities

A comparison of the analytical results for the major modulation formats and detection schemes obtained in Section 12.7.1 to 12.7.4 is provided in Table 12.2. Moreover, to allow a comparison to be made with direct detection, an additional column records the details determined from Eq. (9.7) and Example 9.1. It should be noted, however, that the average number of photons per bit required to maintain a BER of 10^{-9} assumes in the case of ASK that photons arrive over two bit periods because no light is transmitted for a binary zero. Hence the values shown must be doubled if the actual number of photons required to register a binary one with a BER of 10^{-9} is to be recorded (i.e. not the average over the two bit periods). This factor can lead to some confusion in the approaches adopted by different authors (e.g. Refs. 3, 23, 96, 98, 99]. However, no such difficulties occur with the FSK and PSK modulation formats as a constant amplitude carrier signal is transmitted for both the binary one and zero bits.

The average number of received photons per bit at a particular BER as given in Table 12.2 may be determined from the expressions which are provided for the respective error probabilities in the table. An analytical definition for the number of received photons per bit, however, is needed. This may be written down by considering Eq. (8.7) which simply provides the generation rate for electrons from incident photons in an optical detector. This equation therefore represents the received photon rate. To convert this into the photon number per bit N_{p} the expression must be simply multiplied by the signalling interval τ so that:

$$N_{\mathrm{p}} = \frac{\eta P_{\mathrm{S}}\tau}{hf} = \frac{\eta P_{\mathrm{S}}}{hf\,B_{\mathrm{T}}} \tag{12.54}$$

where we have written the peak signal power P_{S} in place of P_{o} in Eq. (8.7) for consistency with the notation in this chapter, and B_{T} is the transmission bit rate. Furthermore, it is useful to note that Eqs. (12.16) and (12.17) which provide the SNRs for heterodyne and homodyne detection, respectively, in their shot noise

Table 12.2 Comparison of optical receiver sensitivities in the quantum or shot noise limit. The upper entry in each detection technique provides the possibility of error determined for the different schemes, whilst the lower entry represents the average number of photons per bit required by an ideal binary receiver ($\eta = 1$) to achieve a BER of 10^{-9}

Receiver / Modulation	Homodyne detection	Heterodyne: Synchronous detection	Heterodyne: Nonsynchronous detection	Direct detection
ASK or OOK	$\frac{1}{2}\text{erfc}\left(\frac{\eta P_s}{2hfB_T}\right)^{1/2}$	$\frac{1}{2}\text{erfc}\left(\frac{\eta P_s}{4hfB_T}\right)^{1/2}$	$\frac{1}{2}\exp-\left(\frac{\eta P_s}{4hfB_T}\right)$	$\frac{1}{2}\exp-\left(\frac{\eta P_s}{hfB_T}\right)$
Av. no. photons per bit*	18	36	40	10.4
FSK	No (only very special receiver)	$\frac{1}{2}\text{erfc}\left(\frac{\eta P_s}{2hfB_T}\right)^{1/2}$	$\frac{1}{2}\exp-\left(\frac{\eta P_s}{2hfB_T}\right)$	No
Av. no. photons per bit	—	36	40	—
PSK	$\frac{1}{2}\text{erfc}\left(\frac{2\eta P_s}{hfB_T}\right)^{1/2}$	$\frac{1}{2}\text{erfc}\left(\frac{\eta P_s}{hfB_T}\right)^{1/2}$	DPSK† $\frac{1}{2}\exp-\left(\frac{\eta P_s}{hfB_T}\right)$	No
Av. no. photons per bit	9	18	20	—

* Values provided assume that the photons arrive over two bit periods.
† Strictly speaking, there is not a heterodyne nonsynchronous demodulation scheme for PSK. Differential PSK (DPSK), however, exhibits a less stringent synchronous detection technique than conventional PSK and is therefore included in the nonsynchronous column for convenience.

limits can also be expressed in terms of the number of received photons per bit by taking $B_T = 2B$ as [Ref. 96]:

$$\left(\frac{S}{N}\right)_{\text{het-lim}} = \frac{\eta P_S}{hf\,B_T} = \eta N_p \tag{12.55}$$

and

$$\left(\frac{S}{N}\right)_{\text{hom-lim}} = \frac{\eta P_S}{hf\,B_T/2} = 2\eta N_p \tag{12.56}$$

At this stage, however, we are concerned with the determination of the average numbers of photons per bit for the various digital modulation schemes listed in Table 12.2. An explanation of how these values are obtained is provided in the following example.

Example 12.3
Calculate the number of received photons per bit in order to maintain a BER of 10^{-9} for:

(a) ASK heterodyne synchronous detection;
(b) ASK heterodyne nonsynchronous detection;
(c) PSK homodyne detection.

An ideal binary receiver may be assumed in all cases.

Solution: (a) Substituting N_p from Eq. (12.54) in Eq. (12.33) gives the probability of error for ASK heterodyne detection as:

$$P(e) = \tfrac{1}{2}\,\text{erfc}\left(\frac{\eta N_p}{4}\right)^{\frac{1}{2}}$$

To maintain a BER of 10^{-9} and with an ideal binary receiver ($\eta = 1$):

$$10^{-9} = \tfrac{1}{2}\,\text{erfc}\left(\frac{N_p}{4}\right)^{\frac{1}{2}}$$

and

$$\left(\frac{N_p}{4}\right)^{\frac{1}{2}} = 4.24$$

Hence:

$$\frac{N_p}{4} \simeq 18 \qquad \text{and} \qquad N_p = 72$$

However, for ASK the 72 photons can arrive over two bit periods, assuming an equal number of ones and zeros. Hence the average number of photons per bit required is 36.

(b) Substituting N_p from Eq. (12.54) into Eq. (12.35) gives:

$$P(e) = \tfrac{1}{2}\exp -\left(\frac{\eta N_p}{4}\right)$$

In this case:

$$\exp -\left(\frac{N_p}{4}\right) = 2 \times 10^{-9}$$

and therefore

$$\frac{N_p}{4} \simeq 20 \qquad \text{and} \qquad N_p = 80$$

Again we are considering ASK modulation so that the average number of received photons per bit is 40. It may be noted that this result is very approximately equal to the number obtained in (a).

(c) For PSK homodyne detection, N_p from Eq. (12.54) may be substituted in Eq. (12.47) to give:

$$P(e) = \tfrac{1}{2}\ \text{erfc}\ (2N_p)^{\frac{1}{2}}$$

Hence

$$(2N_p)^{\frac{1}{2}} = 4.24$$

and

$$N_p \simeq \frac{18}{2} = 9$$

In this case N_p is equal to the average number of photons per bit as 9 photons must be received for zero bit as well as the one bit in order to achieve a BER of 10^{-9}.

As mentioned in Section 12.7.1, the approximation erfc $(u) \simeq \exp - (u)^2$ is used by some authors [Refs. 95, 96] to indicate the rough equivalence between the error probabilities for the synchronous and nonsynchronous detection cases. However, it may be observed from Table 12.2 and Example 12.3 that the nonsynchronous detection in reality requires slightly more photons per bit (slightly higher incoming optical power) than does synchronous detection in the shot noise limit. It should be noted that a more accurate approximation relating the complementary error function to the negative exponential of erfc $(u) \simeq \exp - (u)^2/x\sqrt{\pi}$ for $x > 3$ is used by some authors [Refs. 3, 99].

Nevertheless, aside from these approximations it is possible to write down a generalized expression for the error probabilities of the various modulation formats with synchronous detection schemes recorded in Table 12.2. For fully synchronous heterodyne and homodyne detection the generalized expression therefore takes the form:*

$$P(e) = \tfrac{1}{2}\ \text{erfc}\left(\frac{KZ\eta P_S}{4hf B_T}\right)^{\frac{1}{2}} \tag{12.57}$$

where K is a constant which is equal to 1 for heterodyne detection and 2 for homodyne detection. The constant Z is determined by the modulation scheme as follows: for ASK and FSK, $Z = 1$ and for PSK, $Z = 4$. As with the expressions recorded in Table 12.2, the generalized relationship of Eq. (12.57) can also be used to determine the minimum detectable power level to maintain a particular BER with a specific modulation scheme and synchronous receiver.

* The exception to this is DPSK which, although in reality is synchronous, does not fulfil the necessary continuous phase matching condition.

Example 12.4
Determine the minimum incoming optical power level required to detect a 400 Mbit s^{-1} FSK signal at a BER of 10^{-9} using an ideal heterodyne synchronous receiver operating at a wavelength of 1.55 μm.

Solution: For FSK heterodyne synchronous detection, $K = 1$ and $Z = 1$ in Eq. (12.57); therefore:

$$10^{-9} = \frac{1}{2}\,\mathrm{erfc}\left(\frac{\eta P_S}{2hf\,B_T}\right)^{\frac{1}{2}}$$

Hence for the ideal receiver:

$$\frac{P_S}{2hf\,B_T} = 18$$

and

$$P_S \simeq \frac{36hc\,B_T}{\lambda} = \frac{36 \times 6.63 \times 10^{-34} \times 3 \times 10^8 \times 400 \times 10^6}{1.55 \times 10^{-6}}$$

$$= 1.8 \text{ nW or } -57.4 \text{ dBm}$$

Therefore, the minimum incoming peak power level to maintain a BER of 10^{-9} is -57.4 dBm. Although this level may appear at first sight rather high, it must be remembered that the bit rate is also relatively high at 400 Mbit s^{-1}, a factor which directly contributes to the lowering of the sensitivity for the receiver.

The relative sensitivities of the different modulation formats and detection schemes can be readily deduced from consideration of the average numbers of photons per bit indicated for each detection technique in Table 12.2. However, a more immediate comparison of the relative receiver performances may be obtained from Figure 12.24 [Ref. 4]. It may be observed that the sensitivity improvement of ASK heterodyne coherent detection over IM/DD is 10 to 25 dB in Figure 12.24, whereas in Table 12.2 direct detection would appear to be around 6 dB more sensitive than ASK heterodyne detection. This second statement is true in theory but in practice practical direct detection receivers require in the order of 400 to 4000 photons per bit to maintain a BER of 10^{-9} [Ref. 23]. By contrast, practical ASK heterodyne coherent receivers are far more likely to be operated near the quantum or shot noise limit and thus provide the sensitivity indicated in Table 12.2.

To further illustrate the point concerning the relative receiver sensitivities of IM/DD and coherent systems, ASK heterodyne detection using both synchronous and nonsynchronous detection is compared with direct detection in the theoretical characteristics shown in Figure 12.25 [Ref. 95]. The substantial improvement in receiver sensitivity offered by the coherent detection schemes may be observed even when compared with direct detection with very low photodetector dark current (I_d). Moreover, the reduction in receiver sensitivity with increases in the

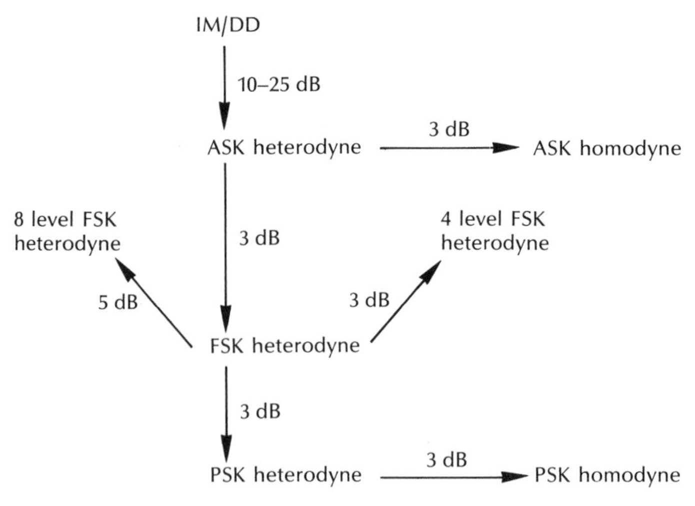

Figure 12.24 Receiver sensitivity improvements using various coherent modulation and demodulation schemes.

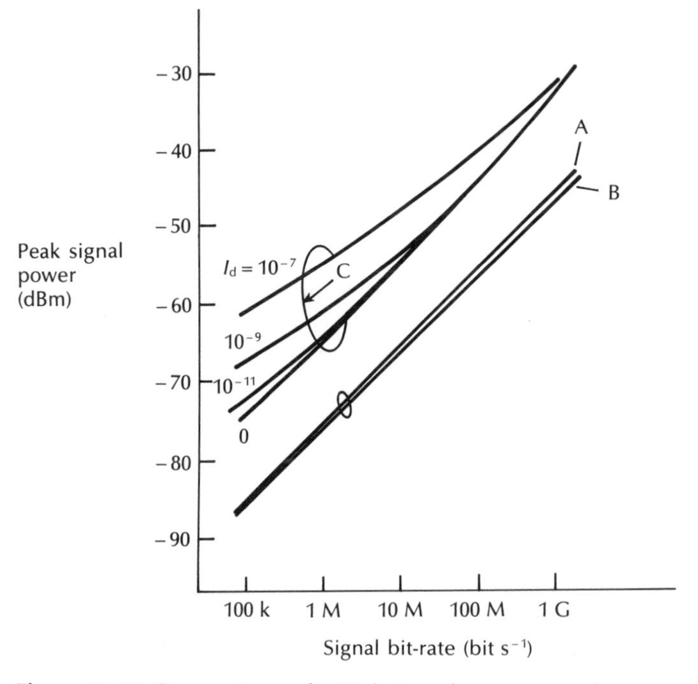

Figure 12.25 Comparison of ASK heterodyne nonsynchronous and synchronous receiver sensitivities with IM/DD for a BER of 10^{-9} using an APD with excess noise factor of 1.0, curve A; ASK heterodyne nonsynchronous detection, curve B: ASK heterodyne synchronous detection; curves C; IM/DD, in which the photodetector dark current (I_d) is a parameter. Reproduced with permission from T. Okoshi, K. Emura, K. K. Kikuchi and R. Th. Kersten, *J. Opt. Commun.*, **2**, p. 89, 1981 [Ref. 95].

transmission bit rate can be noted, as mentioned in Example 12.3. The aforementioned factors are of particular interest when we consider the potential or ultimate repeater spacings that may be afforded by coherent optical fiber transmission [Ref. 20].

Example 12.5

Calculate the absolute maximum repeater spacing that could be provided to maintain a BER of 10^{-9} within a coherent optical fiber system operating at a wavelength of 1.55 μm when the fiber and splice/connector losses average out at 0.2 dB km^{-1}, the optical power launched into the fiber link is 2.5 mW and the transmission rates are 50 Mbit s^{-1} and 1 Gbit s^{-1}. For both bit rates consider the following ideal receiver types:

(a) ASK heterodyne synchronous detection;
(b) PSK homodyne detection.

Solution: (a) Considering the 50 Mbit s^{-1} transmission rate and using the result obtained for ideal ASK heterodyne synchronous detection in Example 12.3 (average photons per bit required 36) from Eq. (12.54)

$$N_{\mathrm{p}} = \frac{P_{\mathrm{S}}}{hf\,B_{\mathrm{T}}} = \frac{P_{\mathrm{S}}}{hf\,B_{\mathrm{T}}} \simeq 36$$

Hence

$$P_{\mathrm{S}} \simeq 36hc\,B_{\mathrm{T}} = \frac{36 \times 6.63 \times 10^{-34} \times 3 \times 10^{8} \times 50 \times 10^{6}}{1.55 \times 10^{-6}}$$

$$= 0.23 \text{ nW or } -66.4 \text{ dBm}$$

The maximum system margin with no overheads is therefore:

$$\textit{Max. system margin} = 4 \text{ dBm} - (-66.4) \text{ dBm} = 70.4 \text{ dB}$$

Moreover, the absolute maximum repeater spacing is:

$$\textit{Max. repeater spacing} = \frac{70.4}{0.2} = 352 \text{ km}$$

For 1 Gbit s^{-1} we have:

$$P_{\mathrm{S}} \simeq \frac{36 \times 6.63 \times 10^{-34} \times 3 \times 10^{8} \times 10^{9}}{1.55 \times 10^{-6}}$$

$$= 4.6 \text{ nW or } -53.4 \text{ dBm}$$

Therefore the maximum system margin is 57.4 dB and

$$\textit{Max. repeater spacing} = \frac{57.4}{0.2} = 287 \text{ km}$$

Again using the result for PSK homodyne detection obtained in example 12.3 and considering first the 50 Mbit s^{-1} rate, then from Eq. (12.54):

$$N_p = \frac{P_S}{hf\,B_T} = \frac{P_S}{hf\,B_T} \simeq 9$$

and

$$P_S \simeq {}^9hc\,B_T = \frac{9 \times 6.63 \times 10^{-34} \times 3 \times 10^8 \times 50 \times 10^6}{1.55 \times 10^{-6}}$$

$$= 58 \text{ pW or } 72.4 \text{ dBm}$$

The maximum system margin is now 76.4 dB so that,

$$\textit{Max. repeater spacing} = \frac{76.4}{0.2} = 382 \text{ km}$$

For 1 Gbit s^{-1}:

$$P_S \simeq \frac{9 \times 6.63 \times 10^{-34} \times 3 \times 10^8 \times 10^9}{1.55 \times 10^{-6}}$$

$$= 1.15 \text{ nW or } -59.4 \text{ dBm}$$

Hence the maximum system margin is 63.4 dB and

$$\textit{Max. repeater spacing} = \frac{63.4}{0.2} = 317 \text{ km}$$

Thus the range of repeater spacings indicated by this example is between 287 and 392 km.

The relative receiver sensitivities of two multilevel FSK modulation schemes are also indicated in Figure 12.24. Hence, as anticipated, these four and eight level FSK heterodyne detection schemes display a reduction in sensitivity of 3 dB and 5 dB, respectively, over binary FSK heterodyne detection. Receiver sensitivity characteristics for other M'ary modulation schemes are shown in Figure 12.26(a) [Ref. 68]. The receiver sensitivities for binary PSK (BPSK), quaternary or four level PSK (QPSK) and sixteen level quadrature amplitude modulation (QAM) against their information transmission capacities are displayed. Furthermore, these characteristics are a function of the receiver thermal noise parameters which are also shown in Figure 12.26(a). It may be observed that at higher transmission rates the receiver sensitivities are degraded by increases in thermal noise, as demonstrated by the corresponding characteristics given in Figure 12.26(b) [Ref. 68].

However, QPSK may provide an M'ary modulation format which will enable the achievement of high capacity coherent optical transmission with no sensitivity penalty in comparison with binary PSK. It may be noted that in Figure 12.26(a) the

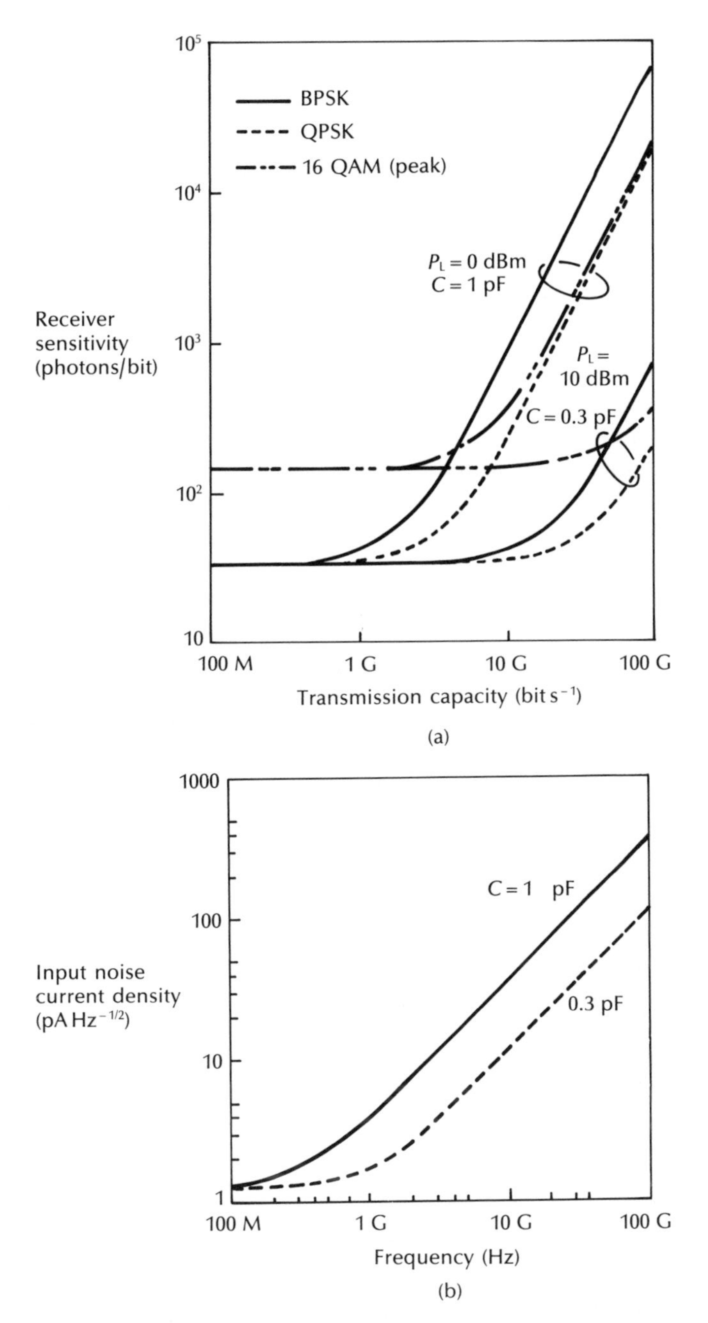

Figure 12.26 (a) Receiver sensitivities for M'ary modulation schemes; (b) Equivalent input noise current density of preamplifier used to determine characteristics provided in (a) where C is the total input capacitance, the transconductance is 50 mS and the numerical factor is 1.1. Reproduced with permission from K. Nosu and K. Iwashita, 'A consideration of factors affecting future coherent lightwave communication systems', *J. Lightwave Technol.*, **6**, p. 686, 1988. Copyright © 1988 IEEE.

sensitivity of QPSK is at least equivalent to (or marginally better than) that of BPSK. Moreover, a recent feasibility study concerned with coherent optical QPSK transmission also supports this observation [Ref. 99]. With higher numbers of levels using PSK or QAM, however, the receiver sensitivity degrades and hence narrower linewidth lasers will be required.

12.8 Single and multicarrier systems

Since the initial measurements of bit error rate on experimental coherent optical fiber systems were reported in 1981, many experimental systems have been developed in the laboratory with research workers' concern shifting from heterodyne ASK to the more sophisticated FSK, CPSK, PSK and DPSK heterodyne demodulation schemes [Refs. 4, 18, 44, 45, 101]. In addition, more recently, interest has grown in ASK using phase diversity reception [Ref. 86] as well as ASK heterodyne detection employing polarization diversity receivers [Ref. 40]. Various of these experimental systems were outlined in Sections 12.4.2.2, 12.5 and 12.6, where their performance characteristics were also indicated. Therefore, in this section we consider the next stage of these developments which is the deployment of coherent optical fiber systems within the telecommunication network. This is approached through a description of the first demonstration of a coherent optical transmission system in an operational network, followed by a brief discussion of potential developments in multicarrier coherent fiber optical system and network strategies.

12.8.1 DPSK field demonstration system

The first demonstration of the deployment of a coherent optical fiber communication system in an operational environment was reported in October 1988 [Ref. 102]. The system configuration for this DPSK demonstrator system is provided in Figure 12.27. Miniaturized LEC lasers were employed as the optical sources at the transmitter and local oscillator, each of which provided a spectral linewidth less than 100 kHz and a launch power of 0 dBm [Ref. 26]. Mechanical adjustment also allowed the output wavelength to be preset with a range of around 50 nm with continuous wavelength tuning over a range of 50 GHz provided by electronic control about the preset wavelength. For the demonstration the operational wavelength was set to 1.534 μm.

A lithium niobate phase modulator was used to apply DPSK modulation at a rate of 565 Mbit s^{-1} to the transmitter laser output. The optical isolator shown in Figure 12.27 was a fiber-coupled 30 dB commercial device [Ref. 103] which when inserted between the LEC laser and the modulator proved sufficient for the suppression of reflection induced linewidth broadening as LEC lasers are relatively insensitive to reflected optical power. The output from the modulator was boosted by a semiconductor laser optical amplifier prior to launching into the installed conventional single-mode fiber cable. This amplifier was introduced to overcome the loss through

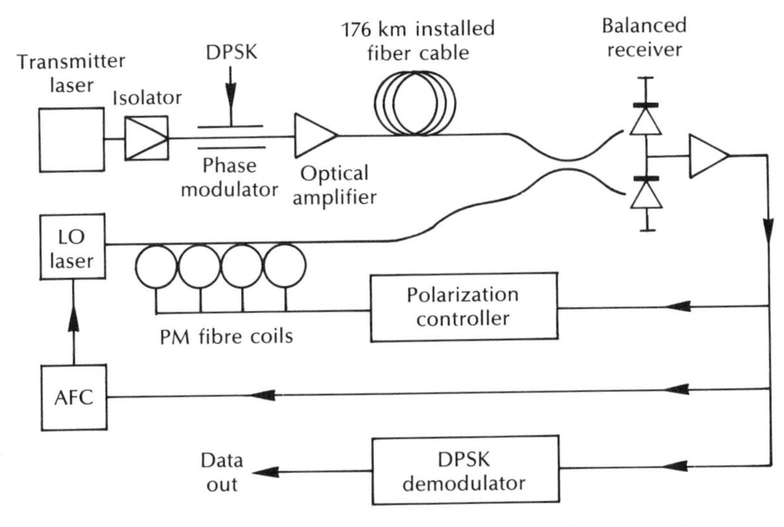

Figure 12.27 Configuration for the DPSK field demonstration system [Ref. 102].

the optical isolator and the modulator, and therefore to increase the launch power from −10 to +1 dBm when transmission over 176 km of installed fiber was required.

At the receive terminal an automated polarization control system was applied to the local oscillator laser output. The device consisted of four transducers formed by winding polarization maintaining fiber onto piezoelectric cylinders, and provided a limitless range of polarization control as described in Section 12.4.2.1. A dual detector balanced optical input was provided at the receiver to minimize the thermal noise penalty. It was also designed to give a low return loss to the local oscillator laser and therefore remove the requirement for an optical isolator. Automatic frequency control applied to the piezoelectric movements in the grating mount of the local oscillator laser allowed the IF between the received carrier and the local oscillator to be stabilized at 847 ± 10 MHz (i.e. $1.5 \times$ bit rate). Finally, the transmitted data were recovered using a delay line demodulator with IF bandpass filtering.

The system which was installed using an eighteen fiber cable in an underground duct between Cambridge and Bedford in the United Kingdom demonstrated a long term measured BER of 5×10^{-9} with a receiver sensitivity of −47.6 dBm (276 photons per bit) over the 176 km route. Error-free operation with no error floor at BER levels above 10^{-13} was also observed using the same system arrangement but operating over 150 km of installed fiber cable. Moreover, subsequent improvements to the system included a hybrid integrated balanced receiver using a GaAs–FET IC preamplifier to improve the sensitivity to −52 dBm as well as the incorporation of an injection laser pumped erbium doped fiber amplifier repeater [Ref. 104].

12.8.2 Multicarrier system and network concepts

It was mentioned in Section 12.1 that a major attribute of coherent optical transmission was its ability to provide wavelength/frequency selectivity with narrow channel spacings for future multicarrier systems and networks. Wavelength division multiplexing technology and techniques have been discussed in relation to IM/DD optical fiber systems (see Sections 5.6.3 and 11.8.3) but it is apparent that far more optical carriers could be employed using coherent optical receivers which may be tuned to specific incoming carrier signals. For such coherent systems, because the channel widths are very narrow (often of the order of the data transmission rate) in comparison with conventional WDM, then the channel spacings are measured in frequency rather than wavelength units. Consequently, multicarrier coherent optical systems and networks are often referred to as frequency division multiplexed (FDM) [Refs. 6, 7, 105 to 108].

It is clear that coherent optical FDM systems provide a powerful strategy for the utilization of the enormous optical bandwidth potential of fibers (over 50 000 GHz between 1.3 and 1.6 μm) whilst avoiding the potential bottleneck created by the speed of the electronics within single carrier systems. Although 10 Gbit s^{-1} may finally be achieved within single channel installed systems, moving to higher rates with electronic TDM looks increasingly doubtful [Ref. 1]. However, far greater capacity could be achieved at more modest transmission rates using coherent optical FDM techniques. For example, even at spacings of 10 GHz several thousand frequency/wavelength channels could be accommodated over the 1.3 to 1.6 μm wavelength band.

At present a favoured technique within coherent multicarrier systems is to use a passive star coupler to distribute or broadcast the optical signals over the network [Refs. 105 to 107]. A block schematic for such an FDM coherent star network is illustrated in Figure 12.28. All the optical carriers (shown as λ_i, λ_j, λ_k, etc.) are generated and modulated individually for transmission over the network. At the network output, tunable, highly selective optical heterodyne receiver local oscillators provide multiple channel access. Alternatively, fixed wavelengths/frequencies could be assigned to the receivers and tunable lasers could be employed at the transmitters. The former strategy has been demonstrated for a small number of channels [Refs. 7, 107]. In the latter demonstration four FSK modulated channels operating at 200 Mbit s^{-1} and spaced 2.45 GHz apart were combined in a 16×16 passive star coupler. Demultiplexing was achieved using a heterodyne FM receiver with an IF of 850 MHz. Moreover, the minimum received optical power required to maintain a BER of 10^{-9} was -55.5 dBm (109 photons per bit) and the degradation due to adjacent channel interference was found to be negligible.

Considering an $N \times N$ coherent passive star network of the form shown in Figure 12.28. The minimum power required at each receiver P_S can be obtained from Eq. (12.55) or Eq. (12.56) as:

$$P_S \geqslant N_p \, hf \, B_T \qquad (12.58)$$

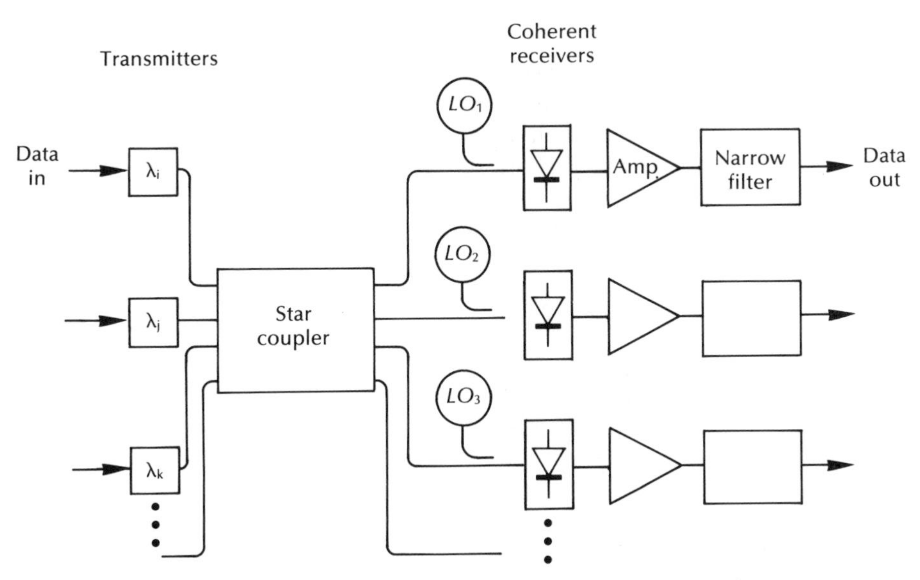

Figure 12.28 Frequency division multiplexed coherent star network.

Then taking account of the star coupler distribution loss and assuming no excess loss through the device, the power required to be launched from each transmitter P_{tx} is:

$$P_{tx} \geqslant N_p \, hf \, B_T \, N \tag{12.59}$$

where N is the number of ports on the coupler.

The relationship given in Eq. (12.59) may be seen to be correct because the division of power on the network is inversely related to N (see Section 5.6.2), as is the portion of the total fiber optical bandwidth available to each terminal. However, since for an optical FDM network the total information throughput capacity $N \times B_T$ is limited by the fiber bandwidth B_{fib} then the transmission capacity (bit rate) available to each terminal is:

$$B_T \leqslant \frac{B_{fib}}{N} \tag{12.60}$$

Substituting Eq. (12.60) into (12.59) gives:

$$P_{tx} \geqslant N_p \, hf \, B_{fib} \tag{12.61}$$

This is an interesting result which indicates that the transmitter power requirement is independent of the number of terminals on the network [Ref. 44].

Example 12.6

Estimate the minimum transmitter power requirement for an optical coherent FDM passive star network using heterodyne synchronous receivers which need an average of 150 photons per bit for reception at the desired BER. It may be assumed that the network is operating from a shortest wavelength of 1.3 μm with an optical bandwidth of 20 THz.

Solution: The minimum required transmitter power P_{tx} may be obtained directly from Eq. (12.61) as the worst case occurs at the shortest wavelength. Hence:

$$P_{tx} \simeq N_p hf B_{fib} = \frac{N_p hc B_{fib}}{\lambda}$$

$$= \frac{150 \times 6.63 \times 10^{-34} \times 3 \times 10^8 \times 20 \times 10^{12}}{1.3 \times 10^{-6}}$$

$$= 0.5 \text{ mW or } -3 \text{ dBm}$$

The result obtained in Example 12.6 indicates that relatively modest transmitter power will, in principle, facilitate the operation of optical coherent FDM star networks with arbitrarily large numbers of terminals. It must be emphasized, however, that significant losses on the network (coupler excess loss, fiber losses, etc.) have not been taken into account and that the restriction on the number of channels will generally be substantially greater than that dictated by Eq. (12.60) in which the channel bandwidth requirement was taken to be approximately equal to the transmission bit rate. In many cases the modulated spectrum can be much wider than this amount (e.g. for wide deviation FSK) and, in addition, guard bands are necessary between channels.

A coherent optical FDM experimental system for cable television (CATV) distribution has recently been demonstrated [Ref. 108]. The experimental ten channel system, which again operated as a broadcast network using a 16×16 passive star coupler, is shown in Figure 12.29. The ten optical FDM transmitters, each incorporating wavelength tunable DBR lasers (see Section 6.6.2), are located on the left of Figure 12.29, together with a channel spacing controller. This latter device provided a channel space locking system based on a reference pulse technique in order to stabilize the channel spacings and therefore suppress the crosstalk penalties [Ref. 110]. The DBR lasers were FSK modulated using a 400 Mbit s^{-1} NRZ format for the TV signal.* The transmitted optical signals were then multiplexed through the star coupler.

At the receive terminals (block on the right of Figure 12.29) balanced polarization diversity receivers were employed with random access channel selection circuits.

* It was assumed that this signal would be high definition TV (HDTV) with its higher bandwidth requirement than conventional TV.

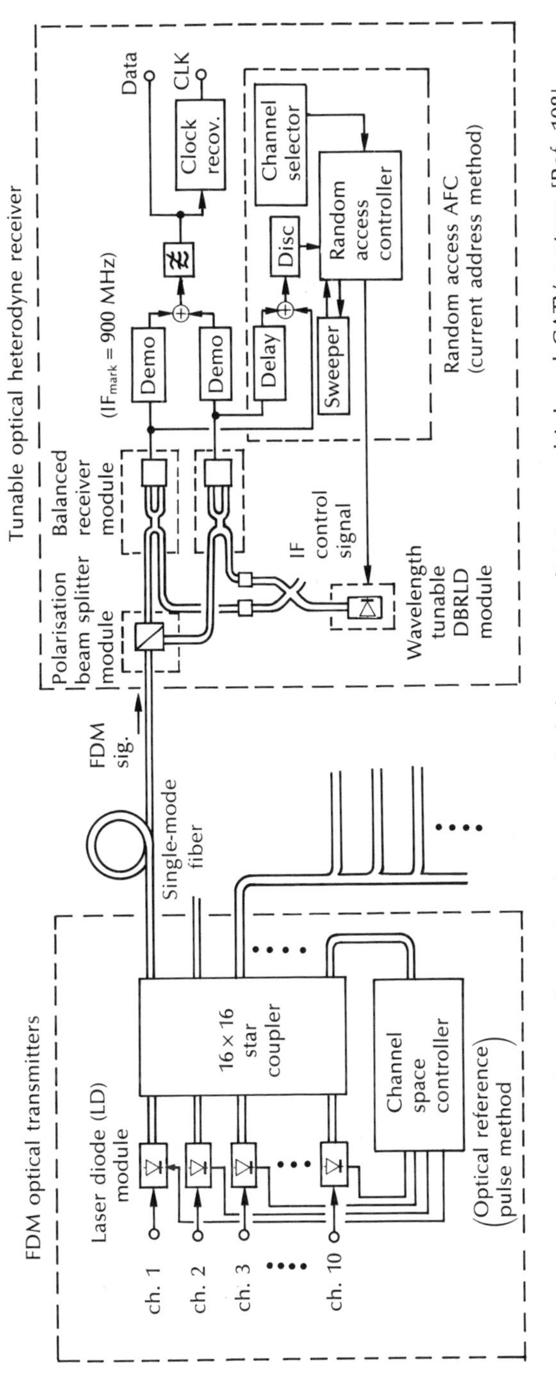

Figure 12.29 Experimental ten channel coherent optical frequency division multiplexed CATV system [Ref. 108].

Hence the receiver could select a particular channel through the control of the frequency of a local oscillator laser which again was a wavelength tunable DBR laser of the same design as the transmitter laser. A single filter detection scheme was used to demodulate the FSK signals which necessitated a large modulation index of between 3 and 4. Although a minimum theoretical channel spacing of 5 GHz was required, the developed system utilized an 8 GHz channel spacing. Receiver sensitivities of -45 dBm at 400 Mbit s^{-1} were obtained for all ten channels. Furthermore, the feasibility of the use of a travelling wave optical amplifier together with a wideband tunable local oscillator laser indicated the potential to distribute over eighty HDTV channels to some 16 000 subscriber terminals [Ref. 108].

Furthermore, coherent optical wavelength/frequency division switching systems are also under investigation for incorporation in future coherent optical networks [Ref. 111]. Such a switching system has been demonstrated for two channels which were wavelength synchronized using 8 GHz spaced optical FSK signals transmitted at a rate of 180 Mbit s^{-1}. The wavelength switch comprised coherent balanced optical receivers using FSK single filter detection connected to wavelength/frequency tunable DBR lasers which provided the wavelength converted output signals. Similar tunable devices were employed for the local oscillator lasers within the coherent receivers and the tuning ranges indicated that thirty-two wavelength channels could be obtained [Ref. 111].

Finally, both single wavelength and multiwavelength coherent carriers have been transmitted through optical amplifiers (see Sections 10.2 to 10.4). In the former case semiconductor laser amplifiers [Ref 112, 113] and erbium doped fiber amplifiers [Ref. 114] have been cascaded within experimental coherent optical fiber systems. However, fiber amplifier devices may prove more successful for use with coherent transmission systems as they have been shown to be polarization insensitive [Ref. 115].

Multicarrier coherent transmission through an erbium doped fiber amplifier has also been demonstrated in the field. For example, British Telecom have reported [Ref. 116] on the operation of a two channel coherent WDM system with 10 GHz channel spacing incorporating an erbium doped fiber amplifier repeater. This system was operated over a distance of 200 km between Edinburgh and Newcastle in the United Kingdom at a rate of 622 Mbit s^{-1} on each channel. Therefore the use of coherent multicarrier systems to provide for very long-haul transmission also appears likely in the future.

Problems

12.1 (a) Outline the rationale behind the pursuit of coherent optical systems. Indicate the major problems encountered in the realization of coherent optical transmission and briefly describe the ways in which they have been overcome.

(b) Discuss, with the aid of a suitable block diagram, a coherent optical fiber communication system. Comment on the differing system requirements to facilitate heterodyne detection in comparison with homodyne detection.

12.2 The frequency stability requirement for a local oscillator laser in an ASK optical heterodyne detection system is 10 MHz. When the laser is emitting at a centre frequency of 1.55 μm and exhibits an output frequency change with temperature of 14 GHz $^{\circ}C^{-1}$, estimate:

(a) the fractional stability necessary for the device;
(b) the maximum temperature change that could be permitted for the device when there is no other form of laser frequency control;
(c) the maximum transmission bandwidth that would be allowed by the laser frequency stability.

12.3 (a) Obtain from first principles the theoretical SNR improvement in the shot noise limit for optical homodyne detection over heterodyne detection. Indicate the primary reason for this improvement.
(b) Briefly discuss the strategies which have been adopted to provide optical homodyne detection.

12.4 A homodyne OOK receiver has a bandwidth of 250 MHz and utilizes a photodiode with a responsivity of 0.6 AW^{-1} at the operating wavelength. The device is shot noise limited and a received SNR of 11 dB is required to provide an acceptable BER. Compute the receiver sensitivity and the photocurrent obtained when the local oscillator laser output power is -3 dBm and the phase difference between this signal and the incoming one is 12°.

12.5 The incoming signal power to an ASK optical heterodyne receiver operating at its shot noise limit is 1.28 nW for a received SNR of 9 dB. Determine the transmission wavelength of the ASK system if the quantum efficiency of the photodetector is 75% at this wavelength and the transmission bandwidth is 400 MHz.

12.6 Outline the major practical constraints associated with coherent optical transmission and discuss the techniques which have been adopted to overcome them.

12.7 To allow nonsynchronous ASK heterodyne detection the linewidths of the signal and local oscillator lasers should be less than 50% of the transmitted bit rate. Estimate the maximum permitted linewidths in nanometres for such ASK system sources:

(a) emitting at a wavelength around 1.30 μm when the transmission rate is 140 Mbit s^{-1};
(b) emitting at a wavelength around 1.55 μm when the transmission rate is 2.4 Gbit s^{-1}.

12.8 Compare and contrast the attributes and drawbacks associated with direct modulation of the laser signal source and indirect modulation of the source in both ASK and FSK coherent optical fiber communication systems.

12.9 (a) Describe what is understood by continuous phase shift keying (CPFSK) modulation within coherent optical transmission. Indicate the benefits of this modulation technique in comparison with FSK.
(b) Discuss the advantages and suggest a drawback associated with coherent optical differential phase shift keying (DPSK) in comparison with synchronous PSK heterodyne detection.

12.10 A coherent PSK optical fiber communication system employing synchronous heterodyne detection requires a minimum input optical power level of -58.2 dBm in order to receive with a BER of 10^{-9}. The system is operated at a transmission rate of 600 Mbit s^{-1} and the quantum efficiency of the receiver photodetector is 80%. Assuming shot noise limited operation at the receiver, obtain the transmission wavelength for the system.

12.11 Outline the major techniques employed to achieve nonsynchronous optical ASK and FSK heterodyne detection. Indicate the benefits of these schemes over the corresponding synchronous demodulation schemes.

12.12 Describe what is meant by phase diversity reception for coherent optical fiber communication systems. Discuss with the aid of a suitable block diagram the salient features of the in-phase and quadrature receiver when used for optical ASK demodulation.

12.13 Verify that in order to obtain a BER of 10^{-9}:

(a) an average of eighteen photons per bit is required within an ideal ASK homodyne detection system;

(b) an average of forty photons per bit is necessary for ideal nonsynchronous FSK heterodyne detection.

12.14 Determine the minimum detectable peak optical power levels for both of the detection schemes in problem 12.13 when the transmission wavelength and bit rate are 1.31 μm and 100 Mbit s^{-1} respectively.

12.15 A coherent DPSK system operating at a wavelength of 1.54 μm uses a photodetector with a quantum efficiency of 83% at this wavelength. In shot noise limited performance a BER of 0.94×10^{-12} is obtained at the coherent optical receiver for a minimum detectable optical power level of 2.1 nW. Calculate both the average number of photons per bit required to maintain the BER and the transmission bit rate of the system under these circumstances.

12.16 An OOK coherent optical fiber system using nonsynchronous heterodyne detection has a transmission wavelength of 1.55 μm. Estimate the number of photons required for a one bit to provide a BER of 10^{-10} when there is shot noise limited detection and the responsivity of the system photodetector at the operating wavelength is 0.7.

12.17 An FSK coherent optical fiber system employing synchronous heterodyne detection has a transmission wavelength of 1.3 μm where the fiber and splice/connector losses average out at 0.4 dB km^{-1}. If 2 mW of optical power is launched into the system link and an ideal photodetector is assumed, determine the absolute maximum repeater spacing to maintain a BER of 10^{-9} at transmission rates of: (a) 140 Mbit s^{-1}; (b) 2.4 Gbit s^{-1}.

12.18 A DPSK coherent optical fiber system operating at a transmission wavelength and bit rate of 1.55 μm and 250 Mbit s^{-1} respectively, has a repeater spacing of 300 km. Assuming a launch power of 0 dBm with shot noise limited detection and average overall transmission losses of 0.2 dB km^{-1} at the operation wavelength, compute the minimum quantum efficiency required for the photodetector to enable the system to function with a BER of 10^{-10}.

12.19 Indicate what is understood by FDM in relation to coherent optical transmission. Describe, with the aid of a suitable diagram, the possible implementation of a coherent multicarrier distribution system based on a passive star coupler.

Estimate the number of photons per bit obtained with an optical coherent FDM passive star network, which is operating from the shortest wavelength of 1.50 μm with an optical bandwidth of 100 nm, when the transmitter powers are 0 dBm. Comment on the result.

12.20 The two spectral transmission regions for coherent multicarrier systems may be considered to be 1.27 μm to 1.35 μm and 1.48 μm to 1.60 μm. Determine the number of FDM channels that could be accommodated in each region when coherent optical PSK transmission at 2 Gbit s^{-1} is to be utilized on each channel. A 25% guard band frequency for filter roll-off should be assumed.

Answers to numerical problems

12.2 (a) 1.93 in 10^7
(b) 7×10^{-4} °C
(c) 50 MHz
12.4 -60.8 dBm; 0.76 μA
12.5 1.32 μm
12.7 (a) 4×10^{-4} nm
(b) 1×10^{-2} nm
12.10 1.57 μm
12.14 (a) 273 pW
(b) 607 pW
12.15 27, 500 MHz
12.16 164
12.17 (a) 160.3 km
(b) 129.5 km
12.18 71.4%
12.19 603
12.20 5590; 6068

References

[1] J. C. Campbell, A. G. Dentai, W. S. Holden and B. L. Casper, 'High performance avalanche photodiode with separate absorption, grading and multiplication regions', *Electron. Lett.*, **19**, pp. 818–819, 1983.
[2] J. Saltz, 'Modulation and detection for coherent lightwave communications', *IEEE Commun. Mag.*, **24**(6), pp. 38–49, 1986.
[3] M. Schwatz, *Information Transmission, Modulation and Noise:* 4th edn, McGraw-Hill, 1990.
[4] T. Okoshi, 'Recent advances in coherent optical fiber communication systems, *J. of Lightwave Technol.*, **LT-5**(1), pp. 44–51, 1987.
[5] T. G. Hodgkinson, D. W. Smith, R. Wyatt and D. J. Malyon, 'Coherent optical fiber transmission systems', *Br. Telecom Technol. J.*, **3**(3), pp. 5–18, 1985.
[6] B. S. Glance, J. Stone, K. J. Pollock, P. J. Fitzgerald, C. A. Burrus Jr, B. L. Kasper and L. W. Stulz, 'Densely spaced FDM coherent star network with optical signals confined to equally spaced frequencies', *J. of Lightwave Technol.*, **6**(11), pp. 1170–1181, 1988.
[7] K. Nosu, H. Toba and K. Iwashita, 'Optical FDM transmission technique', *J. of Lightwave Technol.*, **LT-5**(9), pp. 1301–1308, 1987.
[8] M. C. Brain, 'Coherent optical networks', *Br. Telecom Technol. J.*, **7**(1), pp. 50–57, 1989.
[9] F. E. Goodwin, 'A 3.39 μm infrared optical heterodyne communication system', *IEEE J. Quantum Electron.*, **QE-3**(11), pp. 524–531, 1967.
[10] M. C. Teich, 'Homodyne detection of infrared radiation from a moving diffuse target', *Proc. IEEE*, **57**(5), pp. 789–792, 1969.
[11] O. E. DeLange, 'Wide-band optical communication systems: part II – frequency division multiplexing', *Proc. IRE*, **58**(10), pp. 1683–1690, 1970.
[12] S. Machida, A. Kawana, K. Ishihara and H. Tsuchiya, 'Interference of a AlGaAs

laser diode using 4.15 km single-mode fiber cable', *IEEE J. Quantum Electron.*, **QE-15**(7), pp. 535–537, 1979.
[13] T. Kimura and Y. Yamamoto, 'Progress of coherent optical fibre communication systems', *Opt. Quantum Electron.*, **15**, pp. 1–39, 1983.
[14] F. Favre, L. Jeunhomme, I. Joindot, M. Monerie and J. C. Simon, 'Progress towards heterodyne-type single-mode fibre communication systems', *IEEE J. Quantum Electron.*, **QE-17**(6), pp. 897–905, 1981.
[15] S. Saito, Y. Yamamoto and T. Kimura, 'Optical heterodyne detection of directly frequency modulated semiconductor laser signals', *Electron. Lett.*, **16**, pp. 826–827, 1980.
[16] D. W. Smith, R. A. Harmon and T. G. Hodgkinson, 'Polarisation stability requirements for coherent optical fibre transmission systems', *Br. Telecom Technol. J.*, **1**(2), pp. 12–16, 1983.
[17] I. W. Stanley, 'A tutorial review of techniques for coherent optical fibre transmission systems', *IEEE Commun. Mag.*, **23**(8), pp. 37–53, 1985.
[18] T. Kimura, 'Coherent optical fiber transmission', *J. of Lightwave Technol.*, **LT-5**(4), pp. 414–428, 1987.
[19] D. W. Smith, 'Coherent fiberoptic communications', *Laser Focus*, pp. 92–106, November 1985.
[20] T. Okoshi, 'Ultimate performance of heterodyne/coherent optical fiber communications', *J. of Lightwave Technol.*, **LT-4**(10), pp. 1556–1562, 1986.
[21] R. Lang, 'Injection locking properties of a semiconductor laser', *IEEE J. Quantum Electron.*, **QE-18**(6), pp. 976–983, 1982.
[22] R. Wyatt, D. W. Smith and K. H. Cameron, 'Megahertz linewidth from a 1.5 μm semiconductor laser with HeNe laser injection', *Electron. Lett.*, **18**, pp. 292–293, 1982.
[23] E. Basch and T. Brown, 'Introduction to coherent fiber-optic communication', in E. E. Basch (Ed.), *Optical-Fiber Transmission*, H. W. Sams & Co., pp. 503–542, 1987.
[24] F. G. Walker and J. E. Kaufman, 'Characterization of GaAlAs laser diode frequency noise', *Sixth Top. Mtg. Opt. Fib. Commun.*, New Orleans, USA, paper TUJ5, February 1983.
[25] M. Osinski and J. Buus, 'Linewidth broadening factor in semiconductor lasers – an overview', *IEEE J. Quantum Electron.*, **QE-23**(1), pp. 9–29, 1987.
[26] J. Mellor, S. Al-Chalabi, K. H. Cameron, R. Wyatt, J. C. Regnault, V. W. Devlin and M. C. Brain, 'Performance characteristics of miniature external cavity semiconductor lasers', *Proc. CLEO'89*, Baltimore, USA, paper FP1, April 1989.
[27] T. Myogadani, S. Tanaka and Y. Suetsugu, Polarization fluctuation in single mode fiber cables', *IOOC–ECOC'85*, **1**, pp. 151–154, 1985.
[28] R. Ulrich, 'Polarization stabilization on single-mode fiber', *Appl. Phys. Lett.*, **35**(11), pp. 840–842, 1979.
[29] M. Kubota, T. Oohara, K. Furuya and Y. Suematsu, 'Electrooptical polarization control on single-mode fibres', *Electron. Lett.*, **16**(15), p. 573, 1980.
[30] H. C. Lefevre, 'Single-mode fiber fractional wave devices and polarization controllers', *Electron. Lett.*, **16**(20), pp. 778–780, 1980.
[31] T. Imai, K. Nosu and H. Yamaguchi, 'Optical polarization control utilising an optical heterodyne detection scheme', *Electron. Lett.*, **21**(2), pp. 52–53, 1985.

[32] T. Okoshi, N. Fukaya and K. Kikuchi, 'A new polarization-state control device: rotatable fibre cranks', *Electron. Lett.*, **21**(20), pp. 895–896, 1985.

[33] T. Okoshi, Y. Cheng and K. Kikuchi, 'A new polarization-control scheme for optical heterodyne receivers', *Electron. Lett.*, **21**(18), pp. 787–788, 1985.

[34] N. G. Walker and G. R. Walker, 'Polarization control for coherent optical fibre systems', *Br. Telecom Technol. J.*, **5**(2), pp. 63–76, 1987.

[35] M. J. Creaner, R. C. Steele, G. R. Walker and N. G. Walker, '565 Mbit/s optical PSK transmission system with endless polarization control', *Electron. Lett.*, **24**(5), pp. 270–271, 1988.

[36] N. G. Walker and G. R. Walker, 'Polarization control for coherent communications', *J. of Lightwave Technol.*, **8**(3), pp. 438–458, 1990.

[37] N. G. Walker, G. R. Walker and J. Davidson, 'Endless polarization control using an integrated optic lithium niobate device', *Electron. Lett.*, **24**(5), pp. 266–268, 1988.

[38] T. Okoshi, S. Ryu and K. Kikuchi, 'Polarization diversity receiver for heterodyne/coherent optical-fiber communications', *4th Internat. Conf. Integrated Opt. and Opt. Fiber Commun.*, Tokyo, Japan, paper 30C3-2, June 1983.

[39] T. E. Darcie, B. Glance, K. Gayliard, J. R. Talman, B. L. Kasper and C. A. Burrus, 'Polarization-insensitive operation of coherent FSK transmission system using polarisation diversity', *Electron Lett.*, **23**(25), pp. 1382–1384, 1987.

[40] T. G. Hodgkinson, R. A. Harmon and D. W. Smith, 'Performance comparison of ASK polarization diversity and standard coherent heterodyne receivers', *Electron. Lett.*, **24**(1), pp. 58–59, 1988.

[41] A. D. Kersey, M. J. Marrone and A. Dandridge, 'Adaptive polarisation diversity receiver configuration for coherent optical fibre communications', *Electron. Lett.*, **25**(4), pp. 275–277, 1989.

[42] I. Garrett and G. Jacobsen, 'Theoretical analysis of heterodyne optical receivers for transmission systems using (semiconductor) lasers with non-negligible linewidth', *J. of Lightwave Technol.*, **LT-4**(3), pp. 323–334, 1986.

[43] T. G. Hodgkinson, R. A. Harmon and D. W. Smith, 'Polarisation insensitive heterodyne detection using polarisation scrambling', *Electron. Lett.*, **23**(10), pp. 513–514, 1987.

[44] R. A. Linke and A. H. Gnauck, 'High capacity coherent lightwave systems', *J. of Lightwave Technol.*, **6**(11), pp. 1750–1769, 1988.

[45] D. W. Smith, 'Techniques for multigigabit coherent optical transmission', *J. of Lightwave Technol.*, **LT-5**(10), pp. 1466–1478, 1987.

[46] T. G. Hodgkinson, R. Wyatt and D. W. Smith, 'Experimental assessment of a 140 Mbit/s coherent optical receiver at 1.52 microns', *Electron. Lett.*, **18**, pp. 523–525, 1982.

[47] J. L. Gimlett, 'Low noise 8 GHz *p–i–n*/FET optical receiver', *Electron. Lett.*, **23**, pp. 281–283, 1987.

[48] S. B. Alexander, 'Design of wide-band optical heterodyne balanced mixer receivers', *J. of Lightwave Technol.*, **LT-5**(4), pp. 523–537, 1987.

[49] H. P. Yuen and V. W. S. Chan, 'Noise in homodyne and heterodyne detection', *Opt. Lett.*, **8**(3), pp. 177–179, 1983.

[50] G. L. Abbas, V. W. Chan and T. K. Yee, 'Dual detector optical heterodyne receiver for local oscillator noise suppression', *J. of Lightwave Technol.*, **LT-3**(5), pp. 1110–1122, 1985.

[51] B. L. Kasper, C. A. Burrus, J. R. Talman and K. L. Hall, 'Balanced dual detector receiver for optical heterodyne communication at Gbit/s rates', *Electron. Lett.*, **22**, pp. 413–414, 1986.

[52] A. H. Gnauck, R. A. Linke, B. L. Kasper, K. J. Pollock, K. C. Reichmann, R. Valenzula and R. C. Alferness, 'Coherent lightwave transmission at 2 Gbit/s over 170 km of optical fibre using phase modulation', *Electron. Lett.*, **23**, pp. 286–287, 1987.

[53] A. F. Elrefaie, R. E. Wagner, D. A. Atlas and D. G. Daut, 'Chromatic dispersion limitations in coherent lightwave transmission systems', *J. of Lightwave Technol.*, **6**(5), pp. 704–709, 1988.

[54] K. Iwashita and N. Takachio, 'Chromatic dispersion compensation in coherent optical communications', *J. of Lightwave Technol.*, **8**(3), pp. 367–375, 1990.

[55] A. R. Chraplyvy, 'Limitations on lightwave communications imposed by optical fiber nonlinearities', *Opt. Fiber Commun. Conf. (OFC'88)*, New Orleans, USA, paper TUD3, January 1988.

[56] I. Garrett and G. Jacobson, 'The effect of laser linewidth on coherent optical receivers', *J. of Lightwave Technol.*, **LT-5**(4), pp. 551–560, 1987.

[57] K. Emura, S. Yamazaki, S. Fujita, M. Shikada, I. Mito and K. Minemura, 'Over 300 km transmission experiment on an optical FSK heterodyne dual filter detection system', *Electron. Lett.*, **22**(21), pp. 1096–1097, 1986.

[58] K. Emura, S. Yamazaki, M. Shikada, S. Fujita, M. Yamaguchi, I. Mito and K. Minemura, 'System design and long-span transmission experiments on an optical FSK heterodyne single filter detection system', *J. of Lightwave Technol.*, **LT-5**(4), pp. 469–477, 1987.

[59] R. S. Vodhanel, '1 Gbit/s bipolar optical FSK transmission experiment over 121 km of fibre', *Electron. Lett.*, **24**(3), pp. 163–164, 1988.

[60] R. C. Steele and M. Creaner, '565 Mbit/s AMI FSK coherent system using commercial DFB lasers', *Electron. Lett.*, **25**(11), pp. 732–734, 1989.

[61] K. Iwashita and N. Takachio, '2 Gbit/s optical CPFSK heterodyne transmission through 200 km single-mode fibre', *Electron. Lett.*, **23**(7), pp. 341–342, 1987.

[62] L. G. Kazovsky and G. Jacobsen, 'Multichannel CPFSK coherent optical communications system', *J. of Lightwave Technol.*, **7**(6), pp. 972–982, 1989.

[63] L. L. Jeromin and V. W. S. Chan, 'M'ary FSK performance for coherent optical communication systems using semiconductor lasers', *IEEE Trans. Commun.*, **COM-34**(4), pp. 375–381, 1986.

[64] S. Kobayashi and T. Kimura, 'Optical phase modulation in an injection locked AlGaAs semiconductor laser', *IEEE J. Quantum Electron.*, **QE-18**(10), pp. 1662–1669, 1982.

[65] B. Glance, 'Performance of homodyne detection of binary PSK optical signals', *J. of Lightwave Technol.*, **LT-4**(2), pp. 228–235, 1986.

[66] S. Yamazaki, S Murata, K. Komatsu, Y. Koizumi, S. Fujita and K. Emura, '1.2 Gbit/s optical DPSK heterodyne detection transmission system using monolithic external-cavity DFB LDs', *Electron. Lett.*, **23**(16), pp. 860–862, 1987.

[67] T. Chikama, T. Naitou, H. Onaka, T. Kiyonaga, S. Watanabe, M. Suyama, M. Seino and H. Kuwahara, '1.2 Gbit/s, 201 km optical DPSK heterodyne transmission experiment using a compact, stable external fibre DFB laser module', *Electron. Lett.*, **24**(10), pp. 636–637, 1988.

[68] K. Nosu and K. Iwashita, 'A consideration of factors affecting future coherent

lightwave communication systems', *J. of Lightwave Technol.*, **6**(5), pp. 686–694, 1988.

[69] E. Dietrich, B. Enning, R. Gross and H. Knupke, 'Heterodyne transmission of a 560 Mbit/s optical signal by means of polarization shift keying', *Electron. Lett.*, **23**(8), pp. 421–422, 1987.

[70] R. Calvani, R. Caponi and R. Cisternino, 'Polarization phase-shift keying: a coherent transmission technique with differential heterodyne detection', *Electron. Lett.*, **24**(10), pp. 642–643, 1988.

[71] Y. Imai, K. Iizuka and R. T. B. James, 'Phase-noise-free coherent optical communication system utilizing differential polarization shift keying (DPolSK), *J. of Lightwave Technol.*, **8**(5), pp. 691–698, 1990.

[72] S. Betti, F. Curti, G. De Marchis and E. Iannone, 'Multilevel coherent optical system based on Stokes parameters modulation', *IEEE J. of Lightwave Technol.*, **8**(7), pp. 1127–1136, 1990.

[73] P. Benedelta and P. Poggliolini, 'Performance evaluation of multilevel polarization shift keying modulation schemes, *Electron. Lett.*, **26**(4), pp. 244–246, 1990.

[74] S. Saito, Y. Yamamoto and T. Kimura, 'S/N and error rate evaluation for an optical FSK–heterodyne detection system using semiconductor lasers', *IEEE J. Quantum Electron.*, **QE-19**(2), pp. 180–193, 1983.

[75] Y. Yamamoto, 'AM and FM quantum noise in semiconductor lasers – Part I and II', *IEEE J Quantum Electron.*, **QE-19**(1), pp. 34–58, 1983.

[76] T. G. Hodgkinson, 'Costas loop analysis for coherent optical receivers', *Electron. Lett.*, **22**(7), pp. 394–396, 1986.

[77] S. Watanabe, T. Chikama, T. Naito and H. Kuwahara, '560 Mb/s optical PSK heterodyne detection using carrier recovery', *Electron. Lett.*, **25**, pp. 588–590, 1989.

[78] T. Chirkawa, S. Watanabe, T. Naito, H. Onaka, T. Kiyonaga, Y. Onoda, H. Miyata, M. Suyama, M. Seino and H. Kuwahara, 'Modulation and demodulation techniques in optical heterodyne PSK transmission systems', *J. of Lightwave Technol.*, **8**(3), pp. 309–322, 1990.

[79] J. M. P. Delavaux, L. D. Tzeng, M. Dixon and R. E. Tench, '1.4 Gbit/s optical DPSK heterodyne transmission system experiment', *Fourteenth European Conf. on Opt. Commun., (ECOC'88)*, UK, pp. 475–477, September 1988.

[80] D. J. Malyon, D. W. Smith and R. Wyatt, 'Semiconductor laser homodyne optical phase lock loop', *Electron. Lett.*, **22**, pp. 421–422, 1986.

[81] H. K. Phillip, A. L. Scholtz, E. Bonek and W. Leeb, 'Costas loop experiments for a 10.6 μm communications receiver', *IEEE Trans. Commun.*, **COM-31**(8), pp. 1000–1002, 1983.

[82] L. Kazovsky, L. Curtis, W. Young and N. Cheung, 'All fiber 90° optical hybrid for coherent communications', *Appl. Opt.*, **26**(3), pp. 437–439, 1987.

[83] J. A. Arnaud, 'Enhancement of optical receiver sensitivities by amplification of the carrier', *IEEE J. Quantum Electron.*, **QE-4**, pp. 893–899, 1968.

[84] C. G. Atkins, D. Cotter, D. W. Smith and R. Wyatt, 'Application of Brillouin amplification in coherent optical transmission', *Electron. Lett.*, **22**, pp. 556–558, 1986.

[85] N. G. Walker and J. E. Carroll, 'Simultaneous phase and amplitude measurements on optical signals using a multiport junction', *Electron. Lett.*, **20**, pp. 981–983, 1984.

[86] A. Davis, M. Pettitt, J. King and S. Wright, 'Phase diversity techniques for coherent optical receivers', *J. of Lightwave Technol.*, **LT-5**(4), pp. 561–572, 1987.

[87] L. Kazovsky, P. Meissner and E. Patzak, 'ASK multiport optical homodyne receiver', *J. of Lightwave Technol.*, **LT-5**(6), pp. 770–791, 1987.

[88] K. Emura, R. S. Vodhanel, R. Welter and W. B. Sessa, '5 Gbit/s optical phase diversity homodyne detection experiment', *Electron. Lett.*, **25**(6), pp. 400–401, 1989.

[89] R. Welter and L. G. Kazovsky, '150 Mbit s^{-1} phase diversity ASK homodyne receiver with a DFB laser', *Optical Fiber Commun. Conf. (OFC'88)*, New Orleans, USA, paper TU1, January 1988.

[90] T. G. Hodgkinson, R. A. Harmon and D. W. Smith, 'Demodulation of optical DPSK using in-phase and quadrature detection', *Electron. Lett.*, **21**, pp. 867–868, 1985.

[91] A. W. Davis, S. Wright, M. J. Pettitt, J. P. King and K. Richards, 'Coherent optical receiver for 680 Mbit/s using phase diversity', *Electron. Lett.*, **22**, pp. 9–11, 1986.

[92] M. J. Pettitt, D. Remodios, A. W. Davies, A. Hadjifotiou and S. Wright, 'A coherent transmission system using DFB lasers and phase diversity', *Proc. IEE Colloq.* (UK), pp. 9/1–9/5, 1987.

[93] E. Gottwald and J. Pietzsch, 'Measurement method for determination of optical phase shifts in 3 × 3 fibre couplers', *Electron. Lett.*, **24**, pp. 265–266, 1988.

[94] J. Siuzdak and W. van Etten, 'BER evaluation for phase and polarization diversity optical homodyne receivers using non-coherent ASK and DPSK demodulation', *J. of Lightwave Technol.*, **7**(4), pp. 584–599, 1989.

[95] T. Okoshi, K. Emura, K. Kikuchi and R. Th. Kersten, 'Computation of bit-error rate of various heterodyne and coherent-type optical communication schemes', *J. of Optical Commun.*, **2**(3), pp. 89–96, 1981.

[96] T. Okoshi and K. Kikuchi, *Coherent Optical Fiber Communications*, KTK Scientific Publishers (Tokyo)/Kluwer Academic Publishers (Dordrecht), 1988.

[97] L. G. Kazovsky, 'Optical heterodyning versus optical homodyning: A comparison', *J. of Optical Commun.*, **6**(1), pp. 18–24, 1985.

[98] E. Basch and T. Brown, 'Introduction to coherent optical fiber transmission', *IEEE Commun. Mag.*, **23**(5), pp. 23–30, 1985.

[99] P. S. Henry, R. A. Linke and A. H. Gnauck, 'Introduction to lightwave systems', in S. E. Miller and I. P. Kaminow (Eds.) *Optical Fiber Telecommunications II*, Academic Press, pp. 781–831, 1988.

[100] S. Yamazaki, T. Fujta and K. Emura, 'Feasibility study on optical QPSK heterodyne systems', *Opt. Fiber Commun. Conf. (OFC'88)*, New Orleans, USA, paper WC3, January 1988.

[101] R. A. Linke, 'Optical heterodyne communication systems' *IEEE Commun. Mag.*, pp. 36–41, October 1989.

[102] M. J. Creaner, R. C. Steele, I. Marshall, G. R. Walker, N. G. Walker, J. Mellis, S. Al-Chalabi, I. Sturgess, M. Rutherford, J. Davidson and M. Brain, 'Field demonstration of 565 Mbit/s DPSK coherent transmission system over 176 km of installed fiber', *Electron. Lett.*, **24**(22), pp. 1354–1356, 1988.

[103] J. Mellis, S. Al-Chalabi and L. N. Barker, 'Characteristics of fiber coupled optical isolators at 1.5μm', *Electron. Lett.*, **24**(22), pp. 1353–1354, 1988.

[104] M. C. Brain, M. J. Creaner, R. C. Steele, N. G. Walker, G. R. Walker, J. Mellis, S. Al-Chalabi, J. Davidson, M. Rutherford and I. C. Sturgess, 'Progress towards the field deployment of coherent optical fiber systems', *J. of Lightwave Technol.*, **8**(3), pp. 423–437, 1990.

[105] E. J. Bachus, R. P. Braun, C. Casper, H. M. Foisel, E. Grobmann, B. Strebel and F. J. Westphal, 'Coherent optical multicarrier systems', *J. of Lightwave Technol.*, **7**(2), pp. 375–384, 1989.

[106] R. A. Linke, 'Frequency division multiplexed optical networks using heterodyne detection', *IEEE Network*, pp. 13–20, March 1989.

[107] B. Glance, T. L. Koch, O. Scaramucci, K. C. Reichmann, U. Koren and C. A. Burrus, 'Densely spaced FDM coherent optical star network using monolithic widely frequency-tunable lasers', *Electron. Lett.*, **25**(10), pp. 672–673, 1989.

[108] S. Yamazaki, M. Shibutani, N. Shimoska, S. Murata, T. Ono, M. Kitamura, K. Emura and M. Shikada, 'A coherent optical FDM CATV distribution system', *J. of Lightwave Technol.*, **8**(3), pp. 396–405, 1990.

[109] P. Cochrane, 'Future directions in long haul fibre optic systems', *Br. Telecom Technol. J.*, **8**(2), pp. 5–17, 1990.

[110] N. Shimosaka, K. Kaede and S. Murata, 'Frequency locking of FDM optical sources using widely tunable DBR LDs', *Opt. Fiber Commun. Conf. (OFC'88)*, New Orleans, USA, paper THG3, January 1988.

[111] M. Fujiwara, N. Shimosaka, M. Nishio, S. Suzuki, S. Yamazaki, S. Murata and K. Kaede, 'A coherent photonic wavelength-division switching system for broad-band networks', *J. of Lightwave Technol.*, **8**(3), pp. 416–422, 1990.

[112] N. A. Olsson, M. G. Oberg, L. A. Koszi and G. Przybylek, '400 Mbit/s, 372 km coherent transmission experiment using in-line optical amplifier', *Electron. Lett.*, **24**(1), pp. 36–38, 1988.

[113] D. J. Maylon, R. C. Steele, M. J. Creaner, M. C. Brain and W. A. Stallard, 'Coherent optical transmission at 565 Mbit/s, through five cascaded photonic amplifiers', *Electron. Lett.*, **25**(5), pp. 354–356, 1989.

[114] M. J. Creaner, T. J. Whitley, R. C. Steele, R. A. Garnham, C. A. Millar and M. C. Brain, 'Diode-pumped erbium-doped fibre amplifier repeater in a 565 Mbit/s DPSK coherent optical transmission system', *Proc. European Conf. in Opt. Commun., ECOC*, Gothenburg (Sweden), pp. 82–85, September 1989.

[115] N. H. Taylor and A. Hadjifotiou, 'Optical amplification and its applications,' *Proc. SPIE., Int. Soc. Opt. Eng. (USA),* **1314**, 'Fibre Optics' 90, pp. 64–67, 1990.

[116] M. J. Creaner, D. Spirit, G. R. Walker, N. G. Walker, J. Mellis, S. A. Chalabi, W. Hale, I. Sturgess, M. Rutherford, D. Trivett, M. C. Brain and R. C. Steele, 'Field demonstration of two channel coherent transmission with a diode-pumped fibre amplifier repeater,' *IEEE Opt. Fib. Commun. Conf., OFC'90*, San Francisco, USA, postdealine paper, January 1990.

13

Optical fiber measurements

13.1 Introduction

In this chapter we are primarily concerned with measurements on optical fibers which characterize the fiber. These may be split into three main areas:

(a) transmission characteristics;
(b) geometrical and optical characteristics;
(c) mechanical characteristics.

Data in these three areas are usually provided by the optical fiber manufacturer with regard to specific fibers. Hence fiber measurements are generally performed in the laboratory and techniques have been developed accordingly. This information is essential for the optical communication system designer in order that suitable choices of fibers, materials and devices may be made with regard to the system application. However, although the system designer and system user do not usually need to take fundamental measurements of the fiber characteristics there is a requirement for field measurements in order to evaluate overall system performance, and for functions such as fault location. Therefore, we also include some discussion of field measurements which take into account the effects of cabled fiber, splice and connector losses, etc.

Several organizations have become involved in standardization issues relating to optical fiber measurements. The International Telephone and Telegraph Consultative Committee, the CCITT (the acronym for its name in French) has made a few recommendations, particularly regarding single-mode fiber measurements [Ref. 1], and in the the United States the Electronics Industries Association (EIA) has published over 200 Fiber Optic Test Procedures (FOTPs). Hence there now exists a body of standardized measurement techniques. Such test methods are divided into reference test methods (RTMs) and alternative test methods (ATMs). An RTM provides a measurement of a particular characteristic, strictly according to the definition which usually gives the highest degree of accuracy and reproducibility, whereas an ATM may be more suitable for practical use but can deviate from the strict definition; however, there must be a way to relate such results to those obtained from the reference test method.

The fiber transmission characteristics of greatest interest are those of attenuation and dispersion. For multimode fibers the latter parameter enables the bandwidth to be determined, whereas with single-mode fibers it is the intramodal or chromatic dispersion which is generally provided by manufacturers. Furthermore, the important geometrical and optical characteristics for multimode fibers are size (core and cladding diameters), numerical aperture (see Section 2.2.3) and refractive index profile, but for single-mode fibers they are the effective cutoff wavelength of the second order mode (see Section 2.5.1) and the mode-field diameter (see Section 2.5.2). Measurements of the mechanical characteristics such as tensile strength and durability were outlined in Section 4.7.1 and are therefore pursued no further in this chapter.

When attention is focused on the measurement of the transmission properties of multimode fibers, problems emerge regarding the large number of modes propagating in the fiber. The various modes show individual differences with regard to attenuation and dispersion within the fiber. Moreover, mode coupling occurs giving transfer of energy from one mode to another (see Section 2.4.2). The mode coupling which is associated with perturbations in the fiber composition or geometry, and external factors such as microbends or splices, is for instance responsible for the increased attenuation (due to radiation) of the higher order modes. These multimode propagation effects mean that both the fiber loss and bandwidth are not uniquely defined parameters but depend upon the fiber excitation conditions and environmental factors such as cabling, bending, etc. Also, these transmission parameters may vary along the fiber length (i.e. they are not necessarily linear functions) due to the multimode propagation effects, making extrapolation of measured data to different fiber lengths less than meaningful.

It is therefore important that transmission measurements on multimode fibers are performed in order to minimize these uncertainties. In the laboratory, measurements are usually taken on continuous lengths of uncabled fiber in order to reduce the influence of external factors on the readings (this applies to both multimode and single-mode fibers). However, this does mean that the system designer must be aware of the possible deterioration in the fiber transmission characteristics within the installed system. The multimode propagation effects associated with fiber perturbations may be accounted for by allowing or encouraging the mode distribution to reach a steady state (equilibrium) distribution. This distribution occurs automatically after propagation has taken place over a certain fiber length (coupling length) depending upon the strength of the mode coupling within the particular fiber. At equilibrium the mode distribution propagates unchanged and hence the fiber attenuation and dispersion assume well defined values. These values of the transmission characteristics are considered especially appropriate for the interpretation of measurements to long-haul links and do not depend on particular launch conditions.

The equilibrium mode distribution may be achieved by launching the optical signal through a long (dummy) fiber to the fiber under test. This technique has been used to good effect [Ref. 2] but may require a kilometre of dummy fiber and is therefore not suitable for dispersion measurements. Alternatively there are a number of methods of simulating the equilibrium mode distribution with a much shorter length of fiber. Mode equilibrium may be achieved using an optical source with a mode output which corresponds to the steady state mode distribution of the fiber under test. This technique may be realized experimentally using an optical arrangement which allows the numerical aperture of the launched beam to be varied (using diaphragms) as well as the spot size of the source (using pinholes). In this case the input light beam is given an angular width which is equal to the equilibrium distribution numerical aperture of the fiber and the source spot size on the fiber input face is matched to the optical power distribution in a cross section of the fiber at equilibrium.

Other techniques involve the application of strong mechanical perturbations on a short section of the fiber in order quickly to induce mode coupling and hence equilibrium mode distribution within 1 m. These devices which simulate mode equilibrium over a short length of fiber are known as mode scramblers or mode filters.

A mode scrambler increases the power in the higher order modes relative to the lower order ones, whereas a mode filter typically reduces the power in the higher order modes without seriously affecting the power in the lower order modes. Mode filters are usually required with LED sources because more higher order mode excitation than the equilibrium mode distribution is obtained in most multimode fibers. By contrast, a mode scrambler may be necessary for lens-less excitation with laser sources, as this technique produces more lower order modes than are contained in the equilibrium mode distribution in most multimode fibers. Hence both strategies are employed to achieve equilibrium mode power distributions in multimode fibers.

A simple mode scrambling method [Ref. 3] is to sandwich the fiber between two sheets of abrasive paper (i.e. sandpaper) placed on wooden blocks in order to provide a suitable pressure. Two slightly more sophisticated techniques are illustrated in Figure 13.1 [Refs. 4 and 5]. Figure 13.1(a) shows mechanical perturbations induced by enclosing the fiber with metal wires and applying pressure by use of a surrounding heat shrinkable tube. A method which allows adjustment and therefore an improved probability of repeatable results is shown in Figure 13.1(b). This technique involves inserting the fiber between a row of equally spaced pins, subjecting it to sinusoidal bends. Hence the variables are the number of pins giving the number of periods, the pin diameter d and the pin spacing s.

A common mode filtering technique uses a mandrel wrap applied before the test fiber as illustrated in Figure 13.2(a). In this method four or five turns of fiber are wrapped around a 20 to 30 mm diameter mandrel in order to simulate the equilibrium mode power distribution [Ref. 6]. The other popular mode filtering technique which was mentioned earlier is shown in Figure 13.2(b). This method employs a dummy fiber of length 0.5 to 1 km which is of a similar type to the test fiber. The dummy fiber is spliced before the test fiber such that an equilibrium mode distribution is established after the optical launch.

In order to test that a particular mode scrambler or filter gives an equilibrium mode distribution within the test fiber, it is necessary to check the insensitivity of the far field radiation pattern (this is related to the mode distribution, see Section 2.4.1) from the fiber with regard to changes in the launch conditions. It is also useful to compare the far field patterns from the device and a separate long length

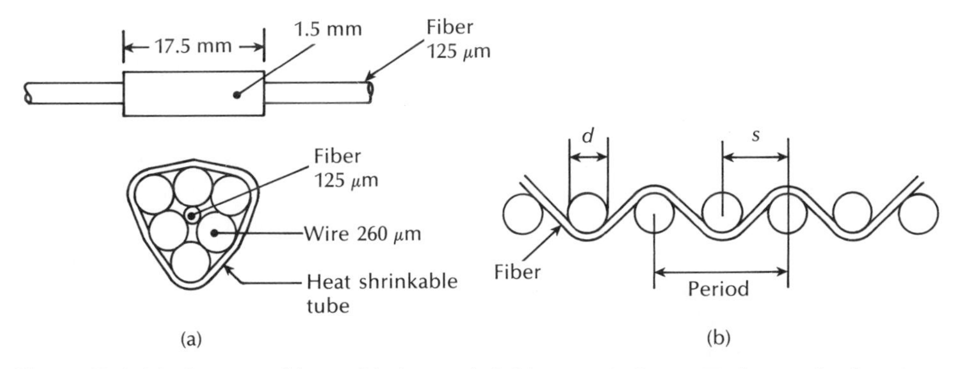

Figure 13.1 Mode scramblers: (a) heat shrinking technique [Ref. 4]; (b) bending technique [Ref. 5].

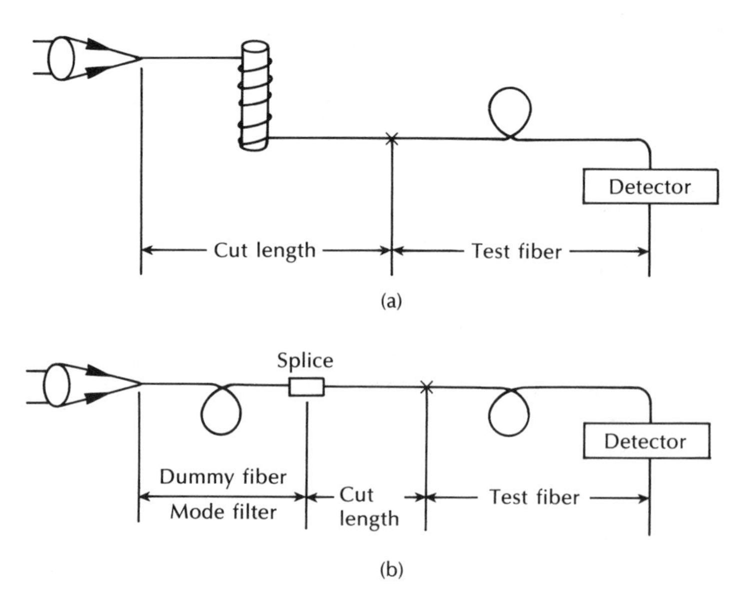

Figure 13.2 Equilibrium mode simulation by mode filtering: (a) mandrel wrap; (b) dummy fiber.

at the test fiber for coincidence [Ref. 2]. However, it must be noted that, at present, mode scramblers or filters tend to give only an approximate equilibrium mode distribution and their effects vary with different fiber types. Hence measurements involving the use of different mode scrambling methods can be subject to discrepancies. Nevertheless, the majority of laboratory measurement techniques to ascertain the transmission characteristics of multimode optical fibers use some form of equilibrium mode simulation in order to give values representative of long transmission lines. Moreover, the current standards agreements regarding equilibrium mode simulation are outlined in FOTP-50 [Ref. 7].

We commence the discussion of optical fiber measurements in Section 13.2 by dealing with the major techniques employed in the measurement of fiber attenuation. These techniques include measurement of both total fiber attenuation and the attenuation resulting from individual mechanisms within the fiber (e.g. material absorption, scattering). In Section 13.3 fiber dispersion measurements in both the time and frequency domains are discussed. Various techniques for the measurement of the fiber refractive index profile are then considered in Section 13.4. The measurement of the fiber cutoff wavelength which has particular relevance to single-mode fibers is then dealt with in Section 13.5. In Section 13.6 we discuss simple methods for measuring fiber numerical aperture. Measurement of the fiber outer and core diameters are then described in Section 13.7.

A far more important parameter than the fiber numerical aperture and core diameter for single-mode fibers is the mode-field diameter. Hence the measurement of this single-mode fiber characteristic is discussed in Section 13.8. This is followed in Section 13.9 with a description of a measurement procedure for reflectance and optical return loss for either a fiber component or an optical link. Finally, field measurements which may be performed on optical fiber links, together with examples of measurement instruments, are discussed in Section 13.10. Particular attention is paid in this concluding section to optical time domain reflectometry (OTDR).

13.2 Fiber attenuation measurements

Fiber attenuation measurement techniques have been developed in order to determine the total fiber attenuation of the relative contributions to this total from both absorption losses and scattering losses. The overall fiber attenuation is of greatest interest to the system designer, but the relative magnitude of the different loss mechanisms is important in the development and fabrication of low loss fibers. Measurement techniques to obtain the total fiber attenuation give either the spectral loss characteristic (see Figure 3.3) or the loss at a single wavelength (spot measurement).

13.2.1 Total fiber attenuation

A commonly used technique for determining the total fiber attenuation per unit length is the cut-back or differential method. Figure 13.3 shows a schematic diagram of the typical experimental set-up for measurement of the spectral loss to obtain the overall attenuation spectrum for the fiber. It consists of a 'white' light source, usually a tungsten halogen or xenon arc lamp. The focused light is mechanically chopped at a low frequency of a few hundred hertz. This enables the lock-in amplifier at the receiver to perform phase-sensitive detection. The chopped light is then fed through a monochromator which utilizes a prism or diffraction grating arrangement to select the required wavelength at which the attenuation is to be measured. Hence the light is filtered before being focused on to the fiber by means of a microscope objective lens. A beam splitter may be incorporated before the fiber to provide light for viewing optics and a reference signal used to compensate for output power fluctuations. As indicated in Section 5.1, when the measurement is performed on multimode fibers it is very dependent on the optical launch conditions. Therefore unless the launch optics are arranged to give the steady state mode distribution at the fiber input, or a dummy fiber is used, then a mode scrambling device is attached to the fiber within the first metre. The fiber is also usually put through a cladding mode stripper, which may consist of an S-shaped groove cut in the Teflon and filled with glycerine. This device removes light launched into the fiber cladding through radiation into the index matched (or slightly higher refractive index) glycerine. A mode stripper can also be included at the fiber output end to remove any optical power which is scattered from the core

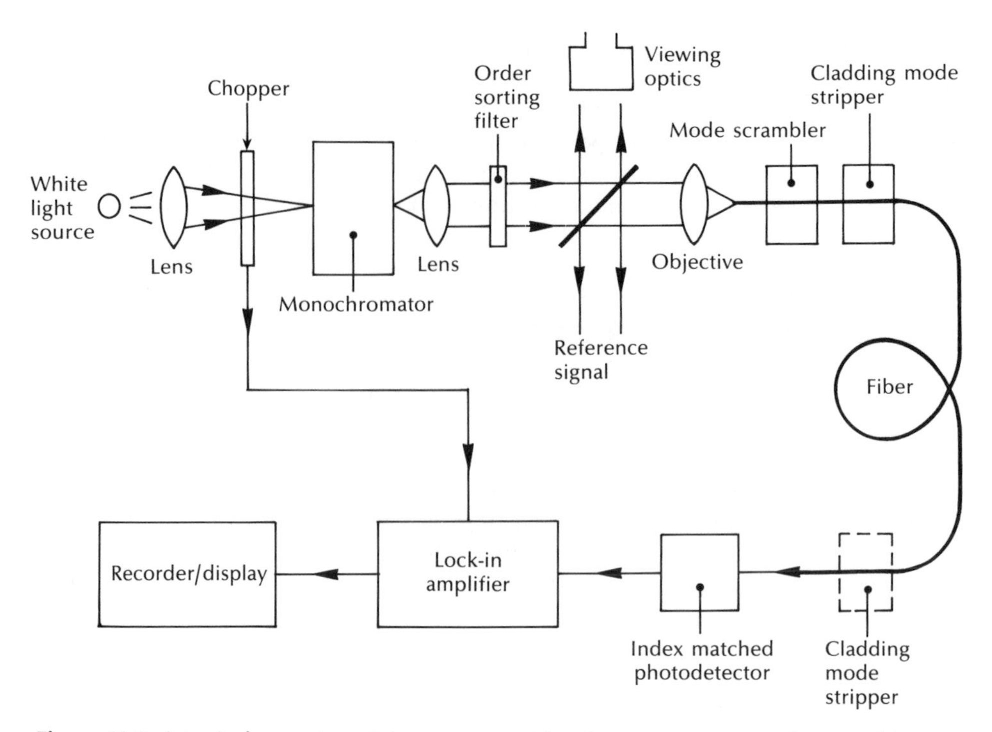

Figure 13.3 A typical experimental arrangement for the measurement of spectral loss in optical fibers using the cut-back technique.

into the cladding down the fiber length. This tends to be pronounced when the fiber cladding consists of a low refractive index silicone resin.

The optical power at the receiving end of the fiber is detected using a *p–i–n* or avalanche photodiode. In order to obtain reproducible results the photodetector surface is usually index matched to the fiber output end face using epoxy resin or an index matching cell [Ref. 8]. Finally, the electrical output from the photodetector is fed to a lock-in amplifier, the output of which is recorded.

The cut-back method* involves taking a set of optical output power measurements over the required spectrum using a long length of fiber (usually at least a kilometre). This fiber is generally uncabled having only a primary protective coating. Increased losses due to cabling (see Section 4.8.1) do not tend to change the shape of the attenuation spectrum as they are entirely radiative, and for multimode fibers are almost wavelength independent. The fiber is then cut back to a point two metres from the input end and, maintaining the same launch

* The cut-back method is outlined in FOTP-46 and FOTP-78 for multimode and single-mode fibers respectively [Refs. 9, 10]. In addition, it is the CCITT reference test method for fiber attenuation [Ref. 1].

conditions, another set of power output measurements are taken. The following relationship for the optical attenuation per unit length α_{dB} for the fiber may be obtained from Eq. (3.3):

$$\alpha_{dB} = \frac{10}{L_1 - L_2} \log_{10} \frac{P_{02}}{P_{01}} \tag{13.1}$$

L_1 and L_2 are the original and cut-back fiber lengths respectively, and P_{01} and P_{02} are the corresponding output optical powers at a specific wavelength from the original and cut-back fiber lengths. Hence when L_1 and L_2 are measured in kilometres, α_{dB} has units of dB km^{-1}.

Furthermore Eq. (13.1) may be written in the form:

$$\alpha_{dB} = \frac{10}{L_1 - L_2} \log_{10} \frac{V_2}{V_1} \tag{13.2}$$

where V_1 and V_2 correspond to output voltage readings from the original fiber length and the cut-back fiber length respectively. The electrical voltages V_1 and V_2 may be directly substituted for the optical powers P_{01} and P_{02} of Eq. (13.1) as they are directly proportional to these optical powers (see Section 7.4.3). The accuracy of the results obtained for α_{dB} using this method is largely dependent on constant optical launch conditions and the achievement of the equilibrium mode distribution within the fiber. In this case only the fiber to detector power coupling changes between measurements and this variation can be made less than 0.01 dB [Ref. 6]. Hence the cut-back technique is regarded as the reference test method (RTM) for attenuation measurements by the EIA as well as the CCITT.

Example 13.1

A 2 km length of multimode fiber is attached to apparatus for spectral loss measurement. The measured output voltage from the photoreceiver using the full 2 km fiber length is 2.1 V at a wavelength of 0.85 μm. When the fiber is then cut back to leave a 2 m length the output voltage increases to 10.7 V. Determine the attenuation per kilometre for the fiber at a wavelength of 0.85 μm and estimate the accuracy of the result.

Solution: The attenuation per kilometre may be obtained from Eq. (13.2) where:

$$\alpha_{dB} = \frac{10}{L_1 - L_2} \log_{10} \frac{V_2}{V_1} = \frac{10}{1.998} \log_{10} \frac{10.7}{2.1}$$

$$= 3.5 \text{ dB km}^{-1}$$

The dynamic range of the measurements that may be taken depends upon the exact configuration of the apparatus utilized, the optical wavelength and the fiber core diameter. However, a typical dynamic range is in the region 30 to 40 dB when using a white light source at a wavelength of 0.85 μm and multimode fiber with a core diameter around 50 μm. This may be increased to around 60 dB by use of a laser source operating at the same wavelength. It must be noted that a laser source is only suitable for making a single wavelength (spot) measurement as it does not emit across a broad band of spectral wavelengths.

Spot measurements may be performed on an experimental set-up similar to that shown in Figure 13.3. However, interference filters are frequently used instead of the monochromator in order to obtain a measurement at a particular optical wavelength. These provide greater dynamic range (10 to 15 dB improvement) than the monochromator but are of limited use for spectral measurements due to the reduced number of wavelengths that are generally available for measurement. A typical optical configuration for spot attenuation measurements is shown in Figure 13.4. The interference filters are located on a wheel to allow measurement at a selection of different wavelengths. In the experimental arrangement shown in Figure 13.4 the source spot size is defined by a pinhole and the beam angular width is varied by using different diaphragms. However, the electronic equipment utilized with this set-up is similar to that used for the spectral loss measurements illustrated

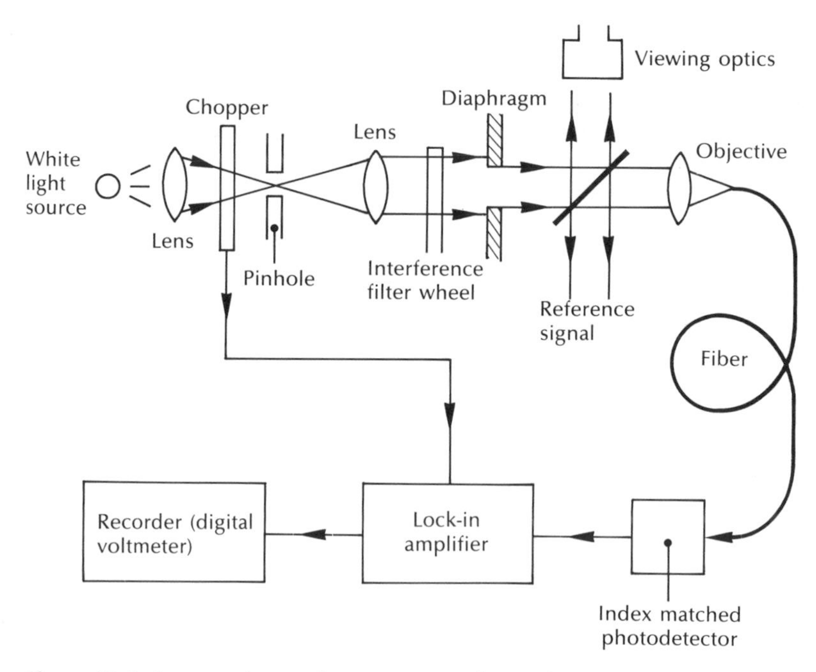

Figure 13.4 An experimental arrangement for making spot (single wavelength) attenuation measurements using interference filters and employing the cut-back technique.

in Figure 13.3. Therefore determination of the optical loss per unit length for the fiber at a particular wavelength is performed in exactly the same manner, using the cut-back method. Spot attenuation measurements are sometimes utilized after fiber cabling in order to obtain information on any degradation in the fiber attenuation resulting from the cabling process.

Although widely used, the cut-back measurement method has the major drawback of being a destructive technique. Therefore, although suitable for laboratory measurement it is far from ideal for attenuation measurements in the field. Several nondestructive techniques exist which allow the fiber losses to be calculated through a single reading of the optical output power at the far end of the fiber after determination of the near end power level. The simplest is the insertion or substitution* technique which utilizes the same experimental configuration as the cut-back method. However, the fiber to be tested is spliced, or connected by means of a demountable connector to a fiber with a known optical output at the wavelength of interest. When all the optical power is completely coupled between the two fibers, or when the insertion loss of the splice or connector is known, then the measurement of the optical output power from the second fiber gives the loss resulting from the insertion of this second fiber into the system. Hence the insertion loss due to the second fiber provides measurement of its attenuation per unit length. Unfortunately, the accuracy of this measurement method is dependent on the coupling between the two fibers and is therefore somewhat uncertain.

The most popular nondestructive attenuation measurement technique for both laboratory and field use only requires access to one end of the fiber. It is the back-scatter measurement method which uses optical time domain reflectometry and also provides measurement of splice and connector losses as well as fault location. Optical time domain reflectometry finds major use in field measurements and is therefore discussed in detail in Section 13.10.1.

13.2.2 Fiber absorption loss measurement

It was indicated in the preceding section that there is a requirement for the optical fiber manufacturer to be able to separate the total fiber attenuation into the contributions from the major loss mechanisms. Material absorption loss measurements allow the level of impurity content within the fiber material to be checked in the manufacturing process. The measurements are based on calorimetric methods which determine the temperature rise in the fiber or bulk material resulting from the absorbed optical energy within the structure.

The apparatus shown in Figure 13.5 [Ref. 12] which is used to measure the absorption loss in optical fibers was modified from an earlier version which measured the absorption losses in bulk glasses [Ref. 13]. This temperature measurement technique, illustrated diagrammatically in Figure 13.5(b), has been

* Description of the substitution method is provided in FOTP-53 [Ref. 11].

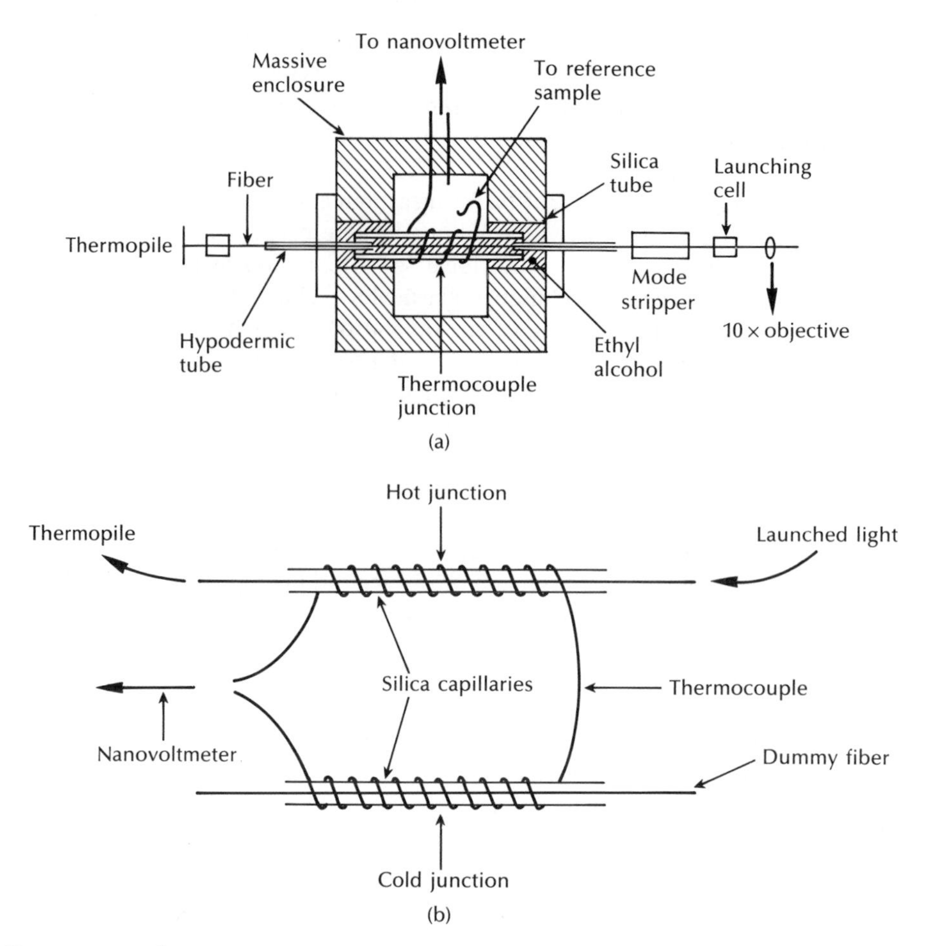

Figure 13.5 Calorimetric measurement of fiber absorption losses: (a) schematic diagram of a version of the apparatus [Ref. 12]; (b) the temperature measurement technique using a thermocouple.

widely adopted for absorption loss measurements. The two fiber samples shown in Figure 13.5(b) are mounted in capillary tubes surrounded by a low refractive index liquid (e.g. methanol) for good electrical contact, within the same enclosure of the apparatus shown in Figure 13.5(a). A thermocouple is wound around the fiber containing capillary tubes using one of them as a reference junction (dummy fiber). Light is launched from a laser source (Nd : YAG or krypton ion depending on the wavelength of interest) through the main fiber (not the dummy), and the temperature rise due to absorption is measured by the thermocouple and indicated on a nanovoltmeter. Electrical calibration may be achieved by replacing the optical fibers with thin resistance wires and by passing known electrical power through one.

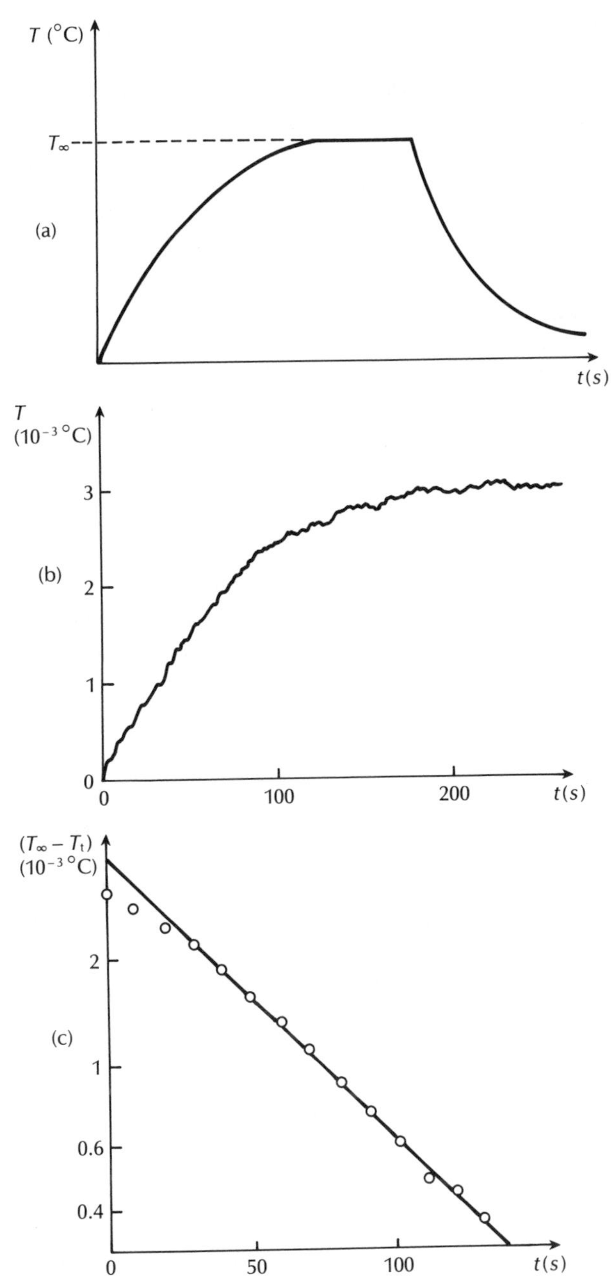

Figure 13.6 (a) A typical heating and cooling curve for a glass fiber sample. (b) A heating curve. (c) The corresponding plot of $(T_\infty - T_t)$ against time for a sample glass rod (bulk material measurement). Reproduced with permission from K. I. White and J. E. Midwinter. *Opto-electronics*, **5**, p. 323, 1973.

Independent measurements can then be made using the calorimetric technique and with electrical measurement instruments.

The calorimetric measurements provide the heating and cooling curve for the fiber sample used. A typical example of this curve is illustrated in Figure 13.6(a). The attenuation of the fiber due to absorption α_{abs} may be determined from this heating and cooling characteristic. A time constant t_c can be obtained from a plot of $(T_\infty - T_t)$ on a logarithmic scale against the time t, an example of which shown in Figure 13.6(c) was obtained from the heating characteristic displayed in Figure 13.6(b) [Ref. 13]. T_∞ corresponds to the maximum temperature rise of the fiber under test and T_t is the temperature rise at a time t. It may be observed from Figure 13.6(a) that T_∞ corresponds to a steady state temperature for the fiber when the heat loss to the surroundings balances the heat generated in the fiber resulting from absorption at a particular optical power level. The time constant t_c may be obtained from the slope of the straight line plotted in Figure 13.6(c) as:

$$t_c = \frac{t_2 - t_1}{\ln (T_\infty - T_{t_1}) - \ln (T_\infty - T_{t_2})} \tag{13.3}$$

where t_1 and t_2 indicate two points in time and t_c is a constant for the calorimeter which is inversely proportional to the rate of heat loss from the device.

From detailed theory it may be shown [Ref. 13] that the fiber attenuation due to absorption is given by:

$$\alpha_{abs} = \frac{CT_\infty}{P_{opt} t_c} \text{ dB km}^{-1} \tag{13.4}$$

where C is proportional to the thermal capacity per unit length of the silica capillary and the low refractive index liquid surrounding the fiber, and P_{opt} is the optical power propagating in the fiber under test. The thermal capacity per unit length may be calculated, or determined by the electrical calibration utilizing the thin resistance wire. Usually the time constant for the calorimeter t_c is obtained using a high absorption fiber which gives large temperature differences and greater accuracy. Once t_c is determined, the absorption losses of low loss test fibers may be calculated from their maximum temperature rise T_∞, using Eq. (13.4). The temperatures are measured directly in terms of the thermocouple output (microvolts), and the optical input to the test fiber is obtained by use of a thermocouple or an optical power meter.

Example 13.2

Measurements are made using a calorimeter and thermocouple experimental arrangement as shown in Figure 13.5 in order to determine the absorption loss of an optical fiber sample. Initially a high absorption fiber is utilized to obtain a plot of $(T_\infty - T_t)$ on a logarithmic scale against t. It is found from the plot that the readings of $(T_\infty - T_t)$ after 10 and 100 seconds are 0.525 and 0.021 μV respectively.

The test fiber is then inserted in the calorimeter and gives a maximum temperature rise of 4.3×10^{-4} °C with a constant measured optical power of 98 mW at a wavelength of 0.75 μm. The thermal capacity per kilometre of the silica capillary and fluid is calculated to be 1.64×10^{4} J °C^{-1}.

Determine the absorption loss in dB km^{-1}, at a wavelength of 0.75 μm, for the fiber under test.

Solution: Initially, the time constant for the calorimeter is determined from the measurements taken on the high absorption fiber using Eq. (13.3) where:

$$t_c = \frac{t_2 - t_1}{\ln (T_\infty - T_{t_1}) - \ln (T_\infty - T_{t_2})}$$

$$= \frac{100 - 10}{\ln (T_\infty - T_{10}) - \ln (T_\infty - T_{100})}$$

$$= \frac{90}{\ln (0.525) - \ln (0.021)}$$

$$= 28.0 \text{ s}$$

Then the absorption loss of the test fiber may be obtained using Eq. (13.4) where:

$$\alpha_{abs} = \frac{CT_\infty}{P_{opt} t_c} = \frac{1.64 \times 10^4 \times 4.3 \times 10^{-4}}{98 \times 10^{-3} \times 28.0}$$

$$= 2.6 \text{ dB km}^{-1}$$

Hence direct measurement of the contribution of absorption losses to the total fiber attenuation may be achieved. However, fiber absorption losses are often obtained indirectly from measurement of the fiber scattering losses (see the next section) by subtraction from the total fiber attenuation, measured by one of the techniques discussed in Section 13.2.1.

13.2.3 Fiber scattering loss measurement

The usual method of measuring the contribution of the losses due to scattering within the total fiber attenuation is to collect the light scattered from a short length of fiber and compare it with the total optical power propagating within the fiber. Light scattered from the fiber may be detected in a scattering cell as illustrated in the experimental arrangement shown in Figure 13.7. This may consist of a cube of six square solar cells (Tynes cell [Ref. 14]) or an integrating sphere and detector [Ref. 15]. The solar cell cube which contains index matching fluid surrounding the fiber gives measurement of the scattered light, but careful balancing of the detectors is required in order to achieve a uniform response. This problem is overcome in the integrating sphere which again usually contains index matching fluid but responds

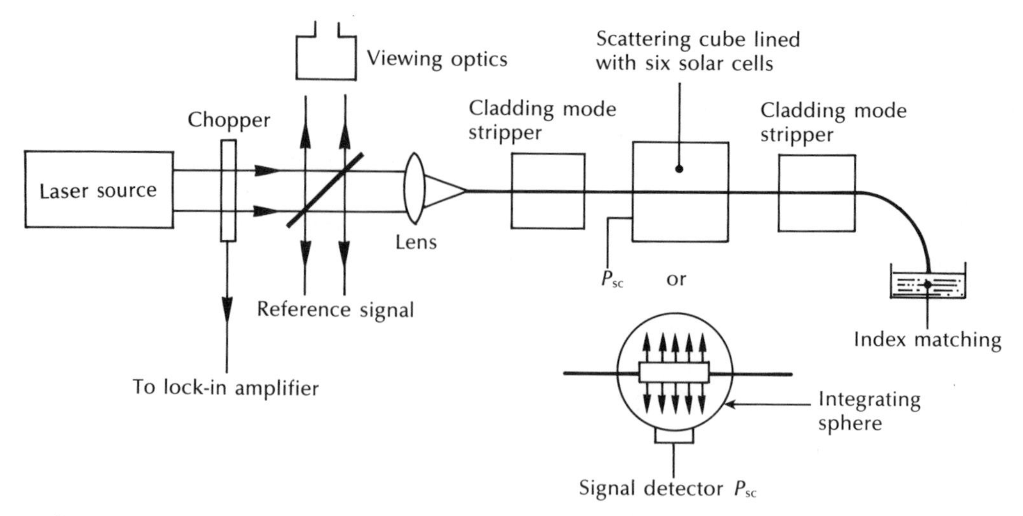

Figure 13.7 An experimental setup for measurement of fiber scattering loss illustrating both the solar cell cube and integrating sphere scattering cells.

uniformly to different distributions of scattered light. However, the integrating sphere does exhibit high losses from internal reflections. Other variations of the scattering cell include the internally reflecting cell [Ref. 16] and the sandwiching of the fiber between two solar cells [Ref. 17].

A laser source (i.e. He–Ne, Nd:YAG, krypton ion) is utilized to provide sufficient optical power at a single wavelength together with a suitable instrument to measure the response from the detector. In order to avoid inaccuracies in the measurement resulting from scattered light which may be trapped in the fiber, cladding mode strippers (see Section 13.2.1) are placed before and after the scattering cell. These devices remove the light propagating in the cladding so that the measurements are taken only using the light guided by the fiber core. Also to avoid reflections contributing to the optical signal within the cell, the output fiber end is index matched using either a fluid or suitable surface.

The loss due to scattering α_{sc} following Eq. (3.3) is given by:

$$\alpha_{sc} = \frac{10}{l(\text{km})} \log_{10}\left(\frac{P_{opt}}{P_{opt} - P_{sc}}\right) \text{ dB km}^{-1} \tag{13.5}$$

where l(km) is the length of the fiber contained within the scattering cell, P_{opt} is the optical power propagating within the fiber at the cell and P_{sc} is the optical power scattered from the short length of fiber l within the cell. As $P_{opt} \gg P_{sc}$, then the logarithm in Eq. (13.5) may be expanded to give:

$$\alpha_{sc} = \frac{4.343}{l(\text{km})} \left(\frac{P_{sc}}{P_{opt}}\right) \text{ dB km}^{-1} \tag{13.6}$$

Since the measurements of length are generally in centimetres and the optical power is normally registered in volts, Eq. (13.6) can be written as:

$$\alpha_{sc} = \frac{4.343 \times 10^5}{l(\text{cm})}\left(\frac{V_{sc}}{V_{opt}}\right) \text{ dB km}^{-1} \tag{13.7}$$

where V_{sc} and V_{opt} are the voltage readings corresponding to the scattered optical power and the total optical power within the fiber at the cell. The relative experimental accuracy (i.e. repeatability) for scatter loss measurements are in the range ± 0.2 dB using the solar cell cube and around 5% with the integrating sphere. However, it must be noted that the absolute accuracy of the measurements is somewhat poorer, being dependent on the calibration of the scattering cell and the mode distribution within a multimode fiber.

Example 13.3

A He–Ne laser operating at a wavelength of 0.63 μm was used with a solar cell cube to measure the scattering loss in a multimode fiber sample. With a constant optical output power the reading from the solar cell cube was 6.14 nV. The optical power measurement at the cube without scattering was 153.38 μV. The length of the fiber in the cube was 2.92 cm. Determine the loss due to scattering in dB km^{-1} for the fiber at a wavelength of 0.63 μm.

Solution: The scattering loss in the fiber at a wavelength of 0.63 μm may be obtained directly using Eq. (13.7) where:

$$\alpha_{sc} = \frac{4.343 \times 10^5}{l(\text{cm})}\left(\frac{V_{sc}}{V_{opt}}\right)$$

$$= \frac{4.343 \times 10^5}{2.92}\left(\frac{6.14 \times 10^{-9}}{153.38 \times 10^{-6}}\right)$$

$$= 6.0 \text{ dB km}^{-1}$$

13.3 Fiber dispersion measurements

Dispersion measurements give an indication of the distortion to optical signals as they propagate down optical fibers. The delay distortion which, for example, leads to the broadening of transmitted light pulses, limits the information-carrying capacity of the fiber. Hence as shown in Section 3.8 the measurement of dispersion allows the bandwidth of the fiber to be determined. Therefore, besides attenuation, dispersion is the most important transmission characteristic of an optical fiber. As discussed in Section 3.8 there are three major mechanisms which produce dispersion in optical fibers (material dispersion, waveguide dispersion and intermodal dispersion). The importance of these different mechanisms to the total fiber dispersion is dictated by the fiber type.

For instance, in multimode fibers (especially step index), intermodal dispersion tends to be the dominant mechanism, whereas in single-mode fibers intermodal dispersion is nonexistent as only a single mode is allowed to propagate. In the single-mode case the dominant dispersion mechanism is intramodal (i.e. chromatic dispersion). The dominance of intermodal dispersion in multimode fibers makes it essential that dispersion measurements on these fibers are performed only when the equilibrium mode distribution has been established within the fiber, otherwise inconsistent results will be obtained. Therefore devices such as mode scramblers or filters must be utilized in order to simulate the steady state mode distribution.

Dispersion effects may be characterized by taking measurements of the impulse response of the fiber in the time domain, or by measuring the baseband frequency response in the frequency domain. If it is assumed that the fiber response is linear with regard to power [Ref. 19], a mathematical description in the time domain for the optical output power $P_o(t)$ from the fiber may be obtained by convoluting the power impulse response $h(t)$ with the optical input power $P_i(t)$ as:

$$P_o(t) = h(t) * P_i(t) \tag{13.8}$$

where the asterisk $*$ denotes convolution. The convolution of $h(t)$ with $P_i(t)$ shown in Eq. (13.8) may be evaluated using the convolution integral where:

$$P_o(t) = \int_{-\infty}^{\infty} P_i(t-x)h(x)\,dx \tag{13.9}$$

In the frequency domain the power transfer function $H(\omega)$ is the Fourier transform of $h(t)$ and therefore by taking the Fourier transforms of all the functions in Eq. (13.8) we obtain,

$$\mathscr{P}_o(\omega) = H(\omega)\mathscr{P}_i(\omega) \tag{13.10}$$

where ω is the baseband angular frequency. The frequency domain representation given in Eq. (13.10) is the least mathematically complex, and by performing the Fourier transformation (or the inverse Fourier transformation) it is possible to switch between the time and frequency domains (or vice versa) by mathematical means. Hence, independent measurement of either $h(t)$ or $H(\omega)$ allows determination of the overall dispersive properties of the optical fiber. Thus fiber dispersion measurements can be made in either the time or frequency domains.

13.3.1 Time domain measurement

The most common method for time domain measurement of pulse dispersion in multimode optical fibers is illustrated in Figure 13.8 [Ref. 20]. Short optical pulses (100 to 400 ps) are launched into the fiber from a suitable source (e.g. AlGaAs injection laser) using fast driving electronics. The pulses travel down the length of fiber under test (around 1 km) and are broadened due to the various dispersion

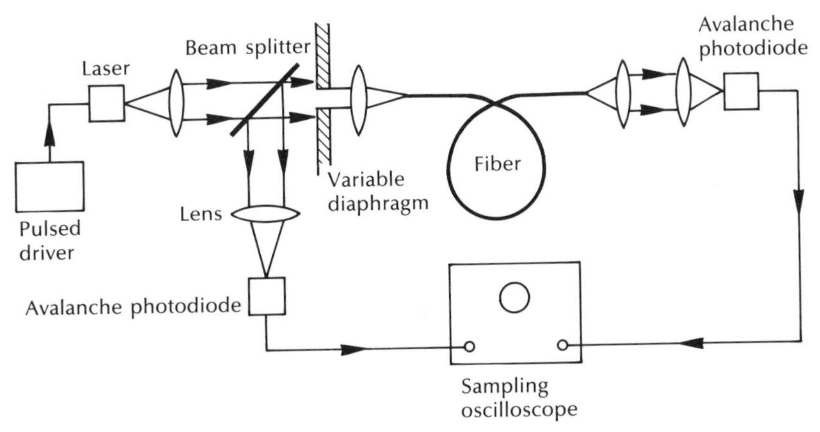

Figure 13.8 Experimental arrangement for making multimode fiber dispersion measurements in the time domain [Ref. 20].

mechanisms. However, it is possible to take measurements of an isolated dispersion mechanism by, for example, using a laser with a narrow spectral width when testing a multimode fiber. In this case the intramodal dispersion is negligible and the measurement thus reflects only intermodal dispersion. The pulses are received by a high speed photodetector (i.e. avalanche photodiode) and are displayed on a fast sampling oscilloscope. A beam splitter is utilized for triggering the oscilloscope and for input pulse measurement.

After the initial measurement of output pulse width, the long fiber length may be cut back to a short length and the measurement repeated in order to obtain the effective input pulse width. The fiber is generally cut back to the lesser of 10 m or 1% of its original length [Ref. 6]. As an alternative to this cut-back technique, the insertion or substitution method similar to that used in fiber loss measurement (see Section 13.2.1) can be employed. This method has the benefit of being nondestructive and only slightly less accurate than the cut-back technique. These time domain measurement methods for multimode fiber are covered in FOTP-51 [Ref. 21].

The fiber dispersion is obtained from the two pulse width measurements which are taken at any convenient fraction of their amplitude. However, unlike the considerations of dispersion in Sections 3.8 to 3.11 where rms pulse widths are used, dispersion measurements are normally made on pulses using the half maximum amplitude or 3 dB points. If $P_i(t)$ and $P_o(t)$ of Eq. (13.8) are assumed to have a Gaussian shape then Eq. (13.8) may be written in the form:

$$\tau_o^2(3\text{ dB}) = \tau^2(3\text{ dB}) + \tau_i^2(3\text{ dB}) \tag{13.11}$$

where $\tau_i(3\text{ dB})$ and $\tau_o(3\text{ dB})$ are the 3 dB pulse widths at the fiber input and output, respectively, and $\tau(3\text{ dB})$ is the width of the fiber impulse response again measured at half the maximum amplitude. Hence the pulse dispersion in the fiber (commonly referred to as the pulse broadening when considering the 3 dB pulse width) in ns km^{-1} is given by:

$$\tau(3\text{ dB}) = \frac{(\tau_o^2(3\text{ dB}) - \tau_i^2(3\text{ dB}))^{\frac{1}{2}}}{L} \text{ ns km}^{-1} \tag{13.12}$$

where $\tau(3\text{ dB})$, $\tau_i(3\text{ dB})$ and $\tau_o(3\text{ dB})$ are measured in ns and L is the fiber length in km. It must be noted that if a long length of fiber is cut back to a short length in order to take the input pulse width measurement then L corresponds to the difference between the two fiber lengths in km. When the launched optical pulses and the fiber impulse response are Gaussian then the 3 dB optical bandwidth for the fiber B_{opt} may be calculated using [Ref. 22]:

$$\begin{aligned} B_{opt} \times \tau(3\text{ dB}) &= 0.44 \text{ GHz ns} \\ &= 0.44 \text{ MHz ps} \end{aligned} \tag{13.13}$$

Hence estimates of the optical bandwidth for the fiber may be obtained from the measurements of pulse broadening without resorting to rigorous mathematical analysis.

Example 13.4

Pulse dispersion measurements are taken over a 1.2 km length of partially graded multimode fiber. The 3 dB widths of the optical input pulses are 300 ps, and the corresponding 3 dB widths for the output pulses are found to be 12.6 ns. Assuming the pulse shapes and fiber impulse response are Gaussian calculate:

(a) the 3 dB pulse broadening for the fiber in ns km^{-1}:
(b) the fiber bandwidth–length product.

Solution: (a) The 3 dB pulse broadening may be obtained using Eq. (13.12) where:

$$\tau(3\text{ dB}) = \frac{(12.6^2 - 0.3^2)^{\frac{1}{2}}}{1.2} = \frac{(158.76 - 0.09)^{\frac{1}{2}}}{1.2}$$

$$= 10.5 \text{ ns km}^{-1}$$

(b) The optical bandwidth for the fiber is given by Eq. (13.13) as:

$$B_{opt} = \frac{0.44}{\tau(3\text{ dB})} = \frac{0.44}{10.5} \text{ GHz km}$$

$$= 41.9 \text{ MHz km}$$

The value obtained for B_{opt} corresponds to the bandwidth–length product for the fiber because the pulse broadening in part (a) was calculated over a 1 km fiber length. Also it may be noted that in this case the narrow input pulse width makes little difference to the calculation of the pulse broadening. The input pulse width becomes significant when measurements are taken on low dispersion fibers (e.g. single-mode).

The above dispersion measurement techniques allow the total dispersion for multimode fibers to be determined. It is clear, however, that intramodal or chromatic dispersion is an important transmission parameter, particularly for single-mode fibers. Moreover, it can also be a significant distortion effect in multimode fibers even though intermodal dispersion is normally dominant. The time domain measurement of chromatic dispersion is outlined in FOTP-168 [Ref. 23]. A typical experimental arrangement is shown in Figure 13.9. The pulse delay versus optical wavelength is measured for both long and short fiber lengths. The source usually comprises multiple injection lasers possibly including wavelength tunable devices (see Section 6.10). When $\Delta T(\lambda)$ is the delay difference for the length difference $L_1 - L_2$, then the specific group delay per unit length $\tau_g(\lambda)$ is given by [Ref. 6]:

$$\tau_g(\lambda) = \frac{\Delta T(\lambda)}{L_1 - L_2} \tag{13.14}$$

Differentiation of Eq. (13.14) provides the chromatic dispersion D_T following Eq. (3.46) where:

$$D_T(\lambda) = \frac{d\tau_g}{d\lambda} \text{ ps nm}^{-1} \text{ km}^{-1} \tag{13.15}$$

and the dispersion slope S from Eq. (3.52):

$$S(\lambda) = \frac{dD_T}{d\lambda} \text{ ps nm}^{-2} \text{ km}^{-1} \tag{13.16}$$

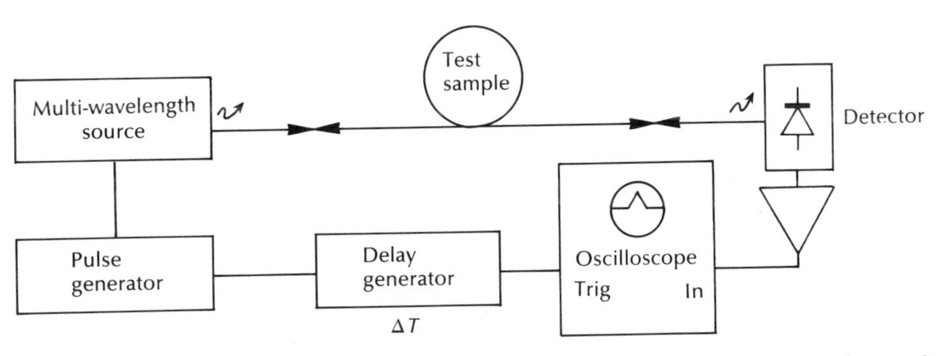

Figure 13.9 Experimental arrangement for the measurement of intramodal or chromatic dispersion by time delay.

This pulse delay method is also one of the two reference test methods to obtain chromatic dispersion in single-mode fibers which are recommended by the CCITT [Ref. 25].

13.3.2 Frequency domain measurement

Frequency domain measurement is the preferred method for acquiring the bandwidth of multimode optical fibers. This is because the baseband frequency response $H(\omega)$ of the fiber may be obtained directly from these measurements using Eq. (13.10) without the need for any assumptions of Gaussian shape, or alternatively, the mathematically complex deconvolution of Eq. (13.8) which is necessary with measurements in the time domain. Thus the optical bandwidth of a multimode fiber is best obtained from frequency domain measurements.

One of two frequency domain measurement techniques is generally used. The first utilizes a similar pulsed source to that employed for the time domain measurements shown in Figure 13.8. However, the sampling oscilloscope is replaced by a spectrum analyser which takes the Fourier transform of the pulse in the time domain and hence displays its constituent frequency components. The experimental arrangement is illustrated in Figure 13.10.

Comparison of the spectrum at the fiber output $\mathscr{P}_o(\omega)$ with the spectrum at the fiber input $\mathscr{P}_i(\omega)$ provides the baseband frequency response for the fiber under test following Eq. (5.10) where:

$$H(\omega) = \frac{\mathscr{P}_o(\omega)}{\mathscr{P}_i(\omega)} \tag{13.17}$$

The second technique involves launching a sinusoidally modulated optical signal at different selected frequencies using a sweep oscillator. Therefore the signal

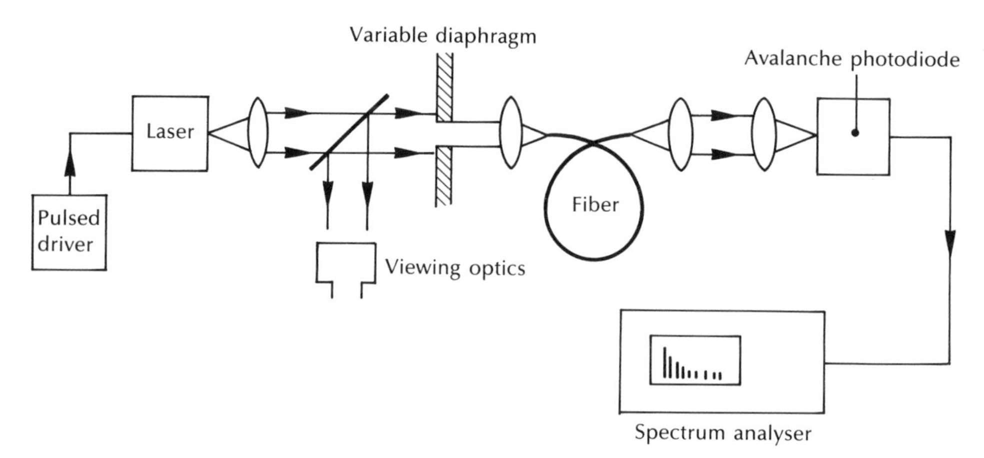

Figure 13.10 Experimental setup for making fiber dispersion measurements in the frequency domain using a pulsed laser source.

energy is concentrated in a very narrow frequency band in the baseband region, unlike the pulse measurement method where the signal energy is spread over the entire baseband region. A possible experimental arrangement for this swept frequency measurement method is shown in Figure 13.11 [Ref. 24]. The optical source is usually an injection laser, which may be directly modulated (see Section 7.5) from the sweep oscillator. A spectrum analyser may be used in order to obtain a continuous display of the swept frequency signal. Again, Eq. (13.17) is utilized to obtain the baseband frequency response, employing either the cut-back or substitution procedure in a similar manner to the time domain measurement (see Section 13.3.1). However, the spectrum analyser provides no information on the phase of the received signal Therefore a vector voltmeter or ideally a network analyser can be employed to give both the frequency and phase information. This multimode fiber frequency domain measurement method is described in FOTP-30 [Ref. 25].

The intramodal or chromatic dispersion for single-mode fibers may also be obtained using frequency domain measurement techniques. The second reference test method recommended by the CCITT [Ref. 1] falls into this category and is known as the phase shift method. This technique is also covered in FOTP-169 [Ref. 26]. To obtain the phase shift $\phi(\lambda)$ versus wavelength, the pulse generator in Figure 13.9 (corresponding to the time domain measurement) is replaced by a high frequency oscillator operating at a constant frequency and the delay generator and oscilloscope are replaced by a phase meter or vector voltmeter. Finally, an electrical or optical reference channel is connected between the oscillator and the meter.

When an optical signal, which is sinusoidally modulated in power with frequency f_m, is transmitted through a single-mode fiber of length L, then the modulation

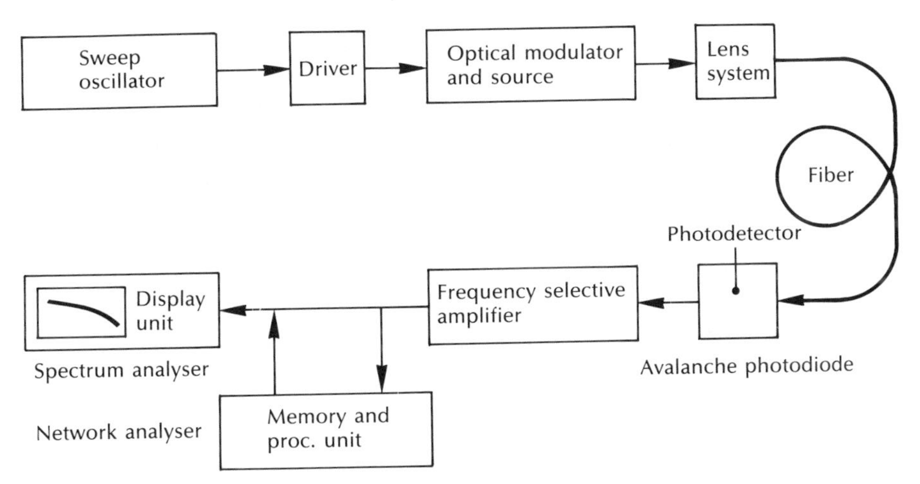

Figure 13.11 Block schematic showing an experimental arrangement for the swept frequency measurement method to provide fiber dispersion measurements in the frequency domain [Ref. 24].

envelope is delayed in time by:

$$\frac{L}{v_g} = \tau_g L \tag{13.18}$$

where v_g is the group velocity which corresponds to the signal velocity. Since a delay of one modulation period T_m or $1/f_m$ corresponds to a phase shift of 2π, then the sinusoidal modulation is phase shifted in the fiber by an angle ϕ_m where:

$$\phi_m = \frac{2\pi\tau_g L}{T_m} = 2\pi f_m \tau_g L \tag{13.19}$$

Hence the specific group delay is given by:

$$\tau_g = \frac{\phi_m}{2\pi f_m L} \tag{13.20}$$

Again the chromatic dispersion and the dispersion slope can be obtained by differentiation following Eqs. (13.15) and (13.16) respectively. Finally, alternative direct measurement techniques using a differential phase shift method are described in FOTP-175.

13.4 Fiber refractive index profile measurements

The refractive index profile of the fiber core plays an important role in characterizing the properties of optical fibers. It allows determination of the fiber's numerical aperture and the number of modes propagating within the fiber core, whilst largely defining any intermodal and/or profile dispersion caused by the fiber. Hence a detailed knowledge of the refractive index profile enables the impulse response of the fiber to be predicted. Also as the impulse response and consequently the information-carrying capacity of the fiber is strongly dependent on the refractive index profile, it is essential that the fiber manufacturer is able to produce particular profiles with great accuracy, especially in the case of graded index fibers (i.e. optimum profile). There is therefore a requirement for accurate measurement of the refractive index profile. These measurements may be performed using a number of different techniques each of which exhibit certain advantages and drawbacks. In this section we will discuss some of the more popular methods which may be relatively easily interpreted theoretically, without attempting to review all the possible techniques which have been developed.

13.4.1 Interferometric methods

Interference microscopes (e.g. Mach–Zehnder, Michelson) have been widely used to determine the refractive index profiles of optical fibers. The technique usually involves the preparation of a thin slice of fiber (slab method) which has both ends accurately polished to obtain square (to the fiber axes) and optically flat surfaces.

The slab is often immersed in an index matching fluid, and the assembly is examined with an interference microscope. Two major methods are then employed, using either a transmitted light interferometer (Mach–Zehnder [Ref. 27]) or a reflected light interferometer (Michelson [Ref. 28]). In both cases light from the microscope travels normal to the prepared fiber slice faces (parallel to the fiber axis), and differences in refractive index result in different optical path lengths. This situation is illustrated in the case of the Mach–Zehnder interferometer in Figure 13.12(a). When the phase of the incident light is compared with the phase of the emerging light, a field of parallel interference fringes is observed. A photograph of the fringe pattern may then be taken, an example of which is shown in Figure 13.12(b) [Ref. 30].

The fringe displacements for the points within the fiber core are then measured using as reference the parallel fringes outside the fiber core (in the fiber cladding). The refractive index difference between a point in the fiber core (e.g. the core axis) and the cladding can be obtained from the fringe shift q, which corresponds to a number of fringe displacements. This difference in refractive index δn is given by [Ref. 6]:

$$\delta n = \frac{q\lambda}{x} \tag{13.21}$$

where x is the thickness of the fiber slab and λ is the incident optical wavelength. The slab method gives an accurate measurement of the refractive index profile, although computation of the individual points, is somewhat tedious unless an

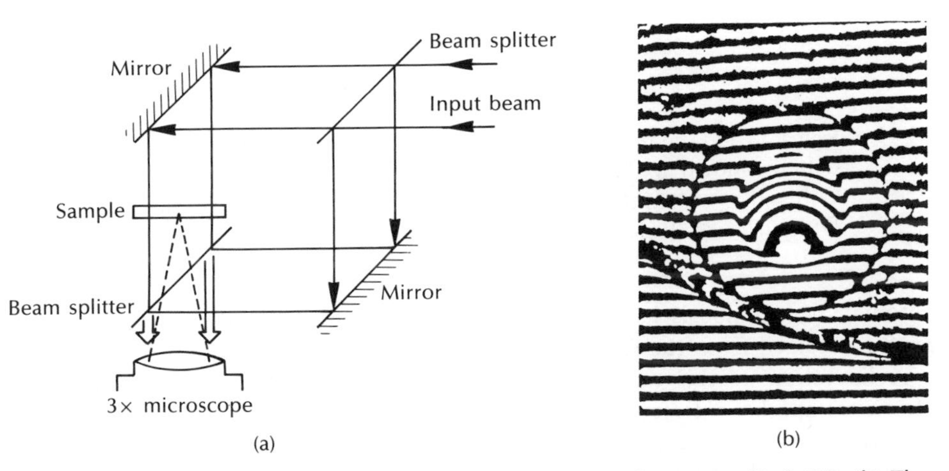

Figure 13.12 (a) The principle of the Mach–Zehnder interferometer [Ref. 27]. (b) The interference fringe pattern obtained with an interference microscope from a graded index fiber. Reproduced with permission from L. G. Cohen, P. Kaiser J. M. MacChesney, P. N. O'Connor and H. M. Presby. *Appl. Phys. Lett.*, **26**, p. 472, 1975.

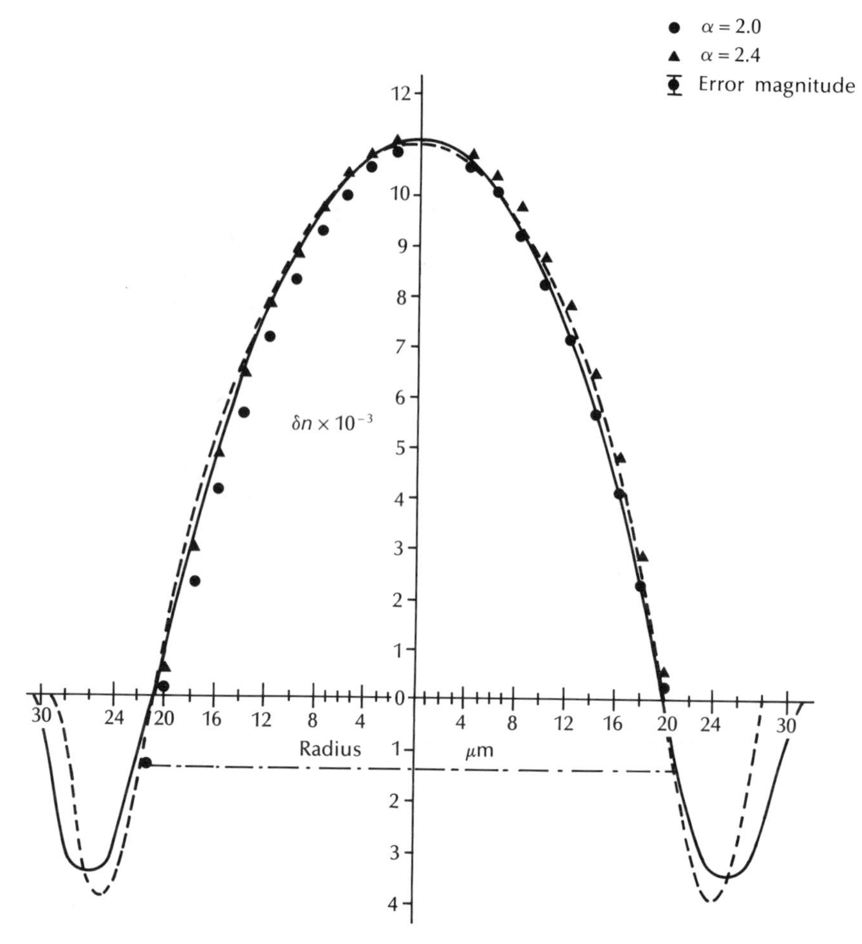

Figure 13.13 The fiber refractive index profile computed from the interference pattern shown in Figure 13.12(b). Reproduced with permission from L. G. Cohen, P. Kaiser, J. M. MacChesney, P. N. O'Connor and H. M. Presby. *Appl. Phys. Lett.*, **26**, p. 472, 1975.

automated technique is used. Figure 13.13 [Ref. 30] shows the refractive index profile obtained from the fringe pattern indicated in Figure 13.12(b).

A limitation of this method is the time required to prepare the fiber slab. However, another interferometric technique has been developed [Ref. 32] which requires no sample preparation. In this method the light beam is incident to the fiber perpendicular to its axis; this is known as transverse shearing interferometry. Again fringes are observed from which the fiber refractive index profile may be obtained.

13.4.2 Near field scanning method

The near field scanning or transmitted near field method utilizes the close resemblance that exists between the near field intensity distribution and the refractive index profile, for a fiber with all the guided modes equally illuminated. It provides a reasonably straightforward and rapid method for acquiring the refractive index profile. When a diffuse Lambertian source (e.g. tungsten filament lamp or LED) is used to excite all the guided modes then the near field optical power density at a radius r from the core axis $P_D(r)$ may be expressed as a fraction of the core axis near field optical power density $P_D(0)$ following [Ref. 33]:

$$\frac{P_D(r)}{P_D(0)} = C(r, z)\left[\frac{n_1^2(r) - n_2^2}{n_1^2(0) - n_2^2}\right] \tag{13.22}$$

where $n_1(0)$ and $n_1(r)$ are the refractive indices at the core axis and at a distance r from the core axis respectively, n_2 is cladding refractive index and $C(r, z)$ is a correction factor. The correction factor which is incorporated to compensate for any leaky modes present in the short test fiber may be determined analytically. A set of normalized correction curves is, for example, given in Ref. 34. For multimode fiber such a transmitted near field method is described in FOTP-43 [Ref. 35]. However, at present there is no transmitted near field FOTP for single-mode fiber.

An experimental configuration is shown in Figure 13.14. The output from a Lambertian source is focused on to the end of the fiber using a microscope objective lens. A magnified image of the fiber output end is displayed in the plane of a small active area photodetector (e.g. silicon p–i–n photodiode). The photodetector which scans the field transversely receives amplification from the phase sensitive

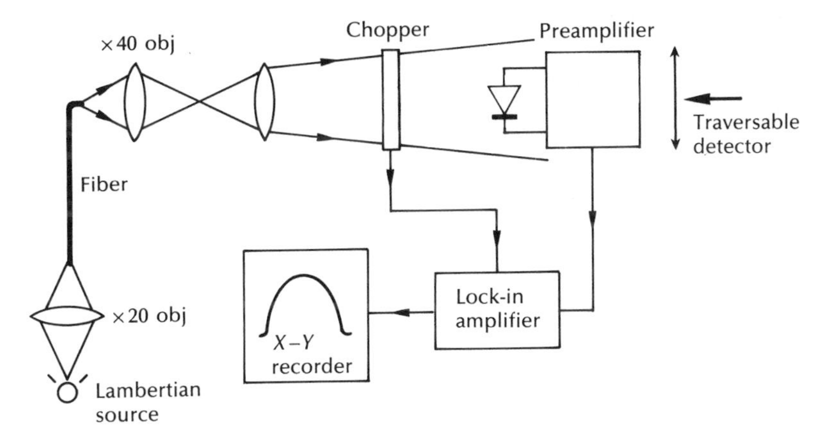

Figure 13.14 Experimental setup for the near field scanning measurement of the refractive index profile [Ref. 37].

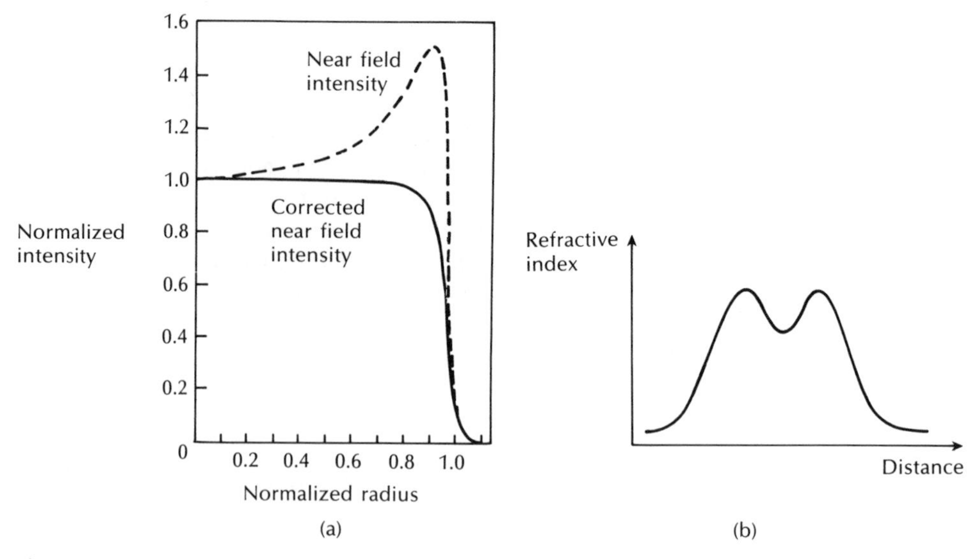

Figure 13.15 (a) The refractive index profile of a step index fiber measured using the near field scanning method, showing the near field intensity and the corrected near field intensity. Reproduced with permission from F. E. M. Sladen, D. N. Payne and M. J. Adams, *Appl. Phys. Lett.*, **28**, p. 225, 1976. (b) The refractive index profile of a practical step index fiber measured by the near field scanning method [Ref. 33].

combination of the optical chopper and lock-in amplifier. Hence the profile may be plotted directly on an X–Y recorder. However, the profile must be corrected with regard to $C(r, z)$ as illustrated in Figure 13.15(a) which is very time consuming. Both the scanning and data acquisition can be automated with the inclusion of a minicomputer [Ref. 34].

The test fiber is generally 2 m in length to eliminate any differential mode attenuation and mode coupling. A typical refractive index profile for a practical step index fiber measured by the near field scanning method is shown in Figure 13.15(b). It may be observed that the profile dips in the centre at the fiber core axis. This dip was originally thought to result from the collapse of the fiber preform before the fiber is drawn in the manufacturing process but has been recently shown to be due to the layer structure inherent at the deposition stage [Ref. 36].

13.4.3 Refracted near field method

The refracted near field (RNF) or refracted ray method is complementary to the transmitted near field technique (see Section 13.4.2) but has the advantage that it does not require a leaky mode correction factor or equal mode excitation. Moreover, it provides the relative refractive index differences directly without recourse to external calibration or reference samples. The RNF method is the most commonly used technique for the determination of the fiber refractive index profile [Ref. 38] and is the EIA reference test method for both multimode and single-mode fibers. Details of the test procedure are provided in FOTP-44 [Ref. 39].

A schematic of an experimental set-up for the RNF method is shown in Figure 13.16. A short length of fiber is immersed in a cell containing a fluid of slightly higher refractive index. A small spot of light typically emitted from a 633 nm helium neon laser for best resolution is scanned across the cross sectional diameter of the fiber. The measurement technique utilizes that light which is not guided by the fiber but escapes from the core into the cladding. However, light escaping from the fiber core partly results from the power leakage from the leaky modes which is an unknown quantity. The effect of this radiated power reaching the detector is undesirable and therefore it is blocked using an opaque circular screen, as shown in Figure 13.16(a). The refracted ray trajectories are illustrated in Figure 13.16(b) where θ' is the angle of incidence in the fiber core, θ is the angle of refraction in the fiber core and θ'' constitutes the angle of the refracted inbound rays external to the fiber core. Any light leaving the fiber core below a minimum angle $\theta''_{\min}$ is prevented from reaching the detector by the opaque screen (Figure 13.16(a)). Moreover, it may be observed from Figure 13.16(b) that this minimum angle corresponds to a minimum angle of incidence $\theta'_{\min}$. However, all light at an angle

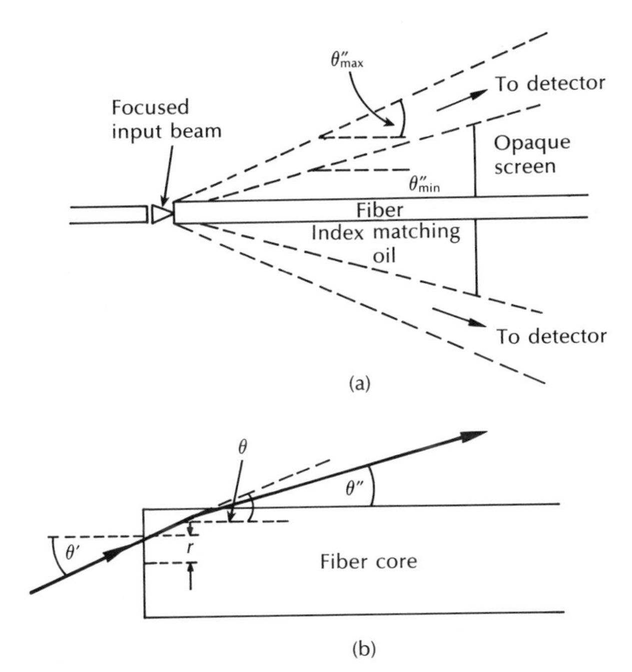

Figure 13.16 Refracted near field method for the measurement of refractive index profile: (a) experimental arrangement; (b) illustration of the ray trajectories [Ref. 40].

of incidence $\theta' > \theta'_{min}$ must be allowed to reach the detector. To ensure that this process occurs it is advisable that input apertures are used to limit the convergence angle of the input beam to a suitable maximum angle θ'_{max} corresponding to a refracted angle θ''_{max}. In addition, the immersion of the fiber in an index matching fluid prevents reflection at the outer cladding boundary. Hence all the refracted light emitted from the fiber at angles over the range θ''_{min} to θ''_{max} may be detected.

The detected optical power as a function of the radial position of the input beam $P(r)$ is measured and a value $P(a)$ corresponding to the input beam being focused into the cladding is also obtained. The refractive index profile $n(r)$ for the fiber core is then given by [Ref. 40]:

$$n(r) = n_2 + n_2 \cos \theta''_{min}(\cos \theta''_{min} - \cos \theta'_{max}) \frac{P(a) - P(r)}{P(a)} \tag{13.23}$$

where n_2 is the cladding refractive index. Furthermore, Eq. (13.23) can be written as:

$$n(r) = k_1 - k_2 P(r) \tag{13.24}$$

It is clear that $n(r)$ is proportional to $P(r)$ and hence the measurement system can be calibrated to obtain the constants k_1 and k_2. For example, a calibration scheme in which the power that passes the opaque screen is monitored as it is translated along the optical axis provided an early strategy [Refs. 41, 42]. Alternative calibration techniques which allow accurate RNF measurements are described in Ref. 38.

13.5 Fiber cutoff wavelength measurements

A multimode fiber has many cutoff wavelengths because the number of bound propagating modes is usually large. For example, considering a parabolic refractive index graded fiber, following Eq. (2.95) the number of guided modes M_g is:

$$M_g = \left(\frac{\pi a}{\lambda}\right)^2 (n_1^2 - n_2^2) \tag{13.25}$$

where a is the core radius and n_1 and n_2 are the core peak and cladding indices respectively. It may be observed from Eq. (13.25) that operation at longer wavelengths yields fewer guided modes. Therefore it is clear that as the wavelength is increased, a growing number of modes are cutoff where the cutoff wavelength of a LP_{lm} mode is the maximum wavelength for which the mode is guided by the fiber.

Usually the cutoff wavelength refers to the operation of single-mode fiber in that it is the cutoff wavelength of the LP_{11} mode (which has the longest cutoff wavelength) which makes the fiber single moded when the fiber diameter is reduced to 8 or 9 μm. Hence the cutoff wavelength of the LP_{11} is the shortest wavelength above which the fiber exhibits single-mode operation and it is therefore an important parameter to measure (see Section 2.5.1). The theoretical value of the cutoff wavelength can be determined from the fiber refractive index profile following Eq. (2.98). Because of the large attenuation of the LP_{11} mode near cutoff, however, the parameter which is experimentally determined is called the effective cutoff wavelength which is always smaller than the theoretical cutoff wavelength by as much as 100 to 200 nm [Ref. 43]. It is this effective cutoff wavelength which limits the wavelength region for which the fiber is 'effectively' single-mode.

The effective cutoff wavelength is normally measured by increasing the signal wavelength in a fixed length of fiber until the LP_{11} mode is undetectable. Since the attenuation of the LP_{11} mode is dependent on the fiber length and its radius of curvature, the effective cutoff wavelength tends to vary with the method of measurement. Moreover, numerous methods of measurement have been investigated [Refs. 6, 43] and because these techniques can give significantly different results, the measurement has caused some problems [Ref. 44]. Nevertheless, three methods were recommended by the CCITT in 1986 [Ref. 1]; two of which, being transmitted power techniques, were recommended as reference test methods. In addition, these two techniques correspond to the EIA standard test method FOTP-80 [Ref. 45].

The effective cutoff wavelength has been defined by the CCITT as the wavelength greater than which the ratio between the total power, including the launched higher order modes, and the fundamental mode power has decreased to less than 0.1 dB in a quasi-straight 2 m fiber length with one single loop of 140 mm radius.* Measurement configurations which enable the determination of fiber cutoff wavelength by the RTMs are shown in Figure 13.17. A single turn configuration is illustrated in Figure 13.17(a), whilst the split mandrel configuration of Figure 13.17(b) proves convenient for fiber handling. The other test apparatus is the same as that employed for the measurement of fiber attenuation by the cut-back method (Figure 13.2). However, the launch conditions used must be sufficient to excite both the fundamental and the LP_{11} modes, and it is important that cladding modes are stripped from the fiber.

In the bending-reference technique the power $P_s(\lambda)$ transmitted through the fiber sample in the configurations shown in Figure 13.17 is measured as a function of wavelength. Thus the quantity $P_s(\lambda)$ corresponds to the total power, including launched higher order modes, of the CCITT definition for cutoff wavelength. Then keeping the launch conditions fixed, at least one additional loop of sufficiently small radius (60 mm or less) is introduced into the test sample to act as a mode filter to suppress the secondary LP_{11} mode without attenuating the fundamental mode at the effective cutoff wavelength. In this case the smaller transmitted spectral power $P_b(\lambda)$ is measured which corresponds to the fundamental mode power referred to in the definition. The bend attenuation $a_b(\lambda)$ comprising the level difference

* It should be noted that the 2 m fiber length corresponds to the length specified in the cut-back attenuation measurements (Section 13.2.1).

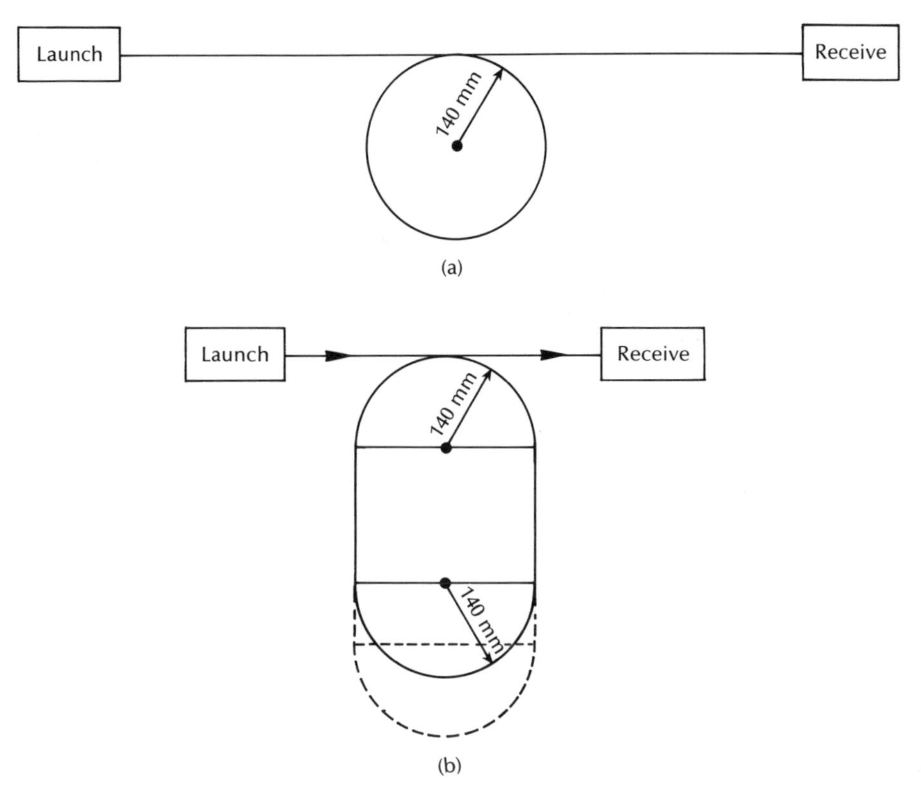

Figure 13.17 Configurations for the measurement of uncabled fiber cutoff wavelength: (a) single turn; (b) split mandrell [Refs. 1, 45].

between the total power and the fundamental power is calculated as:

$$a_b(\lambda) = 10 \log_{10} \frac{P_s(\lambda)}{P_b(\lambda)} \tag{13.26}$$

The bend attenuation characteristic exhibits a peak in the wavelength region where the radiation losses resulting from the small loop are much higher for the LP_{11} mode than for the LP_{01} fundamental mode, as illustrated in Figure 13.18. It should be noted that the shorter wavelength side of the attenuation maximum corresponds to the LP_{11} mode, being well confined in the fiber core, and hence negligible loss is induced by the 60 mm diameter loop, whereas on the longer wavelength side the LP_{11} mode is not guided in the fiber and therefore, assuming that the loop diameter is large enough to avoid any curvature loss to the fundamental mode, there is also no increase in loss. Using the CCITT and EIA definition for the effective cutoff wavelength λ_{ce} it may be determined as the longest wavelength at which the bend attenuation or level difference $a_b(\lambda)$ equals 0.1 dB, as shown in Figure 13.18.

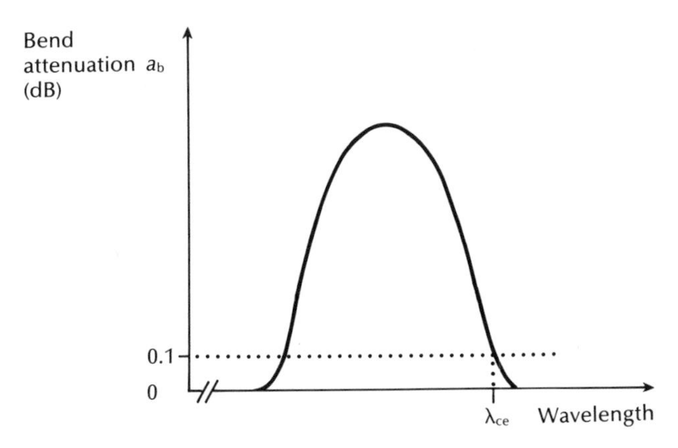

Figure 13.18 Bend attenuation against wavelength in the bending method for the measurement of cutoff wavelength λ_{ce}.

The other RTM is called the power step method [Ref. 46] or the multimode reference technique [Ref. 6]. Again, the fiber configurations shown in Figure 13.17 are employed with the test apparatus the same as that to measure fiber attenuation by the cut-back method. Furthermore, the launch conditions must again be sufficient to excite both the fundamental and LP_{11} modes and, as in the bending method, the transmitted power $P_s(\lambda)$ is measured as a function of wavelength. Next, however, the 2 m length of single-mode fiber is replaced by a short (1 to 2 m) length of multimode fiber and the spectral power $P_m(\lambda)$ emerging from the end of the multimode fiber is measured.

The relative attenuation $a_m(\lambda)$ or level difference between the powers launched into the multimode and single-mode fibers may be computed as:

$$a_m(\lambda) = 10 \log_{10} \frac{P_s(\lambda)}{P_m(\lambda)} \tag{13.27}$$

A typical characteristic showing the level difference as a function of wavelength is provided in Figure 13.19 in which the step reduction of level difference around cutoff may be observed. This results from the increase in power obtained at the output of the single-mode fiber from propagation of the LP_{11} mode, as well as the fundamental LP_{01} mode when going through the cutoff wavelength. To obtain the effective cutoff wavelength, the longest wavelength portion of the characteristic is fitted to a straight line and the intersection of the $a_m(\lambda)$ curve with another parallel straight line displaced by 0.1 dB produces the result. It should be noted, however, that accurate measurement requires an attenuation difference of not less than 2 dB [Ref. 6]. Such a difference may be readily obtained as there are two modes in the primary LP_{01} mode group and four in the secondary LP_{11} mode group. Hence with equal excitation of both groups the maximum attenuation difference is $10 \log_{10} (2 + 4)/2$, or 4.8 dB, when going through cutoff [Ref. 46]. Finally, this method and

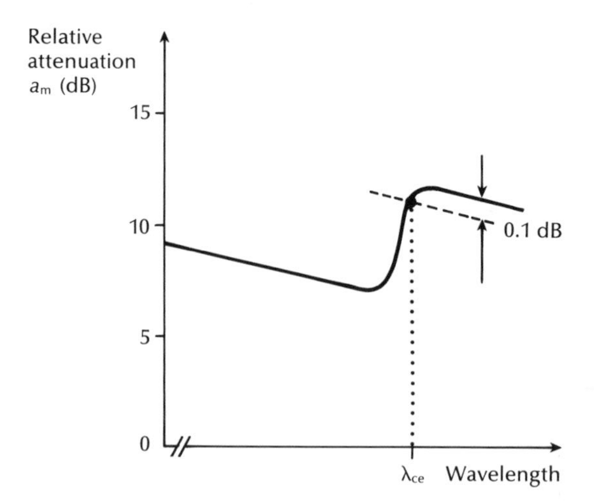

Figure 13.19 Relative attenuation against wavelength in the power step technique for the measurement of cutoff wavelength λ_{ce}.

the bending reference technique have been shown to yield approximately the same values for the effective cutoff wavelength in a round robin test [Ref. 47].

A third method for determination of the effective cutoff wavelength which is recommended by the CCITT as an alternative test method [Ref. 1] is the measurement of the change in spot size with wavelength [Ref. 48]. In this case the spot size is measured as a function of wavelength by the transverse offset method (see Section 13.8) using a 2 m length of fiber on each side of the joint with a single loop of radius 140 mm formed in each 2 m length. When the fiber is operating in

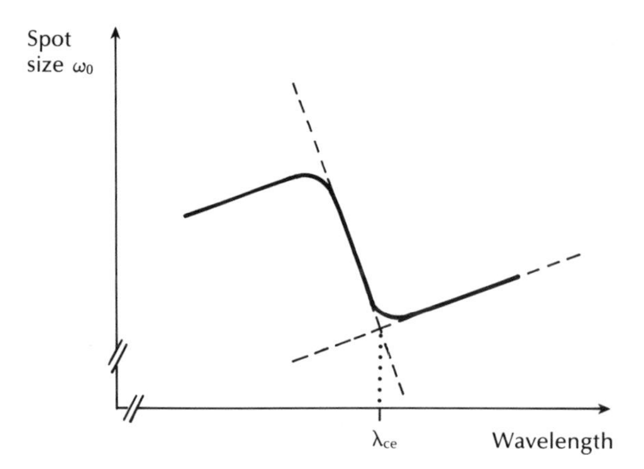

Figure 13.20 Wavelength dependence of the spot size in the spot size technique for the measurement of cutoff wavelength λ_{ce}.

the single-mode region, the spot size increases almost linearly with increasing wavelength [Ref. 43], as may be observed in Figure 13.20. However, as the cutoff wavelength is approached, the contribution from the second order mode creates a significant change in the spot size from the expected single-mode values. At this point two straight lines with a positive and negative slope can be fitted through the measured points, as illustrated in Figure 13.20, and the intersection point corresponds to the effective cutoff wavelength.

The effective cutoff wavelength for a cabled single-mode fiber will generally be smaller than that of the uncabled fiber because of bend effects (both micro- and macrobending). A procedure for this measurement is outlined in FOTP-170 [Ref. 49] which is similar to the transmitted power methods of FOTP-80.

13.6 Fiber numerical aperture measurements

The numerical aperture is an important optical fiber parameter as it affects characteristics such as the light-gathering efficiency and the normalized frequency of the fiber (V). This in turn dictates the number of modes propagating within the fiber (also defining the single-mode region) which has consequent effects on both the fiber dispersion (i.e. intermodal) and, possibly, the fiber attenuation (i.e. differential attenuation of modes). The numerical aperture (NA) is defined for a step index fiber in air by Eq. (2.8) as:

$$NA = \sin \theta_a = (n_1^2 - n_2^2)^{\frac{1}{2}} \tag{13.28}$$

where θ_a is the maximum acceptance angle, n_1 is the core refractive index and n_2 is the cladding refractive index. It is assumed in Eq. (13.28) that the light is incident on the fiber end face from air with a refractive index (n_0) of unity. Although Eq. (13.28) may be employed with graded index fibers, the numerical aperture thus defined represents only the local NA of the fiber on its core axis (the numerical aperture for light incident at the fiber core axis). The graded profile creates a multitude of local NAs as the refractive index changes radially from the core axis. For the general case of a graded index fiber these local numerical apertures $NA(r)$ at different radial distances r from the core axis may be defined by:

$$NA(r) = \sin \theta_a(r) = (n_1^2(r) - n_2^2)^{\frac{1}{2}} \tag{13.29}$$

Therefore, calculations of numerical aperture from refractive index data are likely to be less accurate for graded index fibers than for step index fibers unless the complete refractive index profile is considered. However, if refractive index data is available on either fiber type from the measurements described in Section 13.4, the numerical aperture may be determined by calculation.

Alternatively, a simple commonly used technique for the determination of the fiber numerical aperture is now described by FOTP-177 [Ref. 50] and involves measurement of the far field radiation pattern from the fiber. This measurement may be performed by directly measuring the far field angle from the fiber using a

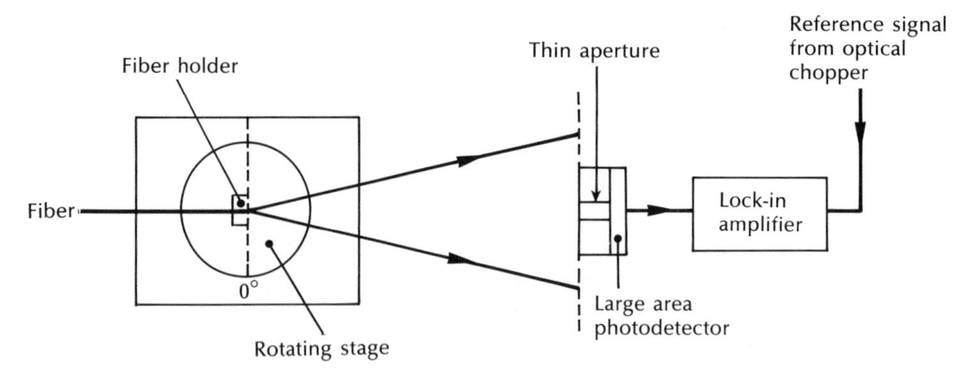

Figure 13.21 Fiber numerical aperture measurement using a scanning photodetector and a rotating stage.

rotating stage, or by calculating the far field angle using trigonometry. An example of an experimental arrangement with a rotating stage is shown in Figure 13.21. A 2 m length of the graded index fiber has its faces prepared in order to ensure square smooth terminations. The fiber output end is then positioned on the rotating stage with its end face parallel to the plane of the photodetector input, and so that its output is perpendicular to the axis of rotation. Light at a wavelength of 0.85 μm is launched into the fiber at all possible angles (overfilling the fiber) using an optical system similar to that used in the spot attenuation measurements (Figure 13.4).

The photodetector, which may be either a small area device or an apertured large area device, is placed 10 to 20 cm from the fiber and positioned in order to obtain a maximum signal with no rotation (0°). Hence when the rotating stage is turned the limits of the far field pattern may be recorded. The output power is monitored and plotted as a function of angle; the maximum acceptance angle being obtained when the power drops to 5% of the maximum intensity [Ref. 51]. Thus the numerical aperture of the fiber can be obtained from Eq. (13.28). This far field scanning measurement may also be performed with the photodetector located on a rotational stage and the fiber positioned at the centre of rotation. Moreover FOTP-177 also outlines a technique to obtain the numerical aperture from the refractive index profile of the fiber.

A less precise measurement of the numerical aperture can be obtained from the far field pattern by trigonometric means. The experimental apparatus is shown in Figure 13.22 where the end prepared fiber is located on an optical base plate or slab. Again light is launched into the fiber under test over the full range of its numerical aperture, and the far field pattern from the fiber is displayed on a screen which is positioned a known distance D from the fiber output end face. The test fiber is then aligned so that the optical intensity on the screen is maximized. Finally, the pattern size on the screen A is measured using a calibrated vernier caliper. The numerical

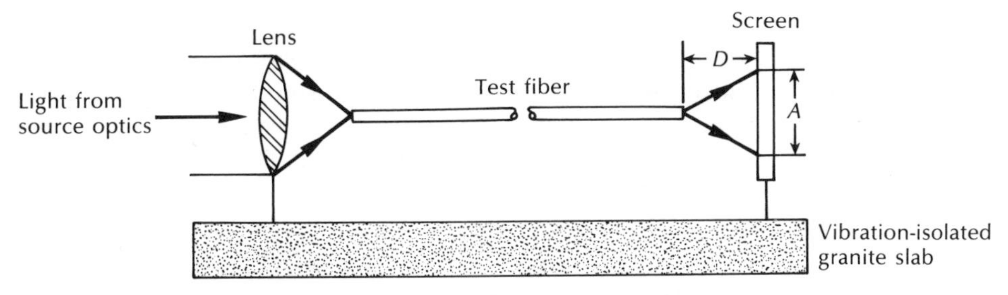

Figure 13.22 Apparatus for trigonometric fiber numerical aperture measurement.

aperture can be obtained from simple trigonometrical relationships where:

$$NA = \sin \theta_a = \frac{A/2}{[(A/2)^2 + D^2]^{\frac{1}{2}}} = \frac{A}{(A^2 + 4D^2)^{\frac{1}{2}}} \tag{13.30}$$

Example 13.5

A trigonometrical measurement is performed in order to determine the numerical aperture of a step index fiber. The screen is positioned 10.0 cm from the fiber end face. When illuminated from a wide angled visible source the measured output pattern size is 6.2 cm. Calculate the approximate numerical aperture of the fiber.

Solution: The numerical aperture may be determined directly, using Eq. (13.30) where:

$$NA = \frac{A}{(A^2 + 4D^2)^{\frac{1}{2}}} = \frac{6.2}{(38.44 + 400)^{\frac{1}{2}}} = 0.30$$

It must be noted that the accuracy of this measurement technique is dependent upon the visual assessment of the far field pattern from the fiber.

The above measurement techniques are generally employed with multimode fibers only, as the far field patterns from single-mode fibers are affected by diffraction phenomena. These are caused by the small core diameters of single-mode fibers which tend to invalidate simple geometric optics measurements. However, more detailed analysis of the far field pattern allows determination of the normalized frequency and core radius for single-mode fibers, from which the numerical aperture may be calculated using Eq. (2.69) [Ref. 52].

Far field pattern measurements with regard to multimode fibers are dependent on the length of the fiber tested. When the measurements are performed on short fiber lengths (around 1 m) the numerical aperture thus obtained corresponds to that defined by Eq. (13.28) or (13.29). However, when a long fiber length is utilized which gives mode coupling and the selective attenuation of the higher order modes, the measurement yields a lower value for the numerical aperture. It must also be noted that the far field measurement techniques give an average (over the local NAs) value for the numerical aperture of graded index fibers. Hence alternative methods must be employed if accurate determination of the fiber's NA is required [Ref. 53].

13.7 Fiber diameter measurements

13.7.1 Outer diameter

It is essential during the fiber manufacturing process (at the fiber drawing stage) that the fiber outer diameter (cladding diameter) is maintained constant to within 1%. Any diameter variations may cause excessive radiation losses and make accurate fiber–fiber connection difficult. Hence on-line diameter measurement systems are required which provide accuracy better than 0.3% at a measurement rate greater than 100 Hz (i.e. a typical fiber drawing velocity is 1 m s^{-1}). Use is therefore made of noncontacting optical methods such as fiber image projection and scattering pattern analysis.

The most common on-line measurement technique uses fiber image projection (shadow method) and is illustrated in Figure 13.23 [Ref. 54]. In this method a laser beam is swept at a constant velocity transversely across the fiber and a measurement

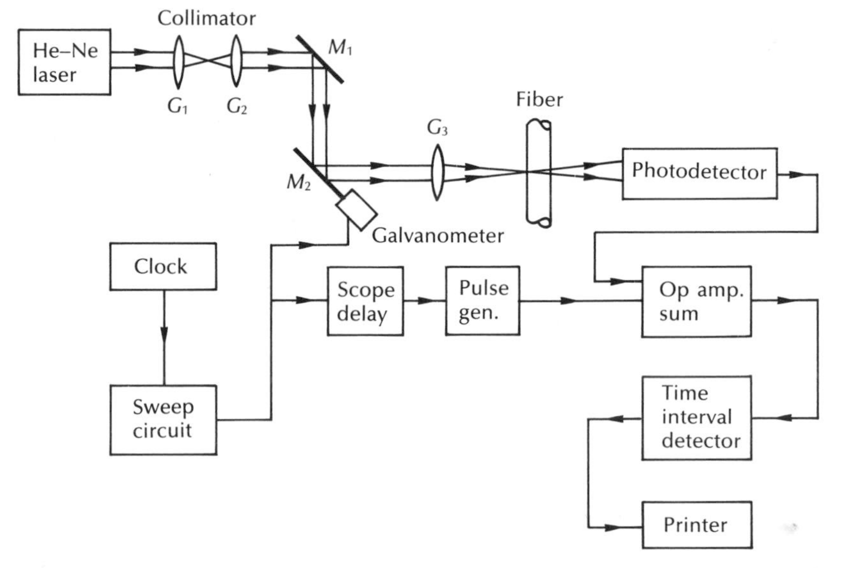

Figure 13.23 The shadow method for the on-line measurement of the fiber outer diameter [Ref. 54].

is made of the time interval during which the fiber intercepts the beam and casts a shadow on a photodetector. In the apparatus shown in Figure 13.23 the beam from a laser operating at a wavelength of 0.6328 μm is collimated using two lenses (G_1 and G_2). It is then reflected off two mirrors (M_1 and M_2), the second of which (M_2) is driven by a galvanometer which makes it rotate through a small angle at a constant angular velocity before returning to its original starting position. Therefore, the laser beam which is focused in the plane of the fiber by a lens (G_3) is swept across the fiber by the oscillating mirror, and is incident on the photodetector unless it is blocked by the fiber. The velocity $\mathrm{d}s/\mathrm{d}t$ of the fiber shadow thus created at the photodetector is directly proportional to the mirror velocity $\mathrm{d}\phi/\mathrm{d}t$ following:

$$\frac{\mathrm{d}s}{\mathrm{d}t} = l\,\frac{d\phi}{\mathrm{d}t} \tag{13.31}$$

where l is the distance between the mirror and the photodetector.

Furthermore, the shadow is registered by the photodetector as an electrical pulse of width W_e which is related to the fiber outer diameter d_o as:

$$d_o = W_e\,\frac{\mathrm{d}s}{\mathrm{d}t} \tag{13.32}$$

Thus the fiber outer diameter may be quickly determined and recorded on the printer. The measurement speed is largely dictated by the inertia of the mirror rotation and its accuracy by the rise time of the shadow pulse.

Example 13.6

The shadow method is used for the on-line measurement of the outer diameter of an optical fiber. The apparatus employs a rotating mirror with an angular velocity of 4 rad s^{-1} which is located 10 cm from the photodetector. At a particular instant in time a shadow pulse of width 300 μs is registered by the photodetector. Determine the outer diameter of the optical fiber in μm at this instant in time.

Solution: The shadow velocity may be obtained from Eq. (13.31) where:

$$\frac{\mathrm{d}s}{\mathrm{d}t} = l\,\frac{\mathrm{d}\phi}{\mathrm{d}t} = 0.1 \times 4 = 0.4\ \mathrm{m\,s^{-1}}$$

$$= 0.4\ \mu\mathrm{m}\,\mu\mathrm{s}^{-1}$$

Hence the fiber outer diameter d_o in μm is given by Eq. (5.24):

$$d_o = W_e\,\frac{\mathrm{d}s}{\mathrm{d}t} = 300\ \mu\mathrm{s} \times 0.4\ \mu\mathrm{m}\,\mu\mathrm{s}^{-1}$$

$$= 120\ \mu\mathrm{m}$$

Other on-line measurement methods, enabling faster diameter measurements, involve the analysis of forward or backward far field patterns which are produced when a plane wave is incident transversely on the fiber. These techniques generally require measurement of the maxima in the centre portion of the scattered pattern from which the diameter can be calculated after detailed mathematical analysis [Refs. 55 to 58]. They tend to give good accuracy (e.g. ± 0.25 μm [Ref. 58]) even though the theory assumes a perfectly circular fiber cross section. Also for step index fibers the analysis allows determination of the core diameter, and core and cladding refractive indices.

Measurements of the fiber outer diameter after manufacture (off-line) may be performed using a micrometer or dial gage. These devices can give accuracies of the order of ± 0.5 μm. Alternatively, off-line diameter measurements can be made with a microscope incorporating a suitable calibrated micrometer eyepiece.

13.7.2 Core diameter

The core diameter for step index fibers is defined by the step change in the refractive index profile at the core–cladding interface. Therefore the techniques employed for determining the refractive index profile (interferometric, near field scanning, refracted ray, etc.) may be utilized to measure the core diameter. Graded index fibers present a more difficult problem as, in general, there is a continuous transition between the core and the cladding. In this case it is necessary to define the core as an area with a refractive index above a certain predetermined value if refractive index profile measurements are used to obtain the core diameter.

Core diameter measurement is also possible from the near field pattern of a suitably illuminated (all guided modes excited) fiber. The measurements may be taken using a microscope equipped with a micrometer eyepiece similar to that employed for off-line outer diameter measurements. However, the core–cladding interface for graded index fibers is again difficult to identify due to fading of the light distribution towards the cladding, rather than the sharp boundary which is exhibited in the step index case. Nevertheless, details of the above measurement procedures are provided in FOTP-58 [Ref. 59].

13.8 Mode-field diameter for single-mode fiber

It was indicated in Section 2.5.2 that for single-mode fiber the geometric distribution of light in the propagating mode rather than the core diameter or numerical aperture is what is important in predicting the operational properties such as waveguide dispersion, launching and jointing losses, and microbending loss. In particular, the mode-field diameter (MFD) which is a measure of the width of the distribution of the electric field intensity is used to predict many of these properties. Alternatively, the spot size which is simply equal to half the MFD, or the mode-field radius, is utilized.

Since the field of the fundamental mode of a circularly symmetric fiber is bell shaped and exhibits circular symmetry (see Figure 2.31), not only is its extent readily described by the MFD, but it can be expressed in terms of both the near field (i.e. the optical field distribution on the output face of the fiber) and the far field (i.e. the radiation pattern at larger distances, typically a few millimetres, from the fiber end face) distributions [Ref. 60]. Hence direct measurement of the MFD may be obtained using either near field or far field scanning techniques. These basic methods are covered for standardization purposes in FOTP-165 [Ref. 61] and FOTP-164 [Ref. 62] respectively.

A typical experimental arrangement for measurement of the near field intensity distribution using a scanning fiber is illustrated in Figure 13.24 [Ref. 60]. As would be expected it has distinct similarities to the experimental setup for the near field scanning of the refractive index profile shown in Figure 13.14. The arrangement utilizes a relatively intense light source (an LED or injection laser) operating at the desired wavelength to inject optical power into the fiber under test. A lens system is required to magnify the fiber output end, the image of which is scanned across a diameter using another fiber on a motor driven translation stage pigtailed to a small area photodiode. The near field MFD, d_n may be obtained using [Ref. 60]:

$$d_n = 2\sqrt{2}\left\{\frac{\int_0^\infty E^2(r)r^3 \, dr}{\int_0^\infty E^2(r)r \, dr}\right\}^{\frac{1}{2}} \tag{13.33}$$

where $|E(r)|^2$ is the local near field intensity at radius r. Equation (13.33) assumes a non-Gaussian field distribution in which the near field MFD is proportional to the rms width of the near field distribution. The numerical integration of the local

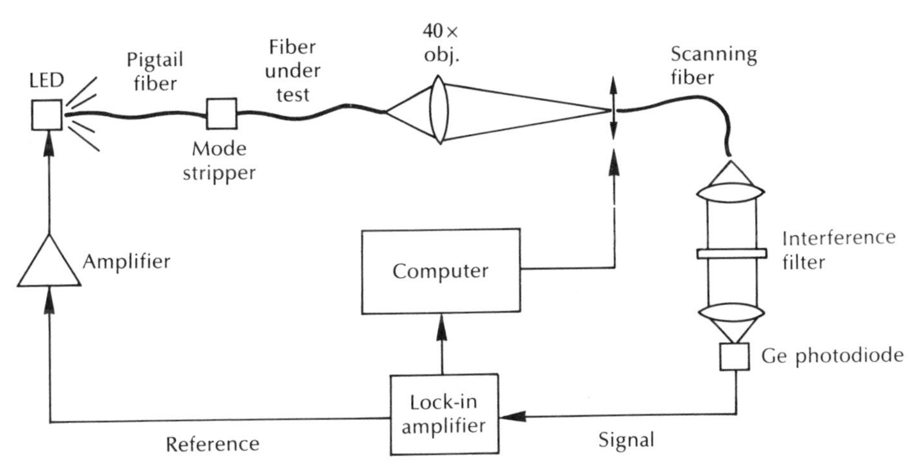

Figure 13.24 Experimental setup for near field intensity distribution measurements (near field scanning) to obtain mode-field diameter.

measured near field intensities at intervals determined by the dynamic range of the setup thus allows d_n to be calculated. Although the near field scanning technique provides a direct way to measure the MFD, the method suffers from inaccuracies resulting from lens distortion, difficulties in locating and stably holding the image plane at the detector, and a limited dynamic range with only a small portion of the optical power reaching the photodetector.

Another direct MFD measurement technique is obtained by scanning the far field intensity distribution. This method is very straightforward to implement, as shown in Figure 13.25. The experimental arrangement required comprises a high intensity light source (an injection laser is normally needed) and a photodetector mounted on a motor driven rotational stage. It is necessary that the far field intensity pattern be detected at a sufficiently large distance from the centre of the fiber output end such that good angular resolution is achieved in detection. When using a pigtailed injection laser source, however, this distance may be as low as a few millimetres. Furthermore, the angular sector scanned in front of the fiber must be sufficiently wide (between ± 20 and 25°) to completely include the main lobe of the radiation pattern. In particular, this aspect is critical when dispersion modified fibers are scanned because they exhibit broad far field distributions.

The far field MFD d_f can be obtained directly by inserting the measured far field intensities into [Ref. 6]:

$$d_f = \frac{\sqrt{2}}{\pi} \left\{ \frac{F(\theta) \sin \theta \cos \theta \, d\theta}{F(\theta) \sin^3 \theta \cos \theta \, d\theta} \right\}^{\frac{1}{2}} \tag{13.34}$$

where $F(\theta)$ corresponds to the measured data. Again, the integration can be performed numerically. It should be noted, however, that in this case the rms far field, or Petermann II [Ref. 63] definition has been adopted by the EIA in FOTP-164. This definition applies to non-Gaussian measurements and is particularly appropriate for dispersion modified fiber operating at a wavelength of 1.55 μm. Other integrative far field methods also include various aperture techniques, two of which are reported in standards, namely; the variable aperture method in FOTP-167 [Ref. 64]; and the knife-edge method in FOTP-174 [Ref. 65].

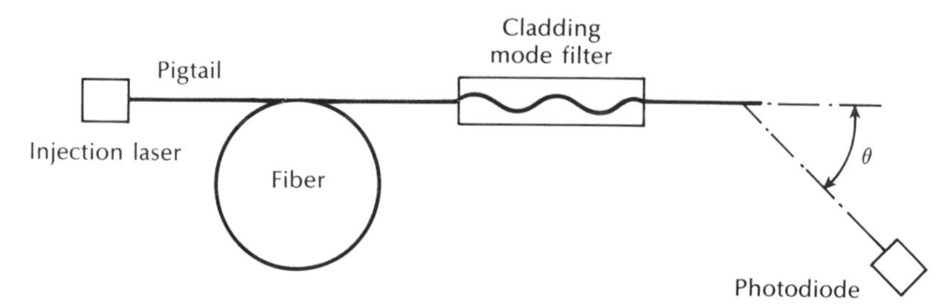

Figure 13.25 Experimental arrangement for far field intensity distribution measurements (far field scanning) to obtain mode-field diameter.

Finally, an indirect method for the measurement of the MFD which has proved popular is the transverse offset technique [Refs. 43, 60, 66, 67]. It overcomes some of the drawbacks associated with the near and far field methods by measurement of the power transmitted through a mechanical butt splice as one of the fibers is swept transversely through the alignment position. The experimental apparatus is shown in Figure 13.26 which employs the same single-mode fiber on either side of the joint. This technique makes use of the dependence of splice loss on spot size for Gaussian modes. Hence the variation of transmitted power with offset, $P(u)$, which is measured on a high precision translation stage, can be fitted to the expected Gaussian dependence. For the case of identical fibers with an MFD of $2\omega_o$ this is given by [Ref. 67]:

$$P(u) = P_o \exp\left(\frac{-u^2}{2\omega_o^2}\right) \tag{13.35}$$

where u is the offset and P_o is the maximum transmitted power. The means of fit to Eq. (13.35) is very important as the pattern departs from the Gaussian distribution. Moreover, it has been found that an unweighted truncated fit with the truncation de-emphasizing the Gaussian tails gives good agreement with near field and far field techniques for circularly symmetric single-mode fiber [Ref. 67].

The transverse offset technique has several advantages; in particular it is efficient in its use of optical power since most of the light is intercepted and transmitted, in contrast to the near field method. Furthermore, it is possible to use a tungsten lamp and monochromatic combination to provide a tunable optical source which allows easy measurement of the MFD as a function of wavelength. The technique therefore lends itself to the determination of the cutoff wavelength as mentioned in Section 13.5. In addition, it is relatively rapid, quite accurate (with less than 2%

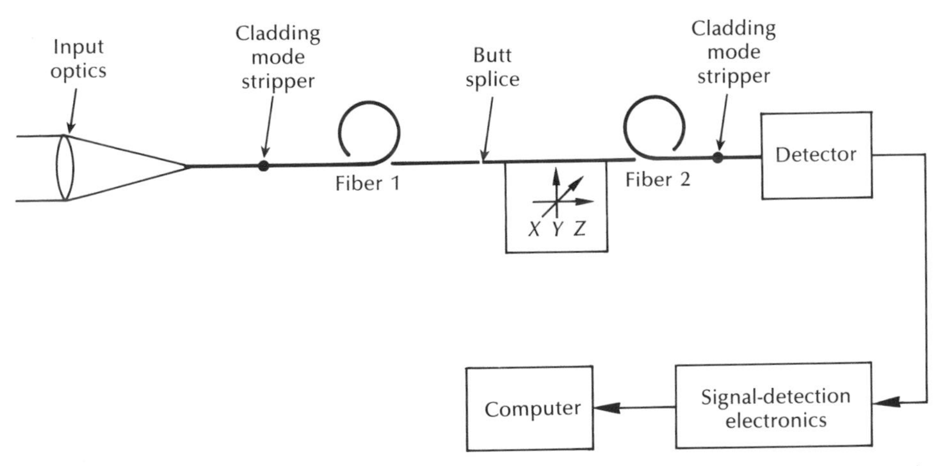

Figure 13.26 Experimental setup for the measurement of mode-field diameter by transverse offset technique.

error in spot size [Ref. 68]) and does not require complex mathematical evaluation. Finally, it is a technique which is described in CCITT G.652 [Ref. 1].

13.9 Reflectance and optical return loss

It was indicated in Section 6.7.4 that reflections along a fiber link (i.e. optical feedback) can adversely affect injection laser stability. Furthermore, multiple reflections can contribute to the noise levels at the optical detector. Fresnel reflection r occurs at a fiber–air interface, as discussed in Section 5.2, giving a reflectance of around 4% or −14 dB. The optical return loss (ORL) is therefore defined as [Ref. 6]:

$$ORL = -10 \log_{10} r \tag{13.36}$$

It should be noted that the term reflectance is sometimes utilized when referring to single components whereas the optical return loss applies to a series of components, including the fiber, along a link.

Low values of reflectance can be obtained with fusion splicing and with carefully designed mechanical joints. For example, the use of index matching gel can substantially reduce reflections. Nevertheless, certain mechanisms can cause larger values of reflectance. These include optical interference produced in the cavity between two fiber end faces as well as reflection from a high index layer formed on the end face of a highly polished fiber. Ideally, the optical return loss needs to be maintained at levels above 40 dB to avoid detrimental effects on the performance of the fiber link [Ref. 6].

Optical return loss measurements can be performed using an optical continuous wave reflectometer (OCWR), as described in FOTP-107 [Ref. 69]. In this arrangement, shown in Figure 13.27, a continuous wave LED or injection laser source is connected to the input port 1 of a four port coupler and a detector is connected to input port 2. Then a jumper cable with the reflecting components to be measured is spliced to output port 4 and output port 3 is made nonreflecting using an index matching gel or a tight fiber loop. The optical power P_r at port 2, which results from reflections caused by the components and the coupler, is thus

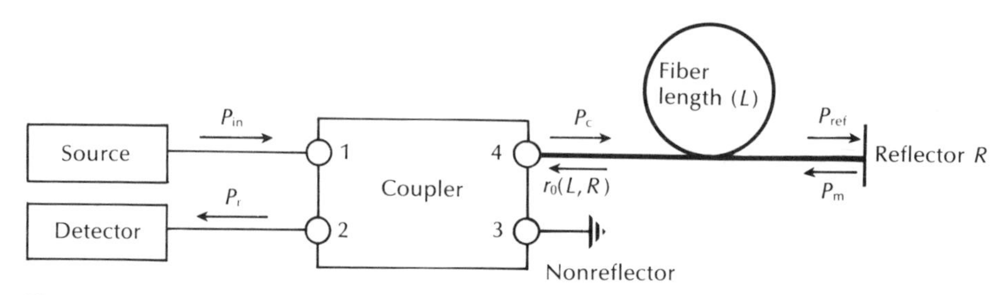

Figure 13.27 Optical return loss measurement using a four port coupler.

measured. Next the jumper cable is removed and replaced by a nonreflecting termination. This allows P_c due only to the coupler to be measured at port 2. The detector is then transferred to port 4 and the power incident upon the reflector P_{ref} is measured.

Apart from the loss in transmission between port 4 and port 2, the fraction of the reflected power from the components under test is $(P_r - P_c)/P_{ref}$. To obtain the ports 4 to 2 loss, the source and detector are connected to ports 4 and 2, respectively, providing a measurement P_{out}. Finally, a power P_{in} is measured by connecting the source directly to the detector such that P_{out}/P_{in} is the fraction of optical power transmitted between port 4 and port 2. Hence the optical return loss is given by:

$$ORL = 10 \log_{10} \left(\frac{P_{out} P_{ref}}{P_{in}(P_r - P_c)} \right) \tag{13.37}$$

The OCWR is a d.c. instrument and only provides a measurement of the overall optical return loss for a component on a link; it does not allow information on the location of a number of reflecting components to be obtained. A device which can, however, provide this information, albeit in a more complex manner, is the optical time domain reflectometer which is described in Section 13.10.1.

13.10 Field measurements

The measurements discussed in the preceding sections are primarily suited to the laboratory environment where quite sophisticated instrumentation may be used. However, there is a requirement for the measurement of the transmission characteristics of optical fibers when they are located in the field within an optical communication system. It is essential that optical fiber attenuation and dispersion measurements, connector and splice loss measurements and fault location be performed on optical fiber links in the field. Although information on fiber attenuation and dispersion is generally provided by the manufacturer, this is not directly applicable to cabled, installed fibers which are connected in series within an optical fiber system. Effects such as microbending (see Section 4.8.1) with the resultant mode coupling (see Section 2.4.2) affect both the fiber attenuation and dispersion. It is also found that the simple summation of the transmission parameters with regard to individual connected lengths of fiber cable does not accurately predict the overall characteristics of the link [Ref. 70]. Hence test equipment has been developed which allows these transmission measurements to be performed in the field.

In general, field test equipment differs from laboratory instrumentation in a number of aspects as it is required to meet the exacting demands of field measurement. Therefore the design criteria for field measurement equipment include:

1. Sturdy and compact encasement which must be portable.
2. The ready availability of electrical power must be ensured by the incorporation of batteries or by connection to a generator. Hence the equipment should maintain accuracy under conditions of varying supply voltage and/or frequency.
3. In the event of battery operation, the equipment must have a low power consumption.
4. The equipment must give reliable and accurate measurements under extreme environmental conditions of temperature, humidity and mechanical load.
5. Complicated and involved fiber connection arrangements should be avoided. The equipment must be connected to the fiber in a simple manner without the need for fine or critical adjustment.
6. The equipment cannot usually make use of external triggering or regulating circuits between the transmitter and receiver due to their wide spacing on the majority of optical links.

Even if the above design criteria are met, it is likely that a certain amount of inaccuracy will have to be accepted with field test equipment. For example, it may not be possible to include adjustable launching conditions (i.e. variation in spot size and numerical aperture) in order to create the optimum. Also, because of the large dynamic range required to provide measurements over long fiber lengths, lossy devices such as mode scramblers may be omitted. Therefore measurement accuracy may be impaired through inadequate simulation of the equilibrium mode distribution.

A number of portable, battery-operated optical power meters are commercially available. Some of these instruments are of small dimension and therefore are designed to be hand-held, whilst others, which generally provide greater accuracy and stability, are slightly larger in size. A typical example of the latter type is shown in Figure 13.28. Such devices usually measure optical power in dBm or dBμ (i.e. 0 dBm is equivalent to 1 mW and 0 dBμ is equivalent to 1 μW; see Example 13.7) over a specified range (e.g. 0.38 to 1.15 μm or 0.75 to 1.7 μm). In most cases the spectral range is altered by the incorporation of different demountable sensor heads (i.e. wide area photodiodes). For example, the optical power meter displayed in Figure 13.28 can be used with five different sensor heads (three of which are shown in the foreground of Figure 13.28) comprising either silicon, germanium or InGaAs photodiodes. The device is also specified to have a measurement range from -100 dBm (0.1 pW) to $+3$ dBm (2 mW) with an accuracy of $\pm 5\%$ when employing the latter sensor head.

It must be noted, however, that although these instruments often take measurements over a certain spectral range this simply implies that they may be adjusted to be compatible with the centre emission frequency of particular optical sources so as to obtain the most accurate reading of optical power. Therefore, the devices do not generally give spectral attenuation measurements unless the source optical output frequency is controlled or filtered to achieve single wavelength operation. Optical power meters may be used for measurement of the absolute

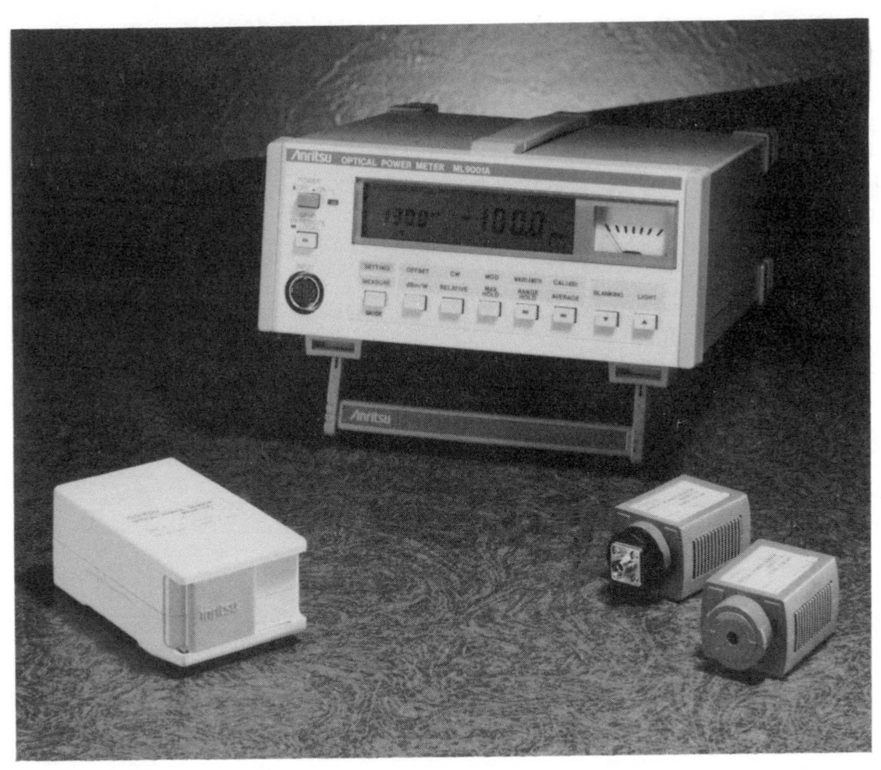

Figure 13.28 The Anritsu ML9001A optical power meter. Courtesy of Anritsu Europe Limited.

optical attenuation on a fiber link by employing the cut-back technique. Other optical system parameters which may also be obtained using such instruments are the measurement of individual splice and connector losses, the determination of the absolute optical output power emitted from the source (see Sections 6.5 and 7.4.1) and the measurement of the responsivity or the absolute photocurrent of the photodetector in response to particular levels of input optical power (see Section 8.6).

Example 13.7

An optical power meter records optical signal power in either dBm or dBμ.

(a) Convert the optical signal powers of 5 mW and 20 μW to dBm.
(b) Convert optical signal powers of 0.3 mW and 80 nW to dBμ.

Solution: The optical signal power can be expressed in decibels using:

$$\text{dB} = 10 \log_{10} \left(\frac{P_o}{P_r}\right)$$

where P_o is the received optical signal power and P_r is a reference power level.

(a) For a 1 mW reference power level:

$$\text{dBm} = 10 \log_{10} \left(\frac{P_o}{1 \text{ mW}}\right)$$

Hence an optical signal power of 5 mW is equivalent to

$$\textit{Optical signal power} = 10 \log_{10} 5 = 6.99 \text{ dBm}$$

and an optical power of 20 μW is equivalent to:

$$\begin{aligned}\textit{Optical signal power} &= 10 \log_{10} 0.02 \\ &= -16.99 \text{ dBm}\end{aligned}$$

(b) For a 1 μW reference power level:

$$\text{dB}\mu = 10 \log_{10} \left(\frac{P_o}{1 \ \mu\text{W}}\right)$$

Therefore an optical signal power of 0.3 mW is equivalent to:

$$\begin{aligned}\textit{Optical signal power} = 10 \log_{10} \left(\frac{P_o}{1 \ \mu\text{W}}\right) &= 10 \log_{10} 30 \\ &= 14.77 \text{ dB}\mu\end{aligned}$$

and an optical signal power of 800 nW is equivalent to:

$$\begin{aligned}\textit{Optical signal power} &= 10 \log_{10} 0.8 \\ &= -0.97 \text{ dB}\mu\end{aligned}$$

There are a number of portable measurement test sets specifically designed for fiber attenuation measurements which require access to both ends of the optical link. These devices tend to use the cut-back measurement technique unless correction is made for any difference in connector losses between the link and a short length of similar reference cable. A block schematic of an optical attenuation meter consisting of a transmitter and receiver unit is shown in Figure 13.29 [Ref. 70]. Reproducible readings may be obtained by keeping the launched optical power from the light source absolutely constant. A constant optical output power is achieved with the equipment illustrated in Figure 13.29 using an injection laser and a regulating circuit which is driven from a reference output of the source derived from a photodiode. Hence any variations in the laser output power are rectified by automatic adjustment of the modulating voltage, and therefore current, from the

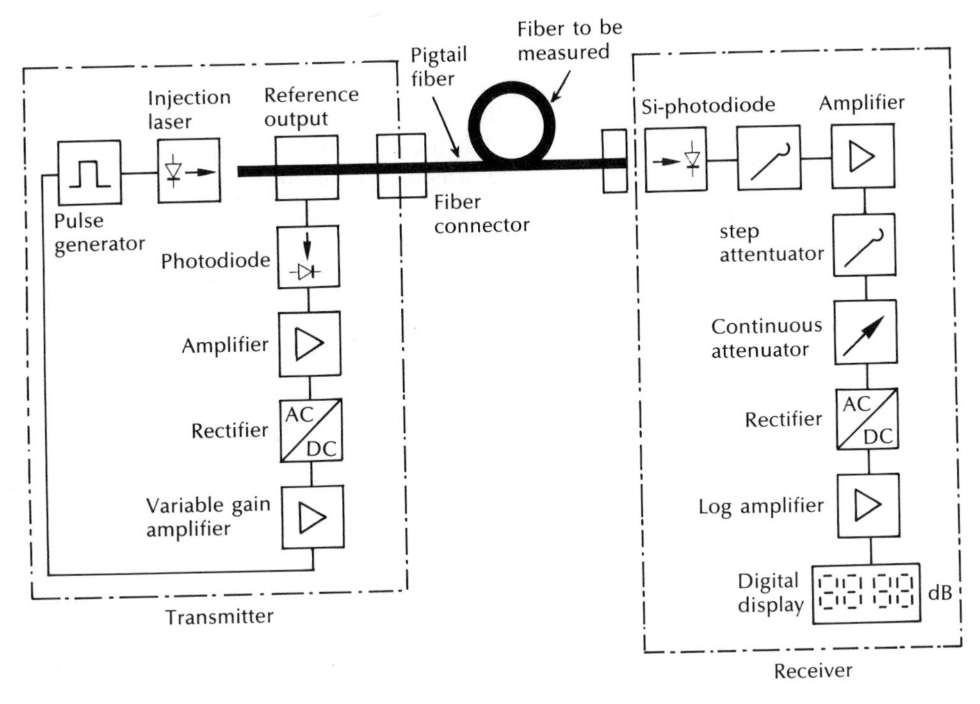

Figure 13.29 An optical attenuation meter [Ref. 70].

pulse generator. A large area photodiode is utilized in the receiver to eliminate any effects from differing fiber end faces. It is generally found that when a measurement is made on multimode fiber a short cut-back reference length of a few metres is insufficient to obtain an equilibrium mode distribution. Hence unless a mode scrambling device together with a mode stripper are used, it is likely that a reference length of around 500 m or more will be required if reasonably accurate measurements are to be made. When measurements are made without a steady state mode distribution in the reference fiber a significantly higher loss value is obtained which may be as much as 1 dB km^{-1} above the steady state attenuation [Refs. 22 and 71].

Several field test sets are available for making dispersion measurements on optical fiber links. These devices generally consist of transmitter and receiver units which take measurements in the time domain. Short light pulses ($\simeq$200 ns) are generated from an injection laser and are broadened by transmission down the optical link before being received by a fast response photodetector (i.e. avalanche photodiode) and displayed on a sampling oscilloscope. This is similar to the dispersion measurements in the time domain discussed in Section 13.3. If it is assumed that the pulses have a near Gaussian shape, Eq. (13.13) may be utilized to determine the pulse broadening on the link, and hence the 3 dB optical bandwidth may be obtained.

13.10.1 Optical time domain reflectometry (OTDR)

A measurement technique which is far more sophisticated and which finds wide application in both the laboratory and the field is the use of optical time domain reflectometry (OTDR). This technique is often called the backscatter measurement method. It provides measurement of the attenuation on an optical link down its entire length giving information on the length dependence of the link loss. In this sense it is superior to the optical attenuation measurement methods discussed previously (Section 13.2) which only tend to provide an averaged loss over the whole length measured in dB km^{-1}. When the attenuation on the link varies with length, the averaged loss information is inadequate. OTDR also allows splice and connector losses to be evaluated as well as the rotation of any faults on the link. It relies upon the measurement and analysis of the fraction of light which is reflected back within the fiber's numerical aperture due to Rayleigh scattering (see Section 3.4.1). Hence the backscattering method which was first described by Barnoski and Jensen [Ref. 72] has the advantages of being nondestructive (i.e. does not require the cutting back of the fiber) and of requiring access to one end of the optical link only.

The backscattered optical power as a function of time $P_{Ra}(t)$ may be obtained from the following relationship [Ref. 73]:

$$P_{Ra}(t) = \tfrac{1}{2} P_i S \gamma_R W_o v_g \exp(-\gamma v_g t) \tag{13.38}$$

where P_i is the optical power launched into the fiber, S is the fraction of captured optical power, γ_R is the Rayleigh scattering coefficient (backscatter loss per unit length), W_o is the input optical pulse width, v_g is the group velocity in the fiber and γ is the attenuation coefficient per unit length for the fiber. The fraction of captured optical power S is given by the ratio of the solid acceptance angle for the fiber to the total solid angle as:

$$S \simeq \frac{\pi(NA)^2}{4\pi n_1^2} = \frac{(NA)^2}{4n_1^2} \tag{13.39}$$

It must be noted that the relationship given in Eq. (13.39) applies to step index fibers and the parameter S for a graded index fiber is generally a factor of 2/3 lower than for a step index fiber with the same numerical aperture [Ref. 74]. Hence using Eqs. (13.38) and (13.39) it is possible to determine the backscattered optical power from a point along the link length in relation to the forward optical power at that point.

Example 13.8

An optical fiber link consists of multimode step index fiber which has a numerical aperture of 0.2 and a core refractive index of 1.5. The Rayleigh scattering coefficient for the fiber is 0.7 km^{-1}. When light pulses of 50 ns duration are launched into the fiber, calculate the ratio in decibels of the backscattered optical power to the forward optical power at the fiber input. The velocity of light in a vacuum is 2.998×10^8 ms^{-1}.

Solution: The backscattered optical power $P_{Ra}(t)$ is given by Eq. (13.38) where:

$$P_{Ra}(t) = \tfrac{1}{2} P_o S \gamma_R W_o v_g \exp(-\gamma v_g t)$$

At the fiber input $t = 0$; hence the power ratio is:

$$\frac{P_{Ra}(0)}{P_i} = \tfrac{1}{2} S \gamma_R W_o v_g$$

Substituting for S from Eq. 5.26) gives:

$$\frac{P_{Ra}(0)}{P_i} = \frac{1}{2}\left[\frac{(NA)^2 \gamma_R W_o v_g}{4n_1^2}\right]$$

The group velocity in the fiber v_g is defined by Eq. (2.40) as:

$$v_g = \frac{c}{N_g} \simeq \frac{c}{n_1}$$

Therefore

$$\frac{P_{Ra}(0)}{P_i} = \frac{1}{2}\left[\frac{NA^2 \gamma_R W_o c}{4n_1^3}\right]$$

$$= \frac{1}{2}\left[\frac{(0.02)^2 0.7 \times 10^{-3} \times 50 \times 10^{-9} \times 2.998 \times 10^8}{4(1.5)^3}\right)$$

$$= 1.555 \times 10^{-5}$$

In decibels

$$\frac{P_{Ra}(0)}{P_i} = 10 \log_{10} 1.555 \times 10^{-5}$$

$$= 48.1 \text{ dB}$$

Hence in Example 13.9 the backscattered optical power at the fiber input is 48.1 dB down on the forward optical power. The backscattered optical power should not be confused with any Fresnel reflection at the fiber input end face resulting from a refractive index mismatch. This could be considerably greater than the backscattered light from the fiber, presenting measurement problems with OTDR if it is allowed to fall on to the receiving photodetector of the equipment described below.

A block schematic of the backscatter measurement method is shown in Figure 13.30 [Ref. 75]. A light pulse is launched into the fiber in the forward direction from an injection laser using either a directional coupler or a system of external

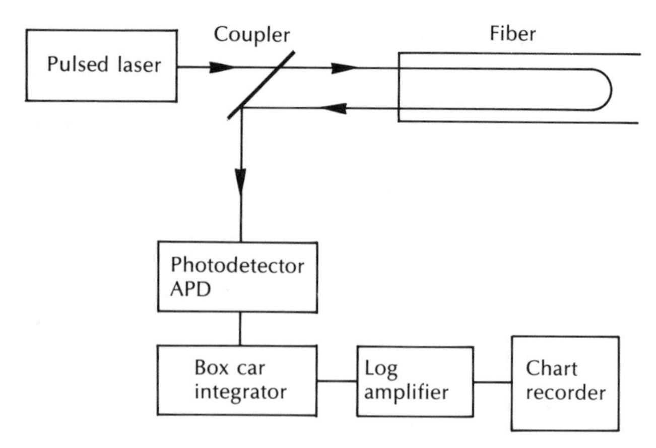

Figure 13.30 Optical time domain reflectometry or the backscatter measurement method.

lenses with a beam splitter (usually only in the laboratory). The backscattered light is detected using an avalanche photodiode receiver which drives an integrator in order to improve the received signal to noise ratio by giving an arithmetic average over a number of measurements taken at one point within the fiber. This is necessary as the received optical signal power from a particular point along the fiber length is at a very low level compared with the forward power at that point by some 45 to 60 dB (see Example 13.9), and is also swamped with noise. The signal from the integrator is fed through a logarithmic amplifier and averaged measurements for successive points within the fiber are plotted on a chart recorder. This provides location-dependent attenuation values which give an overall picture of the optical loss down the link. A possible backscatter plot is illustrated in Figure 13.31 [Ref. 76] which shows the initial pulse caused by reflection and backscatter from the input coupler followed by a long tail caused by the distributed Rayleigh scattering from the input pulse as it travels down the link. Also shown in the plot is a pulse corresponding to the discrete reflection from a fiber joint, as well as a discontinuity due to excessive loss at a fiber imperfection or fault. The end of the fiber link is indicated by a pulse corresponding to the Fresnel reflection incurred at the output end face of the fiber. Such a plot yields the attenuation per unit length for the fiber by simply computing the slope of the curve over the length required. Also the location and insertion losses of joints and/or faults can be obtained from the power drop at their respective positions on the link. Finally the overall link length can be determined from the time difference between reflections from the fiber input and output end faces. Standard methods for these measurements are covered in FOTP-59 to 61 [Refs. 77 to 79] and they provide very powerful techniques for field measurements on optical fiber links. In addition, FOTPs are in process for the measurement of splice or connector loss and the measurement of splice or connector return loss utilizing an OTDR [Ref. 6].

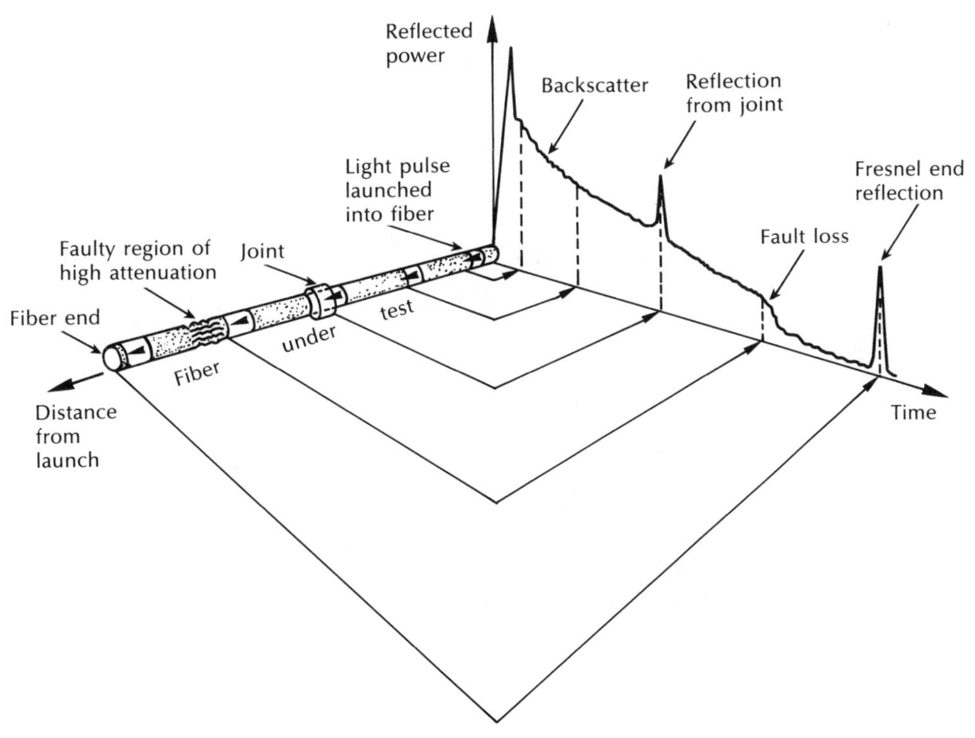

Figure 13.31 An illustration of a possible backscatter plot from a fiber under test [Ref. 76].

A number of optical time domain reflectometers are commercially available for operation both in the shorter and longer wavelength regions. The former instruments emit a series of short (10 to 100 ns), intense optical pulses (100 to 500 mW) from which the backscattered light is received, analysed and displayed on an oscilloscope, or plotted on a chart recorder. A typical example of a high performance OTDR is shown in Figure 13.32. This flexible instrument employs plug-in units to enable it to operate in both the longer wavelength region (at 1.31 μm and 1.55 μm), as well as the shorter wavelength window (0.85μm). The longer wavelength units produce pulse widths in the range 100 ns to 10 μs, the latter pulses providing one-way dynamic ranges of 34 dB and 30 dB at wavelengths of 1.31 μm and 1.55 μm respectively. Hence, when using the device with low loss single-mode fiber, measurements can be made at distance in excess of 150 km with a resolution of up to 10 cm. It must be noted, however, that the accuracy of these distance measurements is around ± 1 m.

A number of themes have been pursued over recent years in relation to OTDR performance. These include the enlargement of the device dynamic range, the enhancement of the device resolution, the reduction of noise levels intrinsic to

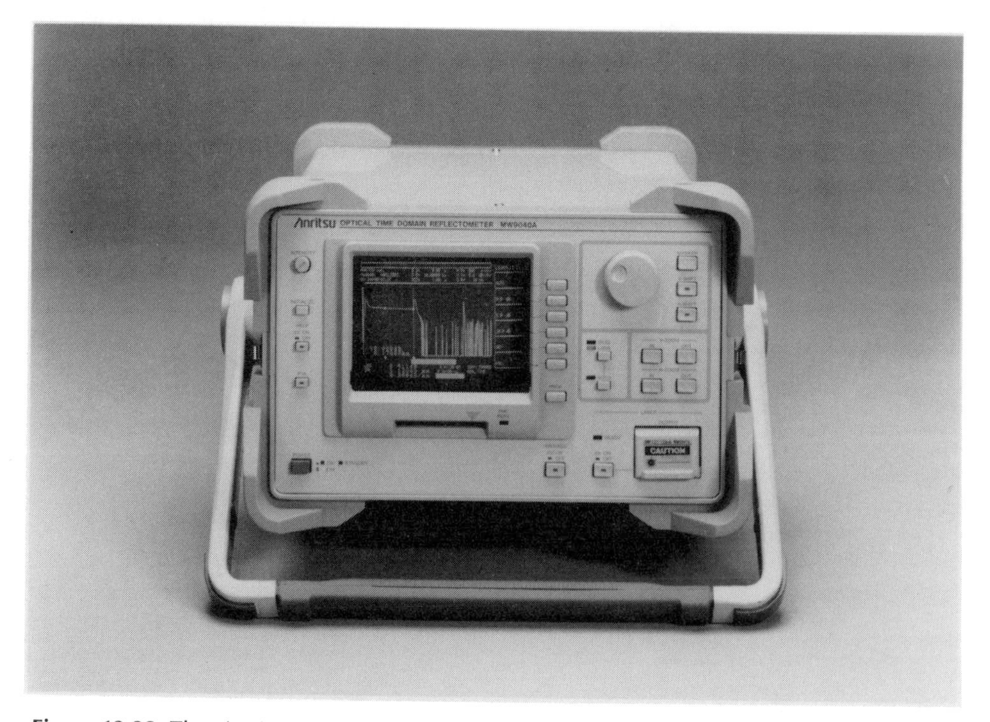

Figure 13.32 The Anritsu MW9040A optical time domain reflectometer. Courtesy of Anritsu Europe Limited.

single-mode fibers and the increase in the user friendliness of the equipment [Ref. 80]. Significant improvements have been obtained in the former two device performance characteristics with a range of strategies including the use of higher input optical power levels, decreasing the minimum detectable optical power and employing narrower pulse widths.

For example, one strategy which has proved successful is the use of a photon counting technique [Ref. 81] in which the backscattered photons are detected digitally. In this method the avalanche photodiode is operated in a Geiger tube breakdown mode [Ref. 82] by biasing the device above its normal operating voltage where it can detect a single photon. The photon counting technique has demonstrated significantly improved receiver sensitivity (i.e. -7 dB) than the best analog system at a wavelength of 1.3 μm. Moreover, a resolution of up to 1.5 cm with high sensitivity (3×10^{-10} W) has been reported when operating at a wavelength of 0.85 μm using a single photon detecting APD at room temperature [Ref. 83].

Single-mode fiber OTDRs exhibit an additional problem over multimode devices, namely, polarization noise. In general, the state of polarization of the backscattered light differs from that of the laser pulse coupled into the fiber at the input end and is dependent on the distance of the backscattering fiber element from the input fiber end. This results in an amplitude fluctuation in the backscattered light known as polarization noise. Interestingly, this same phenomenon can be employed to measure the evolution of the polarization in the fiber with the so-called polarization optical time domain reflectometer (POTDR) [Ref. 43]. However, in a conventional single-mode OTDR reduction of the polarization noise is necessary using a polarization independent acousto-optic deflector (see Section 10.6.2) or, more usually, a polarization scrambler [Ref. 80].

Problems

13.1 Describe what is meant by 'equilibrium mode distribution' and 'cladding mode stripping' with regard to transmission measurements in optical fibers. Briefly outline methods by which these conditions may be achieved when optical fiber measurements are performed.

13.2 Discuss with the aid of a suitable diagram the cut-back technique used for the measurement of the total attenuation in an optical fiber. Indicate the differences in the apparatus utilized for spectral loss and spot attenuation measurement.

A spot measurement of fiber attenuation is performed on a 1.5 km length of optical fiber at a wavelength of 1.1 μm. The measured optical output power from the 1.5 km length of fiber is 50.1 μW. When the fiber is cut back to a 2 m length, the measured optical output power is 385.4 μW. Determine the attenuation per kilometre for the fiber at a wavelength of 1.1 μm.

13.3 Briefly outline the principle behind the calorimetric methods used for the measurement of absorption loss in optical fibers.

A high absorption optical fiber was used to obtain the plot of $(T_\infty - T_t)$ (on a logarithmic scale) against time shown in Figure 13.33(a). The measurements were achieved using a calorimeter and thermocouple experimental arrangement. Subsequently, a different test fiber was passed three times through the same

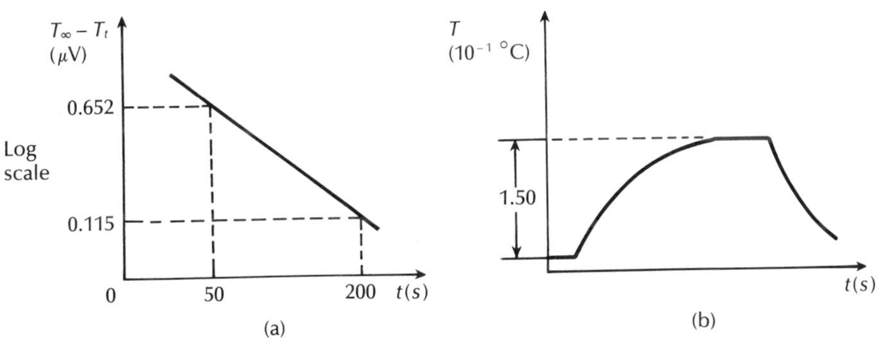

Figure 13.33 Fiber absorption for measurements for Problem 13.3: (a) plot of $(T_\infty - T_t)$ against time for a high absorption fiber; (b) the heating and cooling curve for the test fiber.

calorimeter before further measurements were taken. Measurements on the test fiber produced the heating and cooling curve shown in Figure 13.33(b) when a constant 76 mW of optical power, at a wavelength of 1.06 μm, was passed through it. The constant C for the experimental arrangement was calculated to be 2.32×10^4 J °C^{-1}. Calculate the absorption loss in decibels per kilometre, at a wavelength of 1.06 μm, for the fiber under test.

13.4 Discuss the measurement of fiber scattering loss by describing the use of two common scattering cells.

A Nd : YAG laser operating at a wavelength of 1.064 μm is used with an integrating sphere to measure the scattering loss in an optical fiber sample. The optical power propagating within the fiber at the sphere is 98.45 μW and 5.31 nW of optical power is scattered within the sphere. The length of fiber in the sphere is 5.99 cm. Determine the optical loss due to scattering for the fiber at a wavelength of 1.064 μm in decibels per kilometre.

13.5 Fiber scattering loss measurements are taken at a wavelength of 0.75 μm using a solar cell cube. The reading of the input optical power to the cube is 7.78 V with a gain setting of 10^5. The corresponding reading from the scattering cell which incorporates a 4.12 cm length of fiber is 1.56 V with a gain setting of 10^9. Previous measurements of the total fiber attenuation at a wavelength of 0.75 μm gave a value of 3.21 dB km^{-1}. Calculate the absorption loss for the fiber at a wavelength of 0.75 μm in decibels per kilometre.

13.6 Discuss with the aid of suitable diagrams the measurement of dispersion in optical fibers. Consider both time and frequency domain measurement techniques.

Pulse dispersion measurements are taken on a multimode graded index fiber in the time domain. The 3 dB width of the optical output pulses from a 950 m fiber length is 827 ps. When the fiber is cut back to a 2 m length the 3 dB width of the optical output pulses becomes 234 ps. Determine the optical bandwidth for a kilometre length of the fiber assuming Gaussian pulse shapes.

13.7 Pulse dispersion measurements in the time domain are taken on a multimode and a single-mode step index fiber. The results recorded are:

	Input pulse width (3 dB)	Output pulse width (3 dB)	Fiber length (km)
(a) Multimode fiber	400 ps	31.20 ns	1.13
(b) Single-mode fiber	200 ps	425 ps	2.35

Calculate the optical bandwidth over 1 kilometre for each fiber assuming Gaussian pulse shapes.

13.8 Compare and contrast the major techniques employed to obtain a measurement of the refractive index profile for an optical fiber. In particular suggest reasons why the refracted near field method has been adopted as the reference test method by the EIA.

13.9 The fraction of light reflected at an air–fiber interface r can be obtained from the Fresnel formulae of Eq. (5.1) and for small changes in refractive index:

$$\frac{\delta r}{r} = \left(\frac{4}{n_1^2 - 1}\right)\delta n_1$$

where n_1 is the fiber core refractive index at the point of reflection. Show that the fractional change in the core refractive index $\delta n_1/n_1$ may be expressed in terms of the fractional change in the reflection coefficient $\delta r/r$ following:

$$\frac{\delta n_1}{n_1} = \left(\frac{r^{\frac{1}{2}}}{1 - r}\right)\frac{\delta r}{r}$$

Hence, show that for a step index fiber with n_1 of 1.5, a 5% change in r corresponds to only a 1% change in n_1.

13.10 Describe, with the aid of suitable diagrams, the reference test methods which are utilized to determine the effective cutoff wavelength in single-mode fiber.

13.11 Compare and contrast two simple techniques used for the measurement of the numerical aperture of optical fibers.

Numerical aperture measurements are performed on an optical fiber. The angular limit of the far field pattern is found to be 26.1° when the fiber is rotated from a centre zero point. The far field pattern is then displayed on a screen where its size is measured as 16.7 cm. Determine the numerical aperture for the fiber and the distance of the fiber output end face from the screen.

13.12 Describe, with the aid of a suitable diagram, the shadow method used for the on-line measurement of the outer diameter of an optical fiber.

The shadow method is used for the measurement of the outer diameter of an optical fiber. A fiber outer diameter of 347 μm generates a shadow pulse of 550 μs when the rotating mirror has an angular velocity of 3 rad s^{-1}. Calculate the distance between the rotating mirror and the optical fiber.

13.13 Define the mode-field diameter (MFD) in a single-mode fiber and indicate how this parameter relates to the spot size.

Discuss the techniques which are commonly employed to measure the MFD by either direct or indirect methods. Comment on their relative attributes and drawbacks.

13.14 Outline the major design criteria of an optical fiber power meter for use in the field. Suggest any problems associated with field measurements using such a device.

Convert the following optical power meter readings to numerical values of power: 25 dBm, -5.2 dBm, 3.8 dBμ.

13.15 Describe what is meant by optical time domain reflectometry. Discuss how the technique may be used to take field measurements on optical fibers. Indicate the advantages of this technique over other measurement methods to determine attenuation in optical fibers.

A backscatter plot for an optical fiber link provided by OTDR is shown in Figure 13.34. Determine:

(a) the attenuation of the optical link for the regions indicated A, B and C in decibels per kilometre.

(b) the insertion loss of the joint at the point X.

13.16 Discuss the sensitivity of OTDR in relation to commercial reflectometers. Comment on an approach which may lead to an improvement in the sensitivity of this measurement technique.

The Rayleigh scattering coefficient for a silica single-mode step index fiber at a wavelength of 0.80 μm is 0.46 km^{-1}. The fiber has a refractive index of 1.6 and a numerical aperture of 0.14. When a light pulse of 60 ns duration at a wavelength of 0.80 μm is launched into the fiber, calculate the level in decibels of the backscattered light compared with the Fresnel reflection from a clean break in the fiber. It may be assumed that the fiber is surrounded by air.

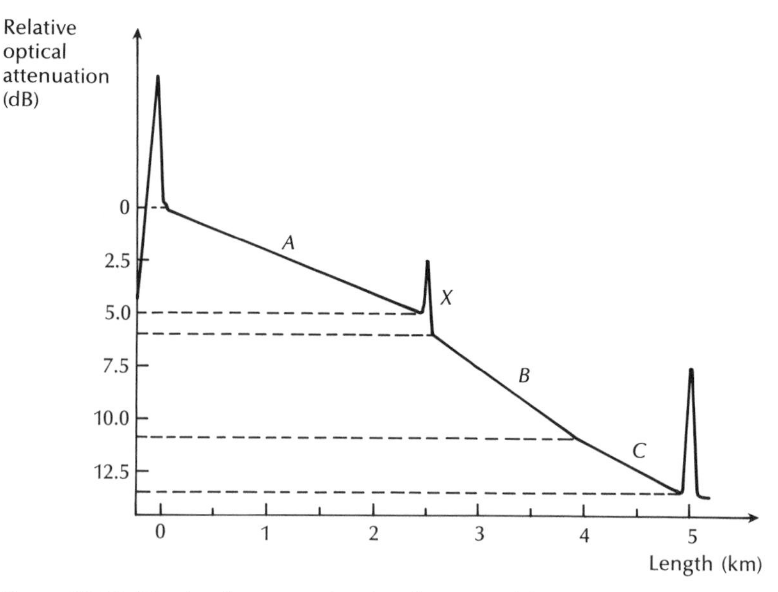

Figure 13.34 The backscatter plot for the optical link of Problem 13.15.

Answers to numerical problems

13.2 5.92 dB km^{-1}
13.3 1.77 dB km^{-1}
13.4 3.91 dB km^{-1}
13.5 1.10 dB km^{-1}
13.6 525.9 MHz km
13.7 (a) 15.9 MHz km;
(b) 7.3 GHz km
13.9 1.0%
13.11 0.44, 17.0 cm
13.12 21.0 cm
13.14 316.2 mW, 302 μW, 2.40 μW
13.15 (a) 2.0 dB km^{-1}, 3.0 dB km^{-1},
2.5 dB km^{-1};
(b) 1.0 dB
13.16 -37.3 dB

References

[1] CCITT. Recommendation G.652, 'Characteristics of a single-mode fiber cable', CCITT document Fascicle III.2, pp. 272–291, 1986.
[2] M. Tateda, T. Horiguchi, M. Tokuda and N. Uchida, 'Optical loss measurement in graded index fiber using a dummy fiber', *Appl. Opt.*, **18**(19), pp. 3272–3275, 1979.
[3] M. Eve, A. M. Hill, D. J. Malyon, J. E. Midwinter, B. P. Nelson, J. R. Stern and J. V. Wright, 'Launching independent measurements of multimode fibres', *2nd European Conference on Optical Fiber Communication* (Paris), pp. 143–146, 1976.
[4] M. Ikeda, Y. Murakami and C. Kitayama. 'Mode scrambler for optical fibres' *Appl. Opt*, **16**(4), pp. 1045–1049, 1977.

[5] S. Seikai. M. Tokuda, K. Yoshida and N. Uchida, 'Measurement of baseband frequency response of multimode fibre by using a new type of mode scrambler', *Electron. Lett.*, **13**(5), pp. 146–147, 1977.

[6] F. P. Kapron, 'Fiber-optic test methods', in *Fiber Optics Handbook for Engineers and Scientists*, F. C. Allard (Ed.), McGraw-Hill, pp. 4.1–4.54, 1990.

[7] FOTP-50. Light launch conditions for long length graded-index optical fiber spectral attenuation measurements.

[8] J. P. Dakin, W. A. Gambling and D. N. Payne, 'Launching into glass–fibre waveguide', *Opt. Commun.*, **4**(5), pp. 354–357, 1972.

[9] FOTP-46. Spectral attenuation measurement for long-length graded-index optical fibers.

[10] FOTP-78. Spectral attenuation cutback measurement for single-mode fibers.

[11] FOTP-53. Attenuation by substitution measurement – for multimode graded-index optical fibers or fiber assemblies used in long length communication systems.

[12] K. I. White, 'A calorimetric method for the measurement of low optical absorption losses in optical communication fibres', *Opt. Quantum Electron.*, **8**, pp. 73–75, 1976.

[13] K. I. White and J. E. Midwinter, 'An improved technique for the measurement of low optical absorption losses in bulk glass', *Opto-electronics*, **5**, pp. 323–334, 1973.

[14] A. R. Tynes, 'Integrating cube scattering detector', *Appl. Opt.*, **9**(12), pp. 2706–2710, 1970.

[15] F. W. Ostermayer and W. A. Benson, 'Integrating sphere for measuring scattering loss in optical fiber waveguides', *Appl. Opt.*, **13**(8), pp. 1900–1905, 1974.

[16] S. de Vito and B. Sordo, 'Misure do attenuazione e diffusione in fibre ottiche multimodo', *LXXV Riuniuone AEI*, Rome, 15–21 Sept. 1974.

[17] J. P. Dakin, 'A simplified photometer for rapid measurement of total scattering attenuation of fibre optical waveguides', *Opt. Commun.*, **12**(1), pp. 83–88, 1974.

[18] L. G. Cohen, P. Kaiser and C. Lin, 'Experimental techniques for evaluation of fiber transmission loss and dispersion', *Proc. IEEE*, **68**(10), pp. 1203–1208, 1980.

[19] S. D. Personick, 'Baseband linearity and equalization in fiber optic digital communication systems', *Bell Syst. Tech. J.*, **52**(7), pp. 1175–1194, 1973.

[20] D. Gloge, E. L. Chinnock and T. P. Lee, 'Self pulsing GaAs laser for fiber dispersion measurement', *IEEE J. Quantum Electron.*, **QE-8**, pp. 844–846, 1972.

[21] FOTP-51. Pulse distortion measurement of multimode glass fiber information transmission capacity.

[22] F. Krahn, W. Meininghaus and D. Rittich, 'Measuring and test equipment for optical cable', *Phillips Telecomm. Rev.*, **37**(4), pp. 241–249, 1979.

[23] FOTP-168. Chromatic dispersion measurement of multimode graded-index and single-mode optical fibers by spectral group delay measurement in the time domain.

[24] I. Kokayashi, M. Koyama and K. Aoyama, 'Measurement of optical fibre transfer functions by swept frequency technique and discussion of fibre characteristics', *Electron. Commun. Jpn*, **60-C**(4), pp. 126–133, 1977.

[25] FOTP-30. Frequency domain measurement of multimode optical fiber information transmission capacity.

[26] FOTP-169. Chromatic dispersion measurement of single-mode fibers by phase-shift method.

[27] W. E. Martin, 'Refractive index profile measurements of diffused optical waveguides', *Appl. Opt.*, **13**(9), pp. 2112–2116, 1974.

[28] H. M. Presby, W. Mammel and R. M. Derosier, 'Refractive index profiling of graded index optical fibers', *Rev. Sci. Instr.*, **47**(3), pp. 348–352, 1976.

[29] B. Costa and G. De Marchis, 'Test methods (optical fibres)', *Telecomm. J. (Engl. Ed.) Switzerland*, **48**(11), pp. 666–673, 1981.

[30] L. C. Cohen, P. Kaiser, J. B. MacChesney, P. B. O'Conner and H. M. Presby, 'Transmission properties of a low-loss near-parabolic-index fiber', *Appl. Phys. Lett.*, **26**(8), pp. 472–474, 1975.

[31] L. C. Cohen, P. Kaiser, P. D. Lazay and H. M. Presby, 'Fiber characterization', in S. E. Miller and A. G. Chynoweth (Eds.), *Optical Fiber Telecommunications*, pp. 343–400, Academic Press, 1979.

[32] M. E. Marhic, P. S. Ho and M. Epstein, 'Nondestructive refractive index profile measurement of clad optical fibers', *Appl. Phys. Lett.*, **26**(10), pp. 574–575, 1975.

[33] P. L. Chu, 'Measurements in optical fibres', *Proc. IEEE Australia*, **40**(4), pp. 102–114, 1979.

[34] M. J. Adams, D. N. Payne and F. M. E. Sladen, 'Correction factors for determination of optical fibre refractive-index profiles by near-field scanning techniques', *Electron. Lett.*, **12**(11), pp. 281–283, 1976.

[35] FOTP-43. Output near-field radiation pattern measurement of optical waveguide fibers.

[36] A. J. Ritger, 'Bandwidth improvement in MCVD multimode fibers by fluorine etching to reduce centre dip', *Eleventh European Conf. on Optical Commun.*, pp. 913–916, 1985.

[37] F. E. M. Sladen, D. N. Payne and M. J. Adams, 'Determination of optical fibre refractive index profile by near field scanning technique', *Appl. Phys. Lett.*, **28**(5), pp. 255–258, 1976.

[38] K. W. Raine, J. G. N. Baines and D. E. Putland, 'Refractive index profiling – state of the art', *J. of Lightwave Technol.*, **7**(8), pp. 1162–1169, 1989.

[39] FOTP-44. Refractive index profile, refracted ray method.

[40] A. H. Cherin, *An Introduction to Optical Fibers*, McGraw-Hill, 1983.

[41] K. I. White 'Practical application of the refracted near-field technique for the measurement of optical fiber refractive index profiles', *Opt. and Quantum Electron.*, **11**(2), pp. 185–196, 1979.

[42] W. J. Stewart, 'Optical fiber and preform profiling technology', *J. of Quantum Electron.*, **QE-18**(10), pp. 1451–1466, 1982.

[43] E.-G. Neuman, *Single-Mode Fibers: Fundamentals*, Springer-Verlag, 1988.

[44] D. B. Payne, M. H. Reeve, C. A. Millar and C. J. Todd, 'Single-mode fiber specification and system performance', *Symp. Opt. Fiber Meas. NBS spec. publ.*, **683**, pp. 1–5, 1984.

[45] FOTP-80. Cutoff wavelength of uncabled single-mode fiber by transmitted power.

[46] R. Srivastava and D. L. Franzen, 'Single-mode optical fiber characterization', *National Bureau of Standards Report*, pp. 1–101, July 1985.

[47] D. L. Franzen, 'Determining the effective cutoff wavelength of single-mode fibers: an interlaboratory comparison', *J. of Lightwave Technol.*, **LT-3**, pp. 128–134, 1985.

[48] C. A. Millar, 'Comment: fundamental mode spot-size measurement in single-mode optical fibers', *Electron. Lett.*, **18**, pp. 395–396, 1982.

[49] FOTP-170. Cable cutoff wavelength of single-mode fiber by transmitted power.

[50] FOTP-177. Numerical aperture of graded-index optical fiber.

[51] D. L. Franzen, M. Young, A. H. Cherin, E. D. Head, M. J. Hackert, K. W. Raine

and J. G. N. Baines, 'Numerical aperture of multimode fibers by several methods: resolving differences', *J. of Lightwave Technol.*, **7**(6), pp. 896–901, 1989.

[52] W. A. Gambling, D. N. Payne and H. Matsumura, 'Propagation studies on single-mode phosphosilicate fibres', *2nd European Conference on Optical Fiber Communication* (Paris), pp. 95–100, 1976.

[53] F. T. Stone, 'Rapid optical fibre delta measurement by refractive index tuning', *Appl. Opt.*, **16**(10), pp. 2738–2742, 1977.

[54] L. G. Cohen and P. Glynn, 'Dynamic measurement of optical fibre diameter', *Rev. Sci. Instrum.*, **44**(12), pp. 1745–1752, 1973.

[55] H. M. Presby, 'Refractive index and diameter measurements of unclad optical fibres', *J. Opt. Soc. Am.*, **64**(3), pp. 280–284, 1974.

[56] P. L. Chu, 'Determination of diameters and refractive indices of step-index optical fibres', *Electron. Lett.*, **12**(7), pp. 150–157, 1976.

[57] H. M. Presby and D. Marcuse, 'Refractive index and diameter determinations of step index optical fibers and preforms', *Appl. Opt.*, **13**(12), pp. 2882–2885, 1974.

[58] D. Smithgall, L. S. Wakins and R. E. Frazee, 'High-speed noncontact fibre-diameter measurement using forward light scattering', *Appl. Opt.*, **16**(9), pp. 2395–2402 1977.

[59] FOTP-58. Core diameter measurement of graded-index optical fibers.

[60] M. Artiglia, G. Coppa, P. DiVita, M. Potenza and A. Sharma, 'Mode field diameter measurements in single-mode optical fibers', *J. of Lightwave Technol.*, **7**(8), pp. 1139–1152, 1989.

[61] FOTP-165. Single-mode fiber, measurement of mode field diameter by near-field scanning.

[62] FOTP-164. Single-mode fiber, measurement of mode field diameter by far-field scanning.

[63] K. Petermann, 'Constraints for the fundamental mode spot size for broadband dispersion-compensated single-mode fibres', *Electron. Lett.*, **19**, pp. 712–714, 1983.

[64] FOTP-167. Mode field diameter measurement-variable aperture in the far field.

[65] FOTP-174. Mode field diameter of single-mode optical fiber by knife-edge scanning in the far field.

[66] J. Streckert, 'New method for measuring the spot size of single-mode fibers', *Opt. Lett.*, **5**, pp. 505–506, 1980.

[67] M. L. Dakss, 'Optical-fiber measurements', in E. E. Basch (Ed.), *Optical-Fiber Transmission*, H. W. Sams & Co., pp. 133–178, 1987.

[68] J. Streckert, 'A new fundamental mode spot size definition usable for non-Gaussian and noncircular field distributions', *J. of Lightwave Technol.*, **LT-3**, pp. 328–331, 1985.

[69] FOTP-107. Return loss.

[70] F. Krahn, W. Meininghaus and D. Rittich, 'Field and test measurement equipment for optical cables', *Acta Electronica*, **23**(3), pp. 269–275, 1979.

[71] R. Olshansky, M. G. Blankenship and D. B. Keck, 'Length-dependent attenuation measurements in graded-index fibres', *Proceedings of 2nd European Conference on Optical Communication* (Paris), pp. 111–113, 1976.

[72] M. K. Barnoski and S. M. Jensen, Fiber waveguides: a novel technique for investigating attenuation characteristics', *Appl. Opt.*, **15**(9), pp. 2112–2115, 1976.

[73] S. D. Personick, 'Photon probe, an optical fibre time-domain reflectometer', *Bell Syst. Tech. J.*, **56**(3), pp. 355–366, 1977.

[74] E. G. Newman, 'Optical time domain reflectometer: comment', *Appl. Opt.*, **17**(11), p. 1675, 1978.

[75] M. K. Barnoski and S. D. Personick, 'Measurements in fiber optics', *Proc. IEEE*, **66**(4), pp. 429–440, 1978.

[76] J. D. Archer, *Manual of Fibre Optics Communication*, STC Components Group, UK, 1981.

[77] FOTP-59. Measurement of fiber point defects using an OTDR.

[78] FOTP-60. Measurement of fiber or cable length using an OTDR.

[79] FOTP-61. Measurement of fiber or cable attenuation using an OTDR.

[80] M. Tateda and T. Huriguchi, 'Advances in optical time-domain reflectometry', *J. of Lightwave Technol.*, **7**(8), pp. 1217–1224, 1989.

[81] P. Healey, 'Optical time domain reflectometry by photon counting', *6th European Conference on Optical Communication* (UK), pp. 156–159, 1980.

[82] P. P. Webb *et al.*, 'Single photon detection with avalanche photodiodes', *Bull. Am. Phys. Soc. II*, **15**, p. 813, 1970.

[83] C. G. Bethea, B. F. Levine, S. Cova and G. Ripamonti, 'High-resolution, high sensitivity optical time-domain reflectometer', *Opt. Lett.*, **13**(3), pp. 233–235, 1988.

14

Applications and future developments

14.1 Introduction

In order to appreciate the many areas in which the application of lightwave transmission via optical fibers may be beneficial, it is useful to review the advantages and special features provided by this method of communication. The primary advantages obtained using optical fibers for line transmission were discussed in Section 1.3 and may be summarized as follows:

(a) enormous potential bandwidth;
(b) small size and weight;
(c) electrical isolation;
(d) immunity to interference and crosstalk;
(e) signal security;

(f) low transmission loss;
(g) ruggedness and flexibility;
(h) system reliability and ease of maintenance;
(i) potential low cost.

Although this list is very impressive, it is not exhaustive and several other attributes associated with optical fiber communications have become apparent as the technology has developed. Perhaps the most significant are the reduced power consumption exhibited by optical fiber systems in comparison with their metallic cable counterparts and their ability to provide for an expansion in the system capability, often without fundamental and costly changes to the system configuration. For instance, a system may be upgraded by simply changing from an LED to an injection laser source, by replacing a *p–i–n* photodiode with an APD detector, or alternatively by operating at a longer wavelength without replacing the fiber cable.

The use of fibers for optical communication does have some drawbacks in practice. Hence to provide a balanced picture these disadvantages must be considered. They are:

(a) the fragility of the bare fibers;
(b) the small size of the fibers and cables which creates some difficulties with splicing and forming connectors;
(c) some problems involved with forming low loss T-couplers;
(d) some doubt in relation to the long-term reliability of optical fibers in the presence of moisture (effects of stress corrosion – see Section 4.7);
(e) an independent electrical power feed is required for any electronic repeaters;
(f) new equipment and field practices are required;
(g) testing procedures tend to be more complex.

A number of these disadvantages are not just inherent in optical fiber systems but are always present at the introduction of a new technology. Furthermore, both continuing developments and experience with optical fiber systems are generally reducing the other problems.

The combination of the numerous attributes and surmountable problems makes optical fiber transmission a very attractive proposition for use within national and international telecommunication networks (PTT applications). To date applications for optical fiber systems in this area have proved the major impetus for technological developments in the field. The technology has progressed from what may be termed first generation systems using multimode step index fiber and operating in the shorter wavelength region (0.8 to 0.9 μm), to second generation systems utilizing multimode graded index fiber operating in both the shorter and longer wavelength regions (0.8 to 1.6 μm). Furthermore, fully engineered third generation systems incorporating single-mode fiber predominantly for operation in the longer wavelength region (1.1 to 1.6 μm) have been installed and are operating in the public telecommunications network. In addition many alternative fiber systems applications have become apparent in other areas of communications where often first and second generation systems provide an ideal solution. Also the growing utilization of optical fiber systems has stimulated tremendous research efforts towards enhanced fiber design. This has resulted in improvement of the associated optoelectronics as well as investigation of 'passive' optics which are likely to provide an advance in the current 'state of the art' of optical fiber communications together with an expansion in its areas of use. Hence, what may be termed fourth generation systems are already close to realization being concerned with both coherent transmission (see Chapter 12) and integrated optics (see Sections 10.5 to 10.9) The potential for fifth generation systems is also apparent. For example, these could involve the development of fluoride glass fibers for very long-haul mid-infrared transmission (see Section 3.7) [e.g. Ref. 1] or the continuing advances in relation to soliton propagation [e.g. Ref. 2]. These latter investigations seek to utilize nonlinear pulse propagation in optical fibers to provide greatly increased channel capacity whilst exhibiting compatibility with integrated optical systems which function in a nonlinear environment.

In this chapter we consider current and potential applications of optical fiber communication systems together with some likely future developments. The discussion is primarily centred around application areas including the public network, military, civil, consumer and industrial which are dealt with in Sections 14.2 to 14.4.

This is followed in Section 14.5 with a description of the continuing developments associated with optical sensor systems which are becoming a major application area. Computer applications are then outlined in Section 14.6 prior to a more detailed discussion in Section 14.7 of the application and developments of optical fiber communications technology within the expanding field of local area networks.

14.2 Public network applications

The public telecommunications network provides a variety of applications for optical fiber communication systems. It was in this general area that the suitability of optical fibers for line transmission first made an impact. The current plans of the major PTT administrations around the world feature the installation of increasing numbers of optical fiber links as an alternative to coaxial and high frequency pair cable systems. In addition it is indicated [Ref. 3] that administrations have largely abandoned plans for millimetric waveguide transmission (see Section 1.1) in favour of optical fiber communications.

14.2.1 Trunk network

The trunk or toll network is used for carrying telephone traffic between major conurbations. Hence there is generally a requirement for the use of transmission systems which have a high capacity in order to minimize costs per circuit. The

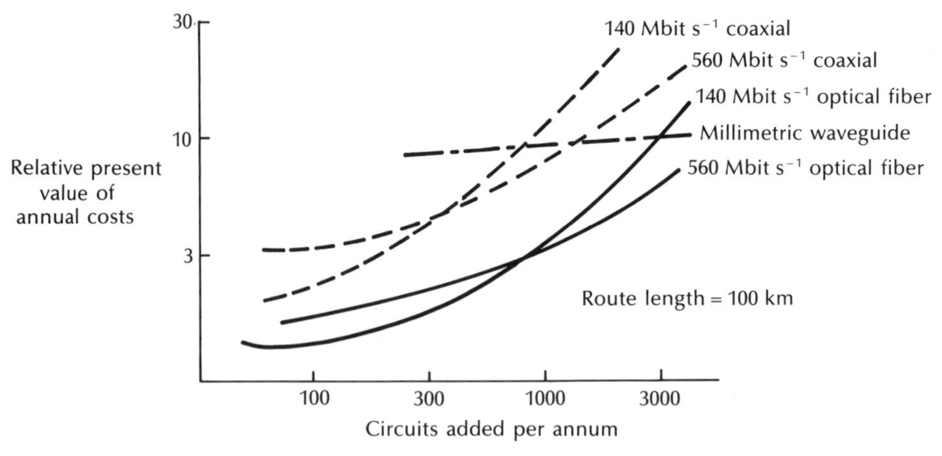

Figure 14.1 Relative present value cost comparison of different high capacity line transmission media. Reproduced with permission of the International Telecommunication Unit (ITU), Geneva, Switzerland, from C. J. Lilly, 'The application of optical fibers in the trunk network', *ITU Telecommunication Journal*, **49-11/1982**, p. 109.

transmission distance for trunk systems can vary enormously from under 20 km to over 300 km, and occasionally to as much as 1000 km. Therefore transmission systems which exhibit low attenuation and hence give a maximum distance of unrepeatered operation are the most economically viable. In this context optical fiber systems with their increased bandwidth and repeater spacings offer a distinct advantage. This may be observed from Figure 14.1 [Ref. 4] which shows a cost comparison of different high capacity line transmission media. It may be observed that optical fiber systems show a significant cost advantage over coaxial cable systems and compete favourably with millimetric waveguide systems at all but the highest capacities. It may also be noted that only digital systems are compared. This is due to the advent of the fully integrated digital public network which invariably means that the majority of trunk routes will employ digital transmission systems. The above media cost observations were confirmed by a more recent study [Ref. 5] which focuses on the interoffice or junction network (see Section 14.2.2). In this application area the use of twisted pair copper cable can prove more expensive than optical fiber.

The speed of operation of most digital trunk optical fiber systems is at present based on the principal digital hierarchies for Europe and North America which were shown in Table 11.1. Proprietary systems (where one contractor supplies the complete system in order to minimize interface problems) operating at 34 Mbit s^{-1} and 140 M bit s^{-1} were installed in the trunk network in the UK on low and high growth rate trunk routes, respectively, in the late 1970s. In the main these systems operate in the 0.85 to 0.9 μm wavelength region using injection laser sources via graded index fiber to silicon APD detectors with repeater spacings of between 8 and

10 km. A typical system power budget for a 140 Mbit s^{-1} system operating over 8 km of multimode graded index fiber at a wavelength of 0.85 μm is shown in Table 14.1 [Ref. 6]. The mean power launched from the laser into the fiber may be improved by over 3 dB using lens coupling rather than the butt launch indicated.

High radiance LED sources emitting at 1.3 μm were also used with multimode graded index fiber in proprietary trunk systems operating at both 34 Mbit s^{-1} and 140 Mbit s^{-1}, most notably in a link between London and Birmingham which is 205 km in length. Field trials of single-mode fiber systems operating in the longer wavelength region demonstrated repeaterless transmission at 565 Mbit s^{-1} over 62 km at 1.3 μm and 140 Mbit s^{-1} over 91 km at 1.5 μm [Ref. 7]. These field trials were followed by the installation of proprietary long wavelength single-mode systems utilizing PIN–FET hybrid receivers between Luton and Milton Keynes, a distance of 27.3 km, and over 52 km between Liverpool and Preston [Ref. 8].

Subsequently the deployment of both 140 and 565 Mbit s^{-1} single-mode fiber systems has taken place in the UK trunk network at an operating wavelength of 1.3 μm and with repeater spacings in the range 30 to 60 km [Ref. 9, 10]. Together with the more recent deployment of 565 Mbit s^{-1} systems operating at 1.55 μm, it meant that by 1990 some 70% of the UK long-haul traffic was conveyed by fiber systems [Ref. 11].

Typical optical power budgets for high performance 565 Mbit s^{-1} single-mode fiber systems operating at the 1.3 and 1.55 μm transmission wavelength are provided in Table 14.2. At the former wavelength a Fabry–Perot injection laser source is utilized, whilst a distributed feedback laser is employed at the 1.55 μm wavelength. In addition, the systems are considered using both *p–i–n* photodiode and APD receivers where in the latter case the receiver sensitivity at both

Table 14.1 A typical optical power budget for a 140 Mbit s^{-1} trunk system operating over 8 km of multimode graded index fiber at a wavelength of 0.85 μm

Mean power launched from the laser transmitter (butt coupling)	−4.5 dBm
APD receiver sensitivity at 140 Mbit s^{-1} (BER 10^{-9})	−48.0 dBm
Total system margin	43.5 dB
Cabled fiber loss (8 × 3 dB km^{-1})	24.0 dB
Splice losses (9 × 0.3 dB each)	2.7 dB
Connector loss (2 × 1 dB each)	2.0 dB
Dispersion–equalization penalty	6.0 dB
Safety margin	7.0 dB
Total system loss	41.7 dB
Excess power margin	1.8 dB

Table 14.2 Optical power budgets for 565 Mbit s^{-1} single-mode fiber trunk systems operating at 1.3 and 1.55 μm

Transmitter type	F–P laser 1.3 μm		DFB laser 1.55 μm	
Receiver type	*p–i–n*	APD	*p–i–n*	APD
Mean power launched from laser transmitter (dBm)	0	0	−3	−3
Receiver sensitivity (dBm)	−32	−39	−33	−40
Total system margin (dB)	32	39	30	37
Connector loss (2 × 1 dB)	2	2	2	2
Penalties (dB)	3	3.5	3	3.5
Ageing and temperature margin (dB)	2	2.5	3	3.5
Design margin (dB)	3	3	3	3
Safety margin (dB)	5	5.5	6	6.5
Maximum repeater spacing (km)	**55**	**70**	**95**	**125**
Cabled fiber loss (dB km^{-1})	0.4	0.4	0.2	0.2
Total cable loss (dB)	22	28	19	25
Total system loss (dB)	32	39	30	37

wavelengths is increased by some 7 dB. The allocation for penalties does not just result from chromatic dispersion but also from laser chirp and extinction ratio penalties, together with reflection noise, jitter and component tolerances. Moreover, it may be observed that the safety margin is separated into an ageing and temperature component as well as a design margin to ensure satisfactory system performance. This latter factor should take account of any cable repairs required over the lifetime of the system operation.

The maximum repeater spacing (or unrepeatered transmission distance) for each of the systems is given in bold in Table 14.2 which enables the total cabled fiber loss to be determined using the loss per kilometre figure. This latter parameter, which represents good quality silica fiber cable, also incorporates the splice losses. It may be observed that the total system losses correspond to the total system margins in all cases and hence excess power margins are not present. However, this is to be expected as the power budgets are calculated using the maximum repeater spacing. It is interesting to note that in the UK public network a 30 km unrepeatered transmission distance is quite sufficient since it is the maximum spacing between existing surface stations and hence power feed points. This removes any requirement for the installation of a metallic conductor for power feed within the system, as well as allowing any repeaters to be installed above ground in a protected internal environment. Benefits gained include significantly reduced system costs along with additional reliability and ease of maintenance.

In the early 1980s the preferred transmission rate for multimode optical fiber trunk systems based on the 1.5 Mbit s^{-1} digital hierarchy (i.e. North America) was 45 Mbit s^{-1}. This was largely due to the fact that much higher growth rates were required for the high speed systems operating at 274 Mbit s^{-1} and above. It was indicated [Ref. 12] that these high speed systems were more appropriate to very long-haul trunk routes (up to 6400 km) where repeater spacings in excess of 25 km were required. However, since 1984 there has been rapid deployment of single-mode fiber systems in the trunk network, typically operating at the 1.3 μm transmission wavelength with repeater spacings around 40 km and bit rates in the range 400 to 600 Mbit s^{-1} [Ref. 13]. This change was stimulated by the nearly two orders in magnitude in bit rate–distance product provided by these single-mode systems over the earlier 45 Mbit s^{-1} multimode fiber systems, together with their potential for substantial upgrade in the future. More recently, however, interest in North America has focused on the synchronous optical network developments discussed in Section 14.2.5.

Work is in progress in Europe, North America and Japan associated with the development of the advanced transmission techniques for IM/DD optical fiber systems described in Section 11.9 for use in trunk networks. Furthermore, the deployment of optical amplifiers in this area of the telecommunication network following the developments discussed in Section 11.10 is likely in the near future. In the longer term the incorporation of coherent optical systems within trunk networks would appear assured, in particular considering the field demonstrations outlined in Section 12.8. Nevertheless it must be noted that the coherent technology is still largely experimental and therefore its deployment may well be gradual rather than rapid, dictated by commercial considerations in relation to the benefits that may be obtained on specific routes.

14.2.2 Junction network

The junction or interoffice network usually consists of routes within major conurbations over distances of typically 5 to 20 km. However, the distribution of distances between switching centres (telephone exchanges) or offices in the junction network of large urban areas varies considerably for various countries as indicated in Figure 14.2 [Ref. 15]. It may be observed from Figure 14.2 that the benefits of long unrepeatered transmission distances offered by optical fiber systems are not as apparent in the junction network due to the generally shorter link lengths. Nevertheless optical fiber junction systems are often able to operate using no intermediate repeaters whilst alleviating duct congestion in urban areas.

In Europe optical fiber systems with transmission rates of 8 Mbit s^{-1}, and for busy routes 34 Mbit s^{-1}, have found favour in the junction network. A number of proprietary systems predominantly operating at 8 Mbit s^{-1} using both injection laser and LED sources via multimode graded index fiber to APD detectors are in operation in the UK with repeater spacings between 7.5 and 12 km. A typical optical power budget for such a system operating at a wavelength of 0.88 μm over 12 km is shown in Table 14.3 [Ref. 6]. It may be noted that the mean power

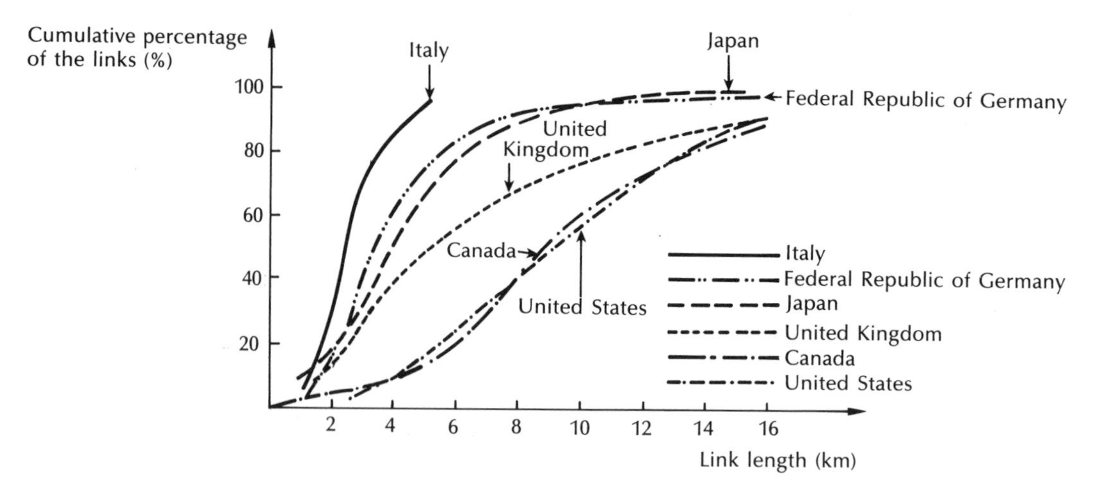

Figure 14.2 Distribution of distances between switching centres in metropolitan areas. Reproduced with permission of the International Telecommunications Union (ITU), Geneva, Switzerland, from O. Cottatelucci, F. Lombardi and G. Pellegrini, 'The Application of optical fibers in the junction network', *ITU Telecommunication Journal*, **49-11/1982**, p. 101.

launched from the laser is reduced below the level obtained with similar dimensioned multimode graded index fiber in the optical power budget shown in Table 14.1. This is due to the lower duty factor when using a 2*B*3*B* code on the 8 Mbit s^{-1} system in comparison with a 7*B*8*B* code used on the 140 Mbit s^{-1} system (see Section 11.6.7).

Table 14.3 Typical optical power budget for a junction system operating at a wavelength of 0.88 μm, a transmission rate of 8 Mbit s^{-1} and an unrepeatered distance of 12 km

Mean power launched from the laser transmitter		−6.0 dBm
Receiver sensitivity at 8 Mbit s^{-1} and a wavelength of 0.88 μm (BER 10^{-9})		−63.0 dBm
Total system margin		57.0 dB
Cabled fiber loss	(12×3.5 dB km^{-1})	42.0 dB
Splice losses	(13×0.3 dB)	3.9 dB
Connector losses	(2×1 dB)	2.0 dB
Dispersion–equalization penalty		0 dB
Safety margin		7.0 dB
Total system loss		54.9 dB
Excess power margin		2.1 dB

In North America, 6 Mbit s^{-1} systems offer flexibility whereas 45 Mbit s^{-1} systems prove suitable for junction traffic requirements of crowded areas. However, economic studies for the United States have indicated that 45 Mbit s^{-1} systems are the most economic choice for the initial service [Ref. 15]. Hence a significant number of commercial 45 Mbit s^{-1} junction systems have been installed. These operate in the shorter wavelength region utilizing injection laser sources, multimode graded index fiber and APD detectors with repeater spacings up to 7.5 km. In addition, several experimental 32 Mbit s^{-1} junction systems have been in operation in Japan since 1980. These systems, which utilize both injection laser and LED sources to APD detectors, have repeater spacings up to 21 km.

Since 1986 PTTs and telephone companies have been installing single-mode fiber in the junction network. This is particularly the case in the United Kingdom and North America [Ref. 16]. These systems using single-mode injection lasers operating at 1.3 μm where the fiber attenuation is in the range 0 4 to 0.5 dB (see Table 14.2) offer the potential for substantial upgrades to cater for the perceived future service requirements of the interoffice networks. In addition, the operation of LEDs with single-mode fiber could prove attractive in these as well as in future local access network (see Section 14.2.3) applications.

An aspect of the interoffice network which has been stimulated by the proliferation of local area networks (LANs) (see Section 14.7) is the requirement for

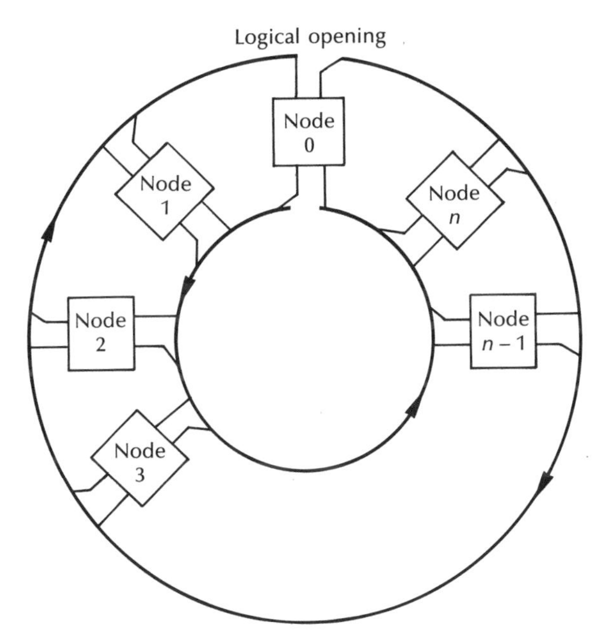

Figure 14.3 Distributed queue dual bus (DQDB) MAN architecture arranged as a physical ring structure.

the interconnection of such networks in the geographically separated area of the junction network. This had led to investigation and standardization discussions associated with packet-based communications over the metropolitan area. Such metropolitan area networks (MANs) are being developed for use in the interconnection of LANs, whilst also potentially providing voice and video services [Refs. 16–18]. Early proposals from the IEEE 802.6 committee suggested a distance optimization of a 50 km diameter network (in order to match the dimensions of typical large metropolitan areas) using a slotted ring architecture.

The emerging standard for MANs, however, is the distributed queue dual bus (DQDB) architecture developed by a subsidiary of Telecom Australia [Refs. 18, 19]. The topology utilizes a dual loop of the transmission medium (nominally optical fiber) that can be arranged in a physical ring but is, in fact, a logical bus. This looped-bus configuration, which is shown in Figure 14.3, eliminates the need to remove data from the medium, as must be done with ring networks. The DQDB MAN employs asynchronous transfer mode (ATM) multiplexing and switching (see Section 14.2.3) in order to provide bandwidth on demand and to efficiently handle bursty traffic. Nevertheless, the MAN standard is compatible with the synchronous optical network developments (see Section 14.2.5) in that it will directly interface with the 155 Mbit s^{-1} level in the synchronous hierarchy. In addition, there is continuing interest in the use of ring topologies operating at high speed with synchronous time division multiplexing for application in the future interoffice networks [Ref. 20].

14.2.3 Local access network

The local access network or subscriber loop connects telephone subscribers to the local switching centre or office. Possible network configurations are shown in Figure 14.4 and include a ring, tree and star topology from the local switching centre. In a ring network (Figure 14.4(a)) any information fed into the network by a subscriber passes through all the network nodes and hence a number of transmission channels must be provided between all nodes. This may be supplied by a time division multiplex system utilizing a broadband transmission medium. In this case only information addressed to a particular subscriber is taken from the network at that subscriber node. The tree network, which consists of several branches as indicated in Figure 14.4(b), must also provide a number of transmission channels on its common links.* However, in comparison with the ring network it has the advantage of greater flexibility in relation to topological enlargement. Nevertheless in common with the ring network, the number of subscribers is limited by the transmission capacity of the links used.

In contrast, the star network (Figure 14.3(c)) provides a separate link for every subscriber to the local switching centre. Hence the amount of cable required is

* The tree network may be considered as a multiple bus network in which access to the local switching centre from a node requires the use of a common bus link. In this context the bus topology is sometimes referred to as the base configuration.

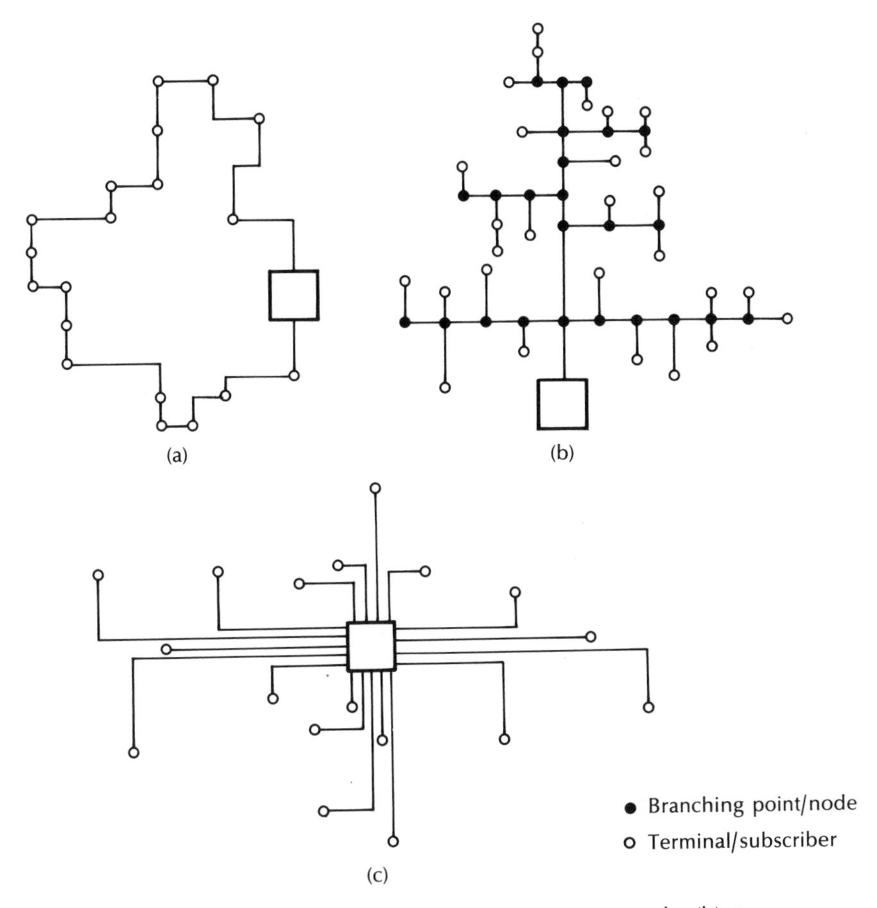

Figure 14.4 Local access network configurations: (a) ring network; (b) tree network; (c) star network.

considerably increased over the ring or tree network, but is offset by enhanced reliability and availability for the subscribers. In addition, simple subscriber equipment is adequate (i.e. no TDM) and network expansion is straightforward. Thus virtually all local and rural telephone networks utilize a star configuration based on copper conductors (twisted pair) for full duplex (bothway) speech transmission. There is substantial interest in the possibility of replacing the existing narrowband local access network twisted pairs with optical fibers. These can also be utilized in the star configuration to provide wideband services (videophone, television, stereo hi-fi, facsimile, data, etc.) to the subscriber together with the narrowband speech channel. Alternatively the enhanced bandwidth offered by optical fibers will allow the use of ring or tree configurations in local and rural networks. This would reduce the quantity of fiber cable required for subscriber loops. However, investigations indicate [Ref. 21] that the cable accounts for only a small fraction of the total network cost. Furthermore, it is predicted that the cost of optical fiber cable may be reduced towards the cost of copper twisted pairs with the large production volume required for local access networks.

Small scale field trials of the use of optical fibers in local access networks have been carried out in several countries including France (the Biarritz project [Ref. 22], Japan (the Yokosuta field trial [Ref. 21]), Canada (the Elie rural field trial [Ref. 23]) and Germany (BIGFON – wideband integrated fiber optic local telecommunications network; a total of ten projects in seven towns [Ref. 21]). These field trials utilized star configurations providing a full range of wideband services to each subscriber through the use of both analog and digital signals on optical fibers.

In the United Kingdom a small Fibrevision (Cable TV) network was installed in Milton Keynes using a switched star configuration [Refs. 24, 25]. This led to the implementation of a cable TV fiber switched star network (SSN) by British Telecom in the Westminster Cable TV franchise area [Ref. 26]. Using the experience gained on that project, British Telecom have designed a modified switched star network which incorporates telephony services to take single-mode fiber through the local access network to the subscriber premises. This active network is known as the broadband integrated distributed star (BIDS), a schematic of which is provided in Figure 14.5 [Ref. 27]. Broadband signals are collected at the headend and then transmitted to a number of hub-sites on 1.3 μm single-mode fiber super-primary links. Each fiber is capable of carrying up to 16 TV channels using analog FM modulation (see Section 11.7.5) to the hub-site which would normally be located at the local telephone exchange. This allows for the interface of the telephony services from the telecommunications network, these being transmitted over a 140 Mbit s^{-1} single-mode fiber primary link to a broadband access point (BAP). Similarly, the broadband services are carried to the BAP on primary fiber links. Finally, secondary single-mode fiber links are employed to transmit the telephony and data services plus two TV channels to the subscriber or customer premises.

An alternative strategy developed by British Telecom involves the use of a passive optical network (PON). Unlike the BIDS network this development is aimed at a staged approach from the initial provision of telephony and other low bit rate services whilst allowing the subsequent upgrading to the distribution of broadband services. The initial concept of telephony on a passive optical network (TPON) is illustrated in Figure 14.6(a) [Ref. 28]. In this case one single-mode fiber operating at 1.3 μm is fed from the local telephone exchange and fanned out via passive optical splitters or tree couplers (see Section 5.6.2) at cabinet and distribution point (DP) positions in order to feed a number of individual subscribers. Each TPON system is designed to provide up to 128 fiber feeds to network terminating equipment (NTE) in customer premises. Time division multiplexing is employed at a rate of 20 Mbit s^{-1} for the telephony services over the broadcast fiber network to the NTEs. The ongoing transmission to the customer premises is provided by a time division multiple access scheme operated at each NTE which associates the appropriate bits of information into assigned time slots.

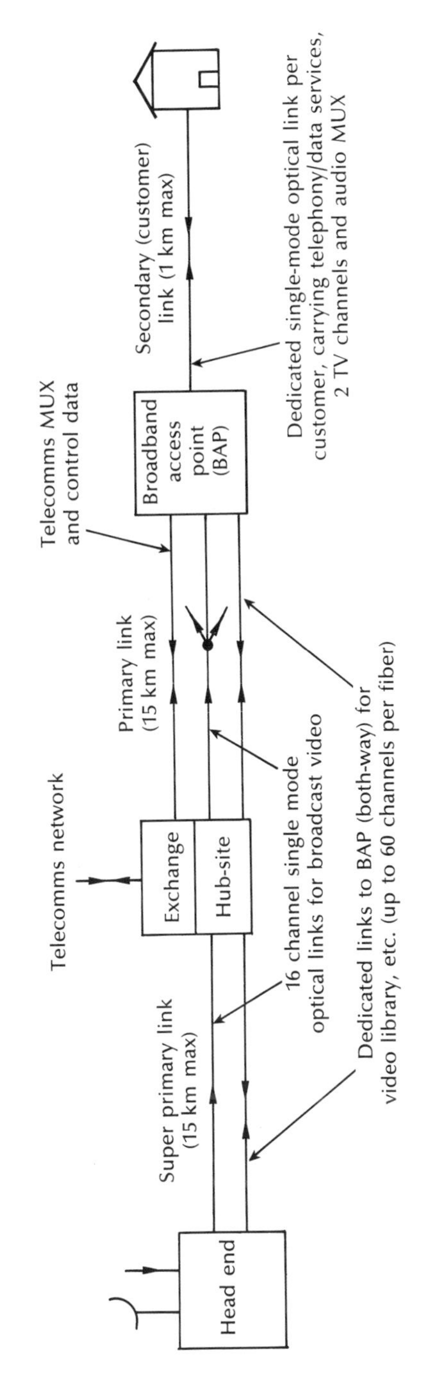

Figure 14.5 Broadband integrated distributed star (BIDS) network.

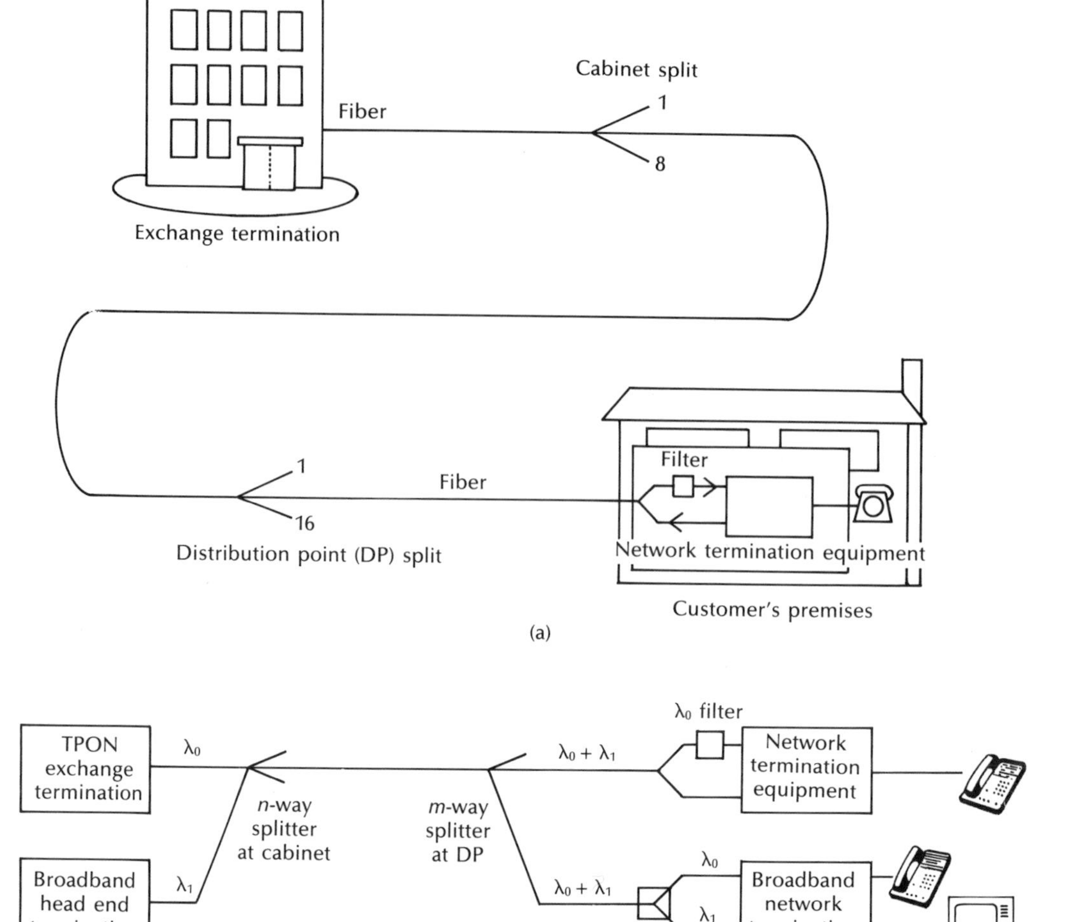

Figure 14.6 Passive optical local access network: (a) telephony on a passive optical network (TPON) configuration; (b) broadband passive optical network (BPON) upgrade.

The enhancement to incorporate broadband services or the broadband passive optical network (BPON) is shown in Figure 14.6(b) [Ref. 29]. It may be observed that the broadband services are carried over the same fiber network architecture but at a different signal wavelength (e.g. 1.55 μm). This use of wavelength division multiplexing (see Section 11.9.3) allows the equivalent of between sixteen and thirty-two TV channels to be transmitted on each additional wavelength signal when analog subcarrier multiplexing (see Section 11.9.2) is employed. For example, a multiplex of thirty-two video modulated carriers can be assembled at the broadband headend in the frequency band 950 MHz to 1.75 GHz for downstream transmission to the customer premises on the 1.55 μm wavelength signal.

British Telecom is conducting a field trial for the above networks at Bishops Stortford, Essex in the United Kingdom over the period 1990 to 1993 [Refs. 30 to 32]. It is envisaged that some 400 residential customers together with twenty-eight businesses will be provided with both telephony and broadband services employing either the BIDS or TPON/BPON systems. Initially, the broadband services provided will be broadcast TV and stereo audio but upgrades to include high-fidelity telephony, video, a video library and high definition TV are planned.

In the United States various trials for the deployment of optical fiber in the local access network have been undertaken. For example, Southern Bell have provided CATV on multimode graded index fiber to just over 250 subscribers at Hunter's Creek, Orlando, Florida [Ref. 33]. This trial has been followed by a more ambitious project to provide basic Integrated Services Digital Network (ISDN), together with CATV, on single-mode fiber to residences at Heathrow, a development north of Orlando. The ISDN terminology describes the end to end digital connectivity with a standard set of interfaces for integrated services delivery. Hence the basic or narrowband ISDN service is intended to provide subscribers with access to two 64 kbit s^{-1} digital circuit switched channels* and a 16 kbit s^{-1} packet switched channel.† Broadband ISDN (BISDN) which can offer transmission rates in the hundreds of Mbit s^{-1} region is also the subject of standardization [Refs. 34, 35] which should facilitate the worldwide compatibility of broadband services such as switched access TV, videophone, high quality audio and multimedia desktop teleconferencing. Nevertheless, it is likely that such services would require optical fiber (and, in particular, single-mode fiber) within the local access network.

A double active star access network topology using fiber is finding application in the United States. It is based on the digital loop carrier (DLC) systems which are often employed to provide gain in the feeder segment of the US local access network, as illustrated in Figure 14.7 [Ref. 35]. The DLC remote node (RN) contains active electronics that provide battery, overvoltage protection, ringing, signalling, coding, hybrid and testing (BURSCHT) functions. Initially, optical fiber was employed in the feeder segment of such systems but more recently the SLC Series 5 Carrier System‡ has used a physical/logical star topology in both the feeder and distribution segments to extend the fiber to the subscriber premises [Ref. 36]. In this case the remote node electronics have been modified such that the BURSCHT functions are removed from the RN to the distant terminal (DT) which may be at or near the subscriber premises.

An alternative local access network architecture which is also finding some application in the United States is supplied by Raynet. It is based on a physical

* Known as B-channels which may be utilized for speech transmission.
† Known as D-channel which may be employed for data transmission.
‡ This system is marketed by AT&T.

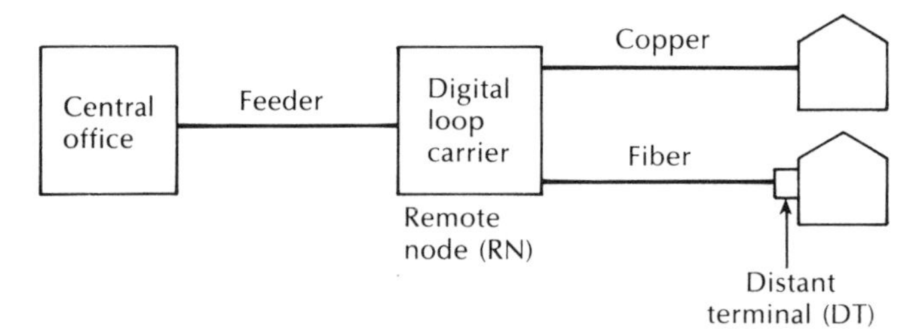

Figure 14.7 AT&T SLC Series 5 local access network architecture.

bus/logical star topology in the distribution segment using active service points, as shown in Figure 14.8 [Ref. 35]. The Raynet fiber bus system utilizes an office interface unit (OIU) at the central office (CO) or remote node (RN) which interfaces with eight 1.5 Mbit s^{-1} digital lines and two protection lines from a central office switch. These are transmitted/received on two-unidiretional fiber buses that carry digitized voice signals, together with signalling information to several subscriber interface units (SIU), each of which supports eight subscriber lines. The SIUs are designed to be placed at or near the subscriber premises such that the copper conductors are only employed over short distances. It may be noted that in comparison with the AT & T SLC Series 5 Carrier System, the Raynet bus architecture reduces the quantity of fiber required in the feeder and distribution segments of the local access network but it increases the number of locations containing active electronics.

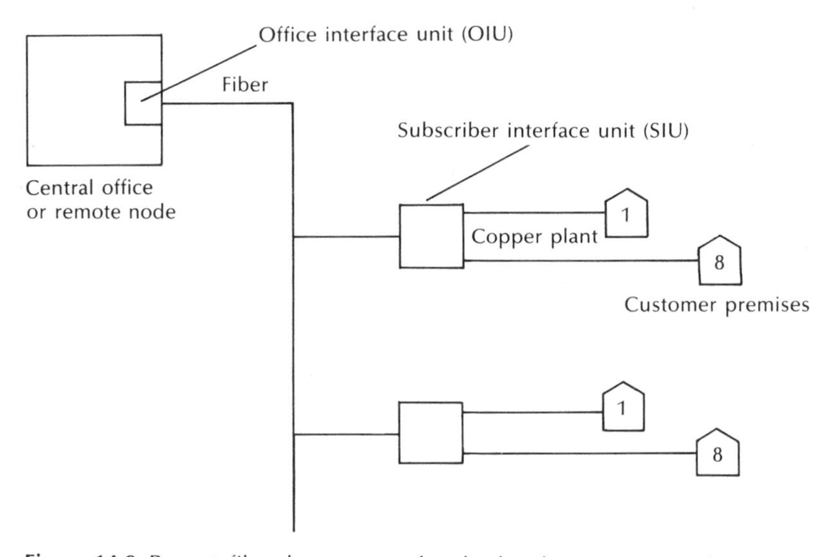

Figure 14.8 Raynet fiber bus system for the local access network.

A passive fiber access network architecture using a double star topology has been demonstrated by Bellcore [Ref. 37]. It is called the Passive Photonic Loop (PPL) and is based on the LAMBDANET system (see Section 11.9.3) which is a very high capacity multiwavelength optical network. The PLL architecture is shown in Figure 14.9 and it employs wavelength division multiplexing techniques to support over twenty wavelength channels. It may be observed from Figure 14.9 that WDM is used between the central office and the remote node (or service access point), which allows a single feed fiber to be shared by a number of channels. Dedicated optical fiber links then connect the RN or SAP to the subscriber premises. Hence the PLL comprises a physical star/logical star topology in both the feeder and distribution segments of the local access network.

Following the LAMBDANET concept, each subscriber is assigned two unique wavelengths, one for upstream transmission and one for downstream transmission. Therefore, information to be transmitted to a particular subscriber is modulated onto the assigned downstream wavelength signal at the central office and then wavelength multiplexed onto the feeder fiber. This wavelength channel is demultiplexed at the remote node and then passed down the required distribution fiber. Thus the subscriber receives the required wavelength signal and no further filtering is necessary at the subscriber premises. The PPL system has been demonstrated experimentally, employing commercially available as well as custom specified components [Ref. 37]. Some twenty channels were operated in the 1.55 μm wavelength band to provide support for ten subscribers. The length of the feeder fiber employed was 9.7 km and the distribution fibers varied in length between 2.2 and 3.3 km. Finally, bidirectional transmission with a bit error rate of less than 10^{-9} when operating at both 600 and 1200 Mbit s^{-1} was obtained.

Although standards relating to the synchronous optical network (SONET) have been adopted (see Section 14.2.5), it is apparent that some broadband services and in particular bursty traffic are not easily accommodated by SONET alone. Hence an alternative packetized multiplexing and switching technique called asynchronous transfer mode (ATM) has been investigated for carrying information within the

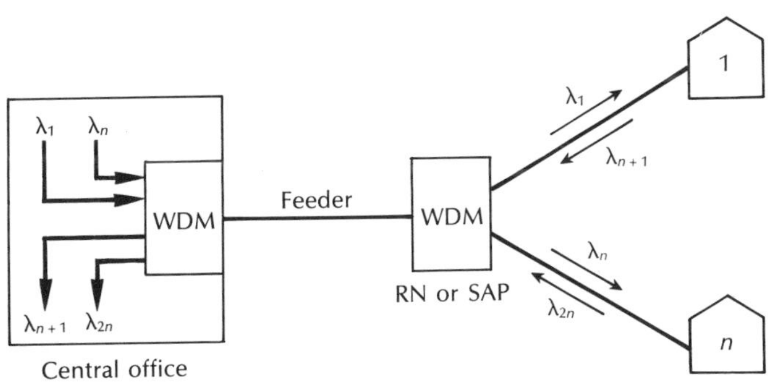

Figure 14.9 Passive photonic loop (PPL) local access network architecture.

SONET payload [Ref. 35]. A particular focus of these activities has concerned the local access network [Ref. 38]. For example, work in this area in the United Kingdom has resulted in the development of ATM operation on a passive optical network [Ref. 39]. Hence, the so called APON system allows a 155 Mbit s^{-1} ATM data stream to be distributed over the shared access PON architectures.

Finally, coherent transmission techniques are also under investigation for the provision of the future optical access network. In particular, coherent optical FDM systems and networks (see Section 12.8.2) offer the potential for many thousands of separate channels within each of the long wavelength windows. Although this powerful attribute of coherent transmission is regarded as offering tremendous potential for future subscriber distribution networks, it is almost certain that initial deployments of optical fiber in the access network will focus on the aforementioned intensity modulation/direct detection technology and that the coherent optical subscriber loop will not be a commercial reality for some time.

14.2.4 Submerged systems

Undersea cable systems are an integral part of the international telecommunications network. They find application on shorter routes especially in Europe. On longer routes, such as across the Atlantic, they provide route diversity in conjunction with satellite links. The number of submerged cable routes and their capacities are steadily increasing and hence there is a desire to minimize the costs per channel. In this context digital optical fiber communication systems appear to offer substantial advantages over current analog FDM and digital PCM coaxial cable systems. High capacity coaxial cable systems require high quality, large diameter cable to overcome attenuation, and still allow repeater spacings of around 5 km. By comparison, it was predicted [Ref. 40] that single-mode optical fiber systems operating at 1.3 or 1.55 μm will provide repeated spacings of 25 to 50 km and eventually even longer.

Research and development of single-mode fiber submerged cable systems is progressing in a number of countries including the United Kingdom, France, the United States and Japan. A successful field trial of a 140 Mbit s^{-1} system was carried out by STC Submarine Systems in Loch Fyne, Scotland in 1980 using a 9.5 km cable length, including a single PCM repeater [Ref. 41]. In the same year a 10 km field trial cable was installed by NTT along the Izu coast in Japan [Ref. 40]. Component reliability together with deep sea cable structure and strength are considered the major problems. These problems were gradually addressed and overcome, however, resulting in the installation of Optican 1 (Tenerife to Gran Canaria), United Kingdom–Belgium 5 (Broadstairs to Ostend), Kynsu to Okinawa, and Honshu to Hokkaido [Ref. 42]. The first two of these short-haul submerged systems, each operating at 280 Mbit s^{-1} per fiber pair, were deployed in 1985 and 1986 respectively. The latter two, which transmit at 400 Mbit s^{-1} on each fiber pair, being installed by NTT.

In 1988 the first generation of repeater transoceanic optical fiber systems was brought into service. This included TAT-8 (United States and France), TPC-3 (Hawaii to Japan) and HAW-4 (California to Japan) [Ref. 43]. These single-mode fiber systems operate at a wavelength of 1.3 μm with a transmission capacity of 560 Mbit s^{-1} (two fibers at 280 Mbit s^{-1} each) and with repeater spacings of approximately 50 km [Ref. 44]. Furthermore, two more 1.3 μm single-mode transoceanic systems have been subsequently installed across the Atlantic (PTAT-1) and Pacific (NPC) oceans [Ref. 45]. Both of these systems contain three fibers (in each direction), each operating at a rate of 420 Mbit s^{-1}.

Second generation submerged systems transmitting at a wavelength of 1.55 μm have been successfully demonstrated. For example, a repeaterless system with a span of 104 km was installed between Tainan and Peng-Hun, off Taiwan in 1988 [Ref. 46]. This system, which was equipped with distributed feedback lasers and avalanche photodiode receivers, operated at a transmission rate of 417 Mbit s^{-1} on each fiber. In addition the UK–Netherlands 12 link using single-mode fiber operating at a wavelength of 1.55 μm was installed in 1989 over an unrepeatered distance of 155 km. However, the transmission rate per fiber was limited to 140 Mbit s^{-1}. The installation of the first, second generation transoceanic system was also completed in 1991. This TAT-9 system between the United States and Europe will also operate at the significantly higher transmission rate of 560 Mbit s^{-1} per fiber with repeater spacings in excess of 100 km [Ref. 45]. Moreover, a further Pacific link (TPC-4) is planned to come into operation in 1992 which will also employ second generation technology at the 565 Mbit s^{-1} data rate. These deployments of systems which are designed for an operational life of twenty-five years provide an indication of the reliability and maturity of the existing 1.55 μm component technology.

The future perspective for submerged optical fiber communications, particularly in the medium- and long-haul range (i.e. distances above 250 km), suggests the use of optical amplifier technology with the present intensity modulation/direct detection systems (see Section 11.10) [Ref. 47]. It is apparent from recent demonstrations that distances up to 1000 km can be achieved using around ten amplifiers without the need for optical isolators or filters [Ref. 48]. In addition, it is likely that one area of early implementation for coherent optical systems (see Chapter 12) will be in undersea applications where the increased transmission distances afforded by this technology can be of great benefit. For example, KDD have already demonstrated 765 bit s^{-1} coherent transmission over 90 km of submerged fiber cable [Ref. 11]. Moreover, in theory, distances of the order of 2000 km could be realized using coherent technology with concatenated optical amplifiers employing optical isolators.

For distances significantly in excess of 2000 km, however, it is suggested that the present second generation systems operating at a wavelength of 1.55 μm and using electronic regenerators may well be continued with because further technology breakthroughs will be required to enable the use of optical amplifiers (either semiconductor or fiber devices) over such distances [Ref. 48]. The limitation is the spontaneous emission at each amplifier, which is both random and signal related, that degrades the signal to noise ratio. Another feature of such systems is the very long interaction length over which dispersion and nonlinear effects can build up and cause distortion. Although these aspects of amplifier operation are under investigation, very-long-haul submarine communications without electronic repeaters may await the deployment of lower loss mid-infrared fiber types (see Section 3.7) within engineered systems. However, it should be noted that coherent transmission over a range of 2200 km at a rate of 2.5 Gbit s^{-1} using erbium doped fiber amplifiers has been demonstrated in the laboratory by NTT in Japan [Ref. 49].

14.2.5 Synchronous networks

The existing baseband digital transmission hierarchies for the telecommunication network in Europe and North America were provided in Table 11.1 (Section 11.5), together with discussion of the time division multiplexing strategy. A schematic of the way in which the European hierarchy is multiplexed up to the 140 Mbit s^{-1} rate from the constituent 2 Mbit s^{-1} (thirty channel) signals is shown in Figure 14.10(a). Difficulties arise, however, with this multiplexing strategy, which is currently adopted throughout the world, in that each 2 Mbit s^{-1} transmission circuit (taking the European example) has its own independent clock to provide for timing and synchronization. This results in slightly different frequencies occurring throughout a network and is referred to as pleisochronous* transmission. Although this strategy is well suited to the transport of bits it suffers a major drawback in that in order to multiplex the different levels (i.e. 2 to 8 to 34 to 140 Mbit s^{-1}) extra bits need to be inserted (bit stuffing) at each intermediate level so as to maintain pleisochronous operation.

The presence of bit stuffing in the existing pleisochronous digital hierarchy makes it virtually impossible to identify and extract an individual channel from within a high bit rate transmission link. Thus to obtain an individual channel the whole demultiplexing procedure through the various levels (Figure 14.10(a)) must be carried out. This process is both complex and uneconomic, particularly when considering future telecommunication networking requirements such as drop and insert where individual channels are extracted or inserted at particular stages. Moreover, a substantial saving in electronic hardware together with increased reliability could be achieved by having a straight, say 2 to 140 Mbit s^{-1} multiplexing/demultiplexing capability as illustrated in Figure 14.10(b) [Ref. 10].

It was therefore clear that a new fully synchronous digital hierarchy was required to enable the international telecommunications network to evolve in the optical fiber era. In particular, this would facilitate the add/drop of lower transmission rate channel groups from much larger higher speed groups without the need for banks of multiplexers and large, unreliable distribution frames.

* Corresponding signals are defined as pleisochronous if their significant instants occur at nominally the same rate, any variation being constrained within specific limits.

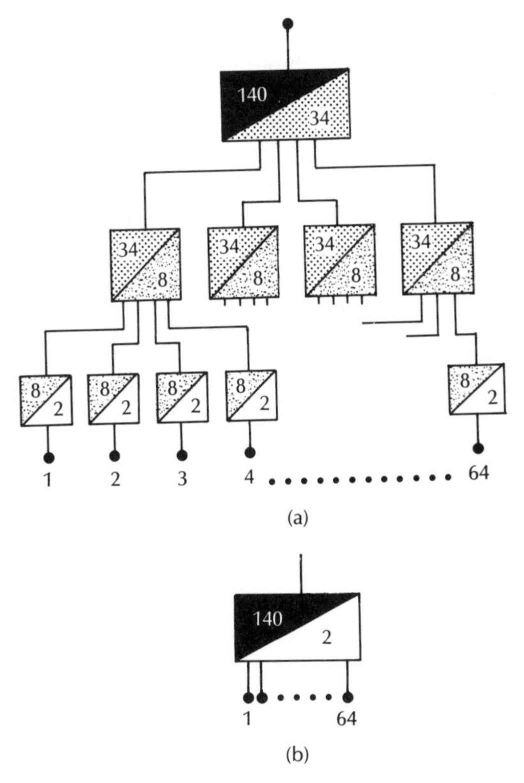

Figure 14.10 European multiplexing hierarchy: (a) existing pleisochronous structure; (b) synchronous multiplexing.

Furthermore, by the mid-1980s the lack of standards for optical networks had led to a proliferation of proprietary interfaces where transmission systems produced by one manufacturer would not necessarily interconnect with those from any other manufacturer such that the ability to mix and match different equipments was restricted. Hence standardization towards a synchronous optical network termed SONET commenced in the United States in 1985 [Ref. 51]. However, two key areas resulted in some modification to the original proposals. These were to make the standard operational in a pleisochronous environment and still retain its synchronous nature and to develop it into an international transmission standard in which the incompatibilities between the existing European and North American signal hierarchies could be resolved. In this latter context the CCITT began deliberation of the SONET concepts in 1986 which resulted in basic recommendations for a new synchronous digital hierarchy (SDH) in November 1988. These recommendations are now published [Refs. 52 to 54]. Prior to these recommendations the American National Standards Institute (ANSI) had issued draft standards relating to SONET [Ref. 55] but as a result of the extensive discussions between the two standards authorities, the two hierarchies are effectively the same. Hence the synchronous optical network recommendations tend to be referred to as SONET in North America and SDH in Europe.

The SONET standard as ultimately developed by ANSI defines a digital hierarchy with a base rate of 51.840 Mbit s^{-1}, as shown in Table 14.4. The OC notation refers to the optical carrier level signal. Hence the base rate signal is OC-1. The STS level in brackets refers to a corresponding synchronous transport signal from which the optical carrier signal is obtained after scrambling (to avoid a long string of ones or zeros and hence enable clock recovery at receivers) and electrical to optical conversion.* Thus STS-1 is the basic building block of the SONET signal hierarchy. Higher level signals in the hierarchy are obtained by byte-interleaving (where a byte is eight bits) an appropriate number of STS-1 signals in a similar manner to that described for the European standard PCM system described in Section 11.5. This differs from the bit interleaving approach utilized in the existing North American digital hierarchy (see Table 11.1). The STS-1 frame structure shown in Figure 14.11 is precisely 125 μs and hence there are 8000 frames per second. This structure enables digital voice signal transport at 64 kbit s^{-1} (1 byte per 125 μs) and the North American DS1-24 channel (1.544 Mbit s^{-1}), as well as the European thirty channel (2.048 Mbit s^{-1}) signals (see Table 11.1) to be accommodated. Other signals in the two hierarchies can also be accommodated. The basic STS-1 frame structure illustrated in Figure 14.11 comprises nine rows, each of 90 bytes which therefore provides a total of 810 bytes or 6480 bits per 125 μs frame. This results in the 51.840 Mbit s^{-1} base rate mentioned above.

The first 3 bytes in each row of the STS-1 frame contain transport overhead bytes, leaving the remaining 783 bytes to be designated as the synchronous payload envelope (SPE). Apart from the first column (9 bytes) which is used for the path overhead, the remaining 774 bytes in the SPE constitute the SONET data payload.

Table 14.4 Levels of the SONET signal hierarchy

Level		Line rate (Mbit s^{-1})
OC-1	(STS-1)	51.840
OC-3	(STS-3)	155.520
OC-9	(STS-9)	466.560
OC-12	(STS-12)	622.080
OC-24	(STS-18)	1244.160
OC-36	(STS-36)	1866.240
OC-48	(STS-48)	2488.320

* STS-1 corresponds to OC-1 which is the lowest level optical signal used at the SONET equipment and network interface.

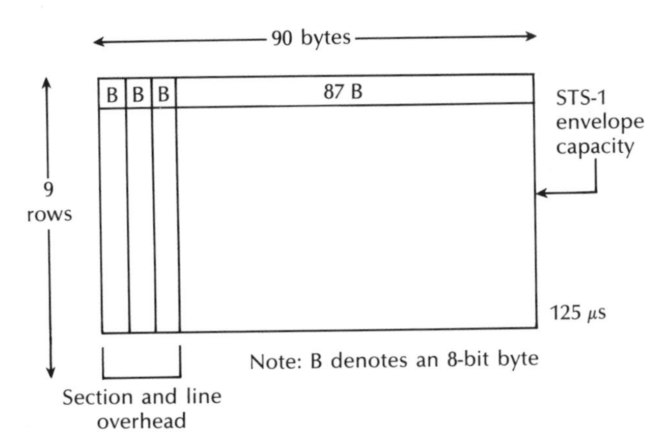

Figure 14.11 STS-1 frame structure.

The transport overhead bytes are utilized for functions such as framing, scrambling, error monitoring, synchronization and multiplexing whilst the path overhead within the SPE is used to provide end to end communication between systems carrying digital voice, video and other signals which are to be multiplexed onto the STS-1 signal. In the latter case a path is defined to end at a point at which the STS-1 signal is created or taken apart (i.e. demultiplexed) into its lower bit rate signals.

The STS-1 SPE does not have to be contained within a single frame; it may commence in one frame and end in another. A 'payload pointer' within the transport overhead is employed to designate the beginning of the SPE within that frame. This provides the flexibility required in order to accommodate different bit rates and a variety of services. Moreover, to accommodate sub-STS-l signal rates a virtual tributary (VT) structure is defined comprising four rates: 1.728 Mbit s^{-1} (VT 1.5); 2.304 Mbit s^{-1} (VT 2); 3.456 Mbit s^{-1} (VT 3); and 6.912 Mbit s^{-1} (VT 6). For example, it may be observed that the 1.544 Mbit s^{-1} and 2.048 Mbit s^{-1} signal streams can each be mapped into a VT 1.5 and VT 2 respectively.

Finally, the higher order multiplexing of a number of STS-1 signals is obviously important in order to achieve the higher bit rates required for wideband services. The format of the STS-N signal frame is shown in Figure 14.12 which, as mentioned previously, is obtained by byte-interleaving N STS-1 signals. In this case the transport overhead bytes of each STS-1 (i.e. the first three single byte columns of each STS-1 signal shown in Figure 14.11) are frame aligned to create the $3N$ bytes of transport overhead, which is illustrated in Figure 14.12. However, the SPEs do not require alignment since the service payload pointers within the associated transport overhead bytes provide the location for the appropriate SPEs.

The synchronous digital hierarchy (SDH), as defined by the CCITT [Refs. 52 to 54], operates in the same manner as described above but differs in some of its terminology. In this case the 125 μs frame structure is referred to as a synchronous transport module (STM) and the base rate STM-1 is 155.520 Mbit s^{-1} which

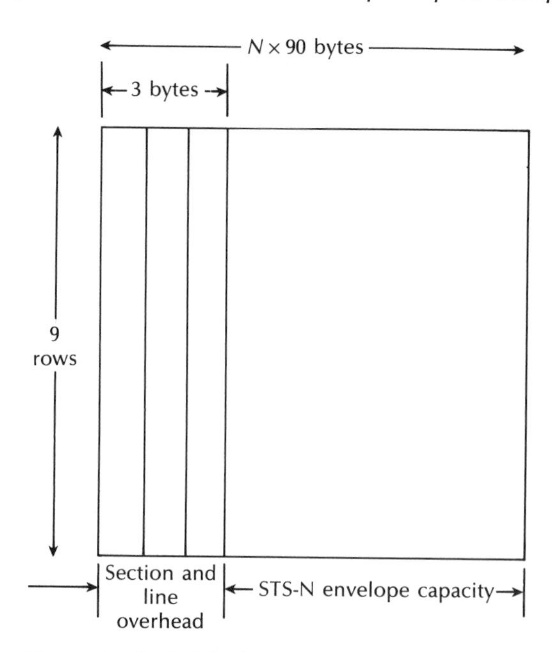

Figure 14.12 STS-N frame structure.

corresponds to OC-3 (STS-3), as may be observed from Table 14.5. Hence the European 140 (i.e. 139.264) Mbit s^{-1} pleisochronous signal can be mapped within an STM-1 signal when including a suitable overhead.

As would be expected, the higher level STM signals also correspond to SONET optical carrier rates as well as providing a match to appropriate multiples in the European pleisochronous hierarchy. Again, these higher levels are formed by simple byte interleaving. Tributaries are used to incorporate the signal rates below the STM-1 rate into the frame format where they may be located by means of pointers. Hence a 2.048 Mbit s^{-1} channel can be readily identified within a 155.520 Mbit s^{-1} stream. As with SONET this will allow a network of high capacity

Table 14.5 Corresponding levels and bit rates for SDH and SONET

SDH level	Line rate (Mbit s^{-1})	SONET level
STM-1	155.520	OC-3
STM-4	622.080	OC-12
STM-8	1244.160	OC-24
STM-12	1866.240	OC-36
STM-16	2488.320	OC-48

cross-connects to be established at nodes throughout the transmission network. Moreover, this facility will enable a managed network to efficiently route and distribute traffic between nodes, dropping off capacity to exchanges for traffic switching [Ref. 56].

14.3 Military applications

In these applications, although economics are important, there are usually other, possibly overriding, considerations such as size, weight, deployability, survivability (in both conventional and nuclear attack [Ref. 57]) and security. The special attributes of optical fiber communication systems therefore often lend themselves to military use.

14.3.1 Mobiles

One of the most promising areas of military application for optical fiber communications is within military mobiles such as aircraft, ships and tanks. The small size and weight of optical fibers provide an attractive solution to space problems in these mobiles which are increasingly equipped with sophisticated electronics. Also the wideband nature of optical fiber transmission will allow the multiplexing of a number of signals on to a common bus. Furthermore, the immunity of optical transmission to electromagnetic interference (EMI) in the often noisy environment of military mobiles is a tremendous advantage. This also applies to the immunity of optical fibers to lightning and electromagnetic pulses (EMP) especially within avionics. The electrical isolation, and therefore safety, aspect of optical fiber communications also proves invaluable in these applications, allowing routeing through both fuel tanks and magazines.

The above advantages were demonstrated with preliminary investigations involving fiber bundles [Ref.3] and design approaches have included multiterminal data systems [Ref. 58] using single fibers, and use of an optical data bus [Ref. 59]. In the former case, the time division multiplex system allows ring or star configurations to be realized, or mixtures of both to create bus networks. The multiple access data highway allows an optical signal injected at any access point to appear at all other other access points. An example is shown in Figure 14.13 [Ref. 6] which illustrates the interconnection of six terminals using two 4×4 transmissive star couplers (see Section 5.6.2).

An experimental optical data bus was installed in the Mirage 4000 aircraft [Ref. 59]. However, significant problems were encountered with optical connection, fiber fragility and low light levels from the LED source. Nevertheless it was concluded that these drawbacks could be reduced by the use of spliced connections, star couplers rather than T-couplers and smaller diameter fibers (100 to 150 μm), which would make it possible to produce cables with smaller radii of curvature, and in which the fiber would be freer.

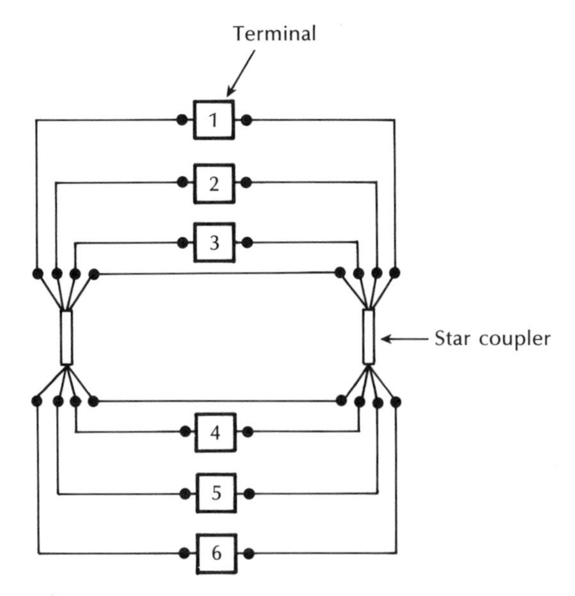

Figure 14.13 Multiple access bus showing the interconnection of six terminals using two four-way transmissive star couplers [Ref. 6].

Studies have also been carried out into the feasibility of using a 1 Mbit s^{-1} optical data bus for flight control, avionic weapons and internal data systems within a helicopter [Ref. 60]. In addition an optical fiber data highway has been installed in the Harrier GR5 aircraft, for operational use. It is incorporated between the communications, navigation and identification data converter located towards the rear of the aircraft and the amplifier control situated beneath the cockpit. The original system specification called for large core diameter plastic fiber for operation around 2.4 kbit s^{-1} [Ref. 60].

Moreover, a flight control system employing optical fibers for an airship has been demonstrated which provided a fly-by-light capability [Ref. 61]. In addition, the focus of investigations in this area has shifted towards the optical fiber realization of the universally accepted military data bus standard MIL-STD-1553B and the ways in which this may be applied within avionics applications [Ref. 62]. Finally, there is substantial interest in the application of optical fiber sensor systems (see Section 14.5) within mobiles (particularly aircraft to provide monitoring functions (e.g. [Ref. 63]). Moreover, these activities are not only confined to the military sphere but are also being pursued within civilian aircraft development.

14.3.2 Communication links

The other major area for the application of optical fiber communications in the military sphere includes both short and long distance communication links. Short distance optical fiber systems may be utilized to connect closely spaced items of electronic equipment in such areas as operations rooms and computer installations. A large number of these systems have already been installed in military installations in the United Kingdom. These operate over distances from several centimetres to a few hundred metres at transmission rates between 50 bauds and 4.8 kbit s^{-1} [Ref. 60]. In addition a small number of 7 MHz video links operating over distances of up to 100 m are in operation. There is also a requirement for long distance communication between military installations which could benefit from the use of optical fibers. In both these cases advantages may be gained in terms of bandwidth, security and immunity to electrical interference and earth loop problems over conventional copper systems.

Other long distance applications include torpedo and missile guidance, information links between military vessels and maritime, and towed sensor arrays. In these areas the available bandwidth and long unrepeatered transmission distances of optical fiber systems provide a solution which is not generally available with conventional technology. A fiber guided weapons system is illustrated in Figure 14.14 whereby a low loss, high tensile strength fiber is used to relay a video signal back to a control station to facilitate targeting by an operator. Such fiber guided anti-tank missiles (ATMS) have been demonstrated using LED sources and multimode fiber over a range of 5 to 10 km, with longer range systems employing single-mode fiber being under development [Ref. 64].

Investigations have also been carried out with regard to the use of optical fibers in tactical communication systems. In order to control sophisticated weapons systems in conjunction with dispersed military units there is a requirement for tactical command and control communications (often termed C^3). These communication systems must be rapidly deployable, highly mobile, reliable and have the ability to survive in military environments. Existing multichannel communication cable links employing coaxial cable or wire pairs do not always meet these requirements [Ref. 65]. They tend to be bulky, difficult to install (requiring long installation times), and are susceptible to damage. In contrast optical fiber cables offer special features which may overcome these operational

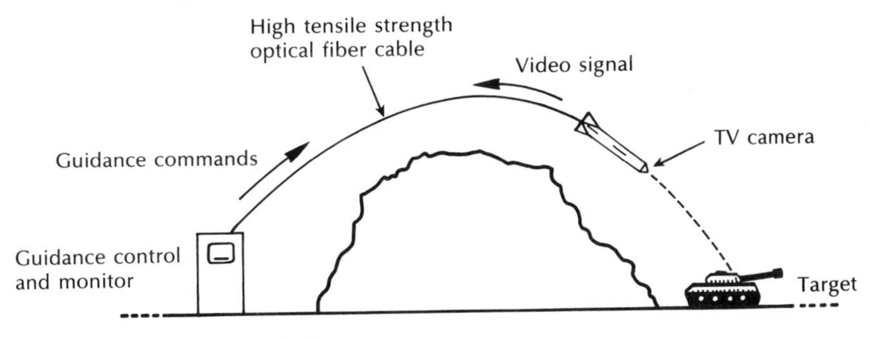

Figure 14.14 A fiber guided weapons system.

deficiencies. These include small size, light weight, increased flexibility, enhanced bandwidth, low attenuation removing the need for intermediate repeaters, and immunity to both EMI and EMP. Furthermore, optical fiber cables generally demonstrate greater ruggedness than conventional deployable cables, making them appear ideally suited for this application.

Optical fiber cables have been installed and tested within the Ptarmigan tactical communication system developed for the British Army [Ref. 66]. They may be utilized as a direct replacement for the HF quad cable system previously employed for the intranodal multichannel cable links within the system [Ref. 6]. The optical fiber element of the system comprises an LED source emitting at a wavelength of 0.9 μm, graded index fiber and a APD detector. It is designed to operate over a range of up to 2 km at data rates of 256, 512 and 2048 kbit s^{-1} without the use of intermediate repeaters. The optical fiber cable assemblies are about half the weight of the HF quad cable, and are quick and easy to deploy in the field. Furthermore, special ruggedized expanded beam optical connectors (see Section 5.5) have been shown to be eminently suitable for use in conditions involving dust, dirt, rough handling and extreme climates. Successful integration of an optical fiber system into a more complex tactical communication system for use in the military environment has demonstrated its substantial operational and technical advantages over HF metallic cable systems.

In summary, it appears that confidence has grown in this technology such that its widescale use in military applications in the future will continue.

14.4 Civil, consumer and industrial applications

14.4.1 Civil

The introduction of optical fiber communication systems into the public network has stimulated investigation and application of these transmission techniques by public utility organizations which provide their own communication facilities over moderately long distances. For example these transmission techniques may be utilized on the railways and along pipe and electrical power lines. In these applications, although high capacity transmission is not usually required, optical fibers may provide a relatively low cost solution, also giving enhanced protection in harsh environments, especially in relation to EMI and EMP. Experimental optical fiber communication systems have been investigated within a number of organizations in Europe, North America and Japan. For instance, British Rail has successfully demonstrated a 2 Mbit s^{-1} system suspended between the electrical power line gantries over a 6 km route in Cheshire [Ref. 3]. Also, the major electric power companies have shown a great deal of interest with regard to the incorporation of optical fibers within the metallic earth of overhead electric power lines [Ref. 67].

It was indicated in Section 14.2.3 that optical fibers are eminently suitable for video transmission. Thus optical fiber systems are starting to find use in commercial television transmission. These applications include short distance links between studio and outside broadcast vans, links between studios and broadcast or receiving aerials, and close circuit television (CCTV) links for security and traffic surveillance. In addition, the implementation of larger networks for cable and common antenna television (CATV) has demonstrated the successful use of optical fiber communications in this area where it provides significant advantages, in terms of bandwidth and unrepeatered transmission distance, over conventional video links.

One of the first commercial optical fiber video systems was installed in Hastings, UK, in 1976 by Rediffusion Limited for the transmission of television signals over a 1.4 km link for distribution to 34,000 customers. Another early optical fiber CATV field trial was the Hi-OVIS project carried out in Japan [Ref. 68]. The project involved the installation of an interactive video system, plus FM audio and digital data to 160 home subscribers and eight local studio terminals in various public premises. The system operated over a 6 km distribution cable consisting of thirty-six fibers plus additional branches to the various destination points; no repeaters were used in this entire network.

Various techniques have been utilized for video transmission including baseband intensity modulation, subcarrier intensity modulation (e.g. FM–IM), pulse analog techniques (e.g. PFM–IM) and digital pulse code modulation (PCM–IM). Generally, digital transmission is preferred on larger CATV networks as it allows time division multiplexing as well as greater unrepeatered transmission distance [Ref. 70]. It also avoids problems associated with the nonlinearities of optical sources. An example of commercial digital video transmission is a 7.8 km optical fiber trunk system operating at 322 Mbit s^{-1} in London, Ontario, Canada [Ref. 71]. This system carries twelve video channels and twelve FM stereo channels along eight fibers installed in a 13 mm cable. A similar digital trunk system has been installed in a CATV network in Denmark [Ref. 72]. This link, using twelve fibers again operating over a distance of 7.8 km, has a capacity of eight channels and twelve FM stereo channels.

However, digital transmission of video signals is not always economic, owing to the cost and complexity of the terminal equipment. Hence, optical fiber systems using direct intensity modulation often provide an adequate performance for a relatively low system cost. For example, a block schematic of a long distance analog baseband video link for monitoring railway line appearances such as road crossings, tunnels and snowfall areas is shown in Figure 14.15 [Ref. 73]. Video signals from TV cameras installed at monitoring points C are gathered to the concentrating equipment through local transmission lines. These signals are then multiplexed in time, frequency or wavelength on to the main transmission line to the monitoring centre. An experimental optical fiber system operating at a wavelength of 1.32 μm using multimode graded index fiber and baseband intensity modulation was installed along a main line of the Japanese national railway [Ref. 73]. Successful video transmission over 16.5 km without intermediate repeaters was achieved, demonstrating the use of fiber systems for this application.

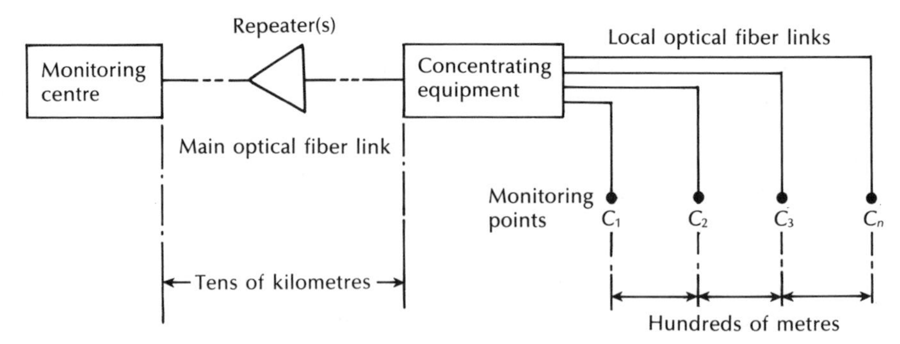

Figure 14.15 Block schematic of an optical fiber baseband video system for railway line monitoring [Ref. 73].

A similar CCTV monitoring system was implemented for the Kobe Mass Transit System, also in Japan [Ref. 74]. This system, which operates over shorter distances of between 300 m and 5 km, also uses analog intensity modulation together with wavelength division multiplexing of four TV channels at wavelengths of 730, 780, 837 and 879 nm on to multimode step index fiber.

The continuing interest in optical fiber video communications has more recently encompassed the use of subcarrier multiplexing (SCM) techniques. These developments have extended the maximum transmission distance without repeaters to over 100 km on single-mode fiber (see Section 11.9.2). Furthermore, the distribution of video signals is under investigation within the telecommunication local access network (see Section 14.2.3) and therefore there is a merging of the optical fiber network strategies that may be adopted in this area with those for the provision of cable TV.

In common with the military applications, other potential civil uses for optical fiber systems include short distance communications within buildings (e.g. broadcast and recording studios) and within mobiles such as aircraft and ships. However, a large market for optical fiber systems may eventually be within consumer applications.

14.4.2 Consumer

A major consumer application for optical fiber systems is within automotive electronics. Work is progressing within the automobile industry towards this end together with the use of microcomputers for engine and transmission control as well as control of convenience features such as power windows and seat controls. Optical fiber communication links in this area provide advantages of reduced size and weight together with the elimination of EMI. Furthermore, it is likely they will reduce costs by allowing for an increased number of control signals in the confined space presented by the steering column and internal transmission paths within the vehicle through multiplexing of signals on to a common optical highway.

Such techniques have been under investigation by General Motors for a number of years and a prototype system was reported [Ref. 75] to have demonstrated the feasibility in 1980. This system utilized a bundle of forty-eight high loss plastic fibers with a simple LED emitting in the visible spectrum. Further developments in the USA and elsewhere suggest that large core diameter (1 mm) single plastic fibers will be utilized in automobile multiplex systems within the passenger compartment, whereas glass fibers may be required to stand the high temperatures (120 °C) encountered in the engine compartment.

More recent work towards the replacement wiring harnesses in automobiles with multiplexed optical fiber data links has also been undertaken in a number of other centres [Refs. 76 to 78]. Several prototype systems have been demonstrated which provide, in addition to the usual control signals (e.g. lights), functions such as centralized locking, power windows and seats, and door lamps. The majority of these systems have employed large core diameter plastic fiber operating at low data rates and have sought to avoid the very high temperatures in the engine compartments. Nevertheless, the provision of optical sensing and monitoring functions (see Section 14.5) within the engine compartment is under investigation [Refs. 79, 80] and work progresses in the development of high temperature polymer fibers [Ref. 81]. In addition, prototype optical sensors for monitoring other automotive parameters such as road adhesion have been implemented [Ref. 82].

Other consumer applications are likely to include home appliances where, together with microprocessor technology, optical fibers may be able to make an impact in the 1990s. However, as with all consumer equipment, progress is very dependent on the instigation of volume production and hence low cost. This is a factor which appears to have delayed wider application of optical fiber systems in this area.

14.4.3 Industrial

Industrial uses for optical fiber communications cover a variety of generally on-premises applications within a single operational site. Hence in the past the majority of industrial applications have tended to fall within the following design criteria [Ref. 83]:

(a) digital transmission at rates from d.c. to 20 Mbit s^{-1}, synchronous or asynchronous, having compatibility with a common logic family (i.e. TTL or CMOS), being independent of the data format and with bit error rates less than 10^{-9};
(b) analog transmission from d.c. to 10 MHz, exhibiting good linearity and low noise;
(c) transmission distances from several metres up to a maximum of kilometres, although generally 1 km will prove sufficient;
(d) a range of environments from benign to harsh, and often exhibiting severe electromagnetic interference from industrial machinery.

Optical fiber systems with performances to meet the above criteria are readily available at a reasonable cost. These systems offer reliable telemetry and control communications for industrial environments where EMI and EMP cause problems for metallic cable links. Furthermore, optical fiber systems provide a far safer solution than conventional electrical monitoring in situations where explosive or corrosive gases are abundant (e.g. chemical processing and petroleum refining plants). Hence the increasing automation of process control, which is making safe, reliable communication in problematical environments essential, is providing an excellent area for the application of optical fiber communication systems.

For example, optical fiber systems were successfully employed in nuclear testing applications in the United States by the Department of Energy. Two plasma diagnostic experiments developed by the Los Alamos Scientific Laboratory [Ref. 84] were carried out at the Nevada Test Site in Mercury, Nevada. These experiments utilized the unique properties of optical fibers to provide diagnostic capabilities which are not possible with coaxial cable systems. In the first experiment a wideband fiber system (1 GHz bandwidth) was used to record the wideband data from gamma ray sources. The second experiment, a neutron imaging system, provided a time and space resolution for a neutron source on a nanosecond time scale. The neutron source was attenuated and imaged through a pinhole on to an array of scintillator filaments, each of which was aligned to a single PCS fiber for transmission via a graded index fiber to a photomultiplier. A pulsed dye laser was used for system calibration. Both amplitude and overall timing calibration were achieved with an optical time domain reflectometer (see Section 13.10.1) being used regularly to record the fiber attenuation. It was estimated [Ref. 84] that this system provided a bandwidth advantage of at least a factor of 10 over coaxial cable, at approximately half the cost, and around one-fiftieth of the weight.

More recently the application of optical fiber communications systems within the industrial environment has grown to encompass two substantive technological areas, namely optical sensor systems and local area networks. These developments are discussed in Sections 14.5 and 14.7.2 respectively

14.5 Optical sensor systems

It was indicated in Section 14.4.3 that optical fiber transmission may be advantageously employed for monitoring and telemetry in industrial environments. The application of optical fiber communications to such sensor systems has stimulated much interest, especially for use in electrically hazardous environments where conventional monitoring is difficult and expensive. There is a requirement for the accurate measurement of parameters such as liquid level, flow rate, position, temperature and pressure in these environments which may be facilitated by optical fiber systems. Early work in this area featured electrical or electro-optical transducers along with optical fiber telemetry systems. A novel approach of this type involved a piezoelectric transducer which was used to apply local deformations to a single fiber highway causing phase modulation of the transmitted signal [Ref. 86]. The unmodulated signal from the same optical source was transmitted via a parallel reference fiber to enable demodulation of the signals from various piezoelectric transducers located on the highway. This technique proved particularly useful when a number of monitoring signals were required at a central control point.

Electro-optical transducers together with optical fiber telemetry systems offer significant benefits over purely electrical systems in terms of immunity to EMI and EMP as well as intrinsic safety in the transmission to and from the transducer. However, they still utilize electrical power at the site of the transducer which is also often in an electrically problematical environment. Therefore much effort is currently being expended in the investigation and development of entirely optical sensor systems. These employ electrically passive optical transducer mechanisms which directly modulate the light for the optical fiber telemetry link.

The fundamental requirement of an electrically passive optical fiber sensor is the modulation of a light beam either directly or indirectly (e.g. with a mechanical linkage) by the measurand (e.g. pressure, temperature, flow velocity, displacement, vibration, strain, etc.) without recourse to electrical interfacing. Modulation of a light beam transmitted via a feed fiber takes place in the modulation zone of the generalized optical fiber sensor system depicted in Figure 14.16. The light beam can be modulated in five of its basic properties: namely, optical intensity, phase, polarization, wavelength and spectral distribution. Most fiber sensors employ the first three, although there is growing interest in the exploitation of the last two. Furthermore, it is possible to categorize the devices into two distinct groups based solely on their use of intensity or phase modulation respectively. This form of classification also largely separates the sensor systems in relation to the type of optical fiber and associated components required.

Optical fiber phase sensors generally utilize an interferometric approach necessitating the use of single-mode fiber and an optical source with spatial coherence (e.g. laser). Therefore, although these devices offer orders of magnitude increased sensitivity over conventional sensing mechanisms, they exhibit a requirement for somewhat sophisticated optical fiber components and optoelectronic devices. Alternatively, optical fiber intensity modulation sensors which may serve in less demanding applications often employ multimode fiber together with an incoherent light source (e.g. LED). These devices therefore tend to be simpler and easier to implement at lower cost.

Optical fiber sensors may be implemented as intrinsic or extrinsic devices. The former type is arranged such that the physical parameter to be sensed acts on the

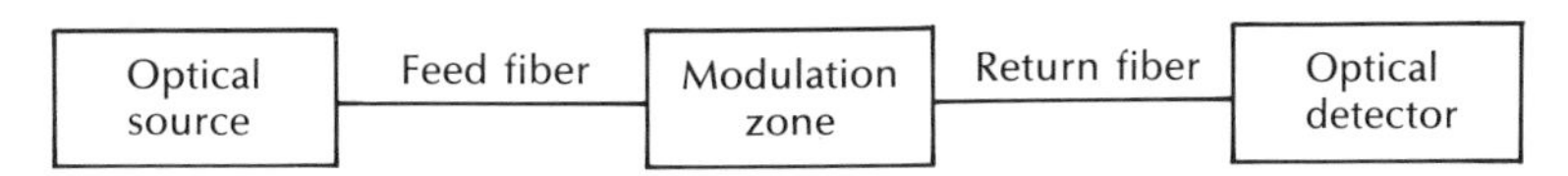

Figure 14.16 Basic optical fiber sensor system.

fiber itself to cause a change in the transmission characteristics. The latter type uses the fiber as a light guide to and from the sensor, which is configured to allow the measurand to change the coupling characteristics between the feed and return fiber (see Figure 14.16).

14.5.1 Phase and polarization fiber sensors

Very sensitive single-mode fiber passive optical phase sensors employ an interferometric approach, as illustrated in Figure 14.17(a). These devices cause interference of coherent monochromatic light propagating in a strained or temperature-varying fiber with light either directly from the laser source, or (as shown in Figure 14.17(a)) guided by a reference fiber isolated from the external influence. The effects of strain, pressure or temperature change give rise to differential optical paths by changing the fiber length, core diameter or refractive index with respect to the reference fiber. This provides a phase difference between the light emitted from the two fibers, giving interference patterns, as illustrated in Figure 14.17(a). Very accurate measurements of pressure or temperature may be obtained from these patterns. For example, using fused silica in such a two arm fiber interferometer, it can be shown that the temperature sensitivity is about 107 rad $^{\circ}C^{-1}$ m^{-1} [Ref. 87]. It should be noted that the interferometric sensor arrangement shown in Figure 14.17(a) is that of the Mach–Zehnder configuration [Ref. 89].

Another common single-mode fiber interferometric sensor which is finding widescale application is the fiber gyroscope [Ref. 90]. This device is based on the classical Sagnac ring interferometer, a fiber version of which is illustrated in Figure 14.17(b). In this device light entering the multiturn fiber coil is divided into two

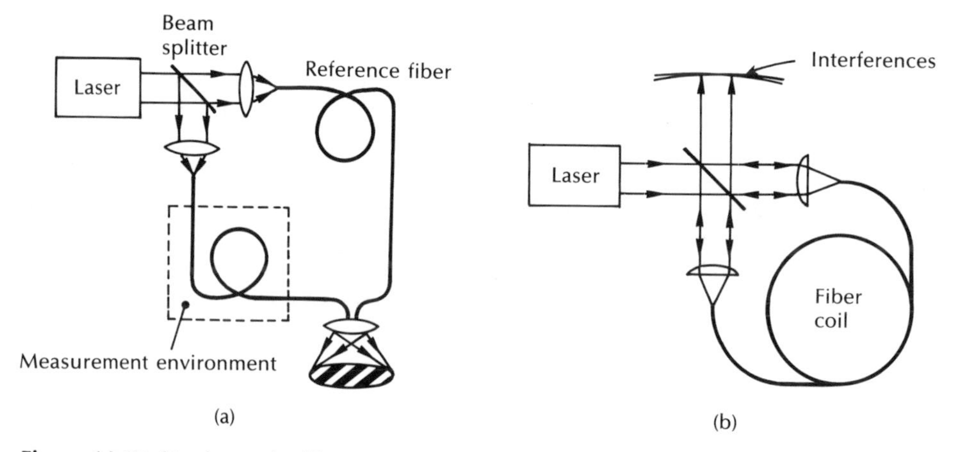

Figure 14.17 Single-mode fiber interferometric sensors: (a) a two arm interferometer (Mach–Zehnder); (b) ring interferometer with multiturn fiber coil (Sagnac).

counterpropagating waves which will return in phase after travelling along the same path in opposite directions. When the fiber coil is rotating about an axis perpendicular to the plane of the coil, however, then the path lengths between the counterpropagating waves differ. This difference produces a phase shift which in turn can be measured by interferometric techniques in order to obtain the rotation. A further interferometric sensor used to measure acoustic pressure which has attracted significant attention is the fiber hydrophone [Ref. 91].

Modulation of the polarization state of light within a fiber may also be utilized to take a physical measurement. A successful implementation of this technique is displayed in the Faraday rotation current monitor shown in Figure 14.18 [Ref. 92]. This device consists of a single polarization maintaining fiber (see Section 3.13.2) which passes up from earth to loop around the current-carrying conductor before passing back to earth. A He–Ne laser beam is linearly polarized and launched into the fiber which is then stripped of any cladding modes. The direction of polarization of the light in the fiber core is rotated by the longitudinal magnetic field around the loop, via the action of the Faraday magneto-optic effect [Ref. 93]. A Wollaston prism is used to sense the resulting rotation and resolves the emerging light into two orthogonal components. These components are separately detected with a photodiode prior to generation of a sum and difference signal of the two intensities (I_1 and I_2). The difference signal normalized to the sum give a parameter which is proportional to the polarization rotation ρ, following [Ref. 93]:

$$K\rho = \frac{I_1 - I_2}{I_1 - I_2} \tag{14.1}$$

where K is a constant which is dependent on the properties of the fiber. Hence a current measurement (either d.c. or a.c.) may be obtained which is independent of

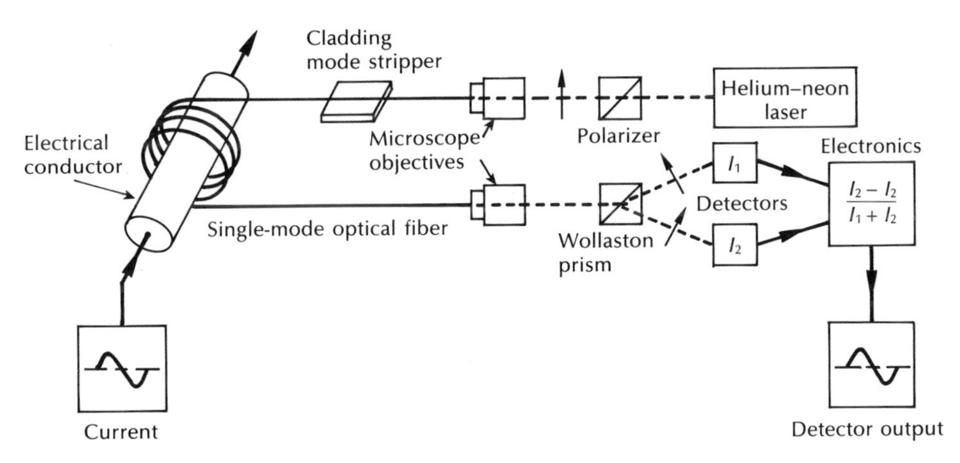

Figure 14.18 Single-mode optical fiber sensor for current measurement [Ref. 92].

the received light power. Finally, it should be noted that the phase and polarization sensors described in this section also constitute intrinsic single-mode fiber devices.

14.5.2 Intrinsic fiber sensors

A popular technique for the realization of an intrinsic multimode fiber sensor involves microbending of the fiber in the modulation region [Refs. 94, 95]. The sensing mechanism for this technique is shown in Figure 14.19(a). Deformation of the fiber on a small scale causes light to be coupled from the guided optical modes propagating in the fiber core into the cladding region where they are lost through radiation into the surrounding region. When the spatial wavelength of the deformation L is correctly chosen, the power coupled from the fiber into radiation modes is high, providing a very high sensitivity to pressure applied to the deformer in a direction perpendicular to the fiber core axis. Furthermore, if the deformation is caused by the measurand (e.g. pressure, vibration, sound, etc.), then the fluctuation in intensity of either the core or cladding light is directly proportional to the measurand for small deformations. Thus monitoring the fiber core or cladding allows detection of the measurand.

The configuration of a microbending fiber sensor for monitoring structural deformation developed by TRW (USA) is shown in Figure 14.19(b) The test data corresponding to this configuration is given in Figure 14.19(c) for a 1.4 mm spatial wavelength of 3 wavelengths [Ref. 94]. It must be noted, however, that a requirement of this sensor type is the removal of the cladding modes (by a cladding mode stripper) both immediately prior to, and after, the modulation zone. This prevents the occurrence of measurement errors from light which may be propagating in the cladding from the source as well as light coupling back from the cladding into the core after the modulation zone. Moreover, although fiber damage resulting from the deformation does not appear to be a severe problem, a settling-in period may be observed prior to the attainment of constant sensor characteristics [Ref. 94].

Also, in common with all fiber intensity modulation sensors, inaccuracies may occur due to source, detector and fiber cable instabilities. These, so-called, common-mode variations usually necessitate the transmission of a separate optical reference signal which is not modulated by the measurand. In this way any optical intensity variations can be removed from the returned measurand signal by comparison with the returned reference signal at the receive terminal. Various strategies have been demonstrated by which referencing for intensity modulation sensors may be achieved. These include: techniques which divide the transmitted optical signal in, or prior to, the modulation region in order to obtain a reference signal [Refs. 96, 97]; methods which employ two separate wavelength optical signals, one of which is modulated by the measurand and the other which acts as the reference signal [Refs. 98, 99]; and a balanced bridge arrangement that produces two return optical signals which allow any intensity variations on the measurand signal to be eradicated [Ref. 100]. It should be noted, however, that the

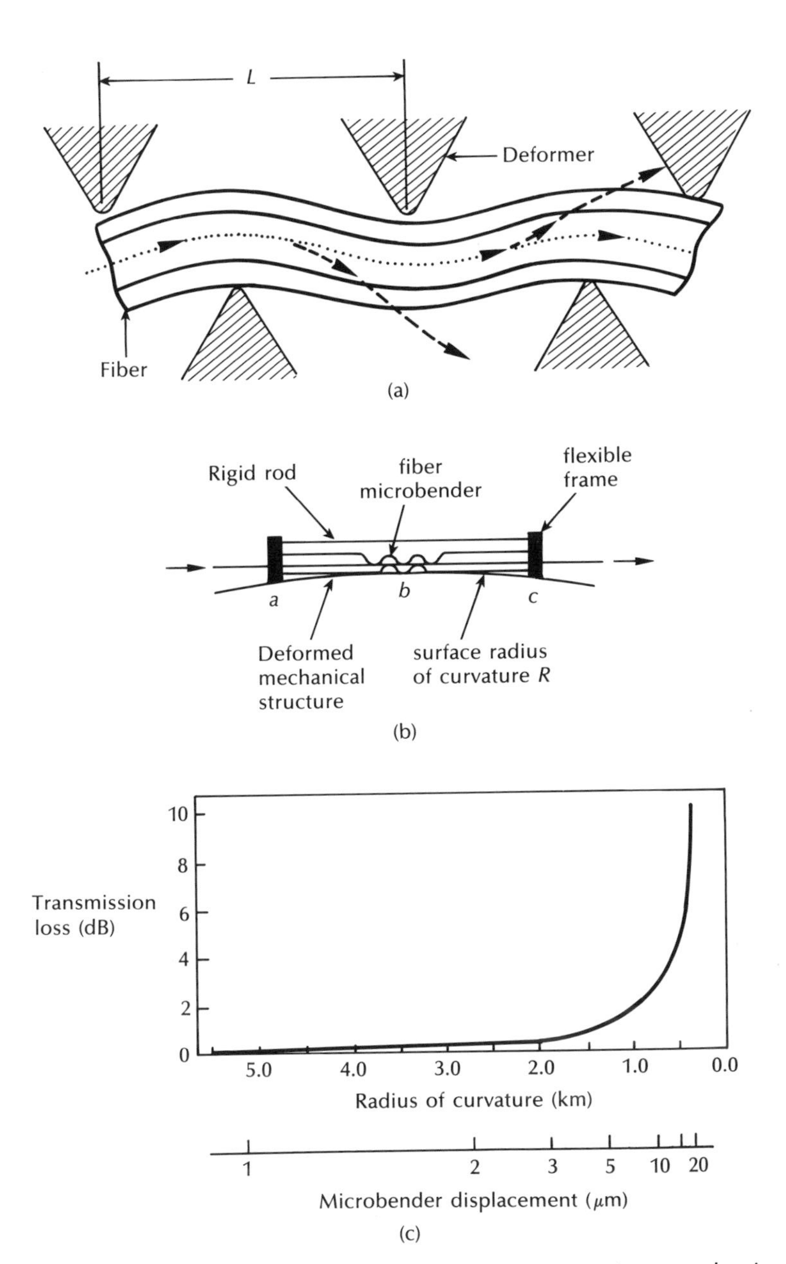

Figure 14.19 Fiber microbending sensor: (a) sensing mechanism; (b) a structure; (c) test data from (b). Reproduced with permission from S. K. Yao and C. K. Asawa, 'Microbending fiber optic sensing', *Proc. SPIE, Int. Soc. Opt. Eng.*, **412**, p. 9, 1983.

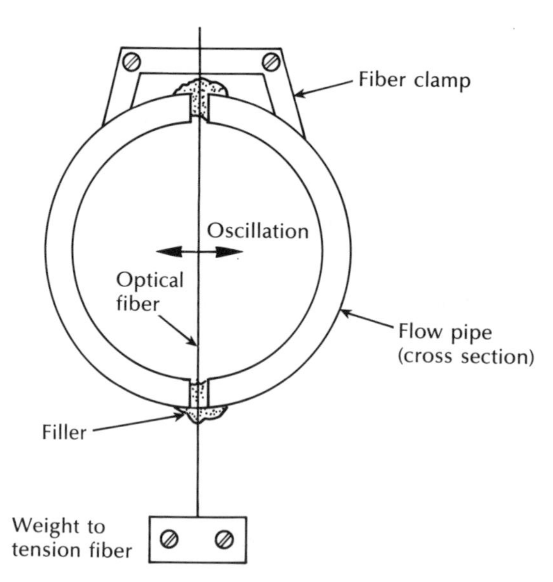

Figure 14.20 An optical fiber flow meter [Ref. 101].

first two techniques have been applied to specific extrinsic sensor mechanisms (see Section 14.5.3), whereas the latter method provides a more universal strategy.

Figure 14.20 shows an intrinsic optical fiber flow sensor mechanism. In this device a multimode optical fiber is inserted across a pipe such that the liquid flows past the transversely stretched fiber. The turbulence resulting from the fiber's presence causes it to oscillate at a frequency roughly proportional to the flow rate. This results in a corresponding oscillation in the mode power distribution within the fiber giving a similarly modulated intensity profile at the optical receiver. The technique has been used to measure flow rates from 0.3 to 3 $\mathrm{m\,s^{-1}}$ [Ref. 101]. However, it cannot measure flow rates below those at which turbulence occurs.

14.5.3 Extrinsic fiber sensors

Numerous extrinsic optical fiber sensor mechanisms have been proposed and investigated [Refs. 91, 95, 102, 103], but to date relatively few practical commercial devices have emerged. A technique which has been realized as a commercial product is illustrated in Figure 14.21 [Ref. 103]. This shows the operation of a simple optical fluid level switch. When the fluid, which has a refractive index greater than the glass forming the optical dipstick, reaches the chamfered end, total internal reflection ceases and the light is transmitted into the fluid. Hence an indication of the fluid level is obtained at the optical detector. Although this system is somewhat crude and will not provide a continuous measurement of a fluid level, it is simple and safe for use with flammable liquids.

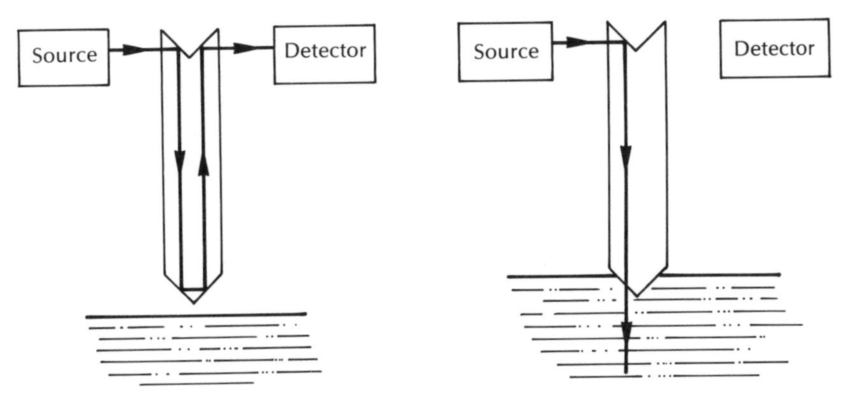

Figure 14.21 Optical fluid level detector.

Intensity modulation of the transmitted light beam is utilized in the extrinsic reflective or fotonic optical sensor shown in Figure 14.22(a) to give a measurement of displacement [Ref. 104]. Light reflected from the target is collected by a return fiber(s) and is a function of the distance between the fiber ends and the target. Hence the position or displacement of the target may be registered at the optical detector. Furthermore, the sensitivity of this sensor may be improved by placing the axes of the feed and return fiber at an angle to one another and to the target [Ref. 105]. Unfortunately, this technique exhibits the drawback mentioned previously with regard to the stability of the optical components which is a feature of intensity

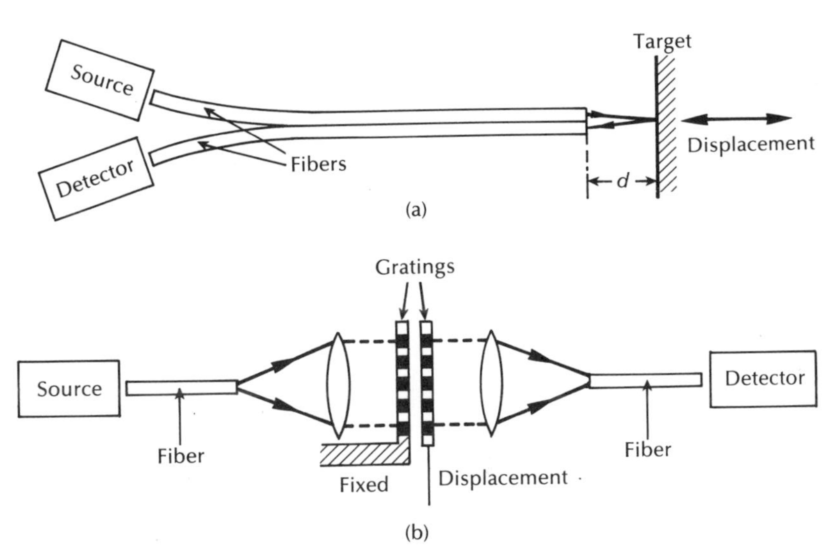

Figure 14.22 Optical displacement sensors: (a) reflective or fotonic sensor; (b) Moiré fringe modulation sensor.

modulated fiber sensors. Nevertheless, it lends itself to the optical sensor referencing techniques described in Section 14.5.2.

A method of overcoming the above drawback with optical fiber intensity modulation sensors is to use a digital measurement technique. Such a technique is shown in Figure 14.22(b) whereby the measurement of displacement is obtained using a Moiré fringe modulator [Ref. 95]. In this case the opaque lined gratings produce dark Moiré fringes. Transverse movement of one grating with respect to the other causes the fringes to move up or down. Thus a count of the fringes as the gratings are moved provides a measurement of displacement. The fringe counting is independent of instabilities within the system components which affect the optical intensity. However, mechanical vibrations may severely affect the measurement accuracy and prove difficult to eradicate. Also there are problems involved with loss of count if, for any reason, optical power to the sensor is interrupted.

A multimode fiber sensor which provides measurement of pressure is illustrated in Figure 14.23(a). In this device the photoelastic effect induced by mechanical stress on a photoelastic material (e.g. piezo-optic glass, polyurethane, epoxy resin)

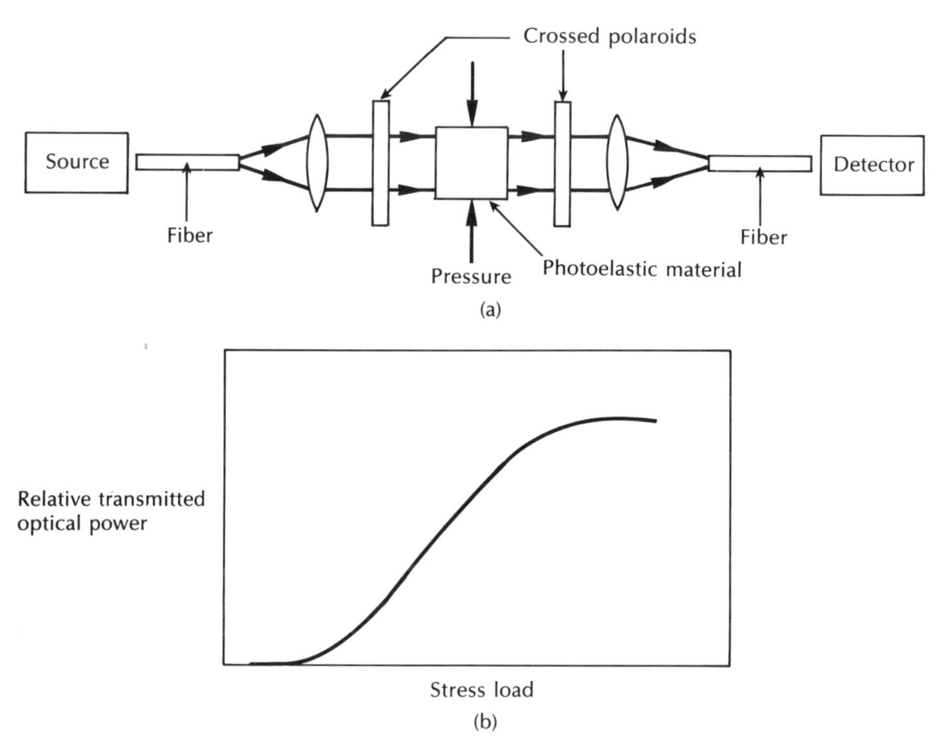

Figure 14.23 Photoelastic pressure sensor: (a) schematic structure; (b) response characteristic.

is utilized to rotate the optical polarization between a pair of crossed polarizers [Refs. 95, 103]. The phenomenon known as birefringence occurs with the application of mechanical stress to the transparent isotropic material, whereby it becomes optically anisotropic giving a variation in transmitted light through the sensor. The response of the sensor is typically sinusoidal, as illustrated in Figure 14.23(b), and a resolution of 10^{-4} of full range has been demonstrated [Ref. 103]. An advantage of this technique is that the stress may be induced directly without the need for an intermediate mechanism (e.g. pressure to displacement). A drawback, however, is that the birefringence exhibited by photoelastic materials is often temperature-dependent, making measurement of a single parameter difficult.

Another technique which has been adopted for commercial exploitation is utilized in the 'Fluoroptic' temperature sensor [Ref. 102]. This device, which is illustrated in Figure 14.24, uses a form of wavelength modulation which occurs when incoherent light is transmitted along a large core (multimode) fiber to a phosphorescent material. The light excites the rare earth phosphor which emits a number of lines at different wavelengths. Two of these spectral lines (in this case at wavelengths of 540 nm and 630 nm) are selected using filters in the receiver. The ratio of intensities of these chosen lines is a single valued function of the temperature of the phosphor. Measurement of this ratio provides an exact measure of the temperature. The resolution of the device is 0.1 °C over the range −50 to +250 °C and is independent of the output light intensity. However, the instrument has rather complex processing electronics and requires calibration during thermal stabilization since the sensitivities of the optical detectors, even when matched, do not track together at the two different emitted wavelengths as the temperature changes.

In general terms, wavelength modulation has the distinct advantage over intensity modulation in that such spectrally encoded sensors are not affected by intensity variations associated with the optical and optoelectronic components in the system. However, the variety of wavelength modulation techniques which have been investigated are significantly fewer in number than those for intensity modulation

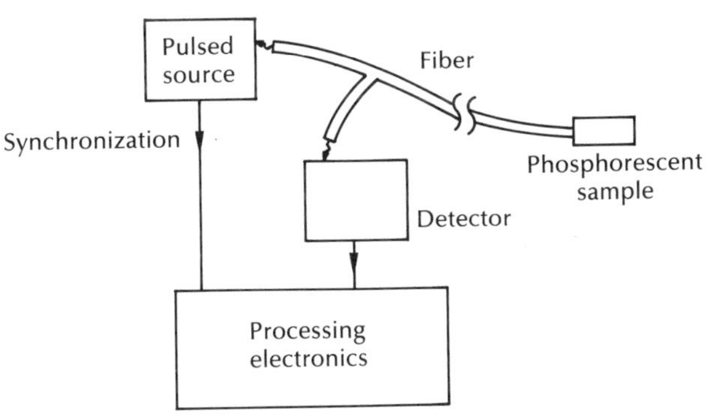

Figure 14.24 Fluoroptic temperature sensor [Ref. 102].

because the demodulation method is inherently more complex, often requiring some form of spectrometer [Ref. 95].

Extrinsic single-mode fiber sensors have been developed to provide noninvasive measurement for several physical measurands (e.g. velocity, fluid surface velocity and vibration). In particular, the all-fiber laser Doppler velocimeter (LDV) illustrated in Figure 14.25 allows the measurement of velocity in gases and fluids, as well as the velocity of objects, to be taken. The arrangement shown in Figure 14.25 employs two single-mode fibers to guide the transmitted beams to and from the probe. A fiber directional coupler is used at the probe to obtain two beams from the single incoming beam. The measurement volume is formed by the region of intersection of these two coherent optical beams which are independently scattered and Doppler shifted. A Doppler difference technique is then employed because the frequency shift is different for each beam as they are travelling in different directions. The two frequency shifts beat together to produce a frequency δf_{Do} which is proportional to the component of velocity of the scattering particle v perpendicular to the mean direction of the incident beams and in their plane, so that [Ref. 89]:

$$\delta f_{\mathrm{Do}} = \left| \frac{2v}{\lambda} n \sin \frac{\theta}{2} \right| \tag{14.2}$$

where λ is the wavelength of the laser source, n is the refractive index of the measurement volume and θ is the angle of convergence between the two input beams.

It may be noted that Eq. (14.2) suggests an ambiguity in the direction of the velocity which may introduce serious errors. This problem can be resolved, however, through the introduction of a frequency shift into one of the transmitted beams. Hence the fiber modulator shown in Figure 14.25 produces the required frequency shift. With such a fiber LDV system, velocity can be measured with high precision in a short period of time. In addition, arrangements based on the classical

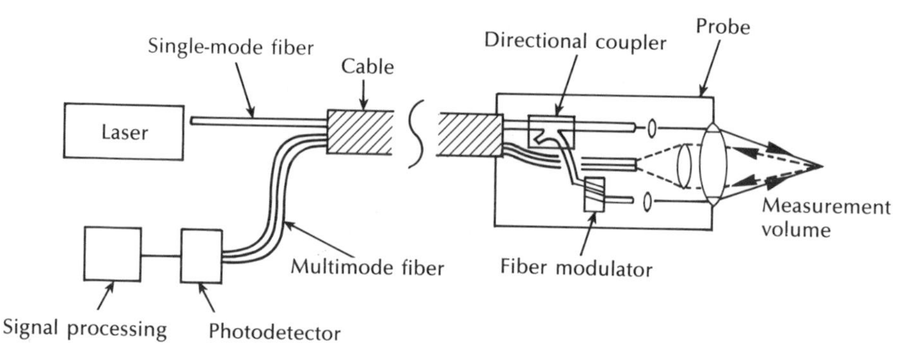

Figure 14.25 Optical fiber laser Doppler velocimeter.

interferometer configurations (e.g. Mach–Zehnder, Fabry–Perot and Michelson) have been employed to produce fiber laser vibrometers for remote vibration measurement [Ref. 89]. Finally, resonant structures, particularly those employing micromachined silicon resonators have attracted increasing interest for application within optical sensor devices in recent years [Refs. 91, 106]. In particular this work has focused upon optically actuated micromachined silicon devices in which a transmitted optical signal is partially absorbed by the resonator structure, creating thermal strain and hence mechanical excitation.* The subsequent detection of the movement of the mechanical system is based on interferometric and/or intensity modulation techniques using the remaining optical signal. For example, such devices have been demonstrated incorporating thermally absorbing surfaces on miniature (10 μm long) silicon bridges [Ref. 91]. In these devices incident optical powers of less than 20 μW (usually delivered via single-mode fiber) are sufficient to produce 10 nm displacements.

Generally, intensity modulation is preferred for use with silicon micromachined devices when the oscillation frequency is relatively low (i.e. less than 20 kHz), whilst the more complex interferometric detection tends to be employed at higher oscillation frequencies. Nevertheless, a steadily increasing number of examples of optical sensor devices using silicon as optically energized microresonators are being reported [Ref. 106], even though the potential ageing and failure mechanisms have still to be fully explored.

14.6 Computer applications

Modern computer systems consist of a large number of interconnections. These range from lengths of a few micrometres (when considering on-chip, very large scale integration (VLSI) connections) to perhaps thousands of kilometres for terrestrial links in computer networks. The transmission rates over these interconnections also cover a wide range from around 100 bit s^{-1} for some teletype terminals to several hundred Mbit s^{-1} for the on-chip connections. Optical fibers are starting to find application in this connection hierarchy where secure, interference-free transmission is required.

Although in its infancy, integrated optics has stimulated interest in connections within equipment, between integrated circuits, and even within hybrid integrated circuits, using optical techniques. Much of this work is still at the research stage and therefore was discussed in Sections 10.5 and 10.6. Moreover optical transmission techniques and optical fibers themselves have found application within data processing equipment. In addition, as indicated in Sections 10.7–10.9 investigations have already taken place into the use of optical fibers for mains isolators and digital data buses within both digital telephone exchanges and computers. Their small size,

* The mechanical excitation of these devices can be explained by photoacoustic effects [Ref. 106].

low loss, low radiation properties and freedom from ground loops provide obvious advantages in these applications.

At present, however, a primary application area for optical fiber communications remains interequipment connections. These provide noise immunity, security and removal of earth loop problems, together with increased bandwidth and reduced cable size in comparison with conventional coaxial cable computer system interconnections. The interequipment connection topology for a typical mainframe computer system (host computer) is illustrated in Figure 14.26. The input/output (I/O) to the host computer is generally handled by a processor, often called a data channel or simply channel, which is attached to the main storage of the host computer. It services all the I/O requirements for the system allowing concurrent instruction processing by the central processing unit (CPU). Each data channel contains an interface to a number of I/O control units. These, in turn, control the I/O devices (e.g. teletypes, visual display units, magnetic disk access mechanisms, magnetic tape drives and printers). When metallic cables are used, the interface between the data channel and the control units comprises a large number (often at least forty-eight) of parallel coaxial lines incorporated into cables. An attractive use of optical fiber interconnection is to serialize this channel interface [Ref. 107] using a multiplex system. This significantly reduces cable and connector bulk and improves connection reliability.

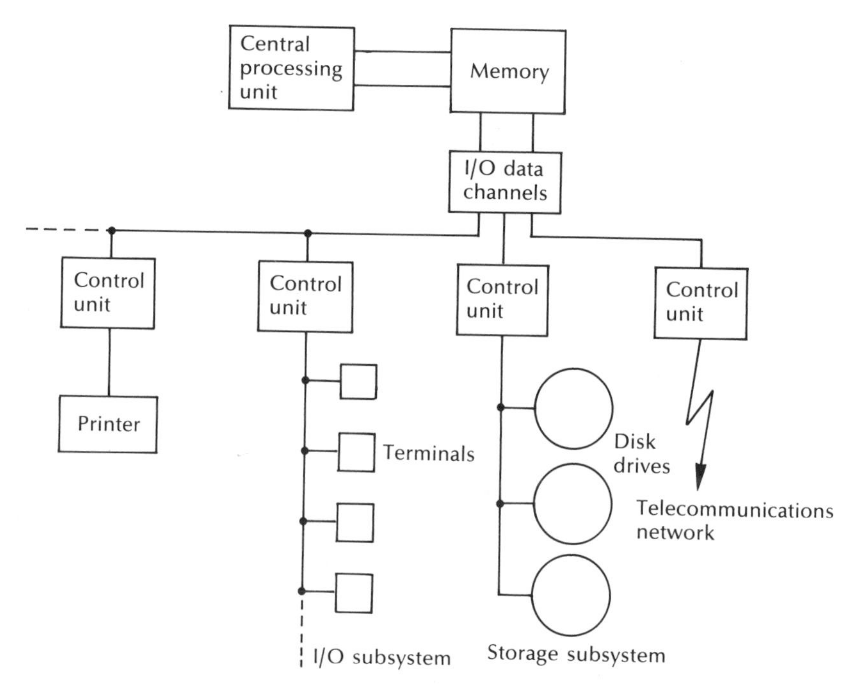

Figure 14.26 Block schematic of a typical mainframe computer system.

Optical fiber links of this type were demonstrated in 1978 by Sperry Univac [Ref. 108], and subsequently Fujitsu developed a product to perform the same function. However, neither of these systems offered enhanced channel performance as measured by the product of bit rate and link length. Developments are therefore continuing with regard to high performance channel links utilizing new protocols for data exchange. An early prototype [Ref. 107] optical fiber serial subsystem designed by IBM Research operates at 200 Mbit s^{-1} over distances of up to 1 km. This system utilized a laser chip mounted on a silicon substrate with the fiber encapsulated in a monolithic dual in line package, and a single chip, high sensitivity, silicon *p–i–n* receiver.

A more recent development to provide high speed communication on a backbone network for use in computer centres is called Datapipe [Ref. 109]. Initially, this system comprised a 275 Mbit s^{-1} single-mode fiber link which could operate over a radius of 30 km. Subsequently, the system was developed to give a 500 Mbit s^{-1} transmission rate providing a site capability of 1 Gbit s^{-1} when a redundant system was installed.

The other interconnection requirement for the mainframe computer system is between the I/O control units and the I/O, terminals. Again, optical fiber systems can provide high speed, multiplexed, secure communication links to replace the multitude of coaxial cables normally required for these interconnections. An example of such a fiber system utilizes a multiplexing system on to a single optical fiber cable for connecting an IBM 3274 controller to its terminals [Ref. 110]. In this case up to thirty-two terminals and printers can be linked to the controller in either a point to point or multidrop* configuration employing a star coupler or beam splitters. This interconnection requirement is often extended due to the trend of connecting numbers of processors together in order to balance the system work load, increase system reliability and share storage and I/O devices. Hence optical fiber systems have been developed for use in local area networks.

14.7 Local area networks

A local area network (LAN), unlike the local telecommunication network, is an interconnection topology which is usually confined to either a single building or group of buildings contained entirely within a confined site or establishment (e.g. industrial, educational, military, etc.). The LAN is therefore operated and controlled by the owning body rather than by a common carrier.† Optical fiber

* The multidrop bus configuration will not allow interconnection of as many as thirty-two terminals due to the insertion losses obtained at the beam splitters or T-couplers.

† Another possible definition of a LAN, based on speed and range of operation, is that a LAN typically operates at a transmission rate of between 100 kbit s^{-1} and 100 Mbit s^{-1} over distances of 500 m to 10 km. Hence a LAN is intermediate between a short range, multiprocessor network (usually data bus) and a wide area network which has historically provided relatively low speed data transmission (up to 100 kbit s^{-1}) over very long distances using conventional communications technology. However, it must be noted that there are always exceptions to these general definitions which are already increasing in number as the technology advances.

communication technology is finding application within LANs to meet the on-site communication requirements of large commercial organizations and to enable access to distributed or centralized computing resources.

Figure 14.27 shows the Open Systems Interconnection (OSI) seven layer network model which was originally developed for wide area networks (WANs). This model, which has been developed by the International Organization for Standardization (ISO), provides a standard architecture that defines communication tasks into a set of hierarchical layers within networks. Each layer performs a related set of functions required in order that one system (nominally a heterogeneous computer) will be able to communicate with another system. The partitioning of these functions following logical methodology was undertaken by an ISO subcommittee, resulting in the reference model comprising seven layers which are listed with a brief definition in Table 14.6.

As there are fundamental differences between LANs and WANs it has been found necessary to redefine the two bottom layers of the OSI model into three layers, as displayed in Figure 14.27. The physical layer, as in the WAN, is responsible for the physical transmission of information.

The functions of data link layer, however, are separated into two layers: namely, the logical link control (LLC) layer which assembles/disassembles data frames or packets and provides the appropriate address and error checking fields, and the medium access control (MAC) layer which organizes communications over the link. The MAC layer embodies the set of logical rules or the access protocol which allow nodes to access the common communication channel, and several MAC options may therefore be provided for the same LLC layer. Furthermore, another major

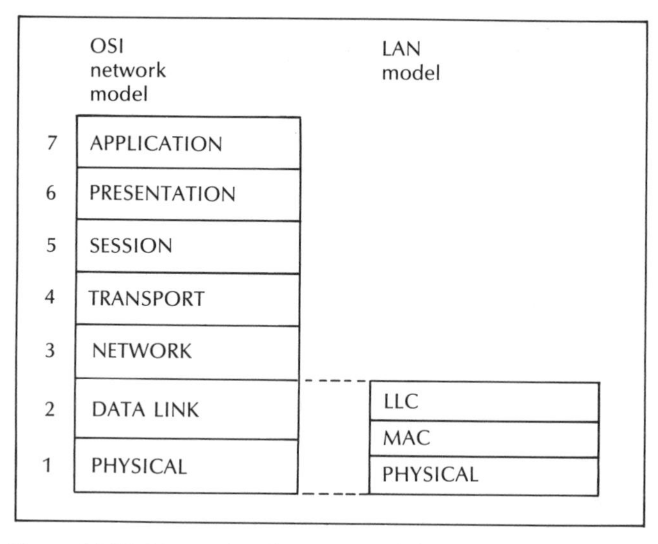

Figure 14.27 Network reference models: OSI and LAN modification.

Table 14.6 Principal functions for the seven levels of the OSI reference model.

Level		Function
1. PHYSICAL	–	For interfacing to the physical transmission medium. Concerned with transmission of unstructured bit stream over the physical link.
2. DATA LINK	–	Provides for the reliable transfer of data across the physical link; controls the transmission and reception of blocks with necessary synchronization, error and flow control.
3. NETWORK	–	For interworking with the underlying telecommunication network. Provides upper layers with independence from data transmission technologies.
4. TRANSPORT	–	Provides reliable transparent transfer between end points; end to end synchronization, flow control and error recovery.
5. SESSION	–	For tying terminals to applications during a session; establishes, manages and terminates connections (sessions).
6. PRESENTATION	–	Provides for manipulation of the data structure. Performs useful transformations on data (eg. encryption, text compression, reformatting).
7. APPLICATION	–	Provides services to the users of the OSI environment (eg. data base management, network management).

difference between WANs and LANs is that each node on the network can be directly connected to all others, thus eliminating the need for routeing via the network layer. As a result of this high degree of connectivity, the specific topologies shown in Figure 14.28 have become commonplace [Ref. 111].

The basic optical fiber LAN topologies are the bus, ring and star [Refs. 112, 113]. In the bus topology, data generally circulate bidirectionally. Data are input and removed from the bus via four port couplers located at the nodes (see Section 5.6.1). In the ring, configuration data usually circulate unidirectionally, being looped through the nodes at each coupling point and hence repeatedly regenerated in phase and amplitude. The optical star forms a central hub to the network which may be either active or passive. In passive operation, the star coupler (see Section 5.6.2) at the hub splits the data in terms of power, so that they are available to all nodes. Line losses are therefore primarily determined by the degree of splitting. With an active hub the data are split electrically in the active star coupler and therefore the only causes of line loss are the fiber and splice/connector losses.

The design of the MAC layer is crucial in LANs because, in general, it is the efficiency of the access protocol which governs the availability of the bandwidth

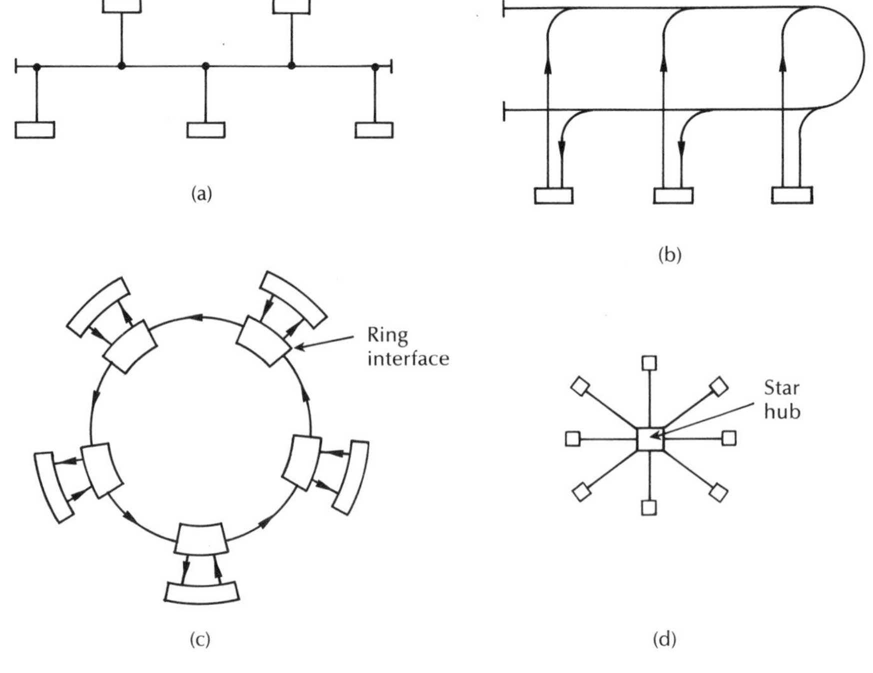

Figure 14.28 Common LAN topologies: (a) linear bus; (b) U-bus; (c) ring; (d) star.

provided by the network for the dual functions of data transmission and channel arbitration. Three specific types of access protocol have gained a fair degree of acceptance, primarily because of their simplicity. These are: (1) random access protocols, the most popular example of which is carrier sense multiple access with collision detection (CSMA/CD) used on the Ethernet LAN; (2) token passing protocols; and (3) time division multiple access (TDMA) protocols.

With CSMA/CD protocols, nodes are allowed to transmit their data as soon as the communication channel is found to be idle (i.e. carrier sense). If the transmissions from two or more nodes collide, the event is detected by the physical layer hardware and the nodes involved terminate transmission before attempting retransmission after a random time interval. Token passing protocols behave as distributed polling systems in which nodes sequentially obtain permission to use the channel. The channel arbitration is determined by the possession of a small distinctive bit packet, known as a token, which can only be held by one node (which is then permitted to use the channel) at any one time. When the data transmission from that node is completed it passes the token packet to the next node in the logical sequence. Finally, with basic (i.e. fixed assignment) TDMA protocols, each node on the network is assigned a fixed length time slot during which it may transmit data. In this way, the protocol operates in a similar manner to that of time division multiplexing on a point to point link (see Section 11.5).

A schematic of an optical fiber regenerative repeater (see Section 11.6.1) which may be used to restore the attenuated signal power in an optical fiber LAN is displayed in Figure 14.29(a). In addition, this device may be employed as an active tap to allow the interconnection of numerous nodes within ring or linear bus topologies. The use of such repeater interfaces allows similar topologies to be realized for optical fiber LANs as their metallic counterparts, but with certain cost penalties. In order to reduce the number of repeaters required, a range of passive tapping components are available (see Section 5.6). Figure 14.29(b) shows a four port tap which allows data to be removed from and added on to a unidirectional optical fiber bus. A pair of three port passive taps which can be used to receive and transmit on to dual linear buses or U-buses are shown in Figure 14.29(c). Unfortunately, as a result of the optical power budget requirements mentioned previously there is an upper limit to the number of passive taps which can be inserted into the linear bus or ring topologies. At present commercial taps incur an excess loss of between 0.1 and 1 dB. In addition to this excess loss the tap must also divert some of the optical signal power to the tapping receiver. Even if the tap only diverts 1% of the transmitted optical signal power, this places a limit of around thirty on the number of taps which may be provided unless active electrical components are included to allow regeneration of the transmitted optical signal.

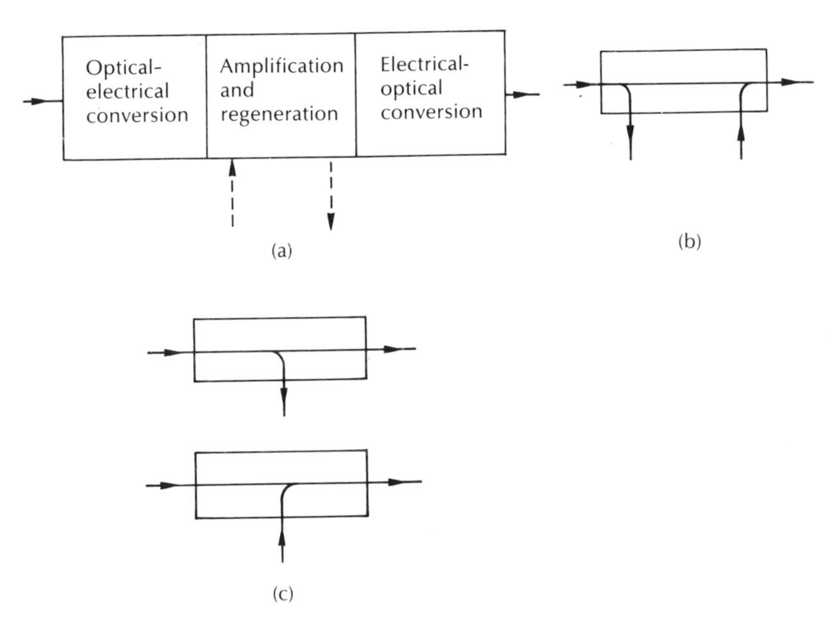

Figure 14.29 Optical fiber LAN tapping devices: (a) regenerative repeater and active tap; (b) four port passive tap; (c) three port passive taps.

Alternatively, the use of optical amplifiers to provide gain and thus enhance the power budgets in such networks is under active consideration (see Section 11.10).

One of the major problems associated with optical fiber buses which use the tapping methods discussed previously is that there is an uneven distribution of the transmitted optical power between the connected receivers. This has two distinct drawbacks, namely that the optical receivers must have a large dynamic range and that detection of the collision events required for CSMA/CD protocols becomes difficult [Refs. 114, 115]. One solution to this problem is to use the passive star couplers discussed in Section 5.6.2. Each node in the star topology has a single input and a single output fiber connected to the hub of the star. Optical power within the input fibers can be distributed more evenly between the output fibers, thus achieving more uniform reception levels. Passive star couplers are currently available using up to 100×100 ports. Typical port to port insertion losses for 32×32 and 64×64 port transmissive passive stars employing multimode fiber are around 19 dB and 23 dB respectively. Variations on the passive star topology include the use of an active star hub [Ref. 116] and the use of cascaded passive stars [Ref. 117].

Activity in the LAN area in general, and more recently in the optical fiber LAN field, has intensified over the last few years as a result of the increasing demand for information technology within commerce and industry. Optical fiber LAN or mixed-media (e.g. fiber and twisted wire pairs) LAN products are available from several equipment suppliers and a good number of networks are at varying stages of development ranging from theoretical studies, sometimes accompanied by experimental investigation, to commercial products [Refs. 118, 119].

An early experiment based on the Ethernet LAN and also developed by Xerox was Fibernet [Ref. 120]. This passive transmissive star network employing nineteen ports and multimode fiber was successfully operated at data rates of 150 and 100 Mbit s^{-1} over distances of 0.5 and 1.1 km, respectively, with zero errors. A more recent development of CSMA/CD network is Fibernet II [Ref. 116] which employs an active star repeater. Collision detection was implemented in the twenty-five port star repeater which was electrically compatible with the standard 10 Mbit s^{-1} Ethernet coaxial cable system. The network span was around 2.5 km, and using eight channel multiplexers the network is capable of handling 200 stations. Alternative passive star-based CSMA/CD networks have been developed by the Codenoll Corporation (Codenet) and Ungermann-Bass Corporation together with the Siecor Corporation to be coupled to their Ethernet compatible Net/One.

Passive bus topologies for optical fiber LANs, as mentioned previously, exhibit drawbacks in relation to optical power losses at the necessary node couplers. However, limited connection bus systems have been demonstrated. For example, Philips have developed a closed optical bus network which can accommodate eight to ten nodes on passive T-couplers. In addition, there are a number of developments by manufacturers (e.g. BICC Data Networks, Siemens) which provide an optical bus interconnection between two coaxial Ethernet segments to form a multimedia network.

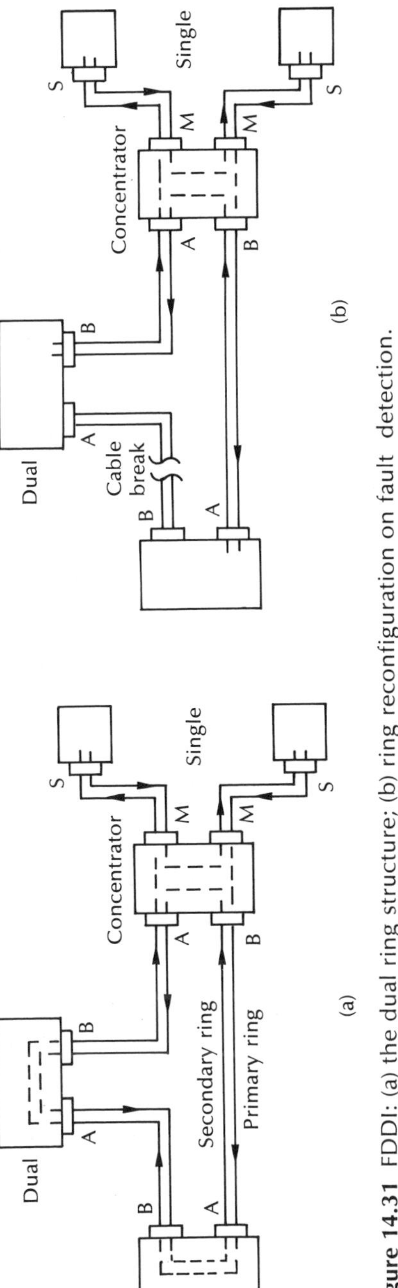

Figure 14.31 FDDI: (a) the dual ring structure; (b) ring reconfiguration on fault detection.

required bit error rate. This results in an overall bit error rate for the network of 10^{-9}.

More recently, a single-mode fiber (SMF) version of the PMD has been defined which utilizes injection laser transmitters [Refs. 123, 129]. This development is shown in Figure 14.30 as SMF–PMD. Two power level categories for the laser transmitters have been specified, the lower level of which retains the same optical receivers as the basic multimode fiber PMD. Nevertheless, it is intended that the SMF–PMD will enable individual links to be extended up to 60 or possibly even 100 km [Ref. 123]. Interestingly, however, the network data rate is set to remain at 100 Mbit s^{-1} (125 Mbaud) under this single-mode fiber standardization proposal [Ref. 129].

The FDDI has in-built recovery mechanisms to enhance its robustness. Network operation is normally designated to the primary ring (Figure 14.31(a)). However, if this ring is broken, then transmission is switched to the secondary ring. The reconfiguration of stations around a cable break, as illustrated in Figure 14.31(b), is controlled by the station management function. In addition, station failure does not affect the network operation due to the provision of optical bypass switches within all the stations. This ensures that the optical path through a station is maintained in the event of a station fault, or power down.

The token passing protocol aspects of the FDDI are somewhat similar to the IEEE 802.5 token passing ring. Each station in the scheme is allowed to transmit only when it is in possession of a token which is circulated sequentially around the stations on the ring. The basic FDDI protocol (i.e. FDDI-1) has provision for handling two priority classes of data traffic: synchronous and asynchronous. At each station, transmission under these classes is controlled by employing timers and by specifying the percentage of the ring bandwidth that can be used by synchronous traffic [Ref. 130].

In order to cater for periodic deterministic traffic, traffic with a single time reference (e.g. speech, possibly video), a circuit-switched mode of operation within the protocol has recently been defined [Refs. 123, 131, 132]. This development, which is referred to as FDDI-11, allocates the 100 Mbit s^{-1} bandwidth of FDDI to circuit-switched data increments of 6.144 Mbit s^{-1} isochronous channels. These 6.144 Mbit s^{-1} portions of circuit-switched data bandwidth are known as wideband channels (WBCs). Up to sixteen WBCs may be assigned using a maximum of 98.304 Mbit s^{-1} of bandwidth. Each of these WBCs offers a full duplex highway which may in turn be flexibly reallocated into a variety of subhighways at modular rates of 8 kbit s^{-1} (e.g. 8 kbit s^{-1}, 64 kbit s^{-1}, 384 kbit s^{-1}, 1.536 Mbit s^{-1} and 2.048 Mbit s^{-1} or combinations thereof). When all sixteen WBCs are allocated, a residual token channel of 768 kbit s^{-1} capacity remains, after allowance for the preamble and cycle header. This packet bandwidth, which consists of twelve bytes every cycle (125 μs), is known as the packet data group (PDG) and is interleaved with the sixteen WBCs. Moreover, WBCs may be assigned on a real-time basis with any of the unassigned bandwidth being made available to the token passing channel. Finally, several WBCs can be combined into aggregates to satisfy the requirements of video, or applications necessitating bandwidths in excess of 6.144 Mbit s^{-1}.

The above implementations of FDDI (i.e. 1 and 11) incorporate the same PMD layer equipments which allow stations spacings of up to 2 km for multimode fiber, or perhaps up to 100 km with single-mode fiber. The implementation of FDDI-11, however, requires an additional standard relating to the hybrid ring control (HRC) [Ref. 133]. The HRC has become the new lowest sublayer of the data link layer, being positioned below and between the MAC and PHY layers, as illustrated in Figure 14.32. It multiplexes data between the (packet) MAC and the isochronous MAC (I-MAC). This function requires that the (packet) MAC be able to transmit data on a noncontinuous basis because packet data are interleaved with isochronous data.

FDDI networks are perceived to have a range of applications within the local area, as illustrated in Figure 14.33. At present a likely application area is as a backbone network for the interconnection of other smaller, lower speed LANs. Some examples are the connection of Ethernet (IEEE 802.3), Token bus (IEEE 802.4) and the Token ring (IEEE 802.5). However, it is also suggested [Ref. 132] that the FDDI could operate as a back-end network for the interconnection of mainframes or minicomputers to peripheral controllers, communication controllers, file servers, database machines, laser printers, etc. Although FDDI networks could possibly provide a replacement for common networks utilized in this area such as Hyperchannel, it is the case that current (nonstandard) developments in this application area are targeted at transmission rates of around 500 Mbit s^{-1} [Ref. 109].

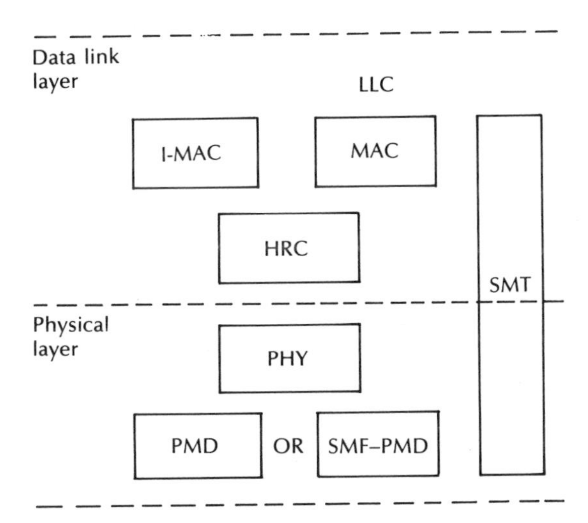

Figure 14.32 FDDI-II relationship to the OSI model.

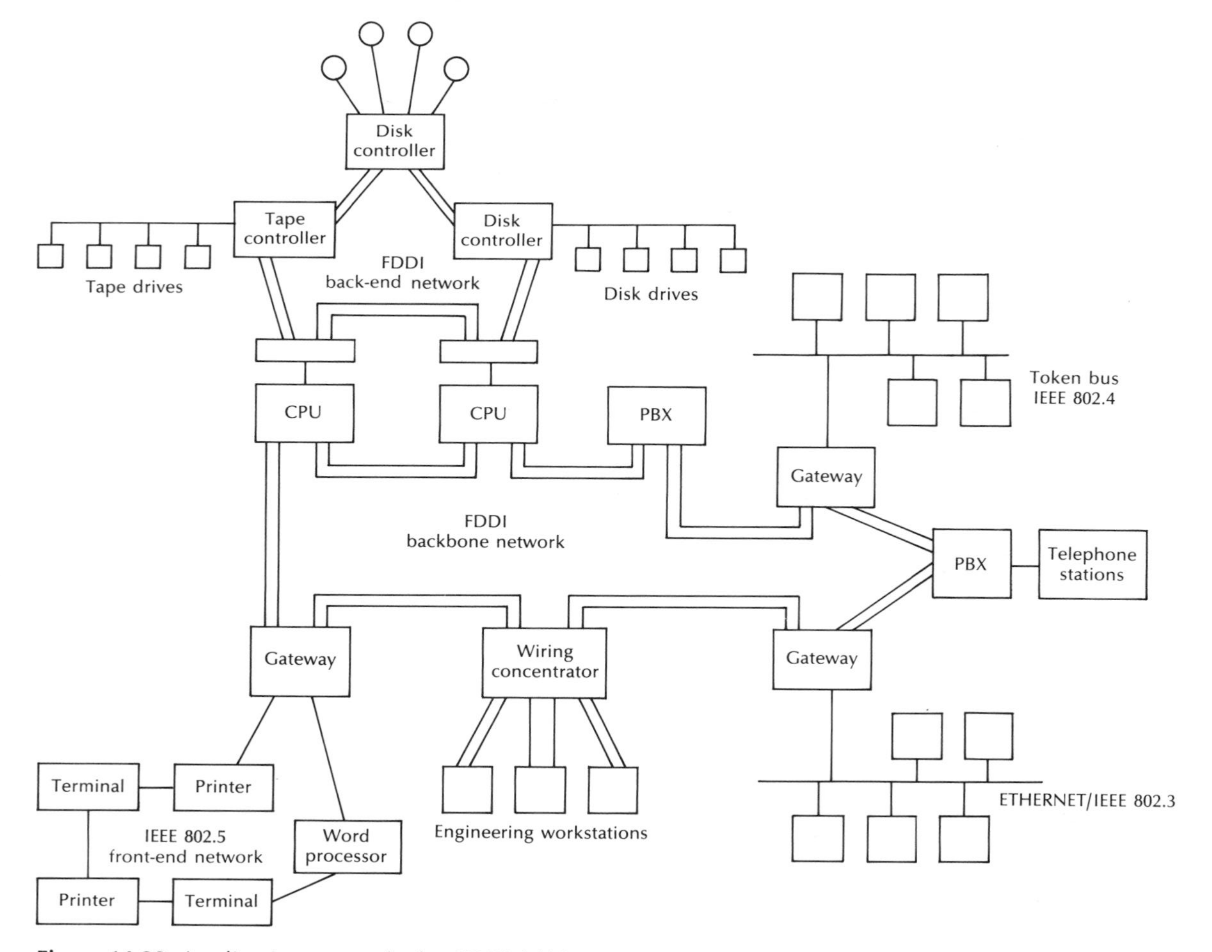

Figure 14.33 Application example for FDDI LANs.

Front-end networks are perhaps the most rapidly expanding application area for FDDI networks [Ref. 132]. Engineering work stations such as computer aided design terminals use a variety of accelerators, processors and LANs. However, it is normally the LAN which ultimately limits the useful performance of such devices. It is here that the bandwidth of FDDI networks should prove sufficient to provide real-time support. In this context it is apparent that FDDI networks (particularly FDDI-11) may well find widespread application within industrial environments. The obvious advantages of the fiber transmission media coupled with much higher bandwidth capacity in comparison with other LAN standardization proposals must assist FDDI LANs for use in such application areas. In addition, the mixed traffic capability to be offered by the FDDI-11 protocol, as well as the fault tolerant dual ring approach, must make these networks an attractive proposition for the provision of backbone and/or front-end networks on industrial sites and within manufacturing premises. It must be noted, however, that to date commercial developments of FDDI LANs have concentrated on the FDDI-1 protocol and it is not clear how quickly FDDI-11 products will become available [Ref. 134].

Notwithstanding the above comment, work is also in progress to define the next generation of LANs which will supersede the FDDI. The so-called 'FDDI follow-on LAN' (FFOL) has been discussed by an ANSI working group since 1990 [Ref. 135]. It is suggested that FFOL would act as a backbone for FDDI networks or as a higher speed front-end network. The proposed data rates for FFOL are intended to integrate with those of the future public telecommunications network under the SDH and SONET developments (see Section 14.2.5), being in the range 150 Mbit s^{-1} to 2.5 Gbit s^{-1}. Furthermore, it is envisaged that the follow-on LAN will also have a dual ring configuration employing optical fiber transmission media. It is suggested that both single-mode and multimode fibers will be put into the FFOL standardization process [Ref. 135], the former fibers providing for the higher speeds over longer distances (e.g. tens of kilometres) whilst the latter graded index fibers would be operated at data rates up to 622 Mbit s^{-1} over distances of around 300 m.

14.7.2 Industrial networks

Various models for industrial communications have been suggested [Refs. 136 to 138], including the four level hierarchical structure illustrated in Figure 14.34. In this structure the factory level (level 1) connects all systems in the plant ranging from administrative to the computer aided design/computer aided manufacturing (CAD/CAM) functions. Office services at level 2 include input/output (I/O) devices and database management. Manufacturing services at this second level include overall planning functions such as numerical-control program distribution together with materials handling and scheduling. At level 3 a cell might typically be a flexible manufacturing system or a warehouse storage unit. At the fourth, or application,

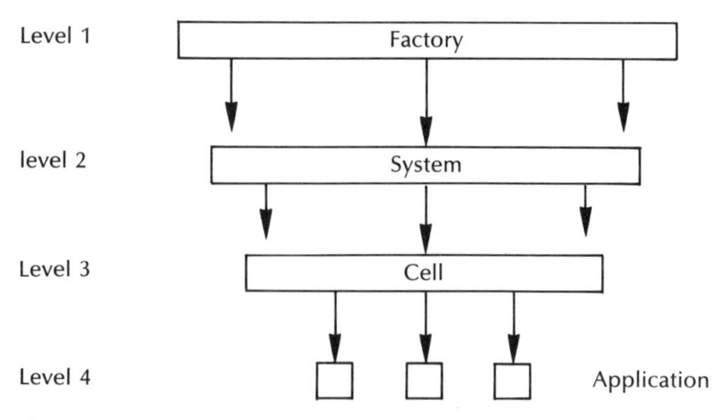

Figure 14.34 Model for industrial communications.

level, controller specific connections occur serving robots, machine tools, computer numerical controllers (CNCs) and, perhaps, advanced guided vehicles (AGVs).

As the service requirements differ from level to level an approach has been suggested [Refs. 136 to 138] whereby each level may adopt its own protocol. For example, application/cell connections generally require a real-time response for automatic monitoring and control, whereas I/O and database transactions at the system level are not critically dependent on message delays. In order to maintain flexibility without increasing complexity a more attractive approach is to employ a single access protocol solution. Hence, in this case all the offered traffic is seen by the network at the same level and the service requirements placed on the protocol for each device may be considered in this context.

A major development over the past few years in industrial local area networking is the Manufacturing Automation Protocol (MAP) activity initiated by General Motors in the United States to fulfil their future factory communication requirements [Ref. 139]. This development has broadened to include many other manufacturing concerns, together with major vendors on an international front. The MAP 3.0 specification which is compatible with the OSI reference model defines a broadband (i.e. frequency division multiplexed) coaxial cable network employing a token passing protocol which originates from the IEEE 802.4 Token bus proposals. Two adjacent 6 MHz channels on the LAN are utilized to provide a data rate for industrial traffic of 10 Mbit s^{-1}. The specification also includes a carrier band option on a single channel operating at 5 Mbit s^{-1}.

Included within the MAP 3.0 specification is an appendix on fiber optics which was written by the European Map User Group (EMUG WG2). The appendix, whose object is to allow the use of optical fiber within MAP where it is considered the appropriate medium, recommends two specific options from the IEEE 802.4 and 802.5 proposals [Ref. 140]. These are:

Token bus (IEEE 802.4H).

(a) use of single passive star with up to thirty-two ports and excess loss less than 3 dB (Figure 14.35);
(b) transmission rate of 10 Mbit s^{-1};
(c) operating wavelength at 0.85 μm;
(d) high sensitivity power budget of 30 dB;
(e) use of dual window multimode fiber with losses of less than 3 dB km^{-1} at 0.85 μm and 1 dB km^{-1} at 1.3 μm.

Token ring (IEEE 802.5C)

(a) use of dual ring configuration;
(b) transmission rate of 16 Mbit s^{-1};
(c) operating wavelength at 0.85 μm;
(d) the use of optical fiber bypass relays as an option;
(e) use of dual window multimode fiber with losses of less than 3 dB km^{-1} at 0.85 μm and 1 dB km^{-1} at 1.3 μm.

The above recommendations within MAP are intended to provide the basis for lower speed industrial optical fiber networks whereas FDDI will provide for some of the higher speed traffic requirements.

The Token ring recommendation within MAP has been suggested in part to cater for networks requiring in excess of thirty-two stations. However, an attractive

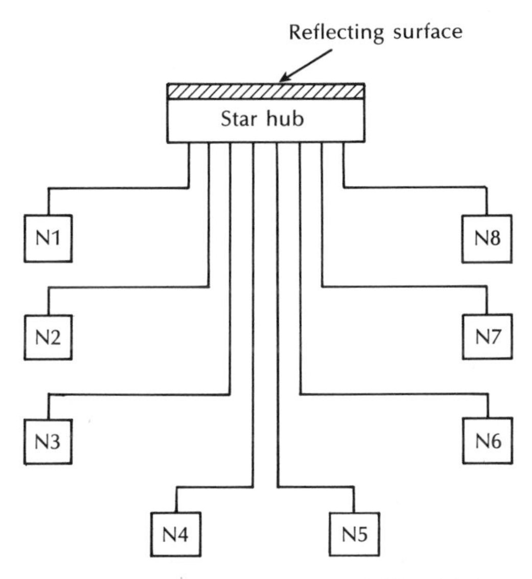

Figure 14.35 Passive star optical fiber LAN topology shown with reflective star coupler hub.

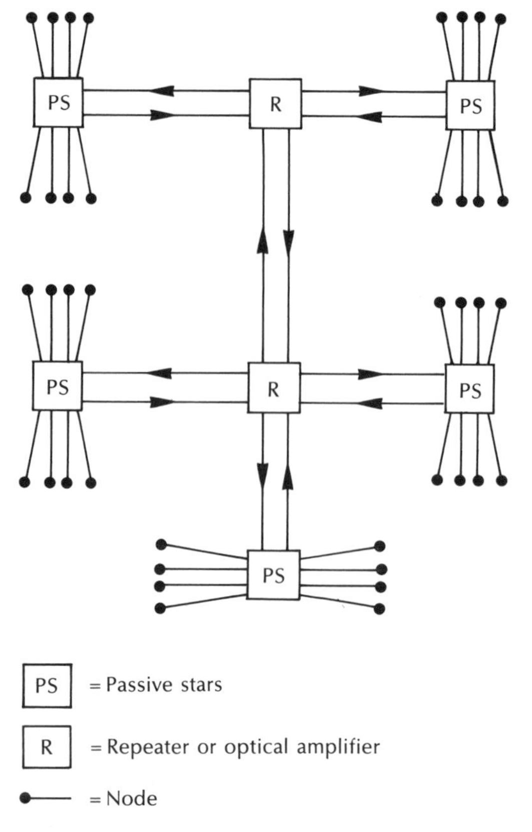

Figure 14.36 Multiple passive star optical fiber LAN topology.

alternative which has not as yet been specified is the use of a multiple star topology [Ref. 141]. Such a configuration is shown in Figure 14.36 in which the smaller passive star coupler devices (see Section 5.6.2) are interconnected by active repeaters. This topology reduces the quantity of fiber required as the large distances on a site are covered by the interstar links. Furthermore, with single-mode fiber networks the active repeaters could be provided by optical amplifiers (see Section 11.9), or even in the future by a derivative of the recently demonstrated optical regenerator [Ref. 142] so that the interstar signals remain in the optical domain. The modular configuration of the multiple star topology is also particularly appropriate to the cellular nature of modern automated production processes and flexible manufacturing systems (FMS) [Ref. 143].

Generally within FMS, a number of automated devices are grouped together to form a flexibly automated cell. A typical industrial plant would then contain a number of these cells, each of which has the capability to perform a variety of automated tasks. Communication is required for the exchange of information between the devices which are associated with individual automated cells as well as between the various cells. Passive star couplers could therefore be located within cells with the intercellular communication being provided on the interstar links.

References

[1] M. C. Brierley, P. W. France, R. A. Garnham, C. A. Millar and W. A. Stallard, 'Long wavelength fluoride fibre system using a 2.7 μm fluoride fiber laser, *Proc. Optical Fiber Commun. Conf., OFC'89* (USA), Postdeadline paper, PD-14, February 1989.

[2] K. Iwatsuki, S. Nishi and M. Saruwatari, '2.8 Gbit/s optical soliton transmission employing all laser diodes', *Electron. Lett.*, **26**(1), pp. 1–2, 1990.

[3] C. P. Sandbank (Ed.), *Optical Fibre Communication Systems*, Chapter X, John Wiley, 1980.

[4] C. J. Lilly, 'The application of optical fibres in the trunk network', *Telecomm. J. (Eng. edn), Switzerland*, **49**(2), pp. 109–117, 1982.

[5] R. S. Bergin Jr, 'Economic analysis of fiber versus alternative media', *IEEE J. on Selected Areas in Commun.*, **SAC-4**(9), pp. 1523–1526, 1986.

[6] P. E. Radley, 'Systems applications of optical fiber transmission', *Radio Electron. Eng.* (*IERE J.*), **51**(7/8), pp. 377–384, 1981.

[7] D. R. Smith, 'Advances in optical fibre communications'. *Physics Bulletin*, **33**, pp. 401–403, 1982.

[8] A. R. Beard. 'High capacity optical systems for trunk networks', *Proc SPIE Int. Soc. Opt. Eng.* (*USA*), **374**, pp. 102–110, 1983.

[9] P. Cockrane, R. Brooks and R. Dawes, 'A high reliability 565 Mbit/s trunk transmission system', *IEEE J. on Selected Areas in Commun.*, **SAC-4**(9), pp. 1396–1403, 1986.

[10] P. Cockrane and M. Brain, 'Future optical fiber transmission technology and networks', *IEEE Commun. Mag.*, pp. 45–60, November 1988.

[11] P. Cockrane, 'Future directions in long haul fibre optic systems', *Br. Telecom Technol. J.*, **8**(2), pp. 5–17, 1990.

[12] A. Javed, F. McAllum and G. Nault, 'Fibre optic transmission systems: the rationale and application'. *Telesis 1981 Two* (Canada), pp. 2–7, 1981.

[13] D. C. Gloge and I. Jacobs, 'Terrestrial intercity transmission systems', in S. E. Miller and I. P. Kaminow (Eds.), *Optical Fiber Telecommunication II*, Academic Press, pp. 855–878, 1988.

[14] P. Matthijsse, 'Essential data on optical fibre systems installed in various countries', *Telecomm. J. (Eng. edn), Switzerland*, **49**(2), pp. 124–130, 1982.

[15] O. Cottatellucci, F. Lombardi and G. Pellegrini, 'The application of optical fibres in the junction network', *Telecomm. J. (Eng. edn), Switzerland*, **49**(2), pp. 101–108, 1982.

[16] S. S. Cheng and E. H. Angell, 'Interoffice transmission systems', in S. E. Miller and I. P. Kaminow (Eds.), *Optical Fiber Telecommunication II*, Academic Press, pp. 833–854, 1988.

[17] R. W. Klessig, 'Overview of metropolitan area networks', *IEEE Commun. Mag.*, **24**(1), pp. 9–15, 1986.

[18] J. F. Mollenauer, 'Standards for metropolitan area networks', *IEEE Commun. Mag.*, **26**(4), pp. 15–19, 1988.

[19] J. O'Sullivan, 'The ICEberg project – a broadband network in Ireland', *Br. Telecommun. Eng.*, **9**, pp. 47–49, August 1990.

[20] T.-H. Wu and M. E. Burrowes, 'Feasibility study of a high-speed SONET self healing ring architecture in future interoffice networks', *IEEE Commun. Mag.*, pp. 33–51, November 1990.

[21] G. Schweiger and H. Middel, 'The application of optical fibres in the local and rural networks', *Telecomm. J. (Eng. edn), Switzerland*, **49**(2), pp. 93–101, 1982.

[22] G. Lentiz, 'The fibering of Biarritz', *Laser Focus (USA)*, **17**(11), pp. 124–128, 1981.

[23] G. A. Trough, K. B. Harris, K. Y. Chang, C. I. Nisbet and J. F. Chalmers, 'An integrated fiber optics distribution field trial in Elie, Manitoba', *IEEE International Conference on Communication*, ICC 82, Philadelphia, PA, USA, 13–17 June 1982, pp. 4D4/1–5, Vol. 2, IEEE, 1982

[24] J. R. Fox, D. I. Fordham, R. Wood and D. J. Ahern, 'Initial experience with the Milton Keynes optical fiber cable TV trial', *IEEE Trans. Comm.*, **COM-30**(9), pp. 2155–2162, 1982.

[25] W. K. Ritchie, 'Multi-service cable-television distribution systems', *British Telecomm. Eng.*, **1**(4), pp. 205–210, 1983.

[26] W. K. Ritchie, 'The BT switched-star cable TV network', *Br. Telecom Technol. J.*, **2**(4), pp, 5–17, 1984.

[27] J. R. Fox and E. J. Boswell, 'Star-structured optical local networks', *Br. Telecom Technol., J.*, **7**(2), pp. 76–88, 1989.

[28] J.R. Stern, J. W. Ballance, D. W. Faulkner, S. Hornung and D. B. Payne, 'Passive optical networks for telephony and beyond', *Electron. Lett.*, **23**(24), pp. 1255–1257, 1987.

[29] D. W. Faulkner and D. I. Fordham, 'Broadband systems on passive optical networks', *Br. Telecom Technol. J.*, **7**(2), pp . 115–122, 1989.

[30] T. R. Rowbotham, 'Plans for a British trial of fibre to the home', *Proc. Second IEE Conf. on Telecommun.*, York , UK, pp. 73–77, April 1989.

[31] D. Clery, 'Trial makes light of cable network', *New Scientist*, p. 28, 27 October 1990.

[32] S. Fenning and P. Rosher, 'A subcarrier multiplexed broadcast video system for the optical field trial at Bishops Stortford', *Br . Telecom Technol. J.*, **8**(4), pp. 26–29, 1990.

[33] R. S. Bergen Jr, 'Southern Bell's fiber-to-the-home projects', *Opt. Fiber Commun. Conf. (OFC'88)*, New Orleans, USA, paper WK2, January 1988.

[34] P. E. White and L. S. Smoot, 'Optical fibers in loop distribution systems', in S. E. Miller and I. P. Kaminow (Eds.) *Optical Fiber Telecommunications II*, Academic Press, pp. 911–931, 1988.

[35] Y.-K. M. Lin, D. R. Spears and M. Yin, 'Fiber-based local access network architectures', *IEEE Commun. Mag.*, pp. 64–73, October 1989.

[36] R. M. Huyler, D. E. McGowan, J. A. Stiles and F. J. Horsey, 'SLC Series 5 Carrier System fiber to the home feature', *FOCLAN'88* (USA), paper 10.7, 1988.

[37] S. S. Wagner, H. Kobrinski, T. J. Robe, H. L. Lemberg and L. S. Smoot, 'Experimental demonstration of a passive optical subscriber loop architecture', *Electron Lett.*, **24**(6), pp. 344–346, 1988.

[38] I. Gallager, J. Ballance and J. Adams, 'The application of ATM techniques to the local network', *Br. Telecom. Technol. J.*, **7**(2), pp. 151–160, 1989.

[39] J. W. Ballance, P. H. Rogers and M. F. Halls, 'ATM access through a passive optical network', *Electron. Lett.*, **26**(9), pp. 560–588, 1990.
[40] I. Yamashita, Y. Negishi, M. Nunokawa and H. Wakabayashi, 'The application of optical fibres in submarine cable systems', *Telecom. J. (Eng. edn), Switzerland*, **49**(2), pp. 118–124, 1982.
[41] P. Worthington, 'Design and manufacture of an optical fibre cable for submarine telecommunication systems', *Proceedings of Sixth European Conference on Optical Communication*, York, UK, pp. 347–356, 1980.
[42] T. R. Rowbotham, 'Submarine telecommunications', *Br. Telecom. Technol. J.*, **5**(1), pp. 5–24, 1987.
[43] P. K. Runge and N. S. Bergano, 'Undersea cable transmission systems', in S. E. Miller and I. P. Kaminow (Eds.), *Optical Fiber Telecommunications II*, Academic Press, pp. 879–909, 1988.
[44] R. L. Williamson, 'Submarine optical telephone cables present status and future prospects', *Proc. SPIE Int. Soc. Opt. Eng. (USA)*, **1120** (*Fibre Optics'89*), pp. 38–42, 1989.
[45] P. W. Black, 'Undersea system design constraints', *Proc. Soc. Photo-Opt and Instrum. Eng., SPIE*, **1314** (*Fibre Optics'90*), pp. 112–115, 1990.
[46] P. R. Trischitta and D. T. S. Chen, 'Repeaterless undersea lightwave systems', *IEEE Commun. Mag.*, pp. 16–21, March 1989.
[47] A. F. Mitchell and W. A. Stallard, 'Application of semiconductor optical amplifiers to submarine systems', *Proc. IEEE ICC'89*, Boston, MA, USA, pp. 1546–1550, June 1989.
[48] P. Cockrane, 'Future directions in undersea fiber optic system technology', *Proc. IOOC'89*, Kobe, Japan, paper 21B1–2, July 1989.
[49] M. Saito, 'An over 2200 km coherent transmission experiment at 2.5 Gbit/s using erbium doped fibre amplifiers', *Proc. Opt. Fiber Commun. Conf. OFC'90*, USA, postdeadline, PD2, January 1990.
[50] T. J. Aprille, 'Introducing SONET into the local exchange carrier network', IEEE Commun. Mag., **28**(8), pp. 34–38, 1990.
[51] R. Ballert and Y.-C. Ching, 'SONET: now It's the standard optical network', *IEEE Commun. Mag.*, pp. 8–15, March 1989.
[52] CCITT Recommendation G.707, 'Synchronous Digital Hierarchy Bit Rates', 1989.
[53] CCITT Recommendation G.708, 'Network Node Interface for the Synchronous Digital Hierarchy', 1989.
[54] CCITT Recommendation G.709, 'Synchronous Multiplexing Structure', 1989.
[55] 'American National Standard for Telecommunications-Digital Hierarchy Optical Interface Rates and Formats Specifications', ANSI T1.105 – 1988, Draft, March 1988.
[56] M. De Block, 'The new synchronous digital hierarchy', *Frost & Sullivan Fibre Optics Conference*, London, UK, section 13, 10–11 April, 1989.
[57] C. S. Grace and R. H. West, 'Nuclear radiation effects on fibre optics', *Proc. No 53, International Conference on Fibre Optics*, London, 1–2 March 1982, pp. 121–128, IERE, 1982.
[58] J. G. Farrington and M. Chown, 'An optical fibre multiterminal data system for aircraft', *Fibre and Integrated Optics*, **2**(2), pp. 173–193, 1979.
[59] J. J. Mayoux, 'Experimental "Gina" optical data bus in the Mirage 4000 aircraft', *Proc. No 53, International Conference on Fibre Optics*, London, 1–2 March 1982, pp. 147–156, IERE, 1982.

[60] D. L. Williams, 'Military applications of fiber optics', *Proc. SPIE Int. Soc. Opt. Eng. (USA)*, **374**, pp. 138–142, 1983.

[61] R. James, 'Fly by light for the Skyship 600', *Proc. SPIE Int. Soc. Opt. Eng. (USA)*, **1120**, pp. 217–223, 1989.

[62] M. J. Kennett and A. E. Perkins, 'The design of a helicopter fibre optic data bus', *Proc. SPIE Int. Soc. Opt. Eng. (USA)*, **1120**, pp. 25–34, 1989.

[63] P. T. Gardiner and R. A. Edwards, 'Fibre optic sensors (FOS) for aircraft flight controls', *Proc. Royal Aeronautical Soc., Applications of Light to Guided Flight*, London, UK, pp. 42–63, January 1987.

[64] A. G. Gleave, 'Optical technology aspects of fibre optic guided weapon duplex links', *Proc. Royal Aeronautical Soc., Applications of Light to Guided Flight*, London, UK, pp. 25–31, January 1987.

[65] M. K. Barnoski, Fiber systems for the military environment', *Proc. IEEE*, **68**(10), pp. 1315–1320, 1980.

[66] P. H. Bourne and D. P. M. Chown, 'The Ptarmigan optical fibre subsystem', *Proc. No 53, International Conference on Fibre Optics*, London, 1–2 March 1982, pp. 129–146, IERE, 1982.

[67] J. Gladenbeck, K. H. Nolting and G. Olejak, 'Optical fiber cable for overhead line systems', *Proceedings of Sixth European Conference on Optical Communication*, York, UK, pp. 359–362, 1980.

[68] S. E. Miller, 'Potential applications', in S. E. Miller and A. G. Chynoweth (Eds.), *Optical Fiber Telecommunications*, pp. 675–683, Academic Press, 1979.

[69] T. Nakahara, H. Kumanaru and S. Takeuchi, 'An optical fiber video system', *IEEE Trans. Commun.*, **COM-26**(7), pp. 955–961, 1978.

[70] A. C. Deichmiller, 'Progress in fiber optics transmission systems for cable television', *IEEE Trans. Cable Television*, **CATV-5**(2), pp. 50–59, 1980.

[71] S. R. Cole, 'Fiber optics for CATV applications', *International Conference on Communications*, pp. 38.3.1–4. IEEE, 1981.

[72] E. A. Lacey, *Fiber Optics*, Chapter 1, Prentice Hall, 1982.

[73] P. Tolstrup Nielsen, B. Scharøe Petersen and H. Steffensen, 'A trunk system for CATV using optical transmission at 140 Mb/s', *Proceedings of Sixth European Conference on Optical Communication*, York, UK, pp. 406–409, 1980.

[74] M. Sekita, T. Kawamura, K. Ito, S. Fujita, M. Ishii and Y. Miyake, 'TV video transmission by analog, baseband modulation of a 1.3 μm-band laser diode', *Proceedings of Sixth European Conference on Optical Communication*, York, UK, pp. 394–397, 1980.

[75] M. Kajino, K. Nishimura, F. Hayashida, T. Otsuka, K. Ito, H. Shiono and T. Yamada, 'A 4-channel WDM baseband video optical fiber system for monitoring of an automated guideway transit', *Proceedings of Sixth European Conference on Optical Communication*, York, UK, pp. 442–445, 1980.

[76] N. Yumoto, H. Ikeda, T. Sugimoto, K. Hayashi and F. Sakamoto, 'Optical data link for automobiles', *Sumitomo Electr. Tech. Rev.*, **23**, pp. 152–158, 1984.

[77] R. P. Page, 'A fibre optic multiplexing system for use on a motor car', *Proc. SPIE Int. Soc. Opt. Eng. (USA)*, **522**, pp. 142–147, 1985.

[78] E. Wolsthoff, 'Ring-main wiring wing light (for cars)', *Funkschau* (Germany), **5**, pp. 34–35, 1986.

[79] P. Extance, R. J. Hazelden, J. W. Birch and C. P. Cockshott, Fibre optics for monitoring internal combustion engines', *Proc. SPIE Int. Soc. Opt. Eng. (USA)*, **734**, pp. 224–230, 1987.

[80] C. P. Cockshott, A. J. Cook, R. J. Hazelden, S. J. Pacand, R. A. Pinnock and I. Sakai, 'Applications of optical sensors in the automobile industry', *Proc. SPIE Int. Soc. Opt. Eng. (USA)*, **1120**, pp. 210–214, 1989.

[81] T. Sasayama and H. Asano, 'Multiplexed optical transmission system for automobiles utilizing polymer fiber with high heat resistance', *Proc. SPIE Int. Soc. Opt. Eng. (USA)*, **989**, pp. 148–155, 1988.

[82] I. Sakai and C. Stenson, 'Optical sensors for infering road adhesion', *Proc. SPIE Int. Soc. Opt. Eng.*, **1314**, pp. 236–243, 1990.

[83] D. A. A. Roworth, 'Fibre optics for industrial applications', *Optics and Laser Technology*, **12**(5), pp. 255–259, 1981.

[84] P. B. Lyons, E. D. Hodson, L. D. Looney, G. Gow, L. P. Hocker, S. Lutz, R. Malone, J. Manning, M. A. Nelson, R. Selk and D. Simmons, 'Fiber optic application in nuclear testing', *Electro Optics/Laser '79 Conf. Exposition*, Anahiem, CA, 23–25 October, 1979.

[85] W. F. Trover, 'Fiber optics for data acquisition and control communications: case histories', *Wire Technology*, **9**(2), pp. 79–87, 1981.

[86] D. E. N. Davies and S. A. Kingsley, 'A novel optical fibre telemetry highway', *Proceedings of First European Conference on Optical Communication*, London, UK, 16–18 September 1975, pp. 165–167, IEE, 1975.

[87] G. B. Hocker, 'Fiber-optic sensing of pressure and temperature', *Appl. Opt,* **18**(9), pp. 1445–1448, 1979.

[88] T. G. Giallorenzi *et al.*, 'Optical fiber sensor technology', *IEEE J. Quantum Electron.*, **18**(4), pp. 626–666, 1982.

[89] D. A. Jackson and J. D. C. Jones, 'Interferometers', in B. Culshaw and J. P. Dakin, (Eds.), *Optical Fiber Sensors: Systems and Applications*, Artech House, pp. 329–380, 1989.

[90] H. C. Lefevre, 'Fibre optic gyroscope', in B. Culshaw and J. Dakin, (Eds.), *Optical Fibre Sensors: Systems and Applications*, Vol. 2, Artech House, pp. 381–429, 1989.

[91] G. D. Pitt, 'Optical fiber sensors', in F. C. Allard, *Fiber Optics Handbook: For Engineers and Scientists*, McGraw-Hill, 1990.

[92] A. J. Rogers, 'Optical fibre current measurement', *Proc. SPIE Int. Soc. Opt. Eng. (USA)*, **374**, pp. 196–201, 1983.

[93] J. Wilson and J. F. B. Hawkes, *Optoelectronics: An introduction*, 2nd edn, Prentice Hall, 1989.

[94] S. K. Yao and C. K. Asawa, 'Microbending fiber optic sensing', *Proc. SPIE Int. Soc. Opt. Eng.*, **412**, pp. 9–13, 1983.

[95] B. E. Jones, R. S. Medlock and R. C. Spooner, 'Intensity and wavelength-based sensors and optical actuators', in B. Culshaw and J. Dakin, (Eds.), *Optical Fibre Sensors: Systems and applications*, Vol.2., Artech House, pp. 431–473, 1989.

[96] Cambridge Consultants, 'Optics challenge electronic sensing', *Eureka*, pp. 78–80, 1983.

[97] J. M. Senior, S. D. Cusworth, N. G. Burrow and A. D. Muirhead, 'An extrinsic optical fibre sensor employing a partially reflecting mirror', *Proc. SPIE Int. Soc. Opt. Eng. (USA)*, **522**, pp. 204–210, 1985.

[98] B. E. Jones and R. C. Spooncer, 'An optical fibre pressure sensor using a holographic shutter modulator with two-wavelength intensity referencing', *Proc. SPIE Int. Soc. Opt. Eng. (USA)*, **514**, pp. 223–226, 1984.

[99] J. M. Senior, G. Murtaza, A. I. Stirling and G. H. Wainwright, 'Dual-wavelength

intensity-modulated optical fibre sensor system', *Proc. SPIE Int. Soc. Opt. Eng. (USA)*, **1120**, pp. 332–337, 1989.

[100] B. Culshaw, J. Foley and I. P. Giles, 'A balancing technique for optical fibre intensity modulated transducers', *Proc. SPIE Int. Soc. Opt. Eng. (USA)*, **574**, pp. 117–120, 1984.

[101] A. Rogers, 'Measurement using fibre optics', *New Electronics* (*GB*), pp. 29–36, 27 October 1981.

[102] B. Culshaw, 'Optical fibre transducers', *The Radio and Electronic Eng.*, **52**(6), pp. 283–290, 1982.

[103] S. K. Yao and C. K. Asawa, 'Fiber optical intensity sensors', *IEEE J. Selected Areas in Commun.*, **SAC-1**(3), pp. 562–573, 1983.

[104] C. Menadier, C. Kissinger and H. Adkins, 'The fotonic sensor', *Instruments and Control Systems*, **40**, pp. 114–120, 1967.

[105] J. A. Powell, 'A simple two-fiber optical displacement sensor', *Rev. Sci. Instrum.*, **45**(2), pp. 302–303, 1974.

[106] B. Culshaw, 'Silicon in optical fiber sensors', in B. Culshaw and J. Dakin (Eds.), *Optical Fiber Sensors: Systems and applications*, Vol. 2, Artech House, pp. 475–509, 1989.

[107] J. D. Crow and M. W. Sachs, 'Optical Fibers for Computer Systems', *Proc. IEEE*, **68**(10), pp. 1275–1280, 1980.

[108] J. A. Eibner, *Fiber Optics for Computer Applications*, Connector Symp. Proc., Vol. 11, Cherry Hill, NJ, 1978.

[109] D. Schibonski, 'A 500 megabit network: the applications and issues', *Proc. European Comp. Commun. Conf.*, London, UK, pp. 257–264, June 1987.

[110] C. P. Wyles, 'Fibre-optic multiplexing system for the IBM 3270 series', *Proc. SPIE Int. Soc. Opt. Eng.* (*USA*), **374**, pp. 78–83, 1983.

[111] D. Clark, K. Pogran and D. Reed, 'An introduction to local area networks', *Proc. IEEE.*, **66**, pp. 1497–1517, 1978.

[112] S. Y. Suh, S. W. Granlund and S. S. Hegde, 'Fiber-optic local area network topologies', *IEEE. Commun. Mag.*, **24**(8), pp. 26–32, 1986.

[113] M. M. Nasseki, F. A. Tobagi and M. E. Marhic, 'Fiber optic configurations for local area networks', *IEEE. J. Selected Areas in Commun.*, **SAC-3**(6), pp. 941–949, 1985.

[114] J. W. Reedy and J. R. Jones, 'Methods of collision detection in fiber optic CSMA/CD networks', *IEEE. J. Selected Areas in Commun.*, **SAC-3**(6), pp. 890–896, 1985.

[115] S. D. Personick, 'Protocols for fiber-optic local area networks', *Journal of Lightwave Technology*, **LT-3**, pp. 426–431, 1985.

[116] R. V. Schmidt, E. G. Rawson, R. E. Norton, S. B. Jackson and M. D. Bailey, 'Fibernet II: a fiber optic ethernet', *IEEE. J. Selected Areas in Commun.*, **SAC-1**(5), pp. 702–710, 1983.

[117] T. Tamura, M. Nakamura, S. Ohshima, T. Ito and T. Ozeki, 'Optical cascade star network – a new configuration for a passive distribution system with optical collision detection capability', *Journal of Lightwave Technology*, **LT-2**(1), pp. 61–66, 1984.

[118] M. R. Finley, 'Optical fibers in local area networks', *IEEE. Commun. Mag.*, **22**(8), pp. 22–23, 1984.

[119] D. Rosenburger and H. H. Witte, 'Optical LAN activities in Europe', *Journal of Lightwave Technology*, **LT-3**(3), pp. 432–437, 1985.

[120] E. G. Rawson and R. M. Metcalfe, 'Fibernet: multimode optical fibers for local computer networks', *IEEE. Trans-Commun.*, **COM-26**(7), pp. 983–990, 1978.

[121] G. W. Lichfield, P. Hensel and D. J. Hunkin, 'Application of optical fibers to the Cambridge ring system', *Proc. 6th Int. Conf. on Computer Commun.*, Sept. 1982, pp. 513–517, 1982.
[122] F. E. Ross, 'FDDI – a tutorial', *IEEE. Commun. Mag.*, **24**(5), pp. 10–17, 1986.
[123] F. E. Ross, 'An overview of FDDI: the Fiber Distributed Data Interface', *IEEE J. Selected Areas in Commun.*, **7**(7), pp. 1043–1051, 1989.
[124] FDDI Token Ring Physical Layer Medium Dependent (PMD), American National Standard, ANSI ×3.166, 1990: ISO Standard 9314-3, 1990.
[125] FDDI Token Ring Physical Layer Protocol (PHY), American National Standard, ANSI ×3.138, 1988: ISO Standard 9314-1, 1989.
[126] FDDI Token Ring Media Access Control (MAC), American National Standard, ANSI ×3.139, 1987: ISO Standard 9314-2, 1989.
[127] FDDI Token Ring Station Management (SMT), Draft proposed American National Standard, Rev. 6.2, 15 May 1990.
[128] T. King, 'Fiber optic components for the Fiber Distributed Data Interface (FDDI) 100 Mbit s^{-1} local area networks'. *Proc. SPIE Int. Soc. Opt. Eng. (USA)*, **949**, pp. 2–13, 1988.
[129] FDDI Token Ring Single-Mode Fiber Physical Layer Medium Dependent (SMF–PMD), Draft proposed American National Standard, Rev. 4, April 1989.
[130] F. E. Ross, 'FDDI – an overview', *IEEE COMPCOM Conf.*, pp. 434–440, 1987.
[131] K. Caves, 'FDDI – 2: a new standard for integrated services high speed LANs', *Proc. European Computer Commun. Conf.*, (London), pp. 245–256, June 1987.
[132] S. D. Rigby, 'FDDI speeds networks', *Commun. Int.*, **15**(4), pp. 67–69, 1988.
[133] FDDI Hybrid Ring Control (HRC), Draft proposed American National Standard, Rev. 6, 11 May 1990.
[134] A. R. Hills, 'FDDI-II – An implementor's perspective', *The Ninth Annual Europ. Fibre Optic Commun. and Local Area Network Conf.* (London), LAN Proc. pp. 194–200, June 1991.
[135] G. E. Mityko, 'The FDDI follow-on LAN', *The Ninth Annual Europ. Fibre Optic Commun. and Local Area Network Conf.* (London), LAN Proc. pp. 201–205, June 1991.
[136] A. Leach, 'Distributing PCs via local area networks', *Control and Instrumentation*, **17**(1), pp. 51–55, 1985.
[137] J. Keogh, 'Real-time control', *Systems Int.*, **14**(9), pp. 79–81, 1986.
[138] J. Pingrie, 'MAP users speak out at Satech', *FMS Magazine*, **5**(1), pp. 26–30, 1987.
[139] P. Cheshire, 'MAP & TOP in perspective: the role of communications', *Proc. European Computer Commun. Conf.* (London), pp. 361–368, June 1987.
[140] N. C. L. Beale, 'Standard fiber optic LANs and options for MAP', *Proc. European Computer Commun. Conf.* (London), pp. 387–400, June 1987.
[141] J. M. Senior, W. M. Walker and A. Ryley, 'Topology and MAC layer access protocol investigation for industrial optical fiber LANs', *Computer Networks and ISDN Systems*, **13**, pp. 275–289, 1987.
[142] C. R. Giles, T. Li, T. H. Wood, C. A. Burrus and D. A. B. Miller, 'An all optical regenerator', *Optical Fiber Commun. Conf., OFC '88*, (USA), Postdeadline PD22, January 1988.
[143] J. Chandler and D. Hobson, 'Connecting for FMS – experiences in building of an educational system', *Computer-Aided Engineering Journal*, pp. 13–20, February 1985.

Appendices

A. The field relations in a planar guide

Let us consider an electromagnetic wave having an angular frequency ω propagating in the z direction with propagation vector (phase constant) β. Then as indicated in Section 2.3.2 the electric and magnetic fields can be expressed as:

$$\mathbf{E} = \mathrm{Re}\{\mathbf{E}_0(x, y) \exp \mathrm{j}(\omega t - \beta z)\} \tag{A1}$$

$$\mathbf{H} = \mathrm{Re}\{\mathbf{H}_0(x, y) \exp \mathrm{j}(\omega t - \beta z)\} \tag{A2}$$

For the planar guide the Cartesian components of $\mathbf{E}_0$ and $\mathbf{H}_0$ become:

$$\frac{\partial E_z}{\partial y} + \mathrm{j}\beta E_y = -\mathrm{j}\mu_r\mu_0\omega H_x \tag{A3}$$

$$\mathrm{j}\beta E_x + \frac{\partial E_z}{\partial x} = \mathrm{j}\mu_r\mu_0\omega H_y \tag{A4}$$

$$\frac{\partial E_y}{\partial x} - \frac{\partial E_x}{\partial y} = -\mathrm{j}\mu_r\mu_0\omega H_z \tag{A5}$$

$$\frac{\partial H_z}{\partial y} + \mathrm{j}\beta H_y = \mathrm{j}\omega\varepsilon_r\varepsilon_0 E_x \tag{A6}$$

$$-\mathrm{j}\beta H_x - \frac{\partial H_z}{\partial x} = \mathrm{j}\omega\varepsilon_r\varepsilon_0 E_y \tag{A7}$$

$$\frac{\partial H_y}{\partial x} - \frac{\partial H_x}{\partial y} = \mathrm{j}\omega\varepsilon_r\varepsilon_0 E_z \tag{A8}$$

If we assume that the planar structure is an infinite film in the y–z plane, then for an infinite plane wave travelling in the z direction the partial derivative with respect to y is zero ($\partial/\partial y = 0$). Employing this assumption we can simplify the above equations to demonstrate fundamental relationships between the fields in such a structure. These are:

$$j\beta E_y = -j\mu_r\mu_0\omega H_x \qquad \text{(TE mode)} \qquad \text{(A9)}$$

$$j\beta E_x + \frac{\partial E_z}{\partial x} = j\mu_r\mu_0\omega H_y \qquad \text{(TM mode)} \qquad \text{(A10)}$$

$$\frac{\partial E_y}{\partial x} = -j\mu_r\mu_0\omega H_z \qquad \text{(TE mode)} \qquad \text{(A11)}$$

$$j\beta H_y = j\omega\varepsilon_r\varepsilon_0 E_x \qquad \text{(TM mode)} \qquad \text{(A12)}$$

$$-j\beta H_x - \frac{\partial H_z}{\partial x} = j\omega\varepsilon_r\varepsilon_0 E_y \qquad \text{(TE mode)} \qquad \text{(A13)}$$

$$\frac{\partial H_y}{\partial x} = j\omega\varepsilon_r\varepsilon_0 E_z \qquad \text{(TM mode)} \qquad \text{(A14)}$$

It may be noted that the fields separate into TE and TM modes corresponding to coupling between E_y, H_x, H_z, ($E_z = 0$) and H_y, E_x, E_z ($H_z = 0$) respectively.

B. Gaussian pulse response

Many optical fibers, and in particular jointed fiber links, exhibit pulse outputs with a temporal variation that is closely approximated by a Gaussian distribution. Hence the variation in the optical output power with time may be described as:

$$P_o(t) = \frac{1}{\sqrt{(2\pi)}} \exp - \left(\frac{t^2}{2\sigma^2}\right) \qquad \text{(B1)}$$

where σ and σ^2 are the standard deviation and the variance of the distribution respectively. If t_e represents the time at which $P_o(t_e)/P_o(0) = 1/e$ (i.e. $1/e$ pulse width), then from Eq. (B1) it follows that:

$$t_e = \sigma\sqrt{2}$$

Moreover, if the full width of the pulse at the $1/e$ points is denoted by τ_e then:

$$\tau_e = 2t_e = 2\sigma\sqrt{2}$$

In the case of the Gaussian response given by Eq. (B1) the standard deviation σ is equivalent to the rms pulse width.

The Fourier transform of Eq. (B1) is given by:

$$\mathscr{P}(\omega) = \frac{1}{\sqrt{(2\pi)}} \exp - \left(\frac{\omega^2\sigma^2}{2}\right) \qquad \text{(B2)}$$

The 3 dB optical bandwidth B_{opt} is defined in Section 7.4.3 as the modulation frequency at which the received optical power has fallen to one half of its constant value. Thus using Eq. (B2):

$$\frac{[\omega(3\text{ dB opt})]^2}{2}\sigma^2 = 0.693$$

and

$$\omega(3\text{ dB opt}) = 2\pi B_{\text{opt}} = \frac{\sqrt{2}\times 0.8326}{\sigma}$$

Hence

$$B_{\text{opt}} = \frac{\sqrt{2}\times 0.8326}{2\pi\sigma} = \frac{0.530}{\tau_e} = \frac{0.187}{\sigma}\text{ Hz}$$

When employing return to zero pulse where the maximum bit rate $B_T(\text{max}) = B_{\text{opt}}$, then:

$$B_T(\text{max}) \simeq \frac{0.2}{\sigma}\text{ bit s}^{-1}$$

Alternatively, the 3 dB electrical bandwidth B occurs when the received optical power has dropped to $1/\sqrt{2}$ of the constant value (see Section 7.4.3) giving:

$$B = \frac{0.530}{\tau_e\sqrt{2}} = \frac{0.375}{\tau_e} = \frac{0.133}{\sigma}\text{ Hz}$$

C. Variance of a random variable

The statistical mean (or average) value of a discrete random variable X is the numerical average of the values which X can assume weighted by their probabilities of occurrence. For example, if we consider the possible numerical values of X to be $x_1, x_2, \ldots x_i$, with probabilities of occurrence $P(x_1), P(x_2)\ldots P(x_i)$, then as the number of measurements N of X goes to infinity, it would be expected that the outcome $X = x_1$ would occur $NP(x_1)$ times, the outcome $X = x_2$ would occur $NP(x_2)$ times, and so on. In this case the arithmetic sum of all N measurements is:

$$x_1P(x_1)N + x_2P(x_2)N + \ldots x_iP(x_i)N = N\sum_i x_iP(x_i) \tag{C1}$$

The mean or average value of all these measurements which is equivalent to the mean value of the random variable may be calculated by dividing the sum in Eq. (C1) by the number of measurements N. Furthermore, the mean value for the random variable X which can be denoted as $\bar{X}$ (or m) is also called the expected value of X and may be represented by $E(X)$. Hence:

$$\bar{X} = m = E(X) = \sum_{i=1}^{N} x_iP(x_i) \tag{C2}$$

Moreover, Eq. (C2) also defines the first moment of X which we denote as M_1. In a similar manner the second moment M_2 is equal to the expected value of X^2 such that:

$$M_2 = \sum_{i=1}^{N} x_i^2 P(x_i) \tag{C3}$$

M_2 is also called the mean square value of X which may be denoted as $\overline{X^2}$.

For a continuous random variable, the summation of Eq. (C2) approaches an integration over the whole range of X so that the expected value of X:

$$M_1 = E(X) = \int_{-\infty}^{\infty} x p_X(x)\, \mathrm{d}x \tag{C4}$$

where $p_X(x)$ is the probability density function of the continuous random variable X. Similarly, the expected value of X^2 is given by:

$$M_2 = E(X^2) = \int_{-\infty}^{\infty} x^2 p_X(x)\, \mathrm{d}x \tag{C5}$$

It is often convenient to subtract the first moment $M_1 = m$ prior to computation of the second moment. This is analogous to moments in mechanics which are referred to the centre of gravity rather than the origin of the coordinate system. Such a moment is generally referred to as a central moment. The second central moment represented by the symbol σ^2 is therefore defined as:

$$\sigma^2 = E[(X-m)^2] = \int_{-\infty}^{\infty} (x-m)^2 p_X(x)\, \mathrm{d}x \tag{C6}$$

where σ^2 is called the variance of the random variable X. Moreover, the quantity σ which is known as the standard deviation is the root mean square (rms) value of $(X-m)$.

Expanding the squared term in Eq. (C6) and integrating term by term we find:

$$\begin{aligned}\sigma^2 &= E[X^2 - 2mX + m^2]\\ &= E(X^2) - 2mE(X) + E(m^2)\\ &= E(X^2) - 2m^2 - m^2\\ &= E(X^2) - m^2\end{aligned} \tag{C7}$$

As $E(X^2) = M_2$ and $m = M_1$, the variance may be written as:

$$\sigma^2 = M_2 - (M_1)^2$$

D. Variance of the sum of independent random variables

If a random variable $W = g(X, Y)$ is a function of two random variables X and Y, then extending the definition in Eq. (C4) for expected values gives the expected value of W as:

$$E(W) = \int_{-\infty}^{\infty} \int_{-\infty}^{\infty} g(x, y) p_{XY}(x, y) \, dx \, dy \tag{D1}$$

where $p_{XY}(x, y)$ is the joint probability density function. Furthermore the two random variables X and Y are statistically independent when:

$$p_{XY}(x, y) = p_X(x) p_Y(y) \tag{D2}$$

Now let X and Y be two statistically independent random variables with variances σ_X^2 and σ_Y^2 respectively. In addition we assume the sum of these random variables to be another random variable denoted by Z such that $Z = X + Y$, where Z has a variance σ_Z^2. If the mean values of X and Y are zero, employing the definition of variance given in Eq. (C6) together with the expected value for a function of two random variables (Eq. (D1)) we can write:

$$\sigma_Z^2 = \int_{-\infty}^{\infty} \int_{-\infty}^{\infty} (x + y)^2 p_{XY}(x, y) \, dx \, dy \tag{D3}$$

As X and Y are statistically independent we can utilize Eq. (D2) to obtain:

$$\begin{aligned} \sigma_Z^2 &= \int_{-\infty}^{\infty} \int_{-\infty}^{\infty} (x + y)^2 p_X(x) p_Y(y) \, dx \, dy \\ &= \int_{-\infty}^{\infty} x^2 p_X(x) \, dx + \int_{-\infty}^{\infty} y^2 p_Y(y) \, dy \\ &\quad + 2 \int_{-\infty}^{\infty} x p_X(x) \, dx \int_{-\infty}^{\infty} y p_Y(y) \, dy \end{aligned} \tag{D4}$$

The two factors in the last term of Eq. (C4) are equal to the mean values of the random variables (X and Y) and hence are zero. Thus:

$$\sigma_Z^2 = \sigma_X^2 + \sigma_Y^2$$

E. Closed loop transfer function for the transimpedance amplifier

The closed loop transfer function $H_{CL}(\omega)$ for the transimpedance amplifier shown in Figure 9.9 may be derived by summing the currents at the amplifier input, remembering that the amplifier input resistance is included in R_{TL}. Hence,

$$i_{det} + \frac{V_{out} - V_{in}}{R_f} = V_{in} \left(\frac{1}{R_{TL}} + j\omega C_T \right) \tag{E1}$$

As $V_{in} = -V_{out}/G$, then

$$i_{det} = -V_{out} \left(\frac{1}{R_f} + \frac{1}{GR_f} + \frac{1}{GR_{TL}} + \frac{j\omega C_T}{G} \right) \tag{E2}$$

Therefore,

$$H_{\mathrm{CL}}(\omega) = \frac{V_{\mathrm{out}}}{i_{\mathrm{det}}} = \frac{-R_{\mathrm{f}}}{1 + (1/G) + (R_{\mathrm{f}}/GR_{\mathrm{TL}}) + (\mathrm{j}\omega C_{\mathrm{T}}R_{\mathrm{f}}/G)}$$

$$= \frac{-R_{\mathrm{f}}/(1 + 1/G + R_{\mathrm{f}}/GR_{\mathrm{TL}})}{[1 + \mathrm{j}\omega C_{\mathrm{T}}R_{\mathrm{f}}/(1 + R_{\mathrm{f}}/R_{\mathrm{TL}} + G)]} \tag{E3}$$

Since,

$$G \gg \left(1 + \frac{R_{\mathrm{f}}}{R_{\mathrm{TL}}}\right) \tag{E4}$$

then Eq. (E3) becomes,

$$H_{\mathrm{CL}}(\omega) \simeq \frac{-R_{\mathrm{f}}}{1 + (\mathrm{j}\omega R_{\mathrm{f}}C_{\mathrm{T}}/G)}\ \mathrm{VA}^{-1}$$

Index